Titan from Cassini-Huygens

Robert H. Brown • Jean-Pierre Lebreton
J. Hunter Waite
Editors

Titan from Cassini-Huygens

Springer

Editors
Robert H. Brown
University of Arizona
1629 E. University Blvd.
Tucson AZ 85721
USA
rhb@lpl.arizona.edu

Jean-Pierre Lebreton
ESTEC
ESA Space Science
Astrophysics Division
2200 AG Noordwijk
Netherlands
Jean-Pierre.Lebreton@esa.int

J. Hunter Waite
Southwest Research Institute (SWRI)
Space Research Division
6220 Clebra Road P.O.Box 28510
San Antonio TX 78228
USA
hwaite@swri.edu

Please note that additional material for this book can be downloaded from
http://extras.springer.com

ISBN 978-94-007-4452-3 ISBN 978-1-4020-9215-2 (eBook)
DOI 10.1007/978-1-4020-9215-2
Springer Dordrecht Heidelberg London New York

Library of Congress Control Number: 2009930523

Front cover figure: This false-color image of Titan is a composite of data taken by the Cassini RADAR instrument and the Cassini VIMS instrument with special emphasis on showing Titan's surface features. *Credit*: U.S. Geological Survey, University of Arizona and NASA JPL, by Laurence Soderblom and Randy Kirk (Cassini RADAR) and Jason W. Barnes (Cassini VIMS)

Back cover figure: [text to be provided by Bob Brown]
Credit: University of Arizona, by Eric Karkoschka (Huygens DISR)

Printed on acid-free paper

Springer is part of Springer Science+Business Media (www.springer.com)

Preface

This book is one of two volumes meant to capture, to the extent practical, the scientific legacy of the Cassini–Huygens prime mission, a landmark in the history of planetary exploration. As the most ambitious and interdisciplinary planetary exploration mission flown to date, it has extended our knowledge of the Saturn system to levels of detail at least an order of magnitude beyond that gained from all previous missions to Saturn.

Nestled in the brilliant light of the new and deep understanding of the Saturn planetary system is the shiny nugget that is the spectacularly successful collaboration of individuals, organizations and governments in the achievement of Cassini–Huygens. In some ways the partnerships formed and lessons learned may be the most enduring legacy of Cassini–Huygens. The broad, international coalition that is Cassini–Huygens is now conducting the Cassini Equinox Mission and planning the Cassini Solstice Mission, and in a major expansion of those fruitful efforts, has extended the collaboration to the study of new flagship missions to both Jupiter and Saturn. Such ventures have and will continue to enrich us all, and evoke a very optimistic vision of the future of international collaboration in planetary exploration.

The two volumes in the series Saturn from Cassini–Huygens and Titan from Cassini–Huygens are the direct products of the efforts of over 200 authors and co-authors. Though each book has a different set of three editors, the group of six editors for the two volumes has worked together through every step of the process to ensure that these two volumes are a set. The books are scholarly works accessible at a graduate-student level that capture the approximate state of knowledge of the Saturn system after the first 4 years of Cassini's tenure in Saturn orbit. The topics covered in each volume range from the state of knowledge of Saturn and Titan before Cassini–Huygens to the ongoing planning for a return to the system with vastly more capable spacecraft.

In something of a departure from the norm for works such as these, we have included an appendix in each of the books featuring the people of Cassini–Huygens who are truly responsible for its success – the people behind the scientific scenes who ensure that everything works as flawlessly as it has. We dedicate the Cassini–Huygens volumes to them and to those who started the journey with us but could not finish it. We hope that all who read the books will share in the new knowledge and gain a deeper appreciation for the tireless efforts of those who made possible its attainment.

The Editors: Bob Brown, Michele Dougherty, Larry Esposito, Stamatios Krimigis, Jean-Pierre Lebreton and J. Hunter Waite

Contents

Chapter 1
Overview

Robert H. Brown, Jean-Pierre Lebreton, and J. Hunter Waite

1.1 Introduction

This book is the one of two volumes meant to capture the main scientific results of the Cassini–Huygens prime mission. The first book, *Saturn from Cassini–Huygens*, contains the material pertinent to Saturn, its satellites and its magnetosphere, except its largest moon Titan. This book, *Titan from Cassini–Huygens* is meant to capture the main scientific results regarding Titan, including its origin and evolution, interior, surface, atmosphere and interaction with Saturn's magnetosphere. The reason we have chosen to treat the results for Titan in a separate volume is that at the time of publication of this book, the Cassini spacecraft has flown by Titan some 62 times, including delivering the Huygens probe, which successfully landed on Titan's surface and survived for over 2 hours, transmitting a wealth of data on Titan's atmosphere and surface. The scientific results obtained for Titan are so voluminous that they cannot be properly captured as part of the "Saturn from Cassini–Huygens" volume and require a separate volume; thus, the two books.

Because there exists a three-volume series of books written prior to Cassini–Huygens' arrival in the Saturn system, that describe in great detail the Cassini Orbiter spacecraft, its science instruments and investigations, the Huygens probe and its science instruments and investigations, and the overall design of the Cassini–Huygens mission to Saturn, we will not attempt to repeat any of that material here. Instead the reader is referred to those volumes (2002, 2004a, b). We do, however, include sections in this book describing the details of the Cassini–Huygens orbital tour as it was executed through the Cassini prime mission, the Cassini Equinox mission, and the plans for the Cassini Solstice mission, because very little of that material was described in the previous volumes.

1.2 Organization

This volume is divided into eight main sections. The first section is introductory material, including this chapter and a chapter describing knowledge of the Saturn system prior to Cassini–Huygens' and as seen from the ground. The second section deals with "bulk" Titan and contains two chapters covering Titan's origin and its interior structure. The third section covers Titan's surface, and contains two chapters describing Titan's geology and the composition of its surface. The fourth section is devoted to Titan's global systemic evolution, and contains three chapters relating to the origin and evolution of Titan's volatiles, the conversion of nitrogen and methane in Titan's atmosphere to complex polymers, and the astrobiological considerations related to Titan's evolution as a chemical system. The fifth section covers Titan's atmosphere and ionosphere, and features five chapters covering the composition and structure of Titan's atmosphere, the composition and structure of Titan's ionosphere and thermosphere, Titan's aerosols, its atmospheric dynamics and meteorology, and its seasonal cycles. The sixth section deals with Titan's magnetospheric interactions and contains two chapters covering mass-loss processes and energy deposition in Titan's upper atmosphere. The seventh section is dedicated to an exposition of the exploratory plan for Titan. It contains two chapters, one covering the past, present and future plans for Cassini's exploration of Titan, and the other discussing plans for returning to the Saturn system with new and more capable spacecraft. The final section contains Cassini-related cartographic products for Titan and a chapter depicting people who worked behind the scenes for Huygens.

Please note that what follows below is our attempt to summarize the main results that are covered in much greater

R.H. Brown (✉)
Department of Planetary Sciences, University of Arizona, Tucson, AZ, USA
e-mail: rhb@arizona.edu

J.P. Lebreton
ESA, Nordwijk, Netherlands

J.H. Waite
Southwest Research Institute, San Antonio, TX, USA

detail in this book. The reader should assume attribution of these synopses to each of the chapters covered as part of each section below.

1.3 Synopses of the Main Results for Titan

1.3.1 Origin, Evolution and Interior (Chapters 3 and 4)

Titan likely formed in a disk of material around Saturn as a regular satellite, but it is very different than its smaller siblings in size, mass, and abundance of rock, possibly suggesting it formed in an accretion disk less massive than the disk around Jupiter from which it is thought the Galilean satellites formed. The outermost layers of primordial Titan probably consisted of water ice mixed with a small amount of ammonia, and heat from Titan's formation likely melted that layer, resulting in a transient surface of water-ammonia liquid in contact with organic molecules. The composition of Titan's atmosphere as determined by Cassini–Huygens suggests that NH_3 liberated from Titan's interior is the source of Titan's atmospheric molecular nitrogen.

Measurements of $^{13}C/^{12}C$ in hydrocarbon molecules in Titan's atmosphere indicate that the bulk of Titan's primordial carbon survived massive escape during Titan's thick-primordial-atmosphere phase, whereas measurements of $^{15}N/^{14}N$ indicate mild to major escape during the same phase. Measurements of $^{40}K/^{40}Ar$ strongly suggest that Titan has outgassed volatiles throughout its history, notably methane and argon.

To date, Cassini measurements have not detected a Titanian magnetic field, thus the actual level of differentiation of its interior is uncertain. Nevertheless, Cassini measurements have spurred the construction of new models of Titan's interior structure and evolutionary history. These models suggest that Titan is at least partially differentiated with a silicate core, a water-rich mantle, and an internal ocean a few tens of kilometers below the surface. Titan's orbit has a relatively high eccentricity, implying low tidal dissipation in its interior, suggesting that its ice crust is cold and that the putative ocean under the ice layer contains ammonia. If this ocean indeed exists and is composed of a mix of water and ammonia, it is probably separated from the core by a layer of high-pressure ices. The obliquity of Titan derived from Cassini images of Titan, however, implies an unreasonable internal structure if Titan is in a Cassini state, and this issue has yet to be resolved.

1.3.2 Surface (Chapters 5 and 6)

Observations of Titan's surface gleaned primarily from measurements by the RADAR, VIMS and ISS instruments reveal that Titan has a complex surface morphology, having many features in common with other icy satellites, many features in common with the Earth and some that seem unique to Titan. Titan has an extremely young surface, constantly modified by aeolian, pluvial, fluvial, lacustrine, cryovolcanic and tectonic processes, all in dynamic interplay, resulting in a surface composition and morphology unique in the solar system.

Displaying a wide variety of geologic features on all size scales sampled by the Cassini data, Titan's surface shows evidence of impact cratering, fluvial erosion resulting from a hydrological cycle similar to that of the Earth but driven by liquid methane, aeolean features driven by a thick and dynamic atmosphere, and tectonic and cryovolcanic features driven by heat escaping from Titan's interior and expressed in a material that is probably water ice mixed with ammonia, carbon dioxide and methane. Titan also has an apparent surficial modification process that is so far unique in planetary exploration experience. Its terrain is apparently being "veneered" by a slow and steady "rain" of complex-organic particles, producing (in most cases) a dramatically different appearance when viewed in radar versus optical wavelengths. Titan thus has a very active surface, constantly being modified by exogenic and endogenic processes.

Titan's surface shows a fair amount of topography. Relief ranges from a few tens to a couple of hundred meters in the region of the Huygens landing site, mountains on Titan range from a few hundred to a couple of thousand meters, and there are a number of regions on Titan interpreted as lakebeds whose depths range from a few tens to a few hundreds of meters. A general relationship seen on Titan is that bright units are topographically high while darker units are topographically low.

Titan shows a number of geological features interpreted as being tectonic or volcanic in origin. Features thought to be tectonic seem to be expressed as long, linear features that are chains of hills and/or tectonic ridges, modified by ongoing geological processes such as erosion and deposition. Features on Titan interpreted as volcanic in origin are globally widely distributed but, so far, small in number. Three major areas interpreted to have volcanic morphologies are Hotei Arcus, Tui Reggio and Ganesha Macula (although the cryovolcanic interpretation for the origin of Ganesha Macula has recently been called into question). There are a number of smaller features on Titan sometimes controversially interpreted as volcanic in origin, among which are clusters of quasi-circular depressions with raised rims that in many cases are thought to be volcanic calderas.

Erosional and depositional features are ubiquitous on Titan. Vast fields of dunes abound in Titan's equatorial regions, and are thought to be composed of millimeter to sub-millimeter complex-organic particles, driven by winds into low-lying areas. Their mere existence implies a large source of organic particles that is poorly understood. Other erosional features, most likely driven by fluid flow, manifest themselves as vast, dendritic channel systems, cutting across almost all types of terrains on Titan. Some of the channel systems have rough, radar-bright floors suggesting that they are dry, while others have smooth, radar-dark floors, suggesting that some may contain liquids. The morphology of the channels suggests that they have been cut through steep terrain by precipitation-driven fluid flow. Liquid methane has been suggested as the fluid responsible for the bulk of the fluvial features on Titan, and studies of its hydrological properties suggest that it could plausibly move enough material under conditions present on Titan to account for most of Titan's fluvial features.

In addition to erosional and depositional features, the idea that Titan has an active hydrological cycle is supported by the widespread presence in Titan's northern, high latitudes of apparent lakes, presumably composed of liquid methane, ethane and other simple hydrocarbons. Though it is thought that lakes on Titan are at least partially seasonal in nature, whose changes are driven by seasonal volatile transport between Titan's polar regions, lake-like features are also seen in Titan's south-polar regions, at least one of which (Ontario Lacus) has been shown to contain liquid ethane.

Impact craters, a central feature of the geology of icy satellites, exist on Titan too, but they are small in number, and only a few have been positively identified. Titan's thick atmosphere may have prevented smaller impactors from reaching Titan's surface, but cannot be responsible for the lack of larger impact craters seen on Titan, suggesting that many craters have been removed by erosion, burial, or relaxation, further arguing that Titan has a young surface, perhaps no older than 500–1000 million years.

For obvious reasons, the geology of areas at and near the Huygens landing site has been the subject of much study. No liquid was seen near the Huygens landing site, but there is abundant evidence of fluvial activity. There are complex systems of channels cut into brighter, presumably highland materials, draining into an area that is darker and smoother, and is perhaps a large, dry lake bed. The lake bed, if that it is what it is, seems much too large to have been fed by the observed, neighboring channels, and perhaps is the geological remnant of a much wetter period in Titan's history or of catastrophic flooding. Cobbles seen in what is thought to be a stream bed near where the Huygens probe landed are strongly suggestive of past fluid flow, and channel systems of various morphologies suggest fluid flow driven by precipitation and/or springs. Overall, the geology of the Huygens landing site is strongly suggestive of fluvial erosion and deposition.

Titan's thick atmosphere, with its complement of methane and nitrogen gases, both spectrally active in the near infrared where surface-composition measurements from reflected light enjoy their greatest efficacy, has made determination of the composition of Titan's surface problematic under the best of conditions. The most detailed and accurate understanding of the composition of one small area on the surface of Titan comes from the measurements of the Huygens probe immediately before and after it landed. It detected small amounts of ethane, and perhaps benzene, cyanogen, and carbon dioxide, possibility liberated by heat transferred to the soil by the probe when it landed. The most plausible interpretation of infrared spectroscopy of the surface near the probe landing site is that Titan's surface is composed of a mix of hydrocarbons, nitriles and water ice. Though controversial, one interpretation of infrared spectroscopy of Titan's surface by the VIMS instrument is that the darker units on Titan are a mixture of water ice and organics, while the bright units are mostly hydrocarbons mixed with little or no water ice. Color measurements using three bands in the infrared centered on Titan's near-infrared atmospheric windows show that there are three main compositional units on Titan, the so-called "dark brown", "dark blue" and "bright" terrains. The dark brown units correlate very strongly with the extensive dune material seen in radar images of Titan's equatorial regions, and a simple interpretation of the difference between the dark-blue and dark-brown units on Titan is the presence of a somewhat greater amount of water ice in the dark-blue materials (although this interpretation has been challenged).

Images of Titan obtained in Titan's 5-μm atmospheric window show a few, large areas with a high reflectance relative to the bulk of Titan's surface, which is otherwise quite dark in this wavelength region. Carbon dioxide or cyanoacetylene have alternatively been proposed as the agent responsible for the high reflectance, but both interpretations have problems when the reflectance of the 5 μm-bright-areas is analyzed in detail in other wavelength regions. Other materials proposed to be widespread on Titan's surface based on possible absorption features seen in infrared spectra are benzene, liquid methane and liquid ethane, possibly arising from atmospheric sources.

That organics are ubiquitous and compositionally dominant on Titan, possibly to a depth of several meters, is supported by both infrared and radar measurements. Passive microwave radiometry indicates that there are few or no exposures of bare water ice on Titan, except perhaps in the bottom of Sinlap crater. Radar radiometry supports the idea that most, if not almost all, of Titan's surface is "veneered" with a layer of organics.

Finally, Titan's lake-like features are arguably some of the most interesting components of the global array of compositional and geological units on Titan. Synthetic aperture radar images and passive microwave radiometry strongly suggest that most of Titan's lake-like features exist in Titan's high northern latitudes, and are filled with an organic liquid. Infrared spectroscopic measurements of Ontario Lacus, the largest lake-like feature known in Titan's apparently lake-poor southern polar regions indicate that the surface of Ontario Lacus is extremely smooth, and at least one component of the material in Ontario Lacus is liquid ethane, probably in solution with liquid methane and other light organics.

1.3.3 Volatiles (Chapters 7–9)

Measurements of the abundance of primordial argon and the abundance ratios of the stable isotopes of nitrogen and carbon are particularly important in understanding the origin and evolution of Titan's atmosphere. If Titan's nitrogen were originally in the form of molecular nitrogen captured from the pre-planetary nebula, the abundance of ^{36}Ar relative to N_2 should be about 11%, but Huygens' measurements show that it is rather about 6 orders of magnitude lower, strongly suggesting that Titan's nitrogen was incorporated originally as ammonia (NH_3) and was subsequently produced by photolysis of NH_3 by solar UV.

Titan's atmospheric $^{14}N/^{15}N$ ratio as measured by Huygens is roughly 2/3 that of the terrestrial value, suggesting preferential escape of ^{14}N from Titan's atmosphere due to the slightly higher average thermal velocity of lighter molecules. Assuming that Titan's $^{14}N/^{15}N$ started near the terrestrial value, and that the observed isotopic ratios are result of massive atmospheric escape, Titan's early compliment of nitrogen could have been 2–10 times larger than present.

As a result of Huygens' measurements of noble gas abundances in Titan's atmosphere, there are two leading hypotheses for the origin of methane on Titan. One hypothesis is that methane was delivered as part of the material out of which Titan formed. Nevertheless, the low Ar abundance and the non-detection of Xe and Kr by Huygens, argue against Titan's methane being delivered as part of the original material out of which Titan formed, unless the low noble-gas abundances are the result of some process that has sequestered most of those gases in the interior of Titan. An alternative hypothesis involves serpentization, a process where hydration of silicate minerals liberates hydrogen which then reacts with carbon compounds such as CO and CO_2 to form methane; which, if either, of these processes is responsible has yet to be resolved.

Regardless of its ultimate origin, methane is the second most abundant constituent of Titan's atmosphere, and without the greenhouse heating provided by the methane, Titan's atmosphere would condense. Methane has a distinct meteorological cycle in Titan's troposphere, and photolysis destroys methane irreversibly in its stratosphere and ionosphere. Thus, the long-term presence of methane in Titan's atmosphere implies a source, probably episodic outgassing from Titan's interior at levels where it can again become a greenhousing agent and participate in photochemistry.

The meteorological role of methane in Titan's atmosphere is similar to that of water vapor on the Earth. Sub-saturated parcels of air convectively ascend to a level in the atmosphere where the condensable species saturate and condense. Clouds then form, and the condensable material rains out of the atmosphere and wets the surface, where the condensable may then be heated, re-evaporate and start the whole process again. Such considerations for Titan indicate that methane rainfall in Titan's equatorial regions would be less than 10 cm per year on average (equivalent to a terrestrial desert). On 100- to 1,000-year timescales, large storms could form, and thus transient, localized rainfall could be much larger. As on the Earth, greater precipitation is expected in the polar regions; the presence of lakes existing mostly above about 80°N on Titan, combined with a 2–4 K equator-to-pole temperature difference supports this. The even larger temperature difference between summer and winter poles, suggests that seasonal changes on Titan should result in pole-to-pole circulation of large amounts of methane on time scales of 10–100 years.

Photolysis of methane combined with rapid escape of the liberated hydrogen causes irreversible loss of Titan's atmospheric methane. Assuming that Titan's atmospheric methane has been periodically replenished over its lifetime, the results of 4.5 billion years of photochemistry on Titan probably have had a profound effect on the composition, appearance and rheology of its surface materials. One major photochemical product is ethane, but its production as an end product occurs at far lower levels than previously thought. Much of the ethane is converted to heavier hydrocarbons, which materials probably compose the many aerosol layers in Titan's atmosphere.

Complex organic chemistry has been occurring in Titan's atmosphere since its formation, producing a wide variety of molecules, including complex macromolecular material similar to laboratory tholins. On Titan's surface exchange processes with the atmosphere and the sub-surface can occur, allowing macromolecular material to further evolve. For example, hydrolysis could produce a wide variety of organics and oxygenated compounds, including many of biological interest, such as amino-acids and urea. Titan's lakes may also play an important astrobiological role because they could accumulate and concentrate important macromolecular species. Additional energy sources such as cosmic rays could allow macromolecules to chemically evolve in a lake. Such chemical evolution may also occur in Titan's sub-surface ocean where early conditions inside Titan could foster the development of prebiotic chemistry and the possible emergence of life.

1.3.4 Atmosphere (Chapters 10–14)

CH_4, the second most abundant molecule in Titan's atmosphere, has an abundance of 5% near the surface, falling to 1.4% in the stratosphere, and increasing again to 12% at Titan's exobase. Titan's troposphere has a well defined tropopause and a stable lower stratosphere with high static stability. The lower stratosphere is cold over the winter pole and warm over the summer pole, while in the middle stratosphere, the temperatures are highest at the equator. Temperature profiles of Titan's atmosphere above 500 km altitude are essentially isothermal at ~170 K with 10 K thermal waves superimposed. It is not clear whether Titan has a mesopause.

CH_4 photolysis in Titan's thermosphere, catalytic reactions in its stratosphere, and dissociation of N_2 by UV and charged particles produce C_2H_6 and HCN as the dominant hydrocarbon and nitrile, respectively. Most photochemical products (except ethylene) increase with altitude at equatorial and southern latitudes, indicative of transport from a high-altitude source to a condensation sink in the lower stratosphere. Northward of 45°N, most atmospheric products are enriched as a consequence of subsidence in the winter polar vortex, particularly for nitriles and more complex hydrocarbons than C_2H_6 and C_2H_2. North of 45°N, most products have smaller increases with altitude than at low latitudes.

Titan has a substantial ionosphere even at altitudes below 1,000 km, and a rich and complex ion-neutral chemistry exists in both Titan's ionosphere and thermosphere. Titan's ionosphere hosts a very large number of ion species, both predicted and unpredicted, including the exciting discovery of negative ions. Reactions driven by solar extreme ultraviolet, x-rays, and energetic magnetospheric electrons drive a complex chemistry that has important effects all the way to Titan's lower atmosphere. Cassini measurements confirm that molecular nitrogen and methane are the major components of Titan's ionosphere and thermosphere, as well detecting molecular hydrogen, acetylene, ethylene, benzene, and propane. Cassini measurements show that electron densities and temperatures significantly exceed the temperature of neutrals in Titan's ionosphere.

Huygens measurements of the average properties of aerosols in Titan's atmosphere show the typical aerosol particle to be an aggregate of much smaller particles (monomers) roughly 50 nm across. Calculations indicate that the average haze particle in Titan's atmosphere is composed of about 3,000 monomers whose radius by equivalent-spherical volume is 0.72 μm, whereas the radius of a spherical particle with the same projected area is 2.03 μm. Titan's aerosols are strongly forward scattering, and their vertical distribution generally falls off with altitude, but there are several "detached" haze layers between roughly 500 and 600 km altitude that are time variable both in latitude and altitude. There is speculation that such variations may be the result of wave phenomena in Titan's atmosphere, but modeling is required to determine whether such speculation is justified. Seasonal variations in Titan's haze distribution, seen in Voyager data are quite similar to what the Cassini instruments found between 2004 and 2008, supporting the idea that Titan's global circulation affects the distribution of aerosols.

Titan's atmosphere is in global cyclostrophic circulation, with a strong seasonal modulation in the middle atmosphere. There are zonal winds mostly in the sense of the satellite's rotation, and a strong, winter, circumpolar vortex with maximum winds of 190 m s^{-1} at mid northern latitudes near 300 km. Seasonal effects may be responsible for stratospheric zonal winds and temperatures being symmetric about an axis that is offset from Titan's rotation axis by about 4°, and subsidence may occur in the north-polar region during winter and early spring. Meridional contrast in tropospheric temperatures at all latitudes is ~5 K at the tropopause and ~3 K at the surface, implying axisymmetric meridional circulations and efficient heat transport. The effect of the methane "hydrological" cycle on the atmospheric circulation is not well understood.

Huygens' probe measurements show, zonal winds on Titan are eastward down to 7 km, then shift westward, and eventually shift eastward again below 1 km and down to the surface. Eastward winds at Titan's surface are consistent with Titan's equatorial dune fields occurring in a mean eastward zonal wind, but unlike those of the Earth which are westward above the surface.

Spacecraft and groundbased observations since the early 1980s suggest that Titan will soon enter a period of rapid seasonal change as its season moves from equinox toward northern summer. As seen from Earth, Titan should start to appear darker as the sub-solar point moves northward. The polar vortex is expected to break up, meridional circulation is expected to reverse, haze should move from Titan's northern hemisphere to its southern hemisphere, and cloud activity, with accompanying vigorous precipitation, all should occur as northern summer approaches. So far Cassini measurements have shown little evidence of the expected change, except perhaps for the disappearance of methane clouds seen over the south polar regions immediately after Cassini's orbital insertion, and the putative loss of some of Ontario Lacus' complement of liquid hydrocarbons. This is possibly not surprising because Cassini arrived during late southern summer on Titan. Nevertheless, Cassini is positioned with its Equinox and Solstice missions to obtain data which, when combined with diligent ground-based observations, will surely reveal exciting and interesting changes on Titan as we move toward its northern summer solstice on 2017.

1.3.5 Magnetospheric Interactions (Chapters 15 and 16)

As mentioned above, Titan's $^{15}N/^{14}N$ ratio suggests a large amount of atmospheric escape. There are several atmospheric escape processes that could have contributed to Titan's low $^{15}N/^{14}N$ ratio, among which are thermal escape, chemical-induced escape, slow hydrodynamic escape, pick-up-ion loss, ionospheric outflow and plasma-ion-induced atmospheric sputtering. Cassini data indicate that hydrogen liberated by photolysis of methane is escaping from Titan. Methane and nitrogen are also escaping from Titan at rate much larger than expected, and much larger than the measured ion loss rates, suggesting that much of the present atmosphere by now should have been lost to space. Slow hydrodynamic escape models seem unable to explain the large escape rates, and neither is the composition of the magnetospheric plasma at Titan's orbit consistent with the suggested maximum loss rates for carbon.

Solar UV, Saturn's magnetosphere, solar wind and galactic cosmic rays drive most of Titan's atmospheric chemistry, aerosol formation, and atmospheric loss, with solar UV dominating at lower altitudes (roughly from about 1,200–400 km), while magnetospheric plasma interactions are more important higher in Titan's atmosphere (~1,400 km). Heavy ions can penetrate below 950 km and cosmic rays (>1 GeV) deposit most of their energy around 70 km. Titan has an induced magnetic field resulting from its interaction with Saturn's magnetosphere which channels energy into Titan's upper atmosphere. Saturn's magnetosphere displays large and systematic changes with local time driving significant changes in the character of the interaction with Titan depending upon Saturn local time. Sub-storms in Saturn's magnetosphere caused by the solar wind can, in turn, modify the magnetospheric interaction with Titan. Energy input from a number of interactions of Titan's atmosphere with Saturn's magnetosphere likely produces large positive and negative ions below ~1,100 km altitudes, some of which have masses exceeding 10,000 Da, and which may act as seed particles for the aerosols observed below 1,000 km altitude. These aerosols likely fall to Titan's surface as polymerized hydrocarbons with trapped free oxygen, possibly participating in a range of, as yet unknown, chemical interactions.

1.4 Open Questions

Chapters 17 and 18 relate both present and future plans for Cassini observations of Titan and concept studies for a return to the Saturn system with in-depth exploration of Titan using an orbiter, balloons and landers. At the time of writing of this chapter, Cassini is executing its Equinox Mission, and the Cassini Solstice Mission, which is in the planning stages, and is likely to be approved for funding soon, carrying with it the future prospect of much deeper and more comprehensive studies of Titan and the Saturn system. Nevertheless, there will be a large number of questions left unanswered, which is as it should be, since diligent scientific inquiry always raises more questions than it answers. Thus, the final section of this chapter will list a few of the major, open and un-answered questions that will likely remain after the Cassini Solstice Mission is finished. The list is not meant to be exhaustive (the reader is referred to the individual chapters for more complete discussions of such), but rather to provide the reader with a taste of the questions to be addressed by future missions and analysis of data.

1.4.1 Titan's Interior

1. How thick is Titan's ice crust?
2. Why are Xenon and Krypton not in Titan's atmosphere? Where are they?
3. Does Titan have a sub-surface ocean and how is its presence expressed in Titan's surface features or composition?
4. Where is the ammonia?

1.4.2 Titan's Geology and Surface Composition

1. What do volcanic, tectonic, and erosional/depositional features look like at optical wavelengths and at spatial resolutions of ~50 m/pixel or better?
2. What is Titan's global topography, especially around volcanoes and tectonic features?
3. What are the details of organic chemistry taking place on Titan's surface, and has Titan's surface hosted pre-biotic chemical evolution?

1.4.3 Titan's Volatiles

1. What is the origin of Titan's methane, how is it destroyed and how does it cycle through Titan's surface, atmosphere and interior? What is D/H in the surface H2O ice, $^{16}O/^{18}O$ in CO, CO2 and H2O, $^{36}Ar/^{38}Ar$, $^{12}C/^{13}C$ in CO, CO2 and surface organics? What are the abundances of Xe and Kr to at least the 10^{-10} mole fraction precision in the atmosphere and in the surface material?
2. What are the processes and pathways producing complex organic molecules in vapor, liquid and solid phases from high in Titan's ionosphere down to its surface?
3. What is detailed composition of the liquids and sediments in the lakes? How have the lakes evolved with time?
4. How do episodic events such as cryovolcanoes, fumaroles and impacts affect Titan's surface and atmosphere?
5. If Titan has a sub-surface ocean, what is its nature and astrobiological potential? Does it communicate with the surface through volcanic or tectonic processes?

1.4.4 Titan's Atmosphere and Ionosphere

1. What is the higher-mass, ion chemistry in Titan's ionosphere and how is it linked to aerosol formation?
2. What are the sources or sinks of negative ions?
3. How are higher order hydrocarbon species such as benzene formed, and what are their roles in aerosol formation, and in both neutral and ion chemistry?
4. What is the composition of Titan's aerosols, how are they formed, and what are their distributions in time and space on Titan?
5. Are stratospheric condensates important in Titan's atmosphere? What physical, chemical, and dynamical processes are important in the mesosphere above 400 km?
6. What happens to the polar vortex when the winter hemisphere moves into summer? What drives the global super-rotation on Titan? Are there planetary-scale waves on Titan? Is there really a methane hydrological cycle on Titan and do the "lakes" migrate seasonally?

1.4.5 Titan's Magnetospheric Interactions

1. What are the dominant processes responsible for atmospheric escape on Titan and what is the true escape rate?
2. As Saturn and Titan progress from southern summer to equinox, how do the various energy input processes evolve?
3. How do Enceladus' plumes affect Saturn's magnetosphere, especially if their water-vapor output varies with time?
4. How does the chemistry being driven by magnetospheric interactions changes in Titan's upper atmosphere with time and season?

The questions above are but a small subset of important questions identified in the various chapters of this book. It is hoped that they will whet the reader's appetite for the feast of knowledge that waits in the pages that follow.

References

C. T. Russell, ed., The Cassini–Huygens mission Volume 1 – Overview, objectives and huygens instrumentation. (2002) Space Sci Rev 114(1–4)

C. T. Russell, ed., The Cassini–Huygens mission Volume 2 – Orbiter in situ investigations. (2004a) Space Sci Rev 114(1)

C. T. Russell, ed., The Cassini–Huygens mission Volume 3 – Orbiter remote sensing investigations. (2004b) Space Sci Rev 115(1–4)

Chapter 2
Earth-Based Perspective and Pre-Cassini–Huygens Knowledge of Titan

Athena Coustenis, Emmanuel Lellouch, Bruno Sicardy, and Henry Roe

Abstract This chapter sets the scene for the current investigation of Titan with Cassini–Huygens, by reviewing the steps that took us there, from the first glimpses through a small telescope to the satellite observations passing by the first hints of an atmosphere about a 100 years ago.

2.1 Context/Introduction

Titan, Saturn's biggest moon, and (by a narrow margin) the second in size among the satellites in our Solar System, was discovered in 1655 and has ever since attracted a lot of attention among scientists and public alike. It has been known for a century now that Titan has a substantial atmosphere. Catalan astronomer José Comas Solà claimed in 1908 to have observed limb darkening on Titan (Comas Solá 1908). Confirmation came from spectroscopic observations by Gerald Kuiper in the 1940s, leading to the discovery of methane (Kuiper 1944), but it was not until the Voyager 1 spacecraft visited Titan in 1980 that the dominantly nitrogen-methane composition and 1.5 bar surface pressure were established. Furthermore, complex organic chemistry is active there, producing multiple layers of orange-colored haze, which render the atmosphere opaque in the optical range of wavelengths. It took a long time and a lot of effort to get a glimpse of the surface, and ground-based observations played a significant role in defining what we know today to be quite a complex landscape.

To address the many questions asked about Titan over the centuries since its discovery, a series of space probes was developed and dispatched towards this intriguing body. Pioneer 11 arrived first, in 1979, followed by Voyager 1 a year later. The scientific understanding of Titan as a planet-like object that emerged from the analysis of Voyager data was improved by ongoing Earth-based observations, using increasingly more powerful optical and spectroscopic techniques, such as radar and adaptive optics, and advanced platforms on Earth-orbiting space observatories. At the same time, new theoretical models were being developed to account for these observations, new theories proposed and debated, and old or repeated measurements re-analyzed.

The latest envoy to Titan, a large and sophisticated international space mission called Cassini–Huygens, was launched in the year 1997. It arrived in July 2004, and started gathering new measurements from an orbit around Saturn that was designed to permit multiple Titan encounters. The Huygens probe descended in Titan's atmosphere on January 14, 2005 and recorded breathtaking data, revealing an intriguing new world, and the most distant one to be landed on by a human-made machine.

2.2 Early History of Titan Observations and First Interpretations

The story of the first detection of Titan is a classic of its kind. On the night of March 25th, 1655, a novice Dutch astronomer, Christiaan Huygens, pointed his telescope at Saturn. According to his notes, Huygens (1659) saw a small star 3 arc minutes away from the planet and almost immediately guessed it was a satellite. He confirmed his guess a few days later when the 'star' had moved. Titan was soon realized to be a sizeable object. Huygens believed Titan to be the biggest satellite of all, and that is the reason for the name it was given, following a proposition by Herschel (Herschel 1847). In Greek mythology, the Titans were brothers and sisters of Kronos, the Greek equivalent of the Roman god Saturn. In that same publication, Herschel named the six other then-known moons of Saturn after individual Titans.

After Solà's claims of having observed an atmosphere around Titan in 1908, James Jeans decided in 1925 to include Titan and the biggest satellites of Jupiter in his theoretical study of escape processes in the atmospheres around solar system objects. His results (Jeans 1931) showed that Titan could have kept an atmosphere, in spite of its small size and weak gravity, if low temperature conditions — that he evaluated as between 60 and 100 K — prevailed. In this case, a

A. Coustenis (✉)
e-mail: athena.coustenis@obspm.fr

R.H. Brown et al. (eds.), *Titan from Cassini-Huygens*,

gas of molecular weight higher than or equal to 16 could not have escaped Titan's atmosphere since the satellite's formation. Jeans' calculations demonstrated that some of the lighter gases such as hydrogen and helium should have escaped, but that several constituents could have been present in non-negligible quantities in the mix of gas and dust particles that condensed to form the Solar System and could have been retained by Titan. The thermodynamical theory of gases then points toward ammonia, argon, neon, molecular nitrogen and methane. Ammonia (NH_3) is solid at the estimated Titan temperature and could therefore not substantially contribute to its atmosphere. The others, however, are gaseous within this same temperature range. Methane (CH_4), unlike argon, neon and molecular nitrogen, exhibits strong absorption bands in the infrared spectrum, which make it relatively easy to detect.

The first formal proof of the existence of an atmosphere around Titan came in 1944, when Gerald Kuiper observed Titan with the new McDonald 82-in. telescope and discovered spectral signatures on Titan at wavelengths longer than 0.6 μm (microns), among which he identified two absorption bands of methane at 6190 and 7250 Å (Kuiper 1944). This discovery was significant not only because it requires a dense atmosphere with a significant fraction of methane, but also because the atmosphere needs to be chemically evolved, since methane requires hydrogen in the presence of carbon, and molecular and atomic hydrogen would have escaped from Titan's weak gravitational field since the formation of the solar system. By comparing his observations with methane spectra taken at low pressures in the laboratory, Kuiper (1952) derived an estimate of the amount of methane on Titan of 200 m-amagat. It was more than a factor 10 less than the current known CH_4 column density, however, not that far off when the one considers that the solar light is reflected by the haze. Kuiper searched for similar behavior in the spectra of other Saturnian satellites, but his data, obtained in 1952, did not show the methane bands. Kuiper concluded that Titan was a unique case in the Saturnian system due to the presence of an atmosphere, of a composition such that it gave the satellite an orange color.

In the years that followed, in spite of much interest, it proved difficult to make significant further progress in exploring or comprehending Titan's atmosphere.

2.3 Pre-Voyager Observations and Predictions

The pre-Voyager and Voyager observations are thoroughly discussed in Hunten et al. (1984) and Owen (1982), and will be summarized here. Until the 1970s, few significant advances were made in the study of Titan, but from 1972 to 1979 a number of scientists concentrated their efforts on seeking a better estimate for the methane abundance and the surface pressure, using observations made in the 1-to-2 μm infrared spectral region. Limb darkening was finally unambiguously observed (Elliot et al. 1975), consistent with an optically thick atmosphere. Radio and infrared measurements indicated ground temperatures that ranged from 165 to 200 K (Low 1965; Gillett et al. 1973; Briggs 1974).

At about this time, Trafton (1972a, 1975) announced the identification of a spectral absorption feature of molecular hydrogen, H_2, on Titan, for which he evaluated an abundance of 5 km-amagat (Fig. 2.1). Kuiper (1952) had already pointed out that a body as small as Titan could not retain hydrogen, and therefore Trafton's discovery clearly pointed to the presence of another gas which would inhibit H_2 escape. Introducing the idea of "limiting flux", Hunten (1973a,b) considered the possibility of methane or nitrogen as the major gas (the latter, which is the correct one, was considered following an idea first voiced by Lewis (1971)).

Trafton (1972b) further conducted observations of the $3\nu_3$ spectral band of methane at 1.1 μm, in which he found an unexpected strong absorption, indicating either a methane abundance at least 10 times higher than that inferred by Kuiper, or a broadening of the CH_4 bands induced by collisions with molecules of another, as yet undetected, but quite abundant, gas in the atmosphere. In either case, the intensity of the absorption band is a function of the methane abundance and of the local atmospheric pressure. By comparing the weak absorption bands of methane in Titan's visible spectrum with spectra of Jupiter and Saturn, in which these bands have almost identical absorption strengths, Lutz et al. (1976) derived from Trafton's measurements a 320 m-amagat abundance for methane, and an estimate for the effective pressure on Titan of about 200 mbar. The immediate consequence of this result was that methane turned into a minor atmospheric component, since even 1.6 km-amagat (Trafton's highest estimate) could only correspond to a partial pressure of methane of 16 mbar.

In spite of much effort directed at the detection of NH_3, the observers of the time failed to produce more than upper limits, which got lower and lower with successive measurements, suggesting that if any of this gas existed on Titan it must either have been photodissociated, with subsequent production of N_2 and H_2, or else trapped on the surface as ammonia ice.

All of the above led to a model (completed by Hunten (1977)), which started with the assumption that dissociation of ammonia should produce molecular nitrogen, which is transparent in the visible and infrared spectrum, in large quantities. In this model the surface temperature and pressure would be quite high (200 K and 20 bar). These high temperatures on the ground could be explained by a pronounced greenhouse effect, resulting essentially from

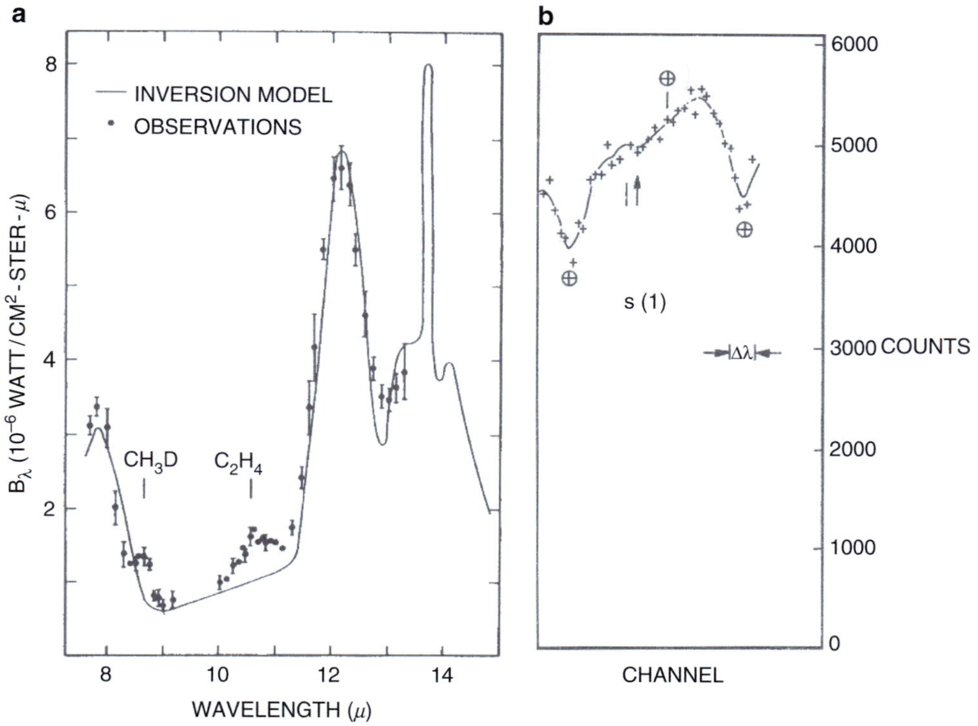

Fig. 2.1 Examples of first pre-Voyager detections of Titan's atmospheric components and models (Gillett et al. 1973 – left – and Trafton 1972a, 1975 – right —). CH_3D and C_2H_4 spectral signatures are identified in the Gillett spectra covering the thermal infrared range. In the example of the spectra obtained by Trafton, the absorption caused by the 3-0 S(1) line of H_2 at 8150.7 A is indicated by an arrow.

pressure-induced opacity in hydrogen at wavelengths longer than 15 μm (Sagan 1973; Pollack 1973).

However, the greenhouse effect was put to the test when Danielson et al. (1973) observed very low UV albedo which implied the presence of an absorbing haze. More detailed photometry (Harris 1961) identified the material composing the haze as red and UV absorbing. Polarization measurements indicated that this material is distributed as a thick layer around Titan (Veverka 1973; Zellner 1973). Indeed, the polarization sense of the reflected radiation was consistent with a deep atmosphere composed of thick haze layers, as opposed to a transparent atmosphere. Caldwell (1977) pointed out that this haze would cause significant heating of the middle atmosphere and lead to a warm stratosphere. These observations of Titan's low albedo and of the positive polarisation of the reflected light, then confirmed the presence of a thick, cloudy atmosphere, with the cloud particles present up to high altitudes. Sagan (1971) was the first to hypothesize that these layers might be composed of complex organic molecules, and this was subsequently backed up by experimental (Sagan 1973; Khare and Sagan 1973) and observational evidence. Indeed, Gillett et al. (1973) and Gillett (1975) found signatures in Titan's thermal emission spectrum not only of methane (CH_4), but also of ethane (C_2H_6) at 12.2 μm, monodeuterated methane (CH_3D, at 9.39 μm), ethylene (C_2H_4, at 10.5 μm) and acetylene (C_2H_2, at 13.7 μm), see Fig. 2.1, thus confirming the emitting stratosphere hypothesis and the chemical richness of this stratosphere. Theoretical considerations suggested that two sorts of aerosols were expected to co-exist in Titan's atmosphere: clouds of condensed CH_4, and a photochemical fog of more complex condensates. The latter would arise as a result of methane photolysis, that is, dissociation by sunlight, mostly at ultraviolet wavelengths. The fragments of methane, CH_2, CH_3 etc., combine, leading to the production of a variety of polymers that condense to form oily droplets. The quantities of organic molecules synthesized over geological time were estimated to be on the order of hundreds of g cm^{-2} (Sagan 1974; Hunten 1977; Chang et al 1979).

Just prior to the Voyager encounter, Jaffe et al. (1980) made radio telescope observations with the newly-completed Very Large Array in New Mexico, and obtained an emission temperature of the surface of 87 ± 9 K, a range that includes the modern value. This brightness temperature implied a surface temperature of about 90 K. They even suggested that conditions on Titan might support oceans of methane, an idea that was ahead of its time, but the paper failed to get the attention it deserved as it was published during the excitement of the Voyager encounter. Hunten (1978) applied the 90 K surface temperature to his model and derived a surface pressure of ~2 bar, which is fairly close to what Voyager found.

Therefore, a second model was in vogue prior to the Voyager encounter (Caldwell 1977), which favored methane as the main component (about 90%) of the atmosphere, and predicted surface conditions of T = 87 K at a pressure of 20 mbar, with a temperature inversion in the higher atmospheric levels, demonstrated by the presence of emission features of hydrocarbon gases in the infrared spectrum of Titan.

The majority of the pre-Voyager chemical models concentrated exclusively on hydrocarbons (Strobel 1974; Allen et al 1980) and, although there were several researchers (Lewis 1971; Hunten 1978; Atreya et al. 1980) considering the effect of a massive nitrogen atmosphere on these reactions, the first serious attempt at modeling Titan's atmospheric chemistry in the presence of N_2 was not made until 1982 (Strobel 1982).

Close examination of Titan's visible and near-infrared spectrum had already revealed at this time that the continuum absorption decreased with frequency, suggesting that the aerosol became more transparent at longer wavelengths. This led to the assumption that it might be possible, at certain frequencies in the near infrared, to probe all the way down to the satellite's surface. Fink and Larson (1979) thus produced the first characterization of the lower atmosphere and the surface of Titan by spectroscopic observations in what is now called the "methane windows" in the near-infrared (between 0.8 and 5 μm). At shorter wavelengths, and down to about 2,200 Å, the brightness stays nearly constant, suggesting that the aerosol is uniformly mixed at high altitudes. The measurements say little about the nature of the aerosols, their composition for example, but the fact that they are present in the atmosphere makes all attempts to interpret spectroscopic observations extremely dependent on assumptions about the haze properties.

2.4 The Voyager Mission to Titan

The previous section describes the situation before Voyager 1 flew by Titan in 1980. Voyager was not the first visitor from Earth to the Saturnian system, as the ringed planet had been visited by a small, unmanned Pioneer probe in 1979. The Pioneer 11 trajectory carried it across the orbit of Titan one day after its closest approach to Saturn, on September 2, 1979, at a distance from the satellite of 363,000 km. This was the first man-made object to enter the realm of Saturn, and it showed the way was safe for Voyager.

The Voyager 1 and 2 missions were launched in 1977. The Voyager 1 encounter with Saturn and Titan took place in November 1980, while Voyager 2 arrived in the Saturnian system in August 1981, some 9 months later. Although not without interest, the data relative to Titan obtained by Voyager 2 are not as extensive as those taken by Voyager 1, because the closest approach distance of Voyager 2 was more than 100 times greater. The closest approach of Voyager 1 to Titan took place on 12 November, 1980, at 6,969 km (4,394 miles) from the satellite's centre. The orbital plane of Titan was crossed from north to south, the spacecraft trajectory inclined with respect to the orbital plane at about 8.7°, at a speed with respect to the satellite of 17.3 km s^{-1}.

A very advanced machine for its era, the Voyager spacecraft carried 105 kg of scientific payload and was capable of operating with a high degree of autonomy at vast distances from the Earth thanks to a 3.7-km antenna used for telecommunication and radio science. For a detailed description of the Voyager missions see Stone and Miner (1981) and http://voyager.jpl.nasa.gov/index.html.

Each Voyager carried the same eleven scientific instruments, four of them mounted on the movable scan platform so they can be pointed at specific targets. The latter were the imaging experiment, consisting of boresighted narrow- and wide-angle cameras; the infrared interferometer spectrometer and radiometer (IRIS); the ultraviolet spectrometer (UVS); and a photopolarimeter-radiometer (PPR). Six other instruments were used to study fields, particles and waves in interplanetary space and near planets, including magnetometers, a plasma detector, a low-energy charged particle detector, a plasma wave detector, a planetary radio astronomy instrument, and a cosmic ray detector. In addition, the spacecraft's radio antenna provided radio scientific investigations (such as radio-occultations with the Radio Science Subsystem (RSS)).

2.4.1 *Visual Appearance and Haze Properties*

The arrival of the Voyager 1 and 2 spacecrafts, respectively in November, 1980 and August 1981, i.e. close to Titan's northern spring equinox, provided definite answers to many of the speculations of that time. In terms of its visual appearance (Smith et al. 1981, 1982), Titan appeared rather disappointing, being entirely enshrouded by a reddish optically thick haze that hindered images of the surface at the time (Fig. 2.2). Total optical depth of the haze was quickly estimated to be >5, and the peak haze brightness was found to occur at about 240 km above Titan's surface. Still, Titan's appearance was not completely bland. It clearly showed a hemispheric asymmetry (Fig. 2.2), with the Southern hemisphere brighter (by about 25% at blue wavelengths), less red, and more uniform than the Northern hemisphere. The latter showed hints for a banded structure and appears surmounted by an even darker polar hood. Early explanations for the hemispheric asymmetries invoked variations in haze density, particle size, and or composition. They were attributed to dynamical effects, caused by seasonal variations in the dis-

Fig. 2.2 Voyager 2 image showing the N/S asymmetry, polar collar and detached haze (Smith et al. 1982).

tribution of solar heating with a ~1 season time lag related to a long radiative time constant of the atmosphere as a whole (Smith et al. 1981). The other feature that was clearly present in the Voyager images was the "upper" (or "detached") haze. In average, the peak of the detached haze was located roughly 100 km above that of the main haze (i.e. at 300–350 km), though at high northern latitudes, the two layers appeared to merge, due to an increase of the altitude of the main haze and a lowering of the upper haze. In high phase-angle observations (Rages and Pollack 1983), haze particles were detected as high as 500 km.

Determining the haze distribution (particle size and number as a function of altitude) was the subject of considerable effort after the Voyager encounters. Data from Pioneer 11 and Voyager photopolarimetry showed large polarization at 90° phase angle, requiring small particles (<0.1 μm for spherical particles), while the Voyager phase-angle observations rather implied particle sizes of 0.2–0.5 μm. Although it was quickly noted that haze particles are not necessarily spherical (e.g. Asano 1979), models based on this idea and later on the possibility that aerosols may be composed of fractal aggregates had to await the development of more powerful computing facilities (e.g. West and Smith 1991, Rannou et al. 1995). These models, which described the haze particles as the aggregation of many tens of ~0.05 μm monomers, provided a solution to the paradox, as the large size of the aggregates permits allows for the reproduction of the strong forward scattering while preserving the large polarization. The Voyager images, in themselves, did not provide much information on the vertical distribution of haze below the optical radius. This was estimated from microphysical models and fits of Titan's geometric albedo in the visible and near-infrared (Neff et al. 1984). Results seemed to indicate a cutoff of the haze below ~70 km. This behavior was deemed reasonable due to the expected coagulation and sedimentation of haze particles in the lower atmosphere, as well as their likely scavenging by organic ices in the lower stratosphere (see below), but ultimately turned out not to be confirmed by Huygens (see Chapter 12).

2.4.2 Atmospheric Bulk Composition and Mean Thermal Structure

Titan's thermal structure was mostly revealed by the Voyager 1 radio-occultation (RSS) data (Lindal et al. 1983). To first order, these data constrained the ratio of atmospheric temperature T to mean molecular mass m and as such, they required an independent knowledge of the gas composition. Using Voyager infrared (IRIS) data, coupled with radiative transfer calculations, it was demonstrated that Titan's surface temperature had to be the range 94 K < T < 97 K (Samuelson et al. 1981), a result that was largely independent of atmospheric composition and opacity sources. In combination with the T/m determination from RSS, this implied that near the surface m was close to 28 amu, indicating N_2 or CO as the major atmospheric gas. In addition, based on solar occultation and airglow measurements (showing emission lines due to molecular and atomic nitrogen), the Voyager ultraviolet experiment (UVS) provided evidence that N_2 was the dominant constituent and methane only a minor constituent, and provided upper limits on argon and carbon monoxide at the several to ten percent level (Broadfoot et al. 1981; Smith et al. 1982; Strobel and Shemansky 1982; Strobel et al. 1993). These latter observations pertained to the upper atmosphere, 800–1,200 km above Titan's surface. These early findings justified that the first analyses of the RSS data were conducted for a pure N_2 atmosphere (Lindal et al. 1983). They indicated a surface pressure, temperature and radius of 1.496 ± 0.02 bar, 94 ± 0.7 K, and 2575 ± 0.5 km, respectively. A troposphere covering the first 42 km of the atmosphere was detected, with a tropopause temperature and pressure of 71.4 ± 0.5 K and 130 mbar, respectively, followed by a well-marked stratosphere extending up to at least ~200 km (~0.75 mbar), the highest level probed by the radio-occultation, where the temperature reached ~170 K. As the presence of other gases modifies the molecular refractivity as well as the mean molecular mass of the atmosphere, all the above figures are sensitive to the precise atmospheric composition. The existence of a methane-rich (a few percent of N_2) stratosphere was confirmed from the detection of 7.7

μm methane emission (Hanel et al. 1981). Lellouch et al. (1989) performed a combined reanalysis of the RSS and of the IRIS observations, accounting for the possible presence of argon as an important atmospheric constituent, and requiring the methane not be supersaturated in the stratosphere. Technically, this analysis allowed for a wide range of argon abundance (0–27%), methane surface abundance (0.5–21%, with a "nominal" surface mixing ratio and humidity of 8% and 80%, respectively), surface temperature (93–101 K), and methane stratospheric abundance (0.–3.4%). Subsequent analyses of the Voyager IRIS spectra (e.g. Courtin et al. 1995, indicating Ar < 6%, and Samuelson et al. 1997a, finding CH_4 ~ 5.7% at the surface) and ground-based observations (e.g. Lemmon et al. 2002, indicating CH_4 ~ 4% at the surface), allowed to considerably narrow this range, before final numbers could be put by Cassini/Huygens (see Chapter 10). Based on an analysis of the IRIS 20–50 μm spectra, some authors (Courtin et al. 1995; Samuelson et al. 1997a) further concluded that methane was supersaturated by a factor 1.5–2 in the upper troposphere, but this was not confirmed by Cassini/Huygens.

Constraints on the atmospheric structure above 200 km were much looser. The IRIS observations, mostly in nadir viewing, contained information on the atmospheric profile up to about 450 km (Coustenis et al. 1989a), but solution thermal profiles were not unique; for example it was not possible to unambiguously determine the altitude and temperature of the stratopause. In the upper atmosphere, temperature constraints from Voyager were even fewer, being (initially) restricted to a single temperature (186 ± 20 K)/density point at 1265 km and some abundance measurements of methane and other hydrocarbons, based on the UVS solar occultation (Smith et al. 1982). Interpolation between conditions in the upper and lower atmosphere indicated a mean ~165 K temperature, but the detailed temperature profile was uncharacterized and had to be largely deduced from theory. The methane abundance was initially measured to be 8% at 1,125 km, indicating a homopause at 925 ± 70 km (Smith et al. 1982). A reanalysis of these data by Vervack et al. (2004) led to drastically different results, with a thermospheric temperature of 153–158 K, CH_4 densities lower than previously inferred by a factor 3–7 and essentially no sign of a homopause up to at least 1,000 km.

2.4.3 Atmospheric Trace Composition and Photochemistry

The Voyager infrared data revealed the richness of Titan's atmospheric composition (Hanel et al. 1981). In addition to N_2, CH_4, and H_2 detected in the troposphere through their collision-induced features, the IRIS spectra confirmed the presence of CH_4, C_2H_2, C_2H_6 and C_2H_4 stratospheric emissions and allowed the discovery of numerous other gases in Titan's stratosphere: the more complex hydrocarbons C_3H_8 (propane), CH_3C_2H (methylacetylene) and C_4H_2 (diacetylene), the nitriles HCN (hydrogen cyanide), HC_3N (cyanoacetylene), and C_2N_2 (cyanogen), as well as carbon dioxide CO_2. The identification of most of these gases was originally made from comparison with laboratory spectra, but the determination of their abundances was based on radiative transfer models. Early values of the molecular abundances (Maguire et al. 1981; Kunde et al. 1981; Samuelson et al. 1981; Samuelson et al. 1983) were superseded by the more comprehensive analyses of Coustenis et al. (1989a,b) which made use of broader datasets and improved spectroscopic databases.

All of these species could be mapped as a function of latitude, and most of them, with the noticeable exception of CO_2 and to a lesser extent of C_2H_6, appeared to be enhanced at high Northern latitudes (Coustenis and Bézard 1995). This study indicated that the least abundant species C_4H_2, HC_3N, and C_2N_2 show the most marked latitudinal variability (the latter two being actually only detected northward of 50°). This is illustrated in Fig. 2.3, where the reported mixing ratios indicate the mean abundance of each compound above their respective expected condensation point in the lower stratosphere at 10-50 mbar (only C_2H_4 does not condense in Titan's conditions). Finally, using a Voyager/IRIS limb sequence recorded over the North Polar region, Coustenis et al. (1991) showed that the abundance of most of the compounds actually increases with altitude, an expected behavior for species formed in the upper atmosphere and diffusing downwards to their condensation sink in the lower stratosphere. Since most species are subject to photolytic losses, an early model (Yung 1987) invoked the accumulation of nitriles in shadow during the preceding winter as the cause of their northern enhancement. It was demonstrated later (Lebonnois et al. 2001) that this explanation does not hold when proper time constants are considered. Most results obtained from Voyager 1 were confirmed by the Voyager 2 /IRIS observations (Letourneur and Coustenis 1993), and no significant variations in composition were observed between Voyager 1 and 2, an expected result given that the two flybys were separated by only 9 months, i.e. about a week of a Titan season.

Another highlight of the Voyager /IRIS observations was the observation of the 8.6 μm feature at high resolution, permitting its proper assignment as the ν_6 band of CH_3D, and allowing the determination of Titan's D/H ratio (Kim and Caldwell 1982; Coustenis et al. 1989b). A clear enrichment (by a factor of 3–12) with respect to the protosolar value was observed. However, the interpretation of this enrichment (nowadays estimated much more accurately to be a factor about 6.5) has remained difficult. Pinto et al. (1986) evaluated several fractionation processes for deuterium, such as differential photolysis and escape, condensation in clouds,

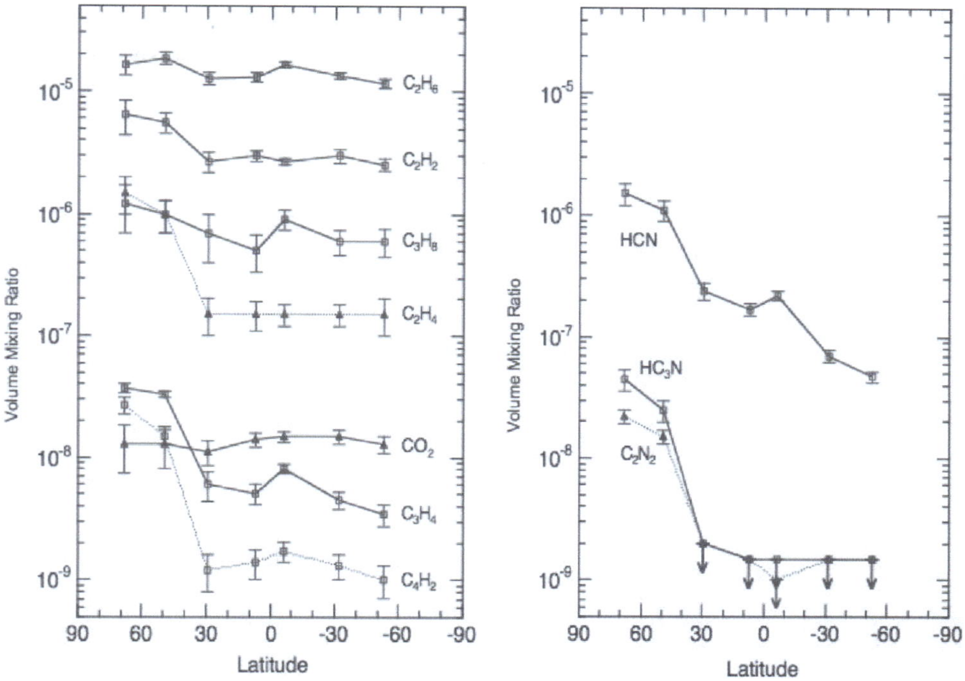

Fig. 2.3 Latitudinal distribution of gases seen by IRIS/ Voyager 1 (Coustenis and Bézard 1995).

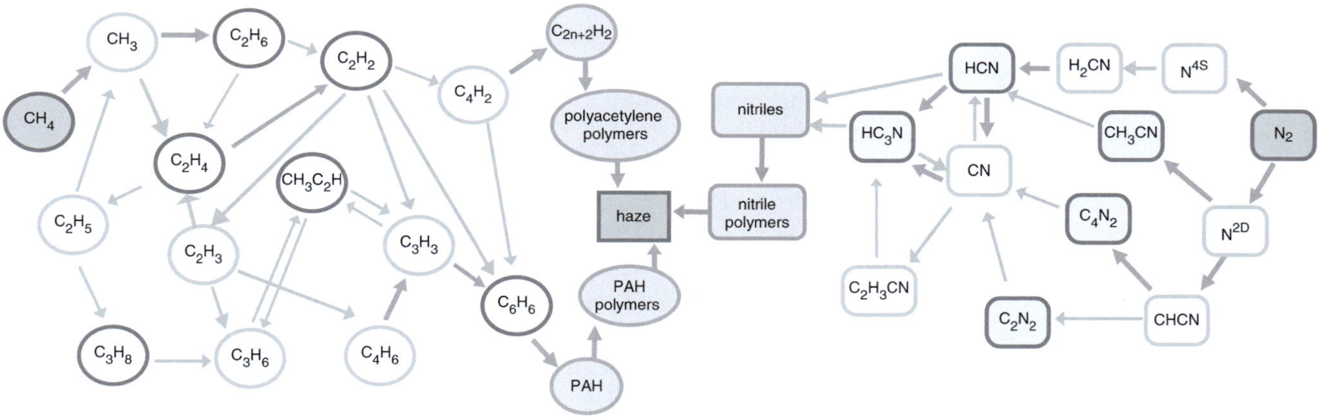

Fig. 2.4 A simplified photochemical scheme of Titan's atmosphere. From Atreya et al. 2006.

over a hydrocarbon ocean and between the ocean and a crust. They concluded that atmospheric evolution was unlikely to entirely explain the D/H enhancement, producing at most a factor ~2.2 enrichment. Lunine et al. (1999) re-examined the problem and concluded to a possible factor of 4 enhancement. This tends to imply that the D/H ratio in Titan's atmosphere was either acquired in the Saturn sub-nebula where Titan had formed, or "upstream" in the cooling solar nebula by isotopic exchange with interstellar-rich ices. More details are given in Chapter 3.

Finally, the IRIS spectra of the North Pole revealed the presence of several additional emission features, not obviously attributable to a particular gas. One of them, at 478 cm^{-1}, was identified as belonging to C_4N_2 ice (Samuelson et al. 1997b), but others, and particularly the very prominent 220 cm^{-1} feature, have remained unidentified so far.

These essential discoveries on the composition of Titan's atmosphere prompted the generation of one-dimensional photochemical models, in which vertical transport was modelled by using the phenomenological approach of eddy diffusion coefficient profile (Strobel 1982; Yung et al. 1984; Toublanc et al. 1995; Lara et al. 1996 for the early ones). The aim was to reproduce the observed composition, based on the expected chemical reactions in a N_2–CH_4 atmosphere. The main reaction schemes were quickly established (see Fig. 2.4 for a simplified but up-to-date understanding of Titan's chemistry). Hydrocarbon chemistry is initiated by the methane dissociation, that produces CH_3, CH_2 (in the $^1CH_2$ and $^3CH_2$ states), and

CH radicals. The methane dissociation occurs directly above 700 km, mostly at Ly α, or through catalytic destruction below 500 km The main photochemical product, ethane, is formed by the self-recombination of the methyl radical (CH_3); high-altitude reactions produce C_2H_4, whose high-altitude photodissociation leads to the formation of C_2H_2, which can be transported downwards to form higher-order hydrocarbons, such as C_4H_2. The C_3 compounds (C_3H_8 and CH_3C_2H) are formed from insertion of a C_1 radical in a C_2 molecule. Nitrogen chemistry is initiated by the dissociation of N_2, either from energetic (<100 nm) UV photons or from magnetospheric electrons in the upper atmosphere, and possibly from galactic cosmic ray in the lower atmosphere, producing $N(^4S)$, $N(^2D)$, N^+, and possibly N^{2+}. The most abundant nitrile, HCN, is formed from reaction of $N(^4S)$ with CH_3 or $^3CH_2$ radicals. Its photolysis leads to CN, which can react with acetylene to form HC_3N; C_2N_2 can be formed in various reactions. Finally, the presence of CO_2 required a source of external oxygen in Titan's atmosphere, in the form of water (not detected at that time), as CO_2 is best formed from CO + OH where OH is produced from the photolysis of water and CO is either of internal or external origin. With few exceptions, photolytic products are expected to condense in Titan's stratosphere and ultimately precipitate to the surface in liquid or solid form. The net results of gas-phase chemistry are thus (i) an irreversible conversion of CH_4 of N_2 into more or less complex hydrocarbons and nitriles, and (ii) the massive production of atomic and molecular hydrogen that can diffuse upwards and easily escape from the atmosphere.

These 1-D photochemical models, which were updated when more species were discovered (namely CO, CH_3CN (acetonitrile), CH_2CCH_2 (allene), H_2O, and C_6H_6 (benzene), see below), were reasonably successful as reproducing the composition of Titan's stratosphere, although for some species (e.g. CH_3CCH, HC_3N) the disagreement between the calculated and predicted abundance could reach an order of magnitude of more. A specific problem was that it turned out difficult to fit all the observed abundances with a single eddy diffusion profile. In particular, the HCN vertical profile seemed to require a larger eddy diffusion than that needed for the major hydrocarbons (Lara et al. 1996). A possible way out was to invoke an additional loss of path of HCN to the haze (Lara et al. 1999). Another important problem, still incompletely solved (see Chapter 10), was to determine the origin of the CO required to form CO_2 (and directly observed from the ground shortly after Voyager). Finally, more recent models, coupling photochemistry and dynamics (Hourdin et al. 2004), have demonstrated the limitations of 1-D photochemical models in their ability to quantitatively reproduce the composition of Titan's stratosphere and even more its latitudinal and seasonal variability.

Titan's photochemistry presumably continues to increasingly complex molecules, eventually proceeding in the solid phase to form the small particles that are the "precursors" of the haze. Evidence for that was provided in part by laboratory experiments (Khare et al. 1984) demonstrating that organic mixtures (named "tholins" after the Greek θωλ ς, meaning "muddy") produced by electron discharge in a gas representative of Titan's atmosphere composition have optical properties similar to those of Titan's aerosols, in particular their reddish colour. The agreement in fact extends over 3 orders of magnitude in wavelength and in absorption coefficient (McKay et al. 1989, 2001). However, photochemical models usually do not track the fate of heavy molecules (with more than ~6 heavy atoms). Thus, Titan's haze was quickly understood as the ultimate product of the coupled hydrocarbon-nitrile photochemistry, but essential chemical pathways are still missing, although several schemes, involving (i) polymers of acetylene and cyanoacetylene (ii) polycyclic aromatic hydrocarbons (iii) polymers of HCN and other nitriles, have been proposed, along with some parameterization (Lebonnois et al. 2002). Notwithstanding these uncertainties, microphysical and photochemical models implied that the total haze production is in the range (0.5–2) × 10^{14} g cm^{-2} s^{-1} (see review in McKay et al. 2001) and laboratory measurements performed under a variety of pressure, temperature and energy source conditions (see Coll et al. 1999 and references therein) indicated that the haze has a C/N ratio of 2–11 and a C/H ratio of 0.7–3.

2.4.4 Thermal Balance

The enormous step forward in characterizing Titan's thermal structure, composition and haze structure justified the study of its thermal balance. Based on a an analytical model, Samuelson (1983) confirmed that the temperature inversion near 40 km was due to stratospheric absorption of UV light and penetration of visible light to near the surface. Building upon these results, McKay et al. (1989) developed a radiative-convective numerical model, including a haze microphysical model, and tuned to fit Titan's geometric albedo (Fig. 2.5). They demonstrated that Titan is in radiative equilibrium in most if not all of the atmosphere, and that the key factors controlling the thermal structure below ~200 km are the absorption of sunlight by haze and the absorption of surface radiation by the far-IR pressure-induced transitions of N_2, CH_4 and H_2 in the troposphere. The haze reflects 30% and absorbs 40% of the solar light. Since it is nearly transparent in the thermal IR, it leads to an "anti-greenhouse" effect (McKay et al. 1991) amounting to −9 K and is also responsible for the well-marked stratosphere, as ozone is on Earth. The tropospheric greenhouse gases, on the other hand, warm Titan's surface by 20 K. Overall, the surface is heated by 11 K over its equilibrium temperature of 84 K. One of the crucial side results of the McKay et al. (1989) study was the demonstration that about 10% of the solar light reaches

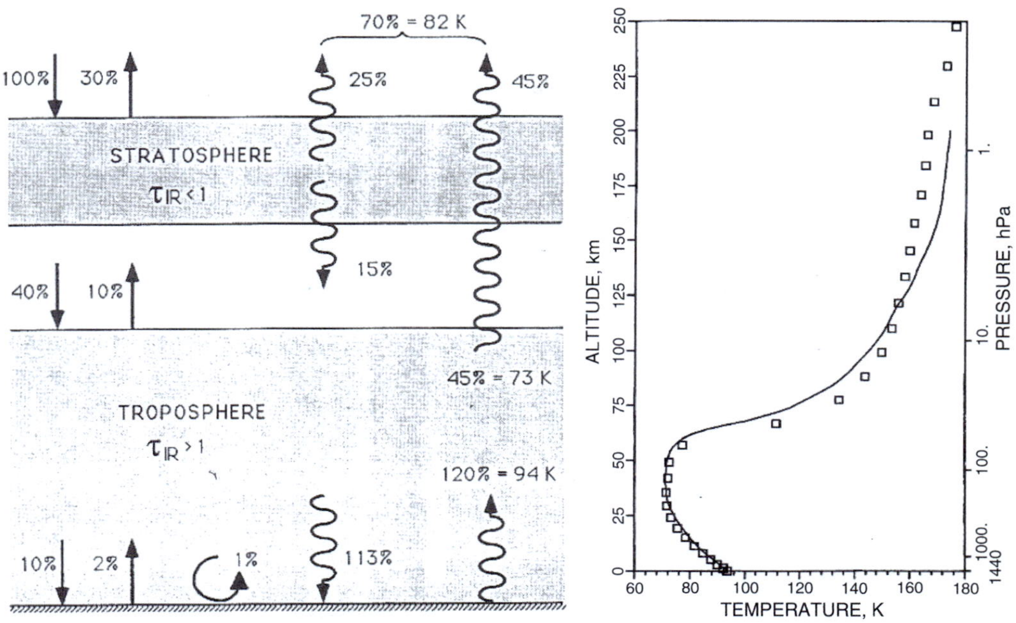

Fig. 2.5 Energy balance (left) and thermal structure (right) of Titan's atmosphere. From McKay et al. (1991) and McKay et al. (1989).

Titan's surface, i.e. that the surface would be observable at some (near-infrared) infrared wavelengths (Fig. 2.5). This turned out to be demonstrated experimentally shortly after.

In Titan's upper atmosphere, heat is transported by conduction and to some extent by radiation. Friedson and Yung (1984) first investigated the heat budget of the upper atmosphere, that appeared governed by UV heating in N_2 and CH_4, non-LTE cooling in C_2H_2, and downward conduction. They were able to reproduce the (then) UVS-derived thermospheric temperature and proposed a temperature profile above 600 km, which included a cold (110 K) mesopause at 736 km. Lellouch et al. (1990) showed, however, that this agreement was illusory due to a large error in the calculation of the heating rates, and that it was in fact difficult to reproduce thermospheric temperatures as cool as observed. The problem was revisited by Yelle (1991). His model included a large number of improvements over the previous studies, such as the inclusion of IR heating and/or cooling by CH_4, C_2H_2, C_2H_6 and the aerosols, a proper treatment of the coupling between vibrational levels, and the rotational cooling by HCN. The latter factor turned out to be key in explaining the ~186 K thermospheric temperature. Based on these models, Yelle predicted a 135–140 K mesopause near 600 km (~0.1 µbar), relatively warm because of C_2H_6 heating in this region.

2.4.5 Circulation and Meteorology

The Voyager IR observations provided indications on the horizontal variability of the temperatures, although with poor latitudinal resolution. Flasar et al. (1981) quickly noted

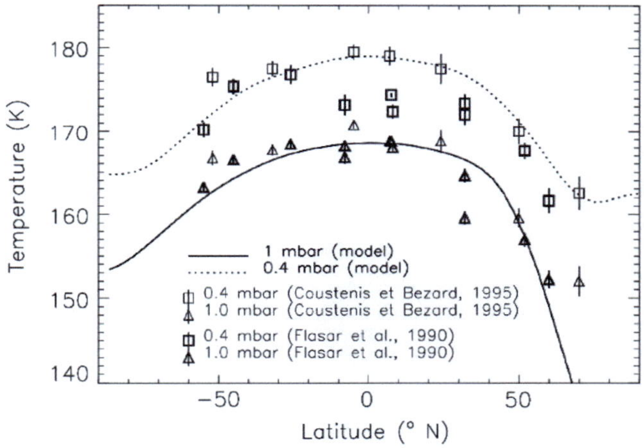

Fig. 2.6 Latitudinal temperature distribution compared to GCM model including coupling with haze. From Rannou et al. (2004).

a hemispheric asymmetry in the stratospheric temperatures, with the Northern hemisphere ~10 K colder than the Southern. This was confirmed in subsequent analyses of the IRIS data (Flasar and Conrath 1990, Coustenis et al. 1995, Bézard et al. 1995), indicating 0.4–1 mbar temperatures relatively uniform – to within ~4 K – in the Southern hemisphere, but decreasing sharply – by more than 15 K – from equator to 60°N latitude (see Fig. 2.6). Applying the thermal wind equation to this limited dataset, Flasar and Conrath (1990) inferred a moderate (60 m/s) zonal jet in the Northern hemisphere. Because the radiative time constant at this level is much shorter than 1 Titan year, a symmetric temperature field, tracking the insolation field, could have been expected for this (spring) season. Flasar and Conrath invoked a case of "dynamical inertia", in which the atmospheric response lags the insolation changes due to the time is takes to redistribute the axial angular momentum associated with temperature changes. Noting that the Northern hemisphere was the place of increased haze and increased abundances of radiatively-active minor species, Bézard et al. (1995) proposed instead that the temperature asymmetry was radiative in origin and related to the asymmetry of opacity sources. This issue was further explored when general circulation models started to be developed in the 1990s. These models, starting from the pioneering work of Hourdin et al. (1995), indicated that Titan's circulation near solstice is essentially characterized by meridional transport from summer pole to winter pole in the stratosphere and above, and in the opposite direction near the surface. This circulation pattern is reversed at the opposite solstice. For short periods around each equinox, the meridional circulation breaks up into two cells, with upwelling near the Equator and downwelling at high latitudes. The models indicated that this meridional circulation must induce strong, prograde, zonal winds. They further showed that a purely dynamical effect was insufficient to explain the temperature asymmetry between the two hemispheres observed by Voyager.

Rannou et al. (2002), Lebonnois et al. (2003) and Luz et al. (2003) demonstrated that strong couplings exist between the thermal field, the wind field, and the altitude-latitude distribution of haze and of the minor compounds (Fig. 2.6). Essentially, the meridional circulation driven by latitudinal temperature contrasts transports the haze to the Northern winter polar region. The accumulation of haze there leads to extra cooling, intensifying the circulation. The same is true for minor compounds, although their feedback on the circulation appears less important than for the haze. Rannou et al. (2002) were able to show that the gross features of the haze structure and temperature field seen by Voyager were consistent with these concepts. In contrast, models in which the haze production varied seasonally but which did not call for dynamical control failed to reproduce the N/S albedo asymmetry, due to too long time constants associated with haze formation and accumulation (Hutzell et al. 1993). Note finally that ground-based observations (Roe et al. 2004) indicated that the thermal field just prior Winter Solstice (Dec. 2000) was then symmetric in latitude, with Equator-to-Pole contrasts of about 10 K in the middle stratosphere.

Although it was immediately realized that methane (but not nitrogen) was a condensable species in Titan's atmosphere, the dynamics and meteorology of Titan's atmosphere remained poorly characterized for a long time after the Voyager mission because the methane vertical profile remained uncertain, temperature constraints were few in the troposphere, and clouds had not been detected yet. For a pure N_2 atmosphere, the lapse rate inferred from RSS in the first few kilometers appeared to follow a dry nitrogen adiabat, suggesting that methane condensation did not occur near the surface (Lindal et al. 1983; Eshleman et al. 1983). By considering the effect of CH_4 on the RSS data, Flasar (1983) showed that the methane surface humidity was below 0.7, otherwise the atmosphere would be unstable to dry convection. This upper limit was refined in the study of McKay et al. (1997), who showed that if allowance was made for the condensation properties of methane/nitrogen mixtures, the methane maximum humidity at the surface was closer to 0.6. Their study further showed that there is no convective zone in the troposphere (i.e. that the lapse rate follows everywhere the radiative lapse rate), and that the lapse rate is stable against dry convection and unstable against moist convection. This suggested that condensation of methane does not occur at all in Titan's atmosphere (methane supersaturation in Titan's troposphere was advocated originally in papers by Courtin et al. 1995 and Samuelson et al. 1997a) or only in patchy form, producing a mean thermal gradient intermediate between the dry and the lapse rates. This latter view was generally confirmed when clouds were discovered from spectroscopy and direct imaging (see below).

Voyager IRIS data indicated that the tropopause temperature, as measured by the 300 cm^{-1} radiance, is uniform to within 1 K (Flasar et al. 1981; Samuelson et al. 1997c; Courtin and Kim 2002), while the 510 cm^{-1} radiance, to which the surface contributes for about 60%, suggests that the surface brightness temperature varies by about 3 K from Equator to both poles. The hemispheric symmetry of the surface temperature is consistent with the very long radiative timescale of the troposphere. Noting that the annually-averaged insolation at 60° is about ½ of its equatorial value, Flasar et al. (1981) estimated that the radiatively forced surface temperatures would show a ~15 K difference between these latitudes, and that meridional winds of ~0.04 cm/s would be sufficient to reduce the contrasts to their observed value. Since then, however, general circulation models (Hourdin et al. 1995; Tokano et al. 1999) have failed to reproduce even this modest temperature contrast, which, given the expected efficiency of meridional transport, remains a challenge to modellers (Tokano and Neubauer 2002).

2.4.6 Speculations About the Surface and the Interior

Immediate post-Voyager photochemical models (Yung et al. 1984) indicated a depletion of atmospheric methane on a ~10^7 year timescale and its irreversible conversion to more complex hydrocarbons, primarily ethane and acetylene. This number and conclusion, which were essentially confirmed in later photochemical models, required that methane is supplied over geologic time from a reservoir more massive than the atmosphere. The natural idea of a pure methane ocean, however, was ruled out by the near-surface sub-saturation of methane. As ethane is liquid at Titan's surface temperature, Lunine et al. (1983) postulated a model in which methane was stored in a surface ocean, mixed with its own photolysis products (ethane and other liquid hydrocarbons such as propane) and with dissolved nitrogen. Since the atmospheric abundance of methane is a function of its mole fraction in the ocean (being lowered by the presence of ethane), it was possible to estimate the oceanic composition. In addition, the ocean depth could be constrained by the quantity of ethane having precipitated to the surface since Titan's formation. In its nominal form, the Lunine et al. (1983) oceanic model was made of 70% C_2H_6 – 25% CH_4 – 5% N_2, and had a ~1 km depth. These figures hold for a global ocean. This was based on considerations by Sagan and Dermott (1982), who argued that Titan's non-zero orbital eccentricity indicates low tidal damping, while discrete and shallow seas would result in a strong tidal erosion and orbit circularization (this argument has since then be revisited, see Sohl et al. 1995; Dermott and Sagan 1995).

A more extensive study of the oceanic composition and depth was performed by Dubouloz et al. (1989), based on the range of atmospheric models explored by Lellouch et al. (1989). This study indicated that the ocean could vary from "shallow and ethane-rich" to "deep and methane-rich", with maximum depths as high as 9 km. Such an ocean could also be a large CO reservoir, and sequester insoluble species at its bottom, in particular acetylene with an estimated ~100 m solid layer. Additional refinements were brought to the model, i.e. regarding the possible presence of other dissolved species and suspended aerosols. Nonetheless, the essential picture of a global, ethane-dominated ocean buffering the atmospheric methane, controlling the lower atmosphere meteorology and providing a sink for the dregs of the atmospheric chemistry remained the accepted paradigm of Titan's surface for about a decade after the Voyager observations. It was ultimately washed out when the first radar echos, and near-infrared lightcurves and images were recorded in the early 1990s.

2.5 Observations of Titan from the Earth and Earth-Orbit in the Post-Voyager Era

In spite of the enormous progress brought by the Voyager investigation, there was still room for other discoveries to be made remotely in the following years. Those were made possible by the advent of more sensitive ground-based instruments and space-borne observatories (e.g. HST and ISO), the development of new techniques such as heterodyne spectroscopy and adaptive optics, the long-awaited opportunity provided by stellar occultations and radar sounding, and the realization that Titan's lower atmosphere and surface were observable in the near-infrared.

2.5.1 Radar Observations

While early photochemical modeling suggested the presence of a multiple kilometer deep global hydrocarbon ocean (see above), the first radar reflectivity measurements were inconsistent with the existence of a deep global ocean. Muhleman et al. (1990) observed Titan with 3.5 cm wavelength radar on three consecutive Earth days (45° of Titan rotation). They found radar reflectivities at least an order of magnitude brighter than what they would have seen for even a shallower few hundred meter deep ocean. Further, the day-to-day measured reflectivities varied by greater than three times their estimated uncertainties, strongly suggesting that the surface was heterogeneous, which was later confirmed by ground- and space-based near-infrared imaging.

By 2001–2002 Titan had moved into the declination range able to be observed with the recently upgraded 13 cm radar at Arecibo Observatory. Campbell et al. (2003) reported the detection of specular reflections at ~75% of the 16 locations on Titan's surface probed with radar, implying materials smooth at the multi-centimeter scale and interpreted to be the sign of small lakes at the sub-Earth latitude of the time (~26°S). Campbell et al. (2003) further noted that the intensity of the specular reflection correlated with the near-infrared surface albedo, although the physical reason for this was unclear. An examination of ground-based near-infrared imaging by West et al. (2005) showed that lakes of the size suggested by the Campbell radar measurements would have been easily detected in the ground-based imaging and that the lack of any detectable specular reflection in the ground-based data placed severe constraints on the presence of open surface liquids at those latitudes accessible from Earth. While it is possible to imagine surface materials that are smooth at centimeter scales and rough at micron scales, a definitive answer that satisfies both the radar and near-infrared measurements remains elusive. As of 2008 the radar tracks acquired by the Cassini mission do not yet overlap with the sampling points of Campbell et al. (2003).

2.5.2 Near-IR and Visible Spectroscopy and Imaging

Major progress toward understanding Titan's lower atmosphere and surface was not made until the exploitation in the late-1980s/early-1990s of the spectral `windows' between strong methane absorption bands at near-infrared wavelengths (McKay et al. 1989; Griffith et al. 1991). While much of the ground-based spectro-imaging work focused on Titan's atmosphere (3.5.2.2), progress was also made during the pre-Cassini era with observations of Titan's surface (3.5.5.1).

2.5.2.1 Surface Composition and Morphology

Titan's atmosphere is a hindrance to near-infrared surface spectroscopy and imaging if the lower atmosphere and the surface are the target. Indeed, the surface is only visible in these narrow "windows" between the strong methane absorption bands and the optical properties of the haze layers varying strongly with wavelength from 'window'-to-'window'. Nonetheless several ground-based spectro-imaging observations of Titan are worth noting for their insights brought into the nature of the surface.

Observations in the methane "windows" led to the detection of Titan's surface rotation light curve (Griffith 1993; Lemmon et al. 1993, 1995; Coustenis et al. 1995; Negrão et al. 2006). In hindsight the surface of Titan was detected in earlier datasets. Cruikshank and Morgan (1980) reported near-infrared flux measurements of Titan in wideband J, H, and K filters over 70 days and at a high confidence level detected variability. This variability is now recognized to be the well-known 16-day rotation period of Titan's high-contrast surface, although this was not fully appreciated at the time. The canonical view that the Voyager flybys could not detect Titan's surface was shown to be wrong by Richardson et al. (2004), who found that careful processing of the Voyager imaging revealed the large-scale surface features.

Following previous studies suggesting the presence of "dirty" water ice on the surface (possibly contaminated with organics or tholins (Griffith et al. 1991; Coustenis et al. 1995)), Griffith et al. (2003) reported the analysis of near-infrared spectra that suggested that water ice was further extensively exposed on Titan's surface. In spectroscopy through the 4.9 μm window of Titan's atmosphere Lellouch et al. (2003) also found that Titan's surface albedo was correlated with the albedo in other windows and consistent with water ice, but that some other surface material with a sloped albedo decreasing over 4.98–5.07 μm must also be present. This result has since been confirmed by Cassini observations and is now thought to be due to the presence of benzene or other higher order aromatic hydrocarbons on Titan's surface. More progress was made toward constraining the nature of Titan's surface with recent spectroscopic observations through those same near-infrared atmospheric windows. Negrão et al. (2006, 2007) compared moderate resolution near-infrared spectroscopy with the improved radiative transfer models of Rannou et al. (2003) and found their data compatible with Titan's surface composition being again a mix of water ice and tholins, but with a third unidentified component.

Hartung et al. (2006) searched for absorption bands of solid CO_2 and constrained the surface of Titan to be covered in less than a few percent CO_2, at least the two locations sampled. At longer wavelengths Coustenis et al. (2006) observed Titan with ISO and detected for the first time the full shape of the 2.85-μm CH_4 window. These authors, once more, found their data compatible with water ice being a major component of Titan's surface, but they additionally suggested a possible CO_2 contribution near 2.74 μm.

The near-infrared 'windows' were also used by several teams of observers to obtain images of Titan's surface by using either the Hubble Space Telescope (Smith et al. 1996), speckle interferometry (Gibbard et al. 1999, 2004) or adaptive optics (Bouchez 2004; Coustenis et al. 2001, 2005; Roe et al. 2004; Gendron et al. 2004; Hirtzig et al. 2007). These efforts revealed, with increasing spatial resolution, the heterogeneous complex nature of Titan's surface, with dark and bright regions sharing the landscape. The presence of different ices and various degrees of elevations was suggested, but none of these theories was unambiguously confirmed and the nature of Titan's surface remained largely mysterious.

The near-infrared maps of Titan's surface using the Hubble Space Telescope and the largest ground-based telescopes (see for instance Smith et al. 1996; Meier et al. 2000; Roe et al. 2004; Coustenis et al. 2005) revealed details on Titan's surface down to the limits of their resolution and inspired much speculation about the nature of the surface (Fig. 2.7). However, at the modest resolution achieved, roughly equivalent to that of the unaided human eye peering up at Earth's moon, no geologic features could be definitively identified.

2.5.2.2 Atmospheric Phenomena

Titan's thick stratospheric haze layer, the end product of a complicated network of methane photochemistry, changes in appearance with season. In particular the north/south brightness ratio varies with both wavelength and seasonal phase. This variation with seasonal phase is best explained by a global circulation model (GCM) coupled with a haze microphysical model (Rannou et al. 2004), in which the haze is advected around the upper atmosphere by winds and also plays a feedback role by heating the atmosphere through solar absorption and cooling it by radiating it in the infrared. The variation in appearance as a function of wavelength is due in part to different altitudes being probed by different wavelengths and in part to the variation in optical properties of the haze particles as a function of wavelength. The haze particles are generally absorbing at visible wavelengths and shorter

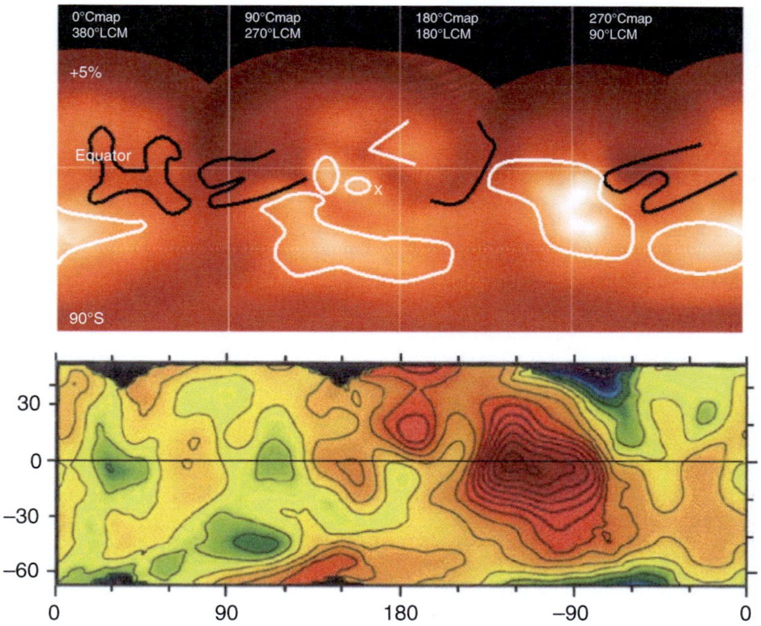

Fig. 2.7 Among the first maps of Titan's surface: two maps taken with the adaptive optics system NAOS at the VLT at 1.28 μm (*upper panel*) and the HST NICMOS at 1.6 μm (*lower panel*). The surface features are coherent from one dataset to the other. The bright areas dominate Titan's leading hemisphere, while the darker ones prevail on the other side. The large bright equatorial region has since then been named "Xanadu Regio" and is observed near 110°LCM. The Huygens landing site is marked with an "X" near 192°LCM and 10°S. Adapted from Fig. 2.7 in Coustenis (2007).

and generally scattering at wavelengths longward of visible. Additionally the haze layers are optically thick at short visible wavelengths and the opacity decreases with increasing wavelength until the haze is optically thin at ~2 μm. Numerous observations have been made at both visible and near-infrared wavelengths of the seasonally varying haze and polar collar or hood that is likely related to a high altitude polar vortex. These include imaging from the Hubble Space Telescope (e.g. Caldwell et al. 1992; Lorenz et al. 1999, 2001) and ground-based adaptive optics systems (e.g. Coustenis et al. 2001; Roe et al. 2002a; Hirtzig et al. 2007). Evidence for the north-south asymmetry reversing, starting first in the visible range, was found by Lorenz et al. (1999) and has since been confirmed in all wavelengths. This is an expected return to the Voyager situation, predicted by seasonal models (Sromovsky et al. 1981; Hutzell et al. 1993), where the south limb was bright in the visible, but dark in the infrared, anti-mirroring the northern limb situation. This situation had reversed in the 90's, showing the famous "Titan's smile" in HST and adaptive optics images (e.g. Caldwell et al. 1992 and Coustenis et al. 2001, respectively).

Initial detections of Titan's tropospheric clouds were controversial, in part due to interpretations of Voyager flyby data suggesting "rain without clouds" in Titan's troposphere (Toon et al. 1988) and methane supersaturation (Courtin et al. 1995; Samuelson et al. 1997b) wherefore tropospheric clouds should not be able to exist in a stable condition. The first published detections of clouds were the whole-disk near-infrared spectra of Griffith et al. (1998). These spectral detections use the variation in atmospheric opacity with wavelength to selectively probe different altitudes and separate surface rotation from tropospheric clouds and from stratospheric haze (Fig. 2.8). The observations revealed an enormous tropospheric storm, covering 5–7% of Titan's disk in September 1995. Griffith et al. (2000) then showed evidence for daily small-scale (covering ~0.1% of Titan's disk) clouds in similar disk-averaged spectra (Fig. 2.8).

Direct imaging of Titan's clouds first showed strong cloud activity in the south polar region in late southern spring (Gibbard et al. 2004; Roe et al. 2002b; Brown et al. 2002; Gendron et al. 2004; Bouchez and Brown 2005; Hirtzig et al. 2006; 2007). These south polar clouds (Fig. 2.9), visible especially at 2.0 μm, persisted for several years as the season progressed into early southern summer. Following a massive south polar storm system in late 2004 (Schaller et al. 2006a) the south polar clouds were observed to dissipate for an extended period of time (Schaller et al. 2006b). A new class of clouds, long, streaky, and confined to near 40°S latitude were discovered in late 2003 and reported in Roe et al. (2005a). This latitudinal clustering of clouds could be indicative of a seasonal region of uplift (see Chapter 14) and this was initially proposed as the probable explanation (Roe et al. 2005a; Griffith et al. 2005). Further observations of these mid-latitude clouds revealed them to be clustered in longitude, suggesting that some aspect of their formation mechanism was linked to the surface and the clouds could be the result of geologic injection of methane to the atmosphere by geysers, cryovolcanoes, etc. (Roe et al. 2005b). Ultimately these clouds most likely result from some combination of surface processes and seasonal circulation.

Other atmospheric phenomena have been observed in Titan's atmosphere from the Earth. A diurnal variation in

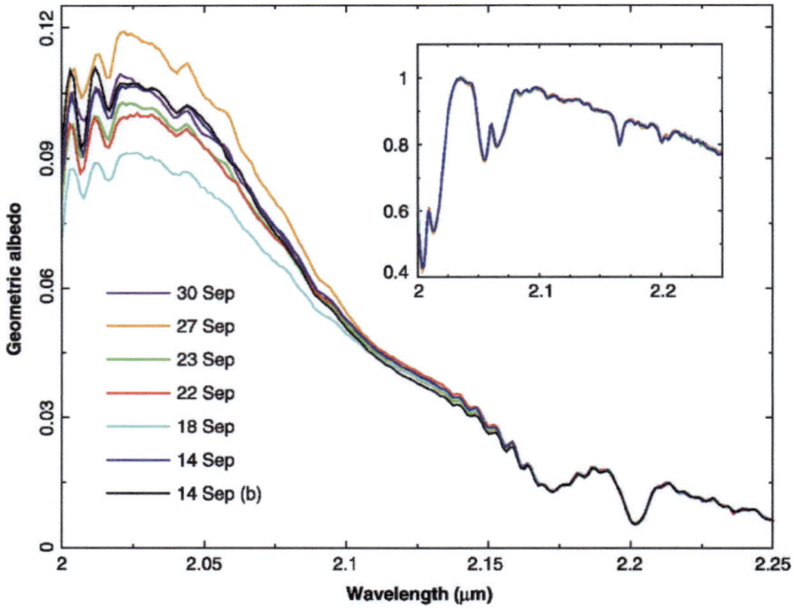

Fig. 2.8 Evidence for clouds on Titan in spectra taken over several days in 1999 (Griffith et al. 2000). Across this wavelength range the observed albedo varies inversely with the atmospheric gas opacity, which is primarily provided by methane and molecular hydrogen. At shorter wavelengths (~2.0–2.08 µm) the opacity is low and the heterogeneity of surface albedo is apparent in the spectral variations during one 16-day Titan rotation. At longer wavelengths (~2.16–2.25 µm) the opacity is high and photons do not penetrate below the stratosphere. Titan's stratosphere varies only on longer time scales and therefore this spectral region remains constant during the period of observations. At intermediate wavelengths (~2.08–2.16 µm) the gas opacity is high enough to block the surface from view but low enough to allow photons to reach the troposphere. Subtle albedo variations in this spectral region are uncorrelated with the surface rotation and indicative of variable tropospheric cloud coverage.

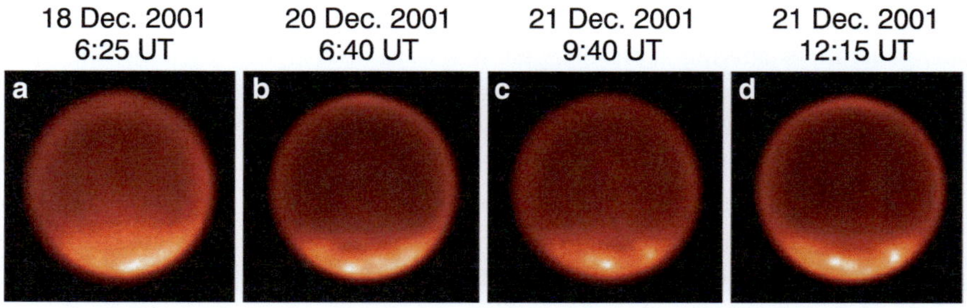

Fig. 2.9 The south polar clouds were the first to be directly imaged. In this sequence, which was typical of Titan during late southern spring, the cloud positions and brightnesses can be seen evolving over several days (Roe et al. 2002b).

the upper troposphere or lower stratosphere is apparent in the consistent brightness enhancement of the haze on the morning limb, first recognized by Coustenis et al. (2001) and confirmed in Hirtzig et al. (2006). In a careful radiative transfer analysis of a spectrally and spatially resolved datacube, Adamkovics et al. (2007) reported spectral evidence for the presence of an optically thin layer of particles at 30 km on the morning limb that was consistent with thin condensed methane clouds and the formation of widespread morning drizzle.

Titan's year is 30 Earth years long, which severely limits the historic record of observations. Few of the techniques used to observe Titan were available 30 years ago. The one long-term study of Titan that has now successfully observed a full Titan year is that of Lockwood et al., at Lowell Observatory (see e.g. Lockwood et al. 1986). In the most recent update (Lockwood and Thompson 2009) they report on 34 years of visible wavelength photometry of Titan and find that the disk-integrated brightness of Titan displays a 10% sinusoidal variation lagging the phase of Titan's seasonal extremes by 1/8 of a Titan year. Of particular interest are the most recent 4 years of observations, overlapping in seasonal phase with their first 4 years of observations. They find that Titan's brightness varied from one Titan year to the next at the same seasonal phase, in this case largely due to the high stratospheric hazes. This is an important reminder to us all that the atmosphere on Titan, much like on Earth, may display significantly different behavior from 1 year to the next.

With this caution in mind observational studies of the seasonal evolution of Titan's clouds continue. Already the shutdown of the south polar cloud system was reported in early southern summer (Schaller et al. 2006b). Global circulation modelers have begun to explore how Titan's tropospheric weather will respond to the changing seasons (see e.g. Tokano 2005; Rannou et al. 2006) and make predictions for when and where clouds should occur. The next one to two decades should see a continuing iterative refinement of these models as the database of cloud observations extends to cover an ever-increasing fraction of a Titan year.

2.5.3 Mid-Infrared, Far-Infrared and Millimeter Spectroscopy

Considerable progress was made over 1980–2000 in characterizing Titan's atmosphere composition and dynamics, especially from observations in the thermal infrared. The advent of cooled grating, echelle, and heterodyne spectrometers permitted to re-explore the thermal infrared range (6–50 μm) with a spectral resolution and sensitivity superior to Voyager /IRIS, in spite of the much larger distance to the target and the modest (at best) spatial resolution available. Such instruments included the echelle spectrometer IRSHELL and its successor TEXES on the IRTF, the Goddard Infrared heterodyne spectrometer (IRHS) and its successor HIPWAC on the same facility or on Subaru, and the Short Wavelength Spectrometer (SWS) of the Infrared Space Observatory (ISO), launched in 1995 and operative until April 1998. Thanks to their very high spectral ($>10^6$) resolution, heterodyne observations, both in the millimeter/submillimeter range and in the 10 μm window, were especially powerful to (i) detect weak and isotopic species (ii) determine the vertical distribution of some gases from line profiling and (iii) obtain direct wind measurements.

Shortly after the discovery of CO_2 by Voyager, a second oxygen-bearing species, CO, was detected in 1982 (Lutz et al. 1983) with a ~60 ppm abundance. CO was expected because the formation of CO_2 from an external source of H_2O involved similar formation of CO from $OH + CH_3 \rightarrow CO + 2H_2$. Although the original discovery of CO was obtained at 1.57 μm, most subsequent studies of CO were performed in the mid-infrared (5 μm – Noll et al. 1996; Lellouch et al. 2003; Lopez-Valverde et al. 2005) and in the mm/submm range (Muhleman et al. 1984; Marten et al. 1988; Gurwell and Muhleman 1995; Gurwell and Muhlemann 2000; Hidayat et al. 1998). The latter set of observations, in particular, generated a lot of controversy as to the precise abundance and vertical profile of CO, as a number of measurements indicated a depletion of the CO stratospheric abundance with respect to its tropospheric value, a behavior that was difficult to explain given the extreme chemical stability of this compound. The overall outcome of all these measurements, however, was that there is no definite evidence that CO is not uniformly mixed in Titan's atmosphere.

Other key compositional studies were performed in the mm/submm range (Bézard et al. 1993; Hidayat et al. 1997; Marten et al. 2002; Gurwell 2004). Highlights include: (i) the first – and so far only – detection of acetonitrile (CH_3CN) – see Fig. 2.10; (ii) the first vertical profiles of nitriles (HCN, HC_3N, CH_3CN) outside of the polar region, confirming the general trend of minor species, especially short-lived ones, to increase with altitude in the stratosphere; (iii) the first evidence of a strongly enhanced $^{15}N/^{14}N$ ratio (by a factor of ~4) in HCN. Initially (Marten et al. 2002), this ratio was implicitly assumed to be representative of Titan's atmosphere, and to reflect its evolution in presence of large escape. The Huygens measurements, however, have indicated that the bulk (i.e. in N_2) $^{15}N/^{14}N$ rate in enriched by a factor 1.5 only (see Chapter 10), so that the higher enhancement in HCN points to an efficient fractionation process, identified as a differential photolysis rate (Liang et al. 2007); (iv) measurements of the $^{18}O/^{16}O$ ratio (Owen 2000; Gurwell 2008), though results haven't been extensively published yet.

A wealth of results on Titan's composition was also achieved in the mid- and far-infrared. New information on hydrocarbon abundances and distributions, particularly ethane and acetylene, was obtained from the ground and from ISO (Kostiuk et al. 2005 and references therein, Coustenis et al. 2003; Roe et al. 2004). Again with IRTF/TEXES, Roe et al. (2003) were able to clearly separate and measure the 748 cm^{-1} feature of propane, partially blended with acetylene

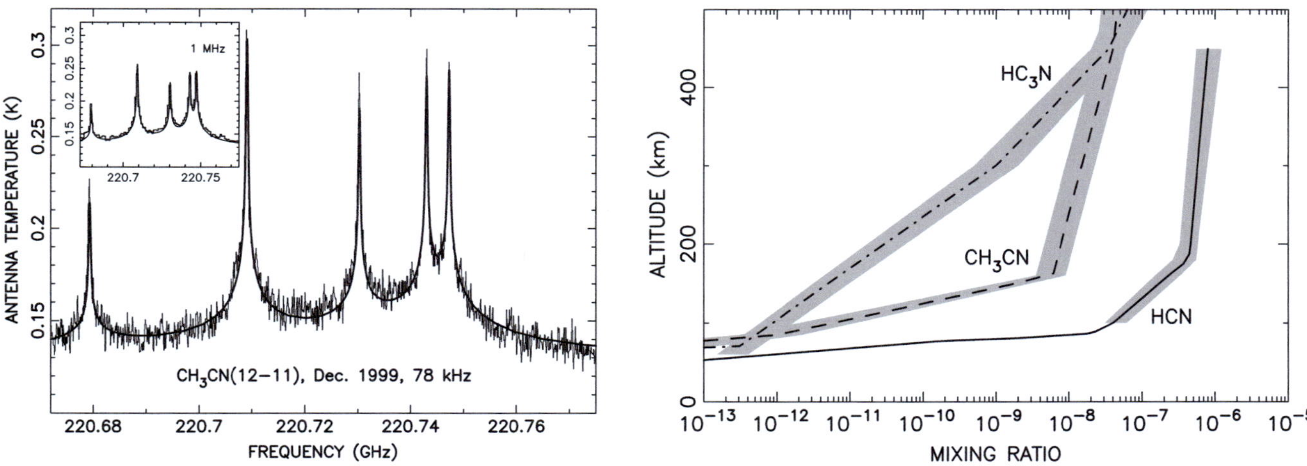

Fig. 2.10 Detection of CH_3CN, (*left*) and vertical profiles of nitriles (*right*) from IRAM-30 m millimeter observations (Marten et al. 2002).

at Voyager/IRIS (and even ISO/SWS) resolution. New measurements of the D/H ratio from CH_3D were also obtained (Orton 1992; Coustenis et al. 2003). These, however, indicated values in the range (5–10) × 10^{-5}, somewhat lower than those obtained from Voyager, than values obtained in the near-IR (de Bergh et al. 1988) and than the "modern", post-Cassini value of ~13 × 10^{-5} (see Chapter 10). In terms of novelty, however, the most important results were the detections of new species, which included water and benzene from ISO (Coustenis et al. 1998, 2003) and allene from IRTF/TEXES (Roe et al. 2004). The evidence for benzene, which followed its detection by ISO in the Giant Planets (Bézard et al. 2001) prompted an examination of the chemical schemes that could produce it at stratospheric levels (Wilson et al. 2003; Wilson and Atreya 2004; Lebonnois 2005); the proposed mechanism involves the recombination of propergyl (C_3H_3 radicals), producing isomers of C_6H_6, and presumably ultimately to benzene through isomerization. The detection of water vapor (Fig. 2.11) was the final (if necessary) proof for an external source of water vapor in Titan's atmosphere (Coustenis et al. 1998). The estimated water mixing ratio (typically 8 ppb at 400 km) and flux – (1.3–4.5) × 10^6 cm^{-2} s^{-1}, referred to flux surface – was generally in line with expectations based on the CO_2 abundance (see Hörst et al. 2008, and the Chapter 10 for an up-to-date discussion on this problem). However, the ultimate origin of water remains elusive, as water can originate either from global sources, such as interplanetary icy dust grains, or from local sources such as the sputtering of icy satellite or ring surfaces (or based on a more recent perspective, Enceladus venting). A surprising aspect is the fact that the above flux is comparable to the supply rate of oxygen into Saturn, which amounts to (4 ± 2) × 10^6 cm^{-2} s^{-1} (Moses et al. 2000), while larger fluxes may be expected at Saturn due to the gravitational focusing it exerts on interplanetary dust particles and the fact that a ring source, interior to Titan's orbit, should also favor Saturn over Titan. The presence of an external oxygen supply in Titan's atmosphere may thus point to the importance of local sources beyond Titan's orbit, but the precise origin of the dominant source (Hyperion, Iapetus, Phoebe…) is still not solved.

The post-Voyager ground-based observations also provided results on the thermal structure and especially the dynamics of Titan's atmosphere. From very high-resolution (70,000 resolving power) observations of the ν_4 band of methane near 8 µm with TEXES, Griffith et al. (2005) obtained the first infrared measurements of the thermal structure in the 300–600 km range, with the first evidence for a mesosphere above about 380 km. Thanks to their unsurpassed spectral resolution, heterodyne observations allowed the first direct Doppler measurements of winds in Titan's atmosphere. The first successful detection of winds was reported as early as 1994, based on measurements acquired in August 1993 with IRTF/IRHS. It was based on differential line position in ethane 12 µm spectra obtained on two (East and West) Titan limb positions. Albeit rather inaccurate and at a spatial resolution essentially equal to a Titan diameter, these measurements, augmented by additional measurements in 1994–1996 and ultimately published in Kostiuk et al. (2001), revealed a wind velocity of order 100 m/s in the prograde direction at 0.1–7 mbar (120–300 km), confirming the speeds inferred from stellar occultation measurements (see below) but uniquely providing their direction. Beyond its scientific impact for the understanding of Titan's super-rotation regime, this breakthrough was also of high importance for planning the Huygens mission. This result was confirmed from improved observations with HIPWAC on Subaru in 2003, indicating an equatorial wind

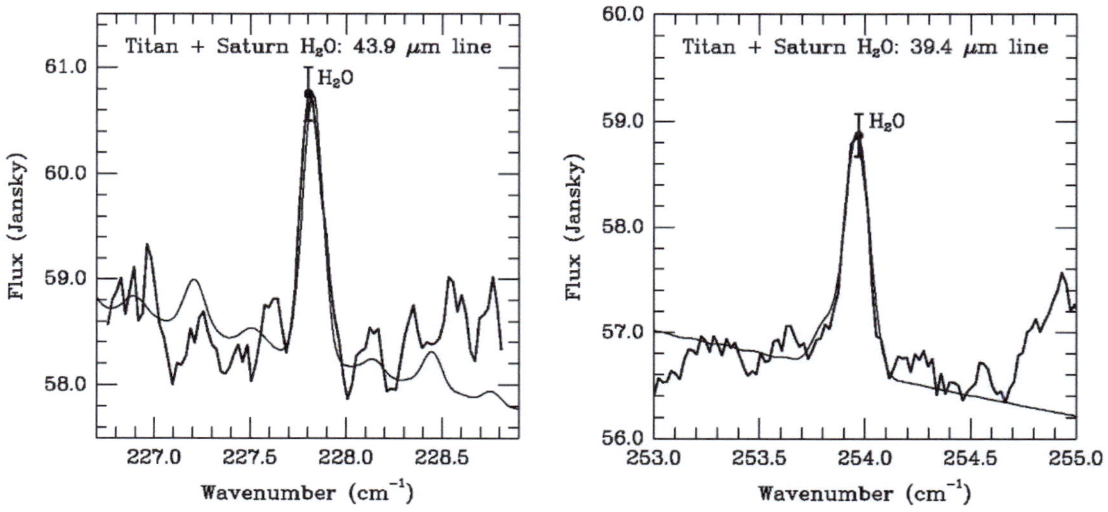

Fig. 2.11 The detection of H_2O from ISO (Coustenis et al. 1998).

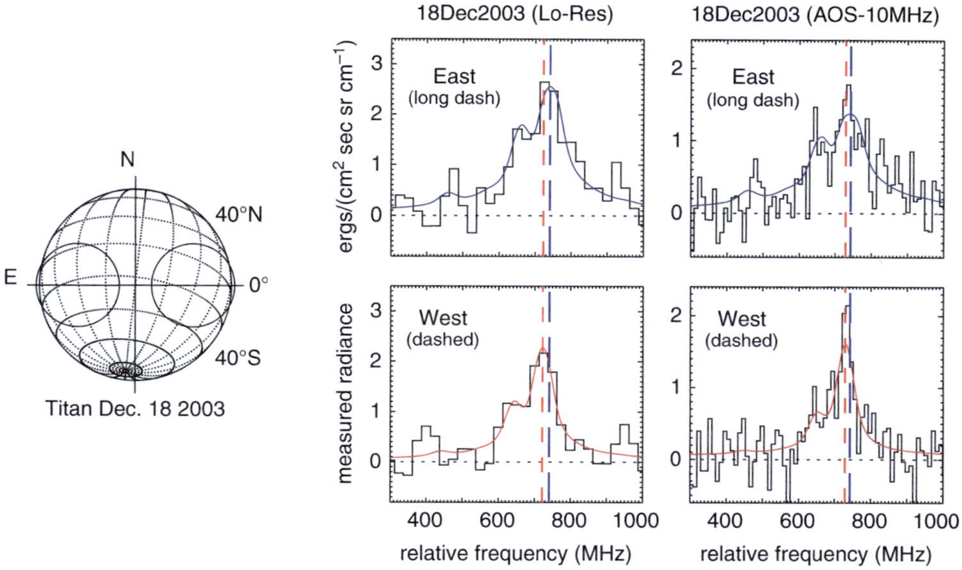

Fig. 2.12 Measurement of winds in Titan's stratosphere using Doppler shift heterodyne spectroscopy of ethane at 12 μm. From Kostiuk et al. (2005)

velocity of 190 ± 90 m/s (Kostiuk et al. 2005, Fig. 2.12). Similar measurements were recorded during the coordinated campaign in support to Huygens on January 15, 2005 (Kostiuk et al. 2006). Wind measurements were also obtained by Moreno et al. (2005), using mm transitions of HC_3N and CH_3CN mapped with the Plateau de Bure interferometer. This permitted the independent determination of the wind speed at two stratospheric altitudes, indicating v = 160 ± 60 m/s at 300 ± 150 km and v = 60 ± 20 m/s at 450 ± 100 km, i.e. a decrease of the zonal winds in the mesosphere. Winds on Titan can also be measured from Doppler-shifted visible solar lines scattered off Titan haze near 200 km (Luz et al. 2005, 2006). Such measurements, performed before and during the Huygens descent with the UVES instrument on the VLT, also unambiguously showed prograde winds, but only lower limits on the wind speed, of order 50–60 m/s, could be determined, due to the partial smearing of the Doppler signals due to seeing effects in the Earth's atmosphere. Ultimately all of these measurements proved highly complementary to the Huygens wind profile and Cassini thermal wind field in constraining the stratospheric dynamics (see Chapter 13).

2.5.4 Stellar Occultations

Ground-based stellar occultations by solar system bodies are powerful tools to probe planetary atmospheres at pressure ranges from one to some hundreds of microbar. Those observations remain rare, however. Titan, for instance, subtends only 1 arcsec or so on the sky, as seen from Earth. So the probability of the satellite passing in front of a star is small. As of today, there have been six documented ground-based stellar occultations by Titan: one on 3 July 1989 (Hubbard et al. 1990, 1993; Sicardy et al. 1990), one on 21 August 1995 (Tracadas et al. 2001), two on 20 December 2001, involving a double star (Bouchez 2004) and two on the same day on 14 November 2003 (Sicardy et al. 2006; Zalucha et al. 2007).

Ground-based stellar occultations probe the stratospheric and mesospheric regions of Titan's atmosphere, between altitude levels ~600 km down to ~250 km, corresponding to pressure levels between one microbar and ~0.25 mbar. These regions are difficult to observe using other techniques (and even with spacecraft, except with an in situ probe), as visible and IR instruments tend to probe deeper layers, while UV instruments are sensitive to higher regions. Occultations allow us to retrieve temperature profiles and especially temperature gradients, detect density, pressure and temperature fluctuations (thought to be caused by gravity waves), measure haze opacities at various wavelengths and in some case – when a so-called "central flash" is observed -, to derive the zonal wind profiles at the ~250 km altitude level (0.25 mbar pressure level).

2.5.4.1 Temperature Profiles

The star disappearance and re-appearance behind Titan provide, via an inversion procedure, the refractivity profile, then the density profile (assuming a given gaseous composition), and finally, temperature and pressure via hydrostatic equilibrium and ideal gas assumptions. Arbitrary boundary conditions make a unique temperature profile retrieval impossible, so that it is difficult to discriminate between various models (given e.g. by Yelle 1991 or Vervack et al. 2004), at least in the probed region (Sicardy et al. 2006). In any case, the retrieved temperature profiles exhibit much more local structures than contained in theoretical models, although the general shape of the modeled temperature profiles agrees with the observations. In particular, two conspicuous inversion layers where detected during the July 1989 occultation near 425 and 450 km (7 and 4 μbar, respectively), where the temperature increases by more than 10 K over an altitude range of less than 10 km (Sicardy et al. 1999). In 2003, only one inversion layer was detected at 515 km (1.5 μbar), where the temperature increases by about 15 K in 6 km. Another remarkable property of this layer is that it is global, i.e. observed essentially at all stations, from latitudes 46°S to 20°N in 1989 and 3°S to 11°N in 2003 (Sicardy et al. 1999 and 2006). This inversion layer was observed again 14 months after the 2003 ground-based observation, when the Huygens probe measured in situ the vertical atmospheric profiles with the HASI instrument (Fulchignoni et al. 2005). The origin of this inversion layer is still debated, but could be linked to tidal waves driven by Titan's orbital eccentricity, through time-dependent components of Saturn's gravitational potential (Strobel 2006). Note also that this inversion has the same altitude as the so-called detached haze layer (Porco et al. 2005).

2.5.4.2 Gravity-Waves

Beyond the inversion layers evoked earlier, the density and temperature profiles exhibit numerous fluctuations that are evidences for internal gravity waves in Titan's stratosphere (Strobel and Sicardy 1997, Sicardy et al. 1999). Those waves are close to saturation, i.e. their negative temperature gradients are close to the adiabatic lapse rate. Thus, near 400–500 km altitudes, the wave amplitudes have increased to reach local convective instability, then breaking and transferring energy and momentum to the atmosphere. A spectral analysis of those fluctuations indicate power spectra with slopes between −2 and −3 for vertical wavelengths between ~5 and ~50 km. The −3 slope is reminiscent of the "universal" gravity wave spectrum observed in Earth's oceans (Bell 1975) and middle atmosphere (Smith et al. 1987). Its origin seems to be linked to the presence of gravity waves propagating upward in a stratified medium, but the detailed mechanisms responsible for such a spectrum are not yet clearly established. Explanations range from breaking of individual wave packets to Doppler-shifted couplings between the horizontal winds forced by the waves or radiation (see the reviews by Gardner 1994 and Zhu 1994).

The amplitude of those waves is consistent with gravity waves freely propagating upwards from lower down, as reported by other authors (e.g. Hinson and Tyler 1983, Friedson 1994). A comparison of the vertical and horizontal coherence distances observed during the 1989 occultation reveals furthermore that the typical aspect ratio of the waves (i.e. the ratio of horizontal over vertical wavelengths) is between ~20 and 100 (Sicardy et al. 1999). Those high values are encountered in other planets like Earth or Neptune (Narayan and Hubbard 1988). Note finally that the same kind of fluctuations have been observed by the HASI instrument during the Huygens descent (Fulchignoni et al. 2005).

2.5.4.3 Winds

During stellar occultations, a central flash may be observed near the middle of the event, when the observer is at less than ~100 km from shadow center. The flash is caused by focusing of light towards the shadow center by the deepest layers probed by the stellar rays. In the case of Titan and for ground-based observations, this layer is situated near 250 km altitude, corresponding to a pressure of about 0.25 mbar. Because this layer is not perfectly spherical, multiple stellar images may appear. A modeling of the flash then allows one to retrieve the shape of the focusing layer. Assuming that this layer is in hydrostatic equilibrium under the effect of gravity and centrifugal acceleration caused by zonal winds, one can eventually deduce the zonal wind profile from the flash shape. Note that it is usually not possible to retrieve a unique solution if only one cut in the central flash is available (as was the case for the second of the two occultations of 14 November 2003, see Zalucha et al. 2007). Thus, multi-station observations are necessary to constrain the zonal wind field. This was achieved in 1989; for the first of the 14 November 2003 occultations, see Fig. 2.13.

The 1989 observations revealed a strong jet near latitude 65°S, with velocities of 180 m/sec, decreasing to about 80 m/s near equator (Hubbard et al. 1993). The 2003 observation revealed, on the other hand, a jet at more than 200 m/s near 55°N, and then a steady decrease to zero as more and more southern latitudes were probed. This profile is actually what is predicted by a Global Circulation Model by Rannou et al. 2004. Note that the 1989 and 2003 flashes were observed half a Titanian year apart. Thus, those observations confirm that a swapping of the wind regime between the two hemispheres occurred between these two dates, as predicted by the GCM. Note in Fig. 2.12 that the 2003 ground-based wind regime agrees well with the one derived by Flasar et al. (2005) from thermal field inferences, based on Cassini/CIRS observations made one year after the the 14 November 2003 occultations.

Figure 2.13 also shows the latitudinal wind profile derived by Bouchez (2004), using an adaptive optics observation of the occultation of a double star by Titan in December 2001. No central flash was observed in that case, but the accuracy of adaptive optics allows one to measure the minute displacements of the stars during the occultation, from which a shape

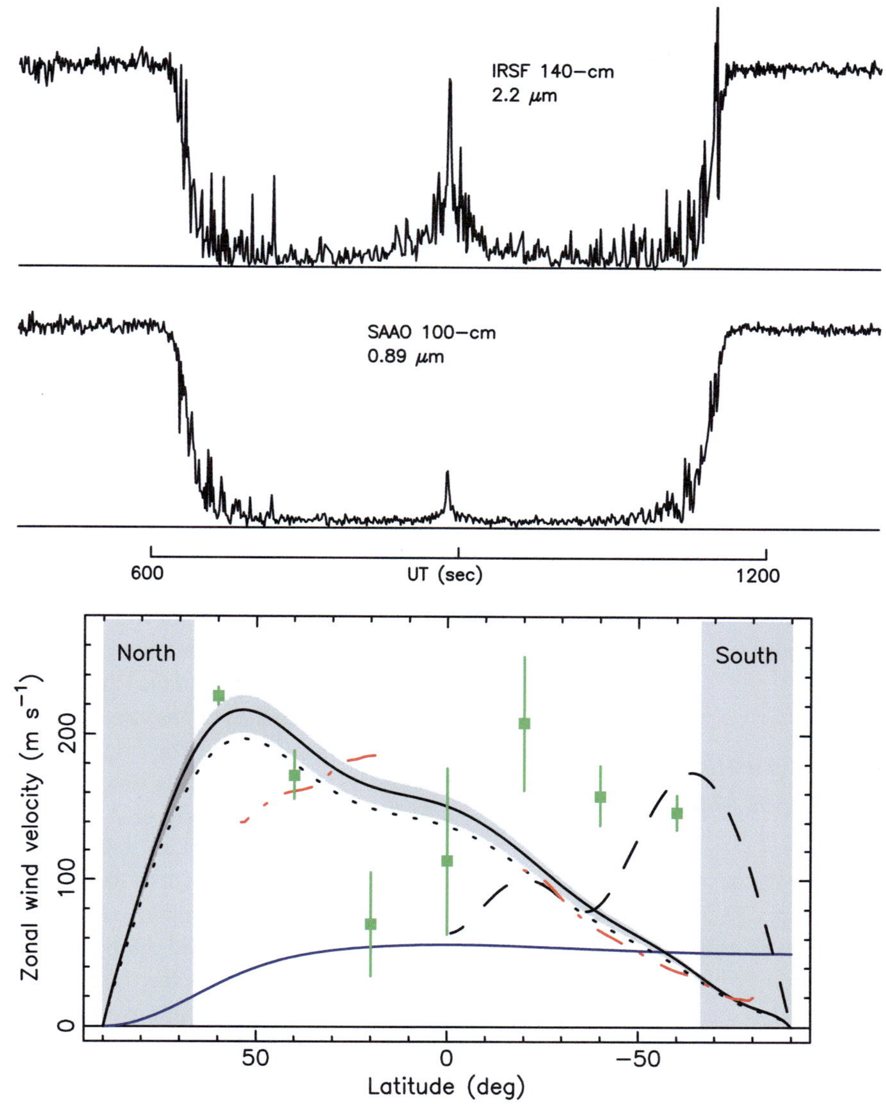

Fig. 2.13 Upper panel – Two Titan occultation light curves observed simultaneously on 14 November 2003 from Sutherland (South Africa), in two different bands: 0.89 μm at the South Africa Astronomical Observatory (SAAO) 1-m telescope and 2.2 μm at the InfraRed Survey Facility of Nagoya University 1.4-m telescope (IRSF), see Sicardy et al. 2006 and Witasse et al. 2006. Note the many local signal fluctuations caused by small density fluctuations in Titan's atmosphere, themselves linked to gravity waves. Note also the central flashes, whose variation with wavelength is solely due to the differential extinctions of the hazes, which become more and more transparent at longer IR wavelengths. Lower panel – zonal wind profile derived from the November 2003 central flash analysis (*solid black curve*, Sicardy et al. 2006), compared to the the Cassini/CIRS profile (*red dash-dots*) obtained one year later (Flasar et al. 2005). The wind profile is also compared with other profiles obtained 1989 (*dashed line, from* Hubbard et al. 1993), and 2001 (*green squares, from* Bouchez 2004). Note that the 1989 wind profile was derived near the northern summer solstice at Titan, and about half a Titanian year before the 2003 profile, taken near the southern summer solstice. The blue line shows the variation of height of the 0.25 mbar isobar (in km, using the same scale as for the zonal wind), taking as arbitrary reference the radius at latitude 90°N (*north pole*). It shows that the radius difference with the north pole value reaches about 50 km at equator and in the entire southern hemisphere.

of isobaric levels can be deduced, and then a zonal wind model, as explained before. The results by Bouchez reveal more complex structure, with a strong jet (225 m/s) at 60°N and another jet at 210 m/sec near 20°S. The northern jet agrees well with what was observed 2 years later, in November 2003. In contrast, the two profiles clearly disagree in the southern hemisphere, where the 2003 winds are weaker. It remains to be seen whether this is evidence for a transient regime, where the southern activity, in terms of winds, damped down between 2001 and 2003, while the high northern latitudes were experiencing a steady strong jet.

2.5.4.4 Haze

The central flash contrast is very sensitive to haze absorption, as optical depths along the line of sight at 250 km reach unity or more in the visible and near IR. This is illustrated in

Fig. 2.13, where the flash observed in K band (2.2 μm) is much more conspicuous than the one in the I band (0.89 μm). The difference is caused by the chromatic dependence of the haze optical depth. Comparison of various central flashes profiles of the 1989 occultation provided a dependence of the form $\tau \propto \lambda^{-q}$, with $q = 1.7 \pm 0.2$ (Hubbard et al. 1993), and the 14 November 2003 central flash analysis provided a consistent value, $q = 1.8 \pm 0.5$ (Sicardy et al. 2006). A lower value of $q = 1.3 \pm 0.2$ was derived by Zalucha et al. (2007) from the analysis of the second occultation observed on 14 November 2003. However, the latter measurement does not apply to the central flash altitude level (250 km), but higher in the atmosphere (typically between 250 and 400 km), so that part of this discrepancy can be due to fact that different regions – with different aerosol properties – are probed by the various experiments. A recent analysis of a solar occultation observed by the *Cassini/VIMS* instrument (Bellucci et al. 2008) shows that the value of q varies from 1.7 to 2.2 between altitude levels of 130 and 300 km, thus confirming the ground-based derived results.

2.5.5 Earth-Based Observations During the Huygens Mission

Campaigns conducted during the descent and landing of the Huygens probe on 14 January 2005 were coordinated by ESA (Witasse et al. 2006) and connected the in situ data with the extended coverage provided by high-technology ground-based observations.

The different approaches include data acquired through Radio telescope tracking of the Huygens signal at 2,040 MHz, studies of the atmosphere and surface of Titan with spectroscopy and imaging, and some attempts to observe the probe's entry signal.

The probe's radio signal was successfully captured by the whole or part of a network of 17 terrestrial radiotelescopes at around 10:19 UTC, thus confirming the overall good state of the mission some 6 h before telemetry data reached Earth relayed by the orbiter (Lebreton et al. 2005). This allowed scientists to furthermore obtain Doppler tracking as an enhancement to the Huygens Doppler Wind Experiment (DWE, Bird et al. 2005), which in the end helped to reconstruct the 2-D horizontal wind field in Titan's atmosphere and permitted to achieve this Huygens goal (Folkner et al. 2006). Another goal consisted in acquiring Very Long Base Interferometry (VLBI) data by the Joint Institute for VLBI in Europe (JIVE) in order to be able to determine the exact position of the probe (within 1 km) during the descent.

In addition, 8 optical observatories participated to this coordinated observations aiming to offer scientific inputs on winds, atmospheric and surface properties, as well as to detect radiation emitted during the Huygens probe entry into Titan's atmosphere. This latter observation goal however, failed (Lorenz et al. 2006; de Pater et al. 2006).

Kostiuk et al. (2006) reported on observations of the Doppler-shift of stratospheric emission lines of ethane around 12 μm performed with HIPWC at the National Astronomical Observatory (NAO), in Hawaii, on January 14 and 15, 2005 and leading to measurements of the magnitude and direction of Titan's stratospheric zonal winds. Zonal winds in Titan's stratosphere were also inferred from observations taken with the UV-Visual Echelle Spectrograph (UVES) of the VLT around January 14, 2005, by Luz et al. (2006), who measured the Doppler-shifted solar spectrum reflected by Titan's atmosphere in the visible range. These Doppler retrievals provided a vertical distribution of the winds between 10 and 450 km in altitude for the first time; the profile showed a prograde wind with an increased velocity in the upper stratosphere.

Adaptive optics spectra and images were taken using three different telescopes in order to characterize Titan's surface and atmosphere at the time of the Huygens descent. Several authors discuss the presence or absence of cloud features at the time of the Huygens descent. The bright tropospheric cloud activity observed in the South pole was not obvious in Jan. 15 and 16 VLT/NACO data (Hirtzig et al. 2006) nor in Gemini Jan. 14 observations (Schaller et al. 2006b), but other cloud-like bright features seem to be present on the western limb. No clouds were observed near the Huygens landing site and it appears that the probe's descent occurred during a rather quiescent period of cloud activity on Titan which lasted for at least 5 months.

Two vertical distributions of atmospheric ethane, fitting equally well high-resolution infrared spectra taken with Subaru, were retrieved by Livengood et al. (2006).

Several investigators focused on surface properties at the Huygens landing site. Hirtzig et al. (2006) exploited their VLT/NACO images to show that the surface spectrum at 1.28, 1.6 and 2.0 μm was compatible with a combination of water ice, possibly mixed CH_4 ice. 2-μm spectra taken on 16 January 2005 also with NACO agreed with this result and indicated that the Huygens landing site was among the darker albedo areas on Titan (Negrão et al. 2007), thus giving a broader context to the probe's measurements.

2.6 Concluding Remarks and Open Questions Before Cassini–Huygens

One consequence of the success of the Voyager encounters was a sharp increase in the level of interest in Titan. Enough was revealed, especially about the atmosphere, to stimulate a considerable number of theoretical and modeling studies, most of which raised further questions which went on to define the goals for a future space mission to the Saturnian System. After initial discussions around a mission called "SOP^2" (for Saturn Orbiter with dual Probe), Cassini–Huygens was born and called to bring answers to these questions. Some of the most intriguing ones included:

- Where did the atmosphere come from, and why is it unique in the outer Solar System?
- What is the degree of complexity achieved by the chemistry on Titan?
- What is the nature of the surface, its composition and topography?
- What is the circulation of the atmosphere? How can best the alkanological cycle be described and where is the methane reservoir?

The scientific objectives advertised by the mission representatives covered the above questions and matched them with precise measurement requirements which included:

- Determine the abundances of atmospheric constituents (including any noble gases)
- Establish isotope ratios for abundant elements, which will help constrain scenarios of formation and evolution of Titan and its atmosphere.
- Observe vertical and horizontal distributions of trace gases
- Detect new molecules
- Investigate energy sources for atmospheric chemistry,
- Study the formation, composition and distribution of aerosols.
- Determine the winds and map the global temperatures, investigate cloud physics, general circulation, and seasonal effects in Titan's atmosphere;
- Search for lightning discharges.
- Determine the physical state, topography, and composition of the surface with in situ measurements at different locations of the disc,
- Infer the internal structure of the satellite.
- Investigate the upper atmosphere, its ionisation, and its role as a source of neutral and ionised material for the magnetosphere of Saturn.

We will be dealing with these and other questions, as well as generally with the progress made with the Cassini–Huygens mission, more in detail in the following chapters.

In the meantime, ground-based measurements are still profitable, even with the Cassini–Huygens mission on the spot since 2004 and for another 2 years at least. We have shown in the previous sections that even a powerful mission like Cassini requires complementary information from the ground (or from Earth-orbiting satellites) as will post-Cassini missions which are already being discussed (see Chapter 18). For example, since the Huygens probe explored the surface at one single area, the ground-based measurements acquired simultaneously allowed for the Cassini–Huygens data to be extrapolated to the whole disk surface. Furthermore, ground-based observations can provide measurements at solar phase angles not attained by Cassini (e.g., at specific small phase angles); complementary observations for regions that are unlit (all flybys include global images and spectra before and after the close encounters; regions that are unlit can be observed within a few days on the ground); data during times the spacecraft is observing other objects to look for time-variable phenomena (cloud formation and decay); and measurements at wavelengths that are not included in spacecraft instrumentation. Furthermore, ground-based stellar occultations, and especially those yielding a central flash, may be valuable as they provide unique insight on the zonal wind regime around 250 km altitude (0.25 mbar level). Monitoring such events might thus tells us how winds seasonally evolve with time.

In addition, future Earth-based studies could bring insights to some of the unanswered questions put forth in this chapter. Indeed, telescopic observations from future large observatories like the Thirty Meter Telescope, the Atacama Large Millimeter/submillimeter Array (ALMA), as well as satellites like Herschel and the SOFIA aircraft, will be able to add to our knowledge of – for instance – Titan's chemical composition, by enhancing the available Titan data in imaging and spectroscopy. Other such large future facilities which could potentially contribute to Titan science include the Expanded Very Large Array (EVLA), the European Extremely Large Telescope (E-ELT), the Giant Segmented Mirror Telescope (GSMT), and the James Webb Space Telescope (JWST). Continuous monitoring from the ground will certainly (beyond any space mission) help to cover the full Titan year and return information on the satellite's complex seasonal phenomena.

References

Adamkovics M, Wong MH, Laver C, de Pater I (2007) Widespread morning drizzle on Titan. Science 318:962–965

Allen M, Pinto IP, Yung YL (1980) Titan: aerosol photochemistry and variations related to the sunspot cycle. Astrophys J 242: L125–L128

Asano S (1979) Light scattering properties of spheroidal particles. Appl Opt 18:712–723

Atreya SK, Kuhn WR, Donahue TM (1980) Saturn – tropospheric ammonia and nitrogen. Geophys Res Let 7:474–476

Atreya SK, Adams EY, Niemann HB, Demick-Montelara JE, Owen TC, Fulchignoni M, Ferri F, Wilson EH (2006) Titan's methane cycle. Plan Space Sci 54:1177–1187

Bell TH (1975) Topographically generated internal waves in the open oceans. J Geophys Res 80:320–327

Bellucci, A.; Sicardy, B.; Drossart, P.; Rannou, P.; Nicholson, P. D.; Hedman, M.; Baines, K. H.; Burrati, B., 2009. Titan solar occultation observed by Cassini/VIMS: Gas absorption and constraints on aerosol composition. Icarus 201, 198-216.

Bézard B, Marten A, Paubert G (1993) Detection of acetonitrile on Titan. Bull Am Astron Soc 25(3):1100
Bézard B, Coustenis A, McKay CP (1995) Titan's stratospheric temperature asymmetry: a radiative origin. Icarus 113:267–276
Bézard B, Drossart P, Encrenaz Th, Feuchtgruber H (2001) Benzene on the giant planets. Icarus 154:492–500
Bird MK, Allison M, Asmar SW, Atkinson DH, Avruch IM, Dutta-Roy R, Dzierma Y, Edenhofer P, Folkner WM, Gurvits LI, Johnston DV, Plettemeier D, Pogrebenko SV, Preston RA, Tyler GL (2005) The vertical profile of winds on Titan. Nature 438:800–802
Bouchez A (2004) Seasonal trends in Titan's atmosphere: haze, wind, and clouds, Ph.D. thesis, California Institute of Technology, Pasadena
Bouchez AH, Brown ME (2005) Statistics of Titan's south polar tropospheric clouds. Astrophys J 618:L53–L56
Briggs FH (1974) The radio brightness of Titan. Icarus 22:48
Broadfoot AL, Belton MJS, Takacs PZ, Sandel BR, Shemansky DE, Holberg JB, Ajello JM, Atreya SK, Donahue TM, Moos HW, Bertaux JL, Blamont JE, Strobel DF, McConnell JC, Dalgarno A, Goody R, McElroy MB (1981) Extreme ultraviolet observations from Voyager 1 encounter with Saturn. Science 212:206–211
Brown ME, Bouchez AH, Griffith CA (2002) Direct detection of variable tropospheric clouds near Titan's south pole. Nature 420:795–797
Caldwell J (1977) Thermal radiation from Titan's atmosphere. In: Burns JA (ed) Planetary satellites, University of Arizona, Tucson, pp 420–437
Caldwell J, Cunningham CC, Anthony D, White HP, Groth EJ, Hasan H, Noll K, Smith PH, Tomasko MG, Weaver HA (1992) Titan: evidence for seasonal change – a comparison of hubble space telescope and voyager images. Icarus 97:1–9
Campbell DB, Black GJ, Carter LM, Ostro SJ (2003) Radar evidence for liquid surfaces on Titan. Science 302:431–434
Chang S, Scattergood T, Aronowitz S, Flores J (1979) Organic chemistry on Titan. Rev Geophys Space Phys 17:1923–1933
Coll P, Coscia D, Smith N, Gazeau M-C, Ramírez SI, Cernogora G, Israël G, Raulin F (1999) Experimental laboratory simulation of Titan's atmosphere: aerosols and gas phase. Plan Space Sci 47:1331–1340
Comas Solá J (1908) Observationes des Satellites Principaux de Jupiter et de Titan. Astron Nachr 179:289–290
Courtin R, Kim SJ (2002) Mapping of Titan's tropopause and surface temperatures from Voyager IRIS spectra. Plan Space Sci 50:309–321
Courtin R, Gautier D, McKay CP (1995) Titan's thermal emission spectrum: re-analysis of the Voyager infrared measurements. Icarus 114:144–163
Coustenis A (2007) «Titan ». In *Encyclopedia of the Solar System*, Second Edition, eds. P. R. Weissman, L.-A. McFadden, T.V. Johnson, Academic Press.
Coustenis A, Bézard B (1995) Titan's atmosphere from Voyager infrared observations IV. Latitudinal variations of temperature and composition. Icarus 115:126–140
Coustenis A, Bézard B, Gautier D (1989a) Titan's atmosphere from Voyager infrared observations: I The gas composition of Titan's equatorial region. Icarus 80:54–76
Coustenis A, Bézard B, Gautier D (1989b) Titan's atmosphere from Voyager infrared observations: II. The CH3D abundance and D/H ratio from the 900–1200 cm^{-1} spectral region. Icarus 82:67–80
Coustenis A, Bézard B, Gautier D, Marten A, Samuelson RE (1991) Titan's atmosphere from Voyager infrared observations III. Vertical distributions of hydrocarbons and nitriles near Titan's north pole. Icarus 89:152–167
Coustenis A, Lellouch E, Maillard J-P, McKay CP (1995) Titan's surface: composition and variability from the near-infrared albedo. Icarus 118:87–104
Coustenis A, Salama A, Lellouch E, Encrenaz Th, Bjoraker G, Samuelson RE, de Graauw Th, Feuchtgruber H, Kessler MF (1998) Evidence for water vapor in Titan's atmosphere from ISO/SWS data. Astron Astrophys 336, L85–L89
Coustenis A, Gendron E, Lai O, Véran J-P, Woillez J, Combes M, Vapillon L, Fusco T, Mugnier L, Rannou P (2001) Images of Titan at 1.3 and 1.6 μm with aqdaptive optics at the CFHT. Icarus 154:501–515
Coustenis A, Salama A, Schulz B, Ott S, Lellouch E, Encrenaz Th, Gautier D, Feuchtgruber H (2003) Titan's atmosphere from ISO mid-infrared spectroscopy. Icarus 161:383–403
Coustenis A, Hirtzig M, Gendron E, Drossart P, Lai O, Combes M, Negrão A (2005) Maps of Titan's surface from 1 to 2.5 μm. Icarus 177:89–105
Coustenis A, Negrao A, Salama A, Schulz B, Lellouch E, Rannou P, Drossart P, Encrenaz Th, Schmitt B, Boudon V, Nikitin A (2006) Titan's 3-μm spectral region from ISO high-resolution spectroscopy. Icarus 180:176–185
Cruikshank DP, Morgan JS (1980) Titan – suspected near-infrared variability. Astrophys J Lett 235:L53+
Danielson RE, Caldwell JJ, Larach DR (1973) An inversion in the atmosphere of Titan. Icarus 20:437–443
de Bergh C, Lutz BL, Owen T, Chauville J (1988) Monodeuterated methane in the outer solar system III. Its abundance on Titan. Astron Astrophys J 329:951–955
De Pater I, Ádámkovics M, Bouchez AH, Brown ME, Gibbard SG, Marchis F, Roe HG, Schaller EL, Young E (2006) Titan imagery with Keck adaptive optics during and after probe entry. J Geophys Res 111: E07S05
Dermott SF, Sagan C (1995) Tidal effects of disconnected hydrocarbon seas on Titan. Nature 374:238
Dubouloz N, Raulin F, Lellouch E, Gautier D (1989) Titan's hypothesized ocean properties:the influence of surface temperature and atmospheric composition uncertainties. Icarus 82:81–94
Elliot JL, Veverka J, Goguen J (1975) Lunar occultation of Saturn. I – the diameters of Tethys, Dione, Rhea, Titan, and Iapetus. Icarus 26:387–407
Eshleman VR, Lindal GF, Tyler GL (1983) Is Titan wet or dry? Science 221:53–55
Fink U, Larson HP (1979) The infrared spectra of Uranus, Neptune, and Titan from 0.8 to 2.5 μm. Astrophys J 233:1021–1040
Flasar FM (1983) Oceans on Titan? Science 221:55–57
Flasar FM, Conrath BJ (1990) Titan's stratospheric temperatures: a case for dynamical inertia? Icarus 85:346–354
Flasar FM, Samuelson RE, Conrath BJ (1981) Titan's atmosphere: temperature and dynamics. Nature 292:693–698
Flasar FM, Achterberg RK, Conrath BJ, Bjoraker GL, Jennings DE, Pearl JC, Romani PN, Simon-Miller AA, Kunde VG, Nixon CA, Bézard B, Orton GS, Spilker LJ, Irwin P, Teanby NA, Spencer JA, Owen TC, Brasunas JC, Segura ME, Carlson R, Matmoukine A, Giearasch PJ, Schinlder PJ, Ferrari C, Showalter MR, Barucci A, Courtin R, Coustenis A, Fouchet T, Gautier D, Lellouch E, Marten A, Prangé R, Strobel DF, Calcutt SB, Read PL, Taylor FW, Bowles N, Samuelson RE, Abbas MM, Raulin F, Ade P, Edgington S, Pilorz S, Wallis B, Wishnow E (2005) Temperatures, winds, and composition in the Saturnian system. Science 307:1247–1251
Folkner WM, Asmar SW, Border JS, Franklin GW, Finley SG, Gorelik J, Johnston DV, Kerzhanovich VV, Lowe ST, Preston RA, Bird MK, Dutta-Roy R, Allison M, Atkinson DH, Edenhofer P, Plettemeier D, Tyler GL (2006) Winds on Titan from ground-based tracking of the Huygens probe. J Geophys Res 111:Issue 7, CiteID E07S02
Friedson AJ (1994) Gravity waves in Titan's atmosphere. Icarus 109: 40–57
Friedson AJ, Yung YL (1984) The thermosphere of Titan. J Quant Spectrosc Radiat Transfer 89:85–90
Fulchignoni M, Ferri F, Angrilli F, Ball AJ, Bar-Nun A, Barucci MA, Bettanini C, Bianchini G, Borucki W, Colombatti G, Coradini M, Coustenis A, Debei S, Falkner P, Fanti G, Flamini E, Gaborit V, Grard R, Hamelin M, Harri AM, Hathi B, Jernej I, Leese MR, Lehto A, Lion Stoppato PF, Lopez-Moreno JJ, Mäkinen T, McDonnell JAM, McKay CP, Molina-Cuberos G, Neubauer FM, Pirronello V,

Rodrigo R, Saggin B, Schwingenschuh K, Seiff A, Simoes F, Svedhem H, Tokano T, Towner MC, Trautner R, Withers P, Zarnecki JC (2005) Titan's physical characteristics measured by the Huygens Atmospheric Instrument (HASI). Nature 438:785–791
Gardner CS (1994) Diffusive filtering theory of gravity wave spectra in the atmosphere. J Geophys Res 99:(20)601–(20)622
Gendron E, Coustenis A, Drossart P, Combes M, Hirtzig M, Lacombe F, Rouan D, Collin C, Pau S, Lagrange A-M, Mouillet D, Rabou P, Fusco Th, Zins S (2004) VLT/NACO adaptive optics imaging of Titan. Astron Astroph 417:L21–L24
Gibbard SG, Macintosh B, Gavel D, Max CE, de Pater I, Ghez AM, Young EF, McKay CP (1999) Titan: high-resolution speckle images from the keck telescope. Icarus 139:189–201
Gibbard SG, Macintosh B, Gavel D, Max CE, de Pater I, Roe HG, Ghez AM, Young EF, McKay CP (2004) Speckle imaging of Titan at 2 μm: surface albedo, haze optical depth, and tropospheric clouds 1996–1998. Icarus 169:429–439
Gillett FC (1975) Further observations of the 8–13 μm spectrum of Titan. Astrophys J 201:L41–L43
Gillett FC, Forrest WJ, Merril KM (1973) 8–13 μm observations of Titan. Astrophys J 184:L93–L95
Griffith CA (1993) Evidence for surface heterogeneity on Titan. Nature 364:511–514
Griffith CA, Owen T, Wagener T (1991) Titan's surface and troposphere, investigated with ground-based, near-infrared observations. Icarus 93:362–378
Griffith CA, Owen T, Miller GA, Geballe T (1998) Transient clouds in Titan's lower atmosphere. Nature 395:575–578
Griffith CA, Hall JL, Geballe TR (2000) Detection of daily clouds on Titan. Science 290:509–513
Griffith CA, Owen T, Geballe TR, Rayner J, Rannou P (2003) Evidence for the exposure of water ice on Titan's surface. Science 300:628–630
Griffith CA, Penteado P, Greathouse TK, Roe HG, Yelle RV (2005) Observations of Titan's mesosphere. Astrophys J 629:L57–L60
Gurwell MA (2004) Submillimeter observations of Titan: global measures of stratospheric temperature, CO, HCN, HC3N, and the isotopic ratios 12C/13C and 14N/15N. Astrophys J 616:L7–L10
Gurwell MA (2008) Carbon, nitrogen and oxygen isotopic ratios in titan's stratosphere. American Astronomical Society, DPS meeting #40, Bull Amer Astron Soc. 20.09 [Abstract]
Gurwell MA, Muhleman DO (1995) CO on Titan: evidence for a well-mixed vertical profile. Icarus 117:375–382
Gurwell MA, Muhlemann DO (2000) CO on Titan: more evidence for a well-mixed vertical profile. Icarus 145:653–656
Hanel R, Conrath B, Flasar FM, Kunde V, Maguire W, Pearl JC, Pirraglia J, Samuelson R, Herath L, Allison M, Cruikshank DP, Gautier D, Gierasch PJ, Horn L, Koppany R, Ponnamperuma C (1981) Infrared observations of the Saturnian system from Voyager 1. Science 212:192–200. Harris, 1961
Harris DL (1961) Photometry and colorimetry of planets and satellites. In: Planets and Satellites. Gerard P. Kuiper and Barbara M. Middlehurst Eds. Chicago, The University of Chicago, p 272
Hartung M, Herbst TM, Dumas C, Coustenis A (2006) Limits to the abundance of CO2 ice on Titan. J Geophys Res Planets 111:E07S09 (7 pages)
Herschel J (1847) Results of Astronomical Observations made during the year 1834-8 at the Cape of Good Hope. London.
Hidayat T, Marten A, Bézard B, Gautier D, Owen T, Matthews HE, Paubert G (1997) Millimetre and submillimetre heterodyne observations of Titan: retrieval of the vertical profile of HCN and the 12C/13C ratio. Icarus 126:170–182
Hidayat T, Marten A, Bézard B, Gautier D, Owen T, Matthews HE, Paubert G (1998) Millimeter and submillimeter heterodyne observations of Titan: the vertical profile of carbon monoxide in its stratosphere. Icarus 133(1):109–133
Hinson DP, Tyler GL (1983) Internal gravity waves in Titan's atmosphere observed by Voyager radio occultation. Icarus 54:337–352
Hirtzig M, Coustenis A, Gendron E, Drossart P, Negrao A, Combes M, Lai O, Rannou P, Lebonnois S, Luz D (2006) Monitoring atmospheric phenomena on Titan. Astron Astrophys 456:761–774
Hirtzig M, Coustenis A, Gendron E, Drossart P, Hartung M, Negrao A, Rannou Combes M (2007) Titan: atmospheric and surface features as observed with NAOS/CONICA at the time of the Huygens' landing. J Geophys Res Planets112:E02S91
Hörst SM, Vuitton V, Yelle RV (2008) Origin of oxygen species in Titan's atmosphere. J Geophys Res 113:CiteID E10006
Hourdin F, Talagrand O, Sadourny R, Courtin R, Gautier D, McKay CP (1995) Numerical simulation of the general circulation of the atmosphere of Titan. Icarus 117:358–374
Hourdin F, Lebonnois S, Luz D, Rannou P (2004) Titan's stratospheric composition driven by condensation and dynamics. J Geophys Res 109:CiteID E12005
Hubbard WB, Hunten DM, Reitsema HJ, Brosch N, Nevo Y, Carreira E, Rossi F, Wasserman LH (1990) Results of Titan's atmosphere from its occultation of 28 Sagittari. Nature 343:353–355
Hubbard WB, Porco CC, Hunten DM, Rieke GH, Rieke MJ, McCarthy DW, Haemmerle V, Clark R, Turtle EP, Haller J, McLeod B, Lebofsky LA, Marcialis R, Holberg JB, Landau R, Carrasco L, Elias J, Buie MW, Persson SE, Boroson T, West S, Mink DJ (1993) The occultation of 28 Sgr by Titan. Astron Astrophys 269:541–563
Hunten DM (1973a) The escape of light gases from planetary atmospheres. J Atmos Sci 30:1481–1494
Hunten DM (1973b) The escape of H2 from Titan. J Atmos Sci 30:726–732
Hunten, DM (1977) Titan's atmosphere and surface. In: Burns JA (ed) Planetary satellites, University of Arizona Press, Tucson, pp 420–437
Hunten DM (1978). A Titan atmosphere with a surface temperature of 200 K. In: JPL (ed) The Saturn system, University of Arizona Press, Tucson. pp 127–140
Hunten DM, Tomasko MG, Flasar FM, Samuelson RE, Strobel DF, Stevenson DJ (1984) Titan. In: Saturn T, Gehrels & Matthews M (eds), University of Arizona Press, pp 671–759
Hutzell WT, McKay CP, Toon OB (1993) Effects of time-varying haze production on Titan's geometric albedo. Icarus 105:162–174
Huygens C (1659) In: Systema saturnium, Hague.
Jaffe W, Caldwell J, Owen T (1980) Radius and brightness temperature observations of Titan at centimetre wavelengths by the very large array. Astrophys J 242:806–811
Jeans JH (1931) The dynamical theory of gases. 3d edition. (Dover reprint).
Khare BN, Sagan C (1973) Red clouds in reducing atmospheres. Icarus 20:311
Khare BN, Sagan C, Arakawa ET, Suits F, CallcottTA, Williams MW (1984). Optical constants of organic tholins produced in a simulated Titanian atmosphere: from soft X-ray to microwave frequencies. Icarus 60:127–137
Kim SJ, Caldwell J (1982) The Abundance of CH3D in the atmosphere of Titan derived from 8 to 14 mm thermal emission. Icarus 52:473–482
Kostiuk T, Fast KE, Livengood TA, Hewagama T, Goldstein JJ, Espenak F, Buhl D (2001) Direct measurement of winds of Titan. Geophys Res Lett 28:2361–2364
Kostiuk T, Livengood TA, Hewagama T, Sonnabend G, Fast KE, Murakawa K, Tokunaga AT, Annen J, Buhl D, Schmlling F (2005) Titan's stratospheric zonal wind, temperature, and ethane abundance a year prior to Huygens insertion. Geophys Res Lett 32:L22205. doi:10.1029/2005GL023897
Kostiuk T, Livengood TA, Sonnabend G, Fast KE, Hewagama T, Murakawa K, Tokunaga AT, Annen J, Buhl D, Schmülling F, Luz D,

Witasse O (2006) Stratospheric global winds on Titan at the time of Huygens descent. J Geophys Res 111:CiteID E07S03

Kuiper GP (1944) Titan: a satellite with an atmosphere. Astrophys J 100:378–383

Kuiper GP (1952) In: G.P. Kuiper, Editor, The Atmospheres of the Earth and Planets, University of Chicago

Kunde VG, Aikin AC, Hanel RA, Jennings DE, Maguire WC, Samuelson RE (1981) C4H2, HC3N and C2N2 in Titan's atmosphere. Nature 292(686–688):1981

Lara LM, Lellouch E, López-Moreno JJ, Rodrigo R (1996) Vertical distribution of Titan's atmospheric neutral constituents. J Geophys Res 101:23262–23283

Lara L-M, Lellouch E, Shematovich V (1999) Titan's atmospheric haze: the case for HCN incorporation. Astron Astrophys 341:312–317

Lebonnois S (2005) Benzene and aerosol production in Titan and Jupiter's atmospheres: a sensitivity study. Plan Space Sci 53:486–497

Lebonnois S, Toublanc D, Hourdin F, Rannou P (2001) Seasonal variations of Titan's atmospheric composition. Icarus 152:384–406

Lebonnois S, Bakes ELO, McKay CP (2002) Transition from gaseous compounds to aerosols in Titan's atmosphere. Icarus 159: 505–517

Lebonnois S, Hourdin F, Rannou P, Luz D, Toublanc D (2003) Impact of the seasonal variations of composition on the temperature field of Titan's stratosphere. Icarus 163:164–174

Lebreton J-P, Witasse O, Sollazzo C, Blancquaert T, Couzin P, Schipper A-M, Jones JB, Matson DL, Gurvits LI, Atkinson DH, Kazeminejad B, Pérez-Ayúcar M (2005) An overview of the descent and landing of the Huygens probe on Titan. Nature 438:758–764

Lellouch E, Coustenis A, Gautier D, Raulin F, Dubouloz N, Frère C (1989) Titan's atmosphere and hypothesized ocean: a reanalysis of the Voyager 1 radio-occultation and IRIS 7.7 μm data. Icarus 79:328–349

Lellouch E, Hunten D, Kockarts G, Coustenis A (1990) Titan's thermosphere profile. Icarus 83:308–324

Lellouch E, Coustenis A, Sebag B, Cuby J-G, Lopez-Valverde M, Schmitt B, Fouchet T, Crovisier J (2003) Titan's 5μm window: observations with the very large telescope. Icarus 162:125–142

Lemmon MT, Karkoschka E, Tomasko M (1993) Titan's rotation – surface feature observed. Icarus 103:329–332

Lemmon MT, Karkoschka E, Tomasko M (1995) Titan's rotational lightcurve. Icarus 113:27–38

Lemmon MT, Smith PH, Lorenz RD (2002) Methane abundance on Titan, measured by the space telescope imaging spectrograph. Icarus 160:375–385

Letourneur B, Coustenis A (1993) Titan's atmospheric structure from Voyager 2 infrared spectra. Planet Space Sci 41:593–602

Lewis JS (1971) Satellites of the outer planets: their physical and chemical nature. Icarus 15:174–185

Liang M-C, Heays AN, Lewis BR, Gibson ST, Yung YL (2007) Source of nitrogen isotope anomaly in HCN in the atmosphere of Titan. Astrophys J 664:L115–L118

Lindal GF, Wood GE, Hotz HB, Sweetnam DN, Eshelman VR, Tyler GL (1983) The atmosphere of Titan: an analysis of the Voyager 1 radio-occultation measurements. Icarus 53:348–363

Livengood TA, Kostiuk T, Sonnabend G, Annen JN, Fast KE, Tokunaga A, Murakawa K, Hewagama T, Schmülling F, Schieder R (2006) High-resolution infrared spectroscopy of ethane in Titan's stratosphere in the Huygens epoch. J Geophys Res 111:CiteID E11S90

Lockwood GW, Lutz BL, Thompson DT, Bus ES (1986) The albedo of Titan. Astrophys J 303:511–520

Lockwood GW, Thompson DT (2009) Seasonal photometric variability of Titan, 1972–2006. Icarus 200:616–626.

Lopez-Valverde MA, Lellouch E, Coustenis A (2005) Carbon monoxide fluorescence from Titan's atmosphere. Icarus 175:503–521

Lorenz RD, Lemmon MT, Smith PH, Lockwood GW (1999) Seasonal change on Titan observed with the hubble space telescope WFPC-2. Icarus 142:391–401

Lorenz RD, Young EF, Lemmon MT (2001) Titan's smile and collar: HST observations of seasonal change 1994–2000. Geophys Res Lett 28:4453–4456

Lorenz, R. D.; Witasse, Olivier Lebreton, Jean-Pierre; Blancquaert, Thierry; de Pater, Imke; Mazoue, Franck; Roe, Henry; Lemmon, Mark T.; Burratti, Bonnie J.; Holmes, Shadrian; Noll, Keith (2006) Huygens entry emission: Observation campaign, results, and lessons learned. J Geophys Res Volume 111, Issue E7, CiteID E07S11

Low FJ (1965) Planetary radiation at infrared and millimeter wavelengths. Lowell Observ Bull 6:184–187

Lunine JI, Stevenson DJ, Yung YL (1983) Ethane ocean on Titan. Science 222:1229+

Lunine JI, Yung YL, Lorenz RD (1999) On the volatile inventory of Titan from isotopic abundances in nitrogen and methane. Plan Space Sci 47:1291–1303

Lutz BL, Owen T, Cess RD (1976) Laboratory band strengths of methane and their application to the atmospheres of Jupiter, Saturn, Uranus, Neptune and Titan. Astron Astrophys J 203:541–551

Lutz BL, de Bergh C, Owen T (1983) Titan: discovery of carbon monoxide in its atmosphere. Science 220:1374–1375

Luz D, Hourdin F, Rannou P, Lebonnois S (2003) Latitudinal transport by barotropic waves in Titan's stratosphere. II. Results from a coupled dynamics-microphysics-photochemistry GCM. Icarus 166: 343–358

Luz D, Civeit T, Courtin R, Lebreton JP, Gautier D, Rannou P, Kaufer A, Witasse O, Lara L, Ferri F (2005) Characterization of zonal winds in the stratosphere of Titan with UVES. Icarus 179: 497–510

Luz D, Civeit T, Courtin R, Lebreton J-P, Gautier D, Witasse O, Kaufer A, Ferri F, Lara L, Livengood T, Kostiuk T (2006) Characterization of zonal winds in the stratosphere of Titan with UVES: 2. Observations coordinated with the huygens probe entry. J Geophys Res 111:CiteID E08S90

Maguire WC, Hanel RA, Jennings DE, Kunde VG, Samuelson RE (1981) C3H8 and C3H4 in Titan's atmosphere. Nature 292:683–686

Marten A, Gautier D, Tanguy L, Lecacheux A, Rosolen C, Paubert G (1988) Abundance of carbon monoxide in the stratosphere of Titan from millimeter heterodyne observations. Icarus 76:558–562

Marten A, Hidayat T, Biraud Y, Moreno R (2002) New millimeter heterodyne observations of titan: vertical distributions of nitriles HCN, HC3N, CH3CN, and the isotopic ratio 15N/14N in its atmosphere. Icarus 158:532–544

McKay CP, Pollack JB, Courtin R (1989) The thermal structure of Titan's atmosphere. Icarus 80:23–53

McKay CP, Pollack JB, Courtin R (1991) The greenhouse and antigreenhouse effects on Titan. Science 253:1118–1121

McKay CP, Martin SC, Griffin CA, Keller RM (1997) Temperature lapse rate and methane in Titan's troposphere. Icarus 129:498–505

McKay CP, Coustenis A, Samuelson RE, Lemmon MT, Lorenz RD, Cabane M, Rannou P, Drossart P (2001) Physical properties of the organic aerosols and clouds on Titan. Plan Space Sci 49:79–99

Meier R, Smith BA, Owen TC, Terrile RJ (2000) The surface of Titan from NICMOS observations with the Hubble Space Telescope. Icarus 145:462–473

Moreno R, Marten A, Hidayat T (2005) Interferometric measurements of zonal winds on Titan. Astron Astrophys 437:319–328

Moses JI, Lellouch E, Bézard B, Gladstone GR, Feuchtgruber H, Allen M (2000) Photochemistry of saturn's atmosphere II. Effects of an influx of external oxygen. Icarus 145:166–202

Muhleman DO, Berge GL, Clancy RT (1984) Microwave measurements of carbon monoxide on Titan. Science 223:393–396

Muhleman DO, Grossman AW, Butler BJ, Slade MA (1990) Radar reflectivity of Titan. Science 248:975–980

Narayan R, Hubbard WB (1988) Theory of anisotropic refractive scintillation: application to stellar occultation by Neptune. Astrophys J 325:503–518

Neff JS, Humm DC, Bergstralh JT, Cochran AL, Cochran WD, Barker ES, Tull RG (1984) Absolute spectrophotometry of Titan, Uranus, and Neptune: 3500–10, 500 angstroms. Icarus 60:221–235

Negrão A, Coustenis A, Lellouch E, Maillard J-P, Rannou Combes M, Schmitt B, McKay CP, Boudon V (2006). Titan's surface albedo from near-infrared CFHT/FTS spectra: modeling dependence on the methane absorption. Plan Space Sci 54:1225–1246

Negrão A, Hirtzig M, Coustenis A, Gendron E, Drossart P, Rannou Combes M, Boudon V (2007). 2-μm spectroscopy of Huygens' landing site on Titan with VLT/NACO. J Geophys Res Planets 112:E02S92

Noll KS, Geballe TR, Knacke RF, Pendleton YJ (1996) Titan's 5 μm spectral window: carbon monoxide and the albedo of the surface. Icarus 126:625–631

Orton G (1992) Ground-based observations of Titan's thermal spectrum. In: Kaldeich B (ed) Symposium on Titan, Toulouse, France, 1991, ESA-SP 338. pp 81–85

Owen T (1982) The composition and origin of Titan's atmosphere. Planet Space Sci 30:833–838

Owen TC (2000) On the origin of Titan's atmosphere. Plan Space Sci 48:747–752

Pinto JP, Lunine JI, Kim SJ, Yung YL (1986) D to H ratio and the origin and evolution of Titan's atmosphere. Nature 319:388–390

Pollack JB (1973) Greenhouse models of the atmosphere of Titan. Icarus 19:43–58

Porco CC, Baker E, Barbara J, Beurle K, Brahic A, Burns JA, Charnoz S, Cooper N, Dawson DD, Del Genio AD, Denk T, Dones L, Dyudina U, Evans MW, Fussner S, Giese B, Grazier K, Helfenstein P, Ingersoll AP, Jacobson RA, Johnson TV, McEwen A, Murray CD, Neukum G, Owen WM, Perry J, Roatsch T, Spitale J, Squyres S, Thomas P, Tiscareno M, Turtle EP, Vasavada AR, Veverka J, Wagner R, West R (2005) Imaging of Titan from the Cassini spacecraft. Nature 434:159–168

Rages K, Pollack JB (1983) Vertical distribution of scattering hazes in Titan's upper atmosphere. Icarus 55:50–62

Rannou P, Cabane M, Chassefière E, Botet R, McKay CP, Courtin R (1995) Titan's geometric albedo: role of the fractal structure of the aerosols. Icarus 118:355–372

Rannou P, Hourdin F, McKay CP (2002) A wind origin for Titan's haze structure. Nature 418:853–856

Rannou P, McKay CP, Lorenz RD (2003) A model of Titan's haze of fractal aerosols constrained by multiple observations. Plan Space Sci 51:963–976

Rannou P, Hourdin F, McKay CP, Luz D (2004) A coupled dynamics-microphysics model of Titan's atmosphere. Icarus 170:443–462

Rannou P, Montmessin F, Hourdin F, Lebonnois S (2006) The latitudinal distribution of clouds on Titan. Science 311:201–205

Richardson J, Lorenz RD, McEwen A (2004) Titan's surface and rotation: new results from Voyager 1 images. Icarus 170:113–124

Roe HG, de Pater I, Macintosh BA, Gibbard SG, Max CE, McKay CP (2002a) NOTE: Titan's atmosphere in late southern spring observed with adaptive optics on the W. M. Keck II 10-meter telescope. Icarus 157:254–258

Roe HG, de Pater I, Macintosh BA, McKay CP (2002b) Titan's clouds from Gemini and Keck adaptive optics imaging. Astrophys J 581: 1399–1406

Roe HG, Greathouse TK, Richter MJ, Lacy JH (2003) Propane on Titan. Astrophys J 597:L65–L68

Roe HG, de Pater I, Gibbard SG, Macintosh BA, Max CE, Young EF, Brown ME, Bouchez AH (2004) A new 1.6-μm map of Titan's surface. Geophys Res Lett 31:7–+

Roe HG, Bouchez AH, Trujillo CA, Schaller EL, Brown ME (2005a) Discovery of temperate latitude clouds on Titan. Astrophys J Lett 618:L49–L52

Roe HG, Brown ME, Schaller EL, Bouchez AH, Trujillo CA (2005b) Geographic control of Titan's mid-latitude clouds. Science 310:477–479

Sagan C (1971) The solar system beyond Mars: an exobiological survey. Space Sci Rev 11:827–866

Sagan C (1973) The greenhouse of Titan. Icarus 18:649–656

Sagan C (1974) The origin of life in a cosmic context. Origins of life 5:497–505

Sagan C, Dermott SF (1982) The tide in the seas of Titan. Nature 300: 731–733

Samuelson RE (1983) Radiative equilibrium model of Titan's atmosphere. Icarus 53:364–387

Samuelson RE, Mayo LA (1997) Steady-state model for methane condensation in Titan's troposphere. Planet Space Sci 45:949–958

Samuelson RE, Hanel RA, Kunde VG, Maguire WC (1981) Mean molecular weight and hydrogen abundance of Titan's atmosphere. Nature 292:688–693

Samuelson RF, Maguire WC, Hand RA, Kunde VG, Jennings DF, Yung YL, Aikin AC (1983) CO2 on Titan. J Geophys Res 88:8709–8715

Samuelson RE, Nath NR, Borysow A (1997a) Gaseous abundances and methane supersaturation in Titan's troposphere. Planet Space Sci 45:959–980

Samuelson RE, Mayo LA, Knuckles MA, Khanna RJ (1997b) C4N2 ice in Titan's north polar stratosphere. Planet Space Sci 45:941–948

Schaller EL, Brown ME, Roe HG, Bouchez AH (2006a) A large cloud outburst at Titan's south pole. Icarus 182:224–229

Schaller EL, Brown ME, Roe HG, Bouchez AH, Trujillo CA (2006b) Dissipation of Titan's south polar clouds. Icarus 184:517–523

Sicardy B, Brahic A, Ferrari C, Gautier D, Lecacheux J, Lellouch E, Roques F, Arlot JE, Colas F, Thuillot W, Sevre F, Vidal JL, Blanco C, Cristaldi S, Buil C, Klotz A, Thouvenot E (1990) Probing Titan's atmosphere by stellar occultation. Nature 343:350–353

Sicardy B, Ferri F, Roques F, Lecacheux J, Pau S, Brosch N, Nevo Y, Hubbard WB, Reitsema HJ, Blanco C, Carreira E, Beisker W, Bittner C, Bode H-J, Bruns M, Denzau H, Nezel M, Riedel E, Struckmann H, Appleby G, Forrest RW, Nicolson IKM, Hollis AJ, Miles R (1999) The structure of Titan's stratosphere from the 28 Sgr occultation. Icarus 142:357–390

Sicardy B, Colas F, Widemann T, Bellucci A, Beisker W, Kretlow M, Ferri F, Lacour S, Lecacheux J, Lellouch E, Pau S, Renner S, Roques F, Fienga A, Etienne C, Martinez C, Glass IS, Baba D, Nagayama T, Nagata T, Itting-Enke S, Bath K-L, Bode H-J, Bode F, Lüdemann H, Lüdemann J, Neubauer D, Tegtmeier A, Tegtmeier C, Thomé B, Hund F, deWitt C, Fraser B, Jansen A, Jones T, Schoenau P, Turk C, Meintjies P, Hernandez M, Fiel D, Frappa E, Peyrot A, Teng JP, Vignand M, Hesler G, Payet T, Howell RR, Kidger M, Ortiz JL, Naranjo O, Rosenzweig P, Rapaport M (2006) The two Titan stellar occultations of 14 November 2003. J Geophys Res 111:E11S91

Smith BA, Soderblom L, Beebe R, Boyce J, Briggs G, Buncer A, Collins SA, Hansen CJ, Johnson TV, Mitchell JL, Terrile RJ, Carr M, Cook Ii AF, Cuzzi J, Pollack JP, Danielson GE, Ingersoll A, Davies ME, Hunt GE, Masursky H, Shoemaker E, Morrison D, Owen T, Sagan C, Veverka J, Strom R, Suomi VE (1981) Encounter with Saturn: Voyager 1 imaging science results. Science 212:163–191

Smith GR, Strobel DF, Broadfoot AL, Sandel BR, Shemansky DF, Holberg JB (1982) Titan's upper atmosphere: composition and temperature from the EUV solar occultation results. J Geophys Res 87:1351–1359

Smith SA, Fritts DC, VanZandt TE (1987) Evidence for a saturated spectrum of atmospheric gravity waves. J Atmos Sci 44:1404–1410

Smith PH, Lemmon MT, Lorenz RD, Sromovsky LA, Caldwell JJ, Allison MD (1996) Titan's surface, revealed by HST imaging. Icarus 119:336–349

Sohl F, Sears WD, Lorenz RD (1995) Tidal dissipation on Titan. Icarus 115:278–294

Sromovsky LA, Suomi VE, Pollack JB, Krauss RJ, Limaye SS, Owen T, Revercomb HE, Sagan C (1981) Implications of Titan's north south brightness asymmetry. Nature 292:698–702

Stone EC, Miner ED (1981) Voyager 1 encounter with the Saturnian system. Science 212:159–162

Strobel DF (1974) The photochemistry of hydrocarbons in the atmosphere of Titan. Icarus 21:466–470

Strobel DF (1982) Chemistry and evolution of Titan's atmosphere. Planet Space Sci 30:839–848
Strobel DF (2006) Gravitational tidal waves in Titan's upper atmosphere. Icarus 182:251–258
Strobel DF, Shemansky DE (1982) EUV emission from Titan's upper atmosphere – Voyager 1 encounter. J Geophys Res 87:1361–1368
Strobel DF, Sicardy B (1997) Gravity wave and wind shear models. In:Lebreton JP (ed) Huygens science, Payload and Mission. ESA SP 1177. European Space Agency, Noordwijk, The Netherlands, pp. 299–311
Strobel DF, Hall DT, Zhu X, Summers MF (1993) Upper limit on Titan's atmospheric argon abundance. Icarus 103:333–336
Tokano T (2005) Meteorological assessment of the surface temperatures on Titan: constraints on the surface type. Icarus 173: 222–242
Tokano T, Neubauer FM (2002) Tidal winds on Titan caused by Saturn. Icarus 158:499–515
Tokano T, Neubauer FM, Laube M, McKay CP (1999) Seasonal variation of Titan's atmospheric structure simulated by a general circulation model. Planet Space Sci 47:493–520
Toon OB, McKay CP, Courtin R, Ackerman TP (1988) Methane rain on Titan. Icarus 75:255–284
Toublanc D, Parisot P, Brillet J, Gautier D, Raulin F, McKay CP (1995) Photochemical modelling of Titan's atmosphere. Icarus 113:2–16
Tracadas PW, Hammel HB, Thomas-Osip JE, Elliot JL, Olkin CB (2001) Probing Titan's atmosphere with the 1995 august stellar occultation. Icarus 153:285–294
Trafton L (1972a) On the possible detection of H2 in Titan's atmosphere. Astrophys J 175:285–293
Trafton L (1972b) The bulk composition of Titan's atmosphere. Astrophys J 175:295–306
Trafton LM (1975) Near-infrared spectrophotometry of Titan. Icarus 24:443–453
Vervack RJ, Sandel BR, Strobel DF (2004) New perspectives on Titan's upper atmosphere from a reanalysis of the Voyager 1 UVS solar occultations. Icarus 170:91–112
Veverka J (1973) Titan: polarimetric evidence for an optically thick atmosphere. Icarus 18:657–660
West RA, Smith PH (1991) Evidence for aggregate particles in the atmospheres of Titan and Jupiter. Icarus 90:330–333
West RA, Brown ME, Salinas SV, Bouchez AH, Roe HG (2005) No oceans on Titan from the absence of a near-infrared specular reflection. Nature 436:670–672
Wilson EH, Atreya SK (2004) Current state of modeling the photochemistry of Titan's mutually dependent atmosphere and ionosphere. J Geophys Res 109:CiteID E06002
Wilson EH, Atreya SK, Coustenis A (2003) Mechanisms for the formation of benzene in the atmosphere of Titan. J Geophys Res – Planets 108(E2):5014–5024
Witasse O, Lebreton J-P, Bird M, Dutta-Roy R et al. (2006) Overview of the coordinated ground-based observations of Titan during the Huygens mission. J Geophys Res Planets 111:E07S01 (12 pages)
Yelle RV (1991) Non-LTE models of Titan's upper atmosphere. Astrophys J 383:380–400
Yung YL (1987) An update of nitrile photochemistry on Titan. Icarus 72:468–472
Yung YL, Allen M, Pinto JP (1984) Photochemistry of the atmosphere of Titan: comparison between model and observation. Astrophys J 55:465–506
Zalucha A, Fitzsimmons A, Elliot JL, Thomas-Osip J, Hammel HB, Dhillon VS, Marsh TR, Taylor FW, Irwin PGJ (2007) The 2003 November 14 occultation by Titan of TYC 1343-1865-1. II. Analysis of light curves. Icarus 192:503–518
Zellner B (1973) The polarization of Titan. Icarus 18:661
Zhu X (1994) A new theory of the saturated gravity wave spectrum for the middle atmosphere. J Atmos Sci 51:3615–3626

Chapter 3
The Origin and Evolution of Titan

Jonathan Lunine, Mathieu Choukroun, David Stevenson, and Gabriel Tobie

Abstract Titan was formed as a regular satellite in a disk that was the outgrowth of the formation of Saturn itself. Unlike the Jovian system, Titan is alone in terms of its size and mass, not part of a system gradational in density and hence rock abundance, perhaps reflecting a smaller disk and greater importance of stochastic events during satellite assembly. Accretional heating of Titan was enough to melt an outer layer of water (a water "magma ocean") and sustain for a short period an environment in which exposed water or water-ammonia liquid was in contact with organic molecules. Initial warm surface conditions are supported by direct samplings of Titan's atmosphere by the mass spectrometers on board Cassini and Huygens, which provide circumstantial evidence that ammonia (NH_3) is the primordial source of Titan's atmospheric molecular nitrogen. Ammonia can be extracted from the liquid phase only if the surface temperature is above the melting point of the mixture, thus implying warm accretion.

The carbon isotopic ratio $^{13}C/^{12}C$ in hydrocarbon molecules measured by the GCMS on Huygens reflects a bulk carbon inventory that did not participate in the massive escape phase of the ancient post-accretional atmosphere, in contrast to nitrogen, whose isotopic ratio $^{15}N/^{14}N$ is modestly enhanced and thus suggests escape, though how much depends on the mechanism. The presence of a significant amount of the ^{40}K decay daughter ^{40}Ar strongly suggests that internal outgassing of volatiles, including methane and argon, has occurred through Titan's history.

Different models of the thermal and structure evolution of Titan's interior have been proposed to explain the persistence of methane at the surface over the age of the solar system (of order 100 times the lifetime of the known reservoirs of methane in the surface and atmosphere), and a modest dearth of impact craters consistent with a surface age of about a billion years. A first post-Cassini–Huygens model suggests that the formation of thin crust enriched in methane clathrate, owing to interactions between the primordial ocean and the primitive atmosphere as well as release of volatiles from the deep interior, could have delayed the crystallization of the internal ocean, favoring outgassing of methane at different epochs.

J. Lunine (✉)
Lunar and Planetary Laboratory, The University of Arizona, Tucson, AZ 85721, USA
e-mail: jlunine@lpl.arizona.edu

M. Choukroun
NASA Jet Propulsion Laboratory, 4800 Oak Grove Dr. Pasadena, CA 91109, USA

D. Stevenson
Division of Geological and Planetary Science, California Institute of Technology, Pasadena, CA 91125, USA

G. Tobie
Laboratoire de Planetologie et Geodynamique, Universite' de Nantes, 44312, Nantes, Cedex 3, France

3.1 Introduction

Titan is a complex body, of essentially the volume and mass of Jupiter's moon Ganymede, with a dense atmosphere and a surface that shows evidence for liquid erosion, high latitudes lakes of (presumably) methane and ethane, and very tentative evidence for episodes of geologic activity. But a great deal remains unknown about Titan, including (at the time of writing) the extent to which rock and ice in the interior have segregated to form a differentiated structure, and whether ammonia – oft-invoked to provide a buoyant low-melting point liquid that could rise through an ice crust – exists within this giant moon's interior. In the present chapter we consider the evidence for the conditions around Saturn within which Titan formed, the energetics of accretion and the subsequent differentiation of the interior, the long-term co-evolution of atmosphere, surface and interior with time, and the origin of the major volatiles N_2 and CH_4 that populate Titan's atmosphere and "hydrosphere". We end with questions raised by the superbly successful Cassini–Huygens mission, the answers to which will likely require a follow-on mission.

3.2 General Constraints on the Environment Around Early Saturn

Saturn, like Jupiter, is composed primarily of hydrogen and helium and this dictates the standard view of the origin of their satellite systems. In this standard picture, the late stages

R.H. Brown et al. (eds.), *Titan from Cassini-Huygens*,

of the formation of these planets involves an inflow of gas from the solar nebula that has too much angular momentum to merge directly with the contracting protogiant planet. As a consequence, a disk is mandatory and it is within this disk that the satellite systems are expected to form (cf. Machida et al. 2008). Mechanisms for the formation of the satellites by giant impact (applicable to our Moon, possibly the Uranian system, Cameron 1975) or capture (applicable to Triton, Goldreich et al. 1989) are unattractive for the Galilean satellites or for Titan and the inner Saturnian satellites, given the great regularity of these satellite systems. In this respect, they are often viewed as "miniature solar systems" and it would be reasonable to suppose that we can understand the environment of the formation of Titan using some of the same conceptual framework that we might use to understand Earth in the context of planet formation around the newly forming Sun. However, there is at least one important difference: The material delivered to the near-Saturn environment has already been much processed by solar nebular processes and is delivered in a manner and on a timescale dictated by the planet formation process rather than merely the gravitational collapse of a cloud of interstellar gas. The other possible major difference is that the formation of Titan may have been mostly in the presence of gas, whereas the formation of the terrestrial planets is thought by most workers to be in the absence of a nebula (cf., the very successful "Nice" model and similar models, Morbidelli and Levison 2008).

The natural dynamic timescales of probable relevance to the Titan-forming region will scale with the orbital timescales around Saturn and are short compared to most of the dynamical timescales in the solar nebula, so the solar nebula can be expected to act as a controlling mechanism on the formation process. In the picture of a "miniature solar system", we must ask about the context provided by the formation of the solar system since this is likely to set much of the composition of the starting material and the timing of accumulation. Inheritance is the crucial issue here: To what extent can we use what we think we understand about the solar system to derive what we expect to find in or on Titan and to what extent must we choose instead to attribute the outcome to processes in the sub-nebula or indeed on the surface of Titan as it formed? To what extent do the satellites (not necessarily Titan itself) tell us the timescales of relevance, by the extent of their accretional or hypothesized ^{26}Al heating? In all likelihood, the inheritance is a mixed one, and to the extent we have data that pertain to this, they do not suggest an extreme dominance of one of the three processes (solar nebula, subnebula or Titan-centric). We begin accordingly with the solar nebula context before consideration of the more local context.

Giant planets must form prior to the dissipation of protoplanetary disks. Optically thick dust disks typically survive for only a few million years (Briceno et al. 2007; Wadhwa et al. 2007), and protoplanetary disks have lost essentially all of their gases by the age of $<10^7$ years (Meyer et al. 2007), implying that giant planets formed on this timescale or less. Recent reviews of models for the formation of gas giant planets include Wuchterl et al. (2000) and Lissauer and Stevenson (2007). Star-like direct quasispherical collapse is not likely, both because of the observed brown dwarf desert and theoretical arguments against the formation of Jupiter-mass objects by fragmentation (Bodenheimer et al. 2000a). The theory of giant planet formation that is favored by most researchers is the *core nucleated accretion model*, in which the planet's initial phase of growth resembles that of a terrestrial planet, but the planet becomes sufficiently massive (typically several Earth masses) that it is able to accumulate substantial amounts of gas from the surrounding protoplanetary disk. The only other giant planet formation scenario receiving significant attention is the *gas instability model*, in which a giant planet forms directly from the contraction of a clump that was produced via a gravitational instability in the protoplanetary disk. Numerical calculations show that Jupiter mass clumps can form in sufficiently gravitationally unstable disks (e.g., Boss 2000; Mayer et al. 2002). However, weak gravitational instabilities excite spiral density waves; density waves transport angular momentum that leads to spreading of a disk, lowering its surface density and making it more gravitationally stable. Rapid cooling or mass accretion is required to make a disk highly unstable. Thus, long-lived clumps can only be produced in protoplanetary disks with highly atypical physical properties (Rafikov 2005). Additionally, gas instabilities would yield massive stellar-composition planets, requiring a separate process to explain the smaller bodies in our Solar System and the heavy element enhancements in Jupiter and Saturn. The existence of intermediate objects like Uranus and Neptune is particularly difficult to account for in such a scenario. Furthermore, metal-rich stars are more likely to host observable extrasolar planets than are metal poor stars (Fischer and Valenti 2005; Udry et al. 2007); this trend is consistent with the requirement of having sufficient condensables to form a massive core, but runs contrary to the requirement of rapid disk cooling needed to form long-lived clumps via gravitational instabilities (Cai et al. 2006).

We focus accordingly on core nucleated accretion. In this model (Pollack et al. 1996; Bodenheimer et al. 2000b; Hubickyj et al. 2005), the formation and evolution of a giant planet is viewed to occur in the following sequence: (1) Dust particles in the solar nebula form planetesimals that accrete one another, resulting in a solid core surrounded by a low mass gaseous envelope. Initially, runaway accretion of solids occurs, and the accretion rate of gas is very slow. As the solid material in the planet's feeding zone is depleted, the rate of solids accretion tapers off. The gas accretion rate steadily increases and eventually exceeds the accretion rate of solids. (2) The protoplanet continues to grow as the gas accretes at a relatively constant rate. The mass of the solid core also increases, but at a slower rate. The term 'solids' is conventionally used to refer to the entire condensed (solid and liquid) portion of the planet. Accretion energy (and radioactive decay) heats a growing planet, and can cause material that was accreted in solid form to melt and vaporize. Vaporization of ices and other heavy compounds can significantly affect the properties of the planet's atmosphere, and its ability to radiate energy and to accrete more gas. Melting has little effect on the overall growth of the planet, except for the capacity of the melt to release or trap gasses. Eventually, the core and envelope masses become equal. (3) Near this point, the rate of gas accretion increases in runaway fashion, and the protoplanet grows at a rapidly accelerating rate. The first three parts of the evolutionary sequence are referred to as the nebular stage, because the outer boundary of the protoplanetary envelope is in contact with the solar nebula, and the density and temperature at this interface are those of the nebula. (4) The gas accretion rate reaches a limiting value defined by the rate at which the nebula can transport gas to the vicinity of the planet. After this point, the equilibrium region of the protoplanet contracts, and gas accretes hydrodynamically into this equilibrium region. This part of the evolution is considered to be the transition stage. (5) Accretion is stopped by either the opening of a gap in the disk as a consequence of the tidal effect of the planet, accumulation of all nearby gas, or by dissipation of the nebula. Once accretion stops, the planet enters the isolation stage. The planet then contracts and cools to the present state at constant mass.

In this picture, the in situ formation of satellite systems is presumed to be during the last stages of giant planet formation (Pollack and Reynolds 1974; Canup and Ward 2002; Mosqueira and Estrada 2003). Although we are interested here in Titan, it is of relevance to review how this picture has been applied to the particularly impressive and massive Galilean satellite system. Indeed, the Galilean satellites exhibit a distribution in mean density with orbital distance that has often been attributed to a thermal environment dictated by the newly formed Jupiter (Pollack and Reynolds 1974). Moreover, Ganymede and Callisto are roughly half water ice and half rock, and the Galileo gravity data have been interpreted to imply that Callisto has most of this ice mixed with rock. It seems that conditions must have been appropriate for the condensation of water ice at the location where Ganymede formed, and the inferred lack of complete differentiation of Callisto requires the formation of that body on a time scale exceeding about 0.1 Myr, so that water ice would not melt and lead to a fully differentiated structure. This timescale argument has figured prominently in the recent models of satellite formation referenced above but is suspect because Callisto may not be hydrostatic, and it is only in hydrostatic conditions that the inversion of gravity to infer moment of inertia can be done with confidence. Titan is a good analog to Callisto in at least one very important respect: They are at similar large distances from the planet they orbit (in units of the planet radius) and accordingly have similar, small tidal and rotational distortions of their gravity, thus exacerbating the concern about hydrostaticity. Titan's gravity is now better characterized than that of the Galilean satellites and recent information (Iess et al. 2008) points to an incompletely differentiated structure, perhaps suggesting an extending accretion time of order 100,000 years or more, but with some lingering concerns about non-hydrostatic effects.

The argument for an imposed temperature gradient in the Galilean satellite forming region is not universally accepted but (more importantly for Titan), it does not gain support from the distribution of the satellite densities around Saturn. The difficulty with the Saturnian system lies, perhaps, in the existence of only one really large satellite. All the others are so small that stochastic effects may play a role in their mean densities, and they are indeed quite varied. It is also possible that a temperature gradient existed in the Saturnian satellite formation environment but that it was less strong than for Jupiter so that ice could condense everywhere.

In addition to a constraint on accretion time, there may be a constraint on the absolute timing of satellite formation, should ^{26}Al heating play a role. The incompatibility of the present spin state and shape of Iapetus suggest that this large moon of Saturn was heated and cooled in its deep interior before a substantial outer rigid layer could be softened, allowing spin-down from a rapid spin state while preserving the corresponding oblate shape. A plausible heat source (perhaps the only one) is then short-lived radioactivities at an abundance a fraction of the full primordial solar values (which would have fully melted the satellite); the constraints imposed by this hypothesis yield a time of formation of 3–5 Myr after the appearance of CAI's in the solar system (Castillo-Rogez et al. 2007). Any substantial gaseous disk and volatile-charged planetesimals by which Titan would have inherited its nitrogen- and carbon-bearing inventory of molecules should have had a lifetime equal to or shorter than this. A similar type of constraint applied to the formation of Rhea yields a formation time of 2–7 Myr (Barr and Canup 2008). This constraint does not violate either the lifetime of the solar nebula, which based on astrophysical observations of equivalent disks around other stars was of order this duration, nor the timescales for satellite assembly (Stevenson et al. 1986).

Bulk composition of solids: The mass-weighted densities of the Saturnian satellites, excluding Titan, is 1,200 $kg{\cdot}m^{-3}$ (Jacobson 2004). Interestingly, although Enceladus's density is close to that of Titan at present, if the current rate of activity

(Tian et al. 2007) had been sustained over the age of the solar system its original density would have been closer to that of the mass-weighted mean for the icy satellites cited above. This mean is consistent with a disk that was much more water-rich than the solar nebula, for example via a highly reduced composition for the carbon-bearing species –CH_4 instead of CO or CO_2 (Johnson and Lunine 2005). If a chemically-reduced composition was an intrinsic property of the circum-Saturnian gas it would imply that the carrier of at least some of the nitrogen was ammonia (Prinn and Fegley 1981), consistent with other considerations discussed below about Titan's noble gas inventory. However, to explain the most ice rich members of the satellite cohort would require a mechanism for the segregation of the rock from the ice across the Saturn system (Wong et al. 2008), or additional enrichment associated with the formation of Saturn itself (Mosqueira and Estrada 2003). It also is not consistent with the bulk density of Titan itself, which however might have increased during the late stages of formation if accretional heating or large impacts drove off water (Section 3.4). Alternatively, the satellite densities may tell us little about the overall disk rock-to-ice ratio if they are in fact pieces of a second (or several) satellites the mass of Titan, which migrated inward by interaction with the gas (Canup and Ward (2006) and was (or were) eventually lost to Saturn after collisions.

Other constraints on the circum-Saturnian nebula are associated more specifically with the composition of Titan's atmosphere, and discussed in the next section.

3.3 Compositional and Physical Constraints on Titan Formation

Understanding the composition of the building blocks that formed Titan requires a good knowledge of the composition of the solar nebula around Saturn. The composition of comets, assessed by ground-based observations and in-situ measurements, has long been inferred as the best indicator of the early Solar nebula composition (e.g. Bockelée-Morvan et al. 2004, and references therein). Henceforth, cometary nuclei, which have been less subject to the further chemical evolution that may have altered the initial composition (Irvine et al. 2000), are believed to be representative of the most pristine materials of the solar nebula. Since the Saturn system was derived from the solar nebula, the models of the formation of giant planets and their satellites must explain volatile enrichments that have been measured by the Galileo probe in C, N, S, and noble gases, at Jupiter (Atreya et al. 1999) relatively to the solar abundances (Anders and Grevesse 1989; Lodders 2003). Such enrichments have been measured on the carbon abundance in Saturn's atmosphere by the Composite InfraRed Spectrometer (CIRS) onboard the Cassini spacecraft (Flasar et al. 2005; Orton et al. 2005), which show an enrichment of ~7 compared to the solar C/H ratio (Hersant et al. 2008). Ground-based measurements have also suggested comparable enrichments in N and S (e.g. Briggs and Sackett 1989) at Saturn. On Titan, in-situ isotopic measurements have been conducted with the Gas Chromatograph and Mass Spectrometer onboard the Huygens probe (e.g. Niemann et al. 2005), and with the Ion Neutral Mass Spectrometer onboard the Cassini spacecraft (e.g. Waite et al. 2006). The fractionation of isotopes currently present in the atmosphere of Titan sheds new light on the processes that have affected the bulk of Titan, from its formation until today. In this section we review the astrophysical models applicable to Titan's formation and compare predicted compositions with the data obtained during the Cassini–Huygens prime mission.

3.3.1 Satellite Formation and the End of the Saturn Subnebula

In the framework of the formation of giant planets, and particularly Saturn, see Section 3.1, two groups of theories aim at explaining the formation of satellites in the framework of this evolution. One group considers that the hydrogen collapse onto Saturn's proto-core is compensated by outward transport of angular momentum, resulting in the formation of a dense subnebula around the giant planet. This scenario has been modeled by Magni and Coradini (2004) for Jupiter's case. Mosqueira and Estrada (2003) used a thick inner region and a thin outer disk to distinguish the conditions at the Galilean satellites. This type of model can be applied at Saturn. The subsequent evolution of the dense Saturn subnebula would have been dominated by turbulent transport. Mousis et al. (2002) developed a turbulent evolution model, which provides some insights into the presence of CH_4 and NH_3 within the subnebula, see Section 3.6.2. However, this model assumed thermal equilibrium at all times, is valid for an optically thin disk, and neglects the intake of gas of Saturn from the subnebula. Another group of theories argues that a dense circumplanetary disk: (1) does not lead to accurate reproduction of the icy compositions at Ganymede and Callisto; and (2) would apply strong torques on the satellites, which may trigger the loss of large satellites to onward decay to the planet. It has thus been suggested that temporary gaps are generated in the solar nebula as a result of giant planet formation (e.g. Lubow et al. 1999). The replenishment of these gaps with solar nebula material would then have formed a gas-starved disk, with a composition close to that of the feeding zone of the giant planet. Such a scenario has been proposed by Canup and Ward (2002, 2006) for Jupiter; it is also applicable to Saturn, although the duration of the events may be different than at Jupiter. The formation of Titan inside the Saturn subnebula occurred later in the evolution of the subnebula.

The condensation of a gas-starved subnebula can be divided into two steps (e.g. Alibert and Mousis 2007): (1) First, the whole subdisk is in equilibrium, with an accretion rate roughly constant. During this phase, the materials falling onto Saturn are resupplied from the surrounding solar nebula in Saturn's feeding zone. (2) Once the solar nebula has disappeared (~0.36 Myr after the accretion of 70% of Saturn's mass, according to Alibert and Mousis 2007), the final evolution corresponds to the accretion of the remaining material inside the subnebula. The accretion rate cannot be constant throughout the disk in such a case. Theoretical calculations from Weidenschilling (1997) suggested that planetesimal growth occurs in a few tens of thousand years. However the observation of disks around young stars provided evidence for micrometer to millimeter scale grains after up to 5 Myr (e.g. Clampin et al. 2003). Planetesimal disruption due to the intense collisions (e.g. Benz 2000) may thus be responsible for the persistence of small-size grains for several millions of years in the solar nebula, and therefore in the Saturn subnebula. Such a setting is very favorable for trapping mechanisms of volatiles to occur in the small particles that formed the planetesimals (e.g. Gautier and Hersant 2005).

The growth of Titan and the other Saturnian satellites in the subnebula was achieved by the collision of particles to form rock-ice planetesimals, which grew within the feeding zone of Saturn. Further growth of planetesimals and the attraction exercised onto the surrounding particles is the most likely mechanism for formation of satellites in the circumplanetary disk. This process probably took place after the hydrodynamic collapse of the solar nebula hydrogen gas and the formation of the subnebula. The stage of evolution of the subnebula at which the satellites were formed remains poorly constrained, whether before the complete depletion of the feeding zone of Saturn or after this event. Nevertheless, the mechanism of planetesimal growth within the subnebula advocated implies a very similar volatile composition between the building blocks that formed Titan and those that formed the envelope of Saturn, accreted in the latest stages of the subnebula.

Little constraint can be brought to the composition of refractory materials on Titan, these materials being mostly the silicates, which are not observable on the surface of Titan. Titan's mean volumic mass is of 1,881 kg·m^{-3} (Burns 1986; Morrison et al. 1986; Schubert et al. 1986). First-order information of the bulk composition of Titan can be derived from this measurement. Stevenson (1992) inferred a ~55:45 rock/ice ratio from this value. If the range of possible silicate densities taken into account varies from ~2,700 kg·m^{-3} (hydrated silicates that may form from reaction during the differentiation, Scott et al. 2002) to ~4,000 kg·m^{-3} (high pressure silicate phases), the ranges of possible silicate and ice contents are ~50–70% and ~30–50%, respectively. These estimates are obtained with a given internal structure model, with two, three, or more layers. The composition of the interior and its state of differentiation still remain uncertain. However, the composition of volatiles on Titan predicted by formation models can be confronted by the Cassini–Huygens data (e.g. Niemann et al. 2005; Waite et al. 2006; Coustenis et al. 2007). Further constraints on formation models of the satellites within the subnebula are also brought by the composition of Saturn's atmosphere and its implication on the accretion mechanism of volatiles.

3.3.2 Trapping of Volatiles

Three different ways to trap volatiles in a condensing nebula have been suggested in the literature: (1) direct condensation of the pure solid of each of the various gases; (2) trapping of gases in amorphous ice (e.g. Owen and Bar-Nun 1995); (3) formation of a sequence of clathrate hydrates on cooling (e.g. Lunine and Stevenson 1985; Gautier and Hersant 2005). One way may be preferred to another, depending on location of the planetesimals in the solar nebula, and the cooling rate. In the case of (3), the efficiency of clathration depends on the availability of crystalline water ice, and that this latter parameter is still poorly constrained. The remaining volatiles that cannot be trapped by crystalline water ice form pure condensates in the nebula. Refractory (or, at least, less volatile) compounds may have formed by chemical reactions in the solar nebula. Specifically, the transformation of CO into CH_4 and higher hydrocarbons by Fischer–Tropsch catalysis, and the formation of NH_4HCO_3 and/or NH_4COONH_2, have been suggested to affect volatile compositions (e.g., Prinn and Fegley 1989). However, such reactions are efficient only at temperatures above 550 K (Fischer–Tropsch), 150 K (NH_4HCO_3 formation), and 130 K (NH_4COONH_2 formation). So their products, although they may be relevant to planet compositions, should not have contributed much to the chemical composition of satellites such as Titan (Prinn and Fegley 1989).

Evolution models of the solar nebula are constrained by the measured compositions in carbon- and nitrogen-bearing species, as well as the abundances in noble gases, in the giant planets and in comets. Trying to reconcile the solar abundances (Anders and Grevesse 1989; Lodders 2003) with measured abundances is difficult. Indeed, comets whose compositions have been measured (e.g. Cochran et al. 2000) display a very low N_2/CO ratio. Also, the measured enrichments in volatiles in Jupiter and Saturn compared to the solar

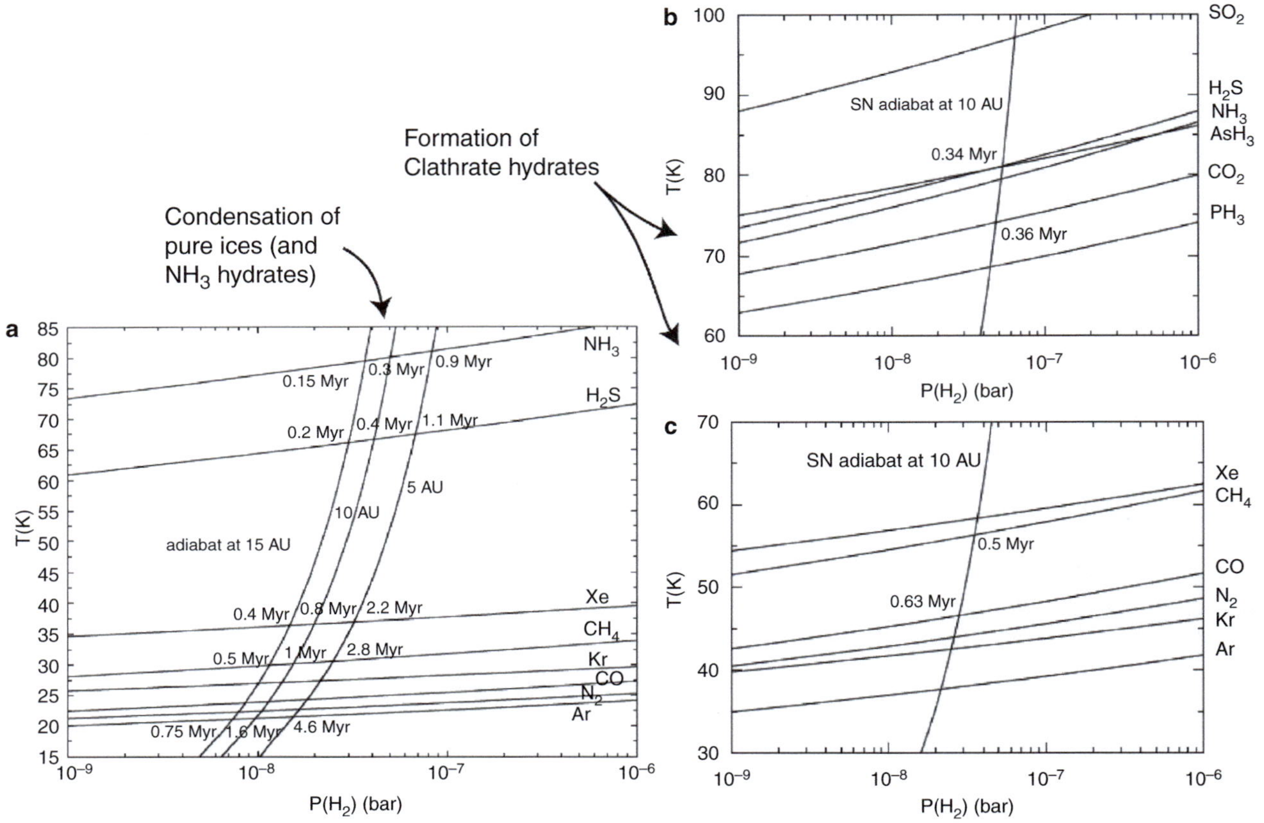

Fig. 3.1 Condensation curves of volatiles and stability curves of clathrate hydrates of these volatiles, and evolutionary track of the cold outer solar nebula. (**a**) Condensation curves of pure ices of H_2S, Xe, CH_4, Kr, CO, N_2, Ar, and of ammonia hydrates, and evolutionary track of the solar nebula at 5, 10, and 15 a.u. from the proto-Sun. The condensation curves are derived from Lodders (2003), and the evolutionary track corresponds to the model of Hersant et al. (2001, 2004). (**b**) Stability curves of CO_2 ice, NH_3 hydrate, and clathrate hydrates of SO_2, H_2S, AsH_3 and PH_3, and evolutionary track of the solar nebula at 10 a.u. (same as *a*). (**c**) Stability curves of clathrate hydrates of Xe, CH_4, CO, N_2, Kr, and Ar, and evolutionary track of the solar nebula at 10 a.u. (same as **a**). (Modified after Hersant et al. (2004) and Hersant et al. (2008).) Actual uptake of gas by the water ice depends on the amount of water ice relative to the amount of gas, and the extent to which the interior of the ice grains are exposed to the gas (hence, kinetics)

nebula remain puzzling. Figure 3.1 shows the evolutionary path of the solar nebula at various distances from the Sun, in temperature versus total nebula pressure diagrams. The evolution tracks correspond to the nominal turbulent model of Hersant et al. (2001). In Fig. 3.1a, the condensation curves of the pure ices of H_2S, Xe, CH_4, Kr, CO, N_2, and Ar are indicated. Ammonia is expected to condense as ammonia hydrate, due to its high freezing point compared to pure NH_3 and to the presence of large amounts of H_2O in the outer solar nebula. At a distance of 5 a.u. from the Sun, i.e. the orbit of Jupiter, temperatures of 20 K are reached ~4.6 Myr after formation of the Sun. This corresponds to the condensation of Ar, which is the last gas to condense. A comparison with the model of Jupiter formation of Pollack et al. (1996) shows that this condensation sequence at 5 a.u. may correspond to phase 2 of the planetary formation. The global enrichment of ~2–3 relatively to the solar abundances at Jupiter (see Gautier and Hersant 2005, and references therein) in the species considered could therefore be explained by this condensation sequence. Applying this scenario to comets leads to quasi-solar N_2/CO ratios, because of the small temperature gap between the condensation curves of these two constituents, hence an almost simultaneous condensation. The upper detection limit of N_2 and the measured CO abundances in the Hyakutake, Hale-Bopp, and Halley comets (Bockelée-Morvan et al. 2004, and references therein) imply a N_2/CO ratio of 10^{-4}, which is inconsistent with a direct condensation hypothesis. Furthermore, the CIRS measurements on C, N, and S abundances in Saturn's atmosphere (Flasar et al. 2005; Orton et al. 2005) also show a differential enrichment in these species, also incompatible with the model of direct condensation of ices.

The second possible medium for volatile trapping is amorphous ice, due to its very high porosity (e.g. Owen and Bar-Nun 1995). This model is derived from experiments conducted at very low pressure and temperature, with a gas condensing simultaneously with water vapor (e.g. Bar-Nun et al. 1988). These experiments have shown that the release of volatiles trapped in equal amounts at 30 K varies with temperature. For instance, upon heating to 100 K, a sample with initially equal amounts of ice, CO, Ar, and N_2 would have released more than 95% of the gases, and the remaining trapped gas would display a N_2/Ar ratio of ~100. This model could therefore explain some differential depletions and enrichments relatively to the solar nebula. However, the cometary N_2/CO ratio cannot be explained by the model of Owen and Bar-Nun (1995). Also, the predicted abundances at Uranus are incompatible with the measurements. It has been advocated by Gautier and Hersant (2005) that this model would likely provide an accurate description of the volatile composition of solids resulting from the condensation of a very tenuous nebular cloud, at a far distance from the star. Nevertheless, the material formed in the outer regions are transported inward, and thus would suffer heating in sufficient amount to trigger the release of the gases, as well as a probable transformation of amorphous ice into crystalline ice at the orbit of the giant planets in the Solar System (Gautier and Hersant 2005). Therefore, it appears that this model is not applicable to the formation and the composition of Titan.

The third model corresponds to the formation of clathrate hydrates of the volatile species within the solar nebula. Clathrate hydrates are non-stoichiometric inclusion compounds, with a hydrogen-bonded H_2O structure forming molecular cages in which a guest gas can be sequestered by van der Waals interactions. Miller (1961) has been the first to suggest that these molecules may be abundant not only on Earth, but also on other objects in the Solar System. Lunine and Stevenson (1985) established a thermodynamic model of the stability of clathrate hydrates for astrophysical applications, and this model has been updated by Iro et al. (2003). The stability of clathrate hydrates depend the nature of the guest gas, its partial pressure and hence its fugacity, and on the degree of occupancy of the cages and the type of structure to a lesser extent (e.g. Sloan 1998). Hersant et al. (2004, 2008) developed a model of the formation of clathrate hydrates within the cooling nebula. Their model considers a nebular elemental abundance, with CO/CH_4 and N_2/NH_3 ratios of 10. In this model, they consider the direct condensation of CO_2 and of ammonia hydrate, and the clathrate of the other species. Fig. 3.1b and c show the stability curves of the solids formed, and the evolutionary track of the solar nebula at 10 a.u. (Hersant et al. 2004, 2008). Gautier and Hersant (2005) have shown that such a model can explain the measured enrichments in volatiles at Jupiter, as well as the nitrogen deficiency in comets relatively to CO (Iro et al. 2003). Hersant et al. (2008) applied this model to the formation of Saturn, and found that the enrichments in carbon, nitrogen, and sulfur measured by CIRS (Flasar et al. 2005; Orton et al. 2005) can be matched by the condensation of CO_2 and the absence of trapping of both CO and N_2 in clathrate hydrates. Such a phenomenon may be due to the complete depletion of the subnebula in water ice, since the evolutionary track allows the formation of these clathrates during the phase 2 of the planet formation scenario. A corollary of the depletion in water ice prior to the formation of N_2 and CO clathrates is that Ar and Kr have been very weakly trapped as clathrates.

The strength of the scenario proposed by Hersant et al. (2008) is that it explains the enrichments measured at Saturn. However, some points remain unclear. First, the model does not take into account the formation of mixed clathrate hydrates, which would incorporate some amounts of a gas that would by itself form a pure clathrate at lower temperatures. The closer to the stability curve of the clathrate hydrate of the second gas, the larger the amounts of gas can be trapped. Therefore, some N_2 and CO should be incorporated in Saturn, even if the available H_2O is not sufficient for trapping these species as pure clathrates. Also, because of the absence of data on the abundance of noble gases at Saturn, the ending point of the condensation of the subnebula is hardly constrained. Hersant et al. (2008) suggested that at 10 a.u. from the Sun, the nebula temperatures remain above 30 K, thus precluding the condensation of the ices of noble gases and of nitrogen. Further data on Saturn's atmospheric composition, as well as experimental measurements on the stability of pure and mixed clathrates at the conditions of the nebula where very little data are available, would help constraining the processes leading to the trapping of volatiles.

3.3.3 Resulting Composition of Titan and Confrontation with Cassini–Huygens Data

The scenario of Titan's formation within Saturn's subnebula presented in Section 3.3.1, and the composition in volatiles resulting from the formation of planetesimals developed in Section 3.3.2, provide constraints on Titan's composition. The mechanism for trapping volatiles in Saturn's subnebula advocated by Hersant et al. (2008) resulted in the growth of large clathrate hydrate grains, by collision of initially micron-scale ice grains that reacted with the gas to form these clathrate hydrates. The scenario suggested by these authors imply that the condensation of the subnebula occurred in a cold, late stage of its evolution. Another turbulent evolution model of Saturn's subnebula has been developed by Alibert and Mousis (2007), which focused on the formation of Titan. This model is also based on the clathration theory of volatiles, but differs from the Hersant et al. (2008) model in that it calculates the rates of conversion of CO and of CO_2 into CH_4, similarly to the model of Mousis et al. (2002). The model of Alibert and Mousis (2007) shows that these conversions may have been very efficient in the early ages of the subnebula, prior to the depletion of the feeding zone. This result suggests that the formation of Titan may have occurred in a "warm" epoch of the subnebula, hence earlier in the process of the agglomeration of clathrate hydrate grains predicted by Hersant et al. (2004, 2008).

The main results obtained with the Gas Chromatograph and Mass Spectrometer onboard the Huygens probe during its descent in the atmosphere of Titan (Niemann et al. 2005) are: (1) presence of 1.4 % to 4.9 % of methane from the upper atmosphere to the surface; (2) presence of Ar, with an unexpectedly large $^{40}Ar/^{36}Ar$; (3) a $^{14}N_2/^{15}N_2$ lower than the terrestrial value; (4) no detection of Xe and Kr ($<10^{-8}$); (5) no direct evidence for CO_2 (present in the instrument itself) or for CO. The measurements conducted in the stratosphere are consistent with measurements of CIRS and INMS onboard Cassini (Coustenis et al. 2007; Waite et al. 2006). Tentative detection of CO at low altitudes, however, has been achieved by Baines et al. (2006). Predicted compositions from the subnebula turbulent evolution models imply: (1) large amounts of methane would have accreted as clathrate hydrates; (2) no primordial Ar is expected; (3) NH_3 would have accreted rather than N_2 in the Hersant et al. (2008) model, whereas the Alibert and Mousis (2007) model infers accretion of both NH_3 and N_2; (4) Kr should not have accreted, whereas Xe is expected in the Hersant et al. (2008) model; (5) CO_2 (and potentially CO for the model of Alibert and Mousis 2007) should be present in significant amounts. Direct comparison between the measurements and the model predictions thus seem to display some discrepancies.

However, the composition of Titan's atmosphere at present-day is related to the long-term evolution of Saturn's largest satellite, which is described in Section 3.5. The formation of Titan inside a subnebula undergoing turbulent evolution, following the scenario of Hersant et al. (2008), is consistent with the presence of large amounts of methane in the bulk Titan, needed to explain the current atmospheric abundance. The low $^{36}Ar/N_2$ ratio is consistent with a non-primordial origin of the atmospheric nitrogen, as predicted by the model; however, the nebular clathration scenario does not account for the presence of Ar, except if this constituent is trapped along with other gases. The large $^{40}Ar/^{36}Ar$ ratio, on the other hand, indicates very significant outgassing from the interior, ^{40}Ar being produced by the decay of ^{4}K. The $^{14}N/^{15}N$ ratio in N_2 has a value lower than terrestrial, which can be related to intense escape of the early atmosphere. However, a recent fractionation model based on the INMS data (Mandt et al. 2008) suggests that over 200 bars of molecular nitrogen are necessary to explain a non-terrestrial value, unless this ratio is primordial. As to the absence of Xe, it could be related to an early clathration at depth within Titan.

These considerations, briefly mentioned here and detailed further in the chapter, thus reconcile the data obtained by Cassini–Huygens with the model of formation of Titan in the framework of the clathration of volatiles in the subnebula (Hersant et al. 2008). Two limitations of this model are that the observed composition can be explained only if CO and N_2 have not been trapped as clathrates due to a lack of H_2O, and if CO_2 is present in large amounts within Titan. Potential evidence for CO_2 on the surface obtained by McCord et al. (2008), although the identification remains controversial, might support this hypothesis. Conversely, the model of Alibert and Mousis (2007) explains the presence of Ar in Titan's atmosphere, but requires the presence of very large amounts of primordial N_2. Xe, and Kr would be expected in this scenario, but they have not been detected so far. The implications of the various formation models as to the origin of nitrogen and of methane are discussed in greater details in Sections 3.6.1 and 3.6.2. So far, none of these models can explain all the signatures distinguished. Further data on the volatile composition of Saturn, and particularly in noble gases, would provide stronger constraints on the evolution of the subnebula. Also, laboratory experiments on the stability of clathrate hydrates at nebular conditions are extremely scarce, as well as data on the reactions that may take place in the subnebula. Future experiments and thermodynamic modeling may provide further constraints on the evolution of the subnebula and on the composition of Titan.

3.4 Accretion of Titan

3.4.1 Range of Accretional Heating Values

During the accretion process, part of the impact energy is converted into heat through two primary processes: shock heating and ejecta blanket deposition (e.g. Senshu et al. 2002). The impact creates a shock wave that compresses the satellite beneath the impact site. Below the impact site, the peak pressure is almost uniform in a quasi-spherical region, called the isobaric core (e.g. Senshu et al. 2002; Monteux et al. 2007). As shock compression is an irreversible process, the entropy below the impact site irremediably increases, leading to the temperature increase. The temperature increase can be determined from the peak pressure within the isobaric core, which is directly related to the impactor velocity, and from the intrinsic properties of the target materials (density, specific heat, thermal expansion, etc.) (Senshu et al. 2002). Furthermore, material near the impact point is excavated and ejected, thus forming the crater and the surrounding ejecta blanket. When the ejecta fall back onto the satellite's surface, their kinetic energy is deposited in a heated ejecta blanket. Excavation as well as the post-impact rebound of the crater have the effect of transferring thermal energy from the vicinity of the impact site to the surrounding zone. This effect reduces the temperature increase just beneath the impact site but induces a further increase of temperature at very shallow depths around the impact site, in particular in the ejecta deposit zone. As the accretion timescale (10^4–10^6 years) is not long enough for most of the accretion energy to escape into space by radiation, the surface temperature irremediably increases.

Despite important progress in our understanding of impacts in the last decades, the amount of energy that can be retained within a growing satellite is still poorly constrained. Most of the studies focused on icy satellites use a simplified approach based on a poorly constrained factor, *h*, which estimates the fraction of accretional energy that is retained at depth and progressively heats the surface of the growing satellite (Kaula 1979; Schubert et al. 1981; Lunine and Stevenson 1987; Mueller and McKinnon 1988; Grasset and Sotin 1996). Following this classical approach, the accretional temperature profile can be approximately calculated as:

$$T_a(r) = \frac{hGM(r)}{c_p r}\left\{1 + ru^2 / 2GM(r)\right\} + T_e,$$

where *r* is the radius, *M*(*r*) the mass contained below the radius *r*, c_p is the specific heat, *u* the velocity of the impactor before being accelerating by the growing satellite, *Te* the temperature of the surrounding environment. Assuming that *h* typically varies between 0.1 and 0.4 and that GM/ru^2 is equal to 4 (Grasset and Sotin 1996), it can be seen from Fig. 3.2 that when the growing proto-satellite reaches a radius of roughly 1,000–1,500 km, melting and vaporization of the surface materials occur (Lunine and Stevenson 1987; Kuramoto and Matsui 1994; Grasset and Sotin 1996). If contaminants such as ammonia are present, melting may occur at lower temperature and therefore at an even smaller radius.

Above this threshold radius, the infalling materials are desegregated, the rocky part sediments at the base of a liquid water layer, which comes from the melting of the ice phase. A proto-atmosphere is also generated, whose composition is determined by the composition of the infalling materials and the stability of the volatile-rich solid phases. The proto-atmosphere once formed limits the radiation into space, resulting in a further increase of the surface temperature. Owing to the blanketing effect of the proto-atmosphere, the surface temperature can reach values higher than 300 K, and as high as 500 K (Kuramoto and Matsui 1994). This leads to the formation of a very massive and hot steam atmosphere in equilibrium with a deep water-rich ocean (Fig. 3.3a). Accordingly, at the end of the accretion process, Titan's interior structure would comprise an inner homogeneous proto-core made of a mixture of rock and icy materials containing a significant fraction of volatiles, overlaid by a silicate layer, resulting from the sedimentation of silicate part of the infalling planetesimals, and a liquid ammonia-enriched water layer (Fig. 3.3a).

The above simple calculation assumes that Titan formed by accreting a multitude of small uniformly distributed impactors (R < 1–10 km). However, it is conceivable that the accretion process first leads to the formation of satellite embryos, a few tens to hundreds kilometers in size, from which Titan has been assembled (e.g. Mosqueira and Estrada 2003). In this condition, it is much more difficult to predict the post-accretional structure. If proto-Titan was the result of slowly assembling hundreds of volatile-rich, cold embryos, melting of the outer layer may have been avoided. The undifferentiated proto-core may extent up to the surface, and the proto-atmosphere would have been relatively tenuous. Nevertheless, the high $^{15}N/^{14}N$ ratio and the low $^{36}Ar/N_2$ ratio measured by the Huygens GCMS (Niemann et al. 2005) implies that a massive ammonia-rich atmosphere was generated during accretion, thus suggesting that a significant portion of the interior has been melted, while the presence of ^{40}Ar suggests as well that substantial outgassing has occurred.

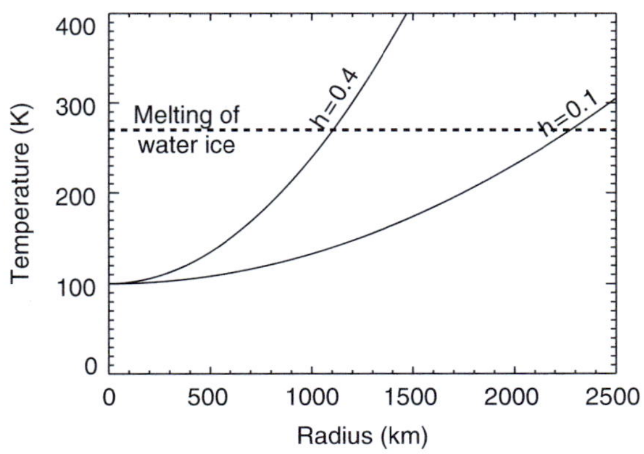

Fig. 3.2 Accretional temperature profile in Titan for different fractions h of the impact energy retained as heat

3.4.2 *Properties of a Primitive Atmosphere Immediately After Accretion*

During accretion and for some time afterwards the surface of Titan is warm enough to support a substantial atmosphere of water and ammonia (assuming this is accreted with the water). For h = 0.1 the final temperature of 300 K corresponds to a pure water saturation vapor pressure of 0.03 bar (Eisenberg and Kauzmann 1969); the corresponding pure ammonia vapor pressure exceeds one atmosphere (Haar and Gallagher 1978). At 176 K the vapor pressure of water is negligible while that of ammonia is of order a millibar (Haar and Gallagher 1978), if the atmosphere is in

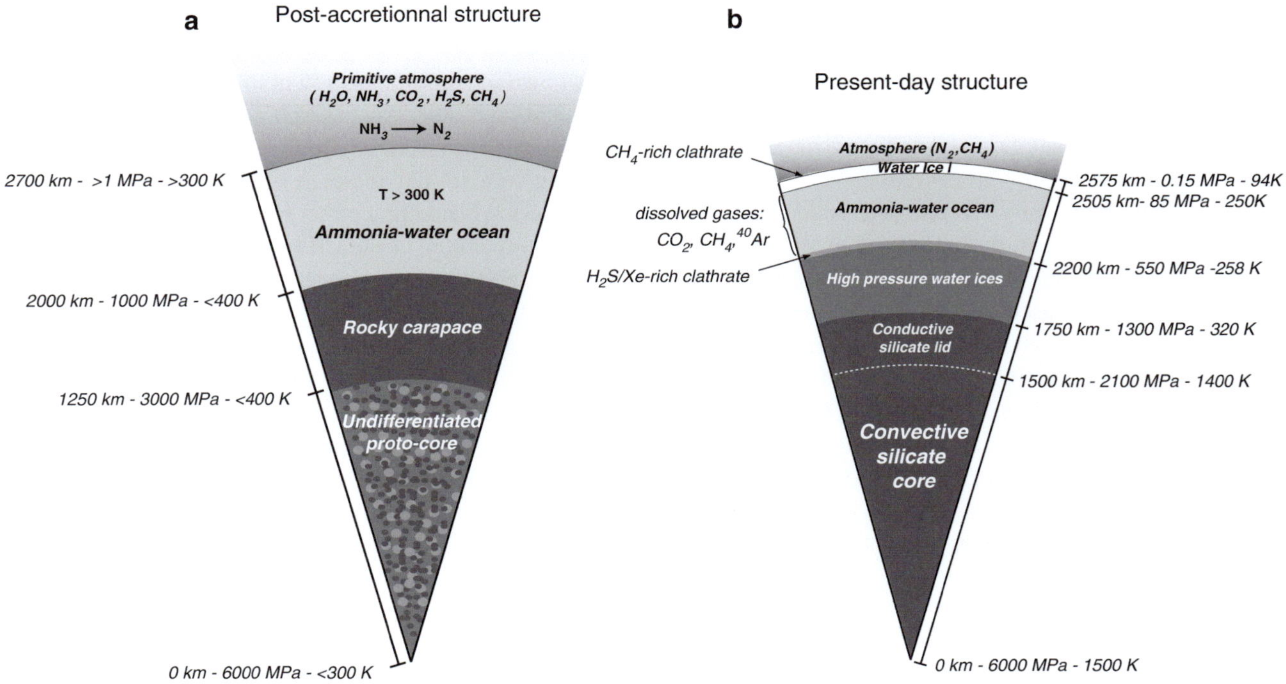

Fig. 3.3 Possible structure and composition of Titan after accretion (**a**) and at present time (**b**). For the post-accretional interior model, the radius of the internal interfaces and the corresponding pressure and temperature are purely indicative. The present-day structure comes from the evolution model of Tobie et al. (2006) assuming an initial ammonia concentration of 5% in the primordial water ocean and a silicate mass fraction of 55% (after Tobie et al. 2009)

equilibrium with a ammonia-water liquid layer having a peritectic composition, which is possible only if large quantities of ammonia have been initially trapped.

However, such temperatures are short-lived after accretion. Cooling of the ammonia-water ocean in the presence of a strong atmospheric greenhouse effect and an insulating ice lid results in complete freezing on a timescale of up to 10^7 years (Lunine 1985; Adams 2006). Methane, on the other hand, has a substantial vapor pressure at 176 K (27 bar), but it would never reach this value in contact with a thick water or water-ammonia ocean. In the presence of a water ocean clathrate formation would occur and the dissociation pressure at 176 K is 0.3 bar, interpolating from the results of Lunine and Stevenson (1985). This is still, however, much more than the ammonia partial pressure (whether we consider pure ammonia or ammonia in solution).

Thus, the era of an accretion-driven water or water-ammonia-rich atmosphere would have existed for at most $\sim 10^7$ years on Titan, supplanted by an atmosphere dominated by methane and possibly nitrogen as ammonia was converted to the latter as suggested by the ratio of argon to nitrogen measured on Titan (Owen 1982; Niemann et al. 2005). Mechanisms include photochemistry in the upper part of, and impacts throughout, the early atmosphere (Atreya et al. 1978; Jones and Lewis 1987; McKay et al. 1988), discussed in more detail in Section 3.6.1 and in Chapter 7.

3.5 Core Formation, Crustal Freezing and Initial Outgassing

3.5.1 Internal Differentiation and Evolution

The bulk density of Titan indicates that the body is roughly 0.5–0.7 by mass silicate (for a silicate density between 3,000 and 4,000 kg·m^{-3}), with the remainder mostly water ice. As illustrated in Section 3.4.1, the proto-core was relatively cold and the complete segregation of ice and rock mixture occurred only once the radiogenic decay of isotopes contained in the silicate minerals raises the temperature above the melting point of water ice. Melting induced a net volume change of the proto-core and a rupture of the overlying silicate layer. This leads to a rapid segregation of the different materials present in the interior (Lunine and Stevenson 1987). According to thermal evolution models, this core overturn would have occurred between 0.1 and 0.5 billion years after accretion (Kirk and Stevenson 1987; Lunine and Stevenson 1987).

The internal differentiation is mainly driven by the density contrast between the different materials composing the interior (silicate, ice, liquid water, hydrate, etc.). Silicate, which is the densest material, migrates toward the center forming a discrete rock core, and the icy materials migrate toward the outer regions, forming a very thick mantle of a mixture of ammonia-water liquid, water ices in different pressure-induced phases, and gas hydrates. The stability of different phases and their density determine the structure of the H_2O thick mantle. Depending on the pressure temperature conditions, H_2O can be in the form of ice I (at $P < 207$ MPa), in the form of high-pressure phase ice (at $P > 207$ MPa), in a liquid form. Accordingly, the cooling and crystallization of the liquid water layer results in the formation of ice I layer at the top of the ocean and of a dense high-pressure ice layer at the bottom. Ammonia, which reduces the temperature of crystallization of the ocean, does not incorporate in the ice phase during crystallization and therefore its concentration in the ocean increases as it crystallizes (Grasset and Sotin 1996; Grasset and Pargamin 2005; Tobie et al. 2005; Mitri et al. 2008; Chapter 4).

The crystallization rate of the ammonia-water ocean is determined by the balance between the heat flow coming out of the silicate core and the heat flow transferred through the outer ice I layer. The energy source within the silicate core is mostly provided by the radioactive decay of long-lived isotopes of K, U, Th with tidal dissipation being negligible there (Sohl et al. 2003; Tobie et al. 2005). Models of thermal evolution indicate that the silicate core should be convective at the present time (Grasset et al. 2000; Sohl et al. 2003; Tobie et al. 2005; Tobie et al. 2006). However, owing to the strong temperature dependence of the silicate viscosity, the convective motions are confined below a thick rigid lid (estimated to 250 km on Fig. 3.3b). The temperature increases quasi-linearly through the lid as the energy is transferred by thermal diffusion, and then it almost follows an adiabat down to the center (Fig. 3.3b). If a sufficient amount of native iron is present, an iron core may eventually form about 1 billion year after differentiation (Grasset et al. 2000). In this case, the deep interior would be comprised of an iron core, probably liquid, and a silicate convective mantle (not shown here, see Grasset et al. 2000). The gravity measurements performed by the Radio Science System on board Cassini (Iess et al. 2008) suggest that the deep interior is not fully differentiated. However, as Titan is not perfectly in hydrostatic equilibrium, the presence of an iron core cannot be definitely ruled out. Additional measurements would be required to determine whether Titan possesses an iron core like its jovian cousin, Ganymede (see Chapter 4).

During the differentiation process, interactions between rock and liquid water may have promoted some chemical reactions. The presence of a few percent of ammonia in solutions with liquid water may have facilitated ionic exchanges between the liquid solutions and the silicate minerals (Engel et al. 1994). In particular, NH_3 may have reacted with sulfate-rich brines leached from the silicate core during its hydration and may have led to the formation of a liquid layer of aqueous ammonium sulfate (Fortes et al. 2007). It is also possible that aqueous alteration of olivine-rich rocks may have produced significant amounts of methane from the conversion of carbon monoxide or dioxide (Atreya et al. 2006). This process, know as serpentinization, is commonly observed on Earth in peridotites dredged from the seafloor and in ophiolites on Earth (see Lowell and Rona 2002, and references therein). However, it is still uncertain how sulfur or ammonia present in the proto-core may have affected the thermochemical equilibrium of the serpentinization reactions. Further laboratory experiments and theoretical models of Titan's interior are needed to better assess the possibility of methane production, notably to determine the kinetics of reactions in presence of NH_3, H_2S and SO_2 and under high pressure conditions.

3.5.2 Primitive Crust Formation

After the accretion epoch, the proto-atmosphere mostly consisted of methane, possibly CO_2 as large amounts of this constituent might have accreted on Titan as it did on Saturn (Hersant et al. 2008), and ammonia and/or nitrogen, the latter being produced by photochemical dissociation of the former (Atreya et al. 1978). The atmosphere was in direct contact with the deep surfacial ammonia-water ocean, a configuration that allowed interactions between these fluid layers. Subsequent evolutions of the primordial atmosphere and of the ocean have therefore been intimately related until the crystallization of an icy shell.

The cooling time of the proto-atmosphere has been calculated to ~10^7 year, from 300 K to the peritectic point of the ammonia-water system at 176 K (Lunine et al. 1989). Over this time span, the temperature drop encompasses the solidification point of several constituents initially present in the atmosphere and/or in the ocean: water ice (267 K for 5% ammonia at a 1-bar pressure, e.g. Kargel 1992), CO_2 ice (~200 K), and clathrate hydrates of CH_4, N_2, CO_2, Ar, and other species (180 to 280 K, depending on gas composition and atmospheric pressure). The temperatures cited here do not take into account the effect of mixed components; for instance, the effect of ammonia on the solidification temperature of clathrates is neglected, the reason being this effect is poorly constrained but expected to be on the same order as the decrease in solidification temperature of ice. In a warm accretion case, which leads to a proto-atmosphere of a few tens of bars, clathrate hydrates would be the very first compounds to crystallize at the atmosphere–ocean interface. Due to this peculiarity, we focus here on the formation of clathrate hydrates on Titan resulting from ocean–atmosphere interactions.

Figure 3.4 schematically describes the coupled evolution of the primordial atmosphere and of the ocean, including the

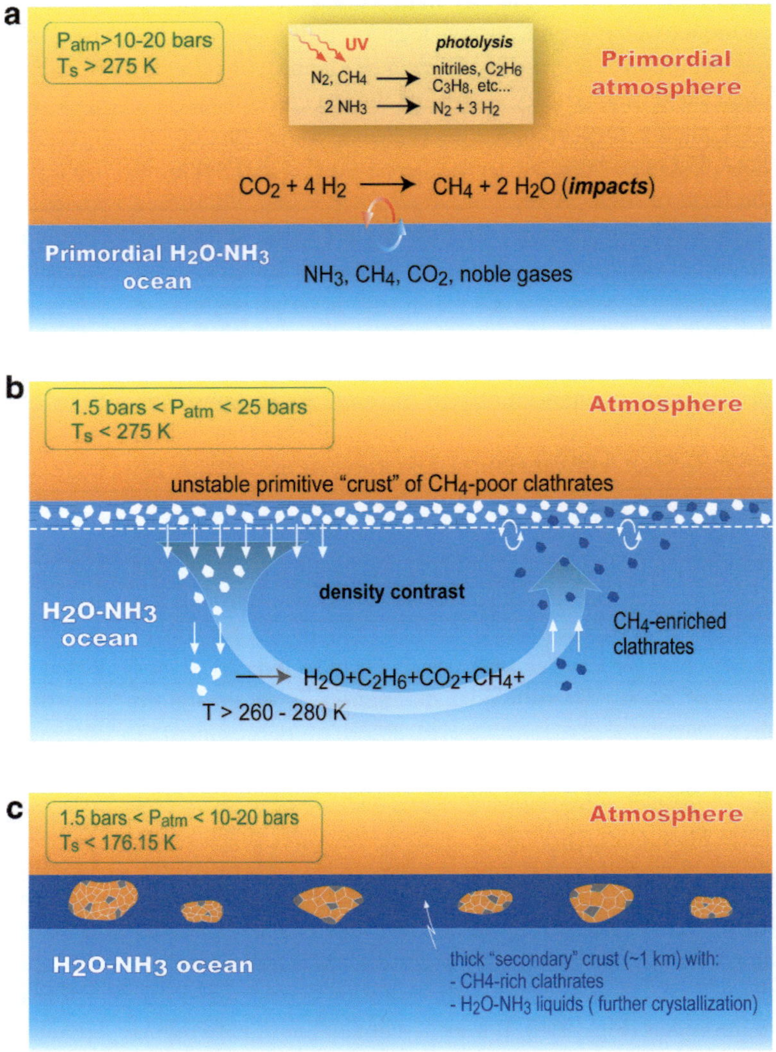

Fig. 3.4 Interactions between the primordial ocean and the early atmosphere of Titan. This scenario leads to the formation of an early, methane clathrate – rich crust. (**a**) Exchanges to regulate vapor pressures and early photochemistry in the atmosphere. (**b**) On cooling, formation of dense methane-poor clathrates at the interface, which sink; within the ocean, formation of methane clathrates and release of atmospheric gases; leads to methane-enriched clathrates. (**c**) Methane-enriched clathrates form an individual layer, isolating the atmosphere from the ocean; during its formation, this crust trapped some amounts of liquid from the ocean, which on cooling formed water ice and ammonia hydrates (Choukroun 2007)

reactions that may have taken place. Figure 3.4a presents the initial stage, right after the end of Titan's accretion. The ocean contained water, ammonia, CO_2, CH_4, and small amounts of noble gases (e.g. Lunine et al. 1989). Leaching of silicates upon accretional melting of ice may have also brought important amounts of ^{40}K, among other alkaline species, to the ocean as well (Engel and Lunine 1994). At temperatures of ~300 K, the atmosphere contained mostly methane, carbon dioxide, and ammonia. UV-driven photochemical processes would have occurred at faster rates than at present-day, owing to both the larger atmospheric pressure and the more energetic solar emissions. These processes, as they do at present-day, have led to the conversion of some NH_3 into N_2 and to the formation of ethane, propane, other hydrocarbons, nitriles, cyanides, etc. Also, the cometary bombardment was probably intense at this epoch (e.g. Zahnle et al. 1992; Griffith and Zahnle 1995), resulting in: (1) addition of some amounts of methane, CO, etc., to the atmosphere, and (2) potential conversion of CO_2 into CH_4.

Figure 3.4b shows the reactions that took place at the interface upon cooling, once the stability domain of clathrate hydrates was reached. The fate of these crystalline compounds formed at the interface was directly dependent on their density relative to the ocean. Densities of clathrate hydrates vary over a wide range, owing to the nature of the guest gas and the cage occupancies: from ~900 kg·m^{-3} for methane clathrates with low cage occupancy, to ~1,150 kg·m^{-3} for carbon dioxide clathrates with full cage occupancy (Sloan 1998). An interesting case is nitrogen clathrate, with a density of 950 to 1,000 kg·m^{-3}. Clathrate hydrates of Kr, H_2S, and Xe, are much denser (up to 1,800 kg·m^{-3} for Xe clathrates). These values need to be compared with the density of a 5 wt% ammonia-water solution at 270 to 300 K, which is on the order of 980–985 kg·m^{-3} (Croft et al. 1988). Therefore, some clathrate species, once formed, would be expected to remain at the ocean–atmosphere interface, while others would conversely sink into the ocean. However, the interaction between a mixture of gases and an aqueous solution does not form clathrates of the individual gases, but rather a mixed clathrate whose composition depends on (1) the partial pressure of gases in the mixture, and (2) the fugacity of each gas (e.g. Sloan 1998). For a primordial atmosphere with high partial pressures of methane and nitrogen, and increasing partial pressures of ethane, the clathrates formed had densities higher than the ammonia-water ocean and thus would have tended to sink into the ocean. Three elements controlled the evolution of these clathrates within the ocean: (1) little to no nitrogen and ethane were initially present in the ocean, (2) the solubility of carbon dioxide is large (3% and more) and increases with pressure, and (3) the solubility of methane, conversely, is very low (less than 0.1% at a few tens of MPa). The implication of these considerations is that the sinking, relatively methane-poor clathrates, likely released some (if not all) of the enclathrated N_2, ethane, and CO_2 at thermodynamic equilibrium. The resulting clathrates therefore contained larger amounts of methane, which probably inverted the density balance between these clathrates and the ocean and induced a buoyant ascent of these methane-enriched clathrates to the surface.

In the final stage of atmosphere cooling, clathrate formation proceeded, with a slight evolution of the chemical composition because of the change in partial pressure of atmospheric gases, owing to the photochemistry. Eventually, more and more methane-rich clathrates accumulated at the ocean–atmosphere interface and formed a solid segregating layer. In laboratory experiments where clathrates are formed by a unique cooling stage of a mixture of gas and water, the texture of the solids is snow-like and therefore highly porous. Genov et al. (2004) studied the growth of porous clathrates and have shown that clathrate porosity exists even at the macroscopic scale. Porosity achieved on clathrate growth has been suggested as a way to trap small quantities of the primordial ocean within the thickening clathrate layer (Fortes et al. 2007; Choukroun et al. 2008; Choukroun et al. 2009). Subsequent evolution of this wet crust upon cooling resulted in the solidification of pure water ice and of ammonia hydrates (Choukroun et al. 2008; Choukroun et al. 2009) and/or ammonium sulfates (Fortes et al. 2007), if these species were present in solution. Although the rates of this crustal formation have not been calculated, it can be inferred that crustal growth had evolving rates: (i) initially slow rates when the temperature was above 270 K, where most of the aforementioned reactions and buoyancy-driven mobilization of the clathrates took place, (ii) reaction rates likely increased once a substantial, thin clathrate layer started to form at the surface, and (iii) eventually slow growth because the chemical exchanges between the atmosphere and the ocean became more and more difficult and mostly controlled by diffusion of the gases through the thickening layer. Therefore, at the end of atmosphere cooling to the freezing points of ammonia hydrates, an insulating layer of methane-rich clathrate hydrates had likely formed (Fig. 3.4c).

Even though a fraction of the atmospheric methane could be incorporated in the primitive crust during the clathration and freezing processes, this fraction remains small. In contrast, Osegovic and Max (2005) showed that during the formation of clathrate from a gas mixture, preferred hydrate-forming materials can be completely consumed, and thus removed from the surrounding environment, as long as water molecules are present in sufficient amount, which was the case when the ocean was in direct contact with the proto-atmosphere. Before a solid crust formed, xenon, which is the preferred hydrate forming material, could have been completely removed from the primitive atmosphere, and would be stored in a heavy compound clathrate (mainly composed of H_2S and Xe) layer at the base of the ocean, (Fig. 3.3b).

3.5.3 Internal Reservoir of Volatiles

At the end of accretion, only the inner undifferentiated portion of Titan's interior was able to hold volatiles in significant amounts. Most of the region outward of this proto-core was probably warm liquid water ($T \geq 300$ K), in which many gaseous compounds have very low solubility, and so potentially very large amounts of gaseous compounds, notably methane, ended up in the primitive atmosphere and on the surface. In contrast, ammonia has a high solubility in water, so that most of the available ammonia remains in the liquid phase and only a relatively small fraction is extracted from the liquid phase and is then converted into nitrogen. Most of the methane initially present in the building blocks ended up in the primitive atmosphere and was lost shortly after accretion owing to the combined effect of strong atmospheric escape and enhanced solar UV photolysis.

After that early epoch, most of the remaining mass of methane is stored in the undifferentiated proto-core. For an undifferentiated proto-core representing 15% to 25% of Titan's total mass ($R_{proto-core}$ = 1,400–1,600 km) and assuming a mass fraction of methane relative to water in Titan's building blocks equal to 1% and a silicate mass fraction of 50%, the total available mass of methane in Titan's interior would range from 1 to 1.7×10^{20} kg, i.e. 360 to 610 times the present-day mass of atmospheric methane (estimated to 2.8×10^{17} kg after Niemann et al. (2005)). Methane in the cold proto-core is stable in the form of clathrate hydrate (e.g. Loveday et al. 2001), but would be released after the proto-core overturn (~0.5 Gyr after accretion in Lunine and Stevenson (1987)). The internal differentiation that leads to the formation of a discrete rocky core is driven by the density contrast between the different materials comprising the interior (silicate, ice, liquid water, hydrate, etc.). The lighter materials rise toward the surface, whereas the denser materials (silicate) converge toward the center.

Figure 3.5a displays the density of different candidate materials computed as a function of pressure from the experimentally determined bulk modulus following Croft et al. (1988):

$$\rho = \rho_0(T_{0,}P_0) \times \left[\frac{K_0^{'}}{K0} + 1\right]^{1/K_0}$$

where ρ_0 is the density at the reference pressure P_0, K_0 is the isothermal bulk modulus and K'_0 is the pressure derivative of the bulk modulus. Table 3.1 summarizes the experimentally-derived parameter values used for each material. For the ammonia-water system, those parameters are computed following Croft et al. (1988). For methane hydrate and ammonia hydrate, the density at reference pressure is calculated from the mean molecular volume V_{mol} and the molecular mixing ratio of water given by Loveday et al. (2001) and Loveday and Nelmes (2003). For gas clathrate, the density depend on the gas molecules trapped in their cages. Following Sloan (1998), clathrates of pure methane have a density of 920 kg·m^{-3} at ambient pressure, whereas clathrates of carbon dioxide have a density of 1,130 kg·m^{-3}. For sake of simplicity, the same pressure dependence is assumed for methane clathrate and carbon dioxide clathrate. Whatever the pressure values, pure methane clathrate is the least dense material, being equal in density to water ice at low pressure (P < 0.2GPa). Even for 15% ammonia-water liquid, methane clathrate

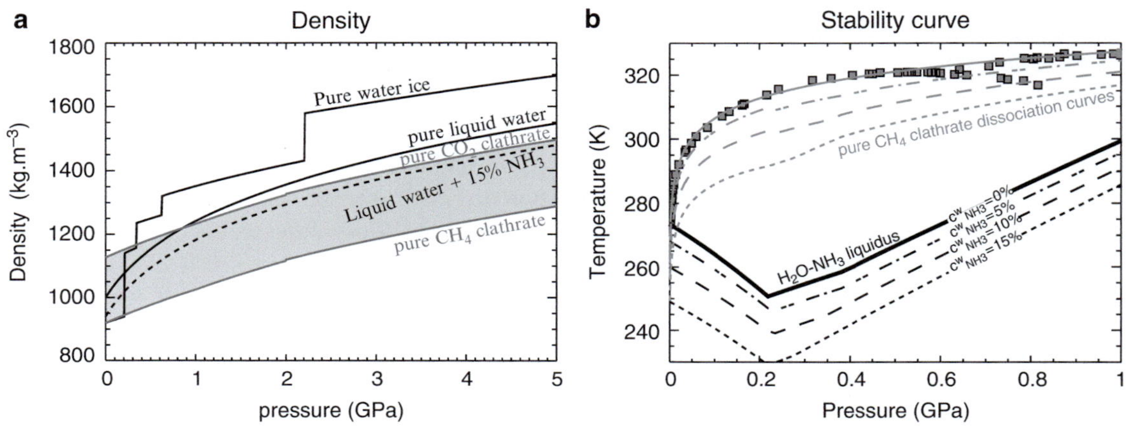

Fig. 3.5 (**a**) Density of different materials potentially present within Titan's interior as a function of pressure (**b**) Stability curve of methane clathrate hydrate compared to the melting curve of ice for different ammonia mass fractions in the liquid phase. *Squares* represent experimental data given by Dyadin et al. (1997). *Curves* with similar line styles correspond to the same ammonia fraction

Table 3.1 Parameters to compute the density as a function of pressure

	Reference pressure P_0 (GPa)	Density at P_0 ρ_0 (kg·m^{-3})	Isothermal bulk modulus K_0 (GPa)	K'_0	Reference
Ammonia-water					
	10^{-4}	Computed at T = 300K	Computed at T = 300K	Computed at T = 300K	Croft et al. (1988)
Ice					
Ice I	10^{-4}	920	9.2	5.5	Sotin et al. (1998)
Ice III	0.207	1,140	8.5	5.7	Sotin et al. (1998)
Ice V	0.34	1,235	13.2	5.2	Sotin et al. (1998)
Ice VI	0.62	1,320	14.9	6.6	Sotin et al. (1998)
Ice VIII	2.21	1,460	24	4.15	Hemley et al. (1987)
Methane hydrate					
MHI	10^{-4}	920	~10[b]	4	Loveday et al. (2001)
MHII	1	1,025	~15[b]	4	Loveday et al. (2001)
MHIII	2	1,120	~25[b]	4	Loveday et al. (2001)
Ammonia monohydrate					
AMHI	10^{-4}	960	8.9	4.2	Loveday and Nelmes (2003)
AMHIV	3	1,355	~100	4	Loveday and Nelmes (2003)

is still 2% less dense. Therefore, only clathrates that are mostly composed of methane would have a density lower than ammonia-water solutions, and would be able to rise through the thick primordial ammonia-water layer during differentiation.

The ascent of methane-rich clathrates and their accumulation at the top of the ammonia-water layer also depend on their stability with respect to the ammonia-water solution. For simplicity, the effect of minor gas compounds on the stability of methane-rich clathrates is neglected. The stability curve of pure methane clathrate in water solution can be parameterised from available experimental data over a wide range of pressure (Sloan 1998) by performing a polynomial fit:

$$T_d^0 = 264.4 + 21.1 \times \log(P)$$

with T^0_d the dissociation temperature in the binary water-methane system in Kelvin, P the pressure in megapascal.

The inhibiting effect of ammonia on methane clathrate has long been suspected, but its net effect on the stability of methane clathrate has only recently been experimentally investigated (Choukroun et al. 2008). Preliminary results indicate that a 7.5% NH_3 concentration could decrease the dissociation curve by up to 20 K. In absence of additional experimental constraints, the effect of ammonia on the methane clathrate stability can be estimated from the binary ammonia-water system, which is well-known (e.g. Grasset and Pargamin 2005; Choukroun and Grasset 2007), by using the parameterization proposed for clathrate stability in salt water solutions (Sloan 1998):

$$T_d^{xNH_3} = \left\{ \frac{1}{T_d^0} - \frac{n\Delta H_m}{\Delta H_d} \left[\frac{1}{T_m^0} - \frac{1}{T_m^{xNH_3}} \right] \right\}^{-1}$$

where T^0_m and T^{xNH3}_m are the melting temperature of ice for the pure water system and ammonia-water system, respectively, computed using the method described by Grasset and Pargamin (2005). ΔH_d is the enthalpy of dissociation for methane clathrate to liquid water and methane gas, n is the hydration number ($\Delta H_d / _n R$ is estimated to 1,090 K). ΔH_m is the enthalpy of melting for pure water to ice (6008 J · mol^{-1} at 273.2 K, Sloan (1998)).

As illustrated in Figure 3.5b, ammonia reduces the stability of methane clathrate to almost the same extent that it decreases the crystallization point of ammonia-water solutions. Most of the methane clathrate released during the overturn would ascend to the top of the outer liquid layer and accumulate there just below the primitive crust. Part of the methane will also be stored as dissolved gas in the water ocean, together with other volatile species such as carbon monoxide and carbon dioxide (Tobie et al. 2009).

3.6 Particular Problems

3.6.1 *Origin of Nitrogen*

Owen's (1982) original suggestion that a very low ratio of $^{36}Ar/N_2$ in Titan's atmosphere would point to ammonia as the source of Titan's nitrogen was based on the strong difference in volatility and affinity with water ice of NH_3 versus N_2, and the rough similarity in vapor pressure of argon with the latter. Thus if the primary carrier of Titan's N_2 were molecular nitrogen itself, conditions cold enough to condense out both Ar and N_2 would lead to a solar composition ratio (about 0.1 Ar/N_2; Anders and Grevesse 1989, for example) for the two in the condensed phase. This will vary somewhat if the gases are adsorbed or enclathrated in water ice, but not by more than an order of magnitude (Lunine and Stevenson 1985; Owen and Bar-Nun 1995). We thus expect an Ar/N_2 ratio in Titan's atmosphere of somewhere between 0.01 and 0.1, some 3 to 4 orders of magnitude higher than what is actually observed. The straightforward conclusion is that nitrogen was delivered to Titan in a much less volatile form, in planetesimals that were formed (or underwent thorough chemical alteration) in a nebula whose temperatures were never low enough to trap most of the argon. Ammonia is most commonly cited as the probable carrier of nitrogen (Hunten 1978) because it is seen in comets (e.g. Kawakita et al. 2007), should have been relatively stable in a circum-Saturnian disk that had pressures elevated above those in the solar nebula (Prinn and Fegley 1981), is involatile relative to N_2 and Ar, and hydrogen-bonds readily to water ice. There should thus be large amounts of ammonia in Titan's interior, from which molecular nitrogen was derived in the atmosphere by photochemistry or impacts.

Atreya (1986) estimates a minimum temperature of 130 K for photochemical conversion of NH_3 to N_2, determined by the condensation of the intermediate product hydrazine N_2H_4, and a maximum temperature of roughly 200 K above which the water vapor pressure is high enough that it would have interfered with the ammonia-photolysis. Under present conditions the ultraviolet flux is such that the conversion would take of order 10^8 years to produce enough N_2 to account for present day abundances, too long given the considerations above. However, the ultraviolet flux was enhanced by several orders of magnitude shortward of 2000 Å during the first few tens of millions of years of the Sun's life (Zahnle and Walker 1982). Provided ammonia could be exposed to the ultraviolet photons, there would seem to have been sufficient flux while the atmosphere was warm enough to permit it to be present in the UV active region. Exposure to UV photons of sufficient energy even with methane present is plausible since methane shielding by its own photolysis occurs at wavelengths much shorter than the 2000 Å where ammonia photolysis occurs. Adams (2006) quantifies these ideas and concludes that many bars of nitrogen could be produced photochemically. Chapter 7 goes into more detail on the photochemical model.

Impacts are potentially able to add to the nitrogen inventory provided atmospheric conditions are appropriate (high ammonia and low water content), leading to production of between several and 20 times the present atmospheric inventory (Jones and Lewis 1987; McKay et al. 1988). However, large impacts would also have led to significant erosion given Titan's low gravity, which in turn implies loss of a significant amount of atmosphere and the need for a relatively high yield of N_2. Impacts of course happened, regardless of whether they were effective or not at converting ammonia to N_2, and so the main question is whether they worked over a primordial atmosphere that no longer exists, or enough of that atmosphere survived to be reflected in the present-day volatile inventory of surface and atmosphere. Because impact loss of atmosphere is an inefficient way to enrich the heavy isotope of nitrogen, the rather modest enrichment in $^{15}N/^{14}N$ of 1.5 times terrestrial (Niemann et al. 2005) might still have been obtained during loss of large amounts of atmosphere.

While the conversion of ammonia to nitrogen in Titan's early atmosphere appears to be possible by at least two mechanisms, a cautionary note is that ammonia has not been seen on Titan at all (except in ACP data – Israël et al. 2005 – where it is plausibly a chemical product in the aerosol). The plumes of Enceladus do show evidence for ammonia (Waite et al. 2009), though no spectroscopic evidence exists for it on Titan's surface (although it would be extremely difficult through the atmospheric methane absorption features for Cassini-VIMS to detect it). Evidence for cryovolcanic features in Cassini Radar data (Lopes et al. 2007) does not directly imply the presence of ammonia since the topography of such features (and hence the material properties of the magma) is unknown.

More troubling for the ammonia hypothesis is the lack of the noble gases krypton and xenon, which were not detected by the Huygens GCMS in the lowermost atmosphere (Niemann et al. 2005), despite their expected presence (Hersant et al. 2008). Clathration in ices at the surface might have removed Kr and Xe but not, substantially, Ar (Lunine and Stevenson 1985; Thomas et al. 2007), and so the lack of argon would still be a signpost for ammonia as the parent molecule of atmospheric molecular nitrogen. However, Jacovi and Bar-Nun (2008) have reported experiments in which adsorption of all three noble gases on photochemical hazes could have, over the age of the solar system, swept out of the atmosphere an amount of argon equivalent to that corresponding to solar abundance relative to atmospheric nitrogen (in fact, several times that value). If the experiments and their interpretation are correct, and if photochemical aerosols have remained preserved at the surface rather than being destroyed by impacts, then the argon abundance may be an artifact of the adsorption on aerosols and not an indication that nitrogen came into Titan as ammonia. Evaluation of the various possibilities for the sequestration of the noble gases would require direct analysis of heated samples of the water ice crust and the dune material – which appears to be composed of the solid organic haze (Paganelli et al. 2008) – an ambitious goal even for a landed mission on Titan.

3.6.2 *Origin of Methane*

A remaining issue about Titan is the origin of atmospheric methane. This problem is independent on the form within which volatiles have been accreted (clathrate hydrates, pure methane ice, trapped in amorphous water ice). Is methane primordial, i.e. has methane been accreted on Titan as such within the Saturn subnebula? Or has carbon been accreted in a different form (CO, CO_2) and has methane been produced later within Titan, owing to internal processes? Another corollary to the present-day atmospheric abundance of methane (up to 5% near the surface, Niemann et al. 2005) is the existence of replenishment processes to counterbalance the irremediable disappearance of CH_4 in photochemical reactions over a short time span (Yung et al. 1984).

3.6.2.1 Accretion of CH_4 or of CO/CO_2

As discussed in Section 3.3, abundant CO and N_2 are reported in comets, compared to small amounts of NH_3 and CH_4 that are merely detectable (Bockelée-Morvan et al. 2004, and references therein). The discrepancy between comet composition, solar nebula bulk composition, and Titan's atmospheric composition, may originate from diverse evolutionary paths of the nebula and/or of Titan. Table 3.2 summarizes the formation models of Saturn and Titan within the subnebula, and compares predicted compositions with the recent Cassini–Huygens measurements.

After the study of Lewis (1972), high concentrations of N_2 and CO relative to NH_3 and CH_4 were inferred for the solar nebula. The chemical model of Lewis and Prinn (1980) showed that the kinetics of the conversion from CO to CH_4 and from N_2 to NH_3 are slow in the solar nebula, relative to the rates of radial transport or of thermal evolution. Accordingly, the dominant carbon- and nitrogen-bearing compounds in the solar nebula are predicted to have been CO, CO_2, and N_2, which is consistent with cometary compositions. Prinn and Fegley (1981) focused on modeling the evolution of a circumplanetary nebula at Jupiter and the resulting chemical composition of the Jovian satellites. With

Table 3.2 Synthetic comparison between selected formation models, associated predicted compositions for Saturn and Titan, and actual Cassini measurements (CIRS at Saturn; CIRS, INMS, and GCMS at Titan)

Stage; comparison	P&F 1981	MGB 2002	A&M 2007	HGTL 2008
Solar Nebula	CO, CH_4, N_2, NH_3	CO, CH_4, N_2, NH_3	CO, CH_4, N_2, NH_3, CO_2, Ar, Kr, Xe, H_2S	CO, CH_4, N_2, NH_3, CO_2, Ar, Kr, Xe, H_2S
Trapping form	Condensation and ammonia hydrates	Clathrate hydrates and NH_3 hydrates	Clathrates, ammonia hydrates and CO_2 ice	Clathrates, ammonia hydrates and CO_2 ice
Preferrential trapping	No	CH_4 and NH_3, no CO	No	No
Saturn subnebula	Conversion of CO and N_2 to CH_4. NH_3	No conversion due to low reaction rates	CO to CH_4 (Fischer–Tropsch catalysis)	H_2O depletion before CO, N_2, Ar, Kr
Saturn atmosphere	CH_4, NH_3	CH_4, NH_3	High CH_4, NH_3, N_2, CO_2, Ar, Kr, Xe, H_2S	High CH_4, NH_3, CO_2, H_2S; low CO, Ar, Kr, Xe
Compatibility with measurements	No enrichments with respect to nebula	No differential enrichments	Large N_2 and noble gases	Good compatibility
Titan (bulk)	CH_4, NH_3	CH_4, NH_3	High CH_4, NH_3, N_2, CO_2, Ar, Kr, Xe, H_2S	High CH_4, NH_3, CO_2, H_2S; Litle to no CO, N_2, Ar, Kr, Xe
Titan atmosphere	CH_4, N_2 from NH_3	CH_4, N_2 derives from NH_3 conversion post-formation	CH_4, high primordial N_2, some N_2 from NH_3, Ar, Kr, Xe	CH_4, N_2 from NH_3 only, little to no CO, primordial N_2, Ar, Kr, Xe
Comparison with measurements	Insufficient components to compare with data	Insufficient components to compare with data	N_2 abundance too high, presence of Kr and Xe, large CO_2	No Ar and CO predicted
Possible explanation for discrepancy with measurements			Trapping of Kr and Xe as clathrates (Thomas et al. 2007)	Some CO, N_2, and noble gases in mixed clathrates or ices

P&F 1981: Prinn and Fegley, 1981; MGB 2002: Mousis et al. 2002; A&M 2007: Alibert and Mousis, 2007; HGTL 2008: Hersant et al. (2008).

their model, these authors calculated the conversion rates of the same compounds and found that: (1) at thermochemical equilibrium, the further away from the proto-Sun the considered location in the nebula is, the smaller the amounts of CO and N_2 are at the end of the condensation process; and (2) iron-catalyzed reactions taking place in the Jovian subnebula also favor the subsistence of NH_3 and of CH_4. This model thus predicts that the Jovian satellites have accreted much larger amounts of ammonia and methane than of carbon monoxide and molecular nitrogen. Applying the model of Prinn and Fegley (1981) to the Saturn subnebula would lead to similar relative abundances. However, this model considers an optically thick circumplanetary disk, with high partial pressures of the various gases, and assumes the cooling to occur at steady-state.

More recent models, such as proposed by Mousis et al. (2002), show that the conversion rates examined by Prinn and Fegley (1981) are much lower in the case of a thin nebular disk. Mousis et al. (2002) computed a turbulent evolution model of the Saturn subnebula, within an optically thin, gas-starved disk. They have suggested that no conversion of CO and N_2, nor of CH_4 and NH_3, could occur within the Saturn subnebula. However, materials supplied by the surrounding solar nebula to the Saturn subnebula would be enriched in CH_4 and NH_3, according to the condensation sequence and the trapping method of volatiles proposed by Mousis et al. (2002). The formation of clathrate hydrates and ammonia hydrates assumed by this model is consistent with the predictions by turbulent evolution models of Gautier et al. (2001); Hersant et al. (2004); Hersant et al. (2008). However, the model of Mousis et al. (2002) does not take into account the large amounts of gases that accrete to Saturn, and it is only valid in the case of a very thin disk. Mousis et al. (2009a) considers the delivery of volatiles to Saturn itself.

Two different models provide an explanation for the enrichment in carbon measured at Saturn, as well as the presence of very large amounts of CH_4 on Titan: the model of Alibert and Mousis (2007), and that of Hersant et al. (2008). The former revisits the chemical reactions taking place in the subnebula and the conversion rates of CO_2 and of CO into CH_4. Alibert and Mousis (2007) consider the effect of the Fischer-Tropsch catalysis at low pressure in the subnebula, based on experimental data obtained at 10^{-2} bars by Sekine et al. (2005). Their calculations suggest that, at a distance from Saturn that corresponds to the location of Titan, CO may be converted very efficiently into CH_4. The rates are 100 times smaller of the conversion of CO_2, this species is thus expected to remain as such and to be present in important amounts in Titan. In this model very large amounts of nitrogen would have also accreted within Titan, and little ammonia would be initially present. The model of Hersant et al. (2008) explains the differential enrichments in carbon, nitrogen, and sulfur in Saturn by a premature end to the clathration process of volatiles because of the complete depletion in H_2O of the nebula prior to the trapping of N_2 and CO. If the same considerations are applied to Titan, the methane would have accreted in large amounts, while CO should be absent or present in very small quantities. This model also predicts that Titan's nitrogen cannot have a primordial origin, therefore it appears consistent with the results of the GCMS (Niemann et al. 2005). Both these models, as well as the very recent model of Mousis et al. (2009b) in which argon, nitrogen and carbon monoxide are depleted from planetesimals upon early migration towards proto-Saturn, propose that a significant part of carbon should be accreted as CH_4 on Titan. This suggests that Titan's methane has a primordial origin.

However, large amounts of CO_2 are also expected to accrete on Titan according to the models of Alibert and Mousis (2007) and of Hersant et al. (2008). During the evolution of Titan subsequent to its formation CH_4 may have also formed by conversion of CO_2 in contact with water and silicates under high pressure in the deep interior (Atreya et al. 2006). CO_2 has not been unambiguously identified on Titan other than in its atmosphere where it is most plausibly a product of the photochemistry of CO with O and Oh ions from the magnetosphere (Horst et al. 2008), and so the test of the hypothesis that methane is principally produced in Titan's interior might have to await a future mission capable of detecting patches of CO_2 localized to regions of cryovolcanic outgassing. It must also be borne in mind that solid or liquid CO_2 at high pressure or trapped as clathrates is more dense than liquid water or ammonia-water. Hence, the potential presence of CO_2 at certainly cannot be ruled out or in at the moment, and this CO_2 might be a source of methane during serpentinization processes. This internally-produced methane would likely rise through the subcrustal liquid layer and induce the icy crust to form methane clathrates. Therefore, although the origin of Titan's methane still remains uncertain, two considerations apply: (1) some fraction of the methane we see on Titan at present has likely been accreted, from the presumed sub-Saturnian nebula, and been trapped in the deep interior; and (2) regardless of methane's origin, outgassing processes must occur throughout Titan's history to bring new methane to the atmosphere via crustal processes that may reflect changes in the internal regime of heat flow

3.6.2.2 Methane Replenishment via Cryovolcanism

Consistent models can explain the accretion of methane with the planetesimals that formed Titan. However, the rates of photochemical dissociation of methane in the atmosphere (Yung et al. 1984) are such that replenishment processes have to be envisaged in order to sustain the abundance of this constituent. Several reservoirs of methane have been foreseen on Titan: a global surface ocean of methane (Sagan and Dermott 1982) or methane + ethane (Lunine et al. 1983); deep reservoir of clathrate hydrates dissociated during cryovolcanic events (Lunine and Stevenson 1987; Tobie et al. 2006); deep-seated methane ocean (Stevenson 1992); porous icy regolith (Kossacki and Lorenz 1996). The results obtained by the Cassini–Huygens mission have ruled out the first two hypotheses, as no global surface ocean could be identified (e.g. Tomasko et al. 2005). The estimated volume of methane contained in Titan's surface lakes (Stofan et al. 2007) would be sufficient to replenish the current methane abundance one time only (Mitri et al. 2007), which does not allow sustaining the presence of methane in the atmosphere throughout Titan's history. The deep-seated global ocean, the interior methane clathrate reservoir, and the porous regolith hypotheses cannot be discarded at present-day. For consistency with the theme of this chapter, we focus here on estimating the likelihood of methane replenishment to occur via cryovolcanism. The reader is referred to Chapter 7 for further descriptions of other hypotheses .

Surface features potentially related to cryovolcanic activity on Titan have been imaged by the Visual and Infrared Mapping Spectrometer (Sotin et al. 2005) and the Cassini Radar Mapper (Lopes et al. 2007) instruments onboard the Cassini spacecraft. Cryovolcanism on icy satellites, conversely to silicate volcanism on Earth, Venus, or Io, is subject to a major conceptual issue: the liquid water produced upon melting of ice is denser than the solid. Therefore, in the pure H_2O system, it is physically impossible to observe extrusion of a liquid on the surface of an icy shell. Henceforth, a cryomagmatic source must fulfill two conditions: (1) by analogy to silicate volcanism, partial melting of ice has to occur in a binary, or more complex, system; (2) the liquid produced upon melting must be less dense than water ice or clathrate hydrates. On top of these restrictions, gaseous methane has a very low solubility in aqueous solutions (~0.1% at a 10 MPa pressure, which corresponds to the present-day interface with the internal liquid layer). Even if a cryomagma saturated in methane ascends up to Titan's surface, the emission of $\sim 3 \times 10^8$ km^3 of cryolavas is necessary in order to supply enough methane to renew the atmospheric methane abundance of $\sim 2.8 \times 10^{17}$ kg (Niemann et al. 2005) one time only. Such a volume would correspond to 3.8 km of cryolava deposited all over Titan's surface over the past 10–100 Ma (Choukroun 2007; Choukroun et al. 2008; Choukroun et al. 2009). Due to the low crater density on Titan, the surface is estimated to be 100 Myr to 1 Gyr old (Lorenz et al. 2007), which is not consistent with the emission of methane-saturated aqueous cryomagma. Therefore, it appears that the cryomagmas need to dissociate clathrate hydrates stored within Titan in order to provide sufficient amounts of methane to replenish the atmospheric methane.

Considering the post-accretional structure of Titan and the thermal evolution of this body described in the previous section, cryovolcanic release of methane throughout Titan's history must be envisaged in two completely different settings. Indeed, the thermal evolution model of Tobie et al. (2006), which is the first to take into account the effect of an insulating, low thermal conductivity lid, suggests that the upper crust rich in methane clathrates has remained in contact with the internal liquid layer for ~3.5 Gyr. It is only after this time span that the internal ocean started to solidify, forming an underlying ice I – rich layer below the clathrate-rich crust. Hence, during the first 4 Gyr, dissociation of clathrate hydrates could have "simply" occurred due to sudden increases in heat flux. Two likely episodes of intense dissociation of the crustal clathrate hydrates have been pointed out by Tobie et al. (2006): (1) after the core overturn and the release of huge amounts of clathrates from the inner undifferentiated core, associated to a peak in heat flux; (2) after ~2 Gyr, once convection had initiated within the silicate core, which also released very large amounts of heat from the rocky core. In both these two episodes, the amounts of heat released from the inner parts of Titan generated an increase in temperature of the adiabatic internal liquid layer, resulting in the dissociation of clathrate hydrates either during transport (core overturn) or from the crust (core overturn + convection initiation in the silicates). In such a scenario, the amounts of methane emitted are up to 8 times larger than the amounts necessary to sustain the presence of methane in the atmosphere (Tobie et al. 2006). However, after the formation of an icy layer underneath at ~3.5 Gyr, the heat released from the core would have somehow been absorbed by the ice prior to reaching the clathrate-rich part of the crust. Nonetheless, large amounts of heat would be brought up after the initiation of convection in the icy layer, which may have induce a third episode of intense outgassing. The emission of cryomagmas and the release of methane during this episode, probably still occuring today, remains challenging for the reasons aforementioned: density and methane content. Also, methane clathrates are stable throughout Titan for an equilibrium thermal profile (e.g. Grasset and Pargamin 2005; Choukroun et al. 2008; Choukroun et al. 2009). Another process thus has to be envisaged to explain present-day cryovolcanic release of methane via the dissociation of clathrate hydrates.

Since the early work of Lewis (1971, 1972), large amounts of ammonia were expected inside icy satellites. As noted above, ammonia dramatically decreases the melting point of ice, by up to 97 K for a eutectic 33 wt% NH_3 concentration (e.g. Rollet and Vuillard 1956). The density of highly concentrated ammonia–water mixtures is actually lower than that of ice Ih, as an example a 30 wt% NH_3 mixture has a volumic mass lower than 0.920 kg·m^{-3} at temperatures higher than 210 K (Croft et al. 1988). This particularity makes ammonia a good candidate cryomagma component. However, interior structure models of Titan predict that the maximum concentration of ammonia in a putative internal liquid layer is on the order of 3-15 wt% NH_3 (Grasset et al. 2000; Sohl et al. 2003; Grasset and Pargamin 2005; Tobie et al. 2006). At no pressure–temperature conditions is such a mixture less dense than water ice or methane clathrate, which is a strong indication that cryomagmas can not originate directly from the internal ocean of Titan. Choukroun et al. (2008) therefore suggested that the cryomagmas are generated within the icy layer of Titan, likely from the melting of ammonia hydrates. Another potential source of cryomagma is ammonium sulfates (Fortes et al. 2007), which also reduces the melting point of ices. Such chemicals might have been formed by reaction of dissolved ammonia with sulfates produced by the early interaction of the primordial ocean with the silicates. However, the model proposed by these authors requires a very large conversion rate of NH_3 into NH_4^+, as well as the complete sulfatation of the silicates. Another obstacle to the presence of ammonium sulfates in Titan's icy crust is their high volumetric mass of 1.77 kg·m^{-3} at atmospheric pressure and 320 K (Material Safety Data Sheet, JT Baker). Nevertheless, whichever the crustal source of cryomagmas is, another issue is the dissociation of clathrates. Indeed, methane clathrates are stable throughout the thermal profile within Titan, even in the case of warm ice intrusions within the icy shell (e.g. Loveday et al. 2001; Choukroun 2007; Choukroun et al. 2008; Choukroun et al. 2009).

Choukroun et al. (2008) have obtained the first experimental data on the dissociation of methane clathrate hydrates under pressure in the H_2O-NH_3-CH_4 system. These data show that: (1) the inhibition of methane clathrates is stronger than expected by Sloan (1998), i.e. a 7.5 wt% NH_3 concentration decreases the dissociation temperature by ~10–20 K at 10 MPa; and (2) the lower the methane content, which implies a lower methane partial pressure, the lower the dissociation temperature is at high pressure. Based on these data and on the buoyancy considerations on cryomagmas compared to ice, a model of cryovolcanism with associated release of methane at present-day has been produced (Choukroun et al. 2008). Figure 3.6 summarizes this model, which is consistent with the internal structure of Titan (Tobie et al. 2006; Chapter 4). In this model, the diffusion of heat from warm ice intrusions associated with upwellings induces the melting of ammonia hydrates that have been previously trapped within the crust. The melts produced, with a high NH_3 concentration on the order of 30 wt%, react with the surrounding methane clathrate hydrates and thus dissolve some methane. Another possibility is that these melts travel from the internal ocean through cracks (Mitri et al. 2008) and then interact with the crustal clathrates. Following this reaction, a cryomagma with large NH_3 and CH_4 concentrations is formed, although the dissociation of clathrates releases

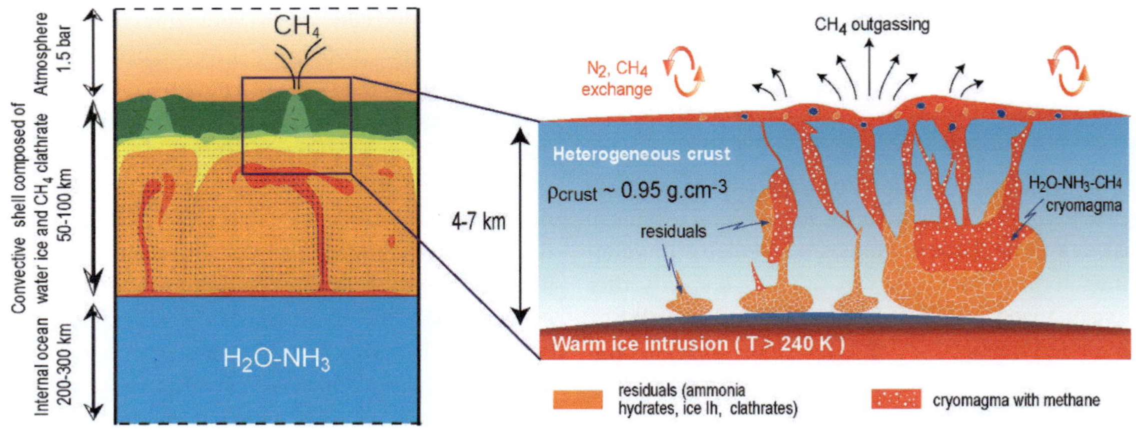

Fig. 3.6 Conceptual model of present-day cryovolcanism and associated release of methane. *Left*: cross-section of Titan's icy shell and computed convection motions within the ice underlaying the clathrate-rich upper crust. *Right*: at the tip of convective plumes, melting of ammonia hydrates generates a low-density, ammonia-rich cryomagma that dissociates the clathrates and may release methane in the atmosphere via direct effusive eruption or explosive eruptions by increase in gas pressure. See text for details (modified after Tobie et al. (2006) and Choukroun et al. (2008))

5.75 times larger H_2O quantities than CH_4. Depending on the degree of reaction, the buoyancy of the cryomagma may remain positive relative to water ice or become negative. On the one hand, effusive cryovolcanic eruption of an H_2O-NH_3-CH_4 cryomagma to the surface of Titan may occur. On the other hand, although the buoyancy becomes negative due to the increase of H_2O concentration, the large amounts of CH_4 released may generate sufficient pressure to break the overlaying ice and thus trigger explosive eruptions. If, instead of NH_3, the main agent is ammonium sulfate (Fortes et al. 2007), explosive cryovolcanism may occur as well. Choukroun et al. (2008a, b) have suggested that cryovolcanism in this fashion could only occur at a very local scale, due to the necessity of high concentrations of ammonia hydrates. However, such a model is consistent with the scarce observations of potential cryovolcanic features on Titan. Also, the amounts of methane released by dissociation of all methane clathrates overlaying an ice plume with the geometry predicted by Tobie et al. (2006b) are such that a broad region, like Tui Reggio or Hotei Reggio that may be cryovolcanic in origin (Barnes et al. 2006), would produce sufficient amounts to sustain the present abundance of methane over a period of ~1 Gyr (Choukroun et al. 2008; Choukroun et al. 2009). According to these results, it seems that cryovolcanism, despite its likely local-scale occurrence on Titan, is a good candidate for a methane replenishment process throughout Titan's history, independently of the other processes that are envisaged. Further experimental data on the stability of clathrate hydrates in presence of inhibitors, as well as thermodynamic models, are necessary to provide better constraints on cryovolcanic processes and on methane outgassing on Titan.

3.7 Presence of an Atmosphere on Titan But Not Ganymede and Callisto

This "meta-question" derives from the close similarities in the bulk properties of the three largest moons in the solar system, and the absence of even a thin equivalent of Titan's atmosphere around Ganymede and Callisto. The possible answers divide naturally into three:

(a) Titan's environment was colder that of Ganymede and Callisto. It is plausible that conditions in the region of Saturn formation were generally colder than at Jupiter. However, this hypothesis runs into difficulty on the basis of the noble gas and volatile pattern in Jupiter, which seems to require that the latter either formed in very cold conditions or accreted planetesimals from a larger distance where the nebular temperature was of order 40 K (Owen and Encrenaz 2006). The contrasting environments for condensation of volatiles or trapping in water ice then must be related to the local circum-planetary nebulae around Jupiter vs Saturn, consistent with the Galilean moons exhibiting a strong gradient in ice to rock ratio with distance from Jupiter suggests a strong temperature gradient there (Lunine and Stevenson 1982).

(b) Titan's atmosphere is impact-generated. Zahnle et al. (2002) point out that impact velocities are systematically lower at Titan than at Ganymede and Callisto, to the extent that Titan might have retained an atmosphere delivered post-accretion by impact of volatile-rich icy bodies (essentially, comets) while Ganymede and Callisto would not. This model is attractive because it appeals to a quantifiable difference in impact velocities between the two systems. However, the presence of nitrogen, methane and other volatiles in the interior of much smaller Enceladus (Waite et al. 2006) is not explained by this mechanism. Enceladus suggests volatile-rich material was accreted at velocities more appropriate to circum-Saturnian material, therefore from a gas disk surrounding Saturn, rather than from solar orbit. Of course, Titan could have received material from both solar and Saturnian sources, though nothing about the compositions of the two bodies requires this.

(c) All three bodies acquired atmospheres but escape driven by magnetospheric and solar UV removed the atmospheres of Ganymede and Callisto. The present nitrogen-dominated atmosphere of Titan is much less stable than that of the Earth's: the rate of loss on Titan corresponds to 10–20% of the nitrogen present in the atmosphere today over the age of the solar system (Strobel 1982). Cassini data suggest that the loss of methane by direct escape is competitive with photochemical destruction (Strobel 2008; Yelle et al. 2008), and therefore the loss rate of methane has been underestimated over Titan's history. Given the fact that Titan's atmosphere really is marginally stable, that the fluences of solar UV, solar wind, and magnetospheric charged particles have been larger at Jupiter than at Saturn, and impact velocities are larger as well for at least Ganymede vs Titan, the long-term stability of Titan-like atmospheres around Ganymede and Callisto should be reevaluated.

Cassini–Huygens data do not allow us to determine which of (a), (b) or (c) was most relevant to determining that Titan would have an atmosphere today while Ganymede and Callisto do not. The answer ultimately may have larger implications. For example, should (a) turn out to be most important – that the Jupiter system was volatile poor compared to the Saturn system – it would in turn raise the issue of whether the ocean beneath Europa's crust might be much poorer in organic molecules than the source region of Enceladus' plumes, with obvious astrobiological implications.

3.8 Questions for Future Missions

The origin and evolution of Titan remains a high priority for future missions that will seek the answers to why Titan is unusually large for the Saturn system and uniquely possesses – among satellites of its size class – a dense atmosphere. Specifically, some of the problems include

1. How thick is the ice crust? Although the asynchronous rotation has been interpreted as indicating a decoupled ice shell of thickness at least 70 km, a more definitive answer will require measurement of the tidal flexure by a fixed lander on Titan's surface.
2. How strongly differentiated is Titan and is there a metal core? This may require a more accurate gravity experiment that, with accelerometry, can remove the noise associated with atmospheric drag on the spacecraft whose acceleration is being measured.
3. Is there ammonia in the interior or on the surface? Higher sensitivity leading to better spatial and spectral resolution than the Cassini VIMS, on an orbiter or aerial platform, will be required to search for deposits of this molecule on Titan's surface. Definitive evidence of the lack of ammonia in the plumes of Enceladus – strongly suggested by Cassini INMS – already suggests that this molecule might have been destroyed in Enceladus's interior by chemical reactions, or perhaps is rarer in the Saturn system than the models described above have assumed.
4. Are there clathrates of methane, ethane, others, on the surface? Because the near-infrared spectra of clathrate is probably similar to water ice, direct sampling of the surface may be required to detect the presence of what should be a fairly common crustal material.
5. Where is the missing Xe and Kr? Is Kr enclathrated at the surface? Are the noble gases trapped in aerosols at the surface? Direct sampling, possibly of heated crustal material including water ice and dune particles will be required to test whether argon, krypton and xenon are trapped in clathrate hydrates, or adsorbed in dune particles.
6. How much recent cryovolcanism has taken place, if any? Circumstantial evidence for cryovolcanism from the appearance of certain features in Cassini radar images will need to be tested with altimetry (topographic information) and high spatial resolution spectra of candidate areas to search for carbon dioxide and ammonia as potential products of cryovolcanism. If active areas are present the serendipitous possibility that it might pass over an area where gases are being emitted mandate the onboard presence of a mass spectrometer.
7. Is there any CO_2? Indications from Cassini data of the potential presence of carbon dioxide, in an area which may also contain cryovolcanic flows (Hayne et al. 2008), suggests this gas is derived from Titan's interior. However, confirmation of the identification of the deposit as CO_2 is required, as well as better evidence that the area is cryovolcanic. Identification of internally-derived CO_2 would support the idea that methane was derived at least partially from "serpentinization" reactions in a liquid layer in Titan's interior that was in contact with hot rock (Atreya et al. 2006). This may require the advanced spectroscopy and topographic information to be provided by a follow-on to Cassini.

Acknowledgements Support from the Cassini Project in the preparation of this chapter is gratefully acknowledged. GT benefited from supports from the French Agence National de Recherche ("Exoclimats" project) and the INSU-Programme National de Planétologie. MC is supported by a NASA Postdoctoral Fellowship, administered by Oak Ridge Associated Universities.

References

Adams EY (2006) Titan's thermal structure and the formation of a nitrogen atmosphere. Ph.D. Dissertation, University of Michigan

Alibert Y, Mousis O (2007) Formation of Titan in Saturn's subnebula: constraints from *Huygens* probe measurements. Astronom Astrophys 465:1051–1060

Anders E, Grevesse N (1989) Abundances of the elements: meteoritic and solar. Geochim Cosmochim Acta 53:197–214

Atreya SK (1986) Atmospheres and ionospheres of the Outer Planets and Their Satellites. Springer, Berlin

Atreya SK, Donahue TM, Kuhn WR (1978) Evolution of a nitrogen atmosphere on Titan. Science 201:611–613

Atreya SK, Wong MH, Owen TC, Mahaffy PR, Niemann HB, de Pater I, Drossart P, Encrenaz T (1999) A comparison of the atmospheres of Jupiter and Saturn: deep atmospheric composition, cloud structure, vertical mixing, and origin. Planetary and Space Sci 47:1243–1262

Atreya SK, Adams EY, Niemann HB, Demick-Montelara JE, Owen TC, Fulchignoni M, Ferri F, Wilson EH (2006) Titan's methane cycle. Planet Space Sci 54:1177–1187

Baines KH, Drossart P, Lopez-Valverde MA, Atreya SK, Sotin C, Momary TW, Brown RH, Buratti BJ, Clark RN, Nicholson PD (2006) On the discovery of CO nighttime emissions on Titan by Cassini/VIMS:derived stratospheric abundances and geological implications. Planet Space Sci 54:1552–1562 (doi: 10.1016/j.pss.2006.06.020)

Barnes JW, Brown RH, Radebaugh J, Burratti BJ, Sotin C, Le Mouëlic S, Rodriguez S, Turtle EP, Perry J, Clark R, Baines KH, Nicholson PD (2006) Cassini observations of flow-like features in western Tui Regio Titan. Geophys Res Lett 33:L16204

Bar-Nun A, Kleinfeld I, Kochavi E (1988) Trapping of gas mixtures by amorphous water ice. Phys Rev B 38:7749–7754

Barr A, Canup R (2008) Constraints on gas giant satellite formation from the interior states of partially differentiated satellites. Icarus 198:163–177

Benz W (2000) Low velocity collisions and the growth of planetesimals. Space Sci Rev 92:279–294

Bockelée-Morvan D, Gautier D, Hersant F, Hure J-M, Robert F (2004) The composition of cometary volatiles. In: Comets II, 391–423

Bodenheimer P, Burket A, Klein R, Boss AP (2000a) In: Mannings V, Boss AP, Russell SS (eds) Michel C, Festou H, Uwe Keller, Harold A. Protostars and planets IV. University of Arizona Press, Tucson, pp 675–701

Bodenheimer P, Hubickyj O, Lissauer JJ (2000b) Models of the in situ formation of detected extrasolar giant planets. Icarus 143:2–14

Boss AP (2000) Possible rapid gas giant planet formation in the solar nebula and other protoplanetary disk. Astrophys J 536:L101–L104

Briceno C, Megeath ST, Gutermuth R et al (2007) The structure and evolution of Young Stellar Clusters. In: Reipurth B, Jewitt D, Keil K(eds) Protostars and planets. University of Arizona Press, Tucson. pp 361–378

Briggs FH, Sackett PD (1989) Radio observations of Saturn as a probe of its atmosphere and cloud structure. Icarus 80(1):77–103

Burns JA (1986) Some background about satellites. In: Burns JA, Matthews MS (eds) Satellites. University of Arizona Press, Tucson, AZ, pp 1–38

Cai K, Durisen RH, Michael S, Boley AC, Mej´ıa AC et al (2006) Astrophys J 636:L149–L152

Cameron AGW (1975) Cosmogonical considerations regarding Uranus. Icarus 24:280–284

Canup RM, Ward WR (2002) Formation of the Galilean satellites: conditions of accretion. Astron J 124:3404–3423

Canup RM, Ward WR (2006) A common mass scaling for satellite systems of gaseous planets. Nature 441:834–839

Castillo-Rogez JC, Matson DL, Sotin C, Johnson TV, Lunine JI, Thomas PC (2007) Iapetus' geophysics: rotation rate, shape and equatorial ridge. Icarus 190:179–202

Choukroun M (2007) Etude expérimentale et thermodynamique des hydrates sous pression: Applications à Titan. Ph.D. Dissertation, Université de Nantes, France

Choukroun M, Grasset O (2007) Thermodynamic model for water and high-pressure ices up to 2.2 GPa and down to the metastable domain. J Chem Phys 127(124):506–511 (doi:10.1063/1.2768957)

Choukroun M, Grasset O, Sotin C, Tobie G (2008) Cryovolcanic release of methane on Titan: experimental constraints from the stability of methane clathrates in presence of ammonia. Lunar Planet. Sci. Conf. XXXIX, abstract #1837, Houston, TX

Choukroun M, Grasset O, Tobie G, Sotin C. Stability of methane clathrate hydrates under pressure: implications for outgassing processes of methane on Titan. *Icarus*, accepted/in press

Clampin M, Krist JE, Ardila DR, Golimowski DA, Hartig GF, Ford HC, Illingworth GD, Bartko F, Benítez N, Blakeslee JP, Bouwens RJ, Broadhurst TJ, Brown RA, Burrows CJ, Cheng ES, Cross NJG, Feldman PD, Franx M, Gronwall C, Infante L, Kimble RA, Lesser MP, Martel AR, Menanteau F, Meurer GR, Miley GK, Postman M, Rosati P, Sirianni M, Sparks WB, Tran HD, Tsvetanov ZI, White RL, Zheng W (2003) Hubble space telescope ACS coronographic imaging of the circumstellar disk around HD 141569A. Astronom J 126:385–392

Cochran AL, Cochran WD, Barker ES (2000) N_2^+ and CO^+ in comets 122P/1995 S1 (deVico) and C/1995 O1 (Hale-Bopp). Icarus 146:583–593

Coustenis A, Achterberg RK, Conrath BJ, Jennings DE, Marten A, Gautier D, Nixon CA, Flasar FM, Teanby NA, Bézard B, Samuelson RE, Carlson RC, Lellouch E, Bjoraker GL, Romani PN, Taylor FW, Irwin PGJ, Fouchet T, Hubert A, Orton GA, Kunde VG, Vinatier S, Mondellini J, Abbas MM, Courtin R (2007) The composition of Titan's stratosphere from Cassini/CIRS mid-infrared spectra. Icarus 189:35–62

Croft SK, Lunine JI, Kargel JS (1988) Equation of state of ammonia-water liquid: derivation and planetological implications. Icarus 73:279–293

Dyadin YA, Aladko EY, Larionov EG (1997) Decomposition of methane hydrates up to 15 kbars. Mendel Comm 7(34–35):69–71

Eisenberg D, Kauzmann W (1969) The structure and properties of water. Oxford University Press, New York

Engel S, Lunine JI, Norton D (1994) Silicate interactions with ammonia-water fluids on early Titan. J Geophys Res 99:3745–3752

Fischer DA, Valenti J (2005) The planet-metallicity correlation. Astrophys J 622:1102–1117

Flasar FM, Achterberg RK, Conrath BJ, Pearl JC, Bjoraker GL, Jennings DE, Romani PN, Simon-Miller AA, Kunde VG, Nixon CA, Bézard B, Orton GS, Spilker LJ, Spencer JR, Irwin PGJ, Teanby NA, Owen TC, Brasunas J, Segura ME, Carlson RC, Mamoutkine A, Gierasch PJ, Schinder PJ, Showalter MR, Ferrari C, Barucci A, Courtin R, Coustenis A, Fouchet T, Gautier D, Lellouch E, Marten A, Prangé R, Strobel DF, Calcutt SB, Read PL, Taylor FW, Bowles N, Samuelson RE, Abbas MM, Raulin F, Ade P, Edgington S, Pilorz S, Wallis B, Wishnow EH (2005) Temperature, winds, and composition in the Saturnian system. Science 307:1247–1251

Fortes AD, Grindrod PM, Trickett SK, Vočadlo L (2007) Ammonium sulfate on Titan: possible origin and role in cryovolcanism. Icarus 188(1):139–153

Gautier D, Hersant F (2005) Formation and composition of planetesimals. Space Sci Rev 116:25–52

Gautier D, Hersant F, Mousis O, Lunine JI (2001) Enrichments in volatiles in Jupiter: a new interpretation of the Galileo measurements. Astrophys J 321:L13–L16

Genov G, Kuhs WF, Staykova DK, Goreshnik E, Salamatin AN (2004) Experimental studies on the formation of porous gas hydrates. Am Mineral 89(8–9):1228–1239

Goldreich P, Murray N, Longaretti PY et al (1989) Neptune's story. Science 245:500–504

Grasset O, Pargamin J (2005) The ammonia-water system at high pressures: implications for the methane of Titan. Planet Space Sci 53:371–384

Grasset O, Sotin C (1996) The cooling rate of a liquid shell in Titan's interior. Icarus 123:101–112

Grasset O, Sotin C, Deschamps F (2000) On the internal structure and dynamics of Titan. Planet Space Sci 48:617–636

Griffith CA, Zahnle K (1995) Influx of cometary volatiles to planetary moons: the atmospheres of 1000 possible Titans. J Geophys Res 100 (E8):16907–16922

Haar L, Gallagher JJ (1978) Thermodynamic properties of ammonia. J Phys Chem Ref. Data 7:635–792

Hayne P, McCord TB, Combe J-Ph, Barnes JW, Hansen GB (2008) Titan: observational Constraints on cryovolcanism. LPSC 39:2010, abstract 1391

Hemley RJ, Jephcoat AP, Mao HK, Zha CS, Finger LW, Cox DE (1987) Static compression of H_2O-ice to 128 Gpa (128 Mbar). Nature 330:737–740

Hersant F, Gautier D, Huré J-M (2001) A two-dimensional model for the primordial nebula constrained by D/H measurements in the solar system: implications for the formation of the giant planets. Astrophys J 554:391–407

Hersant F, Gautier D, Lunine JI (2004) Enrichments I volatiles in the giant planets of the solar system. Planet Space Sci 52:623–641

Hersant F, Gautier D, Tobie G, Lunine JI (2008) Interpretation of the carbon abundance in Saturn measured by Cassini. Planet Space Sci 56:1103–1111

Horst S, Vuitton V, Yelle RV (2008) The origin of oxygen species in Titan's atmosphere. J Geophys Res, 111(E10):E10006

Hunten D (1978) A Titan atmosphere with a surface at 200 K. In: Hunte DM, Morrison D (eds) The Saturn system, NASA Conference Publication 2068, pp 127–140

Iess L, Armstrong JW, Asmar SW, Graziani A, Mackenzie R, Racioppa P, Rappaport N, Tortora P (2008) The gravity field of Titan from the first three Cassini flybys, Geophysical Research Abstracts Vol. 10, EGU2008-A-10849

Iro N, Gautier D, Hersant F, Bockelée-Morvan D, Lunine JI (2003) An interpretation of the nitrogen deficiency in comets. Icarus 127:190–212

Irvine WM, Schloerb FP, Crovisier J, Fegley B, Mumma MM (2000) Comets: a link between interstellar and nebular chemistry. In: Manning V, Boss AP, Russell SS (eds) Protostars and planets IV. University of Arizona Press, Tucson, AZ, pp 1159–1200

Israël G, Szopa C, Raulin F, Cabane M, Niemann HB, Atreya SK, Bauer SJ, Brun J-F, Chassefière E, Coll P, Condé E, Coscia D, Hauchecorne A, Millian P, Nguyen M-J, Owen T, Riedler W, Samuelson RE, Siguier J-M, Steller M, Sternberg R, Vidal-Madjar C (2005) Complex organic matter in Titan's atmospheric aerosols from in situ pyrolysis and analysis. Nature 438:796–799

Jacobson RA (2004) The orbits of the major Saturnian satellites and the gravity field of Saturn from spacecraft and Earth-based observations. Astron J 18:492–501

Jacovi R, Bar-Nun A (2008) Removal of Titan's noble gases by their trapping in its haze. Icarus 196:302–304

Johnson TV, Lunine JI (2005) Saturn's moon Phoebe as a captured body from the outer Solar System. Nature 435:69–71

Jones T, Lewis JS (1987) Estimated impact shock production of N_2 and organic compounds on early Titan. Icarus 72:381–393

Kargel JS (1992) Ammonia-water volcanism on icy satellites: phase relations at 1 atmosphere. Icarus 100:556–574

Kaula WM (1979) Thermal evolution of Earth and Moon growing by planetesimal impacts. J Geophys Res 84:999–1008

Kawakita H, Jehin E, Manfroid J, Hutsemekers D (2007) Nuclear spin temperature of ammonia in Comet 9P/Tempel 1 before and after the Deep Impact event. Icaris 191:513–516

Kirk RL, Stevenson DJ (1987) Thermal evolution of a differentiated Ganymede and implications for surface features. Icarus 69:91–134

Kossacki KJ, Lorenz RD (1996) Hiding Titan's oceans: densification and hydrocarbon storage in an icy regolith. Planet Space Sci 44(9):1029–1037

Kuramoto K, Matsui T (1994) Formation of a hot proto-atmosphere on the accreting giant icy satellite: implications for the origin and evolution of Titan, Ganymede, and Callisto. J Geophys Res 99 (E10):21, 183–21,200

Lewis JS (1971) Satellites of the outer planets: their physical and chemical nature. Icarus 15:174–185

Lewis JS (1972) Low temperature condensation from the solar nebula. Icarus 16:241–252

Lewis JS, Prinn RG (1980) Kinetic inhibition of CO and N_2 reduction in the solar nebula. Astrophys J 238:357

Lissauer JJ, Stevenson DJ (2007) Formation of giant planets. In: Reipurth B, Jewitt D, Keil K (eds) Prostars and planets V. Universityof Arizona Press, Tucson, pp 591–606

Lodders K (2003) Solar system abundances and condensation temperatures of the elements. Astrophys J 591:1220–1247

Lopes RMC, Mitchell KL, Stofan ER, Lunine JI, Lorenz R, Paganelli F, Kirk RL, Wood CA, Wall SD, Robshaw LE, Fortes AD, Neish CD, Radebaugh J, Reffet E, Ostro SJ, Elachi C, Allison MD, Anderson Y, Boehmer R, Boubin G, Callahan P, Encrenaz P, Flamini E, Francescetti G, Gim Y, Hamilton G, Hensley S, Janssen MA, Johnson WTK, Kellehera K, Muhleman DO, Ori G, Orosei R, Picardi G, Posa F, Roth LE, Seu R, Shaffer S, Soderblom LA, Stiles B, Vetrella S, West RD, Wye L, Zebker HA (2007) Cryovolcanic features on Titan's suface as revealed by the Cassini Titan Radar Mapper. Icarus 186:395–412

Lorenz RD, Wood CA, Lunine JI, Wall SD, Lopes RM, Mitchell KL, Paganelli F, Anderson YZ, Wye L, Tsai C, Zebker H, Stofan ER (2007) Titan's young surface: initial impact crater survey by Cassini RADAR and model comparison. Geophys Res Lett 34, L07204. doi:10.1029/2006GL028971

Loveday JS, Nelmes RJ (2003) High-pressure neutron diffraction and models of Titan. High Pressure Res 23:41–47

Loveday JS, Nelmes RJ, Guthrie M, Belmonte SA, Allan DR, Klug DD, Tse JS, Handa YP (2001) Stable methane hydrate above 2GPa and the source of Titan's atmospheric methane. Nature 410: 661–663

Lowell RP, Rona PA (2002) Seafloor hydrothermal systems driven by the serpentinization of peridotite. Geophys Res Lett 29(11):1531, doi:10.1029/2001GLO14411

Lubow SH, Seibert M, Artymowicz P (1999) Disk accretion onto high-mass planets. Astrophys J 526:1001–1012

Lunine JI (1985) Dissertation, California Institute of Technology Dissertation, Calif. Institute of Technology 329 pp

Lunine JI, Stevenson DJ (1982) Formation of the Galilean satellites in a gaseous nebula. Icarus 52:14–39

Lunine JI, Stevenson DJ (1985) Thermodynamics of clathrate hydrate at low and high pressures with application to the outer Solar system. Astrophys J Suppl Ser 58:493–531

Lunine JI, Stevenson DJ (1987) Clathrate and ammonia hydrates at high pressure – application to the origin of methane on Titan. Icarus 70:61–77

Lunine JI, Stevenson DJ, Yung YK (1983) Ethane ocean on Titan. Science 222(4629):1229–1230
Lunine JI, Atreya SK, Pollack JB (1989) Present state and chemical evolution of the atmospheres of Titan, Triton, and Pluto. In: Origin and evolution of planetary and satellite atmospheres. University of Arizona Press, Tucson, AZ
Machida MM, Kokubo E, Inutsuka S-I, Matsumoto T (2008) Angular momentum accretion onto a gas giant planet. Astrophys J 685: 1220–1236
Magni G, Coradini A (2004) Formation of Jupiter by nucleated instability. Planet Space Sci 52:343–360
Mandt KE, Waite JH Jr, Magee BA, Bell JM, Nguyen M-J (2008) Isotopic fractionation in the upper atmosphere of Titan: INMS observations and implications for the atmosphere over geologic time-scales. Titan after Cassini workshop book of abstracts, Corpus Christi, TX, p 29
Mayer L, Quinn T, Wadsley J, Standel J (2002) Formation of giant planets by fragmentation of protoplanetary disks. Science 298:1756–1759
McCord TB, Hayne P, Combe J-P, Hansen GB, Barnes JW, Rodriguez S, Le Mouélic S, Baines KH, Buratti BJ, Sotin C, Nicholson P, Jaumann R, Nelson R (2008) The Cassini VIMS Team. Titan's surface: Search for spectral diversity and composition using the Cassini VIMS investigation. Icarus 194:212–242
McKay CP, Scattergood TW, Pollack JB, Borucki WJ, van Ghyseghem HT (1988) High-temperature shock formation of N_2 and organics on primordial Titan. Nature 322:520–522
Meyer MR, Backman DE, Weinberger AJ, Wyatt MC (2007) Evolution of circumstellar dusks around normal stars. In: Reipurth B, Jewitt D, Keil K (eds) Protostars and planets V. University of Arizona Press, Tucson, pp 573–590
Miller SL (1961) The occurrence of gas hydrates in the solar system. Proc Natl Acad Sci 47(11):1798–1808
Mitri G, Showman AP, Lunine JI, Lorenz RD (2007) Hydrocarbon lakes on Titan. Icarus 186:385–394
Mitri G, Showman AP, Lunine JI, Lopes RMC (2008) Resurfacing of Titan by ammonia-water cryomagma. Icarus 196:216–224
Monteux J, Coltice N, Dubuffet F, Ricard Y (2007) Thermo-mechanical adjustment after impacts during planetary growth. Geophys Res Lett 34, CiteID L24201
Morbidelli A, Levison H (2008) Late evolution of planetary systems. Physica Scripta T130:014028
Morrison D, Owen TC, Soderblom LA (1986) The satellites of Saturn. In: Burns JA, Matthews MS (eds) Satellites. University of Arizona Press, Tucson, AZ, pp 764–801
Mosqueira I, Estrada PR (2003) Formation of regular satellites of giant planets in an extended gaseous nebula I: Subnebula model and accretion of satellites. Icarus 163:198–231
Mousis O, Gautier D, Bockelée-Morvan D (2002) An evolutionary turbulent model of Saturn's subnebula: implications for the origin of the atmosphere of Titan. Icarus 156:162–175
Mousis O, Marboeuf U, Lunine JI, Alibert Y, Fletcher LN, Orton GS, Pauzat F, Ellenger Y (2009a). Determination of the minimum masses of heavy elements in the envelopes of Jupiter and Saturn. Ap J 696:1348–1354
Mousis O, Lunine JI, Thomas C, Pasek M, Marboeuf U, Alibert Y, Ballenegger V, Cordier D, Ellinger Y, Pauzat F, Picaud S (2009b). Clathration of volatiles in the solar nebula and implications for the origin of Titan's atmosphere. Astrophys J 691:1780–1786
Mueller S, McKinnon WB (1988) Three layer Ganymedes and Callistos. Icarus 76(3):437–464
Niemann HB, Atreya SK, Bauer SJ, Carignan GR, Demick JE, Frost RL, Gautier D, Haberman JA, Harpold DN, Hunten DM, Israel G, Lunine JI, Kasprzak WT, Owen TC, Paulkovich M, Raulin F, Raaen E, Way SH (2005) The abundances of constituents of Titan's atmosphere from the GCMS instrument on the Huygens probe. Nature 438:779–784
Orton G, Fletcher L, Irwin P, Bjoracker G, Flasar FM, Wishnow E (2005) The super-solar abundance of methane in Saturn from Cassini CIRS spectra. Geophys Res, Abstracts, 7, European Geophysical Union, p 05823
Osegovic JP, Max MD (2005) Compound clathrate hydrate on Titan's surface. J Geophys Res (Planets) 110:8004
Owen TC (1982) The composition and origin of Titan's atmosphere. Planet Space Sci 30:833–838
Owen TC, Bar-Nun A (1995) Comets, impacts and atmospheres. Icarus 116:215–226
Owen TC, Encrenaz T (2006) Compositional constraints on giant planet formation. Planet Space Sci 54:1188–1196
Paganelli F, Janssen MA, Stiles B, West R, Lorenz RD, Lunine JI, Wall SD, Callahan P, Lopes RM, Stofan E, Kirk RL, Johnson WTK, Roth L, Elachi C (2008) The Radar Team. Titan's surface from Cassini RADAR SAR and high resolution radiometry data of the first five flybys. Icarus 191:211–222
Pollack JB, Hubickyj O, Bodenheimer P, Lissauer JJ, Podolak M, Greenzweig Y (1996) Formation of the giant planets by concurrent accretion of solids and gas. Icarus 124:62–85
Prinn RG, Fegley B (1981) Kinetic inhibition of CO and N_2 reduction in circumplanetary nebulae: implications for satellite composition. Astrophys J 249:308–317
Prinn RG, Fegley B (1989) Solar Nebula chemistry: origin of planetary, satellite, and cometary volatiles. In: Atreya SK, Pollack JB, Matthews MS (eds) Origin and evolution of planetary and satellite atmospheres. University of Arizona Press, Tucson, AZ
Rafikov RR (2005) Can giant planets form by direct gravitational instability? Astrophys J 621:L69–L72
Rollet AP, Vuillard G (1956) Sur un nouvel hydrate de l'ammoniac. C.-R. Acad. Sci. Paris 243:383–386
Sagan C, Dermott SF (1982) Tides in the seas of Titan. Nature 300: 731–733
Schubert G, Stevenson DJ, Ellsworth K (1981) Internal structures of the galilean satellites. Icarus 47:46–59
Schubert G, Spohn T, Reynolds RT (1986) Thermal histories, compositions and internal structure of the moons of the solar system. In: Burns JA, Matthews MS (eds) Satellites. University of Arizona Press, Tucson, AZ, 224–292
Scott HP, Williams Q, Ryerson FJ (2002) Experimental constraints on the chemical evolution of large icy satellites. Earth Planet Sci Lett 203:399–412
Sekine Y, Sugita S, Shido T, Yamamoto T, Iwasawa Y, Kadono T, Matsui T (2005) The role of Fischer–Tropsch catalysis in the origin of methane-rich Titan. Icarus 178:154–164
Senshu H, Kuramoto K, Matsui T (2002) Thermal evolution of a growing Mars, J Geophys Res 107(E12):5118. doi:10.1029/2001JE001819
Sloan EDJ (1998) Clathrate hydrates of natural gases, 2nd edn. Marcel Dekker, New York
Sohl F, Hussmann H, Schwentker B, Spohn T, Lorenz RD (2003) Interior structure models and tidal Love numbers of Titan. J Geophys Res 108(E12). CiteID 5130
Sotin C, Grasset O, Beauchesne S (1998) Thermodynamical properties of high pressure ices. Implications for the dynamics and internal structure of large icy satellites. In: Solar System Ices, Based on reviews presented at the international symposium "Solar system ices" held in Toulouse, France, on March 27–30, 1995, Dordrecht Kluwer Academic, Astrophysics and space science library (ASSL), vol 227. ISBN0792349024, p79
Sotin C, Jaumann R, Buratti BJ, Brown RH, Clark RN, Soderblom LA, Baines KH, Bellucci G, Bibring J-P, Capaccioni P, Combes M, Coradini A, Cruikshank DP, Drossart P, Formisano V, Langevin Y,

Matson DL, McCord TB, Nelson RM, Nicholson PD, Sicardy B, LeMouelic S, Rodriguez S, Stephan K, Scholz CK (2005) Release of volatiles from a possible cryovolcano from near-infrared imaging of Titan. Nature 435:786–789

Stevenson DJ (1992) Interior of Titan. Proceedings of the ESA's Symposium on Titan, European Space Agency Special Publication, SP-338 pp 29–33

Stevenson DJ, Harris AW, Lunine JI (1986) Origins of satellites. In: Burns JA, Matthews MS (eds) Satellites. University of Arizon Press, Tucson, AZ, pp 39–88

Stofan ER, Elachi C, Lunine JI, Lorenz RD, Stiles B, Mitchell KL, Ostro S, Soderblom L, Wood C, Zebker H, Wall S, Janssen M, Kirk R, Lopes R, Paganelli F, Radebaugh J, Wye L, Anderson Y, Allison M, Boehmer R, Callahan P, Encrenaz P, Flamini E, Francescetti G, Gim Y, Hamilton G, Hensley S, Johnson WTK, Kelleher K, Muhleman D, Paillou P, Picardi G, Posa F, Roth L, Seu R, Shaffer S, Vetrella S, West R (2007) The lakes of Titan. Nature 445:61–64

Strobel DF (1982) Chemistry and evolution of Titan's atmosphere. Planet Space Sci 30:839–848

Strobel DF (2008) Titan's hydrodynamically escaping atmosphere. Icarus 198:588–594

Thomas C, Mousis O, Ballenegger V, Picaud S (2007) Clathrate hydrates as a sink of noble gases in Titan's atmosphere. Astron Astrophys 474:L17–L20

Tian F, Stewart AIF, Toon OB, Larsen KW, Esposito LW (2007) Monte Carlo simulations of the water vapor plumes on Enceladus. Icarus 188:154–161

Tobie G, Grasset O, Lunine JI, Mocquet A, Sotin C (2005) Titan's internal structure inferred from a coupled thermal-orbital model. Icarus 175:496–502

Tobie G, Lunine JI, Sotin C (2006) Episodic outgassing as the origin of atmospheric methane on Titan. Nature 440:61–64

Tobie G, Choukroun M, Grasset O, Le Mouelic S, Lunine JI, Sotin C, Bourgeois O, Gautier D, Hirtzig M, Lebonnois S, Le Corre L (2009) Evolution of Titan and implications for its hydrocarbon cycle. Philo Trans R Soc A 367:619–631

Tomasko MG, Archinal B, Becker T, Bézard B, Bushroe M, Combers M, Cook D, Coustenis A, de Bergh C, Dafoe LE, Doose L, Douté S, Eibl A, Engel S, Gliem F, Grieger B, Holso K, Howington-Kraus E, Karkoschka E, Keller HU, Kirk R, Kramm R, Küppers M, Lanagan P, Lellouch E, Lemmon M, Lunine JI, McFarlane E, Moores J, Prout GM, Rizk B, Rosiek M, Rueffer P, Schröder SE, Schmitt B, See C, Smith P, Soderblom LA, Thomas N, West R (2005) Rain, wind and haze during the Huygen's probe descent to Titan's surface. Nature 438:765–778

Udry S, Fischer D, Queloz D (2007) A decade of radial velocity discoveries in the exoplanets domain. In: Reipurth B, Jewitt D, Keil K (eds) Protostars and planets V. University of Arizona Press, Tucson, pp 685–700

Wadhwa M, Amelin Y, Davis AM et al (2007) From dust to planetesimals: implications for the solar protoplanetary disk from short-liver radionuclides. In: Reipurth B, Jewitt D, Keil K (eds) Protostars and planets V. University of Arizona Press, Tucson, pp 836–848

Waite JH, Jr., Lewis WS, Magee BA, Lunine JI, McKinnon WB, 11 others (2009). Liquid water on Enceladus from observations of ammonia and 40Ar in the plume. Nature 460:487–490

Weidenschilling SJ (1997) The origin of comets in the solar nebula: a unified model. Icarus 127:290–306

Wong MH, Lunine JI, Atreya SK, Johnson T, Mahaffy PR, Owen TC, Encrenaz T (2008) Oxygen and other volatiles in the giant planets and their satellites. Rev Mineral Geochem 68:219–246

Wuchterl G, Guillot T, Lissauer JJ (2000). In: Mannings V, Boss AP, Russell SS (eds) Protostars and planets IV. University of Arizona Press, Tucson, pp 1081–1109

Yelle RV, Cui J, Mueller-Wordag I (2008) Methane escape from Titan's atmosphere. J Geophys Res, 113(E10):E10003

Yung YL, Allen M, Pinto JP (1984) Photochemistry of the atmosphere of Titan – comparison between mode and observations. Astrophys J Suppl Ser 55:465–506

Zahnle KJ, Walker JCG (1982) The evolution of solar ultraviolet luminosity. Rev Geophys and Space Phys 20:280–292

Zahnle K, Pollack JB, Grinspoon D, Dones L (1992) Impact-generated atmospheres over Titan, Ganymede, and Callisto. Icarus 95(1):1–23

Chapter 4
Titan's Interior Structure

Christophe Sotin, Giuseppe Mitri, Nicole Rappaport, Gerald Schubert and, David Stevenson

Abstract The goal of this chapter is to give a description of Titan's interior that is consistent with the new constraints provided by the Cassini mission. As the Cassini mission proceeds into its first extended phase, the data obtained during the nominal mission suggest that Titan is at least partially differentiated. An ocean would be present some tens of kilometers below the surface. By comparison with the Galilean icy satellites Ganymede and Callisto, Titan would be composed of a metal/silicate rich core and a H_2O rich outer layer. These conclusions are drawn from the interpretation of the gravity data, the geological data and the presence of a Schumann resonance which has been inferred from the measurement of electric signals during the descent of the Huygens probe into Titan's atmosphere. Titan's high eccentricity implies that the interior has not been very dissipative, there is little tidal heating available for internal dynamics, and the ice layer is cold, which can be achieved if the ocean under the ice layer contains ammonia. This paper also describes observations and interpretations which seem difficult to reconcile with our present understanding of Titan's interior structure and evolution such as the shape of the planet or the obliquity. The last part of the chapter describes heat transfer models which suggest that the lower part of the ice crust could be convective. The NH_3–H_2O phase diagram indicates that the ocean is decoupled from the silicate-rich core by a layer of high-pressure ices. However, the interior model is largely uncertain because the interpretation of the data is still debated at present time. The additional information that will be acquired during the Cassini Solstice Mission should allow us to answer some of the questions.

C. Sotin (✉), G. Mitri, and N. Rappaport
Jet Propulsion Laboratory, California Institute of Technology, 4800 Oak Grove Drive, Mail Stop 183-301, Pasadena, CA, 91109, USA
e-mail: christophe.sotin@jpl.nasa.gov

G. Schubert
University of California at Los Angeles, Los Angeles, CA, USA

D. Stevenson
California Institute of Technology, Pasadena, CA, USA

4.1 Introduction

Titan, the second largest moon in the solar system, is the only one with a thick atmosphere composed mainly of nitrogen (N_2), with methane (CH_4) the next most abundant compound. Several observations suggest that the atmosphere, or at least part of it, is not primordial but has been outgassed during the last tens or hundreds of millions years. First, the large $^{15}N/^{14}N$ ratio compared to the terrestrial value (Niemann et al. 2005) suggests that nitrogen has escaped. The very small amount of ^{36}Ar and the presence of ^{40}Ar (Waite et al. 2005) imply that ^{40}Ar is not primordial, but produced by the decay of ^{40}K, implying some exchange between the silicates in the interior and the atmosphere. The carbon isotopic ratio $^{13}C/^{12}C$ in hydrocarbons is close to the terrestrial value. This is consistent with the idea that hydrocarbons have been recently released from the interior (Tobie et al. 2006). It explains the presence of atmospheric methane although it irreversibly transforms into ethane on time scales of a few tens to a few hundreds of millions years. A better understanding of the physical and chemical processes involved in the outgassing requires knowledge of the interior structure.

Titan is most like the large icy Galilean moons Ganymede and Callisto in terms of mass, radius, and density (Table 4.1). Titan is intermediate in size between Ganymede and Callisto; it is only about 59 km smaller in radius than Ganymede while it is about 172 km larger in radius than Callisto. Titan is also intermediate in density between the large icy Galilean moons; it is slightly denser than Callisto and a little less dense than Ganymede (Table 4.1). With densities between about 1,850 kg m^{-3} and 1,940 kg m^{-3}, all three satellites must be made of about the same mass fractions of rock and ice (Fig. 4.1). It is quite remarkable that these satellites, so similar in size and density, should be so different in almost every other way. Ganymede is differentiated, has a liquid iron core wherein a dipole-like magnetic field is produced and a liquid water ocean deep within its outer ice shell. Callisto is only partially differentiated, with no iron core or intrinsic magnetic field, but with an internal ocean (Schubert et al. 2004).

R.H. Brown et al. (eds.), *Titan from Cassini-Huygens*,

Table 4.1 Comparison between the Galilean satellites and Titan

	ρ (kg/m³)	Radius (km)	C/MR²	k_s [a]	e	Prot (days)
IO	3,528	1,822	0.377	1.3065	0.004	1.77
EUROPA	2,970	1,569	0.347	0.937	0.010	3.55
GANYMEDE	1,940	2,634	0.311	0.8014	0.0015	7.15
CALLISTO	1,851	2,403	0.358	1.13	0.007	16.70
TITAN	1,881	2,575	?	?	0.029	15.95

[a] values come from Davies and Colvin (1998), except for Callisto since their study assumes an undifferentiated Callisto. For Callisto, we take the value of C/MR² in Schubert et al. (2004) and derive the Love number with the Radau formula (Eq. 4.11) which assumes hydrostatic equilibrium.

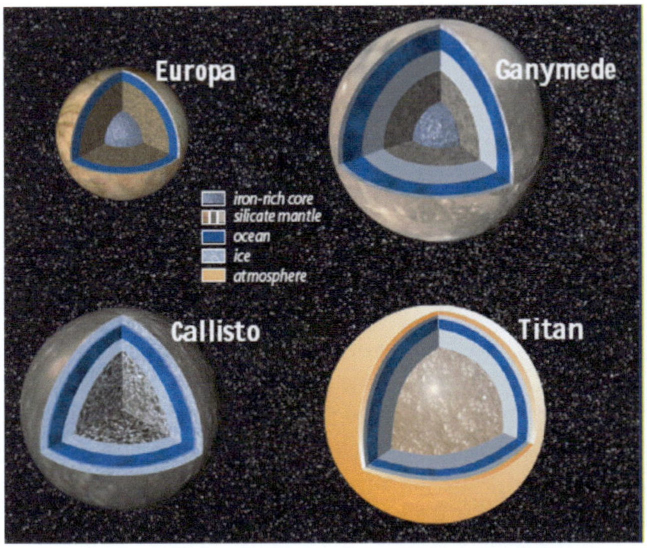

Fig. 4.1 Interior structure of the icy satellites (from Sotin and Tobie 2004)

While the atmosphere provides important data for understanding Titan's evolution, it imposes strong limitations on standard geophysical observations. Measurements of the magnetic and gravity fields have proven to be very powerful for constraining the interior structure of the Galilean satellites. But the intensity of the gravity acceleration varies inversely as the square of the distance. Because of the atmosphere, the Cassini spacecraft cannot approach as near to Titan's surface as the Galileo spacecraft flew past Europa's surface (a few hundreds of kilometers). The signal is therefore much weaker. This paper reports on the results of both the magnetometer (MAG) and the radio science subsystems (RSS) experiments onboard the Cassini spacecraft. Additional information has been deduced by the apparent registration mismatch of radar images, which suggests that Titan's spin rate is not synchronous (Stiles et al. 2008). Information also derives from the electric signal measured by the Permittivity, Wave and Altimetry (PWA) experiment (Grard et al. 1995), a subsystem of the Huygens Atmospheric Structure Instrument (HASI) installed onboard Huygens (Fulchignoni et al. 2005). A quasi-monochromatic signal at around 36 Hz was detected by PWA, visible almost permanently throughout the descent of the Huygens probe (Grard et al. 2006). It was proposed that it could fit the second expected eigenmode of the Schumann resonance (Simões et al. 2007). The inferences of non-synchronous rotation and Schumann resonance are compatible with the presence of an ocean. Other information such as topography will also be discussed and models of the internal dynamics presented.

4.2 Observations that Constrain the Interior Structure

The Cassini prime mission had two main science objectives related to the determination of Titan's interior structure:

- What is the degree of differentiation?
- Is there a liquid layer?

Based on previous information from the Galileo mission (e.g. Schubert et al. 2004) and from the phase diagram of water ice (Fig. 4.2), it is believed that an icy satellite could be composed of several layers which, from the center to the surface, are: an iron-rich core, a silicate shell, a high-pressure ice shell, an ocean, a low-pressure ice layer. All these layers may not be present. In addition, it is possible that other layers exist but only seismic experiments would allow us to know about them. The degree of differentiation is given by the moment of inertia C which is defined by:

$$C = \int_M x^2 dm = \iiint r^4 \rho(r,\theta,\psi) \sin^3(\theta) dr d\theta d\psi \tag{4.1}$$

where ρ(r,θ,ψ) is the density at a given point located by its radius (r), its colatitude (θ) and its longitude (ψ). The parameter x is the perpendicular distance of a mass point to the rotation axis. Assuming that there are n shells, each of them with a constant density, Eq. 4.1 can be written

$$C = \frac{8\pi}{15} \sum_{i=1}^{n} \rho_i \left(R_i^5 - R_{i-1}^5\right) \tag{4.2}$$

where R_{i-1} and R_i are the inner and outer radius of the shell i of density ρ_i, respectively. Note that R_0 is equal to 0. The mass is given by:

$$M = \int_M dm = \frac{4\pi}{3} \sum_{i=1}^{n} \rho_i \left(R_i^3 - R_{i-1}^3\right) \tag{4.3}$$

Since mass and radius are known (Table 4.2), one can calculate the ratio C/MR². If the satellite is undifferentiated and if the density of the material does not vary with depth, there

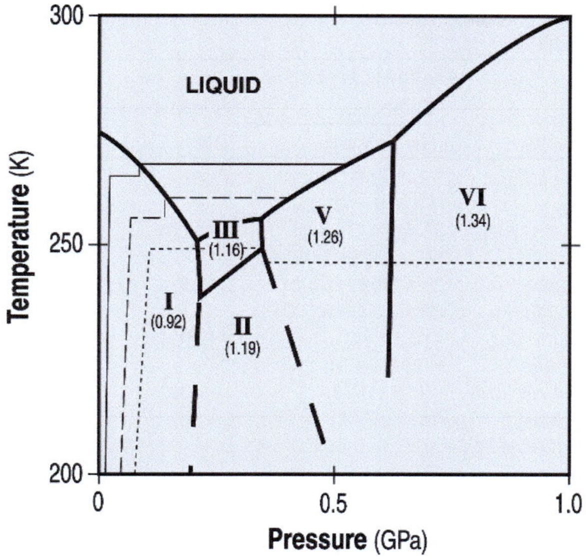

Fig. 4.2 Phase diagram of water ice. The temperature profiles for different heat fluxes have been reported in the ice I stability model. In the liquid layer, the adiabatic gradient is very close to isothermal

Table 4.2 Parameters for Titan. Solution SAT192. The obliquity value comes from Stiles et al. (2008). The values for the different radii comes from Zebker et al. (2009)

Semi-major axis to Saturn (km)	1.2218 10^6 km
Eccentricity	2.846%
Sidereal rotational period (s) if synchronous	1,377,684 ~ 15.945 Earth days
GM (km^3/s^2)	8978.133(4)
Mass (10^{22} kg)/density	13.452 (2)/1.880 (4)
Mean radius (R)	2574.73 (9) km
Mean equatorial radius (a/b)	2574.95 (6) km (2575.15/2574.78 km)
Polar radius c	2574.47 (6) km
$q=\omega^2R^3/GM$	3.96 10^{-5}
Gravity acceleration (m/s^2) at the surface	1.354
Estimated radiogenic power	300–400 GW
Obliquity	0.3°

is only one layer and the moment of inertia is equal to 0.4 MR^2. Values of C/MR^2 for the Galilean icy satellites range from 0.358 for Callisto down to 0.311 for Ganymede (Table 4.1) which has a liquid iron core wherein its intrinsic magnetic field is generated (Anderson et al. 1996; Davies and Colvin 1998; Schubert et al. 2004). Knowledge of the moment of inertia gives the degree of differentiation. Because the moment of inertia can be retrieved from the degree 2 gravity coefficients (see Section 4.2.2), the main investigation for determining the moment of inertia is gravity. But the moment of inertia also influences the orbital characteristics and this chapter also describes the information which can be retrieved from knowledge of the obliquity.

The second important objective related to interior structure is whether an ocean is present at depth. Because Titan's orbit is highly eccentric, the degree 2 gravity coefficients have periodic variations whose amplitude depends on the presence or absence of a liquid layer. The magnetic field investigation might also provide information on the possible existence of an internal ocean as did the magnetometer experiment on Galileo that recorded an induced magnetic field at Callisto and Europa best explained by the presence of an ocean. This chapter explains why gravity and magnetic experiments have so far failed in providing the required measurement. However, investigations not foreseen for this science objective at the time the mission was designed, do give information on the presence of an internal ocean.

4.2.1 Orbital Data (Eccentricity, Obliquity and Rotation Rate)

It was known before Cassini-Huygens that Titan's eccentricity is large. With a value of 2.9% (e.g. Lodders and Fegley 1998), Titan's eccentricity is much larger than those of Europa (1%) and Ganymede (0.15%). The eccentricity of the Galilean satellites is maintained by the Laplace resonance between Io, Europa and Ganymede. For Titan, no such resonance exists and the loss of Titan's orbital energy by internal dissipation should lead to a reduction in orbital eccentricity (Sears 1995). If one neglects the dissipation of tides raised by Titan on Saturn, which is less than 2% of the dissipation within Titan (Tobie et al. 2005), the change in eccentricity (e) due to tidal dissipation is given by (Sohl et al. 1995):

$$\frac{de}{dt}\frac{e}{1-e^2} = -\frac{D}{GM_SM_T}\frac{dE}{dt} \quad (4.4)$$

where M_S and M_T are the mass of Saturn and Titan, respectively, D is the semimajor axis of the orbit and dE/dt is the global dissipation rate. Following the study of Tobie et al. (2005), the maximum amount of dissipation is on the order of the radiogenic heating (400 GW). Such a value is obtained for a viscosity of ice between 10^{14} and 10^{15} Pa·s, which is the expected viscosity for ice above a pure H_2O ocean (Tobie et al. 2005). With this value, the present eccentricity would decrease by about 0.01% per Myr, which means that the orbit would be circular in less than 300 Myr. It suggests that Titan's interior is not dissipative, which poses strong constraints on the viscosity of the interior and on an ocean (see Section 4.3.3). Grasset et al. (2000) showed that the presence of ammonia in the ocean would lower the temperature of the ice crust resulting in larger viscosities and smaller dissipation rate. The presence of ammonia is consistent with the present eccentricity if the initial eccentricity was on the order of 20–30% (Tobie et al. 2005).

Since the Cassini radar started imaging Titan in October 2004, about 150 geological features were mapped twice as new radar paths crossed previous ones during the three first years of the Cassini mission (Stiles et al. 2008). Each landmark was located in Titan-centered inertial, non rotating coordinates. The differences in coordinates of each landmark were inverted to provide Titan pole coordinates (declination and right ascension) and spin rate parameters (Table 4.3 from Stiles et al. 2008). The first and main component is the obliquity of Titan spin axis relative to the normal to its orbital plane. Some landmarks could be imaged at very different values of the true anomaly. These measurements provide a very good constraint on the value of the obliquity, which is found to be 0.3°. This value is much larger than that predicted if both the orbital plane and the equatorial plane precess at the same rate (Fig. 4.3). This equilibrium is known as the Cassini state which is followed by almost all natural satellites. Bills and Nimmo (2008) propose that the presence of an ocean that decouples the icy shell from the silicate core can explain such a large obliquity (see discussion in Section 4.4).

Another parameter, which has been determined by Stiles et al. (2008), is the spin rate. Their study suggests that Titan had a spin rate 1.1 10^{-3} degree/year faster than the synchronous spin rate, during this period of observations. Following the study of Tokano and Neubauer (2005), Lorenz et al. (2008) propose that the variations to synchronous rotation could be the result of exchange of angular momentum between the atmosphere and the icy crust. Such variations are seasonal because wind directions vary with seasons. The amplitude of the spin rate variations also suggests that the

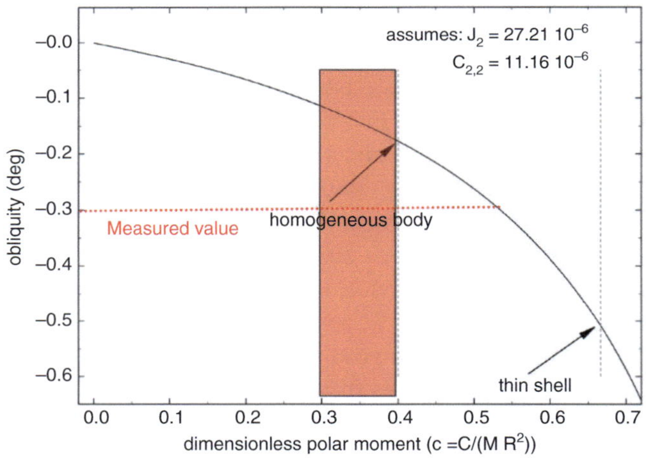

Fig. 4.3 Obliquity versus moment of inertia. This figure, adapted from Bills and Nimmo (2008), shows the value of the obliquity if Titan is in a Cassini state. As described in the text, reasonable values of the moment of inertia (*red area*) range between 0.3 MR^2 for a fully differentiated body and 0.4 MR^2 for a constant density body. The observed value of 0.3° for the obliquity is discussed in Section 4.4

Table 4.3 Orbital parameters from radar misregistration (from Stiles et al. 2008)

Parameter	Estimate +/−1 s N = 50 cases (N = 150 cases)	Nominal value Titan IAU	Estimated–Nominal
Pole RA (deg)	39.483 ± 0.025 (39.470 ± 0.032)	37.589	1.894 (1.881)
Pole DEC (deg)	83.4279 ± 0.0024 (83.4321 ± 0.0031)	83.6710	−0.2431 (−0.2479)
Spin rate (deg/day)	22.57809 ± 0.00011 (22.57790 ± 0.00013)	22.5770	0.0011 (0.0009)
dPRA/dt (deg/century)	−30.1 ± 4.2 (−25.2 ± 4.9)	−2.2031	−27.9 (−23.2)
dPDEC/dt (deg/century)	−0.05 ± 0.36 (0.27 ± 0.44)	−0.1252	0.0752 (0.3952)
dSR/dt (deg/day/century)	0.0523 ± 0.0050 (0.0395 ± 0.0066)	0	0.0523 (0.0395)

icy crust would be decoupled from the silicate core by an ocean. This interpretation is discussed in Section 4.2.5.

4.2.2 Gravity Field

The gravity measurements provide important information on the internal structure of planets and satellites. The values of the degree 2 coefficients provide information on the state of internal differentiation; their variations with true anomaly provide information on the presence of liquid layers at depth and on the amount of tidal dissipation. The gravitational potential can be compared with the shape of the planet. If the topography is known at similar degrees, comparison between topography and gravity anomalies provides quantitative information on the degree of compensation of the crust. Subsurface anomalies like the lunar mascons and the Ganymede gravity anomalies (Palguta et al. 2009) can be detected. However, the ability to reach these conclusions depends on the quality of the data. At the end of the nominal mission, the RSS provided some preliminary information but no definitive answer on the different points described above. After providing the principles of gravity measurements, this paragraph describes the information obtained so far with Cassini.

The gravity potential can be retrieved from accurate measurements of the Line-of-Sight (LoS) Doppler shift when a radio signal is sent from Earth to the spacecraft and from the spacecraft to Earth. It can be written:

$$U = \frac{GM}{r}\sum_{n=0}^{\infty}\left[\left(\frac{R}{r}\right)^n \sum_{m=0}^{n}\begin{pmatrix}\alpha_n^m \cos(m\phi)+\\ \beta_n^m \sin(m\phi)\end{pmatrix} P_n^m(\sin(\lambda))\right]+ \frac{1}{2}\omega^2 r^2 \cos^2(\lambda) \quad (4.5)$$

where GM is the product of the gravitational constant G and the body's mass M (Table 4.2), R is the mean radius, ϕ is longitude, λ is latitude and $P_n^m(\sin(\lambda)$ are the normalized associated Legendre polynomials of degree n and order m (e.g. Blakely 1996). The last term in (5) is the rotational potential where ω is the spin rate in rd/s. The coefficients α and β are determined by the Doppler shift observations and provide the information about the distribution of mass within the body. Since a mass anomaly of degree n provides a gravity signal proportional to $(1/r)^{n+1}$, measurements are limited by the distance of the spacecraft to the center of mass of the planet (Fig. 4.4). If the Cassini spacecraft is too close to the planet, the atmospheric drag creates a signal that is mixed in the observations. It cannot be easily retrieved and the best compromise between the distance to Titan's surface and the atmospheric signal is around 1,400 km altitude (Rappaport et al. 2008a).

First consider the information provided by the degree 2 coefficients. Equation 4.5 can be written:

$$U = \frac{GM}{r}\left[1+\left(\frac{R}{r}\right)^2\left(\begin{array}{l}-J_2P_{2,0}(\sin(\lambda))+\\ J_{2,2}\cos(2\phi)P_{2,2}(\sin(\lambda))\end{array}\right)\right]+ \frac{1}{2}\omega^2r^2\cos^2(\lambda) \quad (4.6)$$

where J_2 and $J_{2,2}$ are constants to be determined and $P_{2,0}(\theta)$ and $P_{2,2}(\theta)$ are associated Legendre polynomials (e.g. Blakely 1996) defined as:

$$P_{2,0}(\sin(\lambda)) = \frac{1}{2}\left[3\cos^2(\theta)-1\right] = \frac{1}{2}\left[3\sin^2(\lambda)-1\right]$$
$$\text{and } P_{2,2}(\sin(\lambda)) = 3\sin^2(\theta) = 3\cos^2(\lambda) \quad (4.7)$$

The values of J_2 and $J_{2,2}$ can be written as the sum of a constant and periodic term

$$J_2 = J_{2h} + \Delta J_2 = \left(\frac{1}{2}\alpha + \frac{1}{3}q\right)k_s + \frac{3}{2}\alpha k_2\, e\cos(v) \quad (4.8)$$

$$J_{2,2} = J_{2,2h} + \Delta J_{2,2} = \left(\frac{1}{4}\alpha k_s + \frac{3}{4}\alpha k_2\right)e\cos(v) \quad (4.9)$$

$$\text{with } \alpha = \frac{M_{Saturn}}{M}\times\left(\frac{R}{D}\right)^3 \text{ and } q = \frac{\omega^2R^3}{GM} \quad (4.10)$$

D is the semimajor axis of Titan's orbit around Saturn. In the case of a synchronous satellite, $\alpha = q$. In Eqs. 4.8 and 4.9, e is the eccentricity, v is the true anomaly, k_s is the secular Love number and k_2 is the periodic Love number. Because the eccentricity is 2.9% and the periodic Love number is smaller than the secular Love number, the periodic term is at most a few percent of the secular term (Fig. 4.4).

If Titan is in hydrostatic equilibrium, $J_2 = \frac{10}{3}J_{2,2}$ and the Radau–Darwin equation can be used to retrieve the moment of inertia about the spin axis (C):

$$\frac{C}{MR^2} = \frac{2}{3}\left[1-\frac{2}{5}\sqrt{\frac{4-k_S}{1+k_S}}\right] \quad (4.11)$$

So far, 4 Titan flybys have been devoted to Titan gravity field measurements. During the nominal mission, three flybys, T11, T22, and T33 occurred on February 27, 2006; on December 28, 2006; and on June 29, 2007, respectively. Another flyby (T38) was initially allocated to the Doppler shift measurements but it was exchanged with the first flyby of the extended mission, T45, which happened on July 31, 2008. The data are still being processed and only the results of the first two flybys (Iess et al. 2007) are reported in Table 4.4. Using only data from T11 and T22, the quadrupole moments were in full agreement despite the fact that the T11 and T22 probed very different regions of Titan and had different geometries: T11 was an equatorial, relatively distant flyby, while T22 was an inclined close flyby with an inclination of 68° (Rappaport et al. 2008a). This agreement suggested that Titan's gravity field could be well represented by a degree 2 model. However, there were two puzzling features in the solution. The first one is that the ratio $J_2/J_{2,2}$ is equal to 2.52 and 2.65 for T11 and T22, respectively. This

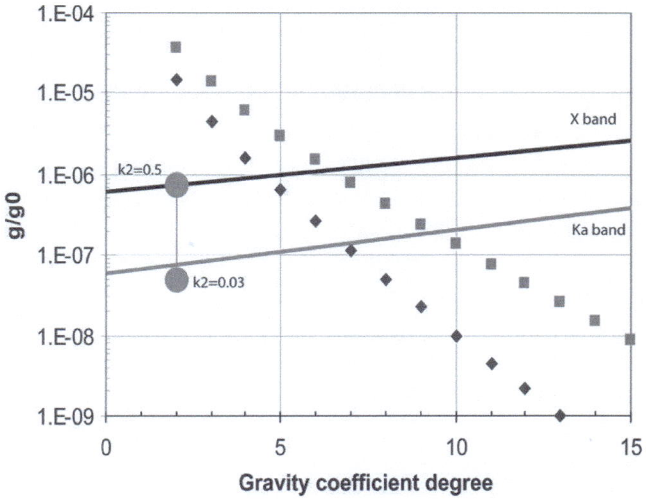

Fig. 4.4 Amplitude of the gravity signal at a given degree as a function of this degree when extrapolating the degree 2 root mean square amplitude to higher degrees with the Kaula's rule (see text for details). Two cases have been assessed: gravity variations in the silicate core only (*filled diamonds*) and gravity variations spread in the whole satellite (*filled squares*). The values can be compared with the accuracy predicted for Doppler-shift measurements in the X band (*black curve*) and K band (*red curve*). Circles indicate the variations of the degree 2 coefficients as a function of eccentricity for two different models of Titan: $k_2 = 0.5$ if there is an internal ocean and $k_2 = 0.03$ if there is no internal liquid layer

Table 4.4 Gravitational parameters for flybys T11 and T22 (Iess et al. 2007). The data for flybys T33 and T45 were not published at the time of writing the present paper

	T11	T22
Time at C/A	27-FEB-2006 08:25:18 UTC	28-DEC-2006 18:05:22 UTC
Lat at C/A	0.04	35.98
Long at C/A (W)	107.367	354.36
Geography	Eastern edge of Xanadu, North of Thui Regio	NE of Fensal – North of Sinlap
R/Alt at C/A (km)	4,388/1,812	3,890/1,297
Subsolar lat/long	−18.106/14.715	−14.12/28.94
Ra/dec	40.98/−6.36	181.49/−30.94
SpaceC vel (km/s)	5.86495	6.0
Mean anomaly	173	197
GM (km^3/s^2)	8.978136 (4) 10^3	8.978130 (4) 10^3
J2	2.81 (33) 10^{-5}	2.86 (6) 10^{-5}
C22	1.117 (5) 10^{-5}	1.079 (22) 10^{-5}

ratio is much smaller than 10/3, suggesting that Titan is not in hydrostatic equilibrium. The second one is related to the signatures in the Doppler residuals at closest approach which suggest that additional information is present. Following these observations, Rappaport et al. (2008b) have included degree 3 coefficients in the inversion of the T11, T22, T33 and T45 flybys. This new inversion allows them to reach two conclusions (Rappaport et al. 2008b). First Titan is in hydrostatic equilibrium. This result contradicts the first analysis performed on T11 and T22 and described above (Iess et al. 2007). It shows that it is important to include higher degree terms in the inversion. Considering the values of J_2 and $J_{2,2}$ given for T11 and T22 and in the absence of published revised values, one can just assess that the value of the Love number k_s calculated with Eqs. 4.8 and 4.9, is between 0.85 and 1.1. It is larger than that of Ganymede which has an iron core and less than the value of 1.5 which corresponds to a constant density body. Implications for the interior structure and dynamics are described in Section 4.3.

The second conclusion is that the degree 3 coefficients account for about 1–3% of the total field. This information allows us to draw a very preliminary comparison with the Moon. If Titan is differentiated, the silicate core is about the size of the Moon whose gravity field is much better known (Konopliv et al. 1998). The root mean-squares amplitude of the gravity signal as a function of the degree n follows a power law which is very close to the $1/n^2$ power law predicted by the Kaula's rule (Kaula 1966). The knowledge of the degree 2 coefficients (Table 4.4) allows us to determine the constant to be used in the Kaula's rule (Fig. 4.4). The values of the contribution of each degree to the total gravity signal at 1,500 km altitude can be determined for variations in the silicate core only (filled diamonds in Fig. 4.4) or in the icy crust (filled squares in Fig. 4.4). One can see that the predicted ratio of the degree 3 contribution to the degree 2 contribution is on the order of 30%, which is one order of magnitude larger than the value reported in Rappaport et al. (2008b). It suggests that lateral density variations on Titan are much less than those on silicate bodies. The stronger slope would mean that the degree 4 variations would be smaller than the periodic contribution if an ocean is present ($k_2 = 0.5$ in Fig. 4.4). Measurements of the gravity field during the extended mission may allow for the determination of the periodic Love number.

4.2.3 Titan's Shape

Topography of a planet and a satellite is determined relative to an equipotential surface which we refer to as the theoretical shape. On Earth, sea level is by definition an equipotential. Since the major deformation is due to the rotation of the planet, the reference surface is a bi-axial ellipsoid that best fits the ocean surface. The difference between the equatorial radius (a) and the polar radius (c) defines the flattening (f) which depends on both the spin rate and the interior structure. It can be expressed as:

$$f = \frac{a-c}{a} = \frac{3}{2}J_2 + \frac{1}{2}q \tag{4.12}$$

A value of 1/298.256 is found for the Earth (e.g. Turcotte and Schubert 1982).

For satellites having synchronous orbit, the hydrostatic shape of the equipotential is that of a tri-axial ellipsoid since a tide-forming potential caused by the planet must be added to the rotational potential (Zharkov et al. 1985). The equator has the shape of an ellipse and is defined by two perpendicular axes with semi-major and semi-minor dimensions a and b, respectively. If the satellite is not on an eccentric orbit, the long axis (a) is aligned with the line joining the satellite and the planet. The rotational and tidal potentials lead to the following expressions for the differences between the radii (Zharkov et al. 1985):

$$a - c = 2\alpha h_s R, 2b - c = \frac{1}{2}\alpha h_s R = \frac{1}{4}(a-c) \tag{4.13}$$

where $h_s = 1 + k_s$ is the degree 2 Love number for deformation and R is the radius of the sphere having a volume equivalent to that of the ellipsoid ($R = \sqrt[3]{abc}$). The differences in radius depend on the spin rate through the value of α (Eq. 4.10), and on the interior structure through the value of the Love number. The value of α is well known and is equal to 3.96×10^{-5} for Titan (Table 4.2). If Titan is undifferentiated ($k_s = 1.5$, $h_s = 2.5$), then a difference of about 500 m is expected between the long equatorial radius and the polar radius. The equatorial radius would vary by 400 m. If Titan is differentiated, the differences would be smaller. If the value of the Love number k_S lies in between those of Ganymede and Callisto ($k_S = 1.0$), the differences become 400 and 300 m, respectively. This simple calculation shows that Titan's shape must be known with an accuracy on the order of 10 m in order to provide information on the interior structure, assuming that the shape is not affected by other processes. It must be noted that neither the gravity data, nor the shape information gives a unique solution for the interior structure (e.g. Moore and Schubert 2003) and that a given value of the Love number can be explained by several models of the interior structure.

The Cassini Radar Altimeter provides information on the topographic height on Titan. Altimetry is performed with the radar's beam pointed at nadir (Elachi et al. 2004). Typical altimeter observations are 300–600 km long. Altimeter data have been acquired on a number of flybys including TA, T3, T8, T13, T16, T19, T21, T23, T28, T28 and T30 (Zebker et al. 2009). The altimeter uses an echo profile (intensity versus time delay referenced to the Cassini spacecraft). The echo profile is defined by the surface radar backscatter cross section as a function of range weighted by the antenna beam pattern. The ephemeris of the Cassini spacecraft determined by the Cassini navigation team has provided the position of the surface relative to Cassini. The distance between the spacecraft and Titan is reconstructed with a precision <100 m. Recently, SAR mono-pulse measurements have been combined with the nadir-looking altimetry (Zebker et al. 2009). It is noticed that there is a much lower density of flybys in the southern hemisphere than in the northern hemisphere. Their preliminary processing suggests that the equatorial bulge (a–c) is equal to 680 m. The variations along the equator (a–b) would be equal to 370 m (Table 4.1). The value of (b–c) is equal to 310 m, which is much more than the value predicted by Eq. 4.13 (b–c = 1/4(a–c) = 170 m). But considering that the uncertainties on the values of (a–c) and (b–c) are on the order of 120 m, one must be careful about the conclusions one can draw from these measurements. The hydrostatic equilibrium solution described by Eq. 4.13 is within the uncertainties of the measurements reported by Zebker et al. (2009).

One can also use Eq. 4.13 to get an idea of the value of the Love numbers k_s and h_s. With (a–c) equal to 680 m, the value of k_s is equal to 2.33 ($h_s = 3.33$), which is much larger than any value expected for Titan with its present spin rate. One interpretation by Zebker et al. (2009) is that Titan acquired its shape at a time when it had a larger spin rate. With a value of $k_s = 1$, it gives a spin rate 30% faster than the present spin rate.

Given the uncertainties on the values describing Titan's shape, Titan may be in hydrostatic equilibrium. The large difference between the equatorial radius and the polar radius suggests that Titan acquired its shape when its spin rate was larger by 15% for $k_s = 1.5$ and 30% for $k_s = 1$.

4.2.4 Titan's Electric and Magnetic Field

The magnetic data acquired during the Galileo mission revealed an induced field inside Callisto and Europa which is best interpreted by the presence of an ocean. These measurements provided one of the major discoveries of the Galileo mission. For Titan, there are several facts that make such measurements difficult. First, Saturn's magnetic field is almost aligned with its spin axis and the inducing field produced as the satellites orbit around Saturn is therefore weak. Second, the flybys are far from the surface of Titan because of the atmosphere. Because the field decreases as $1/x^3$ where x is the distance between the magnetometers on the spacecraft and the location of the magnetic anomalies, the signal is very small and may not be detectable. Third, Titan's ionosphere adds a large external component that must be removed before retrieving the internal part of the magnetic signal.

Another major discovery of the Galileo mission was the detection of Ganymede's intrinsic magnetic field (Kivelson et al. 1996, 2002). It was expected that Titan may have a similar field although its spin rate is about four times smaller. The amplitude of Ganymede's magnetic field at the surface is on the order of 720 nT (Kivelson et al. 1996). If the magnetic dynamo is modeled as a magnetic dipole, then its magnitude is on the order of $1.3\ 10^{20}$ A m^2. If one takes such a value for Titan, it would create a magnetic field on the order of 200 nT at an altitude of 1,500 km. Such a large field would have been observed by the Cassini spacecraft. We can conclude that Titan does not have a field as strong as that of Ganymede. It does not mean that there is no iron core. However, as discussed above, the gravity data suggest that Titan is not as differentiated as Ganymede, which fits well with the idea that Titan does not have an iron core.

As the Huygens probe descended into Titan's atmosphere in January 2005, the horizontal and vertical components of the electric field were recorded by the Permittivity, Wave and Altimetry (PWA) instrument (Béghin et al. 2007). A signal a few Hz wide and centered at 36 Hz is present almost continuously throughout the descent. Different origins of this signal have been investigated, including instrumental effects. A careful analysis of the data (Simões et al. 2007) suggests that this signal is of natural origin and would have been the second eigenmode of a Schumann like resonance in Titan's atmospheric cavity. The Schumann resonance on Earth is the propagation of an electric signal within a cavity limited by two conductive layers which are the ionosphere on the top and the surface (mainly the ocean) at the bottom. The frequency of the resonance is linked to the size of the cavity. On Earth, the Schumann resonance is triggered by powerful atmospheric lightning activity. Such a source has not been detected to date on Titan (Fischer et al. 2007). Another process that could trigger the propagation of the waves is a plasma instability mechanism associated with the corotating Saturnian plasma flow (Béghin et al. 2007). Whatever the triggering process, the electric signal decreases in the atmosphere if there is no electric field. The slope change at 62 km altitude (Fig. 4.5) is due to the presence of a conductive layer formed by the ionization of neutral gas by galactic cosmic rays (Hamelin et al. 2007). Below that layer, the signal decreases linearly. The surface measurements obtained by the Huygens probe suggest that the electric conductivity and permittivity of the crust are so small that the refraction index is close to 1. The Huygens measurements give values of $\sigma \sim 4\ 10^{-10}$ S m^{-1} and $\varepsilon \sim 1.8$ for the ground conductivity and permittivity, respectively (Grard et al. 2006). The value of the permittivity given by the radar experiment is also very small all over the planet (Janssen et al. 2009) but it is at a much higher frequency that may not be relevant to the 36 Hz signal. Provided that the conductivity and the permittivity throughout the crust of ice remain small enough, the profile in Fig. 4.5 can be extrapolated linearly down to the conductive surface with the same slope. The depth of the conductive layer would be at a depth between 15 and 50 km and it is tempting to think that the conductive layer is the ice/ocean interface.

4.2.5 Geological Data – Apparent Misregistration of Geological Features

Three instruments can see surface features on Titan: the radar SAR, the ISS camera and the VIMS. Their spatial resolutions are quite different with 300 to 1,000 m for the radar, a few kilometers for ISS which is limited by light scattering in the atmosphere and between 10 km/pixel and 500 m/pixel for the VIMS in the infrared windows. During the course of the mission, some geological features were observed several times and located assuming synchronous spin rate and zero obliquity. The radar team reported on the apparent misregistration of surface features (Stiles et al. 2008). As written above (Section 4.2.1), most of the misregistration is explained by the obliquity equal to 0.3°. The best solution of the data inversion suggests that Titan's spin rate is 0.001 degree/day faster than the synchronous spin rate. Across the 4 years of the nominal period, this is more than 1.5° (65 km at the equator). However, it must be noted that there is a strong correlation (98%) between the precession and the spin rate.

If one accepts that Titan's spin rate is faster than synchronous, an explanation may be that Titan's Length of Day (LoD) seasonally varies as some angular momentum is exchanged between the dense atmosphere and the icy crust (Tokano and Neubauer 2005). The effect of the atmosphere on the icy crust is much larger if the icy crust is decoupled from the internal solid body by an ocean. Although quite tempting, this model is still debated for two reasons: first, there is no observation of wind direction changes; and second, there is a lag of two years between the predicted spin rate and the observed one (Lorenz et al. 2008). In addition, a recent study (Karatekin et al. 2008) suggests that Titan is likely to rotate almost as a rigid body even when it has an internal ocean, because of a substantial internal gravitational coupling between the crust and the interior. Although the predicted surface rotation rate would be further reduced due to Saturn's torque, Karatekin et al. (2008) acknowledge that other factors, such as larger atmospheric torque, smaller

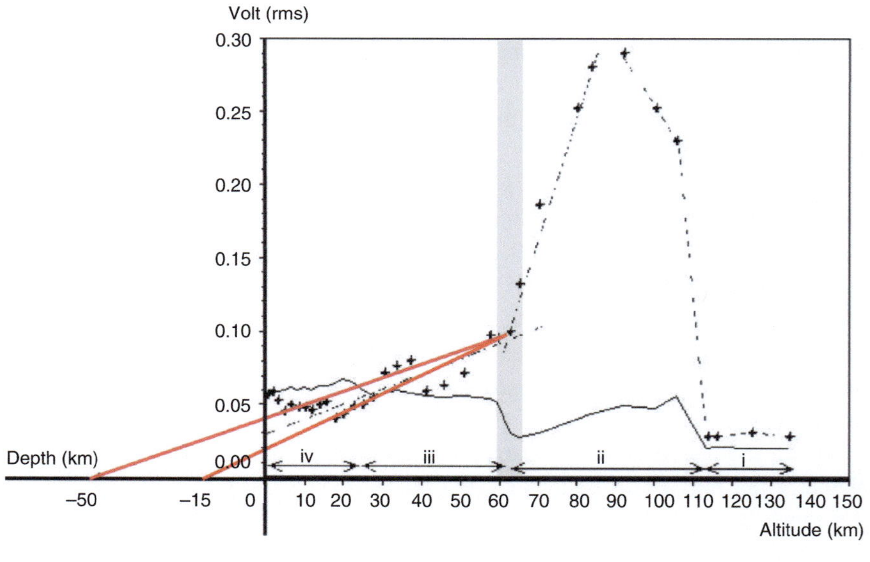

Fig. 4.5 Amplitude of the electric signal at 36 Hz during the descent of the Huygens probe. Between the two conductors, the amplitude must decrease linearly. It suggests that the bottom conductive layer is located between 15 and 50 km below the surface (this figure is adapted from Béghin et al. 2009)

equatorial flattening, and/or viscous relaxation of the icy crust could explain the observed non-synchronous rotation. Future observations in the extended mission should provide a more definite value of Titan's spin rate.

4.3 Modeling the Interior Structure

The goal of that chapter is to give a description of Titan's interior that fits with the constraints provided by the Cassini mission and that is consistent with the ice phase diagram. However, the model is largely uncertain because the interpretation of the data is still debated at present time (see discussions in Section 4.2). This chapter also addresses the question of heat sources and the possibility of convection in the ice crust. It does not deal with Titan's thermal evolution which is described in Chapter 3 of the present book.

4.3.1 Description of the Nominal Model

Based on the comparison with the Jovian icy satellites Ganymede and Callisto, the likelihood of an ocean (electric data, obliquity, cold ice layer to avoid too much dissipation), and the degree of differentiation inferred from the gravity data, the nominal model is a partially differentiated Titan without an iron core and with an ocean. From the interior to the surface, the different shells are:

- A silicate rich core
- A high-pressure ice layer
- An ammonia-rich ocean
- An ice I layer covered by a layer of unknown materials likely to be a mixture of organics and ices

Provided that Titan is in hydrostatic equilibrium, the values of the degree 2 coefficients of the gravity field suggest that the secular Love number is less than 1.5 but larger than that of Ganymede ($k_S = 0.8$). In order to have a range of plausible internal models, the Love number has been calculated for different interior structures where the free parameter is the density of the core. The other parameters that are necessary to compute the interior structure are its radius of 2,575 km and the standard gravitational parameter GM = 8,978.133 $km^3\ s^{-2}$ (Table 4.2). With these values, Titan's mean density is equal to 1,880 ± 4 kg/m^3. If the density of the silicates that formed Titan is equal to 3,500 kg/m^3, the mass fraction of silicates would be a little bit more than 50%. For the curve presented in Fig. 4.8, it has been assumed that there is no porosity and that the H_2O layer is made of two layers: an outer layer of low pressure ice I and an inner layer of high pressure ices with the transition being at the triple point of ice I – liquid – ice III at P = 212.9 MPa (R = 2,404 km). The densities of the ice I layer and high-pressure ice layer are set to 910 and 1,310 kg/m^3, respectively. The value for the density of the high-pressure ice layer is that of ice VI close to its melting point (Hobbs 1974). The density chosen for the ice I outer layer corresponds to an average density taking into account the fact that lower density compounds (clathrates) may be present. This value may vary according to the temperature in the ice crust and also the composition of the ice crust but these variations are of second order for the determination of the Love number k_s. If the value of k_S is higher than 1, it would be difficult to explain the density of the core without invoking porosity. For $k_S = 1$, a fully serpentinized core could explain the density of 2,550 kg/m^3 (Fig. 4.6). It would imply that the silicates have been altered by water. Such a process would have provided a lot of dihydrogen and this effect is discussed in Chapter 3. For smaller values of k_S, a mixture of peridotite (density of 3,500 kg/m^3) and serpentine can explain the observation (Fig. 4.6).

Different models of Titan's interior structure have been proposed and all of them propose that Titan is differentiated (e.g. Grasset and Sotin 1996; Grasset et al. 2000; Sohl et al. 2003; Tobie et al. 2005; Grindrod et al. 2008; Mitri and Showman 2008). The interior structure of the core depends principally on the composition of planetoids that formed Titan. Grasset et al. (2000) discussed that depending on the amount of metallic iron included in the planetoids, the present core of Titan could be composed principally of silicate where convective motions are present or could be composed of two layers with a liquid iron inner core surrounded by a shell of silicate. A schematic representation of the Titan's interior is in Fig. 4.1.

4.3.2 Heat Sources (Radiogenic, Tidal Heating, Latent Heat) at Present

The first heat source that is likely present within Titan's interior today is radiogenic heating due to the decay of the long-lived radiogenic elements such as ^{40}K, ^{232}Th, ^{235}U, and ^{238}U.

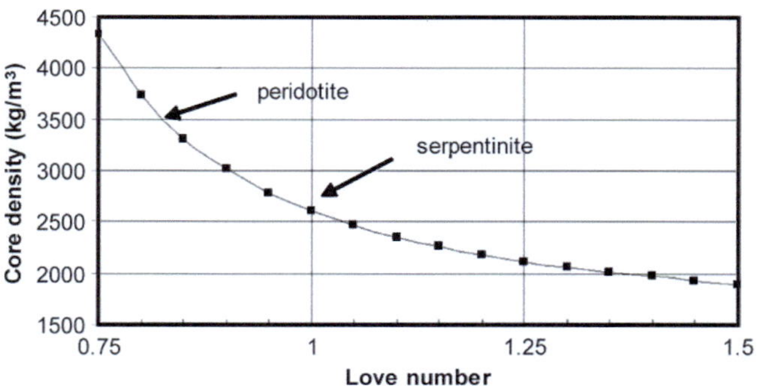

Fig. 4.6 Density of the core for different values of the Love number k_s (see text for values of the parameters)

There are no data on the amount of radiogenic elements that is contained within Titan. The presence of ^{40}Ar and the very low abundance of ^{36}Ar in Titan's atmosphere prove not only that ^{40}K is present and decaying, but also that exchange processes must exist between the interior where ^{40}K is present and the surface. Different composition of planetoids can be assumed (e.g. Grasset et al. 2000). If one takes the example of carbonaceous chondrites, a value of 4 10^{-12} W/kg of silicates would be produced. Depending on the density chosen for the chondrites, the total amount of radiogenic heating ranges between 300 and 400 GW, which gives a surface heat flux between 3.7 and 5 mW/m^2 at present day. The location of the heat sources depends on the differentiation processes. Radiogenic elements are most likely in the silicate core. However, hydrothermal alteration of the silicates may have leached the silicates and concentrated the potassium in the ocean.

The other potential heat source is tidal heating (Peale 1977). A detailed study of tidal heating has been carried out by Tobie et al. (2005). Using this nominal model for Titan and an interior model which is consistent with the gravity and magnetic data (Fig. 4.1), this study shows that the amount of tidal heating is about 100 times smaller for Titan than for Europa. In this study, the physical parameters (viscosity law and shear modulus in particular) are the same for Titan and Europa. The model is based on the deformation of a Maxwell body and the dissipation is maximum for a viscosity which provides a Maxwell time close to the exciting period (Fig. 4.7). Because Titan orbits Saturn at a much larger distance than Europa orbits Jupiter, the viscosity at which maximum tidal heating is modeled, is larger for Titan than for Europa (Fig. 4.7). If the viscosity of ice is optimal (between 10^{14} and 10^{16} Pa s), the amount of tidal heating is slightly larger than radiogenic heating.

4.3.3 Ice Crust: Convection or Conduction in the Ice Crust

The heat produced in the interior of Titan by tidal dissipation and radiogenic decay can be transferred through the ice crust by conduction or by thermal convection (Fig. 4.8). If heat is transferred by conduction, the heat flux (q) is proportional to the temperature gradient (∂T/∂z):

$$q = -k\frac{\partial T}{\partial z} \tag{4.14}$$

where k is the thermal conductivity which is very much temperature dependent in the case of ice (Hobbs 1974). For heat flux between 5 and 10 mW/m^2, the thickness of the conductive layer is a few tens of kilometers (Fig. 4.8).

Convection, which is a much more effective means of heat transfer, can occur if the layer is thick enough. Because of the high viscosity contrast between the surface and the bottom of the ice-I shell, convection must occur in the stagnant-lid regime (McKinnon 1999). Parameterized models have been developed for the case of a layer heated from below with a fixed temperature at the lower interface (Deschamps and Sotin 2001; Mitri and Showman, 2008), which is the temperature of the ocean. In this case, convection is driven by the formation of hot plumes at the ice/ocean interface. The convective heat flux can be written:

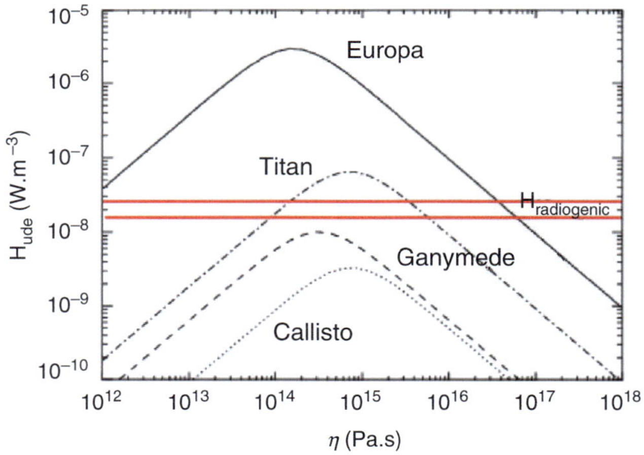

Fig. 4.7 Amount of tidal heating within Europa, Ganymede, Callisto and Titan (Sotin and Tobie 2004). Maximum dissipation occurs when the Maxwell time (viscosity/shear modulus) is on the order of the orbital period. The longer the orbital period, the larger the viscosity. For Titan, tidal heating can be more important than radiogenic heating if the viscosity of ice is in between 10^{14} and 10^{16} Pa s

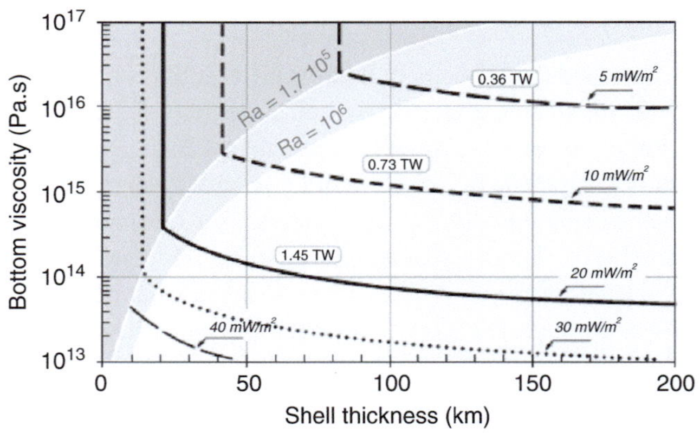

Fig. 4.8 Curves of heat flux in a viscosity – thickness diagram. The viscosity is that of ice at the ocean/ice interface. The dark grey area corresponds to a domain where heat is transferred by conduction. The light grey area is the (viscosity-thickness) domain where convection is marginal. Labels give both the total power coming out of the ice shell and the corresponding heat flux. The curves have been obtained using the equations described in the text

$$q = k\frac{\Delta T_{TBL}}{\delta_{TBL}} = k\frac{\Delta T}{b}\left(\frac{\Delta T_{TBL}}{\Delta T}\right)^{4/3}\left(\frac{Ra}{Ra_{TBL}}\right)^{1/3} \tag{4.15}$$

where Ra_{TBL} is the thermal boundary layer Rayleigh number (see definition below), ΔT_{TBL} is the temperature difference across the thermal boundary layer, b is the thickness of the ice layer, and δ_{TBL} is the thickness of the thermal boundary layer. The temperature difference across the thermal boundary layer is proportional to the viscosity temperature scale ΔT_η (e.g. Davaille and Jaupart 1993; Moresi and Solomatov 1995; Solomatov 1995):

$$\Delta T = 1.43\,\Delta T_\eta \text{ with } \Delta T_\eta = \left|\frac{1}{\partial Ln(\eta)/\partial T}\right| \tag{4.16}$$

where η is the viscosity. Although the viscous behavior of a material depends on several parameters including grain size, stress and temperature, the most important effect comes from temperature (Durham et al. 1997; Goldsby and Kohlstedt 2001). The viscosity can be written as:

$$\eta = A\exp\left(\frac{Q}{RT}\right) \tag{4.17}$$

where A is a constant and Q is the activation energy which is equal to 60 kJ/mole for ice (Durham et al. 1997; Goldsby and Kohlstedt 2001). In Eq. 4.15, the Rayleigh number is the ratio of the thermal buoyancy to the viscous force and can be expressed by:

$$Ra = \frac{\alpha\rho g \Delta T b^3}{\kappa\eta} \tag{4.18}$$

where α is the thermal expansivity of the ice ($1.6\ 10^{-4}\ K^{-1}$), ρ is ice density ($920\ kg\ m^{-3}$), g is gravity ($1.35\ m\ s^{-2}$), ΔT is the temperature difference between the lower interface (ice/ocean) and the surface of the ice-I layer, κ is the thermal diffusivity of the water ice ($1.3\ 10^{-6}\ m^2\ s^{-1}$), and η is the viscosity within the interior of the convective sublayer. The thermal boundary layer Rayleigh number is defined by:

$$Ra_{TBL} = \frac{\alpha\rho g \Delta T_{TBL}\delta_{TBL}^3}{\kappa\eta} \tag{4.19}$$

This number is constant for the case of isoviscous fluids and represents the equilibrium value that leads to stationary convection (e.g. Roberts 1979; Somerscales and Gazda 1986). Deschamps and Sotin (2001) have shown that the thermal boundary layer Rayleigh number slightly depends on the Rayleigh number in models for which viscosity is strongly temperature dependent. These equations allow us to determine the heat flux as a function of the thickness of the layer (Fig. 4.8). The heat flux transferred by convective processes is independent of the thickness of the layer for isoviscous fluids. For more complex viscosities, the stability of the thermal boundary layers depends on the amplitude of the viscosity variations (e.g. Davaille and Jaupart 1993). Therefore a slight dependence of heat flux on the thickness of the layer exists. For a heat flux of 10 mW/m^2, the equilibrium viscosity at the ocean/ice interface is equal to 10^{15} Pa s. This viscosity is one order of magnitude larger for a heat flux of 5 mW/m^2. This approach shows that a layer less than 50 km thick may not be convective. If more complex viscous law are used (e.g. Barr and McKinnon 2007), then the equilibrium viscosity may be shifted. Other processes such as stability of hydrates, partial melting need further investigation to assess the dynamics of the outer ice crust. This study does not investigate the question of the onset of convection which must be discussed in the Chapter 3 on Titan's evolution.

4.4 Discussion and Remaining Questions to be Addressed by Future Mission

At the end of the nominal mission, the data processing of the RSS experiment has not yet provided the degree 2 gravity coefficients. According to the most recent information (Rappaport et al. 2008b), Titan would be in hydrostatic equilibrium and the value of the Love number would lie in between those of Ganymede and Callisto. The data processing shows that higher degree terms must be taken into account in order to explain the observed signal. The future flagship mission to the Jovian system will provide extended coverage with one orbiter around Europa, one orbiter around Ganymede and several close flybys of Callisto. It will be possible to compare the higher degree coefficients of each of these icy moons and infer some important differences between their internal structures.

The comparison between Titan and the two large icy Galilean satellites Ganymede and Callisto shows that satellites similar in radius and mass can have very different degrees of differentiation. Titan has no magnetic field and the state of its internal differentiation is still unconstrained, but it is distinct from Ganymede and Callisto by its thick, mainly nitrogen atmosphere, and by the methane and other hydrocarbons that reside in its atmosphere and on its surface. These differences must be due to conditions experienced by the satellites during their formation and evolution. In particular, tidal heating must have been important in Ganymede to account for its complete differentiation and the formation of its core, but unimportant in the partially differentiated Callisto. The role of tidal heating in the evolution of Titan will remain uncertain until the state of its interior is better constrained.

Titan's eccentricity and obliquity are puzzling. Although some thermal evolution models can explain the present large eccentricity of Titan (e.g. Tobie et al. 2005), their validity can be assessed only if one can obtain an accurate description of the interior structure. If Titan is in a Cassini state, the obliquity found by the radar team (Stiles et al. 2008) implies a moment of inertia of 0.53 MR^2 (Fig. 4.3 modified from Bills and Nimmo 2008). Such a value can be obtained if the outer ice layer is decoupled from the interior. The situation is somehow similar to that for Mercury. Although Mercury has no ocean, the presence of the magnetic field suggests the presence of a liquid iron core. The obliquity of Mercury could be related to the motion of the silicate outer layer and not that of the entire planet. It must be noted that the gravitational coupling between the ice shell and the silicate core is not taken into account by Bills and Nimmo (2008). Karatekin et al. (2008) suggest that the coupling would limit the value of the obliquity. Another interpretation is that Titan is not in a Cassini state.

As discussed in Section 4.2.1 the Titan Length of Day (LoD) may be different from synchronous. Also the explanation of an exchange in angular momentum between the atmosphere and the ice crust is appealing, it is necessary to investigate the effect of the torque exerted by the inner silicate core. The gravitational coupling (Van Hoolst et al. 2008; Szeto and Xu 1997) between the silicate core and the ice layer should prevent the exchange of angular momentum and therefore the change of LoD. Additional data during the extended mission (2010–2017) should allow us to determine how Titan's LoD varies during seasonal variations.

Detecting the presence of an ocean would be a great achievement of the Cassini mission. As already discussed in Section 4.2.2, this can be achieved with either the geoid data or the gravity data. Note that the electric data strongly suggest the presence of a liquid layer some tens of kilometers below the surface. In that case the periodic Love number k_2 could be large enough for the signal contained in the periodic degree 2 term to be larger than the signal contained in the degree 4 or 5, as discussed in Section 4.2.2. It implies that the degree 3 and degree 4 coefficients must be determined in order to be confident in the value of k_2. However, there is another method by which we can approach this problem. This method consists in analyzing flybys that cover the same ground of Titan, while occurring at various mean anomalies of Titan. In this case, the only difference between the two gravity fields would be attributed to the tides and the value of the periodic Love number k_2 could be inferred. This seems however out of the possibility of the Cassini mission and only a dedicated Titan orbiter will be able to obtain this information.

4.5 Conclusions

As the Cassini mission proceeds into its first extended phase, the data obtained during the nominal mission, and published so far, suggest that Titan is at least partially differentiated with an internal ocean some tens of kilometers below the surface. By comparison with the Galilean icy satellites, Titan would be composed of a silicate rich core and a H_2O rich outer layer. These conclusions are drawn from the interpretation of the gravity data, the geological data and the electric signal recorded by the Huygens probe during its descent into Titan's atmosphere. However, there are many observations that still need to be understood. For example, the value of the obliquity which has been inferred from the apparent misregistration of geological features imaged at different times by the Cassini Synthetic Aperture Radar (SAR), is not consistent with a reasonable internal structure if Titan is in a Cassini state. The high eccentricity implies that the interior has not been very dissipative and that there is little tidal heating available for internal dynamics. It seems to imply that the ice layer is cold, which can be achieved if the ocean under the ice layer contains ammonia. Heat transfer models suggest that the lower part of the ice crust would be convective. The NH_3–H_2O phase diagram indicates that this ocean is decoupled from the silicate-rich core by a layer of high-pressure ices. During the Cassini Solstice Mission (2010–2017), additional observations of landmarks will help us constrain Titan's Length of Day as Titan enters a new season. Also, additional gravity paths are required in order to constrain the gravity coefficients and the Love numbers k_s and k_2.

Acknowledgements This work was carried out at the Jet Propulsion Laboratory, California Institute of Technology under contract with the National Aeronautics and Space Administration. The authors thank the anonymous persons who carefully reviewed this paper.

References

Anderson JD et al (1996) Gravitational constraints on the internal structure of Ganymede. Nature 384:541–543

Barr AC, McKinnon WB (2007) Convection in ice I shells and mantles with self-consistent grain size. J Geophys Res (Planet) 112: 2012

Béghin C, Simões F, Krasnoselskikh V, Schwingenschuh K, Berthelier J-J, Besser BP, Bettanini C, Grard R, Hamelin M, López-Moreno JJ, Molina-Cuberos GJ, Tokano T (2007) A Schumann-like resonance on Titan driven by Saturn's magnetosphere possibly revealed by the Huygens Probe. Icarus 191:251–266

Béghin C, Canu P, Karkoschka E, Sotin C, Bertucci C, Kurth WS, Berthelier JJ, Grard R, Hamelin M, Schwingenschuh K, Simões F (2009) New insights on Titan's plasma-driven Schumann resonance inferred from Huygens and Cassini data. Planet Space Sci. doi:10.1016/j.pss.2009.04.006

Bills BG, Nimmo F (2008) Forced obliquity and moments of inertia of Titan. Icarus 196:293–297

Blakely RJ (1996) Potential theory in gravity and magnetic applications. Cambridge University Press, Cambridge, pp 441

Davaille A, Jaupart C (1993) Transient high Rayleigh number thermal convection with large viscosity variations. J Fluid Mech 253:141–166

Davies ME, Colvin TR (1998) The control networks of the Galilean satellites and implications for global shape. Icarus 135:372–376

Deschamps F, Sotin C (2001) Thermal convection in the outer shell of large satellites. J Geophys Res 106:5107–5121
Durham WB, Kirby SH, Stern LA (1997) Creep of water ices at planetary conditions: a compilation. J Geophys Res 102:16293–16302
Elachi C et al (2004) Radar: the Cassini Titan radar mapper. Space Sci Rev 115:71–110
Fischer G, Kurth WS, Dyudina UA, Kaiser ML, Zarka P, Lecacheux A, Ingersoll AP, Gurnett DA (2007) Analysis of a giant lightning storm on Saturn. Icarus 190:528–544
Fulchignoni M et al (2005) In situ measurements of the physical characteristics of Titan's environment. Nature 438(8):785–791
Goldsby DL, Kohlstedt DL (2001) Superplastic deformation of ice: experimental observations. J Geophys Res 106:11017–11030
Grard R, Svedhem H, Brown V, Falkner P, Hamelin M (1995) An experimental investigation of atmospheric electricity and lightning activity to be performed during the descent of the Huygens Probe on Titan. J Atm Terr Phys 57:575–585
Grard R et al (2006) Electric properties and related physical characteristics of the atmosphere and surface of Titan. Planet Space Sci 54:1124–1136
Grasset O, Sotin C (1996) The cooling rate of a liquid shell in Titan's interior. Icarus 123:101–112
Grasset O, Sotin C, Deschamps F (2000) On the internal structure and dynamics of Titan. Planet Space Sci 48:617–636
Grindrod PM, Fortes AD, Nimmo F, Feltham DL, Brodholt JP, Vocadlom L (2008) The long-term stability of a possible aqueous ammonium sulfate ocean inside Titan. Icarus 197:137–151
Hamelin M et al (2007) Electron conductivity and density profiles derived from the Mutual, Impedance Probe measurements performed during the descent of Huygens through the atmosphere of Titan. Planet Space Sci 55:1964–1977
Hobbs PV (1974) Ice physics. Clarendon, Oxford
Iess L, Armstrong JW, Asmar SW, Di Benedetto M, Graziani A, Mackenzie R, Racioppa P, Rappaport N, Tortora P (2007) The determination of Titan gravity field from Doppler Tracking of the Cassini spacecraft. Proc Int Symp Space Flight Dynamics. NASA/CP-2007-214158
Janssen MA, Lorenz RD, West R, Paganelli F, Lopes RM, Kirk RL, Elachi C, Wall SD, Johnson WTK, Anderson Y, Boehmer RA, Callahan P, Gim Y, Hamilton GA, Kelleher KD, Roth L, Stiles B, Le Gall A, and the Cassini Radar Team (2009) Titan's surface at 2.2-cm wavelength imaged by the Cassini RADAR Radiometer: calibration and first results. Icarus 200:222–239
Karatekin O, Van Hoolst T, Tokano T (2008) Effect of internal gravitational coupling on Titan's non-synchronous rotation. Geophys Res Lett 35:L16202. doi:10.1029/2008GL034744
Kaula WM (1966) Theory of satellite geodesy. Blaisdell, Waltham, Massachusetts
Kivelson MG et al (1996) Discovery of Ganymede's magnetic field by the Galileo spacecraft. Nature 384:537–541
Kivelson MG et al (2002) The permanent and inductive magnetic moments of ganymede. Icarus 157:507–522
Konopliv AS, Binder AB, Hood LL, Kucinskas AB, Sjogren WL, Williams JG (1998) Improved gravity field of the moon from lunar prospector. Science 281:1476–1480
Lodders K, Fegley B (1998) The planetary scientist's companion. Oxford University Press, New York, 371 pp
Lorenz RD, Styles BW, Kirk RL, Allison MD, Persi del Marmo P, Iess L, Lunine JI, Ostro SJ, Hensley S (2008) Titan's rotation reveals an internal ocean and changing zonal winds. Science 319:1649–1651
McKinnon WB (1999) Convective instability in Europa's floating ice shell. Geophys Res Lett 26:951–954
Mitri G, Showman AP (2008) Thermal convection in ice-I shells of Titan and Enceladus. Icarus 193:387–396
Moore WB, Schubert G (2003) The tidal response of Ganymede and Callisto with and without liquid water oceans. Icarus 166:223–226
Moresi LN, Solomatov VS (1995) Numerical investigation of 2D convection with extremely large viscosity variations. Phys Fluids 7:2154–2162
Niemann H et al (2005) The abundances of constituents of Titan's atmosphere from the GCMS instrument on the Huygens probe. Nature 438:779–784
Palguta J, Schubert G, Zhang K, Anderson D (2009) Constraints on the location, magnitude, and dimensions of Ganymede's mass anomalies. Icarus 201:615–625
Peale SJ (1977) Rotation histories of the natural satellites. In: Burns JA (ed) Planetary Satellites. Univ. of Arizona Press, Tucson, pp 87–111
Rappaport N, Iess L, Wahr J, Lunine JI, Armstrong JW, Asmar SW, Tortora PF, Di Benedetto M, Racioppa P (2008a) Can Cassini detect a subsurface ocean in Titan from gravity measurements? Icarus 194:711–720
Rappaport NJ, Jacobson RA, Iess L, Racioppa P, Armstrong JW, Asmar SW, Stevenson DJ, Tortora P, Di Benedetto M, Graziani A, Meriggiola R (2008b) The gravity field of Titan from four Cassini flybys. American Geophysical Union, Fall Meeting 2008, Abstract #P21A-1343
Roberts GO (1979) Fast isoviscous Bernard convection. Geophys Astrophys Fluid Dyn 12:235–272
Schubert G, Anderson JD, Spohn T, McKinnon WB (2004) Interior composition, structure and dynamics of the Galilean Satellites. In: Bagenal F, Dowling T, McKinnon WB (eds) Jupiter: the planet, satellites, and magnetosphere. Cambridge University Press, Cambridge, pp 281–306
Sears WD (1995) Tidal dissipation in oceans on Titan. Icarus 113:39–56
Simões F et al (2007) A new numerical model for the simulation of ELF wave propagation and the computation of eigenmodes in the atmosphere of Titan: did Huygens observe any Schumann resonance? Planet Space Sci 55:1978–1989
Sohl F, Sears WD, Lorenz RD (1995) Tidal dissipation on Titan. Icarus 115:278–294
Sohl F, Hussmann H, Schwentker B, Spohn T, Lorenz RD (2003) Interior structure models and tidal Love numbers of Titan. J Geophys Res 108:5130. doi:10.1029/2003JE002044
Solomatov VS (1995) Scaling of temperature- and stress-dependent viscosity convection. Phys Fluids 7:266–274
Somerscales EFG, Gazda IW (1986) Thermal convection in high prandtl number liquids at high Rayleigh numbers. Rensselaer Polytechnic Institute, ME Report HT-5, Troy, NY
Sotin C, Tobie G (2004) Internal structure and dynamics of the large icy satellites. CR Acad Sci Physique 5:769–780
Stiles BW, Kirk RL, Lorenz RD, Hensley S, Lee E, Ostro SJ, Allison MD, Callahan PS, Gim Y, Less L, Perci del Marmo P, Hamilton G, Johnson W-TK, West RD and the Cassini RADAR Team (2008) Determining Titan's spin state from Cassini RADAR images. Astron J 135:1669–1680
Szeto AMK, Xu S (1997) Gravitational coupling in a triaxial ellipsoidal Earth. J Geophys Res – Solid Earth 102:27651–27657
Tobie G, Grasset O, Lunine JI, Mocquet A, Sotin C (2005) Titan's internal structure inferred from a coupled thermal-orbital model. Icarus 175:496–502
Tobie G, Lunine J, Sotin C (2006) Episodic outgassing as the origin of atmospheric methane on Titan. Nature 440:61–64
Tokano T Neubauer F (2005) Wind-induced seasonal angular momentum exchange at Titan's surface and its influence on Titan's length-of-day. Geophys Res Lett 32:L24203. doi:10.1029/2005GL024456
Turcotte DL, Schubert G (1982) Geodynamics: application of continuum physics to geological problems. Wiley, New York, 450 pp
Van Hoolst T, Rambaux N, Karatekin O, Dehant V, Rivoldini A (2008) The librations, shape, and icy shell of Europa. Icarus 195:386–399
Waite H et al (2005) Ion-neutral mass spectrometer results from the first flyby of Titan. Science 308:982–986
Zebker HA, Stiles B, Hensley S, Lorenz R, Kirk RL, Lunine J (2009) Size and shape of Saturn's moon Titan. Science 324:921–923
Zharkov VN, Leontjev VV, Kozenco AV (1985) Models, figures, and gravitational moments of the Galilean satellites of jupiter and icy satellites of saturn. Icarus 61:92–100

Chapter 5
Geology and Surface Processes on Titan

Ralf Jaumann, Randolph L. Kirk, Ralph D. Lorenz, Rosaly M.C. Lopes, Ellen Stofan, Elizabeth P. Turtle, Horst Uwe Keller, Charles A. Wood, Christophe Sotin, Laurence A. Soderblom, and Martin G. Tomasko

Abstract The surface of Titan has been revealed globally, if incompletely, by Cassini observations at infrared and radar wavelengths as well as locally by the instruments on the Huygens probe. Extended dune fields, lakes, mountainous terrain, dendritic erosion patterns and erosional remnants indicate dynamic surface processes. Valleys, small-scale gullies and rounded cobbles such as those observed at the Huygens landing site require erosion by energetic flow of a liquid. There is strong evidence that liquid hydrocarbons are ponded on the surface in high-latitude lakes, predominantly, but not exclusively, at high northern latitudes. A variety of features including extensive flows and caldera-like constructs are interpreted to be cryovolcanic in origin. Chains and isolated blocks of rugged terrain rising from smoother areas are best described as mountains and might be related to tectonic processes. Finally, impact craters are observed but their small numbers indicate that the crater retention age is very young overall. In general, Titan exhibits a geologically active surface indicating significant endogenic and exogenic processes, with diverse geophysical and atmospheric processes reminiscent of those on Earth.

5.1 Introduction

The first detailed discussions about Titan's surface took place at a workshop at NASA Ames in 1973 (Hunten 1974). Although that workshop was dedicated to Titan's atmosphere, two contributions in particular addressed Titan's surface. First, Lewis (1971) considered the formation scenarios for Titan and reasoned that water and ammonia were likely to be the major bulk constituents of Titan. Second, Hunten (1974) recognized that the surface would be the repository for photochemical products of methane photolysis, which could accumulate to a depth of about 1 km over the age of the solar system.

This photochemistry-dominated perspective on Titan prevailed through the Voyager encounters in 1980 and 1981 as Voyager images failed to show details of the surface (although a re-analysis of Voyager images 20 years later did show that some surface contrast can be recovered in these data (Richardson et al. 2004). The Voyager radio-occultation experiment revealed the radius of Titan's solid surface (allowing a more accurate constraint on Titan's bulk density), and also indicated surface atmospheric conditions of 94 K and 1.5 bar (Tyler et al. 1981). Methane humidity was inferred to be well under 100% (e.g. Flasar 1983), which argued against a surface covered in liquid methane (Fig. 5.1).

However, the idea of surface liquids continued to draw interest. Sagan and Dermott (1982) argued that Titan's orbital eccentricity (0.029) requires tidal dissipation to be low. Noting that, as on Earth, tidal dissipation is high in shallow seas, they argued that seas on Titan should therefore be either deep or non-existent. In addition, they concluded (erroneously, in retrospect) that erosion would obliterate topography on Titan, so that, if deep oceans were present, they would be global, with no islands or continents.

R. Jaumann (✉)
DLR, Institute of Planetary Research, Rutherfordstrasse 2, 12489, Berlin, Germany
Freie Universität Berlin, Institute of Geological Sciences, Malteserstr. 74-100, 12249, Berlin, Germany
e-mail: ralf.jaumann@dlr.de

R.L. Kirk and L.A. Soderblom
U.S. Geological Survey, Flagstaff, AZ 86001, USA

R.D. Lorenz and E.P. Turtle
Space Department, John Hopkins University Applied Physics Laboratory, Laurel, MD 20723, USA

R.M.C. Lopes and C. Sotin
Jet Propulsion Laboratory, California Institute of Technology, Pasadena CA 91109, USA

E. Stofan
Proxemy Research, Rectortown VA 20140, USA

H. Uwe Keller
Max-Planck-Institute for Solar System Research, Katlenburg-Lindau, Germany

C.A. Wood
Planetary Science Institute, Tucson, AZ 85719, USA
M. Tomasko
Lunar and Planetary Laboratory, University of Arizona, Tucson, AZ, 85721, USA

R.H. Brown et al. (eds.), *Titan from Cassini-Huygens*,

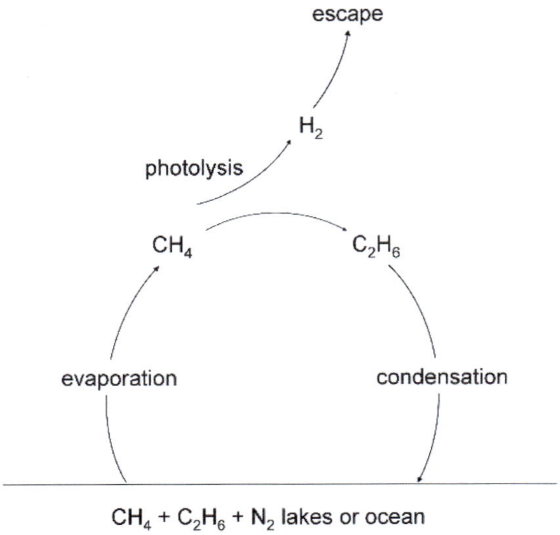

Fig. 5.1 Methane cycle and deposits of photochemical material (Hunten 1974)

These arguments dominated much of the thinking about Titan's surface for the following decade. In fact, the tidal dissipation argument was later shown (Sohl et al. 1995) to be irrelevant as a discriminator between surface types because dissipation in the interior (especially if there were a liquid water-ammonia layer, as thermal models began to predict around this time) would in any case make Titan's eccentricity difficult to understand, whether there were liquid hydrocarbons on the surface or not. Dermott and Sagan (1995) furthermore refuted their own earlier arguments by pointing out that if Titan's liquid hydrocarbons were confined in small basins such as impact craters, tidal dissipation in those surface liquids would be small.

Post-Voyager photochemical models enabled a quantitative treatment of the deposition fluxes of various organic compounds. Notably, the model by Yung et al. (1984) suggested that 100–200 m of solid acetylene and other solids might have been deposited over the age of the solar system, while about 600 m of ethane liquid would be generated. Lunine et al. (1983) suggested that this amount of ethane could resolve the paradox of the low near-surface humidity by the need for a surface reservoir of methane because its inventory in the atmosphere in gaseous form would be depleted in only 10 Myr. The methane vapor pressure above a mixed 'ocean' of methane and ethane, which is a rather less volatile molecule and thus has a low vapor pressure, would be less than that above a pure methane ocean. This scenario was consistent with the radio-occultation data and the photochemical model.

Formerly, the amount of argon in the atmosphere had not been meaningfully constrained by Voyager data, and the resultant uncertainty about mean molecular weight made for some uncertainty about the surface temperature, which could have ranged from 92.5°K to 101°K (Dubouloz et al. 1989). Based on the key assumptions of thermodynamic equilibrium between ocean and atmosphere, and an ethane depth of ~600 m (Yung et al. 1984), this in turn would allow ocean compositions with ethane:methane:nitrogen:argon concentrations of 91:7:2:0 to 5:83:6:6 and ocean depths of 700 m to 9.5 km.

Stevenson and Potter (1986) drew attention to the equator-to-pole temperature gradient on Titan and suggested that, somewhat analogously to the Martian seasonal polar caps of CO_2, there might be seasonal condensation of methane at the winter pole. In the build-up to the Cassini mission, Lunine and Stevenson (1985, 1987) made a number of calculations regarding Titan's origin and evolution and the nature of its surface.

The first data constraining Titan's surface was radar reflectivity measurement by Muhleman et al. (1990). Using the Goldstone 70-m antenna as a transmitter, and the very large array (VLA) as a receiver, they detected a Titan echo during three nights. The published reflectivity values were quite high – certainly ruling out a global ocean – and perhaps indicative of a 'dirty-ice' surface. In fact, the published values were erroneous – a reduction software error leading to their being reported too high by a factor of 2. If the correct value had been reported, the interpretation might have been less challenging.

Soon after radar detection, near-infrared measurements in methane windows began to be made. The realization that Titan's atmosphere would be sufficiently transparent in the near infrared emerged in the late 1980s. Griffith et al. (1991) suggested that a 'dirty ice' composition might be consistent with observations, an interpretation on which 17 years have afforded remarkably little progress, the same conclusion being drawn by McCord et al. (2008) from Cassini VIMS data.

Lemmon et al. (1993) showed that Titan's leading face was brighter than the trailing face, the first detection of surface heterogeneity. Technologically, this discovery could have been made decades before (e.g. McCord et al. 1971; Hunten 1974). Remarkably, Cruikshank and Morgan (1980) searched for spectral variability but failed to detect it although a re-analysis of their data supplemented by new observations did show a light curve (Noll and Knacke 1993 – published shortly before the work by Lemmon et al. (1993)).

Several other workers followed up this result (Griffith 1993; Coustenis et al. 1995). All light curves showed essentially the same shape (although differing in amplitude) at all the wavelengths studied – 0.94, 1.07, 1.28, 1.58 and 2 μm. The more challenging 5 μm light curve would not be measured for almost another decade (Lellouch et al. 2004). It was realized that, in principle, the short-wavelength light curves (e.g. 673, 790, 950 nm) could be measured by appropriately equipped amateur astronomers (Lorenz et al. 2003), and indeed, the surface light curve is sufficiently prominent in the small-telescope cloud-monitoring efforts of Schaller et al. (2005) to be taken into account in their analyses.

Finally, Cassini VIMS measured several light curves during the approach phase in 2004 when Titan was too distant to be resolved (Buratti et al. 2006).

The spatial variability implied by the light curve was more graphically revealed by the first near-infrared maps of Titan made with data from the repaired Hubble space telescope in 1994 (Smith et al. 1996) (Fig. 5.2). The 940 nm map showed leading-edge brightness to be concentrated in a region measuring about 4,000 × 2,500 km, now known as Xanadu.

The map, made from 14 HST observations, had a coverage gap at 0° longitude. This gap was a factor in the decision to publish the Titan maps centered on the anti-Saturn prime meridian (i.e. 180°W) rather than the 0° prime meridian as on Earth: this convention has been largely but not universally followed since. Another set of HST maps (using a longer-wavelength camera) was generated with data acquired in 1997 (Meier et al. 2000).

Ground-based near-infrared imaging with adaptive optics (AO) began to show features in 1993 (Saint-Pe et al. 1993; Combes et al., 1997; Coustenis et al. 2001; Hirtzig et al. (2007); Negraõ et al. (2007)). The same holds true for speckle imaging (Gibbard et al. 1999, 2004). Since then, AO has progressively improved (e.g. Gendron et al. 2004; Roe et al. 2004). By 2004 the 8- and 10-m-class telescopes such as Keck, Gemini and VLT began to surpass the HST resolution. Furthermore, the superior clarity of the 2 μm window used by these telescopes allowed surface features to be detected better. However, the nature of the bright and dark features remained unknown. Attempts to interpret the surface 'spectrum' were hampered by measurement uncertainties and differences in the models used to interpret the spectrum. In particular, given a specific observed albedo, the inferred surface reflectivity was highly dependent on the assumed methane absorption coefficient. These difficulties led to uncertainties of a factor of 2–3 in reflectivity, rendering efforts at compositional interpretation largely futile. The almost universal model, 'dirty ice', still holds (e.g. Soderblom et al. 2007a).

New radar observations of Titan's surface were enabled in 2002 by the upgrade of the Arecibo radio telescope and the changing declination of Titan, which brought it into the view of that facility. This 300-m dish could achieve radar observations of Titan with a much higher signal-to-noise ratio than the Goldstone-VLA work, and allowed the radar spectrum to be measured with more accuracy. The Doppler broadening of the monochromatic transmitted signal was easily seen, allowing the tracking of specific reflecting regions as they moved with Titan's rotation, and on several occasions, striking specular reflections were seen at the sub-telescope point (Campbell et al. 2003). These bright glints required very flat, although actually somewhat dark, surfaces. These results were interpreted as being possibly due to surface liquids, although in retrospect, since these echoes were from latitudes of about −22°, it is possible that flat areas surfaces could have been responsible. It had been noted that no near-infrared specular glints had been seen (West et al. 2005; Fussner 2006) which would have been indicative of liquid surfaces.

In anticipation of Cassini-Huygens's observations, various landforms and surface processes on Titan were considered theoretically, with varying degrees of success (Fig. 5.3).

Rainfall had been suggested as a cleansing mechanism that might render elevated terrain optically brighter than lowlands (Griffith et al. 1991; Smith et al. 1996), especially since elevated terrain might receive more rainfall (Lorenz 1993a). But methane raindrops on Titan would fall slowly and might not reach the ground at all before evaporating in the unsaturated lower atmosphere (Lorenz 1993a). Lorenz and Lunine (1996) initially argued that fluvial erosion would be weak on Titan, especially since the meager sunlight does allow a vigorous hydrological cycle amounting to only ~1 cm of methane rainfall per Earth year. However, subsequent thinking (Lorenz 2000) noted that even though desert regions on Earth receive little rainfall on average, rain and rivers can substantially erode the landscape if that rainfall comes in rare but violent storms. That revised paradigm appears to have been borne out.

It was realized early (e.g. Greeley and Iversen 1985; Allison 1992; Grier and Lunine 1993) that Titan's thick

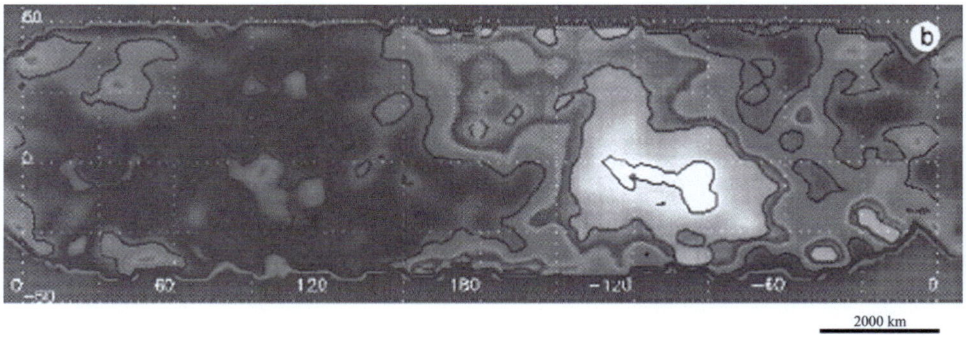

Fig. 5.2 HST Map from Smith et al. (1996), showing brightness heterogeneities

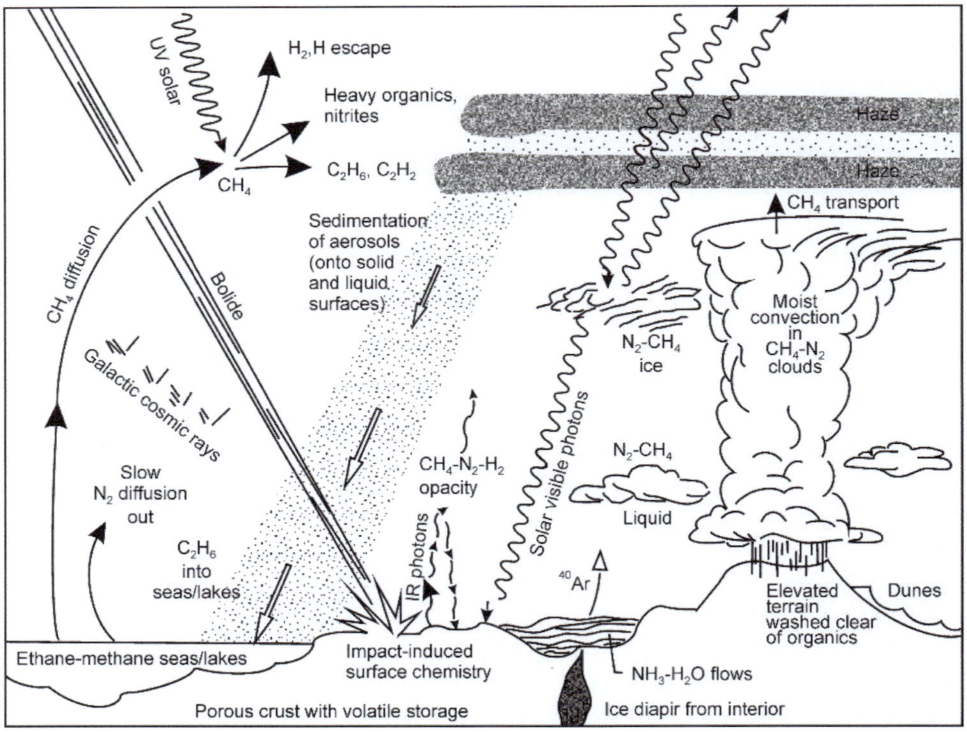

Fig. 5.3 Sketch of Titan surface processes from Lunine (1990)

atmosphere and low gravity should make aeolian transport easy, in the sense that no more than weak winds would be necessary to move fine-grained surface material. Thus, it was expected that dunes might exist on Titan. Conversely, Lorenz et al. (1995) argued that the weak sunlight and the thick atmosphere, which can transport a great deal of heat at low wind speeds, would not permit strong enough winds for even this low of saltation threshold to be exceeded. Furthermore, consideration of sand-generation processes suggested that the budget of sand-sized sediments on Titan would probably be small.

Glacial processes were considered by Lorenz and Lunine (1996) and ruled out for two reasons. First, the accumulation rate of materials would be rather low, so that the driving stresses behind a glacier would be small. Second, the thermodynamic conditions at Titan's surface did not permit solid methane or ethane to be in equilibrium with the observed atmosphere.

Several workers considered possible tides (Sagan and Dermott 1982; Lorenz 1993b; Sears 1995; Ori et al. 1998; Dermott and Sagan 1995) in bodies of liquid: in particular, Lorenz 1994 considered tides in lakes on Titan and noted that the location-dependent tidal tilt could be as high as 2×10^{-5} radians. Lunine (1990) noted that the sedimentation of aerosol particles Tomasko and Smith (1982) or other sediment in Titan liquids would be very slow. In anticipation of possible radar measurements or the floatation dynamics of the Huygens probe, several workers considered waves on the surface of Titan seas. Especially Srokosz et al. (1992) noted that gravity waves could be large and of relatively long period compared to Earth. Ghafoor et al. (2000) suggested that waves of about 0.2 m amplitude could be generated by winds of ~1 m/s, although this estimate was based on scaling terrestrial relations by gravity, without taking air density or other effects into account. Lorenz et al. (2005) showed by wind tunnel experiments that capillary waves in hydrocarbons grow larger than waves in water at the same wind speed, and also demonstrated that wave growth rates had a strong and somewhat nonlinear dependence on air density.

Lorenz (1996b) considered possible cryovolcanism on Titan and noted that the low solubility of methane in water or a water-ammonia solution would be such as to make volatile-enriched cryo-lavas unlikely. Thus, explosive eruptions, plinian or strombolian, for example, could not be expected, and cinder cones and large strato-volcanos resembling Mt. Fuji would not occur, with pancake-dome surface flows appearing more likely instead. Lorenz (1996b) noted that, expressing 10% of the likely geothermal heat flow as latent heat, the approximate fraction of surface eruptions on Earth would yield a global resurfacing rate of 2 mm/year, equivalent to an eruption rate of about 2 km^3 per year. Lorenz (2002) considered the possibility of methane geysers on Titan, noting some interesting similarities with Earth with lower temperature gradients being compensated for by the higher volatility of methane compared to water on Earth.

Several workers investigated impact cratering. Important early considerations were the suggestion of impact shock synthesis in Titan's atmosphere by Jones and Lewis (1987) and the astrobiological role of 'impact oases' containing liquid water created by impact melt (Thompson and Sagan 1992; see also O'Brien et al. 2005). Impact cratering was realized to be a potential tool for detecting variation over time in Titan's atmosphere, since small craters (<20 km diameter) would be relatively rare owing to atmospheric shielding (Engel et al. 1995) effects, which were studied in progressively more detail by Ivanov et al. (1997), Artemieva and Lunine (2003) and Korycansky and Zahnle (2005).

Snapshots illustrating the current thinking about Titan's surface are captured in a number of papers. Lorenz (1993b) discussed the surface in the context of the measurements that might be made by the Huygens probe. Lorenz and Lunine (1997, 2005) summarized results up to Cassini's arrival. Evolving perspectives on Titan's surface, as well as other aspects of Titan, are also described in the books by Lorenz and Mitton (2002, 2008) and Coustenis and Taylor (1999, 2008). It is evident that some predictions were successful (e.g. the overall balance of surface processes; Lorenz and Lunine 2005) and the possibility of strong surface modification by rainfall (Lorenz 2000), while others were grossly wrong, notably the failure to anticipate widespread dune coverage and the significance of latitude in controlling Titan's surface features. This mixed record is paralleled by the remarkable variety in Titan's landscape that Cassini was to reveal.

5.2 Cassini's Exploration of Titan's Surface

Over its 4-year nominal mission Cassini-Huygens has returned a wealth of data about Titan. In addition to the Huygens descent, the Cassini orbiter has observed Titan on 45 close passes to date, steadily increasing spatial coverage at various wavelengths as well as the timespan over which observations were made as spring approaches in the northern hemisphere. Cassini's imaging science subsystem (ISS), RADAR, and visual and infrared mapping spectrometer (VIMS) as well as Huygens' descent imaging spectral radiometer (DISR) have revealed an intriguing surface that is at once familiar and alien, and the combination of this suite of instruments, which complement each other in terms of wavelength, resolution and surface coverage, has proved essential to interpreting Titan's geology.

The ISS began observing Titan in April 2004 as Cassini approached the Saturnian system (Porco et al. 2005). Imaging Titan's surface is a challenge because it is almost completely obscured by the atmosphere at visible wavelengths (Richardson et al. 2004). Therefore, a 0.938 μm narrow band pass filter and infrared polarizer filters were incorporated in the ISS (Porco et al. 2004) to take advantage of (1) a window in methane's absorption spectrum in the near-infrared where the opacity of the atmospheric haze is lower, and (2) the high polarization of the haze at phase angles near 90° (West and Smith 1991). Despite the complications, Cassini has imaged to date almost all (~90%) of Titan's surface at resolutions better than ~10 km (Turtle et al. 2009) and a substantial fraction (~40%) at significantly higher resolutions ranging from a few kilometer down to the limit of ~1 km imposed by atmospheric scattering (Porco et al. 2004). These observations have been combined to produce a 0.938 μm map of the surface (Fig. 5.4).

As the topography expected for Titan is relatively low (Perron and de Pater 2004) and the contrast produced by the solar illumination of slopes is reduced by atmospheric scattering, the brightness variations observed at 0.938 μm are unlikely to be the result of topographic shading. Thus, with the exception of occasional bright tropospheric clouds, the features revealed in the ISS map are due to variations in surface albedo.

The VIMS has obtained spectral data of Titan's surface since its arrival in the Saturnian system in June 2004 (Brown et al. 2006). It operates in a spectral range from 0.35 to 5.2 μm, generating image cubes in which each pixel represents a spectrum consisting of 352 contiguous wavebands. Titan's atmosphere is transparent in atmospheric windows centered at 0.94, 1.08, 1.28, 1.6, 2.0, 2.8 and 5.0 μm (Barnes et al. 2007a), allowing the identification and mapping of surface features. Thus, the VIMS can be used as a multi-spectral camera to analyze the composition, geology and morphology of Titan's surface (Jaumann et al. 2006). The spatial resolution of VIMS data generally averages a few kilometers down to 250 m/pixel. VIMS has primarily mapped the equatorial regions of the satellite. The data with best spatial resolution have been acquired mainly between 30°N and 30°S.

The Cassini RADAR mapper is a multiple-beam sensor operating at four different modes in the 13.8-GHz Ku-band (i.e. 2.2 cm wavelength). The instrument operates in several active modes (synthetic aperture radar – SAR, real aperture scatterometry, and altimetry) and a passive radiometric mode (Elachi et al. 2004). Passive radiometry measures the microwave emission of a surface and complements active imaging. The two remaining modes are altimetry and scatterometry. In the SAR mode, which is unaffected by atmospheric contributions, image data are generated with a ground resolution of up to 300 m (Elachi et al. 2006), and brightness corresponds to surface properties such as roughness, slope, material composition and volume scattering (Elachi et al. 2006). By the end of the nominal Cassini mission, SAR data covered 34.4% of Titan's surface, 27.5% at resolutions <2 km. By early 2009, these areas had increased to 39% and 29.5% respectively (Kirk et al. 2009) (Fig. 5.4). An overview map of the RADAR coverage is also given in Stephan et al. (2009).

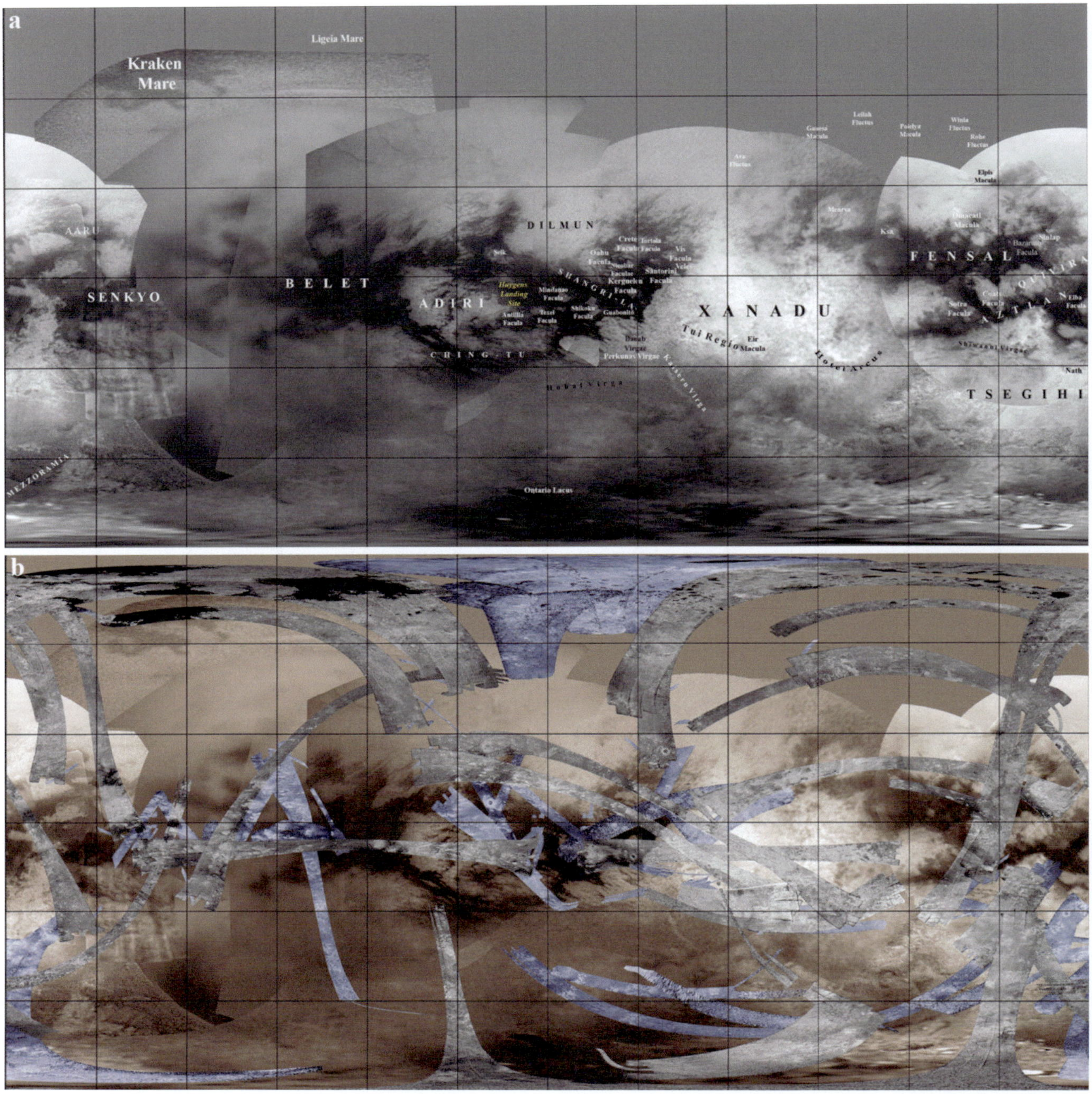

Fig. 5.4 (**a**) Map of Titan's surface compiled from ISS 0.938 μm images acquired between April 2004 and July 2008. The resolution varies according to the viewing geometry. The *dark feature* near the top with the convoluted boundary is Kraken Mare, which extends from ~55°N to ~80°N and covers an area larger than 400,000 km^2. *Narrow bright parallel streaks* are clouds along ~60°N latitude. The majority of 70 names currently approved by the International Astronomical Union are shown. Simple cylindrical projections, north at top. For a variety of other maps of Titan's surface see Stephan et al. (2009). (**b**) RADAR synthetic aperture (SAR) images overlaid onto the ISS map. The maps are displayed in an equidistant projection centered at 0°N and 180°W. Full resolution SAR images obtained through flyby T50 (February 2009) are shown in *gray*, lower resolution high altitude SAR (HiSAR) observations for the same period in *blue*, and the ISS base map in *brown*. Images have been normalized to a constant incidence angle of 35° based on the scattering model fitted by Wye et al. (2007) and scaled logarithmically. *Black* and *dark blue* correspond to −20 db, white to 5 db. Full resolution SAR coverage is 29.5% of Titan; coverage including HiSAR is 39% of Titan

5.3 Titan's Morphology and Topography

5.3.1 Morphology

As described above, radio (Muhleman et al. 1990) and telescopic observations (Smith et al. 1996) of Titan began revealing spatial inhomogeneity on the surface in the 1990s. Optical light curves (Lemmon et al. 1993; Griffith 1993; Coustenis et al. 1995) exhibited similar shapes at multiple wavelengths, although they differ in amplitude as a result of varying contributions from the surface and atmosphere. Maps of the surface reflectance, corrected for atmospheric scattering, were also similar within the limits imposed by noise, viewing geometry, and the possibility of transient cloud activity (e.g. Smith et al. 1996). One major area of bright terrain, subsequently named Xanadu, was known to be centered near the equator at longitudes 80°–140°W. Some variations were visible within the darker areas, but the extent to which they represented real surface features was not well constrained. Thus, the pre-Cassini picture of Titan's surface variability was effectively one-dimensional and based on differences in reflectance that were similar at all wavelengths. Titan was, to put it simply, a world of black and white (Fig. 5.2). Initial observations of Titan made by the ISS on Cassini in the near-infrared atmospheric windows at 0.938 μm (Porco et al. 2005) confirmed the dominance of Xanadu as the largest, brightest region on Titan and filled in a considerable amount of detail (Fig. 5.4). ISS coverage, in particular, quickly provided a low- to moderate-resolution map (5–10 km) of Titan's relative surface brightness from the south pole to the limit of useful illumination (~50°N at the beginning of the Cassini mission). Images obtained during subsequent flybys added detail in particular regions. The latest maps of Titan's surface including most of the feature names currently approved by the IAU (which presently number 70 for all feature categories) are shown in Fig. 5.4. The early ISS observations (Fig. 5.5) revealed the equatorial zone of Titan, apart from Xanadu, to be sharply divided into moderately bright and very dark regions on a scale of tens of degrees.

Many of the dark equatorial regions seemed to have relatively linear boundaries. Higher-resolution ISS imaging (e.g. Fig. 5.5) revealed that the western boundaries of bright areas are often very sharply defined. The combination of sharpness and, in some cases, linearity in these boundaries suggested that at least some of them were tectonically defined (Porco et al. 2005). The eastern boundaries of bright regions are, in general, more diffuse, suggesting that the dark material might have been deposited by west-to-east flow either in the atmosphere or in surface liquids, and/or that bright material was being eroded and re-deposited by west-to-east flow (Porco et al. 2005). Indeed, the equatorial dark regions have been shown to consist of extensive fields of longitudinal dunes (Lorenz et al.

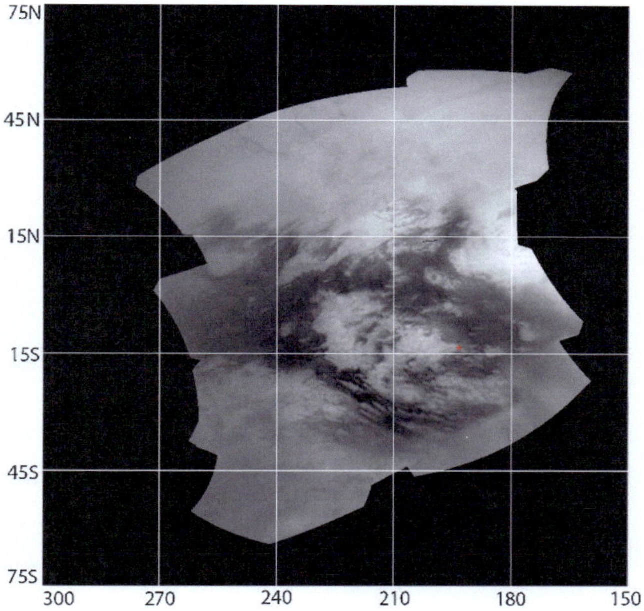

Fig. 5.5 ISS mosaic of the Adiri region, with the approximate location of the Huygens landing site marked by (+). Simple cylindrical projection, north at top. Controlled image cube produced by USGS

2006a). Isolated bright features (termed 'faculae') of all sizes down to the limit of resolution are interspersed in the dark terrain, and smaller dark lanes, designated with the new feature category of 'virgae' by the International Astronomical Union (IAU) Working Group for Planetary System Nomenclature, are present within the bright regions. The branching and meandering morphology of some of these dark streaks suggests channel systems (Porco et al. 2005; Lorenz et al. 2008a; Jaumann et al. 2008). ISS also identified a handful of nearly circular or annular features as candidate impact craters (Porco et al. 2005). Some of these are bright rings within Shangri-la (e.g. Guabonito) and other equatorial dark regions, another is a dark annulus more than 300 km in diameter (Porco et al. 2005) which was subsequently confirmed by RADAR to be a large impact structure now named Menrva (Elachi et al. 2006).

Although ISS observations revealed Titan's mid-latitudes to be relatively featureless, high latitudes show significant albedo variations (Fig. 5.4). The surface of the south polar region exhibits numerous dark features ranging from ~10 km to, in one case, 235 km long. The largest and darkest of these, Ontario Lacus, was hypothesized to be an example of the long-sought hydrocarbon lakes of Titan, based on its simple (but non-circular) closed shape, smooth boundary and low reflectivity compared to its surroundings (McEwen et al. 2005; Turtle et al. 2009). The appearance of new dark features near the South Pole in later ISS observations strengthens the interpretation that they represent liquid ponded on the surface (Turtle et al. 2009). Cassini arrived 5 years before the northern vernal equinox so northern latitudes were poorly illuminated; however, there is sufficient scattered light in Titan's atmosphere

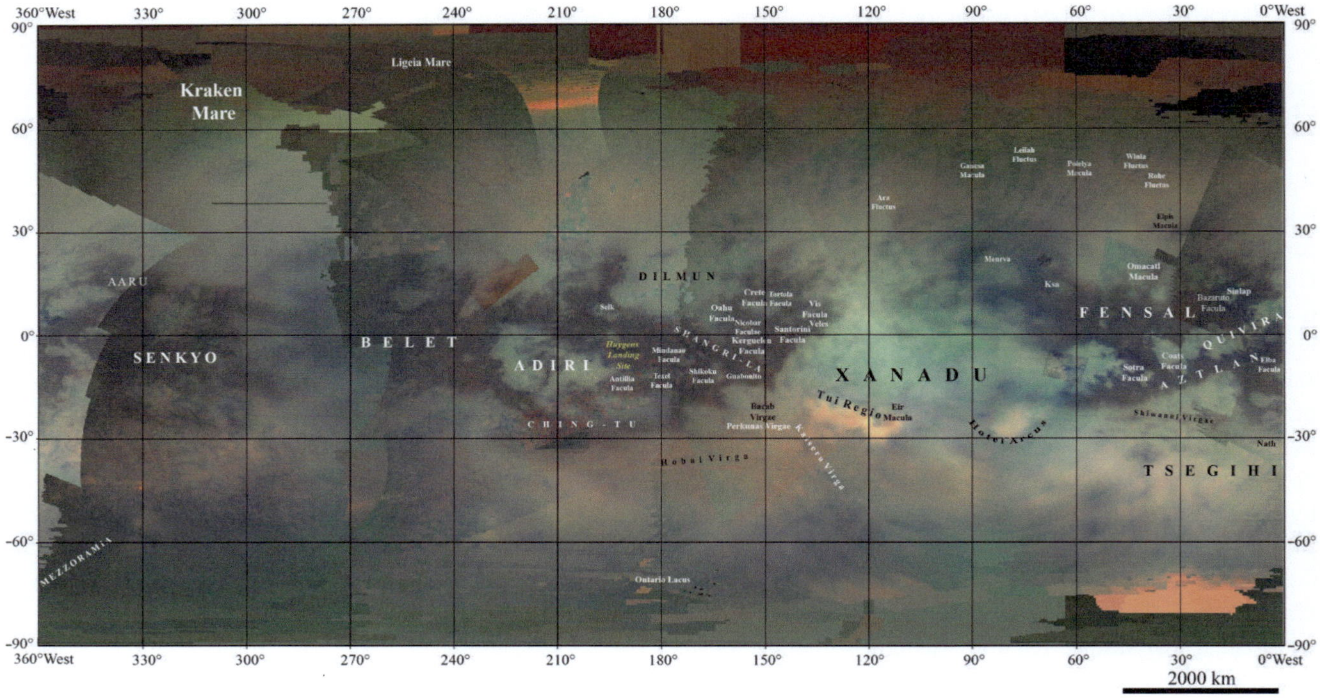

Fig. 5.6 Map of Titan's surface compiled from VIMS images acquired between April 2004 and July 2008, displayed as a false color composite with R = 5.0 μm, G = 2.0 μm, B = 1.3 μm. Simple cylindrical projection, north at top. Adapted from Barnes et al. (2007a). The majority of 70 names currently approved by the International Astronomical Union are shown. Simple cylindrical projections, north at top

that ISS can observe the surface for ~10° beyond the terminator and therefore has also been able to map out the full extents of the much larger northern maria identified by RADAR (Stofan et al. 2007) as well as the distribution of the myriad smaller lakes (Fig. 5.4). The Cassini VIMS spectrometer sees the surface of Titan at the 0.94 μm wavelength used by the ISS as well as through six other atmospheric 'windows' between 1.08 and 5.0 μm (Brown et al. 2004). Global maps of Titan (Jaumann et al. 2006; Barnes et al. 2007a) in all of these bands resemble one another and the ISS images (Porco et al. 2005) to first order, but subtle color differences are apparent, as shown in Fig. 5.6.

Such variations have been studied in a variety of ways, both by grouping surface regions into different spectral classes (Barnes et al. 2007a; Tosi et al. 2008) and by modeling them as mixtures of spectral end members (Soderblom et al. 2007a; McCord et al. 2008; Jaumann et al. 2008). A conclusion common to the majority of these studies is that at least two spectrally distinct types of dark material and two types of brighter material are present. The dark materials are commonly referred to as 'blue' and 'brown' based on their appearance in false-color spectral composites such as Fig. 5.6. The bright materials include (as a minimum) a widespread variety and a unit that is especially bright at 5 μm which is relatively uncommon and quite localized spatially. The relationship between these spectral units and morphological features will be described below. The interpretation of their surface composition is the subject of considerable debate, for a description of which the reader is directed to Soderblom et al. (2009b). In addition to relatively low-resolution (several tens of kilometers) global coverage, the VIMS has also obtained images of local areas with resolutions of a few km (Sotin et al. 2005) and in exceptional cases, a few hundred meters (Jaumann et al. 2008; Barnes et al. 2008). These observations also show albedo and spectral variations at the finest scales resolved. The bright features Tortola Facula (Sotin et al. 2005) and Tui Regio (Barnes et al. 2006) have been hypothesized to be cryovolcanic, based on their morphology in VIMS images. Hotei Arcus, which is exceptionally bright at 5 μm (Barnes et al. 2005) and bears some spectral resemblance to Tui, has also been hypothesized to be volcanic (Soderblom et al. 2009a), and it has been suggested that it has varied in brightness over time (Nelson et al. 2007, 2009a, b). RADAR images of this area reveal lobate features that have been interpreted to be cryovolcanic (Wall et al. 2009). This interpretation is supported by RADAR stereo observations indicating that the lobes are 100–200 m thick.

A series of north–south trending bright and dark streaks to the east of Tsegihi have been interpreted as mountain ranges ~500 km long and 1 m to 1.5 km high (Sotin et al. 2007). The highest-resolution optical observations of Titan's surface are the images obtained by the DISR instrument during and after the descent of the Huygens probe (Tomasko et al. 1997; Tomasko et al. 2005). The DISR images (Fig. 5.7) cover an area ~150 km across at resolutions of tens of meters, improving to a few meters nearest the landing site (images obtained on the surface show the immediate vicinity of the probe with a resolution of millimeters to centimeters).

The DISR data confirmed earlier observations of a sharp-edged boundary between bright and dark regions on the surface of Titan with albedos varying by about a factor of two. The largest of the dark regions near the lander are ~10 km across, but numerous isolated regions of relatively high albedo are distributed within the dark terrain, and dark, dendritic features interpreted to be channels are present in the bright terrain (Soderblom et al. 2007b). Dark linear features ~30 km north of the landing site, overlooked initially due to foreshortening, were later identified as dunes and used to constrain the location of the landing site because they can also be identified in RADAR images of the area (Lunine et al. 2008a).

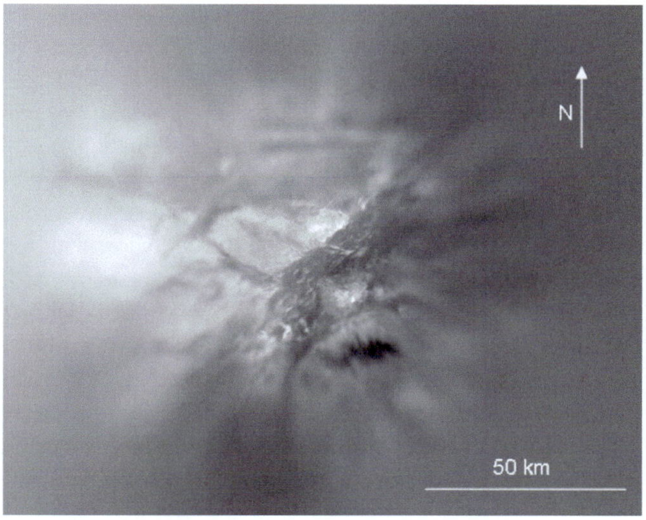

Fig. 5.7 Mosaic of Huygens DISR images centered on the landing site. Simple cylindrical projection with north at top (adapted from Soderblom et al. 2007a)

The Cassini RADAR instrument probes the response of Titan's surface to 2.2-cm microwaves. Like the optical instruments, it provides both low (50–500 km) resolution global coverage and higher-resolution (<1 km) imaging of more limited areas. Low-resolution maps have been made by both active radar scatterometry (Wye et al. 2007) and passive radiometry of thermal emission (Janssen et al. 2009); examples are shown in Fig. 5.8.

The negative correlation between maps (Fig. 5.8) is striking but hardly unexpected. Kirchoff's law states that, in thermal equilibrium, the reflectivity and emissivity averaged over all directions must add up to unity. Scattering and emission measurements obtained at particular viewing geometries need not obey this law but will approximate it to the extent that these processes do not vary greatly in their angular directionality. Perhaps more surprising is the relatively strong correlation between the patterns of reflectivity in the microwave band (Fig. 5.8a) and the near infrared (Fig. 5.6),

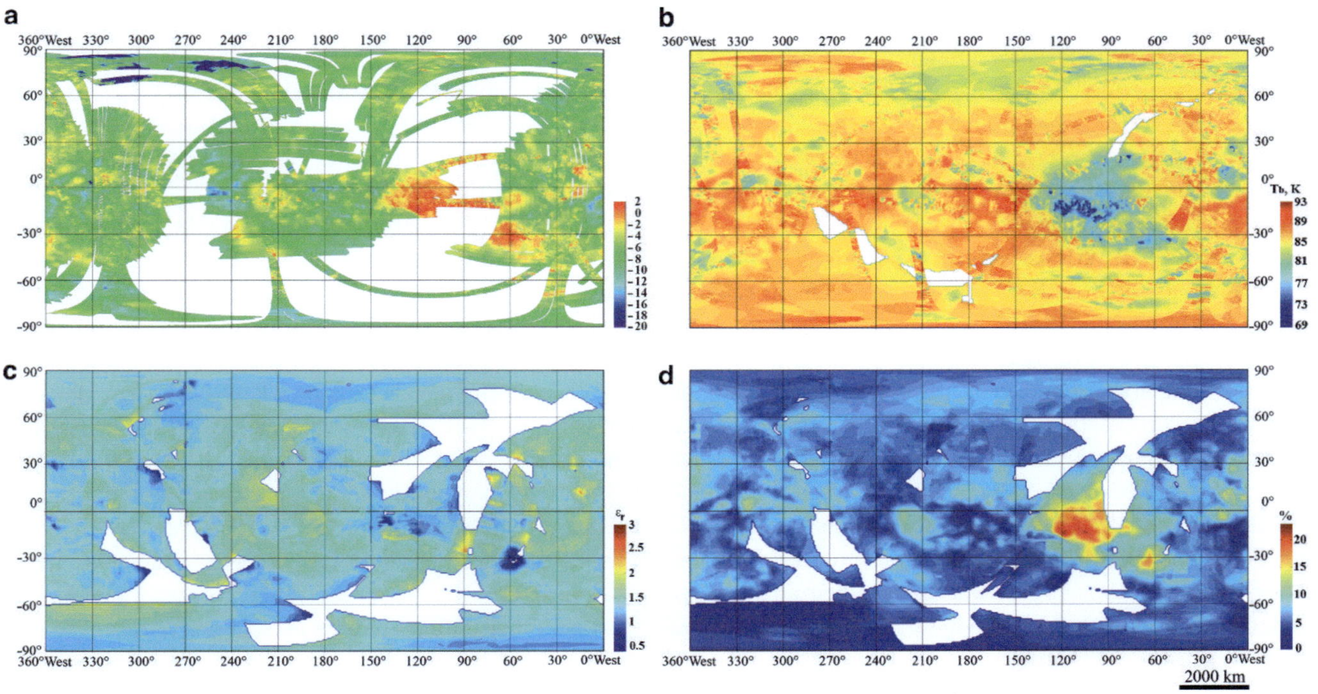

Fig. 5.8 Global maps of Titan's surface properties measured by the RADAR instrument in an equidistant projection centered at 0°N and 180°W. (**a**) Radar backscatter cross-section from scatterometry measurements, normalized to a constant incidence angle of 35° based on the scattering model fitted by Wye et al. (2007) and scaled logarithmically from −20 db to 2 db. Greater backscatter values correspond to rougher or more reflective surfaces or areas of enhanced subsurface volume scattering. (**b**) Surface brightness temperature as measured by passive radiometry, scaled to normal incidence (Janssen et al. 2009). Because physical temperature is nearly constant, this is a proxy for microwave thermal emissivity. (**c**) Surface dielectric constant, inferred from the polarization of thermal emission. (**d**) Fraction of subsurface volume scattering

although Muhleman et al. (1990) also observed the strongest radar echoes from Titan at approximately the longitude of Xanadu. In particular, Xanadu, Tsegihi and other large near-IR bright regions are radar-bright and emit weakly, whereas the darkest equatorial regions such as Shangri-la, Fensal, and Belet are radar-dark and highly emissive. When the data are examined in detail, however, the correlations are less clear and may be entirely absent. The sharp discontinuity in the radar-scattering properties of southwestern Xanadu, within an area that appears relatively uniform in Fig. 5.4a, is perhaps the best example of this phenomenon.

Additional information about the surface comes from more detailed modeling of radiometer measurements. Where overlapping measurements with sufficiently different polarization angles are available, the effective dielectric constant of the surface can be calculated; this, along with the brightness temperature at either polarization, leads, in turn, to an estimate of volume scattering (Janssen et al. 2009). Again, the spatial distributions of dielectric constants and volume scattering correlate roughly but not precisely with one another and with the scatterometry and optical maps. Prominent (optically and radar) bright regions such as Xanadu feature low dielectric constants and high volume scattering. The darkest equatorial areas, conversely, show high dielectric constants and weak volume scattering. However, the maps show subtle variations in detail; for example, all of Xanadu is optically bright, but only part is radar-bright, and an even more limited area exhibits the most greatly enhanced volume scattering. As with the spectral differences seen by the VIMS, these subtle departures from the general correlation between the global RADAR datasets undoubtedly provide clues to the variety and nature of surface units on Titan, but at resolutions of tens to hundreds of km, links to surface features and processes are tantalizingly vague.

Fortunately, RADAR also has the capability of taking synthetic aperture radar (SAR) images with resolutions ranging from ~300 m near the closest approach to a few km at the extreme range limits of this scattering model. Such resolutions are comparable to the best VIMS images and inferior only to the DISR data in spatial resolution, but they cover larger areas amounting to 39% of Titan's surface at present (Fig. 5.4b). Although the SAR images afford only a single-wavelength view of Titan and thus must be considered in conjunction with compositional and textural information in the lower-resolution datasets described above, their high resolution, broad coverage and sensitivity to roughness and topography make them the primary source of information about the morphology of surface features. Not only do radar images delineate the planimetric form of features of different backscattering strength, as described below, they also provide unambiguous evidence about the surface relief in many places through shading, stereo parallax and other clues. Similar information is also present in the near-IR images and might in some cases be interpretable (e.g. Sotin et al. 2007), but relief on Titan is characteristically so subtle that the higher resolution of the SAR provides a tremendous advantage. The following sections in this chapter consider various processes, such as aeolian and fluvial modification, impact cratering, cryovolcanism and tectonics that are believed to be at work on Titan's surface, and describe in detail the reasons for attributing these processes to particular features. Initial attempts have been made to map the TA, T3, T7, T8, and T30 RADAR swaths (Stofan et al. 2006; Lunine et al. 2008a; Lopes et al. 2009). As the brightness variations in the swaths are due to a relatively poorly constrained combination of surface roughness, surface topography, material composition and volume scattering, initial maps reflect radar properties more like geomorphological than geological maps, with the exception of clearly delineated feature types such as cryovolcanic flows and impact crater materials that will be described in detail in subsequent sections. In this section, we will give only a brief overview of the range of surface features and terrains that the RADAR images have revealed.

Because the geographical locations of the first few SAR images resulted, rather fortuitously, in a steady sequence of increasingly dramatic and often surprising feature types, it is reasonable to describe these features in a roughly historical order. The first flyby, TA, included RADAR imaging of the northern mid-latitudes (Elachi et al. 2005). The surface this image revealed was dominated by plains with a variety of radar brightnesses and textures (Stofan et al. 2006). Subsequent flybys, along with the global views of ISS and VIMS, have confirmed that such plains dominate the middle latitudes of both hemispheres. The most prominent feature in TA, a large circular structure named Ganesa Macula, was interpreted as a possible cryovolcanic dome (Lopes et al. 2007a), and the swath included both diffuse flows that might be cryovolcanic and flows with discrete margins and associated caldera-like sources that almost certainly are (Lopes et al. 2007a). It was unclear whether small linear features near Ganesa were associated cryovolcanic features or channels formed by the flow of surface hydrocarbons, but the next RADAR image, obtained during the T3 flyby (Elachi et al. 2006) and covering the same longitudes as TA but closer to the equator, showed a variety of channels not associated with any identified volcanic features and thus certainly fluvial in origin (Lorenz et al. 2008a). These channel systems are one to two orders of magnitude larger than those seen by the DISR at the Huygens landing site (Jaumann et al. 2008; Lorenz et al. 2008a).

The major surprise of T3 was the unambiguous identification of two impact craters, Menrva (450 km in diameter) and Sinlap (80 km), by ISS and RADAR. Only four definite impact craters were found during the entire prime mission, but RADAR, like the ISS, has imaged a much larger number of 'suspicious circular features' that could be modified impact structures (Lorenz et al. 2007; Le Mouélic et al.

2008; Wood et al. 2009a and 2009b). Narrow radar-dark streaks trending roughly east–west were seen in several places, deviating around or terminating at other features such as the Sinlap ejecta blanket. Given the informal descriptive name 'cat scratches', they were tentatively identified as dunes. Subsequent RADAR imaging of the equatorial zone on T8 and later flybys revealed that the 'cat scratches' were ubiquitous, covering large areas that corresponded closely to the patches and bands of lowest near-infrared albedo. As discussed in detail later in this chapter, the identification of the streaks as dunes and the dark equatorial regions as sand seas is among the most secure conclusions about Titan's geology to date (Lorenz et al. 2006a). Studies of subsequent images showed that the sand seas give way to isolated clusters of dunes farther from the equator, with few dunes seen at latitudes greater than 30° north or south (Radebaugh et al. 2008a).

The T3 image also contained relatively clear photogeological evidence of relief features. Topographic shading, identifiable by its orientation relative to the illumination direction and the pairing of bright and dark areas across each feature, was seen not only in the two craters but also in numerous isolated features with an appearance similar to that of terrestrial mountains imaged by radar (Radebaugh et al. 2007). Larger areas of rugged relief were seen in a subsequent RADAR flyby, T7, and the T8 image contained linear chains of these 'mountains' in addition to isolated peaks (Lunine et al. 2008a). The culmination of this sequence was the imaging of Xanadu in the T13 flyby, revealing densely packed mountains over most of the radar-bright areas (Kirk et al. 2006; Wood et al. 2007a; Radebaugh et al. 2009).

The T7 image, which was the first to sample the mid-southern latitudes, also revealed numerous channels, some with clear evidence of being topographically incised, while others were revealed only by the contrast in backscatter with their surroundings. The morphology of the channel systems suggested southward flow, and the southernmost images obtained, at around 65°S, showed a scalloped bright terrain apparently embayed by dark material within Mezzoramia (Fig. 5.4), with little interior texture visible (Lunine et al. 2008a). Unfortunately, the remainder of the image, which was to have continued to even higher latitudes and back toward the equator, was lost as a result of a fault on the spacecraft. The morphology of the dark embaying materials and their location at the terminus of an extensive channel system triggered speculation that the T7 image might have ended at the shore of a hydrocarbon lake, but they are in fact only moderately radar-dark, comparable to dunes. Much stronger evidence of hydrocarbon lakes was seen in subsequent flybys that imaged the north polar area, beginning with T16 (Stofan et al. 2007). Here, numerous dark patches ranging in size from a few to nearly 100 km were seen at latitudes≥70°N, many of them consistent with zero reflected signal, given the noise characteristics of the RADAR instrument. Many of these features are located in apparently steep-sided local depressions with simple (sub-circular to multi-lobed) shapes; others are associated with channel systems, supporting their identification as lakes.

The subsequent north polar RADAR passes T25, T28, T29, and T30 revealed much larger dark features, worthy of the name seas (Lopes et al. 2007b). Indeed, larger-scale but lower-resolution images acquired by the ISS immediately after T25 reveal the full extent of the largest of these, Kraken Mare, to be almost 1,200 km (Turtle et al. 2009). The large seas have more complex coastlines with channels terminating at the shore, indicating they have partially submerged preexisting topography, rather than being confined to individual steep-sided depressions like the smaller lakes (Mitchell et al. 2007; Mitchell et al. 2009). The small lakes and the seas are strongly segregated by longitude (Hayes et al. 2008), which may be a function of the preexisting topography in different regions. Both the steep-sided depressions, some of which contain small lakes, and the rugged, fluvially dissected terrain like that around the seas seem to be characteristic of high latitudes. The rugged high-latitude terrains are distinguished from both Xanadu and other low-latitude mountains and chains in that the latter are typically extremely radar-bright (Radebaugh et al. 2007, 2008a). Images obtained on flybys T36 and T39 late in the prime mission indicated that a few small lakes are present near the south pole, although so far they appear to be much less abundant than at the north pole (Stofan et al. 2008).

In summary, the 'menagerie' of Titan geological features and terrains defined by their radar morphology includes plains, cryovolcanic flows and constructs, mountains, dunes, channels, lakes, and seas (Lopes et al. 2009). An important question is how these morphological entities relate to the patterns of compositional, physical and textural properties seen in lower-resolution datasets from the ISS, the VIMS and RADAR scatterometry and radiometry. Soderblom et al. (2007a) examined coregistered RADAR SAR and VIMS high-resolution spectral data, supplemented by DISR images of the Huygens landing site, to address this issue. Despite the general correlation between optical and radar reflectivity noted previously, the coverage of western Shangri-la and the Huygens landing site (Fig. 5.9) shows only at first glance a few local instances of correspondence between bright and dark features.

Soderblom et al. (2007a) noted, however, that spectral variation in this and the other areas they studied could be interpreted in terms of linear mixing of three end-member materials. Based on their appearance in RGB false-color composites of data from the 2.0, 1.6, and 1.3 μm atmospheric windows, such as that shown in Fig. 5.9 above, these end members can be described as 'bright', 'dark blue' and 'dark brown'. They are consistent with the range of spectral units defined by Barnes et al. (2007a), Jaumann et al. (2008) and

Fig. 5.9 Comparison of VIMS (*top*) and RADAR (*center*) images of the Huygens landing site and setting. VIMS image is a false color composite with R = 2.0 μm, G = 1.6 μm, B = 1.3 μm. DISR mosaic (Fig. 5.7) is shown at the same scale. Simple cylindrical projection with north at top. *Green arrows* connect corresponding bright features around the landing site. *Bottom*: sketch maps at reduced scale showing spectral and morphological units. *Magenta* = VIMS 'blue' spectral end member. *Red* = densely packed longitudinal dune fields as seen in RADAR, *yellow* = sparse dunes, *green* = data quality insufficient for mapping. Dunes correspond closely to VIMS 'brown' spectral end member (adapted from Soderblom et al. 2007a)

McCord et al. (2008) over most of Titan. Soderblom et al. (2007a) showed that, in all the regions they examined, areas dominated by the brown end-member correspond almost exactly to the dune fields identified by their morphology in the RADAR images. Thus, this spectral signature corresponds to the presumably organic material that makes up the dunes. The low reflectivity of the dune material in both the near-infrared and microwave wavelengths accounts for a large part of the correlation that is seen on the global scale between VIMS, ISS and RADAR observations. This correlation is also seen locally, where the resolution of the data is sufficient. Individual dunes visible in RADAR images can be identified both in DISR images and in the highest-resolution VIMS and ISS data (Barnes et al. 2008; Perry et al. 2007c).

With the 'brown' dune material stripped away, relations between the other spectral signatures and the morphology become more complex, and there is no universal correlation between overall optical and radar brightness. The large area of Xanadu which, like other, smaller mountainous regions is very bright in both wavelength ranges, may suggest such a correlation, as the western part of Fig. 5.9 shows, it is more generally the optically bright regions which include both radar-bright and radar-dark features. Soderblom et al. (2007a) noted that blue material is often found peripherally to the bright end member, particularly on the eastern side of bright features; its presence and intermediate albedo at short wavelengths could account for the diffuse appearance of the eastern margins of such features as seen by the ISS. They also showed an example in which local enhancements in the blue end member in Omacatl Macula could be identified as channels and outwash features and in some cases but not consistently, as mountains. Barnes et al. (2007b) extended this result by showing correlations between bluer spectral signatures and mountain ranges and channels in several locations in Xanadu. The two groups of researchers offered a common hypothesis, namely that the bright end member seen by the VIMS is a surface coating that is partially removed as mountains are eroded and is then depleted in the eroded material in channels and outwash. Such blue outwash material would be expected to be deposited peripherally to high-standing bright terrains on all sides, but would probably be covered by dunes at the western margins. If the bright material is an atmospherically deposited coating, it is likely to be organic. The comparatively high reflectance of the blue material at 0.9–1.0 μm could be explained by water ice, which is expected to be the most important constituent of Titan's crust (Stevenson 1992) and thus might be the main constituent of sedimentary deposits as well. Consensus has yet to be achieved, however, on whether the spectral information in the full set of atmospheric windows is consistent with the presence of H_2O, and the reader is referred to Chapter 6 for a discussion of this issue.

5.3.2 Topography

Our knowledge of Titan's size, global shape and topography was extremely limited prior to the Cassini mission. Voyager radio occultations at 6°N, 102°W and 8°S, 284°W (Lindal et al. 1983) indicated a radius of 2,575 ± 0.5 km. This was sufficient to show that the solid body of Titan was smaller than Ganymede, so that it is not the largest satellite in the solar system, and also placed a limit of about 1 km on the relative topography between the two points probed, which were located in northern Xanadu and the dark region of Senkyo, respectively. The polar flattening that was expected on theoretical grounds if Titan were in hydrostatic equilibrium and had uniform density, is even smaller, on the order of 100 m (Lorenz and Lunine 2005).

The Cassini data now available provide information about Titan's topography in a wide variety of ways. As noted above, photogeological interpretation of the optical and especially the RADAR images supplies evidence of the presence of local relief and the direction of slopes. Thus, for example, the direction of apparent cryovolcanic flow features (Elachi et al. 2005; Lopes et al. 2007a) and fluvial channels (Lorenz et al. 2008a) indicates the direction of surface dip at the time the features formed. The pattern of flow directions seems to be primarily eastward in the region north of Xanadu, but southward within Xanadu itself and to the south. The tendency of dunes to end at or divert around 'islands' of brighter terrain suggests that the latter are elevated and act as obstacles (Radebaugh et al. 2008b), even where 'mountainous' relief is not directly visible as it is in other places (Radebaugh et al. 2007). If the interpretation of the dark features at the north pole as lakes (Stofan et al. 2007) is correct, they must be situated in at least local topographical lows.

Such geologic evidence for topography is important but ultimately rather frustrating because the quantitative element is lacking. Knowledge of the magnitudes as well as directions of slopes is needed to answer questions such as these: What are the viscosities, and hence the possible compositions, of cryolavas? What are the discharges of the channel systems? How much material is in the dunes, and how high must the obstacles be that divert them? How deep are the lakes? Direct topographical measurements are also needed to determine whether deformation has altered the downslope direction since flows or channels formed. The methods available for quantifying the topography of Titan include stereo photogrammetry, shape-from-shading applied to both optical and RADAR images, radar altimetry, a novel type of radar processing known as SARTopo (Styles et al., 2009) provide only local and relative topographical information, but the latter three are capable of covering a substantial portion of Titan and providing information about absolute as well as relative elevations.

Photogrammetry, the geometrical analysis of stereo pairs of optical images has provided much of our knowledge about the quantitative topography of Earth and other planets. The local relief on Titan is never more than a few kilometers and often much less. The Huygens DISR cameras obtained images of the landing site with resolutions of tens of meters and with view angles ranging from 6.5° off nadir to the horizon (Tomasko et al. 2005). Although the locations and pointing of the exposures could not be controlled to ensure stereo imaging, and roughly half the images were lost as a result of the failure of one of the two downlink channels, useful stereo pairs of images with similar resolutions but different viewing geometries were obtained (Fig. 5.10a).

In particular, Soderblom et al. (2007b) were able to analyze a cluster of six overlapping images in the bright terrain north of the landing site and an additional stereo pair in the darker terrain closer to the landing point. Because these two regions were not connected by additional data, it is impossible to determine their relative elevations, but the digital topographical model (DTM) of the bright region, Fig. 5.10b, shows that peaks in the interior of the bright terrain are elevated ~180 m above the 'coastline' with nearby dark terrain and≤250 m above the lowest points in the latter. The horizontal resolution of the DTM is clearly sufficient to show that the dark, branching channels seen in the images are flowing downhill, and in some cases to distinguish the channels themselves. Where the slopes of the channel walls can be measured, they are on the order of 30°, indicating very rapid downward cutting. The second DTM (Fig. 5.10c) shows that the somewhat brighter lineations within the dark terrain stand high, and that the dark basins in between are flat but differ slightly in elevation. Again, the range of elevations is ≤50 m. Soderblom et al. (2007b) interpreted these patterns of relief as a consistent pattern of erosive fluid flow out of the bright highlands and eastward through the dark terrain on which Huygens landed. Other questions of interest, such as the relief of the landing site over longer distances and whether the 'stubby' lower-order drainage systems to the west of those seen in Fig. 5.10b differ in their transverse or longitudinal slopes, might yet be addressed if the DISR dataset can be 'mined' to find additional useful stereo pairs for mapping.

Shape-from-shading, also known as photoclinometry, is an alternative approach to extracting topographical information from images. It can be applied to unpaired images and can achieve single-pixel resolution because it is based on interpreting brightness (shading) variations in slopes rather than measuring geometrical parallax, it is more model-dependent. Variations in intrinsic brightness (albedo) can be misinterpreted as shading and cause artifacts. Atmospheric scattering must also be taken into account, and this is a severe problem on Titan. Any amount of scattering will dilute the contrast of the surface features with a uniform 'haze' that must be accounted for in order to recover the amplitude of topography correctly, but a dense atmosphere will also change the directionality of surface illumination and thus the relation between brightness and topography. In the extreme case of an atmosphere so dense that illumination becomes non-directional, valley floors (which see the least amount of sky) rather than slopes away from the sun would be the darkest topographic features (Grieger 2005; Keller et al. 2008). Despite these challenges, several researchers have tried to exploit the strengths of the VIMS instrument to reduce uncertainties and thus permit photoclinometry. Comparing the surface contrasts in multiple atmospheric windows highlights variations in color that may imply variations in surface albedo, and allows the contrast of surface features to

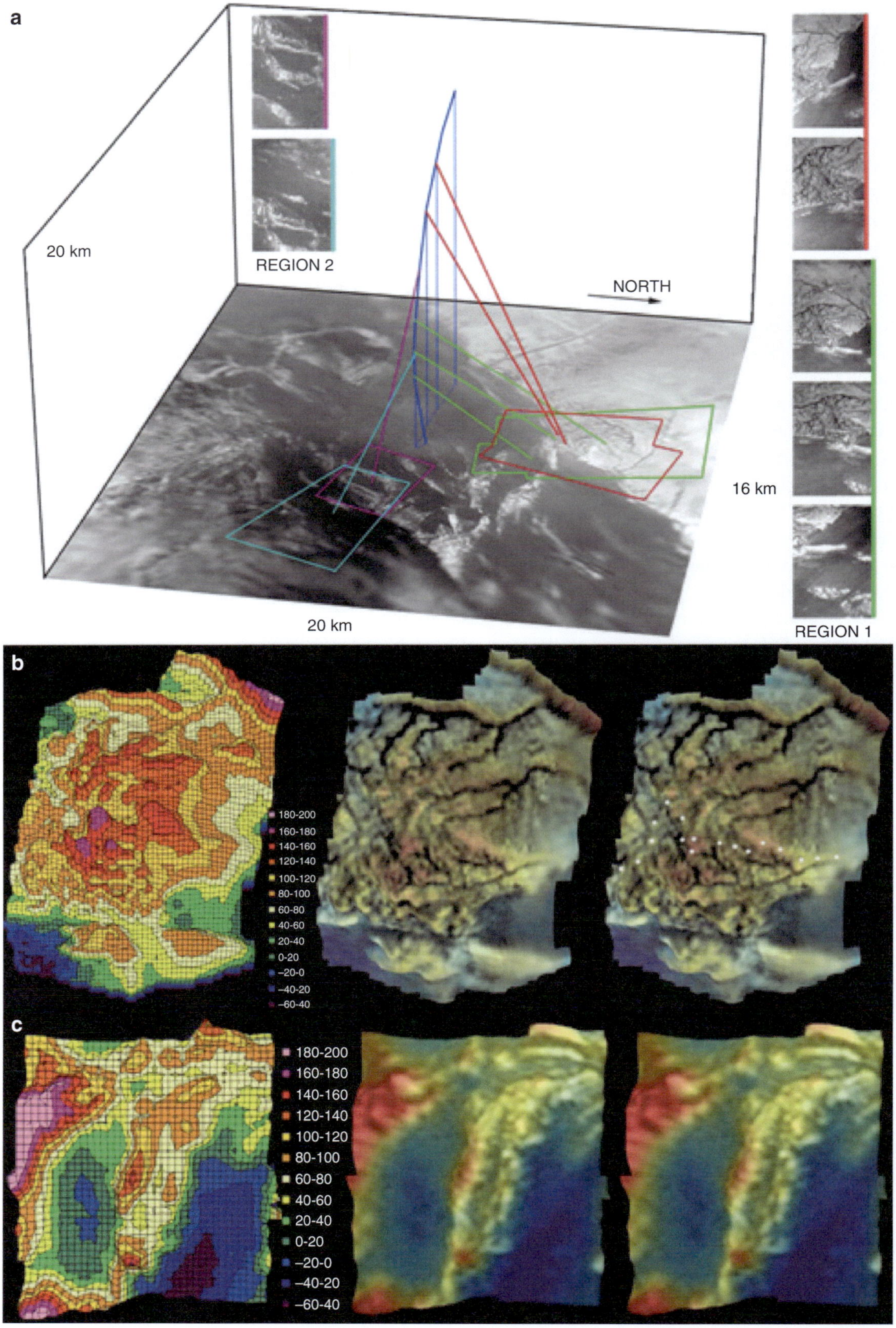

Fig. 5.10 Topographic models of bright and dark areas near the Huygens landing site, based on stereo analysis of DISR images. View is from the east. (**a**) Descent trajectory showing the acquisition geometry of the images used. (**b**) Region 1 in bright terrain. (**c**) Region 2 in dark terrain. For each region, a color-coded topographic map is shown, accompanied by a synthetic stereo pair showing an oblique view of the images with color-coding according to elevation added (adapted from Soderblom et al. 2007b)

be extrapolated to zero atmospheric opacity. Restricting the application of the method to features that show the characteristics of slope shading, with bright and dark paired on the up-sun and down-sun sides of features, is an additional guard against error. Sotin et al. (2005) employed all these safeguards to construct photoclinometric profiles across east–west trending linear features, typically 5–10 km apart, near Tortola Facula, arriving at heights of a few hundred meters. The spectral signature of this area strongly suggests that it is a dune field, however, and RADAR images of such areas indicate a typical dune separation of only 1–4 km (Radebaugh et al. 2008b). Barnes et al. (2008) subsequently constructed a photoclinometric profile across dunes from the much higher resolution VIMS data from T20. They obtained peak-to-trough heights of 30–70 m for the best-resolved dunes with separations of ~2 km, though this result is based on an empirical haze correction that is not explained in detail. A much larger set of features, interpreted to be north–south trending mountains to the east of Tsegihi, has been modeled by Sotin et al. (2007). The most distinct portions of these paired bright-dark lineations are ~200 km long, though they can be traced considerably farther. The estimated height based on photoclinometry is 1–1.5 km.

Radarclinometry refers to the application of shape-from-shading to SAR images. On Titan, radarclinometry is advantageous because higher-resolution images are available over a broad area, and the atmosphere is transparent and hence does not interfere with image radiometry. Like photoclinometry, radarclinometry yields results that are model-dependent and will be distorted if the intrinsic reflectivity of the surface is not constant over the region being mapped. As we have seen, the radar backscatter cross-section of Titan varies dramatically on a global, regional and local scale, so it is necessary to be very selective and apply the technique only to areas where slope shading clearly outweighs any intrinsic variations in brightness. Where, as is usual on Titan, areas of apparent uniformity are small, one-dimensional or profiling techniques are superior to two-dimensional radarclinometry methods that attempt to produce a DTM of a rectangular image area. Kirk et al. (2005) constructed such profiles across some of the putative cryovolcanic flows in the TA image but also quoted a cautionary example in which a seemingly plausible profile showing a smooth variation in elevation across a flow probably resulted from gradational backscatter variations rather than actual topography. Sharply bounded flows, which could be modeled with greater confidence, had heights ≤300 m and bounding slopes of ≤10°. The mountains seen in T3 and later images were clearly more appropriate targets for radarclinometry. Not only do they exhibit slope shading in the form of closely paired radar-facing bright slopes and away-facing dark slopes, the bright faces appear foreshortened as a result of the geometry of image formation (Fig. 5.11), just as in RADAR images of terrestrial mountains.

Preliminary profiles of isolated mountains in T3 yielded low heights averaging ~300 m (Radebaugh et al. 2006), but these values were later roughly doubled by Radebaugh et al. (2008b) who argued that a diffuse scattering law fits both Xanadu and the isolated mountains on Titan better than the more slope-dependent function that they had used initially; the weaker dependence of backscatter on incidence angle necessitates steeper slopes and greater heights to account for the same image data. Most mountains imaged in subsequent flybys also turned out to be larger, horizontally as well as vertically, than the isolated peaks in the T3 image, though still typically measuring <20 km across the base. Finally, Kirk and Radebaugh (2007) used simulated data to demonstrate that the radarclinometry results for many of Titan's mountains are affected by the limited resolution of the RADAR images. Because RADAR's resolution limits the apparent foreshortening of the bright slope, it distorts the modeled photoclinometric profile, but Kirk and Radebaugh (2007) showed that the away-facing (dark) slope is nonetheless recovered accurately and can be used to calculate the corrected height.

The influence of radar speckle noise on reconstructed topographic profiles can be estimated by applying the radarclinometry algorithm to a profile extracted from a featureless plain containing no detectable geological features, merely speckle. For a typical case in which data are drawn from the T3 RADAR image, this process yields a profile with ~60 m RMS variations (~200 m peak-to-peak). Such noise-driven fluctuations could be mistaken for small mountains (typically <2 km across) but would significantly affect conclusions about heights only for the smallest mountains modeled. Radebaugh et al. (2007) examined adjacent profiles across a few mountains and drew similar conclusions about the reliability of the height results from the variability of such profiles. The accuracy with which the absolute scale of the mountains is determined is a separate issue. Radarclinometry results are model-dependent in the sense that the amplitude of the recovered topography depends on the scattering law assumed to apply to the surface. To quote an extreme example of the size of error that this may introduce: when a scattering law appropriate to Titan as a whole (Radebaugh et al. 2006) was used early on, it led to heights a factor of 2 smaller than those obtained later by re-analysis based on a diffuse cosine (incidence) law (Radebaugh et al. 2007). The diffuse scattering shown by Xanadu and some isolated mountains imaged at a variety of incidence angles is well established, so realistic limits of the scale error for radarclinometry results are considerably tighter. For example, changing the diffuse scattering law from cosine (incidence) to cosine2 (incidence) (the maximum range of diffuse scattering characteristics estimated by Wye et al. 2007) reduces the inferred height of mountains by only 10%.

Taking such resolution effects into account, Radebaugh et al. (2008b) found mountain heights ranging from 100 to

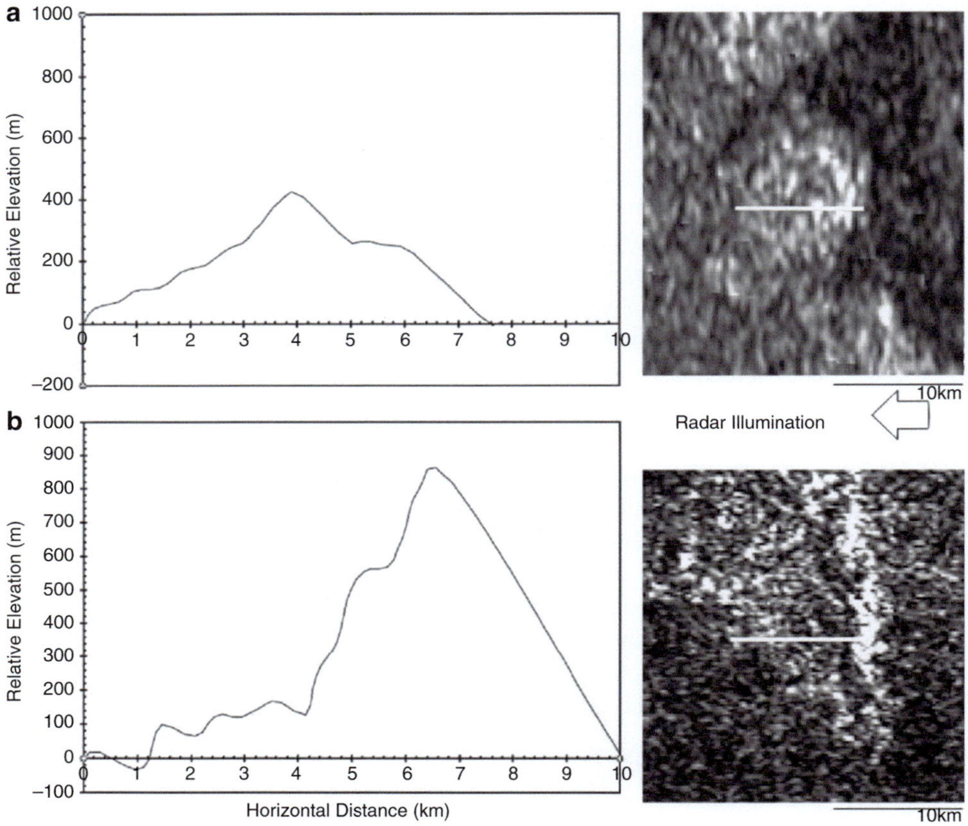

Fig. 5.11 Radarclinometric trace across two mountains in the T3 RADAR swath. *Line* through the mountain image denotes the pixel trace where data was obtained. *Graphs* show mountain heights to the same scale in the form of a cross section through the mountain, with vertical exaggeration. The higher mountain (b) shows marked 'leaning' towards the *left*, due to understeepening of the bright-side slope from resolution effects (see text for discussion) (adapted from Radebaugh et al. 2007)

2,000 m (Fig. 5.12), with values near the upper end of this range apparently common in the rugged terrain of Xanadu.

Maximum slope inclinations on the mountains ranged widely, from <10° to 60° in a few cases, with a mean around 30°. Although their relief is small by terrestrial standards, these were the tallest and steepest features found on Titan at the time, justifying the use of the term 'mountains' rather than merely 'hills'.

Lorenz et al. (2006a) used the same approach to radarclinometry to construct profiles across dunes in the Belet sand sea. This application is potentially problematic because the underlying surface is visible between the dunes in many places (e.g. the dunes and interdunes show as dark and bright stripes in images in which the direction of illumination is along the dune crest, so that no topographic shading is possible), and the difference between its backscatter cross-section and that of the dunes themselves could be misinterpreted as a topographic slope. Lorenz et al. (2006a) addressed this issue by selecting an image area where the dunes were especially well resolved, and the bright facing slope and dark back slope, both narrow, could be distinguished from the uniform interdune areas, and where the latter were distinctly darker than the surrounding terrain, suggesting they were at least partially covered by dune material. Dunes meeting these criteria, with a typical separation of 2.5–3 km, were found to have trough-to-crest heights of ~150 m. Thus, these dunes are close in both their horizontal and vertical dimensions to similar forms on Earth (Bagnold 1941; Lancaster 1982, 1995; Lorenz et al. 2006b). Careful selection of dunes with prominent bright slopes and minimal interdune contrast should minimize systematic errors in a profile by the same token, it may result in the selected dunes being atypically large.

An entirely different approach to radarclinometry has been applied to Ganesa Macula by Neish et al. (2008). Ganesa is circularly symmetric in outline and has been compared to the (considerably smaller) steep-sided domes on Venus (e.g. McKenzie et al. 1992). Unfortunately, the intrinsic backscatter cross-section of the feature is highly variable, so that no single profile with even approximately constant backscatter properties can be found. Neish et al. (2008) therefore attempted to model Ganesa by postulating a normalized shape

Fig. 5.12 (**a**) Maximum mountain heights as estimated by radarclinometry (using a diffuse backscatter law) on the vertical axis plotted against the heights of the same mountains corrected for the limited resolution of the images based on assumed symmetry (Kirk and Radebaugh 2007). Mountains in theT3 crater region are lower than those found in other regions, and because they have smaller slopes, they also plot closer to the diagonal line on which the raw resolution-corrected heights are equal. (**b**) Numbers of mountains found at those preferred resolution-corrected heights are shown below on a similar horizontal scale (from Radebaugh et al. 2007)

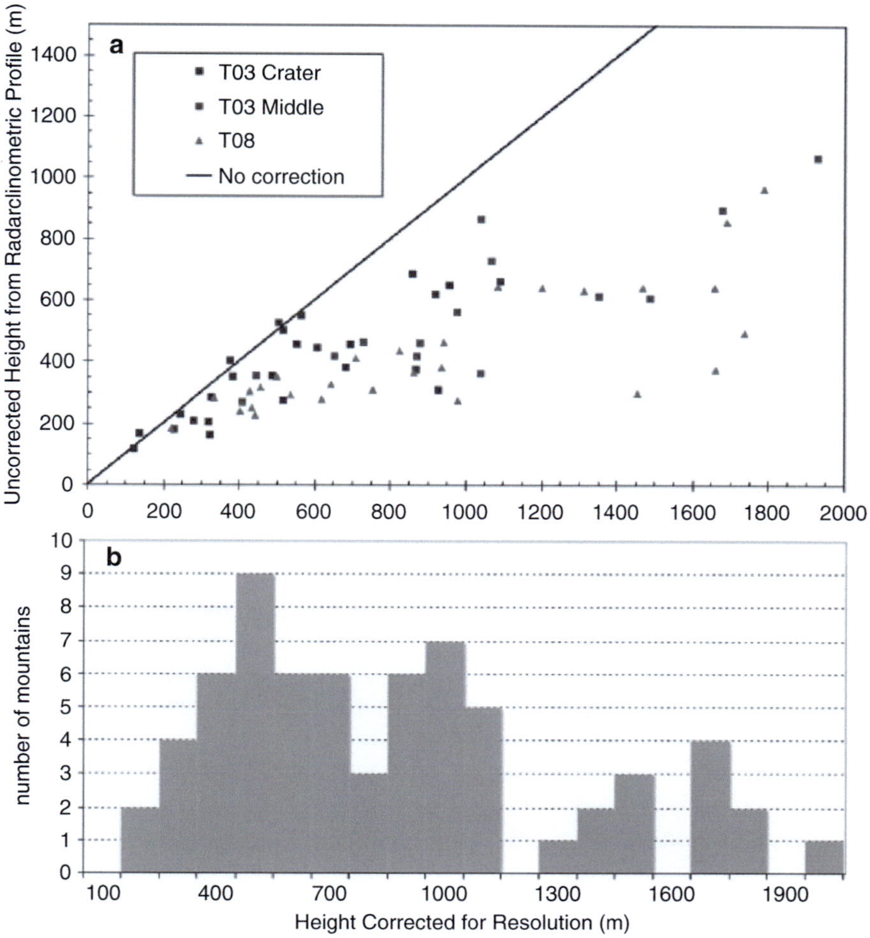

(i.e. a profile of height relative to the maximum as a function of distance from the center) and identifying the height that best fitted the TA image data. Best-fit heights varied from 2.0 to 4.9 km, depending on the shape that was assumed, with a conical shield model giving both the largest height and the best fit, though not so much better as to rule out other, more steep-sided shapes.

Altimetry, the techniques discussed so far provide, at best, information about the relative heights of features within a local area. The RADAR instrument obtains measurements of absolute height in altimeter mode, typically at ranges of 4,000 –10,000 km before or after closest approach, yielding an elevation profile ~600 km long in each case (Elachi et al. 2004). The intrinsic uncertainty in range measurements is ~35 m, and ranges are relative to the spacecraft trajectory, which is believed to be reconstructed with an accuracy of ~100 m. Samples are obtained as often as every km along track, but each sample is influenced by the distribution of elevations in a footprint that ranges from 20 to 50 km in diameter. Zebker et al. (2009) described how systematic errors also arise from the curvature of Titan, off-nadir pointing, and the interaction of these effects with the local distribution of elevations, so that bias corrections based on detailed simulations of the altimetry process are necessary in order to get consistent elevation estimates from different observation geometries. The first 18 altimetry profiles reveal a remarkably small range of elevations, remaining almost everywhere within 500 m of the reference radius of 2,575 km. The global distribution of high and low elevations will be described below in conjunction with the 'SARTopo' data-set. Considered individually, some profiles are remarkably flat, with <100 m of relief over hundreds of kilometers, while others include both flat segments and elevated regions several hundred meters high and typically 100–200 km across. Figure 5.13 shows 'radargrams' or plots of the echo strength as a function of delay and along-track distance that illustrate these extremes of flat and rugged topography.

The much greater echo dispersion toward the end of the T19 profile (Fig. 5.13b) is noteworthy; this is evidently rugged terrain that contains depressions below the level of the rest of the profile as well as peaks rising above the plain. It corresponds to bright terrain in ISS images, whereas the flatter early part of the profile is in infrared-dark terrain. Relating smaller topographic variations to geomorphic features

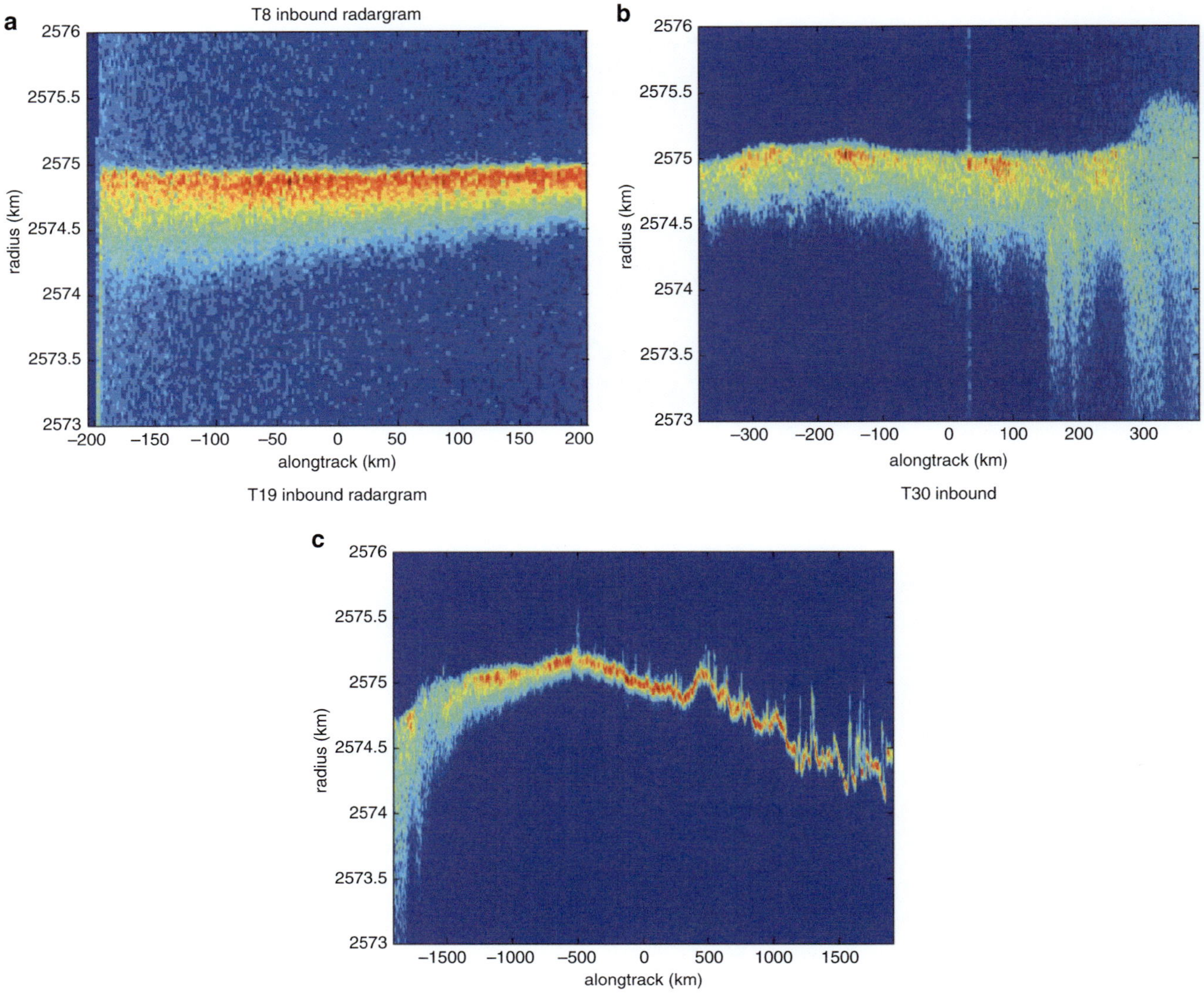

Fig. 5.13 'Radargrams' (Zebker et al. 2009) illustrating results of the Cassini RADAR altimeter mode operation from three representative Titan flybys. *Horizontal coordinate* is along-track distance, *vertical coordinate* corresponds to elevation (absolute planetary radius, with 2,575.0 km the reference radius for Titan), and *color-coding* indicates relative signal strength from *blue* (weak) to *red* (strongest). (**a**) Data for inbound segment of T8 flyby show an extremely flat surface with almost no variation in elevation and a mean near 2,575.0. Apparent narrowing of the echo to the right is an artifact of decreasing range to target, which shrinks the altimetry footprint and decreases the dispersion in heights. (**b**) Outbound portion of T19 flyby, for which range and apparent dispersion increase to the right. Topographic variations of several hundred meters are evident, particularly near the right end of the profile. The increase in echo depth and height is noteworthy and probably indicates very rugged terrain, with lows below the level of the surrounding plains, as well as elevated peaks. (**c**) Inbound portion of T30 flyby. Note different horizontal scale. In this instance, altimetry was obtained up to closest approach, resulting in a long profile with high resolution that resolves numerous local topographic features with amplitudes of several hundred meters. Many of these local features are visible in the T28 RADAR image which overlaps the profile, but the long-wavelength elevation changes do not correspond to identifiable variations in the images (adapted from Zebker et al. 2009)

is difficult because most of the altimetry profiles do not overlap high-resolution SAR images. An important exception is the T30 dataset, which was deliberately extended through closest approach to provide a ~4,000-km-long profile with footprints as small as 6 km (Fig. 5.13c), most of which overlaps the T28 SAR image. Many individual peaks in the elevation profile can be identified with ridges in the image, but there are no evident changes in image tone or texture that correspond to the long-wavelength elevation changes (Zebker et al. 2009).

'SARTopo', two main shortcomings of the altimetry dataset, the shortness of most profiles and the lack of SAR image context, have been addressed by developing a novel topographic mapping technique during the mission which produces elevation profiles from unpaired Cassini SAR images (Stiles et al. 2008). All SAR imaging, including that of Cassini,

works by locating features along-track by their Doppler shift time history, and across-track by their range to spacecraft, which is influenced by both their cross-track ground coordinates and their elevation. Cassini's RADAR obtains images that are quite wide compared to the antenna beam pattern, so features can also be located in angle by using the decreasing sensitivity on each edge of the beam as a reference. With only a single beam, this approach would only work for a target of uniform brightness. Fortunately, RADAR uses five beams to obtain parallel image swaths and build up its full-width image. Where two adjacent beams overlap, it is possible to compare the signal strength of a feature in each of them, and thus determine its angular position regardless of its intrinsic brightness. From the angle and the range, horizontal and vertical coordinates can then be calculated. A simplified way of describing this process is that the non-uniform beams 'paint' a pattern of stripes onto the surface of Titan at fixed angles relative to the spacecraft trajectory, and the range-measuring function of SAR imaging effectively 'views' these stripes from a vantage point on the other side of the image swath, which makes their distortion by surface topography visible. Thus, the technique is directly analogous to the 'structured light' methods (e.g. laser striping) used in three-dimensional modeling (DePiero and Trivedi 1996). In practice, only 1 to 3 of the overlapping beam pairs yield useful information, but they each produce a topographic profile over the full length of the image (excluding extremely dark areas such as lakes), with a horizontal resolution of ≤10 km and a vertical precision of ~100 m. Absolute accuracy depends on the accuracy of both the spacecraft trajectory and the pointing of the antenna; the pointing errors are dominant but are at least partially corrected for by a global bundle adjustment in which pointing biases for each flyby are selected so as to minimize the mismatch where different profiles cross. No special conditions apply to data acquisition, so 'SARTopo' profiles can be constructed from images from the entire mission. However, confidence in the method was greatly increased by the availability of the long T30 inbound altimetry segment which coincides with the T28 SAR image. Where the T30 altimetry and T28 SAR topography ground tracks were close together compared to their footprint sizes, the elevations obtained by the two methods were found to be extremely well correlated (Stiles et al. 2008).

As shown in Fig. 5.14, the 'SARTopo' dataset enormously increases the amount of information available about absolute elevations on Titan compared to altimetry, although the distribution of both kinds of data is dictated by the tour design and is less uniform than would be ideal.

Figure 5.14 also shows reasonably good agreement between altimetry and 'SARTopo' elevations and excellent agreement (at the 100–200 m level) where pairs of 'SARTopo' profiles cross. Perhaps the most noteworthy point is that this global dataset contains variations of less than ±1,500 m relative to the nominal radius of 2,575 km. This is similar to the range of local relief described above, but it is important to note that measurements of local relief do not constrain the long-wavelength departures from a spherical shape to be small, as they apparently are. The north polar region does appear to be systematically several hundred meters lower than the equatorial zone, but the nature of this discrepancy and its implications for the distribution of lakes on Titan are only gradually becoming clear as additional SARTopo profiles covering more of the body are obtained. Present results are clearly indicative of global flattening but inconclusive as to the relative elevations of the northern and southern hemispheres (Zebker et al. 2009). Figure 5.15 shows higher-resolution 'SARTopo' profiles across several features of particular interest.

These results were unexpected and somewhat puzzling, in that surprisingly little relief is associated with some of the most dramatic features on Titan. In part, this reflects the overall low relief of Titan, but in several cases the pattern as well as the magnitude of the topography was surprising. Ganesa Macula does not obviously display a domical or shield shape as expected; it appears to slope upward gently from west to east, with a deep trough outside the eastern edge. A similar pattern of gentle western and steep eastern slopes seems to be repeated several times across this area. Menrva, which is unequivocally an impact crater, also has surprisingly little relief, given its ~450 km diameter. The rim is only ~300 m high. The center of the crater and surrounding ring are elevated above the terrain outside the crater, and are surrounded by a shallow moat that lies below the exterior terrain, but the total range of elevations is <700 m. This pattern of relief is consistent with the theory that the long-wavelength topography has been removed by viscous relaxation. Finally, Xanadu, which before the Cassini mission had been postulated by Griffith et al. (1991)), Smith et al. (1996) and others to be an elevated 'continent,' although rugged and sloping regionally toward the east, is also remarkably flat, with only a few hundred meters of relief. There is no abrupt elevation transition at the sharp western boundary that would explain the termination of the Shangri La dune field here. Within Xanadu, individual peaks in the 'SARTopo' profiles can be identified with major mountain ranges, but the apparent range of topography is much less than the 2 km locally obtained by radarclinometry. This is most likely a resolution effect; with its footprint of ~1 0 km, the 'SARTopo' averages over much of the local relief on a scale of 2–10 km. Thus, Xanadu appears not to be elevated on average, but it is nonetheless rugged and contains extremes of both high and low topography compared to the surrounding dune fields. Additional 'SARTopo' profiles across Xanadu in a northwest-southeast direction, based on images obtained at the end of the prime mission, suggest a modest southward slope for at least the western portion of Xanadu (Stiles et al. 2008; Radebaugh et al. 2009). The apparent pattern of southward-draining channels seen in the

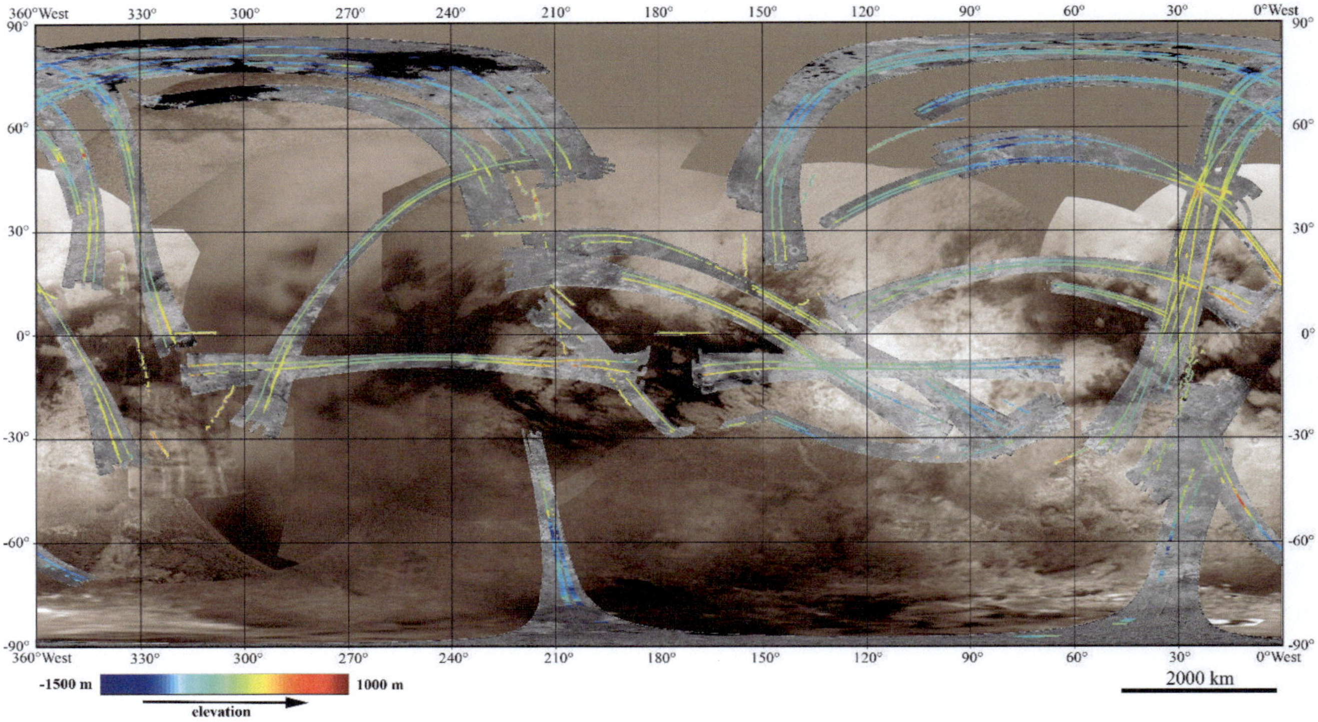

Fig. 5.14 Global mosaic of RADAR images through T29 with color-coded elevation profiles obtained by 'SARTopo' processing (Stiles et al. 2009) as described in text. Simple cylindrical projection, north at top, centered on 180° longitude

T13 image suggests that the topographic lows (valley floors) dip to the south, and 'SARTopo' unfortunately can shed no light on this question because it averages over the local valleys and peaks. In many other locations, however, the 'SARTopo' data are more readily interpretable (Stiles et al. 2008). Patches of rugged, dissected terrain outside Xanadu, for example, are shown to be elevated by hundreds of meters. Both small lakes and large seas are associated with local topographic lows, and the lakes appear to sit ~300 m above the seas in absolute elevation.

Radargrammetry, the geometric analysis of stereo SAR images by methods analogous to photogrammetry, is the final source of topographic information to be discussed. The typical geometry of these images leads to a local precision of relative heights on the order of 100 m if features can be matched with a precision of one pixel (175 m), which seems to be possible in the highest-resolution parts of the images. Because the RADAR image formation process does not depend on the orientation of the spacecraft, a major source of error in photogrammetry, absolute elevations should also be obtainable, with accuracies comparable to the spacecraft's position accuracy of ~100 m. Before the T18 flyby late in 2006, no overlapping SAR images were available, but a quasi-stereoscopic measurement of the depth of the Sinlap crater was possible under the assumption of symmetry. Under this assumption, the apparent foreshortening of the front and back walls of the depression was compared, yielding an estimated rim-to-floor depth of 1,300 ± 200 m for the 81 km diameter crater (Elachi et al. 2006). After T18, nearly all SAR images overlapped one or more previous swaths, and the amount of potentially useful stereo imagery accumulated rapidly, as shown in Fig. 5.16.

Not all of these overlaps are equally useful for topographic mapping, both because of the varying resolution of the images as a function of the distance from the closest approach, and because it is a fundamental characteristic of RADAR imaging that the viewing and lighting geometries are linked. Image pairs with parallel ground tracks and illumination from the same side are thus the easiest to find feature correspondences in but provide the least amount of stereo parallax (sometimes none). Opposite-side viewing provides very strong stereo convergence, but matching features can be difficult because, while intrinsic scattering variations are similar in the two images, slope shading is reversed (Plaut 1993). Images that cross at a wide angle are intermediate in their properties but generally more similar to the opposite-side case. Even viewing opposite-side or crossing stereo pairs for qualitative geological interpretation can be quite difficult unless the terrain is so subdued that shading effects are minimal; fortunately, given the prevalence of high-angle and opposite-side pairs, this is quite often the case on Titan. An additional obstacle to stereo mapping, and

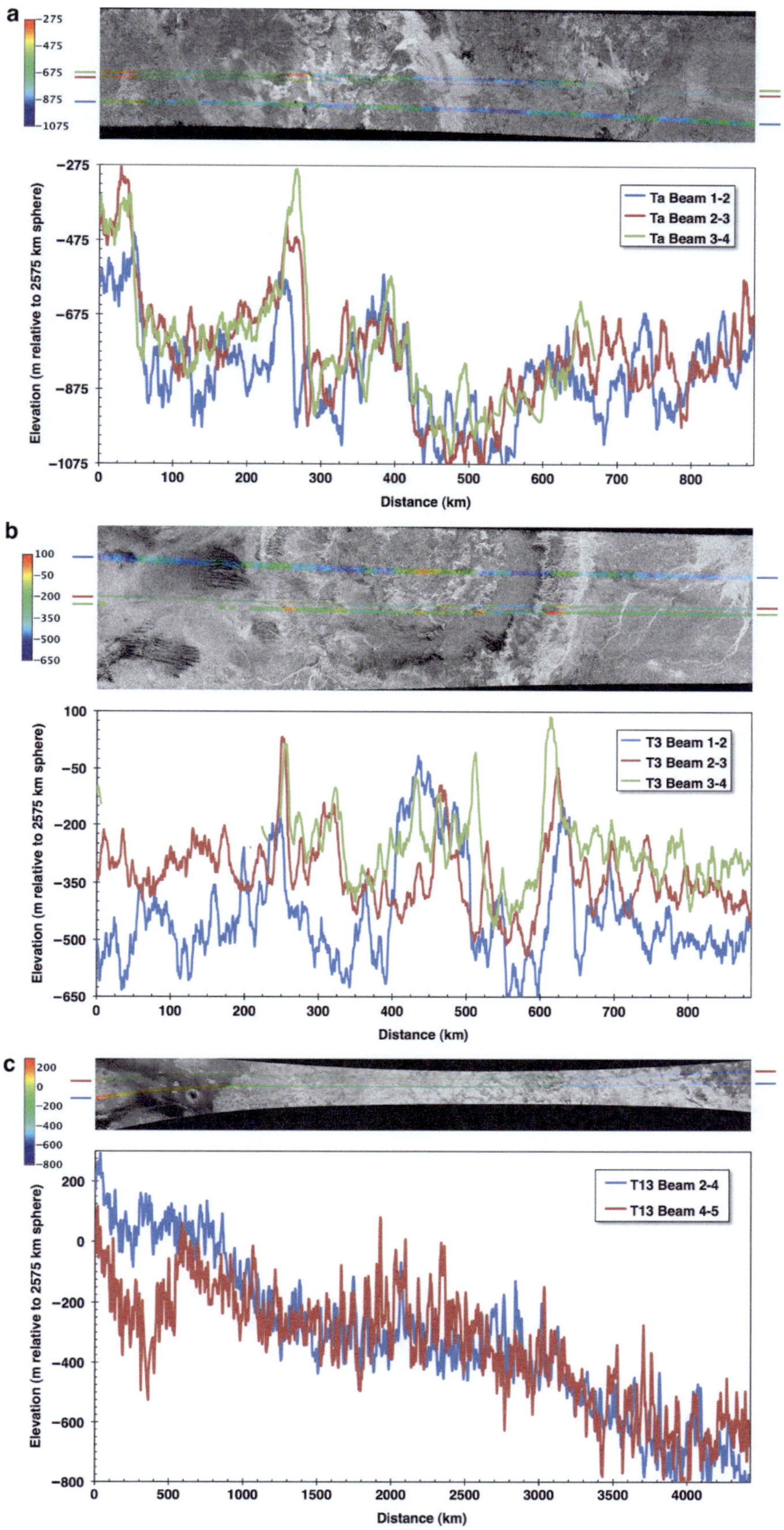

Fig. 5.15 Examples of 'SARTopo' profiles over specific features of interest. (**a**) Ganesa Macula TA. (**b**) Menrva crater, T3. (**c**) Xanadu Regio, T13. The *blue*, *red*, and *green elevation profiles* correspond to the color-coded elevations overlaid on the images as indicated by the matching colored tick-marks in the margin (data from Stiles et al. 2009)

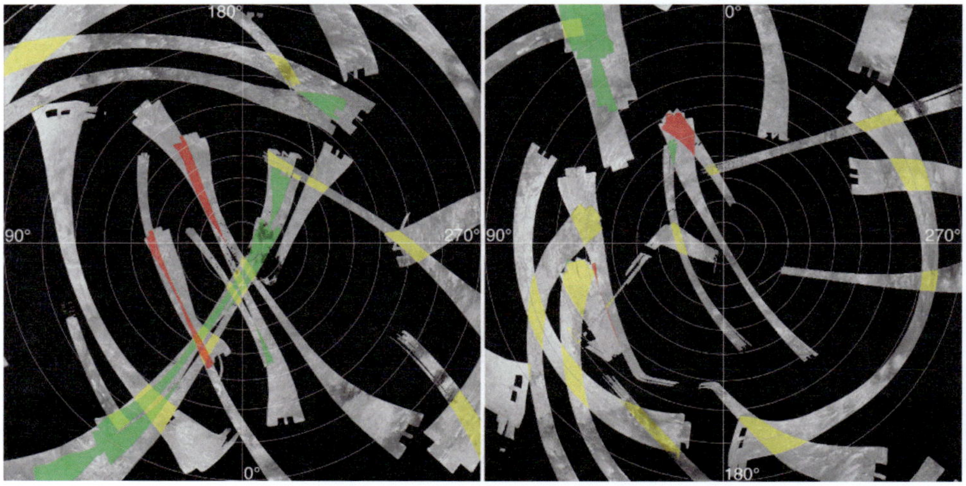

Fig. 5.16 Global mosaic of RADAR images through T50 with stereo overlap areas color-coded according to geometry: *green* for same-side viewing, *yellow* for high-angle crossing, *red* for opposite-side viewing. Total coverage by full-resolution SAR images to date is 29.5%, yielding overlapping coverage for 2.1% of Titan. Additional coverage of ~10% of Titan by lower resolution images obtained at high altitude is not shown. Polar stereographic projections of the northern (left, 0° longitude at bottom) and southern hemisphere (right, 0° longitude at top), with parallels of longitude drawn every 10° (from Kirk et al. 2009)

particularly to obtaining absolute elevations, was the discovery that the pre-Cassini model of Titan's spin-axis orientation and rotation rate was considerably in error, leading to misregistrations of up to 30 km between overlapping SAR images (Stiles et al. 2008). Initial work with RADAR stereo therefore focused on sets of images (T16–T18–T19 and T25–T28–T29) obtained within a short interval at the same orbital phase so that the errors induced by the rotation model would be minimized.

Initial radar stereo topographic results for Titan were based on automated and manual measurements of the relative parallax between corresponding features, followed by converting parallax to relative height using relatively simple formulae based on the incidence angles of the two images (Plaut 1993). A DTM of the T16–T19 overlap, produced by adapting matching software from the Magellan mission (Hensley and Schafer 1994), showed total relief of ~1,000 m, mostly associated with an arcuate set of scarps in the southern part of the overlap that may be remnants of a highly degraded impact crater (Kirk et al. 2007). The DTM did not fully resolve details of the small lakes in apparently steep-sided depressions (Stofan et al. 2007) at its northern end, so the authors supplemented it with manual measurements of lake shores, scarps, and other discrete features. With a few exceptions, the shorelines of lakes in this area were found to be within <100 m of the same elevation, consistent with the hypothesis that they are coupled by subsurface flow in an 'alkanofer' (the hydrocarbon equivalent of an aquifer; Hayes et al. 2008). The basins in which lakes were located were found to be a few hundred meters deep, 600 m at most. Unlike the lake levels, the rim elevations of different depressions were not found to be uniform. These results agree generally with those from altimetry (Hayes et al. 2008) and 'SARTopo' (Stiles et al. 2008).

Both Hensley at JPL (Kirk et al. 2008) and the USGS group (Kirk et al. 2009) subsequently developed rigorous sensor models for the Cassini RADAR, which can be used to calculate absolute horizontal and vertical coordinates from matched image features, taking proper account of the full 3D geometry of the images so that accurate results can be obtained even for crossing image pairs. The USGS sensor model is incorporated in a commercial stereo-mapping software system which allows spacecraft trajectories to be adjusted so as to bring stereo data into better agreement with other data, such as altimetry or 'SARTopo', and permits editing DTMs interactively based on a stereoscopic display in addition to automated image matching. Kirk et al. (2009) describe a total of six DTMs produced by their approach. The largest, from the T25 and T28 images (Fig. 5.17), covers part of the north polar region, with several of the seas in the eastern half and numerous steep-sided depressions not filled by lakes in the west. A basic but crucial result was that the putative dark seas have the lowest elevations, consistent with the interpretation that they are seas of liquid hydrocarbons. Shoreline elevations are constant within the uncertainty of the measurements, and the highest points are ~1,200 m above the shore. This strongly suggests that the lakes are at most a few hundred meters deep, and probably shallower (Lorenz et al. 2008b). The depressions in the western half of the DTM are typically 300–500 m deep, and thus are comparable

Fig. 5.17 Topographic map of part of Titan's north polar seas, from T25 and T28 RADAR images, in polar stereographic projection with center longitude 295°W. The north pole is located above the large island, Mayda Insula, at center right. Dark areas along the southern edge of the map are believed to be arms of a single large sea, Kraken Mare. A second sea, Ligeia Mare, is located to the northeast of the right end of the map. (**a**) Orthorectified T28 image. Radar illumination direction is approximately from the bottom. (**b**) Orthoimage color-coded with elevation relative to the 2,575-km reference sphere. (**c**) Orthoimage color-coded with adirectional (downhill) slope in degrees over a 5-km baseline (from Kirk et al. 2009)

to the lake-filled basins measured by Kirk et al. (2007). Slopes are notably low, <5° over 5 km and less over longer distances, even in this rugged-appearing part of Titan.

Kirk et al. (2009) spent considerable effort on validating this DTM, and found good agreement (<100 m in absolute height and RMS difference) between it and collocated SARTopo data with the exception of a few areas where the SARTopo measurement is biased toward brighter (and possibly higher or lower) terrain on one side of the nominal profile location. The stereo results also agree closely with radarclinometry in areas of uniform scattering properties; elsewhere, backscatter variations distort the radarclinometry profiles significantly. The DTM also corresponds well to the morphology revealed in the images, and appears to resolve features as small as 5 km across.

A surprising result of this investigation was the apparent success of the automatic image-matching algorithm in very dark portions of the seas, though not in the darkest and completely featureless areas. The DTM indicates a few hundred meters of relief in areas believed to be liquid-covered. This is inconsistent with laboratory measurements of the microwave absorption coefficient of liquid hydrocarbons by Paillou et al. (2008), which imply that bottom features would not be visible through lake areas deeper than about 7 m. Some of the DTM results in the darker parts of the seas may be the result of spurious matching between speckle noise in the images, but portions with backscatter cross-sections between −10 and −15 db (e.g., around the northern sides of Mayda Insula) clearly show channels and other features common to both images of the pair, and manual measurements of these features confirm that their elevations vary by several hundred meters. This apparent contradiction has not yet been explained satisfactorily. One possibility is that the moderately-dark areas are not in fact liquid-filled seas of varying depth, but are relatively smooth, perhaps swampy, exposed surfaces of varying texture. Another possibility is that the laboratory results represent only an upper limit on the absorptivity of liquid hydrocarbon, and that Titan's seas are sufficiently transparent that the RADAR is indeed seeing the bottom through a few hundred meters of liquid.

Kirk et al. (2009) also report that topographic rises of only about 200 m are sufficient to stop or divert dunes in the equatorial "sand seas." This result suggests the dunes themselves are no more than 200 m high, and direct measurements of a few of the largest dunes yield heights of ~150 m. Heights of hills near the Huygens landing site agree well with the DISR stereo results described above, and the RADAR DTM appears to show a gentle (<0.1°) eastward slope in the valley in which the probe landed. This is consistent with the fluvial geology described in the following section. A RADAR DTM of Ganesa Macula (Fig. 5.18) is entirely consistent with the SARTopo profiles (Fig. 5.15) but provides a much clearer picture of the overall pattern of relief. Ganesa Macula is not consistently or centrally elevated as would be expected for a volcanic dome or shield; points on the eastern rim are high but the western rim is low. The overall pattern of alternating highs and lows with north–south trends is suggestive of tectonic activity. Leilah Fluctus and other radar-bright channels and fans to the east of Ganesa are probably rough fluvial deposits, and are consistently lower

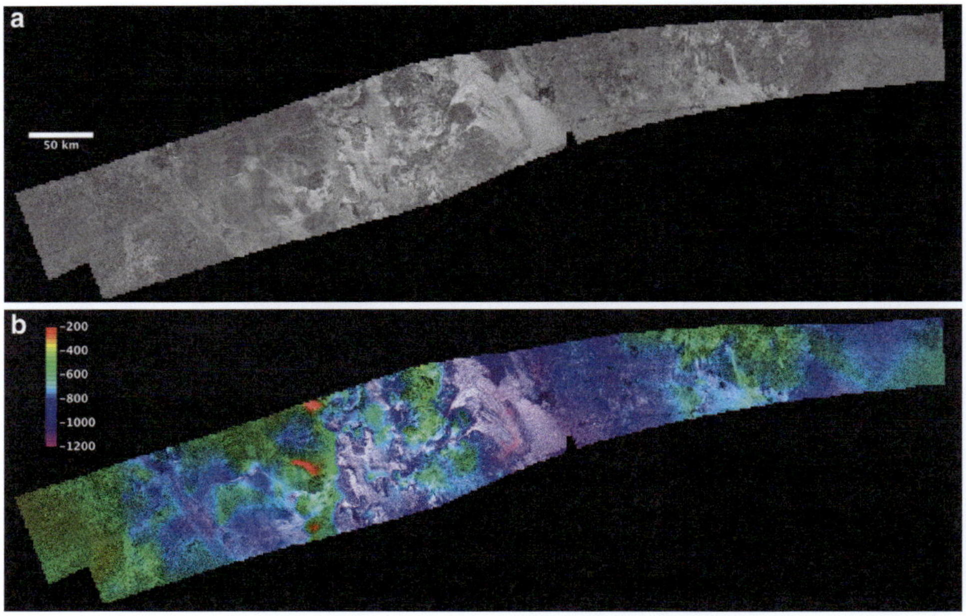

Fig. 5.18 Topographic map of Ganesa Macula (*circular feature at left*) and Leilah Fluctus (*bright fans at center*) from TA and T23 SAR images. Equirectangular projection, center latitude 51.8°N, north at top. (**a**) Orthorectified TA SAR image. Radar illumination is approximately from bottom. (**b**) Orthoimage color coded with elevation (from Kirk et al. 2009)

in elevation than radar-dark materials, indicating significant erosion and redeposition have taken place. Thus, whatever its origin, Ganesa Macula seems clearly to have been extensively modified.

Stereo topography in the vicinity of Hotei Arcus (Fig. 5.19) lends support to the interpretation by Wall et al. (2009) that the lobate features seen here in the SAR images are cryovolcanic flows. These features are clearly elevated above the radar-bright plain on which they sit. Kirk et al. (2009) infer a viscosity on the order of 10^4 Pa s from the width and 100–200 m thickness of these flows, based on the model of Hulme (1974). In contrast, the narrow, radar-bright channels emerging from the rugged mountains on the southeastern border of the DTM show little elevation change along their length, consistent with a much less viscous working fluid such as pluvial/fluvial liquid methane.

Both the USGS and JPL groups are in the process of extending their stereo DTM production to as many of the image overlaps as possible. Accurate georeferencing of the images, based on improved models of Titan's rotation, are crucial to this process. Kirk et al. (2009) report that the interim rotation model, used for flybys after T30, is incompatible with the Stiles et al. (2008) model used for TA-T30. Fortunately, the accumulating set of SAR image overlaps is being used to improve the models of Titan's rotation so they are consistent over the full mission, as well as to make stereo DTMs. Important additional results concerning the topography of Titan can thus be expected in the near future.

5.4 Geology at the Huygens Landing Site

Little was known about the surface before the arrival of Cassini/Huygens in 2005. Multispectral and RADAR observations on approach and during the first few flybys provided tantalizing and in some cases, puzzling views of the surface (Porco et al. 2005; Barnes et al. 2007a; Elachi et al. 2005). Cassini provides global coverage but cannot resolve objects and structures smaller than some hundreds of meters in size. Therefore, the close-up views that Huygens' descent imager/spectral radiometer (DISR) would provide were eagerly anticipated.

Huygens was projected to land on Titan's anti-Saturnian hemisphere just below the equator on a boundary between bright and dark terrain. The DISR three-camera combination (high resolution imager HRI, medium resolution imager MRI, and side looking imager SLI (Tomasko et al. 2002) pictured the surface from the near-nadir angle (8°) to above the horizon (96°), providing a view not unlike that of a passenger flying on an airplane. The DISR revealed a landscape that is amazingly Earth-like. The panorama in Fig. 5.20 shows the landscape

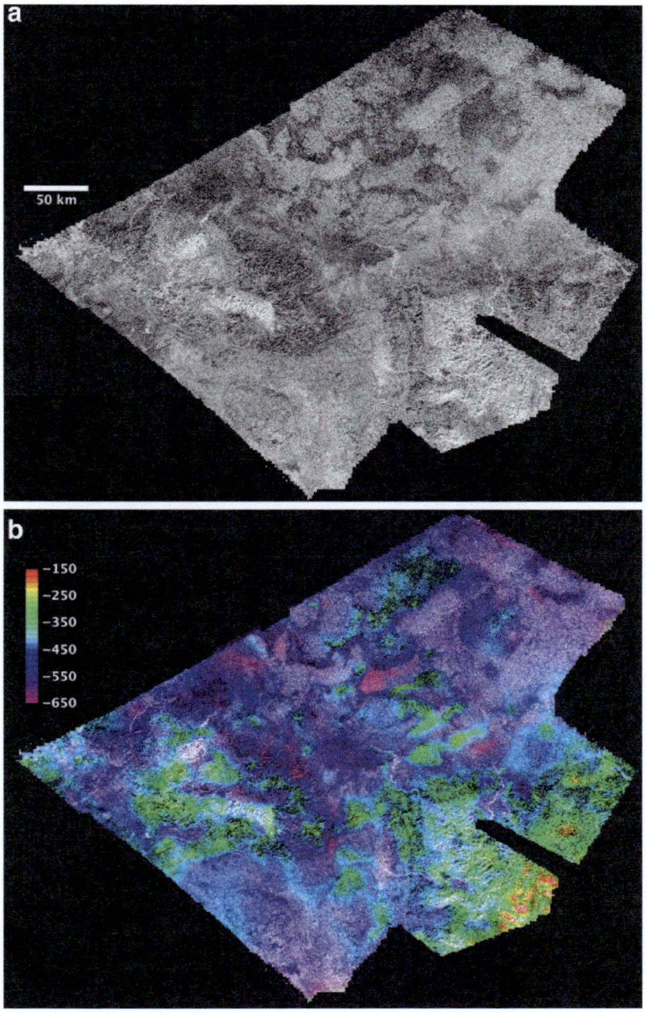

Fig. 5.19 Topographic map of part of Hotei Arcus, from T41 and T43 SAR images. Equirectangular projection, center latitude 29.9°S, *north at top*. (**a**) Orthorectified T41 SAR image. Radar illumination direction is approximately from the *upper left*. (**b**) Orthoimage color coded with elevation. (from Kirk et al. 2009)

north of the Huygens landing site (HLS) from an altitude of 8 km (based on comparison with http://saturn.jpl.nasa.gov/multimedia/images/image details.cfm?imageID=1309).

A long 'coastline' separates a dark, dry 'lake bed' in the foreground from bright, rugged highland terrain that is incised by a dendritic system of channels. Emplaced on the lake bed are bright features with the appearance of sandbars. This is an image that could have been taken on Earth or in other words, one that shows a geologically very active surface. Huygens was clearly very fortunate to land on the border of two completely different terrains. Which are the geological processes that formed this exceptional landscape? And how does the HLS fit into the global geological scenario as derived from Cassini observations?

Before we address these questions, we will briefly discuss the available DISR data to explain their quality and limitations. An overview of the complexity of the data reduction is provided by Karkoschka et al. (2007). DISR mosaics are composed of individual HRI, MRI and SLI images that are mapped into a mosaic using a model for probe drift, spin, yaw, and pitch. A complicating factor is the influence of the atmosphere. First, due to the haze extending down to the surface, the latter was only visible from altitudes below 45 km (Tomasko et al. 2005), which limited the area mapped by the DISR to about 2,500 km^2. Second, atmospheric scattering increases image brightness and decreases contrast with increasing altitude. All images displayed in this chapter were highly stretched in contrast to make features visible. In addition, a brightness gradient was introduced from the nadir towards the horizon, which needs to be removed. Proper correction requires a detailed characterization of the optical properties of Titan's atmosphere (Tomasko et al. 2005). The maximum image contrast found after correction is less than 20%, indicating remarkably small variations in surface reflectance. Unfortunately, half of the images recorded by the DISR were lost due to the failure of one of Cassini's communication channels (Lebreton 2005). This limited the availability of images for topography reconstruction by the stereo method (Soderblom et al. 2007b). The DISR images were broad-band and sensitive primarily in the near-infrared, but surface color information can be retrieved using data from the downward looking visible spectrometer (DLVS) which covers the wavelength range from 0.4 μm to 1 μm, and the downward looking infrared spectrometer (DLIS), which operates between 0.85 and 1.6 μm.

5.4.1 The Huygens Landing Site Terrain

First, we will address the question of where on Titan's surface Huygens landed. The answer requires a topographic and geomorphological analysis of the HLS and an interpretation of images taken on the surface.

Given the DISR's limited field of vision, the difference in scale between the Cassini and DISR observations complicated the identification of the HLS. The DISR cameras provided clear images of the landing site and its surroundings over an area barely large enough to permit a comparison with the views offered by the instruments on board Cassini. ISS and VIMS have viewed the landing site many times, while the RADAR SAR has done two close and three more distant views of the LS: T8 and T41 full resolution and T13, T36, T39 at high altitude. The first VIMS images of the landing site are of relatively low resolution (Rodriguez et al. 2006) and only permit comparing large-scale features, while recent VIMS observations of the landing site (T47) provide resolutions of about 500 m/pixel (Jaumann et al. 2009). The resolution of the SAR and ISS images is sufficient for a more

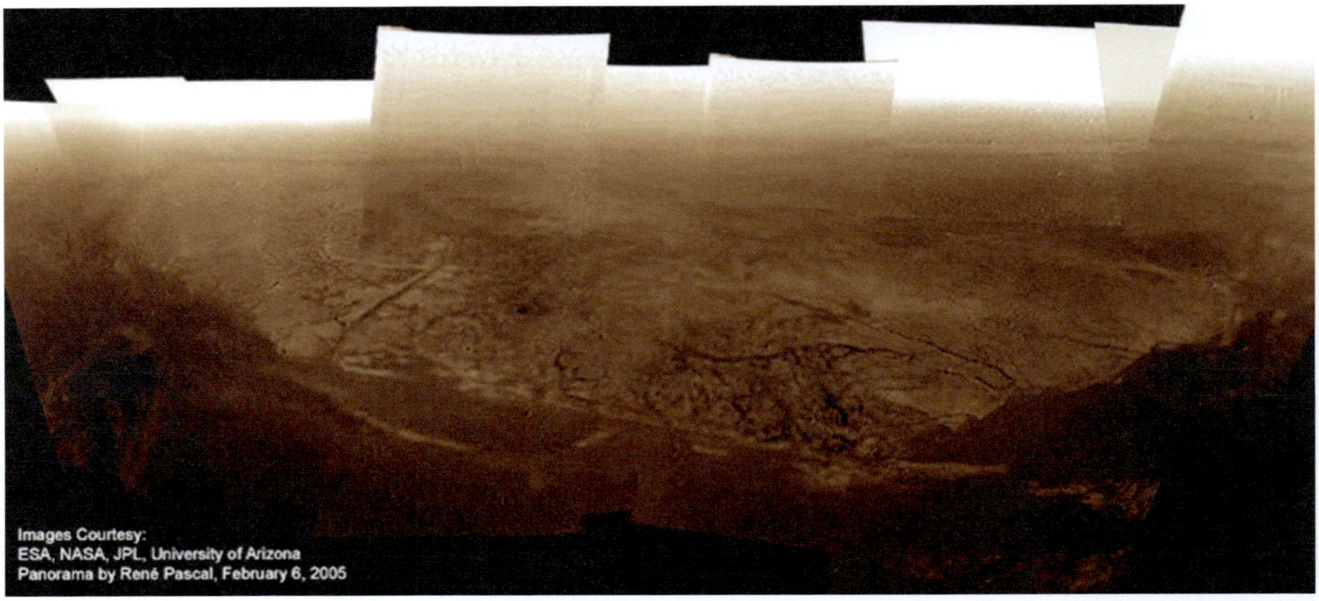

Fig. 5.20 Panorama view taken by the DISR cameras during decent. Viewing direction is towards the north

detailed comparison with the DISR mosaic provided by Karkoschka et al. (2007). Lunine et al. (2008a) located the landing site by matching the dark lines seen in the northernmost part of the DISR mosaic with the 'cat scratches', or dunes (Lorenz et al. 2006b) visible in the T8 SAR image. Huygens landed at longitude 167.6°E and latitude 10.2°S, with 0.1° accuracy (Karkoschka et al. 2007). Kirk et al. (2009) have updated these coordinates to 167.58°E 10.58°S by repeating the comparison but using a newly reprocessed version of the T8 SAR image that is consistent with a revised model of Titan's rotation (Stiles et al. 2008). The relation between the landing site and surrounding features seen in the SAR image is unchanged, but the coordinates are shifted by ~0.4°, mainly in latitude, because of the previously unsuspected systematic error in georeferencing the images. Figure 5.21 shows how the DISR mosaic fits in with the SAR and ISS images.

A detailed comparison between the SAR image and the DISR mosaic is provided in Fig. 5.22.

The good visual match with the ISS image and the dunes in the SAR image permits an accurate placement of the DISR mosaic. To match orientation, the DISR mosaic needs to be rotated 5° counterclockwise. However, this does not imply that the whole mosaic is off by 5°–10°. Whereas the outer parts of the mosaic may contain significant distortions with orientation off by as much as 5°–10°, the orientation in the center (within 5 km of the landing site) is probably good to 1°–2° (Keller et al. 2008).

The DISR mosaic ought to agree well with the ISS observations since the ISS filter wavelength (938 nm) is included in the DISR band pass, but there are some notable exceptions. For example, one would expect the terrain in the north of the DISR mosaic to be darker (Fig. 5.19). Most probably this is due to increased atmospheric scattering at higher off-nadir angles. The most noticeable difference between DISR and ISS data becomes visible in an area southeast of the landing site that appears dark to the DISR but bright to the ISS. This Dark Spot (DS; Fig. 5.22) is an area of high SAR brightness, as is the rugged terrain covered by rivers north of the landing site that was seen in detail by the DISR. A natural explanation for this high SAR brightness is that the river area and the DS area are roughly on the scale of the SAR wavelength (2.2 cm) and/or exhibit strong volume scattering.

The reason why these two river areas appear so different to the DISR lies in the difference in viewing and solar-phase angles. This becomes immediately evident when one takes a close look at the river area north of the landing site. Figure 5.23 shows how it appeared bright relative to the lake early in the descent, when it was observed at a low solar phase angle (22°).

Just before landing, when observed at higher phase angles (typically 65°), it becomes darker relative to the lake bed, most notably along the coastline and the rivers. This darkening appears to be restricted to the river area, and is not seen in the bright 'islands' off the coast, whose nature seems to be fundamentally different. The fact that the brightness of river and lake terrains varies differently with the phase angle means that they exhibit different degrees of non-Lambertian reflectance. This difference can be interpreted as a variation in surface roughness and shadow hid-

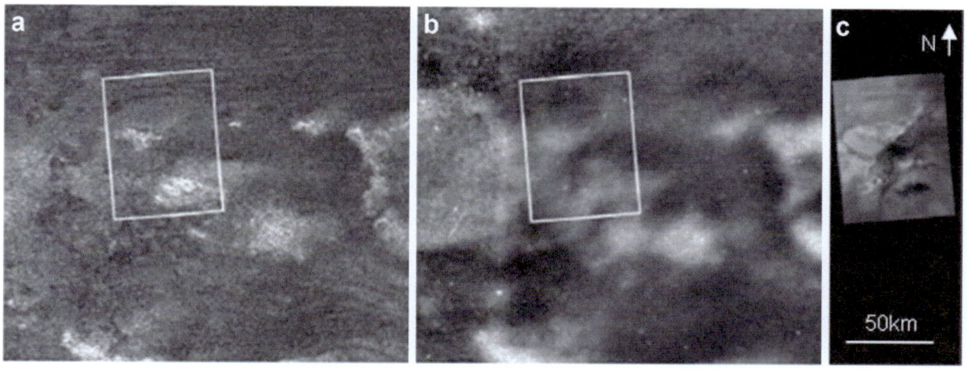

Fig. 5.21 Huygens landing site identified in the Cassini RADAR (**a**) and ISS (**b**) images. The rectangle outlines the location of the Huygens DISR mosaic (**c**), shown enlarged in Fig. 5.20b (Keller et al. 2008)

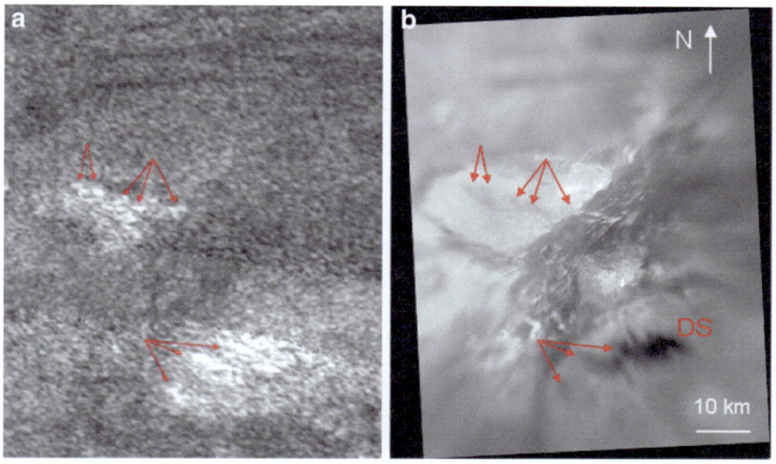

Fig. 5.22 Detailed comparison between the Cassini RADAR (**a**) and Huygens DISR (**b**) images of the landing site. The top arrows point at the river terrain seen in detail by the DISR, the bottom arrows point at the suspected river terrain seen by the DISR as a dark spot (labeled DS). The DISR mosaic (Karkoschka et al. 2007) is centered on the Huygens landing site. The mosaic is rotated by 5° counterclockwise to align the dunes in the north and to better match the dark spot with the radar-bright features in the south (Keller et al. 2008)

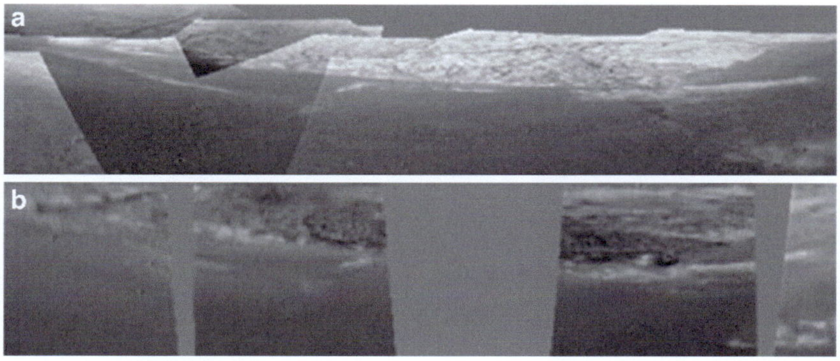

Fig. 5.23 The brightness of the river area relative to the lake bed depends on the solar phase angle. These two panoramas show the same stretch of coastline in perspective view (Mercator projection), but one (**a**) was recorded at high altitude (low phase angle) and the other (**b**) at low altitude (high phase angle). They are reprojected to the same viewpoint and a resolution of 1 km (adapted from Keller et al. 2008)

ing (e.g. Hapke 1981, 1984). Higher surface roughness causes brightness to decrease more swiftly as the phase angle increases, the implication being that the river terrain is rougher than the lake terrain, which is consistent with the RADAR observations (albeit on a different spatial scale: microns for DISR and ~2 cm for RADAR). The fact that the deepest darkening is associated with the coastline and the rivers themselves suggests that steep slopes also play a role. This hypothesis is consistent with the idea that the river terrain is an old and eroded (perhaps original) part of the crust, and that the lake bed is sedimentary in nature. The fact that the DS is also dark in the visible band when observed at solar phase angles between 90° and 105° but bright at the radar wavelength strongly suggests it is another river area. A river can be discerned in one of the SLI images. In the same way, Keller et al. (2008) identified the nature of the SAR-bright terrain in the vicinity of the landing site: it represents a hilly terrain crisscrossed by rivers. It is also bright at visible wavelengths, but only at low solar phase angles. SAR-dark terrain is flat and lies below the bright terrain. The conclusion is that Huygens landed on a floodplain between two river systems, one close by to the north, the second larger and further away to the southeast.

5.4.2 Topography and Geomorphology at the Huygens Landing Site

For interpreting the topography of the HLS, stereo images are essential. Due to the loss of half the DISR data, only a few images were left that could be used to create digital terrain models (DTMs). Luckily, the most robust set of images straddles the major bright–dark boundary between the bright rugged highlands with dark dendritic drainages that lie 5 km north of the HLS and the dark lower plains where Huygens landed, which exhibit evidence of fluvial scour (Tomasko et al. 2005). Figure 5.10 provides a quantitative scale for the digital elevations in the river terrain. The area of the region measures roughly 3×5 km. The integrated DTM is arbitrarily leveled because no surface datum exists. What is noteworthy is the ruggedness of the topography; in many places the slopes are near 30°, or about the angle of repose for most unconsolidated natural materials (e.g. Soderblom et al. 2007b).

There are two distinctly different kinds of drainage networks in the bright highlands. Both show elements of tectonic control to varying degrees, as does the coastline between the bright and dark regions, characterized in places by sharp, angular junctions between fairly linear segments. The first type of drainage is the intermediate-order integrated dendritic network mapped in Fig. 5.10. Perron et al. (2006) discussed these networks and compared them to analogous terrestrial systems. This type is prevalent in the right-hand side of the upper panel in Fig. 5.24.

These branching networks are at least of the fourth order (the number of tributary branches counted upstream from the boundary with the dark plain). In terrestrial systems, such dendritic patterns result from a distributed source, i.e. rainfall. In the case of Titan, rainfall is presumed to be either precipitating methane or a methane–ethane mixture in which, by virtue of the vapor pressures, methane dominates (Lorenz 1993b). Another interesting characteristic of these networks is that the DTM suggests a topographic divide (white dots in Fig. 5.10b) that separates two drainage systems: one that flows north and eastward and another south and eastward. In places the divide has been breached and the drainage flow of one has evidently been captured by the other.

In contrast, the drainage system on the left side of the same panel (Fig. 5.24, top) shows a very different pattern.

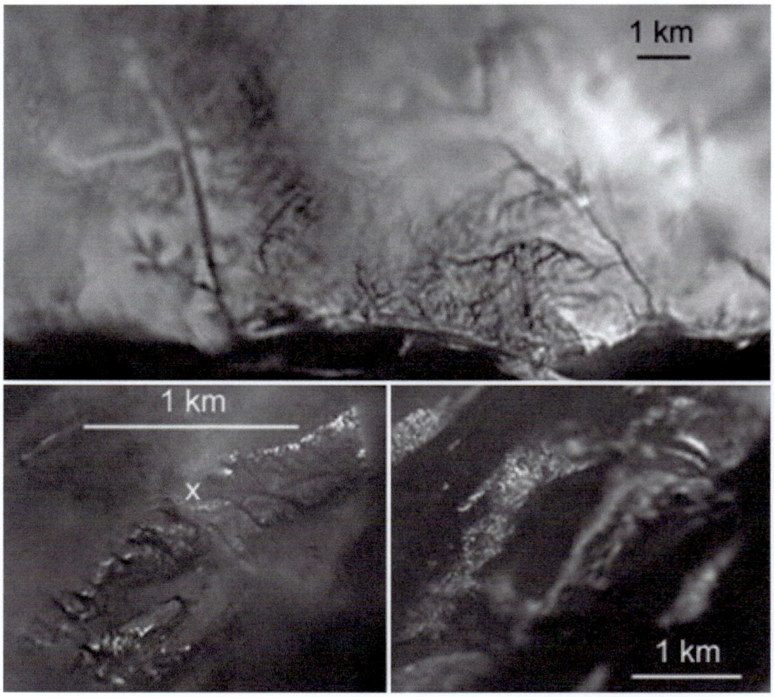

Fig. 5.24 Patterns of integrated drainage, flow, and erosional scour in the Huygens landing site region. *Upper*: two styles of drainage networks in the bright region 5–10 km north of the Huygens landing site (16 m/pixel). *Lower left*: high-resolution (2 m/pixel) view of the erosional channels around the Huygens landing site (marked with white x). *Lower right*: medium-resolution view (8 m/pixel) of bright ridges standing above a dark plain evidently carved by surface flow based on the stereo model of Fig. 5.10c (center ~4 km ESE of the landing site). Mosaics are from Karkoschka (2006)

This type exhibits lower-order branching (second and third order only) with short, "stubby" upper tributaries (Tomasko et al. 2005). Rather than rainfall, this is more characteristic of springs sapping at the heads of closed canyons, as seen in many arid climates on Earth (e.g. the southwestern United States). Although the eastern drainage basins exhibit several linear segments suggestive of tectonic influence, the lower-order kind on the west side of the panel appears much more tectonically controlled, evidently by linear fault patterns; these are delineated in Fig. 5.25.

Soderblom et al. (2007b) further speculated that a bright deposit of material may have been extruded into and onto this canyon system; in places, this bright unit forms linear patterns as if fault valley walls confined it. Perhaps this represents an eruption of fresh (i.e. cleaner) material from ammonia-water cryovolcanoes, a process believed to be probably operative on Titan's surface (Lewis 1971; Stevenson 1992; Lopes et al. 2007a). In this interpretation, the faults were later reactivated, forming zones of weakness and thus allowing the methane fluid to erode the channel floors.

The peak-to-peak relative topography variation derived the DTM covers about 200 m. Assuming a constant surface albedo in the valley, the derived river bed topography with a depth of 27 m and a width of 100 m is reasonable and agrees with the DTM model. Whether the river beds are intrinsically dark (Soderblom et al. 2007b) remains an open question. Keller et al. (2008) investigated the influence of the (diffuse) illumination in Titan's atmosphere. They demonstrated that the limited visibility of the sky from the channel bottoms (shading and reflectivity variation due to slopes (shape-from-shading)) suffices to produce the slight darkening of the rivers. The overall contrast of the river terrain images is only 6% even after correction for the atmosphere. In fact, the river beds are slightly brighter than the lake bed area (Keller et al. 2008). If the river beds were of the same material as the lake area one would expect the river beds to be darker because some shading will occur irrespective of the model details, assuming that the rivers are dry. This condition could be inferred from the fact that the Huygens landing site was not covered by liquid although it did seem damp (Niemann et al. 2005), and no or only very little precipitation was observed. In fact, the river systems are so small (the total length of the main river appears to be less than 15 km) that their drainage probably cannot flood the relatively large lake bed. According to estimates, the discharge of the valley system at the landing site amounts to about 1 m^3/s at short recurrence intervals (Jaumann et al. 2008), which translates into a run-off production rate of 0.6 mm/h. The discharge of larger valley systems on Titan that feed supposed lakes of smaller size than the lake bed at the landing site is greater by about four orders of magnitude (Jaumann et al. 2008). On the other hand, modeled methane convective storms on Titan might produce rainfalls of up to 50 mm/h (Hueso and Sanchez-Lavega 2006), so that the discharge might be greater by two orders of magnitude (Jaumann et al. 2008) but would still not suffice to flood the lake bed.

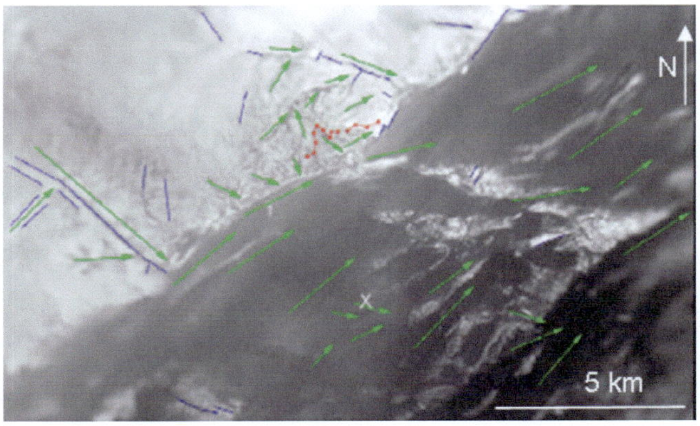

Fig. 5.25 Inferred tectonic and fluid-flow patterns: tectonic patterns are *blue*; drainage divide is *red*; flow directions are shown as *green arrows*; and the Huygens landing site is marked by the *white "x"*. The area covered is 18.3 × 10.9 km (10.2°–10.51°S, 192.2°–192.61°W); the map projection is simple cylindrical (16 m/pixel) (Soderblom et al. 2007b)

The lower left panel in Fig. 5.24 illustrates erosive flow across the dark lake bed unit that surrounds the Huygens landing site. Flow here is generally from WSW to ENE and appears to be cutting down from one level to another. DISR images of the surface acquired after landing show the dark surface strewn with what appear to be rounded cobbles (Tomasko et al. 2005) ranging from 5 to 20 cm in size, placing constraints on the ability of methane flows to transport material (Keller et al. 2008). The lower right of Fig. 5.24 covers the second area mapped stereoscopically (Fig. 5.10c). The error in the relief may be as great as 30–50 m; but it is clear that the bright units protrude as ridges above the dark plain. The landing site shows surface scour flowing to the east, leaving a coherent pattern of bright ridges around 100 m high. Being limited, stereo coverage does not provide reliable relative height scaling between the two regions. Nevertheless, these ridges seem to be an extension of the river terrain (Fig. 5.10b) stretching from the north of the HLS to the dark spot in the southeast. Lower parts of the bright rugged terrain appear to have been flooded and covered with dark sediments. This leads to the interpretation that the dark material (tholins formed in the atmosphere?) has collected in the lower parts of Titan's surface, possibly swept down from the higher brighter (washed-clean) terrain by precipitating methane.

Figure 5.25 is an interpretive map of inferred flow directions and structural lineaments that appear to partly control the patterns of both drainage systems as well as parts of the coastline between the bright highlands and the lower dark plains. The flow patterns are interpreted from composite information (1) taken directly from the DTM, (2) mapped from the dark channel floors that stand out against the bright highland material, (3) mapped from higher-resolution images in which erosive channels can clearly be seen in the dark plain (cf. Fig. 5.24), and (4) inferred from patterns of bright markings in the dark plains (e.g. along-shore shoals and sandbar-like features, patterns that taper off downstream). It appears that, although there clearly is some flow through and from the highlands down into the lower darker plains, the dominant flow direction across the plain is parallel to the coastline, with strong flows from west to east across the region. This suggests major sources of liquid methane/ethane in some region(s) to the west. A careful analysis of the reflectance of the lake bed area reveals a brightness gradient to lower (darker) values with the distance from the coast (Fig. 5.26).

This is consistent with the assumption that bright material also was swept into the lake bed by the river systems (see also Jaumann et al. 2009). The discrete levels between the streamlined islands (sandbars) fit in with the idea of a major flooding.

It is interesting to speculate whether there is liquid exposed at the surface in those areas; the Huygens probe landed on a dry plain, but liquid methane appeared to be present just below the surface (Niemann et al. 2005; Lorenz et al. 2006b). Perron et al. (2006) modeled the methane precipitation rates required to erode the dendritic valleys in the highlands to the north of the HLS. They pointed out that modeled precipitation rates are highly dependent on the grain size of the material entrained in transport. They used size ranges taken from the landing-site images (see Fig. 5.27) of 1–10 cm and inferred precipitation rates of 0.5–15 mm/h. However, the cobbles seen at the HLS are probably not at all representative of the grain size of the material transported in the dendritic highland channels to the north.

These cobbles (5–20 cm after Keller et al. 2008) can only be related to the floods that scoured the dark HLS plain flowing from west to east.

The images taken after landing provide additional evidence that liquid once flowed at the Huygens landing site (Fig. 5.27, Keller et al. (2008)). In addition to the rounded cobbles and pebbles revealed on the surface in the SLI image, 'matrix' material appears in between them and flow-like features are visible. An analysis of the results from the surface science package (SSP) on Huygens showed that the 'matrix' is a granular material which is either non-cohesive or has some liquid component mixed in (Zarnecki et al. 2005). The small-scale structure seen in the MRI image (lower panel in Fig. 5.27) may show the granularity of the surface suggested by Zarnecki et al. (2005). The MRI image also shows some pebbles measuring a few centimeters on the surface. The overall intensity gradient is caused by light from the DISR surface lamp.

The derived size distribution of the cobbles (Keller et al. 2008) markedly differs from that on small solar system bodies. For example, surface coverage by boulders sharply decreases with size on the asteroid Eros (Chapman et al. 2002). On the other hand, the increase in surface coverage with size is comparable to that seen on Mars, where fluvial processes are expected to have been important (Smith et al. 1997). Also, terrestrial flooding events tend to leave large clasts on the surface whereas smaller ones will be buried (Hassan and Church 1994). The effect smears out over multiple events, so that the degree to which the cobbles seen on Titan's surface are sorted by size suggests that they were transported by prolonged flows. On the other hand, a single large flooding event cannot be ruled out. Flow speeds on the order of 1 m s^{-1} can be derived, assuming that the cobbles consist of water ice (Burr et al. 2006; Keller et al. 2008); however, the nature of the cobbles is unknown. If they are made from water ice, their spectroscopic signature (Clark, 1981) has not been detected, nor has the high emissivity of water ice been registered in the RADAR observations. A layer of tholin sediments suppressing the water ice absorption bands might cover

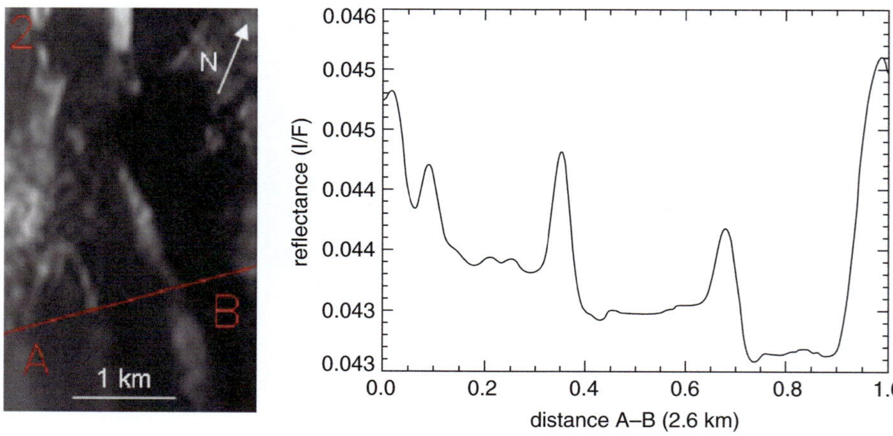

Fig. 5.26 Brightness profile across the track shown on the *left image*

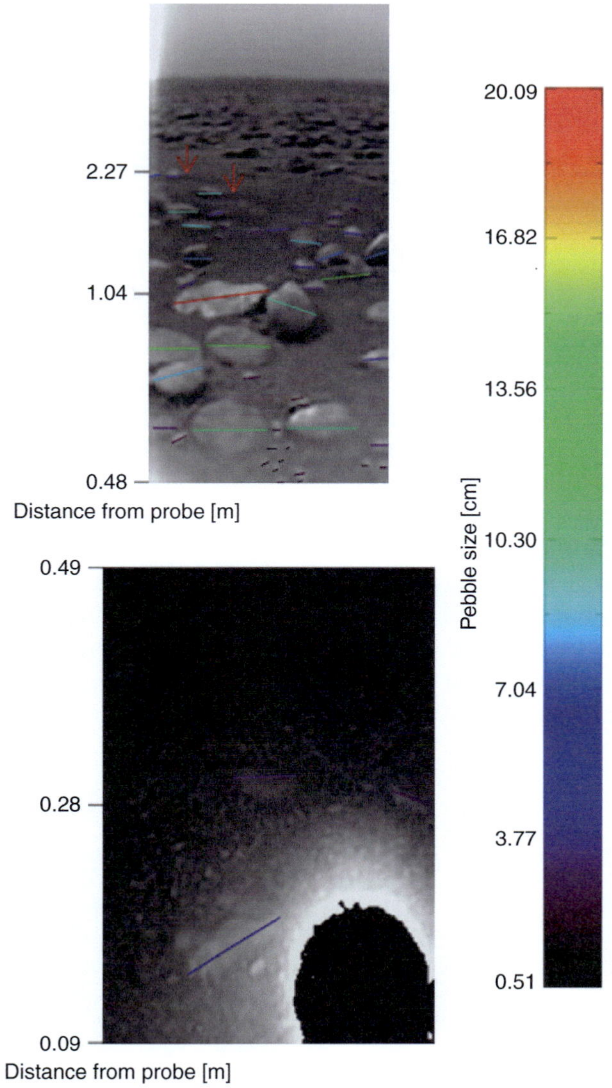

Fig. 5.27 Cobbles identified in the MRI and SLI images. The size refers to the pebble diameter in the horizontal direction. Distance from the probe is measured along the surface. The two red arrows indicate possible flow features. The *stray light* at the left edge of the SLI image is from the science surface lamp. The *dark patch* at the lower right corner of the MRI image is overexposed in all images (Keller et al. 2007)

the cobbles. To mask water-ice emissivity, the cobbles would have to appear rough at the RADAR wavelength of 2.2 cm. It is conceivable that the cobbles consist of solidified sediments, such as benzene and other aliphatic hydrocarbons or nitriles (Kargel 2007; Clark et al. 2006; Soderblom et al. 2009b). In this case, their density would be slightly (about 10%) lower than that of water ice and their strength weaker, which is consistent with their rounded appearance.

There may be substantial surface coverage by larger boulders even though none are seen in the surface images as the area imaged after landing is relatively small. In the last HRI images taken from altitudes between 200 and 300 m, the resolution is sufficient to detect meter-sized objects. None, however, were found (Keller et al. 2008).

The DISR spectrometers observed the surface in near-IR atmospheric methane windows at 751 nm, 827 nm, 0.93, 1.07, 1.28, and 1.59 μm. Generally, observed intensity increased with decreasing solar phase angle due to a combination of surface and atmospheric backscattering. Superposed on this trend were comparatively small variations resulting from reflectance (and hence material or roughness) variability intrinsic to the surface. Haze becomes increasingly dominant towards the lower wavelengths, contributing only 10% to the observed intensity at 1.59 μm but almost half the light at 751 nm. The highland/lake bed reflectance ratio in the methane windows steadily increases from around 1.10 at 751 nm to 1.25 at 1.28 μm, beyond which it jumps to 1.7 at 1.59 μm. The highland is redder than the lake bed over the whole near-IR wavelength range. The contrast between highland and lake bed found in the DLVS methane windows (10-15%) is consistent with the 13% found in the imager band pass.

The observed reflectance spectrum of the dark lake bed terrain at the HLS covers wavelength ranges from the UV to the near IR and was derived using the DISR surface lamp to compensate for the methane absorption bands. It is important to realize that the phase angle is very small, which is why the opposition effect enhances the I/F above the average derived by Tomasko et al. (2008). The only significant spectral feature is the absorption at 1.5 μm. In Fig. 5.28, the shape of the highland spectrum is based on a rough estimate arrived at by multiplying the ratios with the reflectance reconstructed for the lake (Schröder and Keller 2008).

It is tempting to interpret this reddening as representing tholins, implying that they are more abundant on the highland than in the lake bed. Different types of tholins have been

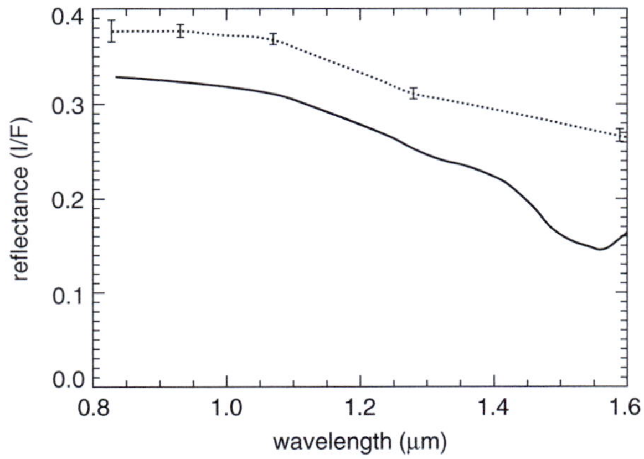

Fig. 5.28 Reconstructed spectrum of the highlands near the HLS (*dotted curve*), obtained by multiplying the lake bed spectrum from Schröder and Keller (2008) by the land/lake ratios in the methane windows. Note that the lake bed reflectance is that for phase angle 0° whereas the ratios were determined at phase angle 35° (Keller et al. 2008)

synthesized in the laboratory. The black tholins of Bernard et al. (2006) are red over the whole DISR wavelength range, whereas the reflectance spectrum of yellow tholin features an absorption line at 1.5 µm. The presence of both types of tholin cannot explain the jump in reflectance at 1.59 µm. It was suggested that the 1.5 µm absorption is caused by water ice (Soderblom et al. 2007a; Rodriguez et al. 2006). The absence of weaker absorption lines around 1.2 µm requires the water ice crystals to be smaller than 10 µm. Why larger water ice particles should not be present is not obvious. Therefore, it seems more probable that the 1.5-µm absorption is due to organic material. Figure 5.28 shows that the jump in the highland spectrum at 1.59 µm completely erases the shallow absorption line of the lake bed spectrum. It still features a blue near-IR slope. Therefore, water ice is not exposed in the bright terrain of the highland area, as suggested by Tomasko et al. (2005) and Rodriguez et al. (2006) on the basis of VIMS spectral data.

The spectrophotometric maps that were generated (Keller et al. 2008) are ideal for constructing false-color maps from the ratio of the intensity measured in different methane windows. Figure 5.29 shows DLVS spectrophotometric maps overlaid on panoramas of the Huygens landing site, with their footprints colored according to the ratio of the intensities in the 0.83/0.75 µm near-IR methane windows.

While the influence of atmospheric scattering below 1 µm is strong, similar maps using DLIS in the wavelength region beyond 1 µm show that the bright highlands are clearly redder than the dark lake bed terrain (Fig. 5.30).

There seems to be a correlation with the phase angle of backscattering (NW direction). Models of surface reflectance suggest that the phase-angle contribution cannot explain these observations (Schröder and Keller 2009). There is no obvious correlation between color variations and geological features within the lake bed. The reddish terrain in the lake bed south of the landing site (marked with an asterisk in Fig. 5.30a) is peculiar. The reddest DLIS footprints are closest to the very dark terrain towards the south (dark spot in Fig. 5.22), which is interpreted as rugged ground incised by rivers, like that in the north. They appear to be located within an outflow channel of the dark spot. Keller et al. (2008) speculate that the red color may be associated with this outflow, e.g. with the sediment it once transported.

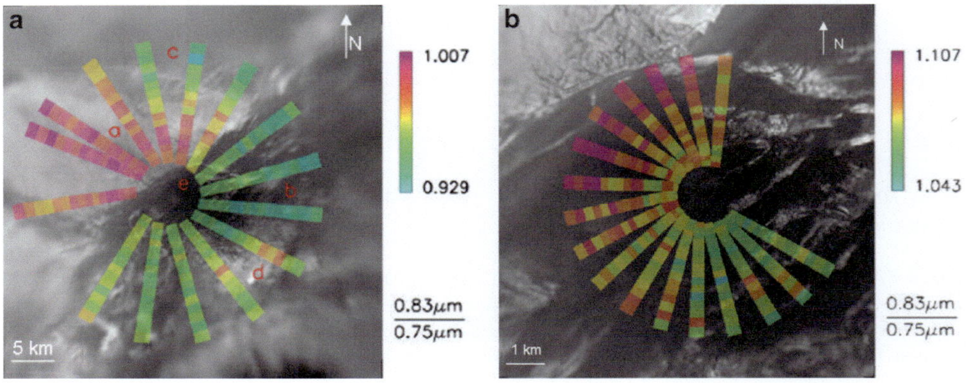

Fig. 5.29 Ratio of the intensity in the methane windows at 0.827 µm and 0.751 µm, displayed in false colors for two DLVS spectrophotometric maps with different resolutions (**a**) and (**b**). (Keller et al. 2008) Projection is gnomonic with the landing site exactly at the center of each panorama (Karkoschka et al. 20087

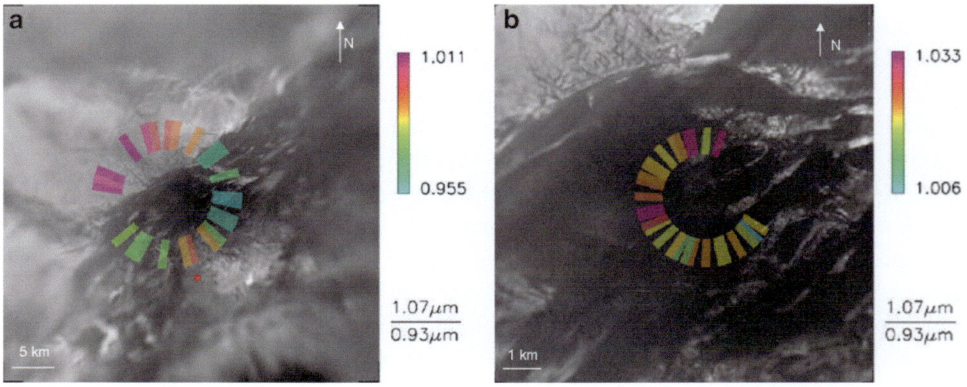

Fig. 5.30 Ratio of the intensity in the methane windows at 1.07 and 0.93 µm, displayed in false colors for two DLIS spectrophotometric maps at different resolutions (**a**) and (**b**). Labeled with a *red asterisk* is an anomalously red area south of the landing site in (**a**)

5.4.3 Regional Context, Connection with Orbiter Observations

Optical images show bright and dark areas distributed along the equator. The Huygens probe landing site is located on a boundary line between bright and dark surface areas. Classification of surface units based on VIMS observations through methane windows yields 3 major classes: bright terrain correlated with highlands, dark brownish terrain correlated with lowlands and often covered by dunes, and bluish dark terrain generally found around white terrain (Soderblom et al. 2007a). In this system, the Huygens landing site is on dark blue terrain. Blue terrain comprises less than 10% of the equatorial dark region, which by itself covers about 25% of the equatorial belt. Bluish terrain is generally found at the transition from bright (elevated) terrain (Soderblom et al. 2007a) to dark terrain. It is thought to be a mixture of bright and dark material, possibly swept down into the dark lowlands. This interpretation agrees with the discovery by Keller et al. (2008) of a brightness gradient perpendicular to the 'coastline' tapering out in the dark lake bed. Therefore, the landing site may not be fully representative of Titan's low- and middle-latitude surface. However, the Huygens landing scenario luckily provides insight into two major terrains, the white, rugged and hilly river terrain and the dark flat lake bed. Huygens' observations show that the large channels and river bed terrains identified by the SAR (Lunine et al. 2008a) also show up in the small-scale DISR images. The HLS is shaped by large-scale fluvial erosion, even though surface liquids have not yet been found in the equatorial belt either by orbiter- or Earth-based observations (Fussner 2006; West et al. 2005). The effect may be seasonal.

The region near the HLS is free of the pervasive dune fields found elsewhere, the nearest dunes being about 30 km north. The geomorphology of the landing-site region is that of bright rugged higher terrain and lower dark plain, in which the major ongoing erosional and depositional processes, often appearing to reflect tectonic control, are fluvial and pluvial. Terrain dominated by marked fluvial erosion may well be typical of most of the highland surfaces of Titan, possibly even more so at higher latitudes where lakes have been detected.

The model of Soderblom et al. (2007b) suggests that the immediate environment of the HLS consists of a dark substrate that is mantled with bright deposits derived from the highland regions a few kilometers to the north. If that model is valid, the dark channels of the highlands to the north and the plains of the HLS itself represent exposed dark (water-ice-rich? (Soderblom et al. 2007b)) substrate. The evidence of water ice provided by orbiter observations through the methane windows and in the far infrared (Rodriguez et al. 2006) is tenuous and not corroborated by the DISR spectra taken at the surface (Keller et al. 2008). The observed brightness gradient near the coastline suggests that bright material has been swept onto the dark lake bed plain. Water ice is a major constituent of Titan (Fortes and Grindrod 2006; Tobie et al. 2005), and it was expected that it would be detected in DISR spectra acquired near the surface. It seems that all of Titan's surface is covered with layers of organic material that are thick enough to obscure the water ice underneath, except perhaps in cases of enhanced tectonic activity or perturbation by impactors (O'Brien et al. 2005; Korycansky ad Zahnle 2005).

Even though dunes are only tens of kilometers apart from channels in certain places (Elachi et al. 2005), their presence implies the prevalence of dry conditions. This is also the case at the HLS where the 'cat scratches' are about 30 km north of the lake bed (fluvial channel). In apparent contrast, the HLS is damp with methane and ethane (Niemann et al. 2005; Lorenz et al. 2006b). Climatic differences (humid at the HLS and dry at the 'cat scratches') are not plausible across such a short distance without significant elevations in between. It is more probable that dunes are located in higher terrains (plateaus) where the methane table does not intersect the surface. However, RADAR DTM data do not indicate the presence of such elevation differences. Perhaps they are too small to measure, much less than the few hundred m between the lowlands and bright faculae (Kirk et al. 2009). In some areas, dunes may be absent because of "blocking" by high topography to the west. Dune fields are only found within ±30° of the equatorial belt (Barnes et al. 2007b; Radebaugh et al. 2008a) where precipitation may not be as strong as near the poles, where lakes are found.

The DISR found a rugged topography, occasionally approaching the angle of repose, near level plains. Fluvial systems drain into adjacent relatively flat, dark lowland terrains. In addition to flows from highland drainages, the lowland area shows evidence of additional flows parallel to the highland–lowland boundary, leaving bright outliers that resemble terrestrial sandbars, which implies major west-to-east flooding across the plains. Consistent with this hypothesis, images taken after landing on the surface show abundant 5–20-cm-sized smooth rounded cobbles strewn about an apparent flood plain. A stereo model for part of the dark plains region east of the landing site shows surface scour flowing to the east, leaving a pattern of bright ridges that are on the order of 100 m high. Tectonic patterns are also evident in (1) the rectilinear, low-order, stubby drainages and (2) the numerous straight and angular margins along the apparent coastline at the highland–lowland boundary. Because the features appear to be relatively young, they suggest that the equatorial atmosphere occasionally becomes saturated with enough methane to create violent rainfall; hence the erosion seen in the DISR images.

5.5 Tectonic and Volcanic Surface Features and Processes

Tectonic and volcanic features are surface expressions of internal geological activity. Considering the state of the icy Galilean satellites and Titan's large eccentricity, faulting systems are to be expected. The low density of impact craters suggests that resurfacing has been taking place in the geologically recent past, and that cryovolcanic features may be visible if not buried by dunes or obscured by atmospheric debris. In addition, the presence in the atmosphere of ^{40}Ar (Niemann et al. 2005), a product of the decay of ^{40}K, suggests that degassing has occurred from the interior, where silicates are present, to the icy surface.

Since RADAR and optical observations (ISS and VIMS) only overlap in a few places, many of the features described in this chapter have been observed by only one of these instruments. It must be noted that the resolution of RADAR images is on the order of 300 m whereas the near-infrared cameras have resolutions around 1–10 km in the mapping phase. During the nominal mission, the VIMS was able to acquire 25 cubes at resolutions lower than 5 km, half of them during the dedicated T20 flyby. During this flyby, the VIMS also tested a push-broom mode which provided 'noodles' several hundreds of kilometers long. These long stripes with a resolution as low as 500 m per pixel cross some of the radar swaths, so that a few places have now been observed at comparable resolutions by both instruments (Soderblom et al. 2007a; Barnes et al. 2007b). However more studies on correlating radar and optical images have to be performed and thus, it is one major goal of the extended mission to obtain RADAR observations of places where optical instruments have found key geological features, and vice versa.

5.5.1 *Tectonic Features*

The driving forces of tectonics on Titan are not understood at this point; most possibly tectonic features appear to be at least partially degraded and embayed by surrounding plains units. Possible tectonic features include linear and ridge-like formations that may be chains of hills (Lopes et al. 2007a; Barnes et al. 2007b) and dark sub-parallel and branching lineaments (called virgae) that may represent fractures which, as on Earth, may subsequently have served as fluvial channels (Perry et al. 2007a, b).

Although lineaments are rare on Titan, the optical instruments have found very few elongated features, mostly linear boundaries, that can be interpreted as tectonic. One of the most evident sets of linear features has been observed during four flybys: T17, T20, T21, and T34 by VIMS (Fig. 5.31).

It is located at about 30°S latitude and 330°W longitude. One north–south feature about 500 km long seems to cross both dark and bright terrains. The two images taken by VIMS were obtained at an incidence angle around 60° and an

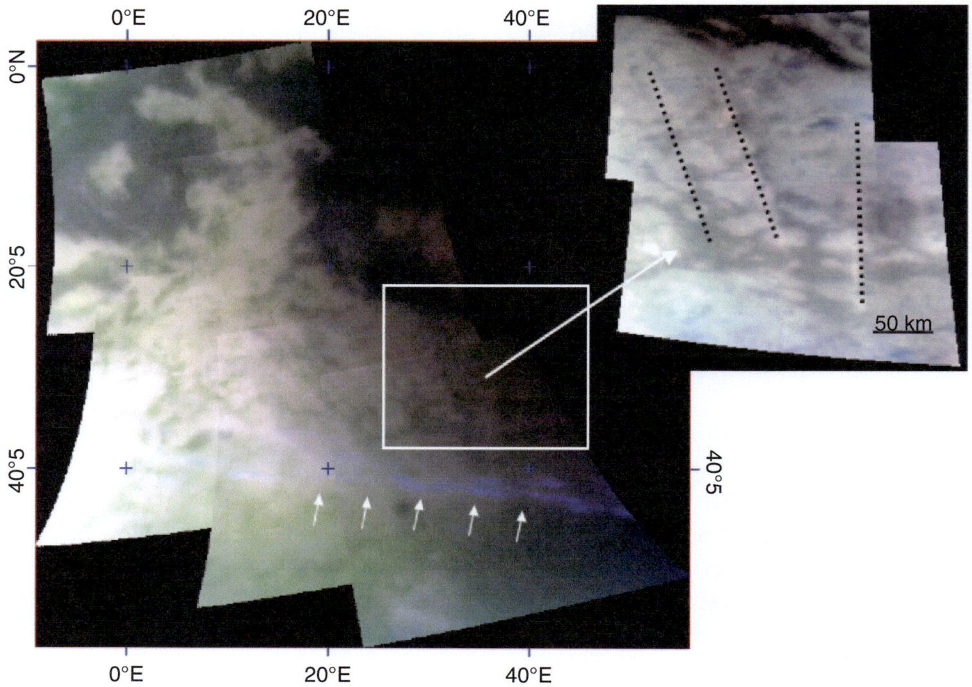

Fig. 5.31 Lineaments in the southern hemisphere which cross-cut bright and dark areas (dotted lines in inserted picture). *Arrows* indicate cloud. False color VIMS image is mapped with 5.2 μm as *red*, 2.00 μm as *green*, and 1.28 μm as *blue*

emission angle close to vertical. Analysis of the Sun and anti-Sun facing slopes yields estimates of 6°–9°, suggesting that topography ranges between 1 and 2 km (Sotin et al. 2007). Southeast of the first is another north–south-trending linear feature. Such structures could have been created by either compression (fold) or extension (rift), but in the absence of additional data it is difficult to determine the origin of such linear features. In addition, clouds oriented in an east–west direction were observed during three of the four observations, which can be interpreted as orographic.

Two types of possibly tectonic terrain have been observed by RADAR: linear chains of radar-bright raised topography (Radebaugh et al. 2007; Lunine et al. 2008a) and more irregularly shaped fragments or larger expanses of rugged radar-bright terrain, such as Xanadu (Wood et al. 2007a; Radebaugh et al. 2009). The linear mountain chains were identified in the T8, T21 and T25 swaths (Radebaugh et al. 2007; Lunine et al. 2008a; Lopes et al. 2009). Radebaugh et al. (2007) measured mean slopes close to 10° and elevations not higher than 650 m for features in the T3 and T8 swaths. These mountainous features are locally surrounded by diffuse deposits that are interpreted as erosional aprons and dune material. Covering the surrounding plains, they stop at the mountains, suggesting that the latter are older. Radebaugh et al. (2007) tentatively suggested that these mountain chains were formed by crustal compressional tectonics and the up-thrusting of blocks. Radebaugh et al. (2007) compared mountain and blanket volumes to arrive at an estimate of the original mountain height. Assuming 10–25% of the typical terrestrial erosion rate of 0.5–1 mm/year (Summerfield and Hulton 1994) for the erosion on Titan because precipitation is rarer (Lorenz and Lunine 2005), Radebaugh et al. (2007) suggested that the typical mountain age may be as young as 20–100 million years, which is intriguing as the mountains do appear to be old relative to their surroundings.

Smaller fragments of radar-bright terrain are seen in many RADAR swaths, and have been described as isolated hills or blocks (Stofan et al. 2006; Radebaugh et al. 2007; Lopes et al. 2009). These may either be eroded fragments of the linear mountain chains and thus originally compressional or extensional in origin, or they may be blocks of impact ejecta, or formed by the dissection and erosion of a preexisting layer of material (Radebaugh et al. 2007). Larger regions of rugged terrain make up Xanadu (Radebaugh et al. 2009) and are also seen at the south (Stofan et al. 2008) and north pole. Further analysis is required to determine whether these regions are tectonic in origin or were formed by dissection and erosion.

5.5.2 Volcanic Features

Radar evidence of cryovolcanism appears in both the northern and the southern hemispheres, but it is not ubiquitous. The most easily identified features are bright flow fronts and circular features which appear to be the sources of some flows. The largest structure speculatively identified as cryovolcanic is Ganesa Macula (Elachi et al. 2005; Lopes et al. 2007a), a 180-km-wide circular feature with an apparent depression at its center and radial drainage on its sides (Fig. 5.32).

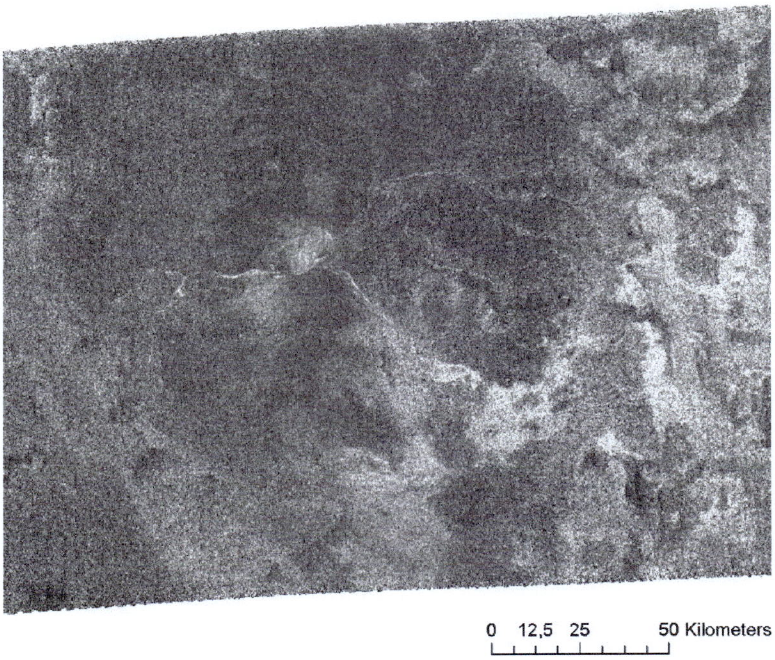

Fig. 5.32 RADAR image of Ganesa Macula centered at (49.7°N; 87.3°W) (after Lopes et al. 2007a), for explanation see text

It is centered at approximately 49.7°N and 87.3°S. The feature is radar-bright on the edges facing the instrument, which suggests a positive relief. The dark top is consistent with a flat surface similar to the steep-sided domes on Venus. In the center, a bright circular feature 20 km in diameter has originally interpreted as a caldera with two bright flows towards the west and southeast (Lopes et al. 2007a). However, 'SARTopo' and RADAR stereo are not consistent with this interpretation (Kirk et al. 2009), so it seems probable that the bright margin and dark top are areas of differing radar backscatter cross-section rather than different slopes. The RADAR instrument has also mapped several flow complexes. The largest one, Winia Fluctus, is located 1,340 km east of Ganesa's eastern edge (Lopes et al. 2007a). It extends from 50°W, 52°N to 44°W, 47°N. This large flow forms a series of lobate deposits consistent with cryovolcanic or sediment-laden flows. This extensive complex has an area of no less than 23,700 km^2. Although simple thermodynamic arguments suggest that flows could be composed of ammonia hydrates, the same arguments can be used to explain the lack of ammonia in the crust if the crust has grown at the expense of an internal ocean.

Several other features observed by the RADAR instrument have been interpreted as flows associated with calderas about ten to several tens of kilometers in diameter. One of these flows was also observed by the VIMS during the T20 observation (Fig. 5.33). Other features are Rohe Fluctus and Ara Fluctus, observed in the RADAR Ta image, which have flowlike features emanating from possible calderas (Lopes et al. 2007a).

The radar-bright flow seems to originate at a quasi-circular feature with rough rims (Stofan et al. 2006). There is an excellent correlation between the radar-bright flow and the VIMS-bright terrain (Le Corre et al. 2009). Compared to most of Titan's surface, the flow shows a low 2.69/2.78 μm ratio, which is compatible with the presence of fine-grained CO_2 ice (Le Corre et al. 2009). Note that the presence of CO_2 cannot be confirmed by the absorption at 4.92 μm because of the low integration time used for this kind of observation. On the other hand, ammonia seems unlikely because it absorbs at 2.03 μm, which does not seem compatible with the brightness of the flow at this wavelength. However, no conclusion can be drawn at present on the basis of this single example of a flow feature seen by both the RADAR instrument and the VIMS. Additional opportunities are planned in the extended mission, and observations then should clarify the composition.

South of Xanadu, VIMS has observed two large areas of very high reflectivity at 5 μm: near Hotei Arcus and Tui Regio (Figs. 5.4 and 5.6) (Barnes et al. 2005, 2006). Tui Regio is located southwest of Xanadu. Centered near 125°W/24°S, it is ~150 km wide and extends 1,500 km in an east–west direction (Fig. 5.34).

Barnes et al. (2006) describes at least three long, thin, lobate, spectrally self-similar tendrils emanating from a central point that is the brightest VIMS pixel in the immediate vicinity (located at 134°W/22°S). The tendrils extend north, southwest, and west-northwest from this central point. The total surface area of the Tui region is 2.5 105 km^2. Tui Regio is an area for which no SAR images have been acquired so far.

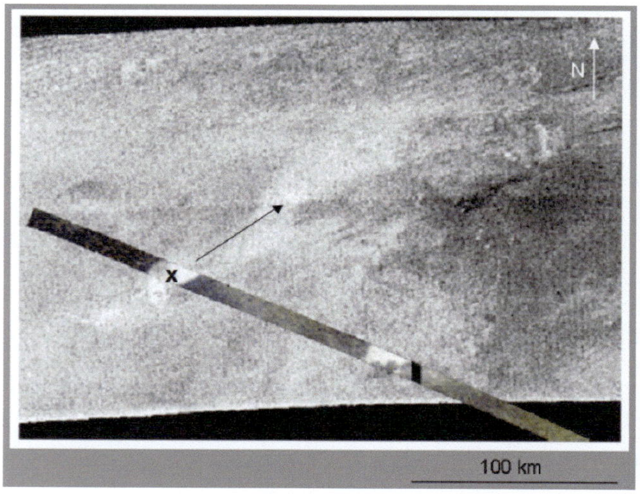

Fig. 5.33 Correlation between RADAR images (T3 flyby) and VIMS infrared image obtained during T20 flyby at 68°–73°W, 18°–22°N (Le Corre et al. 2009). The (x) shows the effusion point and the *arrow* the flow direction (Lopes et al. 2007a)

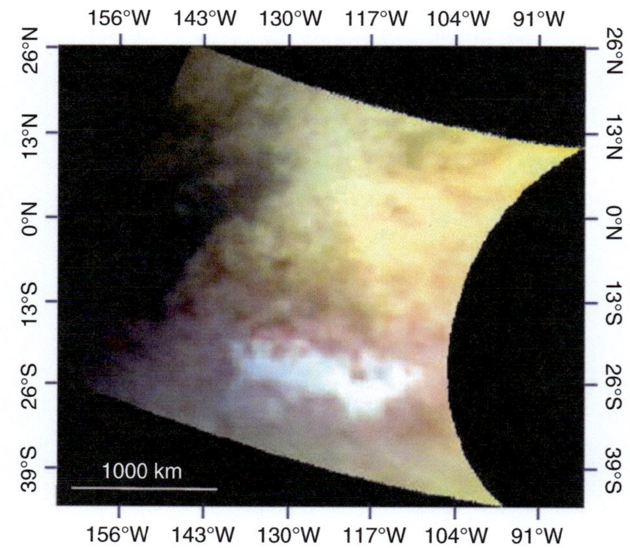

Fig. 5.34 Tui Regio as observed by VIMS with a very bright 5-μm surface feature

The second area near Hotei Arcus (Barnes et al. 2005). This 5-μm-bright spot is located at 80°W, 25°S. It has been observed on several occasions since the Cassini orbit insertion in 2004 (Barnes et al. 2005). It extends 450 km from north to south and 400 km from east to west, covering an area about 2×10^5 km^2, similar in size to Tui Regio. No high-resolution VIMS images of this area are acquired so far. It should be noted that Barnes et al. (2005) already discussed the possibility that the surface might be composed of CO2. RADAR has obtained images showing lobate flowlike features of possible cryovolcanic origin (Wall et al. 2009; Soderblom et al. 2009b). Nelson et al. (2009b) have reported that variability and NH3 rather than CO2 could be present. Digital terrain modeling of Kirk et al. (2009) shows that the flowlike features are quite thick and tower above the narrow fluvial channels also seen in the area (Fig. 5.19).

A last feature that can be described is Tortola Facula (Fig. 5.35), which was observed during the first Titan flyby (TA). Note that the location of this feature is where the Huygens probe was originally intended to land if a telecommunications problem had not forced the tour to be modified.

This first image demonstrates that the VIMS is capable of observing Titan's surface at a high resolution of 2.03 μm, where scattering by aerosols is low and the amount of reflected light quite large (Sotin et al. 2005). For this observation, the resolution was about 3 km/pixel. In the center there is a bright circular feature about 30 km in diameter, with a dark dot in the center. Two elongated wings extend westward. The structure resembles a volcanic edifice with lobate flows (Sotin et al. 2005). Although there is a correlation between brightness temperature, a radar-bright surface and the location of Tortola Facula as seen by VIMS, the geomorphologic interpretation of RADAR data is not consistent with this feature being cryovolcanic. Tortola Facula is now interpreted as mountainous material (Lopes et al. 2009).

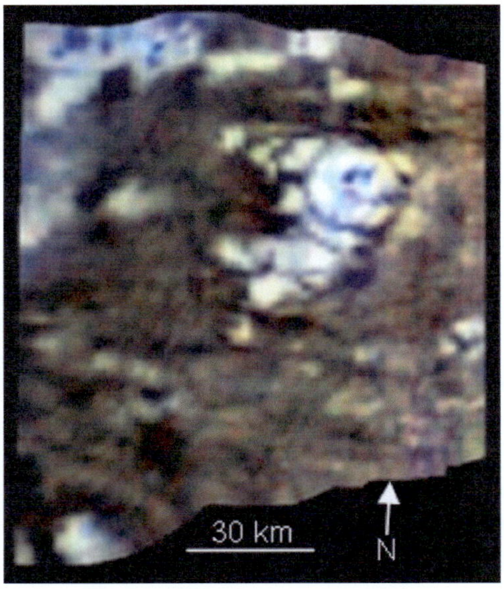

Fig. 5.35 VIMS observation of Tortola Facula (8.8°N, 143.1°N)

Several features observed by the RADAR instrument and the VIMS suggest the presence of cryovolcanic edifices and flows on Titan's surface. The best proof that cryovolcanism is indeed occurring would be to witness an eruption by detecting the thermal anomaly due to erupted material either by VIMS or RADAR in its radiometry mode, or changes in appearance of surface features, as on Enceladus or Io. The low density of impact craters on Titan suggests global resurfacing some hundreds of millions of years ago (Lorenz et al. 2007). The situation is similar on Venus where volcanic edifices and large volcanic flows are present but no evidence of volcanic eruptions was observed while spacecraft (Pioneer Venus, Venera, Magellan, Venus Express) were orbiting the planet.

The description of geological features as being probably of cryovolcanic origin underlines the need for further observation of these features by both radar and optical instruments. Such work will be carried out during the Cassini Equinox Mission and any further extensions of the Cassini mission, possibly until 2017, resources permitting.

5.5.3 Theoretical Considerations for Endogenic Processes

On the basis of thermal-orbital coupled models and numerical simulations of thermal convection, Tobie et al. (2006) showed that methane might have been released after interior differentiation (~4 Gyr ago) had formed a discrete rock core, and stored in the form of clathrate hydrate in the outer layer above an ammonia-water ocean. The remaining clathrate reservoir would have been dissociated when the outer layer was destabilized by thermal convection. Tobie et al. (2006) suggested that methane was released into Titan's atmosphere in three periods: during accretion; at about 2.5 Gyr. when convection processes started in the silicate core and generated greater heat flux in the outer ice layers; and more recently (a few hundred Myr ago) when the thickness of the outer ice crust became large enough for convection processes to take place. This last episode would explain both resurfacing and the presence of methane and ^{40}Ar in the atmosphere, assuming that methane can be destabilized in the crust as hot ice moves to the surface (Sotin et al. 2009). The stability of methane is described in Fig. 5.36.

Given a surface temperature of 94 K and a reasonable temperature gradient of between 2 and 5 K/km, methane exists in its liquid phase. However, water ice and liquid methane react to form clathrate hydrate, which is stable in Titan's subsurface. There are about six molecules of H_2O for one molecule

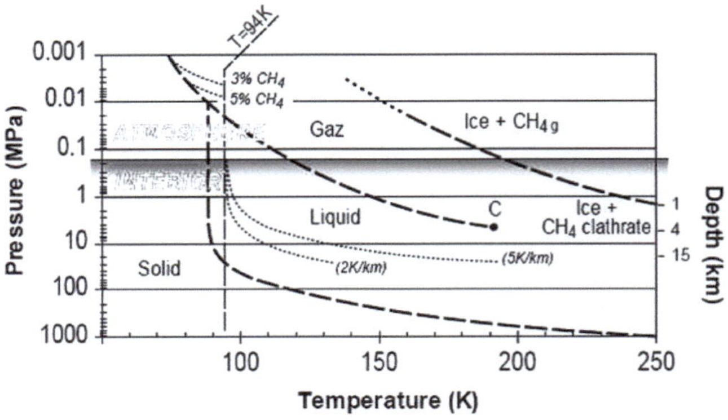

Fig. 5.36 Phase diagram of methane and methane clathrate with temperature profiles in the interior

of CH_4 in this compound. As long as there is enough H_2O, methane clathrate should be the preferred stable phase.

Methane clathrate can be destabilized if hot material (T ~ 250 K) can reach a depth of 1 km (Fig. 5.36). This is not possible in a simple stagnant lid convection regime (Sotin et al. 2009). However, if the lithosphere is weakened by faults generated by impacts or tectonic activity, upwelling plumes could move much closer to the surface and destabilize methane clathrates there. Another process that may be considered is tidal heating that might enable upwelling plumes to reach the depth necessary for the destabilization of methane clathrates. It must be noted that this kind of destabilization implies the formation of methane as and is likely to cause explosive eruptions due to the great pressure created by the release of methane gas (Sotin et al. 2005).

In the description of cryovolcanic features (above), CO_2 has been mentioned as a possible constituent of the 5-μm-bright features. CO_2 is solid under Titan's subsurface conditions, but the lack of CO_2 in the atmosphere makes it unstable at the surface. It is interesting to note that CO_2 has been recorded by the VIMS and interpreted as a cryovolcanic gas (Baines et al. 2006).

Ammonia hydrate is the favored component by which to explain the flow features on Titan's surface (Mitri et al. 2007). The reason is that the melting temperature of ammonia is about 100 K lower than that of H_2O. If the crust is made of ammonia hydrate, any increase in heat flux can lead to partial melting, the liquid phase being composed of ammonia-water that may eventually reach the surface. However, the presence of ammonia in the crust is difficult to reconcile with the crystallization of the crust above an ocean. During the freezing of an ammonia-rich ocean, the solid would be composed of pure water and the ocean would be enriched with ammonia.

The physical and chemical processes that can explain cryovolcanic edifices and flows are being studied. To understand these processes, laboratory experiments on complex systems (H_2O, NH_3, CH_4) are required. However, observations and interpretations of the surface composition of these cryovolcanic features are needed as well. The combination of radar and optical images should allow us to make progress on this topic during the extension of the Cassini mission. The relative paucity of tectonic features identified on Titan to date makes it unlike other icy satellites in the Solar System and demonstrates the need for complete, high-resolution coverage of the surface, as the distribution and nature of tectonic features is critical to understanding the interior and surface evolution of planetary bodies. In addition, the origin of expanses of rugged terrain such as Xanadu, have formed through a poorly constrained combination of tectonic and erosional processes which requires not only further analysis and modeling but also higher-resolution data. Modeling of Titan's rotation by Stiles et al. (2008), interpreted by Lorenz et al. (2008c) indicates a decoupled ice shell and a possible liquid subsurface layer (see also, Sotin et al. 2009).

5.6 Erosional/Depositional Surface Features, Processes and Redistribution of Material

5.6.1 *Aeolian Features and Processes*

The fraction of Titan's surface covered by dunes is larger than on any terrestrial planet. This comes as something of a surprise, because before Cassini, dunes were thought unlikely on the basis of two lines of reasoning (Lorenz et al. 1995; Lorenz 1996a). These assumptions have meanwhile been proven wrong (Lorenz 2006) for rather interesting reasons:

First, expected speeds of thermally driven winds near Titan's surface were extremely low due to the low solar flux, the large column mass of the atmosphere and the small radius of Titan. This kind of energy-flux argument, which correctly predicts wind speeds of a few meters per second on Earth, suggests wind speeds on Titan of only about 1 cm/s. In contrast, despite Titan's thick atmosphere and low gravity, which both favor the transport of material by wind, the threshold wind speeds required to move sand-sized particles on Titan, which are defined to range between 62 and 2,000 μm, are on the order of 0.5–1 m/s.

The second issue is that of the supply of sand-sized particles. Because of the weak sunlight, the hydrological cycle was expected to be weak overall, corresponding to about 1 cm of liquid methane per Earth year. Furthermore, raindrops would fall relatively slowly, weakening their erosive effect. The freeze-thaw action that serves to break down bedrock on Earth would not occur on Titan since the heat capacity of the atmosphere buffers the surface against large diurnal temperature changes. It was further conjectured that seas of liquids on Titan's surface might act as traps for any sand that was formed (for example the sand-sized fraction of ejecta from impact craters – this is believed to be the principal source of sand on Venus).

The expectation that near-surface winds would be gentle was attributable to the theory that Titan's winds are likely to be driven by the same mechanism (differential solar heating) that drives winds on Earth, on Mars and in every other planetary atmosphere in the solar system, a not altogether unreasonable expectation. However, Titan has the additional factor of a significant gravitational tide in its atmosphere due to its eccentric orbit around the massive planet Saturn: while unique in this solar system, this could in fact be an important phenomenon on many tidally locked extrasolar planets. The potential height difference associated with the tide is several hundred times larger than that exerted by the Moon on the Earth.

It was noted in Lorenz et al. (1995) that measures of average wind speeds may be of limited utility in assessing sand transport processes in that winds above the transport threshold may in fact be quite rare. Thus, the controlling factor in aeolian feature formation would be the probability distribution of various wind speeds and the length and width of the high-speed 'tail' of the skewed distribution. Wind speeds are characterized by non-Gaussian statistics, with log-normal or more typically, Weibull functions being used to describe them, at least in the wind energy literature (Lorenz 1996a).

Whether the winds on Titan are made powerful enough to move sand by statistical fluctuation exceeding the threshold or perhaps more likely, by atmospheric tides remains to be determined in model studies (and by lander in-situ measurements on a future mission). However, important constraints on the patterns of dune-forming winds may be derived from the orientation of dunes seen in RADAR images (Fig. 5.43) (Lorenz et al. 2006a).

As for the supply of sand, various perceptions have changed since the assessment by Lorenz and Lunine (1996). First, the low latitudes appear devoid of bodies of liquid which might act as sand traps. The production of sand-sized materials by impacts seems even less effective than previously thought. Views of the freeze-thaw cycle have not changed.

Porco et al. (2005) and Perry et al. (2006) noted some 'streaky' boundaries between light and dark terrain in Cassini's initial optical reconnaissance of Titan. Some streaks were identified, and the eastern boundaries of several faculae (bright spots) were noted to be diffuse, while the western boundaries were sharp (Shangri-la and Fensal-Aztlan, Fig. 5.4), suggesting eastward surface transport. It could not be determined unambiguously whether these features were the result of aeolian or fluvial transport, although the consistent pattern of well-defined margins to the west and diffuse margins to the east suggested the former and implied net west-to-east transport (Porco et al. 2005; Perry et al. 2006).

The second SAR swath on flyby T3 showed many long and narrow radar-dark sub-parallel features which, nicknamed 'cat scratches' (Fig. 5.37), were tentatively interpreted (Elachi et al. 2006) as being aeolian in origin, although other processes could not be ruled out.

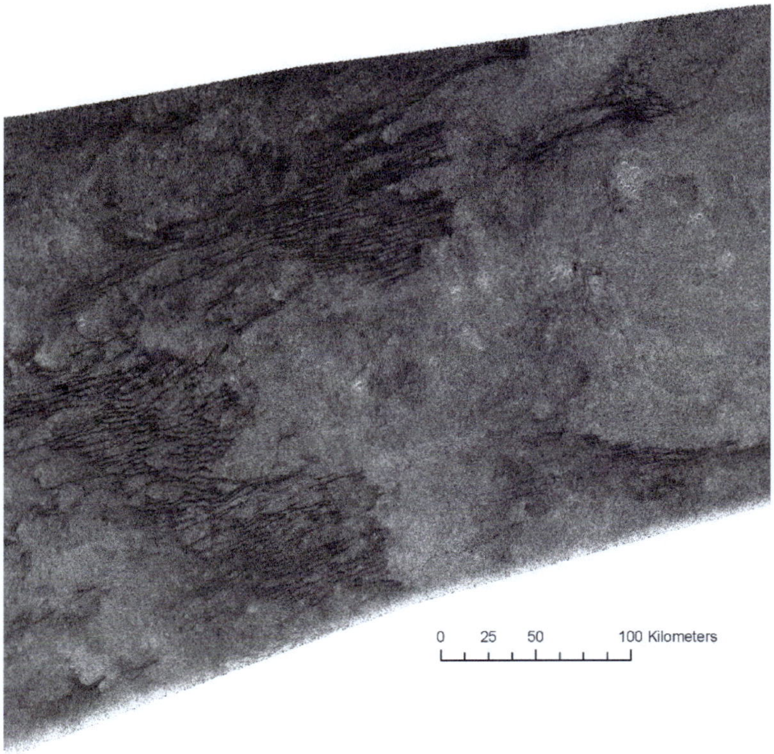

Fig. 5.37 'Cat Scratches' as first seen in RADAR orbit T3 (Elachi et al. 2006)

It was the near-equatorial flyby T8 which cleared up the interpretation (Lorenz et al. 2006a). First, this flyby passed over terrain that was completely covered with sand in many places (Fig. 5.38).

While in the T3 data, the cat scratches were seen simply as dark material deposited above a brighter substrate in the T8 data the topographic shape of the sand surface was the major control on the RADAR appearance. Second, the equatorial ground track of the spacecraft meant that the dunes, oriented E–W, were illuminated broadside on, so that there were many opportunities to detect topographic glints (Fig. 5.39), which provided a clear, positive-relief impression of the nature of the dunes.

Finally, the dunes in this dark region appear larger than those in the T3 data. As described in Radebaugh et al. (2008a), the T3 dunes had a mean length of 16 km while the mean length in T8 was 38 km. Based on RADAR data, Radebaugh et al. (2008a) counted several thousand individual dunes. Some network dunes were detected, suggesting complex wind regimes in specific locations, but predominantly the dunes are linear. Figure 5.40 shows a segment of the T25 swath featuring dunes in Aztlan. The bright interdune areas can be seen as well as the abrupt termination of dunes where they reach the western edge of topographic obstacles.

The presence of dunes in the T8 swath (Figs. 5.38, and 5.39), during which the Huygens landing site was imaged as well, was in fact instrumental in co-registering the 2 cm RADAR image with the near-IR DISR image (e.g. Lunine et al. 2008a). The datasets do not always correlate well, especially when they are small-scale, but the dunes (seen only in the distance as horizontal dark streaks in the DISR side-looking images) can be identified in both (Fig. 5.41).

The dunes are typically aligned close to the E–W direction (Radebaugh et al. 2008a), and while in some cases the direction along the dune axis is ambiguous, the overall pattern is pre-eminently one of net sand transport from west to east. Dunes terminate abruptly at the western edge of obstacles, and pick up gradually again further on. Some obstacles

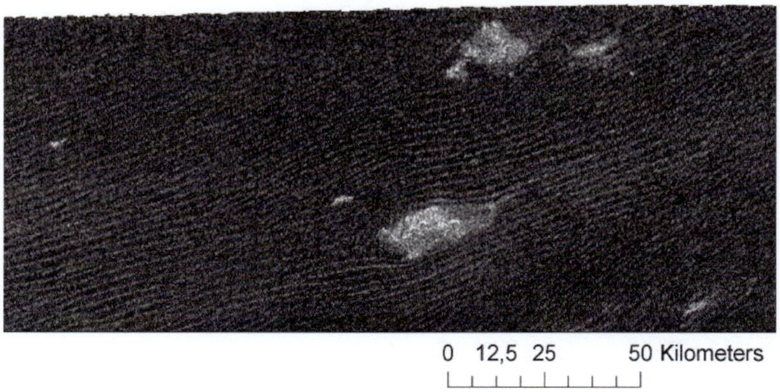

Fig. 5.38 Dunes in Belet observed by RADAR in T8. The longitudinal nature of the dunes is evident from their interaction with bright topographic obstacles

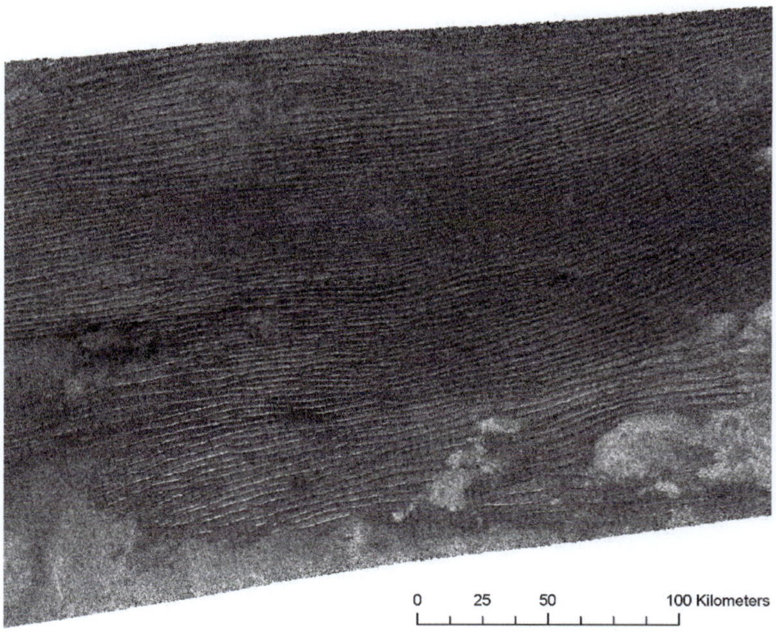

Fig. 5.39 Section of the T8 RADAR-swath in Belet (illumination from North/Top of image). Although the dunes are seen as dark and overlie brighter material at lower left, the sharp uprange glints seen throughout, and particularly prominent at *lower left*, show the dunes to have positive relief of some 100–150 m

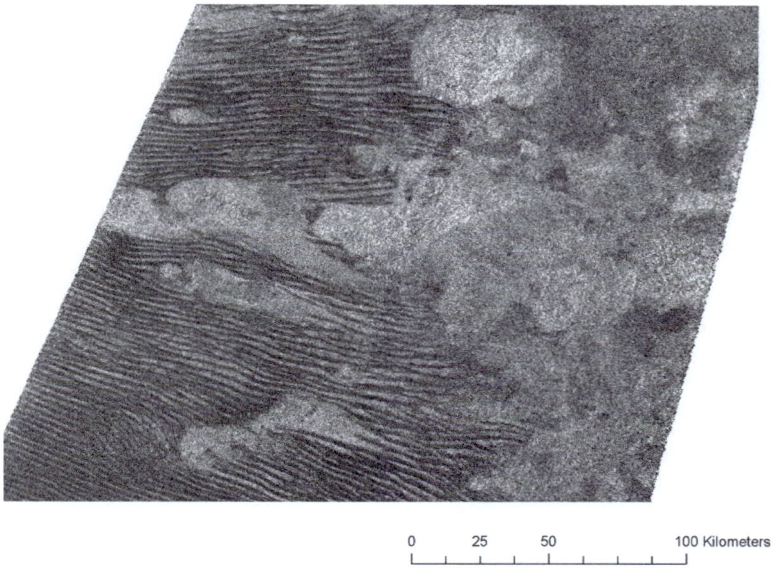

Fig. 5.40 A segment (*north up*) of the T25 RADAR-swath, showing dunes in Aztlan. The bright interdune areas can be seen as well as the abrupt termination of dunes when they reach the western edge of topographic obstacles

Fig. 5.41 Huygens landing site identification in T8 was facilitated by two dunes seen near the horizon in DISR SLI images north of the landing site indicated by (+). DISR mosaic overlaid on RADAR data. The image is about 100 km across

have 'tails' in the downstream direction. These characteristics are consistent with the ISS observation that equatorial faculae typically have sharp western and diffuse eastern margins (Porco et al. 2005; Perry et al. 2006).

Dunes were detected in the T20 high-resolution VIMS 'noodle' at Fensal (Barnes et al. 2008) ~50°W, 10°N where dunes had already been identified in T17 RADAR data. Spectroscopically, the dune material appears to contain less water ice than other units on Titan, and various organic materials would be consistent with the data. Barnes et al. (2008) measured a typical dune spacing of 2 km and determined by photoclinometry that the dunes in this area are 30–70 m high.

The optimum particle size for saltation in Titan's atmosphere, assuming that interparticle cohesion is similar to that of terrestrial materials, is about 250 μm (Greeley and Iversen 1985). Such material may be produced by coarser materials such as impact ejecta or fluvial sediments breaking down, or by agglomeration of finer material, such as particles created in the atmospheric haze. At present, the latter origin appears to be favored: the optically dark appearance of the material and its spectral characteristics support organic composition, suggesting that the sand formed from haze particles. <1-μm haze particles may be converted into 250-μm and grains by sintering over long timescales or perhaps more likely, by cycles of wetting and drying in Titan's lakes. The latter scenario would require the transport of sand from the lakes at the poles to the equatorial regions where the dunes are found.

An initial estimate of the volume of river channels indicated that it is insufficient to account for the volume of sand needed to construct the dunes. More heavily eroded areas have since been found, so that this calculation may need to be revised. The observed distribution of impact craters – originally thought to be a likely source for sand-sized material (they are believed to be the dominant sand source on Venus (Garvin 1990) – is unable to provide the required volume, unless some other process has broken down larger ejecta.

Estimating the total volume of sand, Lorenz et al. (2008b) noted that about 40% of the low-latitude half of Titan appears covered in dunes. Based on radarclinometry, radiometry and similarity, they estimated the average to be between 200,000 and 800,000 km³ of material, corresponding to a thickness of several meters over the whole planet.

The dunes represent an important set of constraints on Titan's meteorology by virtue of their extent, their shape, and their orientation. There are also implications for photochemistry in that the dunes represent the largest known reservoir of organic material (Lorenz et al. 2008b), as described above. First, their distribution, confined to the tropics, defines the latitudes equatorward of 30°N and S as having, at least sometimes, the conditions required for dune formation (namely available and transportable (i.e. dry) sediment, and winds strong enough to move the material), whereas other latitudes appear not to satisfy these requirements. Since the sand source has not yet been unambiguously identified, this constraint pertains to wind and humidity. With regard to humidity, models (e.g. Rannou et al. 2006) had already suggested that the low latitudes on Titan would eventually dry out unless resupplied by a surface methane source. Mitchell et al. (2006) explored this question further and estimated that some 1–2 m of liquid methane per year might be evaporating from the lower latitudes. They found that the latitudinal extent of the dry region depends on the total methane inventory, with between 7 and 20 m agreeing best with observations (such as dune extent and the humidity recorded by Huygens).

Second, the predominance of the longitudinal (linear) dune shape requires a modestly changing (typically bidirectional) wind regime (e.g. Lancaster 1995). Factors producing this variation include seasonal changes (the usual reason for this wind regime on Earth) and possibly the gravitational tide in the atmosphere.

Finally, the dune orientation pattern is important for diagnosing the tropospheric winds, which feature few clouds which could act as tracers. Tokano (2008) has explored the winds in a global circulation model (GCM), finding that, indeed, surface winds should exceed the saltation threshold of 0.5–1 m/s not infrequently and that, furthermore, bidirectional winds are encountered in the course of a Titan year due to seasonal changes in the hemisphere-to-hemisphere Hadley circulation. However, this model (indeed, all present models of Titan's general circulation) predicts that near-surface winds at low latitudes are predominantly easterly (i.e. blowing westwards), which clashes with the appearance of the dunes. Tokano (2008) reported on GCM experiments in which he introduced Xanadu as a large positive relief feature (i.e. a hill) and changed its albedo in an attempt to modify the winds but was unable to generate low-latitude westerlies.

This is a fundamental issue that is not understood – a simple consideration of the overall planetary angular momentum balance suggests low-latitude easterly winds at the surface. The answer could lie in a non-uniform distribution of topography (e.g. more mountains at middle latitudes) that modifies the surface drag balance. Another possible factor is mountain torque, where higher pressure on one side of a mountain range than on the other produces a torque, even without winds flowing. The 1-mbar pressure bulge due to the gravitational tide in Titan's atmosphere (Tokano and Neubauer 2002) could be significant in this regard. Measurements of Titan's apparently changing spin state by correlated RADAR images (Lorenz et al. 2008b) may help understand what controls the surprising surface wind pattern.

While the dunes almost exclusively indicate eastward sand transport, there are regional deviations of up to 45° (see Fig. 5.42).

The pattern remains to be fully interpreted, although one immediate impression is of deviation around the 2,500-km continental-scale feature Xanadu – much as dunes deviate around bright obstacles only a few km across. There may also be some convergence of flow towards the center of large dune fields like Belet, which might be a 'sea-breeze' effect by which daytime solar heating of a dark dune field causes updrafts and convergence.

5.6.2 Fluvial Features and Processes

Surface conditions on Titan differ from those on Earth, yet the similarity of Earth's and Titan's fluvial features, especially at the Huygens site, suggests comparable physical processes. Titan's surface temperature of 94°K is too cold for liquid water to be stable, so that methane and ethane are the only substances that will remain liquid on the surface. Only methane has a vapor pressure high enough to participate in a hydrological cycle (Lorenz et al. 2008a). Its viscosity at 95 K is $1.8 \cdot 10^{-4}$ Pa s, approximately five times lower than that of water at 298°K ($9 \cdot 10^{-4}$ Pa s) (Hanley et al. 1977; National Institute of Standards and Technology, NIST (2005). Liquid methane (CH_4) had been suggested as the major fluid on Titan (Lunine 1993) long before the Huygens results confirmed that parts of Titan's surface have been modified (Tomasko et al. 2005; Soderblom et al. 2007a) by fluvial processes. Based on the fundamental hydraulic formulae for surficial flow, Burr et al. (2006) demonstrated that the required conditions for sediment entrainment and transport by liquid methane on Titan are within a factor of a few (for organic sediment) or several (for water ice) of those required by liquid water flow on Earth or Mars. On a lower-gravity body, the same discharge requires greater width and depth than on a higher-gravity body as flow velocity is lower due

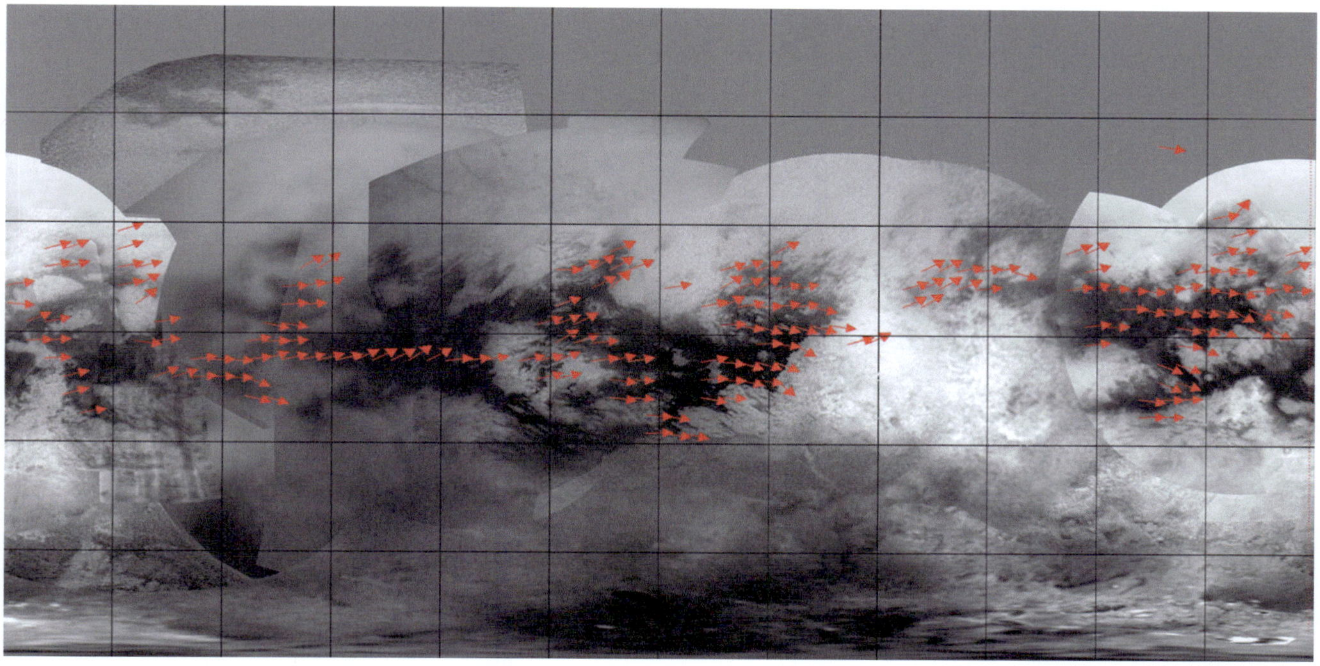

Fig. 5.42 Compilation map of radar-determined dune wind directions overlain on an ISS base map. The Radar coverage is quite incomplete, but dunes are seen on essentially all the optically dark low-latitude terrain

to weaker driving forces (Moore et al. 2003). Assuming similar discharges for comparable slopes and a channel-floor roughness similar to unconfined beds on Earth, flow velocity, depth and width can be adjusted to the lower gravity on Titan (Jaumann et al. 2008). Thus, channels on Titan should have a depth, width and velocity of 1.39, 1.61 and 0.46 times those of unconfined erosional channels on Earth. Although the necessary flow velocity is generally lower for Titan than for Earth or Mars, the required flow depth for high-density organic sediment on Titan is higher, suggesting that similar overland flow processes occur on Titan as on Earth and Mars (Burr et al. 2006). According to the scaling of sediment transport relations to conditions on Titan by Burr et al. (2006), frictional shear velocities of 0.5–7 cm/s, unit discharges (mass flow rate per unit area or flow velocity times depth) of 0.005–8 m^2/s and shallow flow depths of 0.2–6 m at a slope of 0.001 seem to be sufficient to move grains 0.1 mm – 15 cm in size. Rounded material of this size is exposed at the Huygens landing site (Tomasko et al. 2005).

Fluvial systems on Earth are characterized by overland flow in the interfluve areas and by channalized flow forming dendritic and branching patterns of surface erosion (Collins, 2005). The sediments entrained and transported by surficial flow on Earth and Mars are materials derived from the planets' crust as well as from secondary weathering. Sediment on Titan would consist of water ice derived from Titan's outer water-ice crust that is broken up by mass wasting and fluvial processes (Lorenz and Lunine 1996; Griffith et al. 2003) as well as of organic material which, having been formed by photochemical reactions of hydrocarbons, settles from the atmosphere (Khare et al. 1978; Tran et al. 2003; Lorenz and Lunine 2005). At Titan's surface temperatures the brittle behavior of water ice on a small scale is similar to that of rocks on Earth, reaching the compressive and tensile strength of granite (Durham et al. 1983; Cuda and Ash 1984), whereas organic material is weaker and can easily be moved.

Images collected by the Huygens descent imager/spectral radiometer (DISR) during the probe's descent through Titan's atmosphere have revealed the presence of branching networks. At the Huygens site, two types of branching networks are exposed: one pattern of highly dendritic dark networks trending primarily from west to east and a second linear depression oriented northwest to southeast and fed from both sides by many isolated dendritic networks. The systems differ in the sinuosity and complexity of their tributaries, which indicates different maturity, with the NW–SE trending system appearing to be tectonically controlled. Stereoscopic analysis shows that these dark networks are, indeed, elongated topographic depressions comparable to 'valleys' (Tomasko et al. 2005), with the dark material covering only their inner part ('floor') (Jaumann et al. 2008). As fluvial landforms (channels) on Earth and Mars (Irwin et al. 2005; Jaumann et al. 2005) are known to be significantly narrower and shallower than the valleys they occupy; the same was expected on Titan. Another indication that the branching features are valleys formed by fluid is their obvious order of tributaries, which grow in size from the smallest ones at the highest-order tributaries to the largest. The fact that the highest-order segments are terminated by a sharp transition between the bright terrain that harbors the networks and the dark plains-like terrain where Huygens landed is also consistent with this interpretation. Given these similarities to terrestrial river networks (Leopold et al. 1964) and the fact that fluvial channels require sediment entrainment and transport for their formation, the networks are consistent with the suggestion that the dark curvilinear branching features on Titan are, in fact, fluvial valleys (Porco et al. 2005; Elachi et al. 2005; Lorenz et al. 2008a; Jaumann et al. 2008), necessitating surficial and channelized overland flow. We used the term 'valley' throughout this paper to refer to the elongated topographic depressions visible in the optical and RADAR data, and the term 'channel' to describe that part within the valley through which fluid flows and which is much smaller than the dark branching features.

Although methane has a short lifetime (~10^7 years (Strobel 1982)) its high concentration in the atmosphere implies that the atmosphere is recharged by reservoirs at the surface and subsurface (Sotin et al. 2005; Tobie et al. 2006), in which case these fluvial surface features are part of this cycle and should be common on Titan. From RADAR and VIMS as well as ISS observations, numerous examples of these valley networks have been identified (Porco et al. 2005; Tomasko et al. 2005; Stofan et al. 2006; Perron et al. 2006; Barnes et al. 2007a, b; Lorenz et al. 2008a; Jaumann et al. 2008). Fluvial features and valley networks are exposed at all latitudes from the equator to the poles (Table 5.1; Fig. 5.43) (Jaumann et al. 2008).

Table 5.1 So far published valley-like features as identified in VIMS and RADAR data (Barnes et al. 2007b; Porco et al. 2005; Stofan et al. 2006; Tomasko et al. 2005; Perron et al. 2006; Lorenz et al. 2008a; Jaumann et al. 2008). Adapted from Jaumann et al. 2008

	Length	Width	Position	References
A	~100–200 km	800–2,000 m	50°W/12°S	Barnes et al. (2007)
B	<300 km	800–1,000 m	123°W/9°S	Barnes et al. (2007)
C	~50–100 km	800–2,000 m	66°W/10°S	Barnes et al. (2007), Lorenz et al. (2008), T13, Fig. 5.7
D	~1,000 km	10–20 km	Various, mainly south polar region	Porco et al. (2005)
E	10–200 km	500–1,000 m	75°W/23°N	Stofan et al. (2006), Lorenz et al. (2008), T3, Fig. 5.3
F	<5 km	17–100 m	192°W/10°S	Tomasko et al. (2005), Perron et al. (2006)
G	140 km	350–1,400 m	140°W/8°S	Jaumann et al. (2008)
H	1,200 km	1,400–3,000 m	192°W/10°S	Jaumann et al. (2008)
I	<50 km	<500 m	85°W/50°N	Lorenz et al. (2008), TA, Fig. 5.2
J	<100 km	<3,000 m	15°W/50°S	Lorenz et al. (2008), T7, Fig. 5.5
K	<150 km	1–2 km	10°W/60°S	Lorenz et al. (2008), T7, Fig. 5.6, speculative
L	>100 km	500–1,000 m	345°W/75°N	Lorenz et al. (2008), T25, Fig. 5.8

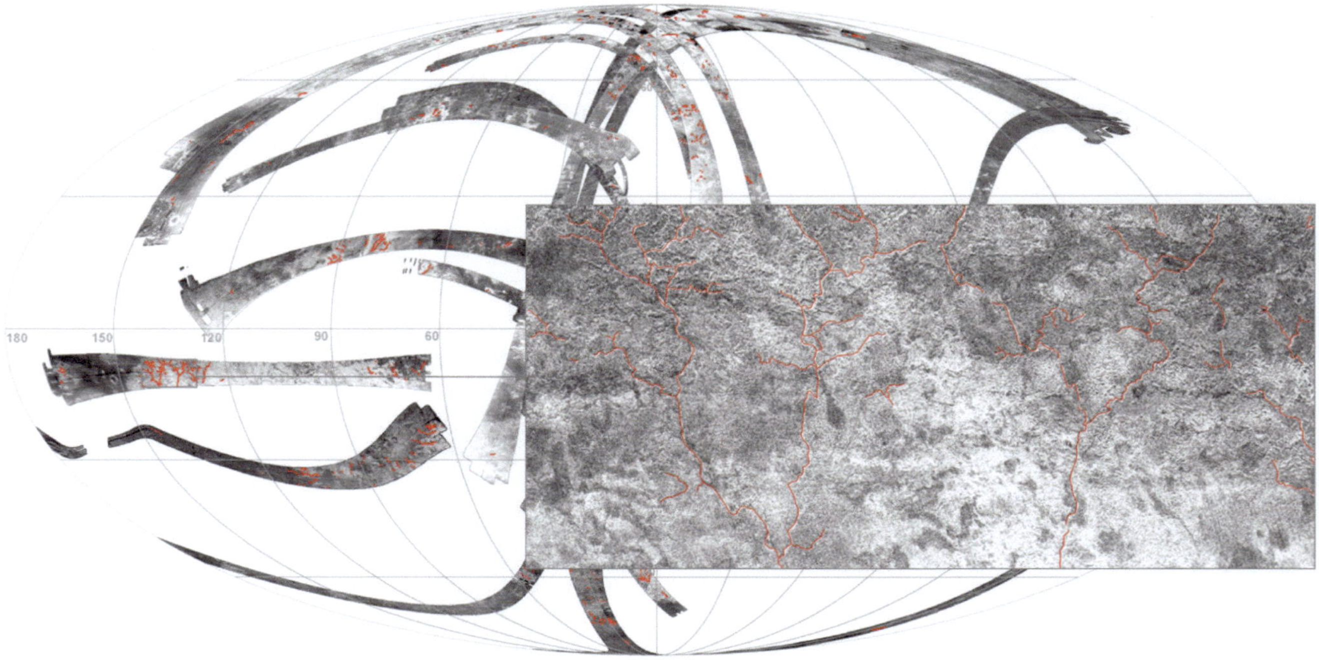

Fig. 5.43 Global RADAR-map of Titan (Mollweide projection). Distribution of putative fluvial channels are marked in *red*. Cassini RADAR: T00A, T003, T007, T008, T013, T016, T017, T018, T019, T021, T023, T025, T028 (Oct. 2004 – Apr. 2007) (after Langhans et al. 2009)

However, due to restricted resolution of image data available (RADAR >300 m; VIMS >500 m and ISS >1 km), the distribution of valley networks appear biased to large systems, while Huygens-like locally developed networks are 'hidden' on a global scale. Yet the global distribution of the larger systems and the presence of small-scale systems at the landing site suggest a global distribution of all sizes of valley networks on Titan.

Within the three major geological systems on Titan – high-standing bright terrains, darker plains, and dunes – valley networks correlate with bright terrain and, in some places, also with plains, but dunes are free of channel-like features (Barnes et al. 2007a, b; Lorenz et al. 2008a; Jaumann et al. 2008). The morphometric properties of valley networks vary widely (Fig. 5.44).

The longest channels identified so far have not been observed end-to-end, but they have a characteristic length of a few hundred kilometers, some reaching more than 1000km (e.g. Fig. 5.45) (Porco et al. 2005; Barnes et al. 2007a, b; Lorenz et al. 2008a; Jaumann et al. 2008).

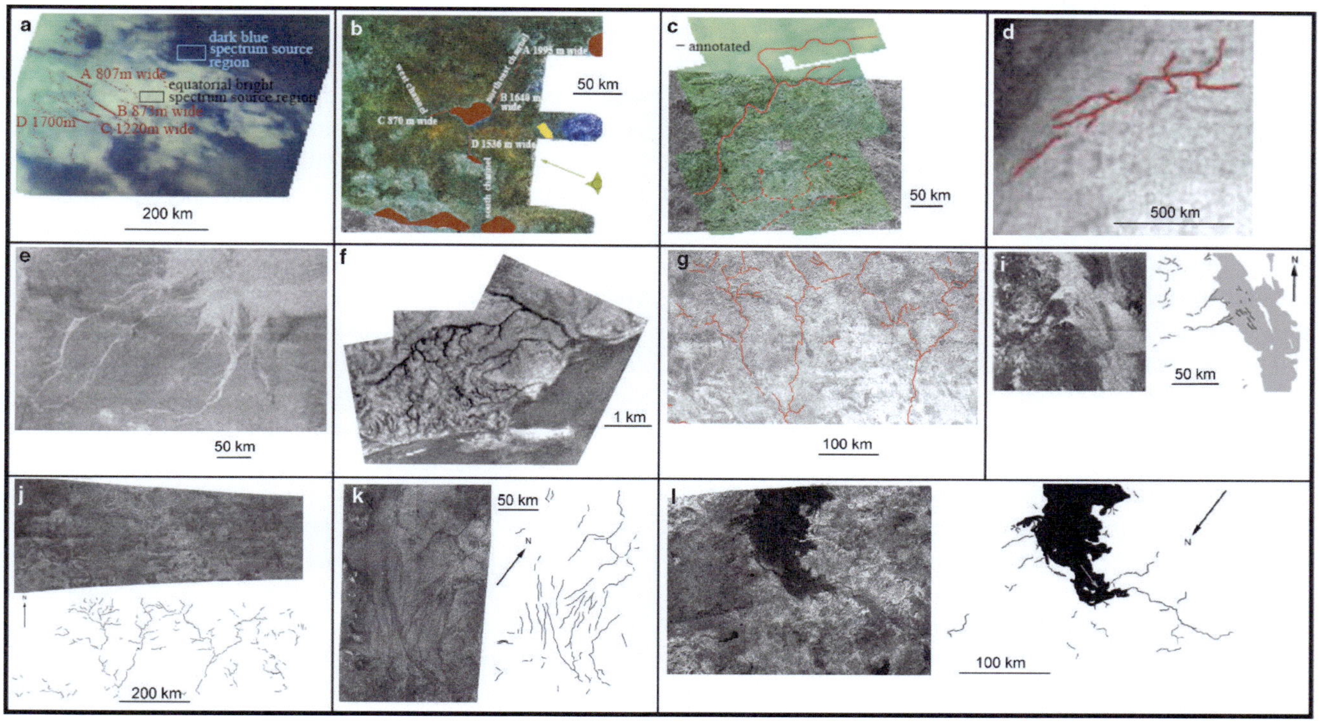

Fig. 5.44 Channel-like features so far published (***a–c*** Barnes et al. 2007b; ***d*** Porco et al. 2005; ***e*** Stofan et al. 2006; ***f*** Tomasko et al. 2005; Perron et al. 2006; ***g*** Jaumann et al. 2008; ***i-l*** Lorenz et al. 2008a; see also Table 5.1)

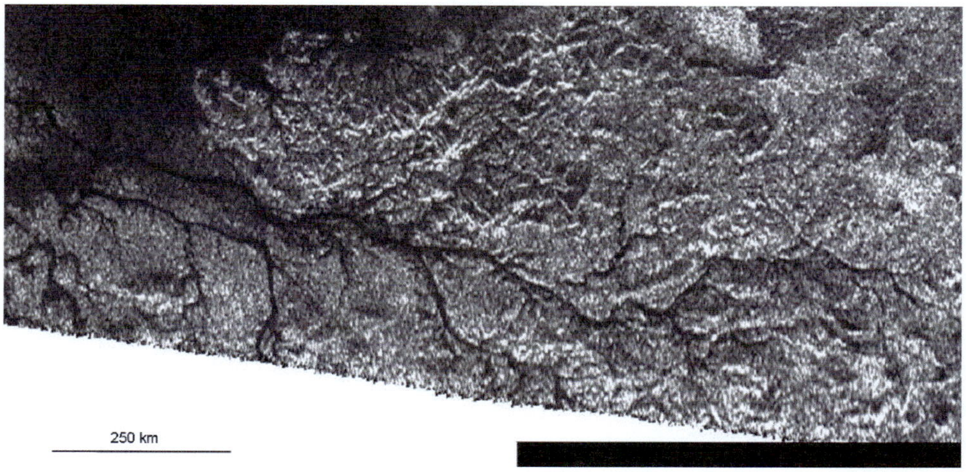

Fig. 5.45 Channel system at 255°W, 75°S (H in Table 5.1) as identified by RADAR observations during orbit T28 (1.4 km/pixel resolution)

Valley widths vary substantially from the resolution limit of ~300 m for RADAR and ~ 500m for VIMS to a few kilometers (Table 5.1). RADAR data indicate very shallow channels with almost no apparent depth that suggest an easily erodable substrate as well as systems cutting more than 100 m deep into more resistant bedrock (Lorenz et al. 2008a). RADAR measurements of some channels indicate negative relief with walls sloping ~20°, implying an incised depth of over 500 m (Lorenz et al. 2008a). Valley networks on Titan are mostly dendritic with at least six traceable stream orders. However, considering resolution restrictions and the large width of the highest-order channel segments identifiable, the stream order of the larger systems could be much higher. Also, there is no widespread evidence of subsurface control of stream patterns, although a few straight segments joining at wider angles (c.f. virgae) may constitute isolated examples of surface/subsurface control (e.g. Figs. 5.25 and 5.44d), although they may also be due to surface slope effects (Lorenz et al. 2008a). However, the well-developed dendritic systems and the relative lack of rectangular tectonic control indicate amounts of fluid sufficient to carve out the channels, at least at the time of their formation.

The hierarchical arrangement of tributaries and their dendritic behavior in combination with the sinuous shape of their reaches appear similar to that of valley networks on Earth, which is indicative of a distributed source of liquid. They clearly differ from the short and box-shaped tributary systems on Mars that primarily originate from groundwater sapping or seepage erosion. Meandering and braiding or anabranching indicate low gradients, at least at the time when valley networks formed (Lorenz et al. 2008a). The sinuosity of meandering rivers on Earth ranges from about 1.2 to 2.5 (Leopold et al. 1964), and Lorenz et al. (2008a) found similar values on Titan. The morphometric parameters of Titan's valley networks are an interesting result of fluvial processes involving a fluid of about 1/2 the density and about 1/5 the viscosity of water on a body with only 1/7 the gravity of Earth. Although quantitative analyses of Titan's valley networks still lack topographic information and better resolution, observations demonstrate that, independently of the fluid involved, erosion happens as a fundamental geological process wherever liquids are available.

Some channels are radar-bright (high radar backscatter), which suggests they are filled with rough boulders; others are radar-dark (low radar backscatter) and may either contain liquids or deposits of ice or organic sediments on their floors, with particle sizes smaller than the radar wavelength (Lorenz et al. 2008). In some places, valley networks are covered by RADAR as well as VIMS data (Barnes et al. 2007b), which facilitates corroborating the theory that these features are channels and investigating their near-infrared spectral properties. The sharpest color images were obtained from a false RGB color composite of band ratios using VIMS channels at 1.29/1.08 μm, 2.03/1.27 μm and 1.59/1.27 μm (Barnes et al. 2007a; Soderblom et al. 2007a; Jaumann et al. 2008). This band ratio imaging technique permits detecting spectral heterogeneities. Based on these ratios, three spectral units can be distinguished: Whitish material mainly distributed in the topographically high areas; bluish material adjacent to the bright-to-dark boundaries; and brownish material that correlates with sand seas (Soderblom et al. 2007a; Lorenz et al. 2006a; Barnes et al. 2007a). Although the spectral units are distinct, their composition is not known at this time (McCord et al. 2006). Bright materials may consist of precipitated aerosol dust composed of methane-derived organics (Soderblom et al. 2007a) superimposed on water-ice bedrock. The bluish component might contain some water ice as its defining feature (Soderblom et al. 2007a). It is no simple matter, however, to distinguish between specific organics and ices because all these molecules have comparable absorptions, resulting in similar spectral slopes. In addition, differences in particle size will have an effect on the depth of absorption bands and related spectral slopes. Nevertheless, the spectral signatures are real and indicate composition differences and/or changes in particle size. In VIMS data, bluish material is exposed where RADAR indicates valleys and channels (Barnes et al. 2007a; Jaumann et al. 2008). Thus, the material of the bluish unit may be deposits on valley floors.

Although these features in Fig. 5.46 are at the resolution limit of the VIMS, the pixels covering the valleys are distinct from surrounding pixels: the latter resemble bright material, whereas the valley pixels are mostly the same as those of dark bluish boundary regions bordering bright terrain (Fig. 5.44) (Barnes et al. 2007b; Jaumann et al. 2008).

Burr et al. (2006) estimated that unit discharges (mass flow rate per unit area or flow velocity times depth) of 0.005–8 m^2/s are required to move 0.1-mm–15-cm particles; these values translate into discharges of 0.05–80 m^3/s for 10 m channel width, and 5–8,000 m^3/s discharge for 1,000 m channel width, respectively (Jaumann et al. 2008).

On Earth, active channel width and discharge can be correlated empirically (Osterkamp and Hedman 1982). Adapting this correlation to Martian channels yields reasonable results (Irwin et al. 2005). Adapting the active channel width-to-discharge correlation to Titan's surface conditions (Jaumann et al. 2008) yields discharges for short recurrence intervals on Titan that range from <1 m^3/s for the smallest identified channel segments (about 10 m wide) at the Huygens landing site to 1,600 m^3/s for the widest active channel segments (1,000 m wide) of valley systems so far observed in RADAR and VIMS data. At longer recurrence intervals, these values increase depending on channel width from 66 to 7,950 m^3/s for 10 years recurrence, and to 178 m^3/s peaking at 13,510 m^3/s for 100 years recurrence, respectively (Jaumann et al. 2008). Discharges for recurrence intervals >10 years are

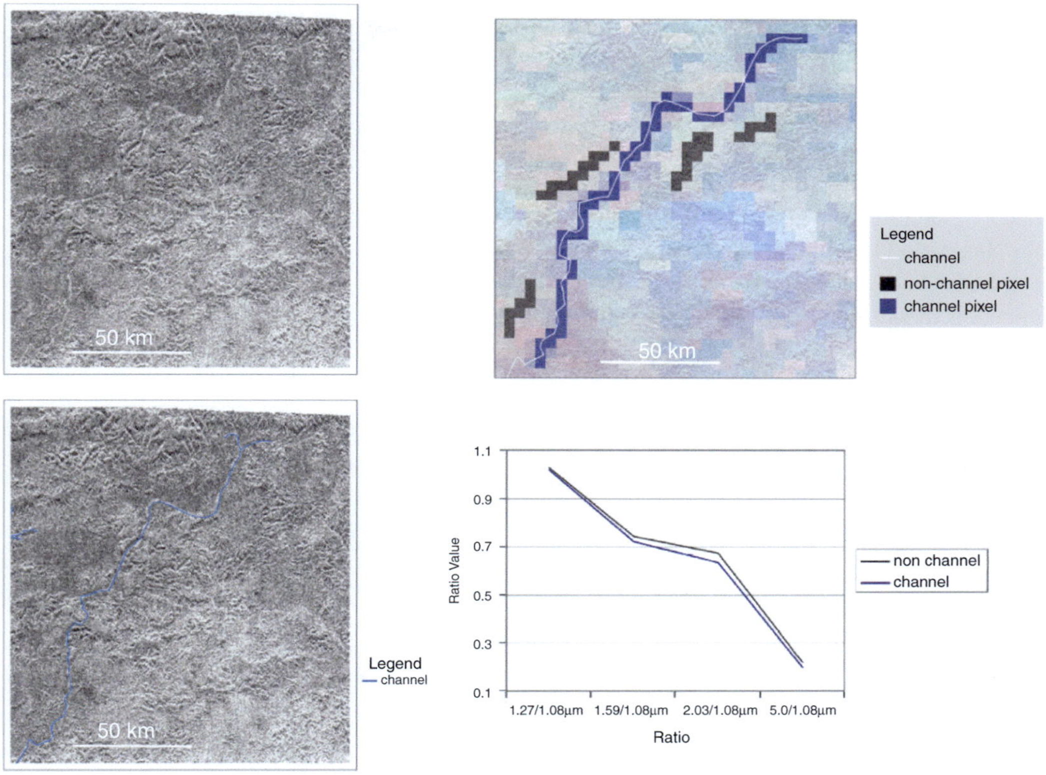

Fig. 5.46 Valleys in Xanadu at 12°S, 125°W as identified in RADAR (30/04/2006, T13) (*top and bottom left*) with 350 m/pixel resolution and VIMS (13/12/2004, TB) with 4.5 km/pixel (*top right*). The spectral characteristics within the valley are distinct from those of adjacent areas and are comparable with the bluish unit identified in false color images

most consistent with the unit discharge values expected for sediment transport under Titan's conditions (Burr et al. 2006), indicating at least periodical erosion events.

Although the available resolution and the lack of detailed topographic information prevent the exact identification of watersheds, connecting the starting points of the smallest recognizable tributaries defines an area which corresponds to the drainage area of a channel network. Based on this approach, Jaumann et al. (2008) estimated runoff production rates for small as well as large channel segments. Runoff production rates are highest in small, low-order channel networks like that observed at the Huygens landing site, with small catchment areas and steep slopes. The cobbles identified at the Huygens landing site can be easily moved within a recurrence interval of about 10 years, corresponding to a runoff production rate of about 60 mm/hr (Jaumann et al. 2008). In higher-order systems, runoff production rates are significantly lower at short recurrence intervals (<1 mm/h) and reach about 50 mm/hr at a 1,000 years recurrence interval. However, in the wider, higher-order channels, materials tend to be more crushed and segmented due to the longer transport history. Thus, even at the mouth of an extended valley system, sand and small cobbles of centimeter size can be transported in reasonable times (<100 years) (Jaumann et al. 2008). This is consistent with the rates (0.5–15 mm/h) needed to move sediment of 1–15 mm (Perron et al. 2006). Modeled convective methane storms on Titan, producing a rainfall of 110 kg/m^2 within 5–8 h (20–50 mm/h) (Hueso and Sanchez-Lavega 2006), are sufficient to trigger such flash floods. Even moderate storms with a rainfall of 30 kg/m^2 within 5–10 h (6–12 mm/h) will explain the observed runoff rates. On the other hand, precipitation rates as estimated from current atmospheric conditions might not exceed 1 cm/year (Lorenz 2000; Rannou et al. 2006; Tokano et al. 2006), which is comparable to terrestrial deserts. To satisfy the overall energy balance of the hydrological cycle, the heavy rainstorms indicated above must be infrequent – perhaps centuries apart, which is consistent with the estimated runoff production rates. Even if we assume slow drizzle (Tokano et al. 2006), long-term accumulation of liquid in the subsurface might entail the formation of large underground reservoirs. Such accumulations could be subject to cryovolcanic and tectonically-induced melting and/or release (Sotin et al. 2005) of large amounts of liquid that form in combination with unconfined material, as in lahar-like mud flows (Fortes and Grindrod 2006) or hyper-concentrated flows (Burr et al. 2006).

The observed valley and channel morphologies and the erosional parameters are at least consistent with fluid surface flow and marked variations in stream flow (Lorenz et al. 2008a;

Jaumann et al. 2008). Applying a global circulation model, Rannou et al. (2006) found a variety of cloud types prevailing at different latitudes and seasons. In particular, they noted that low latitudes would dry out unless methane was resupplied from the surface or subsurface. In contrast, high latitudes act as sinks for methane, and thus become more or less permanently saturated. More specifically, they predicted 'sporadic' large clouds at middle latitudes in summer, and similarly sporadic but smaller-scale clouds at low latitudes. An independent model has been developed by Mitchell et al. (2006) which similarly predicts sporadic seasonal rainfall at all latitudes, but persistent humidity near the poles. Rain therefore appears to occur everywhere, but methane will evaporate more rapidly at low latitudes. This paradigm seems broadly consistent with the observations of Jaumann et al. (2008) and Lorenz et al. (2008), with erosion and removal of fines at low latitudes and generally depositional channels at higher latitudes. The evidence that rivers can flow at low latitudes for more than 400 km before evaporating will serve as a valuable joint constraint on the hydraulic and meteorological setting for rivers on Titan (Lorenz et al. 2008a). However, because erosional and depositional processes are found in the equatorial and lakes in the polar regions, and since recent large-scale storm activity was concentrated near the South Pole (Porco et al. 2005; Schaller et al. 2006a, b), we cannot determine at this time whether the observed erosion (1) was mainly driven by paleoclimatic processes, (2) is a seasonal effect or (3) is due to volcanic outgassing, and to what extent these processes are responsible for increases or decreases in surface erosion. More global repeated observations of erosional features are needed to further constrain the global circulation and the related methane cycle.

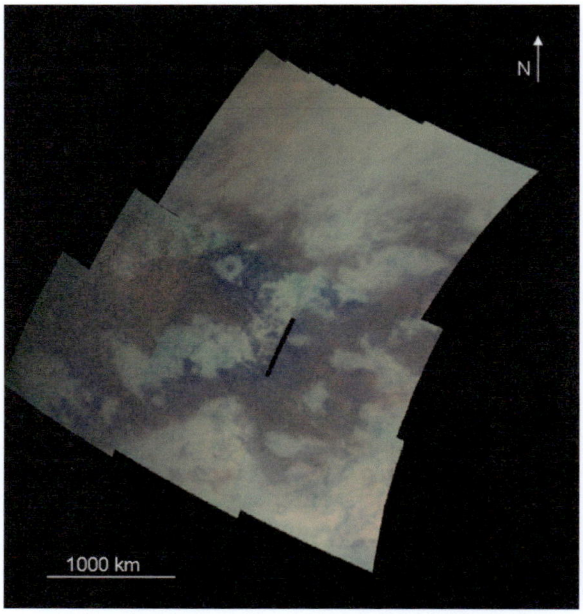

Fig. 5.47 Mosaic of VIMS data covering the Quivira/Aztlan region (0°–40°W, 10°S–10°N. Wavelength ratios at 1.29/1.08 μm, 2.03/1.27 μm and 1.59/1.27 μm were composed to generate *blue*, *green* and *red false-color images* in order to enhance the overall spectral contrast

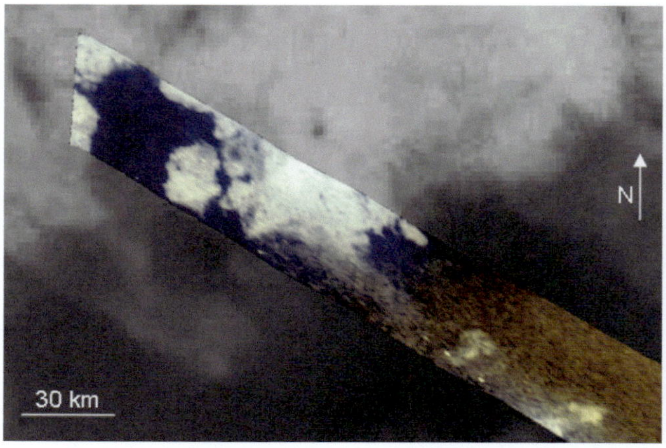

Fig. 5.48 VIMS observed an area at 30°W and 7°S in the Quivira/Aztlan region that marks the bright-dark boundary at a prominent indentation. Wavelength ratios at 1.29/1.08 μm, 2.03/1.27 μm and 1.59/1.27 μm were used to generate blue, green and red false-color images in order to enhance the overall spectral contrast. The transition from bright high-standing material through bluish material to brownish dunes is clearly indicated. VIMS observation overlain on ISS data (orbit T20) (adapted from Jaumann et al. 2008)

5.6.3 Depositional Features

The main optical characteristics of Titan's surface are bright and dark regions. Comparisons with RADAR observations (Elachi et al. 2006) and DISR measurements (Tomasko et al. 2005) at the Huygens landing site indicate that bright material is topographically high at lower latitudes, whereas dark colored material tends to form lower-lying plains (Soderblom et al. 2007a). Boundaries between different spectral units are interfingered (Fig. 5.47).

As mentioned above, bluish equatorial areas are dune-free and adjacent to bright high-standing material. So far, the highest-resolved VIMS observations were acquired during Cassini's 20th Titan flyby (October 25, 2006), showing a bright-dark boundary region at about 30°W and 7°S (Fig. 5.48), with a dark indentation merging with bright material.

The northernmost part of this area is covered by the highest-resolution spectral image cube with 500 m/pixel (Fig. 5.49) (Jaumann et al. 2008).

The bay-like area is located between high-standing bright material in the north, west and east and dune material in the south (Fig. 5.48) (Jaumann et al. 2008). Spectra of this area decrease in albedo and spectral slope from northeast to

Fig. 5.49 The northernmost part of the *dark indentation* shows a 25-km-wide protrusion with details at a resolution of about 500 m/pixel. Bright material is separated from dark material by a distinct boundary. Within the dark materials, absorptions and spectral slopes increase from the north to south indicating a change in material composition, density and/or particle size (Jaumann et al. 2008)

southwest (Jaumann et al. 2008). This variation might be due to selective separation of materials caused by density differences (e.g. the density of liquid methane (National Institute of Standards and Technology, NIST (2005), water ice (National Institute of Standards and Technology, NIST (2005; Burr et al. 2006) and organics (Hanley et al. 1977; Khare et al. 1994) at 95 K and 1.6 bar is 450 kg/m^3, 992 kg/m^3 and 1,500 kg/m^3, respectively) and/or by differences in particle size, as observed at the Huygens landing site. Thus, the bluish zone resembles an area where material is arranged spectrally in a systematic way, clearly separated from the bright high-standing units and dune-covered areas, suggesting that this region may covered with eroded and transported sediment.

Processes that can displace icy bedrock materials and rearrange them in a systematic manner might cause these specific spectral correlations. The major processes for moving surface material are aeolian, glacial and fluvial erosion and transport.

The freedom from dunes and the systematic albedo and spectral-slope patterns, which differ from the usually random distribution of windblown deposits, do not support the aeolian origin of these materials. In particular, the clear spectral distinction between bluish and brownish materials in the color ratios (Fig. 5.48) suggests that these are two separate surface units produced by different formation processes. Soderblom et al. (2007a) demonstrated, based on RADAR and VIMS data, that bluish areas are completely uncovered by dune material, which abruptly terminates at this specific spectral boundary.

A glacial deposit might be consistent with the albedo features seen in the bay-like area south of Quivira (Jaumann et al. 2008). Morphologically, they resemble moraine-like deposits such as those formed by the movement of glaciers. However, similar features at the Huygens landing site do not show glacial morphology (Tomasko et al. 2005) and are not interpreted to be glacial in origin. In addition, the usually poorly sorted appearance characteristic of glacial deposits does not agree with the systematic spectral slope pattern (Jaumann et al. 2008). Moreover, no glacial features have been reported in RADAR or other VIMS observations so far, nor are they expected, given the thermodynamics of the present climate (Lorenz and Lunine 1996). Lastly, the surroundings of northern bay-like area do not show any evidence of glacial movement.

On the other hand, there is evidence of fluvial processes at the Huygens landing site and elsewhere on Titan (see Section 6.4 and 6.6.2). The systematic spectral patterns and albedo variations of the units at the bay-like area, which seem to be due to systematic changes of composition and/or particle size, as well as their position between the high-standing lands in the north and the dunes in the south (Fig. 5.48) suggest that materials accumulated there during the low-velocity phase at the end of a flood event. In total, the bluish accumulation zone extending south from the bay-like area to the brownish dune area covers over 20,000 km^2, marking a huge sediment accumulation zone.

The material covering the bluish plains apparently originated from the bright Quivira region (Fig. 5.47) surrounding the bay-like area (Jaumann et al. 2008).

Cryovolcanism-induced melting (Sotin et al. 2005) can produce a sudden release of large amounts of liquids which combine with unconfined material to form lahar-like mudflows (Fortes and Grindrod 2006). Within such flows, material settles out as flow velocity decreases, providing systematic sediment characteristics like those observed in the bluish plains. Nevertheless, the major argument against the cryovolcanic-tectonic explanation is the lack of large volcanic constructs in the regions adjacent to the bay-like area.

To account for the widespread plain deposits of fine-grained material, a release of large fluid volumes is required. High-energy flow will cause mechanical weathering and large accumulations of fine material in alluvial fans that can contribute towards building dunes, at least in the transition zone between bluish plains and vast dune fields. Violent equatorial storms with heavy rainfalls of methane could explain such erosion (Jaumann et al. 2008).

Fluvial surface runoff of sediment-loaded liquid and sediment settling due to decreasing flow velocity is compatible with the spectral observations in northern bay-like area, and liquid release by precipitation and surface runoff seems to explain these dune-free plains plausibly as accumulation zones.

5.6.4 *Lakes*

Prior to Cassini, speculation that Titan had hydrocarbon seas was based on the detection of methane in the atmosphere by Voyager (Sagan and Dermott 1982; Lunine et al. 1983), and Earth-based radar observations did find some evidence of specular reflection (Campbell et al. 2003) that could have been consistent with equatorial surface liquids. Images from both the Huygens probe and the initial passes of the Cassini furnished evidence that liquids had carved channels into the surface (as described above) (Tomasko et al. 2005; Elachi et al. 2005, 2006; Porco et al. 2005) but no evidence of liquids currently on the surface. Extensive sand seas in the equatorial region (Lorenz et al. 2006a) indicated that surface liquids might not be globally distributed, or might be ephemeral.

These findings were consistent with the fact that relative methane humidity is less than 100% at all but the highest latitudes, indicating that any standing bodies of liquid would evaporate unless constantly replenished. From a theoretical standpoint, methane precipitation near the poles, the preferential deposition of ethane in polar regions (e.g. Samuelson et al. 1997), and the possibility of a subsurface liquid methane table indicated that the polar regions are the most likely place on Titan to find hydrocarbon lakes or seas (Lunine et al. 1983; Rannou et al. 2006; Tobie et al. 2006; Mitri et al. 2007).

ISS observations of the South Pole in July 2004 (T0) and June 2005 (Fig. 5.50) revealed more than 50 dark features, tens to hundreds of kilometers long, poleward of ~70°S (Porco et al. 2005; McEwen et al. 2005; Turtle et al. 2009).

Low albedos and observed morphologies suggest that these features could be hydrocarbon lakes or lake beds. Indeed, the rounded boundaries were sufficiently reminiscent of shorelines so that one of the most prominent features was given the official name Ontario Lacus by the IAU. Observations of nearby channels and south-polar convective methane clouds (Porco et al. 2005; Schaller et al. 2006a, b), as well as changes documented between the images acquired in 2004 and 2005 (Turtle et al. 2009), support the interpretation that these features are, or have once been, filled with liquid; however, these images alone cannot provide conclusive evidence for lakes. Near-infrared specular reflections, which would demonstrate the presence of surface liquids, cannot be observed by the ISS at high latitudes due to illumination geometry limitations and have not been detected at low latitudes (Fussner 2006).

The north polar T16 RADAR pass in July 2006 detected over 75 circular, radar-dark features thought to be hydrocarbon lakes (Stofan et al. 2007) (Fig. 5.51).

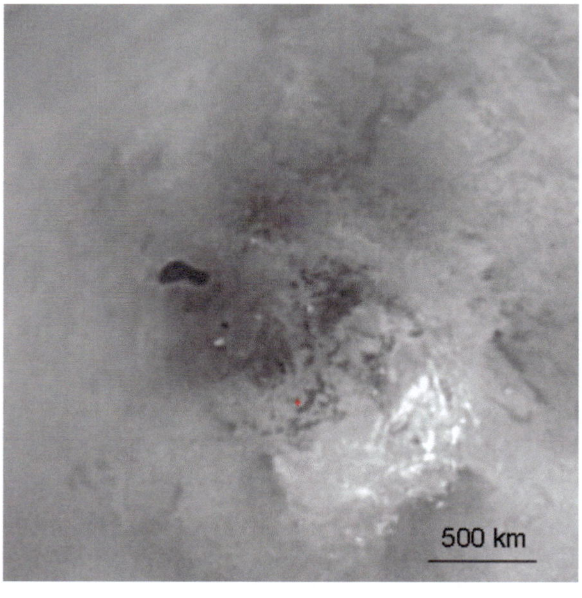

Fig. 5.50 ISS South Pole observation, including the 235-km-long Ontario Lacus to the *left of center*. + marks the South Pole

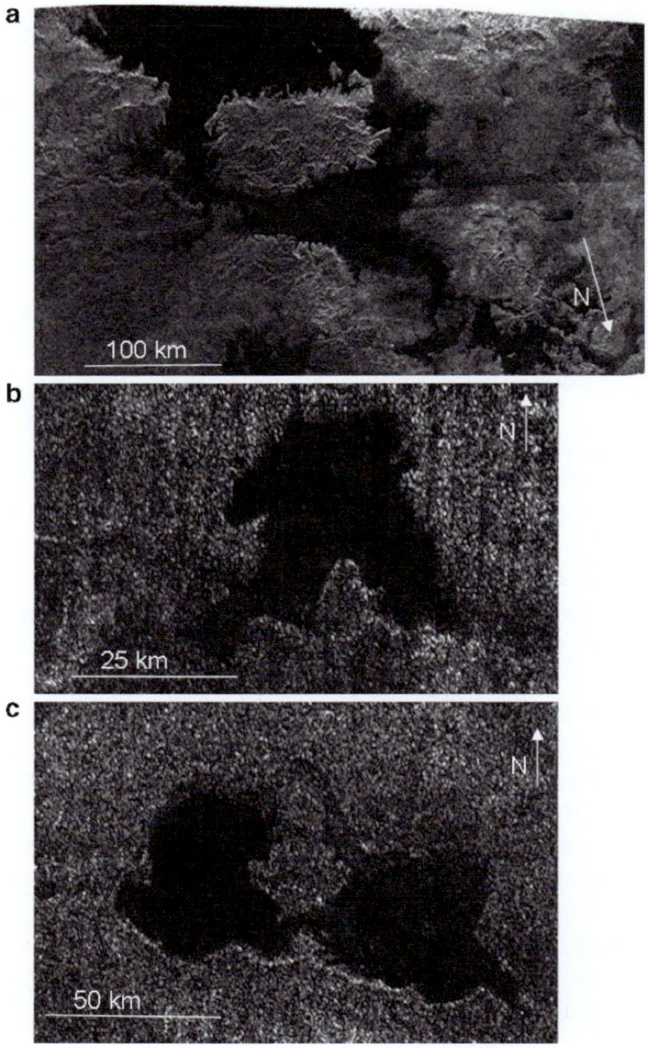

Fig. 5.51 RADAR images of Titan lakes. (**a**) 90 × 150 km island in large sea at 79N, 310W longitude. The shorelines of the large seas resemble drowned river valleys on Earth. NASA/JPL PIA09810. (**b**) Lake with multiple bays at 74N, 65W longitude. The lake is 20 × 25 km. NASA/JPL PIA08741. (**c**) Two lakes at 73N, 46W longitude with lobate shorelines. The lakes appear to have multiple shorelines, and the lake on the right is fed by a now apparently dry channel. NASA/JPL PIA08740

They are located poleward of about 70°N latitude and vary in shape from irregular to highly circular. The radar signature is strongly suggestive of these features being lakes; the return is extremely low – at the noise floor of the data. Such a low radar return is consistent only with a very smooth surface at wavelengths comparable to that of the Cassini RADAR (2.2 cm) alternatively, with a non-reflecting, absorbing surface. Given the association of the dark patches with channels, the location of many of them in topographic lows, and the fact that some of the depressions are partially filled led to the interpretation that the morphology and the radar return are most consistent with the dark material being liquid hydrocarbons (a mixture of methane and ethane that can be liquid under Titan conditions), although the interpretation that these features are playas cannot be ruled out, either. However, based on VIMS spectra, Brown et al. (2008) showed that Ontario Lacus in the southern hemisphere contains liquid ethane. The dark lake interior is surrounded by two separate annuli that follow the lake interior's contours. The inner annulus is uniformly dark, but not so much as the interior lake, and is generally 5–10 km wide at the lake's southeastern margin. Barnes et al. (2009) propose that it represents wet lakebed sediments exposed by either tidal sloshing of the lake or seasonal methaneloss leading to lower lake-volume.

Subsequent passes by RADAR (T18, T19, T25, T28, T29, T30) and more recent ISS mapping over the north polar regions have revealed over 650 dark depressions between about 55°N and the pole (Hayes et al. 2008; Turtle et al. 2009; Mitchell et al. 2009) (Figs. 5.4 and 5.52), with the same radar characteristics as measured before (Stofan et al. 2007; Hayes et al. 2008). From their albedo radar backscatter, their morphology and their association with channels, these features are most probably liquid although deposits of dry material cannot be ruled out.

Mitchell et al. (2009) classified lakes by shape (circular, irregular, nested, canyon-like, and diffuse), location of depressions, size, and nature of fill (unfilled, partially filled, nature of radar backscatter). They found five major types of lakes: circular rimmed, irregular, irregular rimmed, unfilled and seas. Lakes range in size from <10 km^2 to Kraken Mare the area of which exceeds 400,000 km^2. Circular rimmed lakes tend to be small, located in depressions and reminiscent of terrestrial crater lakes. Irregular lakes have scalloped edges, tend to become more radar-dark towards the interior, and are sometimes only partially filled. Some irregular lakes resemble flooded terrestrial river valleys. Irregular rimmed lakes are similar to irregular lakes but have prominent topographic rims. Unfilled lakes are topographic depressions with a lake-like morphology whose radar backscatter is similar to that of the surrounding terrain. Seas are the largest (>45,000 km^2), some with islands and many with an estuary/fjord-like shoreline morphology. ISS observations of high northern latitudes, acquired as the north pole emerged from winter, have filled in gaps between the RADAR swaths and illustrated the full extent of Mare Kraken (>1,100 km long) and other northern lakes and seas (Fig. 5.53).

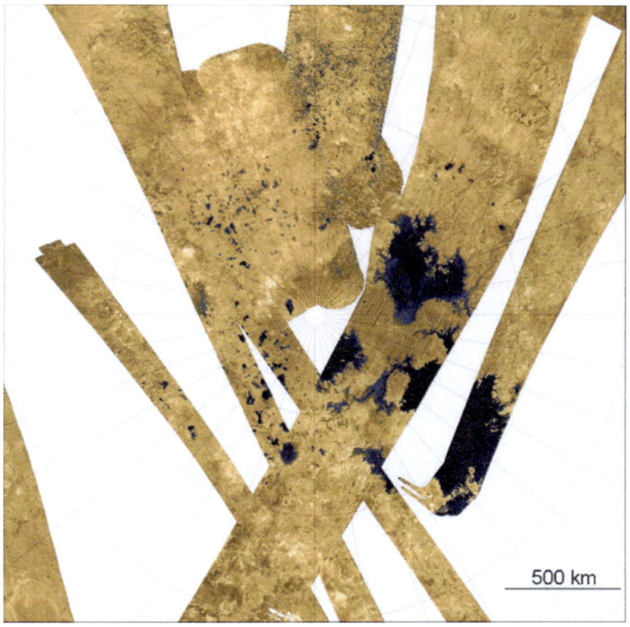

Fig. 5.52 This false-color RADAR mosaic of Titan's north polar region shows lakes in blue and black, and the areas likely to be solid surface in light brown. The terrain in the *upper left* of this *mosaic* is imaged at lower resolution than the remainder of the image. The lakes and larger seas are likely composed predominantly of liquid methane and ethane. The large sea in the upper right center has a minimum area of 100,000 km^2 (NASA/JPL PIA1008)

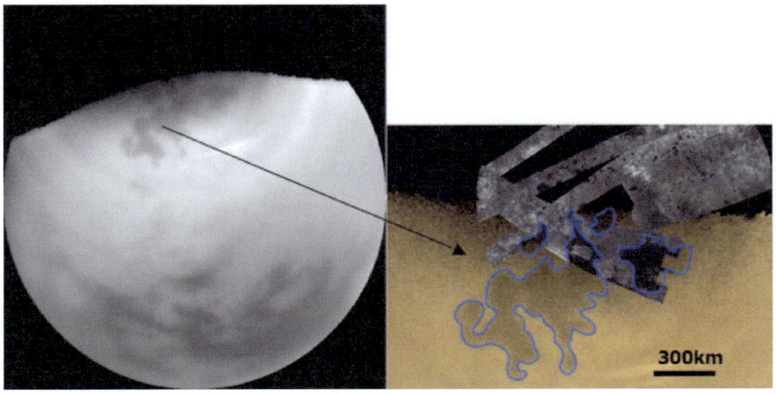

Fig. 5.53 Kraken and Ligeia Maria as seen by RADAR (*right*) (T25) and ISS (*left*) from February 2007. North is up. *Blue lines* illustrate the margins of lakes and seas

Mitchell et al. (2009) divided the north polar region into five areas, each with a distinctive lake morphology and distribution. Region 1 consists of an approximately circular cluster of scattered, small, irregular rimmed lakes, none of which appear to be fed by visible surface channels. Region 2 has all morphological types except seas, some channels and a relatively low density of lakes. Seas and the largest lakes are located in region 3, with channels feeding and connecting lakes. Regions 4 and 5 display low lake density, circular and irregular lakes and a high proportion of unfilled lakes. The cause of this morphological clustering is under investigation (Mitchell et al. 2009). Stofan et al. (2007) interpreted the fact that some lakes are associated with dark channels as indicating that they are drainage or groundwater drainage lakes, while the steep edges and lack of channels in other lakes suggests that they are seepage or groundwater drainage lakes. Of course, in the case of Titan, it is not groundwater but underground sinks of liquid hydrocarbons that feed lakes. Mitchell et al. (2009) observed that unfilled lakes occur in proximity to filled lakes at all northern latitudes where lakes are found, suggesting that lake fill levels are controlled by precipitation rates as well as subsurface and surface drainage.

Hayes et al. (2008) classified lakes by their radar backscatter into three classes: dark, granular (lakes with intermediate backscatter) and bright. The 394 dark lakes Hayes et al. (2008) studied are located above 65°N, with lakes closer to the pole showing less backscatter, suggesting they are possibly deeper. Granular lakes are located below 77°N and are found in proximity to bright lakes. Hayes et al. (2008) interpreted their brightening as consistent with the degree of radar penetration to the lake bottom. Bright lakes are the same as the unfilled class of Mitchell et al. (2009) and are 200–300-m deep depressions; Hayes et al. (2008) mapped 152 bright lakes. They modeled the rate of outflow from lakes using material properties from Huygens probe data, considering the case of a lake emptied by subsurface transport and evaporation and the case of a lake fed by subsurface liquid hydrocarbons until depleted by evaporation. These models indicate that the timescales of lake drainage are on the order of tens of years, consistent with evaporative and seasonal cycles.

To further constrain the nature of the lakes, Paillou et al. (2008) measured the dielectric constant of LNG (liquefied natural gas) mainly composed of methane, in the Ku-band, obtaining $\varepsilon=1.75$ - j0.002j. This value was then utilized to model the radar backscatter of some of the T16 lakes in a two-layer scattering model, with a layer of liquid methane overlying a rougher tholin substrate. They found that for a detection limit of about –27 dB, the Cassini radar can penetrate approximately 7 m of liquid methane at a mean incidence angle of about 33°. In 77 lakes studied, two groups were identified: 'bright lakes' ($\sigma 0 > -19$ dB), which corresponds to a likely depth of less than 2 m, and 'dark' lakes ($\sigma 0 < -22$ dB) that are deeper than 3 m. Paillou et al. (2008) found that 'dark lakes' correspond to those lakes that are larger than 100 km^2. The darkness of these large lakes supports the expectation that winds should be quite low on Titan (Tokano et al. 2008; Lorenz et al. 2005), as larger lakes would be the most likely to have surfaces roughened by wind.

The T39 pass by the Cassini spacecraft on December 20, 2007 marked the first RADAR imaging of Titan's south polar region. The swath contains two very dark features (Fig. 5.54), both associated with channels.

The very low RADAR return indicates that the features are liquid-filled lakes, consistent with the interpretation of the north polar features (Stofan et al. 2008). Both the south polar lakes have very diffuse margins and are highly circular. The T39 swath also contains a few other features that can be seen as an indication of unfilled or drained lakes. While the T39 swath represents a limited view of the south polar region, the density of lake features is much lower than that typically seen in the northern polar region. The broader view of ISS (Fig. 5.4) also demonstrates a difference in the amount of dark material at each pole (Turtle et al. 2009). Lunine et al. (2008b) proposed that the lower density of lakes most likely is the result of seasonal evaporation from the southern to the northern hemisphere, although there may be elevation differences between

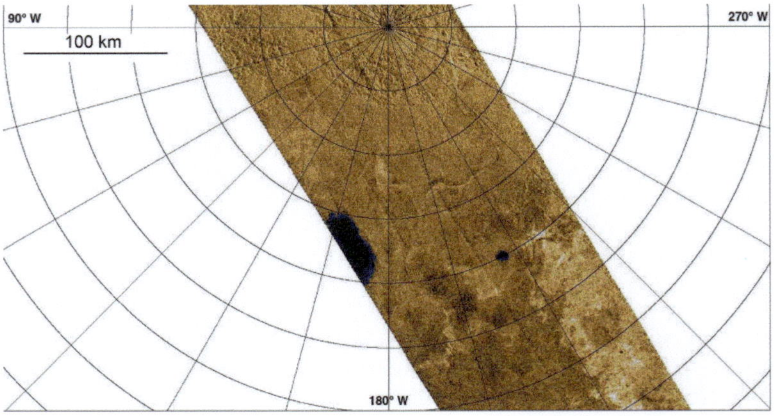

Fig. 5.54 RADAR image of south pole taken on the T39 pass on Dec. 20, 2007. The false color in this image is used to highlight two possible lakes (*blue*) located in the mountainous terrain near the south pole. As indicated by this swath, there are fewer lakes near the south pole compared to a comparable size region at the north pole. The polar projection of this image is approximately 351 m/pixel, and the radar illumination is from the *lower left* with incidence angles from about 22 to 38°. NASA/JPL PIA10218

the two poles that would result in defective retention of surface liquids in the south, or alternatively, differences in surface properties or geology that would favor subsurface retention in the south polar region. As it is currently the end of summer in the southern hemisphere, methane would have evaporated from the southern polar region and been transported north, which may explain the difference between the number of lakes observed by RADAR in December 2007 and the dark features observed by ISS 2.5 years earlier, in June 2005 (Turtle et al. 2009). Lunine et al. (2008b) calculated that lake depths in the south could not have been >20 m if they dried out completely (assuming pure methane). For ethane/methane mixtures, shallower lake depths are possible, but ethane would be left behind, allowing the lake to survive as such.

Three models of origin have been considered for the topographic depressions in which the lakes lie: impact cratering, volcanism and karstic processes (Stofan et al. 2007; Wood et al. 2007a, b; Mitchell et al. 2009). Under the first two mechanisms, lake liquids are not related to the origin of the depressions, while under the latter mechanism, depressions form by dissolution of surface materials by liquids.

Impact cratering was suggested as a possible origin for lake depressions owing to the highly circular shape of some lakes. Nonetheless, the size distribution of lake basins and their preferential location at high latitudes is not consistent with an impact origin (Mitchell et al. 2009). However, we cannot rule out that a few of the lakes may lie in impact craters, although it is probably not a dominant process in the formation of lake depressions.

The raised rims and bright haloes surrounding some lakes as well as their nested appearance led to the suggestion that some may be volcanic craters, with lakes occupying them like volcanic crater lakes on Earth (Stofan et al. 2007; Wood et al. 2007b). Possible calderas have been observed elsewhere on Titan (e.g. Lopes et al. 2007b), though most have associated lobate deposits rather than bright haloes. However, the concentration of lake depressions in the polar regions and the irregular shape of most lakes argue against a volcanic origin of most depressions. As with the impact origin, we cannot rule out that a limited number of lakes may be located in cryovolcanic calderas.

On Earth, water can dissolve soluble rock (e.g. limestone, dolomites, salt) at or near the surface to form characteristic karstic landforms including sinkholes (dolines) and caverns (e.g. Blair 1987). On Titan, lakes are much larger than the typical 1–2-km diameter of terrestrial dolines, but lake morphologies and clustering are consistent with a karstic origin (Mitchell et al. 2009). Consisting predominantly of water ice, the crust of Titan is not soluble in methane or ethane (Lorenz and Lunine 1996), so that if dissolution processes are operating, some other methane/ethane-soluble material layer must be present, at least in the polar regions. The organic materials that form in the atmosphere of Titan and are deposited on the surface (e.g. Yung et al. 1984) could form thick enough layers (100 s of meters) and may be composed of methane/ethane-soluble material (e.g. acetylene) (Mitchell et al. 2009). However, better knowledge of crustal compositions and methane/ethane solubility rates is needed.

A variation on the karst hypothesis suggests that lakes result from thermokarstic processes. On Earth, sinkholes and depressions can also form in periglacial areas when permafrost melts. However, the low insolation at Titan's poles makes this mechanism unlikely. The seas of Titan are larger than any suggested impact, volcanic or karst structures; they must be filling regional low spots.

Not all lake depressions have to be of the same origin: impact, volcanic and dissolution processes may all play a role in the formation of lakes on Titan. While a karst-like origin is preferred for the majority of lakes at this time (e.g. Mitchell et al. 2009), uncertainties about the nature of dissolution processes on Titan call for further observation and analysis.

Assuming a conservative average depth of 20 m for the lakes, Lorenz et al. (2008b) estimated that they contain approximately 8,000 km^3 of liquid hydrocarbons. Considering the unimaged portions of the north and south polar regions, they estimated that the total lake hydrocarbon inventory amounts to 10^4–10^5 km^3. However, Lorenz et al. (2008b) do not believe that this volume is able to buffer the amount of methane in the atmosphere of Titan over long timescales, suggesting that episodic injection of methane into the atmosphere by cryovolcanism is necessary (or climate change is likely) and probably has occurred in the past (e.g. Lorenz et al. 1999).

5.7 Cratering and Surface Ages

Impact craters are found on all solid bodies in the solar system, and have been detected on Titan, too. However, their number is so small that two major conclusions are immediately obvious. The surface of Titan has been very dynamic, as indicated by aeolian and erosional features as well as the presence of liquid lakes, and there are so few impact craters that most parts of the surface seem to be relatively young (Porco et al. 2005; Wood et al. 2006; Lorenz et al. 2007; Jaumann et al. 2008; Brown et al. 2008; Jaumann and Neukum 2009; Wood et al. 2009; Turtle et al. 2009). Removal of impact craters by burial and erosion is likely, given the evidence of fluvial, aeolian and cryovolcanic processes and the relatively degraded appearance of hills and ridges. However, the existence of the large Menrva impact structure suggests that larger (and thus potentially older) craters have been preserved in some places (Porco et al. 2005; Wood et al. 2006; Lorenz et al. 2007; Le Mouélic et al. 2008).

Although suspiciously circular features have been detected with optical sensors (Porco et al. 2005), only RADAR has the resolution and the topographic information to identify impact craters convincingly. Within the roughly 39% of Titan's surface imaged by RADAR, only a handful of impact features have been identified with any certainty, along with a few dozen others that may be older and eroded. By the standard of Rhea and other satellites of Saturn, tens of thousands of craters could be expected.

The largest impact feature yet identified on Titan is the 440-km-wide impact basin Menrva (Fig. 5.55).

This two-ring basin is circular, with a well-defined raised outer rim that drops down to a broad inner moat (Fig. 5.15). This annular area is relatively featureless as seen by RADAR and is visible as a dark ring in ISS images (Fig. 5.4, 20°N, 87°W). Within the moat is a rough-textured inner ring about 125 km in diameter that encircles a smoother inner zone (Wood et al. 2009). The rough interior differs from that of two-ring basins on the Moon, Mars and Mercury that typically have a flat interior. Erosion appears to be relatively minor in Menrva, with only a few streams cutting its outer rim and although there are numerous radial grooves on the inside of the rim, they have not breached it. The next largest positively identified impact crater is Sinlap (Figs. 5.55 and 5.56), with an 80-km-wide raised rim and flat floor.

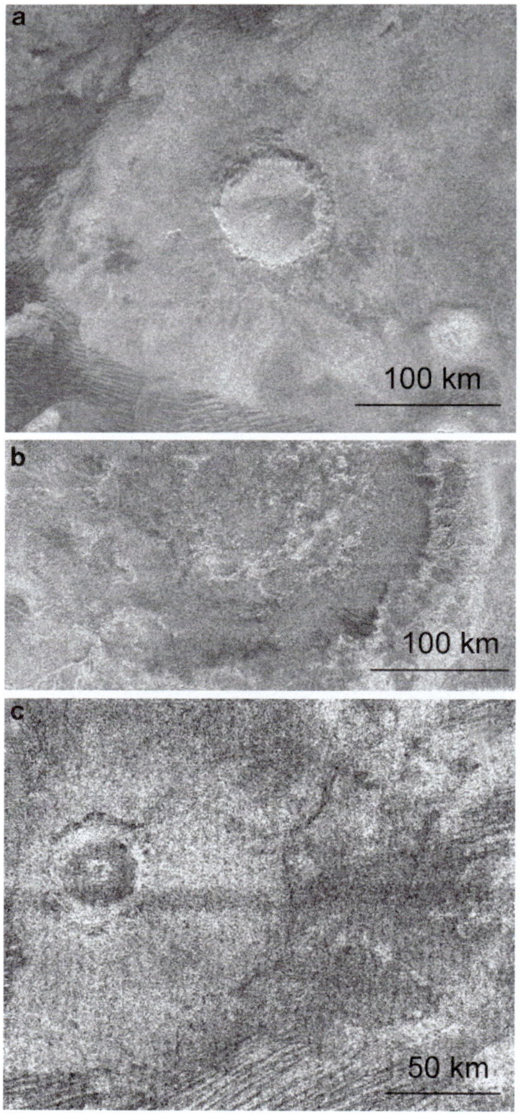

Fig. 5.55 Impact craters on Titan: (**a**) Sinlap: 16°W, 11°N, 80 km in diameter; (**b**) Menrva:~19°N, 87°W; diameter of 450 km; (**c**) Ksa ~15°N, 65°W near Menrva; 29 km in diameter (Lorenz et al. 2007)

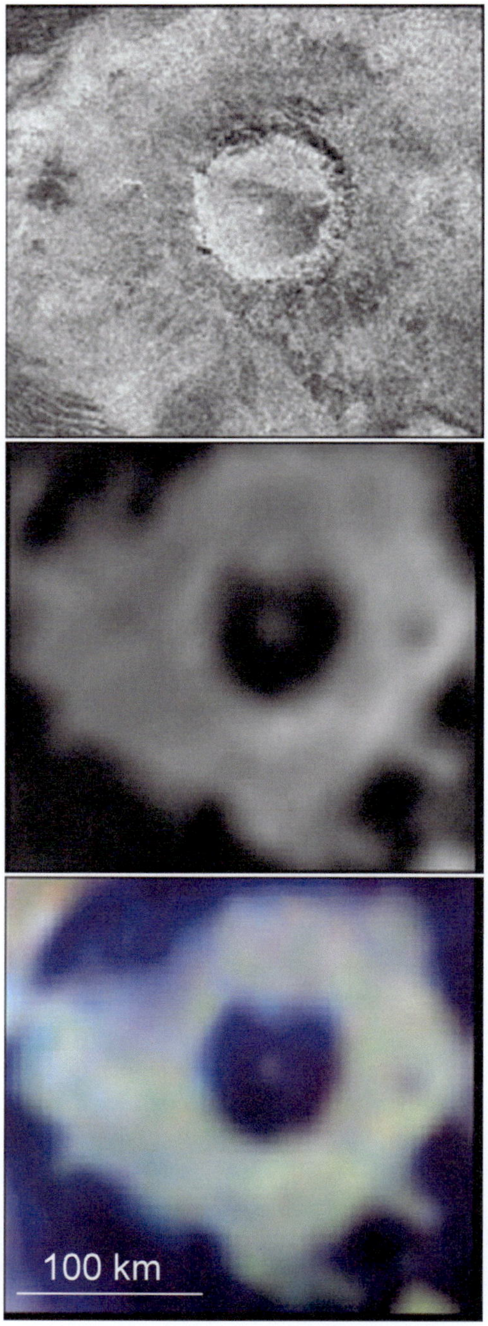

Fig. 5.56 SAR image (*top*) and VIMS T13 2 μm image (*middle*) and VIMS false color ratio composite images (*bottom*) of the Sinlap crater, showing a possible central peak (adapted from Le Mouélic et al., 2008)

A crater as large as Sinlap would be expected to be complex, with a central peak and terraced walls. In the RADAR image, Sinlap lacks both, although a small off-center dark spot is visible in the VIMS image (Le Mouélic et al. 2008). Sinlap appears to be a normal impact crater that has been modified by geological processes that smoothed its inner rim walls and largely buried its central peak. This degree of modification is odd, considering that two different units of apparent ejecta surround the crater. A slightly dark unit with a diameter of ~40 km is cut by shallow radial lineations. This dark unit sits on a broader bright one that appears to taper towards the southeast.

Surprisingly, Titan's impact craters are similar to craters formed on silicate planets rather than icy satellites (Wood et al. 2007a). Viscous creep has shallowed medium- to large-diameter craters on Ganymede and other icy satellites (Schenk 1993), but on Titan only the large basin Menrva (Fig. 5.15), whose inner ring is higher than the surrounding terrain, shows morphological evidence of topographic relaxation. Similarly, Titan's craters lack another characteristic of impact craters on other icy moons – central pits rather than central peaks (Wood et al. 2009).

Ksa is a 29-km-wide impact crater (Fig. 5.57) that looks remarkably like a fresh complex crater on a rocky world. Ksa has a mountainous circular rim, a dark floor and a bright central peak. Its well-preserved ejecta blanket with radial lineations dramatically illustrates the youth of Ksa.

A number of other less certain but probable impact craters have been cataloged by Wood et al. (2009). Two of the best examples are very similar bright-rimmed structures with radar-dark (presumably smooth) floors and short bright exterior deposits (Fig. 5.57).

One, situated at 23°N/140°W, is 75 km in diameter, and the second at 26°N, 10°W is 170 km across. Both these features resemble impact craters on Venus. The two are quite probably of impact origin, but the jagged textures of their rims are distinct from the smoother rims of the proven impact craters on Titan. Perhaps they are younger and very little modified, or perhaps they formed on a substrate with different characteristics.

About five dozen other possible impact craters have been identified so far on Titan (Wood et al. 2009). In general, they are circular and appear to have elevated rims and interiors whose radar reflectivity contrasts to that of the rims. Many of the larger of these features are found within Xanadu and show evidence of having been significantly eroded. Others are partially or nearly completely covered by dunes, a few are cut by fluvial channels, and many are surrounded by what appears to be talus. Clearly, as on Earth, there are multiple processes at work on Titan that can modify and presumably erase craters.

An unexpected finding is that about half of the putative craters on Titan have diameters smaller than 20 km (Wood et al. 2009). The dense atmosphere of Titan led to model predictions that few small projectiles would reach the surface, resulting in a dearth of craters smaller than 20 km (e.g. Engel et al. 1995). A number of the small features detected on Titan are quite likely to be of impact origin, but for many the uncertainty remains.

The paucity of impact craters means that the surface of Titan must be very young; however, there are many uncertainty factors in crater generation models. Based on early Cassini observations, it was estimated (Porco et al. 2005; Lorenz et al. 2007) that the average age was less than a billion years, as on Earth. A more recent study (Wood et al. 2009) of possibly eroded impact craters is consistent with that estimate but demonstrates that different surfaces may have significantly different crater retention ages. In particular, no impact craters have been found that cut into dunes, suggesting that the latter are the youngest geological features on Titan. At the other end of the age spectrum is Xanadu, whose eastern half has the highest concentration of possible impact craters yet seen on Titan. The distribution of supposed impact craters is still being investigated, but they seem to be much more common on terrain that is optically bright, speculatively suggesting that bright terrains are remnants of an older surface of Titan (Wood et al. 2007c).

5.8 Geological Evolution

One hypothesis based on an evolutionary model for Titan (Tobie et al. 2006) proposes that about 0.5 billion years ago the onset of convection created high heat flows within Titan,

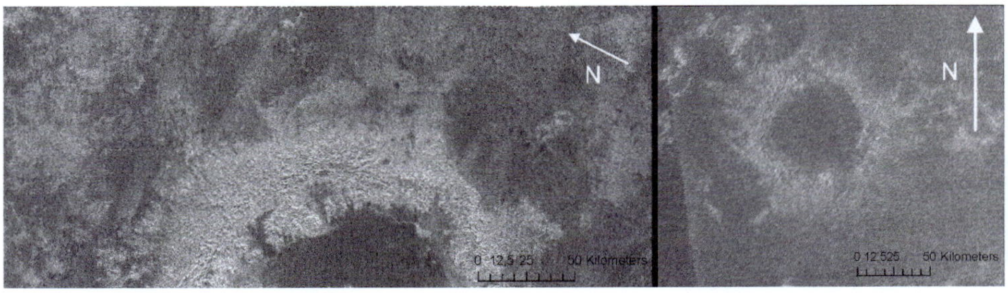

Fig. 5.57 Two bright-rimmed, dark-floored features likely to be impact craters have been imaged by Cassini RADAR. *Right*: 75-km-diameter feature at 23°N, 140°W; *left*: 170-km-diameter object on edge of the RADAR image at 26°N, 10°W

leading to a high outgassing rate for methane. Based on this assumption, Wood et al. (2007a) speculate that the earlier terrain did not survive the tempestuous tectonics of the preceding thin-crust era, and that only a few parts of the current surface are remnants of an older crust. In this scenario, most of the surface has formed more recently, with the dune fields and fluvial features being apparently the youngest.

Lopes et al. (2009) mapped different terrain types and morphological features from SAR data and inferred timelines for the emplacement of various features. A picture of Titan's geological evolution is emerging from the stratigraphic relationships and the distribution of geomorphological units. Geological surface features indicate that Titan has been and still might be a very active world, and that erosional processes, both from fluvial and aeolian activity, play a major and ongoing role in modifying Titan's surface.

Titan's surface is young, even assuming that all crateriform structures ('suspiciously circular features') are due to impacts (Wood et al. 2009). Given the abundant evidence for fluvial and aeolian materials modifying craters, removal of impact craters by burial and erosion has clearly happened. The terrain surrounding the Menrva impact basin indicates that erosional processes have degraded the SW outer rim in particular. Several other tentatively identified impact craters have radar-bright rims and radar-dark interiors. It is likely that numerous craters have been buried or partially buried by aeolian deposits; in some cases, only the rims are left exposed. Fluvial erosion also seems to have played a major role, particularly in areas such as Xanadu. Lorenz et al. (2007) pointed out that the size-density distribution of craters on Titan is similar to Earth's, indicating rapid obliteration by erosion and burial. On Titan, erosion might even play a more important role than on Earth because a major reason for the paucity of impact craters on Earth is tectonic activity, including plate tectonics. Plate tectonics is the major reason Earth as a whole has a dearth of craters because the majority of the surface is oceanic crust which is quite young, but it's only one of the players on the continental crust, which is quite old and appears more heavily cratered than Titan. It is not yet clear what role cryovolcanic and tectonic modification have played in crater obliteration on Titan as, so far, evidence of these processes has not been as abundant as that of erosional processes.

Bright, hummocky, and mountainous terrain, found in numerous patches including Xanadu (Lopes et al. 2009), appear to be the oldest units on Titan. There are crateriform structures in Xanadu, but it is not clear from the available data whether Xanadu and other areas of bright, hummocky, and mountainous terrain are older than the few clearly identified impact craters on Titan. It is clear, however, that patches of bright, hummocky, and mountainous terrain are scattered all over the surface, and that nowhere do they appear uneroded or stratigraphically younger than adjacent local terrain types. It is likely that these, and perhaps Menrva, are the oldest geological formations on Titan that are still preserved.

The bright, hummocky, and mountainous terrains are likely tectonic in origin, and one possibility is that all these areas were formed at the same time by the same mechanism, during an early phase of Titan's history when tectonic activity was taking place. Another possibility is that patches of this terrain were formed at different times by different mechanisms, but in an earlier era when tectonics was more persistent. We also have to consider the possibility that some of patches of bright, hummocky, and mountainous terrain may not be tectonic in origin but purely erosional. Erosion has certainly played a role in the morphology of mountainous terrain. Radebaugh et al. (2007) pointed out that peak morphologies and surrounding diffuse blankets suggest erosion. Most of the possible tectonic features appear to be at least partially degraded and embayed by plains. Radebaugh et al. (2007) argued that fluvial runoff has played a significant role.

Plains units, including mottled plains as discussed by Stofan et al. (2006), may be younger than patches of bright, hummocky, and mountainous terrain, although plains units may have been deformed in an earlier phase and eroded into bright, hummocky and mountainous units. The processes of deposition and cryovolcanism probably contributed to their formation, but the featureless nature of the plains provides few clues to their origin. An exception to this are brighter plains flanking mountainous units, which are interpreted to be erosional aprons (Radebaugh et al. 2007), although they are much smaller.

Cryovolcanism may be a relatively young process, although it cannot be excluded that older eroded cryovolcanic features exist but are hard to identify based on the currently available resolution. The cryovolcanic features identified to date are mostly located at high northern latitudes. It is unclear whether Ganesa Macula has suffered fluvial erosion, as the origin of channel-like features on its flanks could be either volcanic or fluvial. The large flow fields so far identified do not show any evidence of fluvial erosion, perhaps implying that they are quite young. In agreement with the observation that cryovolcanic features are not widely distributed on Titan's surface and appear to be young, Tobie et al. (2006) and Mitri et al. (2008) showed that cryovolcanic activity is probably a marginal geological process with episodic activity. Mitri et al. (2008) proposed a model of cryovolcanism that involves cracks at the base of the ice shell and the formation of ammonia-water pockets in the ice. Large-scale tectonic stress patterns (tides, non-synchronous rotation, satellite volume changes, solid-state convection, or subsurface pressure gradients associated with topography) would enable these ammonia-water pockets to erupt effusively onto the surface. Tobie et al. (2006) suggested that episodic outgassing associated with cryovolcanic activity of methane stored as clathrate hydrates within an ice-I shell might explain the present abundance of methane in the atmosphere.

Fluvial and aeolian deposits appear to be very young features on Titan, while liquid filled lakes are probably the youngest. Mapping of the currently available RADAR data indicates that fluvial features are widespread over both latitude and longitude and appear on many different scales. At high latitudes, fluvial erosion appears to be the dominant modification process. Aeolian activity may dominate at lower latitudes, and dunes overlay fluvial deposits in places. Fluvial as well as lacustrine activity most likely varies with the seasons, as has been observed at high southern latitudes (Turtle et al. 2009).

5.9 Summary and Conclusion

Titan, a complex moon with organic substances, shares features with other large icy satellites as well as the terrestrial planets. It is subject to tidal stresses, and its surface appears to have been modified tectonically to form mountains. It is likely that cryovolcanism exists where liquid water, perhaps in concert with ammonia, methane and carbon dioxide, makes its way to the surface from the interior. Cassini has revealed that Titan has the largest known abundance of organic material in the solar system apart from Earth, and that its active hydrological cycle is analogous to that of Earth, but with methane replacing water.

The surface of Titan exhibits morphological features of different sizes and origins created by geological processes that span the entire dynamic range of aeolian, fluvial, tectonic and cryovolcanic activities (Tomasko et al. 2005; Porco et al. 2005; Sotin et al. 2005; Elachi et al. 2005; Lorenz et al. 2006a; Wood et al. 2006; Stofan et al. 2007; Lopes et al. 2007a; Radebaugh et al. 2007; Barnes et al. 2007; Lorenz et al. 2007; Lorenz et al. 2008; Jaumann et al. 2008; Turtle et al. 2008; Lunine et al. 2008;). Titan is characterized by a geologically active surface indicating endogenic and exogenic processes (Fig. 5.58).

Titan's Topography Stereo analysis of the dendritic region at the Huygens landing site (Tomasko et al. 2005) indicates an elevation of 50–200 m relative to the large darker plain. It suggests that the brighter areas within the darker terrain are higher as well. Based on radar shape-from-shading, isolated mountains and chains range from subdued ones, with heights ≤300 m, to structures as high as 2,000 m, with slopes approaching 45° in a few cases (Radebaugh et al. 2007).

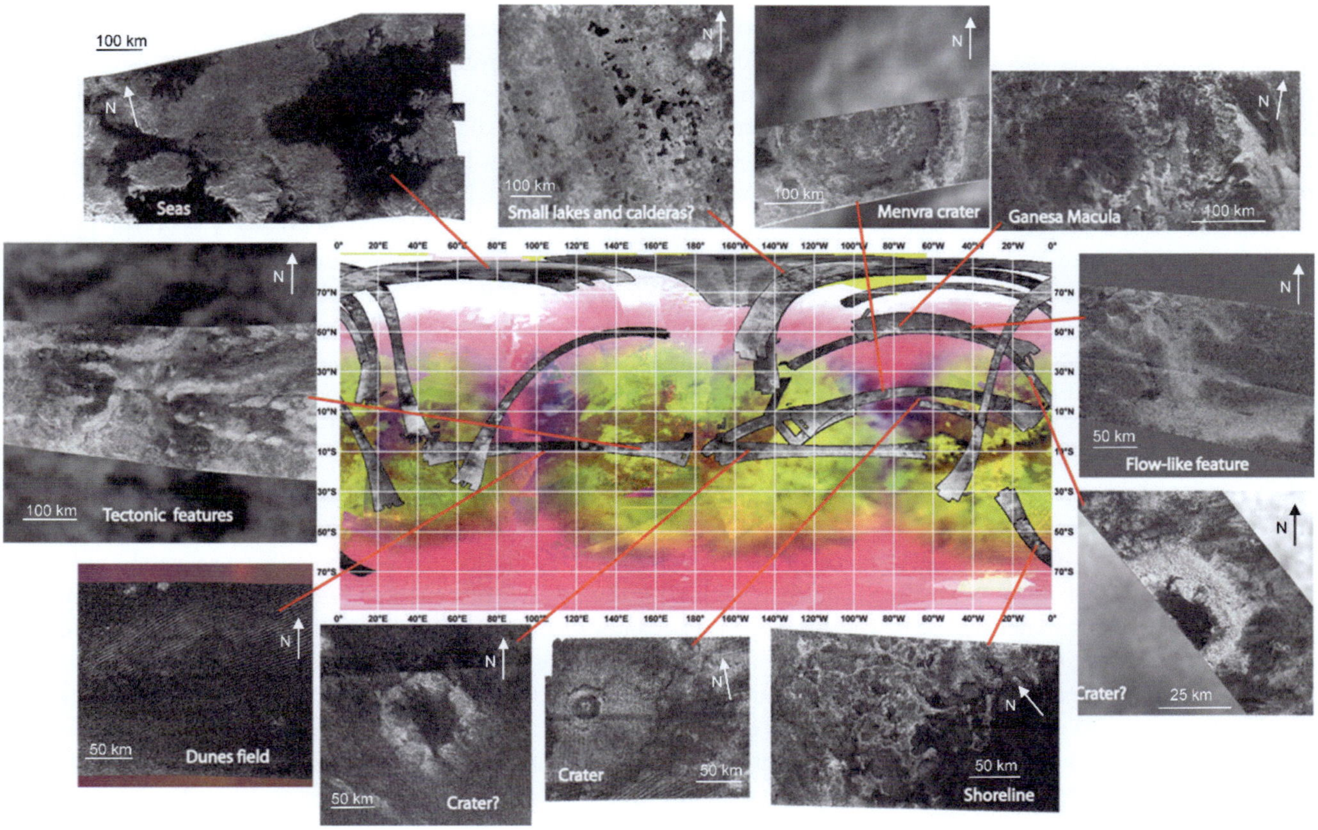

Fig. 5.58 VIMS color-coded map of Titan's surface overlain by RADAR swath. A few features were chosen to illustrate the geological diversity of Titan's surface. Colors are coded with 1.59 µm/1.27 µm as red, 2.03 µm /1.27 µm as green and 1.27 µm /1.08 µm as blue

Stereo analysis of RADAR images taken near the largest north polar lakes indicates elevation variations ≤1,200 m and slopes over 10-km baselines of generally less than 5° (Kirk et al. 2009). Steep-sided depressions, some of which contain smaller lakes (Stofan et al. 2007), have depths as great as 600 m (Kirk et al. 2007, 2008, 2009)

Geological and Geomorphological Units Based on the spectral signature in the infrared methane windows as well as on VIMS color composites, three major units can be distinguished: whitish material mainly distributed in topographically high areas; bluish material adjacent to bright-to-dark boundaries; and brownish material that correlates with dunes (Lorenz et al. 2006a; Soderblom et al. 2007a; Jaumann et al. 2008; Le Mouélic et al. 2008). Although the spectral units are distinct, their compositions are not known at this time. Bright materials may consist of precipitated aerosol dust composed of methane-derived organics (Soderblom et al. 2007a) superimposed on water-ice bedrock. The bluish component might contain water ice as its defining feature (Soderblom et al. 2007a; Le Mouélic et al. 2008). DISR spectra do not confirm water ice (Keller et al. 2008) at the Huygens landing site, which lies in a bluish transition zone (Soderblom et al. 2007a). It is no simple matter to distinguish between specific organics and ices because all these molecules have comparable absorptions, resulting in similar spectral slopes. In addition, different particle sizes will have an effect on the depths of absorption bands and related spectral slopes. Nevertheless, the spectral signature variations are real and indicate differences in composition and/or particle size that are related to geological processes (Soderblom et al. 2007a; Jaumann et al. 2008). In terms of identified geomorphological units, Titan's surface has mountainous, rugged terrains that are interpreted as tectonic in origin, cryovolcanic units, dunes formed by aeolian processes, fluvial channels, lakes and impact craters.

Geology at the Huygens Landing Site Although there are no liquid hydrocarbon pools at the Huygens site, traces of once flowing liquid are evident (Tomasko et al. 2005). Surprisingly like Earth, the brighter highland regions show complex systems draining into flat, dark lowlands. Images taken after landing appear to be of a dry river bed. If the darker region is interpreted as a dry lake bed, it is too large to have been formed by the creeks and channels in the neighboring bright areas. It might have been created by other larger river systems or some large-scale catastrophic event, which predates deposition by the rivers seen in these images. The major elements of Titan's surface albedo variations can be interpreted as being controlled by flows of low-viscosity fluids driven by topographic variation, whether caused by precipitation as suggested by the dendritic networks or spring-fed flows as suggested by the stubby networks. Rounded cobbles at the site vary in diameter between 3 mm, the resolution limit of the Huygens imager, and a maximum of 20 cm (Keller et al. 2008). The geology at the Huygens landing site strongly implies intense resurfacing by erosion and deposition.

Tectonics and Volcanism What forces are driving tectonics on Titan is not clear at this point; as is also common on Earth, most of the possible tectonic features appear to be at least partially modified by other processes and embayed by surrounding plains units. Among the possible tectonic features are linear and ridge-like features in the form of chains of hills. Evidence of cryovolcanism appears in both the northern and the southern hemispheres, but it is not ubiquitous (Lopes et al. 2007a). The most easily identified features are bright flow fronts and craters which appear to be the sources of some flows. The largest structures identified as cryovolcanic are flow fields near Hotei Arcus, Tui Regio and Ganesa Macula (Sotin et al. 2005; Wall et al. 2009; Barnes et al. 2006; Elachi et al. 2005). North of 70°N latitude is a zone of many roundish to irregular depressions that contain lakes (Stofan et al. 2007). These depressions have raised rims and deep centers and are often nested. One possible interpretation is that they are volcanic calderas (Wood et al. 2007b, c).

Erosional Surface Features Titan exhibits erosional as well as depositional areas. *Aeolian*: The most pervasive landforms on Titan are large dune fields that extend, mostly in equatorial regions, over distances of up to a few thousand kilometers (Lorenz et al. 2006a). The dunes are dark at radar and visible wavelengths. Their existence implies a source of particles that can be distributed by winds, and a lack of nearby surface liquids to entrap them. *Fluvial:* Cutting across many different types of terrains are river channel systems (Tomasko et al. 2005; Porco et al. 2005; Lorenz 2006; Stofan et al. 2007; Barnes et al. 2007b; Lorenz et al. 2008a; Jaumann et al. 2008). Some channels are radar-bright, which suggests that they are filled with rough boulders; others are radar-dark and may either contain liquids or have smooth deposits on their floors. Tributaries suggest that rainfall feeds the rivers, and gentle bends, rather than tortuous sinuosity, suggest that the downhill slopes are at least somewhat steep. Hence, while a cryovolcanic origin of the observed channel-like features cannot be ruled out, the sharp bends and branched networks of Titan's valleys are more consistent with distributed than with the localized sources that one would expect in cases of cryovolcanic liquid release. Cryovolcanic flow features identified so far are distinct from the observed branching channels (Lopes et al. 2007a; Lunine et al. 2008a). Liquid methane (CH_4) is suggested to be the main fluid on Titan. Its viscosity at surface temperature (95 K) is 1.8×10^{-4} Pa s, which is approximately five times lower than that of water at 298 K (9×10^{-4} Pa s) (Hanley et al. 1977; National Institute of Standards and Technology, NIST (2005). Thus, liquid methane will produce turbulent flows on Titan's surface with significant erosional power, generating discharge and runoff sufficient to move sediment (Perron et al. 2006; Jaumann et al. 2008). *Depositional:* Besides dunes, there are dune-free areas surrounding bright topographically high material that might be comparable to plains such as those seen at the Huygens landing site (Soderblom et al. 2007a; Jaumann et al. 2008). *Lakes:* The presence of abundant lakes of methane/ethane (Stofan et al. 2007; Lunine et al. 2008b; Turtle et al. 2009) in the northern polar region and of some in the southern polar region (Brown et al. 2008; Turtle et al. 2009) implies the existence of an active hydrological cycle and also, perhaps, a wetter climate at the poles (Stofan et al. 2007; Brown et al. 2008).

Cratering and Surface Ages With the exception of Io impact craters exist on all solid bodies in the solar system, and have been detected on Titan. However, since very few have been observed, they must be rapidly destroyed or buried by other geological processes (Porco et al. 2005; Wood et al. 2006; Wood et al. 2007c; Lorenz et al. 2007; Le Mouélic et al. 2008; Jaumann and Neukum 2009; Wood et al. 2009). The morphologies of the impact craters are more similar to those seen on silicate planets than on icy satellites (Wood et al. 2007c). Removal of impact craters by burial and erosion is likely, given the evidence of fluvial, aeolian and cryovolcanic processes and the relatively degraded appearance of other topographic features, e.g. hills and ridges. The dearth of craters compared with other icy satellites indicates that the surface of Titan is young and modified by other geologic processes. However, the existence of the large Menrva impact structure (>400 km in diameter) suggests that in some places, larger (and thus potentially older) craters may have been preserved (Porco et al. 2005; Wood et al. 2006; Lorenz et al. 2007; Le Mouélic et al. 2008; Jaumann and Neukum 2009).

*Geological Evolution*One hypothesis based on an evolutionary model for Titan (Tobie et al. 2006) proposes that about 0.5 billion years ago, the onset of convection created intense heat flows within Titan, leading to a high outgassing rate for methane. Based on this assumption, Wood et al. (2009a and 2009b) speculated that the earlier terrain did not survive the tempestuous tectonics of the thin-crust era and that only a few parts of the current surface are remnants of the oldest crust. In this scenario, most of the surface has formed more recently, with dune fields and some fluvial channels being the youngest features.

Our geological exploration of Titan has revealed a world with an extremely young surface which has been modified by a complex interplay of aeolian, pluvial, fluvial, lacustrine, cryovolcanic and tectonic processes. Only Earth rivals Titan in the dynamic interplay of geological processes. However, as discussed above, numerous questions remain regarding the evolution of Titan's surface and interior. We do not understand how Titan's geological processes evolve over time or how they interact at system level, and how this compares to Earth, and we need to constrain the nature of the organic inventory on Titan more precisely, including the interchange between atmosphere, surface and subsurface. Some of these questions can be at least partially addressed during Cassini's extended mission. Others will require future missions to Titan.

Determining whether Titan is still geologically active today and constraining the history of its surface calls for a number of new observations. The internal density structure and the ice shell thickness need to be determined, and the global distribution and morphology of volcanic and tectonic features must be established at resolutions on the order of 50 m/pixel or better. High-resolution altimetry data of geological features such as volcanoes and tectonic features are also required to constrain surface and interior evolution. Subsurface sounding to provide a three-dimensional view of these features would also be extremely revealing. Determining how active Titan is requires measurements of seismic activity and the frequency of cryovolcanic eruptions, which would also help to constrain Titan's internal structure.

Understanding the organic chemistry of Titan including the composition of its lakes, its meteorological cycles, and its fluvial processes also will be aided by high-resolution imaging and altimetry data. The few instances we have in which RADAR and VIMS data can be combined illustrate the necessity of obtaining high-resolution compositional coverage of key types of surface features. While a limited number of features can be studied during Cassini's extended mission, a future mission with a highly capable near-infrared mapping spectrometer extending to 6 μm, as well as a lander equipped with mass spectrometers for investigating the elemental chemistry are critical to understanding atmosphere-surface-subsurface interactions.

These measurements require orbital and lander as well as mobile atmospheric platforms (Coustenis et al. 2008); simultaneity of measurements from multiple vantage points would greatly enhance our ability to quantify how Titan operates as a system. Possible high-priority landing sites on Titan include lakes or seas, the equatorial dune fields and cryovolcanic features. This chapter illustrates that Titan is a complex, active world, with only Earth having a similar diversity of landforms and processes that have shaped and continue to shape its surface. Future exploration of Titan is necessary to expand our basic understanding of Titan and to constrain its geological evolution.

Acknowledgements We gratefully acknowledge the long years of work by the entire Cassini team that allowed these data of Titan to be obtained. We also acknowledge NASA, ESA, ASI, DLR, CNES, and JPL that provide support for the international Cassini team. We thank K. Stephan, M. Langhans and J. Perry, for data processing support. H. U. Keller thanks S. Schröder for his assistance. The review and discussions of Ron Greeley are highly appreciated. Part of this work was performed at DLR, Institute of Planetary Research with support from the Helmholtz Alliance 'Planetary Evolution and Life' and at the Jet Propulsion Laboratory under contract to the National Aeronautics and Space Administration Outer Planets Research and Planetary Geology & Geophysics program.

References

Allison M (1992) A preliminary assessment of the titan planetary boundary layer, pp113–118. In: Proceedings symposium on titan toulouse, France, September 1991, European Space Agency SP-338, Noordwijk, The Netherlands

Artemieva N, Lunine JI (2003) Cratering on Titan: impact melt, ejecta, and the fate of surface organics. Icarus 164:471–480

Bagnold R (1941) Physics of wind-blown sand and desert dunes. London:Methuen

Baines K, Drossard P, Lopez-Valverde MA, Atreya SK, Sotin C, Momary TW, Brown RH, Buratti BJ, Clark RN, Nicholson PD (2006) On the discovery of CO nighttime emissions on Titan by Cassini/Vims: derived stratospheric abundances and geological implications. Planet Space Sci 54:1552–1562

Barnes J, Brown RH, Turtle EP, McEwen AS, Lorenz RD, Janssen M, Schaller EL, Brown ME, Buratti BJ, Sotin C, Griffith C, Clark R, Perry J, Fussner S, Barbara J, West R, Elachi C, Bouchez AH, Roe HG, Baines KH, Belluchi G, Bibring J-P, Capaccioni F, Cerroni P, Combes M, Corradini A, Cruikshank DP, Drossart P, Formisano V, Jaumann R, Langevin Y, Matson DL, McCord TB, Nicholson PD, Sicardy B (2005) A 05-micron-bright spot on Titan: evidence for surface diversity. Science 210, 92–95. doi:10.1126/science.1117075

Barnes J, Brown RH, Radebaugh J, Buratti BJ, Sotin C, Le Mouèlic S, Rodriguez S, Turtle EP, Perry J, Clark R, Baines KH, Nicholson P (2006) Cassini observations of flow-like features in western Tui Regio, Titan. Geophys Res Lett 33: L16204. doi:10.1029/2006GL026843

Barnes J, Brown RH, Soderblom L, Buratti BJ, Sotin C, Rodriguez S, Le Mouèlic S, Baines KH, Clark R, Nicholson P (2007a) Global-scale surface spectral variations on Titan seen from Cassini/VIMS. Icarus 186:242–258. doi:10.1016/j.icarus.2006.08.021

Barnes JW, Radebaugh J, Brown RH, Wall S, Soderblom L, Lunine J, Burr D, Sotin C, Le Mouèlic S, Buratti BJ, Clark R, Baines KH, Jaumann R, Nicholson PD, Kirk RL, Lopes R, Lorenz R, Mitchell K, Wood CA, the Cassini RADAR Team (2007) Near-infrared spectral mapping of Titan's mountains and channels. J Geophys Res 112(E11): CiteID E11006. doi:10.1029/2007JE02932

Barnes J, Brown RH, Soderblom L, Sotin C, Le Mouèlic S, Rodriguez S, Jaumann R, Beyer RA, Buratti BJ, Pitman K, Baines KH, Clark R, Nicholson P (2008) Spectroscopy, morphometry, and photoclinometry of Titan's dunefields from Cassini/VIMS. Icarus 195:400–414. doi:10.1016/j.icarus.2007.12.006

Barnes J, Brown RH, Soderblom JM, Soderblom LA, Jaumann R, Jackson B, Le Mouèlic S, Sotin C, Buratti BJ, Pitman KM, Baines KH, Clark R, Nicholson PD, Turtle EP, Perry J (2009) Shorelike features of of Titan's Ontarion Lacus from Cassini/VIMS observations. Icarus 201: 217–225. doi:10.1016/j.icarus.2008.12.028

Bernard J-M, Quirico E, Brissaud O, Montagnac G, Reynard B, McMillan P, Coll P, Nguyen M-J, Raulin F, Schmitt B (November 2006) Reflectance spectra and chemical structure of Titan's tholins: application to the analysis of Cassini-Huygens observations. Icarus 185:301–307

Blair RW(1987) Karst Landforms and Lakes. In: Short NM, Blair RW (eds) Geomorphology from Space: a global overview of regional landform, chapter 7. NASA SP-486, Washington, DC

Brown RH, Baines KH, Bellucci G, Bibring J-P, Buratti BJ, Bussoletti E, Capaccioni F, Cerroni P, Clark RN, Coradini A, Cruikshank DP, Drossart P, Formisano V, Jaumann R, Langevin Y, Matson DL, McCord TB, Miller E, Nelson RM, Nicholson PD, Sicardy B, Sotin C (2004) The Cassini visual and infrared mapping spectrometer investigation. Space Sci Rev 115:111–168

Brown RH, Baines KH, Bellucci G, Buratti BJ, Capaccioni F, Cerroni P, Clark RN, Coradini A, Cruikshank DP, Drossart P, Formisano V, Jaumann R, Langevin Y, Matson DL, McCord TB, Mennella V, Nelson RM, Nicholson PD, Sicardy B, Sotin C, Baugh N, Griffth CA, Hansen GB, Hibbitts CA, Momary TW, Showalter MR (2006) Observations in the Saturn system during approach and orbital insertion, with Cassini's visual and infrared mapping spectrometer (VIMS). Astron Astrophys 446:707–716

Brown RH, Soderblom LA, Soderblom JM, Clark RN, Jaumann R, Barnes JW, Sotin C, Buratti B, Baines KH, Nicholson PD (2008) The identification of liquid ethane in Titan's Ontario Lacus. Nature 454:607–610. doi:10.1038/nature07100

Buratti BJ, Sotin C, Brown RH, Hicks MD, Clark RN, Mosher JA, McCord TB, Jaumann R, Baines KH, Nicholson PD, Momary T, Simonelli DP, Sicardy B (2006) Titan, preliminary results on surface properties and photometry from VIMS observations of the early flybys. Planet Space Sci 54:1498–1502

Burr DM, Emery JP, Lorenz RD, Collins GC, Carling PA (2006) Sediment transport by liquid surficial flow: application to Titan. Icarus 181:235–242

Campbell DB, Black GJ, Carter LM, Ostro SJ (2003) Radar evidence for liquid surfaces on Titan. Science 302:431–434

Chapman CR, Merline WJ, Thomas PC, Joseph J, Cheng AF, Izenberg N (2002) Impact history of eros: craters and boulders. Icarus 155: 104–118

Clark RN (1981) Water frost and ice: the near-infrared spectral reflectance 0.65–2.5 μm. J Geophys Res 86:3087–3096

Clark RN, Curchin JM, Brown RH, Waite JH, Cruikshank DP, Jaumann R, Lunine J, Hoefen TM, Cravens TE, Yelle RV, Vuitton V, Baines KH, Buratti BJ, Barnes J, McCord TB, Nicholson PD (2006) Detection of widespread aromatic and aliphatic hydrocarbon deposits on Titan's surface observed by VIMS and excess benzene observed in Titan's thermosphere observed by INMS. Bull Am Astron Soc 38:574

Collins GC (2005) Relative rates of fluvial bedrock incision on Titan and Earth. Geophys Res Let 32(22, L22202):1–4

Combes M, Vapillon L, Gendron E, Coustenis A, Lai O, Wittemberg R, Sirdey R (1997) Spatially-resolved images of Titan by means of adaptive optics. Icarus 129:482–497

Coustenis A, Taylor F (1999) Titan: the earth-like moon. World Scientific, Singapore

Coustenis A, Taylor F (2008) Titan: exploring an earth-like world. World Scientific, Singapore

Coustenis A, Lellouch E, Maillard J-P, McKay CP (1995) Titan's surface: composition and variability from the near infrared albedo. Icarus 118:87–104

Coustenis A, Gendron E, Lai O, Véran J-P, Woillez J, Combes M, Vapillon L, Fusco T, Mugnier L, Rannou P (2001) Images of Titan at 1.3 and 1.6 μm with adaptive optics at the CFHT. Icarus 154:501–515

Coustenis A, Atreya SK, Balint T, Brown RH, Dougherty MK, Ferri F, Fulchignoni M, Gautier D, Gowen RA, Griffith CA, Gurvits LI, Jaumann R, Langevin Y, Leese MR, Lunine JI, McKay CP, Moussas X, Müller-Wodarg I, Neubauer F, Owen TC, Raulin F, Sittler EC, Sohl F, Sotin C, Tobie G, Tokano T, Turtle EP, Wahlund J-E, Waite JH, Baines KH, Blamont J, Coates AJ, Dandouras I, Krimigis T, Lellouch E, Lorenz RD, Morse A, Porco CC, Hirtzig M, Saur J, Spilker T, Zarnecki JC, Choi E, Achilleos N, Amils R, Annan P, Atkinson DH, Bénilan Y, Bertucci C, Bézard B, Bjoraker GL, Blanc M, Boireau L, Bouman J, Cabane M, Capria MT, Chassefière E, Coll P, Combes M, Cooper JF, Coradini A, Crary F, Cravens T, Daglis IA, de Angelis E, de Bergh C, de Pater I, Dunford C, Durry G, Dutuit O, Fairbrother D, Flasar FM, Fortes AD, Frampton R, Fujimoto M, Galand M, Grasset O, Grott M, Haltigin T, Herique A, Hersant F, Hussmann H, Ip W, Johnson R, Kallio E, Kempf S, Knapmeyer M, Kofman W, Koop R, Kostiuk T, Krupp N, Küppers M, Lammer H, Lara L-M, Lavvas P, Le Mouélic S, Lebonnois S, Ledvina S, Li J, Livengood TA, Lopes RM, Lopez-Moreno J-J, Luz D, Mahaffy PR, Mall U, Martinez-Frias J, Marty B, McCord T, Menor Salvan C, Milillo A, Mitchell DG, Modolo R, Mousis O, Nakamura M, Neish CD, Nixon CA, Nna Mvondo D, Orton G,

Paetzold M, Pitman J, Pogrebenko S, Pollard W, Prieto-Ballesteros O, Rannou P, Reh K, Richter L, Robb FT, Rodrigo R, Rodriguez S, Romani P, Ruiz Bermejo M, Sarris ET, Schenk P, Schmitt B, Schmitz N, Schulze-Makuch D, Schwingenschuh K, Selig A, Sicardy B, Soderblom L, Spilker LJ, Stam D, Steele A, Stephan K, Strobel DF, Szego K, Szopa C, Thissen R, Tomasko MG, Toublanc D, Vali H, Vardavas I, Vuitton V, West RA, Yelle R, Young EF (2008) TandEM: Titan and Enceladus Mission. Exp Astron 23:893–946. doi:10.1007/s10686-008-9103-z

Cruikshank D, Morgan JS (1980) Titan – suspected near-infrared variability. Astrophys J 235:L53–L54

Cuda V, Ash RL (1984) Development of an uniaxial ice tensile specimen for low temperature testing. Cold Reg Sci Technol 9:47–52

de Pater I, da´mkovics A, Bouchez M, Brown AH, Gibbard ME, Marchis SG, Roe F, Schaller HG, Young EL (2006) Titan imagery with Keck adaptive optics during and after probe entry. J Geophys Res 111(E7, E07S05):1–16

DePiero FW, Trivedi MM (1996) 3-D computer vision using structured light: design, calibration and implementation issues. Adv Comp 43:243–278

Dermott SF, Sagan C (1995) Tidal effects of disconnected hydrocarbon seas on Titan. Nature 374:238–240

Dubouloz N, Raulin F, Lellouch E, Gautier D (1989) Titan's hypothesized ocean properties – the influence of surface temperature and atmospheric composition uncertainties. Icarus 82:81–96

Durham WB, Heard HC, Kirby SC (1983) Experimental deformation of polycristallyne H_2O ice at high pressure and high temperature: preliminary results. J Geophys Res 88:377–292

Elachi C, Allison M, Borgarelli L, Encrenaz P, Im E, Janssen MA, Johnson WTK, Kirk RL, Lorenz RD, Lunine JI, Muhleman DO, Ostro SJ, Picardi G, Posa F, Roth LE, Seu R, Soderblom LA, Vetrella S, Wall SD, Wood CA, Zebker HA (2004) RADAR: the Cassini Titan radar mapper. Space Sci Rev 115:71–110

Elachi C, Wall S, Allison M, Anderson Y, Boehmer R, Callahan P, Encrenaz P, Flamini E, Franceschetti G, Gim Y, Hamilton G, Hensley S, Janssen M, Johnson W, Kelleher K, Kirk R, Lopes R, Lorenz R, Lunine J, Muhleman D, Ostro S, Paganelli F, Picardi G, Posa F, Roth L, Seu R, Shaffer S, Soderblom L, Stiles B, Stofan E, Vetrella S, West R, Wood C, Wye L, Zebker H (2005) First views of the surface of Titan from the Cassini radar. Science 308:970–974

Elachi C, Wall S, Janssen M, Stofan E, Lopes R, Kirk R, Lorenz R, Lunine J, Paganelli F, Soderblom L, Wood C, Wye L, Zebker H, Anderson Y, Ostro S, Allison M, Boehmer R, Callahan P, Encrenez P, Flameni E, Francescetti G, Gim Y, Hamilton G, Hensley S, Johnson W, Kelleher K, Muhleman D, Picardi G, Posa F, Roth L, Seu R, Shaffer S, Stiles B, Vetrella S, West R (2006) Titan radar mapper observations from Cassini's T_a and T_3 Flybys. Nature 441:709–713. doi:10.1038/nature04786

Engel S, Lunine JI, Hartmann WK (1995) Cratering on Titan and implications for Titan's atmospheric history. Planet Space Sci 43:131–154

Flasar FM (1983) Oceans on Titan? Science 221:55–57

Fortes AD, Grindrod PM (2006) Modeling of possible mud volcanism on Titan. Icarus 182:550–558

Fussner S (2006) Search for 938 nm Specular Enhancement on Titan. M.S. thesis, University of Arizona, Tucson, AZ

Garvin JB (1990) The global budget of impact-derived sediments on Venus. Earth Moon Planets 50–51:175–190

Gendron E, Coustenis A, Drossart P, Combes M, Hirtzig M, Lacombe F, Rouan D, Collin C, Pau S, Lagrange A-M, Mouillet D, Rabou P, Fusco Th, Zins G (2004) VLT/NACO adaptive optics imaging of Titan. Astron Astroph 417:L21–L24

Ghafoor NA-L, Zarnecki JC, Challenor P, Srokosz MA (2000) Wind-driven surface waves on Titan. J Geophys Res105(E5):12,077–12,091

Gibbard SG, Macintosh B, Gavel D, Max CE, de Pater I, Ghez AM, Young EF, McKay CP (1999) Titan: high-resolution speckle images from the Keck telescope. Icarus 139:189–201

Gibbard SG, Macintosh B, Gavel D, Max CE, de Pater I, Roe HG, Ghez AM, Young EF, McKay CP (2004) Speckle imaging of Titan at 2 microns: surface albedo, haze optical depth and tropospheric clouds 1996–1998. Icarus 169:429–439

Greeley R, Iversen J (1985) Wind as a geological process. Cambridge University Press, Cambridge

Grieger B (2005) Shading under Titan's sky. Planet Space Sci 53(5): 577–585

Grier JA, Lunine JI (1993) Speculation into possible aeolian and fluvial dune deposits on Titan. Bull Am Astron Soc 25(3):1105–1106

Griffith CA (1993) Evidence for surface heterogeneity on Titan. Nature 364:511–514

Griffith CA, Owen TC, Wagener R (1991) Titan's surface and troposphere, investigated with ground-based, near-infrared observations. Icarus 93:362–378

Griffith CA, Owen T, Geballe TR, Rayner J, Rannou P (2003) Evidence for exposure of water ice on Titan's surface. Science 300:628–630

Hanley HJM, Haynes WM, McCarty RD (1977) The viscosity and thermal conductivity coefficient for dense gaseous and liquid methane. J Phys Chem Ref Data 6:597–601

Hapke B (1981) Bidirectional reflectance spectroscopy: I – theory. J Geophys Res 86:3039–3054

Hapke B (1984) Bidirectional reflectance spectroscopy, III – correction for macroscopic roughness. Icarus 59:41–59

Hassan MA, Church M (1994) Vertical mixing of coarse particles in gravel bed rivers: a kinematic model. Water Resour Res 30:1173–1186

Hayes A, Aharaonson O, Lewis K, Mitchell K, Lorenz R, Callahan P, Lunine J, Lopes R, Wall S, Elachi C, Mitri G, Kirk R, Gim Y, Stofan E, the Cassini RADAR Team (2008) Hydrocarbon lakes on Titan: distribution and interaction with a porous regolith. Geophys Res Lett 35: L09204. doi:10.1029/2008GL033409

Hulme G (1974) The interpretation of lava flow morphology, Geophys. F. R. Ast. Soc. 39:361–383

Hensley S, Schafer S (1994) Automatic DEM generation using Magellan stereo data. Geoscience and remote sensing symposium, IGARSS 94. 'Surface and atmospheric remote sensing: technologies, data analysis, and interpretation 3:1470–1472

Hirtzig M, Coustenis A, Gendron E, Drossart P, Hartung M, Negra˜o A, Rannou P, Combes M (2007) Titan: atmospheric and surface features as observed with Nasmyth adaptive optics system near-infrared imager and spectrograph at the time of the Huygens mission. J Geophys Res 112(E2, E02S91):1–12

Hueso R, Sanchez-Lavega A (2006) Methane storms on Saturn's moon Titan. Nature 442:428–431

Hulme G (1974) The interpretation of lava flow morphology, Geophys. FR Ast Soc 39:361–383

Hunten DM (ed) (1974) The atmosphere of Titan. Proceedings of a workshop held 25–27 July, 1973 at Ames Research Center, NASA SP-340

Irwin RPR, Craddock A, Howard HD (2005) Interior channels in Martian valley networks: discharge and runoff production. Geology 33:489–492

Ivanov BA, Basilevsky AT, Neukum G (1997) Atmospheric entry of large meteoroids: implication to Titan. Planet Space Sci 45:993–1007

Janssen MA, Lorenz RD, West R, Paganelli F, Stiles B, Wall SD, Callahan P, Kirk RL, Roth L, Anderson Y, the Cassini RADAR Team (2009) Titan's surface at 2.2-cm wavelength imaged by the Cassini RADAR radiometer: calibration and first results. Icarus 200:222–239

Jaumann R, Neukum G (2009) The surface age of Titan. 40th Lunar and planetary lunar and planetary science conference, Houston, abstract #1641

Jaumann R, Reiss D, Frei S, Neukum G, Scholten F, Gwinner K, Roatsch T, Matz K-M, Hauber E, Köhler U, Head JW, Hiesinger H, Carr M (2005) Interior channels in Martian Valleys: constraints on fluvial erosion by measurements of the Mars express high resolution stereo camera. Geophys Res Lett 32:L16203. doi:10.1029/2002GL023415

Jaumann R, Stephan K, Brown RH, Buratti BJ, Clark RN, McCord TB, Coradini A, Capaccioni P, Filacchione G, Cerroni P, Baines KH, Bellucci G, Bibring JP, Combes M, Cruikshank DP, Drossart P, Formisano V, Langevin Y, Matson DL, Nelson RM, Nicholson PD, Sicardy B, Sotin C, Soderblom LA, Griffith C, Matz KD, Roatsch T, Scholten F, Porco CC (2006) High-resolution Cassini-VIMS mosaics of Titan and the icy Saturnian satellites. Planet Space Sci 54:1146–1155

Jaumann R, Brown RH, Stephan K, Barnes JW, Soderblom LA, Sotin C, Le Mouélic S, Clark RN, Soderblom J, Buratti BJ, Wagner R, McCord TB, Rodriguez S, Baines KH, Cruikshank DP, Nicholson PD, Griffith CA, Langhans M, Lorenz RD (2008) Fluvial erosion and post-erosional processes on Titan. Icarus 197:526–538

Jaumann R, Stephan K, Sotin C, Brown RH, Langhans M, Soderblom J, Soderblom LA, Le Mouélic S, Clark RN, Buratti BJ, Baines KH, Cruikshank DP, Nicholson PD, Filacchione G, Wagner R, Barnes J, Nelson RM (2009) Erosion and stratigraphic relations on Titan. 40th Lunar and planetary lunar and planetary science conference, Houston, abstract #1599

Jones TD, Lewis JS (1987) Estimated impact shock production of N2 and organic compounds on early Titan. Icarus 72:381–393

Kargel JS (2007) Theory of geochemical/geological homology applied to hydrocarbon and organic substances on icy satellites and other solid planetary objects. Workshop on Ices, Oceans, and Fire: Satellites of the Outer Solar System, held August 13–15, 2007, Boulder, Colorado, LPI-Contribution #1357:72–73

Karkoschka E (2006) http://webgis.wr.usgs.gov/pigwad/down/titan_dl.htm.S

Karkoschka E, Tomasko MG, Doose LR, See C, McFarlane EA, Schröder SE, Rizk B (2007) DISR imaging and the geometry of the descent of the Huygens probe within Titan's atmosphere. Planet Space Sci 55(13):1896–1935

Keller HU, Grieger B, Küppers M, Schröder SE, Skorov YV, Tomasko MG (2008) The properties of Titan's surface at the Huygens landing site from DISR observations. Planet Space Sci 56:728–752

Khare BN, Sagan C, Bandurski EL, Nagy B (1978) Ultraviolet-photoproduced organic solids synthesized under simulated jovian conditions: Molecular analysis. Science 199:1199–1201

Khare BN, Sagan C, Thompson WR, Arakawa ET, Meisse C, Tuminello PS (1994) Optical properties of poly-HCN and their astronomical applications. Can J Chem 72:678–694

Kirk RL, Radebaugh J (2007) Resolution effects in radarclinometry. ISPRS Working Group IV/7 Workshop "Advances in Planetary Mapping 2007", Houston, March, 2007. Available online at http://www.dlr.de/pf/Portaldata/6/Resources/dokumente/isprs_2007/Kirk_3__ISPRS_2007.pdf http://www.dlr.de/pf/desktopdefault.aspx/tabid-656/5144_read-7836/

Kirk RL, Callahan P, Seu R, Lorenz RD, Paganelli F, Lopes R, Elachi C, the Cassini RADAR Team (2005) RADAR reveals Titan topography. 36th Lunar and planetary lunar and planetary science conference, Houston, abstract #2

Kirk RL, Wall SD, Lorenz RD, Lunine JI, Radebaugh J, Soderblom LA, Stiles BW, Jansssen MA, Paganelli R, Lopes R, the Cassini RADAR Team (2006) A high resolution view of the Xanadu region of Titan from the Cassini RADAR. Amer Astron Soc Bull 38(3):52.03, p 579

Kirk RL, Howington-Kraus E, Mitchell KL, Hensley S, Stiles BW, the Cassini RADAR Team (2007) First stereoscopic radar images of Titan. 38th Lunar and planetary lunar and planetary science conference, Houston, abstract #1427

Kirk RL, Howington-Kraus E, Stiles BW, Hensley S, the Cassini RADAR Team (2008) Digital topographic models of Titan produced by radar stereogrammetry with a rigorous sensor model. 39th Lunar and planetary lunar and planetary science conference, Houston, abstract #2320

Kirk RL, Howington-Kraus E, Redding BL, Becker TL, Lee EM, Stiles BW, Hensley S, Hayes AG, Lopes RMC, Lorenz RD, Mitchell KL, Radebaugh J, Paganelli F, Soderblom LA, Stofan ER, Wall SD, Wood CA, Elachi C, the Cassini RADAR Team (2009) Three-dimensional views of Titan's diverse surface features from Cassini RADAR stereogrammetry. Icarus (submitted)

Korycansky DG, Zahnle KJ (2005) Modeling crater populations on Venus and Titan. Planet Space Sci 52:695–710

Lancaster N (1982) Linear dunes. Prog Phys Geog 6:476–504

Lancaster N (1995) Geomorphology of desert dunes. Routeledge, London

Langhans M, Jaumann R, Stephan K, Brown RH, Buratti BJ, Clark R, Baines KH, Nicholson PD, Lorenz RD (2009) Fluvial valleys on Titan: a global perspective. 40th Lunar and planetary science conference, Houston, abstract #1681

Leberton J-P, Witasse O, Sollazzo C, Blancquaert T, Cousin P, Schipper A-M, Jones JB, Matson DL, Gurvits LI, Atkinson DH, Kazeminejad B, Perez-Ayucar M (2005) An overview of the descent and landing of the Huygens probe on Titan, Nature 438:758–764. doi:10.1038/nature04347

Le Corre L, Le Mouélic S, Sotin C, Barnes J, Brown RH, Buratti B, Jaumann R, Rodriguez S, Soderblom J, Soderblom LA, Clark R, Baines KH, Nicholson PD (2009) Analysis of a cryolava flow on Titan Planet Space Sci 54:870–879: doi:10.1016/j.pss.2009.03.005

Le Mouélic S, Paillou P, Janssen MA, Barnes JW, Rodriguez S, Sotin C, Brown RH, Baines K, Buratti BJ, Clark RN, Crapeau M, Encernaz PJ, Jaumann R, Geudtner D, Paganelli F, Soderblom L, Tobie G, Wall S (2008) Mapping and interpretation of Sinlap crater on Titan using Cassini VIMS and RADAR data. J Geophys Res 113:E04003. doi:10.1029/2007JE002965

Lellouch E, Schmitt B, Coustenis A, Cuby J-G (2004) Titan's 5-micron lightcurve. Icarus 168:209–214

Lemmon MT, Karkoshka E, Tomasko M (1993) Titans Rotation: Surface Feature Observed. Icarus 103:329–332

Leopold LB, Wolman MG, Miller JP (1964) Fluvial processes in geomorphology. W.H. Freeman, San Francisco

Lewis JS (1971) Satellites of the outer planets: their physical and chemical nature. Icarus 15:174–185

Lindal GF, Wood GE, Hotz HB, Sweetnam DN, Eshleman VR, Tyler GL (1983) The atmosphere of Titan – an analysis of the Voyager 1 radio occultation measurements. Icarus 53:348–363

Lopes RMC, Mitchell KL, Stofan ER, Lunine JI, Lorenz R, Paganelli F, Kirk RL, Wood CA, Wall SD, Robshaw LE, Fortes AD, Neish CD, Radebaugh J, Reffet E, Ostro SJ, Elachi C, Allison MD, Anderson Y, Boehmer R, Boubin G, Callahan P, Encrenaz P, Flamini E, Francescetti G, Gim Y, Hamilton G, Hensley S, Janssen MA, Johnson WTK, Kelleher K, Muhleman DO, Ori G, Orosei R, Picardi G, Posa F, Roth LE, Seu R, Shaffer S, Soderblom LA, Stiles B, Vetrella S, West RD, Wye L, Zebker HA (2007a) Cryovolcanic features on Titan's surface as revealed by the Cassini Titan Radar mapper. Icarus 186:395–412. doi:10.1016/j.icarus.2006.09.006

Lopes R, Mitchell KL, Wall SD, Mitri G, Janssen M, Ostro S, Kirk RL, Hayes AG, Stofan ER, Lunine JI, Lorenz RD, Wood C, Radebaugh J, Paillou P, Zebker H, Paganelli F, the Cassini RADAR Team (2007) The lakes and seas of Titan. Eos 88, No. 51:569–576

Lopes RMC, Stofan ER, Peckyno R, Radebaugh J, Mitchell KL, Mitri G, Wood CA, Kirk RL, Wall SD, Lorenz RD, Lunine JI, Turtle EP, Craig J, Ollerenshaw RJ, Hayes A, Paganelli F, West R, Styles BW, Callahan PS, Anderson Y, Valora P, Elachi C, the Cassini RADAR Team (2009) Distribution and interplay of geologic processes in Titan from Cassini RADAR data. Icarus (accepted)

Lorenz RD (1993a) The life death and afterlife of a raindrop on Titan. Planet Space Sci 41:647–655

Lorenz RD (1993) The surface of Titan in the context of ESA's Huygens Probe. ESA J 17:275–292

Lorenz RD (1994) Crater lakes on Titan: rings, horseshoes and bullseyes. Planet Space Sci 42:1–4

Lorenz RD (1996a) Martian surface windspeeds, described by the Weibull distribution. J Spacecraft Rockets 33:754–756

Lorenz RD (1996b) Pillow lava on Titan: expectations and constraints on Cryovolcanic Processes. Planet Space Sci 44:1021–1028

Lorenz RD (2000) The weather on Titan. Science 290:467–468

Lorenz RD (2002) Thermodynamics of Geysers: application to Titan. Icarus 156:176–183

Lorenz RD (2006) Thermal interactions of the Huygens probe with the Titan environment: constraint on near-surface wind. Icarus 182:559–566. doi:10.1016/j.icarus.2006.01.009

Lorenz RD, Lunine JI (1996) Erosion on Titan: past and present. Icarus 122:79–91

Lorenz RD, Lunine JI (1997) Titan's surface reviewed: the nature of bright and dark terrain. Planet Space Sci 45:981–992

Lorenz RD, Lunine JI (2005) Titan's surface before Cassini. Planet Space Sci 53:557–576

Lorenz RD, Mitton J (2002) Lifting Titan's veil. Cambridge University Press, Cambridge

Lorenz RD, Mitton J (2008) Titan unveiled. Princeton University Press Princeton, NY

Lorenz RD, Lunine JI, Grier JA, Fisher MA (1995) Prediction of Aeolian features on planets: application to Titan paleoclimatology. J Geophys Res 88:26,377–26,386

Lorenz RD, McKay CP, Lunine JI (1999) Analytic stability of Titan's climate: sensitivity to volatile inventory. Planet Space Sci 47:1503–1515

Lorenz RD, Dooley JM, West JD, Fujii M (2003) Backyard spectroscopy and photometry of Titan, Uranus and Neptune. Planet Space Sci 51:113–125

Lorenz RD, Kraal E, Eddlemon E, Cheney J, Greeley R (2005) Sea-surface wave growth under extraterrestrial tmospheres – preliminary wind tunnel experiments with application to Mars and Titan. Icarus 175:556–560

Lorenz RD, Wall S, Radebaugh J, Boubin G, Reffet E, Janssen M, Stofan E, Lopes R, Kirk R, Elachi C, Lunine J, Paganelli F, Soderblom L, Wood C, Wye L, Zebker H, Anderson Y, Ostro S, Allison M, Boehmer R, Callahan P, Encrenaz P, Ori GG, Francescetti G, Gim Y, Hamilton G, Hensley S, Johnson W, Kelleher K, Mitchell K, Muhleman D, Picardi G, Posa F, Roth L, Seu R, Shaffer S, Stiles B, Vetrella S, Flamini E, West R (2006a) The sand seas of Titan: Cassini RADAR observations of equatorial fields of longitudinal dunes. Science 312:724–727. doi:10.1126/science.1123257

Lorenz RD, Niemann HB, Harpold DN, Way SH, Zarnecki JC (2006b) Titan's damp ground: constraints on Titan surface thermal properties from the temperature evolution of the Huygens GCMS inlet. Meteorit Planet Sci 41:1705–1714

Lorenz RD, Wood CA, Lunine JI, Wall SD, Lopes RM, Mitchell KL, Paganelli F, Anderson YZ, Wye L, Tsai C, Zebker H, Stofan ER (2007) Titan's young surface: initial impact crater survey by Cassini RADAR and model comparison. Geophys Res Lett 34:L07204. doi:10.1029/2006GL028971

Lorenz RD, Lopes RMC, Paganelli F, Lunine JI, Kirk RL, Mitchell KL, Soderblom LA, Stofan ER, Ori G, Myers M, Miyamoto H, Radebaugh J, Stiles B, Wall SD, Wood CA, the Cassini RADAR Team (2008a) Fluvial channels on Titan: initial Cassini RADAR observations. Planet Space Sci 56(8):1132–1144. doi:10.016/j.pss.2008.02.009

Lorenz RD, Mitchell KL, Kirk RL, Hayes AG, Zebker HA, Paillou P, Radebaugh J, Lunine JI, Janssen M-A, Wall SD, Lopes RM, Stiles B, Ostro S, Mitri G, Stofan ER, the Cassini RADAR Team (2008b) Titan's inventory of organic surface materials. Geophys Res Lett 35:L02206. doi:10.1029/2007GL032118

Lorenz RD, Stiles B, Kirk RL, Allison M, Persi del Marmo P, Iess L, Lunine JI, Ostro SJ, Hensley S (2008c) Titan's rotation reveals an internal ocean and changing zonal winds. Science 319:1649–1651

Lunine JI (1990) Evolution of the atmosphere and surface of Titan. ESA SP-315:159–165

Lunine JI (1993) Does Titan have an ocean? Review of current understanding of Titan's surface. Rev Geophys 31:133–149

Lunine JI, Stevenson DJ (1985) Evolution of Titan's coupled ocean-atmosphere system and interaction of ocean with bedrock, pp 741–757. In: Ices in the solar system. Proceedings of the advanced research workshop, Nice, France, January 16–19, 1984, Reidel, Dordrecht

Lunine JI, Stevenson DJ (1987) Clathrate and ammonia hydrates at high pressure – application to the origin of methane on Titan. Icarus 70:61–77

Lunine JI, Stevenson DJ, Yung YL (1983) Ethane ocean on Titan. Science 222:1229–1230

Lunine JI, Elachi C, Wall SD, Janssen MA, Allison MD, Anderson Y, Boehmer R, Callahan P, Encrenaz P, Flamini E, Francischetti G, Gim Y, Hamilton G, Hensley S, Johnson WTK, Kelleher K, Kirk RL, Lopes RM, Lorenz R, Muhlemen DO, Orosei R, Ostro SJ, Paganelli F, Paillou P, Picardi G, Posa F, Radebaugh J, Roth LE, Seu R, Shaffer S, Soderblom LA, Stiles B, Stofan ER, Vetrella S, West R, Wood CA, Wye L, Zebker H, Alberti G, Karkoschka E, Rizk B, McFarlane E, See C, Kazeminejad B (2008a) Titan's diverse landscapes as evidenced by Cassini RADAR's third and fourth looks at Titan. Icarus 195:415–433. doi:10.1016/j.icarus.2007.12.022

Lunine JI, Mitri G, Elachi C, Stofan E, Lorenz R, Kirk RL, Mitchell K, Lopes R, Wood CA, Radebaugh J, Wall SD, Soderblom LA, Paillou Ph, Farr T, Stiles B, Callahan P, RADAR Science Team (2008b) Lack of South polar lakes on Titan. 39th Lunar and planetary lunar and planetary science conference, Houston, abstract #1637

McCord T, Johnson TV, Elias JH (1971) Saturn and its satellites: narrow band spectrophotometry (0.3–1.1 μm). Astrophys J 165:413–424

McCord TB, Hansen GB, Buratti BJ, Clark RN, Cruikshank DP, D'Aversa E, Griffith CA, Baines EKH, Brown RH, Dalle Ore CM, Filacchione G, Formisano V, Hibbitts CA, Jaumann R, Lunine JI, Nelson RM, Sotin C, the Cassini VIMS Team (2006) Composition of Titan's surface from Cassini VIMS. Planet Space Sci 54(15):1524–1539

McCord TB, Hayne P, Combe J-P, Hansen GB, Barnes JW, Rodriguez S, Le Mouèlic S, Baines KH, Buratti BJ, Sotin C, Nicholson P, Jaumann R, Nelson R, the Cassini VIMS Team (2008) Titan's surface: search for spectral diversity and composition using the Cassini VIMS investigation. Icarus 194:212–242. doi:10.1016/j.icarus.2007.08.039

McEwen A, Turtle E, Perry J, Dawson D, Fussner S, Collins G, Porco C, Johnson T, Soderblom L (2005) Mapping and monitoring the surface of Titan. Bull Am Astron Soc 37, abstract 53.04

McKenzie D, Ford PG, Liu F, Pettengill GH (1992) Pancakelike domes on Venus. J Geophys Res 97:15967–15976

Meier R, Smith BA, Owen TC, Terrile RJ (2000) The surface of Titan from NICMOS observations with the Hubble space telescope. Icarus 145:462–473

Mitchell JL, Pierrehumbert RT, Frierson DMW, Caballero R (2006) The dynamics behind Titan's methane clouds. Proc Natl Acad Sci 103:18421–18426

Mitchell KL, Wall SD, Stofan ER, Lopes RMC, Janssen M, Stiles B, Paillou P, Mitri G, Lunine J, Ostro SJ, Lorenz RD, Farr TG, Kirk RL, Radebaugh J, the Cassini RADAR Science Team (2007) Titan's north polar lakes as observed by Cassini RADAR: an update. Workshop on ices, oceans, and fire: satellites of the outer solar system, boulder, CO, 13–15 August 2007, Abstract #6042, Lunar and Planetary Institute, Houston

Mitchell KL, Stofan ER, Radebaugh J, Lopes RMC, Paillou Ph, Lunine JI, Wall SD, Lorenz RD, Hayes A, Kargel JS, Ventura B, Casarano D, Notarnicola C, Mitri G, Farr TG, Stiles BW, the Cassini Radar Science Team (2009) Titan's north polar lake district: insights from the Cassini Titan Radar Mapper. Icarus (submitted)

Mitri G, Showman AP, Lunine JI, Lorenz RD (2007) Hydrocarbon lakes on Titan. Icarus 186:385–394. doi:10.1016/j.icarus.2006.09.004

Mitri G, Lunine JI, Showman AP, Lopes R (2008) Resurfacing of Titan by Ammonia-water Cryomagma. Icarus 196:216–224. doi:10.1016/j.Icarus.2008.02.024

Moore JM, Howard AD, Dietrich WE, Schenk PM (2003) Martian layered fluvial deposits: implications for Noachien climate scenarios. Geophys Res Lett 30(24):2292. doi:10.1029/2003GL019002

Moroz VI (2002) Estimates of visibility of the surface of Venus from descent probes or balloons. Planet Space Sci 50:287–297

Muhleman DO, Grossman AW, Butler BJ, Slade MA (1990) Radar reflectivity of Titan. Science 248:975–980

National Institute of Standards and Technology (NIST) (2005) Chemistry web book. Standard reference data base, No. 69, June 2005. http:webbook.nist.gov./chemistry

Negraõ A, Hirtzig M, Coustenis A, Gendron E, Drossart P, Rannou P, Combes M, Boudon V (2007) The 2-mm spectroscopy of Huygens probe landing site on Titan with very large Telescope/Nasmyth Adaptive Optics System Near-Infrared Imager and Spectrograph. J Geophys Res 112(E2, E02s92): 1–14

Neish CD, Lorenz RD, Kirk RL (2008) Radar Topography of domes on Venus, Earth, and Titan. Icarus 196:554–564

Nelson RM, Kamp L, Matson DL, Irwin PGJ, Baines KH, Boryta MD, Leader FE, Jaumann R, Smythe, WD, Sotin C, Clark RN, Cruikshank DP, Drossart P, Pearl JC, Hapke BW, Lunine J, Combes M, Bellucci G (2007)Saturn's Titan: Cassini VIMS reports regional reflectance change consistent with surface activity. 38th Lunar and planetary lunar and planetary science conference, Houston, abstract #2158

Nelson RM, Kamp LW, Lopes RMC, Matson DL, Kirk RL, Hapke BW, Wall SD, Boryta MD, Leader FE, Smythe WD, Mitchell KL, Baines KH, Jaumann R, Sotin C, Clark CN, Cruikshank DP, Drossart P, Lunine JI, Combes M, Bellucci G, Bibring, JP, Capaccioni F, Cerroni P, Coradini A, Formisano V, Filacchione G, Langevin Y, McCord TB, Mennella V, Nicholson PD, Sicardy B, Irwin PGJ, Pearl JC (2009a) Photometric changes on Saturn's Titan: evidence for active cryovolcanism. Geopgys Res Lett 36:L04202. doi:10.1029/2008GL036206

Nelson RM, Kamp LW, Matson DL, Irwin PGJ, Baines KH, Boryta MD, Leader FE, Jaumann R, Smythe WD, Sotin C, Clark CN, Cruikshank DP, Drossart P, Pearl JC, Habke BW, Lunine JI, Combes M, Bellucci G, Bibring J-P, Capaccioni F, Cerroni P, Coradini A, Formisano V, Filacchione G, Langevin RY, McCord TB, Mennella V, Nicholson PD, Sicardy B (2009b) Saturn's Titan: surface change, ammonia, and implications for atmospheric and tectonic activity. Icarus 199:429–441

Niemann HB, Atreya SK, Bauer SJ, Carignan GR, Demick JE, Frost RL, Gautier D, Haberman JA, Harpold DN, Hunten DM, Israel G, Lunine JI, Kasprzak WT, Owen TC, Paulkovich M, Raulin F, Raaen E, Way SH (2005) The composition of Titan's atmosphere from the GCMS on the Huygen probe, and implications for the origin of nitrogen and methane. Nature 438:779–784

Noll KS, Knacke RF (1993) Titan: 1–5 μm photometry and spectrophotometry and a search for variability. Icarus 101:272–281

O'Brien DP, Lorenz RD, Lunine JI (2005) Numerical calculations of the longevity of impact oases on Titan. Icarus 173:243–253

Ori G, Marinangeli L, Baliva A, Bressan M, Strom R (1998) Dynamics of liquids on Titan's surface. Planet Space Sci 46:1417–1421

Osterkamp WR, Hedman ER (1982) Perenial-streamflow characteristics related channel geometry and sediment in Missouri River Basin. U.S. Geological Survey Professional Paper 1242

Paillou P, Mitchell K, Wall S, Ruffie G, Wood C, Lorenz R, Stofan E, Lunine J, Lopes R, Encrenaz P (2008) Microwave dielectric constant of liquid hydrocarbons: application to the depth estimation of Titan's lakes. Geophys Res Lett 35:L05202. doi:10.1029/2007GL032515

Perron JT, de Pater I (2004) Dynamics of an ice continent on Titan. Geophys Res Lett 31(17):L17S04. doi:10.1029/2004GL019802

Perron JT, Lamb MP, Koven CD, Fung IY, Yager E, Ádámkovics M (2006) Valley formation and methane precipitation rates on Titan. J Geophys Res 111:E11001. doi:10.1029/2005JE002602

Perry JE, McEwen AS, Turtle EP, Fussner S, Cassini ISS Team (2006) Equatorial faculae on Titan: distribution and orientation. 37th Lunar and planetary lunar and planetary science conference, Houston, abstract #2170

Perry JE, Turtle EP, McEwen AS, Dawson DD (2007a) Cassini ISS observations of Titan: the Titan-20 Flyby. 38th Lunar and planetary lunar and planetary science conference, Houston, abstract #2219

Perry JE, Turtle EP, McEwen AS, Dawson DD, Porco CC (2007) Cassini ISS observations of Titan's trailing hemisphere. Workshop on ices, oceans, and fire: satellites of the outer solar system, abstract #6064

Perry JE, Turtle EP, McEwen AS, Dawson DD, Porco CC (2007) Cassini ISS observations of Titan's trailing hemisphere. Bull Am Astron Soc 39, abstract 44.04

Plaut JJ (1993) Stereo imaging. In: Guide to Magellan image interpretation. JPL Pub 932-24, 33

Porco CC, West RA, Squyres S, McEwen A, Thomas P, Murray CD, Delgenio A, Ingersoll AP, Johnson TV, Neukum G, Veverka J, Dones L, Brahic A, Burns JA, Haemmerle V, Knowles B, Dawson D, Roatsch T, Beurle K, Owen W (2004) Cassini imaging science: instrument, characteristics and anticipated scientific investigations at Saturn. Space Sci Rev 115:363–497

Porco CC, Baker E, Barbara J, Beurle K, Brahic A, Burns JA, Charnoz S, Cooper N, Dawson DD, Del Genio AD, Denk T, Dones L, Dyudina U, Evans MW, Fussner S, Giese B, Grazier K, Helfenstein P, Ingersol AP, Jacobson RA, Johnson TV, McEwen A, Murray CD, Neukum G, Owen WM, Perry J, Roatsch T, Spitale J, Squyres S, Thomas P, Tiscareno L, Turtle EP, Vasavada AR, Veverka J, Wagner R, West R (2005) Imaging of Titan from the Cassini spacecraft. Nature 434:159–168

Radebaugh J, Lorenz R, Kirk R, Lunine J (2006) Mountains on Titan observed by Cassini RADAR. 37th Lunar and planetary lunar and planetary science conference, Houston, abstract #1007

Radebaugh J, Lorenz R, Kirk R, Lunine J, Stofan E, Lopes R, Wall S, the Cassini RADAR Team (2007) Mountains on Titan observed by Cassini RADAR. Icarus 192:77–92. doi:10.1016/j.icarus.2007.06.020

Radebaugh J, Lorenz R, Wall S, Lunine J, Kirk R, Lopes R, Paganelli F, Stofan E, the Cassini RADAR Team (2008a) Dunes on Titan from Cassini RADAR. Icarus 194:690–703. doi:10.1016/j.icarus.2007.10.015

Radebaugh J, Kirk RL, Lopes RM, Stofan ER, Valora P, Lunine JI, Lorenz RD, the Cassini RADAR Team (2008) Mountains on Titan as evidence of global tectonism and erosion. 39th Lunar and planetary lunar and planetary science conference, Houston, abstract #2206

Radebaugh J, Lorenz RD, Kirk RL, Wall SD, Wood CA, Lunine JI, Stofan ER, n Lopes RMC, Valora P, Farr TG, Hayes AG, Styles BW, Mitri G, Zebker H, Janssen M, Wye L, Le Gall A, Mitchel KL, the Cassini RADAR Team (2009) Regional geomorphology and history of of Titan's Xanadu province. Icarus (submitted)

Rannou P, Montmessin F, Hourdin F, Lebonnois S (2006) The latitudinal distribution of clouds on titan. Science 311(5758):201–205

Richardson J, McEwen A, Lorenz RD (2004) Titan's surface and rotation – new insights from voyager images. Icarus 170:113–124

Rodriguez S, Le Moue´ lic S, Sotin C, Cle´ net H, Clark RN, Buratti B, Brown RH, McCord TB, Nicholson D, Baines KH, theVIMS Science Team (2006) Cassini/VIMS hyperspectral observations of the HUYGENS landing site on Titan. Planet Space Sci 54:1510–1523

Roe HG, de Pater I, Gibbard SG, Macintosh BA, Max CE, Young EF, Brown ME, Bouchez AH (2004) A new 1.6 micron map of Titan's surface. Geophys Res Lett 31:L17S03. doi:10.1029/2004GL019871

Sagan C, Dermott SF (1982) The tides in the seas of Titan. Nature 300:731–733

Saint-Pe O, Combes M, Rigaut F, Tomasko M, Fulchignoni M (1993) Demonstration of adaptive optics for resolved imagery of solar sys-

tem objects: preliminary results on Pallas and Titan. Icarus 105:263–270
Samuelson RE, May LA, Knuckles MA, Khanna RJ (1997) C4N2 ice in Titan's north polar atmosphere. Planet Space Sci 45:941–948
Schaller EL, Brown ME, Roe HG, Bouchez AH, Trujillo CA (2005) Cloud activity on Titan during the Cassini mission. 36th Lunar and planetary lunar and planetary science conference, Houston, abstract #1989
Schaller EL, Brown ME, Roe HG, Bouchez AH (2006a) A large cloud outburst at Titan's south pole. Icarus 182:224–229
Schaller EL, Brown ME, Roe HG, Bouchez AH, Trujillo CA (2006b) Dissipation of Titan's south polar clouds. Icarus 184:517–523
Schenk PM (1993) Central pit and dome craters – exposing the interiors of Ganymede and Callisto. J Geophys Res 98:7475–7498
Schröder SE, Keller HU (2008) The reflectance spectrum of Titan's surface at the Huygens landing site determined by the descent imager/spectral radiometer. Planet Space Sci 56:753–769
Schröder SE, Keller HU (2009) The unusual phase curve of Titan's surface observed by Huygens' descent imager/spectral radiometer. Planet Space Sci. in press:doi:10.1016/j.pss.2009.03.012
Sears WD (1995) Tidal dissipation in oceans on Titan. Icarus 113:39–56
Smith BA, Soderblom L, Beebe RF, Boyce JM, Briggs G, Bunker A, Collins SA, Hansen C, Johnson TV, Mitchell JL, Terrile RJ, Carr MH, Cook AF, Cuzzi JN, Pollack JB, Danielson GE, Ingersoll AP, Davies ME, Hunt GE, Masursky H, Shoemaker M, Morrison D, Owen T, Sagan C, Veverka J, Strom R, Suomi VE (1981) Encounter with Saturn – voyager 1 imaging science results. Science 212:163–191
Smith PH, Lemmon MT, Lorenz RD, Sromovsky LA, Caldwell JJ, Allison MD (1996) Titan's surface, revealed by HST imaging. Icarus 119:336–339
Smith PH, Bell JF III, Bridges NT, Britt DT, Gaddis L, Greeley R, Keller HU, Herkenhof KE, Jaumann R, Johnson JR, Kirk RL, Lemmon M, Maki JN, Malin MC, Murchie SL, Oberst J, Parker TJ, Reid RJ, Sablotny R, Soderblom LA, Stoker C, Sullivan R, Thomas N, Tomasko MG, Ward W, Wegryn E (1997) Results from the Mars pathfinder camera. Science 278:1758–1761
Soderblom L, Anderson J, Baines K, Barnes J, Barrett J, Brown R, Buratti B, Clark R, Cruikshank D, Elachi C, Janssen M, Jaumann R, Kirk R, Karkoschka E, Le Mouélic S, Lopes R, Lorenz R, Lunine J, McCord T, Nicholson P, Radebaugh J, Rizk B, Sotin C, Stofan E, Sucharski T, Tomasko M, Wall S (2007a) Correlations between Cassini VIMS spectra and RADAR SAR images: Implications for Titan's surface composition and the character of the Huygens probe landing site. Planet Space Sci 55:2025–2036. doi:10.1016/j.pss.2007.04.014
Soderblom L, Tomasko M, Archinal B, Becker T, Bushroe M, Cook D, Doose L, Galuszka L, Hare T, Howington-Kraus A, Karkoschka E, Kirk R, Lunine J, McFarlane E, Redding B, Rizk B, Rosiek M, See C, Smith P (2007b) Topography and geomorphology of the Huygens landing site on Titan. Planet Space Sci 55:2015–2024. doi:10.1016/j.pss.2007.04.015
Soderblom LA, Brown RH, Soderblom JM, Barnes JW, Kirk RL, Jaumann R, Mackinnon J, Mackowski DW, Baines KH, Buratti BJ, Clark RN, Nicholson PD, Sotin C (2009a) The geology of Hotei Regio, Titan: correlation of Cassini VIMS and RADAR. Icarus (submitted)
Soderblom LA, Clark RN, McCord TB, Tomasko MG, Janssen MA, Nieman HB, Barnes J, Brown RH (2009b) Composition. In: Brown RH, Lebreton J-P, Waite H (eds) Titan from Cassini-Huygens. Springer, NY
Sohl F, Sears WD, Lorenz RD (1995) Tidal dissipation on Titan. Icarus 115:278–294
Sotin C, Jaumann R, Buratti BJ, Brown RB, Clark RN, Soderblom LA, Baines KH, Bellucci G, Bibring J-P, Capaccioni F, Cerroni P, Coradini A, Cruikshank DP, Drossart P, Formisano V, Langevin Y, Matson DL, McCord TB, Nelson RM, Nicholson PD, Sicardy B, Le Mouélic L, Rodriguez S, Stephan K, Scholz CK (2005) Release of volatiles from a possible cryovolcano from near-infrared imaging of Titan. Nature 435:786–789
Sotin C, Le Mouèlic S, Brown RH, Barnes J, Soderblom L, Jaumann R, Burattin BJ, Clark RN, Baines KH, Nelson RM, Nicolson P, the Cassini VIMS Team (2007) Cassini/VIMS observations of titan during the T20 flyby. 38th Lunar and planetary lunar and planetary science conference, Houston, abstract #2444
Sotin C, Stevenson DJ, Rappaport NJ, Mitri GA, Schubert G (2009) Interior structure. In: Brown RH, Lebreton J-P, Waite H (eds) Titan from Cassini-Huygens. Springer, NY
Srokosz MA, Challenor PG, Zarnecki JC, Green SF (1992) Waves on Titan, European space agency. (ESA) SP-338, Noordwijk, The Netherlands, pp 321–323
Stephan K, Jaumann R, Karkoschka E, Barnes JW, Kirk R, Tomasko MG, Turtle EP, Le Corre L, Langshans M, Le Mouelic S, Lorenz R, Perry, J. (2009) Mapping products of Titan's surface. In: Brown RH, Lebreton J-P, Waite H (eds) Titan from Cassini-Huygens. Springer, NY
Stevenson DJ (1992) The interior of Titan. In: Proceedings of the symposium on Titan. ESA SP-338, ESA, Noordwijk, The Netherlands, 23–33
Stevenson DJ, Potter AE (1986) Titan's latitudinal temperature distribution and seasonal cycle. Geophys Res Lett 13:93–96
Stiles B, Kirk RL, Lorenz R, Hensley S, Lee E, Ostro S, Gim Y, Hamilton G, Johnson WTK, West R, the Cassini RADAR Team (2008) Estimating Titan's spin state from Cassini SAR data. Astronom J 135:1669–1680. doi:10.1088/0004-6256/135/5/1669
Stofan ER, Lunine JI, Lopes R, Paganelli F, Lorenz RD, Wood CA, Kirk R, Wall S, Elachi C, Soderblom LA, Ostro S, Janssen M, Wye L, Zebker H, Anderson Y, Allison M, Boehmer R, Callahan P, Encrenaz P, Flamini E, Francescetti G, Gim Y, Hamilton G, Hensley S, Johnson W, Kelleher K, Muhleman D, Picardi G, Posa F, Roth L, Seu R, Shaffer S, Stiles B, Vetrella S, West R (2006) Mapping of Titan: results from the first Titan radar passes. Icarus 185(2):443–456. doi:10.1016/j.icarus.2006.07.015
Stofan ER, Elachi C, Lunine JI, Lorenz RD, Stiles B, Mitchell KL, Ostro S, Soderblom L, Wood C, Zebker H, Wall S, Janssen M, Kirk R, Lopes R, Paganelli F, Radebaugh J, Wye L, Anderson, Y, Allison M, Boehmer R, Callahan P, Encrenaz P, Flamini E, Francescetti G, Gim Y, Hamilton G, Hensley S, Johnson WTK, Kelleher K, Muhleman D, Paillou P, Picardi G, Posa F, Roth L, Seu R, Shaffer S, Vetrella S, West R (2007) The lakes of Titan. Nature 445. doi:10.1038/nature05438
Stofan ER, Elachi C, Lunine JI, Lorenz RD, Kirk RL, Lopes RM, Wood CA, Radebaugh J, Wall SD, Mitchell KL, Soderblom LA, Paillou P, Farr T, Stiles B, Callahan P, the RADAR Science Team (2008) Varied geologic terrains at Ttian's south pole: first results from T39. 39th Lunar and planetary lunar and planetary science conference, Houston, abstract #1492
Strobel DF (1982) Chemistry and evolution of Titan's atmosphere. Planet Space Sci 30:839–848
Styles BW, Hensley S, Gim Y, Bates DM, Kirk RL, Hayes A, Radebaugh J, Lorenz RD, Mitchell KL, Callahan PS, Zebker H, Johnson WTK, Wall SD, Lunine JI, Wood CA, Janssen M, Pelletier F, West RD, Veeramacheneni C, and the Cassini RADAR Team (2009) Determining Titan surface topography from Cassini SAR data, Icarus, in press.
Summerfield MA, Hulton NJ (1994) Natural controls of fluvial denudation rates in major world drainage basins. J Geophys Res 99: 13871–13883
Thompson WR, Sagan C (1992) Organic chemistry on Titan – surface interactions. Symposium on Titan, Toulouse, September 1991, ESA SP-338, 167–176
Tobie G, Mocquet A, Sotin C (2005) Tidal dissipation within large icy satellites: Europa and Titan. Icarus 177:534–549
Tobie G, Lunine JI, Sotin C (2006) Episodic outgassing as the origin of atmospheric methane on Titan. Nature 440:61–64

Tokano T (2008) Dune-forming winds on Titan and the influence of topography. Icarus 194:243–262
Tokano T, Neubauer F (2002) Tidal winds on Titan caused by Saturn. Icarus 158:499–515
Tokano T, McKay CP, Neubauer FM, Atreya SK, Ferri F, Fulchignoni M, Niemann HB (2006) Methane drizzle on Titan, Nature 442:432–435
Tomasko MG, Smith PH (1982) Photometry and polarimetry of Titan Pioneer 11 observations and their implications for aerosol properties. Icarus 51:65–95
Tomasko MG, Doose LR, Smith PH, West RA, Soderblom LA, Combes M, Bezard B, Coustenis A, deBergh C, Lellouch E, Rosen-qvist J, Saint-Pe O, Schmitt B, Keller HU, Thomas N, Gliem F (1997a) The descent imager/spectral radiometer (DISR) aboard Huygens. In: Wilson A (ed) Huygens – science, payload and mission, Vol SP-1177. ESA Publications Division, ESTEC, Noordwijk, The Netherlands, pp 109–138
Tomasko MG, Buchhauser D, Bushroe M, Dafoe LE, Doose LR, Eibl A, Fellows C, McFarlane E, Prout GM, Pringle MJ, Rizk B, See C, Smith PH, Tsetsenekos K (2002) The descent imager/spectral radiometer (DISR) experiment on the Huygens entry probe of Titan. Space Sci Rev 104:469–551
Tomasko MG, Archinal B, Becker T, Bézard B, Bushroe M, Combes M, Cook D, Coustenis A, de Bergh C, Dafoe LE, Doose L, Douté S, Eibl A, Engel S, Gliem F, Grieger B, Holso K, Howington-Kraus A, Karkoschka E, Keller HU, Kirk R, Kramm R, Küppers M, Lellouch E, Lemmon M, Lunine J, McFarlane E, Moores J, Prout M, Rizk B, Rosiek M, Rüffer P, Schröder SE, Schmitt B, Smith P, Soderblom L, Thomas N, West R (2005) Rain, winds and haze during the Huygens probe's descent to Titan's surface. Nature 438(8):765–778. doi:10.1038/nature04126
Tomasko MG, Doose L, Engel S, Dafoe LE, West R, Lemmon M, Karkoschka E (2008) A model of the aerosols in Titan's atmosphere. Planet Space Sci 56(5):669–707
Tosi F, Orosei R, Seu R, Filacchione G, Coradini A, Lunine JI, Capaccioni F, Cerroni P, Adriani A, Moriconi ML, the Cassini VIMS and RADAR Teams (2008) Analysis of selected Cassini VIMS and RADAR data over the surface of Titan through multivariate statistical methods, 39th Lunar and planetary lunar and planetary science conference, Houston, abstract #1357.
Tran BN, Jones JC, Ferris JP, Persans PD, Chera JJ (2003) Simulation of Titan haze formation using a photochemical flow reactor: the optical constants of the polymer. Icarus 165:379–390
Turtle EP, Perry JE, McEwen AS, DelGenio AD, Barbara J, West RA, Dawson DD, Porco CC (2009) Cassini imaging of Titan's high-latitude lakes, clouds, and south-polar surface changes. Geophys Res Lett 36:L02204. doi:10.1029/2008GL036186
Tyler GL et al (1981) Radio science investigations of the Saturn system with Voyager 1 – preliminary results. Science 212:201–206
Wall SD, Lopes RM, Stofan ER, Wood CA, Radebaugh JL, Stiles BW, Nelson RM, Kamp LW, Janssen MA, Lorenz RL, Lunine JI, Farr TG, Mitri G, Paillou P, Mitchell KL (2009) Cassini RADAR images at Hotei Arcus and western Xanadu, Titan: evidence for recent cryovolcanic activity. Geophys Res Lett 36:L04203. doi:10.1029/2008 GL036415
West RA, Smith P-H- (1991) Evidence for aggregate particles in the atmospheres of Titan and Jupiter. Icarus 90:330–333
West RA, Brown ME, Salinas SV, Bouchez AH, Roe HG (2005) No oceans on Titan from the absence of a near-infrared specular reflection. Nature 436:670–672
Wood C, Lunine JI, Lopes RM, Stofan ER, Mitchel K, Radebaugh (2006) Crateriform structures on Titan. 37th Lunar and planetary lunar and planetary science conference, Houston, abstract #1659
Wood C, Kirk RL, Stofan E, Stiles B, Zebker H, Ostro S, Radebaugh J, Lorenz RD, Callahan P, Wall S (2007a) Xanadu is old, rugged, and low-lying. Am Astron Soc Bull 39(3):44.05
Wood CA, Mitchell KL, Lopes RMC, Radebaugh J, Stofan E, Lunine J (2007b) Volcanic calderas in the north polar region of Titan. 38th Lunar and planetary lunar and planetary science conference, Houston, abstract #1453
Wood CA, Lorenz R, Radabaugh J (2007c) How Titan works, a RADAR perspective. 38th Lunar and planetary lunar and planetary science conference, Houston, abstract #1338
Wood CA, Kirk R, Lorenz RD (2009a) Numbers, distribution and morphologies of impact craters on Titan. 40th Lunar and planetary lunar and planetary science conference, Houston, abstract #2242
Wood CA, Lorenz RD, Kirk RL, Lopes RMC, Mitchell KL, Stofan EM (2009b) Impact craters on Titan. Icarus (submitted)
Wye LC, Zebker HA, Ostro SJ, West RD, Gim Y, Lorenz RD, the Cassini RADAR Team (2007) Electrical properties of Titan's surface from Cassini RADAR scatterometer measurements. Icarus 188:367–385. doi:10.1016/j.icarus.200
Yung YL, Allen M, Pinto JP (1984) Photochemistry of the atmosphere of Titan: comparison between model and observations. Astrophys J Suppl 55:465–506
Zarnecki JC, Leese MR, Hathi B, Ball AJ, Hagermann A, Towner MC, Lorenz RD, McDonnell JAM, Green SF, Patel MR, Ringrose TJ, Rosenberg PD, Atkinson KR, Paton MD, Banaszkiewicz M, Clark BC, Ferri F, Fulchignoni M, Ghafoor NAL, Kargl G, Svedhem H, Delderfeld J, Grande M, Parker DJ, Challenor PG, Geake JE (December 2005) A soft solid surface on Titan as revealed by the Huygens surface science package. Nature 438:792–795
Zebker HA, Gim Y, Callahan P, Hensley S, Lorenz R, the Cassini RADAR Team (2009) Analysis and interpretation of Cassini Titan radar altimeter echos. Icarus (submitted) Icarus 200:240–255

Chapter 6
Composition of Titan's Surface

L.A. Soderblom, J.W. Barnes, R.H. Brown, R.N. Clark, M.A. Janssen, T.B. McCord, H.B. Niemann, and M.G. Tomasko

Abstract The Huygens Probe returned the first *in situ* data on Titan's surface composition in January 2005. Although Huygens landed on a dry plain, the Gas Chromatograph Mass Spectrometer (GCMS) showed evidence of methane moisture in the near subsurface suggesting methane precipitation at some time in the past. Heavier organic molecules were not found to be abundant in the atmosphere or at the surface, but the GCMS surface results did show ethane to be present and tentatively identified cyanogen, benzene, and carbon dioxide. During descent, aerosol particles were processed with the Aerosol Collector and Pyroliser; results suggested that the aerosols contain both nitriles and hydrocarbons. The Descent Imager/Spectral Radiometer (DISR also carried by the probe) measured the visible and near-infrared spectral reflectance of the dark plain surface at the landing site. Those data suggest a mixture of water ice, tholin-like materials, and dark neutral material with a blue slope in the near infrared; identification of water ice is suggested but inconclusive. Most remarkably DISR did not detect spectral features, beyond those for methane, for a wide range of spectrally active hydrocarbon and nitrile compounds that had been expected to be present on the surface.

The Cassini Visual and Infrared Mapping Spectrometer (VIMS) observes the spectral properties of Titan's surface through atmospheric windows between intense methane absorption bands. VIMS data show Titan's dark blue units (in RGB composites of 2.0, 1.6, and 1.3 μm) to exhibit lower relative albedos in the 1.6, 2.0, and 5 μm windows interpreted (though not unambiguously) to result from enhancement in water ice. Spectra for bright units do not exhibit depressed albedo in these windows. This gives strong evidence that the bright units are bright organic solids and not exposed water ice. The other dark equatorial unit, the dark brown unit, correlates with the vast seas of dunes discovered in the Cassini RADAR SAR (Synthetic Aperture Radar) images, suggesting that the dunes are composed of dark organic grains. If the bright materials and dark dunes are both largely organics, then they appear to consist of physically and/or chemically different hydrocarbons and/or nitriles. The VIMS and RADAR data together lead to a model where a dark blue substrate is mantled by the seas of dark organic dunes seen in SAR images and by thinner units of bright organic solids that are invisible to SAR.

Carbon dioxide has been suggested as a reasonable compositional component of Titan's surface. The GCMS did tentatively identify CO_2 at the surface. VIMS observations of south mid-latitude 5 μm bright spots Hotei Regio and Tui Regio have been suggested as attributable to carbon dioxide. CO_2 might explain both an unusual spectral slope in the 2.7–2.8 μm spectral region and an absorption band near 4.92 μm. However the VIMS 4.92 μm band is shifted significantly in wavelength from the position observed in the laboratory rendering the CO_2 identification in VIMS Tui Regio spectra inconclusive. An alternate suggestion for the source of the 4.92 μm feature in the VIMS Tui Regio spectrum is the nitrile cyanoacetylene (HC_3N); it offers a better spectral match than does CO_2. Cyanoacetylene is a known thermospheric product detected by both the Composite Infrared Spectrometer (CIRS) and the Ion and Neutral Mass Spectrometer (INMS) Cassini instruments. If Tui Regio in fact shows a high abundance of cyanoacetylene it raises questions as to by what processes such materials are concentrated.

L.A. Soderblom(✉)
Astrogeology Team, United States Geological Survey, Flagstaff, AZ, 86004, USA
e-mail: lsoderblom@usgs.gov

J.W. Barnes
University of Idaho, Moscow, ID, 83844, USA

R.H. Brown and M.G. Tomasko
University of Arizona, Tucson, AZ, 85721, USA

R.N. Clark
United States Geological Survey, Denver, CO, 80225, USA

M.A. Janssen
Jet Propulsion Laboratory, California Institute of Technology, Pasadena, CA, 91109, USA

T.B. McCord
The Bear Fight Center, P.O. Box 66722 Fiddler's Road, Winthrop, WA, 98862, USA

H.B. Niemann
National Aeronautics and Space Administration, Greenbelt, MD, 20771, USA

R.H. Brown et al. (eds.), *Titan from Cassini-Huygens*,

Other surface absorption features in the 4.8–5.2 μm spectral region have been attributed to various aromatic and aliphatic hydrocarbons including benzene. Because of the low signal precision of VIMS data at these wavelengths, these features are difficult to detect, particularly in Titan's dark regions. As a result there is a debate over the certainty of their existence. One such argued absorption feature near 5.05 μm most closely matches laboratory spectra of benzene, a compound detected both at the surface by the GCMS and at high altitude by INMS in greater abundance than expected. Another absorption feature at 4.97 μm, also in debate, is best matched by spectra of the low-molecular-weight alkanes, methane and ethane, suggestive of moist surfaces wetted with such liquids consistent with GCMS observations of subsurface methane moisture.

The Cassini RADAR measurements constrain electrical properties related to Titan's surface composition in its scatterometry and radiometry modes. Analysis of the scatterometry observations yields an average dielectric constant of $\varepsilon \sim 2.2$. The global passive microwave radiometry map shows the effective ε to be quite uniform over the globe; >95% of the surface shows a narrow range of $\varepsilon \sim 1.5 \pm 0.3$. Both data sets suggest a high degree of volume scattering indicating substantial porosity making higher-ε materials including fractured, porous water ice, possible. At the same time, these data preclude substantial exposures of solid sheets of water ice ($\varepsilon \sim 3.1$) in the near surface except perhaps as local outcrops as at Sinlap crater ($\varepsilon \sim 2.5$). The radiometry analysis also yields global maps of thermal emissivity and of volume scattering. These properties show Titan's surface on the global scale to be consistent with fluffy blankets and veneers of organics, perhaps with graded density increasing with depth. The higher emissivity of the radar-dark dunes is consistent with grains having hydrocarbon and/or nitrile rich materials.

Cassini SAR images showed the north-polar region (>70°N) to exhibit a plethora of features resembling terrestrial lakes and seas. Further support for their being liquid is provided from analysis of high-resolution microwave radiometry that shows the north polar lakes to have high emissivity (~0.985) and low equivalent dielectric constant (~1.6) consistent with methane-ethane liquid. Most significant VIMS found absorption bands in south polar lake Ontario Lacus that evidence the presence of ethane, probably in liquid solution with methane, nitrogen, and other low-molecular-weight hydrocarbons.

6.1 Introduction: Sources of Titan's Surface Composition

Based on its bulk density (1,880 kg/m^3) Titan is presumed to consist mostly of water ice and silicates (cf. Tyler et al. 1981 and Tobie et al. 2005). Its interior is also hypothesized to contain shells of methane clathrate hydrate and ammonia hydrate that could enable a variety of cryovolcanic fluids to erupt to the surface (cf. Chapters 3 and 4). Titan's dense cold atmosphere consists largely of nitrogen with a few percent of methane. It has long been known that energetic processes operating in the upper atmosphere, UV photolysis and impinging energetic particles, generate a vast array of hydrocarbon and nitrile by-products ranging in complexity from simple molecules (e.g. ethane) to large complex organic aerosols (e.g. tholins) (cf. Chapters 7, 8, and 12). So large might be this atmospheric contribution to the surface that thick accumulations of hydrocarbons and nitriles and even oceans of ethane and methane were hypothesized (Sagan and Dermott 1982; Lunine et al. 1983). Unlike any other terrestrial planet or satellite, Titan's surface composition then derives from massive amounts of material delivered to it by active processes ongoing in both in its atmosphere and its interior. Like the Earth and Mars these surface materials are subsequently altered both chemically and physically and eroded, transported and re-deposited by a wide range of geologic and atmospheric processes (cf. Chapter 5). Here we review the history of theoretical, laboratory, and observational work on which was based our pre-Cassini concepts for the major sources and the processes that led to the composition of Titan's modern surface (also see Chapter 2).

6.1.1 *Atmospheric Organics and Nitrile Compounds*

Kuiper (1944) discovered Titan's atmosphere and its methane absorption bands but it was not until the 1970s that the full complexity of the atmospheric photochemistry began to unfold. After Trafton (1972) identified H_2 and Danielson et al. (1973) detected ethane (C_2H_6), work rapidly progressed in understanding the methane photochemistry. Strobel (1974) showed H_2, C_2H_2 (acetylene), and C_2H_6 to be among the expected by-products and that ethane should be the primary photolytic product. Hunten (1974) posited a km-thick surface layer of photochemical debris. In 1980 data from the Voyager 1 Radio Science, IRIS, and UVS instruments showed N_2 to be the dominant specie (as had been suspected; cf. Hunten 1974) of a thick, cold atmosphere with a surface pressure of 1.5 bar at T ~ 95 K (Tyler et al. 1981; Samuelson et al. 1981; Hanel et al. 1981; Broadfoot et al. 1981: Lindal et al. 1983). From the IRIS data Hanel et al. (1981) identified CH_4, C_2H_2, C_2H_4 (ethylene), C_2H_6, HCN (hydrogen cyanide) and suggested C_3H_4 (methylacetylene) and C_3H_8 (propane). Allen et al. (1980) showed that ethane is actually generated from methane in the lower stratosphere wherein photosensitized dissociation of acetylene provides a catalyst. Refined models for organic and nitrile photochemistry led to estimated deposition over 4.5 Gy of ~600 m of ethane, 100 m of acetylene,

20 m of propane, 20 m of hydrogen cyanide, and lesser amounts of propyne, cyanoacetylene, cyanogen and carbon dioxide (Yung et al. 1984; Raulin 1987). Relative to methane all of these by-products have very low vapor pressures (lower by more than four orders of magnitude). Subsequent modeling by Lara et al. (1994) indicated only ~200 m of ethane, still an ample amount to produce surface lakes and seas. Ethane and methane are stable as liquids at Titan's surface temperature. Acetylene, on the other hand, is a solid at 95 K, with the tendency to explosively transform to polyacetylenes (Lorenz and Lunine 1997).

6.1.2 Titan Aerosols: Tholins

In addition to the specific hydrocarbon and nitrile molecules just discussed another key contributor to the surface composition is the formation of aerosols that are deposited on the surface. Sagan (1974) suggested that a range of energetic chemical processes could generate dark organic polymers that would explain Titan's dark orange haze. Khare et al. (1984) simulated these "tholin" particles in the laboratory and published optical properties (from X-ray to microwave wavelengths covering 5 orders of magnitude); these data have enjoyed wide use for over 25 years. Khare and colleagues synthesized these heavy polymers by placing mixtures of N_2-CH_4 in an electric discharge ionization chamber. Tholin production is thought to occur as a fraction of the acetylene and nitriles is converted to heavy polymers (Sagan et al. 1992; Lunine 1993; McKay et al. 2001). Subsequent work has expanded the synthesis of tholins under various energy sources (including UV photolysis), varying composition, and over a range of pressure (Ramirez et al. 2002; Tran et al. 2003; Imanaka et al. 2004, 2005; Bernard et al. 2006). These simulations may be more representative of Titan aerosols and reveal a range of aerosols with widely varying composition, optical properties, colors, and albedos could be forming at different levels in Titan's atmosphere. The tholins produced by ultraviolet photolysis are similar to those produced by plasma discharge in the real part of the refractive index but the imaginary part is a factor of ~10 lower in the red and short-wavelength IR (Tran et al. 2003). Using a 90:10 mixture of N_2/CH_4 with a cold plasma source, Imanaka et al. (2004) found that varying the chamber pressure produces aerosol particles with a range of carbon/nitrogen ratio. Tholins created at low pressures (13–26 Pa, equivalent to 200–300 km Titan altitude) exhibit C/N~1.5–2 relative whereas those formed at higher pressure (160–2,300 Pa, equivalent to 80–180 km altitude) have C/N~3. Imanaka and colleagues showed that the increase in the aromatic compounds (lower C/N ratio) results in formation of nitrogen-containing polycyclic aromatic compounds in lower pressure cases, whereas tholin formed at high pressure exhibits very few aromatic compounds. These studies show that we might expect a wide range in tholins composition, color, albedo and optical properties and are important in consideration of the nature of the tholins analyzed by the Huygens probe and in the variety of aerosols being continually deposited at the surface.

6.1.3 Hydrocarbon Oceans and Seas

Because estimated photochemical dissociation of CH_4 is so rapid and irreversible (due to atmospheric escape of H_2) methane would be depleted given its current atmospheric abundance in a few times 10^7 years (cf. Lunine et al. 1983; Yung et al. 1984; Lara et al. 1994). A global ocean-worth of methane would be required to sustain CH_4 over geologic time. But global methane oceans were not consistent the Voyager 1 radio occultation data that showed the methane abundance to be well below that expected to be in equilibrium with a surface ocean (Eshelman et al. 1983; Flasar 1983; Lunine et al. 1983). Flasar noted however that liquid methane mixed with its photochemical by-products like ethane or propane would substantially lower the methane partial pressure. Lunine et al. (1983) developed a comprehensive model of a mostly-ethane ocean a kilometer or more deep and showed the Voyager 1 data (abundance of 3 mole percent methane) could be consistent with an ocean of 75% ethane, 20% methane, and 5% nitrogen. Lunine and colleagues further predicted that because the photochemical product acetylene (C_2H_2) would be insoluble, it should form a hundred meter deep layer on the ocean floor (cf. Raulin 1985). Lunine (1993) and Lorenz and Lunine (1997) integrated new observational constraints on global hydrocarbon oceans and from earth-based radar reflectivity (Muhleman et al. 1990, 1995), radiometry (Grossman and Muhleman 1992), and near-infrared spectroscopy (Griffith et al. 1991) and concluded the data were consistent with neither a global ocean of light hydrocarbons nor a surface of pure water ice. Near infrared light-curve observations revealed substantial surface heterogeneity on the global scale (Lemmon et al. 1993, 1995; Griffith 1993; Coustenis et al. 1995) and 0.940 μm images from HST resolved a continental-scale discrete bright region (Smith et al. 1996) subsequently named Xanadu, further discounting the likelihood of a uniform global ocean. Rather a mixture of surface types including dirty water ice, hydrocarbon lakes and seas, a wide range of hydrocarbon and nitrile deposits, subsurface reservoirs of hydrocarbon liquids, seemed more plausible; this is rather like the picture that has emerged from Cassini-Huygens.

6.1.4 *Subsurface Oceans, H_2O Ice, Clathrates, Hydrates – Sources of Cryovolcanism*

As discussed above, the short photochemical lifetime (10^{7-8} yrs) of methane in the atmosphere means that unless the current CH_4 abundance is not typical of the geologic past, there must be a large reservoir on or below Titan's surface. Cometary impact as a continuous source for methane over the last 4 Gy is estimated to be insignificant (cf. Lunine 1993). Because global-scale open bodies of liquid methane were precluded, the next hypothesis was that the reservoir must lie in the subsurface. As a consequence Titan's atmospheric history would be closely coupled with that of its interior. In an early prescient paper Lewis (1971) suggested that the methane source could lie in the subsurface as a layer of methane clathrate hydrate.

In addition to methane and water the third principal ice-forming component in the outer solar system is ammonia. In fact the origin of the nitrogen atmosphere has long been suggested as photolysis of NH_3 in an early hot atmosphere (cf. Hunten et al. 1984). Lunine and Stevenson (1987) and Stevenson (1992) evolved a model for the Titan's interior that after differentiation and cooling results in a (1) central rocky core (radius ~ 1,900 km), (2) overlain by shells of high pressure polymorphs of water ice and water-ammonia hydrate ice (~250 km), (3) overlain by a thick (~200 km) ammonia-water ocean, (4) overlain by a layers of solid methane-clathrate-hydrate, and (5) finally overlain by a porous ice I crust possibly filled with liquid hydrocarbons. A near-surface methane-ethane ocean could provide the source for replenishing the atmospheric methane. Such a subsurface global-scale aquifer could also provide a sink for the 200-to-600 m of liquid ethane generated over geologic time. Kossaki and Lorenz (1996) modeled the porosity that the ice crust could sustain and found it to be realistic to provide such an oceanic-scale reservoir for methane and ethane. The final question relative to Titan's surface composition is the possibility that these materials have erupted from the subsurface. Lunine and Stevenson (1987) discussed methane-driven explosive cryovolcanism that likely brought with it ice to the surface. Ammonia-water-hydrate volcanism might also be profuse as this material has a density close to, and a much lower melting point (~176 K) than, pure water ice (Kargel et al. 1991; Yarger et al. 1993; Lorenz 1996). Ammonia-water slurries could erupt as viscous gelatinous fluids depending on the proximity of their composition to the water-rich peritectic (Kargel et al. 1991). Although in general it is difficult for ammonia hydrate volcanic fluid to reach the surface today, owing its slightly higher density compared to ice I or methane clathrate, Stevenson (1992) noted that during early history things were different. The density of the NH_3-H_2O system is strongly temperature dependent and early in Titan's evolution when the interior was much hotter, the ammonia hydrate was more buoyant (even more so than methane clathrate hydrate) and ammonia hydrate cryolavas could have more easily erupted onto the primitive crust. Clearly there were and still are ample prospects for eruption of water ice and ammonia-water slurries to erupt to and construct the surface we see today.

6.2 In Situ Observations of Surface Composition from the Huygens Probe

6.2.1 *Gas Chromatograph Mass Spectrometer/Aerosol Collector Pyrolyzer*

The Huygens probe, released from the Cassini spacecraft on Christmas Day 2004, descended to Titan's surface on January 14, 2005 and made the first *in situ* observations of Titan's thick lower atmosphere and of the surface. A key integrated instrument that measured the composition of the atmosphere, near-surface gasses, and aerosols was the Gas Chromatograph Mass Spectrometer (GCMS) and Aerosol Collector Pyrolyzer (ACP). The GCMS used a quadrupole mass analyzer with an electron multiplier detection system. The gas sampling system provided continuous atmospheric composition measurements and batch processing through three gas chromatographic columns. The mass spectrometer used five electron impact ion sources (of either 70 or 25 eV); three ion sources served as detectors for the three gas chromatographic columns, the other two used for direct atmosphere sampling and for analyzing the products of the ACP that are generated by heating to 600°C. More complete descriptions of the instruments can be found in Niemann et al. (2002) and Israël et al. (2002).

A broad variety of heavy hydrocarbons and nitriles are photochemically generated above 500 km altitude (see Chapters 7, 8, and 12) although most condense at the low temperatures of the lower stratosphere and the tropopause. It is therefore not unexpected that the GCMS measurements during descent, all done below 146 km, show only H_2, CH_4, N_2, and Ar and did not show well-defined signatures of other hydrocarbon or nitrile molecules (Niemann et al. 2005). In addition during descent two aerosol samples were collected by the ACP (one from 135-to-35 km and a second from 25-to-20 km) and were pyrolized at 600°C and the products then analyzed by the GCMS (Israël et al. 2005). The two samples gave virtually identical results and Israël and colleagues concluded that the most likely components were ammonia (NH_3) and hydrogen cyanide (HCN). They noted that these molecules are not necessarily parent molecules prior to pyrolysis but interpreted the results to show that the aerosols include a solid organic refractory core rich in nitriles. That Titan's aerosols contain substantial nitriles would be an important finding. However, owing to the very poor statistics in the data and potential confusion by an array of other species present at the same *m/z*, Biemann (2006) argued that the identification of NH_3 and HCN is highly unreliable.

Although methane was found by the GCMS to be undersaturated as the probe neared the surface with a relative humidity ~45%, after landing the methane level was seen to increase rapidly. On impact of the Huygens probe the GCMS inlet port, which protruded several cm below the probe shield, was evidently driven into the surface. Evidence indicates the inlet line heater also heated the surface materials in the vicinity of the inlet by direct contact, gas convection and conductance, and/or radiation. The detailed nature of the thermal contact of the GCMS inlet port with the surface material is not known, nor is the temperature to which the surface materials were heated. However a model extrapolating the measured temperatures of the heater and flange (Lorenz et al. 2006a) suggests the surface at the inlet was heated to ≥140 K (Fig. 6.1). During this heating, the methane count rate increased by 40% while the nitrogen count rate remained constant (Fig. 6.2). This increased value for methane remained nearly constant for about 3,000 s, and then gradually decreased to about 25% above the pre-impact value at ~4,000 s after impact, near the time of the last data transmission. Niemann and colleagues further suggested that the slight decrease in CH_4 after 3,000 s might be due to the gradual depletion of liquid methane in the soil around the inlet. An explanation involving CH_4 that had been condensed or trapped during the descent was considered highly unlikely because of the high temperature in the inlet line and the observed sustained elevated rate of evaporation after landing. Niemann et al. (2005) concluded that the GCMS surface data strongly indicate the presence of liquid methane mixed with the subsurface materials. Taken with the observation that the methane humidity at the surface was only 45% this could evidence recent methane precipitation. Recent models (cf. Tokano et al. 2006: Griffith et al. 2008) suggest that the atmospheric conditions are in fact close to those needed for low-level precipitation.

Niemann et al. (2005) reported that the Huygens GCMS spectra taken after landing on Titan's surface exhibit mass peaks attributable to several heavier molecules (Fig. 6.3).

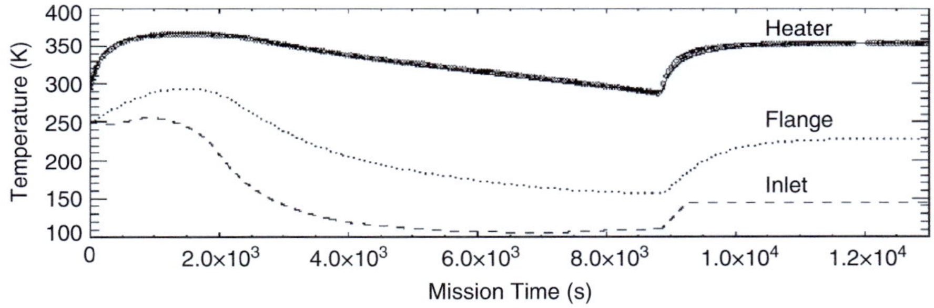

Fig. 6.1 GCMS inlet heater and flange temperatures and modeled inlet temperature (Lorenz et al. 2006a).

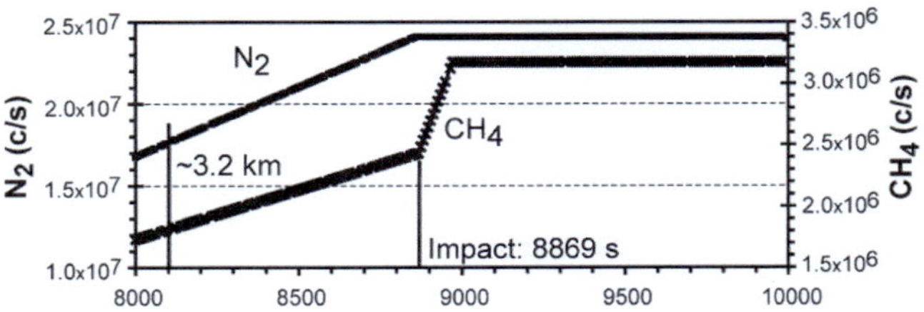

Fig. 6.2 GCMS count rates for nitrogen and methane after landing versus mission time in seconds. Whereas N_2 remained constant after landing (mission time ~8870 s), the CH_4 rate rose by 40% in the first 2 min after landing (Niemann et al. 2005).

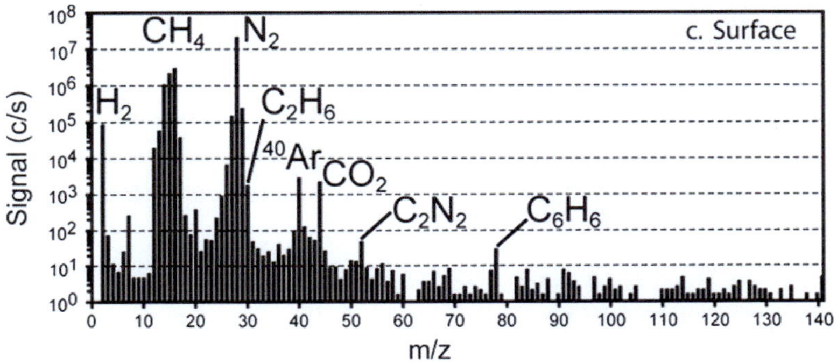

Fig. 6.3 Averaged GCMS mass spectrum acquired at the surface showing ion count rates versus mass per charge (m/z). Beyond increased methane, compounds found at the surface included firm detection of ethane and tentative detection of cyanogen, benzene, and carbon dioxide (Niemann et al. 2005).

Their results included firm identification of ethane (C_2H_6) and tentative detections of cyanogen (C_2N_2), benzene (C_6H_6), and carbon dioxide. Cyanogen, benzene, and carbon dioxide are far less volatile than methane and would therefore show a smaller mass peak, regardless of their surface abundances. Water ice could not be detected by the instrument, even if in large abundance, because of the low vapor pressure of the H_2O ice. Carbon monoxide could not be detected directly by the GCMS because of strong interference in the mass spectra with molecular nitrogen. The gas chromatograph subsystem had a dedicated column for CO but none was detected above the detection threshold.

6.2.2 Descent Imager/Spectral Radiometer

The Huygens probe also carried the multi-channel optical instrument, DISR or Descent Imager/Spectral Radiometer, operating in the violet to short-wavelength infrared spectral region. Among its 14 channels this instrument included upward- and downward-looking visible-wavelength spectrometers (ULVS and DLVS; 0.48–0.96 µm; Δλ ~ 0.006 µm), upward- and downward-looking infrared spectrometers (ULIS and DLIS; 0.87–1.70 µm; Δλ ~ 0.015–0.022 µm), and three visible-wavelength CCD framing cameras (spectral bandpass ~ 0.66–1.00 µm; IFOV ~ 1, 2, and 3 mrad) that provided imaging from horizon to near-nadir (Tomasko et al. 2002). DISR also carried a lamp that was sequenced to turn on at an altitude of 700 m to directly illuminate the surface. Its primary purpose was to measure the spectral reflectance of the surface without interference from deep methane absorption bands or from the strongly scattering aerosol haze. Surface detail became increasingly visible in the DISR camera frames at altitudes below ~40 km from which ~240 images were returned. These images provided resolution on the surface ranging from ~50 m/pixel down to a few mm/pixel. The DISR images revealed two general classes of terrain: (1) brighter, rugged highland units showing dissection by drainage networks and (2) darker lowland plains where the probe landed that showed pervasive evidence of fluid erosion (see Chapter 5).

Throughout the descent, but in particular below 20 km during the final stages of approach to the surface, multiple data sets were collected with the DLVS and DLIS spectrometers. The solar zenith angle at the landing site during this part of the descent was ~33°. Correlation of these spectral data with camera images provide geologic context for interpretation of the visible and infrared spectra. Figure 6.4 shows a DLVS map of ratio of the spectral radiance (I/F) measured in two visible-wavelength methane windows (0.827 and 0.751 µm) acquired from an altitude of ~18 km (Tomasko et al. 2005). These maps are overlain on a mosaic of higher resolution DISR camera images. The phase angles range from ~15°–55°. Even without

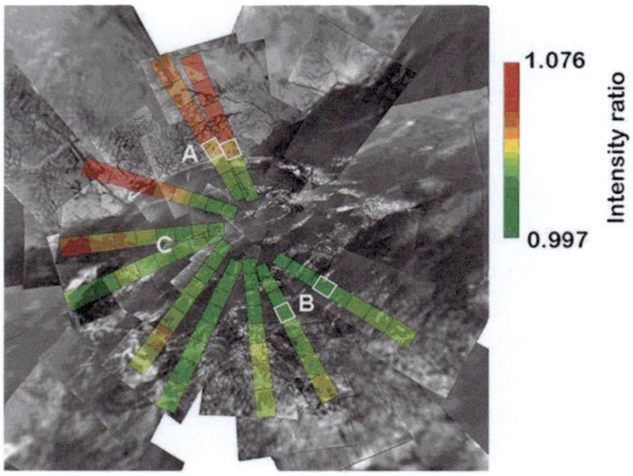

Fig. 6.4 DISR maps of the DLVS spectral ratio in two methane windows (827/751 nm) exhibit variations in visible-wavelength reflectivity around the Huygens landing site. DLVS footprints are overlain on a 23 × 23 km mosaic of DISR images and color-coded by ratio intensity. Spectra for footprints A and B are shown in Fig. 6.5 (Tomasko et al. 2005).

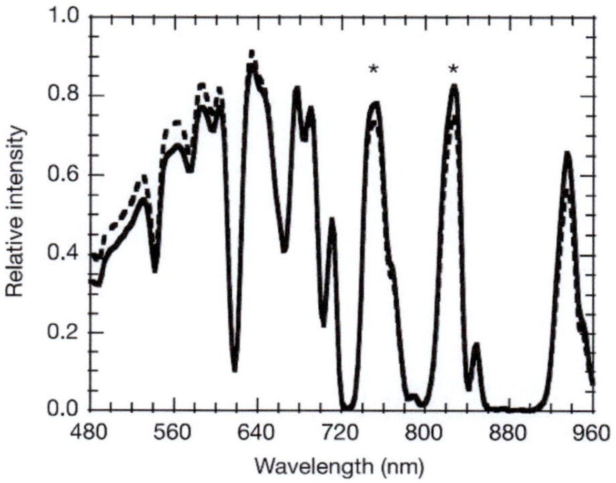

Fig. 6.5 Comparison of DISR DLVS spectra of brighter dendritic highlands (solid line, A in Fig. 6.4) and darker lowland plain at the Huygens landing site (dashed line, location B). Asterisks show the methane windows chosen for the spectral maps of Fig. 6.4. These DISR spectra have been normalized to their total intensity to emphasize the difference in spectral slope (Tomasko et al. 2005).

modeling the effects of atmospheric absorption and aerosol scattering (most difficult in the visible wavelengths), the DLVS spectral radiance data show that the brighter highlands, that lie a few kilometers north of the landing site and exhibit extensive dendritic networks, are apparently redder in the visible than are the darker fluvial-scoured plains of the Huygens site itself. Figure 6.5 shows that this reddening extends over the entire spectral range of the DISR visible spectrometer. Figure 6.6 displays maps of 3 spectral ratios of radiance measured with the DLIS from an altitude of ~4 km through four methane windows (0.93, 1.07, 1.28, and 1.59 µm; Keller et al. 2008). The phase angles in these maps range from ~20°–50°. As in the case of the visible wavelengths, these maps illustrate that

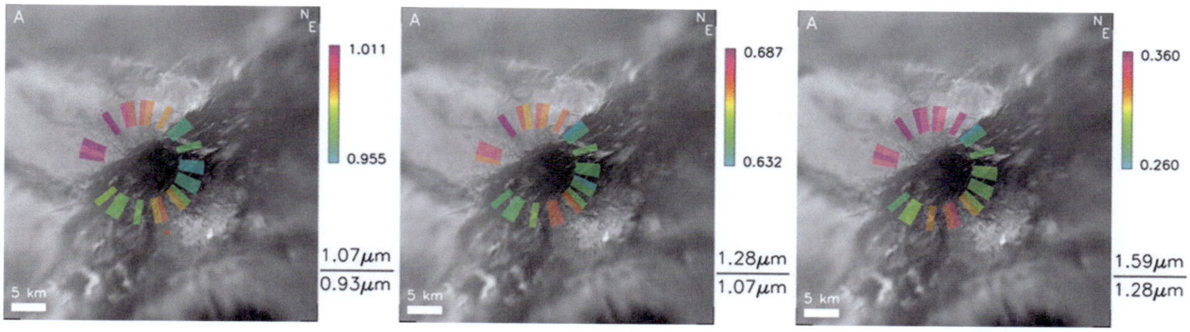

Fig. 6.6 DISR DLIS maps of infrared spectral variations around the Huygens landing region acquired from an altitude of ~4 km. Ratios of observed spectral radiance (I/F) are shown for DLIS observations acquired at the centers of four methane windows (Keller et al. 2008). As in the case of visible wavelengths, highlands are redder than the dark plain of the landing site in the near infrared.

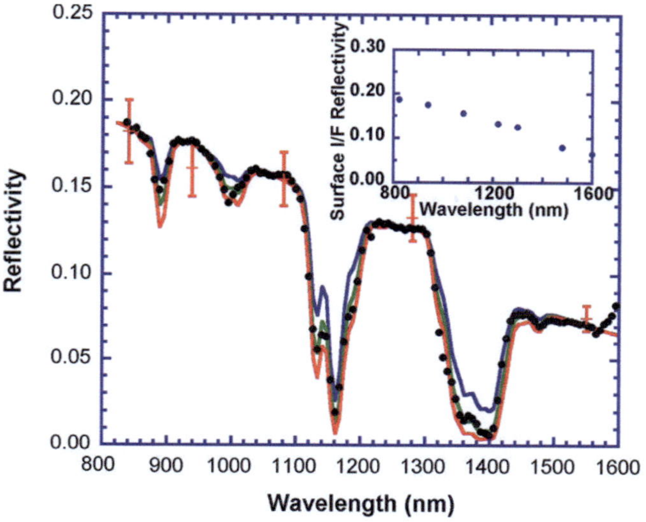

Fig. 6.7 Near-surface methane mole fraction derived from a model of the DISR DLIS lamp-only spectrum with data taken from altitude of ~25 m (*black dots*). Estimated absolute surface reflectance derived from DLIS/ULIS intensity for the centers of the methane windows are shown by the red error bars. Three models of CH_4 mole fraction are shown for comparison: 3% (*blue*), 5% (*green*), and 7% (*red*) assuming surface reflectivity interpolated from the 7 wavelengths shown in inset. The 5% model provides the best fit (Tomasko et al. 2005).

the systematic redness of the bright dendritic highlands, relative to the dark plains of the landing site region, also continues into the near infrared. These workers point out that although it is natural to attribute the difference in redness to composition (e.g. different tholins, other hydrocarbons, or water ice content) or particle properties (e.g. reworking, sorting, compaction, aggregation in fluvial or eolian processes), it is possible that contributions to tholin aerosol scattering over bright versus dark terrains differ sufficiently to cause such effects.

Useful surface signals from DISR lamp illumination were acquired from altitudes ~50 m down to the surface. The last DLIS spectrum, taken with the lamp from ~25 m just before probe touchdown, allowed Tomasko et al. (2005) to estimate both the near-infrared surface reflectance and the methane mole fraction near the surface. This model lamp-only spectrum (shown by the black dots in Fig. 6.7) represents the product of the surface reflectance and the two-way CH_4 transmission and was derived as follows. First surface reflectance was estimated directly from the ratio of the DLIS to ULIS intensities for the centers of the windows where methane absorption is negligible; 3-σ red arrow bars show these estimates for 5 wavelengths. Spectra acquired between ~750 m and ~50 m were used; at these altitudes surface-reflected lamp flux was negligible. Next the solar contribution to the 25-m DLIS spectrum was subtracted using an average of DLIS spectra from ~100 m, again where the surface-reflected lamp flux was negligible. This difference spectrum was then divided by the spectral response of the lamp and scaled to fit the surface reflectance derived from DLIS/ULIS for the centers of the windows. Four methane absorption bands are seen in this lamp-only spectrum. Model fits for 3 methane concentrations (3%, 5%, 7%) were made using an interpolation of selected surface reflectance values (shown in inset) and methane absorption coefficients from E. Karkoschka, University of Arizona. A methane mole fraction of 5% (equivalent to a methane relative humidity of ~50%) provided the best fit to the observed spectrum (green curve in Fig. 6.7) in close agreement with 4.9% from 8 km to the surface as derived by the GCMS (Niemann et al. 2005).

To extract the best estimate of the surface reflectance from the 25-m DLIS lamp-only spectrum it was divided by the modeled methane transmission; spectral regions where the transmission was <90% were excluded. Fig. 6.8 compares this result (+'s) with the spectral shape of the near-infrared surface reflectance measured after landing where the lamp and spectrometer were only a few tens of centimeters from the surface. The post-landing DLIS spectrum was divided by the spectral response of the lamp and also scaled to match the absolute surface reflectance derived for the centers of the CH_4 windows as the described in the paragraph above. The infrared spectrum of the surface is characterized by a gradual drop in reflectance from 0.8 to 1.6 μm of from ~0.18 to ~0.07. The excellent agreement between the shapes of the two independent determinations of the ground reflec-

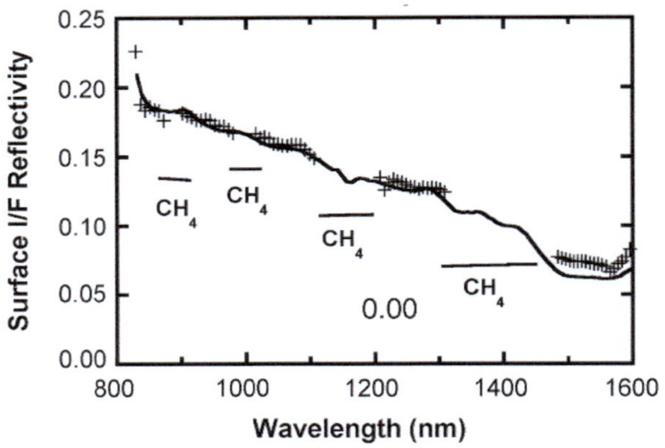

Fig. 6.8 DISR DLIS near infrared reflectance of the Huygens landing site. Spectral reflectivity (+'s) at wavelengths for which methane transmission is <90% (estimated from lamp-only model shown of Fig. 6.7) is compared to the reflectance measured with the DISR lamp after landing (*solid line*). The post-landing spectrum was scaled to match the DLIS/ULIS ratios (*red error bars*) shown in Fig. 6.7. Spectral regions of strong methane absorption are indicated (Tomasko et al. 2005).

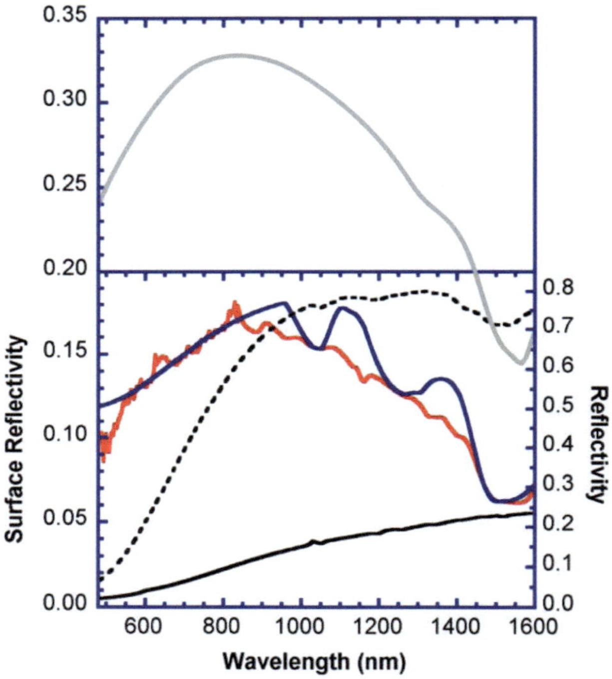

Fig. 6.9 Surface reflectance for the Huygens landing site derived from a combination of the DISR visible and near infrared spectrometers (*see text: red line* Tomasko et al. 2005; *thick gray curve from* Schröder and Keller 2008). Blue curve is a synthetic spectrum of a mixture of coarse low-temperature water ice, yellow tholins, and an unknown component with featureless blue slope dropping falling between 850 and 1500 nm. Spectra of a yellow tholin (*dashed line*) and a dark tholin (*solid black line*) are from Bernard et al. (2006).

tivity give confidence to both the estimated methane abundance and to the derived infrared reflectance of the surface. Tomasko and colleagues constructed the complete DISR spectrum of the landing site surface (0.48–1.70 μm, red curve in Fig. 6.9) by combining the DLIS surface spectrum with the post-landing lamp-on DLVS spectrum, after dividing it by the lamp spectral response and scaling it to match the DLIS surface spectrum at 0.8 μm.

Schröder and Keller (2008) independently derived the surface reflectance also from a combination of the lowest altitude and post-landing lamp-on DLVS and DLIS spectra. Their result (Fig. 6.9 broad grey curve) exhibits the same shape as that of Tomasko and colleagues but shows a higher overall reflectance by ~1.7. Their higher reflectance values could arise partly from differences in approaches: whereas Tomasko et al. (2005) used the ratio of the upward and downward fluxes from low altitude DLIS and ULIS spectra (confirmed in a reanalysis by Jacquemart et al. 2008), Schröder and Keller derived the reflectance by using absolute calibration of the spectral radiance measured in the laboratory for the flight lamp prelaunch. It is certainly possible that the absolute lamp calibration could have changed and could be in error by ~1.7, but it seems unreasonably large. Schröder and Keller contend, rather, that their higher overall result can be explained as an opposition surge. The phase angles of the lamp in the last DLIS and DLVS spectra acquired just prior to landing, and that were used to scale the absolute reflectance spectra, were ~0.07° and 0.23°, respectively. The authors point out that such an opposition surge could only be detected from the absolute value of the reflected lamp flux from these near-zero phase angle observations. In visible and infrared Titan observations taken from the Earth or Cassini at very low phase angles, shadow hiding that causes an opposition effect would be diluted by the pervasive aerosol scattering. It is entirely possible that the two sets of surface reflectance spectra are entirely compatible, the lower values from Tomasko et al. (2005) representing reflectance from diffuse illumination over a wide range of phase angles. If Schröder and Keller are correct the landing site surface has remarkable textural properties. Surfaces that exhibit such large opposition effects are typically quite pitted, porous, and fluffy. For the dark surface at the Huygens landing site plain to have maintained such a porous, fluffy state having experienced episodes of fluvial erosion and transport in its most recent geologic past (see Chapter 5) seems odd. Conceivably a thin layer of fluffy aerosol tholin coats the plain.

Although there is uncertainty in the overall magnitude of the surface reflectance depending on analysis approach used and phase angle considerations, all analyses thus far agree on the fundamental shape of the spectrum. The basic characteristics are: (1) an albedo peaking at 0.83 μm of about 0.18 or 0.33 (whether from Tomasko et al. 2005 and Jacquemart et al. 2008 or from Keller et al. 2008 and Schröder and Keller 2008, respectively); (2) a red slope in the visible <0.83 μm; (3) a blue slope with an overall drop in reflectance between 0.83–1.42 μm; and (4) a broad ~30% absorption apparently centered ~1.54 μm. Most remarkable is the absence of absorption features in the DISR surface spectrum other than the 1.54 μm band. This is at odds with pre-Cassini-Huygens predictions that spectrally active organic species, such as ethane, acetylene, propane, ethylene, hydrogen cyanide might well emerge in the DISR surface spectrum. Also shown in Fig. 6.9 are reflectance spectra for two laboratory tholins (from Bernard et al. 2006) representative of a wide range of optical properties such samples exhibit: one black-brown in color and a second that is a brighter yellow-orange. As noted earlier DLVS data suggest that the bright dendritic-channeled highlands several km north of the landing site are redder in the visible than the materials at the dark plain of the landing site (see Figs. 6.4 and 6.5). And although aerosol scattering could contribute to this color difference, a concentration of brighter, redder tholins (such as the illustrated yellow-orange tholin of Bernard et al. 2006) in the bright highlands may be the correct conclusion.

As discussed more fully in sections that follow, Earth-based spectroscopic observations have suggested evidence that Titan's dark regions contain water ice mixed with dark contaminants: organics or possibly silicates (Coustenis et al. 1995; Griffith et al. 2003; Lellouch et al. 2004). Tomasko et al. (2005) showed water ice to be a plausible candidate for the 1.54 μm band as illustrated by the blue curve in Fig. 6.9 that is a synthetic spectrum of low-temperature coarse-grained (750 μm) water ice, a yellow tholin, and an unknown component exhibiting a featureless blue slope >0.83 μm. However absorption bands at 1.04 and 1.25 μm expected for coarse-grained water ice are not evident. Tomasko and colleagues argue that fine-grained water ice alone cannot be the explanation of both the blue slope and the absence of these bands; fine-grained water ice would suppress these bands but the blue slope in the near-IR is more consistent with large-grained ice. Intimate mixing of fine-grained ice with some material with a strong blue slope could explain the absence of the weaker H_2O bands. Nonetheless the close fit of this 1.54 μm feature to a strong absorption of such an expected component as H_2O seems unlikely to be coincidental. The identification of water ice as a component is reasonable but not conclusive, as various organics exhibit similar absorptions in this wavelength region. In fact the yellow-orange tholin from Bernard et al. (2006) (dashed curve in Fig. 6.9) shows a feature in the 1.5 μm but as these workers note this sample was contaminated by adsorbed water (strongly apparent at 3 μm) and this feature could be also be partly due to adsorbed water.

From DLIS datasets like those shown in Fig. 6.6 Keller et al. (2008) were able to estimate the ratio between bright and dark terrains for discreet wavelengths centered in the methane windows (error bars in Fig. 6.10). Accurate extension of these ratios to visible-light windows ($\lambda < 0.75$ μm) is not feasible owing to the dominance of aerosol scattering at shorter wavelengths at altitudes above a few km. Applying these ratios to the reflectance at the landing site from Schröder and Keller (bottom curve in Fig. 6.10) leads to an estimate of the near-infrared spectrum of the bright highland region (top curve in Fig. 6.10). Although the data in this spectrum are sparse it suggests that the brighter highlands exhibit less of a blue slope in the near infrared and a weaker or absent 1.5 μm absorption feature. Both of these factors are consistent with the notion that the highlands have greater concentration of brighter, redder tholin and if water ice is responsible for the 1.54 μm band at the landing site, the highlands have much less water ice.

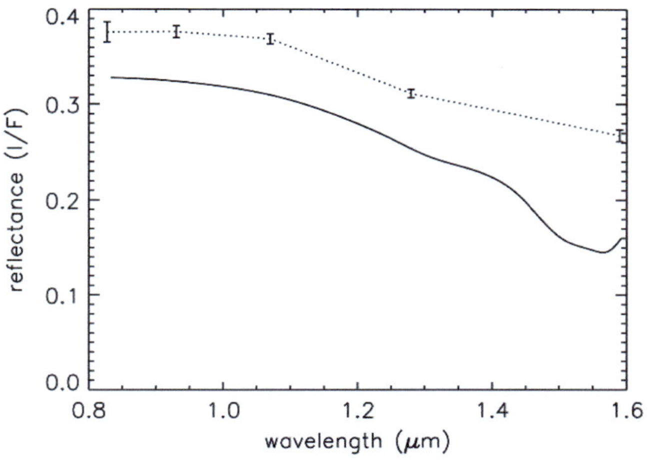

Fig. 6.10 DISR near-infrared reflectance (error bars at centers of 5 methane windows) for bright dendritic highlands a few km north of the Huygens landing site (Keller et al. 2008). Lower curve (*solid*) is same data as the thick grey curve of Fig. 6.9.

6.3 Surface Composition from Short-Wavelength Infrared Spectroscopy

The Visual and Infrared Mapping Spectrometer (VIMS) aboard the Cassini spacecraft is the primary orbital instrument for global studies and mapping of Titan's surface composition. VIMS is an imaging spectrometer that operates in the 0.35–5.17 μm spectral region and consists of two instrumental subsystems covering different spectral ranges (Brown et al. 2004). The VIMS-VIS channel is a multispectral imager that covers the spectral range from 0.35–1.05 μm and uses a frame transfer CCD detector on which spatial and spectral information is simultaneously stored. The VIMS-IR channel covers a wavelength range from 0.85–5.17 μm and is able to penetrate the haze to observe Titan's surface. On approach to Titan, VIMS observations typically commence several hours before closest approach permitting hemispheric mapping with full spectral cubes with spatial resolution ≤50 km/pixel. About 4 h before closest approach four-image mosaics provide coverage of the disk. Resolutions ~10 km/pixel are achieved ~0.5 h before closest approach and spectral cubes can be collected in regional mosaics. Rarely collected isolated cubes acquired during close flybys can yield resolutions as high as 250 m/pixel.

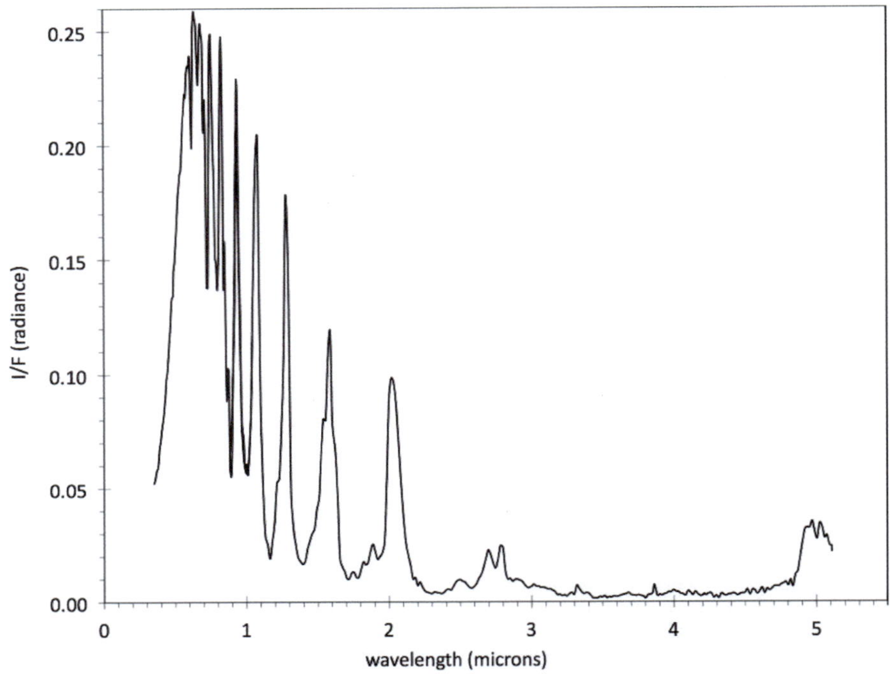

Fig. 6.11 Typical VIMS Titan spectrum. Methane is the principal absorbing gas but additional features due to CO absorption and to H_2 and N_2 collisional bands contribute in the 2-to-5 μm region. Scattering by aerosol particles dominates in the visible and rapidly falls off with increasing λ beyond ~1 μm.

Figure 6.11 shows a typical complete VIMS radiance [I/F(λ)] spectrum for Titan, combining data from the VIMS-VIS and VIMS-IR channels. The spectrum is dominated by deep methane absorption bands. The rapid rise in brightness from violet-to-red wavelengths is due to pervasive red-orange tholin particles in the atmosphere. Between 2.1–2.6 μm and again between 2.9–4.8 μm CO absorption and absorption in H_2 and N_2 collisional bands significantly augment the methane absorptions, as known from pre-Cassini models of atmospheric composition. Radiation reflected from the surface is detectable by VIMS only within the atmospheric windows (i.e. the peaks in Fig. 6.11). Even within atmospheric windows, at wavelengths shorter than ~0.9 μm, surface signals are overwhelmed by multiple scattering from atmospheric aerosols. Consequently only the VIMS data for the eight atmospheric windows ≥0.9 μm (centered at 0.94, 1.08, 1.28, 1.6, 2.0, 2.7, 2.8, and 5 μm) are useful to collect spectral information and resolve surface detail. All of these windows are within the spectral range of VIMS-IR channel on which this section of the chapter focuses. Figure 6.12 (McCord et al. 2008) illustrates the typical variations seen in VIMS spectra and compares several examples of common ices found in the outer solar system whose spectral properties might be distinguished by VIMS in these windows. For the most part, the narrow spectral windows through which VIMS must work provides incomplete spectral information for the surface. Even in the 5 μm window, spectrally the widest, the analysis is hampered by low signal precision.

6.3.1 The Case for H_2O Ice

The inference that water ice is an abundant compositional component making up Titan's crust and surface materials is certainly reasonable based the bulk density (1,880 kg/m^3) and the general abundance of water ice in the outer solar system. Interior models suggest a liquefied ammonia-water ice mantle below which are layers of high pressure water ice and above which is a shell of methane clathrate hydrate, all surrounding a rocky core (Stevenson 1992; Lorenz and Lunine 2005; Tobie et al. 2006). Ammonia-water-hydrate volcanism could be profuse as this material has a density close to and a much lower melting point (~176 K) than, pure I_h water ice (Kargel et al. 1991, Yarger et al. 1993, Lorenz 1996). Cryovolcanic eruptions driven by methane clathrate hydrate are even more likely given the lower density of such material (Lunine and Stevenson 1987; Tobie et al. 2006). The Cassini RADAR investigation has produced evidence of cryovolcanic flows and domes (cf. Lopes et al. 2006 and Chapter 5). Thus both impact cratering and cryovolcanic eruption of water ice throughout geologic time may have brought water-rich materials to the surface. In all likelihood water ice is the fundamental structural component of the topographic forms of Titan's terrains. Whether spectral features of this ice are visible to remote sensing spectrometers, against the tendency for burial by precipitating aerosols, remains ambiguous.

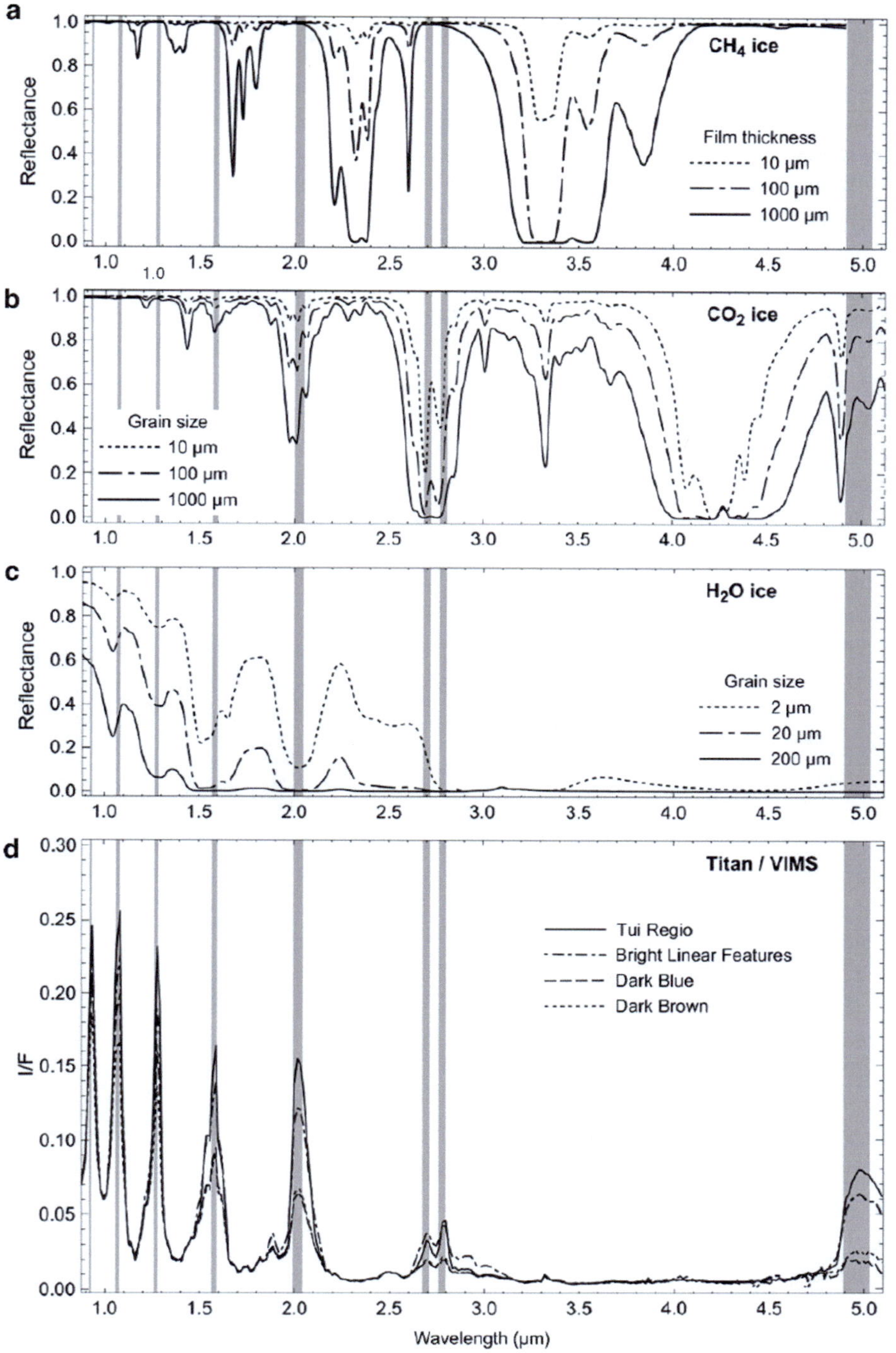

Fig. 6.12 Representative calibrated VIMS spectra (*bottom*) and sample laboratory spectra for various candidate materials: (**a**) CH_4 data from Grundy et al. (2002). (**b**) CO_2 data from Hansen (2005). (**c**) H_2O ice at 110 K from G. Hansen (from McCord et al. 2008).

Using a Fourier transform spectrometer (FTS) on the Canada-France-Hawaii (CFH) telescope Griffith et al. (1991) estimated Titan's surface albedo in the 1.3, 1.6, and 2.0 μm methane windows and found that at 1.6 and 2.0 μm (where water ice strongly absorbs) the albedo was significantly lower relative to that at 1.3 μm. They concluded the data to be consistent with a surface of water ice mixed with some dark component. Telescopic observations of Titan's light curve in several methane windows showed the surface to be globally heterogeneous with a brighter leading hemisphere and darker trailing hemispheres (Lemmon et al. 1993; Griffith 1993). This global heterogeneity was further documented by Smith et al. (1996) who imaged Titan with Hubble Space Telescope in the 0.94 and 1.08 μm windows and showed the leading hemisphere to contain a large, bright equatorial region (subsequently named Xanadu). Griffith (1993) suggested that although water ice might be in evidence, active surface processes would have to be continually removing the rain of organic debris. Lemmon et al. (1995) argued against water ice noting that the 2 μm albedo of the bright hemisphere was higher than would be expected for H_2O ice. Coustenis et al. (1995) also observed Titan with the FTS on the CFH telescope through four methane windows (1.1, 1.3, 1.6, 2.0 μm) and also concluded that the evidence was consistent with H_2O ice. More recently, Griffith et al. (2003) derived surface albedo in six atmospheric windows (1.07, 1.28, 1.58, 2.0, 2.9, and 5.0 μm) and again concluded that water ice was the best match over this extended spectral range. Coustenis et al. (2005) using data collected at the CFH telescope and the VLT (Very Large Telescope in Chile) generated global maps in methane windows from 1 to 2.5 μm and concluded that while the dark regions were consistent with the "dirty water ice" of Griffith and colleagues, the bright regions were definitely not and suggested bright hydrocarbon ices such as ethane, a subject we will return to later in this chapter. It should be noted that in offering that dirty water ice was consistent with the dark regions, these workers did not include spectra of organics and nitriles that might mimic the spectral behavior of water ice (cf. Clark et al. 2009b).

McCord et al. (2006) analyzed an early set of medium resolution VIMS-IR observations (acquired on TA) consisting of a 2×3 mosaic covering the entire disk with a resolution of ~50 km/pixel (Fig. 6.13). Using a range of modeling approaches (including radiative transfer codes following Griffith et al. 2003) McCord and colleagues estimated the spectral albedo of the surface within the methane windows between 0.85 and 5.2 μm for five different locations (two dark and three bright). Two compositional classes emerged associated with the lower albedo and the higher albedo materials (with significant variation among the brighter areas). These results showed again that the spectrum of water ice mixed with dark contaminants best matches the reflectance of the lower albedo regions, whereas the spectra for brighter regions are not matched water ice mixtures (Fig. 6.14). Rodriguez et al. (2006) analyzed VIMS spectral cubes that cover the Huygens landing site and based on low relative reflectance at 1.6 and 2.0 μm also concluded that dark units in the vicinity of the landing site might be enriched in water ice (Fig. 6.15). This hypothesis was consistent with the spectral evidence from DISR suggesting water-ice absorption at 1.54 μm as a possible component of the dark plain of the landing site (Tomasko et al. 2005).

McCord et al. (2006) noted a puzzle in the peculiar reversal in slope in the dark-area spectrum between the sub-windows at 2.7 and 2.8 μm (see Fig. 6.14). If actually a characteristic of the shape of the surface spectrum, this slope reversal would rule out H_2O (and NH-bearing compounds). Coustenis et al. (2003 and 2006) obtained a surface spectrum for the Titan 2.7-μm window using ISO data that also showed a 2.74 μm absorption. Clark et al. (2009b) see all of this as strong evidence against water ice because water ice is strongly absorbing near 2.8 μm, lying in the wing of the 3 μm OH fundamental. Clark and colleagues emphasize that various hydrocarbons and nitriles exhibit spectra that could mimic the water ice in the

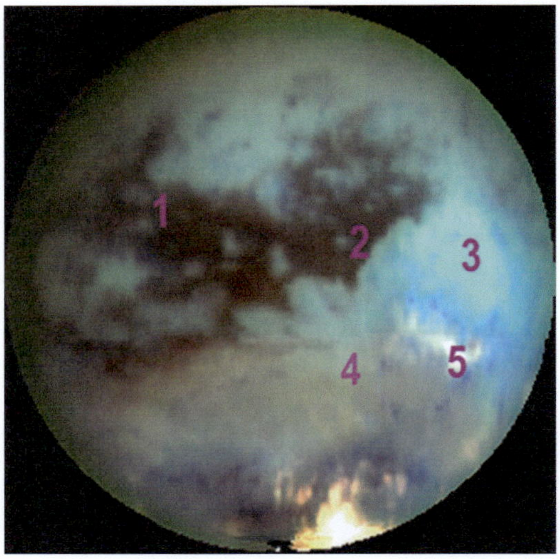

Fig. 6.13 Cassini VIMS full disk view acquired on the first Titan flyby (TA). Spectral unit areas studied by McCord et al. (2006) are labeled: (1) ddark, (2) dark, (3) nominal bright, (4) 5 μm-bright, and (5) anomalously bright. This 4-frame mosaic has a resolution of ~50 m/pixel and is an RGB color composite through atmospheric windows at 5.0, 2.0, and 1.3 μm respectively.

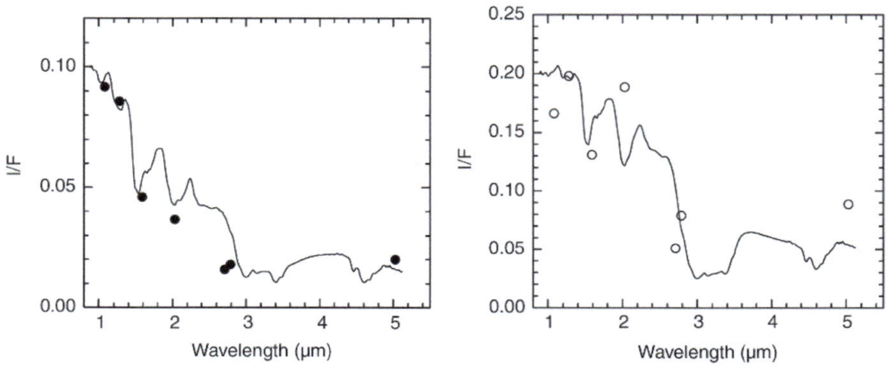

Fig. 6.14 Surface reflectance for dark (location 1 in Fig. 6.13) and bright areas (*location* 5) derived from VIMS data. Dark units appear consistent with water ice mixed with organics whereas bright units are generally not consistent with water ice (from McCord et al. 2006)..

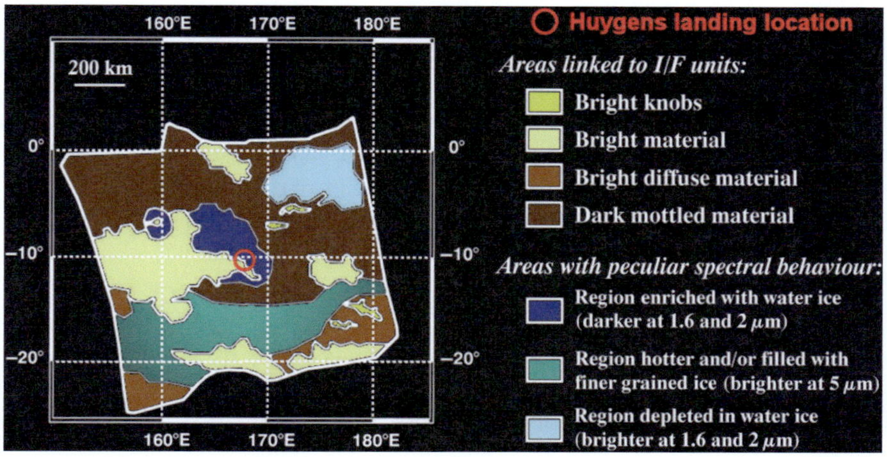

Fig. 6.15 VIMS spectral units derived for the Huygens landing region. Based on low relative reflectance at 1.6 and 2.0 μm, Rodriguez et al. (2006) concluded that the dark plains unit at the landing site to possibly be enriched in water ice.

Fig. 6.16 Orthographic projections of VIMS global mosaics centered at 180°W, 90°W, and 0° from left to right (from Barnes et al. 2007a). VIMS spectral images acquired through 7 atmospheric windows were used to assemble the mosaics. Nine units were identified; two show as dark brown and dark blue in equatorial regions. The views shown here were composited with 4.8–5.2 μm as red, 2.0 μm as green and 1.28 μm as blue.

windows in the 1–3 μm region. It should be noted that the observed reversal in slope between 2.7 and 2.8 μm for dark regions may not be at all representative of the surface spectrum. It is extremely difficult to extract reliable surface spectra for these two atmospheric windows for the following reasons: (1) the methane absorption coefficients are not well determined here and (2) the signals are extremely weak because the surface reflectance of the dark units at the wavelengths of these two windows is likely extremely low. In summary even though it is extremely likely that water ice is a major component in Titan's surface terrains, direct spectroscopic evidence is not without significant ambiguity.

6.3.2 Mapping Global Compositional Units – Spectral and Geomorphological Correlations

Barnes et al. (2007a) published the first global maps of Titan's near-IR spectral units based on VIMS spectral images acquired in seven atmospheric windows (0.94, 108, 1.28, 1.6, 2.0, 2.8, 5 μm, Fig. 6.16). These maps show a strong latitudinal organization in the distribution of spectral types of Titan's surface units. Barnes and colleagues identified 9 units: (1) four equatorial, constrained roughly in the ± 30° latitude belt, (2) four at mid-latitudes, sharing the attribute of being brighter at 5 μm, and (3) a spectrally gray region in a south polar zone (roughly pole-ward of 55S°). The mid-latitude zones extend from ~25°S to ~55°S and from ~25°N to the northern terminator that marked the edge of polar winter early during the nominal Cassini mission. Barnes et al. (2007a) found the large Tsegihi region (see Fig. 6.13) to be typical of the 5 μm-bright units that dominate the mid-latitudes. A region similar to Tsegihi in the south mid-latitude belt (location 4 in Fig. 6.13) was also recognized by McCord et al. (2006) to be much brighter at 5 μm than the equatorial bright regions (location 3). Barnes et al. (2005, 2006) and McCord et al. (2006) described an extremely bright 5 μm feature in the southern mid-latitudes (location 5 in Fig. 6.13) that later was formally named Tui Regio (Fig. 6.16). McCord et al. (2006) also reported Tui Regio feature to exhibit an excessively high 2.8/2.7 μm ratio, making it stand out from all other features on the disk. The global maps of Fig. 6.16 also show a very similar 5 μm-bright feature known as

Hotei Regio. Later in this chapter we return to discussions of compositional interpretations for these unusual spectral units.

According to Barnes et al. (2007a) four spectral units make up Titan's equatorial zone (~30°N to ~30°S); they consist of two bright and two dark units and are found only in this equatorial zone. The equatorial bright units break into Xanadu and a general equatorial bright unit (basically all bright units in the equatorial band other than Xanadu). The dark units split into two units dubbed "dark blue" and "dark brown" as first suggested by Soderblom et al. (2007). The first of these stands out as blue in color composites where the 1.3 μm brightness is used as the blue component and longer wavelengths are used as green and red. This includes both color composites like Fig. 6.16 in which RGB are assigned to 5, 2.0, 1.3 μm as well as like Fig. 6.17 where RGB are assigned to 2.0, 1.6 and 1.3 μm. The difference between the dark blue and dark brown arises because the dark blue unit is much more reflective at 1.3 μm relative to longer wavelength windows (e.g. 1.6, 2.0, and 5 μm) whereas the dark brown unit is more uniformly dark at all wavelengths, including 1.3 μm.

Soderblom et al. (2007) explored the correlations between the VIMS equatorial spectral units and terrain units imaged by the Cassini RADAR instrument in SAR mode (Synthetic Aperture Radar). Figure 6.17 compares the two for typical equatorial regions. In general the patterns seen in the VIMS optical images and in the SAR images bear very little resemblance to one another. The right panel of Fig. 6.17 provides as a typical example; the boundary of the large bright region in the VIMS color composite image shows almost no correspondence to the patterns seen in SAR image. Soderblom et al. (2007) did, however, recognize one strong correlation: the VIMS dark brown unit consistently maps with the vast fields of radar-dark dunes discovered in SAR images (Lorenz et al. 2006b). Although isolated correlations do exist between radar-bright and VIMS-bright units (cf. Fig. 6.17), many of these correspondences are secondary effects arising from fields of dark brown, radar-dark dunes that surround other terrains. Hence most of these correlations arise also from the distribution of the dune fields that *are* strongly correlated between two data types.

Figure 6.18 illustrates typical spectra of VIMS dark blue and dark brown units. They are compared to synthetic spectra for water ice computed from optical constants of Grundy and Schmitt (1998). The VIMS dark blue spectrum crosses over the dark brown one exhibiting higher reflectivity at 1.3 μm and lower reflectivity at 2.0 μm. This difference must be a real surface property because such a crossover cannot arise from unaccounted atmospheric effects. If varying content of water ice were responsible for the difference, then the dark blue unit would be more water-ice rich and the dark brown relatively poor in water ice. Soderblom et al. (2007) suggested that the coarse-grained grains of the dunes to have less water ice than the dark blue units and to be richer in, or even exclusively composed of, hydrocarbons and nitriles. This association was further born out by subsequent VIMS observations that actually resolved the dunes in the brown spectral end-member (Barnes et al. 2008).

Following earlier work (Griffith et al. 2003; Coustenis et al. 2005; McCord et al. 2006; de Pater et al. 2006; Negrão et al. 2006; Hirtzig et al. 2007), Soderblom and colleagues further concluded that the bright equatorial regions exhibit little evidence of water ice and inferred them to be deposits of fine, bright aerosol dust. Figure 6.19 shows a simple 3-D cluster (plotting I/F(λ) for 3 methane windows, 1.3, 1.6, and 2.0 μm) for an equatorial region containing roughly equal proportions of dark blue, dark brown, and bright units. This 3-D cluster is remarkably flat, effectively two-dimensional. An interpretation is that this flatness signals 3 end-members

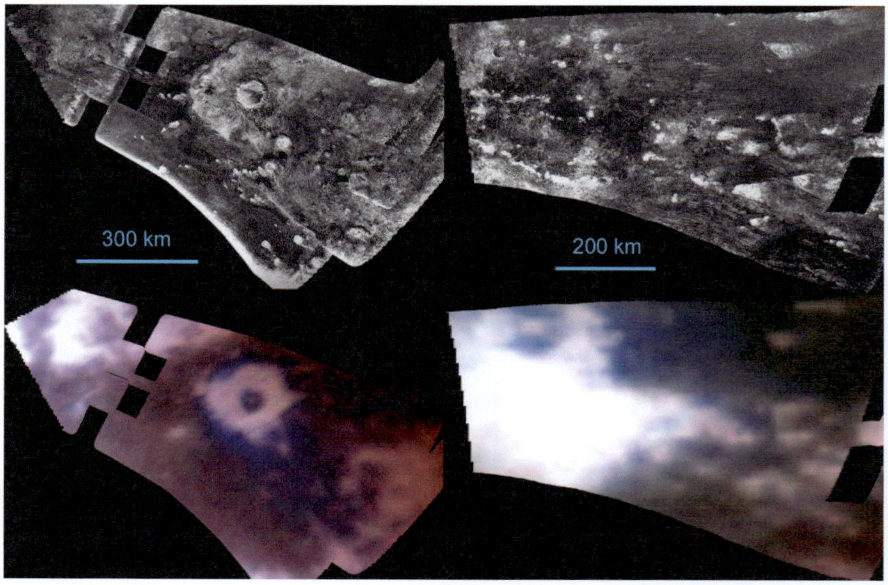

Fig. 6.17 Correlations between VIMS spectral units (*lower set*) and terrain units seen in SAR images (*upper*) from Soderblom et al. (2007). The image pair on the left shows the region of Sinlap crater; to the right is the region of the Huygens landing site. Dark brown units correlate with dune fields imaged by the SAR; other correlations are only weakly apparent (VIMS color composites with RGB = 2.0, 1.6, 1.3 μm).

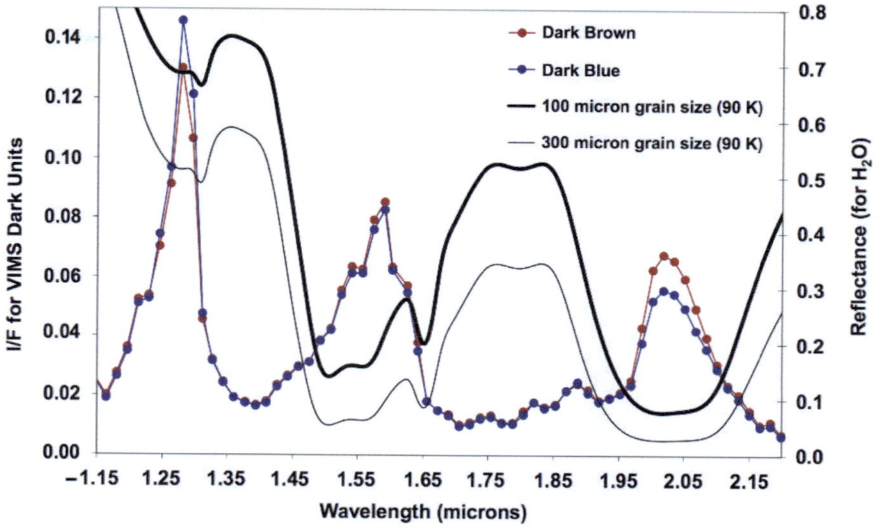

Fig. 6.18 Typical VIMS spectra for dark brown and dark blue equatorial units near Sinlap crater (from Soderblom et al. 2007). They are compared with model spectra of water ice derived from optical constants of Grundy and Schmitt (1998).

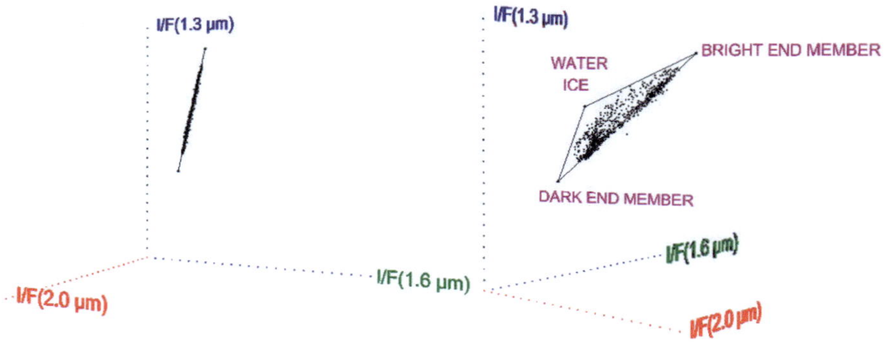

Fig. 6.19 Three-D Plot of I/F in each of 3 methane windows (1.3, 1.6, and 2.0 μm). The very flat, sheet-like nature of the cluster suggests a simple three-component mixture is involved (Soderblom et al. 2007).

forming a simple mixing plane. Soderblom et al. (2007) proposed the end-members to consist of: (1) bright hydrocarbon/nitrile dust, (2) dark coarse-grained hydrocarbon/nitrile grains of the dunes, and (3) water ice. If water ice were an end-member, its variation would most strongly affect the positions in the cluster by deflecting points in the cluster away from and toward the z-axis. The least water ice-rich part of the cluster would then be that part farthest from the z-axis. As can be seen in perspective view of the cluster in the right panel of this figure there is a linear distribution of points along its edge, the ends of that are labeled 'bright end member' and 'dark end member.' Soderblom et al. (2007) hypothesized that either (1) some combination of chemical and/or mechanical processes is converting bright fine-grained tholin airfall deposits into the dark saltating grains of organics or (2) that the material of dark dune materials may be derived from two different types of photochemical aerosol products.

Whereas a strong correlation exists between the VIMS dark brown units and the SAR dunes, correlations between the VIMS dark blue and bright units and the SAR units are largely lacking. Figure 6.20 may give some hint of correlation. Here dunes are generally rare with only one patch of dark brown dunes (lower right inset). In the rest of the region the dark blue units are often correlated with features in the SAR that resemble sinuous flow features; both SAR-bright and SAR-dark cases are observed. Likewise the bright, rough hills seen in SAR appear to be correlated with both bright and dark blue VIMS units. Based on these relationships, Soderblom et al. (2007) postulated that the bright equatorial materials constitute thin mantles that are effectively transparent to the SAR leading to this general lack of correlation. In their model for the VIMS units, both the bright aerosol and dark brown hydrocarbon dune deposits mantle a dark blue substrate posited to be water ice-rich. They hypothesize that a reasonable candidate for the bright end member is a thin mantle of bright aerosol dust that contains acetylene and other simple hydrocarbon solids whereas the dark dunes are made of more complex hydrocarbons and/or nitriles. In another VIMS-SAR correlation study of mountains and

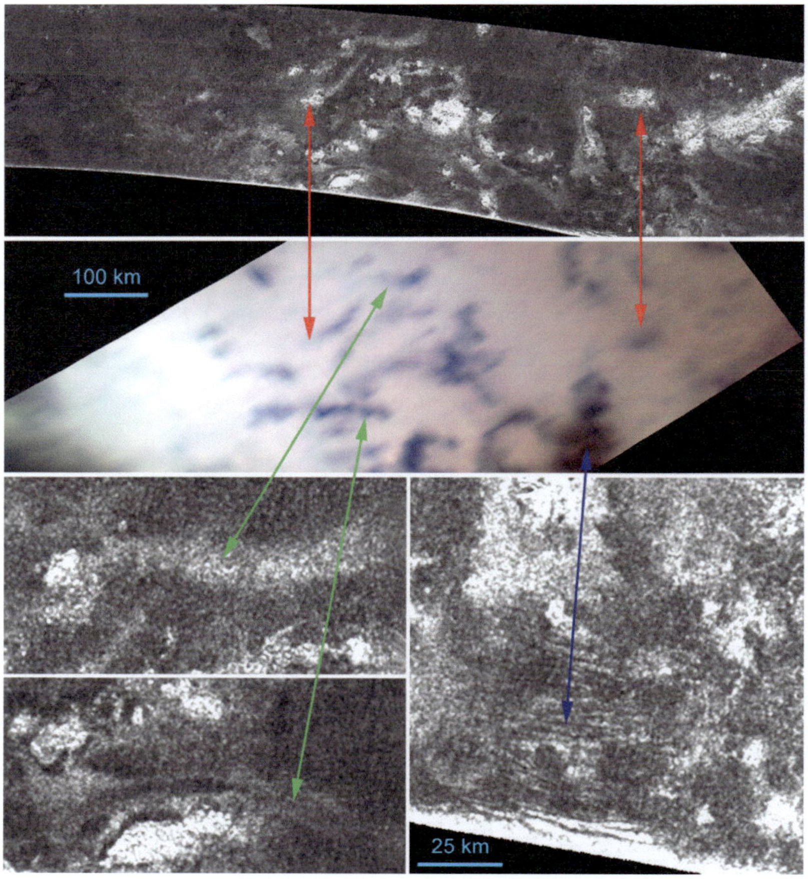

Fig. 6.20 Correlations between VIMS dark blue and VIMS bright units with SAR units. Top two panels cover a 300 × 1,000 km area in Omacatl Macula (20°N, 45°W). Top: SAR image from RADAR T3 swath. Middle: T9 VIMS image (color composite same as Fig. 6.17). Lower: 4× enlargements of SAR image. Blue arrow denotes VIMS dark brown dune patch. Red and green arrows indicate bright hills and flow-like features seen in SAR that sometimes correlate with VIMS dark blue units (Soderblom et al. 2007).

channels in the western part of Xanadu, Barnes et al. (2007b) also conclude that thin mantles of bright material, etched off of the mountains and channels floors exposing the darker blue units is a plausible explanation of the surface stratigraphy.

6.3.3 The Search for CO_2 Ice

Carbon dioxide is another common ice in the outer solar system that could well be expected at the surface of Titan. In fact VIMS has found CO_2 on several of Saturn's satellites including Hyperion, Phoebe, Iapetus, and Enceladus (Buratti et al. 2005; Clark et al. 2005, Brown et al. 2006a,b; Cruikshank et al. 2007; Filacchione et al. 2007). Kress and McKay (2004) predicted that CO_2 might be abundant at Titan's surface, supplied by comet impacts early in its history. Solid carbon dioxide has been suggested for bright regions from Earth-based spectroscopy (Griffith et al. 2003; Coustenis et al. 2006) and from VIMS spectra of regions near the Huygens landing site (Rodriguez et al. 2006). Carbon dioxide ice reflects highly in the short-wavelength atmospheric windows ($\lambda < 1.4$ µm), as does water ice, but is more highly reflective than is water ice in the 5 µm window for small grain sizes. However the global concentration may be quite low as the Huygens GCMS tentatively detected CO_2 in the surface mass spectra but only at ~10^{-3} the concentration of CH_4 (Niemann et al. 2005). Additionally, from high-resolution Fabry-Perot spectroscopy using the VLT/NACO, Hartung et al. (2006) found no evidence of CO_2 in the 2 µm window and placed an upper limit of 7% CO_2 in bright regions centered around 300°W. Nonetheless CO_2 ice is an interesting possibility for the materials of the 5 µm-bright spots, Tui Regio and Hotei Regio. In fact Barnes et al. (2005) speculated that the materials in Hotei Regio could be eroded layers or recent deposits of CO_2.

As mentioned earlier, McCord et al. (2006) noted a further anomalous spectral behavior of Tui Regio. In addition to its extreme 5 µm brightness, it exhibits a very high 2.8/2.7 µm ratio relative to the rest of the Titan disk seen in the TA mosaic (Fig. 6.21). Characterizing this behavior

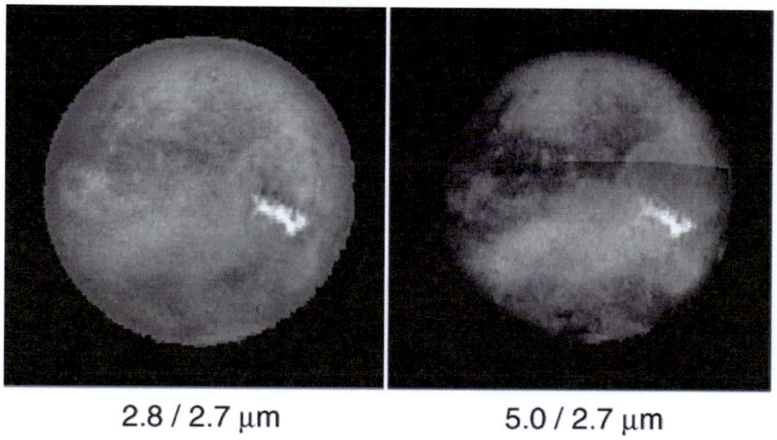

Fig. 6.21 Anomalous spectral behavior of Tui Regio (McCord et al. 2006). Shown are ratios from a mosaic of VIMS cubes acquired on the TA flyby. Tui Regio, one of two prominent 5-μm-bright regions (see Fig. 6.19), stands out from all other features in both ratios but most particular in the 2.8/2.7 μm ratio.

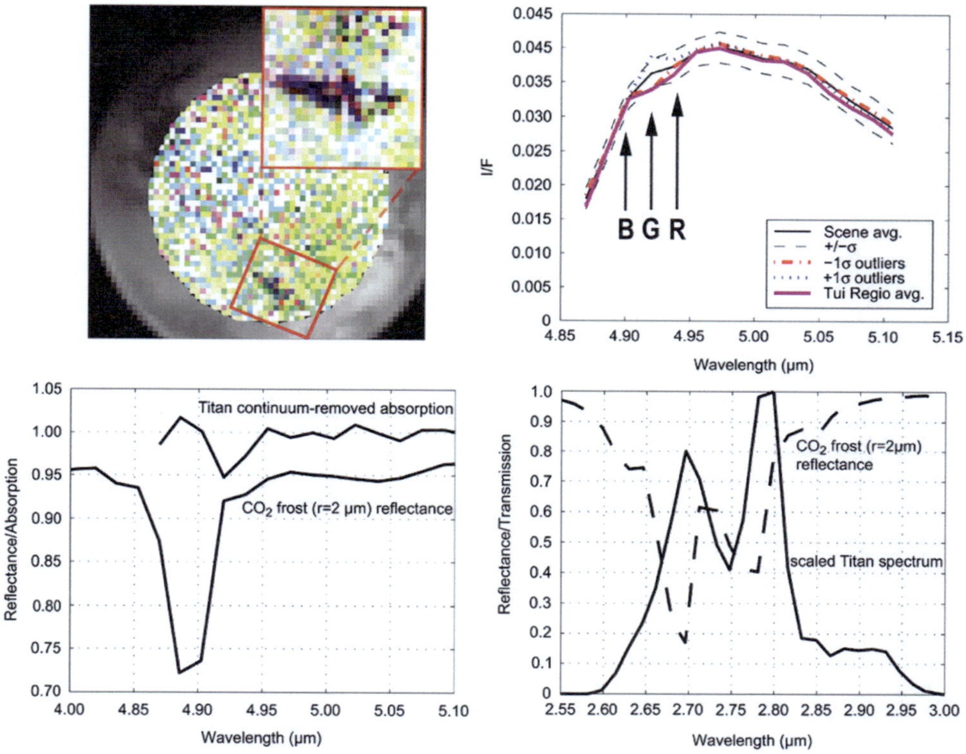

Fig. 6.22 Possible identification of CO_2 in Tui Regio (McCord et al. 2008). Upper left: RGB color composite of spectral channels denoted in upper right. Upper right: VIMS I/F spectra of Tui Regio and the frame-wide average. Lower left: Continuum-removed Tui spectrum in the 5 μm window compared to spectrum of fine-grained CO_2. Lower right: Typical VIMS spectrum overlaid on spectrum of fine-grained CO_2 showing the absorption that could lower the I/F at 2.7 μm and raise the 2.8/2.7 μm ratio.

(high 2.8/2.7 μm ratio and bright at 5 μm) as "CO_2-like ice," McCord et al. (2008) used it as a spectral end-member in mapping global spectral types. These workers concluded Titan's equatorial and mid-latitudes could be mapped by five spectral units: water frost, CO_2-like frost, an unknown material bright at 2 μm, a dark spectrally neutral material, and a generalized spectrum for atmospheric scattering. They found the CO_2-like unit to be ubiquitous in the bright regions but highly concentrated in Tui Regio. Hotei Regio was not included as their mapping was restricted to only about half of the equatorial and mid-latitude regions. Coustenis et al. (2006) had suggested CO_2 to explain a spectral feature seen in ISO spectra at 2.74 μm and McCord et al. (2008) likewise suggested that CO_2 ice might explain the anomalous Tui Regio spectral behavior in the same wavelength region. They suggested that the high 2.8/2.7 μm ratio could be explained by greater absorption of CO_2 at 2.7 μm compared to 2.8 μm (Fig. 6.22, lower right).

Compared to the other atmospheric windows, the 5 μm window is spectrally much broader, covering ~4.8–5.2 μm in the VIMS spectral range. Consequently it offers an important opportunity to search VIMS spectra for discrete absorptions that might be diagnostic of surface composition. Unfortunately this is also a wavelength region for which VIMS spectra have low signal precision. In searching this spectral region McCord et al. (2008) found one absorption feature at 4.92 μm that exhibits sufficient strength to be easily mapped (Fig. 6.22). They offered fine-grained CO_2 frost (grain size <10 μm) as a candidate for the 4.92 μm absorption. They found this feature to be strongest in Tui Regio as illustrated in the color composite in the upper left inset (RGB=4.94, 4.92, 4.90 μm). Also shown in the upper right are the average spectra for Tui Regio and for the scene-wide average. The scene-wide average was used as a continuum and fitted to and removed from the Tui Regio spectrum; the result is shown in the lower left along with a laboratory spectrum of CO_2 ice with a 2-μm grain size.

Unsatisfying in the tentative identification of CO_2 is that the positions of the absorption feature seen in the CO_2 laboratory spectrum compared to that seen in the Tui Regio continuum-removed spectrum (bottom left, Fig. 6.22) are substantially shifted from one another, by as much as 25 nm. McCord and colleagues suggest that the position of the CO_2 absorption could be shifted to match the VIMS feature by other contaminants that could alter vibrational frequencies of bonds. They note that Bernstein et al. (2005) observed wavelength shifts but only as high as 6 nm for H_2O/CO_2 mixtures. In summary while CO_2 remains a tantalizing and potentially even a major component of Titan's surface, VIMS spectral evidence for its presence remains unconvincing.

6.3.4 Spectral Evidence of Organics and Nitriles in the 5 μm Window

As mentioned earlier, although the SNR is quite low in the 5 μm window, it is by far the widest spectrally of the atmospheric windows available to VIMS that covers the 4.8–5.2 μm part of the window. As such this region offers excellent opportunities to identify and map absorption features specific to many expected compositional surface components. Many hydrocarbons and nitriles known or anticipated to exist in Titan's surface materials are spectrally active in this region. In addition to methane at the surface the GCMS confirmed ethane to be present and tentatively identified cyanogen, carbon dioxide, and benzene. A great many more complex organic and nitrile molecules are also likely to be present at the surface. In Titan's thermosphere, the Cassini Ion and Neutral Mass Spectrometer (INMS) instrument detected high abundances of benzene and throughout the stratosphere CIRS has detected trace abundances of a broad spectrum of organic compounds (see Chapters 7 and 8). Clark et al. (2009a,b) have begun to compile a comprehensive database of optical constants and spectral reflectance (including over the VIMS wavelength region) for a large number of simple-to-complex hydrocarbons and nitriles. An example of that data for the alkanes (that includes methane and ethane) is shown in Fig. 6.23. Many of the expected surface molecular components have diagnostic absorption features accessible to VIMS in the 5 μm window.

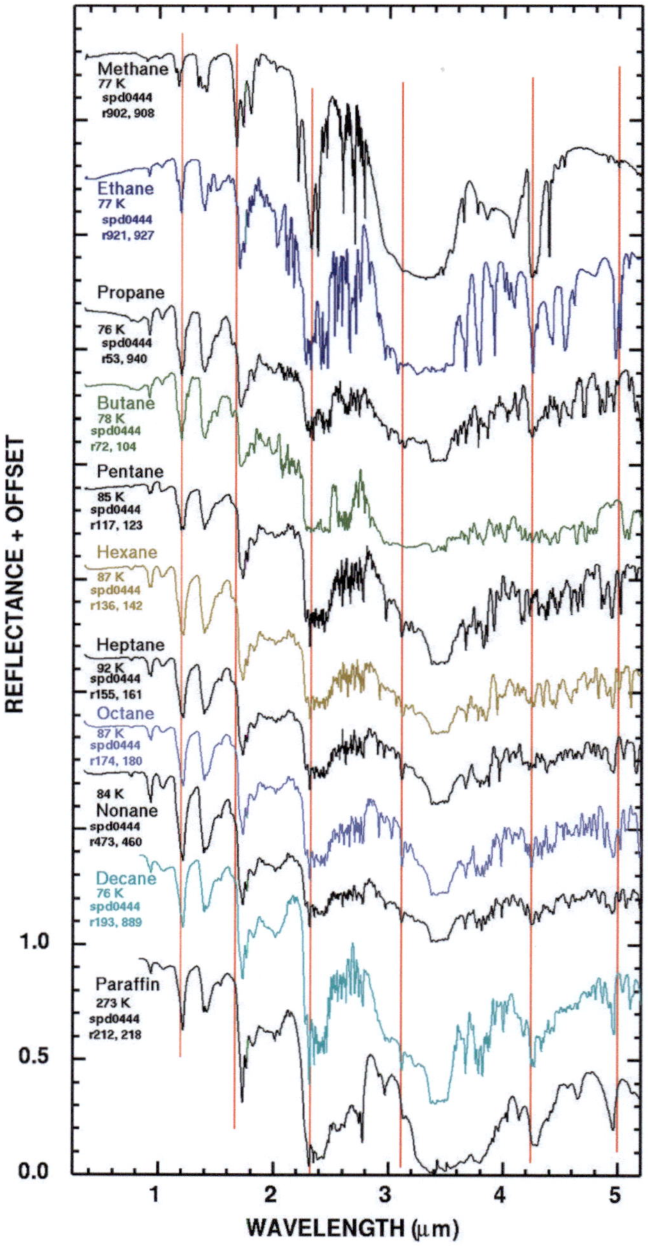

Fig. 6.23 Reflectance spectra of the first ten alkane family members plus paraffin in the VIMS spectral range. Vertical lines show how methane's absorption features correspond to those across the alkane family. Spectra have been offset as follows: methane, 4.1; ethane, 3.6; propane, 3.3; butane, 2.8; pentane. 2.3; hexane, 2.0; heptane, 1.5; octane, 1.2; nonane, 0.85; decane, 0.4; and paraffin, 0.05 (from Clark et al. 2009a).

As described in the previous section McCord et al. (2008) reported an absorption in VIMS spectra of Tui Regio at 4.92 µm that they suggested as due to carbon dioxide. Clark et al. (2009b) find another compound, cyanoacetylene (HC_3N), to provide a better spectral match. This compound is a known photochemical product in the thermosphere and lower atmosphere and was detected by CIRS and INMS (see Chapters 7 and 8). The HC_3N reflectance spectrum in Fig. 6.24 was convolved to VIMS spectral resolution and shows a feature at the wavelength of the shallow absorption in Tui Regio. Clark et al. (2009b) note that as in the case of CO_2, HC_3N exhibits absorption in the two 2.7 and 2.8 µm sub-windows that could explain the higher reflectance at 2.8 µm (Fig. 6.25). But as cautioned earlier the VIMS-observed ratio is strongly affected by scattering and absorption uncertainties that make its interpretation dubious for any spectral fit. Nonetheless the spectral match in the 5 µm window makes cyanoacetylene a leading candidate for the 4.92 µm feature. Higher spectral resolution observations will be needed to confirm this identity.

To search VIMS data for absorption signatures of other organic species in the noisy 5 µm region, Clark et al. (2009b) employed a technique developed in terrestrial remote sensing applications known as the Tetracorder algorithm (Clark et al. 2003; Swayze et al. 2003). VIMS cubes were searched for features using data using mathematically constructed absorption bands derived from the spectral libraries of organic species (Clark et al. 2009a,b; Curchin et al. 2009). In addition to the 4.92 µm band found by McCord and colleagues, Clark et al. (2009b) found three additional features: at 5.05 µm assigned to benzene (C_6H_6), at 4.97 µm consistent with moist surfaces coated with small amounts liquid methane and/or ethane, and at 5.01 µm that remains unidentified.

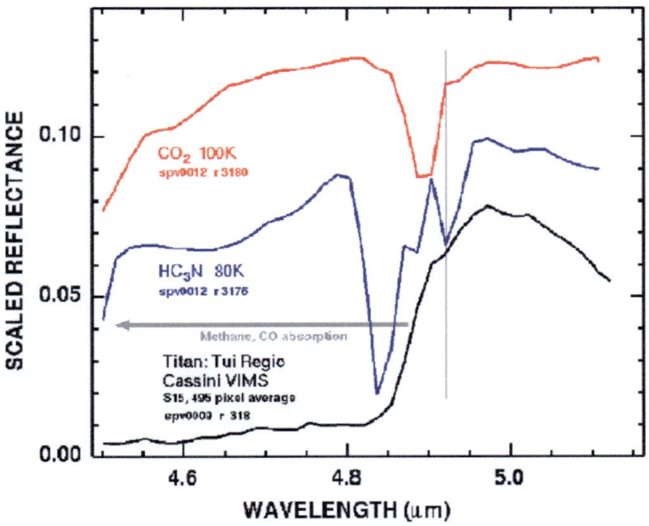

Fig. 6.24 VIMS spectrum of Tui Regio with potential compositional candidates. A feature in cyanoacetylene (HC_3N) offers a closer spectral match to the 4.92 µm feature found at Tui Regio than does CO_2 (Clark et al. 2009b).

Benzene, an aromatic hydrocarbon, was found in higher abundance than expected by INMS (Chapter 8). Clark and colleagues did not find evidence for acetylene (C_2H_2), expected to be in Titan's surface materials in higher abundances than benzene (cf. Chapters 7 and 8). The VIMS 5.05-µm feature matches the position, width, and shape of a unique spectral feature due to benzene (Fig. 6.26). The benzene absorption arises from an overtone of an out-of-plane bending mode whose fundamental lies near 10 µm. Clark et al. (2009b) note that the 5-µm-region absorptions for higher-weight aromatic hydrocarbons are shifted to longer wavelengths outside the VIMS spectral range leaving benzene with a unique feature at 5.05 µm. Of all available spectral libraries for solid and liquid organic, nitrile, and mineral species, Clark et al. (2009b) find that only the spectrum of

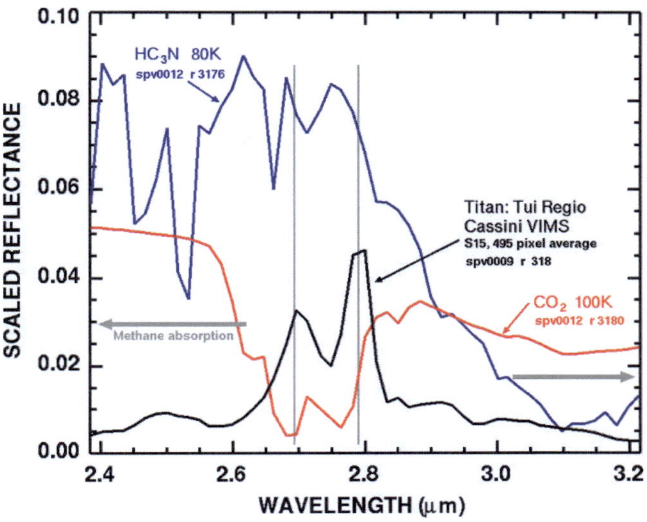

Fig. 6.25 Absorption features in CO_2 and HC_3N in the 2.6–2.9 µm spectral region. Both compounds are spectrally active in this region and either could explain the much higher VIMS signal at 2.8 µm compared to 2.7 µm (Clark et al. 2009b).

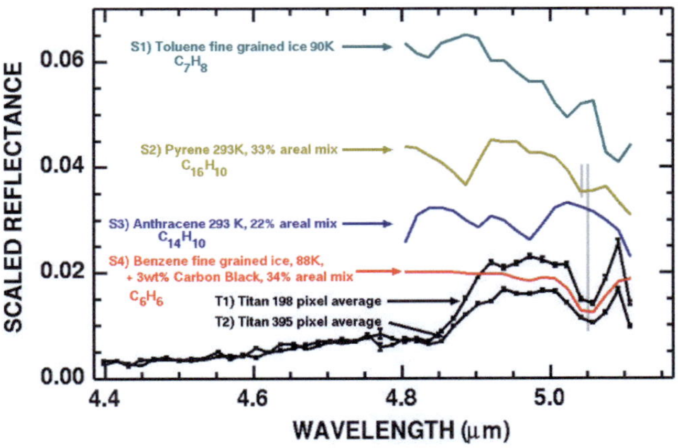

Fig. 6.26 Absorption band seen in VIMS spectra of Titan's dark equatorial regions that correlates with a spectral feature in solid benzene (Clark et al. 2009b).

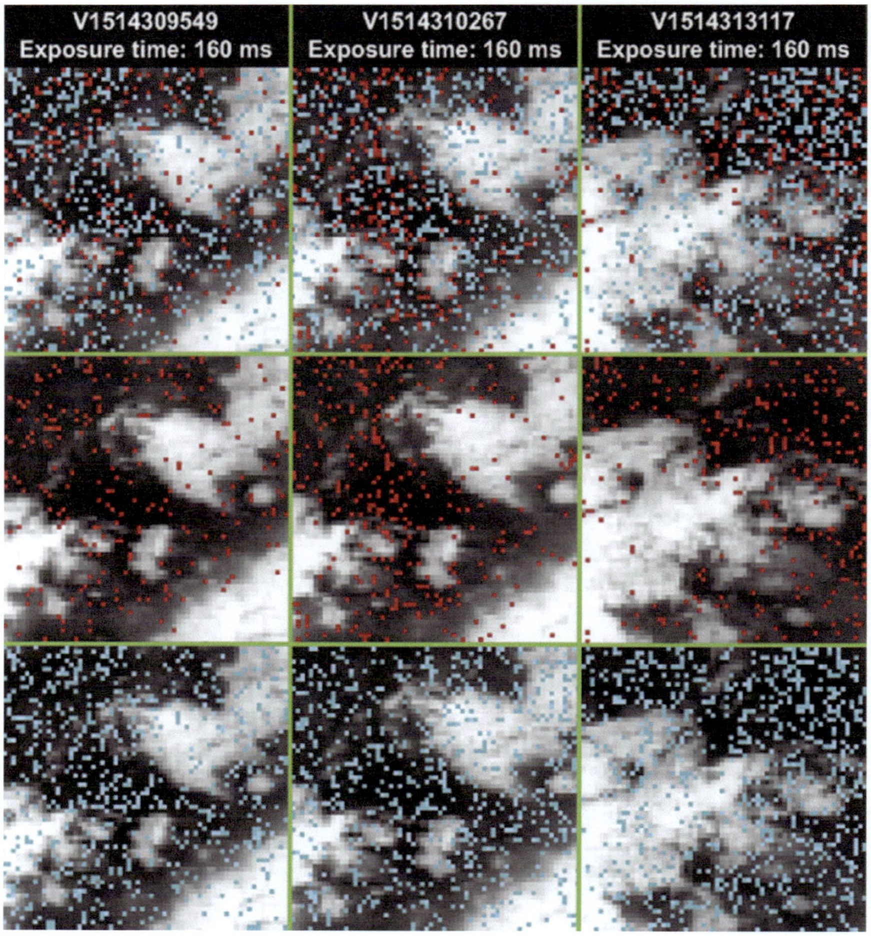

Fig. 6.27 Band-fitting results for individual data cubes from McCord et al. (2008) using the same benzene laboratory spectrum used by Clark and colleagues. Red pixels are matches to the laboratory absorption spectrum whereas cyan pixels show correlations with its inverse (excerpted from McCord et al. 2008).

benzene matches the VIMS feature. The posited 5.05 μm benzene absorption appears preferentially associated with Titan's darker terrains, whereas the 4.92 μm attributable to cyanoacetylene appears in bright terrain, with its best exposure in Tui Regio, the brightest region on Titan in the 5-μm window.

McCord et al. (2008) contested the work of Clark et al. (2006a,b) identifying widespread surface exposures of benzene; they concluded that the 5.05 μm feature was an artifact of analysis arising from the low signal precision of VIMS data in this region. McCord and colleagues analyzed VIMS data using the same Tetracorder algorithm and the same laboratory spectrum used by Clark and colleagues. But in addition they searched for matches using both the laboratory absorption band and then with its inverse. Comparison with maps of inverse correlations they posited would provide a test for the rigor of the identifications. Other than a tendency for both positive and negative hits to be associated with lower signal strength in darker regions (Fig. 6.27), they found no systematic patterns in the alarmed pixels that seemed to correlate with Titan's surface features. But there is a different interpretation of their results shown in Fig. 6.27. The middle tier shows that the red pixels (that correlate with the shape of the laboratory benzene band) are clearly concentrated in the dark regions and rarely appear in the bright regions. It appears that the evidence, although not conclusive, suggests the concentration of a 5.05 μm absorber in the dark regions.

A third feature weak feature at 4.97 μm was identified and mapped by Clark et al. (2009b) with VIMS spectra for the 5 μm window. This occurs in rare, isolated dark equatorial regions. The absorption feature best matches the spectra of the low-molecular-weight alkanes, methane and ethane (Fig. 6.28). Such materials would occur as liquids because as discussed below, ices of these compounds are not stable at the surface, certainly not in the dark equatorial areas where

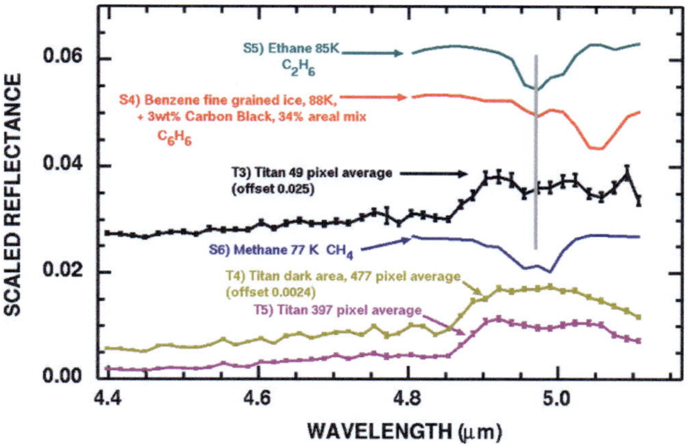

Fig. 6.28 Average VIMS spectrum for isolated equatorial dark regions that exhibit a weak 4.97 µm absorption (*black curve*). This feature matches similar features that may be seen in liquid or solid ethane and methane (Clark et al. 2009b).

this feature is concentrated. The feature is absent in most of the dark regions suggesting they are dry. In the rare places where it does occur, the strength of the observed absorption indicates optical path lengths of less than a few millimeters, suggestive to Clark and colleagues of a surface wetted with liquid methane and ethane, possibly as wet sand, muddy soil, or slush. This hypothesis is consistent with surface observations at the Huygens landing site in which the GCMS detected excess methane to be exhausted from the subsurface on heating post-landing (Niemann et al. 2005) and with the moist sand-like physical nature of the surface as detected on impact with the Surface Science Package (Zarnecki et al. 2005). As discussed later in this chapter, Brown et al. (2008) found further spectral evidence for liquid ethane in a south polar lake, Ontario Lacus.

6.3.5 The Case for Methane and Ethane Surface Ice

The general nature of the optically bright materials such as found in Xanadu is broadly debated. Following the early suggestion of Coustenis et al. (2001, 2005), Negrão et al. (2006, 2007) and Hirtzig et al. (2007) have proposed methane and ethane ices for the bright materials. Although it is true that such ices have high reflectance in the 1.6 µm and 2.0 µm windows as observed, they are not stable as solids at Titan's equatorial surface temperature. Lindal et al. 1983 derived equatorial surface temperatures of 94.0 ± 0.7 K from Voyager Radio Occultation measurements whereas these ices all have melting points well below that (90.7 K and 90.4 K for methane and ethane, respectively). For these ices to be stable at the equatorial latitudes would require the bright terrains to be elevated at least 3-to-4 km above the mean surface of Titan (Lorenz and Lunine 2005). So far Cassini RADAR altimetry measurements have shown no deviations of more than a few hundred meters from a hydrostatic figure (Lorenz et al. 2007). Likewise radarclinometry of Titan's mountains implies that the highest peaks are no greater than 2 km above the surrounding terrain (Radebaugh et al. 2008). In addition the Huygens Atmospheric Structure Instrument measured an HLS surface temperature of 93.65 ± 0.25 K (Fulchignoni et al. 2005). The bright highland region 5 km north of the landing site is no more than 250 m above the adjacent plain (Tomasko et al. 2005, Soderblom et al. 2007). Thus these light alkanes would not be stable as solid "ices" at the equator or mid-latitudes. It is conceivable that they could be stable in the polar regions given a pole-to-equator temperature gradient of a few degrees (cf. Janssen et al. 2009 and Chapter 13). But to even form polar deposits of methane or ethane ice materials would have to be very pure because ethane/methane mixtures have melting points that are considerably lower than for either (Mitri et al 2007). As predicted from photochemistry (cf. Lunine et al. 1983) liquid methane and ethane very probably coexist in any surface liquids.

6.3.6 The Case for Ammonia Ice

Nelson et al. (2009) reported significant, apparent temporal excursions in the reflectance of a region known as Hotei Arcus that were derived from VIMS observations made between mid-2004 to mid-2006. The estimated changes were greatest in the 2 and 5-µm spectral windows, reaching roughly a factor of 2. They hypothesized that these changes reflect current geological activity and are caused by ground fogs or surface condensations of some bright material that subsequently dissipates. They suggested that release of ammonia by ongoing, active cryovolcanism could be the explanation for the apparent brightness variations. Many of these VIMS observations were made when Hotei Arcus was quite close to Titan's bright limb resulting in large emission angles (>60°) and large optical path lengths making derivation of surface reflectance variation at wavelengths <3 µm both difficult and dubious. That the expected form of ammonia in a Titan cryovolcanic lava is a 1:6 solution of ammonia and water as a monohydrate (Stevenson 1992), casts doubt on the idea that the effluent responsible for the apparent brightness changes is pure ammonia. Rather the effluent would be ammonia hydrate, whose reflectance spectrum is only marginally different than that of water ice (cf. Brown et al. 1988). Nevertheless, if their hypothesis were confirmed, it would represent a remarkable advance in our understanding of Titan's geologic activity.

6.4 Compositional Constraints from Microwave Observations

Over the last 20 years a mix of active and passive microwave experiments have been made with radars and radio systems, both Earth-based and aboard Cassini, to observe the microwave reflectance and emission of Titan's surface. Such experiments have included monostatic radar (transmit and receive from about the same vantage point) and passive radiometry; most of these include dual polarization measurements. At microwave wavelengths (millimeters-to-decimeters) Titan's atmosphere is effectively transparent and its dielectric constant (ε) and physical state (temperature, porosity, fracturing, layering) dominate the signal. In the simple case of a flat solid half-space obeying Fresnel reflection and Kirchhoff's law, the measured dielectric constant (ε) can be related directly to composition given the known range of ε values for materials expected at Titan's surface (Table 6.1). But as for most planetary objects, Titan's surface is far from simple and its physical state (largely porosity, fracturing, layering, and surface texture) strongly influences the effective dielectric constant that is observed (i.e., ε obtained from the measurements assuming that the Fresnel's equations are applicable for the surface). Nonetheless, microwave observations provide crucial constraints on the combinations of composition ingredients and physical state that can be allowed in models of Titan's surface.

Muhleman et al. (1990) first measured Titan's radar reflectivity at 3.5 cm by transmitting from the 70 m Goldstone radio telescope in California and receiving at the Very Large Array in New Mexico. These early data indicated Titan's surface to have a very high radar cross-section (σ_0 ~ 0.35) and suggested longitudinal variation, both of which discounted a deep global ocean of liquid hydrocarbons as hypothesized by Lunine et al. (1983). Subsequent observations of the same type revised the average value downward to 0.125 ± 0.02 (Muhleman et al. 1995); this was consistent with that found by Goldstein and Jurgens (1992) of 0.14 ± 0.03 using the Goldstone radar antenna in monostatic mode. Although these values still precluded a global ethane-methane ocean, models with dirty water ice and layers of solid hydrocarbons and nitriles were allowed. Muhleman et al. (1995) also reported longitudinal variations that, assuming synchronous rotation, gave higher values for the leading hemisphere (σ_0 ~ 0.15) relative to the trailing (σ_0 ~0.10). Comparison with the near-IR light curve (Lemmon et al. 1993, 1995; Griffith 1993; Coustenis et al. 1995) and near-IR HST images (Smith et al. 1996) revealed the radar bright units and optically bright units to be correlated (Lorenz and Lunine 1997). The leading hemisphere is the site of the continental scale bright region formally named Xanadu that is also exceptionally bright as seen in Cassini radar and optical data.

Table 6.1 Dielectric properties of possible Titan surface materials

Material	Dielectric constant
Liquid hydrocarbons	1.6–1.9
Solid hydrocarbons	2.0–2.4
CO_2 ice	2.2
Water ice	3.1
Water-ammonia ice	4.5
Organic heteropolymers	4.5–5.5
Meteoritic material	8.6

From Thompson and Squyres (1990); Lorenz (1998); Lorenz et al. (2003); and Wye et al. (2007)

About three years prior to the arrival of Cassini-Huygens, Campbell et al. (2003) used the 305-m Arecibo radio telescope (sometimes in combination with the Green Bank Telescope as the receiver) to map 13-cm radar returns from a southern latitude band on Titan at ~26°S. Data were collected in two circularized polarizations, the same-sense and the opposite-sense (SC and OC) from the transmitted beam. If surfaces were mirror-like they would show up as strong specular components in the OC; surfaces that were rough and highly scattering would show up in both. Most striking in the Arecibo data was the discovery of strong specular reflections seen in the OC echoes for ~75% of the regions observed. The offered hypothesis was that numerous extremely flat surfaces were scattered across Titan's surface, possibly signaling liquid hydrocarbon lakes or flat icy solids. Campbell and colleagues also derived a range of dielectric constant for this band of 1.5–2.2 with a mean of ~1.8. This range overlaps ε values for liquid and solid hydrocarbons though is inconsistent with solid water ice (Table 6.1). Campbell and colleagues also confirmed that the leading hemisphere, seen to be brighter in the near-IR, was brighter in 13-cm radar reflection and found that most of the returned echo showed strong diffuse scattering with roughly equal polarizations. They suggested that volume scattering in radar-transparent water ice could explain the high return from the bright region. A fractured, porous, highly scattering layer of subsurface water ice would exhibit a much lower dielectric constant than solid water ice (ε~3) consistent with the observed dielectric values.

6.4.1 *Compositional Constraints from the Cassini RADAR Scatterometer*

The Cassini RADAR is a 2.18-cm microwave system has four operational modes: a synthetic aperture radar imager (SAR), a low-resolution scatterometer, an altimeter, and a passive radiometer (Elachi et al. 2004; West et al. 2009). The scatterometer is a real-aperture active radar mode that allows raster scanning of large regions (~10^6 km^2) at low resolution (~100 km) out to a range of ~25,000 km. This allows repetitive coverage and mapping of individual features over large range of incidence angles (~0°–80°). The observed angular dependence of the radar cross-section (σ_0) yields information on surface composition (dielectric constant, ε) and physical properties (e.g. surface roughness and slope). Figure 6.29 illustrates two sets of overlapping scatterometry results for part of Xanadu extending westward across the southern equatorial zone (100°W to 220°W, Wye et al. 2007). These maps (normalized by the Titan's average backscatter behavior) illustrate remarkably good agreement from data collected on two separate Titan flybys, TA and T8. To the right are scatter plots for σ_0 versus incidence angle, in which Xanadu points are separately color-coded from other regions illustrating the highest backscatter there. In the left middle and lower panels are maps of residuals from an average Hagfors-based scattering law fitted to backscatter data for each scan (shown on the right). The top left map is from a 0.94 μm Cassini Imaging Science Subsystem mosaic (cf. Porco et al. 2005) and illustrates in detail the systematic correlation between radar-bright and optically bright regions as was seen at hemispheric scale in Earth-based observations.

The scatter plot illustrates the existence of discrete specular reflections near zero-incidence angle as well as a general diffuse-scattering component, both as seen in the Arecibo 13-cm data (Campbell et al. 2003). The scatterometer mode is measured in a single polarization, same-sense linear (SL) so polarization ratios cannot be used to derive electrical properties. However scattering models can be fitted to the σ_0-incidence angle distributions to derive information on surface properties (ε and scattering factors). The specular component, presumably arising from mirror-like flat surfaces with a distribution of tilt angles relative to local normal, can be fitted with simple models to derive ε and mean slope. The high level and gradual fall-off of the diffuse backscatter with incidence angle imply a combination of low-loss volume scattering and surface roughness. Wye et al. (2007) fitted these data with two models: a model composed of a quasi-specular Hagfors term plus a diffuse cosine term and a Gaussian quasi-specular model (cf. Fig. 6.29). The two models give similar results. For the entire scatterometry data set available, with the Hagfors model Wye et al. (2007) report mean dielectric constant, slope, and radar albedo of 2.2 ± 0.05, 7.71° ± 0.5°, and 0.34 ± 0.001, respectively. Their estimate of the total-power albedo of 0.34, is much higher than the total-power albedo measured by Campbell and colleagues of 0.21 at 13 cm. They suggest the discrepancy can be attributed to greater small-scale roughness at 2 cm than at 13 cm. The total variance in dielectric constant (ε~1.9–3.6) shows that Titan to exhibit a globally heterogeneous surface. The mean scatterometry value (ε ~ 2.2) is significantly higher than the range expected for liquid hydrocarbons (1.6–1.9; Table 6.1) but is in the range for solid simple hydrocarbon solids (2.0–2.4). Although solid sheets of water ice over major parts of the surface are precluded, loose porous mixtures with water ice could also fit the observed data.

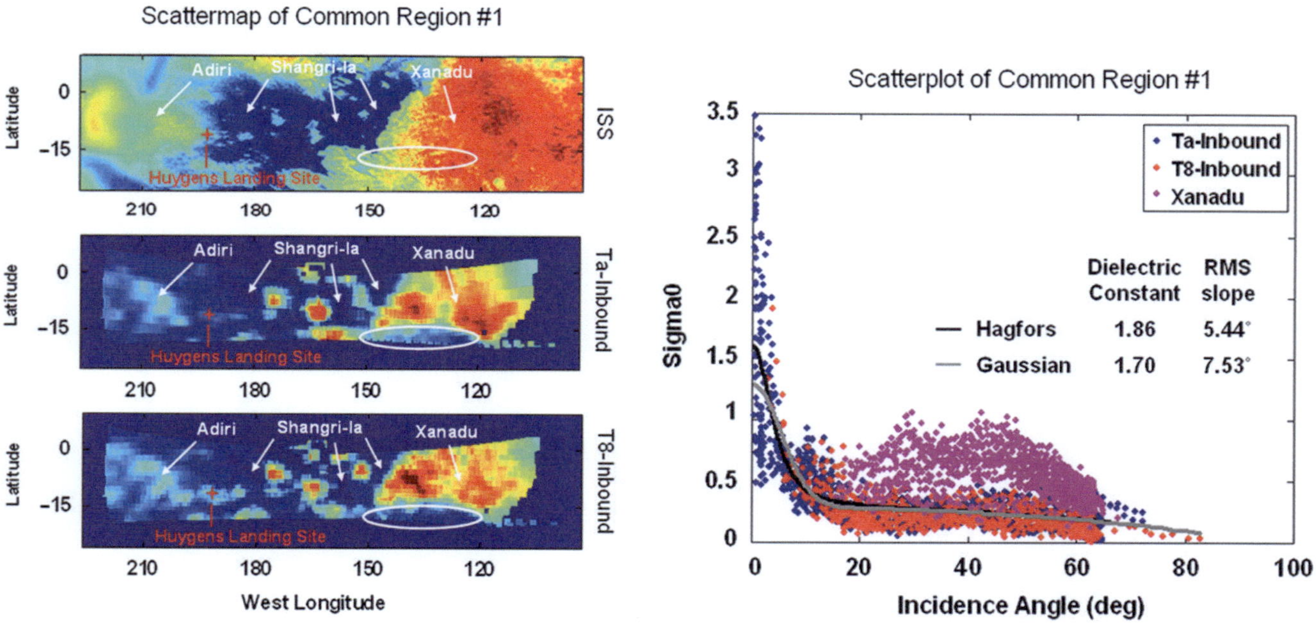

Fig. 6.29 Comparison of RADAR scatterometer maps to 0.94 μm ISS image brightness and model fits to scatterometer data (from Wye et al. 2007). In the left middle and lower panels the backscatter data for TA-inbound and T8-inbound data agree well and correlate with the 0.94 μm ISS image brightness (*upper, red = brighter and blue = darker*). In the right two model fits are shown: a quasi-specular Hagfors model (ε ~ 1.86) and a quasi-specular Gaussian scattering model (ε ~ 1.70).

6.4.2 Constraints from the Cassini RADAR Radiometer

The Cassini RADAR system includes a passive microwave radiometer mode that, as do all other modes, operates at 2.18 cm (Elachi et al. 2004, West et al. 2009). Janssen et al. (2009) describe calibration and mapping of the thermal microwave emission from Titan's surface with the radiometric data from 69 segments of Titan flybys from TA (October 2004) through T30 (May 2007) covering ~94% of Titan's surface. Radiometric data are obtained in three operational modes: radiometer-only, with scatterometry, and with SAR (~ranges of 25,000–100,000 km, 10,000–25,000 km, 1,000–5,000 km that yield ~radiometer resolutions of 150–650, 60–150, and 6–30 km, respectively). In radiometry-only mode most of the disk was scanned twice in orthogonal polarizations. As the radiometer operates in a single linear polarization, orthogonal polarizations were obtained by rotating the spacecraft between scans. During scatterometer mode only a single polarization could be collected. So higher resolution polarization data (higher than the radiometer-only data) are not available except for a case where the nearly identical geometry of the TA and T8 could be used to cover the same region in orthogonal polarizations. Fortuitously, that region included the Huygens landing site and Xanadu. In SAR mode the radiometer obtains its highest resolution data with all five radar beams.

Polarization pairs from low-resolution radiometry-only segments, combined with the unique TA and T8 polarization pair, were first used by Janssen et al. (2009) to derive a low-resolution (~50–500 km) global map of the dielectric constant covering ~78% of Titan's surface (Fig. 6.30). A model from White and Cogdell (1973) for thermal emission from a moderately rough dielectric surface was used to estimate ε from the polarimetry pairs. In general thermal emission from a surface is polarized in proportion to its effective dielectric constant as determined by Kirchoff's law and the Fresnel coefficients for reflection from a dielectric interface. So the map in Fig. 6.30 is of an effective dielectric constant, effective in the sense that the Fresnel coefficients are assumed applicable.

Most remarkable about the ε map is its global uniformity. With the exception of a handful of higher values in the equatorial region on the right side of the map (see 0°–60°W, 10°S–30°N), the effective ε is narrowly constrained (~1.6 ± 0.5). The rare higher ε values include Sinlap crater (~11°N, 13°W) that might be a water-ice rich region; it maps as a VIMS dark blue unit (Soderblom et al. 2007 and LeMouélic et al. 2008). Xanadu exhibits an extremely low ε (~1.0) that suggests high porosity and fracturing consistent with the strong backscatter seen by Wye et al. (2007). The global average of ε ~ 1.7 is very close to the average for the south-equatorial band (~26°S) observed by Campbell et al. (2003) that yielded an average ε~1.8. Although these values are consistent with liquid hydrocarbons, SAR imaging so far reveals no lakes in the equatorial region. Most likely these values point to porous layers of solid hydrocarbons perhaps mixed with water ice.

An important application of the global dielectric mosaic of Fig. 6.30 is in derivation of normal-incidence brightness temperature maps and thereby emissivity maps to the extent that the physical temperature of the surface can be considered known. The dependence of brightness temperature on emission angle is a function of the dielectric constant. Thereby the global map of ε can be used to reduce the radar

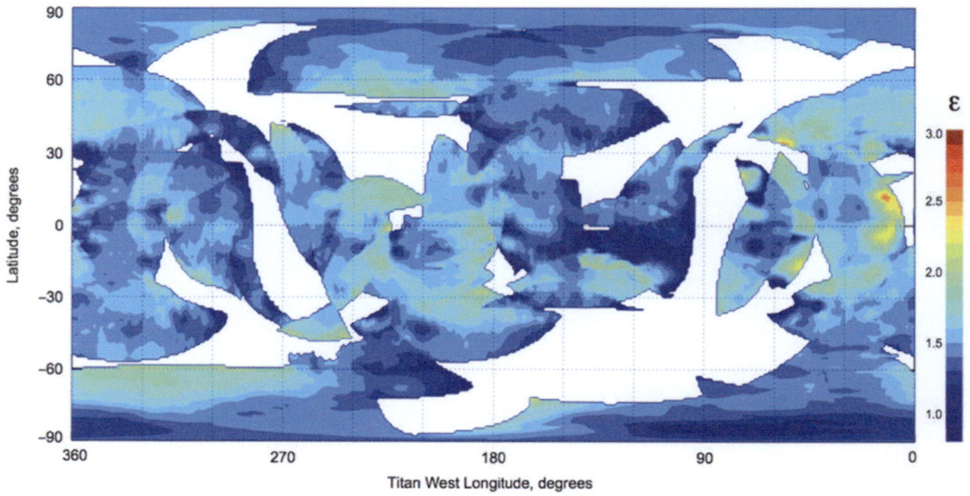

Fig. 6.30 Map of effective dielectric constant from 2.2-cm Cassini RADAR passive radiometry polarization data. Globally Titan exhibits a low average value: ε ~ 1.6. Higher and lower values rare: for example at Sinlap crater and in Xanadu. (Janssen et al. 2009).

microwave thermal emission measurements, acquired over a wide range of emission angle, polarization, and resolution, to equivalent normal-incidence brightness temperature (T_b) that could be assembled into meaningful maps. Janssen et al. (2009) first assembled a low-resolution base map from data collected at ranges > 70,000 km, where sidelobe complications were minimal. Successively closer range layers were adjusted to fit and added to the accumulating global mosaic.

The absolute level of the brightness temperature maps was controlled using two Titan surfaces, the dunes and the northern seas (discussed in the next section), that Janssen and colleagues posit to obey Fresnel reflection principles and Kirchoff's law. In addition the Huygens probe surface temperature measurement (93.65 ± 0.25, Fulchignoni et al. 2005) and the assumed uniformity of equatorial temperatures gives a precise value for the equatorial physical temperature. From the Fresnell equations that at normal incidence lead to emissivity=$[(\sqrt{\varepsilon} - 1)/(\sqrt{\varepsilon} + 1)]^2$ and the estimated dielectric constant of the dunes ($\varepsilon \sim 1.6$) yields an emissivity of ~0.98 and, from the physical temperature, a brightness temperature $T_b \sim 92$ K, thus providing an accurate scale for the T_b map. By comparison, Muhleman et al. (1990) provided the first estimate of microwave thermal emission from Titan's surface giving an average brightness temperature of 83.8 ± 6.4 K. Voyager data gave a physical temperature for the surface to be ~94 K, varying by only a few degrees from equator to pole (Hanel et al. 1981; Tyler et al. 1981, Lindal et al. 1983) and Muhleman derived an emissivity of ~0.9. The T_b map of Fig. 6.31 exhibits an equivalent emissivity in the range of 0.85–0.95, consistent with the early Earth-based estimate. The T_b final map in Fig. 6.31 varies by two orders of magnitude in resolution, covers ~94% of Titan's surface, and has an estimated precision of ~1 K.

Whereas the map of effective dielectric constant is globally quite uniform, the normal incidence brightness temperature map shows a surface that is complex and highly variegated at many scales. The global VIMS map (from Barnes et al. 2007a; shown in hemispheric projections in Fig. 6.16) is provided for comparison with the radiometry-based maps (Fig. 6.31, middle). Correlations between the VIMS and T_b maps, particularly in equatorial regions where they exhibit highest resolution, are remarkable. The dark brown dune fields correlate in detail with regions of highest T_b and therefore emissivity throughout the equatorial band. Xanadu and a number of the VIMS dark blue units, including Sinlap, stand out as areas of anomalously low T_b. The Sinlap ejecta blanket has been investigated using RADAR and VIMS data with a conclusion that regions of exposed water ice may be responsible for its unusually high effective dielectric constant (Le Mouélic et al. 2008). Were these simple Fresnel surfaces obeying Kirchhoff's law they would exhibit much higher T_b consistent with the observed values of ε in Fig. 6.31. The obvious factor to consider in reconciling this inconsistency is volume scattering caused by inhomogeneities in the subsurface. Wye et al. (2007) developed a model to describe the effects volume scattering in lowering the effective emissivity. Janssen et al. (2009) applied such a model to map the volume scattering component (Fig. 6.31, bottom). In this model the subsurface scattering is assumed to be perfectly isotropic; the fraction of the transmitted wave that is scattered back out of the surface is a constant fraction (f_{vol}) that is independent of emission angle. Two regional-scale features with high volume scattering and low emissivity stand out in this map, one in Xanadu and another at Tui Regio southeast of Xanadu (~60°W, 30°S). These features are also distinguished by their low and non-physical effective dielectric constant ($\varepsilon \sim 1$). Janssen and colleagues note that the radiometric properties of these features are similar to those of the icy satellites, the only other regions in the solar system that display such a strong combination of low emissivity, low specular reflection, and high scattering.

In summary radar radiometry maps of brightness temperature, emissivity, and volume scattering describe a surface that is globally diverse with properties that are lunar-like in most regions and like the icy satellites in rare locales. The low average effective dielectric constant is consistent with, but not necessarily definitive of a surface covered with a loose blanket of solid organic deposits. Titan may have a graded-density surface like that of the Moon that exhibits low effective ε but is composed of material of much higher intrinsic ε. Effective dielectric constants approaching unity in regions like Xanadu signal extreme wavelength-scale roughness and subsurface inhomogeneities (fractured, porous, snow-like, fluffy layers). Subsurface (volume) scattering is ubiquitous and a major contributor to the surface emissivity except for regions of dune fields and the northern seas. While the thermal emission depends on both compositional and physical parameters of the surface, the latter appear to dominate in the interpretations. The strong indication of a widespread graded-density surface covering and the low dielectric constant of surfaces without such covering (e.g., dune fields and Huygens landing site) are consistent with a surface composition resulting from the slow deposition and processing of organic compounds from the atmosphere, although there may be a few regions with some fraction of exposed solid ice at the surface. The presence of high ε materials like water ice or ammonia hydrate remains a possibility if they exist as extremely porous layers.

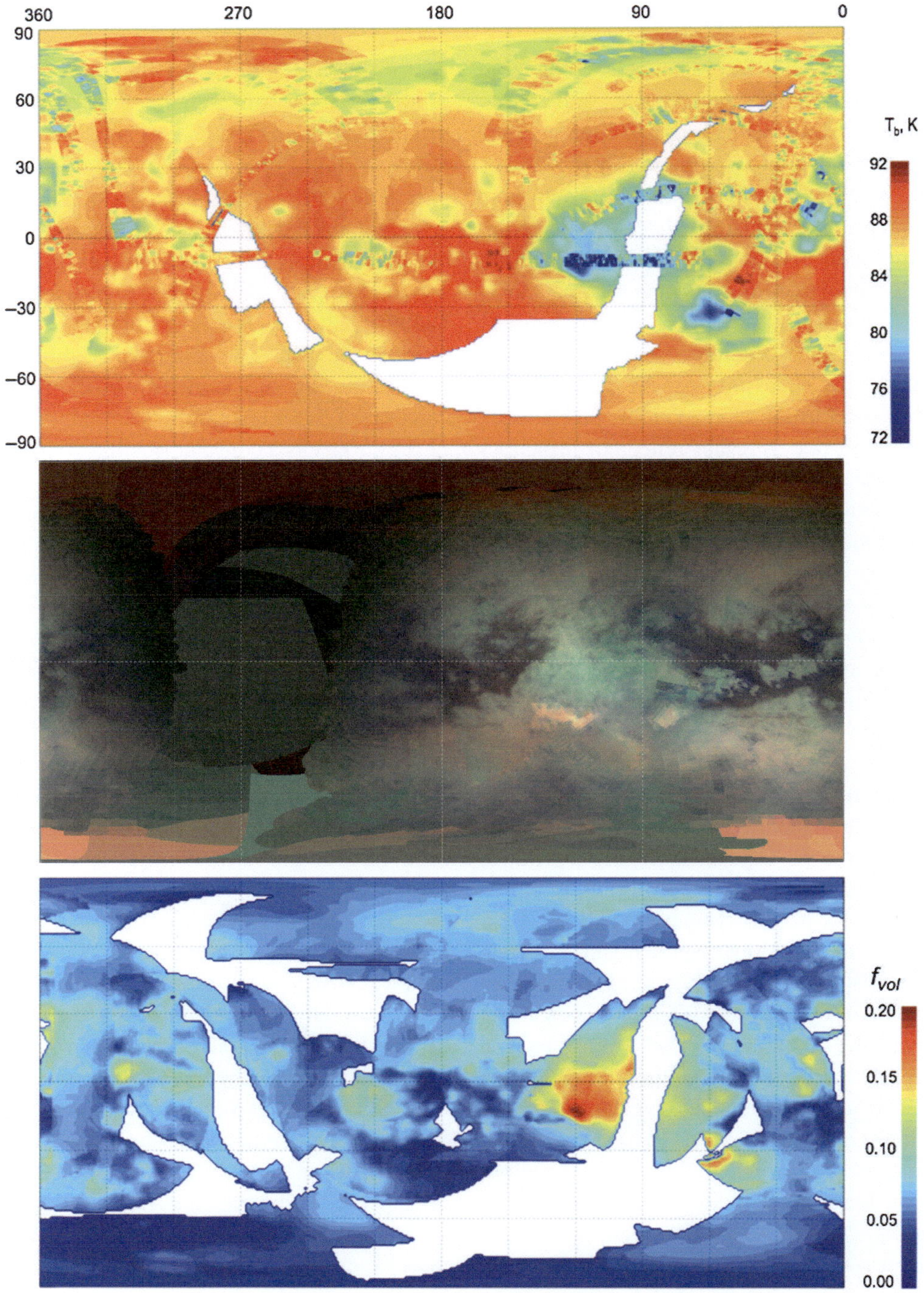

Fig. 6.31 Comparison of RADAR radiometer and VIMS global maps. Top: Microwave emission at 2.2 cm was reduced to brightness temperature at normal incidence (T_b) with a relative error ~1 K (spatial resolution ~6–600 km). Middle: Mosaic of VIMS image cubes taken through three atmospheric windows are composited with R = 4.8–5.2 μm, G = 2.0 μm, and B = 1.3 μm (from Barnes et al. 2007a). Bottom: Volume scattering fraction was derived from the measured brightness temperature and effective dielectric constant using a global model for the physical temperature of the surface. Xanadu stands out as a region dominated by volume scattering along with a second region to the southeast. Top and bottom panels are from Janssen et al. (2009).

6.5 Compositional Constraints on Polar Lakes from RADAR Radiometry and VIMS

Although global hydrocarbon oceans covering the surface were largely ruled out before Cassini-Huygens arrived at Saturn, there still was the tantalizing expectation of discovering lakes or even regional seas composed of such liquids. It was conceivable that the Huygens probe might in fact land in one. An early 0.94 μm image acquired in mid-2005 on T5 with the Cassini Imaging Subsystem (Turtle et al. 2009) revealed a large dark feature in the south polar region (now named Ontario Lacus) that was suggestive of a lake in its dark pattern, but it was not until SAR images were acquired in mid-2006 on T16 of the north polar region did we have convincing evidence for a vast array of lake-like features north of 75° (Stofan et al. 2007 and Chapter 5). But the evidence for their being liquid was circumstantial, largely based morphological pattern and extremely low radar cross-section in SAR images. Two additional lines of evidence from the Cassini Orbiter have emerged to further characterize the composition and physical properties of the polar lakes.

Although dual-polarization RADAR radiometry is not yet available for the polar features, the microwave radiometer data can be used to establish their consistency with liquid hydrocarbons. Figure 6.32 (Janssen et al. 2009) shows a SAR mosaic of several of the largest north polar seas alongside the polar radiometer normal-incidence brightness temperature map. If liquid, the northern seas should provide the clearest case (even beyond the dunes) for the use of the Fresnell equations because their radar reflectivity is so low that their backscatter is undetectable in SAR images (Paganelli et al. 2007). The high-resolution radiometry map in Fig. 6.32 shows the radar-dark northern seas to be emissively bright (T_b ~ 88.5 K), consistent with the low dielectric constant of liquid ethane/methane mixtures (ε ~ 1.6) surrounded by terrain of lower emissivity. This dielectric constant translates to an emissivity~0.985 so their measured brightness temperature of 88.5 K to implies a physical temperature ~90 K, corresponding to an equator-to-pole temperature difference of ~2 K. More precisely, allowing for a 500-m depression of the sea surface consistent with recent Cassini RADAR altimetry and a temperature lapse rate of 1 K/km, Janssen et al. (2009) estimate an equator-to-pole difference of 2 ± 0.8 K, consistent with the 2–3 K estimates from CIRS observations (see Chapter 13). Hence the brightness temperatures of these features are perfectly consistent with open bodies of liquid ethane-methane-nitrogen as has been long conjectured to exist.

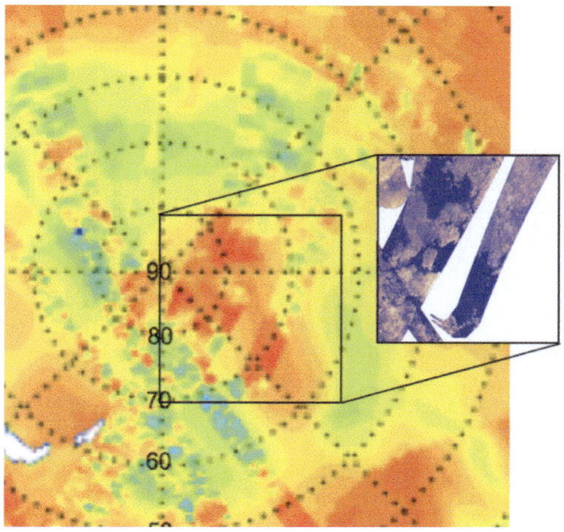

Fig. 6.32 North polar normal-incidence brightness temperature map from RADAR radiometry shown with an inset of high-resolution SAR images (Janssen et al. 2009). The northern seas exhibit higher brightness temperatures (and therefore emissivity) than their surroundings consistent with a low dielectric constant (ε ~ 1.6) expected for liquid methane/ethane seas.

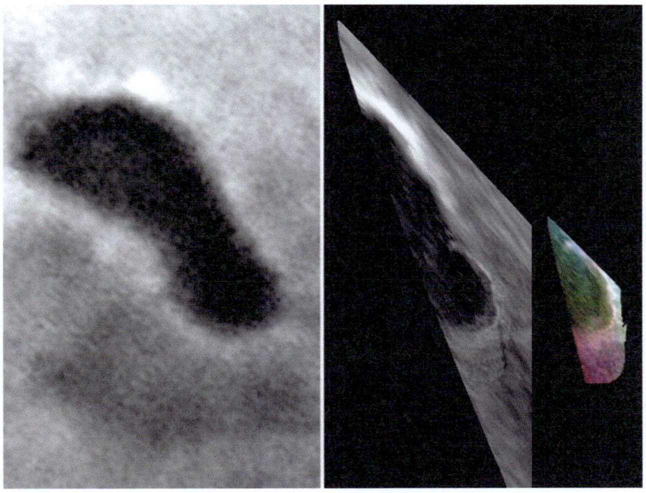

Fig. 6.33 ISS and VIMS images of the south polar lake, Ontario Lacus. The ISS image (*left, acquired mid 2005*) is compared with two VIMS images (*right, from T38*). The middle B/W image is a mosaic of four VIMS cubes and is the sum of 11 VIMS channels in the 5 μm window. To its right is a color composite of the highest resolution VIMS cube in the set of four in which R = 5 μm, G = 2.8 μm, B = 2.0 μm (Brown et al. 2008).

During a close Titan flyby (~1,100 km) on T38 VIMS observed Ontario Lacus in the south-polar region and detected spectral absorption features attributed to liquid ethane (Brown et al. 2008). Also observed with the ISS narrow angle camera in June 2005, this lake-like feature was named after Lake Ontario in North America that it resembles in both size and shape. The VIMS observations consisted of four spectral cubes (B/W 5 μm mosaic in Fig. 6.33) with emission angles varying between ~40°–80° and with a best resolution of ~500 m/pixel (color composite on far right in Fig. 6.33). The "shoreline" of Ontario Lacus is quite bright in all atmospheric windows from 2-to-5 μm relative to the very dark interior of the lake. A narrow annular area intermediate in albedo (referred to by Brown et al. as the "beach" and as the inner annulus by Barnes et al. 2009) lies just inside the bright shoreline and roughly follows its shape suggesting it was emplaced in an earlier era of a higher lake level (Barnes et al. 2009).

Calibrated I/F spectra from inside and outside the lake show strong absorption features in Titan's atmosphere (red and black I/F spectra plotted on left axis in Fig. 6.34). Subtle spectral differences do exist however within the 2-, 2.7- and 5-μm windows arising from photons scattered at the surface. To isolate these subtle spectral differences between the lake's interior and terrain around its shoreline, ratios were made between spectra from the interior of Ontario Lacus and from a nearby area outside of the lake to suppress effects of the strong atmospheric absorptions. Care was exercised to select spectra with nearly identical path length and viewing geometry to minimize the introduction of residuals in the ratios from strong atmospheric absorptions. The ratios are shown in blue for the dark lake interior and in green for the narrow annulus or beach just inside the bright shoreline (plotted on right axes). Residual features in the ratios appear in the 2, 2.8, and 5 μm atmospheric windows.

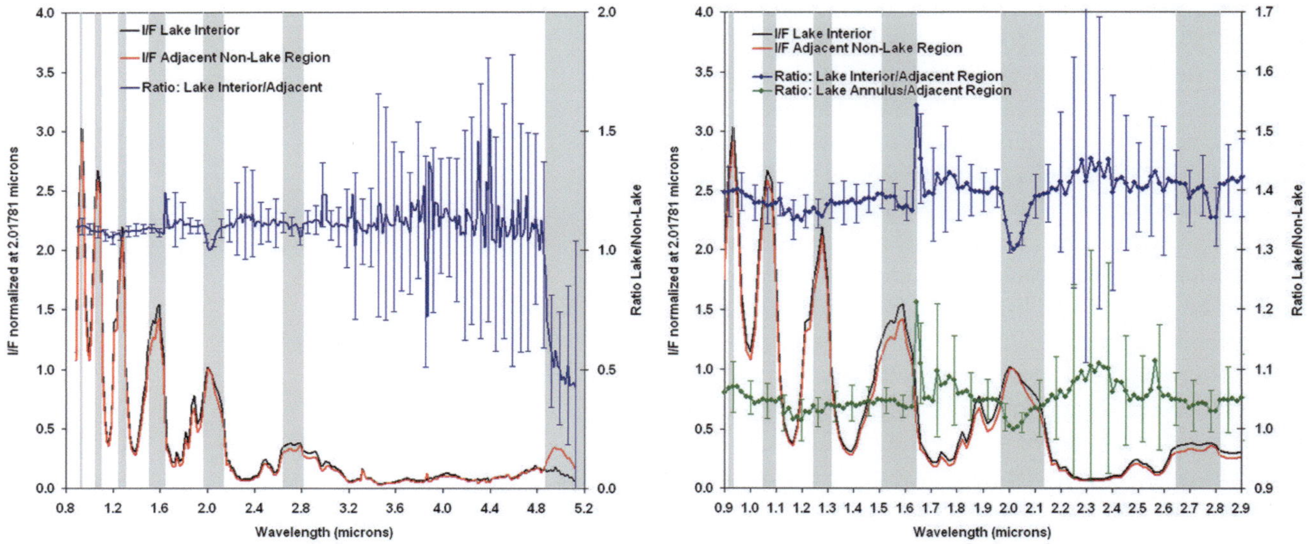

Fig. 6.34 Spectra of regions in and adjacent to Ontario Lacus. I/F spectra were normalized at 2.0178 μm (*red and black, left axis*). Ratio spectra are shown with 1-σ error bars (*blue and green, right axis*). The effectiveness of suppressing strong atmospheric bands is confirmed because the methane absorptions are effectively cancelled in the spectral ratio. The vertical, gray bands denote the spectral channels within atmospheric windows through which VIMS can resolve surface detail (Brown et al. 2008).

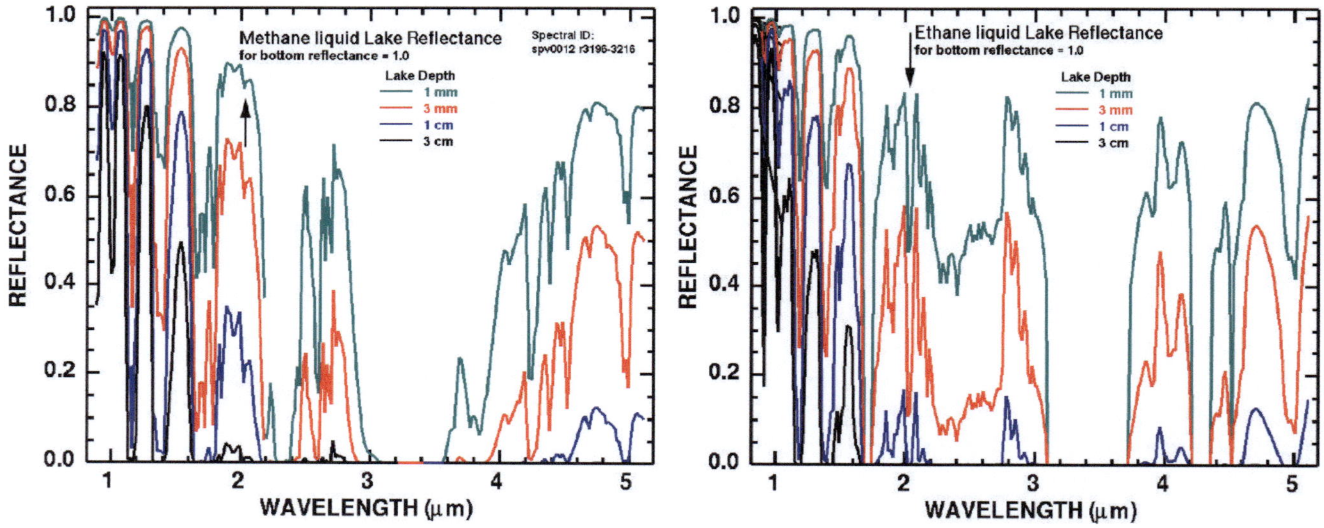

Fig. 6.35 Reflectance models for thin liquid layers of methane and ethane from laboratory absorption coefficients illustrate spectral features expected in VIMS spectra. The sharp 2.018-μm ethane absorption feature, not seen in methane, is called out by the arrows (from Clark et al. 2009b).

The spike at ~1.64 μm is due to noise. That the ratio spectra are nearly flat across the rest of the windows (shown in gray) that lie between the strong methane absorptions illustrates that the atmospheric absorptions were mostly canceled out.

Brown et al. (2008) identified four candidate features associated with the surface of the lake. In increasing wavelength, the first is a well-developed feature that occurs at 2.018 μm. It exhibits a roughly Gaussian profile, is slightly offset from the center of the 2 μm window shifted toward shorter wavelength, another factor that supports it not being a residual atmospheric feature. Relative to the lake interior, this feature shows at about half strength in the spectrum for the annulus or beach (green curve). A very weak feature centered near 2.11 μm was also tentatively identified; it shows as a depression in the long-wavelength wing of the 2.018 μm feature and forms a shallow slope in the right side of the 2 μm atmospheric window. A third possible feature, also weak, is centered near 2.79 μm. The fourth feature is a very steep drop in reflectance near 4.8 μm, that continues downward out to 5.17 μm at the limit of the VIMS spectral range.

An alkane expected to be abundant in Titan lakes and seas is liquid ethane (Lunine et al. 1983). Figure 6.35 shows two

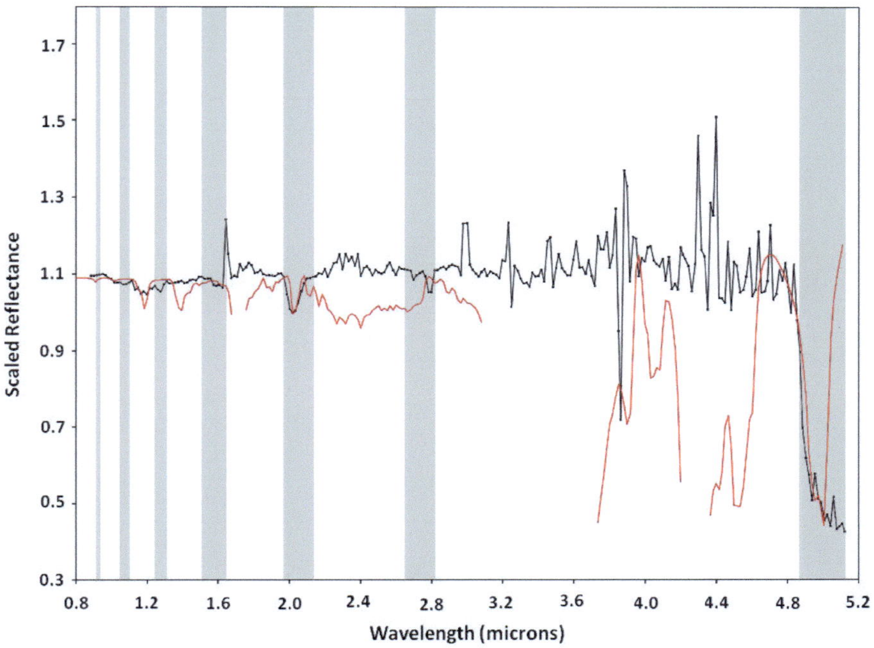

Fig. 6.36 VIMS spectral ratio of the interior of Lake Ontario to adjacent terrain compared with a model liquid ethane spectrum. The liquid ethane spectrum was calculated for a path length of 300 μm from absorption coefficients measured by Clark et al. (2009a) used to model spectra in Fig. 6.34. The vertical, gray bands denote the spectral channels within atmospheric windows through which VIMS can resolve surface detail (Brown et al. 2008).

sets of model spectra for thin liquid layers of methane and ethane that were derived from optical constants measured in the laboratory by Clark et al. (2009b). Most notable are (1) the absorption at 2.018 μm in ethane that is absent in the methane spectrum and (2) the deep absorption slope at 4.8 μm for ethane that is much weaker for methane. Both of these features coincide with the prominent features seen in the Ontario Lacus ratio spectrum (Fig. 6.34). These optical constants were used by Brown et al. (2008) to derive a synthetic ethane spectrum that is compared with the observed ratio spectrum in Figure 6.36. As evident in that figure, most of ethane's absorption bands do not fall in atmospheric windows. However three laboratory-measured bands lie in the 2 μm atmospheric window. At VIMS resolution two are unresolved and merge into the 2.018 μm feature and the third very weak band at 2.11 μm, near the long-wavelength edge of the 2 μm window, and as described above, may be evident as a shoulder on the wing of the 2.018 μm feature. The synthetic spectrum of Fig. 6.36 required a path length of only 300 μm to fit the 2.018 μm absorption. Brown et al. (2008) do not conclude this to be the depth of the lake but rather the band is simply filled in by strong atmospheric scattering in the 2-μm window. Clark et al. (2009b) suggest alternatively that the short path length points to a liquid-saturated mud flat. This seems inconsistent with SAR images that show near-zero backscatter for the north polar lakes (Stofan et al. 2007).

The other strong indicator of lake composition is the steep drop in reflectance beyond 4.8 μm. This is characteristic of alkanes in general, and ethane, propane and butane in particular (Grundy et al. 2002 and Clark et al. 2009a,b). The spectrum of liquid ethane matches the short wavelength side seen in the ratio spectrum in Fig. 6.36 and additional contributions from propane, butane and higher-order alkanes could easily explain the continued drop in Titan's reflectance with increasing wavelength. The unique collection of four VIMS cubes acquired during the T38 flyby have emission angles ranging from 40° to 80° all acquired at an incidence angle of ~65°. This allowed extrapolation of the 5-μm I/F to zero airmass and thereby estimation of the relative albedo inside and outside the lake. Although the signal precision in this window is low, scattering at 5 μm is <10% and a thin-atmosphere approximation applies. Figure 6.37 illustrates that the I/F at 5 μm for the central, dark interior of Ontario Lacus approaches zero when extrapolated to zero airmass. This analysis shows that almost all photons incident on the lake interior at this wavelength are absorbed. Brown and colleagues posited that for the lake to have a net reflectivity so close to zero, it must have a surface that is extremely smooth and free of scatterers. To eliminate any specular glint from waves, they suggested that Ontario Lacus is filled with a quiescent liquid with few scattering particles larger than a few microns. In conclusion the VIMS data provides strong evidence for the presence of liquid ethane in Ontario Lacus, with suggestion of other compounds such as propane and butane. Although the VIMS data do not speak to the presence or absence of liquid methane in Ontario Lacus, work on

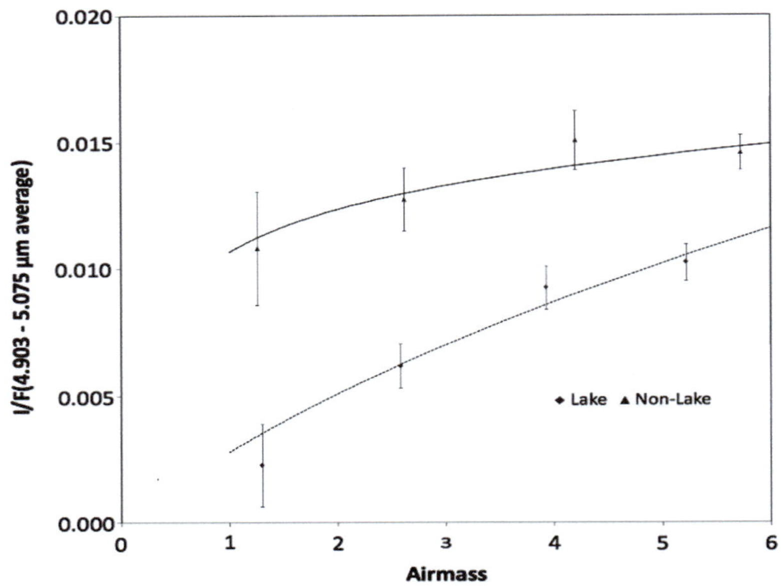

Fig. 6.37 Average brightness in the 5-µm window (I/F) versus airmass for Ontario Lacus interior (*lower*) and an adjacent region outside the lake (*upper*). Airmass or atmospheric path length is approximated by sec(z), where z is zenith or emission angle through the atmosphere (Brown et al. 2008).

the seasonal evolution of hydrocarbon lakes on Titan shows that if liquid ethane exists in these lakes, in all probability so does liquid methane (Mitri et al. 2007).

6.6 Summary

Huygens landed on what appeared to be a dry fluvial-scoured plain with a methane humidity of ~50% at the surface. However subsequent to landing the GCMS detected a high level of subsurface moisture perhaps signaling recent methane precipitation. The GCMS surface data also showed traces of ethane and possibly cyanogen, benzene, and carbon dioxide. Aerosol pyrolitic products analyzed by the ACP/GCMS suggest the aerosols may contain nitriles in addition to organics. The DISR visible-near-IR reflectance spectrum of the surface suggests a mixture of water ice and organics, although evidence for H_2O is uncertain. Huygens data suggest a surface of mostly of complex hydrocarbon and nitrile molecules forming involatile solids, possibly mixed with fine-grained water ice.

As seen in telescopic spectra, the VIMS data suggest that if water ice is mixed with surface materials it is concentrated in the dark units and is in lower abundance or even absent in bright units. This gives strong evidence that the bright units are bright hydrocarbon solids and not exposed water ice. Spectral evidence for any water ice has been challenged as spectral ratios between 2.8 µm and 2.7 µm are higher than would be expected for ice and other organics could mimic water ice in other atmospheric windows. VIMS spectral data show further that dark spectral units break into "dark blue" and "dark brown" as seen in composites with RGB = 2.0, 1.6, 1.3 µm. The dark brown units show less evidence of water ice than do the blue units and therefore appear richer in dark organics and nitriles. These brown units also correlate directly with the vast seas of SAR-dark dunes that belt the equatorial zone. If the VIMS bright materials and dark brown dune fields are both mostly hydrocarbons and/or nitriles, they must consist of physically and/or chemically different organics. SAR images do not distinguish boundaries between VIMS dark blue and bright units suggesting a model in which organic-rich thin mantles and dunes overly a dark blue substrate that may be water-rich.

Carbon dioxide is a probable component of the surface material and the GCMS data suggest trace amounts. VIMS spectra for 5 µm bright spots have been suggested as attributable to CO_2 and telescopic and VIMS studies have offered that CO_2 might explain the peculiar spectral features in the 2.7–2.8 µm region. VIMS Tui Regio spectra show an absorption feature at 4.92 µm that has been suggested at due to CO_2. However the VIMS 4.92 µm feature is significantly shifted from spectral position observed for the CO_2 in the laboratory. This shift appears larger than that expected for contaminants altering bond vibrational frequencies making the identification inconclusive. A feature in the spectrum of cyanoacetylene offers a closer spectral match to the 4.92-µm feature than does CO_2. Cyanoacetylene is a thermospheric molecule seen by both CIRS and INMS. Other VIMS absorption features have been suggested in the 4.8–5.2 µm region.

Because of the low signal precision of these features, they are very difficult to detect (unless locally concentrated as for the Tui Regio feature); a debate continues over the certainty of their existence. One such debated feature at 5.05 μm appears to occur across dark equatorial regions and most matches an absorption in benzene, a molecule tentatively identified in trace amounts at the surface by the GCMS and at altitude in greater abundance than expected by INMS. Another such feature at 4.97 μm occurs in isolated dark equatorial locals and is matched by a feature in methane and ethane. This suggests moist surfaces wetted with such liquids, perhaps consistent with GCMS detection of subsurface methane moisture.

Cassini's RADAR investigation measures microwave scattering, emission, and dielectric properties related to surface composition and physical state. Modeling of radar scatterometry for equatorial regions gives an average dielectric constant ($\varepsilon \sim 2.2$) higher than for liquid hydrocarbons and in the range for solid hydrocarbons. However the high degree of volume scattering permits models of higher-ε materials, like porous water ice and ammonia hydrate, that are mixed with organics and nitriles. The passive microwave radiometry acquired by the Cassini RADAR investigation in orthogonal polarizations yields accurate global maps of effective dielectric constant. The radiometry ε map is globally very uniform; >95% of the surface exhibits $\varepsilon \sim 1.5 \pm 0.3$. All of the microwave data preclude ubiquitous exposures of solid sheets of water ice ($\varepsilon \sim 3.1$). Sinlap crater ($\varepsilon \sim 2.5$) is a rare case that might be a concentration of water ice. The Cassini radiometry data also yield maps of brightness temperature (a proxy for emissivity) and volume scattering. The high emissivity of the dunes is consistent with uniform, coarse grains of organic material. On the global average these data suggest Titan's surface is consistent with fluffy blankets and veneers of organics. Mixtures of solid organics with porous, fractured water ice remain a possible model for the average global surface.

VIMS spectra and RADAR passive radiometry provide evidence that the polar lakes are actually liquid hydrocarbons. Microwave radiometry shows that the north polar lakes to have high brightness temperatures and therefore emissivity and low dielectric constant ($\varepsilon \sim 1.6$) relative to their surroundings; these properties are all consistent with the north polar lakes containing methane-ethane liquids. VIMS measured the surface spectrum of Ontario Lacus, a large south-polar feature first seen in an ISS image that resembles a lake in shape and albedo. Derivation of surface brightness of Ontario Lacus from VIMS at 5 μm yielded a near-zero I/F consistent with a mirror-like non-scattering surface. An important finding by VIMS was spectral absorption features most pronounced in Ontario Lacus' dark interior that match laboratory-modeled spectra for and strongly indicate the presence of ethane, probably in liquid solution with methane and nitrogen.

Acknowledgements This research was carried out under funding from the Cassini Flight Project managed by the Jet Propulsion Laboratory, Caltech for NASA. We thank Brent Archinal, Marc Buie, Dale Cruikshank, and Randy Kirk for their valuable critical input.

References

Allen M, Pinto JP, Yung YL (1980) Titan-Aerosol photochemistry and variations related to the sunspot cycle. Astrophys J Lett 242:L125

Barnes JW, Brown RH, Turtle EP, McEwen AS, Lorenz RD, Janssen M, Schaller EL, Brown ME, Buratti BJ, Sotin C, Griffith C, Clark R, Perry J, Fussner S, Barbara J, West R, Elachi C, Bouchez AH, Roe HG, Baines KH, Bellucci G, Bibring J-P, Capaccioni F, Cerroni P, Combes M, Coradini A, Cruikshank DP, Drossart P, Formisano V, Jaumann R, Langevin Y, Matson DL, McCord TB, Nicholson PD, Sicardy B (2005) A 5-micron-bright spot on Titan: evidence for surface diversity. Science 310:92–95

Barnes JW, Brown RH, Radebaugh J, Buratti BJ, Sotin C, Le Mouelic S, Rodriguez S, Turtle EP, Perry J, Clark R, Baines KH, Nicholson PD (2006) Cassini observations of flow-like features in western Tui Regio, Titan. Geophys Res Lett 33:doi:10.1029/2006GL026843

Barnes JW, Brown RH, Soderblom L, Buratti BJ, Sotin C, Rodriguez S, Le Mouèlic S, Baines KH, Clark R, Nicholson P (2007a) Global-scale surface spectral variations on Titan seen from Cassini/VIMS. Icarus 186:242–258

Barnes JW, Radebaugh J, Brown RH, Wall S, Soderblom L, Burr D, Sotin C, Le Mouèlic S, Rodriguez S, Buratti BJ, Clark R, Baines KH, Jaumann R, Nicholson PD, Kirk RL, Lopes R, Lorenz RD, Mitchell K, Wood CA, and the Cassini RADAR Team (2007b) Near infrared spectral mapping of Titan's mountains and channels. J Geophys Res 112:doi:10.1029/2007JE002932. E11006

Barnes JW, Brown RH, Soderblom L, Sotin C, Mouèlic Le, Stèphane R, Sebastien J, Ralf B, Ross A, Buratti BJ, Pitman K, Baines KH, Clark R, Nicholson P (2008) Spectroscopy, morphometry, and photoclinometry of Titan's dune fields from Cassini/VIMS. Icarus 195:400–414

Barnes JW, Brown RH, Soderblom JM, Soderblom LA, Jaumann R, Jackson B, Le Mouélic S, Sotin C, Buratti BJ, Pitman KM, Baines KH, Clark RN, Nicholson PD, Turtle EP, Perry J (2009) Shoreline features of Titan's Ontario Lacus from Cassini/VIMS observations. Icarus 201:217–225

Bernard J-M, Quirico E, Brissaud O, Montagnac G, Reynard B, McMillan P, Coll P, Nguyen M-J, Raulin F, Schmitt B (2006) Reflectance spectra and chemical structure of Titan's tholins: application to the analysis of Cassini Huygens observations. Icarus 185:301–307

Bernstein MP, Cruikshank DP, Sandford SA (2005) Near-infrared laboratory spectra of solid H2O/CO2 and CH3OH/CO2 ice mixtures. Icarus 179:527–534

Biemann K (2006) Complex organic matter in Titan's aerosols? Nature 444:E6

Broadfoot AL, Sandel BR, Shemansky DE, Holberg JB, Smith GR, Strobel DF, McConnell JC, Kumar S, Hunten DM, Atreya SK, Donahue TM, Moos HW, Bertaux JL, Blamont JE, Pomphrey RB, Linick S (1981) Extreme ultraviolet observations from Voyager 1 encounter with Saturn. Science 212:206–211

Brown RH, Cruikshank DP, Tokunaga AT, Smith RG, Clark RN (1988) Search for volatiles on icy satellites. I–Europa. Icarus 74:262–271

Brown RH, Baines KH, Bellucci G, Bibring J-P, Buratti BJ, Capaccioni F, Cerroni P, Clark RN, Coradini A, Cruikshank DP, Drossart P, Formisano V, Jaumann R, Langevin Y, Matson DL, McCord TB, Mennella V, Miller E, Nelson RM, Nicholson D, Sicardy B, Sotin C

(2004) The Cassini Visual and Infrared Mapping Spectrometer (VIMS) investigation. Space Sci Rev 115:111–168

Brown RH, Baines KH, Bellucci G, Buratti BJ, Capaccioni F, Cerroni P, Clark RN, Coradini A, Cruikshank DP, Drossart P, Formisano V, Jaumann R, Langevin Y, Matson DL, McCord TB, Mennella V, Nelson RM, Nicholson PD, Sicardy B, Sotin C, Baugh N, Griffith CA, Hansen GB, Hibbitts CA, Momary TW, Showalter MR (2006a) Observations in the Saturn system during approach and orbital insertion, with Cassini's visual and infrared mapping spectrometer (VIMS). Astron Astrophys 446:707–716

Brown RH, Clark RN, Buratti BJ, Cruikshank DP, Barnes JW, Mastrapa RME, Bauer J, Newman S, Momary T, Baines KH, Bellucci G, Capaccioni F, Cerroni P, Combes M, Coradini A, Drossart P, Formisano V, Jaumann R, Langevin Y, Matson DL, McCord TB, Nelson RM, Nicholson PD, Sicardy B, Sotin C (2006b) Composition and physical properties of enceladus' surface. Science 311:1425–1428

Brown RH, Soderblom LA, Soderblom JM, Clark RN, Jaumann R, Barnes JW, Sotin C, Buratti B, Baines KH, Nicholson PD (2008) The identification of liquid ethane in Titan's Ontario Lacus. Nature 454:607–610

Buratti BJ, Cruikshank DP, Brown RH, Clark RN, Bauer JM, Jaumann R, McCord TB, Simonelli DP, Hibbitts CA, Hansen GB, Owen TC, Baines KH, Bellucci G, Bibring J-P, Capaccioni F, Cerroni P, Coradini A, Drossart P, Formisano V, Langevin Y, Matson DL, Mennella V, Nelson RM, Nicholson PD, Sicardy B, Sotin C, Roush TL, Soderlund K, Muradyan A (2005) Cassini visual and infrared mapping spectrometer observations of Iapetus: detection of CO2. Astrophys J 622:L149–L152

Campbell D, Black G, Carter L, Ostro S (2003) Radar evidence for liquid surfaces on Titan. Science 302:431–434

Clark RN, Swayze GA, Livo KE, Kokaly RF, Sutley SJ, Dalton JB, McDougal RR, Gent CA (2003) Imaging spectroscopy: earth and planetary remote sensing with the USGS Tetracorder and expert systems. J Geophys Res 108:5-1–5-43

Clark RN, Brown RH, Jaumann R, Cruikshank DP, Nelson RM, Buratti BJ, McCord TB, Lunine J, Baines KH, Bellucci G, Bibring J-P, Capaccioni F, Cerroni P, Coradini A, Formisano V, Langevin Y, Matson DL, Mennella V, Nicholson PD, Sicardy B, Sotin C, Hoefen TM, Curchin JM, Hansen G, Hibbits K, Matz K-D (2005) Compositional maps of Saturn's moon Phoebe from imaging spectroscopy. Nature 435:66–69

Clark RN, Curchin JM, Brown RH, Cruikshank DP, Jaumann R, Lunine J, Hoefen TM, Baines KH, Buratti BJ, Barnes J, Nicholson PD (2006b) Detection of widespread aromatic and aliphatic hydrocarbon deposits on Titan's surface observed by VIMS, AGU Fall Meeting Proceedings, Abstract P11A-03

Clark RN, Curchin JM, Brown RH, Waite JH, Cruikshank DP, Jaumann R, Lunine J, Hoefen TM, Cravens TE, Yelle RV, Vuitton V, Baines KH, Buratti BJ, Barnes J, McCord TB, Nicholson PD (2006a) Detection of widespread aromatic and aliphatic hydrocarbon deposits on Titan's surface observed by VIMS and excess benzene observed in Titan's thermosphere observed by INMS. Bull Am Astron Soc 38:574. Abstract 48-04

Clark RN, Curchin JM, Barnes J, Jaumann R, Soderblom L, Cruikshank DP, Lunine J, Stephan K, Hoefen TH, Lemouélic S, Sotin C, Baines KH, Buratti B, Nicholson P (2009b) Detection and mapping of hydrocarbon deposits on Titan, submitted to Icarus

Clark RN, Curchin JM, Hoefen TM, Swayze GA (2009b) Reflectance spectroscopy of organic compounds i: alkanes. J Geophys Res 114:E03001

Coustenis A, Lellouch E, Maillard JP, McKay CP (1995) Titan's surface, composition and variability from the near-infrared albedo. Icarus 118:87–104

Coustenis A, Gendron E, Lai O, Véran J-P, Woillez J, Combes M, Vapillon L, Fusco T, Mugnier L, Rannou (2001) Images of Titan at 1.3 and 1.6 μm with adaptive optics at the CFHT. Icarus 154:501–515

Coustenis A, Salama A, Schultz B, Ott S, Lellouch E, Encrenaz T, Gautier D, Feuchtgruber H (2003) Titan's atmosphere from ISO mid-infrared spectroscopy. Icarus 161:383–403

Coustenis A, Hirtzig M, Gendron E, Drossart P, Lai O, Combes M, Negrão A (2005) Maps of Titan's surface from 1 to 2.5 μm. Icarus 177:1, 89–105

Coustenis A, Negrão A, Salama A, Schulz B, Lellouch E, Rannou P, Drossart P, Encrenaz T, Schmitt B, Boudon V, Nikitin A (2006) Titan's 3-micron spectral region from ISO high-resolution spectroscopy. Icarus 180:176–185. doi:10.1016/j.icarus.2005.08.007

Cruikshank DP, Dalton JB, Ore CM, Dalle BJ, Stephan K, Filacchione G, Hendrix AR, Hansen CJ, Coradini A, Cerroni P, Tosi F, Capaccioni F, Jaumann R, Buratti BJ, Clark RN, Brown RH, Nelson RM, McCord TB, Baines KH, Nicholson PD, Sotin C, Meyer AW, Bellucci G, Combes M, Bibring J-P, Langevin Y, Sicardy B, Matson DL, Formisano V, Drossart P, Mennella V (2007) Surface composition of Hyperion. Nature 448:54–56

Curchin JM, Shaffer CJ, Clark RN, Mc Mahon RJ, Hoefen TM (2009) Reflectance spectroscopy of cyanoacetylene (HC_3N), submitted to Icarus

Danielson RE, Caldwell J, Larach DR (1973) An inversion in the atmosphere of Titan. Icarus 20:437

de Pater I, Ádámkovics M, Bouchez AH, Brown ME, Gibbard SG, Marchis F, Roe HG, Schaller EL, Young E (2006) Titan imagery with Keck adaptive optics during and after probe entry. J Geophys Res 111:E7, E07S05, 1–16

Elachi C, Allison MD, Borgarelli L, Encrenaz P, Im E, Janssen MA, Johnson WTK, Kirk RL, Lorenz RD, Lunine JI, Muhleman DO, Ostro SJ, Picardi G, Posa F, Rapley CG, Roth LE, Seu R, Soderblom LA, Vetrella S, Wall SD, Wood CA, Zebker HA (2004) The Cassini Titan RADAR Mapper. Space Sci Rev 115:71–110

Eshelman VR, Lindal GF, Tyler GL (1983) Is Titan wet or dry? Science 221:53–55

Filacchione G, Capaccioni F, McCord TB, Coradini A, Cerroni P, Bellucci G, Tosi F, DíAversa E, Formisano V, Brown RH, Baines KH, Bibring JP, Buratti BJ, Clark RN, Combes M, Cruikshank DP, Drossart P, Jaumann R, Langevin Y, Matson DL, Mennella V, Nelson RM, Nicholson PD, Sicardy B, Sotin C, Hansen G, Hibbitts K, Showalter M, Newman S (2007) Saturn's icy satellites investigated by Cassini-VIMS I. Full-disk properties: 350–5100 nm reflectance spectra and phase curves. Icarus 186:259–290

Flasar FM (1983) Oceans on Titan? Science 221:55–57

Fulchignoni M, Ferri F, Angrilli F, Ball AJ, Bar-Nun A, Barucci MA, Bettanini C, Bianchini G, Borucki W, Colombatti G, Coradini M, Coustenis A, Debei S, Falkner P, Fanti G, Flamini E, Gaborit V, Grard R, Hamelin M, Harri AM, Hathi B, Jernej I, Leese MR, Lehto A, Lion Stoppato F, López-Moreno JJ, Mäkinen T, McDonnell JAM, McKay CP, Molina-Cuberos G, Neubauer FM, Pirronello V, Rodrigo R, Saggin B, Schwingenschuh K, Seiff A, Simões F, Svedhem H, Tokano T, Towner MC, Trautner R, Withers P, Zarnecki JC (2005) In situ measurements of the physical characteristics of Titan's environment. Nature 438:785–791

Goldstein RM, Jurgens RF (1992) DSN observations of Titan TDA PR 42–109. Jet Propulsion Laboratory, Pasadena CA

Griffith CA (1993) Evidence for surface heterogeneity on Titan. Nature 364:511–514

Griffith, CA, McKay, CP, Ferri, F (2008) Titan's Tropical Storms in an Evolving Atmosphere. Astrophys. J. 687: L41–L44.

Griffith CA, Owen T, Wagener R (1991) Titans surface and troposphere, investigated with ground-based, near-infrared observations. Icarus 93:362–378

Griffith CA, Owen T, Geballe TR, Rayner J, Rannou (2003) Evidence for the exposure of water ice on Titan's surface. Science 300:628–630

Grossman AW, Muhleman DO (1992) Observations of Titan's radio light-curve at 3.5 cm, Bull Amer Astron Soc 24:954

Grundy WM, Schmitt B (1998) The temperature-dependent near-infrared absorption spectrum of hexagonal H_2O ice. J Geophys Res 103:E11, 25809–25822

Grundy WM, Schmitt B, Quirico E (2002) The temperature-dependent spectrum of methane ice I between 0.7 and 5 mm and opportunities for near-infrared remote thermometry. Icarus 155:486–496

Hanel R, Conrath B, Flasar FM, Kunde V, Maguire W, Pearl J, Pirraglia J, Samuelson R, Herath L, Allison M, Cruikshank D, Gautier D, Gierasch P, Horn L, Koppany R, Ponnamperuma C (1981) Infrared observations of the saturnian system from Voyager 1. Science 212:192–200

Hansen GB (2005) Ultraviolet to near-infrared absorption spectrum of carbon dioxide ice from 0.174 to 1.8 μm. J Geophys Res 110:doi:10.1029/2005JE002531. E11003

Hartung M, Herbst TM, Dumas C, Coustenis A (2006) Limits to the abundance of surface CO2 ice on Titan. J Geophys Res 111:E07S09

Hirtzig M, Coustenis A, Gendron E, Drossart P, Hartung M, Negrão A, Rannou P, Combes M (2007) Titan, atmospheric and surface features as observed with Nasmyth adaptive optics system near-infrared imager and spectrograph at the time of the Huygens mission. J Geophys Res 112:E2, E02S91, 1–12

Hunten DM (1974) Scientific summary. In: Hunten DM (ed) The atmosphere of Titan. Proc. NASA AMES Conf., Moffett Field, CA, NASA Spec. Publ., SP-340, pp 4–8

Hunten DM, Tomasko MG, Flasar FM, Samuelson RE, Strobel DF, Stevenson DJ (1984) In: Gehrels T, Matthews MS (eds) Titan, in Saturn. University of Arizona Press, Tucson, pp 671–787

Imanaka H, Khare BN, Elsila JE, Bakes ELO, McKay CP, Cruikshank DP, Sugita S, Matsui T, Zare RN (2004) Laboratory experiments of Titan tholin formed in cold plasma at various pressures, implications for nitrogen-containing polycyclic aromatic compounds in Titan haze. Icarus 168:344–366

Imanaka H, Khare BN, McKay CP, Cruikshank DP (2005) Complex refractive indices of tholins produced from various initial gas mixtures and formation pressures, Implications for Titan, the early Earth, and the outer solar system bodies. Bull Amer Astron Soc 37:772

Israël G, Cabane J-F, Brun G, Niemann S, Way H, Riedler W, Steller M, Raulin F, Coscia D (2002) Huygens probe aerosol collector pyrolyser experiment. Space Sci Rev 104:435–466

Israël G, Szopa C, Raulin F, Cabane M, Niemann HB, Atreya SK, Bauer SJ, Brun J-F, Chassefière E, Coll P, Condé E, Coscia D, Hauchecorne A, Millian P, Nguyen M-J, Owen T, Riedler W, Samuelson RE, Siguier J-M, Steller M, Sternberg R, Vidal-Madjar C (2005) Complex organic matter in Titan's atmospheric aerosols from in situ pyrolysis and analysis. Nature 438:796–799

Jacquemart D, Lellouch E, Bézard B, de Bergh C, Coustenis A, Lacome N, Schmitt B, Tomasko M (2008) New laboratory measurements of CH_4 in Titan's conditions and a reanalysis of the DISR near-surface spectra at the Huygens landing site. Planetary Space Sci 56:613–623

Janssen MA, Lorenz RD, West R, Paganelli F, Lopes RM, Kirk RL, Elachi C, Wall SD, Johnson WTK, Anderson Y, Boehmer RA, Callahan P, Gim Y, Hamilton GA, Kelleher KD, Roth L, Stiles B, Le Gall A, the Cassini RADAR Team (2009) Titan's surface at 2.2-cm wavelength imaged by the cassini radar radiometer: calibration and first results. Icarus 200:222–239

Kargel JS, Croft SK, Lunine JI, Lewis JS (1991) Rheological properties of ammonia-water liquids and crystal-liquid slurries – planetological applications. Icarus 89:93–112

Keller HU, Grieger B, Küppers M, Schröder SE, Skorov YV, Tomasko MG (2008) The properties of Titan's surface at the Huygens landing site from DISR observations. Planetary Space Sci 56:728–752

Khare BN, Sagan C, Arakawa ET, Suits F, Callcott TA, Williams MW (1984) Optical constants of organic tholins produced in a simulated Titanian atmosphere – from soft x-ray to microwave frequencies. Icarus 60:127–137

Kossaki KJ, Lorenz RD (1996) Hiding Titan's ocean: densification and hydrocarbon storage in an icy regolith, Planet. Space Sci 44:1029–1037

Kress ME, McKay CP (2004) Formation of methane in comet impacts: implications for Earth, Mars, and Titan. Icarus 168:475–483

Kuiper GP (1944) A satellite with an atmosphere. Astrophys J 100:378–383

Lara LM, Lorenz RD, Rodrigo R (1994) Liquids and solids on the surface of Titan, results of a new photochemical model. Planet Space Sci 42:5–14

Lellouch E, Schmitt B, Coustenis A, Cuby J-G (2004) Titan's 5-micron lightcurve. Icarus 168:209–214

Lemmon MT, Karkoschka E, Tomasko M (1993) Titan's rotation: surface feature observed. Icarus 103:329–332

Lemmon MT, Karkoschka K, Tomasko M (1995) Titan's rotational lightcurve. Icarus 113:27–38

Lewis JS (1971) Satellites of the outer planets: their physical and chemical nature. Icarus 15:174–185

Lindal GF, Wood GE, Hotz HB, Sweetnam DN, Eshleman VR, Tyler GL (1983) The atmosphere of Titan – an analysis of the Voyager 1 radio occultation measurements. Icarus 53:348–363

Lopes R, Mitchell KL, Stofan ER, Lunine JI, Lorenz R, Paganelli F, Kirk RL, Wood CA, Wall SD, Robshaw LE, Fortes AD, Neish CD, Radebaugh J, Reffet E, Ostro SJ, Elachi C, Allison MD, Anderson Y, Boehmer R, Boubin G, Callahan P, Encrenaz P, Flamini E, Francescetti G, Gim Y, Hamilton G, Hensley S, Janssen MA, Johnson WT, Kelleher K, Muhleman DO, Ori G, Orosei R, Picardi G, Posa F, Roth LE, Seu R, Shaffer S, Soderblom LA, Stiles B, Vetrella S, West RD, Wye L, Zebker HA (2006) Cryovolcanic features on Titan's surface as revealed by the Cassini Titan radar mapper. Icarus 186:395–412

Lorenz RD (1996) Pillow lava on Titan, expectations and constraints on cryovolcanic processes. Planet Space Sci 44:1021–1028

Lorenz RD (1998) Preliminary measurements of the cryogenic dielectric properties of water–ammonia ices: implications for radar observations of icy satellites. Icarus 136:344–348

Lorenz RD, Lunine JI (1997) Titan's surface reviewed: the nature of bright and dark terrain. Planet Space Sci 45:981–992

Lorenz RD, Lunine JI (2005) Titan's surface before Cassini. Planet Space Sci 53:557–576

Lorenz RD, Biolluz G, Encrenaz P, Janssen MA, West RD, Muhleman DO (2003) Cassini RADAR: prospects for Titan surface investigations using the microwave radiometer. Planet Space Sci 51:353–364

Lorenz RD, Wall S, Radebaugh J, Boubin G, Reffet E, Janssen M, Stofan E, Lopes R, Kirk R, Elachi C, Lunine J, Mitchell K, Paganelli F, Soderblom L, Wood C, Wye L, Zebker H, Anderson Y, Ostro S, Allison M, Boehmer R, Callahan P, Encrenaz P, Ori GG, Francescetti G, Gim Y, Hamilton G, Hensley S, Johnson W, Kelleher K, Muhleman D, Picardi G, Posa F, Roth L, Seu R, Shaffer S, Stiles B, Vetrella S, Flamini E, West R (2006a) The sand seas of Titan Cassini RADAR observations of longitudinal dunes. Science 312:724–727

Lorenz RD, Niemann HB, Harpold DN, Way SH, Zarnecki JC (2006b) Titan's damp ground: constraints on Titan surface thermal properties from the temperature evolution of the Huygens GCMS inlet. Meteoritics Planet Sci 41:1705–1714

Lorenz RD, Callahan S, Gim Y, Alberti G, Flamini E, Seu R, Picardi G, Orosei R, Zebker H, Lunine J, Hamilton G, Hensley S, Johnson WTK, Schaffer S, Wall S, West R, Francescetti G (2007) Titan's shape, radius and landscape from Cassini radar altimetry. Lunar Planetary Sci XXXVIII LPI Cont No 1338:1329

Lunine JI (1993) Does Titan have an ocean – a review of current understanding of Titans surface. Rev Geophys 31:133–149

Lunine JI, Stevenson DJ (1987) Clathrate and ammonia hydrates at high pressure: application to the origin of methane on Titan. Icarus 70:61–77

Lunine JI, Stevenson DJ, Yung YL (1983) Ethane ocean on Titan. Science 222:1229–1230

McCord TB, Hansen GB, Buratti BJ, Clark RN, Cruikshank DP, D'Aversa E, Griffith CA, Baines EKH, Brown RH, Dalle Ore CM, Filacchione G, Formisano V, Hibbitts CA, Jaumann R, Lunine JI, Nelson RM, Sotin C, the Cassini VIMS Team (2006) Composition of Titan's surface from Cassini VIMS. Planet Space Sci 54: 1524–1539

McCord TB, Hayne P, Combe J-P, Hansen GB, Barnes JW, Rodriguez S, Le Mouélic S, Baines EKH, Buratti BJ, Sotin C, Nicholson P, Jaumann R, Nelson R, The Cassini Vims Team (2008) Titan's surface: search for spectral diversity and composition using the Cassini VIMS investigation. Icarus 194:212–242

McKay CP, Coustenis A, Samuelson RE, Lemmon MT, Lorenz RD, Cabane M, Rannou P, Drossart P (2001) Physical properties of the organic aerosols and clouds on Titan. Planet Space Sci 49:79–99

Mitri G, Showman AP, Lunine JI, Lorenz RD (2007) Hydrocarbon lakes on Titan. Icarus 186:385–394

Mouélic Le, Stéphane P, Philippe J, Michael A, Barnes JW, Rodriguez S, Sotin C, Brown RH, Baines KH, Buratti BJ, Clark RN, Crapeau M, Encrenaz PJ, Jaumann R, Geudtner D, Paganelli F, Soderblom L, Tobie G, Wall S (2008) Mapping and interpretation of Sinlap crater on Titan using Cassini VIMS and RADAR data. J Geophys Res 113:4003

Muhleman DO, Grossman AO, Butler BJ, Slade MA (1990) Radar reflectivity of Titan. Science 248:975–980

Muhleman DO, Grossman AO, Butler BJ (1995) Radar investigation of Mars, Mercury, and Titan. Annu Rev Earth Planet Sci 23:337–374

Negrão A, Coustenis A, Lellouch E, Maillard J-P, Rannou P, Schmitt B, McKay CP, Boudon V (2006) Titan's surface albedo variations over a Titan season from near-infrared CFHT/FTS spectra. Planet Space Sci 54:1225–1246

Negrão A, Hirtzig M, Coustenis A, Gendron E, Drossart P, Rannou P, Combes M, Boudon V (2007) The 2-μm spectroscopy of Huygens probe landing site on Titan with very large telescope/nasmyth adaptive optics system near-infrared imager and spectrograph. J Geophys Res 112:E2, E02S92, 1–14

Nelson RM, Kamp LW, Matson DL, Irwin PGJ, Baines KH, Boryta MD, Leader FE, Jaumann R, Smythe WD, Sotin C, Clark RN, Cruikshank DP, Drossart P, Pearl JC, Hapke BW, Lunine J, Combes M, Bellucci G, Bibring J-P, Capaccioni F, Cerroni P, Coradini A, Formisano V, Filacchione G, Langevin RY, McCord TB, Mennella V, Nicholson PD, Sicardy B (2009) Saturn's Titan: surface change, ammonia, and implications for atmospheric and tectonic activity. Icarus 199:429–441

Niemann HB, Atreya SK, Bauer SJ, Biemann K, Block B, Carignan GR, Donahue TM, Frost RL, Gautier D, Haberman JA, Harpold D, Hunten DM, Israël G, Lunine JI, Mauersberger K, Owen TC, Raulin F, Richards JE, Way SH (2002) The gas chromatograph mass spectrometer for the Huygens probe. Space Sci Rev 104:553–591

Niemann HB, Atreya SK, Bauer SJ, Carignan GR, Demick JE, Frost RL, Gautier D, Haberman JA, Harpold DN, Hunten DM, Israël G, Lunine JI, Kasprzak WT, Owen TC, Paulkovich M, Raulin F, Raaen E, Way SH (2005) The composition of Titan's atmosphere from the GCMS on the Huygens probe, and implications for the origin of nitrogen and methane. Nature 438:779–784

Paganelli F, Janssen MA, Stiles B, West R, Lorenz RD, Lunine JI, Wall SD, Callahan P, Lopes RM, Stofan E, Kirk RL, Johnson WTK, Roth L, Elachi C, Team TR (2007) Titan's surface from the Cassini Radar SAR and high resolution radiometry data of the first five flybys. Icarus 191:211–222

Porco CC, Baker E, Barbara J, Beurle K, Brahic A, Burns JA, Charnoz S, Cooper N, Dawson DD, Genio D, Anthony D, Denk T, Dones L, Dyudina U, Evans MW, Fussner S, Giese B, Grazier K, Helfenstein P, Ingersoll AP, Jacobson RA, Johnson TV, McEwen A, Murray CD, Neukum G, Owen WM, Perry J, Roatsch T, Spitale J, Squyres S, Thomas P, Tiscareno M, Turtle EP, Vasavada AR, Veverka J, Wagner R, West R (2005) Imaging of Titan from the Cassini spacecraft. Nature 434:159–168

Radebaugh J, Lorenz RD, Kirk RL, Lunine JI, Stofan ER, Lopes RMC, Wall SD, the Cassini Radar Team (2008) Mountains on Titan observed by Cassini Radar. Icarus 192:77–91

Ramirez SI, Coll P, da Silva A, Navarro-González R, Lafait J, Raulin F (2002) Complex refractive index of Titan's aerosol analogues in the 200–900 nm domain. Icarus 156:515–529

Raulin F (1985) Chimie organique dans Titan: experiences de simulation en laboratoire et speculations. In The Atmospheres of Saturn and Titan. Eur. Space Agency Spec. Publ., SP-241, pp 161–173

Raulin F (1987) Organic chemistry in the oceans of Titan. Adv Space Res 7:71–81

Rodriguez S, Le Mouélic S, Sotin C, Clénet H, Clark RN, Buratti B, Brown RH, McCord TB, Nicholson D, Baines KH, the VIMS Science Team (2006) Cassini/VIMS hyperspectral observations of the HUYGENS landing site on Titan. Planet Space Sci 54:1510–1523

Sagan C (1974) Organic chemistry in the atmosphere. In Hunten DM (ed) The Atmosphere of Titan. NASA Spec. Publ., SP-340, pp 134–141

Sagan C, Dermott SF (1982) The tide in the seas of Titan. Nature 300:731–733

Sagan C, Thompson WR, Khare BN (1992) Titan: a laboratory for prebiological organic chemistry. Account Chem Res 25:286–292

Samuelson RE, Hanel RA, Kunde VG, Maguire WC (1981) Mean molecular weight and hydrogen abundance of Titan's atmosphere. Nature 292:688–693

Schröder SE, Keller HU (2008) The reflectance spectrum of Titan's surface at the Huygens landing site determined by the descent imager/spectral radiometer. Planetary Space Sci 56:753–769

Smith H, Lemmon MT, Lorenz RD, Sromovsky LA, Caldwell JJ, Allison MD (1996) Titan's surface revealed by HST imaging. Icarus 119:336–349

Soderblom LA, Kirk RL, Lunine JI, Anderson JA, Baines KH, Barnes JW, Barrett JM, Brown RH, Buratti BJ, Clark RN, Cruikshank DP, Elachi C, Janssen MA, Jaumann R, Karkoschka E, Mouélic SLe, Lopes RM, Lorenz RD, McCord TB, Nicholson PD, Radebaugh J, Rizk B, Sotin C, Stofan ER, Sucharski TL, Tomasko MG, Wall SD (2007) Correlations between Cassini VIMS spectra and RADAR SAR images: implications for Titan's surface composition and the character of the Huygens Probe landing site. Planetary Space Sci 55:2025–2036

Stevenson DJ (1992) The interior of Titan. Proceedings of the Symposium on Titan, ESA SP-338. ESA, Noordwijk, The Netherlands, pp 29–33

Stofan ER, Elachi C, Lunine JI, Lorenz RD, Stiles B, Mitchell KL, Ostro S, Soderblom L, Wood C, Zebker H, Wall S, Janssen M, Kirk R, Lopes R, Paganelli F, Radebaugh J, Wye L, Anderson Y, Allison M, Boehmer R, Callahan P, Encrenaz P, Flamini E, Francescetti G, Gim Y, Hamilton G, Hensley S, Johnson WTK, Kelleher K, Muhleman D, Paillou P, Picardi G, Posa F, Roth L, Seu R, Shaffer S, Vetrella S, West R (2007) The lakes of Titan. Nature 445:61–64

Strobel DF (1974) The photochemistry in the atmosphere of Titan. Icarus 21:466–470

Swayze GA, Clark RN, Goetz AFH, Chrien TG, Gorelick NS (2003) Effects of spectrometer band pass, sampling, and signal-to-noise ratio on spectral identification using the Tetracorder algorithm. J Geophys Res 108:9-1–9-30

Thompson WR, Squyres SW (1990) Titan and other icy satellites: dielectric properties of constituent materials and implications for radar sounding. Icarus 86:336–354

Tobie G, Grasset O, Lunine JI, Mocquet A, Sotin C (2005) Titan's internal structure inferred from a coupled thermal-orbital model. Icarus 175:496–502

Tobie G, Lunine J, Sotin C (2006) Episodic outgassing as the origin of atmospheric methane on Titan. Nature 440:61–64

Tokano T, McKay CP, Neubauer FM, Atreya SK, Ferri F, Fulchignoni M, Niemann HB (2006) Methane drizzle on Titan. Nature 442:432–435

Tomasko MG, Buchhauser D, Bushroe M, Dafoe LE, Doose LR, Eibl A, Fellows C, Farlane EM, Prout GM, Pringle MJ, Rizk B, See C, Smith H, Tsetsenekos K (2002) The Descent Imager/Spectral Radiometer (DISR) experiment on the Huygens entry probe of Titan. Space Sci Rev 104:469–551

Tomasko MG, Archinal B, Becker T, Bézard B, Bushroe M, Combes M, Cook D, Coustenis A, de Bergh C, Dafoe LE, Doose L, Douté S, Eibl A, Engel S, Gliem F, Grieger B, Holso K, Howington-Kraus E, Karkoschka E, Keller HU, Kirk R, Kramm R, Küppers M, Lanagan P, Lellouch E, Lemmon M, Lunine J, McFarlane E, Moores J, Prout GM, Rizk B, Rosiek M, Rueffer P, Schröder SE, Schmitt B, See C, Smith P, Soderblom L, Thomas N, West R (2005) Rain, winds and haze during the Huygens probe's descent to Titan's surface. Nature 438:765–778

Trafton L (1972) On the possible detection of H_2 in Titan's atmosphere. Ap J 175:285–293

Tran BN, Joseph JC, Ferris JP, Persans PD, Chera JJ (2003) Simulation of Titan haze formation using a photochemical flow reactor – the optical constants of the polymer. Icarus 165:379–390

Turtle EP, Perry JE, McEwen AS, DelGenio AD, Barbara J, West RA, Dawson DD, Porco CC (2009) Cassini imaging of Titan's high-latitude lakes, clouds, and south-polar surface changes. Geophys Res Lett 36:L02204

Tyler GL, Eshleman VR, Anderson JD, Levy GS, Lindal GF, Wood GE, Croft TA (1981) Radio Science investigations of the Saturn system with Voyager 1: preliminary results. Science 212:201–206

West R, Anderson Y, Boehmer R, Borgarelli L, Callahan P, Elachi C, Gim Y, Hamilton G, Hensley S, Janssen M, Johnson WTK, Kelleher K, Lorenz R, Ostro S, Roth L, Shaffer S, Stiles B, Wall S, Wye L, Zebker H (2009) Cassini RADAR sequence planning and instrument performance, IEEE Trans Geosci Remote Sensing. 47:1775–1795

White TL, Cogdell JR (1973) Lunar polarization studies at 3.1 mm wavelength. Moon 6:235–249

Wye LC, Zebker HA, Ostro SJ, West RD, Gim Y, Lorenz RD, the Cassini RADAR Team (2007) Electrical properties of Titan's surface from Cassini RADAR scatterometer measurements. Icarus 188:367–385

Yarger J, Lunine JI, Burke M (1993) Calorimetric studies of the ammonia-water system with application to the outer solar system. J Geophys Res 98:no. E7, 13109–13117

Yung YL, Allen M, Pinto JP (1984) Photochemistry of the atmosphere of Titan – comparison between model and observations. Astrophys J Suppl Ser 55:465–506

Zarnecki JC, Leese MR, Hathi B, Ball AJ, Hagermann A, Towner MC, Lorenz RD, McDonnell JAM, Green SF, Patel MR, Ringrose TJ, Rosenberg PD, Atkinson KR, Paton MD, Banaszkiewicz M, Clark BC, Ferri F, Fulchignoni M, Ghafoor NAL, Kargl G, Svedhem H, Delderfield J, Grande M, Parker DJ, Challenor PG, Geake JE (2005 A soft solid surface on Titan as revealed by the Huygens Surface Science Package. Nature 438:792–795

Chapter 7
Volatile Origin and Cycles: Nitrogen and Methane

Sushil K. Atreya, Ralph D. Lorenz, and J. Hunter Waite

Abstract The story of Titan's two most abundant volatile constituents, nitrogen and methane, is intertwined. The focus of this paper is the origin and evolution of Titan's nitrogen atmosphere and the cycle of methane from its production to destruction to replenishment. Relevant observational results from Cassini–Huygens, Voyager and the Earth as well as various hypotheses and models are reviewed. The origin of nitrogen by direct capture, and from dissociation of primordial nitrogen-bearing molecules, especially ammonia, by impact, photolysis, thermal and other processes is evaluated. Similarly, the origin of methane from Saturn's sub-nebula or by water-rock reactions in Titan's interior is reviewed. The role of methane in regulating Titan's climate is noted, and similarities and differences between the methane cycle in Titan's troposphere and the hydrological cycle on Earth are discussed. The fate of methane in the stratosphere and the upper atmosphere/ionosphere is examined in order to evaluate the possibility and extent of an ocean of ethane, requirement of methane replenishment, and the role of product aerosols in maintaining Titan's nitrogen atmosphere.

7.1 Historical Perspective: From Christiaan Huygens to Cassini–Huygens

Titan is unique in the solar system. Whereas satellites in general are not known for atmospheres, Titan is not only endowed with an atmosphere, it has a massive atmosphere with surface pressure exceeding that on Earth by 50%. Moreover, its atmosphere is largely made up of nitrogen (N_2). Its second most abundant constituent, methane (CH_4), displays a meteorological cycle similar to the hydrological cycle on Earth, complete with surface reservoirs, evaporation, clouds, rain, and dendritic channels presumably carved by the rain of methane. In Titan's stratosphere and the ionosphere, the two volatiles, nitrogen and methane, together create a complex web of chemical reactions. The resulting trace species condense or otherwise produce multiple layers of aerosols, filling Titan's atmosphere with photochemical smog and haze. The aerosols play a crucial role on Titan, however. They provide stratospheric warming, which is critical in preventing nitrogen from condensing.

The above modern-day view of Titan had its origins in 1655 when Christiaan Huygens discovered Titan – a large satellite, in geosynchronous orbit at 20.25 Saturn radii from the planet Saturn. First indication of an atmosphere around Titan began to emerge in 1908. José Comas Solá recorded limb darkening, which he attributed to the "existence of a strongly absorbing atmosphere around Titan". This was a remarkable finding for the time, considering Titan's angular size is only 0.1 arc sec, which is at the limit of the seeing capability of best ground-based telescopes. Yet Comas Solá was gifted with exceptional acuity, and his sketches (e.g. Lorenz 1997) of the Galilean satellites are broadly accurate, and he knew that Mars had no canals.

The first quantitative spectroscopic observations of Titan began in earnest in 1944 with Gerard Kuiper's detection of methane at 6190 Å in the visible and in the near-infrared. Kuiper (1944) derived 0.2 km-am for the methane abundance. Trafton (1972) detected methane in the $3\nu_3$ band of CH_4 at 1.1 μm, and arrived at a much greater abundance of 1.6 km-am, or 16 mb, considering self-broadening in a pure CH_4 atmosphere. An even greater surface pressure of 200 mb was suggested by Lutz et al. (1976) from their analysis of the weak methane bands in the visible where pressure broadening is not a complicating factor. However, they concluded that methane is a minor constituent of the atmosphere, with any number of cosmogonically abundant and stable constituents such as Ar, Ne or N_2 as possible candidates for the background gas. Lewis (1971) had previously speculated on the possibility of a nitrogen atmosphere on Titan, result-

S.K. Atreya (✉)
University of Michigan, Ann Arbor, MI, 48109-2143, USA
e-mail: atreya@umich.edu

R.D. Lorenz
Johns Hopkins University, Applied Physics Laboratory, Laurel, MD, 20723, USA

J. Hunter Waite
South West Research Institute, San Antonio, TX, 78228, USA

R.H. Brown et al. (eds.), *Titan from Cassini-Huygens*,

ing from the photolysis of ammonia outgassed from the ices that were needed to explain Titan's low density. Danielson et al. (1973) analyzed the spectroscopic observations also, and concluded that the atmosphere was thin, comprising only a few millibars of CH_4. In the late 1970s, results of Titan's surface temperature began to emerge. Conklin et al. (1977) interpreted their 3 mm observations to imply a surface temperature of 200 K, which was surprisingly greater than Titan's black body radiation temperature of 82 K. This led Hunten (1978) to propose an atmosphere of 20 bar of N_2, whose pressure-induced opacity could explain the high surface temperature. In their pre-Voyager model, Atreya et al. (1978) calculated that approximately 20 bar of N_2 could be produced *initially* on primordial Titan, from the photolysis of ammonia (NH_3) during the satellite's accretionary heating phase. Ground-based observations could not detect nitrogen, as N_2 has no rotational-vibrational spectral lines due to dipole transitions. Thus, by the end of the 1970s the situation with Titan's basic characteristics was less than satisfactory – surface temperature ranged from 82 K to 200 K, surface pressure ranged from a few millibars to 20 bar, and the make up of the atmosphere from pure CH_4 to N_2 with traces of CH_4 and/or H_2 thrown in. This broad range defined the envelope of the engineering models used for guiding the Voyager 1 encounter with Titan.

On 12 November 1980, Voyager 1 flew past Titan at a distance of 6490 km (3915 km from surface). For the first time, it detected nitrogen, found trace quantities of methane in the stratosphere, a thick haze blanketing the satellite – which prevented composition and most other measurements below the lower stratosphere – a pressure of 1496 ± 20 mb and a temperature of 94 ± 0.7 K at the surface, as well as numerous minor constituents in the stratosphere, including C_2-hydrocarbons and nitriles (for a complete summary of the Voyager 1 and Voyager 2 findings, see Atreya 1986; Coustenis and Taylor 1999; Lorenz and Mitton 2008). N_2 was identified in its Rydberg, Lyman-Birge, and Lyman-Birge-Hopfield bands by the ultraviolet spectrometer (Broadfoot et al. 1981; Strobel and Shemansky 1982) but placed only an upper limit on the argon abundance (see Section 7.2.1).

The Voyager radio occultation experiment derived a mean molecular weight that was "consistent" with a pure nitrogen atmosphere, but the possibility of a small non-zero argon component could not be ruled out either (Eshleman 1982). This is because the radio occultation experiment derives the mean molecular weight, not composition. By itself, it yields only a *relative* pressure–temperature profile from a determination of the atmospheric refractivity. The refractivity of the atmosphere as a function of time (height) during occultation is derived from the angle of refraction of the received frequency of the radio signal (S and X bands, respectively at 2.293 GHz or 13 cm, and 8.6 GHz or 3.6 cm on Voyager). The relative pressure–temperature profile can then be obtained from refractivity, which is related to the atmospheric composition (mean molecular weight). In order to obtain the absolute pressure–temperature profile, an *independent* knowledge of either the temperature or the composition is required (Eshleman et al. 1977). The composition of the troposphere was not constrained at the time of the Voyager measurements. Therefore, the radio occultation experiments relied on the temperature measured by the Voyager infrared spectrometer at 1 mb in the stratosphere as a reference temperature. Any uncertainty in the reference value would propagate to the lower levels. This resulted in a slight uncertainty in the tropospheric and the surface temperatures of Titan. The resulting surface temperature had a range from 93 K to 95 K. The corresponding mean molecular weights were 27.8 and 28.3, respectively (Eshleman et al. 1983; Lindal et al. 1983). The average of the two values led Eshleman (1982) to conclude that the mean molecular weight was consistent with a pure N_2 atmosphere. On the other hand, the higher value (28.3) could accommodate up to 4% Ar, whereas the smaller value (27.8) could be explained by including 2% CH_4. Since the composition of the troposphere was unknown, even the average value of the mean molecular weight (28) could accommodate within a small range various proportions of Ar (36 AMU) and CH_4 (16 AMU). Less than a year after the Voyager 1 flyby of Titan, Voyager 2 observed Titan on 26 August 1981, but from a much greater distance of 666,190 km. Despite its remoteness, Voyager 2 yielded valuable data on Titan's multiple haze layers and the photochemical species.

The Voyager flybys gave a fleeting but tantalizing glimpse into the unusual and complex world of Titan. Their findings were instrumental in establishing the importance of studying Titan in detail, for it may even provide clues to the way Earth was before life emerged. The detailed composition of the gases and the aerosols, what lay beneath the thick haze, and the nature and composition of the surface were some of the mysteries of the time that could be revealed only by returning to Titan with an orbiter and an entry probe. A grass-roots effort by scientists in the United States and Europe in the mid 1980s culminated in what we know now as the Cassini–Huygens mission, a joint undertaking of NASA, ESA and ASI (Italian Space Agency). The first ever in situ exploration of Titan was carried out by the ESA-furnished Huygens probe. After its release from the Cassini orbiter on 25 December 2004, the Huygens probe reached Titan on 14 January 2005, and began a historic 148-min descent through the moon's atmosphere and landed on the surface. Although not designed as a lander mission, the probe continued to send the composition data to the Cassini orbiter for another 72 min upon reaching the surface, and the carrier signal from the probe was recorded by ground-based radio telescopes for yet another 2 h and 2 min. This was a remarkable feat, considering the Huygens probe was expected to send data from the surface for only 3 min. Titan is also being explored remotely by the Cassini orbiter in the multiple close flybys of the moon since the arrival of the spacecraft at the Saturn System on 1 July 2004. As a consequence of these in situ and remote observations, new and vastly improved data have become available which form much of the basis of the discussions in this chapter.

7.2 Origin and Evolution of Titan's Nitrogen Atmosphere

Two distinct possibilities existed for the origin of Titan's nitrogen before the Voyager observations: (1) N_2 is primordial, i.e. it was captured directly as N_2, or (2) N_2 is secondary, formed from the dissociation of ammonia, i.e. NH_3, not N_2, is primordial. The Voyager observations did not yield the data that could unambiguously answer the question of the origin of nitrogen on Titan. We had to wait another 25 years for Cassini–Huygens to settle the issue. In the following paragraphs, we review the various possibilities and present the current status of the origin of nitrogen on Titan.

7.2.1 *Direct Capture of N_2*

Lewis and Prinn (1980) suggested that the dominant form of nitrogen in the solar nebula was N_2. Owen (1982) initially proposed that Titan's atmosphere resulted from the direct capture of the nebular material. In that case, its Ne/N_2 should be solar, or 6.2 (using $N/H = 6.76 \times 10^{-5}$ for the solar nitrogen from Grevesse et al. 2005, but scaling the neon value up by 2.7 to account for the newer x-ray stellar data of Drake and Testa 2005, so that solar $Ne/H = 2.10 \times 10^{-4}$, as discussed in Table 1 of Atreya 2007b). However, neon has never been detected in Titan's atmosphere. The Voyager ultraviolet spectrometer (UVS) observations of Ne at 736 Å yielded a very low *upper limit* of 0.01 for Ne/N_2 above Titan's homopause (Strobel and Shemansky 1982). In the well-mixed homosphere, the Ne/N_2 would be even smaller, not greater, than this upper limit of 0.01. This is because the concentration of species in the diffusive separation region above the homopause is controlled by their individual molecular weights (or, scale heights). This would result in greater mixing ratio of a lighter species, such as Ne (20 AMU, compared to the heavier background of ~28 AMU) above the homopause than in the well-mixed atmosphere below. Thus the Voyager UVS data imply that the Ne/N_2 mixing ratio on Titan is well below the solar value of 6.2. The same conclusion is reached by examining the tropospheric mean molecular weight, M. As determined by the Voyager radio occultation experiment, M is very close to 28 (Eshleman 1982), which also does not permit any significant quantities of neon to be present in the atmosphere. Owen (1982) surmised that the way out of the neon dilemma is that N_2 was initially trapped in a hydrate, together with CH_4 and Ar. This idea was based on an earlier work of Miller (1961) who proposed that the ice in Saturn's moons is largely in the form of a methane hydrate ($CH_4 \cdot 6H_2O$), and of Lewis (1973) who suggested that the methane hydrate in the cooling solar nebula goes to completion at 60 K. Neon is highly volatile, and even at that temperature its vapor pressure is 40 bar, so that it could not have been trapped in the hydrate. Thus the non-detection of neon was not surprising, according to Owen (1982).

If the above scenario of Titan's nitrogen atmosphere forming from the degassing of N_2-CH_4-Ar containing hydrate is correct, it would imply that solar proportions of argon and carbon must be present. The solar $^{36}Ar/N_2$ ratio is 0.11 (solar $N/H = 6.76 \times 10^{-5}$ from Grevesse et al. (2005), but $Ar/H = 3.62 \times 10^{-6}$ from Anders and Grevesse (1989), as the former used oxygen as proxy to derive Ar, and Ne whose correction is given in the previous paragraph). The Voyager UVS observations near a resonance feature of argon at 1048 Å yield an upper limit of 0.06 for $^{36}Ar/N_2$ above the homopause (Strobel and Shemansky 1982). Unlike neon, the argon mole fraction could be larger in the homosphere. This is because ^{36}Ar is heavier than the background atmosphere of N_2, which would lead to a drop in the mole fraction of ^{36}Ar above the homopause due to molecular diffusive separation. Thus, the Voyager data on argon could neither support nor discard the direct capture of N_2 hypothesis of the origin of Titan's atmosphere. The interpretation based on CH_4/N_2 was similarly ambiguous. The solar C/N is 4.07 (based on solar C/H and N/H from Grevesse et al. (2005)). The primordial carbon could be in the form of CO, CO_2, carbon grains or organic material. CO and CO_2 were found to be quite small in the atmosphere by Voyager. If most of the carbon were tied up in methane, then CH_4/N_2 would be solar, or ~8. If indeed methane was so abundant on Titan, a huge reservoir of liquid methane would be present on the surface (the methane triple point of 90.67 K is close to Titan's surface temperature of 94 K). This would lead to the collapse of the atmosphere, as nitrogen gas readily dissolves in liquid methane. The Voyager Infrared Interferometer Spectrometer Radiometer (IRIS) observations yielded $CH_4/N_2 = 0.02$ at 100 mbar in the lower stratosphere (Hanel et al. 1981). Although no direct measurements of methane could be made below the tropopause, using the IRIS data and the Voyager radio refractivity data Samuelson et al. (1997) derived a 50% supersaturation of CH_4 in the upper troposphere and a CH_4 mole fraction close to 6% near the surface.

Lacking the critical data on the distribution of methane in the troposphere and the mole fraction of primordial argon in the homosphere, the Voyager observations were not able to settle the question of the origin of nitrogen on Titan. There were still two competing hypotheses (a) the pre-Voyager model of Atreya et al. (1978) and Hunten (1978) where nitrogen was acquired from the Saturnian subnebula in the form of NH_3, whose subsequent photolysis led to the formation of an N_2 atmosphere in Titan's warmer past, and (b) the post-Voyager model of Owen (1982) according to which nitrogen was captured directly. In a later publication, Owen (2000), however, agreed with the model of the formation of N_2 from NH_3 photolysis, considering that direct trapping of N_2 requires the temperature of planetesimals to be below 30 K, which is far too low for the Saturn subnebula where Titan formed.

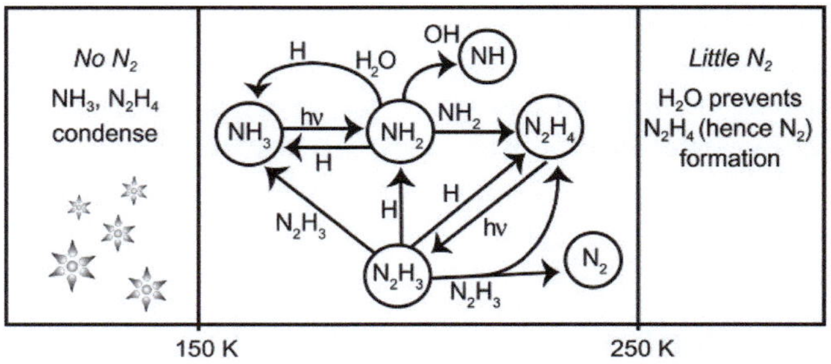

Fig. 7.1 Photochemical production of N_2 from NH_3 in Titan's primordial atmosphere (updated from Atreya et al. 1978). Below 150 K, little NH_3 is in vapor phase, and whatever small intermediate product hydrazine (N_2H_4) forms from it condenses, thus preventing N_2 formation. Above 250 K, water vapor has sufficiently large vapor pressure, so that OH from its photolysis reacts with NH_2 and NH_3 drastically decreasing the subsequent yield of N_2, despite some recycling of NH_3 by H from H_2O

7.2.2 N_2 as a Secondary Atmosphere from Primordial NH_3

Speculations of a nitrogen atmosphere resulting from the photolysis of ammonia were made as early as 1971 (Lewis 1971). Considering Titan's density of 1.8 g/cm^3, it is believed that Titan is made of roughly 40% ice by mass, and the rest rock. A small fraction of the ice is expected to have been in the form of ammonia ice initially. Prinn and Fegley (1981) argued that kinetic inhibition to the conversion of N_2 to NH_3 in the nebula and the subnebulae of the outer planets was minimal. Thus, NH_3, not N_2, was the dominant form of nitrogen in the Saturnian subnebula where Titan formed. Even before this reaffirmation, the idea of Titan's nitrogen as being a secondary atmosphere had caught on (Section 7.1). Three possibilities exist for producing N_2 from the dissociation of NH_3: photolysis, impact, and endogenic.

7.2.2.1 N_2 from NH_3 Photolysis

During the accretionary heating phase the volatiles were released from the ices to the atmosphere of Titan. Lunine and Stevenson (1987) suggested an atmosphere that contained copious quantities of methane, ammonia and water vapor. Atreya et al. (1978) developed a photochemical model, according to which the photodissociation of NH_3 takes place at wavelengths below 300 nm, and peaking at ~195 nm. Their photochemical scheme for ammonia is shown in Fig. 7.1. The photolysis of ammonia produces amidogen radicals (NH_2). About a third of ammonia so destroyed is recycled by the reaction of these radicals with hydrogen atoms. Nearly two-thirds of the NH_2 radicals recombine to form hydrazine (N_2H_4) molecules (Fig. 7.1). Hydrazine acts essentially as the rate limiting step in the subsequent production of N_2. If the temperature is sufficiently warm, N_2H_4 remains in the vapor phase and is dissociated by the solar ultraviolet flux to form hydrazyl radicals (N_2H_3). The self-recombination reaction of N_2H_3 leads to the production of N_2 on Titan (Fig. 7.1). If the temperature is too low, little NH_3 would be present in the form of vapor, and whatever small amount of hydrazine is formed from it would condense and snow out of the atmosphere, preventing further chemistry and production of N_2 (impacts are ineffective in converting NH_3 or N_2H_4 in any phase to N_2, as discussed in the next section). Atreya et al. (1978) found that the ideal temperature range is determined not just by the thermodynamic phase of N_2H_4, but also the relative humidities of water and ammonia. At relatively high temperatures, a large quantity of water vapor coexists with the ammonia vapor. The photolysis of H_2O vapor results in the highly reactive hydroxyl molecules (OH). The latter react with NH_2, severely limiting the formation of N_2H_4 and the subsequent production of N_2. At low temperatures, hydrazine condenses and N_2 does not form, as discussed above. With these considerations, Atreya et al. (1978) determined that the ideal range of temperatures is 150–200 K for the production of nitrogen from ammonia on Titan. Allowing for the increased solar ultraviolet flux in the pre-main sequence of the Sun (Zahnle and Walker 1982), Atreya et al. (1978) and Atreya (1986) calculated that up to 20 bar of N_2 could be produced in less than 2 Myr on primordial Titan initially, i.e. not accounting for subsequent N_2 escape.

Atreya et al. (1978) assumed an isothermal primordial atmosphere in their nitrogen evolution model. For more realistic time scales for the formation of nitrogen, possible vertical gradients in the atmospheric temperature should be taken into account. Considering a range of possible abundances of H_2O, CH_4 and NH_3 on primordial Titan (Lunine and Stevenson 1987), Adams (2006) developed a radiative transfer model for primordial Titan. Additional constraints on the pre-main sequence solar flux from x-ray to the infrared were included, based on the work of Ribas et al. (2005) and Zahnle and Walker (1982). The NH_3 photochemical scheme used was similar to the one used by Atreya et al. (1978). The use of the atmospheric temperature structure derived from radiative transfer led to longer time scales for the generation of nitrogen than the isothermal model of Atreya et al. (1978). The longer time scales are a consequence of the much smaller abundance of ammonia available in the photochemical regime of Titan's stratosphere because of cold trapping of the gas at the tropopause, which was not an issue for the isothermal models with temperatures far greater than the tropopause values. For example, current models require a little over 30 Myr to produce 10 bar of N_2 on primordial Titan. This will be discussed further in Section 7.2.2.4.

7.2.2.2 N_2 from Impacts

High velocity impacts were suggested by Jones and Lewis (1987) and McKay et al. (1988) as a probable cause of Titan's atmosphere. Considering the possibility of shock-induced dissociation of ammonia from high velocity impacts during Titan's accretion, McKay et al. (1988) carried out a laboratory simulation experiment using a 1.06-μm Nd-YAG laser on a mixture of NH_3 and CH_4. This process generated relatively large quantities of N_2. For a simplified accretion scenario, they could produce up to 25 bar on N_2 in the late stage of Titan's accretion. Although an attractive idea, the shock-induced production of N_2 encounters several difficulties. Nitrogen was not the only molecule produced in their experiment. The total yield of numerous hydrocarbons was even greater than N_2, i.e. up to 100 bar. H_2 production was nearly 4 bar. The H_2 amounts are so enormous that they exceed the present H_2 abundance a thousand-fold, unless the hydrogen escape rate throughout Titan's geologic history was also improbably large, exceeding 1000 times the present value (Waite et al. 2005) of 6×10^9 cm^{-2} s^{-1} from the Ion and Neutral Mass Spectrometer (INMS). We are not aware of any mechanisms that could sustain such a rapid loss rate of H_2 from Titan. The large hydrocarbon production may also pose a further challenge, as most of them would end on the surface, depositing a layer 15 km thick. There is little direct evidence of such a thick aerosol layer on Titan's surface, although a substantial organic component to the surface material on Titan cannot be ruled out. Finally, the shock experiment did not include water vapor. As we discussed in the context of NH_3 photolysis, presence of water vapor would inhibit N_2 production (Section 7.2.1.1), especially in the cooling phase of the shock.

A variation on the impact-induced delivery of Titan's atmosphere was proposed by Griffith and Zahnle (1995). These authors suggested that cometary impacts resulted in Titan's nitrogen atmosphere. In this scenario, the source of nitrogen lies in the comets that condense out of the solar nebula, not in the planetesimals that condensed from the Saturnian subnebula to accrete Titan. In the cometary hypothesis, nitrogen is believed to have been supplied originally in the form of complex organic molecules or molecular nitrogen. If comets were indeed the source of Titan's atmosphere, the D/H ratio on Titan should resemble that in comets. Although the D/H ratio has been determined in the ice of only a handful of comets (Halley – Eberhardt et al. 1995; Balsiger et al. 1995; Hyakutake – Bockelee-Morvan et al. 1998; and Hale-Bopp – Meier et al. 1998a), in each case its value was found to be larger than that in Titan's hydrogen reservoirs. The average value is found to be $(3.16 \pm 0.34) \times 10^{-4}$, which is greater than the D/H ratio either from CH_3D ($1.3\ (+0.15, -0.11) \times 10^{-4}$; Bézard et al. 2007; Coustenis et al. 2007) or H_2 ($(2.3 \pm 0.5) \times 10^{-4}$, Niemann et al. 2005) in Titan's troposphere. The cometary D/H has also been determined in one organic compound, HCN, and the value $(2.3 \pm 0.4) \times 10^{-3}$ (Meier et al. 1998b) is more than ten times greater than the above values on Titan. Thus, comets do not appear to be the main source of Titan's atmosphere. Finally, direct infusion of N_2 by comets is unlikely, as comets are deficient in N_2 since the temperature in the region of their condensation between the orbits of Uranus and Neptune was too high (50–60 K) for trapping nitrogen. We discuss the cometary hypothesis further in Section 7.2.2.4.

7.2.2.3 Endogenic N_2

Matson et al. (2007) have surmised that the dissociation of ammonia in Titan's interior may have produced some, if not all, of Titan's nitrogen, especially if catalyzed by clay or metals. If Titan formed 2.5 to 5.0 Myr after the calcium–aluminum inclusions (CAIs) were created (as suggested for Iapetus, Castillo-Rogez et al. 2007) then differentiation occurred early and a stable core formed. During differentiation, heat from short-lived radioisotope decay and gravitational energy enabled rapid serpentinization of most of the silicate phase. (In serpentinization, water-rock reactions comprising oxidation and hydrolysis of low-silica and ultramafic silicate rocks such as olivine or pyroxene occurs, resulting in large quantities of heat, together with molecular hydrogen, serpentine, brucite and other minerals, as discussed in more detail in Section 7.3.5.) Water and chemical reactants were trapped by the hydrated silicate that accumulates into the core. After a few hundred million years, temperatures become high enough for ammonia decomposition to take place, producing molecular nitrogen and hydrogen. As Titan cooled, nitrogen was trapped in ice as a clathrate-hydrate. The outgassing of nitrogen occurred from time to time when the clathrates were destabilized. This scenario predicts that nitrogen is still being released from the interior. This would lead to a replenishment of nitrogen lost to space, and resetting of the $^{14}N/^{15}N$ isotope ratio to terrestrial or near-terrestrial value, as is the case with $^{12}C/^{13}C$ in the CH_4

that is also believed to be released from Titan's interior to replenish the methane destroyed photochemically. The present data yield near-terrestrial $^{12}C/^{13}C$, whereas $^{14}N/^{15}N$ is severely depleted. At the moment, there is no simple solution to the isotope dilemma. In view of the above, the endogenic production of N_2 should be considered as tentative for the time being until further research is carried out. It is important to recognize also that any endogenic source would only supplement, not replace, the photolytic origin of nitrogen. This is because warmer Titan in the past would have certainly put large quantities of NH_3 vapor into the atmosphere, which could not have escaped photochemical destruction to produce nitrogen.

7.2.2.4 Origin and Evolution of Titan's Nitrogen Atmosphere: The Cassini–Huygens Perspective

The Cassini–Huygens Mission collected the data that finally helped settle the question of the origin and evolution of the atmosphere on Titan. Two sets of measurements are particularly relevant: the primordial argon abundance, and the abundance of the nitrogen and carbon isotopes. None were available until they were done by Cassini–Huygens in 2005. The abundance of primordial argon, ^{36}Ar, can shed light on the form of nitrogen acquired by Titan originally. The nitrogen isotope ratio is important for determining the mass of N_2 that must be formed originally to explain the present N_2 amount in the atmosphere. And, the comparison of the nitrogen isotope abundances along with the carbon isotope abundances provides clues to the evolutionary history of the gas. The Gas Chromatograph Mass Spectrometer (GCMS) on the Huygens probe measured the $^{14}N/^{15}N$ ratio in the troposphere of Titan by analyzing the N_2 data at 28 and 29 Da, i.e. $^{14}N^{14}N$ and $^{15}N^{14}N$ (Niemann et al. 2005). Similarly, the Ion and Neutral Mass Spectrometer on the Cassini orbiter measured the ratio at ~1000 km altitude in the thermosphere (Waite et al. 2005). Likewise, the $^{12}C/^{13}C$ was measured by both instruments from $^{12}CH_4$ and $^{13}CH_4$. Primordial argon, ^{36}Ar, as well as radiogenic argon, ^{40}Ar, produced from the radioactive decay of potassium, ^{40}K, in the rocks in the core, were also measured by the INMS and the GCMS. Table 7.1 lists these values. A comparison to the terrestrial values is given also, using the GCMS data. We use the GCMS data for the mixed atmosphere for this comparison instead of the INMS data, as the latter were measured in the thermosphere (~1000 km) where they are diffusively separated, so their extrapolation to well-mixed atmosphere is somewhat model-dependent.

The mole fraction of ^{36}Ar is 2.8×10^{-7} (Table 7.1), which is well below the solar $^{36}Ar/N_2$ value of 0.11 (Section 7.2.1.1), and ^{38}Ar was not even detected. This implies that nitrogen was not captured directly as N_2, nor did it arrive as N_2 from the subnebula, but as another nitrogen compound, most likely ammonia (NH_3). Had Titan acquired its nitrogen in the form of N_2, it would have also accreted large (i.e., solar) quantities of primordial argon, i.e. nearly 11% by volume, or over 300,000 times greater than measured, and over 1 million times greater when N_2 escape is included – see below. During the accretionary heating phase, ammonia should have been present as vapor in the atmosphere of Titan. Subsequent photolysis of this ammonia then led to the nitrogen we find on Titan today, as discussed in Section 7.2.2.1. The $^{14}N/^{15}N$ is found to be 183 in Titan's troposphere, which is ~0.67 times the value in the terrestrial atmosphere (Table 7.1). This implies loss of nitrogen from the atmosphere to space

Table 7.1 $^{12}C/^{13}C$, $^{14}N/^{15}N$, D/H ratios, and the ^{36}Ar, ^{40}Ar, Kr, and Xe mole fractions (f)

Ratio	GCMS[a] in homosphere	INMS[b] in heterosphere	Titan/Earth[(c)] using GCMS
$^{14}N/^{15}N$	183 ± 5 (from N_2)	240 ± 20 (1,450 km)	0.67
		155 ± 5 (1,000 km)	
		143 ± 5 (extrapolated to surface)	
$^{12}C/^{13}C$	82.3 ± 1	118 ± 10 (1,450 km)	0.92
		83 ± 5 (1,000 km)	
		81.0 (extrapolated to surface)	
D/H	$(2.3 \pm 0.5) \times 10^{-4}$		1.44
$f\,^{36}Ar$	$(2.8 \pm 0.3) \times 10^{-7}$	$<(4 \pm 1) \times 10^{-7}$	7.0×10^{-3}
$f\,^{40}Ar$	$(4.3 \pm 0.1) \times 10^{-5}$	$(1.26 \pm 0.05) \times 10^{-5}$ (1050 km)	3.6×10^{-3}
fKr, and fXe	$<10^{-8}$	Not measured	

[a] GCMS measurements are from the Huygens probe from 146 km to the surface. Errors quoted represent 1 standard deviation (Niemann et al. 2005). The GCMS determined the D/H from HD and H_2. The value derived from the Cassini CIRS data using CH_3D and CH_4 is D/H = 1.3×10^{-4} (Bézard et al. 2007; Coustenis et al. 2007). The $^{14}N/^{15}N$ from ground-based submillimeter observation of HCN is 60 ± 6 (Marten et al. 2002), confirmed by Cassini/CIRS HCN data (Vinatier et al. 2007). This is or about a factor of four less than the terrestrial value of 272, most likely due to the fractionation in photochemical production of HCN (Niemann et al. 2005).

[b] INMS values are from Mandt et al. (2009), except for ^{40}Ar (Magee et al. 2009), revised from the values published by Waite et al. (2005) from the first flyby of the Cassini orbiter between 1174 km and 3000 km above Titan's surface.

[c] The terrestrial inorganic $^{12}C/^{13}C$ = 89.45, is the Vienna Pee Dee Belemnite Standard.

in Titan's past. Lammer et al. (2000) found that non-thermal escape due to solar wind ion pick-up and atmospheric sputtering could have resulted in a loss of up to 40 bar of N_2. However, this conclusion was based on a pre-Cassini–Huygens $^{14}N/^{15}N = 60 \pm 6$ derived from HCN (Marten et al. 2002). That nitrogen isotope ratio is a factor of three smaller than that in the principal reservoir of nitrogen, N_2 (Table 7.1), most likely due to fractionation in the process of photochemical formation of HCN.

From a comparison of the current $^{14}N/^{15}N = 183$ in N_2 in Titan's atmosphere measured by the Huygens GCMS with the terrestrial $^{14}N/^{15}N$, Niemann et al. (2005) estimate that the mass of primitive atmosphere of Titan was between two and ten times today's value and perhaps several times the present mass was lost over geologic time, based on a model of diffusive separation of ^{15}N from ^{14}N by Lunine et al. (1999). Lammer et al. (2000) concur that diffusive separation is an important process, but ion pick-up and atmospheric sputtering caused by a high solar wind outflow during a Post T-Tauri phase of the Sun is required to explain the large nitrogen isotope anomaly. They estimate a loss of roughly 10 bar of atmosphere from Titan due to this process in the first 500 Myr, in general agreement with the earlier finding (Lunine et al. 1999). (Note that the mass loss estimates in both Lunine et al. and Lammer et al. were based on the then available smaller $^{14}N/^{15}N = 60 \pm 6$ from HCN, but the mass loss numbers given here have been scaled to reflect the larger GCMS value of 183.) Comparing the GCMS nitrogen isotope ratio with the terrestrial value implies that Titan would have started out with a nitrogen pressure of between 5 and 10 bar in its primordial atmosphere, in order to explain the present surface pressure of 1.5 bar. Considering a radiative transfer model for the primordial atmosphere of Titan, prevalent solar fluxes, and the photochemical scheme of Atreya et al. (1978) and Atreya (1986) for the production of N_2 from NH_3 (Section 7.2.2.1), Adams (2006) calculated that it would take, respectively, 17 and 33 Myr to accumulate 5 and 10 bar of N_2 on primordial Titan. The time scale could be as low as 13–25 Myr for 5–10 bar, if the methane abundance was 30% lower, because of the role of CH_4 in radiative transfer. As we discussed in Section 7.2.2.1, the ideal temperature range for photochemical conversion of NH_3 to N_2 is between 150 and 250 K. Post-accretionary evolution models show that Titan's temperature was in this range for 100–200 Myr (Lunine and Stevenson 1985), which is more than adequate for producing the above initial inventory of nitrogen, even allowing for uncertainties in the primordial atmosphere models. If needed, the ammonia photolysis model is also capable of accommodating even much greater loss of nitrogen than the above 3–8 bar, considering the long duration of ideal temperatures for photolysis on primordial Titan.

The nitrogen isotope ratio in Titan's atmosphere is also remarkable from another point of view. It seems to place Titan in the same family of solar system objects as the terrestrial planets, in so far as the origin of nitrogen is concerned. Later we will see the similarities extend beyond nitrogen. In the right panel of Fig. 7.2, we illustrate this point by showing a comparison between Titan's $^{15}N/^{14}N$ isotope ratio with that of Venus, Mars and the Earth. The $^{15}N/^{14}N$ ratio in the atmosphere of Mars is larger than the solid Mars value as measured in the AL84001 meteorite, but is consistent with it after escape of nitrogen from the atmosphere is accounted for. Thus, the three terrestrial planets, Venus, Earth and (solid) Mars, have very similar $^{15}N/^{14}N$ ratios. The $^{15}N/^{14}N$ ratio of Titan would also be similar to the terrestrial planet value if Titan lost some nitrogen as discussed above. The nitrogen in the atmospheres of Venus (3.5% by volume), Mars (2.7%) and the Earth (78%) is also secondary, i.e. it formed from the dissociation of a nitrogen molecule, most likely ammonia, in the early stages of planetary evolution. The Cassini–Huygens observations show that similar to the terrestrial planets, Titan's nitrogen atmosphere is also secondary.

Finally, a comparison of Titan's nitrogen isotope abundance with that on comets, solar wind and other primordial sources (left panel of Fig. 7.2) provides further constraints to the origin of nitrogen. Very few measurements exist for the comets, however. Furthermore, the nitrogen isotopes in more

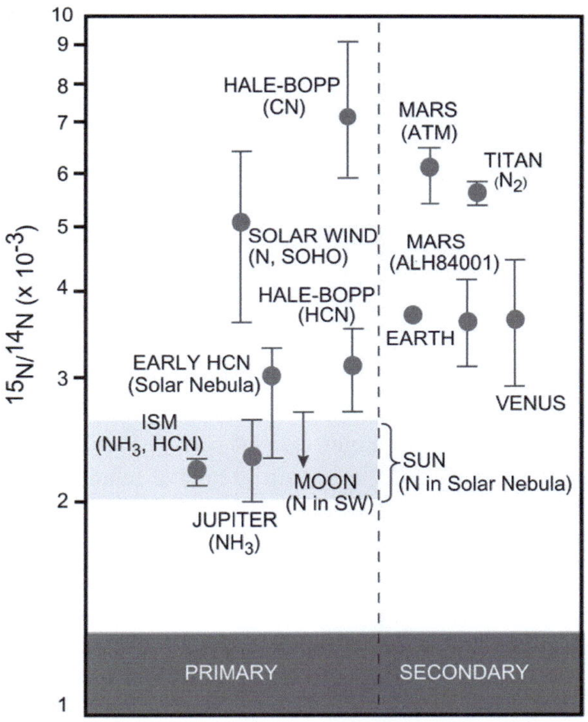

Fig. 7.2 Nitrogen isotope ratio in Titan's atmosphere compared with those on the terrestrial planets, and primary sources, including the interstellar medium (ISM), Sun (SW: solar wind, SOHO: Solar and Heliospheric Observatory), Jupiter (which represents the protosolar value) and comet Hale-Bopp. For Mars, two values are shown, in the atmosphere (ATM) and Allan Hills meteorite (ALH84001); the latter presumably represents the solid Mars value (prepared in consultation with Paul R. Mahaffy)

than one molecule (two) nitrogen containing molecules have been determined for only one comet so far. Submillimeter observations of HCN in comet Hale-Bopp yield $^{14}N/^{15}N$ = 323 ± 46, or $^{15}N/^{14}N = 3.10 \times 10^{-3}$ for the central value (Jewitt et al. 1997). Another observation of the same comet gives a nearly identical value of 330 for $^{14}N/^{15}N$ (or, 3.03×10^{-3} for $^{15}N/^{14}N$) from HCN (Ziurys et al. 1999), which is close to the terrestrial value of $^{14}N/^{15}N = 272$ (or, 3.68×10^{-3} for $^{15}N/^{14}N$). On the other hand, optical data yield an $^{14}N/^{15}N = 140 \pm 30$ from CN in comets Hale-Bopp, LINEAR and a dozen other comets (Arpigny et al. 2003; Hutsemékers et al. 2005). Waite et al. (2009) and Mandt et al. (2009) propose a cometary, rather than the subnebular (NH_3), origin for Titan's N_2, considering that the revised INMS value of $^{14}N/^{15}N = 143 \pm 5$ in the homosphere of Titan matches the value of this ratio in the CN of a dozen comets. This is an interesting alternative, but several caveats should be noted. First, the INMS value of $^{14}N/^{15}N$ in the homosphere (143) is model-dependent, as it is derived by extrapolating the value measured in the heterosphere (above 900 km, where $^{14}N/^{15}N = 188$), where diffusive separation dominates. Second, the derived INMS value for the homosphere (143) differs from the in situ determination of $^{14}N/^{15}N = 183$ directly in the homosphere by the Huygens GCMS. The GCMS value does not resemble the cometary $^{14}N/^{15}N$, being greater than that from CN and less than that from HCN. Third, the cometary $^{14}N/^{15}N$ ratio used for comparing the INMS derived $^{14}N/^{15}N$ for Titan's homosphere is not from the principal reservoir of nitrogen on comets, ammonia, or even the next most abundant one, HCN, but from a tertiary product, CN. The $^{14}N/^{15}N$ in HCN is more than twice that from CN in comet Hale-Bopp, the only comet for which the ratio has been determined in both HCN and CN ($^{14}N/^{15}N$ from NH_3 is not yet available, but may be greater than that in HCN due to photofractionation). Fourth, it would take an unrealistically large number of impacts of comets to deliver the amount of nitrogen present on Titan. Finally, such impacts, or impacts of smaller number of larger comets delivering equivalent amount of N_2, would introduce excessive amounts of H_2 into Titan's atmosphere and result in a D/H value different from that observed on Titan, as discussed in Section 7.2.2.2. The cometary nitrogen reservoir and its potential for delivering N_2 to Titan (third and fourth points above) are further elaborated in the following paragraphs.

The dominant nitrogen bearing molecule on comets is ammonia, whose abundance is 0.5–1.5% relative to water (Bockelee-Morvan et al. 2004). The nitrogen isotope ratio in NH_3 is not measured. However, experience shows that $^{14}N/^{15}N$ fractionation occurs in photochemistry, so that the ratio is smaller in the product than in the parent. This would imply that the $^{14}N/^{15}N$ in NH_3 of comets might be greater than that in HCN, a product of the chemistry between NH_3 and CH_4 of comets. $^{14}N/^{15}N = 323$ from HCN in comet Hale-Bopp, the only comet in which the ratio has been determined. The cometary $^{14}N/^{15}N$ ratio from HCN is a factor of 2.3 *greater* than the INMS value for Titan's homosphere (143). On the other hand, $^{14}N/^{15}N = 140 \pm 30$ in CN of certain comets is similar to the extrapolated INMS value for the homosphere of Titan. Incidentally, photochemical fractionation of nitrogen isotopes is evident in comets as it is on Titan. The $^{14}N/^{15}N$ ratio in the cometary CN is a factor of 2.3 smaller than that in HCN; CN being a photoproduct of HCN (Hutsemékers et al. 2005; Ziurys et al. 1999). Similarly, the $^{14}N/^{15}N$ ratio in Titan's main (parent) nitrogen reservoir, N_2 (143, using the revised INMS value), is also 2.3 times the value derived from HCN (60), a photochemical product of the N_2-CH_4 chemistry. If comets were indeed the source of Titan's nitrogen, the nitrogen isotope ratio in its atmosphere should reflect that in the dominant nitrogen bearing molecules in the comet, NH_3 and HCN, not CN. The INMS $^{14}N/^{15}N$ ratio is a factor of 2.3 smaller than that in the cometary HCN, and probably even less compared to that in the cometary NH_3. If, for some mysterious reason, CN of comets were still responsible for producing Titan's N_2, the paragraph below gives an estimate of the magnitude of required cometary impacts.

In comets, the abundance of HCN is 0.1% relative to H_2O (Bockelee-Morvan et al. 2004). The abundance of CN has not been determined, but is expected to be much smaller than this, being a product of HCN. For example, Wilson and Atreya (2004) find CN/HCN = 10^{-3} from their photochemical model of Titan's atmosphere. In the most optimistic, but clearly unrealistic, scenario where *all* of the cometary HCN was converted to CN, the abundance of CN would be same as HCN, or 0.1% relative to water. A typical comet of 2 km diameter would then deliver 4×10^9 kg of CN to Titan, taking the comet density to be same as that of water ice. The amount of N_2 in Titan's atmosphere today is 9×10^{18} kg, based on a 1500 mbar atmosphere containing 95% N_2 and 5% CH_4. Assuming that *all* of the cometary CN was converted to N_2 upon the comet's impact, again a stretch, it would take a staggering two billion 2 km diameter comets to deliver the amount of nitrogen currently present on Titan. In reality the number would be far greater. Since CN is expected to be only a small fraction of HCN, that not all of the CN is going to end up in N_2, and that the starting inventory of nitrogen was perhaps three to five times greater than that present on Titan today, the number of comets would easily exceed a trillion comets. Fewer, but bigger (than 2-km), comets delivering an equivalent amount of N_2 could be envisioned also. However, cometary impacts, whether in the form of large comets or smaller ones, whether recent or in the past, including those in the purported Late Heavy Bombardment between 3.8 and 4.1 Gyr ago, etc., would also deliver excessive H_2 and produce too large a D/H ratio, etc., as mentioned previously in this section and discussed in Section 7.2.2.2. It should be noted also that the probability of a comet striking

Titan is low. For example, the probability of a comet striking Mars, a somewhat larger object, is only once in 62 Myr.

In summary, photolysis of ammonia on primordial Titan is found to be the most robust mechanism for producing Titan's nitrogen atmosphere. Nevertheless, further measurement of the chemical composition and isotope abundances of nitrogen and other elements in a multitude of comets together with modeling will benefit our understanding of the contribution of comets to the make-up of the planets and satellites in the solar system.

7.3 The Cycle of Methane on Titan

The maintenance of a stable atmosphere of nitrogen on Titan is critically dependent on the warming resulting from the presence of its second most abundant atmospheric constituent, methane. In the absence of methane, Titan's nitrogen would condense and the atmosphere would collapse (Lorenz et al. 1997a, 1999). The photochemistry of methane produces hazes and molecular hydrogen, amongst other products. The hazes absorb the incoming solar infrared radiation and heat the stratosphere. The collision-induced opacity due to the H_2-N_2, CH_4-N_2 and N_2-N_2 collisions raises the temperature below. The resulting methane initiated warming of the atmosphere is crucial for preventing nitrogen from condensing out from the atmosphere as droplets. On the other hand, the stability of methane itself is not a given. Methane has a meteorological cycle in the troposphere that could lead to its loss to the interior, if only temporarily. In the stratosphere and the ionosphere, photochemistry destroys methane irreversibly. Episodic resupply of methane from its possible storage in the interior of Titan is necessary to restore it to levels where it can participate in the atmospheric radiative balance and photochemistry, although the amount available to condense on to the surface in reservoirs such as the polar lakes may be highly variable. In this section, we will review this complex meteorological–photochemical–hydrogeochemical cycle of methane on Titan that is somewhat similar to the terrestrial hydrological cycle, and was referred to as the "methalogical cycle" by Atreya et al. (2006).

7.3.1 The Meteorology of Methane

For the first time, the vertical distribution of methane in Titan's troposphere was available following the in situ measurements made by the Huygens Gas Chromatograph Mass Spectrometer (Niemann et al. 2005). The GCMS determined the CH_4 mole fraction at a high spatial resolution starting at an altitude of 146 km altitude above the surface, and on the surface after the probe's landing. The distribution, shown in Fig. 7.3, is remarkable in many respects. First, the CH_4 mole fraction is uniform, with a value of $1.4 \pm 0.07 \times 10^{-2}$ from the lower stratosphere (146 km) to ~32 km altitude which is just below the tropopause (40 km). Second, the mole fraction gradually increased in the troposphere until it reached a maximum value of $4.9 \pm 0.25 \times 10^{-2}$ at 8 km, below which it remained constant down to a level just above the surface. Third, the methane signal suddenly increased as the probe descended through the 16 km altitude and remained high for a couple of km (see inset in Fig. 7.3). Finally, the methane mole fraction began to increase again after the probe landed on Titan, reaching a plateau within 2 min to a value that was ~50% greater than that just above the surface (not shown). The methane mole fraction then remained constant essen-

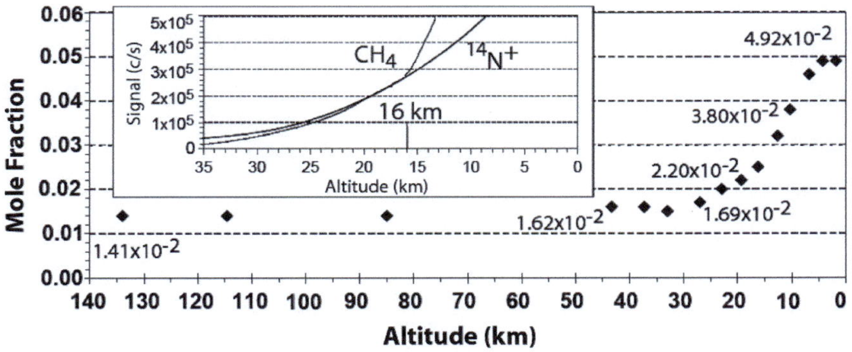

Fig. 7.3 Methane mole fraction in Titan's atmosphere from the GCMS. The CH_4 mole fraction is $1.4 \pm 0.07 \times 10^{-2}$ in the stratosphere. It increased below 32 km, reaching a plateau of $4.9 \pm 0.25 \times 10^{-2}$ at about 8 km. The inset shows the behavior of CH_4 as the probe descended through 16–14 km region. A sudden increase in the methane counting rate relative to N_2 (in this case $^{14}N^+$, which is used as a proxy for N_2 to get around the problem of saturation in the N_2 counts) was detected starting at 16 km and lasting over about 2 km. This was presumably caused by a droplet (or frozen droplet) entering the heated inlet of the GCMS as the probe descended through a cloudy/hazy region (Niemann et al. 2005). The haze apparently extended to at least 21 km where it was detected by the Huygens Descent Imager (DISR), but was first registered by the GCMS at ~16 km where its concentration was large enough for detection by the GCMS

tially to the end of data transmission from the probe 1 h and 12 min later.

The uniform mixing ratio of methane in the lower stratosphere is due to the fact that the unit optical depth, where photolysis destroys most of the methane, is much higher, i.e. between 700 and 800 km. The increase in the methane signal at 16 km indicates that the probe descended through a cloud or haze of methane that evaporated in the heated inlet of the GCMS. The increase of signal at 16 km does not imply either the beginning or end of the cloud/haze layer; it is just where the mass of CH_4 condensate – most likely, a liquid or frozen droplet – was large enough for it to be noticed by the GCMS. The increase in the CH_4 mole fraction after landing indicates moist soil from which methane was evaporated by the heated underbelly of the probe. A slight decrease in the methane mole fraction toward the end of the surface measurements was noticed also, indicating the pool of methane in the Huygens landing site was limited. The above behavior of methane is similar to the familiar behavior of water on the Earth. In both cases, convective ascent of a subsaturated air parcel from the surface to the lifting condensation level (LCL) results in saturation and condensation of the condensible species – methane on Titan and water vapor on the Earth. Clouds form. Precipitation takes place. The surface becomes wet. Above the LCL, the mole fraction of the condensible species is limited to its saturation value, reaching the lowest point at the tropopause, and not rising above it in the warmer stratosphere.

The relative humidity of methane in the Huygens landing site (10.3° S, 192.37° W) was found to be ~50% just above the surface (temperature 93.8 K, CH_4 triple point 90.67 K). The measured CH_4 mole fraction was constant to ~8 km altitude, as is expected in equilibrium. If the LCL were indeed at 8 km, the methane relative humidity should reach 100% at ~8 km. However, the relative humidity turns out to be only 80% considering CH_4 in equilibrium above pure liquid methane, as shown by solid line in Fig. 7.4. On the other hand, since nitrogen gas dissolves in liquid methane, the CH_4 saturation vapor pressure is lower over this binary mixture (Kouvaris and Flasar 1991; Thompson et al. 1992). As a consequence, methane in fact reaches 100% relative humidity at 8 km (dashed line in Fig. 7.4, Atreya et al. 2006), not 80%. This is where the condensation of methane is expected to begin. Above this altitude, methane would remain at 100% relative humidity, if equilibrium prevailed. This is just what is found up to ~15 km altitude (Fig. 7.4). The departure of the CH_4 data from the saturated CH_4 mole fraction above ~15–16 km is due to the transition of methane condensate from the liquid to the solid phase above this altitude, to which the formulation of Kouvaris and Flasar does not apply.

The above discussion illustrates classic textbook example of equilibrium thermodynamics at work in Titan's atmosphere. The methane cloud/haze could actually extend well beyond the 16 km altitude where the GCMS first detected it. In fact, the Descent Imager (DISR) on the probe detected haze at ~21 km level (Tomasko et al. 2005), which is most likely the upper extent of the solid methane particles. Such a transition from liquid methane droplets below ~15–16 km to solid particles above is entirely consistent with the principles of equilibrium thermodynamics (Atreya et al 2006; Tokano et al. 2006). The detection of both an upper ice cloud of methane and a drizzle of liquid methane below has been reported by Adamkovics (2007) from their infrared obser-

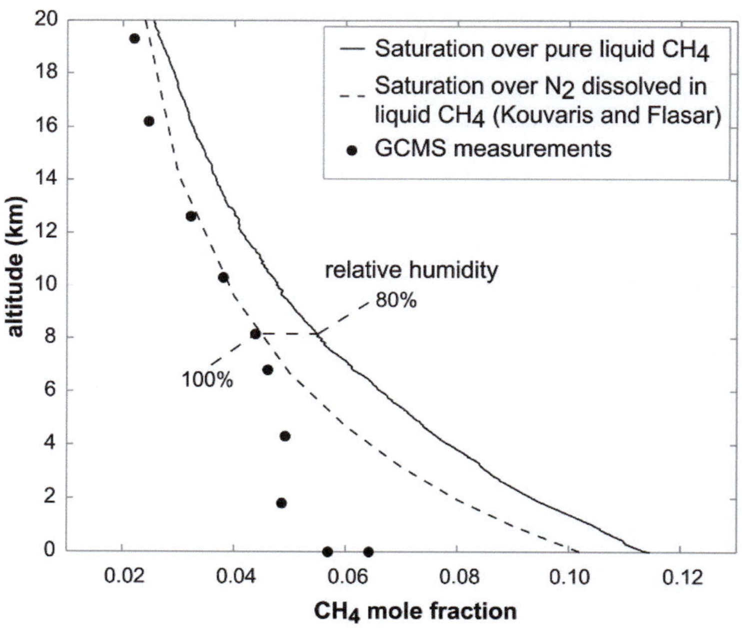

Fig. 7.4 GCMS measurements of the CH_4 mole fraction (*dots*) are shown against the saturation mole fraction curves for the CH_4 vapor over pure CH_4 liquid (*solid line*) and for the N_2 dissolved in the CH_4 liquid (*dashed line*; Kouvaris and Flasar 1991). The relative humidity of CH_4 is ~50% just above the surface. CH_4 is subsaturated up to 8 km, but reaches 100% relative humidity at this altitude, indicating the onset of condensation. The discrepancy between the data and model (*dashed curve*) above ~15 km is because of the change to solid phase of methane (after Atreya et al. 2006)

vations in the equatorial region covering the continent of Xanadu.

Based on the above cycle the average annual rainfall of methane in the equatorial region, such as the landing site of the Huygens probe, would amount to less than 10 cm per year, similar to the annual rainfall in terrestrial deserts. From time to time, convective storms could take place. Lorenz (2000) poses a global average energy limit on re-saturation of the atmosphere if a storm locally depletes it. The time interval between storms is estimated to be 100–1000 years, and could be much longer at a given location. Greater and more frequent precipitation is expected in the polar regions. Lakes have been inferred from the Cassini radar observations, largely at high northern latitudes of 78°N and higher (Stofan et al. 2007). The larger surface reservoir of liquid methane and the 2-4 K colder temperatures in the polar regions than the equatorial region (Flasar et al. 1981) result in ideal thermodynamic conditions for greater condensation and subsequent rainfall at the polar latitudes. The pole-to-pole circulation of methane is expected to occur on time scales of 10–100 years (Lunine and Atreya 2008).

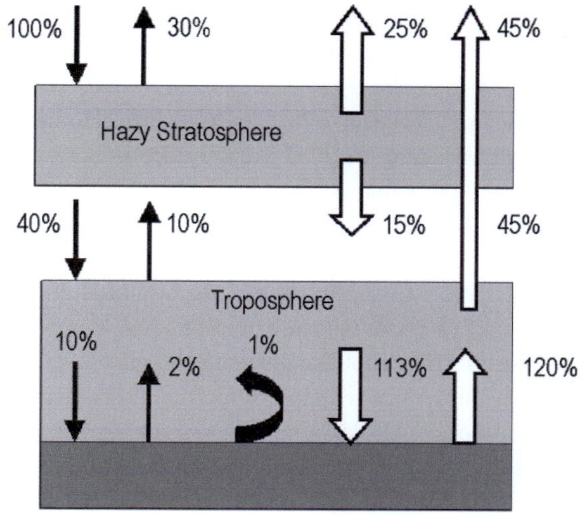

Fig. 7.5 Globally-averaged, annually-averaged energy balance of Titan's surface and atmosphere (original data from McKay et al. 1991). Fluxes are expressed as a percentage of the top-of-atmosphere insolation of 3.7 Wm^{-2}. Solid arrows are shortwave (solar) fluxes; open arrows are the longwave (thermal) fluxes. Curved solid arrow indicates convective flux driving weather and clouds of ~0.04 Wm^{-2}

7.3.1.1 Clouds

While it was speculated, largely on the near-coincidence of the Voyager-determined surface temperature with the triple point of methane, that methane might participate in a hydrological cycle as water does on Earth, there was no evidence to suggest an ongoing hydrometeorology on Titan until the late 1990s. Griffith et al. (1998) inferred the existence of clouds in Titan's upper troposphere from near-infrared ground-based spectroscopy acquired in 1995. This cloud system covered some 10% of Titan's disk, and may well be related to a cloud observed at nearly the same time at around 40° N by the Hubble Space Telescope (Lorenz and Mitton 2008; Lorenz 2008). No direct compositional determination of the cloud was made, but its altitude was consistent with that at which methane convective clouds might be expected to form. Further observations (Griffith et al. 2000) showed that the cloud cover varied on hourly timescales, suggesting that the clouds were actively convecting and/or precipitating.

In Titan's thick atmosphere and low gravity, even the largest raindrops fall gently. Lorenz (1993) showed that drops – controlled by the balance of surface tension and aerodynamic forces – could grow to about 9.5 mm in diameter (compared with about 6.5 mm for water drops on Earth; this is a statistical limit, occasional giants can exist for short periods) but would fall at only 1.6 m s^{-1}, compared with 9 m s^{-1} on Earth. Furthermore, the heat flux available to drive convection and evaporation from the surface in McKay et al.'s (1989, 1991) radiative–convective model is only ~1% of the incident sunlight, or about 0.04 Wm^{-2} on average (Fig. 7.5). Dividing by the latent heat of methane yields an average rainfall rate of only 1 cm per Earth year (the corresponding calculation for Earth yields, correctly, the typical rainfall of about 100 cm per year). These factors initially suggested (Lorenz and Lunine 1996) that rainfall would not be a major agent of erosion on Titan. However, Lorenz (2000) cautioned that, as with desert regions on Earth, landscapes can be fluvially dominated even with low average rates of precipitation, if that average rate is manifested by rare, but violent, rainstorms. Griffith et al. (1998) had noted that the Titan atmosphere holds several meters liquid equivalent of precipitable methane.

Clouds began to be observed consistently on Titan from about 2001, as a result of several factors. Not least was the increased intensity of observation – several large telescopes able to image Titan became available around this time, and the performance of their adaptive optics systems got progressively better. Compounding the telescopes' availability was the establishment that Titan had time-variable weather, making it worthwhile to observe. Another factor was Titan itself, which cooperated by providing regular displays of large clouds around the south pole, which was at that time approaching midsummer. The association of the clouds with those latitudes where solar heating was strongest suggested that these clouds are at least driven by convection.

Griffith et al. (2005) observed the cloud tops of evolving clouds at southern midlatitudes (41°–61° S) in December 2005 (TB) using VIMS. Spectral data (the fit of the 2.1–2.2 μm spectral region) indicate cloud-top heights and overall opacity. The evolution of the cloud-top heights suggest that small centers of vigorous updraft existed in the clouds, in some cases observed to be rising at 2–4 m s^{-1} in larger cloud systems, and 8–10 m s^{-1} in more compact (perhaps less well-mixed or younger) clouds.

Spectra from Cassini's Visual and Infrared Mapping Spectrometer in 2005 revealed the presence of a vast tropospheric cloud on Titan at latitudes 51° to 68° N (Griffith et al. 2006) and the feature likely extended to the pole, although winter darkness prevented its observation. The derived characteristics indicate that this cloud is composed of ethane and forms as a result of stratospheric subsidence and the particularly cool conditions near the moon's north pole during the winter season. Preferential condensation of ethane, perhaps as ice, at Titan's poles during the winters may partially explain the lack of liquid ethane oceans on Titan's surface at middle and lower latitudes.

Tokano et al. (2001) modeled the methane transport in the troposphere with a global circulation model including condensation. They suggest that the latitudinal distribution of methane humidity indicated by Voyager (2% mole fraction at high latitude, about 5% at low latitude, Samuelson et al. 1997) could not be reproduced by purely atmospheric processes, and that the surface must somehow influence the humidity. Cloud data were not given in this paper.

A similar conclusion was reached by Rannou et al. (2006) who noted that the circulation would tend to desiccate low latitudes, yet the Huygens data had by this time confirmed the Voyager indication of ~5% methane mixing ratio at low latitude and thus a humidity of ~50%. Rannou et al. (2006) suggested that a low-latitude methane source was required. This model predicted cloud opacity as a function of latitude and season, notably suggesting persistent clouds at both poles in all seasons, and a seasonal midlatitude cloud in fall (e.g. at 40°N between 1985 and 1995).

Ongoing Cassini and ground-based observations show that the south polar convective clouds seem to have ceased towards the end of 2005. Sporadic clouds have been seen since at a variety of latitudes – these are discussed in the companion chapter by Lorenz et al. (2009).

7.3.2 Methane and the Climate on Titan

The climate of a planetary body reflects the balance between the absorption of sunlight, and the emission of thermal radiation to space. In the absence of atmospheric effects, these in turn depend on the body's distance from the Sun, and the optical reflectivity or albedo of the surface; the deviation from unity of the emissivity of most planetary surfaces is small and usually ignored. An atmosphere complicates the picture considerably, typically by reflecting and absorbing sunlight, and by blocking thermal emission. These factors are all significant at Titan, and were first quantified, analytically, by Samuelson (1983), who showed that the main roles of Titan's haze are to absorb blue light, while red and near-infrared light is scattered and penetrates to the surface, and the gas in the dense atmosphere acts as a powerful greenhouse, absorbing thermal infrared radiation. A further analytic investigation of these effects for Titan and for the early Earth was made by McKay et al. (1999).

The radiative balance was evaluated in much more detail with a numerical radiative–convective model constructed by McKay et al. (1989) – this considered 24 spectral intervals in the visible and near-infrared and some 46 intervals in the thermal infrared, including opacities for nitrogen, methane and hydrogen. The atmosphere was divided into 30 plane-parallel layers, and a microphysics code calculated the optical properties of spherical tholin haze particles in each level. The effects of the haze profile, and methane absorption, allowed the model to be adjusted to fit the visible/near-IR spectrum of Titan measured from the ground, and the temperature profile was adjusted to achieve radiative balance. Convective adjustment was applied where the profile exceeded the local adiabat. The resultant profile could be compared with the temperature profile measured by the Voyager radio-occultation experiment. This model was used to show that condensation clouds of methane played a minimal role in Titan's overall radiative balance, a conclusion that has been borne out with time.

This model, albeit a globally averaged one, allowed the competing effects (McKay et al. 1991) of antigreenhouse cooling by stratospheric haze (analogous to the hypothetical 'nuclear winter' scenario for the Earth) and greenhouse warming in the troposphere to be evaluated in familiar terms. In essence, the effective temperature of Titan, which is determined by the distance from the Sun and the planetary albedo, is 82 K – the antigreenhouse effect drops the temperature by 9 K, whereas the greenhouse increases it by 21 K, giving a net overall greenhouse warming of 12 K.

Meanwhile it was recognized that several effects would drive Titan's climate with time. Not only did the solar luminosity change, forcing the climate directly, but also since methane is a condensible gas at Titan's conditions, a feedback exists wherein the amount of gas (and thus the strength of the greenhouse thermal opacity) depends on the temperature, and this feedback can magnify the effects of the solar forcing. Furthermore, over time, the relative amounts of methane and ethane in a surface-atmosphere reservoir would change as photolysis progressively converted methane to ethane. Additional radiative effects, depending on the photolysis rate, would include the amount of haze in the atmosphere, and the small but significant opacity due to molecular hydrogen, also produced by photolysis.

An initial attempt to explore the solar forcing and volatile greenhouse effects was made by Lunine and Rizk (1989) who used a semianalytic grey radiative model. This effort found that temperatures on Titan several gigayears ago could have been rather lower, by 10–15 K, than present, largely due to the photochemical evolution of the greenhouse inventory and slightly amplified by the solar luminosity variation. These effects were explored in more detail by McKay et al. (1993) who noted that the sensitivity of Titan's climate to solar luminosity and possible surface heating was considerably amplified by the existence of a large surface volatile reservoir. In other words, if Titan just had a few surface lakes, the surface temperature and pressure would change only modestly over the age of the solar system, while if there was a massive ocean, the pressure could have been much lower in the past (~0.2 bar 4 Gyr ago) than at the present (Fig. 7.6). Under these conditions, the ocean would have been frozen at about 74 K. McKay et al. (1993) kept present-day hydrogen greenhouse and haze production rates, and noted, but did not study, the further complication that under these cold early conditions, nitrogen ice clouds would form in the atmosphere and affect the radiative balance.

The possibility that Titan's atmosphere may have been appreciably thinner in the past prompted consideration of possible geomorphological signatures of such history (e.g. Lorenz et al. 1995), in particular the increased abundance of small craters whose formation would be inhibited by atmospheric shielding at the present epoch (Engel et al. 1995). Further studies with the McKay et al. radiative–convective model explored a couple of additional scenarios. The situation of Titan around a red-giant star (either our own, some 5 Gyr from now, or perhaps some Titan-like body around a Saturn-like planet around a red giant elsewhere in the galaxy) was explored

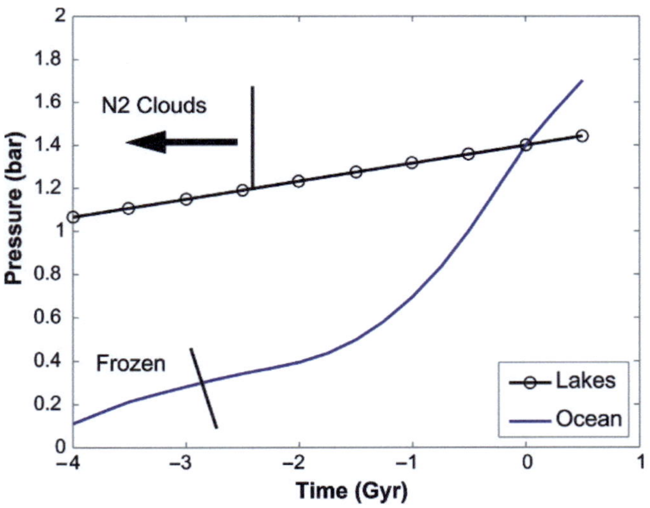

Fig. 7.6 Possible history of Titan's atmospheric pressure over the age of the solar system, assuming photolytic conversion of a methane ocean and the production of haze and hydrogen at present rates (data from McKay et al. 1993). If there are only lakes of methane (with insufficient volume to dissolve much of the atmospheric inventory of N_2) the evolution is modest, but if a large ocean of methane existed, the pressure could have been much lower when the Sun was fainter several Gyr ago

in Lorenz et al. (1997a). Here, strong heating causes the atmosphere to 'puff up' – higher temperatures give a large scale height, and since haze is assumed to form at a fixed low pressure the haze column becomes very large and surface temperatures are not substantially enhanced. However, as the solar spectrum changes to admit more penetrating red light, and to produce less UV and thus produce less haze, surface temperatures can increase profoundly – well above the ammonia-water melting point of 176 K – even without invoking the enhanced greenhouse due to a large volatile reservoir.

Lorenz et al. (1997b) also explored the climate in another direction – with present-day and reduced solar luminosities, but with the methane abundance as an adjustable parameter to allow for the possibility that methane supply to the surface-atmosphere system might be episodic, while its photochemical destruction is continuous. Thus there may have been episodes, perhaps many and sometimes long, when there was no methane available. This not only removes the CH_4–CH_4 and CH_4–N_2 collision-induced thermal opacity, but also (as hydrogen escapes over a subsequent period) would remove the N_2–H_2 greenhouse. These cooling effects are only partly offset by the loss of near-infrared CH_4 solar absorption and the clearing of the atmospheric haze: removal of methane drops the present-day surface temperature by about 5 K, although stratospheric temperatures fall dramatically from about 180 K today to around 60 K.

Lorenz et al. (1999) made a set of convenient–analytic fits to an ensemble of several hundred runs of the McKay et al. radiative-convective model, and reported the joint sensitivity of surface temperature to methane humidity and surface albedo: under present conditions, equilibrium surface temperatures rise by about 1 K for every 0.1 decrease in surface reflectivity, which might lead to some variation of surface temperatures across Titan's variegated surface. A drop of 0.1 in surface relative humidity from the present value of ~0.5 similarly results in a 1 K surface temperature drop. Also, temperatures drop by about 1 K for every 1 km increase in elevation. These analytic fits allowed further exploration of the evolution of the coupled surface-atmosphere climate system. Such a system (and the effect applies also to the Martian climate, where a substantial fraction of the total CO_2 inventory seasonally freezes onto the surface, as well as the water greenhouses on Earth and paleo-Venus) will have an equilibrium where the surface temperature and the resultant atmosphere are self-consistent, in that the surface-atmosphere thermodynamic equilibrium $P = f(T)$ and the radiative equilibrium $T = f(P)$ curves cross, where T represents the surface temperature and P represents the partial pressures of the various gases. As first noted by McKay et al. (1993) a small inventory ('lakes') is well behaved, having only one equilibrium (Fig. 7.7). However, if the volatile inventory of methane, ethane and nitrogen is large, there is the potential for strong positive feedback and the two curves form three equilibria of

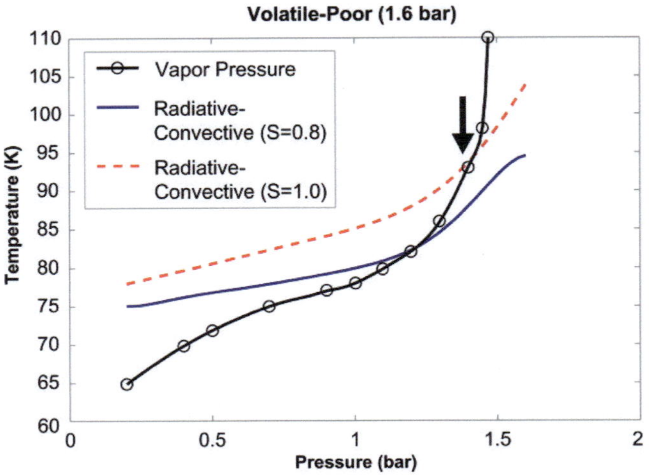

Fig. 7.7 Well-behaved surface-atmosphere equilibrium for a volatile-poor Titan (with ~0.1 bar of liquid at the present day, plus 1.5 bar atmosphere). The solid black curve with circles represents the surface-atmosphere thermodynamic equilibrium, in effect the vapor pressure of the ocean as a function of temperature. The red dashed curve represents the radiative–convective equilibrium, i.e. the temperature as a function of the greenhouse effect in the atmosphere for the present solar constant. Where the red and black curves cross (*marked with an arrow*) indicates a stable self-consistent equilibrium. The stippled blue curve shows the same for a solar constant of 80% the present value – because the greenhouse curves are much shallower than the thermodynamic equilibrium curve, the crossing point moves only gradually as the solar constant increases (data from Lorenz et al. 1999)

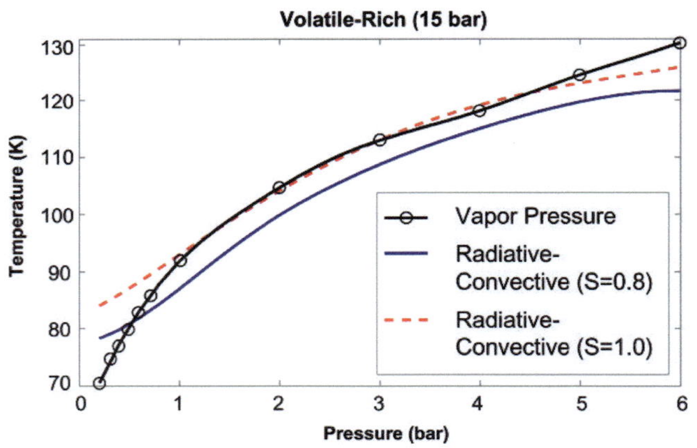

Fig. 7.8 Extreme climate sensitivity in a volatile-rich Titan (with 16 bar of volatiles). This figure is the same as figure 7.7, but with a larger volatile inventory. The corresponding vapor pressure curve (*black*) yields a thicker atmosphere for a given temperature, and the slope of the curve is very similar to the greenhouse curves. For a low solar constant (*blue*) there is a single stable self-consistent equilibrium. However, for the present solar constant (*red*) there are three equilibria. The central crossing is unstable, but two stable equilibria (*arrowed*) exist – one at present conditions, the other with a much higher temperature and pressure. Small perturbations may cause the climate to flip between these conditions (data from Lorenz et al. 1999)

which only the outer two are stable (Fig. 7.8). The choice of which of these two climates will Titan be in would depend on the history of the system, and indeed discontinuous jumps between them can occur – a runaway greenhouse, or equivalently in the opposite direction, atmospheric collapse. This hysteresis makes climate prediction forward or backwards a significant challenge.

Up to this point we have not considered variations of temperature with latitude. Such variations are important in the surface-atmosphere equilibrium, in that the planetwide strength of the greenhouse effect depends on the partial pressure of the various greenhouse gases, which will be close to thermodynamic equilibrium with the coldest part of the planet. Therefore, it is the coldest parts of the planet, and thus to a large extent the effectiveness of equator-pole heat transport that controls the planetary average temperature.

In simple 1-D energy balance climate models used in the 1970s to study the stability of the terrestrial climate to lower insolation (so-called Budyko–Sellers models) the heat transport – embodying ocean as well as atmospheric motions – is represented by a single heat transfer parameter D, which empirically today has a value of ~1 Wm^{-2} K^{-1}. For Martian paleoclimate studies, the terrestrial value has been scaled by atmospheric pressure. For Titan, however, this approach yields a transport that is too high (Lorenz et al. 2001), and results in an equator-to-pole temperature gradient that is orders of magnitude lower than the 3–4 K suggested by Voyager observations. A much lower value, D~0.02 Wm^{-2} K^{-1} is required, and appears consistent with the idea that some climates, including the Earth's, may select combinations of heat transport modes in order to maximize the entropy production by the turbulent heat flux. While this idea is controversial, it is of interest that both the dimensions and the value itself of the heat transport parameter D are the same as the planetary entropy production dS/dt – absorbed solar flux divided by absolute temperature. For the Earth, this is ~300 Wm^{-2}/300 K = 1 Wm^{-2} K^{-1}, while for Titan the flux is ~3 Wm^{-2} and the temperature ~100 K, yielding dS/dt~0.03 Wm^{-2} K^{-1}.

Even though its thermal inertia is such that the bulk of Titan's atmosphere does not warm and cool with the seasons, the surface temperature may see seasonal variations. These were first explored by Stevenson and Potter (1986) who suggested an analogy with the seasonal frost cycle on Mars. Under their scenario, some meters of methane might condense onto the cold winter pole (i.e. forming a transient liquid polar cap – or seasonal lakes) before re-evaporating in spring. In reality, the seasonal variations (see chapter by Lorenz) are rather more complex, involving varying winds and humidity, as well as the availability of condensation nuclei. Nonetheless, the possibility of seasonal evaporation of surface methane lakes has been underscored by Mitri et al. (2007).

Global circulation models of growing sophistication are being applied to understanding Titan's winds and, in particular, the seasonal and latitudinal variations in methane amount, clouds and precipitation. While these models are able to produce some basic features of the Titan climate such as dry conditions at low latitudes where dunes are seen (e.g. Mitchell 2008) and more abundant clouds at high latitudes (e.g. Rannou et al. 2006), many details remain to be resolved. Further, to date, there has been no latitudinally-resolved modeling of Titan paleoclimates with conditions different from the present.

7.3.3 Photochemical Destruction of Methane in the Stratosphere: The Ethane Ocean Dilemma

The cycle of methane discussed in the above subsections represents at best a closed cycle of surface reservoirs, volatile evaporation, cloud formation, followed by precipitation, similar to the hydrological cycle on the Earth. In the worst case scenario, it may represent a net loss of methane due to surface runoff and loss to the interior of Titan through cracks and fissures in the surface. Any such loss is expected to be temporary, however, as various geological processes are likely to pump the methane back out to the surface over time. On the other hand, photochemistry destroys atmospheric methane irreversibly due to the escape of its product, hydrogen, as shown first by Strobel (1974). Consequently, the lifetime of methane in the atmosphere is ~30 Myr (Wilson and Atreya 2004; Yung et al. 1984). The UV photolysis occurs primarily above ~600 km in the stratosphere and is responsible for about one-third of the total CH_4 destruction rate of 4.8×10^9 cm^{-2} s^{-1} (8.7×10^9 cm^{-2} s^{-1} when referenced to the surface, E.H. Wilson, personal communication 2008), whereas the catalytic destruction by acetylene (C_2H_2) – itself a product of the CH_4 photochemistry – in the lower atmosphere accounts for the rest (Wilson and Atreya 2004). In each case, ethane (C_2H_6) is the principal product, as seen in the chemical scheme below. (About 10–15% of the methane is destroyed in the ionosphere above 900 km, but little ethane is produced.)

$$2(CH_4 + h\nu \rightarrow CH_3 + H) \quad \text{and}$$
$$CH_3 + CH_3 + M \rightarrow C_2H_6 + M$$
$$\text{Net: } 2CH_4 \rightarrow C_2H_6 + 2H$$

$$2(C_2H_2 + h\nu \rightarrow C_2H + H)$$
$$2(C_2H + CH_4 \rightarrow C_2H_2 + CH_3)$$
$$CH_3 + CH_3 + M \rightarrow C_2H_6 + M$$
$$\text{Net: } 2CH_4 \rightarrow C_2H_6 + 2H$$

[M is the concentration of the background atmosphere]

Based on the above photochemical fate of methane, Lunine et al. (1983) predicted that over geologic time, a kilometer-deep ocean of ethane is expected to cover Titan's surface. This was a bold but reasonable prediction for its time, considering the limited state of knowledge about the composition and structure of Titan's atmosphere. A quarter century later, observations made from the Cassini–Huygens give no evidence of a global ethane ocean, however. The lack of ethane oceans may be attributed to a much more complex cycle of methane including photochemistry and geology than was previously envisioned and a better understanding of atmospheric vertical mixing. Current photochemical models show that many other products either compete with or are formed from ethane itself. Of special significance to the question of ethane oceans is the detection of benzene (C_6H_6), which was not included in the model of Yung et al. (1984), which was the basis of Lunine et al.'s prediction of global ethane oceans.

Benzene was identified tentatively by the Infrared Space Observatory (ISO, Coustenis et al. 2003). Its presence at 1–4 ppbv levels in the stratosphere was confirmed by the Cassini Composite Infrared Spectrometer (CIRS, Coustenis et al. 2007), and at much higher mixing ratio of 1–5 ppmv above 900 km by the Ion and Neutral Mass Spectrometer (Waite et al. 2005, 2007). Despite these differences in the mixing ratio, the column abundance of C_6H_6 in these two parts of the atmosphere is similar (Atreya 2007a), indicating potent production mechanisms throughout the atmosphere. Several pathways have been proposed for the formation of benzene in the neutral atmosphere and the upper ionosphere, but all can eventually be traced back to ethane. Benzene is the precursor to the formation of polycyclic aromatic hydrocarbons, PAH's, whose polymerization may result in some of Titan's most abundant haze layers. Both the ionosphere and the stratosphere contribute significantly to the haze.

Thus, contrary to the earlier notions, ethane is not the end product (or even close to it) of the methane photochemistry, but it goes on to produce heavier hydrocarbons that subsequently form the multitude of haze and soot layers in Titan's stratosphere and higher (Wilson and Atreya 2004; Waite et al. 2007). Wilson and Atreya (2004) found that the inclusion of benzene and other chemical pathways in a coupled atmosphere-ionosphere photochemical model led to a reduction in the downward flux of ethane at the tropopause by a factor of 4–6 from a value of $\sim 6 \times 10^9$ cm^{-2} s^{-1} in the model that did not include them (Yung et al. 1984) and was the basis of earlier ethane ocean predictions. The new models thus predict much less condensation of ethane, reducing the depth of a global ethane ocean to 150–250 m from the previous predictions of a kilometer.

Atreya et al. (2006) found that the vertical distribution of ethane is also significant factor in determining the amount of ethane condensation. They found the best fit to the CIRS data on ethane requires substantially smaller eddy diffusion below 200 km than that in the previous models. The smaller stratospheric mixing seems consistent also with the conclusion of Yelle et al. (2008) who found a value of $2–5 \times 10^7$ cm^2 s^{-1} for the eddy diffusion coefficient at the homopause, using

^{40}Ar as the observational constraint, or about an order of magnitude smaller than in pre-Cassini models where the analysis was based on the CH_4 distribution alone (see chapter by Strobel et al. (2009) for a comprehensive discussion of composition and structure). Since the eddy diffusion coefficient varies inversely with some power of the atmospheric density, the lower values assumed by Atreya et al. (2006) for the lower atmosphere are reasonable. The reduced eddy mixing results in a further reduction in the downward condensation flux of ethane, by 30–50%. The reduction factor is probably larger, as the Atreya et al. (2006) model was done prior to the complete analysis of the upper atmospheric mixing from the INMS data (Yelle et al. 2008) that indicated a more severe reduction of eddy mixing in the upper atmosphere. Thus, with the above new constraints on atmospheric mixing and the composition from Cassini, the depth of a global ethane ocean is expected to be no greater than 100–150 m, not a kilometer.

Finally, geologic considerations indicate that the depth could be even smaller. Models of the interior show that methane on Titan may be outgassed only episodically, with the last such episode occurring 600 Myr ago (Tobie et al. 2006). This would further reduce the possible depth of the ethane ocean to less than 10 m over the geologic time. Since any ethane condensation would be gradual, it is quite likely that even a good fraction of this last bit may be largely sequestered as ethane clathrate in Titan's regolith (Atreya et al. 2008; Lunine and Atreya 2008). Mousis and Schmitt (2008) estimate that a cryovolcanic icy crust of less than 2.3 km thick is required to bury all the ethane (and other less abundant hydrocarbon liquids) produced over the geologic time. Since the ethane rain ultimately originates in the stratosphere (Sections 7.3.1.1 and 7.3.3), its sequestration as ethane-clathrate is expected to be at shallow depths just below the surface, unlike the methane clathrates that are predicted to be present in the high pressure ice some 50–100 km below Titan's surface (Section 7.3.4).

In conclusion, although roughly 40% of Titan's methane is expected to be converted to ethane over geologic time, global oceans of ethane are no longer predicted by current photochemical–geological models, nor are they seen in the Cassini data. On smaller time scales, ethane may still condense in the atmosphere, especially at mid-high latitudes, and form lakes. Thermodynamic considerations predict a relative humidity of ethane ranging from 80% to 100% in the equatorial region, as shown in Fig. 7.9 (Atreya et al. 2008). Thus the condensation of ethane there would be at best marginal, and moreover any ethane drizzle may be sequestered in shallow subsurface clathrates. This seems to be the case in the Huygens landing site (10.3° S, 192.37° W), where the probe mass spectrometer detected a slow evaporation of ethane from Titan's moist surface (Niemann et al. 2005). On the other hand, at polar latitudes, where the temperatures are 2–4°cooler, ethane would almost certainly condense (Fig. 7.9). Since ethane liquid is fully miscible in the methane liquid, Titan's polar lakes are expected to be composed largely of a methane–ethane mixture (Atreya et al. 2008; Lunine and Atreya 2008). Brown et al. (2008) have, in fact, identified spectral features of ethane in Ontario Lacus (72° S, 183° W), which was observed by VIMS through Titan's atmospheric windows at 2.0, 2.7 and 5.0 μm during Cassini's 38th close flyby of the satellite. The authors conclude that "ethane, probably in liquid solution with methane, nitrogen and other low-molecular mass hydrocarbons, is contained in Ontario Lacus". (The very large abundance of methane gas in the atmosphere prevents the identification of methane liquid on the surface.) Since the stratospheric mole fraction of ethane vapor is about a factor of 10^4 smaller than the CH_4 mole fraction, ethane liquid may constitute a small fraction of the methane–ethane lakes at any given time. However, a realistic assessment of eventual fraction of the ethane liquid in the lakes is a complicated matter. Lorenz (1993) suggested that pure methane droplets may not survive their descent through Titan's troposphere. Even the survival of a ternary mixture of CH_4-N_2-C_2H_6 droplets against evaporation depends on the relative humidity of C_2H_6 (Graves et al. 2008). Until the distribution of ethane is determined from the lower stratosphere down to the surface, the fractional amount of ethane in the polar lakes will continue to remain highly uncertain.

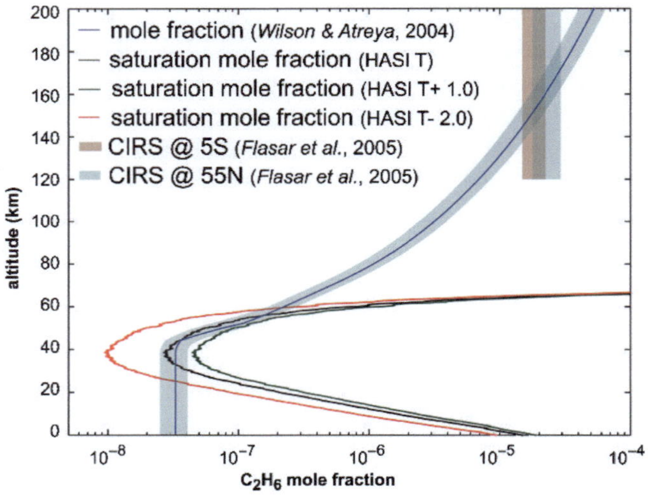

Fig. 7.9 Photochemical profile of ethane with uncertainty envelope (*blue curve*) and its comparison with (**a**) the CIRS data at 5°S and 55°N (*vertical bars, with uncertainty envelopes*), and (**b**) C_2H_6 saturation vapor pressures, using the central HASI temperatures near the equator (*black curve*), HASI temperatures increased by 1° to accommodate the uncertainty in the temperature determination (*green curve*), and temperatures reduced by 2° relative to the HASI equatorial temperatures, to mimic high polar latitude conditions (*red curve*). The resulting C_2H_6 relative humidity at the tropopause is 75–100% in the equatorial region, and over 200% at the polar latitudes (after Atreya et al. 2008)

7.3.4 Methane Replenishment – The Source of Methane

The time scale for the photochemical conversion of the present atmospheric abundance of methane to heavier hydrocarbons, nitriles and hazes on Titan is ~30 Myr, as discussed above. In the absence of recycling or replenishment, Titan would have lost its methane, and therefore much of its nitrogen atmosphere, a long time ago. It is conceivable that the level of methane in the atmosphere has fluctuated over geologic time. However, it seems unlikely that Titan could lose all of its methane at some point in time and then some process would trigger a re-supply of the entire atmosphere at another epoch, considering the sheer amount of methane and nitrogen involved. The most plausible scenario is that the methane destroyed in the atmosphere is somehow replenished from time to time. The $^{12}C/^{13}C$ isotope ratio provides evidence that this must be happening. On Titan, the measured $^{12}C/^{13}C$ ratio of 82.3 (Table 7.1) is only slightly below the terrestrial inorganic value (89.45), or 92% terrestrial, whereas the $^{14}N/^{15}N$ ratio is only 67% terrestrial (Table 7.1), or less if the INMS extrapolation from thermosphere to the homosphere is used (the slight excess of the ^{13}C isotope on Titan could result from methane escape or geology in Titan's interior). This implies irreversible escape of nitrogen from the atmosphere, but replenishment of methane. Unlike the giant planets where the products of the methane chemistry can be recycled back to methane by thermochemical reactions in the planets' H_2-rich, hot and dense interiors, the hydrocarbon and nitrile aerosols cannot be turned back into methane in Titan's cold interior even if the aerosols could all wash into the interior of this solid object. Therefore, a direct source of methane itself is required, irrespective of where the methane came from in the first place. No significant source of methane exterior to the satellite can be identified, so the resupply needs to occur from within. Indeed, if methane is stored as a clathrate-hydrate ($CH_4.6H_2O$) in the high pressure ice in Titan's interior, as has been suggested by Lunine and Stevenson (1985, 1987), Tobie et al. (2006) and others, it could be the source of atmospheric methane, irrespective of its ultimate origin. Any number of processes, including impacts and cryovolcanism (Lunine et al. 2009; Tobie et al. 2006; Sotin et al. 2005) can destabilize the clathrates and release the stored methane to the atmosphere. It is important to ask whether Titan's methane cache is large enough to replenish the amount destroyed by photochemistry in the atmosphere. We address this question below.

Originally, Titan should have accreted approximately four times more carbon (C) than nitrogen (N) (using the solar elemental abundances of Grevesse et al. 2005), considering that the inter-elemental abundance ratios in Titan should reflect the solar composition of the Saturnian subnebula where Titan was formed. Assuming that most of the carbon ended up as methane, and that the primordial atmosphere of Titan contained 5–10 bar of N_2 (Section 7.2.2.4), the partial pressure of methane then should have been 30–80 bar. Thus, the average partial pressure of CH_4 on primordial Titan is expected to be ~55 bar, or 750 times its present partial pressure using the present CH_4 mole fraction of ~5%. Using the current photochemical destruction rate of 4.8 × 10^9 cm^{-2} s^{-1} for CH_4 from Wilson and Atreya (2004), we find that only 4%, or roughly 2.5 bar, of the original methane has been destroyed over geologic time. Today, the atmosphere contains ~75 mbar of CH_4. Therefore, most of the original methane should still be present on Titan, presumably stored as clathrate in Titan's interior with a small fraction as liquid in the lakes. This implies that Titan should have plenty of methane storage to replenish the gas destroyed in the atmosphere. It would take up to 100 billion years to destroy the entire methane inventory of Titan!

The above conclusion is based on several simplifying assumptions, including a constant rate of photochemical destruction throughout the geologic time, and that all of the carbon was sequestered in methane. Nevertheless, it is fairly robust, considering that the solar UV flux has remained nearly uniform after the Sun entered its main sequence and methane is by far the dominant carbon compound formed, irrespective of whether it was produced in the Saturn subnebula or in the interior of Titan. Somewhat larger or smaller amounts than given above could have been destroyed depending on the past history of methane, but it is not likely to change the basic conclusion, considering the wide margin between the amount destroyed and the potential reservoir. As discussed later in this section, ~2% of the original ^{40}Ar has escaped (Waite et al. 2005), which is comparable to the 4% of the CH_4 destroyed (and replaced). This indicates comparable rates of release considering that ^{40}Ar is released from the core whereas CH_4 is released from only 50–100 km below the surface. Related to the question of the replenishment of methane is whether the liquid contained in the lakes alone is sufficient to replace the methane destroyed in the atmosphere. An estimate of methane in the lakes can be obtained by considering the available radar observations and using terrestrial analogy.

Above 55° N latitude, imaging radar observations have covered 55.4% of the terrain through Titan's 30th close encounter in May 2007. A variety of lake types covering almost 10% of that terrain are seen (Hayes et al. 2008; Lorenz et al. 2008a) ranging in size from a few kilometers, which is the smallest recognizable in data of ~300 m/pixel, to several hundred kilometers. The combination of low radar reflectivity and high microwave brightness, thus implying high emissivity, and the feature morphology and association with channels all point to present-day liquid, presumably methane and ethane. Lorenz et al. (2008b) attempted to estimate the liquid volume in the lakes. Using terrestrial analogs as a guide (e.g. typical lake depth in meters equals its dimension in kilometers), the typical 20-km wide lake may be expected to have a depth on the order of 10–20 m. Multiplying by the radar-observed lake area of 400,000 km^2 and assuming similar contributions from the rest of the Northern hemisphere as yet unobserved by radar yields an inventory of ~2 × 10^4 km^3 of liquid – some hundreds of times the known oil and gas reserves on Earth (BP 2007). However, if depth scales with size, the inventory (e.g. Lorenz 1998) is dominated by the single largest lake, and such an average approach may underestimate the inventory – e.g. Ligeia Mare and Kraken Mare are on the order of 200 and 400 km across, and thus may have average depths ten times higher than that above, with a correspondingly higher total inventory, in other words >2 × 10^4 km^3 of liquid.

The discussion above pertains only to the North. While radar coverage of the south polar regions is presently very limited with only the 39th close flyby observing beyond 70° south latitude so far, southern coverage in the Cassini Equinox Mission will become comparable with the present north polar coverage. From this small southern sampling a couple of lake features are evident, and a couple of years earlier a likely lake candidate Lacus Ontario (~235 km across) was identified in ISS images (Turtle et al. 2009). However, the overall impression (Lunine et al. 2008) is of rather less lake coverage than in the north, and thus a complete inventory of Titan's bulk liquid reservoir as more data emerge is likely to increase by no more than a few tens of percent from the estimate given above for the north only. Therefore, for the time being we take the total volume of methane liquid in all of Titan's lakes to be ~3 × 10^4 km^3, keeping in mind that it could be ten times higher as discussed above.

As discussed above, ~2.5 bar, or 1.4 × 10^{19} kg of methane is estimated to have been destroyed over geologic time, whereas the total CH_4 inventory in the lakes is 1.3 × 10^{16} kg using the lower limit of the lake volume and 1.3 × 10^{17} kg if the lake volume were ten times higher. Thus, the lakes are capable of replenishing no more than 1% of the methane destroyed by photochemistry in the atmosphere. This means that if methane was present in Titan's atmosphere throughout most of its geologic history, it had to be supplied from Titan's interior from time to time. Episodic outgassing of methane was proposed by Tobie et al. (2006) whose model indicates that the last such episode took place 600 Myr ago. Evidence of outgassing from the interior was provided by the Huygens GCMS that measured smaller than expected abundance of ^{40}Ar (Table 7.1) in the atmosphere (Niemann et al. 2005) and from INMS (Waite et al. 2005, 2007). ^{40}Ar is a product of radioactive decay of ^{40}K whose half-life of 1.3 Gyr is much shorter than the geologic time of 4.5 Gyr, yet not all of the ^{40}Ar has apparently leaked out to the atmosphere. Waite et al. (2005) estimate that ~2% of the ^{40}Ar that was produced originally has escaped to date. Since ^{40}K is a component of the rocks in the core of Titan, detection of its product, ^{40}Ar in the atmosphere, indicates that outgassing is taking place from even a couple of thousand kilometers below the surface. Hence methane, which is believed to be stored as clathrate in the ice shell above a water-ammonia ocean presumed to be present at a depth of perhaps 50–100 km below the surface, should have no difficulty getting out when the clathrates are destabilized. When it does, it would be in the form of liquid at the surface, re-charging the lakes and the atmosphere in turn with methane.

7.3.5 Origin of Methane

Though methane is believed to be stored in Titan's interior, it requires an explanation as to where it came from in the first place. Kinetic models of the Saturnian subnebula have wavered between methane and non-methane carbon-bearing compounds including CO_2, CO, etc. as the principal form of carbon in the subnebula. As a result, two distinct possibilities for the origin of methane were envisioned (a) methane was delivered *to* Titan by the planetesimals during accretion, or (b) methane formed in situ *on* Titan, from non-methane molecules, such as CO, CO_2, etc. A possible test of the hypotheses was expected to be provided by the measurement of the abundance of the heavy noble gases, argon (^{36}Ar), krypton (Kr) and xenon (Xe). Mousis et al. (2002) predicted nearly solar ratios of $^{36}Ar/C$, Kr/C and Xe/C in Titan's atmosphere if the conversion of CO to CH_4 (and N_2 to NH_3) occurred in the subnebula, as in the model of Prinn and Fegley (1981). In this scenario, CH_4 was produced in the subnebula and delivered as CH_4 to Titan during accretion. Mousis et al. (2002) argued, however, that instead of forming in the subnebula, CH_4 is trapped, along with NH_3 and Xe as clathrate-hydrates from the feeding zone of Saturn. Their pre-Cassini–Huygens model predicted six times solar Xe/C ratio in Titan's atmosphere, and four and five times solar $^{36}Ar/C$ and Kr/C, respectively, if these latter gases were fully trapped (Ar and Kr require 10–15 K lower trapping temperatures than Xe). In this scenario also, methane was delivered to Titan during accretion.

The Huygens GCMS made the first measurements of the noble gases, and found $^{36}Ar/CH_4$ to be a factor of 10,000 times less than the predictions of the Mousis et al. model. Moreover, Xe and Kr were not even detected, with their mixing ratios below the 10^{-8} detection limit of the GCMS. This seems to imply that methane was not delivered to Titan directly as CH_4, otherwise the heavy noble gases, especially Xe which forms clathrates at similar temperatures as CH_4, would certainly be present in Titan's atmosphere. The unexpected non-detection of Xe and Kr and the extremely low abundance of ^{36}Ar has spawned a number of possible explanations, including gradual absorption of Xe and Kr, but not argon, into clathrates near the surface (Thomas et al. 2007), burial of Xe and Kr as clathrates in Titan's interior below the water-ammonia ocean (Lunine et al. 2009; Tobie et al. 2008), and sequestration of Xe and Kr in aerosols (Jacovi and Bar-Nun 2008). The surface clathrate model fails to explain the highly subsolar argon in the atmosphere. The aerosol hypothesis fails to explain why the heavy noble gases were not detected by the GCMS when the surface material was vaporized. The trapping below the ocean hypothesis is somewhat selective and does not allow for the possibility of any destabilization of noble gas clathrates at all during the entire geologic time, but is arguably one of two current leading hypotheses for the origin of Titan's methane, the other being serpentinization as discussed below. In the final section of this chapter, we discuss future measurements that could help discriminate between the various hypotheses for the origin of Titan's methane.

An alternative hypothesis for the origin of methane is that methane did not arrive at Titan as CH_4, but was actually formed on Titan. The process involves water-rock reactions – serpentinization – in which hydration of Fe, Mg or Cr-rich ultramafic silicate minerals, such as olivine [(Mg, Fe)$_2$SiO$_4$] or pyroxene [(Mg, Fe)SiO$_3$], first liberates hydrogen ($H_{2(aq)}$), which reacts in turn with the primordial carbon in the form of CO_2, CO, carbon grains or organic material to produce CH_4, as below:

$$CO_{2(aq)} + [2 + (m/2n)]H_{2(aq)} \rightarrow (1/n)\, C_nH_m + 2H_2O$$
$$CO_{2(aq)} + 4H_{2(aq)} \rightarrow CH_4 + 2H_2O$$
$$C + 2H_{2(aq)} \rightarrow CH_4$$

Atreya et al. (2006) have estimated that the yield of methane from the above process is adequate to explain Titan's methane, and the most plausible period for the production of methane is in the early stages of the formation of Titan when the accretionary heating and heating due to decay of short-lived radioactive elements allowed the water-ammonia ocean to extend all the way to the rocky core. Black Smokers (hydrothermal vents) in terrestrial oceans are known to produce methane and other hydrocarbons by serpentinization, but at relatively high temperatures of 350–400°C. On the other hand, copious quantities of methane are also produced by serpentinization in Lost City, some 15–20 km from spreading centers, where the temperatures are 40–90°C, much lower than in Black Smokers (Kelley et al. 2005). Such low-temperature serpentinization may have occurred over a long period in Titan's interior in the past. The methane produced in the interior could then be stored in the high pressure ice in Titan's interior as Titan cooled. The separation of the rocky core from the water ocean by intervening ice on present day Titan prevents current production of methane.

Finally, McKay and Smith (2005) and Schulze-Makuch and Grinspoon (2005) have proposed microbial production of methane on Titan. In their scenario, acetylene and hydrogen, resulting from the action of sunlight on methane in Titan's atmosphere, serve as nutrients, and methane is produced in turn as a byproduct of metabolism: $C_2H_2 + 3H_2 \rightarrow 2CH_4$. Unlike life on Earth that depends on (liquid) water as solvent or medium, the non-aqueous, non-polar methane liquid in Titan's surface serves as solvent or medium for Titan's methanogens. If indeed Titan's methane were produced by such microbes, one would expect the $^{12}C/^{13}C$ to be greater than the inorganic value of 89.5, as is the case for all living things on Earth. However, the $^{12}C/^{13}C$ ratio on Titan was measured to be 82.3, which is smaller, not larger, than the terrestrial inorganic standard. This argues against a biological origin for Titan's methane. One could imagine that Titan's microbes are "exotic", different from life as we know it (see, e.g. NRC 2007), but at this time we have no evidence that is the case. Moreover, if methanogens were responsible for producing even a fraction of Titan's two hundred and fifty trillion tons of methane in the atmosphere today, they would have put a serious dent in Titan's hydrogen and acetylene inventory. We have no evidence of that either. Nevertheless, life as we *don't* know it is an intriguing thought that deserves further inquiry in any future exploration of the satellite.

7.4 Summary and Future Observations

In many ways, Saturn's moon, Titan, seems so similar to the Earth – its cycle of methane is akin to Earth's hydrological cycle, its dominant atmospheric constituent is nitrogen, as on Earth, its nitrogen atmosphere is secondary, most likely a product of ammonia, as on Earth, and in its past, it had all the right conditions –liquid water, methane and ammonia in a warm environment – necessary for forming pre-biotic, perhaps even biogenic species, as on Earth. Yet there are notable differences. Being ten times farther from the Sun than the Earth, Titan receives only one percent of the solar energy compared to Earth, so that the satellite has remained perpetually frozen for 4 billion years. A purported ocean beneath the frozen ice shell can remain liquid only due to an admixture of 5–10% ammonia, an antifreeze, with water. Ammonia is toxic to life as we know it. The average annual rainfall of methane pales in comparison with the annual average rainfall of water on Earth. Titan's atmosphere is full of smog created by chemistry between its two most abundant constituents, nitrogen and methane, and the smog is persistent unlike the Earth. It is these differences and similarities between Titan and the Earth that make Titan such a unique and intriguing object in the solar system. The Cassini–Huygens Mission has opened new vistas into Titan's past and present, but the view is limited. As is the nature of every new endeavor, Cassini–Huygens' tantalizing findings have only whetted our appetites to return to this fascinating object to investigate more probing issues that are beyond the scope of even the extended Cassini mission.

In particular, the origin, fate, and cycle of Titan's methane require further exploration. The missing isotope ratios, particularly D/H in the surface H_2O ice, $^{16}O/^{18}O$ in CO, CO_2 and H_2O, $^{36}Ar/^{38}Ar$, $^{12}C/^{13}C$ in CO, CO_2 and abundant surface organics, determination of Xe and Kr to at least the 10^{-10} mole fraction level in the atmosphere, together with a thorough analysis of the surface material are important for discriminating between various hypotheses of the origin of methane. Long term, synoptic observations of the atmospheric phenomena including clouds and precipitation are essential supporting data. Identification and quantification of complex organic molecules in the vapor, condensed and solid forms from the ionosphere to the surface will be crucial. The composition and evolution of the lakes and the dune material is critical. Monitoring for signs of sudden or episodic events such as cryovolcanoes, fumaroles and impacts will provide an insight into the existence and nature of the purported subsurface ocean, its potential for habitability, as well as the origin of Titan's nitrogen. These and other outstanding questions can best be addressed by exploring Titan as a coupled system of the interior, surface and the atmosphere-ionosphere, employing diverse platforms including balloons, orbiters and landers. In future missions to Titan, a concerted effort should be made to fully characterize Titan's surface, since this is where the atmospheric products reside, but in higher concentration, and this is where the material from Titan's interior would end up in any cryovolcanic eruptions (Atreya 2007a). The surface is key to Titan's mysteries.

Acknowledgments SKA and JHW acknowledge support by the NASA Planetary Atmospheres Program and the Cassini–Huygens Mission. RDL acknowledges the support of Cassini. Patricia Egeler Dowd provided expert help with editing and figures.

References

Adamkovics M, Wong MH, Laver C, DePater I (2007) Widespread morning drizzle on Titan. Science 318:962–965

Adams EY (2006) Titan's thermal structure and the formation of a nitrogen atmosphere. Thesis, The University of Michigan, Ann Arbor

Anders E, Grevesse N (1989) Abundances of the elements – meteoritic and solar. Geochim Cosmochim Acta 53:197–214

Arpigny C, Jehin E, Manfroid J, Hutsemekers D, Schulz R, Stuwe JA, Zucconi J-M, Ilyin I (2003) Anomalous nitrogen isotope ratio in comets. Science 301:1522–1524

Atreya SK (1986) Atmospheres and ionospheres of the outer planets and their satellites. Springer, New York-Berlin

Atreya SK (2007a) Titan's organic factory. Science 316:843–845

Atreya SK (2007b) Saturn probes: why, where, how? Proceedings of the International Planetary Probe Workshop, NASA Jet Propulsion Laboratory Document, Table 1, Chapter 4_6ATREY in http://ippw.jpl.nasa.gov/20070607_doc/

Atreya SK, Donahue TM, Kuhn WR (1978) Evolution of a nitrogen atmosphere on Titan. Science 201:611–613

Atreya SK, Adams EY, Niemann HB, Demick-Montelara JE, Owen TC, Fulchignoni M, Ferri F, Wilson EH (2006) Titan's methane cycle. Planet Space Sci 54:1177–1187

Atreya SK, Lunine JI, Niemann HB, Wilson EH (2008) Where have Titan's ethane oceans gone? Geophys Res Abst 10, EGU2008-A-04876

BP (2007) BP statistical review of world energy. BP, London, UK

Balsiger H, Altwegg K, Geiss J (1995) D/H and $^{18}O/^{16}O$ ratio in the hydronium ions and neutral water from in situ measurements in comet Halley. J Geophys Res 100:5827–5834

Bézard B, Nixon CA, Kleiner I, Jennings DE (2007) Detection of $^{13}CH_3D$ on Titan. Icarus 191:397–400

Bockelee-Morvan D, Gautier D, Lis DC, Young K, Keene J, Phillips T, Owen T, Crovisier J, Goldsmith PF, Bergin EA, Despois D, Wootten A (1998) Deuterated water in Comet C/1996 B2 (Hyakutake) and its implication for the origin of comets. Icarus 133:147–162

Bockelee-Morvan D, Mumma MJ, Weaver HA (2004) The composition of cometary volatiles. In: Festou MC, Keller HU, Weaver HA (eds) Comets II. The University of Arizona Press, Tucson, pp 391–423

Broadfoot AL, Sandel BR, Shemansky DE, Holberg JB, Smith GR, Strobel DF, McConnell JC, Kumar S, Hunten DM, Atreya SK, Donahue TM, Moos HW, Bertaux JL, Blamont JE, Pomphrey RB, Linick S (1981) Extreme ultraviolet observations from Voyager 1 encounter with Saturn. Science 212:206–211

Brown RH, Soderblom LA, Soderblom JM, Clark RN, Jaumann R, Barnes JW, Sotin C, Buratti B, Baines KH, Nicholson PD (2008) The identification of liquid ethane in Titan's Ontario Lacus. Nature 454:607–610. doi:10.1038/nature07100

Castillo-Rogez JC, Matson DL, Sotin C, Johnson TV, Lunine JE, Thomas PC (2007) Iapetus' geophysics: Rotation rate, shape, and equatorial ridge. Icarus 190:179–202

Conklin EK, Ulich BL, Dickel JR (1977) 3mm observations of Titan. Bull Am Astron Soc 9:471

Coustenis A, Taylor F (1999) Titan: the earth-like moon. World Scientific, Singapore

Coustenis A, Salama A, Schultz B, Ott S, Lellouch E, Encrenaz Th, Gautier D, Feuchtgruber H (2003) Titan's atmosphere from ISO mid-infrared spectroscopy. Icarus 161:383–403

Coustenis A, Achterberg RK, Conrath BJ, Jennings DE, Marten A, Gautier D, Nixon CA, Flasar FM, Teanby NA, Bézard B, Samuelson RE, Carlson RC, Lellouch E, Bjoraker GL, Romani PN, Taylor FW, Irwin PGJ, Fouchet T, Hubert A, Orton GS, Kunde VG, Vinatier S, Mondellini J, Abbas MM, Courtin R (2007) The composition of Titan's stratosphere from Cassini/CIRS mid-infrared spectra. Icarus 189:35–62

Danielson RE, Caldwell JJ, Larach DR (1973) An inversion in the atmosphere of Titan. Icarus 20:437–443

Drake JJ, Testa P (2005) The 'solar model problem' solved by the abundance of neon in nearby stars. Nature 436:525–528

Eberhardt P, Reber M, Krankowsky D, Hodges R (1995) The D/H and $^{18}O/^{16}O$ ratios in water from comet P/Halley. Astron Astrophys 302:301–316

Engel S, Lunine JI, Hartmann WK (1995) Cratering on Titan and implications for Titan's atmospheric history. Planet Space Sci 43: 1059–1066

Eshleman VR (1982) Argon in Titan's atmosphere? Science 217:200

Eshleman VR, Tyler GL, Anderson JD, Fjeldbo G, Levy GS, Wood GE (1977) Radio science investigation with Voyager. Space Sci Rev 21:207–232

Eshleman VR, Lindal GF, Tyler GL (1983) Is Titan wet or dry? Science 221:53–55

Flasar FM, Samuelson RE, Conrath BJ (1981) Titan's atmosphere: temperature and dynamics. Nature 292:693–698

Flasar FM and the CIRS Team (2005) Titan's atmospheric temperatures, winds, and composition. Science 308:975–978

Graves SD, McKay CP, Griffith CA, Ferri F (2008) Rain and hail can reach the surface of Titan. Planet Space Sci 56:346–357

Grevesse N, Asplund M, Sauval J (2005) The new solar composition. In: Alecian G, Richard O, Vauclair S (eds) Element stratification in stars: 40 years of atomic diffusion, EAS Publications Series pp 21–30

Griffith CA, Zahnle K (1995) Influx of cometary volatiles to planetary moons: the atmospheres of 1000 possible Titan's. J Geophys Res 100:16907–16922

Griffith CA, Owen T, Miller GA, Geballe T (1998) Transient clouds in Titan's lower atmosphere. Nature 395:575–578

Griffith CA, Hall JL, Geballe TR (2000) Detection of daily clouds on Titan. Science 290:509–513

Griffith CA, Penteado P, Baines K, Drossart P, Barnes J, Bellucci G, Bibring J, Brown R, Buratti B, Capaccioni F, Cerroni P, Clark R, Combes M, Coradini A, Cruikshank D, Formisano V, Jaumann R, Langevin Y, Matson D, McCord T, Mennella V, Nelson R, Nicholson P, Sicardy B, Sotin C, Soderblom LA, Kursinski R (2005) The evolution of Titan's mid-latitude clouds. Science 310:474–477

Griffith CA, Penteado P, Rannou P, Brown R, Boudon V, Baines KH, Clark R, Drossart P, Buratti B, Nicholson P, McKay CP, Coustenis A, Negrao A, Jaumann R (2006) Evidence for a polar ethane cloud on Titan. Science 313:1620–1622

Hanel RA, Conrath B, Flasar FM, Kunde V, Maguire W, Pearl JC, Pirraglia J, Samuelson R, Herath L, Allison M, Cruikshank DP, Gautier D, Gierasch PJ, Horn L, Koppany R, Ponnamperuma C (1981) Infrared observations of the Saturnian system from Voyager 1. Science 212:192–200

Hayes A and the Cassini RADAR Team (2008) Hydrocarbon lakes on Titan: distribution and interaction with a porous regolith. Geophys Res Lett 35:L09204

Hunten DM (1978) A Titan atmosphere with a surface of 200 K. In: Hunten DM, Morrison D (eds) The Saturn System, NASA Conference Publication 2068, pp. 127–140

Hutsemékers D, Manfroid J, Jehin E, Arpigny C, Cochran A, Schulz R, Stüwe JA, Zucconi J-M (2005) Isotopic abundances of carbon and nitrogen in Jupiter-family and Oort Cloud comets. Astron Astrophys 440:L21–L24

Jacovi R, Bar-Nun A (2008) Removal of Titan's noble gases by their trapping in its haze. Icarus 196:302–304

Jewitt DC, Matthews HE, Owen T, Meier R (1997) Measurements of $^{12}C/^{13}C$, $^{14}N/^{15}N$ and $^{32}S/^{34}S$ ratios in comet Hale-Bopp (C/1995 01). Science 278:90–93

Jones TD, Lewis JS (1987) Estimated shock production of N_2 and organic compounds on early Titan. Icarus 72:381–393

Kelley DS, Karson JA, Früh-Green GL, Yoerger DR, Shank TM, Butterfield DA, Hayes JM, Schrenk MO, Olson EJ, Proskurowski G, Jakuba M, Bradley A, Larson B, Ludwig K, Glickson D, Buckman K, Bradley AS, Brazelton WJ, Roe K, Elend MJ, Delacour A, Bernasconi SM, Lilley MD, Baross JA, Summons RE, Sylva SP (2005) A serpentinite-hosted ecosystem: the lost city hydrothermal field. Science 307:1428–1434

Kouvaris LC, Flasar FM (1991) Phase equilibrium of methane and nitrogen at low temperatures – application to Titan. Icarus 91: 112–124

Kuiper GP (1944) Titan: a satellite with an atmosphere. Astrophys J 100:378–383

Lammer H, Stumptner W, Molina-Cuberos GJ, Bauer SJ, Owen T (2000) Nitrogen isotope fractionation and its consequence for Titan's atmospheric evolution. Planet Space Sci 48:529–543

Lewis JS (1971) Satellites of the outer planets: their physical and chemical nature. Icarus 15:174–185

Lewis JS (1973) Chemistry of the outer solar system. Space Sci Rev 14:401–410

Lewis JS, Prinn RG (1980) Kinetic inhibition of CO and N_2 reduction in the solar nebula. Astrophys J 238:357–364

Lindal GF, Wood GE, Hotz HB, Sweetnam DN, Eshleman VR, Tyler GL (1983) The atmosphere of Titan – an analysis of the Voyager 1 radio occultation measurements. Icarus 53:348–363

Lorenz RD (1993) The life, death and afterlife of a raindrop on Titan. Planet Space Sci 41:647–655

Lorenz RD (1997) Did Comas Solá discover Titan's atmosphere? Astron Geophys 8:16–18

Lorenz RD (1998) Many lakes make an ocean. Abstract #1689, XIXth Lunar and Planetary Science Conference, Houston, TX

Lorenz RD (2000) The weather on Titan. Science 290:467–468

Lorenz RD (2008) The changing face of Titan. Physics Today 61:34–39

Lorenz RD, Lunine JI (1996) Erosion on Titan: past and present. Icarus 122:79–91

Lorenz R, Mitton J (2008) Titan unveiled: Saturn's mysterious moon explored. Princeton University Press, Princeton

Lorenz RD, Lunine JI, Grier JA, Fisher MA (1995) Prediction of aeolian features on planets: application to Titan paleoclimatology. J Geophys Res (Planets) 88:26377–26386

Lorenz RD, Lunine JI, McKay CP (1997a) Titan under a red giant sun: a new kind of "habitable" moon. Geophys Res Lett 24(22): 2905–2908

Lorenz RD, McKay CP, Lunine JI (1997b) Photochemically driven collapse of Titan's atmosphere. Science 275:642–644

Lorenz RD, McKay CP, Lunine JI (1999) Analytical investigation of climate stability on Titan: sensitivity to volatile inventory. Planet Space Sci 47:1503–1515

Lorenz RD and Lunine JI, McKay CP, Withers PG (2001) Entropy production by latitudinal heat flow on Titan. Mars and Earth. Geophys Res Lett 28:415–418

Lorenz RD and the Cassini RADAR Team (2008a) Fluvial channels on Titan: initial Cassini radar observations. Planet Space Sci 56: 1132–1144

Lorenz RD and the Cassini RADAR Team (2008b) Titan's inventory of organic surface materials. Geophys Res Lett 35:L02206

Lorenz RD, Flasar FM, Brown ME (2009) Seasonal cycles. In: Brown RH, Lebreton J-P, Waite JH (eds) Titan From Cassini–Huygens. Springer, Dordrecht

Lunine JI, Atreya SK (2008) The methane cycle on Titan. Nat Geosci 1:159–164

Lunine JI, Rizk B (1989) Thermal evolution of Titan's atmosphere. Icarus 80:370–389

Lunine JI, Stevenson DJ (1985) Thermodynamics of clathrate hydrate at low and high pressures with application to the outer solar system. Astrophys J Suppl 58:493–531

Lunine JI, Stevenson DJ (1987) Clathrate and ammonia hydrates at high pressure – application to the origin of methane on Titan. Icarus 70:61–77

Lunine JI, Stevenson DJ, Yung YL (1983) Ethane ocean on Titan. Science 222:1229–1230

Lunine JI, Yung YL, Lorenz RD (1999) On the volatile inventory of Titan from isotopic abundances in nitrogen and methane. Planet Space Sci 47:1291–1303

Lunine JI and the RADAR Science Team (2008) Lack of south polar methane lakes on Titan. Abstract #1637, XXIXth Lunar and Planetary Society Conference, Houston, TX

Lunine JI Choukroun M, Stevenson DJ, Tobie G (2009) The origin and evolution of Titan. In: Brown RH, Lebreton J-P, Waite JH (eds) Titan from Cassini–Huygens. Springer, Dordrecht

Lutz BL, Owen T, Cess RD (1976) Laboratory band strengths of methane and their applications to the atmospheres of Jupiter, Saturn, Uranus, Neptune and Titan. Astrophys J 203:541–551

Magee BA, Mandt KE, Waite JH, Westlake J, Gell DA, De La Haye V (2009) INMS derived composition of Titan's upper atmosphere: analysis methods and model. Planet Space Sci in press

Mandt KE, Waite JH, Lewis W, Magee BA, Bell JM, Lunine J, Mousis O, Cordier D (2009) Isotopic evolution of the major constituents of Titan's atmosphere based on Cassini data. Planet Space Sci in press

Marten A, Hidayat T, Biraud Y, Moreno R (2002) New millimeter heterodyne observations of Titan: vertical distributions of nitriles HCN, HC_3N, CH_3CN, and the isotopic ratio $^{15}N/^{14}N$ in its atmosphere. Icarus 158:532–544

Matson DL, Atreya SK, Castillo-Rogez J, Johnson TV, Adams EY, Lunine JI (2007) Endogenic origin of Titan's N_2. Abstract 2007AGUFM.P21D.04M, American Geophysical Union Fall Meeting, San Francisco

McKay CP, Smith HD (2005) Possibilities for methanogenic life in liquid methane on the surface of Titan. Icarus 178:274–276

McKay CP, Scattergood TW, Pollack JB, Borucky WJ, van Ghyseghem HT (1988) High-temperature shock formation of N_2 and organics on primordial Titan. Nature 332:520–522

McKay CP, Pollack JB, Courtin R (1989) The thermal structure of Titan's atmosphere. Icarus 80:23–53

McKay CP, Pollack JB, Courtin R (1991) The greenhouse and anti-greenhouse effect on Titan. Science 253:1118–1121

McKay CP, Pollack JB, Lunine JI, Courtin R (1993) Coupled atmosphere-ocean models of Titan's past. Icarus 102:88–98

McKay CP, Lorenz RD, Lunine JI (1999) Analytic models of the greenhouse and anti-greenhouse effects: application to Titan and the early Earth. Icarus 137:57–61

Meier R, Owen TC, Matthews HE, Jewitt DC, Bockelee-Morvan D, Biver N, Crovisier J, Gautier D (1998a) A determination of the HDO/H_2O ratio in Comet C/1995 01 (Hale-Bopp). Science 279: 842–844

Meier R, Owen TC, Jewitt DC, Matthews HE, Senay M, Biver N, Bockelee-Morvan D, Crovisier J, Gautier D (1998b) Deuterium in Comet C/1995 01 (Hale-Bopp): detection of DCN. Science 279:1707–1710

Miller SL (1961) The occurrence of gas hydrates in the solar system. Proc Natl Acad Sci USA 47:1798–1808

Mitchell J (2008) The drying of Titan's dunes: Titan's methane hydrology and its impact on atmospheric circulation. J Geophys Res 113:E08015, doi:10.1029/2007JE003017

Mitri GL, Showman AP, Lunine JI, Lorenz RD (2007) Hydrocarbon lakes on Titan. Icarus 186:385–394

Mousis O, Schmitt B (2008) Sequestration of ethane in the cryovolcanic subsurface of Titan. Astrophys J 677:L67–L70

Mousis O, Gautier D, Bockelée-Morvan D (2002) Turbulent model of the Saturn subnebula: implications for the origin of methane in Titan's atmosphere. Icarus 156:162–175

NRC (2007) The limits of organic life in planetary environments, National Research Council Publication, Washington DC

Niemann HB and the GCMS Team (2005) The abundances of constituents of Titan's atmosphere from the GCMS instrument on the Huygens probe. Nature 438:779–784

Owen TC (1982) The composition and origin of Titan's atmosphere. Planet Space Sci 30:833–838

Owen TC (2000) On the origin of Titan's atmosphere. Planet Space Sci 48:747–752

Prinn RG, Fegley B (1981) Kinetic inhibition of CO and N_2 reduction in circumplanetary nebulae: implications for satellite composition. Astrophys J 249:308–317

Rannou P, Montmessin F, Hourdin F, Lebonnois S (2006) The latitudinal distribution of clouds on Titan. Science 311:201–205

Ribas I, Guinan EF, Gudel M, Audard M (2005) Evolution of the solar activity over time and effects on planetary atmospheres. I. High-energy irradiances (1–1700 Å). Astrophys J 622:680–694

Samuelson RE (1983) Radiative equilibrium model of Titan's atmosphere. Icarus 53:364–387

Samuelson RE, Nath NR, Borysow A (1997) Gaseous abundances and methane supersaturation in Titan's troposphere. Planet Space Sci 45:959–980

Schulze-Makuch D, Grinspoon DH (2005) Biologically enhanced energy and carbon cycling on Titan? Astrobiology 5:560–567

Sotin C, Jaumann R, Buratti BJ, Brown RH, Clark RN, Soderblom LH, Baines KH, Bellucci G, Bibring JP, Capaccioni F, Cerroni P, Combes M, Coradini A, Cruikshank DP, Drossart P, Formisano V, Langevin Y, Matson DL, McCord TB, Nelson RM, Nicholson PD, Sicardy B, LeMouelic S, Rodriguez S, Stephan K, Scholz CK (2005) Release of volatiles from a possible cryovolcano from near-infrared imaging of Titan. Nature 435:786–789, doi:10.1038/nature03596

Stevenson DJ, Potter BE (1986) Titan's latitudinal temperature distribution on seasonal cycles. Geophys Res Lett 13:93–96

Stofan ER, Elachi C, Lunine JI, Lorenz RD, Stiles B, Mitchell KL, Ostro S, Soderblom L, Wood C, Zebker H, Wall S, Janssen M, Kirk R, Lopes R, Paganelli F, Radebaugh J, Wye L, Anderson Y, Allison M, Boehmer R, Callahan P, Encrenaz P, Flamini E, Franscescetti G, Gim Y, Hamilton G, Hensley S, Johnson WTK, Kelleher K, Muhleman D, Paillou P, Picardi G, Posa F, Roth L, Seu R, Shaffer S, Vetrella S, West R (2007) The lakes of Titan. Nature 445:61–64

Strobel DF (1974) The photochemistry of hydrocarbons in the atmosphere of Titan. Icarus 21:466–470

Strobel DF, Shemansky DE (1982) EUV emission from Titan's upper atmosphere: Voyager 1 encounter. J Geophys Res 87:1361–1368

Strobel DF, Atreya SK, Bézard B, Ferri F, Flasar FM, Fulchignoni M, Lellouch E, Muller-Wodarg I (2009) Atmospheric composition and structure. In: Brown RH, Lebreton J-P, Waite JH (eds) Titan from Cassini–Huygens. Springer, Dordrecht

Thomas C, Mousis O, Ballenegger V, Picaud S (2007) Clathrate hydrates as a sink of noble gases in Titan's atmosphere. Astron Astrophys 474:L17–L20

Thompson WR, Zollweg JA, Gabis DH (1992) Vapor-liquid equilibrium thermodynamics of $N_2 + CH_4$: model and Titan applications. Icarus 97:187–199

Tobie G, Lunine JI, Sotin C (2006) Episodic outgassing as the origin of atmospheric methane on Titan. Nature 440:61–64

Tobie G, Gautier D, Hersant F, Lunine JI (2008) Interpretation of Titan's atmospheric composition measured by Cassini–Huygens. EPSC Abstract, 3, EPSC2008-A-00325

Tokano T, Molina-Cuberos GL, Lammer H, Stumptner W (2001) Modelling of thunderclouds and lightning generation on Titan. Planet Space Sci 49:539–560

Tokano T, McKay CP, Neubauer FM, Atreya SK, Ferri F, Fulchignoni M, Niemann HB (2006) Methane drizzle on Titan. Nature 442: 432–435

Tomasko MG and the DISR Team (2005) Rain, winds and haze during the Huygens probe's descent to Titan's surface. Nature 438:765–778

Trafton LM (1972) The bulk composition of Titan's atmosphere. Astrophys J 175:295–306

Turtle EP, Perry JE, McEwen AS, DelGenio AD, Barbara J, West RA, Dawson DD, Porco CC (2009) Cassini imaging of Titan's high latitude lakes, clouds, and south polar surface changes. Geophys Res Lett 36:L02204, doi:10.1029/2008GL036186

Vinatier S, Bézard B, Nixon CA (2007) The Titan $^{14}N/^{15}N$ and $^{12}C/^{13}C$ isotopic ratios in HCN from Cassini/CIRS. Icarus 191:712–721

Waite JH, Niemann H, Yelle RV, Kasprzak WT, Cravens TE, Luhman JG, McNutt RL, Ip W-H, Gell D, De La Haye V, Müller-Wordag I, Magee B, Borggren N, Ledvina S, Fletcher G, Walter E, Miller R, Scherer S, Thorpe R, Xu J, Block B, Arnett K (2005) Ion neutral mass spectrometer results from the first flyby of Titan. Science 308:982–986

Waite JH, Young DT, Cravens TE, Coates AJ, Crary FJ, Magee B, Westlake J (2007) The process of tholin formation in Titan's upper atmosphere. Science 316:870–872

Waite HW, Lunine JI, Owen TC, Gautier D, Niemann HB, Strobel DF (2009) Conversion of methane and nitrogen to complex polymers on Titan. In: Brown RH, Lebreton J-P, Waite JH (eds) Titan from Cassini–Huygens. Springer, Dordrecht

Wilson EH, Atreya SK (2004) Current state of modeling the photochemistry of Titan's mutually dependent atmosphere and ionosphere. J Geophys Res 109:E06002, doi:10.1029/2003JE002181

Yelle RV, Cui J, Müller-Wodarg ICF (2008) Methane escape from Titan's atmosphere. J Geophys Res 113:E10003, doi:10.1029/2007JE003031

Yung YL, Allen MA, Pinto JP (1984) Photochemistry of the atmosphere of Titan: comparison between model and observations. Astrophys J Supp 55:465–506

Zahnle KJ, Walker JCG (1982) The evolution of solar luminosity. Rev Geophys Space Phys 20:280–292

Ziurys LM, Savage C, Brewster MA, Apponi AJ, Pesch TC, Wyckoff S (1999) Cyanide chemistry in comet Hale-Bopp (C/1995 O1). Astrophys J 527:L67–L71

Chapter 8
High-Altitude Production of Titan's Aerosols

J.H. Waite, Jr., D.T. Young, J.H. Westlake, J.I. Lunine, C.P. McKay, and W.S. Lewis

Abstract Measurements with the Cassini Ion and Neutral Mass Spectrometer (INMS) and two Cassini Plasma Spectrometer (CAPS) sensors, the Ion beam Spectrometer (IBS) and the Electron Spectrometer (ELS), have revealed the presence of a significant population of heavy hydrocarbon and nitrile species well above the homopause, with masses as large as several thousand Daltons (Da). The INMS ion and neutral spectra cover the mass range 1–100 Da. The IBS has measured positive ions up to 350 Da, while the ELS has detected concentrations of negative ions as high as 20% of the total negatively charged ionosphere component extending to over 13,000 Da. These measurements have motivated the development of new atmospheric models and have significant implications for our knowledge and understanding of Titan's haze layers.

The existence of a thick haze obscuring Titan's surface was inferred from remote-sensing observations at infrared and ultraviolet wavelengths during the mid-1970s (Danielson et al. 1973; Veverka 1973; Zellner 1973; Trafton 1975) and confirmed by Voyager 1 and 2 imaging, which revealed the existence of two principal haze layers, a main layer and a thin detached layer ~100 km above it, both merging at high northern latitudes (Smith et al. 1981, 1982). It was recognized early on (e.g., Danielson et al. 1973) that photochemistry occurring in the upper atmosphere of Titan was the likely source of the haze-forming aerosols, and in the years leading up to the Voyager encounters several laboratory experiments were performed in an attempt to synthesize materials whose properties were similar to those of the postulated hazes (see reviews by Chang et al. 1979 and Cabane and Chassefière 1995). Substances investigated as possible candidates for the haze-forming aerosols included polymers of acetylene, ethylene, and HCN (Scattergood and Owen 1977; Podolak and Bar-Nun 1979) and "tholins," complex organic solids, brownish in color, produced in a simulated reducing planetary atmosphere through UV irradiation and electric discharge (Khare and Sagan 1973; Sagan and Khare 1979).

Prior to the Voyager encounters, the only species known with certainty to be present in Titan's atmosphere were CH_4 and C_2H_6, although there was evidence for the presence of C_2H_2 and C_2H_4 as well (Gillett 1975). The presence of N_2, predicted by Hunten (1977) and Atreya et al. (1978), had not yet been established, although Titan's reddish-brown albedo suggested that nitrogen-bearing species (and/or sulfur-bearing ones) should be present in the haze aerosols (Scattergood and Owen 1977; Chang et al. 1979). The Voyagers revealed that Titan's atmosphere consists predominantly (>90%) of molecular nitrogen (Broadfoot et al. 1981; Tyler et al. 1981) with methane as the next most abundant species and provided positive identifications of several hydrocarbons including C_2H_2, C_2H_4, and C_3H_8 as well as of the nitriles HCN, HC_3N, and C_2N_2 (Hanel et al. 1981, 1982; Kunde et al. 1981; Maguire et al. 1981).

During the interval between the Voyager encounters and the arrival of Cassini in the Saturn system, several photochemical models were developed to describe the production of hydrocarbons and nitriles resulting from the dissociation of N_2 and CH_4 in Titan's upper atmosphere by electron impact (N_2) and UV irradiation (CH_4) (e.g., Yung et al. 1984; Toublanc et al. 1995; Wilson and Atreya 2004). More a number of laboratory, modeling, and theoretical studies were undertaken to investigate the formation of the haze layers and the physical, optical, and chemical properties of the aerosols in light of both the Voyager data and new remote-sensing

J.H. Waite, Jr.,(✉), D.T. Young, J.H. Westlake, and W.S. Lewis
Southwest Research Institute, P.O. Drawer 28510 San Antonio, TX 78228, USA
e-mail: hwaite@swri.edu

J.H. Waite, Jr., J.H. Westlake
University of Texas at San Antonio, One UTSA Blvd. San Antonio, TX 78249, USA

J.I. Lunine
Lunar and Planetary Laboratory, University of Arizona, Tucson, AZ 85721, USA

C.P. Mckay
NASA Ames Research Center, Mail Stop 245–3, Moffett Field, CA 94035, USA

R.H. Brown et al. (eds.), *Titan from Cassini-Huygens*,

observations (see reviews by Cabane and Chassefière 1995 and McKay et al. 2001). Post-Voyager experiments to synthesize aerosol analogs in the laboratory involved both the production of tholins in a simulated Titan N_2-CH_4 atmosphere (e.g., Thompson et al. 1994; Coll et al. 1999) and the creation of the photopolymers of C_2H_2, C_2H_4, and HCN (Bar-Nun et al. 1988; Scattergood et al. 1992) as well as of HC_3N and HC_3N/C_2H_2 (Clarke and Ferris 1997). The spectral and optical properties of tholins were found to be consistent with Titan's albedo and with the refractive properties of Titan's haze particles, suggesting that tholins are good analogs for Titan's aerosols (Khare et al. 1984).

The ultimate sources of Titan's aerosols are the gas-phase dissociation products of CH_4 and N_2. However, as noted by Lebonnois et al. (2002), the transition from gas-phase compounds to solid-phase aerosols is poorly understood. They suggested three possible chemical pathways that could polymerize simple molecules to macromolecules, which are the presumed precursors to aerosol particles, producing: (1) polymers of acetylene and cyanoacetylene, (2) polycyclic aromatics, and (3) polymers of HCN and other nitriles, and polyynes. Their model suggested a total production rate of 4 $\times 10^{-14}$ g cm^{-2} s^{-1} and a C/N ratio of 4, in a production zone slightly lower than 200 km altitude. Wilson and Atreya (2003) considered similar pathways and concluded that the growth of polycyclic aromatic hydrocarbons (PAH) throughout the lower stratosphere could play an important role in haze formation. They suggested that the peak chemical production of haze would lie near 220 km, with a column integrated production rate of 3.2×10^{-14} g cm^{-2} s^{-1}. Wilson and Atreya (2003) pointed out that the discovery of benzene in Titan's atmosphere by ISO (Coustenis et al. 2003) favored the PAH pathway. Trainer et al. (2004) found that for particles produced from a mixture of 10% CH_4 in N_2 the results were consistent with a large fraction of aromatics, including specific mass spectral peaks likely due to PAHs. However, at lower concentrations of CH_4 (1% and lower), the mass fraction of PAHs greatly diminished, and an aliphatic pathway dominated.

Laboratory simulations also indicate a possible key role for PAHs. Khare et al. (2002) reported on an analysis of the time-dependent chemical evolution of gas phase products in a Titan simulation. They found an early dominance of aromatic ring structures that led in the later stages of the experiment to the appearance of nitrile and amine compounds. Thompson et al. (1991) reported the yields of gaseous hydrocarbons and nitriles produced at pressures (1,700 Pa and 24 Pa) in a continuous-flow, low-dose, cold plasma discharge excited in an atmosphere consisting of 10% CH_4 and 90% N_2 at 295 K. At 1,700 Pa, 59 gaseous species including 27 nitriles were detected while at 24 Pa, 19 species are detected, including six nitriles and three other unidentified N-bearing compounds. The types of molecules formed changed even more markedly, with high degrees of multiple bonding at 24 Pa prevailing over more H-saturated molecules at 1,700 Pa. Imanaka et al. (2004) conducted a series of experiments from high (2,300 Pa) to low (13 Pa) pressure. They found an increase in the aromatic compounds and a decrease in C/N ratio in tholins formed at low pressures, indicating the presence of the nitrogen-containing polycyclic aromatic compounds in tholins formed at low pressures. They concluded that the haze layers at various altitudes might have different chemical and optical properties, but most importantly they found that there is a fundamental change in the nature of haze production between pressures above and below roughly 100 Pa.

8.1 Cassini Observations of Heavy Hydrocarbons in Titan's Upper Atmosphere

Earlier models of the photochemistry responsible for initiating the production of complex acetylene polymers and polyaromatic hydrocarbons (PAHs) suggested that the formation of heavy hydrocarbons such as benzene occurs primarily in the well-mixed portion of Titan's atmosphere below the homopause (pressure ~2×10^{-3} Pa near 750 km) (e.g., Wilson and Atreya 2003, 2004). Ion-neutral reactions near the ionospheric peak (pressure ~5×10^{-6} Pa at 1,100 km) were thought to be an additional, although much weaker source of complex hydrocarbons. Thus, prior to the arrival of Cassini, it was expected that there would be little benzene or other complex hydrocarbons, and thus they would be only marginally detectable at altitudes above ~950 km, the region sampled by the Cassini orbiter during its passes through Titan's upper atmosphere. However, measurements with the Cassini Ion and Neutral Mass Spectrometer (INMS) and two Cassini Plasma Spectrometer (CAPS) sensors, the Ion beam Spectrometer (IBS) and the Electron Spectrometer (ELS), have revealed the presence of a significant population of heavy hydrocarbon and nitrile species well above the homopause, with masses as large as several thousand Daltons (Da). The INMS ion and neutral spectra cover the mass range 1–100 Da (Fig. 8.1; Tables 8.1 and 8.2) (Waite et al. 2005; Magee et al. 2009). The IBS has measured positive ions up to 350 Da (Fig. 8.2; Crary et al. 2009), while the ELS has detected a concentration of negative ions as high as 20% of the total negatively charged ionospheric component extending to over 13,000 Da (Fig. 8.3; Coates et al. 2009).

The INMS is a true (quadrupole) mass spectrometer designed to measure the abundance of ion and neutral species in Titan's upper atmosphere (Waite et al. 2004). The IBS and ELS sensors, on the other hand, measure ion flux as a funatic of ion energy/charge from which pseudo-mass spectra can be derived. IBS, designed for the supersonic solar wind and the

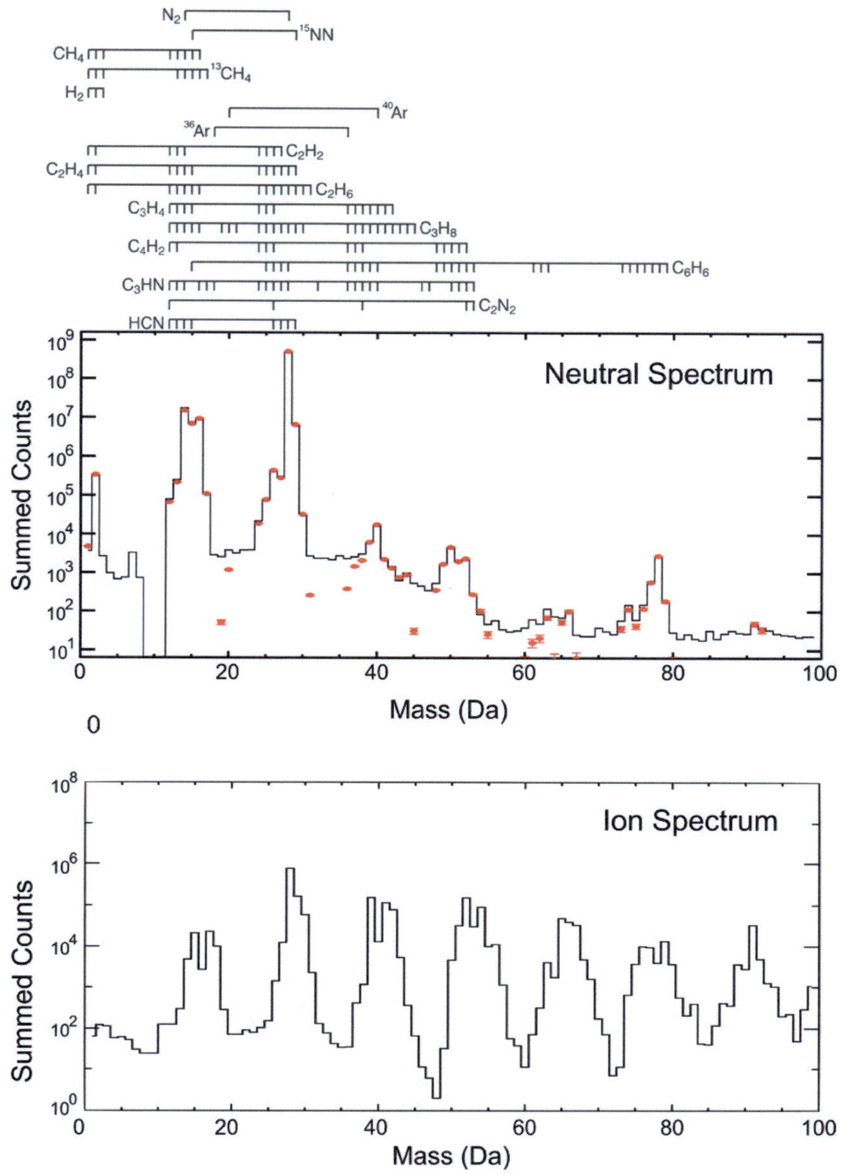

Fig. 8.1 Composite mass spectra for neutrals (*top*) and ions (*bottom*) based on Cassini INMS data (*black line*) acquired during 17 flybys of Titan. Data were taken between 1,000 and 1,100 km. The mass deconvolution used to produce Table 8.1 is indicated above the spectrum and the totals are marked on the spectrum (*top panel*) with red dots, the top panel (from Waite et al. (2007) and is reprinted with permission from AAAS)

cold ionosphere of Titan, has relatively high energy resolution ($\Delta E/E = 1.7\%$). ELS, which is designed to measure hot plasma electrons, has lower resolution ($\Delta E/E = 17\%$) but is in addition sensitive to negative ions (Waite et al. 2007; Coates et al. 2007, 2009).

As a consequence of the low temperatures of ions in Titan's ionosphere (120–250 K between ~950 and ~1,600 km respectively), ion thermal velocities are small (100's of m/s, depending on ion mass). During encounters with Titan Cassini travels at supersonic velocities (~6 km/s) relative to the cold ionosphere, which allows the use of IBS and ELS energy/charge spectra to infer ion mass/charge regardless of charge state or polarity. Crary et al. (2009) used INMS data combined with IBS to extract mass spectra and estimate ion temperatures and flow speeds. Similarly Coates et al. (2007, 2009) and Waite et al. (2007) analyzed ELS data to produce negative ion mass spectra.

The peak flux measured by IBS or ELS can be identified with ion mass by the relationship $E_i = m_i V_{s/c}^2/2 + 8kT$ where $V_{s/c}$ is spacecraft velocity and m_i is the mass of the ith species (Crary et al. 2009). Taking T = 150 K as a characteristic temperature, $E_i = 0.188\,m_i + 0.013\,eV$. Carbon is the smallest mass of interest here, so $m_i \geq 12$ Da and $m_iV_{s/c}^2/2 >>$; 8 kT. Thus $E_i = m_iV_{s/c}^2/2$ to a very good approximation

Table 8.1 Neutral species mixing ratios measured by INMS in the altitude region between 1100 and 1000 km. Values are globally averaged over 20 flybys as reported by Magee et al. (2009)

Major species	
	Mixing ratio
N_2	$0.963 \pm 0.44 \times 10^{-3}$
$^{14}N^{15}N$	$1.08 \times 10^{-2} \pm 0.06 \times 10^{-2}$
CH_4	$2.17 \times 10^{-2} \pm 0.44 \times 10^{-2}$
$^{13}CH_4$	$2.52 \times 10^{-4} \pm 0.46 \times 10^{-4}$
H_2	$3.38 \times 10^{-3} \pm 0.23 \times 10^{-3}$
Minor species	
	Mixing ratios
C_2H_2	$3.42–3.43 \times 10^{-4}$
C_2H_4	$3.91–3.97 \times 10^{-4}$
C_2H_6	$6.05–4.57 \times 10^{-5}$
HCN	$2.40–2.44 \times 10^{-4}$
^{40}Ar	$2.14–2.26 \times 10^{-5}$
CH_3CCH	$0.92–1.13 \times 10^{-5}$
C_3H_6	$2.33–3.45 \times 10^{-6}$
C_3H_8	$2.87–4.38 \times 10^{-6}$
C_4H_2	$5.55–5.65 \times 10^{-6}$
C_2N_2	$2.14–2.20 \times 10^{-6}$
C_6H_6	$2.48–2.50 \times 10^{-6}$
C_2HCN	$1.48–1.54 \times 10^{-6}$
C_2H_3CN	$3.46–4.39 \times 10^{-7}$
C_2H_5CN	$1.54–2.87 \times 10^{-7}$
C_7H_8	$2.51–5.37 \times 10^{-8}$

Table 8.2 Ion densities measured by INMS for five passes

	T16	T17	T18	T21	T23
CH_5^+	1.7	19.7	4.85	0.27	12.97
	(0.1)	(0.3)	(0.18)	(0.03)	(0.25)
$C_2H_5^+$	1.3	95.8	11.06	0.55	61.16
	(0.2)	(0.6)	(0.21)	(0.04)	(0.87)
$HCNH^+$	18.5	499.5	49.4	2.47	242.8
	(0.3)	(1.3)	(0.48)	(0.10)	(2.74)
$C_3H_3^+$	30.1	69	44.19	16.45	159.9
	(0.4)	(1.4)	(1.10)	(0.87)	(4.48)
$C_4H_3^+$	3.1	9.5	3.33	0.48	21.66
	(0.3)	(0.5)	(0.27)	(0.12)	(1.13)
$C_4H_5^+$	3.8	6.3	4.06	0.75	26.98
	(0.3)	(0.4)	(0.30)	(0.15)	(1.38)
$C_6H_5^+$	3.8	0.4	2.67	1.68	20.67
	(0.4)	(0.1)	(0.23)	(0.28)	(1.47)
$C_6H_7^+$	3.4	0.6	3.72	0.90	33.51
	(0.4)	(0.1)	(0.27)	(0.20)	(2.26)
$C_7H_7^+$	9.2	1.2	11.35	5.88	83.97
	(0.9)	(0.2)	(0.50)	(0.72)	(5.79)
Altitude (km)	950	1000	960	1000	1000
LST	17.4	10.5	4.8	20.4	14.1

and mass (in Daltons) can be inferred using $m_i = 5.32E_i$. Under these circumstances the energy resolution of IBS is equivalent to an effective mass resolution of 30 at 28 Da. Peaks in the IBS energy spectra can be resolved up to ~200 Da (Fig. 8.2), while the maximum mass observed so far is ~350 Da. By assuming maxwellian velocity distributions and comparing IBS with INMS, Crary et al. (2009) were able to correct IBS data for spacecraft potential and ion winds along the spacecraft track to obtain pseudo-mass spectra (Fig. 8.4).

Figure 8.5 shows an example of ion density calculated by Crary et al. using this technique. The agreement between the two instruments below ~1,600 km is very good as is agreement with the Cassini Langmuir Probe measurements of total plasma density (Wahlund et al. 2009). The lack of agreement above this altitude is caused by ion heating, which invalidates the cold ion assumption used to interpret IBS data. Heavy ions >100 Da become a major constituent below ~1,200 km and tend to increase deeper in the ionosphere, becoming as much as ~50% of the total at 950 km, the lowest altitudes visited by Cassini (Crary et al. 2009). Altitude profiles strongly suggest that the abundance of ions >100 Da continues to increase rapidly with depth in the atmosphere (Fig. 5; also Wahlund et al. 2009).

Detailed knowledge of the abundance of heavy ions is important for understanding the chemistry of ion formation. Using data from all 14 Titan encounters studied so far, Crary et al. were able to obtain 130 mass spectra. (The number of spectra is limited by the need to swing the CAPS sensors across the Cassini ram direction every ~60 s as is evidenced by the spacing of peaks in Fig. 8.6, which shows the percent occurrence calculated for each mass bin from 100 to 200 Da. On more recent passes such as T55, motion of the sensors has been halted to yield a much larger number of spectra in the ram direction. These have not yet been analyzed.) When taken together and examined statistically the spectra show a consistent mass peak spacing of 12–14 Da as expected for compounds consisting of carbon and nitrogen. Analysis of the abundant groups >100 Da, together with INMS data <100 Da and inferences of possible chemical reactions, suggests that ions in this range are most likely aromatic hydrocarbons (Waite et al. 2007). Crary et al. (Fig. 8.7) also infer that acetylene and nitrile polymers, particularly naphthalene, are relatively common, appearing in >70% of the spectra. In addition they conclude that aromatic hydrocarbons, particularly PAHs, are the most likely component of the positive ion spectra. Sittler et al. (2009) have advanced arguments to the effect that the negative ions observed by ELs could be fullerenes (C60), a suggestion based on analogies to observations of fullerenes formed under laboratory conditions similar to those in Titan's atmosphere. Since neutrals and ions are cold, energy would come from hot electrons (0.1 ~ few eV) that are abundant in the ionosphere (Coates et al., 2009). Sittler et al. go on to suggest that because there are a number of sources of oxygen present in the atmosphere, particularly ~ke V oxygen and water group ions driven in to the atmosphere by corotation, oxygen might in fact be trapped in the fullerenes, a phenomenon also observed in the laboratory. Once formed the fullerenes would settle into haze layers as discussed above, eventually reaching the surface where they would represent a source of oxygen that might contribute to

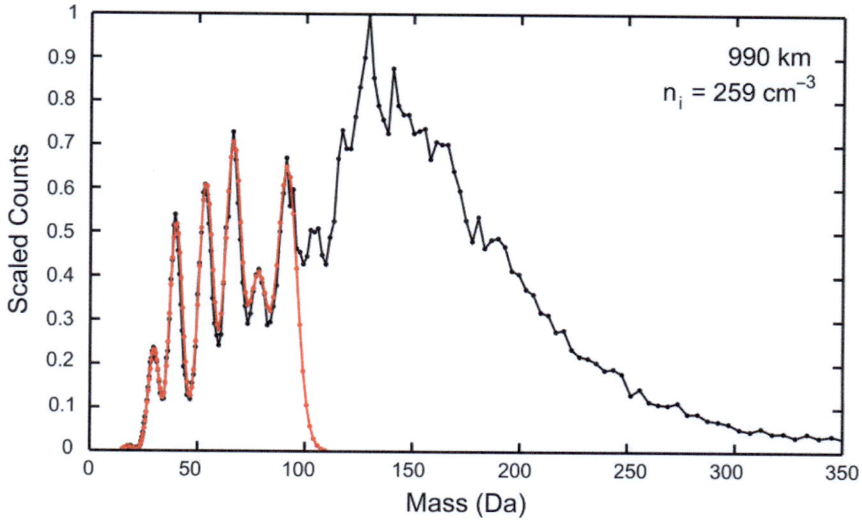

Fig. 8.2 IBS spectrum from flyby T26. The match to the Cassini INMS ion spectrum below 100 Da is marked in red. Note the significant ion densities above 100 Da

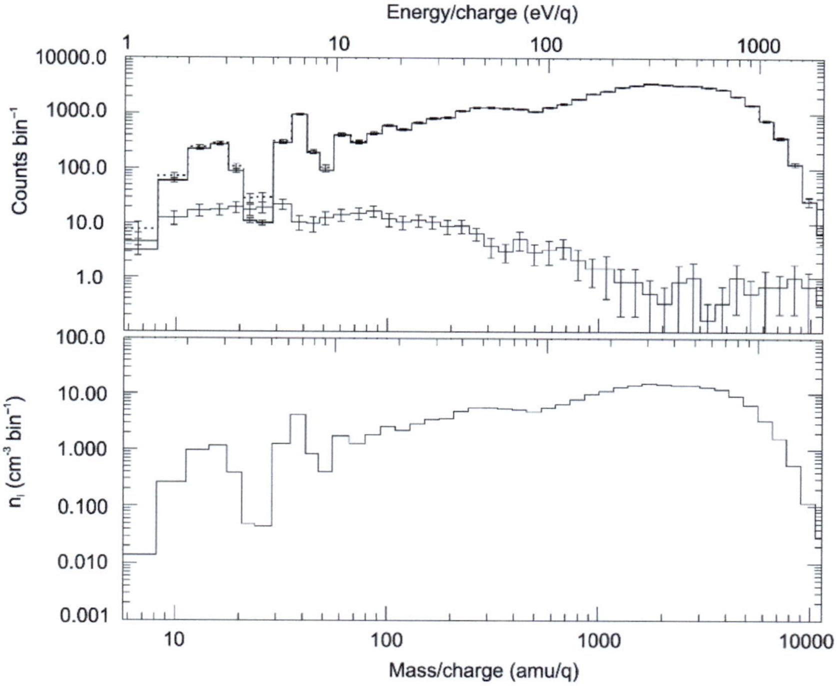

Fig. 8.3 Inferred mass/charge spectrum of negatively charged ions using ELS energy/charge data (energy scale is shown at top of figure). Spectra were taken at 953 km during the T16 pass. Upper trace in top panel shows count rate spectrum corrected for photoelectron contribution. Lower panel shows spectrum converted to differential number density (from Coates et al. (2007))

pre-biological chemistry. Although somewhat speculative in nature (there are many steps involved, not all of which are well established) the possibility of fullerene formation is certainly worth further investigation.

Figure 8.8 shows ELS energy-time spectrograms for four encounters. The sharp spikes in all four panels indicate the presence of cold negative ions that are seen only when ELS sweeps through the ram direction. Although ELS has much lower energy resolution than IBS, the same principles of analysis can be applied, enabling energy spectra to be converted into pseudo-mass spectra (Fig. 8.3). Mass resolution is limited to ~5 at ~16 Da, and, as for IBS, drops at higher masses. Because of the low instrument resolution and high ion masses observed by ELS, corrections for spacecraft potential and winds can be neglected.

The finding of abundant heavy negative ions using ELS is one of the truly surprising "discoveries" made with Cassini. Although photoelectron peaks are seen during daylight encounters, it is very clear that the peaks in the spectra identified as negative ions are far too narrow and unidirectional to be misidentified as electrons, which are both isotropic and hot (>>1 eV, equivalent to >> 10,000 K). As Fig. 8.3 shows, the mass of negative ions extends from 17 Da to >10,000 Da. Negative ions are a permanent feature of the ionosphere, having been observed on all 23 encounters thus far during which spacecraft pointing was favorable for observations. Two very clear negative ion peaks can typically be identified in the spectra at 22 ± 4 Da and 44 ± 8 Da with a possible third peak at 82 ± 14 Da (Figs. 8 and 2 of Vuitton et al. 2007).

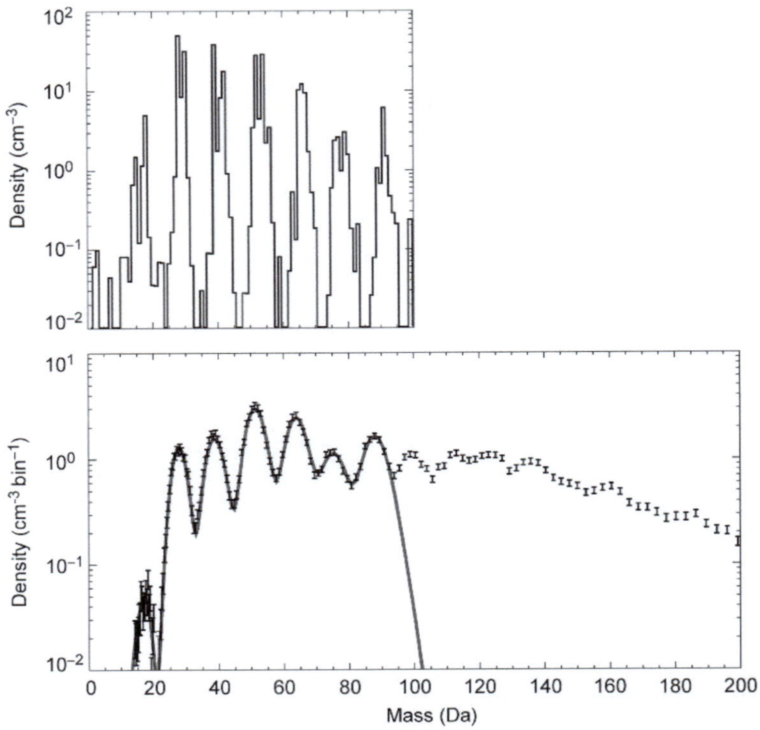

Fig. 8.4 Mass spectra from the INMS (*upper panel*) and the IBS (*lower panel*) from 1,025 km during the ingress leg of the T26 encounter. The lower panel shows a best fit to the IBS data below 100 Da using the method described in the text. Note that, although poorly resolved, mass peaks are still visible above 100 Da. The bottom panel (from Crary et al. (2009) with permission from Elsevier)

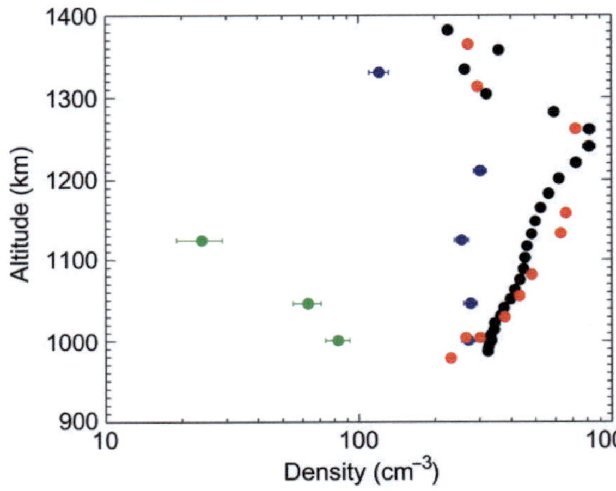

Fig. 8.5 Comparison of INMS and IBS ion densities during the T26 encounter (Crary et al. 2009). INMS total ion density is shown in black. Using data from the IBS, Crary et al. calculated the total ion density (*red*), density of ions below 100 Da (*blue*) and density of ions heavier than 100 Da (*green*) (reprinted with permission from Elsevier)

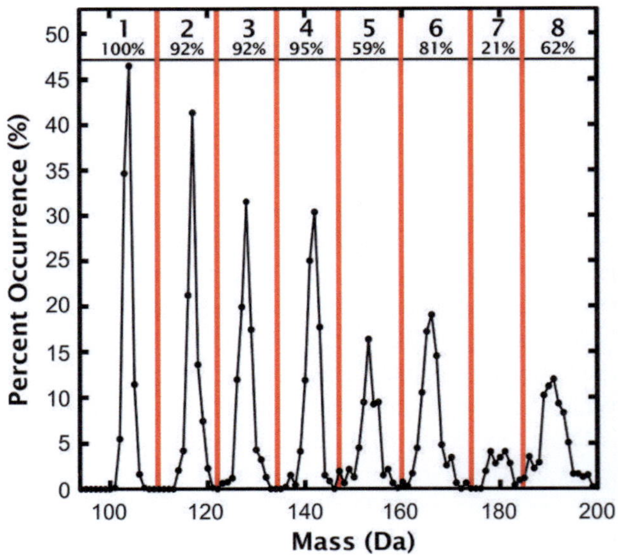

Fig. 8.6 The percent occurrence calculated for each mass bin from 100 to 200 Da (Crary et al. 2009). Total percent occurrence for each group is shown in the box above each peak (reprinted with permission from Elsevier)

GROUP	ALIPHATIC COMPOUNDS						AROMATIC COMPOUNDS	
	ACETYLENE POLYMERS	NITRILE POLYMERS				ALIPHATIC COPOLYMERS	PAH POLYMERS	NITRILE AROMATIC POLYMERS
1 (99 Da. - 110 Da.) Peak at 103 Da.		104 ($C_4N_4^+$)	100 ($C_6N_2^+$)	105 ($C_5H_3N_3^+$)	98-101 (C_7N^+)	103 ($C_5H_3N^+$)	103 ($C_8H_7^+$)	104 ($C_7H_6N^+$)
2 (111 Da. - 122 Da.) Peak at 117 Da.	122 ($C_{10}H_2^+$)		112 ($C_7N_2^+$)		110-112 (C_8N^+)		117 ($C_9H_9^+$)	
3 (123 Da. - 134 Da.) Peak at 127 Da.		130 ($C_5N_5^+$)	124 ($C_8N_2^+$)	131 ($C_6H_3N_4^+$)	122-124 (C_9N^+)	129 ($C_7H_5N^+$)	128 ($C_{10}H_8^+$)	129-130 ($C_9H_{7\text{-}8}N^+$)
4 (135 Da. - 147 Da.) Peak at 141 Da.	146 ($C_{12}H_2^+$)		136 ($C_9N_2^+$)		134-136 146-148 ($C_{10\text{-}11}N^+$)		141 ($C_{11}H_9^+$)	
5 (148 Da. - 158 Da.) Peak at 153 Da.		156 ($C_6N_6^+$)	148 ($C_{10}N_2^+$)	157 ($C_7H_3N_5^+$)		155 ($C_9H_7N^+$)	153 ($C_{12}H_9^+$)	156 ($C_{11}H_{10}N^+$)
6 (159 Da. - 169 Da.) Peak at 165 Da.	170 ($C_{14}H_2^+$)		160 ($C_{11}N_2^+$)		158-160 ($C_{12}N^+$)		165 ($C_{13}H_9^+$)	

LEGEND				
Percent compared to group probability	0% - 12%	13% - 24%	25% - 36%	37% - 48%

Fig. 8.7 Color-coded figure showing the likelihood of chemical groups to be present in the high mass ion population. Probabilities are determined from the percent occurrence spectrum shown in Fig. 8.6 (reprinted from Crary et al. (2009) with permission from Elsevier)

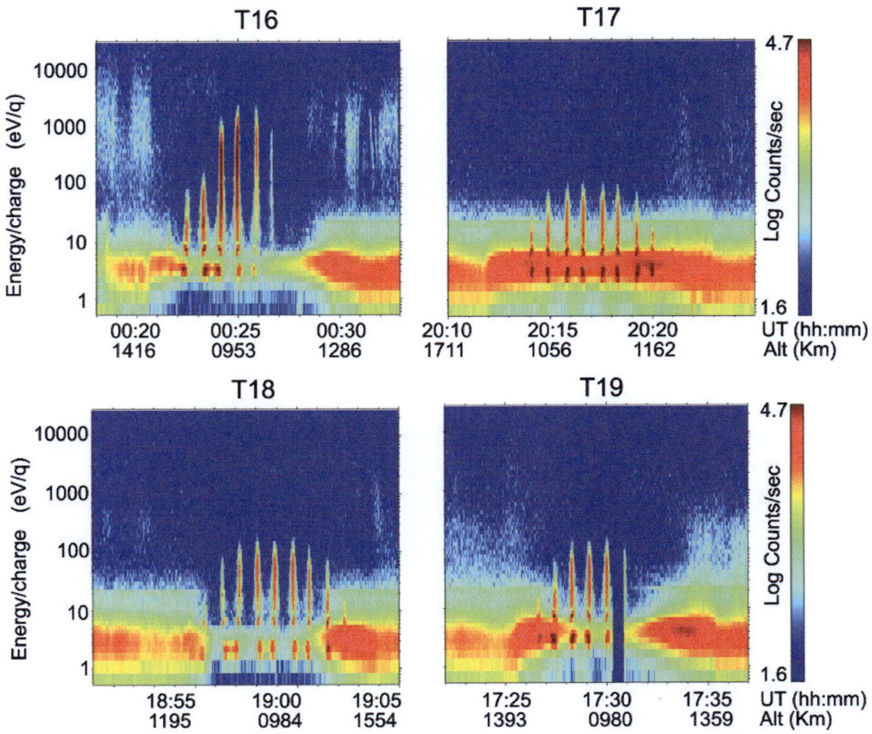

Fig. 8.8 Energy-time spectrograms of ELS data taken during T16 to T19 and centered on closest approach to Titan (Coates et al. 2007). The prominent peaks in each panel are due to cold negative ions rammed into the instrument. Intense fluxes below ~10 eV correspond to photoelectrons from the spacecraft which disappear when spacecraft potential becomes negative in the ionosphere. Fluxes between ~10 and 30 eV correspond to ionospheric photoelectrons

CAPS observations combining IBS and ELS data show that heavy positive and negative ions are present on every pass during the primary mission, 14 in all, where pointing was appropriate. (cf. Table 1 of Crary et al. 2009 and Table 1 of Coates et al. 2009; Wahlund et al. 2009. Analysis has not yet been completed for passes in the extended mission, which began in summer 2008.) Heavy positive ions become a significant component of the positive ion component of the ionosphere below 1,200 km, while at the same altitude heavy negative ions begin to become a prominent fraction (~20%) of the total negatively charged ionospheric component, presumably electrons (Coates et al. 2007, 2009; Waite et al. 2007).

Total negative ion density increases with decreasing altitude as does the maximum negative ion mass (Fig. 8.9). The latter is strongly dependent on altitude, varying from approximately few 100 Da at ~1,400 km to as much as 13,800 Da at 950 km (Coates et al. 2007, 2009). At that altitude negative ions can reach densities up ~200 cm^{-3} and make up as much as 20% of all the negative charge present (Wahlund et al. 2009).

Although a particle with mass of 13,800 Da is a very large molecule indeed, it is doubtful that these are true molecules. Rather they are most likely aerosols formed by the clumping of smaller molecules. Since only mass/charge can be inferred from ELS measurements, if ion charge is >1 electron, then ion mass and size would be proportionately larger. Although simple to estimate in principle, calculating particle size depends on an assumption of density. Solid particles might have a characteristic density ~1 g/cm^3 whereas fractal particles might conceivably have densities as low as 0.001 g/cm^3. The estimated radius for maximum observed mass (13,800 Da) is then in the range 3.8 to 38 nm, the size of small aerosols which, it has been suggested, could be precursors of larger aerosols seen at lower altitudes (see Chapter 12 and references therein).

Since size and electrical charge are important microphysical properties of the aerosols, estimations of size are critical to developing theories of formation and growth (Tomasko et al. 2008; Lavvas et al. 2009). At ~1,000 km altitude the size of heavy negative ions is much smaller than the local Debye length (approximately few centimeters) leading to an estimated particle potential $\varphi \sim -2.5\,kT_e/e$ (Goertz 1989). Since only mass/charge is known, in order to estimate particle size we need to establish ion charge. The electron energy spectrum measured by ELS is variable, at times in sunlight dominated by a photoelectron peak at a few electronvolts (Vuitton et al. 2009). In shadow, the characteristic equilibrium electron temperature is ~1,000 K. On the other hand, the assumption of thermal equilibrium between electrons and ions would result in a temperature of ~150 K. These values lead to a wide range of particle potentials ranging from ~−5 to ~−0.01 V depending on conditions and location.

With an estimate of potential, particle charge can be estimated from $Q = 4\pi\varepsilon_0 a\varphi \exp(-a/\lambda_D)$ where a is particle radius which we earlier estimated to be in the range of 3.8 to 38 nm depending on density. Since $a << \lambda_D$ we neglect the exponent. Substituting values for the constants and units gives the simple equation: $Q \sim 0.7\,a\varphi$ (electrons) where a is in nm and φ in V. The total particle mass $M = km = m/Q = 1.4\,m/a\varphi$ in units of amu, where k is a factor for converting pseudo-mass to total mass. For example a 10,000 Da particle with 1 nm radius and a potential of 1 V would have a total mass of 14,000 amu (Da is a measure of mass/charge).

Given the resolution of ELS the identity of the ions is necessarily somewhat speculative. The high electron affinity of

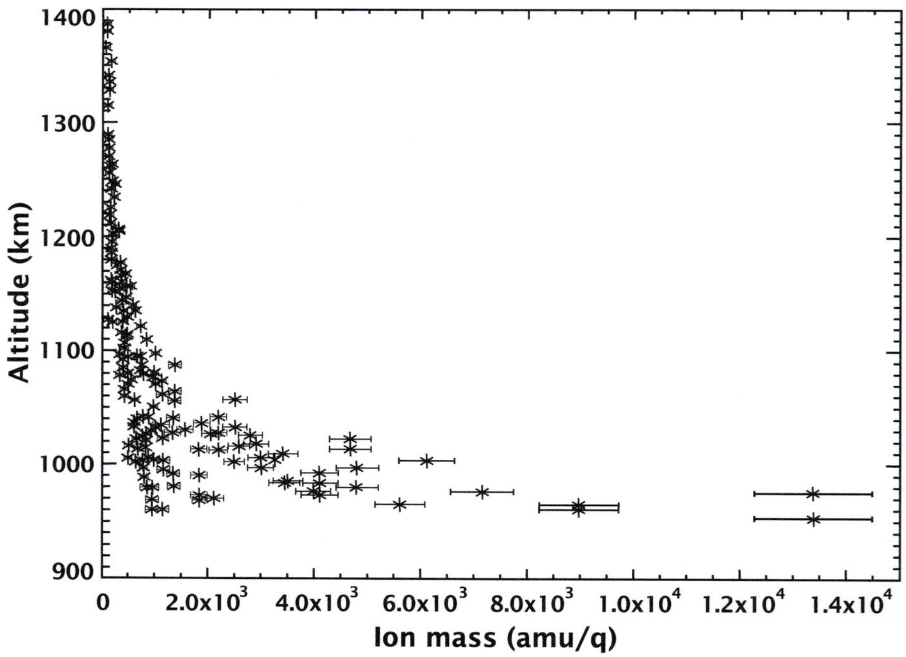

Fig. 8.9 Heavy negative ion densities as a function of altitude showing the rapid increase in density in the aerosol formation layer (below 1,000 km) (reprinted from Coates et al. (2009) with permission from Elsevier)

NH_2^- (0.77 eV) and CN^- (3.86 eV) makes them candidates for the components of the first peak at ~22 Da. (Although ion masses are 16 and 26 respectively both could be present in varying amounts.) Similar considerations suggest that the second peak might be NCN^-, $HNCN^-$ or C_3H^- (see Coates et al. 2007) or C_4H^- (Vuitton et al. 2009). Vuitton et al. also suggest C_5N^- as a possible candidate for third peak. Candidates for heavier ions involve polyynes, nitriles, PAHs, Fullerenes (Sittler et al., 2009) and cyano-nitriles. There is no shortage of possibilities, but at present there is not enough information to narrow the selection. However details of composition matter less than the fact that there is a rich soup of heavy organic compounds conducive to forming aerosols.

Table 8.3 Ion neutral reactions deemed important for the upper atmosphere as assessed by Carrasco et al. (2008)

Reaction	Global rate constant, *k* × 10^{-9} cm^{-3} s^{-1}	Branching ratio (*br*)	Δk/k
$CH^+ + CH_4 \rightarrow C_2H_3^+ + H_2$	1.30	0.84	0.20
$CH^+_2 + CH_4 \rightarrow C_2H_4^+ + H_2$	1.30	0.7	0.15
$CH^+_2 + HCN \rightarrow C_2H_2N^+ + H$	1.80	1	0.20
$CH^+_3 + CH_4 \rightarrow C_2H_5^+$	1.10	1	0.20
$CH^+_4 + CH_4 \rightarrow CH_5^+ + CH_3$	1.14	1	0.15
$CH^+_5 + C_2H_2 \rightarrow C_2H_3^+ + CH_4$	1.48	1	0.20
$CH^+_5 + C_2H_4 \rightarrow C_2H_5^+ + CH_4$	1.50	1	0.20
$CH^+_5 + C_2H_6 \rightarrow C_2H_7^+ + CH_4$	1.35	0.85	0.15
$C_2H^+_2 + CH_4 \rightarrow C_3H_5^+ + H$	0.89	0.79	0.10
$C_2H^+_3 + CH_4 \rightarrow C_3H_5^+ + H_2$	0.19	1	0.20
$C_2H^+_4 + C_2H_2 \rightarrow c\text{-}C_3H_3^+ + CH_3$	0.84	0.77	0.10
$C_2H^+_5 + C_2H_2 \rightarrow c\text{-}C_3H_3^+ + CH_4$	0.19	0.36	0.10
$C_2H^+_5 + C_2H_4 \rightarrow C_3H_5^+ + CH_4$	0.35	1	0.15
$C_2H^+_5 + C_3H_8 \rightarrow C_3H_7^+ + C_2H_6$	0.63	1	0.15
$C_2H^+_5 + H_2O \rightarrow H_3O^+ + C_2H_4$	1.86	1	0.65
$C_2H^+_5 + HCN \rightarrow HCNH^+ + C_2H_4$	2.70	1	0.20
$C_3H^+ + CH_4 \rightarrow C_2H_3^+ + C_2H_2$	0.87	0.9	0.20
$C_3H^+_5 + C_2H_2 \rightarrow C_5H_5^+ + H_2$	0.38	1	0.15
$N(^3P)^+ + CH_4 \rightarrow CH_3^+ + NH$	1.15	0.53	0.35
$N(^3P)^+ + CH_4 \rightarrow CH_4^+ + N$	1.15	0.05	0.35
$N(^3P)^+ + CH_4 \rightarrow HCNH^+ + H_2$	1.15	0.32	0.35
$N(^3P)^+ + C_2H_2 \rightarrow C_2H_2^+ + N$	1.50	0.70	0.15
$N(^3P)^+ + C_2H_2 \rightarrow CNC^+ + H_2$	1.50	0.15	0.15
$N(^3P)^+ + C_2H_4 \rightarrow C_2H_2N^+ + H_2$	1.45	0.05	0.24
$N^+_2 + H_2 \rightarrow N_2H^+ + H$	1.70	1	0.24
$N^+_2 + CH_4 \rightarrow CH_2^+ + H_2 + N_2$	1.14	0.08	0.15
$N^+_2 + CH_4 \rightarrow CH_3^+ + H + N_2$	1.14	0.89	0.15
$N^+_2 + C_2H_2 \rightarrow C_2H_2^+ + N_2$	0.46	1	0.25
$N^+_2 + C_2H_4 \rightarrow C_2H_3^+ + H + N_2$	1.30	0.63	0.15
$N_2H^+ + CH_4 \rightarrow CH_5^+ + N_2$	0.89	1	0.30
$H_3O^+ + HCN \rightarrow HCNH^+ + H_2O$	3.80	1	0.15
$HCNH^+ + CH_3CN \rightarrow C_2H_4N^+ + HCN$	3.80	1	0.20
$HCNH^+ + NH_3 \rightarrow NH_4^+ + HCN$	2.30	1	0.30
$CNC^+ + C_2H_2 \rightarrow C_3H^+ + HCN$	0.80	0.91	0.40
$CNC^+ + C_2H_4 \rightarrow C_2H_2N^+ + C_2H_2$	1.00	0.25	0.20

Source: Adapted from Carrasco et al. (2008), with permission from Wiley Interscience.

8.2 New Chemical Models Based on the Cassini Results

The measurements made by Cassini have motivated a new round of modeling and analysis, which has contributed to our present level of understanding of organic formation in Titan's atmosphere. Waite et al. (2007) were the first to highlight the correspondence of the ion and neutral mass spectra in the upper atmosphere and suggested that this demonstrated a strong degree of coupling in the ion-neutral chemistry that resulted in the observed complexity of the organic compounds. Furthermore, they showed that ion neutral chemistry involving $C_4H_3^+$ was the likely formation path of the observed benzene and that benzene was in chemical equilibrium with its protonated ion counterpart, $C_6H_7^+$. Vuitton et al. (2006) had earlier postulated that most of the unexpected ion peaks measured by INMS in the T5 flyby were the result of protonated nitrile compounds. They used this basic premise to develop a complex zero-dimensional model of the ion neutral chemistry and to infer much of the trace nitrile composition in the neutral atmosphere. Carrasco et al. (2008) examined this same ion neutral data set using two chemical reaction pruning methods coupled with a Bayesian statistical analysis of the reaction rate uncertainty to conclude that only 35 key ion molecule reactions were needed to describe the basic ion-neutral coupling and that they were not dominated by proton transfer as Vuitton et al. had implied. Rather they consisted of 32 growth reactions leading to chemical complexity (see Table 8.3) including 22 condensation (bond rearrangement reactions), 5 protonations, and 5 charge transfer reactions.

Using a method that accounts for diurnal variations in the energy input into the upper atmosphere, De La Haye et al. (2008) have developed a coupled ion-neutral composition model for Titan's atmosphere over the altitude range from 600 to 2,000 km. The model demonstrates the important role played by ion-neutral chemistry in the formation of both hydrocarbons and nitriles, although the relative importance of ion-neutral vs neutral chemistry varies with local time, altitude, and species.

More comprehensive one-dimensional models of the atmosphere that extend from the surface to the exobase and include both gases and particulate chemistry have been recently presented by Lavvas et al. (2008a,b) and Krasnopolsky (2009). Krasnopolsky assumed the temperature structure measured by HASI, whereas Lavvas et al. calculated the thermal structure self consistently, which in fact agrees with the HASI derived profile. Both models generate the haze structure from the gaseous species photochemistry and the authors compare their results to the Cassini INMS and CIRS composition measurements in the thermosphere and stratosphere, respectively, and to the DISR measurements of the haze from the Huygens probe. Lavvas et al. identified a new source of haze particles from 500 to 900 km based on nitrile-hydrocarbon copolymer chemistry, but did not include ion-neutral chemistry and thus failed to identify the source of the high latitude macromolecules observed by CAPS and INMS. Haze formation in their model is dominated by nitrile and aromatic chemistry below 300 km. The total rate of precipitation they calculate is 1.27×10^{-14} g cm^{-2} s^{-1} (4 kg cm^{-2} Gy^{-1}), which is in the range estimated by McKay et al. (2001), $0.5\text{–}2 \times 10^{-14}$ g cm^{-2} s^{-1} (1.6–6.3 kg cm^{-2} Gy^{-1}). Krasnopolsky (2009) included ion-neutral chemistry, ambipolar diffusion, and atmospheric escape as well as both positive and negative ion chemistry that to first order agrees with the CAPS and INMS results. The major haze production in this case is via the reactions $C_6H + C_4H_2$ (polyyne chemistry), $C_3N + C_4H_2$ (copolymer chemistry), and the condensation of hydrocarbons below 100 km first suggested by Hunten (2006). Overall, the estimated precipitation rate is equal to $1.2\text{–}2.2 \times 10^{-14}$ g cm^{-2} s^{-1} (4–7 kg cm^{-2} Gy^{-1}).

Tables 8.4 and 8.5 compare the models discussed above to INMS measurements at 1,050 km and to CIRS measurements at 300 km, respectively. The modeling community before 2008 in general over-predicted the abundance of the C2 group, possibly due to overproduction of $C_2H_5^+$ and $C_3H_5^+$ in the case of the primarily ionospheric models of Vuitton et al. (2006, 2007), or to the lack of ion-neutral chemistry in the case of earlier models. Lavvas et al. (2008 a, b); De La Haye et al. (2008); Lara et al. (1996); and Yung et al. (1984) all underestimate the abundance of the C3 group. Moreover the estimates of benzene mixing ratios vary significantly. Lavvas et al. (2008a,b) and Wilson and Atreya (2004) find mixing ratios from a few times 10^{-10} to 1×10^{-9}, while Krasnopolsky (2009) and De La Haye (2008) find values in the 10^{-7} range. Vuitton et al. (2007) calculate a ratio of 3.0×10^{-6}, which comes closest to observations, possibly because of their inclusion of a currently undefined 2-body reaction process.

Two recent papers have addressed the structure of the haze particles based on Cassini–Huygens observations. Liang et al. (2007) used UVIS ultraviolet observations of the continuum near 190 nm to infer the altitude distribution and size of the haze scattering the incident solar flux. Their analysis indicated that the upper atmosphere near 1,000 km contained up to 10^4 cm^{-3} macromolecules with sizes of the order of 10–20 nm. The photochemical calculations of Liang et al. suggest that polyyne polymers play a major role in the haze formation process. Lavvas et al. (2009) combined the UVIS ultraviolet observations with the ISS visible scattering observations to address the relationship between the detached haze layer and the haze layers extending to the surface. They found that the detached haze layer is formed as a result of changes in the particle sedimentation rate with altitude. They assert that the observed mass flux 2.7 to 4.6×10^{-14} g cm^{-2} s^{-1}, which is approximately what McKay et al. (2001), needed to explain the main haze layer, originates from production mechanisms dominated by upper atmospheric processes first identified by Waite et al. (2007). They therefore suggest that the main haze layer in Titan's stratosphere is formed primarily by sedimentation and coagulation of particles in the detached layer. More recently Sittler et al. (2009) suggest that sufficient fullerenes are formed to contribute ~7% to the total aerosol infall.

Figure 8.10 summarizes the post Cassini–Huygens understanding of the conversion of methane and nitrogen to organic macromolecules, then to organic aerosols, and eventually to organic materials on the surface of Titan. The process begins with the dissociation and ionization of methane and molecular nitrogen by solar ultraviolet radiation and energetic particles in the upper atmosphere of Titan. This process sets in motion a rich ion neutral chemistry that produces heavy positive ions and neutrals that eventually form macromolecules, many of which are negatively charged. Macromolecules precipitate rapidly from the formation layer near 1,000 to 550 km where they reach a size of 40 nm. They then slow through atmospheric viscosity to form the detached haze layer (Lavvas et al. 2009, 2008a,b). Some additional radical chemistry occurs during this descent, but most of the formation chemistry has already taken place in the low-pressure upper atmosphere as a result of ion neutral chemistry. Below 550 km the particles began to coagulate into the main aerosol layer and some additional chemistry likely takes place in this growth process. The cold temperatures of the troposphere lead to condensation of ethane and other organics onto the aerosols (Hunten 2006), before they finally precipitate onto the surface. The loss of hydrogen to space guarantees that the process will irreversibly convert methane in the atmosphere into organic residue on times scales from 10 to 70 million years (Mandt et al. 2009).

8.3 Conclusions: Laboratory Simulations and the Future of Titan Exploration

No laboratory simulations have been published that explicitly try to match the Cassini data. However, it is instructive to look at published results from Titan simulations and compare them to the INMS spectra. For example, comparison of the results of Thompson et al. (1991) to the INMS results of Waite et al. (2007) shows that there is a broad similarity in the two results and suggests that properly conducted laboratory simulations will be able to reproduce the INMS spectrum (Fig. 8.11).

The laboratory mass spectrometer measurements of Imanaka et al. (2004) are compared to the mass peaks identi-

Table 8.4 Comparison of mixing ratios from the INMS with several models

Species		Magee et al. (2009)	Krasnopolsky (2009)	Carrasco et al. (2008)	Lavvas (2008a)	De La Haye et al. (2008)	Vuitton et al. (2007)	Vuitton et al. (2006)	Wilson and Atreya (2004)	Toublanc et al. (1995)	Lara et al. (1996)	Yung et al. (1984)
C_2H_2	Max	3.43×10^{-4}	4.2×10^{-4}	3.2×10^{-3}	1.4×10^{-4}	2.2×10^{-4}		3.0×10^{-4}	5.4×10^{-4}	3.6×10^{-3}	1.1×10^{-3}	7.4×10^{-3}
	Min	3.42×10^{-4}	3.8×10^{-4}	1.5×10^{-3}	1.4×10^{-4}	9.1×10^{-5}			4.2×10^{-4}	3.1×10^{-3}		6.1×10^{-3}
C_2H_4	Max	3.97×10^{-4}	7.0×10^{-4}	2.3×10^{-3}	4.5×10^{-4}	6.8×10^{-4}	1.0×10^{-3}	6.0×10^{-3}	1.5×10^{-3}	2.8×10^{-3}	1.5×10^{-3}	5.4×10^{-3}
	Min	3.91×10^{-4}	6.0×10^{-4}	1.1×10^{-3}	4.4×10^{-4}	9.1×10^{-5}			1.3×10^{-3}	2.8×10^{-3}		4.2×10^{-3}
C_2H_6	Max	6.05×10^{-5}	1.2×10^{-4}	2.2×10^{-4}		9.5×10^{-5}		1.0×10^{-4}	2.1×10^{-4}	1.1×10^{-5}	3.5×10^{-4}	1.1×10^{-3}
	Min	4.57×10^{-5}	1.1×10^{-4}	5.5×10^{-5}		2.4×10^{-5}				8.8×10^{-6}	2.9×10^{-4}	9.6×10^{-4}
HCN	Max	2.44×10^{-4}	6.8×10^{-4}	1.5×10^{-3}	9.3×10^{-4}	2.1×10^{-4}	2.0×10^{-4}	2.0×10^{-4}	7.1×10^{-4}	4.6×10^{-4}		2.3×10^{-3}
	Min	2.40×10^{-4}	5.3×10^{-4}	7.1×10^{-4}	7.5×10^{-4}	5.8×10^{-5}			5.0×10^{-4}	4.7×10^{-4}		1.5×10^{-3}
C_3H_6	Max	3.45×10^{-6}								1.3×10^{-6}		6.8×10^{-6}
	Min	2.33×10^{-6}								6.9×10^{-7}		4.8×10^{-6}
C_3H_8	Max	4.38×10^{-6}	5.2×10^{-6}		4.9×10^{-8}	1.5×10^{-7}				7.2×10^{-6}	6.7×10^{-7}	7.6×10^{-7}
	Min	2.87×10^{-6}	2.4×10^{-6}		1.2×10^{-8}	6.5×10^{-8}				3.1×10^{-6}	3.4×10^{-7}	4.1×10^{-7}
C_4H_2	Max	5.65×10^{-6}	2.4×10^{-5}		2.5×10^{-6}	1.6×10^{-6}	1.0×10^{-5}	6.0×10^{-5}	3.7×10^{-6}	1.4×10^{-5}	1.6×10^{-5}	1.5×10^{-4}
	Min	5.55×10^{-6}	1.1×10^{-5}		8.8×10^{-7}	7.3×10^{-7}			3.1×10^{-6}	6.4×10^{-6}	7.9×10^{-6}	7.8×10^{-5}
C_2N_2	Max	2.20×10^{-6}				2.2×10^{-9}			2.4×10^{-8}	2.1×10^{-7}		3.4×10^{-5}
	Min	2.14×10^{-6}				1.4×10^{-9}			2.1×10^{-8}	1.0×10^{-7}		1.9×10^{-5}
HC_3N	Max	1.54×10^{-6}			9.8×10^{-6}	2.3×10^{-7}	4.0×10^{-5}	2.0×10^{-5}	2.3×10^{-6}	1.6×10^{-5}		2.3×10^{-4}
	Min	1.48×10^{-6}			4.1×10^{-6}	1.8×10^{-7}			1.9×10^{-6}	5.2×10^{-6}		1.2×10^{-4}
C_2H_3CN	Max	4.39×10^{-7}			8.0×10^{-6}		1.0×10^{-5}	1.0×10^{-5}	1.1×10^{-6}			
	Min	3.46×10^{-7}			4.3×10^{-6}				1.4×10^{-6}			
C_2H_5CN	Max	2.87×10^{-7}	3.6×10^{-7}				5.0×10^{-7}	5.0×10^{-7}				
	Min	1.54×10^{-7}	1.9×10^{-7}									
C_6H_6	Max	2.50×10^{-6}	2.7×10^{-7}		4.6×10^{-10}	6.1×10^{-7}	3.0×10^{-6}		1.4×10^{-9}			
	Min	2.48×10^{-6}	1.4×10^{-7}		5.7×10^{-9}	1.3×10^{-7}			2.2×10^{-10}			
C_7H_8	Max	5.37×10^{-8}	2.7×10^{-7}				2.0×10^{-7}					
	Min	2.51×10^{-8}	1.3×10^{-7}									

Source: From Magee et al. (2009), with permission from Elsevier.

Table 8.5. Comparison of CIRS data with several models.

		CIRS data		Models					
		Vinatier (2007)	Teanby (2007)	Krasnopolsky (2009)	Lavvas (2008)	Wilson and Atreya (2004)	Lara (1996)	Toublanc (1995)	Yung (1984)
C_2H_2	Max	4.9×10^{-6}	4.9×10^{-6}	1.2×10^{-5}	2.9×10^{-6}	7.4×10^{-6}	3.1×10^{-5}	2.1×10^{-5}	7.4×10^{-5}
	Min	3.6×10^{-6}	1.5×10^{-5}						
C_2H_4	Max	2.0×10^{-7}		1.3×10^{-7}	3.3×10^{-8}	4.6×10^{-8}	1.5×10^{-8}	2.3×10^{-8}	1.5×10^{-7}
	Min	2.2×10^{-8}							
C_2H_6	Max	2.3×10^{-5}		3.1×10^{-5}	2.2×10^{-5}	1.9×10^{-5}	6.5×10^{-5}	4.4×10^{-5}	2.3×10^{-4}
	Min	2.2×10^{-5}							
HCN	Max	1.3×10^{-6}	6.2×10^{-6}	3.7×10^{-6}	2.0×10^{-6}	1.3×10^{-6}		2.2×10^{-6}	9.7×10^{-6}
	Min	5.9×10^{-7}	1.4×10^{-6}						
C_3H_8	Max	5.5×10^{-7}		5.6×10^{-6}	1.5×10^{-6}		9.6×10^{-7}	1.5×10^{-5}	9.6×10^{-6}
	Min	6.2×10^{-7}							
C_4H_2	Max	4.5×10^{-9}		1.6×10^{-7}	1.0×10^{-8}	2.9×10^{-8}	1.2×10^{-6}	8.0×10^{-8}	6.1×10^{-8}
	Min	2.8×10^{-8}							
HC_3N	Max		5.4×10^{-7}	3.4×10^{-6}	8.7×10^{-9}	2.2×10^{-8}		2.2×10^{-8}	9.8×10^{-7}
	Min		1.9×10^{-8}						
C_6H_6		3.6×10^{-9}				2.1×10^{-10}	8.4×10^{-11}	8.9×10^{-10}	

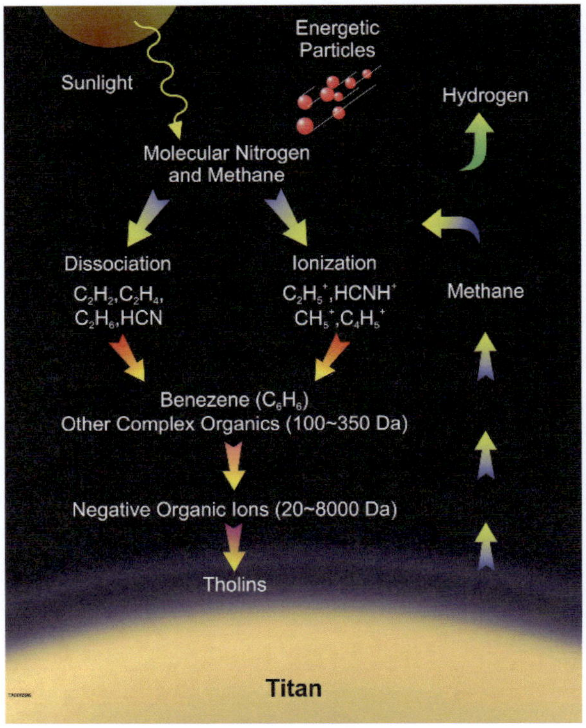

Fig. 8.10 Cartoon illustrating tholin formation as result of high-altitude ion-neutral chemistry. Over tens of millions of years methane is, with the loss of hydrogen to space, irreversibly converted to the complex hydrocarbon-nitrile compounds that are the precursors of the aerosols (adapted from Waite et al. (2007) with permission from AAAS)

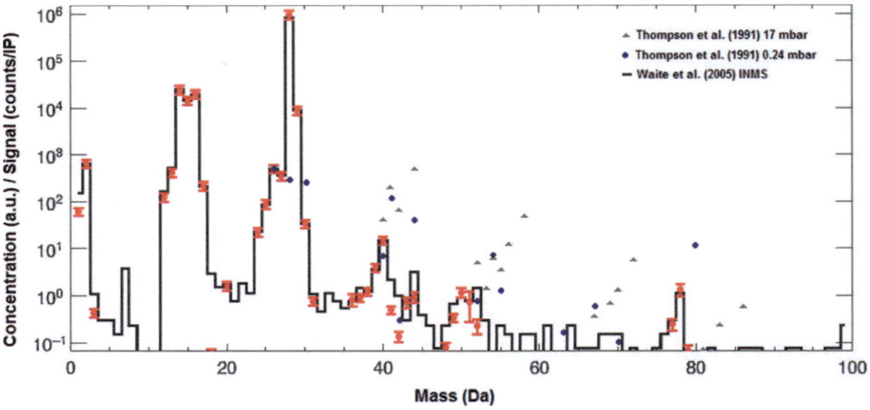

Fig. 8.11 Comparison of INMS spectrum and laboratory simulations. Solid line and red symbols are INMS observations of Titan's atmosphere at an altitude of ~1,000 km (Waite et al. 2005). Blue and gray symbols are laboratory results at 24 and 1,700 Pa respectively. The general comparability of the different results suggests that properly conducted laboratory simulations will be able to assist in understanding the INMS spectrum.

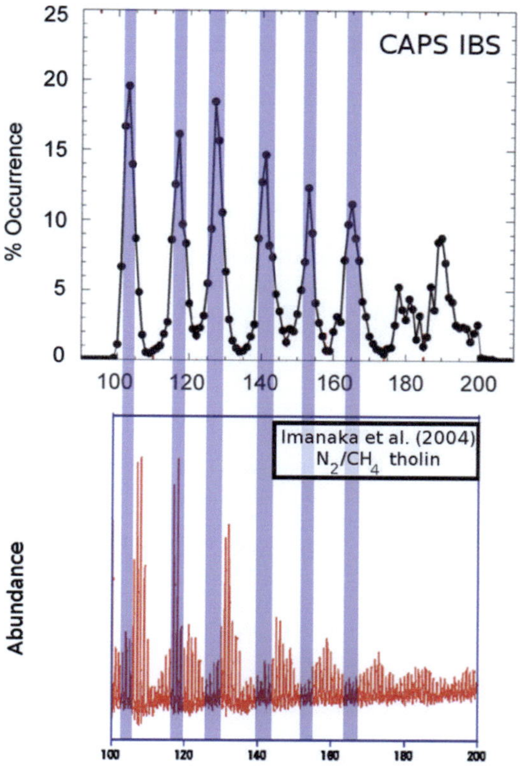

Fig. 8.12 The correspondence between the peaks observed in the high-mass positive ions from the IBS/INMS cross correlation and a mass spectrum of laboratory tholins. The second peak, which would correspond to the C9 group, is similar in both cases. The top panel (reprinted from Crary et al. (2009)), and the bottom panel (adapted from Imanaka et al. (2004), with permission from Elsevier)

fied in the IBS ion mass spectrum in Fig. 8.12. While they are suggestive, the results indicate that the very low pressure ion-neutral chemistry present in Titan's upper atmosphere has not been adequately explored in the laboratory and should be the focus of future investigations. Sittler et al. (2009) have carefully examined laboratory processes that lead to fullerence formation and their results suggest further lines of inquiry. The richness of Titan's chemical environment discovered by Cassini, and the complexity of chemical processes being explored in the laboratory together point to the need for further exploration in both arenas. In the case of Cassini this will take the form of an additional ~60 encounters with the fascinating atmosphere of Titan during the Cassini Solstice Mission.

References

Atreya S, Donahue T, Kuhn W (1978) Evolution of a nitrogen atmosphere on Titan. Science 201:611–613

Bar-Nun A, Kleinfeld I, Ganor E (1988) Shape and optical properties of aerosols formed by photolysis of acetylene, ethylene, and hydrogen cyanide. J Geophys Res 93:115–122

Broadfoot A, Sandel B, Shemansky D, Holberg J, Smith G et al (1981) Extreme ultraviolet observations from Voyager 1 encounter with Saturn. Science 212:206–211

Cabane M, Chassefière E (1995) Laboratory simulations of Titan's atmosphere: organic gases and aerosols. Planet Space Sci 43:47–65

Carrasco N, Plessis S, Dobrijevic M, Pernot P (2008) Toward a reduction of the bimolecular reaction model for titan's ionosphere. Int J Chem Kinet 40:699–709

Chang S, Scattergood T, Aronowitz S, Flores J (1979) Organic chemistry on Titan. *Rev Geophys Space Phys 17(8):1923–1933*

Clarke D, Ferris J (1997) Titan haze: structure and properties of cyanoacetylene and cyanoacetylene–acetylene photopolymers. Icarus 127:158–172

Coates AJ, Crary FJ, Lewis GR, Young DT, Waite Jr JH, Sittler Jr EC (2007) Discovery of heavy negative ions in Titan's ionosphere. Geophys Res Lett 34:L22103

Coates AJ, Lewis G, Wellbrock A, Jones G, Young DT et al (2009) Heavy negative ions in titan's ionosphere: altitude and latitude dependence. Planet Space Sci doi:10.1016/j.pss.2009.05.009

Coll P, Coscia D, Smith N, Gazeau M, Ramrez S et al (1999) Experimental laboratory simulation of Titans atmosphere: aerosols and gas phase. Planet Space Sci 47:1331–1340

Coustenis A, Salama A, Schulz B, Ott S, Lellouch E et al (2003) Titan's atmosphere from ISO mid-infrared spectroscopy. Icarus 161:383–403.

Crary FJ, Magee BA, Mandt KE, Waite JH, Westlake JH, Young DT (2009) Heavy ions, temperatures and winds in titan's ionosphere: Combined cassini caps and inms observations. Planet Space Sci. In press

Danielson RE, Caldwell J, Larach DR (1973) An inversion in the atmosphere of Titan. Icarus 20:437–443

De La Haye V, Waite JH, Cravens TE, Robertson IP, Lebonnois, S (2008) Coupled ion and neutral rotating model of Titan's upper atmosphere. Icarus 197:110–136

Gillett F (1975) Further observations of the 8–13 micron spectrum of Titan. Astrophys J 201:L41

Goertz CK (1989) Dusty plasmas in the solar system. Rev Geophys 27:271–292

Hanel R, Conrath B, Flasar F, Kunde V, Maguire W et al (1981) Infrared observations of the Saturnian system from Voyager 1. Science 212:192–200

Hanel R, Conrath B, Flasar F, Kunde V, Maguire W et al (1982) Infrared observations of the Saturnian system from Voyager 2. Science 215:544–548

Hunten DM (1977) Titan's atmosphere and surface. In: IAU Colloq.#28: Planetary satellites. University of Arizona Press, Tucson, AZ

Hunten, DM (2006) The sequestration of ethane on Titan in smog particles. Nature 443:669–670.

Imanaka H, Khare B, Elsila J, Bakes E, McKay C et al (2004) Laboratory experiments of Titan tholin formed in cold plasma at various pressures: implications for nitrogen-containing polycyclic aromatic compounds in Titan haze. Icarus 168:344–366

Khare BN, Sagan C (1973) Red clouds in reducing atmospheres. Icarus 20:311–321

Khare BN, Sagan C, Arakawa ET, Suits F, Callcott TA, Williams MW (1984) Optical constants of organic tholins produced in a simulated Titanian atmosphere: from soft X-ray to microwave frequencies. Icarus 60:127–137

Khare B, Bakes ELO, Imanaka H, McKay C, Cruikshank D, Arakawa E (2002) Analysis of the time-dependent chemical evolution of Titan haze tholin. Icarus 160:172–182.

Krasnopolsky V (2009) A photochemical model of Titan's atmosphere and ionosphere. Icarus *201:226–256*

Kunde VG, Aikin AC, Hanel RA, Jennings DE, Maguire WC, Samuelson RE (1981) C_4H_2, HC_3N and C_2N_2 in Titan's Atmosphere. Nature 292:686–688

Lara LM, Lellouch E, Lopez-Moreno JJ, Rodrigo R (1996) Vertical distribution of Titan's atmospheric neutral constituents. J Geophys Res 101:23261–23283

Lavvas PP, Coustenis A, Vardavas IM (2008a) Coupling photochemistry with haze formation in Titan's atmosphere, Part I: Model description. Planet Space Sci 56:27–66

Lavvas PP, Coustenis A, Vardavas IM (2008b) Coupling photochemistry with haze formation in Titan's atmosphere, Part II: Results and validation with Cassini/Huygens data. Planet Space Sci 56:67–99

Lavvas P, Yelle R, Vuitton, V (2009) The detached haze layer in Titan's mesosphere. Icarus *201:626–633*

Lebonnois S, Bakes ELO, McKay C (2002) Transition from gaseous compounds to aerosols in Titan's atmosphere. Icarus 159:505–517

Liang M, Yung Y, Shemansky D (2007) Photolytically generated aerosols in the mesosphere and thermosphere of Titan. Astrophys J Lett 661:199–202

McKay CP, Coustenis A, Samuelson RE, Lemmon MT, Lorenz RD et al (2001) Physical properties of the organic aerosols and clouds on Titan. Planet Space Sci 49:79–99

Magee BA, Waite JH, Mandt KE, Bell J, Westlake J, Gell DA, De La Haye V (2009) INMS derived composition of titan's upper atmosphere: Analysis methods and model comparison. Planet Space Sci. In press doi:10.1016/j.pss.2009.06.016

Maguire WC, Hanel RA, Jennings DE, Kunde VG, Samuelson RE (1981) C_3H_8 ad C_3H_4 in Titan's Atmosphere. Nature 292:683–686

Mandt KE, Waite JH, Lewis W, Lunine J, Magee B (2009) Isotopic evolution of titan's main constituents. Planet Space Sci. In press doi:10.1016/j.pss.2009.06.005

Podolak M, Bar-Nun A (1979) A constraint on the distribution of Titan's atmospheric aerosol. Icarus 39:272–276

Sagan C, Khare BN (1979) Tholins – organic chemistry of interstellar grains and gas. Nature 277:102–107

Scattergood T, Owen T (1977) On the sources of ultraviolet absorption in spectra of Titan and the outer planets. Icarus 30:780–788

Scattergood TW, Lau EY, Stone BM (1992) Titan's aerosols. I – Laboratory investigations of shapes, size distributions, and aggregation of particles produced by UV photolysis of model Titan atmospheres. Icarus 99:98–105

Sittler EC, Jr, Ali A, Cooper JF, Hartle RE, Johnson RE, Coates AJ, Young DT (2009) Heavy ion formation in Titan's ionosphere: Magnetospheric introduction of free oxygen and a source of Titan's aerosols? Planet. Space Sci. (2009) doi:10.1016/j.pss.2009.07.017

Smith B, Soderblom L, Beebe R, Boyce J, Briggs G et al (1981) Encounter with Saturn: Voyager 1 imaging science results. Science 212:163–191

Smith G, Strobel D, Broadfoot A, Sandel B, Shemansky D, Holberg J (1982) Titan's upper atmosphere: Composition and temperature from the EUV solar occultation results. J Geophys Res 87: 1351–1359

Teanby NA, Irwin PG, de Kok R, Vinatier S, Bézard B et al (2007) Vertical profiles of HCN, HC_3N, and C_2H_2 in Titan's atmosphere derived from Cassini/CIRS data. Icarus 186:364–384

Thompson RW, Henry T, Schwartz J, Khare B, Sagan C (1991) Plasma discharge in N_2 + CH_4 at low pressures: experimental results and applications to Titan. Icarus 90:57–73

Thompson RW, McDonald G, Sagan C (1994) The Titan haze revisited: magnetospheric energy sources and quantitative tholin yields. Icarus 112:376–381

Tomasko MG, Doose L, Engel S, Dafoe LE, West R et al (2008) A model of Titan's aerosols based on measurements made inside the atmosphere. Planet Space Sci 56:669–707

Toublanc D, Parisot JP, Brillet J, Gautier D, Raulin F, McKay CP (1995) Photochemical modeling of Titan's atmosphere. Icarus 113:2–26

Trafton L (1975) Near-infrared spectrophotometry of Titan. Icarus 24:443–453

Trainer M, Pavlov A, Jimenez J, McKay C, Worsnop D et al (2004) Chemical composition of Titan's haze: are PAHs present? Geophys Res Lett 31:17–17

Tyler GL, Eshleman VR, Anderson JD, Levy GS, Lindal GF et al (1981) Radio science investigations of the Saturn system with Voyager 1 – Preliminary results. Science 212:201–206

Veverka J (1973) Titan: Polarimetric evidence for an optically thick atmosphere? Icarus 18:657–660

Vinatier S, Bezard B, Fouchet T, Teanby N, Kok R et al (2007) Vertical abundance profiles of hydrocarbons in Titan's atmosphere at 15 S and 80 N retrieved from Cassini/CIRS spectra. Icarus 188:120–138

Vuitton V, Yelle RV, Anicich VG (2006) The nitrogen chemistry of Titan's upper atmosphere revealed. Astrophys J Lett 647:175–178

Vuitton V, Yelle RV, McEwan MJ (2007) Ion chemistry and N-containing molecules in Titan's upper atmosphere. Icarus 191:722–742

Vuitton V, Lavvas P, Yelle R, Galand M, Wellbrock A et al (2009) Negative ion chemistry in titan's upper atmosphere. Planet Space Sci. In press doi:10.1016/j.pss.2009.04.004

Wahlund J-E, Galand M, Mueller-Wodarg I, Cui J, Yelle RV et al (2009) On the amount of heavy molecular ions in Titan's ionosphere. Planet Space Sci. In press

Waite JH, Lewis WS, Kasprzak WT, Anicich VG, Block BP et al (2004) The Cassini Ion and Neutral Mass Spectrometer (INMS) investigation. Space Sci Rev 114:113–231

Waite JH, Niemann H, Yelle RV, Kasprzak W, Cravens T et al (2005) Ion neutral mass spectrometer results from the first flyby of Titan. Science 308:982–986

Waite JH, Young DT, Cravens TE, Coates AJ, Crary FJ, Magee B, Westlake, J (2007) The process of tholin formation in Titan's upper atmosphere. Science 316:870

Wilson EH, Atreya SK (2003) Chemical sources of haze formation in Titan's atmosphere. Planet Space Sci 51:1017–1033

Wilson EH, Atreya SK (2004) Current state of modeling the photochemistry of Titan's mutually dependent atmosphere and ionosphere. J Geophys Res 109:E06002

Yung YL, Allen M, Pinto JP (1984) Photochemistry of the atmosphere of Titan – comparison between model and observations. Astrophys J Suppl. S. 55:465–506

Zellner B (1973) The polarization of Titan. *Icarus* 18:661–664

Chapter 9
Titan's Astrobiology

F Raulin, C McKay, J Lunine, and T Owen

Abstract This chapter describes the aspects of Saturn's moon Titan of astrobiological interest. Titan's prebiotic-like chemistry is reviewed, from the high atomosphere to the surface and subsurface, using the Cassini-Huygens data, with the help of theoretical modeling and experimental simulations. Similarities with and differences from the environment of the pre-biotic Earth are presented, and the lessons to be learned for Earth's organic chemical evolution on the prebiotic Earth discussed. The question of habitability and life on and in Titan is then considered, including the possibility of an exotic type of life that might exist in the liquid methane/ethane lakes. Finally, the relation between Titan and the destiny of life on Earth is discussed.

9.1 From Astrobiology to Titan

Since our technology allows us to leave our planet and send spacecraft to explore other worlds, we have a direct tool to try to answer the question "is there life elsewhere", including primitive and microscopic life. This is one of the main questions of Exobiology, the name of a new scientific field introduced in the 1960s by the microbiologist Nobel Laureate Joshua Lederberg, when NASA was preparing the Apollo program. Exobiology was initially the study of extraterrestrial life and its potential interactions with terrestrial life. This field has increased its territory and is now the science which studies life in the universe: its origins, in particular on Earth, its evolution and its distribution in the entire universe, as well as the study of structures and processes which are related to life. Astrobiology has been introduced more recently, in the mid 1990s, by NASA and the opening of the NAI (NASA Astrobiology Institute) centers. Almost a synonym of exobiology, astrobiology covers also the destiny of Life.

Thus astrobiology includes many different approaches. Several are related to terrestrial life, which remains the only example of life we know, and provides an essential reference for our quest for life elsewhere. It includes the study of the origin of life on Earth and the study of terrestrial life in extreme environments. It also includes the study of extraterrestrial organic chemistry (of potential prebiotic interest), and the search for biomarkers and signs for biological activity in extraterrestrial environments. From what we think we understand about the origin of life on our planet, the emergence of living systems requires three main conditions: the presence of organic matter, the presence of liquid water and the availability of energy. The prime planetary targets for astrobiology are those where these three conditions are fulfilled. However, there are other targets also of astrobiological importance where liquid water is not present, but where a complex organic chemistry is going on. The study of such bodies may provide important insights for our understanding of terrestrial prebiotic chemistry.

Mars, likely to have been very similar to the Earth during the first billion of years of the histories of both planets, when life was originating on Earth, remains the preferred target for many astrobiologists to search for (either extinct or extant) extraterrestrial life. Comets are also of large astrobiological interest, because of the complex organic chemistry which may be evolving in them, from the nucleus to the coma. Such cometary chemistry may even have participated in the prebiotic chemistry on Earth, through early impacts. The Voyager mission has revealed two other astrobiologically interesting places in the outer solar system. Europa, with the likely presence of an internal water ocean is a potentially habitable satellite. Titan, with its dense N_2 atmosphere similar to that of the Earth, and its active atmospheric organic chemistry, appeared like a prebiotic planetary environment. It was considered as a prebiotic reactor at the planetary scale, offering the possibility of looking at some of the chemical processes

F. Raulin (✉)
LISA, University of Paris, Paris, 12and 7, France
e-mail: raulin@lisa.univ-paris12.fr

C. McKay
NASA-Ames

J. Lunine
LPL, University of Arizona, Tucson, Arizona

T. Owen
University of Hawaii

R.H. Brown et al. (eds.), *Titan from Cassini-Huygens*,

which occurred on the primitive Earth, in spite of the absence of permanent bodies of liquid water on its surface. However, because of its low temperature conditions Titan was often considered as a frozen primitive Earth, kept in a cosmic refrigerator for 4 billion years.

After the arrival of the Cassini spacecraft in the Saturn system, and the many observations by the Cassini–Huygens mission of the giant planet, its rings and its satellites, many discoveries of tremendous importance for astrobiology have been achieved. The observations of Enceladus have made this small Saturn satellite a new target for astrobiological studies, while the observations of Titan have strongly strengthened its astrobiological importance, with four main characteristics. (1) The many similarities which can be found when comparing Titan and the early Earth. (2) The presence of an active and complex organic chemistry in Titan's environment, from the high atmosphere to the surface and very likely in the subsurface. This is also one of the similarities with the early Earth. (3) The presence of an internal water-ocean, considered very likely from the Cassini-Radar observations. This makes Titan a potential habitable environment, of obvious astrobiological importance. (4) Lastly, Titan can also be considered as a model for understanding the future of the Earth, and of terrestrial life. Several of the aspects which are presented here are quite speculative, and still under discussion, without a full consensus between the three contributors of the present chapter. Thus this chapter includes four complementary parts, each one being the personal view of its author, and not necessarily endorsed by the other contributors of the chapter. Section 9.2 by T. Owen, covers the question of similarities and differences between Titan and the primitive Earth. Section 9.3, by F. Raulin, describes the organic chemistry in the different parts of Titan's environment and its astrobiological consequences. Section 9.4, by C.P. Mckay, is devoted to the question of habitability and the possibility of life on and in Titan. Section 9.5, by J.I. Lunine, presents what we can learn from Titan to foresee the future of life on Earth.

9.2 Titan: A Fiercely Frozen Echo of the Early Earth?

9.2.1 Introduction

How and where life originated on the Earth are both unknown. We have good reasons to believe that the early environment at the surface of the planet included energy from the sun, a relatively thick atmosphere enabling the presence of open bodies of water and all the elements essential to the one example of life we know. Some as yet unknown fraction of both volatiles and refractories must have been delivered by impacting planetesimals. A probable pathway from this random mixture to the origin of life can be described in a consensus model for any type of carbon/water life. In its essence, this model proposes the formation and subsequent evolution of organic compounds through a series of intermediate steps. The increasing complexity of the resulting compounds ultimately leads to the occurrence of that astonishing transition from ordinary matter to matter that is alive.

How this transition occurred is still a mystery to us. As the French astrobiologist André Brack has put it: "The discovery of ethanol in the Interstellar Medium does not mean burgundy is also there – and it's a hell of a lot easier to make an exquisite burgundy than to make life."

It is the nature of the primitive environment that enabled this improbable chain of events that concerns us here. We want to examine present conditions on Titan to see what, if any, insights Titan can give us to explain the origin and characteristics of the life-giving circumstances on the early Earth. Such an inquiry cannot be carried out in the necessary detail on the Earth itself because of the near absence of the planet's early geological record. Erosion and subduction have almost completely eliminated the first 500 million years of history. This is the time frame within which life on Earth must have begun. Thus scientists investigating this problem have been forced to rely mainly on theoretical arguments, laboratory experiments, and whatever they could glean from studying the meteorites or the ancient, desiccated surface of the moon.

Because of its 94 K surface temperature, Titan offers us a volatile-rich, chemically primitive world trapped in time. Titan has an atmosphere of nitrogen, methane and hydrogen corresponding to conditions in the very early phases of solar system history. Hence studying this satellite might be expected to produce some insights into certain features of the early Earth. These insights will obviously be seriously limited by the extreme differences in surface temperature and composition that we find today.

9.2.2 The Origin of the Earth's Atmosphere

Historically there have been three principal models for the origin of the Earth's atmosphere, two exogenous, one endogenous.

Exogenous (A): Capture of gases from the solar nebula (Urey 1952). Dominant gases: CH_4, NH_3, H_2, and H_2O

Exogenous (B): Delivery by rocky and icy planetesimals. Dominant gases: *ROCKY* – CO_2, NH_3, H_2O; *ICY* – CH_4, CO_2, NH_3, CO (Turekian and Clark 1975; Owen and Bar-Nun 1995; Delsemme 1998).

Endogenous (C): Gases stored in the rocks making up the bulk of the Earth were released by volcanism to form the atmosphere (Rubey 1951). Dominant gases H_2O, CO_2, N_2.

Let us examine the pros and cons of each approach.

(A) Has the striking advantage of producing a highly reducing atmosphere and providing an ideal mixture of gases for the production of HCN, amino acids and other monomers representing very early steps in the chemical evolution that preceded biological evolution on Earth. The watershed experiment supporting this pathway was carried out by Stanley Miller (1953). Miller subjected a mixture of methane, ammonia, hydrogen, and water vapor to a spark discharge between two electrodes in a closed flask. The spark simulated lightning on the early Earth, supplying energy to drive chemical reactions. Among the organic compounds produced were several amino acids, the building blocks of proteins. This experiment was so convincing it has been illustrated repeatedly in thousands of textbooks. The implications of Miller's results have been discussed by numerous authors following Miller and Urey (1959). The bad aspect of this approach is its assumption that the Earth owed the composition of its atmosphere to conditions in the surrounding solar nebula. This is now seen to be impossible. Among other arguments, the formation of the moon by a giant impact renders this idea obsolete.

(B) Exogenous: In this variation of (A) the gases are brought to the Earth by the impacts of meteorites, micrometeorites, IDPs and comets. These impacts have the advantage of delivering the elements essential to biology – CHON – in various simple compounds. The way is then open for chemical evolution in the atmosphere and on the ground.

This model gains support from the discovery of indigenous amino acids in some carbonaceous chondrites. In particular, Cronin et al. (1988) found a total of 74 amino acids in the Murchison meteorite. These include the ones produced in the Miller–Urey experiment.

Lest our readers become too excited by these results, we refer them to Brack's cautionary statement in the introduction. Amino acids, whether produced in Miller's experiment or found in meteorites, are a very long way from life.

Comets are appealing as an extraterrestrial source of volatiles as they carry a solar abundance of carbon relative to H_2O. They also contain a higher relative abundance of nitrogen than any of the meteorites (Geiss 1988). This complement includes ~1% each of NH_3 and CH_4 and as much as ~20% CO (Bockelée-Morvan et al. 2004). There is clearly a potential here for producing a reducing atmosphere by impacts of comets on the early Earth (Oro 1961; Delsemme 1998, 2006).

(C) The major advantage of this endogenous scenario is that it duplicates the CO_2, N_2 and H_2O in the present atmosphere. A huge amount of CO_2 that was once in the atmosphere is now found in the form of carbonate rocks such as limestone. In this sense (C) conforms to the powerful hypothesis of Lyell: geological changes in the physical world result from geological processes that are acting at the present time. Lyell's hypothesis freed geology from a reliance on cataclysms to explain the current state of the Earth's surface. (However we now know that early bombardment of the Earth produced a series of cataclysms that are no longer occurring). The disadvantage of this model is that it does not provide the highly reducing atmosphere so beneficial to prebiological chemistry. This can be remedied to some extent by the admixture of CH_4 released from mid-ocean ridges that would produce a weakly reducing atmosphere (Kasting and Brown 1998).

9.2.3 General Remarks

A weakness of all of these models is that they do not reproduce the relative abundances and isotope ratios of the noble gases that we find in our atmosphere. Comets may be an exception; unfortunately we have no information about noble gases in comets yet. A future comparison of cometary and terrestrial abundances of these chemically inert markers should tell us how much matter comets have delivered to the Earth.

Both scenarios (B) and (C) benefit from new work on the lifetime of H_2 in Earth's early atmosphere. Recent calculations of the rate at which hydrogen could escape from the Earth indicate that the residence time for H_2 in the early atmosphere was far longer than had been thought (Tian et al. 2005). This means that the lifetime of any CH_4 or NH_3 in the atmosphere of the early Earth was also longer.than the previously calculated duration. This lengthens the time available for prebiotic chemical reactions in a favorable hydrogen-rich environment.

We conclude that each of these three generalized scenarios has its strong and weak points. Can Titan help us discriminate among these alternative models?

Titan offers us an atmosphere of 1.5 bar consisting of 98.4% N_2 and 1.6% CH_4, plus many trace constituents such as C_2H_6, C_2H_2, C_2N_2, HCN etc. (Niemann et al. 2005; Waite et al. 2005, 2007; Coustenis et al. 2007). Water vapor is almost absent owing to the 94 K surface temperature (the sublimation pressure of H_2O ice is about 10^{-14} Pascal at T = 100 K (Feistel and Wagner 2007). Thus OH does not oxidize the radicals produced by the photolysis of CH_4 and N_2, as happens on Earth. Instead these radicals are free to interact with each other and produce new species (see Raulin, next section). A minor exception is proffered in the uppermost atmosphere. Here H_2O from the vaporization of impacting icy grains from Saturn's icy satellites and rings makes OH available (Coustenis et al. 2007). This chemistry has no effect on the troposphere.

Surprisingly, the photochemistry on Titan occurs in the high upper atmosphere (~1,000 km) rather than in the middle atmosphere as originally predicted and assumed in photochemical models (see Strobel 1982; Yung et al. 1984; Hébrard et al. 2007b, and references included). High molecular weight (~8,000 Da) compounds made here are probably the major progenitors of the aerosols making up Titan's smog (Waite et al. 2007). The result is a thick haze layer that is almost opaque at visible wavelengths and completely impenetrable in the UV (Tomasko et al. 2005). The aerosols ultimately precipitate forming thick deposits and dunes on Titan's surface (Lorenz et al. 2006, 2008).

9.2.4 Lessons for Earth

The low surface temperature on Titan could be considered to be a substitute for the long lifetime of free hydrogen in the Earth's atmosphere. Both of these conditions act to keep the production of OH suppressed allowing reducing conditions to persist.

The existence of the global "smog" layer on Titan led Sagan and Chyba (1997) to explore the significance of a similar layer in the atmosphere of the early Earth.

One of the criticisms of models invoking early hydrogen-rich atmospheres ((A) and (B)), has been that CH_4 and NH_3 as well as their immediate photochemical products would be rapidly dissociated by UV radiation from the sun. The Sagan-Chyba Titan-analog hypothesis provides a smog layer that shields the lower atmosphere from this radiation. The photochemistry in such an atmosphere on the early Earth would occur above the aerosol layer, just as it does on Titan. Precipitating aerosols would filter down through the protected $CH_4 + NH_3$ atmosphere below the smog. In this scenario, these precipitating aerosols would land in a warm ocean instead of on the rock-hard 94 K ice surface of Titan.

This imaginative model is attractive, but how close is it to reality? Would this lower atmosphere on the early Earth contain NH_3 as Sagan and Chyba suggested? Titan tells us that this part of the story should be correct for scenarios (B) or (C). We will show in Section 9.2.4 why NH_3 must have delivered nitrogen to Titan. The evidence for this conclusion strongly supports the presence of NH_3 on the early Earth as well. Titan thus leads us to favor scenario (C), as we have already discarded (B).

(A) does not provide for the delivery of NH_3. However, Trainer et al. (2006) introduced CO_2 to a comparable amount of CH_4 in their model for the early Earth's atmosphere and found that an aerosol layer would still be produced. The history of nitrogen and the emergence of so much methane remain undefined.

So far, so good. But Titan has only taken us as far as validating the idea of a UV shielding smog layer with precipitating organic aerosols. We have not yet satisfied the boundary condition on the early terrestrial environment set by the origin of life. What molecular complexity is sufficient to lead to life? The compounds produced must be able to store the information needed for reproduction and evolution (cf. McKay, Section 9.4).

It will take a future Titan lander to see whether anything remotely near this complexity has been or is being produced there.

The information available at the present time suggest that it is simply impossible that comparisons of this aspect of Titan's present 94 K environment with the early Earth will be of any use (but see Sections 9.3 and 9.4). It is much more fruitful to examine a more basic question: Can Titan tell us how the materials necessary for biopoesis reached the Earth?

9.2.5 Original Sources of Volatiles: Titan and Earth

Titan can only provide a useful comparison to the source of volatiles on the early Earth if they were delivered by icy planetesimals (Section 9.2.2 (C)). If Titan's thick N_2 atmosphere originally came from a solar mixture of gases that was either captured or degassed with no subsequent losses as Urey (1952) postulated (A), one would expect the ratio of Ar/N_2 to be 1/15–1/60 and Ne ≈ N (Owen 1982; Owen and Bar-Nun 1995; Grevesse et al. 2005). Instead the mixing ratio of Ne is $< 10^{-8}$ and Ar/N_2 is 2.8×10^{-7} (Niemann et al. 2005). This again defeats model (A) in its original form. As Ar and N_2 have similar volatilities. To produce the observed element ratios the nitrogen was evidently delivered as NH_3 or some other easily condensed or captured nitrogen compound. This compound or mixture of compounds was subsequently dissociated, thereby allowing the formation of the N_2 we see today. Calculations demonstrate that this is an entirely reasonable hypothesis (Atreya et al. 1978).

The value of $^{36}Ar/N_2$ in our atmosphere is 3×10^{-5}, indicating that our nitrogen also arrived as a nitrogen compound, not as N_2 (Owen and Bar-Nun 1995). The ratio of $^{15}N/^{14}N$ in our nitrogen is 3.7×10^{-3}, notably higher than the value in protosolar N^2, $2.3 \pm 0.3 \times 10^{-3}$, (Owen et al. 2001; Meibom et al. 2007). This disparity again implicates species such as NH_3 as the nitrogen carrier(s). (We cannot make this isotope test on Titan because the fractionation through atmospheric escape has greatly depleted ^{14}N to the point that $^{15}N/^{14}N$ is 1.5 X terrestrial (Niemann et al. 2005)).

A priori, NH_3 is a more likely source of N than N_2 as it is far more easily trapped in rocks and ice. N_2 is both much more volatile and less chemically active. This conclusion is substantiated by the absence of N_2 in comets whereas NH_3 is widely prevalent (e.g., Cochran et al. 2008, Bockelée-Morvan et al. 2004).

Thus there seems to be a strong link between the sources of nitrogen on Titan and the Earth, in fact it appears highly probable that they are identical.

Why is there 12 times as much N_2 in the atmosphere of Titan as in the atmosphere of our 10X more massive planet? The answer seems to lie in the huge amount of ice on Titan, ~50% of the satellite's mass. If we again use the nuclei of comets as an analogue of the icy planetesimals that made Titan, we can estimate the amount of nitrogen that was delivered to Titan by this ice. In cometary comae $NH_3/H_2O \approx 1\%$ (Bockelée-Morvan et al. 2004). This leads to about ten times as much nitrogen in the ice as see in the atmosphere.

The same pathway exists for the delivery of methane to Titan as the abundance of methane in comets is also ~1%. At the present time, methane is only 1.6% of Titan's atmosphere. This amount of methane will be destroyed by photolysis in approximately 20×10^6 years (Strobel 1982; Yung et al. 1984). Thus there must be a much larger reservoir of methane beneath the surface. A ~ 2 km methane clathrate cap on top of a water ocean beneath the crust would provide this reservoir (Tobie et al. 2006). Evidence for such an ocean may come from observations of Titan's rotation (Lorenz et al. 2008).

We now come back to Earth. Clearly cometary bombardment of the early Earth could have supplied the carbon and nitrogen on our planet if we include the carbon containing solids quantified in Comet Halley (Geiss 1988; Owen and Bar-Nun 1995, 2000). These tentative conclusions are actually independent of each other – Titan's gases coming from local icy planetesimals and Earth's gases from cometary bombardment. But they may be telling us that the comets we have studied so far may have come from the same family of icy planetesimals as those that built Titan. H_2O obviously was by definition an integral component of the icy planetesimals. Early cometary bombardment ensures that the Earth would have received water from this same source. However, the differing values of D/H in the examples of cometary water analyzed thus far and the D/H in our oceans clearly shows that comets could not be the *only* source of water on Earth (Eberhardt et al. 1995; Balsiger et al. 1995). The additional source of telluric water was probably the rocks making up the bulk of the planet's mass (Owen and Bar-Nun 1995, 2000; Drake 2005). Other types of comets whose H_2O has not yet been analyzed may also have played a role in carrying the water (Hsieh and Jewitt 2006).

The critical test of the hypothesis that comets are genetically related to the building blocks of Titan will be a determination of the D/H of Titan's ice. We don't yet know this value.

These various complications introduce sufficient uncertainty that it is impossible to conclude this section with a crisp comparison of volatile delivery to Titan and Earth. Titan was obviously built by icy planetesimals. On Earth such objects must have played some role through the early bombardment by comets, but just how important this delivery system was compared to volatiles released from the overwhelming mass of rock making up our planet remains to be determined.

9.2.6 Summary

Titan as we find it today provides an interesting model for some ideas about conditions on the primitive Earth.

Examining the sources of Titan's H_2O, N_2, CH_4 and H_2 we find a plausible source for the C and N and some of the H_2O on the early Earth through bombardment by icy planetesimals. In particular, the origin of N_2 on Titan is demonstrably similar to its origin on Earth.

Perhaps most intriguing is the possible role of an organic smog. If the route for the production of organic molecules on Titan – photolysis in the upper atmosphere, production of aerosols forming a smog layer, precipitation of aerosols to the surface – was followed on Earth, it would have been an excellent way of making and delivering pre-biologically interesting compounds to the early terrestrial oceans (Miller et al. 1998). Titan therefore tantalizes us with visions of the Miller–Urey scenario for the early Earth.

These appear to be the only two hints that Titan presently offers us to aid our efforts to understand the environment at the surface of the early Earth.

Our best hope for moving forward lies in a future mission that can determine abundances and isotope ratios of all the heavy noble gases, the values of D/H and $^{16}O/^{17}O/^{18}O$ in Titan's ice, and $^{15}N/^{14}N$ in any ammonia that has not cycled though Titan's atmosphere. Comparing these results with their counterparts in comets and in the terrestrial atmosphere will surely constrain models for atmospheric origin and early evolution on both worlds. Whatever these new data reveal, the low temperature of Titan's surface absolutely forbids the complex chemistry that ultimately led to life on Earth.

After H_2, H_2O is the most abundant molecule in the universe. It is virtually impossible to imagine a planetary system forming without a plethora of icy planetesimals. Hence the same basic icy delivery system we have found to apply to both Titan and Earth could easily be a common feature in any planetary system.

The Kepler Mission to be launched in 2009 (Borucki et al. 2003) will soon give us a statistically meaningful search for Earth-like planets around other stars. If they exist, it seems virtually certain that these planets will have had access to the same sweet suite of volatiles we find on Earth. Titan has strengthened our confidence that this will be so.

9.3 Prebiotic-like Organic Chemistry

9.3.1 *Laboratory Simulations, Modeling and Observation*

Since the now historical experiment by Stanley Miller (Miller 1953), many experimental as well theoretical works have been achieved which demonstrate that a reducing atmosphere is very favorable for gas phase organic syntheses. Moreover, a N_2–CH_4 atmosphere appears to be one of the most interesting ones for prebiotic chemistry. It allows the formation of a wide variety of organics. These include compounds that are supposed to have played a key role in (terrestrial) prebiotic chemistry, such as hydrogen cyanide, cyanoacetylene and cyanogen (Toupance et al. 1975). Consequently, with the detection of its dense atmosphere, dominated by dinitrogen, N_2, and the confirmation of the presence of noticeable mole fraction of methane, CH_4, by the Voyager instruments in the early 1980s, Titan suddenly became a planetary body of tremendous interest for the study of extraterrestrial organic chemistry.

Many research works were then more specifically carried out on the chemical processes involved in Titan's atmosphere. Simulation experiments on N_2–CH_4 gas mixtures at temperature and pressure close to those encountered in Titan's atmosphere started to be developed in 1982 (Raulin et al. 1982) and photochemical models of the N_2–CH_4 atmosphere as early as 1982 (Strobel 1982) and 1984 (Yung et al. 1984). These two approaches have been extensively followed since that time and have provided a tremendous amount of results which can now be coupled to the exceptional abundance of observational data provided by the Cassini–Huygens mission. The latter has shown that organic chemistry is indeed present on a global scale, from Titan's high atmosphere to its surface. Together, all the parts constitute what could be called, by analogy with our planet, the 'geofluids' of Titan: air (gas atmosphere), aerosols (solid atmosphere) and surface (solid as well as liquid). Organic chemistry should also be present in the subsurface and in the internal water ocean.

9.3.2 *Organic Chemistry in the Atmosphere*

The coupling of CH_4 chemistry with N_2 chemistry in Titan's atmosphere induces the formation of many organic compounds in the gas and particulate phase. Those are mainly hydrocarbons, nitriles and complex refractory organics. In addition to observation and laboratory simulation, this atmospheric chemistry has been widely studied through photochemical modeling.

9.3.2.1 Photochemical Modeling

Several photochemical models describing the chemical and physical pathways driving the chemical evolution of Titan's atmosphere have been published since the first detailed model by Yung et al (1984). Recent examples include Wilson and Atreya (2004), Hébrard et al. (2007a, b), Sekine et al. (2008a, b); and refs therein. Until now, the data (mainly the vertical concentration profiles) produced by these models were relatively poorly constrained by observational data, since only a few observations were providing reliable concentration values at a given altitude. Furthermore, the models usually involve the eddy diffusion coefficient, the profile of which is poorly known.

Nevertheless, these models already provide a quite good understanding of the chemical processes involved in Titan's complex atmospheric chemistry. The primary processes are the dissociation of CH_4 and N_2 by the solar UV photons and the electrons from Saturn's magnetosphere. The resulting species formed in the high atmosphere yield to the production of simple hydrocarbons and N-compounds, specially C_2H_2 and HCN, which play a key role in the general chemical scheme. Once they are formed, they diffuse down to the lower atmospheric regions, where the available UV photons are less energetic, but are absorbed by the two compounds and are capable of their dissociation. The radicals produced by these primary processes allow the formation of more complex hydrocarbons and nitriles. Additional CH_4 dissociation probably occurs in the low stratosphere through photocatalytic processes involving C_2H_2 and polyynes.

Volatile hydrocarbons, from C_2 (ethane, ethylene and acetylene) to C_6 (mainly benzene), and volatile nitriles, mainly from C_1 (HCN) to C_4 (C_4N_2) are formed from these radical reactions in the mid atmosphere. The end product of this chemistry are macromolecular organic compounds made of C, H and N, which are the main constituents of Titan's atmospheric aerosols. Their molecular structure is strongly linked to the invoked chemical pathways.

Several schemes have been published to describe these processes that increase the complexity of the carbon matter from simple starting material (methane molecules) to high molecular weight products. The potential role of polyynes, such as C_4H_2 and C_6H_2, and cyanopolyynes, such as HC_5N, in the chemical processes of organic complexification has been described (Scemama et al. 2002; Vuitton et al. 2006; Jolly and Bénilan 2008; and references included). The photopolymerization of simple unsaturated hydrocarbons in the presence of HCN and HC_3N has also been proposed (Clarke et al. 2000; Tran et al. 2005, 2008 and refs included) to explain the formation of high molecular weight compounds.

After a specific study on the mechanisms of formation of benzene (Wilson et al. 2003) Wilson and Atreya (2003) have developed a 1-D photochemical model to investigate the chemical mechanism responsible for the haze formation. Their results suggest the potentially important role of aromatic compounds in the formation of the hazes, from benzene to polycyclic aromatic hydrocarbons (PAH), with a peak production around 220 km altitude. The authors indicate that the nitrile pathway based on HCN polymerization is another possibility. The important role of benzene and the production of PAHs is also considered in Lebonnois's study (2005). More recently, Sekine et al (2008a, b) have studied heterogeneous reactions on the surface of Titan's aerosols. The results of their calculations show that the aerosols may remove a large fraction of atomic hydrogen in the stratosphere and mesosphere and favor the formation of unsaturated species in the gas phase. New modeling have also been developed to couple the photochemical processes with the haze formation (Lavvas et al. 2008a, b)

Kinetic predictions for these atmospheric processes may strongly depend on the nature of the assumed mechanism and the omission of an important reaction may render the result invalid. Furthermore, many of the data feeding the photochemical models are not well known. This is the case, for instance with kinetic constants, especially at low temperatures and for complex species, but also UV and IR spectra, quantum yields and branching ratios of photodissociation reactions. This is also the case with most of the heterogeneous reactions in spite of their potentially important role in the chemical evolution of Titan's atmosphere. This may introduce strong uncertainties in the photochemical models, which then propagate. As shown in a recent quantitative study (Hébrard et al. 2007a, b) such propagation induces strong uncertainties in the vertical concentration profiles, as shown on Fig. 9.1 for HC_3N. These uncertainties have strong consequences on the conclusions which can be inferred from the models. With the resulting error bars the modeled vertical profile fits with most of the observations: this suggests that it is difficult to really constrain the chemical scheme of the model without having better precision on the input parameters. In any case, the information provided by such work can be used for identifying the reactions for which kinetic data and other associated parameters are essential and need to be experimentally determined.

9.3.2.2 Laboratory Simulation Experiments

The development of so-called "simulation experiments" in the laboratory is another approach to study Titan's organic chemistry that is very complementary to the photochemical modeling. These experiments can integrate the many different physical and chemical processes involved in the chemical evolution of Titan's atmosphere. In this atmosphere the main energy sources are solar UV radiation and mid-energy electrons coming from Saturn's magnetosphere. Only UV photons of wavelengths shorter than about 150 nm can dissociate methane. The most abundant solar photons in this range of wavelengths are those at Lyman alpha (121.6 nm). This radiation can easily be simulated in the laboratory by using monochromatic lamps. But the Lyman α line does not allow the photodissociation of N_2, which requires photons of wavelength shorter than about 100 nm. UV radiation of such short wavelengths is difficult to obtain in the laboratory with irradiation systems compatible with closed or even flow

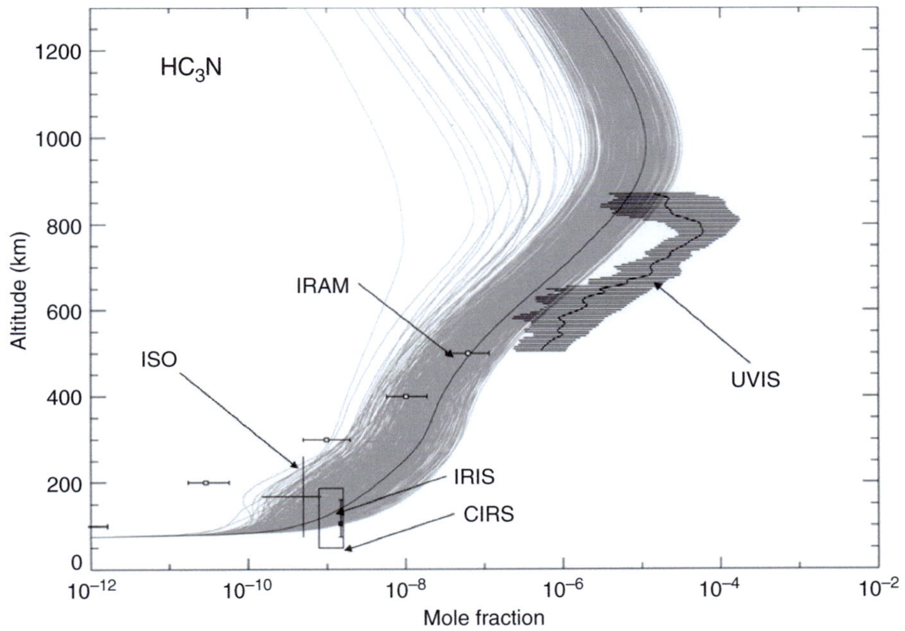

Fig. 9.1 Influence of uncertainties in rate constants and photolysis rates on the outputs of 1-D photochemical modeling of Titan's atmosphere: example of the vertical distribution of HC_3N and associated uncertainties derived from the input uncertainties. Observational data are plotted for comparison (Courtesy: Eric Hébrard. Credit: F. Raulin in G. Horneck & P. Rettberg Eds: Complete Course in Astrobiology. Page 232. 2007. Copyright Wiley-VCH Verlag GmbH & Co. KGaA. Reproduced with permission)

reactors. For that reason, all laboratory experiments simulating the chemical evolution of Titan's atmosphere, which have been carried out so far on N_2–CH_4 initial gas mixtures, use an electron impact energy source, which mimics Titan's atmospheric electrons impinging from Saturn magnetosphere. Another kind of laboratory experiment compatible with a short wavelength UV source is to use an initial gas mixture with another N-compound, which can be photo-dissociated by UV photons of longer wavelengths and which is also present in Titan's atmosphere, although at trace level. Such compounds include HCN and HC_3N (Clarke et al. 2000; Tran et al. 2003a, b, 2005, 2008).

Many of the simulation experiments were able to use the availability of data on the physical conditions of Titan's atmosphere. The fly-bys of Titan by the two Voyager spacecraft in the early 1980s provided for the first time the vertical temperature and pressure structure of Titan atmosphere in addition to its main composition. Many experiments have then been carried out on the chemical evolution of N_2–CH_4 gas mixtures to simulate Titan's atmosphere chemistry in the laboratory (Raulin et al. 1982; Khare et al. 1984, 1986; Thompson et al. 1991; Coll et al. 1998, 1999a, 2003; Ramirez et al. 2001; Bernard et al. 2002, 2003; 2006; Imanaka et al. 2004; Somogyi et al. 2005; Szopa et al. 2006; Nguyen 2007; Nguyen et al. 2007; and many other references included). Different energy sources have been employed in these experiments: corona and arc discharge (Ramirez et al. 2001, and references included), cold plasma at room temperature (Khare et al. 1984; Thompson et al. 1991; Imanaka et al. 2004; Szopa et al. 2006) and at room and low temperature (Coll et al. 1999a, b, 2003; Bernard et al. 2002, 2003). Different conditions of total pressure and the CH_4/N_2 ratio of the initial gas mixture and energy level (from about 1 to several electronvolt) have been used. These global simulations of Titan's atmospheric chemistry usually employed an open reactor with a flow of a low pressure N_2–CH_4 gas mixture. In many cases, the gas phase end products (molecules) are analyzed by GC-MS (Gas Chromatography and Mass Spectrometry) techniques; the transient species (radicals and ions) can be determined by on line UV-visible spectroscopy; in that case, the evolution of the system can also be theoretically described using coupled physical and chemical (ions and neutrals) models (Bernard et al. 2002).

Gas Phase Products

In most of these experiments, all the volatile organics that were detected in Titan's atmosphere before the Cassini–Huygens mission have been obtained. The results were within the correct order of magnitude of their relative concentration in most cases. Even C_4N_2, which was detected from the Voyager-IRIS data in the condensed phase in Titan's atmosphere, has been detected in some experiments. This detection was difficult because of the thermal degradation of this species, the lack of mass spectrometry data, and its gas-chromatographic co-elution with another organic compound. Thanks to a preliminary study (Aflalaye et al. 1995) of its main physical–chemical characteristics (mass spectrum, gas chromatographic behavior, thermal stability), it was specifically searched for and at last identified (Coll et al. 1999b).

Consequently, these experiments seem to mimic the real chemical processes which are going on in Titan's atmosphere. In fact, these experiments produce many other organics that can be assumed to be also present in Titan's atmosphere. For instance the systematic study by Coll et al (1998, 1999a) shows the presence of more than 150 compounds identified in the gas phase. The identified organic products are mainly hydrocarbons and nitriles. Among the other organics formed in these experiments and not yet detected in Titan's atmosphere, one should note the presence of polyynes (C_4H_2, C_6H_2, C_8H_2) and probably cyanopolyyne HC_4CN. These compounds are also included in photochemical models of Titan's atmosphere, where they could play a key role in the chemical schemes allowing the transition from the gas phase products to the aerosols, as already discussed above.

Also of astrobiological interest is the detection in these simulation experiments of organic compounds with chiral carbon atom (C*) such as

$$CH_3C^*H(C_2H_5)CH{=}CH_2 \text{ and } CH_3C^*H(CN)CH{=}CH_2$$

These compounds can exist on two stereoisomeric forms, D and L, The presence of an enantiomeric excess ($D/L \neq 1$) is linked to the question of the origin of homochirality.

The absence at a detectable level of molecules carrying amino groups, like amines, must be highlighted.

Experiments on N_2–CH_4 mixtures that include CO at the 100 ppm level (Bernard et al. 2003; Coll et al. 2003) show the incorporation of O atoms in the produced organics, with an increasing diversity of the products (more than 200 were identified). The main O-containing organic compound is oxirane (also named ethylene oxide), $(CH_2)_2O$, now a good candidate to be searched for in Titan's atmosphere. These studies also show the formation of ammonia (Bernard et al. 2003) at noticeable concentration, opening new avenues in the chemical schemes of Titan's atmosphere. A very recent photochemical study by Tran et al. (2008) on gas mixtures including N_2, CH_4, and H_2 at Titan stratosphere mixing ratios, plus C_2H_2, C_2H_4, HC_3N, HCN and CO at various mixing ratios, including those of Titan's stratosphere, shows the formation of several O-organics, mainly C_1–C_7 aldehydes and ketones. This study indicates that different pathways are involved in the photochemical experiments than in experiments energized by plasma discharge.

Nevertheless, these simulation experiments appear to be very useful guides for further searches for trace species in Titan's atmosphere, both by remote sensing and in situ observations. They allow the definition of new candidates likely to be present in the atmosphere, to tune and even calibrate the instruments used for their detection, and to add new species and related chemical schemes in chemical models. In addition, these experiments concern not only the atmospheric gas phase, but also the aerosol phase, since they generally produce solid material assumed to be a laboratory analogue of Titan's haze particles, usually called Titan's "tholins".

Titan's Tholins

The word "tholins" was invented by Carl Sagan in the late 1970s from the Greek word "tolos", meaning muddy. It is used as a name for the solid organic product obtained after irradiation of a variety of gas mixtures of cosmically abundant molecules by a variety of energy sources of astrophysical importance (Sagan and Khare 1979). The solid materials produced during laboratory experiments simulating the chemical evolution of Titan's atmosphere are thus called Titan's tholins. It has commonly been assumed that these tholins were good laboratory analogues of Titan's aerosols, offering the possibility of using their properties (spectral characteristics in the visible, IR and UV, index of refraction, size and morphology, etc.) to interpret and model the observational data obtained for Titan's aerosols. Since the first work by Sagan and Khare more than 20 years ago (Sagan et al. 1984; Khare et al. 1984), they have been extensively studied (Khare et al. 1986; McDonald et al. 1994; Ehrenfreund et al. 1995; McKay 1996; Coll et al. 1998, 1999a; Ramirez et al. 2002; Bernard et al. 2002, 2006; Tran et al. 2003a, b, 2008; Cruickshank et al. 2005; McGuigan et al. 2006; Szopa et al. 2006; Nguyen. 2007; Nguyen et al. 2007; and included references). These laboratory analogues show very different properties depending on the experimental conditions (temperature, pressure, CH_4/N_2 etc) used for their production (Coll et al. 1998, 1999a; Bernard et al. 2002, 2006; Imanaka et al. 2004; Cruickshank et al. 2005). For instance, the average C/N ratio of the product varies from less than 1 to more than 11, as shown in Table 9.1. Experimental protocols using low pressure and low temperature closer to Titan's conditions (Coll et al. 1998, 1999a; Ramirez et al. 2001, 2002; Bernard et al. 2002, 2006) and recovering Titan's tholins without oxygen contamination in a glove box purged with pure N_2 have been developed to provide more representative laboratory analogues of Titan's aerosols. Systematic studies have been carried out on the influence of the pressure of the starting gas mixture on the elemental composition of the tholins. They show that two different chemical–physical regimes are involved in the processes, depending on the pressure, with a transition pressure around 1 mbar (Bernard et al. 2002; Imanaka et al. 2004).

The molecular structure of the Titan tholins is still poorly known. Several possibilities have been considered such as HCN polymers or oligomers, HCN–C_2H_2 co-oligomers, HC_3N polymers, HC_3N–HCN co-oligomers (Tran et al. 2003a, b, 2008). They are probably made of macromolecules of largely irregular structure. Gel filtration chromatography of the water soluble fraction of Titan tholins shows an average molecular mass of about 500–1,000 Da (McDonald et al. 1994). We note in passing that the CAPS instrument on Cassini has detected ions with molecular masses in this range in Titan's upper atmosphere (Waite et al. 2007). Information on the chemical groups included in their structure has been obtained from their infrared and ultraviolet and X-ray photoelectron spectra (Tran et al. 2003a; Imanaka et al. 2004; Bernard et al. 2006; and references included) and from analysis by pyrolysis-GC-MS techniques (Ehrenfreund et al. 1995; Coll et al. 1998; Somogyi et al. 2005; McGuigan et al. 2006) and high resolution mass spectrometry (Sarker et al. 2003). The data show the presence of aliphatic and benzenic hydrocarbon groups, of CN, NH_2, and C=NH groups. Direct analysis by chemical derivatization techniques before and after hydrolysis allowed the identification

Table 9.1 C/N and C/H values of Titan's tholins observed from various simulation experiments

Reference	Reactor	Solid sampling	T° (K)	% CH4	P (mbar)	C/N	C/H
Sagan et al. (1984)	Open		Ambient	10	<10	1.9	0.63
	Open		Ambient	0.1	1.3	0.75	0.6
McDonald et al. (1994)	Open		ambient	10	2.6	1.5	0.6
Coll et al. (1995)	Closed	Glove Box	100	10	900	10.5	1.1
McKay (1996)	Open		Ambient	10	1150	5.1	1.0
Coll et al. (1997)	Open	Glove Box	Ambient	2	2	1.8	0.69
Coll et al. (1999a)	Open	Glove Box	100–150	2	2	2.8	0.8
Tran et al. (2003a&b, 2008)	Open		Ambient	1	900	9.7–11.4	1.3–1.4

of amino acids or their precursors (Khare et al. 1986), even if the hydrolysis is carried out at neutral pH (Nguyen 2007; Raulin et al. 2007). Their optical properties have been determined (Khare et al. 1984; McKay, 1996; Ramirez et al. 2002; Tran et al. 2003b; Imanaka et al. 2004; Cruickshank et al. 2005). In particular their complex refractive indices have been obtained (Khare et al. 1984; Ramirez et al. 2002; Tran et al. 2003b), including evaluations with error bars (Ramirez et al. 2002). These data can be used to interpret observational data related to Titan, in particular by modellers who compute the properties of Titan's aerosols. Let us also mention that the potential nutritious properties of Titan tholins have also been studied and it has been show that they can be metabolized by microorganisms (Stoker et al. 1990).

Recently, the possible isotopic fractionation of carbon in tholins, compared to the starting methane has been studied (Nguyen et al. 2007). The results show no important enrichment of light or heavy carbon. This suggests that the chemical scheme to the building of tholins does not involve a large number of reactions. This laboratory data can also be used to interpret the results from in situ analysis of Titan's aerosols.

9.3.2.3 Observational Data

Before the exploration of Titan by Cassini–Huygens, several organic compounds were already detected in Titan's stratosphere by Voyager, ISO and ground-based observations. These include C_1–C_4 hydrocarbons and nitriles (both with saturated and unsaturated chains) and benzene as expected from laboratory simulation experiments. Inorganics were also detected: H_2, CO, CO_2 and H_2O. Since the Cassini arrival in the Saturn system, the presence of water vapour and benzene in the stratosphere has been unambiguously confirmed by the CIRS instrument (de Kok et al. 2007; Coustenis et al. 2007). CIRS has also been able to provide the vertical and latitudinal variation (Teanby et al. 2006, 2007; Vinatier et al. 2007a, b), including in the polar region (Vinatier et al. 2007a; Teanby et al. 2008), of many species. But so far, with the exception of isotopomer species, such as N isotopes nitriles (Vinatier et al. 2007b) and deuterated acetylene (Coustenis et al. 2008), no molecule other than those detected by Voyager and other pre-Cassini observations has been identified by remote sensing.

On board the Huygens probe, the GC-MS did not detect a large variety of organic compounds in the atmosphere along the descent trajectory followed by the probe (~150–0 km altitude). The mass spectra show that the mid- and low stratosphere and the troposphere are poor in volatile organic species, with the exception of methane (Niemann et al. 2005). This is probably due to the condensation of these species on the aerosol particles which seem to play a key role in chemical evolution of Titan's geofluid.

An essential discovery of astrobiological importance from Cassini observations is the detection of many organic species in spite of the very high altitude (1,100–1,300 km) by the direct analysis of the ionosphere by the INMS instrument during the low altitude Cassini fly-by's of Titan (Waite et al. 2007). The interpretation of the INMS data indicates the possible presence of a large variety of hydrocarbons (up to C_8) and nitriles (up to C_6) at such a high altitude (Vuitton et al. 2007). The mass range of INMS is limited to 100 Da, however, the extrapolation of its data and of CAPS (Cassini Plasma Spectrometer) strongly suggests that much higher molecular weight species may be present in the ionosphere (Waite et al. 2007). In particular the CAPS Ion Beam Spectrometer detected complex hydrocarbon-nitrile species of molecular weight up to approximately 350 Da and its Electron Spectrometer data suggest the presence of very heavy negative ions (up to approximately 8,000 Da). The data indicate the presence of naphthalene and anthracene in the ionosphere and the key role of benzene and PAH chemistry at these very high altitudes. These discoveries could drastically modify our understanding of the organic processes involved in Titan's atmosphere, and could demonstrate that ionospheric chemistry plays a key role in the formation of the complex organic compounds and aerosol nuclei, which was not considered before. However, we also need to understand if and how the ionospheric chemistry is coupled with the lower atmosphere chemistry and what is the real relationship between the high molecular weight compounds detected in the ionosphere and the stratospheric aerosols.

Another essential astrobiological discovery from Cassini–Huygens is the first direct data on the chemical composition of the particles in the low atmosphere. They have been collected and treated by the ACP instrument. ACP was designed to collect the aerosols during the descent of the Huygens probe on a filter in two different regions of the atmosphere. Then the filter was heated in a closed oven at different temperatures and the gases produced were analysed by the GC-MS instrument. The ACP data suggest that the aerosol particles include refractory organics which release HCN and NH_3 during pyrolysis (Israël et al. 2005). This strongly supports the hypothesis that Titan's laboratory tholins are similar to Titan's aerosols, since several tholins also release HCN and NH_3 during pyrolysis (Israël et al. 2005). These new and first in situ measurement data are compatible with an aerosol particles made of a refractory organic nucleus, covered with condensed volatile compounds (Fig. 9.2). The potential presence of nitrile groups (–CN), amino groups (–NH_2, –NH– and –N<) and /or imino groups (–C=N–) in the refractory complex organics of the aerosol nucleus, derived from the chemical nature of the compounds produced by pyrolysis, also supports the "tholins hypothesis". Moreover, comparison of the data obtained for the first (mainly stratospheric particles) and second (mid troposphere) samplings indicate

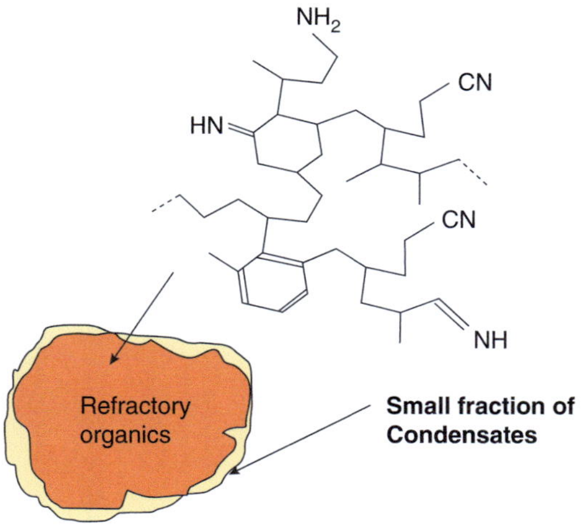

Fig.9.2 Main structure and model of the chemical composition of Titan's aerosols derived from the Cassini–Huygens ACP data (Credit: F. Raulin et al, Titan: an astrobiological laboratory in the solar system. in Instruments, Methods, and Missions for Astrobiology, SPIE 2007 -X – 6694–20, 2007)

that the aerosol composition is homogeneous (Israël et al. 2005). This also fits with some of the data obtained by DISR relative to the aerosol particle which indicates a relatively constant size distribution of the particles with altitude (with a mean dimension of the order of one micron). However, it must be pointed out that the ACP results are controversial as they are not as specific or extensive as the experimenters had hoped (Biemann 2006; Israël et al. 2006). Further laboratory calibrations are needed to solve this problem.

9.3.3 Organic Chemistry on the Surface and Sub-surface

9.3.3.1 Surface Chemistry

After sedimentation down to the surface, the atmospheric aerosols should form a deposit of complex refractory organics and frozen volatiles. But this is probably not the end of their chemical evolution. They may interact with the water ice of the surface, or even with liquid water which may be present episodically, produced by cryovolcanic activity or by the energy released by large impacts. Modeling of cometary impacts on Titan's surface shows that it may melt surface water ice, offering possible episodes of liquid water as long as several thousand years (Artemevia and Lunine 2003, 2005; O'Brien et al. 2005 and references included). This could provide, although for short time periods, conditions allowing terrestrial-like prebiotic syntheses, in spite of low temperatures of the environment.

Laboratory study of the behavior of Titan's tholins in the presence of water is a way to predict the possible chemical evolution of the aerosols on Titan's surface. As mentioned above, the hydrolysis of tholins has already been studied: under very acidic conditions (HCl 6N), it releases many amino acids (Khare et al. 1986). However such conditions of pH are far from that of Titan. Recently systematic studies have been carried out, using different conditions of pH for the hydrolysis, from very acidic (HCl 6N) to neutral (pure water) (Nguyen 2007; Raulin et al. 2007). The results demonstrate that the formation of amino acids is still observed at neutral pH, as well as that of several other organics, specially urea, carboxylic acids and hydroxyl-carboxylic acids as shown on Fig. 9.3. Moreover, studies on the kinetics of the reaction between tholins and water (Neish et al. 2008; 2009) indicate that such hydrolysis processes are possible on Titan since they occur on timescales less than those available in impact melts), Thus additional organics evolved through exposure of the aerosols to liquid water could be present on Titan's surface.

The observational data related to the chemical composition of Titan's surface are still limited. The GC-MS was able to analyze the atmosphere near the surface for more than one hour after the touchdown. The corresponding mass spectra show the signature of several organics, including cyanogen and benzene, indicating that the surface is much richer in volatile organics than the low stratosphere and the troposphere (Niemann et al. 2005). These observations are in agreement with the hypothesis that in the low atmosphere of Titan, most of the organic compounds are in the condensed phase.

Data of the SSP instrument on Huygens suggest the presence of water ice (Zarnecki et al. 2005). Its accelerometer measurements can be interpreted as evidence for the presence of small water ice pebbles on the surface where Huygens has landed, in agreement with the DISR surface pictures.

On the other hand, DISR collected the infrared reflectance spectra of the surface with the help of a lamp, illuminating the surface before the Huygens probe touched down. The retrieving of these infrared data (Tomasko et al. 2005) shows the presence of water ice, but no clear evidence – so far – of tholins.

These new data show the diversity of the locations where organic chemistry is taking place on Titan. Surprisingly the high atmosphere looks very active, with neutral and ion organic processes. In the lower atmosphere this chemistry seems "frozen" into the solid aerosols. These aerosols precipitate, covering Titan's surface with frozen volatile organics: hydrocarbons such as benzene and nitriles, such as cyanogen, together with refractory organic materials, including the hydrolysis products of the aerosol organic nucleus.

Irradiating effects of cosmic rays reaching Titan's surface may induce additional organic syntheses, particularly in the liquid bodies seen on Titan polar regions, likely to be made

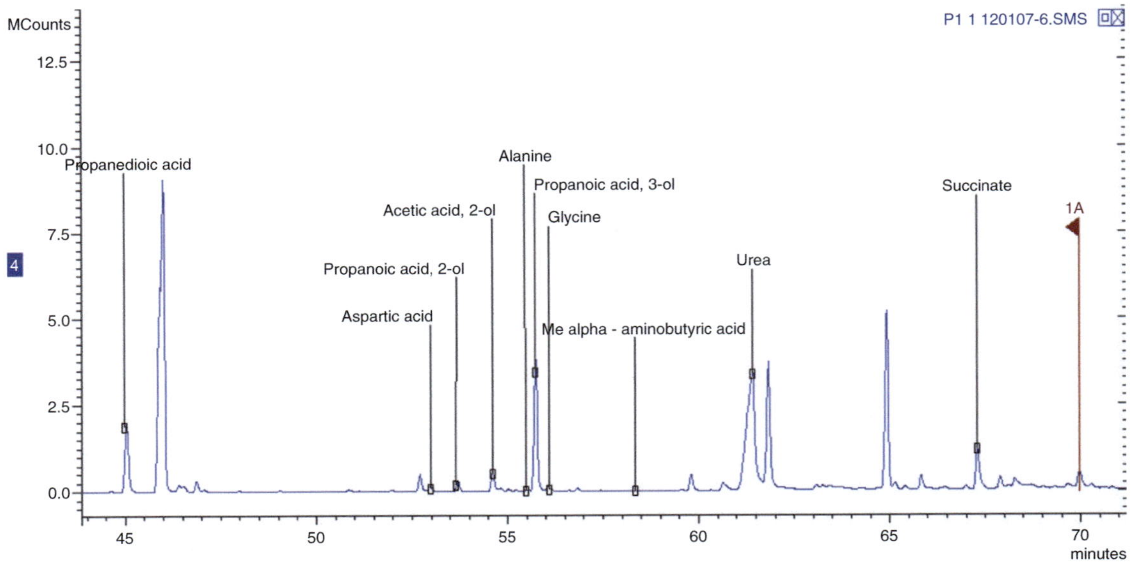

Fig.9.3 GC-MS analysis of products formed after hydrolysis at neutral pH of Titan's tholins (Courtesy: Mai-Julie Nguyen. Credit: F. Raulin et al, Titan: an astrobiological laboratory in the solar system. in Instruments, Methods, and Missions for Astrobiology, SPIE 2007 -X – 6694–20, 2007)

of low molecular weight hydrocarbons, mainly methane and ethane (Brown et al. 2008). This could indeed allow the additional formation of reactive compounds such as azides as well as the polymerization of HCN (Raulin et al. 1995). Moreover, the interface between the liquid phase and the solid icy deposits at the surface may include sites of catalytic activity favorable to these additional chemical reactions.

Titan's lakes represent a very interesting organic reservoir (Raulin 2008a). Assuming thermodynamic equilibrium between the liquid of the lake and the near-surface atmosphere, an atmospheric abundance of methane of 5% would correspond to a 65% C_2H_6– ~31% CH_4 and ~4% N_2 main composition of the lake (Dubouloz et al. 1989). Possible flux to the surface and solubility in such a solvent of many expected or detected atmospheric species have been previously estimated (Raulin 1987; Dubouloz et al. 1989). The results show that many organic compounds should have a concentration in the lakes drastically increased compared to their concentration in the atmosphere (Table 9.2).

9.3.3.2 Sub-surface

Models of the internal structure of Titan suggest the possible presence of a water-ammonia ocean (with up to 15% ammonia) below a several 10 km thick water ice layer. Cassini radar observations of the behavior of Titan's surface features strongly support this hypothesis (Lorenz et al. 2008). This liquid water body may provide an efficient way to convert simple organics into complex molecules, and to reprocess chondritic organic matter into prebiotic compounds.

As on the Earth, an important fraction of organic material on Titan may have been incorporated in the planetary body during its formation or imported by impactors after its accretion. The icy planetesimals from which Titan was built in the Saturn sub-nebula would have included not only water ice, and trapped gases which allow the formation of the primordial Titan's atmosphere (Owen, 2000; this chapter, previous section), but also organic materials, similar to the complex organic matter found in carbonaceous chondrites. The same material may have been an important exogenous source of organic matter after the accretion of Titan's solid body, especially at the close of accretion, when models assume that Titan had a primitive surface water-ammonia ocean (Lunine and Stevenson 1987; Fortes 2000).

The density of Titan (1.9 g/cm^3) suggests that it is made of a combination of water ice and denser materials, mainly silicates, but also including materials from carbonaceous meteorites, such as CII chondrites, whose density is 2.5–2.9 relative to water. CII chondrites include about 2.5% C. The most abundant fraction of organic carbon in carbonaceous chondrites is an unextractable complex organic material, which releases many compounds of biological interest after hydrolysis. Carbonaceous chondrites and micrometeorites also include several free amino acids or their soluble precursors. Purines and pyrimidines bases as well as carbohydrates, including simple sugars have also been identified in carbonaceous chondrites and may be also present in micrometeorites.

Table 9.2 Order of magnitude of concentration of minor constituents in Titan's lakes (T: 92.5 K; 65% ethane), adapted from Raulin (1987). If no other indication, concentration is given in ppm = part per million in volume. (s): indicates that the limiting factor is the solubility, otherwise it is the atmospheric flux or (?) it is not known. The atmospheric concentration of minor species is a stratospheric value (adapted from Raulin 2008b and refs included)

	Concentration (ppm if not %)			Concentration (ppm if not %)	
Solutes	Lake	Atmosphere	Solutes	Lake	Atmosphere
[Hydrocarbons]			**[Nitriles]**		
Alkanes			**Alkanitriles**		
Ethane C_2H_6	65%	10	Methanenitrile HCN	3 s	0.1
Propane C_3H_8	2%	0.5	Ethanenitrile CH_3CN	30 s	0.02
Butane C_4H_{10}	4000		Ethanedinitrile (cyanogen) C_2N_2	0.6 s	1×10^{-3}
2-Methylpropane $(CH_3)_3CH$	4000		Propanenitrile C_2H_5CN	50 s	
Pentane C_5H_{12}	400		Butanenitrile C_3H_7CN	1	
2-Methylbutane $(CH_3)_2CC_2H_5$	400		2-Methylpropanenitrile $(CH_3)_2CHCN$	1 s	
Dimethylpropane $(CH_3)_4C$	400		2-Methylbutanenitrile $(CH_3)_2CHCH_2CN$	1	
Hexane C_6H_{14}	40		**Alkenenitriles**		
2-Methylpentane $(CH_3)_2CHC_3H_7$	40		Propenenitrile (acrylonitrile) CH_2CHCN	10 s	
2,2-Dimethylbutane $(CH_3)_3C_3H_7$	40		2-Butenenitrile $CH_3CHCHCN$	1 s	
2,3-Dimethylbutane $(CH_3)_2CC(CH_3)_2$	40		2-Methylpropenenitrile $CH_2C(CH_3)CN$	1 s	
3-Methylhexane $C_2H_5C_H(CH)_3C_3H_7$	4		2-Butenenitrile CH_2CHCH_2CN	1 s	
Alkenes, Benzene			**Alkynenitriles**		
Ehene (ethylene) C_2H_4	4%	0.4	Propynenitriles (cynoacetylene) HC_3N	3 s	1×10^{-3}
Propene C_3H_6	x		2-Butynenitrile CH_3CCCN	2 s	
2-Methylpropene $CH_2C(CH_3)_2$	5		**[Other compounds]**		
1-Butene $CH_2CHC_2H_5$	5		**N-heterocycles**		
2-Butene (cis and trans) $CH_3CHCHCH_3$	5		Pyrimidine $C_4C_4N_2$	2 ?	
1,3-Butadiene $CH_2CHCHCH_2$	5		Adenine $C_5H_5N_5$	0.01 ?	
Benzene C_6H_6	5 s	3×10^{-4}			
Alkynes, Allene					
Ethyne (acetylene) C_2H_2	400 s	2	**Inorganics**		
Propyne (methylacetylene) CH_3CCH	30 s	8×10^{-3}	Carbon dioxide CO_2	10 s	1.6×10^{-3}
1-Butyne $CHCC_2H_5$	400 s		Carbon monoxide CO	4 s	45
1,3-Butadiyne (diacetylene) C_4H_2	0.5 s	1×10^{-3}	Water H_2O	2×10^{-7} s	4×10^{-4}
Allene CH_2CCH_2	100 s		Ammonia NH_3	5 ?	

Thus a large variety of organic compounds of prebiotic and biological interest are present in these chondritic samples and may have been imported on Titan, as on the Earth.

Fortes (2000) estimated that the chondritic influx on Titan could have produced a concentration of dissolved organic matter of about 1 g L^{-1} in Titan's ocean. To this direct feeding, one should add the possible prebiotic sysntheses occurring in the vicinity of the hypothetical hydrothermal systems present in the primordial oceans on Titan. A concentration of several 0.1% by mass of organics may be sufficient for prebiotic evolution, depending on the chemical nature of the organics. For instance, a concentration of 0.1 mol L^{-1} of HCN in aqueous solution (equivalent to 0.3 % by mass) allows the polymerization of HCN toward its tetramer and higher oligomers. This is considered as a key step in the prebiotic processes to the formation of many molecules of biological interest (Ferris et al. 1978; Ferris and Hagan 1984). Low temperatures reduce the rate constants of prebiotic chemical reactions, but may increase the concentration of reacting organics as the solution cools to its eutectic, which increases the rate of the reaction. In addition, high pressure conditions such as those present in subsurface oceans, may also induce chemical condensation reactions, essential for the formation of biological macromolecules starting from their monomers, such as polypeptides and polynucleotides from their building blocks (amino acids and nucleotides).

Finally, hydrothermal vents, if they are present on the floor of Titan's oceans, are also favorable locations for increasing chemical complexity, thanks to the heterogeneous processes which can occur at the interface between the hot gases and liquid and solid phases. These processes can be favored by the potential catalytic properties of the mineral phases and by the high thermal gradients, which protect the products from thermal degradation.

These processes may have been very efficient at the beginning of Titan's history since models assume that this subsurface water-ammonia ocean may have been in direct contact with the internal bedrock, and with the atmosphere, offering another important similarity with the primitive Earth and the potential implication of deep sea hydrothermal vents in terrestrial prebiotic chemistry. It is thus possible that an organic chemistry in liquid water may have occurred in Titan. It would have, allowed a CHNO prebiotic chemistry, which may have slowly evolved to complex organic systems, including self-replicating macromolecules. Thus, although all the discussion above is highly speculative, even the possibility on Life in Titan cannot be excluded.

9.3.4 Summary

A complex organic chemistry is occurring on Titan. It is present in the atmosphere, from the ionosphere to the surface, as evidenced by many observations, especially from Cassini–Huygens, and by modeling and experimental laboratory simulations. The main complex products of the atmospheric chemistry is the macromolecular material included in the atmospheric aerosols, likely to have a molecular composition close to laboratory Titan's tholins. These materials can evolve on Titan's surface through many exchange processes with the atmosphere and the sub-surface. In particular once in contact with water (water ice or even episodically liquid water) these compounds could follow hydrolysis processes, and produce a wide variety of organics, including oxygenated compounds. Products of such chemical evolution could include many compounds of biological interest, such as amino-acids and urea. Of particular astrobiological importance are the Titan's lakes, where many atmospheric organic compounds could accumulate and reach much higher concentration than in the atmosphere. Additional exchanges processes with the atmosphere induced by high energy cosmic rays could allow those solutes to chemically evolve in the lake.

Complex organic processes may be also present in Titan's surface ocean where conditions during the early history of Titan could have been compatible with the development of an efficient prebiotic chemistry and the emergence of Life.

Altogether, Titan shows many analogies with the early Earth and its organic chemistry with the terrestrial prebiotic chemistry. The important role of Titan's aerosols in Titan's organic chemistry should provide many insights on the potential role of aerosols which may have been present in the primitive atmosphere of the Earth, on terrestrial prebiotic chemistry.

9.4 Habitability and Life

The primary Astrobiology interest in Titan concerns the complex organic chemistry there and its possible relationship to prebiotic organic synthesis on the early Earth. However, several authors have pointed out that it would be premature to dismiss the possibility that biological as well as chemical processes are occurring on Titan.

It is useful to begin the discussion of life on Titan with a review of the ecological requirements for life on Earth. These are simply summarized as: (1) a source of energy, (2) suitable elements including C, H, N, O, P, S, and (3) liquid water. Life, like any self-organizing open system, requires a source of energy. On Earth life uses two sources of energy for primary production; light energy and chemical redox pairs. Photosynthetic organisms can use sunlight at levels as low as 10^{-4} the solar constant at 1 AU (Raven et al. 2000). Life can also use a wide range of redox couples (e.g. CO_2 and H_2). No other energy source (e.g. thermal gradients, electrical and magnetic fields, gravitational) are used now for primary productivity by life on Earth. The main elements that compose life on Earth are C, H, N, O, P, and S – with the central role being that of carbon. Life on Earth is properly called carbon-based and liquid water mediated. Liquid water is perhaps the most important ecological requirement for life on Earth. Of the three requirements it is the one that is most limiting for considerations of life on other worlds. Hence the strategy often stated for searching for life is to first "follow the water". While water is pervasive in its role in biochemistry on Earth, there have been suggestions that other solvents may also form the medium of carbon-based life (Benner et al. 2004; NRC 2007). These suggestions have important implications for possible life on Titan but it is useful to begin with Earth-like life in liquid water.

The surface temperature of Titan (94 K) is much too cold for water to exist as a liquid. This would seem to rule out Earth-like life. However, Thompson et al. (1992) suggested that impacts on the surface would create transient pools of water that could allow for growth. Thompson et al. (1992) calculated that more than 70% of Titan's organic inventory has been exposed to impact melted water for mean periods of around 1,000 years. Presumably spores from one such pool would spread over the surface awaiting the next impact.

As discussed in Chapter 4, there is considerable evidence that Titan has a subsurface layer composed of an ammonia–water mixture (Tobie et al. 2005; Sotin and Tobie 2008). Sites of cryovolcanisms on the surface, where this ammonia and water mixture reaches the surface, are possible locations for life that lives in liquid water but can tolerate ammonia. An advantage of cryovolcanisms sites over impact site is that the former might occur repeatable in certain locations. Fortes (2000) suggested that if Titan has such an ocean below the surface then this could be a suitable habitat for the origin and persistence of life. Organisms in such an ocean would have to survive high pressures, low temperatures and high concentrations of ammonia. The conditions would be extreme but not enough to preclude the survival of some hardy microorganisms known on Earth.

Fortes (2000) suggested that any such micro-organisms would be chemolithotrophs, possibly deriving energy from the reduction of hydrocarbons to CH_4, or the oxidation of NH_3 to N_2. Methanogens have been shown to survive in high concentrations of ammonia at neutral pH (McKay et al. 2008).

If life on Titan is based on liquid water it is therefore confined to relatively small and transient impact craters. Similarly if life on Titan is based on ammonia–water mixtures it is confined to the subsurface and must rely on chemical energy available underground. In either case life would not be expected to be widespread on the surface and it is unlikely to have any global effects that could be detected on a future mission.

For this hypothetical life to be widespread on Titan and have a global environmental effect would require that it grow in a liquid medium that is widespread on the surface: liquid CH_4/C_2H_6. This has been suggested by McKay and Smith (2005) and Schulze-Makuch and Grinspoon (2005). McKay and Smith (2005) suggested that life in liquid methane on the surface of Titan could derive energy by the reduction of hydrocarbons to CH_4. For example, the reaction

$$C_2H_2 + H_2 \rightarrow CH_4$$

yields 334 kJ/mole under Titan surface conditions. This is much larger than the minimum energy required to power methanogen growth on Earth of ~10 kcal/mole (42 kJ/mole) determined by Kral et al. (1998). McKay and Smith (2005) pointed out that if there was widespread life on the surface of Titan living in liquid CH_4 and consuming atmospheric H_2, then it could have a clearly observable effect on the atmosphere through the depletion of H_2 near the surface.

There are no photochemical source or sinks of H_2 in the lower atmosphere of Titan so its mixing ratio should be constant. McKay and Smith (2005) computed that there would be a measurable reduction if the biological consumption exceeded ~10^8 molecules cm/s. In principle this effect could be measured by the Huygens Probe GCMS. However, after the work of McKay and Smith (2005) it has been shown by Sekine et al. (2008b) that the atmospheric haze can also absorb and desorb H_2. This could represent an alternative source or sink of H_2 in the lower atmosphere to explain any observed deviation from a constant mixing ratio. If there are methanogens living in liquid methane on Titan they are producing CH_4 but they are not a net source of CH_4 to the atmosphere. The organisms would be merely recycling photolysis products of CH_4 back into CH_4 and thereby indirectly deriving energy from sunlight. This type of biology is not a source of CH_4 that can be invoked to explain how Titan's atmosphere maintains CH_4 despite constant photochemical destruction and loss of H_2 to space.

9.5 Titan and the Destiny of Life on Earth

In what ways might Titan be a particular analog for some aspect of the processes that have shaped our own planet over geologic time, different from that of Venus or Mars? Titan is distinguished among the bodies of the solar system in two fundamental respects. First, it has an active organic chemistry that operates only occasionally in the presence of liquid water, and therefore any aqueous chemistry proceeds slowly and is interrupted before it proceeds too far (although the extent to which such aqueous chemistry proceeds is an important goal beyond the Cassini–Huygens mission). Second, the methane hydrological cycle described elsewhere (Lunine and Atreya 2008) lacks an ocean because the methane is rapidly escaping into the upper atmosphere where it is destroyed, limiting its ability to accumulate in the surface-atmosphere system.

The first of these distinctions leads us to consider how Titan might inform us of the prebiology of the early Earth. That the early Earth had oxygen in the form of carbon dioxide (including carbo-nates) and water in its atmosphere and on its surface is in little dis-pute, and so Titan's environment today does not resemble that of the early Earth before life. However, the presence of organics on Titan's surface that might come into transient contact with liquid water on large spatial and long temporal scales makes Titan a unique environment for observing the steps leading toward the origin of life.

In one sense the presence of liquid water as an episodic rather than continuous feature adds to Titan's attraction as an astrobiological target. A key question in the origin of life problem is the timescale over which abiotic aqueous chemistry evolves into biochemistry. That is, assuming all of the essential components (organic, mineralogical, energetic) are present, how long does it take for organic chemistry to become dominated by information carrying molecules, to become selective in the kinds of molecules synthesized, and to achieve enantiopurity. Is it millions of years, thousands, hundreds? None of these timescales is accessible in the laboratory. On Titan, it is possible to calculate the time-scale that liquid water survives in impact-generated or volcanically-generated thermally-elevated environments to within an order of magnitude for a given volume and geometry. For large impact craters, timescales range from 10^2 to 10^3 years (O'Brien et al. 2005). Timescales in cryovolcanic zones are more difficult to estimate because they depend on the composition, viscosity, gas-content, and other parameters of the flow, and on the nature of the surroundings (Neish, et al. 2006).

Not only timescales, but spatial scales are important. Systems require not only a flow of free energy, but also sufficiently large dimensionality in order to exhibit spontaneous development of complexity (Kauffman and Clayton 2006). For chiral symmetry breaking leading to strong dominance of one "handed" form over another – enantiopurification –

this corresponds to chemistry over a large area. Hence a planetary surface is qualitatively different, in this view, than a laboratory reaction vessel. Titan clearly provides a variety of geologic environments for the staging of organic chemistry, in which there exists free energy for reactions but not the destructive amounts of particle radiation that, on Europa for example, would quickly reduce any exposed organic molecules to unreactive and uninteresting kerogens.

Sampling of promising Titan environments, then, to determine the types of compounds present, their role in storing information for synthesis of catalytic agents, and the extent to which isomeric or stereo-chemical (chiral, e.g.) selectivity exists, provides a potential for developing an empirical timescale for the origin of life. Such would be realizable only with future missions of exploration to Titan and sophisticated sampling of organic phases on the surface.

The second distinction that Titan possesses is that it may tell us something about how the far future climate of the Earth will be-have. This too, is in the purview of the field of astrobiology, which includes as "Goal 6" of the NASA Astrobiology roadmap, (Des Marais et al. 2008) "Understand the principles that will shape the future of life, both on Earth and beyond". It is remarkable that Titan continues to possess enough methane for a hydrological cycle that includes surface liquids in spite of the fact that the depletion timescale of the methane in the atmosphere is only tens of millions of years. The source of resupply remains uncertain though plausible models for internal sources have been published (Tobie et al. 2006). The point of comparison to the future Earth is in the rapid escape of the methane from the upper atmosphere, which is thanks to the very weak temperature drop from surface to tropopause compared with that for the Earth. From Titan's surface to its tropopause there is a drop of 15°C corresponding to a change in the mixing ratio of methane of a factor of three. For the Earth the corresponding drop in temperature is over 100°C and in mixing ratio a factor of 100–1,000 from the surface value. Hence while the Earth's water is tightly bound to the troposphere, Titan's methane is not. Methane is lost by both UV radiation which breaks apart the methane, followed by loss of hydrogen, and by direct escape of methane (Yelle et al. 2008).

While Earth's surface water is so tightly bound by the cold tropopause that one can consider the terrestrial hydrological cycle almost a closed system, this will not be the case in the far future. As the Sun's luminosity increased in the past, surface temperature increases were buffered by the carbon-silicate cycle (Kasting 1988). However, the CO_2 available for buffering has declined over geologic time to the point that the feedback provided by the carbon-silicate cycle will eventually fail to prevent an increase in the surface temperature as the Sun's luminosity continues to rise. At some point in the far future the surface temperature increase, which involves a positive feedback through evaporation of ocean water, will lead to a large increase in the tropopause temperature. The timing and severity of the tropopause temperature increase is difficult to predict because of the non-linearity of the equations governing radiative-convective climates (Pujol and North 2002), but the general trend will be toward steeply increasing amounts of water vapor in the Earth's stratosphere, where the molecule is subject to breakup by ultraviolet light followed by escape of hydrogen. What happens to the Earth's hydrosphere during rapid escape of water over hundreds of millions of years might be replicated to some extent by the methane cycle we observe on Titan today, further informed by studies of Titan's recent climate and geologic history provided by Cassini–Huygens data.

That Titan's hydrologic cycle might be a methane analog to the Earth's far future hydrologic cycle is obviously somewhat speculative, but that it is active today and different from that of our present home world is inarguable. By studying the methane hydrologic cycle on Titan we expand the range of planetary climates we can observe at close range, providing not only insight into the Earth of today and perhaps the far future but also into possible types of climates that might obtain on planets around other stars. Understanding how an active hydrologic cycle works when the primary volatile is in rapid escape is an opportunity that only Titan affords us in our solar system.

Acknowledgments The authors wish to thank Catherin Neish for her careful reading of the initial manuscript and her several suggestions to improve it. F. Raulin thanks CNES and ESA for supports in the preparation of this paper.

References

Aflalaye A, Andrieux D, Bénilan Y et al (1995) Thermally instable polyynes and n-organics of planetological interest: new laboratory data and implications for their detection by in situ and remote sensing techniques. Adv Space Res 15(10):5–11

Artemevia N, Lunine JI (2003) Cratering on Titan: impact melt, ejecta, and the fate of surface organics. Icarus 164:471–480

Artemevia N, Lunine JI (2005) Impact cratering on Titan. II. Global melt, escaping ejecta, and aqueous alteration of surface organics. Icarus 175:522–533

Atreya SK, Donahue TM, Kuhn WR (1978) Evolution of a nitrogen atmosphere on Titan. Science 201:611–613

Balsiger H, Altwegg K, Geiss J (1995) D/H and 18O/16O ratio in the hydronium ion and in neutral water from in situ ion measurements in comet Halley. J Geophys Res 100:5827–5834

Benner SA, Ricardo A, Carrigan MA (2004) Is there a common chemical model for life in the universe? Curr Opin Chem Biol 8:672–689

Bernard JM, Coll P, and Raulin F (2002) Variation of C/N and C/H ratios of Titans aerosols analogues. Proceedings of the 2nd European Workshop on Exo-/Astro-Biology, ESA SP 518:623–625

Bernard JM, Coll P, Coustenis A, Raulin F (2003) Experimental simulation of Titan's atmosphere detection of ammonia and ethylene oxide. Planet Space Sci 51:1003–1011

Bernard JM, Quirico E, Brissaud O et al (2006) Reflectance spectra and chemical structure of Titan's tholins: application to the analysis of Cassini–Huygens observations. Icarus 185:301–307

Biemann K (2006) Astrochemistry: complex organic matter in Titan's aerosols? Nature 444:E6

Bockelée-Morvan S, CrovisierMJ, Mumma MJ, Weaver HJ (2004) The composition of cometary volatiles. In:, Festou MC, Keller HU, Weaver HA (eds) COMETS II, University of Arizona Press, Tucson, pp 381–414

Borucki WD et al. (2003) Kepler Mission: A mission to find Earth-size planets in the habitable zone. Proceedings of the Toward Other Earths: DARWIN/TPF and the Search for Extrasolar Terrestrial Planets, Heidelberg, Germany, 22–25 April 2003, ESA SP-539, October 2003.

Brown RH, Soderblom LA, Soderblom LA, Soderblom JM, Clark RN et al (2008) The identification of liquid ethane in Titan's Ontario Lacus. Nature 454:607–610

Clarke DW, Joseph JC, Ferris JP (2000) The design and use of a photochemical flow reactor: a laboratory study of the atmospheric chemistry of cyanoacetylene on Titan. Icarus 147:282–291

Cochran AL, Jehin E, Manfroid J, Hutsemkers D et al (2008) Nitrogen isotope ratios in comets. In: Santos NC, Pasquini N, Correia ACM & Romaniello M (eds) Precision Spectroscopy in Astrophysics, ESO Astrophysics symposia, Proceedings of the ESO/Lisbon/Aveiro Conference Portugal, 11–15 September 2006, pp 263–265, Springer, Berlin

Coll P, Coscia D, Gazeau et al. (1995) Organic chemistry in Titan's atmosphere: new data from laboratory simulations at low temperature. Adv Space Res 16(2):93–104

Coll P, Coscia D, Gazeau et al. (1997) New planetary atmosphere simulations: application to the organic aerosols of Titan. Adv Space Res 19(7):1113–1119

Coll P, Coscia D, Gazeau MC, Raulin F (1998) Review and latest results of laboratory investigation of Titan's aerosols. Orig Life Evol Biosph 28:195–213

Coll P, Coscia D, Smith N et al (1999a) Experimental laboratory simulation of Titan's atmosphere: aerosols and gas phase. Planet Space Sci 47(10–11):1331–1340

Coll P, Guillemin JC, Gazeau MC, Raulin F (1999b) Report and implications of the first observation of C_4N_2 in laboratory simulations of Titan's atmosphere. Planet Space Sci 47(12):1433–1440

Coll P, Bernard JM, Navarro-González R, Raulin F (2003) Oxirane: an exotic oxygenated organic compound in Titan? Astrophys J. 598: 700–703

Coustenis A, Achterberg RK, Conrath BJ et al (2007) The composition of Titan's stratosphere from Cassini/CIRS mid-infrared spectra. Icarus 189:35–62

Coustenis A, Jennings DE, Jolly A, Bénilan Y et al (2008) Detection of C_2HD and the D/H ratio on Titan. Icarus 197(2):539–548

Cronin JR, Pizzarello S, Cruickshank DP (1988) Organic matter in carbonaceous chondrites, planetary satellites, asteroids and comets. In: Kerridge JF, Matthews MS (eds) Meteorites and the Early Solar System, University of Arizona Press, Tucson

Cruickshank DP, Imanaka H, Dalle Ore CM (2005) Tholins as coloring agents on outer Solar System bodies. Adv Space Res 36(2): 178–183

Delsemme A (1998) Our cosmic origins – from the Big Bang to the emergence of life and intelligence. Cambridge University Press, Cambridge

Delsemme A (2006) in Comets and the origin and evolution of life. In: Thomas PJ, Hicks RD, Chyba CF, Mckay CP Springer, Berlin, pp 29–68

Des Marais DJ, Nuth JA, Allamandola LJ, Boss AP, Farmer J, Hoehler TM, Jakosky BM, Meadows VS, Pohorille A, Runnegar B, Spormann AM (2008) NASA Astrobiology Roadmap. Astrobiology August 8(4):715–730

Drake MJ (2005) Origin of water in the terrestrial planets. Meteoritics Planet Sci 40:1–9

Dubouloz N, Raulin F, Lellouch E, Gautier D (1989) Titan's hypothesized ocean properties: the influence of surface temperature and atmospheric composition uncertainties. Icarus 82:81–96

Eberhardt P, Reber M, Krankowsky D, Hodges RR (1995) The D/H and $^{18}O/^{16}O$ ratios in water from comet P/Halley. Astron Astrophys 302:301–316

Ehrenfreund P, Boon JP, Commandeur J, Sagan C et al (1995) Analytical pyrolysis experiments of Titan aerosol analogues in preparation for the Cassini–Huygens mission. Adv Space Res 15(3):335–342

Feistel R, Wagner W (2007) Sublimation pressure and subliumation enthalpy of H2O ice Ih between 0 and 273.16K. Geochem. Cosmochim. Acta 71:36–45

Ferris JP; Joshi PC; Edelson EH; Lawless JG (1978) HCN: a plausible source of purines, pyrimidines and amino acids on the primitive earth. J Mol Evol 11(4):293–311

Ferris JP, Hagan WJ Jr (1984) HCN and chemical evolution: the possible role of cyano compounds in prebiotic synthesis. Tetrahedron 40(7): 1093–1120

Fortes AD (2000) Exobiological implications of a possible ammonia-water ocean inside Titan. Icarus 146:444–452

Geiss J (1988) Composition in Halley's Comet: clues to origin and history of cometary matter. Rev Mod Astron 1:1–27

Grevesse N, Asplund M, Sauval J (2005) The new solar chemical composition. In Alecian G, Richard O, Vauclair S (eds) Element stratification in stars: 40 years of atomic diffusion, EAS Publications Series 17, pp 21–30. doi:10.1051/eas

Hébrard E, Dobrijevic M, Bénilan Y, Raulin F (2007a) Photochemical kinetics uncertainties in modeling Titan's atmosphere: a review. J Photochem Photobiol C: Photochem Rev 7:211–230

Hébrard E, Dobrijevic M, Bénilan Y, Raulin F (2007b) Photochemical kinetics uncertainties in modeling Titan's atmosphere:First consequence. Planet Space Sci 55:1470–1489

Hsieh H, Jewitt D (2006) A population of comets in the main asteroid belt. Science 312:561–563

Imanaka H, Khare BN, Elsila JE et al (2004) Laboratory experiments of Titan tholin formed in cold plasma at various pressures: implications for nitrogen-containing polycyclic aromatic compounds in Titan haze. Icarus 168:344–366

Israël G, Szopa C, Raulin F et al (2005) Evidence for the presence of complex organic matter in Titan's aerosols by in situ analysis. Nature 438:796–799

Israël G, Szopa C, Raulin F et al (2006) Astrochemistry: complex organic matter in Titan's aerosols? (Reply). Nature 444:E6–E7

Jolly A, Bénilan Y (2008) Review of quantitative spectroscopy of polyynes. J Quant Spectro Rad Transfer 109:963–973

Kasting JF (1988) Runaway and moist greenhouse atmospheres and the evolution of Earth and Venus. Icarus 74:472–494

Kasting J and Brown LL (1998) The early atmosphere as a source of biogenic compounds. In: Brack A (ed) The molecular origins of life: assembling the pieces of the puzzle, Cambridge University Press, Cambridge

Kauffman SA, Clayton P (2006) On emergence, agency, and organization. Biol Philos 21:501–521

Khare BN, Sagan C, Arakawa ET et al (1984) Optical constants of organic tholins produced in a simulated Titanian atmosphere: from soft X-rays to microwave frequencies. Icarus 60:127–137

Khare BN, Sagan C, Ogino H et al (1986) Amino acids derived from Titan tholins. Icarus 68:176–184

de Kok R, Irwin PGJ, Teanby NA et al (2007) Oxygen compounds in Titan's stratosphere as observed by Cassini CIRS. Icarus 186: 354–363

Kral TA, Brink KM, Miller SL, McKay CP (1998) Hydrogen consumption by methanogens on the early Earth. Orig Life Evol Biosph 28:311–319

Lavvas PP, Coustenis A, Vardavas IM (2008a) Coupling photochemistry with haze formation in Titan's atmosphere, Part I: Model description. Planet Space Sci 56:27–66

Lavvas PP, Coustenis A, Vardavas IM (2008b) Coupling photochemistry with haze formation in Titan's atmosphere, Part II: Results and validation with Cassini/Huygens data. Planet Space Sci 56:67–99

Lebonnois S (2005) Benzene and aerosol production in Titan and Jupiter's atmospheres: a sensitivity study. Planet Space Sci 53:486–497

Lorenz RD, Wall S, Radebaugh J, Boubin G et al (2006) The Sand Seas of Titan: Cassini RADAR Observations of Longitudinal Dunes. Science 312:724–727

Lorenz RD, Stiles BW, Kirk RL, Allison MD et al (2008) Titan's rotation reveals an internal ocean and changing zonal winds. Science 319:1649–1651

Lunine JI, Stevenson DJ (1987) Clathrates and ammonia hydrates at high pressure: application to the origin of methane on Titan. Icarus 70:61–77

Lunine JI, Atreya SK (2008) The methane cycle on Titan. Nat Geosci 1:159–164

McDonald GD, Thompson WR, Heinrich M, Khare BN, Sagan C (1994) Chemical investigation of Titan and Triton tholins. Icarus 108:137–145

McGuigan M, Waite JH, Imanaka H, Sacks RD (2006) Analysis of tholin pyrolysis products by comprehensive two-dimensional gas chromatography-time-of-flight mass spectrometry. J Chromatogr A 1132:280–288

McKay CP (1996) Elemental composition, solubility, and optical properties of Titan's organic haze. Planet Space Sci 44(8):741–747

McKay CP, Smith HD (2005) Possibilities for methanogenic life in liquid methane on the surface of Titan. Icarus 178:274–276

McKay CP, Porco CC, Altheide T et al (2008) The possible origin and persistence of life on Enceladus and detection of biomarkers in the plume. Astrobiology 8(5):909–919

Meibom A, Krot AN, Robert F, Mostefaoui S, Russell SS, Petaev MI, Gounelle M (2007) Nitrogen and carbon isotopic composition of the Sun inferred from a high-temperature solar nebular condensate. Astrophys J 656:L33–L36. doi 10.1086/512052

Miller SL (1953) A production of amino acids under possible primitive Earth conditions. Science 117:528–529

Miller SL, Urey HC (1959) Organic compound synthesis on the primitive Earth. Science 130:245–51

Miller SL, Lyons JR, Chyba C (1998) Organic shielding of greenhouse gases on early Earth. Science 279:779

National Research Council (2007) The limits of organic life in planetary systems. Space Studies Board

Neish CD, Lorenz RD, O'Brien DP (2006) The potential for prebiotic chemistry in the possible cryovolcanic dome Ganesa Macula on Titan. Int J Astrobiol 5:57–65

Neish CD, Somogyi A, Lunine JI, Smith MA (2008) Rate measurements of the hydrolysis of complex organic macromolecules in cold aqueous solutions: implications for prebiotic chemistry on the early Earth and Titan. Astrobiology 8(2):273–287

Neish CD, Somogyi A, Lunine JI, Smith MA (2009) Low temperature hydrolysis of laboratory tholins in ammonia-water solutions: implications for prebiotic chemistry on Titan. Icarus 201(1):412–421

Nguyen MJ (2007) PhD thesis, University Paris 12, December 2007.

Nguyen MJ, Raulin F, Coll P et al (2007) Carbon isotopic enrichment in Titan's tholins? Implications for Titan's aerosols. Planet Space Sci 55:2010–2014

Niemann HB, Atreya SK, Bauer SJ et al (2005) The abundances of constituents of Titans' atmosphere from the GCMS instrument on the Huygens probe. Nature 438:779–784

O'Brien DP, Lorenz RD, Lunine JI (2005) Numerical calculations of the longevity of impact oases on Titan. Icarus 173:243–253

Oro J (1961) Comets and the formation of biochemical compounds on the primitive earth. Nature 190:389–390

Owen T (1982) The composition and origin of Titan's atmosphere. Planet Space Sci 30:833–838

Owen T (2000) On the origin of Titan's atmosphere. Planet Space Sci 48:747–52

Owen T, Bar-Nun A (1995) Comets, impacts and atmospheres. Icarus 116:215–226. doi:10.1006/icar.1995.1122

Owen T, Bar-Nun A (2000) Volatile contributions from icy planetesimals. In Canup RM, Righter K (eds) Origin of the Earth and Moon, University of Arizona Press, Tucson, pp 459–475

Owen T, Mahaffy PR, Niemann HB, Atreya S, Wong M (2001) Protosolar nitrogen. Astrophys J 553:L77–L79. doi:10.1086/320501

Pujol T, North GR (2002) Runaway greenhouse effect in a semi-gray radiative-convective model. J. Atmos Sci 59:2801–2810

Ramirez SI, Coll P, Da Silva A et al (2002) Complex refractive index of Titan's aerosol analogues in the 200 900 nm domain. Icarus 156:515–530

Ramirez SI, Navarro-Gonzalez R, Coll P, Raulin F (2001) Possible contribution of different energy sources to the production of organics in Titan's atmosphere. Adv Space Res 27(2):261–270

Raulin F, Mourey D, Toupance G (1982) Organic syntheses from CH4–N2 atmospheres: implications for Titan. Origin Life 12:267–279

Raulin F (1987) Organic chemistry in the oceans of Titan. Adv Space Res 7(5):71–81

Raulin F, Bruston P, Paillous P, Sternberg R (1995) The low temperature organic chemistry of Titan's geofluid. Adv Space Res 15(3):321–333

Raulin F, Nguyen MJ and Coll P (2007) Titan: an astrobiological laboratory in the solar system. In: Richard B Hoover, Gilbert V Levin, Alexei Y Rozanov, Paul CW Davies (eds) Proceedings of the SPIE 6694, Instruments, Methods, and Missions for Astrobiology X, Editors, SPIE Publisher, Washington DC, USA, DOI: 10.1117/12.732883, SPIE Publisher, Washington DC

Raulin F (2008a) Planetary sciences. Organic lakes on Titan. Nature 454:587–589

Raulin F (2008b) Astrobiology and habitability of Titan. Space Sci Rev 135(1–4):37–48

Raven JA, Kübler JE, Beardall J (2000) Put out the light, and then put out the light. J Mar Biol Assoc UK 80:1–25

Rubey WW (1951) Geologic history of sea water. An attempt to state the problem. Bull Geol Soc Am 62:1111–48

Sagan C, Khare BN (1979) Tholins: Organic chemistry of interstellar grains and gas. Nature 277:102–107

Sagan C, Khare BN, Lewis J (1984) Organic matter in the Saturn system. In "Saturn" edited by T. Gehrels & M.S. Matthews, University of Arizona Press, Tucson, p 788

Sagan C, Chyba C (1997) The early faint sun paradox: organic shielding of ultraviolet-labile greenhouse gases. Science 276:1217–1221

Sarker N, Somogyi A, Lunine JI, Smith MA (2003) Titan aerosol analogues: analysis of the nonvolatile tholins. Astrobiology 3(4):719–726

Scemama A, Chaquin P, Gazeau MC, Bénilan Y (2002) Theoretical study of the struture and properties of polyynnes and monocyano- and dicyanopolyynes: prediction for long chain compounds. J Phys Chem A 106:3828–3837

Schulze-Makuch D, Grinspoon DH (2005) Biologically enhanced energy and carbon cycling on Titan? Astrobiology 5:560–567

Sekine Y, Imanaka H, Matsui T, Khare BN et al (2008a) The role of organic haze in Titan's atmospheric chemistry I. Laboratory investigation on heterogeneous reaction of atomic hydrogen with Titan tholin. Icarus 194:186–200

Sekine Y, Lebonnois S, Imanaka H, Matsui T et al (2008b) The role of organic haze in Titan's atmospheric chemistry II. Effect of heterogeneous reaction to the hydrogen budget and chemical composition of the atmosphere. Icarus 194:201–211

Somogyi A, Oh CH, Smith MA, Lunine JI (2005) Organic environments on Satuirn's moon, Titan: simulating chemical reactions and analyzing products by FT-ICR and ion-trap mass spectrometry. J Am Soc Mass Spectrom 16:850–859

Sotin C, Tobie G (2008) PLANETARY SCIENCE: Titan's Hidden Ocean. Science 319:1629–1630

Stoker CR, Boston PJ, Mancinelli RL et al (1990) Microbial metabolism of tholins. Icarus 85:241–256

Strobel D (1982) Chemistry and evolution of Titan's atmosphere. Planet Space Sci 30:839–848

Szopa C, Cernogora G, Boufendi L et al (2006) PAMPRE: a dusty plasma experiment for Titan's tholins production and study. Planet Space Sci 54:394–404

Teanby NA, Irwin PGJ, de Kok R et al (2006) Latitudinal variations of HCN, HC_3N and C_2N_2 in Titan's stratosphere derived from Cassini/CIRS data. Icarus 181:243–255

Teanby NA, Irwin PGJ, de Kok R et al (2007) Vertical profiles of HCN, HC_3N and C_2N_2 in Titan's atmosphere derived from Cassini CIRS data. Icarus 186:364–384

Teanby NA, Irwin PGJ, de Kok R et al (2008) Global and temporal variations in hydrocarbons and nitriles in Titan's stratosphere for northern winter observed by Cassini/CIRS. Icarus 193:595–611

Thompson WR., Sagan C, Stevenson D, Wing M (1992) Impact mediated chemical evolution on Titan. Bull Am Astron Soc 24. [Poster]

Thompson WR, Todd H, Schwartz J, Khare BN, Sagan C (1991) Plasma discharge in N_2 + CH_4 at low pressures: experimental results and applications to Titan. Icarus 90:57–73

Tian F, Toon OB, Pavlov AA, De Sterck H (2005) A hydrogen-rich early earth atmosphere. Science 308:1014–1017

Tobie G, Lunine JI, Sotin C (2006) Episodic outgassing as the origin of atmospheric methane on Saturn's moon Titan. Nature 440:61–64

Tobie G, Grasset O, Lunine JI, Mocquet A, Sotin C (2005) Titan's internal structure inferred from a coupled thermal-orbital model. Icarus 175(2):496–502

Tomasko MG, Archinal B, Becker T, Bézard B et al (2005) Rain, winds and haze during the Huygens probe's descent to Titan's surface. Nature 438:765–778

Toupance G, Raulin F, Buvet R (1975) Formation of prebiological compounds in models of the primitive Earth's atmosphere. I : CH4-NH3 and CH4–N2 atmospheres. Origin f Life 6:83–90

Trainer MG, Pavlov AA, DeWitt HL, Jimenez JL et al (2006) Organic haze on Titan and the early Earth. Proc Natl Acad Sci 103(48):18035–18042

Tran BU, Ferris JP, Chera JJ (2003a) The photochemical formation of a Titan haze analog. Structural analysis by X-ray photoelectron and infrared spectroscopy. Icarus 162:114–124

Tran BN, Joseph JC, Ferris JP et al (2003b) Simulation of Titan haze formaiton using a photochemical flow reactor. The optical constants of the polymer. Icarus 165:379–390

Tran BN, Joseph JC, Force M et al (2005) Photochemical processes on Titan: irradiation of mixtures of gases that simulate Titan's atmosphere. Icarus 177:106–115

Tran BN, Force M, Briggs RG et al (2008) Titans' atmospheric chemistry: photolysis of gas mixtures containing hydrogen cyanide and carbon monoxide at 185 and 254 nm. Icarus 193:224–232

Turekian KK, Clark SP Jr (1975) The nonhomogeneous accumulation model for terrestrial planet formation and the consequences for the atmoshere of venus. J Atmos Sci 32:1257–1261

Urey HC (1952) The Planets: their origin and development, Yale University press, UK

Vinatier S, Bézard B, Fouchet T et al (2007a) Vertical abundance profiles of hydrocarbons in Titan's atmosphere at 15°S and 80°N retrieved from Cassini/CIRS spectra. Icarus 188:120–138

Vinatier S, Bézard B, Nixon CA (2007b) The Titan 14N/15N and 12C/13C isotopic ratios in HCN from Cassini/CIRS. Icarus 191:712–721

Vuitton V, Doussin JF, Bénilan Y, Raulin F, Gazeau MC (2006) Experimental and theoretical study of hydrocarbon photochemistry applied to Titan stratosphere. Icarus 185:287–300

Vuitton V, Yelle RV, McEwan MJ (2007) Ion chemistry and N-containing molecules in Titan's upper atmosphere. Icarus 191:722–742

Waite H Jr, Niemann H, Yelle RV, Kasprzak WT et al (2005) Ion neutral mass spectrometer results from the first flyby of Titan. Science 308:982–986

Waite JH Jr, Young DT, Cravens TE et al (2007) The process of tholin formation in Titan's upper atmosphere. Science 316:870–875

Wilson EH, Atreya SK (2003) Chemical sources of haze formation in Titan's atmosphere. Planet Space Sci 51:1017–1033

Wilson EH, Atreya SK and Coustenis A (2003) Mechanisms for the formation of benzene in the atmosphere of Titan. J Geophys Res 108(E2):5014

Wilson EH, Atreya SK (2004) Current state of modeling the photochemistry of Titan's mutually dependent atmosphere and ionosphere. J Geophys Res Planets 109(E6):E06002

Yelle RV, Cui J, Müller-Wodarg ICF (2008). Methane escape from Titan's atmosphere. J Geophys Res Planets 113(E10):CiteID E10003

Yung YL, Allen M, Pinto JP (1984) Photochemistry of the atmosphere of Titan: comparison between model and observations. Astrophys J Suppl Sert 55:465–506

Zarnecki JC, Leese MR, Hathi B et al (2005) A soft solid surface on Titan as revealed by the Huygens surface science package. Nature 438:792–795

Chapter 10
Atmospheric Structure and Composition

Darrell F. Strobel, Sushil K. Atreya, Bruno Bézard, Francesca Ferri, F. Michael Flasar, Marcello Fulchignoni, Emmanuel Lellouch, and Ingo Müller-Wodarg

Abstract Titan's atmosphere is predominantly N_2 with CH_4 the next most abundant molecule. It has a mole fraction of 0.05 just above the surface decreasing to 0.014 in the stratosphere. Above the homopause (~800–850 km), it increases to 0.12 at the exobase. The third abundant molecule is H_2 with a tropospheric mole fraction of 0.001 increasing to 0.004 at ~1000 km and ~0.02 at the exobase (~1500–1600 km). This chapter reviews the various measurements acquired by the Voyager flybys, Huygens Probe, orbiting Cassini spacecraft, ground-based and orbiting telescopes of the large suite of hydrocarbons, nitriles, other nitrogen and also oxygen bearing compounds. Titan possesses a mostly stable troposphere with a well defined tropopause ($T \sim 70$ K at ~44 km) and a lower stratosphere with a high static stability, which is extremely cold over the winter polar region (currently northern hemisphere) and warm over the summer pole. Remarkably in the middle stratosphere, the warmest temperatures occur at the equator and the largest meridional temperature gradients are found in the winter hemisphere. The stratopause from the summer pole to about 45° N remains at a relatively constant pressure of 0.1 mbar/300 km and then it rises rapidly upward to ~0.01 mbar/400 km at the winter north pole, where it is the warmest region in the entire atmosphere. One possible interpretation of the Huygens Atmospheric Structure Instrument (HASI) temperature profile is that Titan's atmosphere is essentially isothermal ~170 K from 500–1100 km, with large amplitude thermal waves (10 K) superimposed. The existence and location of a well defined mesopause is an open question.

The chemistry of Titan's atmosphere is driven by CH_4 photolysis in the thermosphere and catalytic reactions in the stratosphere, and by N_2 dissociation due to both UV photons and energetic electrons. Ethane is the most abundant gas product and HCN is the dominant nitrile. The mixing ratios of all photochemical species, except C_2H_4, increase with altitude at equatorial and southern latitudes, indicative of transport from a high-altitude source to a condensation sink in the lower stratosphere. Northward of 45° N, most product compounds are enriched as a consequence of subsidence in the winter polar vortex, particularly for nitriles and more complex hydrocarbons than C_2H_6 and C_2H_2. North of 45° N, most products have lower increases with altitude than at low latitudes.

D.F. Strobel (✉)
The Johns Hopkins University, Baltimore, MD, 21218–2687, USA
e-mail: strobel@jhu.edu

S.K. Atreya
University of Michigan, Ann Arbor, MI, 48109–2143, USA
e-mail: atreya@umich.edu

B. Bézard, M. Fulchignoni and E. Lellouch
LESIA, Observatoire de Paris-Meudon, 92195, Meudon Cedex, France
e-mail: Bruno.Bezard@obspm.fr
e-mail: marcello.fulchignoni@obspm.fr
e-mail: emmanuel.lellouch@obspm.fr

F. Ferri
CISAS G.Colombo, University of Padova, 35131, Padova, Italy
e-mail: francesca.ferri@unipd.it

F. Michael Flasar
NASA Goddard Space Flight Center, Greenbelt, MD, 20771, USA
e-mail: f.m.flasar@nasa.gov

I. Müller-Wodarg
Space and Atmospheric Physics Group, Imperial College London, Prince Consort Road, London, SW7 2BW, UK
e-mail: i.mueller-wodarg@imperial.ac.uk

10.1 Historical Introduction

The first definitive detection of an atmosphere on Titan was Kuiper's (1944) discovery of near-IR CH_4 bands in absorption. But it was not until the 1970s, that Titan became an object of intense study. Lewis (1971) noted that Titan's low density implied an interior composition rich in ices and suggested photolysis of outgassed ammonia would lead to a nitrogen atmosphere. Trafton (1972) reported the possible detection of H_2 and interpreted the 20 μm spectral region in terms of a strong greenhouse. Gillett et al. (1973) detected pronounced peaks at 12 (C_2H_6) and 8 (CH_4) μm in narrow band spectra, which Danielson et al. (1973) interpreted as emission from a thermal inversion layer and substantial amounts of C_2H_6 (~3 × 10^{19} cm^{-2}) in the atmosphere. Two NASA Workshops and their reports (Hunten 1974; Hunten and Morrison 1978) enhanced

R.H. Brown et al. (eds.), *Titan from Cassini-Huygens*,

and focused the intense interest in Titan in anticipation of the Voyager flybys years later. Lewis' (1971) ideas for a nitrogen atmosphere from ammonia photolysis were pursued by Hunten (1978) and Atreya et al. (1978). The ambiguous (at the time) spectroscopic evidence on what are the major atmospheric species and surface pressure was interpreted by Hunten (1978) to imply a very thick background atmosphere (~20 bar of nitrogen) in addition to the observed CH_4. Danielson et al. (1973), in contrast, interpreted the spectroscopic evidence in terms of a very thin few mbar atmosphere of mostly CH_4. With these two divergent working models, the Voyager Missions' scientific objectives for Titan's atmosphere were well focused toward a definitive determination of its composition and structure. More pre-Voyager history and background may be found in the companion Chapter 2.

The Voyager spacecrafts yielded definitive measurements and answers to the outstanding questions at the time about Titan's atmosphere. The most noteworthy results were (1) composition, 97% N_2, except in the lowest 15 km, with 1.5–3% CH_4, <10% Ar, <5% CO (Broadfoot et al. 1981; Hanel et al. 1981; Strobel and Shemansky 1982; Strobel et al. 1993; Vervack et al. 2004), (2) the density, pressure, and temperature profiles at equatorial latitudes from the surface up to 200 km from the Voyager radio (refractive) occultation experiment (Lindal et al. 1983; Lellouch et al. 1989), (3) thermal and chemical structure from the Voyager IRIS Instrument at the surface, tropopause, and stratosphere at latitudes between 60° S to 70° N and detection of a large suite of hydrocarbons and nitriles in Titan's stratosphere (Hanel et al. 1981), and (4) the equatorial thermospheric structure from the Voyager Ultraviolet Spectrometer solar (absorptive) occultation experiment for N_2 and CH_4 between 900 and 1400 km (Broadfoot et al. 1981; Vervack et al. 2004). Additional Voyager background perspective can be found in Chapter 2.

The Cassini–Huygens Mission visited the Saturnian System some 20 years later with well-instrumented orbiting spacecraft and Huygens Probe, which descended to Titan's surface, to study in-depth Titan's atmosphere. In the following discussion the scientific discoveries and findings from this Mission on the structure and composition are reviewed and integrated with scientific knowledge derived from ground-based and orbiting observatories and Voyager.

10.2 Vertical Structure of the Atmosphere: Mass Density, Pressure, and Temperature

To understand the structure of Titan's atmosphere one must keep in mind certain basic facts from solar system dynamics (cf. Fig. 10.1). First, the axial tilt of Saturn and Titan is 26.73° and hence seasonal effects are important. Second, Saturn's orbital eccentricity is 0.05415, which yields a variation in the distance from the Sun of slightly greater than 1 AU and in the solar flux of ~20 %. Perihelion last occurred in 2003.56 and summer solstice in Titan's southern hemisphere was 2002.76, as shown in Fig. 10.1. The variation in solar distance and thus solar flux will be most important in the troposphere and stratosphere, whereas the much larger solar cycle variations in UV and EUV solar radiation will overwhelm the smaller eccentricity effects in the thermosphere. The Voyager spacecraft flybys were at high solar activity, whereas orbit insertion of the Cassini spacecraft in July 2004 was during the descending phase from peak activity, which occurred in April 2000, to solar minimum conditions in the August-November 2007 timeframe. Accordingly, the nominal Cassini–Huygens Mission was mostly at low solar activity.

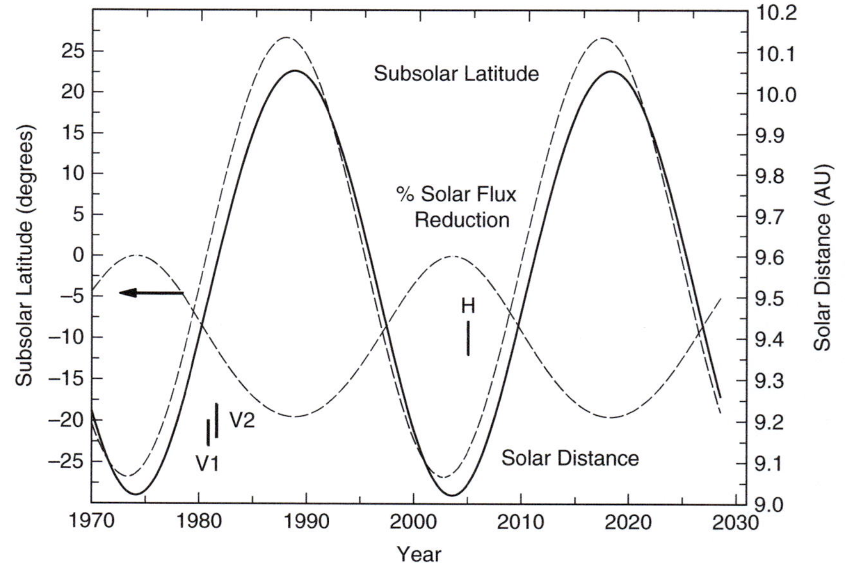

Fig. 10.1 The latitude of the subsolar point (dashed line), the distance from the Sun in AU (*solid line*), and the percentage variation in solar flux relative to perihelion as a function of time in years (*dashed line, read ordinate as %*). Last perihelion was 2003.56; last summer solstice in southern hemisphere was 2002.76. The Voyager flybys (V1, V2) and the Huygens Probe entry (H) are indicated

The most significant spatial variation in a planetary atmosphere is the vertical stratification due to gravity and characterized by an e-folding pressure scale height H, which for Titan varies from 15 to 100 km. As a consequence it is natural to take the distribution of the three thermodynamic variables that describe planetary atmospheres: mass (ρ) or number (n) density, pressure (p), and temperature (T) and write them, e. g. the latter as

$$T(\phi,\theta,z,t) = \langle T\rangle(z)+\overline{T}(\theta,z,t)+T'(\phi,\theta,z,t).$$

The first term is the vertical temperature profile with T averaged over the globe on each height or pressure surface if $z = -\int H(d\ln p)$, depending on the altitude variable. The second term is temperature departure from the first term averaged over longitude, ϕ, on a latitudinal circle at latitude θ and height z. The last term is the longitudinally varying departure from the first two terms and characterizes waves in the atmosphere. Seasonal variations are presumed to be present in the latter two quantities and absent from the first term in the troposphere and lower stratosphere, because of the long radiative time constants. Since most pronounced seasonal time variations are expected at high latitudes, 1D vertical profiles obtained at equatorial latitudes should be approximately equal to rigorously calculated $\langle T\rangle(z)$ from global data.

In Fig. 10.2, various 1D profiles obtained at equatorial latitudes are compared. For the Voyager radio occultation data acquired at 6.2°N and 8.5°S, the Lellouch et al. (1989) analysis is adopted. For the UVS solar occultation data taken at 4°N and 16°S, the Vervack et al. (2004) profile is shown and would imply T ~ 155 K atmosphere, if isothermal. From these data sets, the Yelle et al. (1997) engineering model was constructed for Cassini–Huygens Mission planning. Note that the agreement of this model with the Huygens Atmospheric Structure Instrument (HASI, Fulchignoni et al. 2005) density profile was excellent below 600 km. HASI measured entry deceleration from which density was derived directly down to an altitude of ~175 km, but pressure and temperature were inferred with the assumption of hydrostatic equilibrium. After ejection of the aeroshell and parachute deployment at z ~ 150–175 km, the pressure and temperature were independently measured and density derived from the equation of state for a real gas.

The lack of any change in Titan's lower atmosphere between the Voyager flybys and Huygens Probe is easily understood by the very long radiative time constants (~300 year at 10–20 km, ~60 year at 50 km) in comparison to a Saturnian year (= 29.5 year). In the thermosphere, the mass density measured by HASI is higher by up to a factor of 2 than inferred by Vervack et al. (2004) from the Voyager solar occultation measurements. The Voyager data were taken during solar maximum activity, whereas the HASI data were obtained when solar activity was between medium and minimum solar activity. Hence one would have expected Titan's upper atmosphere to have contracted relative to the Voyager epoch. Note that the Vervack et al. (2004) inferred temperature ~155 K, is about 20 K colder than HASI, so that extrapolating their densities to lower altitudes would merge with the HASI densities at ~1000 km. However due to Saturn's elliptical orbit around

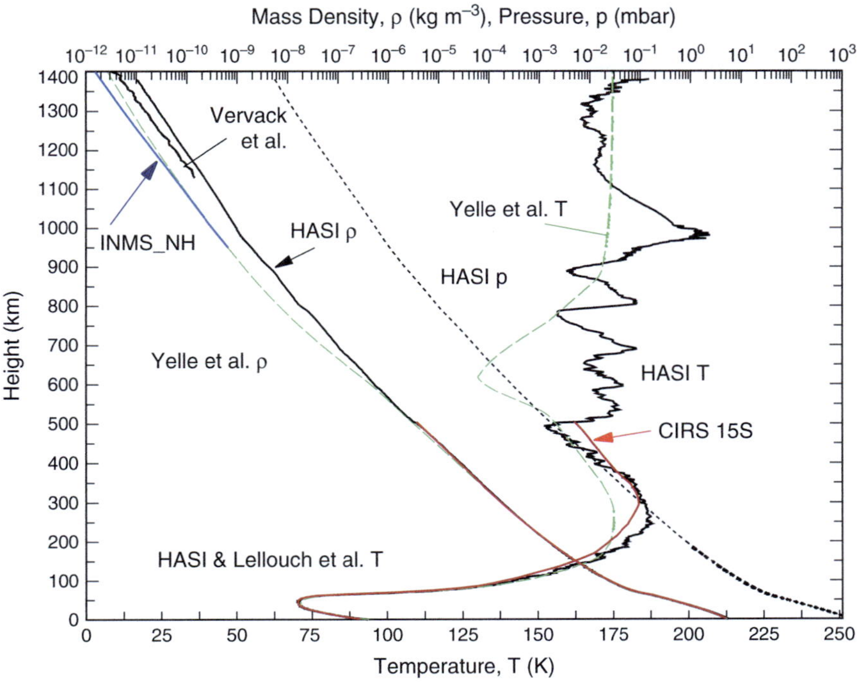

Fig. 10.2 Comparison of Cassini–Huygens HASI density, pressure, and temperature profiles (*black lines*, Fulchignoni et al. 2005), CIRS density and temperature profiles at 15 S (*red lines*, Vinatier et al. 2007a), average INMS Northern Hemisphere density profile (*blue line*, Müller-Wodarg et al. 2008) with the Voyager Lellouch et al. (1989) and Vervack et al. (2004) results, and the Yelle et al. (1997) engineering model (*green lines*), all as functions of altitude above surface

the Sun, Titan was at 9.44 AU during the Voyager 1 flyby and 9.04 AU, at the time of Huygens Probe entry. This could explain the expanded atmosphere, but not the colder temperatures in the thermosphere as the solar EUV and UV fluxes at solar maximum activity for Voyager flybys ratioed to solar fluxes pertinent to Probe descent would be ~2, more than enough to offset the $(9.44/9.04)^2$ ~1.09 factor due to Titan being closer to the Sun during the Probe descent. There is presently no explanation for the temperature difference between Vervack et al. and HASI.

Also shown in Fig. 10.2 is the Ion Neutral Mass Spectrometer (INMS) in situ data averaged for the northern hemisphere (Müller-Wodarg et al. 2008), which is about a factor of 2.4 lower than the HASI densities at 10.3° S. As discussed in more detail below and in Chapter 11, the drag and torque on the Cassini spacecraft allow determination of densities, which are systematically higher than measured INMS densities by a factor of ~2.6 on each pass through the upper atmosphere when INMS is making measurements of the neutral atmosphere. The drag determined densities from HASI support the Cassini spacecraft derived values.

With reference to Fig. 10.2, Titan possesses a well defined equatorial tropopause with $T = 70.43 \pm 0.25$ K at 44 km. Based on the mean temperature profile the troposphere is statically stable. The low temperatures and high densities in the troposphere require that the atmosphere be treated as a real rather than an ideal gas. Titan's lower stratosphere has a high static stability with dT/dz ~1 K km^{-1}, from strong solar heating by its haze absorbing visible and UV radiation and near-IR CH_4 absorption (cf. Sec. 10.4.1). The stratopause according to HASI data is located at 260 km where $T = 187$ K. However, the CIRS stratospheric temperature retrievals do not agree with the HASI location and instead place it at ~312 km, where T = 183 K (Vinatier et al. 2007a). At this altitude, the HASI temperature is 184 K, but more important the CIRS temperature is 181 K at 265 km, and computation of the IR radiance with the HASI derived atmosphere is inconsistent with the observed CIRS CH_4 7.7 μm band radiance. In spite of these differences in temperature, there is remarkable agreement in mass densities up to 500 km, the upper boundary for CIRS retrievals.

Fulchignoni et al. (2005) identify the temperature minimum, 153 K at 494 km, as the Titan mesopause. The Yelle et al. (1997) engineering model required a well defined mesopause somewhere to satisfy Voyager data obtained in the lower/middle atmosphere and the thermosphere, but its location and minimum temperature were less constrained. Just above this HASI minimum is a pronounced temperature inversion with peak temperature of 169 K and temperature gradient of ~3.5 K km^{-1}. This feature was previously detected in stellar occultation measurements, (e. g., Sicardy et al. 2006) and is the location of a detached haze layer (Porco et al. 2005). Only a very small optical depth would suffice to absorb sufficient solar radiation to balance the implied substantial heat loss by thermal conduction (Lavvas et al. 2009).

The HASI temperature profile contains oscillations growing in amplitude above 200 km, which is consistent with propagating waves in a gas with exponential decreasing density, a possibility anticipated by Strobel and Sicardy (1997) in their engineering study. One possible interpretation of the HASI temperature profile is that Titan's atmosphere is essentially isothermal ~170 K from 500–1100 km, with large amplitude thermal waves (10 K) superimposed on an isothermal basic state. Calculation of the wave temperature lapse rate (dT/dz) yields values that exceed the adiabatic lapse rate ($= g/c_p$, which varies from 1.24 at the surface to 0.56 K km^{-1} at 1400 km) at multiple altitudes throughout the atmosphere with the conclusion that the waves attain breaking/saturation amplitudes (cf. Fig. 3 in Fulchignoni et al. 2005). The large vertical wavelengths associated with the largest amplitudes imply either strong horizontal winds and/or high phase speeds for propagating waves.

The mean temperature profile above 1000 km does not have to be isothermal. Strobel (2008) has theorized that the large escape rates deduced from Cassini INMS data for CH_4 and H_2 (Yelle et al. 2008; Cui et al. 2008) can be explained by slow hydrodynamic escape, with no requirement for large non-thermal escape. The bulk outflow produces adiabatic cooling and approximately $T \propto r^{-1}$. Superposition of waves on this slowly decreasing mean temperature profile could also account for the derived HASI temperature (and density) data above ~1000 km. Potential wave sources are gravitational tides, large wind shear creating Kelvin–Helmholtz instability, and gravity waves forced by topography. Further discussion on wave dynamics of the atmosphere is found in Chapter 13.

10.3 The Height and Latitude Structure of the Atmosphere

Below the relevant Cassini observations performed during the nominal mission (RSS radio occultations, CIRS remote sensing, UVIS solar and stellar occultations, and INMS in situ measurements) are discussed with an emphasis on the height and latitude variations of density, pressure, and temperature.

10.3.1 Radio Occultations

There was only one Titan earth occultation of the Voyager spacecraft (Vgr 1), with ingress at 6°N and egress at 9°S. The retrieved temperatures were nearly identical in the troposphere, with a lapse rate close to the dry adiabat in the lowest 3–4 km. Cassini had four occultations by Titan, with ingress and egress soundings at 74°S, 69°S, 53°S, 34°S, 33°S, 32°S, 53°N, 74°N. Although the data have not been fully analyzed, the occultations agree with the Voyager IRIS observations in indicating a small meridional variation of temperature in the troposphere. At the tropopause, temperatures retrieved from the occultations vary by 4–5 K. Near the surface they differ by ~3 K or less. The coldest temperatures are at higher northern latitudes. Temperature profiles near 30°S are nearly adiabatic above the surface. At higher latitudes they become more stably stratified above the surface and in the winter, northern latitudes, temperatures are isothermal or have an inversion above the surface. Data for the lower northern polar stratosphere is of particular interest, because accurate retrievals from CIRS far-IR limb soundings are exceedingly difficult and there is no information from CIRS mid-IR soundings of the polar lower stratosphere as it is too cold.

10.3.2 Remote Sensing

In Fig. 10.3, the temperature of Titan's stratosphere and mesosphere is displayed as a function of latitude (75° S to 75° N) and pressure (0.001 to 10 mbar) and based on CIRS data acquired over the period from July 2004 to September 2006 (Achterberg et al. 2008). Referring to Fig. 10.1, this period is mostly solstitial. While no altitude scale is shown, one can refer to Fig. 10.2 and use the altitude scale for the CIRS 15S profile, which should be reasonable for latitudes spanning 65° S to 30° N. Remarkably in the stratosphere, the warmest temperatures occur at the equator and the largest meridional temperature gradients are found in the winter (currently, northern) hemisphere rather than in the summer (southern) hemisphere. The other noteworthy feature is the extremely cold lower polar stratosphere in the northern winter hemisphere, where the temperatures decrease by ~25–30 K from 30° N to 75° N on constant pressure surfaces between 1 and 10 mbar. The stratopause remains at a relatively constant altitude and pressure of 300 km and 0.1 mbar, respectively from the summer pole to about 45° N, and then it rises rapidly upward to ~400 km at the winter north pole. In fact, the warmest region in the entire atmosphere is the winter, northern polar "stratopause" at ~0.01 mbar/400 km and about 20 K warmer than the equatorial and summer, southern regions. The polar night only extends to 315 km, so it's possible that this warm region may have a diabatic origin. However, the enhanced mixing ratios of nitriles and hydrocarbons at winter, northern polar latitudes (cf. Fig. 10.10) suggests that subsidence and adiabatic heating is certainly occurring.

At 19 μm, CIRS detects thermal radiation from the surface because the atmospheric opacity is small and the latitudinal variation of surface temperature can be inferred from brightness temperatures. Jennings et al. (2009) find that the north pole surface is 3 K lower and the south pole surface 2 K lower than the equatorial surface temperature of 93.7 ± 0.6 K, equal to the HASI value at 10° S.

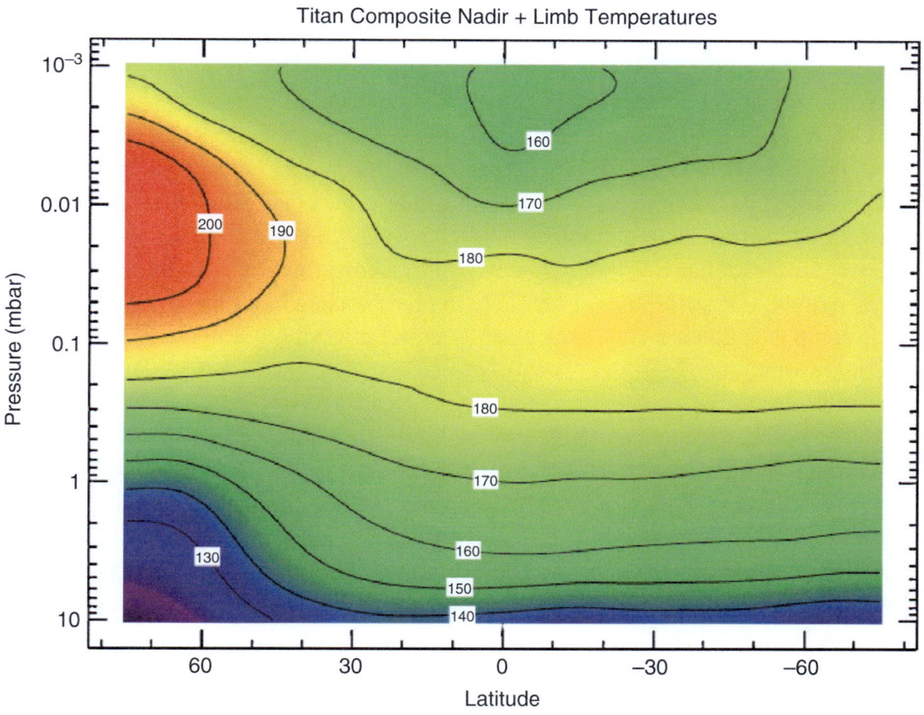

Fig. 10.3 CIRS zonal mean temperatures (K) from limb and nadir spectra recorded between July 2004 and September 2006. Retrieved temperatures were averaged over 5° latitudinal intervals and smoothed three times with 10° boxcar function (after Achterberg et al. (2008))

While HASI and refractive stellar occultations (discussed below) have detected wave structure in the atmosphere, CIRS has not to date and can only place an upper limit on wave temperature amplitudes of ~1 K. But it must be kept in mind that CIRS can only probe the atmosphere up to ~500 km, and then in the limb mode. The total data coverage of the atmosphere to date is limited in extent and it is possible that waves may be detectable by CIRS eventually. However, it is necessary to average CIRS spectra with a consequent loss of spatial resolution.

10.3.3 Solar and Stellar Occultations

The Voyager UVS solar occultations were equatorial and provided the extraordinary measurement that CH_4 was well mixed up to at least 1000 km (Vervack et al. 2004), since confirmed by the INMS (Waite et al. 2005; Yelle et al. 2008). The Cassini Ultraviolet Imaging Spectrometer (UVIS) has performed two solar occultations during the nominal mission, but no published reports are available on the results. Shemansky et al. (2005) have reported on the first UVIS stellar occultations. Their inferred temperature profile above 400 km contains (1) a convective region around 450 km where the temperature gradient is adiabatic and (2) a mesopause at 615 km with T = 114 K. By comparison, the HASI temperature was ~175 K there. When the UVIS retrieved temperature profile is used with the hydrostatic equation and lower boundary condition at 400 km from CIRS and HASI data to calculate the N_2 density profile, one obtains at 1200 km an N_2 density that is a factor of ~8 and ~20 times lower than measured by INMS and HASI, respectively. Also the UVIS inferred CH_4 densities at 1200 km when coupled with this calculated N_2 density implies that CH_4/N_2 ratio is ~0.5, also at variance with INMS values of ~0.035 at 1200 km. The INMS mixing ratios for major species should be far more accurate than absolute values of individual densities. Note that interstellar hydrogen ionizes and cuts off the stellar UV flux below 91.1 nm. Above this wavelength N_2 UV absorption occurs only in predissociated electronic bands that are very difficult to interpret at UVIS spectral resolution.

Beside absorptive occultations, there have been a few refractive stellar occultations (July 1989; 21 August 1995; 20 December 2001; two on 14 November 2003) mostly at visible wavelengths. These occultations were measured with ground-based telescopes, while the previously discussed ones were measured by onboard Cassini instruments. For the November 2003 events as representative, Sicardy et al. (2006) probed the 0.1–250 μbar region of Titan's atmosphere. However, the 0.1–1 μbar region is subject to initial conditions in their onion peeling analysis procedure and for pressures greater than 12 μbar (400 km), the stellar fluxes are attenuated by haze absorption in addition to the usual refraction effects. The retrieved temperature profile is in essential agreement with the HASI temperature profile and, in particular, the presence of abundant wave signatures and temperature gradients that exceed the adiabatic lapse rate. As noted above, there is a pronounced thermal inversion layer at 1.5 μbar (~507–515 km). From the central flash analysis which extracts latitudinal information, Sicardy et al. (2006) found an isopycnic surface equal to approximately the 250 μbar surface which rises about 50 km in altitude above an arbitrary north pole reference height when 45° N is reached and remains relatively constant all the way to the south pole. This is consistent with very strong stratospheric zonal winds in the winter northern hemisphere and much weaker zonal winds in the summer hemisphere, as found by CIRS (Achterberg et al. 2008)

10.3.4 In Situ Measurements

The first in-situ measurements of Titan's upper atmosphere were carried out by the Cassini INMS during the first targeted Titan flyby (TA) on 26 October 2004. This initial dataset has been considerably expanded with subsequent targeted flybys, of which a total of 19 offered the INMS adequate observing conditions and produced data of sufficient quality. The INMS measured gas densities during 13 flybys primarily in the northern hemisphere and during the remaining 6 flybys in the southern hemisphere. Altitudes of closest approach ranged from 950 km (T16) to 1175 km (TA). The INMS can measure either in the Closed Source Neutral (CSN) or Open Source Ion (OSI) mode to retrieve neutral or ion densities, respectively. Ion composition was measured only during 9 of the 19 flybys and neutral densities during all, though not always on both inbound and outbound parts of the flybys.

Using the neutral gas density measurements of N_2 and CH_4 in the northern hemisphere, Müller-Wodarg et al. (2008) constructed a simple empirical model of gas densities and temperatures in the northern hemisphere based on flybys TA-T32. During these flybys the solar declination angle on Titan changed from -23° to -11°, so the season was in transition between southern hemisphere summer conditions and equinox. The model by Müller-Wodarg et al. (2008) can hence be regarded as an average representation of Titan's winter hemisphere. The vertical range is from 1000 to 1600 km, covering altitudes that were sufficiently sampled by the INMS. In their statistical analysis of the INMS N_2 and CH_4 densities, Müller-Wodarg et al. (2008) found no consistent trends with longitude or local time, which however may partly be due to uneven sampling which favored polar nightside or low latitude dayside measurements, primarily in Titan's Saturn-facing longitude sector (Cui et al., 2008). A clear latitudinal trend was detected in the INMS densities at fixed height levels above around 1100 km (Fig. 10.4), with densities increasing towards the equator by around 70%. The density model was used to infer thermospheric temperatures in Fig. 10.4 which decrease with height from 149 ± 10 K at 1000 km to 140 ± 13 K near 1600 km, and with latitude below 1200 km with values near 1000 km from 164 ± 6 K at 20°N to 131 ± 6 K near 80°N.

While in-situ observations by the INMS have been made in the southern hemisphere during 9 passes, the statistical variation between these has been too large to derive a consistent latitudinal trend. Preliminary analysis of the structure of the southern hemisphere suggests temperatures to be warmer by ~10–15 K there than average values in the northern hemisphere, as expected from solar heating.

While the INMS has provided the most comprehensive set of observations yet in Titan's thermosphere, one uncertainty remains in the absolute calibration of densities. Comparison of INMS densities with those by the HASI observations at equatorial latitudes near 1000 km latitude shows INMS values to be 2.4 times smaller than the HASI values. The Cassini Attitude and Atriculation Control Subsystem (AACS) detects torques on the spacecraft as it enters Titan's upper atmosphere on each flyby, providing an additional independent measurement of total density. Comparison of AACS-derived densities at 1000 km and those from the INMS show INMS densities to again be smaller by an average factor of 2.6 than those from the AACS, very similar to the discrepancy factor between HASI and INMS at that altitude. This unresolved discrepancy however does not affect the inferred temperatures.

10.4 Interpretation of Atmospheric Temperature Structure

10.4.1 Radiative Budget of Troposphere and Stratosphere

Solar energy is absorbed in Titan's atmosphere through methane and haze absorption and at the surface. The Descent Imager/Spectral Radiometer (DISR) aboard the Huygens probe measured the downward and upward fluxes of sunlight at wavelengths from 350 to 1600 nm and altitudes from 150 to 0 km. These measurements have been analyzed to derive the vertical distribution and optical properties of haze aerosols (see Chapter 12), the absorption coefficients of methane in Titan's conditions, and the solar energy deposition profile at the latitude of the Huygens landing (10°S) (Tomasko et al. 2008). Averaged over a Titan day at the location and season of the Huygens landing, about 78% of the incoming sunlight is overall absorbed: ~11% at the surface, ~48% in the atmosphere below 150 km and ~19% above. The disk-averaged solar flux profile, calculated with the Huygens haze model, is shown in Fig. 10.5. It is in remarkable agreement with the earlier estimate of McKay et al. (1989) based on ground-based albedo and Voyager data.

Radiative cooling of the atmosphere occurs through thermal emission by haze particles and gases beyond ~7 µm (molecular bands and collision-induced opacity). Tomasko et al. (2008) calculated the radiative cooling rate profile at latitudes and time close to the Huygens landing using Cassini/CIRS nadir and limb measurements. As shown in Fig. 10.5, collision-induced opacity is the main source of atmospheric cooling below 80 km while, above, molecular

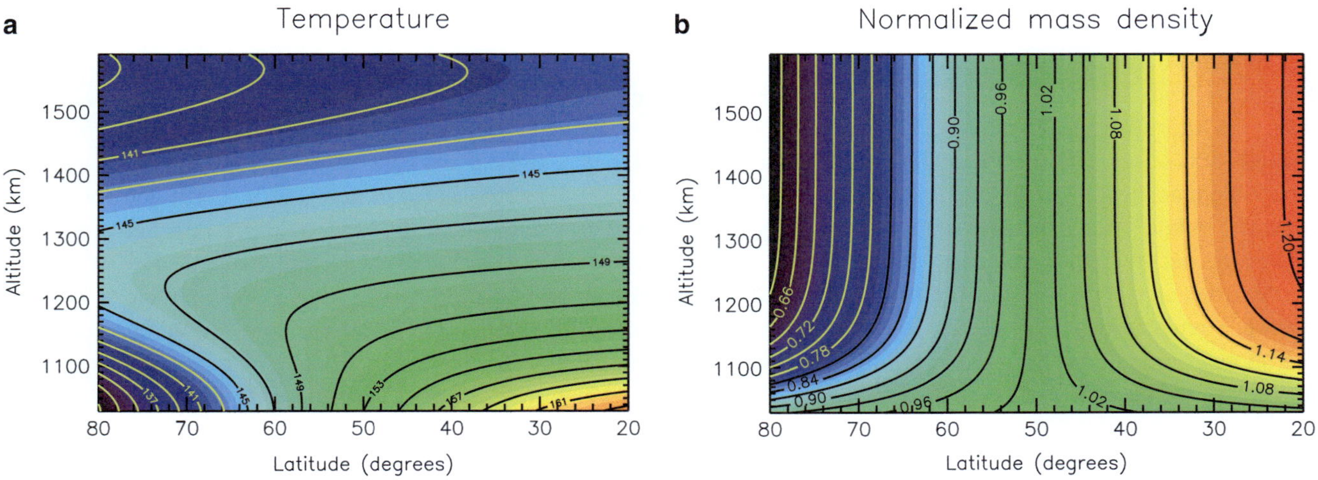

Fig. 10.4 (**a**) Temperatures in Titan's thermosphere inferred from an empirical model of atmosphere densities measured by INMS in the Northern Hemisphere. While the atmosphere is nearly isothermal above around 1200 km altitude, temperatures below 1100 km increase towards the equator. (**b**) Normalized Mass Densities at each height to the value at 50N from the empirical model. Note the bulge at the equator (after Müller-Wodarg et al. (2008))

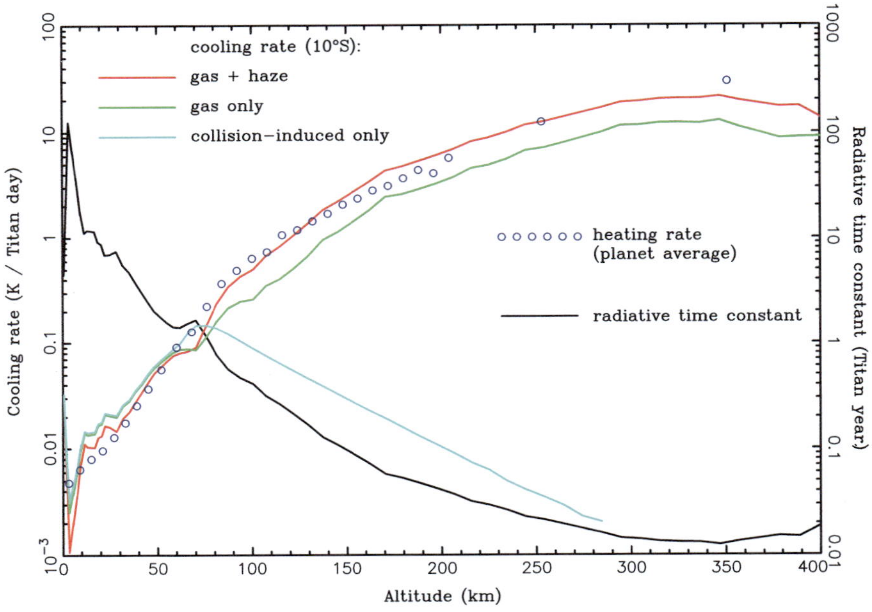

Fig. 10.5 Cooling rate profiles calculated around 10°S from Cassini/CIRS and Huygens data are compared with the solar heating rate profile, averaged over the planet, calculated from Huygens/DISR measurements (Tomasko et al. 2008). Cooling rates are best constrained in the ranges 10–65 km and 90–400 km and heating rates below 200 km. The radiative time constant (defined here as the temperature divided by the cooling rate) is also shown (*right y-scale*)

band and haze opacities dominate and are of comparable importance. The cooling rate reaches 10–20 K per Titan day at altitudes 250–400 km and decreases rapidly at lower levels. Dividing the temperature by the cooling rate provides a radiative time constant, also displayed in Fig. 10.5. It is short in the upper stratosphere and mesosphere (~0.5 Earth year, i.e. 0.015 Titan year, in the range 300–400 km), reaches a Titan year (29.5 Earth years) around 80 km, and is about 10 Titan years at 20 km. The cooling rate profile is close to that determined by Bézard et al. (1995) at 53°S from Voyager data at northern spring equinox. In the region best constrained by the Voyager/IRIS measurements (140–250 km), the two profiles agree within 20%.

The CIRS-derived cooling rate profile around 10°S is very close to the *disk-averaged* heating rate profile based on the Huygens haze model. Both decrease by three orders of magnitude from 350 to 10 km and agree at all levels within 50%. On the other hand, the solar heating rate at 10°S exceeds the cooling rate at all altitudes below 160 km, with a maximum net heating of about 0.5 K per Titan day located around 120 km. The general circulation likely redistributes this excess heat to higher latitudes.

Bézard et al. (1995) have shown that the latitudinal variations of composition strongly affect the radiative forcing in the stratosphere. At 50°N, the cooling rates derived from Voyager in the range 120–270 km (5–0.15 mbar) were some 20–40% larger than at 53°S despite the colder temperatures (7 to 15 K), a consequence of larger concentrations of infrared emitters (gas and particles). But, as the heating rate was also larger, radiative balance was still approximately achieved at this location within error bars. Using a general circulation model, Lebonnois et al. (2003b) also pointed out the strong enhancement of radiative cooling at high winter latitudes induced by the larger concentrations of some gases (C_2H_6, C_2H_2 and HCN). This thermal effect was found similar to that due to the coupling between aerosols and the general circulation (Rannou et al. 2002) in the mesosphere but smaller in the stratosphere around 1 mbar (190 km).

10.4.2 Radiative Processes in the Upper Atmosphere

Above 400 km (~0.03 mbar) non-LTE effects become important for hydrocarbon cooling rates as shown by Yelle (1991). Above 700 km (~0.1 μbar) he showed that HCN LTE rotational line cooling dominates radiative cooling to the exobase. The dominant heating processes above 400 km are near-IR CH_4 heating by absorption of solar radiation in the 1.7, 2.3, and 3.3 μm bands. Above ~650 km solar UV CH_4 heating emerges as the principal heat source and above 900 km, the thermosphere is, to zeroth order, in radiative equilibrium where solar EUV/UV heating is balanced by HCN rotational line cooling (Yelle 1991). Without HCN cooling calculated thermospheric temperatures would far exceed the observed values.

Yelle's models consistently predicted a pronounced mesopause at 600 km (~0.5 μbar) with ~25 K temperature drop from the 300 km temperature, whereas HASI data suggested the mesopause is at 494 km (Fulchignoni et al. 2005) with a temperature minimum of 153 K to be compared with ~185 K at 300 km. In Fig. 10.2 the Yelle et al. engineering model is representative of Yelle (1991) models and can be compared with the HASI profiles. As noted above the steep temperature gradient in the HASI temperature profile above 494 km and inferred by Sicardy et al. (2006) from stellar occultations requires solar heating of the detached haze layer detected by Porco et al. (2005). Above this haze layer, Liang et al. (2007b) present evidence that the aerosols extend up to 1000 km. Heating associated with discreet haze layers and the broad distribution of aerosols above the visible disk of Titan were not included in the Yelle (1991) model, but were included in Lavvas et al. (2009), who used the Yelle computer code and found that the net effect was to raise the mesopause by ~50 km without changing its temperature, in farther disagreement with the HASI inferred mesopause at 494 km.

10.5 Composition

This section emphasizes the observational facts and divides the composition into four major categories. The major constituents, N_2, CH_4, H_2, and the inert, noble gases all have long chemical time constants and hence are expected to be well mixed throughout the homosphere, which extends from the surface to about 800–850 km (Yelle et al. 2008). The one exception to being well-mixed is CH_4, which is condensable in the troposphere. Minor constituents are subdivided into three main classes: pure hydrocarbons, nitriles, and molecules containing oxygen.

10.5.1 Major Constituents and Inert, Noble Gases

From the Voyager Mission, the major constituents of the atmosphere were known. But there was considerably uncertainty in the CH_4 mixing ratio: its variation in the troposphere, its magnitude in the stratosphere, and whether the tropopause region was an effective cold trap. The Huygens Gas Chromatograph Mass Spectrometer (GCMS) provided definite answers to these questions (Niemann et al. 2005). The tropospheric CH_4 mixing ratio (mole fraction) is variable with a maximum value just above the surface of 0.0492 and decreases gradually to initially an asymptotic value of 0.0162 at 32 km, and then more slowly to its measured stratospheric value of 0.0141 as shown in Fig. 10.6. In addition to GCMS, the Huygens Descent Imager/Spectral Radiometer (DISR) measured the surface CH_4 mixing ratio by near-IR spectroscopy to be 0.051 ± 0.008, in agreement with GCMS (Jacquemart et al. 2008). The CIRS inferred stratospheric value of 0.016 ± 0.005 (Flasar et al. 2005) is consistent with the in situ measurement.

The third most abundant species is H_2, for which GCMS has not reported a value. From the Voyager IRIS measurements the H_2 inferred mixing ratio was 0.00112 ± 0.00016 by Samuelson et al. (1997) and 0.001 ± 0.0004 by Courtin et al. (1995). CIRS measurements yield the same mixing ratio: 0.00096 ± 0.00024 southward of 40° N (Courtin et al. 2008). These values are appropriate for the troposphere/tropopause region. At very high altitudes INMS finds the H_2 mixing ratio to be 0.00405 ± 0.00003 (Waite et al. 2005). Given the large separation distance (~1000 km) between these two independent measurements, the factor of 4 difference would not seem significant except that H_2 tends to be essentially uniformly mixed throughout the atmosphere due to its long chemical lifetime and its enormous escape rate out of the atmosphere (1.1×10^{28} s^{-1}, Cui et al.; 2008) which is essentially equal to the maximum possible rate. This small positive gradient in the H_2 mixing ratio is not produced in photochemical models to date and remains a challenge to modelers.

The ^{40}Ar isotope is an important tracer used to determine vertical mixing in the atmosphere and the location of the homopause (cf. Yelle et al. 2008). The measured tropospheric mixing ratio by Huygens GCMS is $(4.32 \pm 0.1) \times 10^{-5}$

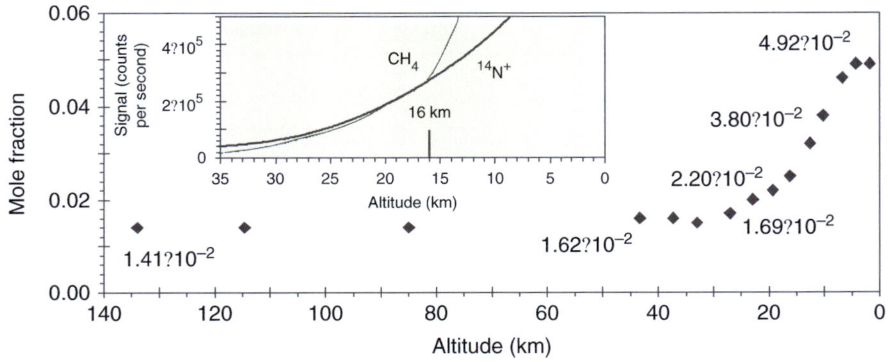

Fig. 10.6 The mole fraction of CH_4 to N_2 versus altitude. The CH4 mole fraction is 0.0141 in the stratosphere., at ~8 km, it reached a plateau of about 0.049. The inset shows an increase of CH_4 at 16 m/z, when compared to N_2 (in this case $^{14}N^+$) at m/z = 14, near 16 km. This is probably due to condensates evaporating in the inlet system of the mass spectrometer as the Huygens probe passed through the methane haze (after Niemann et al. (2005))

(Niemann et al. 2005), whereas in the thermosphere INMS measured $(7.1 \pm 0.1) \times 10^{-6}$ (Waite et al. 2005). Primordial ^{36}Ar is much less abundant, only $(2.8 \pm 0.3) \times 10^{-7}$ (Niemann et al. 2005). Other noble gases, Ne, Kr, and Xe, were not detected by the Huygens GCMS. The Huygens GCMS placed an upper limit of 10^{-8} on Kr and Xe, well below their cosmogonic abundance, as was also the very low, constraining value for ^{36}Ar. The $^{14}N/^{15}N$ isotope ratio measured in situ in N_2 by the GCMS (183; Niemann et al. 2005) is a factor three larger than that inferred from HCN (Vinatier et al. 2007b). The implications of these isotopes for the evolution of Titan's nitrogen atmosphere are discussed in detail in Chapter 7.

10.5.2 Minor Constituents – Hydrocarbons Other than Methane

Voyager/IRIS observed a suite of hydrocarbons produced from methane photochemistry and mapped their abundances in the lower stratosphere between 50°S and 70°N (Coustenis and Bézard 1995). Cassini/CIRS improved over the IRIS observations thanks to a higher spectral resolution (up to 0.5 cm^{-1}), a broader spectral range (10–1500 cm^{-1}), the use of limb-viewing geometry, and a much more extended spatial and temporal coverage (Fig. 10.7). Nadir measurements provide information on the mean gas abundances in a broad region usually centered around 120 km while limb-viewing measurements probe the atmosphere up to ~500 km.

The hydrocarbon mixing ratios in the lower stratosphere have been derived by Coustenis and Bézard (1995), Flasar et al. (2005), and Coustenis et al. (2007) from IRIS and CIRS nadir spectra respectively (Fig. 10.8). Ethane (C_2H_6) is the most abundant photochemical product (~10 ppm at 120 km), followed by acetylene (C_2H_2), propane (C_3H_8), ethylene (C_2H_4), methyl acetylene (CH_3C_2H), diacetylene (C_4H_2) and benzene (C_6H_6, which was not detected in IRIS spectra). The mixing ratio decreases with the complexity of the molecule as would be expected and, for a given number of C-atoms, saturated species are more abundant that unsaturated ones. All hydrocarbons are more abundant northward of 45°N. This enrichment is larger for C_4H_2 and CH_3C_2H than for the more stable species C_2H_6 and C_2H_2. It is more pronounced in the Voyager data recorded in 1980, a season close to northern spring equinox, than observed by Cassini in 2005 shortly after winter solstice (Figs. 10.1 and 10.8). In contrast, images taken in the period 1999–2002 with the Keck telescope at 8–13 μm suggest an accumulation of C_2H_4 in the polar stratosphere south of 60°S (Roe et al. 2004). If real, this polar enhancement, persisting through late southern Spring, has rapidly disappeared as it was not seen by Cassini three years later, i.e. one tenth of a Titan year later.

Vertical profiles have been inferred from two CIRS sequences combining nadir and limb spectra at 15°S and 80°N (Vinatier et al. 2007a) and subsequently at other latitudes (Vinatier 2007; Teanby et al. 2008b). Information is available from about 500 km down to 100–150 km. At southern and mid northern latitudes, all mixing ratio profiles, except ethylene, increase with height (Fig. 10.9). This results

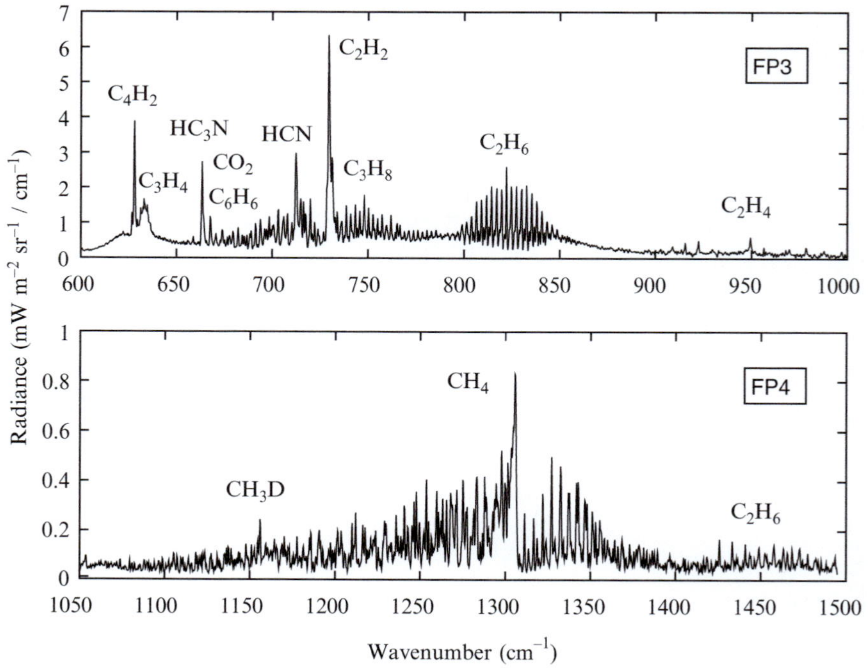

Fig. 10.7 An average of Cassini/CIRS limb spectra recorded at 100–200 km altitude and high northern latitudes, showing emission features from various hydrocarbons and nitriles. To analyze the observations, the temperature profile is first retrieved from the 1305 cm^{-1} (7.7 μm) methane band and the gas mixing ratio profiles are derived from the corresponding emission bands (from Bézard 2009))

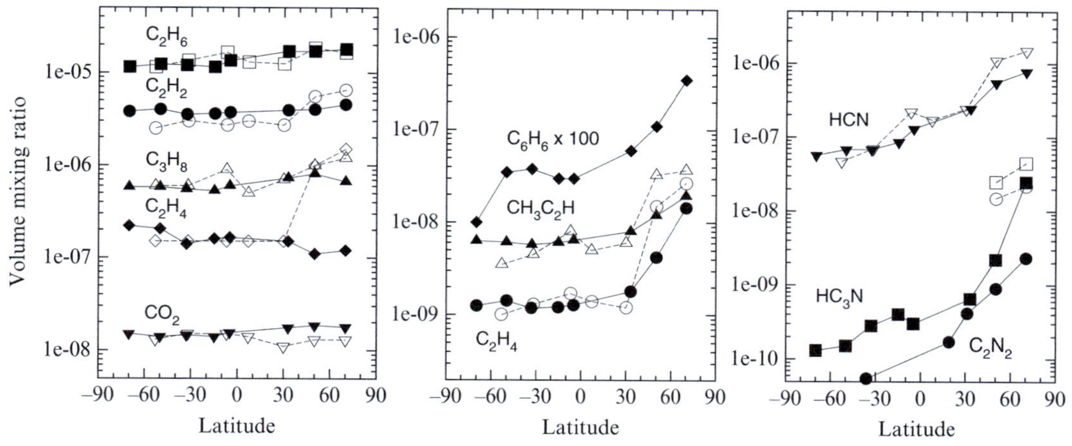

Fig. 10.8 Meridional variation of composition in the lower stratosphere around 120 km. Filled symbols and solid lines represent data from Cassini/CIRS spectra in 2005. Empty symbols and dashed lines correspond to mixing ratios derived by Voyager 1 in November 1980 (from Bézard 2009)

from photochemical production in the upper atmosphere (>500 km) and loss by condensation in the lower stratosphere (60 to 100 km depending on the species), which together maintain a positive concentration gradient. This gradient is moderate for C_2H_6 and C_2H_2 and steepest for C_4H_2 which undergoes chemical losses in the stratosphere (see Fig. 10.9). Ethylene is the only species whose mixing ratio decreases with height between 120 and 250 km, a behavior that may result from the fact that it does not condense and can be enriched in the lower atmosphere by equatorward transport of polar air (Crespin et al. 2008). At high northern latitudes, beyond 45°N, the vertical profiles change drastically and exhibit larger concentrations in the lower stratosphere at least below 250–300 km (Figs. 10.9 and 10.10). At 50–60°N, concentration minima are seen at 350–400 km for C_4H_2 and C_2H_2. These minima also exist at ~80°N but seem to occur at somewhat lower altitudes (~300 km) and be less pronounced. At this latitude (80°N), minima are also observed for CH_3C_2H, C_2H_4 and C_2H_6 as shown in Fig. 10.9. A similar pattern exists for nitriles and likely results from dynamical processes associated with the polar vortex as discussed in Sec. 10.5.3. Little is known on the vertical profile of benzene, the least abundant hydrocarbon detected: at 79°N, its mixing ratio is constant within error bars (3–4 ppb) between 180 and 320 km (Vinatier 2007), i.e. about 10 times larger than at mid-latitudes and ~120 km (Coustenis et al. 2007).

Above 500 km, the atmospheric composition can be probed by stellar or solar occultations observed from spacecraft. A reanalysis of the Voyager 1 UVS solar occultations at latitudes of 4°N (ingress) and 16°S (egress) provided density profiles of two hydrocarbons besides methane: C_2H_2 and C_2H_4 (Vervack et al. 2004). The C_2H_2 mixing ratio agrees very well with that derived from Cassini/CIRS at 15°S (Fig. 10.9) while the C_2H_4 value is about 2 orders of magnitude larger than the CIRS value at 240 km, suggesting that a minimum exists in its mixing ratio profile somewhere between 250 and ~450 km. Above 500 km, the mixing ratios of C_2H_2 and C_2H_4 increase with height to reach respectively $(0.6–2) \times 10^{-3}$ and $(0.3–2) \times 10^{-3}$ at 950 km, pointing to a source at higher altitudes. A stellar occultation observed by Cassini/UVIS at 36°S provided density profiles of several hydrocarbons between ~1000 km and a lower limit ranging from 615 to 845 km (Shemansky et al. 2005). The C_2H_4 number densities inferred between 845 and 1000 km agree within error bars with those from Voyager 1 UVS. On the other hand, the C_2H_2 number density at 615 km is about 20 times larger than the UVS value at the same altitude and the corresponding mixing ratio (assuming a CH_4 mole fraction of 0.014) is about 50 times larger than the CIRS value at 500 km, which casts doubts on the reported UVIS results.

At higher altitudes, the composition of the thermosphere and ionosphere has been probed in situ by the INMS aboard Cassini. During the first low-altitude pass through Titan's atmosphere at ~39°N latitude (Ta), INMS detected the neutrals C_2H_2, C_2H_4, C_2H_6, C_3H_4, C_4H_2 and C_6H_6 around 1200 km. Benzene has been detected by INMS in many passes through Titan's atmosphere. An analysis of the signals recorded during T16 indicates a mole fraction of 1.3×10^{-6} at 950 km near the North Pole (Vuitton et al. 2008). This implies an ionospheric source of about 10^7 molecules cm^{-2} s^{-1}, of the same magnitude as the neutral production rate in the stratosphere needed to explain the CIRS abundance. Measurements of ion densities coupled with simple chemical models provide an additional probe of the composition of the neutral atmosphere. This allowed the detection of C_2H_4, polyynes (C_4H_2, C_6H_2, C_8H_2) and possibly methylpolyynes (CH_3C_4H, CH_3C_6H) and a determination of their mole fractions around 1100 km at 74°N during the T5 flyby (Vuitton et al. 2006, 2007). A review of the composition of Titan's thermosphere and ionosphere can be found in Chapter 11.

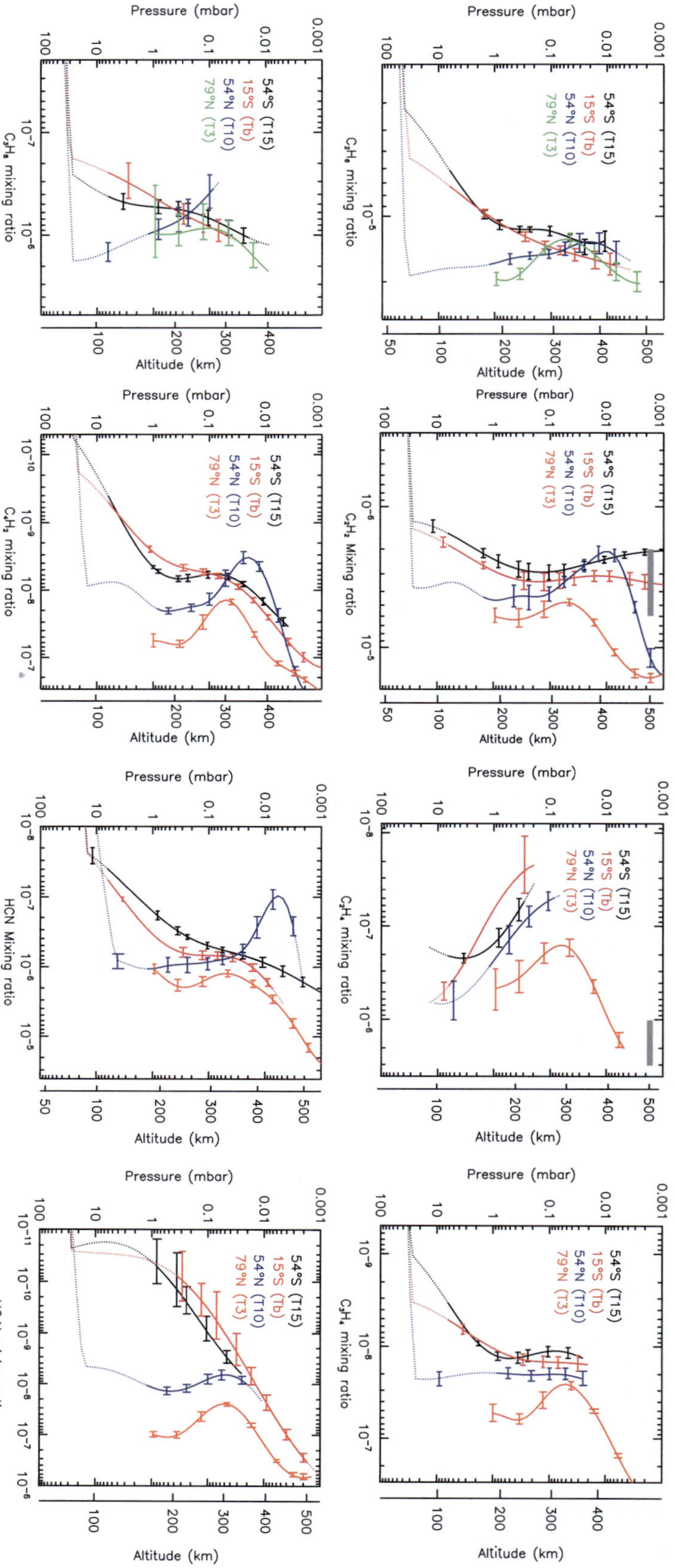

Fig. 10.9 Vertical profiles of hydrocarbons and nitriles obtained from inversion of Cassini/CIRS nadir and limb spectra recorded at four latitudes (54°S, 15°S, 54°N, and 79°N). The error bars are indicated at the maxima of the inversion kernels, which is the level of maximum information in the spectral sequences analyzed. The thick grey segment represents the Voyager 1 UVS measurement at 4°N (adapted from Vinatier 2007)

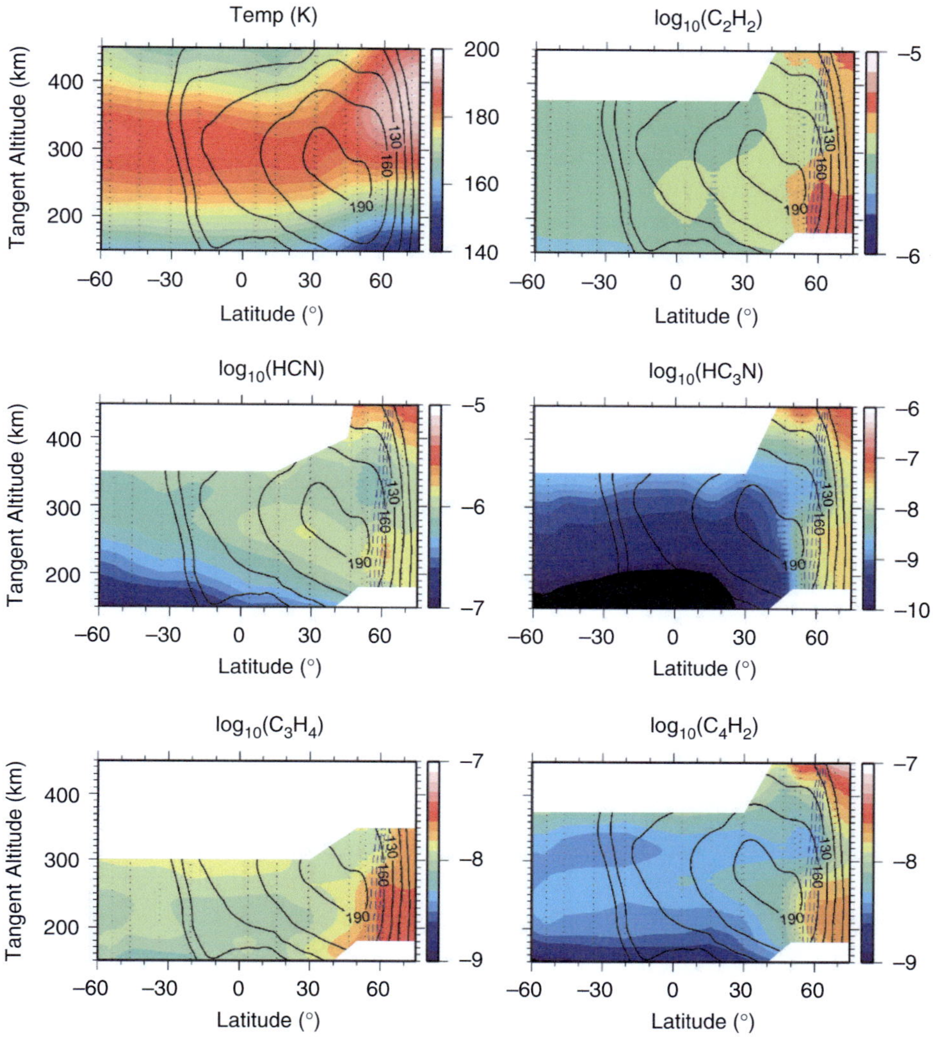

Fig. 10.10 Cross sections of temperature and composition through Titan's atmosphere constructed from Cassini/CIRS observations in 2006–2007. Composition is given as a volume mixing ratio and the position of the observed profiles are denoted by dotted lines. Contours indicate the vortex zonal wind speeds (in m s^{-1}) and blue dashed lines show the region with the steepest horizontal potential vorticity gradient, which indicates a dynamical mixing barrier (from Teanby et al. 2008)

10.5.3 Minor Constituents – N-Bearing Species and Nitriles

Voyager/IRIS and Cassini/CIRS detected three nitriles in Titan's stratosphere: hydrogen cyanide (HCN), cyanoacetylene (HC_3N) and cyanogen (C_2N_2) (Coustenis and Bézard 1995; Teanby et al. 2006). Mixing ratios derived from nadir spectra are shown in Fig. 10.8 as a function of latitude, assuming uniform profiles above the condensation levels. HCN, the simplest and most abundant nitrile, is detected at all latitudes. Its mixing ratio is fairly constant southward of 20°S (~1 × 10^{-7} around 150 km) and gradually increases towards the north. The north-to-south asymmetry is more dramatic for the two other species. In particular, the C_2N_2 signature is extremely weak in CIRS spectra of the southern hemisphere and equatorial region and indicates a mixing ratio as low as 6 × 10^{-11} (Teanby et al. 2009). On the other hand, it is clearly detected at high northern latitudes with a mixing ratio of 9 × 10^{-10} at 50°N and around 3 × 10^{-9} beyond 70°N. As for hydrocarbons, the enhancement of nitriles north of ~45°N was more pronounced at the Voyager encounter time than at the beginning of the Cassini mission.

Vertical profiles of HCN and HC_3N have been derived using CIRS limb observational sequences in the range 100–500 km at a variety of latitudes (Teanby et al. 2007; Vinatier 2007). From southern to mid-northern latitudes (<40°N), the HCN mole fraction increases regularly with height (by a factor of ~30 from 100 to 400 km) while HC_3N exhibits a much steeper gradient, increasing by at least 3 orders of magnitude between 200 and 500 km. The HC_3N profile is consistent with those derived from ground-based millimeter and submillimeter heterodyne observations (Marten et al. 2002; Gurwell 2004) that do not resolve Titan's disk and are thus more heavily weighted towards mid-latitudes (30°S–30°N). At latitudes northward of 45°N, the profiles are more uniform below 250–300 km while concentration minima are observed around 350–400 km for HCN and 320 km for HC_3N. The HCN minimum is more pronounced (and slightly higher) at 50–60°N than at 80°N, leading to an HCN-depleted zone limited in latitude and altitude, as shown in Fig. 10.10 (Teanby et al. 2008b). A CIRS limb sequence at 82°N with high vertical resolution (10 km) has revealed a layered structure in the profiles of HCN and HC_3N, with variations of about an order of magnitude over a few tens of km (Teanby et al. 2007). This structure is reminiscent of the discrete haze layers seen in Cassini images (Porco et al. 2005) and might imply a close connection between hazes and nitriles.

Acetonitrile (CH_3CN) has only been observed from the ground (Marten et al. 2002). Its disk-averaged mole fraction is about 20–50 times lower than that of HCN in the 200–500 km range, slightly increasing with height from (0.8 to 4) × 10^{-8} in this interval. Below 150 km, it decreases sharply down to the condensation level around 80 km with a vertical gradient even steeper than for HC_3N.

The enrichment of hydrocarbons and nitriles in the stratosphere at winter latitudes seems to be roughly correlated with their vertical concentration gradient at low latitudes and anti-correlated with their lifetime in the stratosphere (e.g. Teanby et al. 2009; Bézard 2009). This is consistent with subsidence in the winter polar vortex, bringing air rich in photochemical compounds from their formation region in the upper atmosphere down to the stratosphere (see Sections 10.6 and 10.7 and Chapter 13). The polar vortex acts to isolate high winter latitudes but details of the hydrocarbon and nitrile distributions, such as the concentration minima observed at 350–400 km, provide evidence for dynamics more complex than predicted by general circulation models (Teanby et al. 2008b).

Above the region probed by thermal emission measurements, information is scarce. The Voyager UVS occultation data saw evidence for absorption by HCN and/or HC_3N but could not disentangle the two. The Cassini UVIS occultation at 36°S indicates a HCN mole fraction around 3×10^{-5} at 615 km increasing to 1×10^{-3} at ~1000 km (assuming a CH_4 mole fraction of 0.014) but these data may need to be reanalyzed (see Section 10.3.3). Cassini INMS detected in situ HCN, HC_3N and C_2N_2 around 1200 km at ~39°N with a mole fraction less than 5 ppm (Waite et al. 2005). Ion measurements conducted during a low-altitude pass at 1027 km near 74°N latitude were analyzed to determine indirectly the abundances of several N-bearing molecules (Vuitton et al. 2006, 2007). The HCN, HC_3N and CH_3CN mole fractions derived from INMS data are much larger than the CIRS or ground-based measurements at 500 km (about 10, 30 and 200 times respectively), which implies production in the upper atmosphere and transport or diffusion to lower levels. INMS data also provide the first evidence for the presence of cyanobutadiyne (HC_5N) at the ppm level, acrylonitrile (C_2H_3CN), propionitrile (C_2H_5CN), ammonia (NH_3), methyleneimine/methanimine (H_2CNH), probably methylcyanopolyynes (CH_3C_3N, CH_3C_5N) and two unidentified N-bearing species (C_5H_5N, C_6H_7N), which reveals the complexity of the chemistry taking place in the upper atmosphere (cf. Ch. 12).

10.5.4 Minor Constituents – Oxygen Compounds

Although Titan's atmosphere is a reducing environment, it contains oxygen compounds, three of which have been detected so far. Carbon dioxide, the first to be discovered (by *Voyager* 1 in 1980; Samuelson et al. 1983), is also the best characterized, as observations by *Voyager* 1, *Voyager* 2, *ISO*, and *Cassini* all give consistent results (see de Kok et al. 2007; Coustenis et al. 2007, and references therein). To within error bars, CO_2 appears to be uniformly mixed with latitude and altitude above its condensation level in the lower stratosphere and up to at least 200 km, with a $(1.5 \pm 0.4) \times 10^{-8}$ mixing ratio. CO_2 can be easily formed through reaction of CO and OH, itself produced from the photolysis of water. CO was first discovered from ground-based observations at 1.6 μm (Lutz et al. 1983), and extensively re-observed since then, especially at 5-μm (in reflectance spectroscopy, thermal emission, and non-thermal emission) and through its mm/submm rotational lines observed from the ground and recently from *Cassini* (see Hörst et al. 2008 for a review of all measurements). Although the early observations generated a large body of controversy as to the abundance and stratospheric vertical distribution of CO, the most recent measurements point to a mean, latitude-independent, mixing ratio of about 40 ppm, with no clear evidence for altitude variation from the surface to the upper atmosphere. For both CO_2 and CO, the spatial and vertical uniformity can be understood in terms of their long chemical lifetimes (typically a few thousand and 500 million years, respectively) with respect to the horizontal and vertical transport timescales. The third oxygen species, water vapor, has been detected by ISO (Coustenis et al. 1998). The H_2O column density is $2.6^{+1.9}_{-1.6} \times 10^{14}$ mol cm^{-2}, and the best-determined mixing ratio is $8^{+6}_{-4} \times 10^{-9}$ at 400 km, although it most likely increases with altitude. A preliminary assessment of the CIRS/*Cassini* spectra indicates that water is detected in these data (Bjoraker et al. 2008), but no estimates on the amount and distribution of water are available yet. Finally, upper limits on CH_3OH and CH_3C_2O are available from INMS/*Cassini* (Vuitton et al. 2007).

The presence of H_2O and CO_2 is a direct proof of an external source of water in Titan's atmosphere, and photochemical models designed to reproduce the observed abundance of all three oxygen species require reassuringly similar OH (H_2O) fluxes (Table 10.1). The essential unsolved question is the ultimate origin of this water influx, which possibly includes distant (micrometeorite ablation) and/or

Table 10.1 Water (OH), CO, and O^+ fluxes in several chemical models. Fluxes, φ, ($10^6\ cm^{-2}\ s^{-1}$) are referred to the surface

Reference	Target-molecule(s)	φ (OH)	φ (CO)	φ (O(³P))
Lara et al. (1996); Lara (1998)	CO_2, CO	3	0.83	–
Coustenis et al. 1998[a]	H_2O	1.3–4.5	–	–
Wilson and Atreya 2004	CO_2, H_2O	5	$(5x10^{-5})$[b]	–
Hörst et al. 2008[c]	CO_2, CO, H_2O	2–9	–	1–4

[a] From rescaling of Lara et al. (1996) model.
[b] Imposed mixing ratio.
[c] Including sensitivity to eddy diffusion profile.

local (sputtering of icy satellite or ring surfaces, Enceladus venting) sources.

The origin of CO is more uncertain. Early models (Samuelson et al. 1983; Yung et al., 1984; Toublanc et al. 1995; Lara et al. 1996) assumed that the formation of CO proceeds through $OH + CH_3 \rightarrow CO + 2\ H_2$ (i.e. that a flux of water *alone* could explain all three O-bearing compounds). However, even under this assumption, the OH fluxes required to sustain the observed amounts of H_2O and CO_2 fail to produce a ~ 40 ppm CO mixing ratio and the current chemical loss of CO exceeds its steady-state production by typically $(1–2) \times 10^6\ cm^{-2}\ s^{-1}$. The problem was exacerbated with the realization (Wong et al. 2002) that the $OH + CH_3$ reaction essentially recycles water, leading to only ~2 ppm of CO in equilibrium. Suggestions to solve this dilemma included: (i) alternate external sources, such as CO contained in micrometeorites (Lara et al. 1996) or delivered by cometary impacts (Lellouch et al. 2003), (ii) surface or internal sources such as oceanic evaporation (Dubouloz et al. 1989) or volcanic outgassing (Baines et al. 2006), and (iii) a CO abundance not in equilibrium and reflecting a larger primordial abundance (Wong et al. 2002; Wilson and Atreya 2004).

The recent discovery of O^+ ions precipitating into Titan's atmosphere (Hartle et al. 2006) provides a new source of oxygen and especially CO because through collisions with the atmosphere O^+ is ultimately converted into ground-state $O(^3P)$, which can react with CH_3 to produce CO, either directly ($O(^3P) + CH_3 \rightarrow CO + H_2 + H$) or through HCHO (Hörst et al. 2008). These authors showed that essentially all of the incoming O^+ is converted to CO, so that the observed O^+ influx rate of $\sim 10^6\ cm^{-2}\ s^{-1}$ can provide the observed CO abundance in ~300 million years and approximately balance the chemical loss of CO to CO_2. Studying the sensitivity of their model calculations to eddy diffusion profiles, Hörst et al. (2008) demonstrated that the abundances and profiles of CO, H_2O, and CO_2 can be fit simultaneously with realistic fluxes of OH (H_2O) and $O(^3P)$ (O^+) .

In Tables 10.2 and 10.3, a brief summary of Titan's stratospheric composition is given at two latitudes (15°S, 54°N) for the 5 mbar pressure level, where the altitude is 115–120 km. It should be kept in mind that the stratospheric composition varies considerably with altitude and latitude with few exceptions such as CH_4 and CO. The reader is strongly encouraged to consult Figs. 10.8–10.10 in conjunction with Tables 10.2 and 10.3.

Table 10.3 Stratospheric composition 54°N: mixing ratios (by volume) at approximately 115 km/5 mbar

Hydrocarbonspe		Nitriles		Oxygen compounds	
CH_4		HCN	0.9 ppm (1)	CO	
C_2H_6	16 ppm (1)	CH_3CN		CO_2	25 ppb (1)
C_2H_2	4 ppm[a] (1)	HC_3N	15 ppb[a] (1)	H_2O	
C_2H_4	0.6 ppm (1)	C_2N_2	1 ppb (3)		
C_3H_8	1.5 ppm (1)				
CH_3C_2H	20 ppb (1)				
C_4H_2	20 ppb[a] (1)				
C_6H_6	4 ppb[a] (1)				
H_2	0.0012 (2)				

[a] At 170 km (1 mbar).Source: (1) Vinatier (2007); (2) Courtin et al. (2008); (3) Teanby et al. (2009)

Table 10.2 Stratospheric composition 15°S: mixing ratios (by volume) at approximately 120 km/5 mbar.

Hydrocarbonspe		Nitriles		Oxygen compounds	
CH_4	0.014 (1)	HCN	0.1 ppm (2)	CO	47 ppm (7)
C_2H_6	10 ppm (2)	CH_3CN	20 ppb[a] (5)	CO_2	16 ppb (7)
C_2H_2	2 ppm (2)	HC_3N	1 ppb[a] (5)	H_2O	0.4 ppb (3)
C_2H_4	0.4 ppm (2)	C_2N_2	0.06 ppb (6)		
C_3H_8	0.5 ppm (2)				
CH_3C_2H	8 ppb (2)				
C_4H_2	1 ppb (2)			Noble	gases
C_6H_6	0.4 ppb (3)			^{36}Ar	0.28 ppm (1)
H_2	0.00096 (4)			^{40}Ar	43 ppm (1)

[a] At 300 km (disc-averaged).
Source: After Bézard 2009 with additional data points from Teanby et al. 2009; (1) Niemann et al. (2005); (2) Vinatier et al. (2007a); (3) Coustenis et al. (2007); (4) Courtin et al. (2008); (5) Marten et al. (2002); (6) Teanby et al. (2009); (7) de Kok et al. (2007)

10.5.5 Isotope Ratios

Numerous isotopic ratios have now been determined in Titan's atmosphere, either from in situ measurements (Huygens GCMS, Cassini INMS) or from spectroscopic observations at IR and mm wavelengths obtained from the ground and especially from Cassini/CIRS (see Table 10.4). Most isotopic ratios have been determined separately in several molecules, allowing one to study (i) formation and evolution processes of the atmosphere as a whole, as traced by major species and (ii) chemical fractionation effects in minor species. The interpretation of isotopic ratios in CH_4, N_2 and CO is discussed in detail in the Chapters 3 and 7. Briefly, the enhanced D/H in Titan's methane compared to the protosolar value – typically by a factor of 6 – is thought to reflect a combination of (i) an initially high D/H in the primordial icy material that formed Titan and (ii) fractionation processes at work in the atmosphere, such as methane photolysis and escape (Pinto et al. 1986; Lunine et al. 1999; Mousis et al. 2002; Cordier et al. 2008). As regards N_2, the factor ~1.5 enhancement in $^{15}N/^{14}N$ has been interpreted as due to non-thermal escape of N_2 (Lunine et al. 1999), and suggests that Titan's primitive atmosphere was 2–10 times thicker than nowadays (Niemann et al. 2005). A similar explanation, involving a massive loss of CO in Titan's early history, was put forward by Wong et al. (2002) to explain the factor of ~2 higher than terrestrial $^{18}O/^{16}O$ ratio reported by Owen et al. (1999). There is recent evidence, however, that the enrichment in ^{18}O is considerably milder (see Table 10.4). Finally, atmospheric escape is also qualitatively consistent with the fact that the $^{12}C/^{13}C$ ratio, as measured in CH_4, is slightly (by 8 %), but unambiguously, smaller than the terrestrial value of 89.

The $^{12}C/^{13}C$ ratio has been measured in many molecules besides CH_4, including CH_3D, C_2H_2, C_2H_6, HCN, HC_3N, CO,

Table 10.4 Isotopic ratios determined from indicated parent molecules and techniques for Titan's atmosphere compared with values for the Earth and the Sun

Isotope	Molecule	Technique	Titan value	Reference	Earth	Solar[e]
D/H	H_2	GCMS	$(2.3\pm0.5)\times10^{-4}$	Niemann et al. 2005	1.56×10^{-4}	1.9×10^{-5}
	CH_4	IR, ground	$(1.25\pm0.25)\times10^{-4}$	Penteado et al. 2005		
	CH_4	CIRS	$(1.17^{+0.23}_{-0.28})\times10^{-4}$	Coustenis et al. 2007		
	CH_4	CIRS	$(1.32^{+0.15}_{-0.11})\times10^{-4}$	Bézard et al. 2007		
	C_2H_2	CIRS	$(2.09\pm0.45)\times10^{-4}$ [a]	Coustenis et al. 2008		
$^{12}C/^{13}C$	CH_4	GCMS	82.3±1	Niemann et al. 2005	89	89
	CH_4	INMS	~81 [b]	Waite et al. 2005		
	CH_4	CIRS	76.6±2.7	Nixon et al. 2008a		
	CH_3D	CIRS	82^{+27}_{-18}	Bézard et al. 2007		
	C_2H_2	CIRS	84.8±3.2	Nixon et al. 2008a		
	C_2H_6	CIRS	89.8±7.3	Nixon et al. 2008a		
	$CH_4+C_2H_2+C_2H_6$ [c]	CIRS	80.8±2.0	Nixon et al. 2008a		
	HCN	mm, ground	108±20 [d]	Gurwell 2004		
	HCN	CIRS	75±12	Vinatier et al. 2007b		
	HC_3N	CIRS	79±17	Jennings et al. 2008		
	CO_2	CIRS	84±17	Nixon et al. 2008b		
$^{14}N/^{15}N$	N_2	GCMS	183±5	Niemann et al. 2005	272	440±60 [f]
	N_2	INMS	168–211 [b]	Waite et al. 2005		
	HCN	mm, ground	60–70	Marten et al. 2002		
	HCN	mm, ground	72±9 [d]	Gurwell 2004		
	HCN	CIRS	56±8	Vinatier et al. 2007b		
$^{16}O/^{18}O$	CO	mm, ground	~250	Owen et al., 1999	499	499
	CO_2	CIRS	346±110	Nixon et al. 2008b		
$^{36}Ar/^{40}Ar$	atoms	GCMS	$(6.5\pm0.8)\times10^{-3}$	Niemann et al. 2005	0.0034	3440

[a] From nadir observations. A slightly lower value $(1.63\pm0.27)\times10^{-4}$ is obtained from limb observations.

[b] Estimated in lower atmosphere from extrapolation of values measured in thermosphere (95.6 ± 0.1 for $^{12}C/^{13}C$ and 172–215 for $^{14}N/^{15}N$).

[c] Weighted-mean average.

[d] Assuming thermal profile from Lellouch (1990). Higher values (132 ± 25 for $^{12}C/^{13}C$ and 94 ± 13 for $^{14}N/^{15}N$) are found when the profile from Coustenis and Bézard (1995) is used.

[e] Lodders (2003).

[f] Proto-solar value, based on the Jupiter value (Owen et al. 2001).

and CO_2. To within error bars, all measurements are consistent with the value in CH_4, although for hydrocarbons Nixon et al. (2008a) report a possible trend for an increasing value with molecular mass. In contrast, the D/H ratio measured in H_2 (Niemann et al. 2005) is somewhat higher than in CH_4, which might result from fractionation at atmospheric escape. Fractionation by photolysis may also occur. Given the higher binding energy of the C-D bond compared to C-H, this would tend to enhance the D/H ratio in C_2H_2, as marginally observed (Coustenis et al. 2008), but decrease it in H_2 . Evidence for such an effect is best seen in HCN, in which $^{15}N/^{14}N$ is about three times higher than in N_2. As demonstrated by Liang et al. (2007a), this results from the self-shielding of $^{14}N^{14}N$ against photolysis, while $^{14}N^{15}N$, whose predissociation states are shifted from those of $^{14}N^{14}N$, remains optically thin to greater depths. For pure N_2 photolysis, this would in fact lead to a $HC^{14}N/HC^{15}N$ ratio as low as 23, considerably overpredicting the observed value of ~60. This implies an additional source of non-fractionated atomic N, attributed to ion/electron impact on N_2.

10.6 Sources, Sinks, and Photochemistry of Atmospheric Composition

The atmosphere of Titan contains many trace constituents in the form of hydrocarbons (C_xH_y), nitriles ($C_xH_yN_z$) and aerosols, which are products of the photolysis of methane and nitrogen in the upper atmosphere. Even before Voyager discovered this suite of constituents and that the atmosphere was primarily nitrogen, Strobel (1974) developed the basic framework of methane photochemistry in Titan's atmosphere and concluded that, in the absence of escaping hydrogen atoms and molecules being recaptured, methane would be irreversibly destroyed on a time scale short compared to Titan's age. Yung et al. (1984) developed the first comprehensive post-Voyager model of Titan photochemistry that explained satisfactorily the limited composition data from Voyager.

The photochemistry of CH_4 occurs in two distinct, broad altitude/pressure regions. When coupled with the high efficiency for hydrogen escape, it leads to irreversible loss of CH_4. In addition, CH_4 may be escaping the atmosphere at large rates (cf. Yelle et al. 2008; Strobel 2008, 2009). The lifetime for atmospheric CH_4 (for column density ~7.5×10^{24} cm^{-2}) is only ~45 MY when its loss rate is constrained by its limiting flux through the lower stratosphere and ~15 MY when it is constrained by the largest model chemical loss rate (see Table 10.5 and discussion below) and implies that it must be continually resupplied from the interior. (Table 10.5 below provides a convenient list of all loss rates and fluxes discussed in this section for easy reference.) Direct CH_4 photolysis is driven principally by the intense solar Lyman α line at 121.6 nm, because UV radiation is only absorbed by CH_4 with a significant cross section below 145 nm, even though only 4.5 eV (= 285 nm) is needed to break CH_4 apart. This photolysis is centered at pressures ~7 nbar (~825 km) and in photochemical models contributes a net destruction rate of ~2.9×10^9 cm^{-2} s^{-1} (e. g. Yung et al. 1984). The radicals released in CH_4 photolysis react to form initially C_2H_y hydrocarbons (C_2H_6, C_2H_4, C_2H_2). Methane can also be destroyed by indirect catalytic C_2H_2 dissociation, whereby the radicals C_2H and C_2 dissociate CH_4 and recycle C_2H_2. This process is most important in the 0.1–1 mbar region (190–300 km) and contributes a loss rate of ~1×10^{10} cm^{-2} s^{-1}, significantly larger than the direct rate. To a lesser extent, methane is also destroyed by ion chemistry, with two CH_4 molecules dissociated per ion pair created. From Wilson and Atreya (2004) their globally averaged, integrated electron production rate times 2 is ~3.6×10^8 cm^{-2} s^{-1} or when referenced to the surface ~7×10^8 CH_4 cm^{-2} s^{-1}, are destroyed.

Thus an upward flux of CH_4 driven by photochemistry, must be balanced by downward fluxes of more complex, C_2, C_3, C_4, etc., hydrocarbons carrying the same total number of

Table 10.5 Comparison of various rates for CH_4 photochemistry, escape fluxes from, and limiting diffusion fluxes through Titan's atmosphere

Process	Rates referenced to surface (cm^{-2} s^{-1})
H_2 escape flux at exobase	Y: 7.2×10^9; WA: 5.0×10^9; L: 4.2×10^9; C: 1.4×10^{10}; S9: 1.1×10^{10}
H escape flux at exobase	Y: 5.5×10^9; WA: 3.0×10^9; L: 9.3×10^8
CH_4 escape flux at exobase	Yelle08: $(2.5–3) \times 10^9$; S9: 2.1×10^9
Ionospheric chemical CH_4 loss[a]	WA: ~7×10^8
Direct (Ly α) CH_4 dissociation[a]	Y: 2.9×10^9
C_2H_2 catalytic CH_4 dissociation[a]	Y: 1.1×10^{10}
Total net CH_4 chemical loss[a] and gas phase C atom flux	Y: 1.5×10^{10}; WA: 8.7×10^9; L: 1.3×10^{10}
Downward C_2H_6 flux at tropopause	Y: 5.8×10^9; WA: 2.2×10^9; L: 4.1×10^9
Downward C_2H_2 flux at tropopause	Y: 1.2×10^9; WA: 5.9×10^8; L: 3.2×10^8
Downward C_3H_8 flux at tropopause	Y: 1.4×10^8; WA: 2.9×10^8; L: 3.7×10^8
Downward mass flux of aerosols	L: 1.5×10^{-14}; B: 1.5×10^{-14}; M: $(0.5–2) \times 10^{-14}$ g cm^{-2} s^{-1}
Downward C atom flux in aerosols	L: 3.8×10^8
Downward N atom flux in aerosols	L: 3.0×10^8
Downward H atom flux in aerosols	L: 2.9×10^8
Limiting flux of H_2 (@0.4%)	S08: 1.3×10^{10}

Y: Yung et al. (1984); WA: Wilson and Atreya (2004); C: Cui et al. (2008); L: Lavvas et al. (2008); S8: Strobel (2008); S9: Strobel (2009); Yelle08: Yelle et al. (2008); B: Bézard (2009); M: McKay et al. (2001).

[a] Rate = $\int_{r_0}^{r_{ex}} L(r)(r^2/r_0^2)dr$, r_0, r_{ex} = Titan's radius, exobase.

carbon atoms. All of these less saturated hydrocarbons condense as liquids or solids in the lower stratosphere and vicinity of Titan's cold tropopause (~70 K), with the exception of C_2H_4, to form a pervasive haze layer. Eventually they precipitate from the atmosphere and accumulate on the surface, although to date the precise composition, location, and depth of these end products of CH_4 photolysis is still under intense study by Cassini scientists (cf. Chapter 6). The most abundant photolysis product is C_2H_6 (cf. Figs. 10.8–10.10), which is expected to undergo condensation in the stratosphere 40–50 km above the surface.

The tentative detection of benzene (C_6H_6) at ppbv levels by ISO (Coustenis et al. 2003), added a new dimension to Titan photochemistry, as benzene is the first ring molecule that is generally a precursor to a whole host of heavier hydrocarbons including polycyclic aromatic hydrocarbons (PAH's). Aerosols can form upon polymerization of the PAH's. The only aerosols predicted prior to the detection of benzene were from polyyne ($C_{2n}H_2$) polymers, HCN polymers, and the condensation of stable C_2–C_4 hydrocarbons in the lower stratosphere. Wilson and Atreya (2004) developed a comprehensive post-ISO model of the coupled chemistry of the neutral atmosphere and the ionosphere. A simplified chemical scheme for the neutral atmosphere from Atreya et al. 2006 is shown in Fig. 10.11. Waite et al. (2007) reported molecules as large as tens of thousand daltons may be present at ionospheric heights, which are probably precursors to aerosol formation in the thermosphere. Cassini UVIS has detected aerosols from 1000 km down to the main haze layer (Liang et al. 2007b).

Synthesis of complex hydrocarbons can be defeated by H-atom cracking reactions: e. g., $H + C_{2n}H_2 + M \rightarrow C_{2n}H_3 + M$, $H + C_{2n}H_3 \rightarrow C_2H_2 + C_{2n-2}H_2$. Since the main Titan haze layer is composed of large, condensed molecules with most of the mass contributed by hydrocarbons, Titan must somehow limit the efficiency of these H-atom cracking reactions. The haze particles may play a fundamental role in this process by suppressing the H atom concentration via heterogeneous reactions which recombine H atoms on aerosol surfaces and release H_2 gas (cf. Lebonnois et al. 2003a; Sekine et al. 2008a, b). These heterogeneous reactions are more than competitive with the cracking reactions that are slow at Titan's pressures and temperatures.

The key observational facts constraining CH_4 photochemistry are (1) the INMS inferred H_2 escape rate of (1.1–1.4) × 10^{10} cm^{-2} s^{-1} (Cui et al. 2008; Strobel 2009), (2) the most abundant photochemically produced gas phase hydrocarbon molecule is C_2H_6 followed by C_2H_2 (cf. Figs. 10.8, 10.10), (3) the stratospheric CH_4 mole fraction is 0.0141 from GCMS (Niemann et al. 2005), and (4) the thermospheric CH_4 mole fraction obtained by INMS at the lowest altitudes sampled is slightly lower than the stratospheric value (Yelle et al. 2008) and indicative of CH_4 destruction. Bézard (2009) inferred the mass production rates of C_2H_6 and haze from CIRS derived C_2H_6 and DISR derived haze profiles under the assumption that diffusive transport dominates below their respective formation regions. He obtained a production ratio of 15 times more C_2H_6 than haze.

From a large number of photochemical model calculations, three were selected to illustrate mass balance and partitioning of photochemical products: (1) Yung et al. (1984) for comprehensive post-Voyager analysis, (2) Wilson and Atreya (2004), and (3) Lavvas et al. (2008) for contemporary Cassini–Huygens analysis. Table 10.5 gives relevant magnitudes of CH_4 destruction rates, and downward, escape, and limiting fluxes that pertain to CH_4 photochemistry. In spite of the 24 year time span and improvement in input quantities and more observational constraints, Yung et al. (1984) and Lavvas et al. (2008) are in essential agreement on these important rates. In addition to these gas phase production rates, Lavvas et al. (2008) calculated individual production rates of C, N, H atoms that are incorporated into haze particles

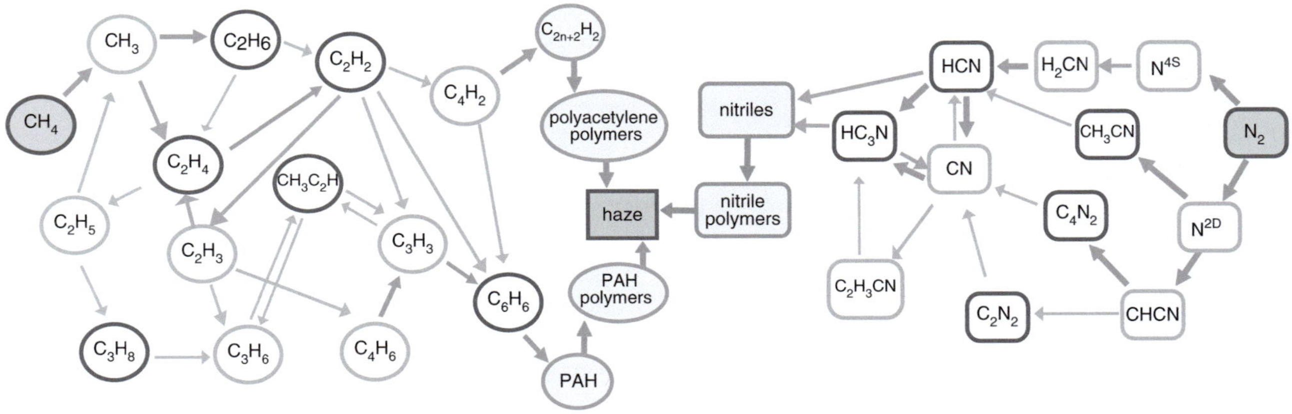

Fig. 10.11 Simplified schematic of the photochemistry of N_2 and CH_4 in Titan's atmosphere through the formation of polymers (after Atreya et al. (2006))

and the total haze production rate (cf. Table 10.5). This latter value is 1/14 of the gas phase mass production rate of C_2H_6, as Bézard (2009) concluded from an empirical analysis of Cassini and Huygens data. But Atreya et al. (2006) argued that this ratio may be as large as ½, based on Wilson and Atreya (2004) whose gas phase production rates are systematically lower than Yung et al. (1984) and Lavvas et al. (2008).

The chemical loss and escape of CH_4 must be balanced by upward transport from the lower atmosphere. With the further hypothesis that CH_4 is the source of all other carbon and hydrogen bearing molecules in the atmosphere and escaping from the atmosphere, then stoichiometrically $2\ CH_4 \rightarrow C_2H_{2n} + (4\text{-}n)\ H_2$ must hold. Note for n = 0, the measured INMS H_2 escape rate places a lower limit on the net CH_4 destruction rate of (5.5–7) × 10^9 cm^{-2} s^{-1}. From Table 10.5 a comparison of the INMS inferred H_2 escape rates with total net CH_4 model loss rates suggests that n = 2 would be the preferred value. But observations indicate that C_2H_6 is the dominant product which implies n = 3. From the Yung et al. (1984) the effective H_2 (H_2 + ½ H) escape rate is ~1 × 10^{10} cm^{-2} s^{-1}, the lower end of INMS H_2 values, whereas Wilson and Atreya (2004) and Lavvas et al. (2008), calculated less, 0.65 × 10^{10} and 0.47 × 10^{10} cm^{-2} s^{-1}, respectively. But certainly some hydrogen is escaping in atomic form, as ionospheric chemistry leads to production of H atoms high in the atmosphere. Thus models to date under predict the H_2 escape rate, because prior to the INMS measurements it was thought that H_2 escape occurred at the Jeans rate, a factor of ~3 lower. To achieve the larger INMS H_2 escape rate and still have C_2H_6 the dominant product of CH_4 chemistry, the CH_4 chemical destruction rate must be also larger than the values given in Table 10.5.

The photochemistry of N_2 occurs mostly in Titan's thermosphere and ionosphere, because the N_2 bond strength of 9.7 eV requires energetic photons, electrons, and ion to break the molecule apart. With no optically allowed excitation paths into repulsive electronic states, dissociation occurs preferentially by indirect paths. Solar radiation between 80 and 100 nm can excite predissociating electronic states which are ≤ 25 % of the total source of N atoms (Liang et al. 2007a). The ion chemistry of N_2^+ preserves the N_2 bond, but dissociative ionization of N_2 by either electron impact or solar EUV radiation will produce one N atom and one N^+ ion, which will react with CH_4 to yield either an N atom or an ion (H_2CN^+ or HCN^+). The latter ion reacts with CH_4 to form the former ion and recombination of H_2CN^+ produces the nitrile HCN and net loss of CH_4. Energetic electrons of photolytic and magnetospheric origins can also directly dissociate N_2. The net efficiency for producing nitriles depends on the extent of self destruction of odd nitrogen by the reaction $N + NH \rightarrow N_2 + H$. Photochemical models estimate the net production/downward flux of N atoms from the thermosphere ~(0.5–1) × 10^9 cm^{-2} s^{-1}, whereas self destruction by the above reaction limits the downward HCN flux to only ~3 × 10^8 cm^{-2} s^{-1} and an equal number incorporated into haze particles (Lavvas et al. 2008). Once the CN bond is formed, it is preserved down through the stratosphere.

10.7 Chemistry and Transport of Atmospheric Constituents

Both Voyager IRIS and Cassini CIRS spectra imply that there is enrichment of several organic compounds at high northern latitude during northern winter and early northern spring. Because most of these species are formed high up in the atmosphere from the photo- and electron-impact dissociation of CH_4 and N_2, their mixing ratio surfaces increase with altitude. A meridional circulation tilts these mixing ratio surfaces in the following sense (cf. Holton and Schoeberl 1988; Bacmeister et al. 1995; Dire 2000). Downward motion (subsidence) brings down higher mixing ratios from above while upward motion brings lower mixing ratios from below. Hence enrichment with latitude implies subsidence over Titan's North Pole in its current winter season during the Cassini Nominal Mission. In Fig. 10.12 is a schematic diagram of what is thought to be the relevant dynamical processes that generate constituent transport and based on analogy with the much better understanding of these processes in the Earth's middle atmosphere. In Fig. 10.10 there are blue dashed lines that show the region where horizontal potential vorticity gradients are the steepest (~55–65° N) and location of dynamical mixing barriers. In this figure the more abundant C_2H_2 and HCN clearly show enhanced mixing ratios in the northern polar region at high altitudes expected from subsidence. In the case of HCN, most of its formation occurs at higher altitudes, whereas for C_2H_2 there is also significant formation in the stratosphere. However, C_2H_2 has a local maximum in mixing ratio at ~200 km in the polar night suggesting a local source and certainly not consistent with simple explanation from subsidence associated with a meridional circulation. On the other hand HCN has a local minimum in its mixing ratio center at ~350 km and 60° N. This cannot be due to gas phase conversion to other gas phase nitriles, e. g. H_3CN in Fig 10.10. But it could be due to heterogeneous reactions removing nitriles and/or condensation, although the temperature is still warm by Titan's standards. Chapter 13 discusses dynamics in much more detail.

10.8 Concluding Remarks

At the end of the Cassini–Huygens Nominal Mission, we have a good first order knowledge of the composition and thermal structure of the atmosphere with the exception of the

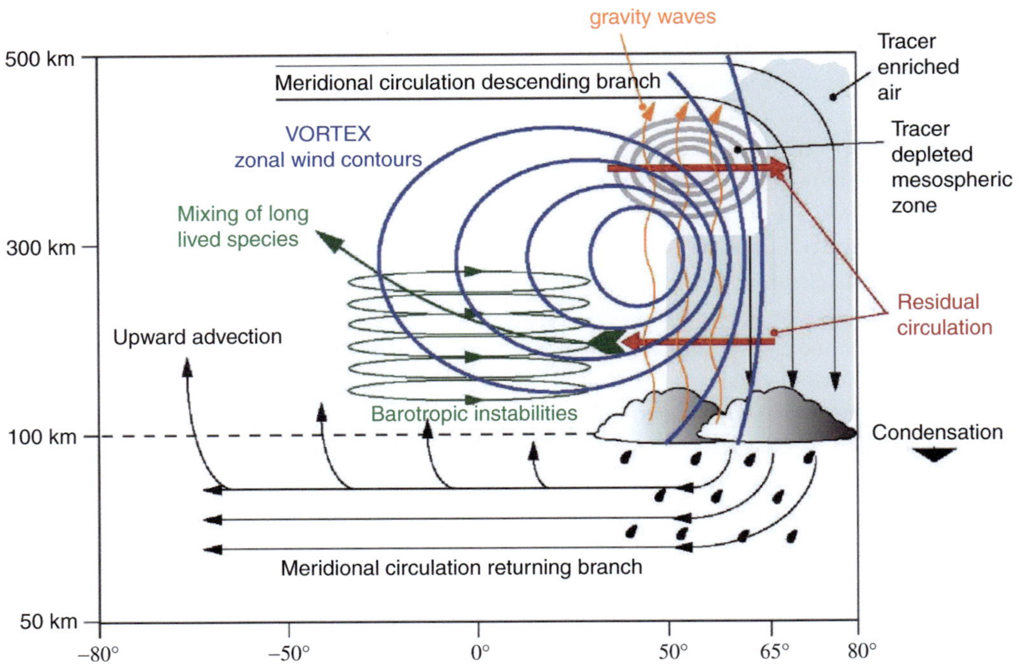

Fig. 10.12 Schematic of dynamical processes thought to be important in Titan's winter north polar region: a single cell meridional circulation, a polar vortex of strong zonal winds, wave induced mixing by inferred equatorial barotropic and gravity waves (after Teanby et al. (2008))

~500–950 km region, variously called the ignorosphere, agnostosphere, etc. Although HASI inferred the thermal structure at equatorial latitudes through this region, it did not yield a pronounced mesopause widely expected from theory. Ground-based millimeter-wavelength provided limited, globally averaged, composition measurements, in comparison to the detailed altitudinal and latitudinal composition and temperature measurements by CIRS for the stratosphere. The UVIS stellar occultation could also provide composition and thermal structure in the ignorosphere region, but the only published analysis (Shemansky et al. 2005) has deficiencies noted above. Better knowledge of the ignorosphere structure would be helpful to resolve the vexing factor of ~2.5 density discrepancy among HASI, INMS, and AACS.

The thermosphere is a chemical factory that initiates the formation of complex positive and negative ions in the high thermosphere as a consequence of magnetospheric–ionospheric–atmospheric interaction involving solar EUV and UV radiation, energetic ions and electrons. This factory produces very heavy positive and negative ions and large molecules, e.g., benzene, naphthalene, nitriles, which apparently condense out and are detectable in solar and stellar UV occultations at ~1000 km (Liang et al. 2007a), and initiate the haze formation process. As these particles fall through the ignorosphere and grow, they become detectable by remote sensing: UVIS at ~1000 km, ISS at ~500 km and eventual become ubiquitous throughout the stratosphere. These haze particles are strong absorbers of solar UV and visible radiation and play a fundamental role in heating Titan's stratosphere and perhaps the ignorosphere (Lavvas et al. 2009). The differential heating with latitude drives wind systems in Titan's middle atmosphere, much as ozone does in the Earth's middle atmosphere. How intimately coupled is the upper thermosphere with the stratosphere and what is the role of the ignorosphere? It is certainly the region of direct photolysis of CH_4. Are complex positive and negative ions and large molecules from above necessary for formation of the main haze layer? What is the role of heterogeneous reactions in determining the partitioning of gas phase molecules and the composition of haze particles?

The Cassini Mission has just entered the first extended mission known as the Cassini Equinox Mission to understand the transition for solstitial conditions to equinoctial conditions, especially as the northern polar region emerges out of the polar night and the polar vortex breaks up. In the planning stages is a proposal to extend the Cassini Mission into a northern, summer solstice mission to explore that transition from equinox into solstitial conditions. Thus the entire duration of the Cassini Mission would be a half of Saturnian year and would permit observations of a full seasonal cycle on Titan.

Acknowledgments DFS acknowledges support by the Cassini–Huygens Mission through JPL Contract No. 1280253. SKA acknowledges support by the Cassini–Huygens Program and Planetary Atmospheres Program of NASA. BB, EL, and M Fulchignoni acknowledge financial support from CNRS and CNES. The authors are indebted to Dr. Sandrine Vinatier for providing Fig. 10.9 and Dr. Eric Wilson supplying his rates referenced to the surface in Table 10.5.

References

Achterberg RK, Conrath BJ, Gierasch PJ, Flasar FM, Nixon CA (2008) Titan's middle-atmospheric temperatures and dynamics observed by the Cassini Composite Infrared Spectrometer. Icarus 194:263–277, doi:10.1016/j.icarus.2007.09

Atreya SK, Donahue TM, Kuhn WR (1978) Evolution of a nitrogen atmosphere on Titan. Science 201:611–613

Atreya SK, Adams EY, Niemann HB, Demick-Montelara JE, Owen TC, Fulchignoni M, Ferri F, Wilson EH (2006) Titan's methane cycle. Planet Space Sci 54:1177–1187. doi:10.1016/j.pss.2006.05.028

Bacmeister JT, Schoeberl MR, Summers ME, Rosenfield JR, Zhu X (1995) Descent of long-lived trace gases in the winter polar vortex. J Geophys Res 100:11669–11684

Baines KH, Drossart P, Lopez-Valverde MA, Atreya SK, Sotin C, Momary TW, Brown RH, Buratti BJ, Clark RN, Nicholson PD (2006) On the discovery of CO nighttime emissions on Titan by Cassini/VIMS: derived stratospheric abundances and geological implications. Planet Space Sci 54:1552–1562. doi:10.1016/j.pss.2006.06.020

Bézard B (2009) Composition and chemistry of Titan's stratosphere. Phil Trans Royal Soc A 367:683–695. doi:10.1098/rsta.2008.0186

Bézard B, Coustenis A, McKay CP (1995) Titan's atmospheric temperature asymmetry: a radiative origin? Icarus 111:267–276

Bézard B, Nixon CA, Kleiner I, Jennings DE (2007) Detection of $^{13}CH_3D$ on Titan. Icarus 191:397–400

Bjoraker G, Achterberg R, Anderson C, Samuelson R, Carlson R, Jennings D (2008) Cassini/CIRS observations of water vapor in Titan's stratosphere, Bull Amer Astron Soc 40(31.12):446

Broadfoot AL et al (1981) Extreme ultraviolet observations from Voyager 1 encounter with Saturn. Science 212:206

Cordier D, Mousis O, Lunine JI, Moudens A, Vuitton V (2008) Photochemical enrichment of deuterium in Titan's atmosphere: new insights from Cassini–Huygens. Astrophys J Lett 689:L61–L64

Courtin R, Gautier D, McKay CP (1995) Titan's thermal emission spectrum: reanalysis of the Voyager infrared measurements. Icarus 114:114–162

Courtin RD, Sim C, Kim S, GautierD, Jennings DE (2008) Latitudinal variations of tropospheric H_2 on Titan from the Cassini CIRS investigation. 40th DPS meeting Abstract. 31.01, Ithaca, New York, Bull Amer Astron Soc 40(31.01): 446

Coustenis A, Bézard B (1995) Titan's atmosphere from Voyager infrared observations. IV. Latitudinal variations of temperature and composition. Icarus 115:126–140

Coustenis A, Salama A, Lellouch E, Encrenaz T, Bjoraker GL, Samuelson RE, de Graauw T, Feuchtgruber H, Kessler MF (1998) Evidence for water vapor in Titan's atmosphere from ISO/SWS data. Astron Astrophys 336:L85–L89

Coustenis A, Salama A, Schulz B, Ott S, Lellouch E, Encrenaz TH, Gautier D, Feuchtgruber H (2003) Titan's atmosphere from ISO mid-infrared spectroscopy. Icarus 161:383–403. doi:10.1016/S0019-1035(02)00028-3

Coustenis A, Achterberg RK, Conrath BJ, Jennings DE, Marten A, Gautier D, Nixon CA, Flasar FM, Teanby NA, Bézard B, Samuelson RE, Carlson RC, Lellouch E, Bjoraker GL, Romani PN, Taylor FW, Irwin PGJ, Fouchet T, Hubert A, Orton GS, Kunde VG, Vinatier S, Mondellini J, Abbas MM, Courtin R (2007) The composition of Titan's stratosphere from Cassini/CIRS mid-infrared spectra. Icarus 189:35–62. doi:10.1016/j.icarus.2006.12.022

Coustenis A, Jennings DE, Jolly A, Bénilan Y, Nixon CA, Vinatier S, Gautier D, Bjoraker G, Romani P, Carlson R, Flasar FM (2008) Detection of C_2HD and the D/H ratio on Titan. Icarus 197:539–548. doi:10.1016/j.icarus.2008.06.003

Crespin A, Lebonnois S, Vinatier S, Bézard B, Coustenis A, Teanby NA (2008) Diagnostics of Titan's stratospheric dynamics using Cassini/CIRS data and the 2-dimensional IPSL Circulation Model. Icarus 197:556–57. doi:10.1016/j.icarus.2008.05.010

Cui J, Yelle RV, Volk K (2008) Distribution and escape of molecular hydrogen in Titan's thermosphere and exosphere. J Geophys Res 113:E10004. doi:10.1029/2007JE003032

Danielson RE, Caldwell J, Larach DR (1973) An inversion in the atmosphere of Titan. Icarus 20:437–443

de Kok R, Irwin PGJ, Teanby NA, Lellouch E, Bezard B, Vinatier S, Nixon CA, Fletcher L, Howett C, Calcutt SB, Bowles NE, Flasar FM, Taylor FW (2007) Oxygen compounds in Titan's stratosphere as observed by Cassini CIRS. Icarus 186:354–363. doi:10.1016/j.icarus.2006.09.016

Dire JR (2000) Seasonal photochemical and meridional transport model for the stratosphere of Titan, Icarus, 145:428–444 (cf. Bull.A. A. S., 24:955–956, 1992)

Dubouloz N, Raulin F, Lellouch E, Gautier D (1989) Titan's hypothesized ocean properties – The influence of surface temperature and atmospheric composition uncertainties. Icarus 82:81–96. doi:10.1016/0019-1035(89)90025-0

Flasar FM et al (2005) Titan's atmospheric temperatures, winds, and composition. Science 308:975–978

Fulchignoni M, Ferri F, Angrilliet F et al (2005) Titan's physical characteristics measured by the Huygens Atmospheric Structure Instrument (HASI). Nature 438:785–791. doi:10.1038/nature04314

Gillett FC, Forrest WJ, Merrill KM (1973) 8–13 micron observations of Titan. Astrophys J 187:L37–L38

Gurwell MA (2004) Submillimeter observations of Titan: Global measures of stratospheric temperature, CO, HCN, HC3N, and the isotopic 12C/13C and 14N/15N. Astrophys J 616:L7–L10

Hanel RA et al (1981) Infrared observations of the Saturnian system from Voyager 1. Science 212:192–200

Hartle RE, Sittler EC, Neubauer FM, Johnson RE, Smith HT, Crary F, McComas DJ, Young DT, Coates AJ, Simpson D, Bolton S, Reisenfeld D, Szego K, Berthelier JJ, Rymer A, Vilppola J, Steinberg JT, Andre N (2006) Preliminary interpretation of Titan plasma interaction as observed by the Cassini Plasma Spectrometer: comparisons with Voyager 1. Geophys Res Lett 33:8201. doi:10.1029/2005GL024817

Holton JR, Schoeberl MR (1988) The role of gravity wave generated advection and diffusion in transport of tracers in the mesosphere. J Geophys Res 93:11075–11082

Hörst S, Vuitton V, Yelle RV (2008) The origin of oyxgen species in Titan's atmosphere. J Geophys Res 113:E10006. doi:10.1029/2008JE003135

Hunten D M (1974) The atmosphere of Titan. Washington, NASA SP-340

Hunten DM (1978) A Titan atmosphere with a surface temperature of 200 K. In Hunten DM, Morrison D (eds) The Saturn System, NASA Conf. Publ. 2068, Washington DC, pp 127–140

Hunten DM, Morrison D (eds) (1978) The Saturn System, NASA Conf. Publ. 2068, WashingtonDC, pp 127–140

Jacquemart D, Lellouch E, Bézard B, de Bergh C, Coustenis A, Lacome N, Schmitt B, Tomasko M (2008) New laboratory measurements of CH_4 in Titan's conditions and a reanalysis of the DISR near-surface spectra at the Huygens landing site. Planet Space Sci 56:613–623

Jennings DE, Nixon CA, Jolly A, Bézard B, Coustenis A, Vinatier S, Irwin PGJ, Teanby N, Romani PN, Achterberg RK, Flasar FM (2008) Isotopic ratios in Titan's atmosphere from Cassini CIRS limb sounding: HC_3N in the north. Astrophys J Lett 681:L109–L111

Jennings DE, Flasar FM, Kunde VG, Samuelson RE, Pearl JC, Nixon CA, Carlson RC (2009) Titan's surface brightness temperatures, Astrophys J, 691:L103–L105

Kuiper GP (1944) Titan: A satellite with an atmosphere. Astrophys J 100:378–383

Lara L (1998) Correction to "Vertical distribution of Titan's atmospheric neutral constituents" by L M Lara et al., J Geophys Res 103:(E11), 25775–25775

Lara LM, Lellouch E, L´opez-Moreno JJ, Rodrigo R (1996) Vertical distribution of Titan's atmospheric neutral constituents. J Geophys Res 101:23261–23283, doi:10.1029/96JE02036

Lavvas PP, Coustenis A, Vardavas IM (2008) Coupling photochemistry with haze formation in Titan's atmosphere, Part II: Results and validation with Cassini/Huygens data. Planet Space Sci 56:67–99

Lavvas PP, Yelle RV, Vuitton V (2009) The detached haze layer in Titan's mesosphere, Icarus, in press

Lebonnois S, Bakes ELO, McKay CP (2003a) Atomic and molecular hydrogen budget in Titan's atmosphere. Icarus 161:474–485. doi:10.1016/S0019-1035(02)00039-8

Lebonnois S, Hourdin F, Rannou P, Luz D, Toublanc D (2003b) Impact of the seasonal variations of composition on the temperature field of Titan's stratosphere. Icarus 163:164–174

Lellouch E (1990) Atmospheric models of Titan and Triton. Ann Geophys 8:653–660

Lellouch E, Coustenis A, Gauthier D, Raulin F, Dubouloz N, Frere C (1989) Titan's atmosphere and hypothesized ocean: a reanalysis of the Voyager 1 radio-occultation and IRIS 7.7 μm Data, Icarus, 79:328–349

Lellouch E, Coustenis A, Sebag B, Cuby J-G, L'opez-Valverde M, Schmitt B, Fouchet T, Crovisier J (2003) Titan's 5-μm window observations with the Very Large Telescope. Icarus 162:125–142. doi:10.1016/S0019-1035(02)00079-9

Lewis JS (1971) Satellites of the outer planets: Their physical and chemical nature. Icarus 15:174–185

Liang M-C, Heays AN, Lewis BR, Gibson ST, Yung YL (2007a) Source of nitrogen isotope anomaly in HCN in the atmosphere of Titan. Astrophys. J. 664:L115–L118

Liang M-C, Yung YL, Shemansky DE (2007b) Photolytically generated aerosols in the mesosphere and thermosphere of Titan. Astrophys J 661:L199–L202

Lindal GF, Wood GE, Hotz HB, Sweetnam DN (1983) The atmosphere of Titan: an analysis of the Voyager radio occultation measurements. Icarus 53:348–363

Lodders K (2003) Solar system abundances and condensation temperatures of the elements. Astrophys J 591:220–1247

Lunine JI, Yung YL, Lorenz RD (1999) On the volatile inventory of Titan from isotopic abundances in nitrogen and methane. Planet Space Sci 47:1291–1303

Lutz BL, de Bergh C, Owen T (1983) Titan–Discovery of carbon monoxide in its atmosphere. Science 220:1374–1376

Marten A, Hidayat T, Biraud Y, Moreno R (2002) New millimeter heterodyne observations of Titan: Vertical distributions of nitriles HCN, HC_3N, CH_3CN, and the isotopic ratio $^{15}N/^{14}N$ in its atmosphere. Icarus 158:532–544

McKay CP, Pollack JB, Courtin R (1989) The thermal structure of Titan's atmosphere. Icarus 80:23–53

McKay CP, Coustenis A, Samuelson RE, Lemmon MT, Lorenz RD, Cabane M, Rannou P, Drossart P (2001) Physical properties of the organic aerosols and clouds on Titan. Planet Space Sci 49:79–99

Mousis O, Gautier D, Coustenis A (2002) The D/H Ratio in methane in Titan: origin and history. Icarus 159:156–165

Müller-Wodarg ICF, Yelle RV, Cui J, Waite HJ Jr (2008) Horizontal structure and dynamics of Titan's thermosphere. J Geophys Res 113:E10005. doi:10.1029/2007JE003033

Niemann HB et al (2005) The abundances of constituents of Titan's atmosphere from the GCMS instrument on the Huygens probe. Nature 438:779–784. doi:10.1038/nature.04122

Nixon CA, Achterberg RK, Vinatier S, Bézard B, Coustenis A, Irwin PGJ, Teanby NA, de Kok R, Romani PN, Jennings DE, Bjoraker GL, Flasar FM (2008a) The 12C/13C isotopic ratio in Titan hydrocarbons from Cassini/CIRS infrared spectra. Icarus 195:778–791. doi:10.1016/j.icarus.2008.01.012

Nixon CA, Jennings DE, Bézard B, Teanby NA, Achterberg RK, Coustenis A, Vinatier S, Irwin PGJ, Romani PN, Hewagama T, Flasar FM (2008b) Isotopic ratios in Titan's atmosphere from Cassini CIRS limb sounding: CO2 in the equator and south. Astrophys J Lett 681:L101–L103

Owen T, Biver N, Marten A, Matthews H Meier R (1999) Saturn VI (Titan). IAU Circ 7306

Owen T, Mahaffy PR, Niemann HB, Atreya S, Wong M (2001) Protosolar Nitrogen. Astrophys. J. 553:L77–L79

Penteado PF, Griffith CA, Greathouse K, de Bergh C (2005) Measurements of CH3D and CH4 in Titan from infrared spectroscopy, Astrophys J 629:L53–L56

Pinto JP, Lunine JI, Kim S-J, Yung YL (1986) The D to H ratio and the origin and evolution of Titan's atmosphere. Nature 319:388–390

Porco CC et al (2005) Imaging of Titan from the Cassini spacecraft. Nature 434:159–168

Rannou P, Hourdin F, McKay CP (2002) A wind origin for Titan's haze structure. Nature 418:853–856

Roe HG, de Pater I, McKay CP (2004) Seasonal variation of Titan's stratospheric ethylene (C_2H_4) observed. Icarus 169:440–461

Samuelson RE, Maguire WC, Hanel RA, Kunde VG, Jennings DE, Yung YL, Aikin AC (1983) CO_2 on Titan. J Geophys Res 88:8709–8715

Samuelson RE, Nath NR, Borysow A (1997) Gaseous abundances and methane supersaturation in Titan's troposphere. Planet Space Sci 45:959–980

Sekine Y, Lebonnois S, Imanaka H, Matsui T, Bakes ELO, McKay CP, Khare BN, Sugita S (2008a) The role of organic haze in Titan's atmospheric chemistry: II. Effect of heterogeneous reaction to the hydrogen budget and chemical composition of the atmosphere. Icarus 194:201–211. doi:10.1016/j.icarus.2007.08.030

Sekine Y, Imanaka H, Matsui T, Khare BN, Bakes ELO, McKay CP, Sugita S (2008b) The role of organic haze in Titan's atmospheric chemistry: I. Laboratory investigation on heterogeneous reaction of atomic hydrogen with Titan tholin. Icarus 194:186–200. doi:10.1016/j.icarus.2007.08.031

Shemansky DE, Stewart AIF, West RA, Esposito LW, Hallett JT, Liu X (2005) The Cassini UVIS Stellar Probe of the Titan Atmosphere. Science 308:978–982. doi:10.1126/science.1111790

Sicardy B, Colas F, Widemann T, et al. (2006) The two stellar occultations of November 14, 2003, J Geophys Res 111, doi:10.1029/2005JE002624

Strobel DF (1974) The Photochemistry of Hydrocarbons in the atmosphere of Titan. Icarus 21:466–470

Strobel, DF (2008) Titan's hydrodynamically escaping atmosphere, Icarus, 193:588–594, doi:10.1016/j.icarus.2007.08.014

Strobel, DF (2009) Titan's hydrodynamically escaping atmosphere: escape rates and the structure of the exobase region, Icarus 202: 632–641, doi:10.1016/j.icarus.2009.03.007

Strobel DF, Shemansky DE (1982) EUV Emission from Titan's upper atmosphere: Voyager 1 encounter. J Geophys Res 87:1361

Strobel DF, Sicardy B (1997) Gravity wave and wind shear models. In: Lebreton JP (ed) HUYGENS Science, Payload and Mission, ESA SP, vol 1177, pp 299–311

Strobel DF, Hall DT, Zhu X, Summers ME (1993) Upper limit on Titan's atmospheric argon abundance. Icarus 103:333–336

Teanby NA et al (2006) Latitudinal variations of HCN, HC_3N and C_2N_2 in Titan's stratosphere derived from Cassini CIRS data. Icarus 181:243–255

Teanby NA et al (2007) Vertical profiles of HCN, HC_3N and C_2H_2 in Titan's atmosphere derived from Cassini/CIRS data. Icarus 186:364–384

Teanby NA, et al. (2008) Titan's polar vortex structure revealed by chemical tracers. J Geophys Res, 113:E12003, doi:10.1029/2008JE003218

Teanby NA et al. (2009) Titan's stratospheric C_2N_2, C_3H_4, and C_4H_2 abundance from Cassini/CIRS far-infrared spectra. Icarus, 202, 620–631, doi:10.1016/j.icarus.2009.03.022

Teanby NA, Irwin PGJ, de Kok R, Nixon CA (2009) Dynamical implications of seasonal and spatial variations in Titan's stratospheric composition. Phil Trans Royal Soc A 367:697–711, doi10.1098/rsta.2008.0164

Tomasko M, Bézard B, Doose L, Engel S, Karkoschka E, Vinatier S (2008) Heat balance in Titan's atmosphere. Planet Space Sci 56: 648–659. doi:10.1016/j.pss.2007.10.012

Toublanc D, Parisot JP, Brillet J, Gautier D, Raulin F, McKay CP (1995) Photochemical modeling of Titan's atmosphere. Icarus, 113:2–26. doi:10.1006/icar.1995.1002

Trafton LM (1972) On the possible detection of H_2 in Titan's atmosphere. Astrophys J 175:285–293

Vervack RJ Jr, Sandel BR, Strobel DF (2004) New perspectives on Titan's upper atmosphere from a reanalysis of the Voyager 1 UVS solar occultations. Icarus 170:91–112

Vinatier S (2007) Analyse des spectres infrarouges thermiques émis par l'atmosphère de Titan enregistrés par l'instrument Cassini/CIRS. Ph.D. Thesis. University of Denis Diderot, Paris VII

Vinatier S et al (2007a) Vertical abundance profiles of hydrocarbons in Titan's atmosphere at 15°S and 80°N retrieved from Cassini/CIRS spectra. Icarus 188:120–138

Vinatier S, Bézard B, Nixon CA (2007b) The Titan 14N/15N and 12C/13C isotopic ratios in HCN from Cassini/CIRS, Icarus, 191:712–721. doi:10.1016/j.icarus.2007.06.001

Vuitton V, Yelle RV, Anicich VG (2006) The nitrogen chemistry of Titan's upper atmosphere revealed. Astrophys J 647:L175–L178

Vuitton V, Yelle RV, McEwanMJ (2007) Ion-chemistry and N-containing molecules in Titan's upper atmosphere. Icarus 191:722–742. doi:10.1016/j.icarus.2007.06.023

Vuitton V, Yelle RV, Cui J (2008) Formation and distribution of benzene on Titan. J Geophys Res, 113:E05007. doi:10.1029/2007JE002997

Waite Jr JW et al. (2005) Ion neutral mass spectrometer results from the first flyby of Titan. Science, 308:982–986

Waite JH, Young DT, Cravens TE, Coates AJ, Crary FJ, Magee B, Westlake J (2007) The process of tholin formation in Titan's upper atmosphere. Science 316:870–875. doi10.1126/science.1139727

Wilson EH, Atreya SK (2004) Current state of modeling the photochemistry of Titan's mutually dependent atmosphere and ionosphere. J Geophys Res 109:E06002. doi:10.1029/2003JE002181

Wong A-S, Morgan CG, Yung YL, Owen T (2002) Evolution of CO on Titan. Icarus 155:382–392. doi:10.1006/icar.2001.6720

Yelle RV (1991) Non-LTE models of Titan's upper atmosphere. Astrophys. J. 383:380–400

Yelle RV, Strobel DF, Lellouch E, Gautier D (1997) Engineering models for Titan's atmosphere. In: Lebreton JP (ed) HUYGENS Science, Payload and Mission, ESA SP, vol 1177, pp 243–256

Yelle RV, Cui J, Müller-Wodarg ICF (2008) Methane escape from Titan's atmosphere. J Geophys Res 113:E10003. doi:10.1029/2007JE003031

Yung YL, Allen M, Pinto JP (1984) Photochemistry of the atmosphere of Titan – comparison between model and observations. Astrophys J 55:465–506. doi:10.1086/190963

Chapter 11
Composition and Structure of the Ionosphere and Thermosphere

T. E. Cravens, R. V. Yelle, J.-E. Wahlund, D. E. Shemansky, and A. F. Nagy

Abstract Airglow emissions, radio and solar occultation data from the Voyager mission over a quarter of a century ago provided the main source of information on the composition and structure of Titan's upper atmosphere and ionosphere until October 2004, when the Cassini Orbiter first encountered Titan during the "Ta" fly-by. During this encounter, in situ measurements were made by many instruments onboard the Orbiter, including the Ion and Neutral Mass Spectrometer (INMS), the Radio Wave and Plasma Wave Spectrometer (RPWS), the Magnetometer (MAG), and the Cassini Plasma Spectrometer (CAPS). For example, INMS measurements confirmed that the major neutral species were molecular nitrogen and methane. Other species detected included molecular hydrogen, acetylene, ethylene, benzene, and propane. The Langmuir probe part of the RPWS experiment observed substantial ionospheric electron densities and measured electron temperatures significantly exceeding the neutral temperature. A large set of data on the upper atmosphere and ionosphere has been collected during the many Titan encounters following Ta. The first composition measurements for the ionosphere were made by INMS during the outbound leg of the T5 pass in April 2005. A rich and complex ion-neutral chemistry scheme was predicted prior to the Cassini mission and the INMS composition data indeed revealed the presence of a very large number of ion species, both predicted and unpredicted. Stellar occultation measurements made by the Cassini Ultraviolet Spectrometer (UVIS) provided important information on the structure and composition of Titan's upper atmosphere, and radio occultation measurements made by the Radio Science Subsystem (RSS) revealed the existence of a substantial ionosphere even for altitudes below 1000 km. The discovery of negative ions in the ionosphere was also very exciting. A vigorous modeling effort aimed at explaining the structure and composition of the upper atmosphere and ionosphere is helping to put the data into a broader theoretical context. For example, solar extreme ultraviolet and x-ray radiation and energetic electrons from Saturn's magnetosphere interact with the upper atmosphere producing the ionosphere and initiating a complex neutral and ion chemistry that has important effects extending deep into the atmosphere.

11.1 Introduction

11.1.1 Brief Overview of How Titan's Upper Atmosphere and Ionosphere Fit into a Comparative Picture of Atmospheres and Ionospheres in the Solar System

Our understanding of the upper atmosphere and ionosphere began to develop in the 1950s with sounding rockets and satellite measurements made at Earth (Schunk and Nagy 2000; Banks and Kockarts 1973). Planetary upper atmospheric and ionospheric studies ("aeronomy") started in the early 1960s when spacecraft were sent to other planets (c.f., Nagy and Cravens 2002). In the Earth's atmosphere the mesosphere is where the temperature profile decreases from its stratospheric maximum (due to absorption of solar radiation by ozone), and its upper boundary is the mesopause where, by definition, the temperature profile has a minimum (typically, T (Earth) ≈ 180 K). At Earth, the temperature increases with altitude above the mesopause to values of about T (Earth) ≈ 1000 K in an atmospheric region called the thermosphere.

T.E. Cravens (✉)
University of Kansas, Department of Physics & Astronomy, Malott Hall, 1251 Wescoe Hall Dr. Lawrence, KS 66046 USA
e-mail: cravens@ku.edu

R.V. Yelle
Lunar and Planetary Laboratory, University of Arizona, P.O. Box 210092, Tucson, AZ 85721, USA

J. Wahlund
Department of Physics and Astronomy, Uppsala University, SE-75120 Uppsala, Sweden

D.E. Shemansky
Aerospace and Mechanical Engineering, University of Southern California, Los Angeles, CA 90089 USA

A.F. Nagy
Space Physics Research Laboratory, University of Michigan, 2455 Hayward, Ann Arbor, MI 48109 USA

R.H. Brown et al. (eds.), Titan from Cassini-Huygens,

Absorption of EUV radiation is a key feature of all planetary thermospheres including Titan's. At high enough altitudes in all atmospheres, the neutral density is low enough such that the collisional mean free path exceeds a typical scale length (usually taken to be the neutral scale height). This atmospheric region is called the exosphere with a lower boundary called the exobase. Atoms and molecules residing above the exobase with speeds exceeding the escape speed can contribute to atmospheric loss.

Soft x-ray and EUV photons with energies exceeding the ionization potential of a neutral constituent (e.g., N_2) create ions and photoelectrons when they are absorbed. Following Schunk and Nagy (2000), the "ionosphere is considered to be that region of an atmosphere where significant numbers of free thermal (<1 eV) electrons and ions are present." The presence of free electrons and ions, combined with relatively low neutral densities, drive processes not operating in the lower atmosphere. For example, the electrical conductivity in the ionosphere is large, particularly in directions along the magnetic field, so that large electrical currents can flow, allowing the magnetosphere to couple to the ionosphere. Ionospheric layers have been defined at Earth, originally based on characteristics of the vertical electron density profiles, each layer with different controlling processes (e.g., chemical control of densities, with molecular ions O_2^+ and NO^+ dominating, in the E-region), but this terrestrial nomenclature is not always useful for other planets (Fox and Yeager 2006). The concept of "Chapman layer" (Schunk and Nagy 2000) in which the vertical density structure is determined by the atmospheric opacity to ionizing radiation is another useful concept that applies to ionospheres in general, including Titan's.

How well are planetary thermospheres and ionospheres, and Titan's in particular, explained by the concepts originally developed for Earth? Clearly there are differences in heliospheric distances, neutral compositions (set by lower atmosphere and surface properties), and intrinsic magnetic field strengths (which affect how the solar wind interacts with the upper atmosphere and ionosphere). The neutral thermospheric temperature on the dayside increases with altitude at almost all planets and satellites, but the vertical increase of the temperature above the mesopause for Venus and Mars ($\Delta T \approx 300$ K) is much less than at Earth due to strong CO_2 cooling at Venus and Mars. And at Titan, due to a lower solar EUV flux (Titan and Saturn are located at a heliospheric distance of 9.5 AU yielding a solar flux only about 1% that at Earth) and due to efficient cooling by HCN (Yelle 1991), the temperature gradient is quite small or even negative (Müller-Wodarg et al. 2006). Earth and Titan are similar in that molecular nitrogen is the major species, but they differ in that the next most abundant atmospheric species at Earth is molecular oxygen and at Titan is methane. The atmospheric and ionospheric chemistry is quite different for the two bodies. At all planets and satellites the ionosphere plays an important role in how that body interacts with the surrounding magnetospheric and/or interplanetary environments. In particular, Titan's upper atmosphere and ionosphere strongly interact with Saturn's magnetospheric plasma.

11.1.2 Brief Review of Our Knowledge of Titan's Thermosphere and Ionosphere Prior to the Cassini Mission

Radio and stellar occultation data obtained from the Voyager mission over a quarter of a century ago plus theoretical models (c.f., Hunten et al. 1984; Neubauer et al. 1984; Bird et al. 1997; Vervack et al. 2004; Wilson and Atreya 2004; Cravens et al. 2004) provided the main source of information on Titan's thermosphere and ionosphere until October 2004, when the Cassini Orbiter first encountered Titan during the "Ta" fly-by.

The thermal and density structure of the thermosphere was derived from solar occultation measurements made by the Ultraviolet Spectrometer (UVS) during the Voyager 1 encounter with Titan during November 1980 (Smith et al. 1982). Smith et al. (1982) derived an exospheric temperature of 185 ± 20 K and also found a methane mole fraction of 8% ± 3% and an acetylene mole fraction of roughly 1%. Vervack et al. (2004) re-examined the Voyager 1 UVS occultation data and found a lower exospheric temperature of 150 ± 3 K and a CH_4 mole fraction of 6% ± 1% (Smith et al. 1982). The thermal balance of the upper atmosphere was modeled by Yelle (1991) and Yelle et al. (1997) who recognized that heating above the mesopause due to absorption of solar EUV radiation is closely balanced by cooling due to infrared emission by hydrogen cyanide (HCN). Müller-Wodarg et al. (2000, 2003) carried out the first global three-dimensional general circulation simulation of Titan's thermosphere, using these Voyager inputs. Day-to-night temperature gradients were found to be ≈10–20 K in the upper thermosphere, depending on the solar conditions adopted, and the typical wind speeds were about 30–60 m/s.

Dayglow emissions associated with N_2 molecular bands and N transition lines were measured by the Voyager UVS experiment and also by the Cassini UVIS experiment (Broadfoot et al. 1981; Ajello et al. 2007, 2008; Liu and Shemansky 2008) confirming that N_2 was the dominant neutral species and that energy deposition processes due to solar radiation and/or particle precipitation were operating (see Chapter 16). The EUV airglow emission observed by the UVIS experiment appears to be mainly present on the dayside with only weak or sporadic nightside emission, but work has not yet been carried out on correlating the observed emissions with the ionospheric structure.

Photochemical models of Titan's stratosphere and lower thermosphere were developed prior to Cassini with most of these models focused on explaining the distributions of hydrocarbon and nitrogen-bearing species (including nitriles) that were observed in the stratosphere (c.f., Atreya 1986; Wilson and Atreya 2004; Hunten et al. 1984; Strobel 1985; Hidayat et al. 1997, 1998), but with a few models considering the thermospheric minor neutral composition (e.g., Yung et al. 1984; Yung 1987; Toublanc et al. 1995; Lara et al. 1996; Lebonnois et al. 2001; Wilson and Atreya 2004). A common theme of these models was that solar (or other) radiation dissociates N_2 and CH_4 producing chemical radicals that initiate a complex chemistry that leads to the formation of more complex hydrocarbon species (C_2H_2, C_2H_4, C_2H_6, C_3H_8 ...) and nitrogen-bearing species (HCN, HC_3N ...). Fox and Yelle (1997) and Wilson and Atreya (2004) pointed out that ionospheric processes play a key role in the nitrile chemistry due to the difficulty of breaking the N_2 bond with anything other than EUV photons or energetic particles.

An ionosphere on Titan was first detected in 1980 by the Voyager 1 radio occultation experiment (Bird et al. 1997) and in situ plasma measurements were made a distance of 2.7 Titan radii down the magnetotail. The first in situ Cassini measurements of Titan's main ionosphere were made in 2004 (Wahlund et al. 2005), although many theoretical studies were carried out prior to Cassini (Ip 1990; Keller et al. 1992, 1994, 1998; Keller and Cravens 1994; Gan et al. 1992, 1993; Fox 1996a, b; Fox and Yelle 1997; Galand et al. 1999; Banaskiewicz et al. 2000; Lilensten et al. 2005; Wilson and Atreya 2004; Cravens et al. 2004). These early studies suggested that both solar photons and electrons from Saturn's outer magnetosphere act as ionization sources (Strobel 1985; Atreya 1986; Bauer 1987; Gan et al. 1992, 1993; Ip 1990). Ionospheric layers near 600 km due to meteor ablation (Molina-Cuberos et al. 2001) and in the lower atmosphere near 60 km due to cosmic ray ionization (e.g., Capone et al. 1976) were also postulated.

The presence of methane in the neutral atmosphere made it clear from the start that the ion chemistry was probably quite complex (Strobel 1985; Atreya 1986; Fox 1996a, b; Fox and Yelle 1997; Keller et al. 1992, 1998; Wilson and Atreya 2004). Pre-Cassini theoretical studies used data-bases of ion-neutral reactions and reaction rates (Anicich and McEwan 1997; Anicich et al. 2004) to outline an ion-neutral scheme that has in many respects (but certainly not all) been confirmed by the Cassini ion composition measurements made by the Ion and Neutral Mass Spectrometer (INMS) (Cravens et al. 2006). Primary ionization processes create N_2^+, N^+, CH_4^+, CH_3^+, and other "primary" ion species, but ion-neutral reactions with methane largely convert these initial ion species into CH_5^+ and $C_2H_5^+$, which then react with minor neutral species like C_2H_2 and HCN to form ion species such as $HCNH^+$, $C_3H_5^+$, as well as much higher-mass species. Dissociative recombination reactions (c.f., Adams and Smith 1988; McLain et al. 2004, 2009) ultimately act as the chemical loss process in the main ionosphere, although pre-Cassini studies also suggested that transport processes are probably important at high altitudes (e.g., Ip 1990; Keller and Cravens 1994; Cravens et al. 1998).

Plasma and field data in Saturn's outer magnetosphere and in Titan's magnetotail was also acquired during Voyager's passage 2.7 Titan radii downstream (Neubauer et al. 1984; Ness et al. 1982), and this data also prompted many modeling studies of the plasma dynamics associated with the magnetospheric interaction with Titan's upper atmosphere and ionosphere (see Sittler et al. Chapter 16).

11.1.3 A Brief Overview of the Current Chapter

Much has been learned about the thermosphere and ionosphere of Titan since the Cassini fly-by in October 2004, and this chapter will summarize this new information. Section 11.2 will focus on the upper neutral atmosphere and Section 11.3 on the ionosphere. By "upper atmosphere" we will mean the region above about 900 km where solar EUV radiation is absorbed (i.e., the thermosphere). Other chapters in this book will cover related topics such as the overall structure and composition of the atmosphere (Chapter 10), atmospheric dynamics (Chapter 13), atmospheric escape (Chapter 16), energy deposition and magnetospheric interaction (Chapter 16), and aerosol formation (Chapters 8 and 12).

11.2 Structure and Composition of the Upper Neutral Atmosphere

11.2.1 Brief Review of Basic Processes Relevant to the Neutral Upper Atmosphere

The structure and composition of a neutral atmosphere (i.e., densities, flow velocities, and temperatures) can in principle be determined by solving appropriate coupled fluid conservation equations for each atmospheric constituent, subject to external inputs of momentum, particles, and energy. The operative equations are the continuity, momentum (Newton's second law of motion), and energy equations. In practice, these equations have been approximated in different ways. For example, the equation of motion can be written for the neutral gas (e.g., see Müller-Wodarg et al. 2000) as a whole and this equation equates the mass per unit volume (or mass density ρ) multiplied by the "convective" derivative of the velocity (**u**) of a fluid "parcel" with respect to time to the total force per unit volume on a fluid parcel. The force includes gravity and the pressure gradient force (i.e., a force from high pressure to low pressure regions). The pressure (p) can be related to the density and the temperature with the equation of state (p = ρ R T where R is the gas constant and T is temperature).

Hydrostatic balance is usually a good approximation in the vertical direction for most planetary atmospheres. A full solution of the momentum equation, including horizontal terms, is essentially what 3D global general circulation codes accomplish, thus predicting wind speeds. Horizontal pressure gradient forces are particularly important in driving winds and these pressure gradients are affected by temperature variations, which in turn depend on the distribution of heating and cooling processes. The topic of atmospheric dynamics at Titan is mainly covered in Chapter 13, although some aspects of thermospheric dynamics will be touched upon in the current chapter.

Titan's atmosphere is quite spherical and has a large scale height relative to its radius (H/R ≈ .05, whereas for Venus H/R ≈ .001). The gravitational acceleration, $g(r) = 1.35\ \mathrm{m\ s^{-2}}\ (R_T/r)^2$ varies considerably with radial distance, r, over the extent of the atmosphere. The radius of Titan is $R_T = 2575$ km. The atmospheric scale height is given by (c.f., Banks and Kockarts 1973):

$$H(r) = k_B T(r) / mg(r) \tag{11.1}$$

where m is the molecular mass (either the average mass below the homopause, or the species-specific mass at higher altitudes) and T(r) is the temperature. The neutral density profile for the major species N_2 for an isothermal atmosphere is given by the following expression (Lindal et al. 1983):

$$n(r) = n(r_1)\exp\{-C_{N2}(r_1^{-1} - r^{-1})\} \tag{11.2}$$

where $n(r_1)$ is a reference density at the bottom of the thermosphere and $C_{N2} = (m_{N2}g(R_T)R_T^2)/k_BT$.

Minor neutral species are described by appropriate fluid conservation equations that include the effects of collisions with major neutral species, as well as chemical sources and sinks as appropriate. The distributions of minor species with short chemical lifetimes are mainly determined by chemical processes, whereas transport/dynamical processes determine how longer-lived species, such as methane and hydrogen cyanide, are distributed. The vertical momentum equation for a species can, by neglecting inertial terms, be converted into a diffusion equation. The vertical flux of species, s, is given by (Schunk and Nagy 2000; Banks and Kockarts 1973):

$$\Phi_s = -D_s\left[\frac{dn_s}{dz} + \frac{n_s}{H_s} + (1+\alpha_s)\frac{n_s}{T}\frac{dT}{dz}\right] - K\left[\frac{dn_s}{dz} + \frac{n_s}{\langle H\rangle} + \frac{n_s}{T}\frac{dT}{dz}\right] \tag{11.3}$$

where H_s is the scale height for species s (Eq. 11.1 but with mass given by m_s) and <H> is the scale height using the average mass. α_s is an effective thermal diffusion coefficient (see De La Haye et al. 2007a). The molecular diffusion coefficient (D_s) for species s diffusing through the major species (N_2) varies inversely as the neutral density (for values see Cui et al. 2008; De La Haye et al. 2007a; Mason and Marrero 1970). The eddy diffusion coefficient, K, is meant to account for, in strictly 1D vertical models, the vertical mixing of species by "eddies" generated both by small-scale dynamics (e.g., gravity wave breaking) and by large-scale dynamics. The homopause is located where D = K, and below this altitude (really a transition region) the density profiles of all chemically long-lived species are expected to have the same scale height.

The collisional mean free path increases inversely with neutral density (and hence with altitude). The uppermost atmospheric region (called the exosphere) is where the mean free path exceeds the scale height (the exobase is the transition altitude and is located near 1450 km at Titan), and particle trajectories are largely unaltered by collisions. Atoms or molecules with speeds at the exobase greater than the escape speed ($v_{esc} = \sqrt{2GM/r}$) and moving upward will escape. The exobase escape speed at Titan is very low at ≈1.5 km/s. The escape associated with the thermal tail of the Maxwellian distribution is known as Jeans escape and at Titan, both H and H_2 have large escape fluxes (Yelle et al. 2006; De La Haye et al. 2007a; Cui et al. 2008; Lebonnais et al. 2003). A large escape flux can also affect the vertical distribution of a species according to Eq. 11.3.

11.2.2 Observed Variations of the Total Neutral Density and the Major Neutral Species and the Thermospheric Temperature

11.2.2.1 Total Neutral Density

The Cassini INMS measures both neutral and ion species with mass to charge ratios (m/q) ranging from 0.5 to 8.5 Da and from 11.5 to 99.5 Da using a radio-frequency quadrupole mass analyzer. In its closed source neutral (csn) mode ambient neutrals enter into an antechamber from which thermalized neutrals are guided past an electron filament at which a fraction of the molecules are ionized. The resulting ions are then guided to the mass analyzer (see the instrument descriptions in Kasprzak et al. 1996, and Waite et al. 2004). The first in situ mass spectrometer measurements of Titan's upper atmosphere were made by the INMS during the Ta flyby in October 2004 (Waite et al. 2005). Typically, a given neutral species contributes to the measured signal at several mass peaks as a consequence of dissociative ionization by the filament electrons and due to isotope effects (c.f., Waite et al. 2005; Cui et al. 2008).

INMS measurements of the main neutral constituents during the early encounters have been analyzed by Waite et al. (2005), Yelle et al. (2006), Müller-Wodarg et al. (2006) and De La Haye et al. (2007a). These analyses interpreted the measurements at varying altitudes along the spacecraft trajectory as vertical profiles, assumed hydrostatic equilibrium and derived temperatures that ranged from 149 to 158 K, with some indication that the temperature varied between the Ta (mainly dayside) and T5 (nightside and high latitude) flybys. Fig. 11.1 shows nitrogen and methane densities from 4 early flybys (De La Haye et al. 2007a). The N_2 densities for Ta measured at an altitude of 1200 km were 6.84×10^8 cm^{-3} and 6.48×10^8 cm^{-3} for ingress and egress respectively, whereas the T5 densities were only about half these values, indicating the presence of some latitudinal or day-night variations. De La Haye et al. fit the N_2 density profiles with a hydrostatic equilibrium expression using an isothermal temperature profile with T=149 ± 3 K, a value close to the 152.9 K (evening) and 157.7 K (morning) values found by Vervack et al. (2004) in their analysis of the Voyager UVS data. This similarity is surprising because the Cassini measurements were made for solar minimum conditions and the Voyager 1 encounter occurred near solar maximum. Differences between inbound and outbound neutral densities are evident in INMS data on many flybys, suggesting that horizontal as well as vertical atmospheric density variations are present in the thermosphere. This is not surprising given that the spacecraft travels about 10^3 km horizontally for altitudes below 1,500 km (e.g., Müller-Wodarg et al. 2006). Smaller-scale structures superimposed on the overall altitude variations are also evident in the density profiles, and these have been interpreted by Müller-Wodarg et al. (2006) as gravity waves.

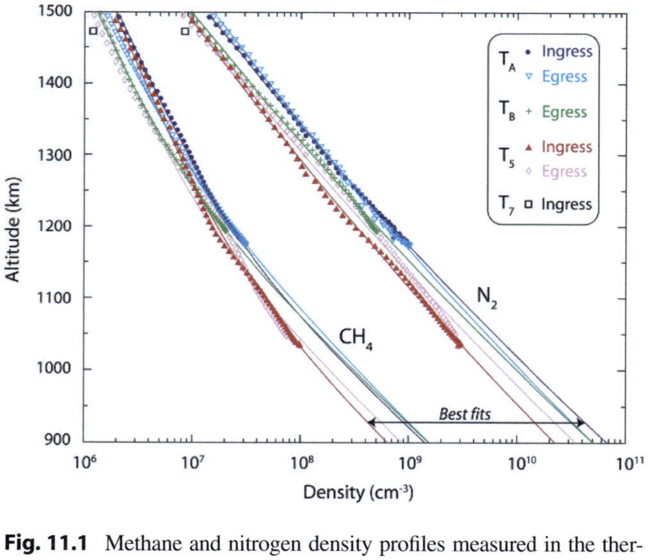

Fig. 11.1 Methane and nitrogen density profiles measured in the thermosphere by the Cassini INMS experiment during the Ta, Tb, T5, and T7 flybys. At closest approach (CA) the local times of these flybys were 16.7, 16.8, 0.7, and 17.4 hours, respectively. The latitudes at CA were 38.8°N, 59.1°N, 73.9°N , and 53.1°S. The solid lines are model fits to the data as discussed in the text. Ingress are red lines and egress are blue lines (from De La Haye et al. (2007a) copyright 2007 American Geophysical Union–further reproduction or electronic distribution is not permitted).

CH_4 mole fractions estimated from this early data for 950 km ranged from 2.55% to 2.99% and are in good agreement with results from the Voyager UVS occultation experiment (Vervack et al. 2004), suggesting that the thermospheric structure was similar to that at the time of the Voyger 1 encounter. This similarity is somewhat surprising because the Cassini measurements were made for solar minimum conditions and the Voyager 1 encounter occurred near solar maximum.

More recent studies have considered the larger data sets acquired during the prime Cassini mission. Müller-Wodarg et al. (2008) examined the distribution of N_2 and CH_4 using data from 13 targeted flybys. Details of the geometrical and solar activity conditions can be found in that paper, but these flybys all occurred in the northern hemisphere and the solar F10.7 flux was always less than 100×10^{-22} W m^{-2} Hz^{-1} (i.e., solar minimum). Potentially, the thermospheric structure may vary with latitude and local time, because of the variation in solar insolation, and with longitude or Titan orbital position, because of possible variations in energy input from Saturn's magnetosphere or the solar wind. Secular and/or seasonal variations are also possible. Müller-Wodarg et al. (2008) showed that significant variability exists and must be taken into account in the analysis of individual profiles.

Yelle et al. (2008) and Cui et al. (2008) avoided the complexities of horizontal variations, and the pitfalls inherent in analyzing individual profiles, by constructing average density profiles for the northern hemisphere from the data set described above. Some results are shown in Fig. 11.2. The distribution

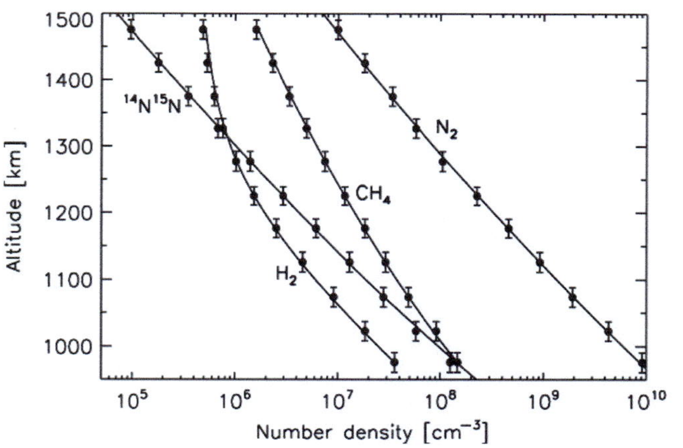

Fig. 11.2 Globally averaged INMS density profiles for N_2, CH_4, N_2, and $^{14}N^{15}N$ for data through early 2008. The solid line fit shown for N_2 was obtained with a thermospheric temperature of 154 K. The best fit to the methane and hydrogen data were obtained with diffusion models with escape fluxes of 3.0×10^9 cm^{-2} s^{-1} and 1.3×10^{10} cm^{-2} s^{-1}, respectively (from Cui et al. (2009a))

of mass density (including only contributions from N_2 and CH_4) for this average profile implies a temperature of approximately 150 K, in accord with the earlier results. The CH_4 mole fraction at 1,000 km is 2.88%. Note that the CH_4 mole fraction varies strongly with altitude in this region because of diffusive separation, as will be elaborated on below.

In order to deal with the problem of horizontal and vertical density variations being confounded, Müller-Wodarg et al. (2008) employed a combined analysis of all the available data. There was insufficient data to constrain variations with all important geophysical variables so it was assumed that the variations were primarily vertical and latitudinal. The measurements were interpolated onto a uniform altitude grid and the latitudinal variations at each altitude were fit with Legendre polynomial expansions. The derived coefficients for the N_2 and CH_4 distributions are provided in Müller-Wodarg et al. (2008). This empirical model allows a determination of true altitude profiles for the N_2 and CH_4 distributions that may in turn be analyzed to determine the temperature profile by assuming hydrostatic equilibrium, as shown in Fig. 11.3. The derived temperature is fairly uniform at high altitude with a value of 142 ± 2 K, but develops pronounced structure below 1,200 km. Near 1,000 km, the base of the model, the temperature varies from 165 K near the equator to 140 K near the north pole. The variations in CH_4 mole fraction behave differently. The mole fraction at 1,030 km is nearly constant with latitude at a value of 1.85%, but at 1,200 km the mole fraction varies from 3.3% near the equator to 5.9% at the poles. This variation is partly due to the shifting of the constant pressure altitude as latitude varies.

Cui et al. (2009a) took a different approach than Müller-Wodarg et al. by sorting data into various bins defined by ranges of geophysical variables. By considering 3 bins in latitude from 0–30°, 30–60°, and 60–90° and fitting the density distribution derived for each bin, Cui et al. (2009a) determined best-fit isothermal temperatures of 153, 151, and 146 K. The trend is quite consistent with that found in Müller-Wodarg et al. (2008). Cui et al. (2009a) also divided the data into longitude bins and found that the Saturn side of Titan is warmer than the anti-Saturn side and the magnetospheric ram side is warmer than the wake side. Dividing the data into day and night bins, Cui et al. show that the nightside is warmer than the dayside, confirming a result first suggested by De La Haye et al. (2007a). Cui et al.'s analysis shows what horizontal variations may be present and provides the sense of the variations, but with the small number of samples presently available it is impossible to determine which geophysical variations are the most important.

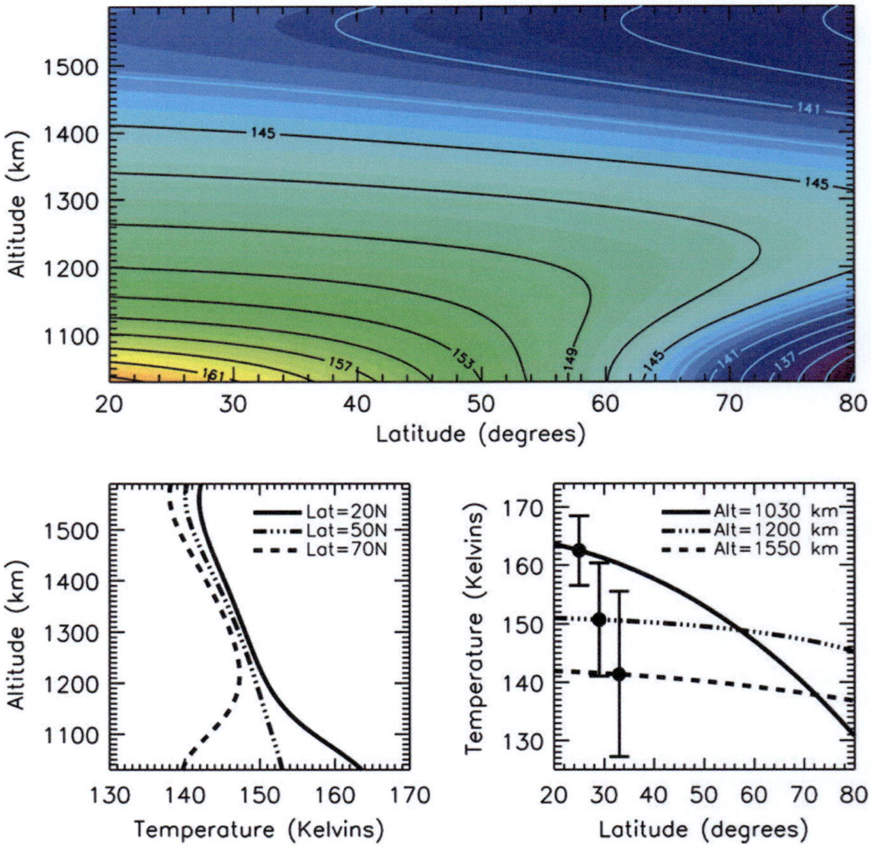

Fig. 11.3 Temperatures in the thermosphere of Titan from an empirical model based on densities of major neutral species by the Cassini INMS. Top panel: temperature contours versus latitude and altitude. Bottom panels: Temperatures versus altitude and versus latitude from the empirical model (from Müller-Wodarg et al. (2008); copyright 2008 American Geophysical Union–further reproduction or electronic distribution is not permitted).

11.2.2.2 Dynamical Implications of Measured Neutral Density Structure

Pre-Cassini predictions from a Thermospheric General Circulation Model (TGCM) (Müller-Wodarg et al. 2000, 2003) suggested thermospheric wind speeds up to 60 m/s. These models solved the continuity, energy and momentum equations, including viscosity and thermal conduction terms, on a latitude, local time, pressure grid including the variation of gravity with height and the spherical form for the divergence terms. Both factors are important because of the large extent of Titan's thermosphere. The lower boundary of the model was set at 600 km and the upper boundary was at 1,500 km. Solar heating parameterized by a heating efficiency was included and radiative cooling by HCN after Yelle (1991). The model calculated the distribution of N_2, CH_4, and HCN including advection and diffusion but not chemistry. The production of HCN was parameterized with a Chapman production function adjusted to produce densities consistent with more elaborate photochemical models.

Assuming a stationary lower boundary, Müller-Wodarg et al. (2003) predicted a fairly simple day/night circulation pattern, but the authors also point out that strong winds at the lower boundary could dominate the upper atmosphere. Strong winds at the 600 km are not unexpected because the stratosphere is known to super-rotate with wind speeds on the order of 100 m/s. These winds extend at least up to 500 km (Achterberg et al. 2008), and it is unlikely that they die away completely by 600 km.

The Cassini data are obviously inconsistent with a simple day/night circulation pattern. In fact, the available evidence suggests that the dayside is cooler than the nightside (De La Haye et al. 2007a; Cui et al. 2009a). At the present time, the energy sources for the upper atmosphere are unclear, leading Müller-Wodarg et al. (2008) to consider the circulation pattern implied by the empirical temperature field. Both a stationary and super-rotating lower boundary, with the Achterberg et al. (2008) wind field, were considered. With a stationary lower boundary, maximum zonal and meridional wind speeds of 50 and 150 m/s, respectively, were predicted, and with the Achterberg et al. (2008) winds at the lower boundary the maximum zonal and meridional winds were 120 and 80 m/s, respectively. The winds are driven by the strong latitudinal temperature gradient and so blow from the equator to the north pole and, by assumption, from west to east. The different circulation patterns in these two cases can be understood as a manifestation of different coriolis forces.

Bell et al. (2009) also presented results from a thermospheric general circulation model, incorporating essentially the same physics as Müller-Wodarg et al (2003), but in addition they included the possibility of non-hydrostatic vertical distributions. The circulation during the height of southern summer was studied assuming the Achterberg et al. (2008) winds at the lower boundary. They calculated a strong latitudinal temperature gradient, from 142 K at the southern pole to 187 at the northern pole, at an altitude of 600 km. The temperature gradient vanishes by 1,000 km, however, and the atmosphere in this region is essentially isothermal at a temperature of 160 K, which is inconsistent with the observations, perhaps indicating that some heating and/or cooling processes, or a momentum source, has not been accounted for.

11.2.2.3 Diffusive Separation – Methane Structure in the Thermosphere

The effects of diffusive separation in Titan's upper atmosphere are clearly seen in scale heights for different species (Fig. 11.2), but quantitative analysis reveals some surprises. Fits of a diffusion model (i.e., Eq. 11.3) to the average data shown in Fig. 11.2, yield an eddy diffusion coefficient of K $= 4 \times 10^8$ cm^2 s^{-1}. This value is consistent with the range of 4–8 $\times 10^8$ cm^2 s^{-1} determined by Vervack et al. (2004) from Voyager UVS occultation measurements, but smaller than the values of 5.2×10^9 cm^2 s^{-1}, 10×10^9 cm^2 s^{-1} and 3.9×10^9 cm^2 s^{-1} determined by De La Haye et al. (2007a) for the Ta, Tb, and T5 passes. These differences could be due to the analysis of individual passes rather than vertical profiles inferred from the entire data set. However, the differences in any case do not provide too much insight into vertical transport because CH_4 does not appear to be a good tracer.

The best tracer for vertical transport in Titan's atmosphere is the heavy species, argon-40 (^{40}Ar), which is not likely to escape rapidly, and which is chemically inert. The ^{40}Ar altitude distribution should be determined entirely by molecular and eddy diffusion. Moreover, the mole fraction of this species has also been measured in the lower atmosphere by the GCMS experiment at a value of 4.3×10^{-5} (Niemann et al. 2005). Yelle et al. (2008) showed that the combination of the INMS and GCMS measurements imply an eddy diffusion coefficient in the upper atmosphere of 2–5 $\times 10^7$ cm^2 s^{-1}, with a best-fit value of 3×10^7 cm^2 s^{-1}. This value has been confirmed in separate analyses by Cui et al. (2008, 2009a), using a slightly expanded data set (2 additional passes), and by Bell et al. (2009). The homopause is located at 850 km for this value of the eddy diffusion coefficient.

The discrepancy between the eddy diffusion coefficients derived from CH_4 and ^{40}Ar can be explained as the consequence of rapid escape of CH_4 (Yelle et al. 2006, 2008). A large escape flux affects the density profile of a minor constituent in a manner similar to a large eddy coefficient. To see this, we rewrite the diffusion Eq. 11.3 (Yelle et al. 2008) as:

$$\frac{dX_i}{dr} = \frac{D_i}{D_i + K}\left(\frac{1}{H_a} - \frac{1}{H_i}\right)\left(1 - \frac{\Phi_i}{\Phi_l}\right) \tag{11.4}$$

where Φ_i is the flux and Φ_l is the limiting flux, defined by

$$\Phi_l = D_l N_a \left(\frac{1}{H_a} - \frac{1}{H_i} \right) X_i .$$

The observations indicate that there has been little diffusive separation of CH_4, implying a low value for dX_i/dr. H_a is the average scale height. From Eq. 11.4 we see that a small dX_i/dr can be the result of either K » D or $\Phi_i \approx \Phi_l$. The ^{40}Ar data imply that K « D near 1,000 km, therefore Φ_i must be close to Φ_l. Fits of a diffusion model to the CH_4 data using the eddy coefficient determined from the ^{40}Ar data imply a CH_4 escape flux of 2.5–3.0 × 10^9 cm^{-2} s^{-1}, a very large escape value. The implications of this value are discussed further in Chapter 16 on escape.

Bell et al. (2009) offer an alternative explanation for the well-mixed CH_4 profile. These authors propose that the CH_4 altitude distribution is affected by chemical loss of CH_4. Photolysis and gas phase chemical reactions are too slow compared with diffusion to significantly affect the CH_4 distribution, so instead Bell et al. suggested that CH_4 is lost through incorporation into aerosols. The idea is based on laboratory experiments by Jacovi and Bar-Nun et al. (2008) that show that noble gases can be incorporated into laboratory analogs of Titan aerosols. This interesting hypothesis will require further scrutiny, including why chemical loss to aerosols would force the CH_4 distribution to be parallel to the N_2 distribution and whether the aerosol density could be high enough to accommodate the required amount of CH_4.

Measurements made by the Cassini UVIS experiment using the stellar occultation technique have also provided information on the structure of the upper atmosphere (Shemansky et al. 2005). The attenuation of the radiation from a star or the Sun as the ray path cuts deeper through the atmosphere depends on the slant column densities of the absorbing species and on the relevant photoabsorption cross sections at the wavelengths of interest, and this allows slant column density profiles to be derived from the variation of the intensity. Using different features in the observed ultraviolet spectra the UVIS experiment obtained column densities for methane and several minor species for altitudes between 400 and 1,700 km, as shown in Fig. 11.4 for the Tb flyby.

Absolute density comparisons require inversion of these profiles to convert column densities (N(r)) to number densities (n(r)), where r is the radial distance, and this was carried out for the UVIS observations of methane (Shemansky et al. 2005; methane abundance listed in Table 11.1). The slant column density is approximately given by $N(r) \approx 2n(r)H(r)Ch(\chi = 90°, r)$ where H is the scale height of the species of interest, and $Ch(\chi = 90°, r)$ is the Chapman function (Banks and Kockarts 1973) which accounts for geometrical effects. The effective slant path in the lower thermosphere of Titan is approximately $N(r)/n(r) \approx 0.8 \times 10^8$ cm. The methane mixing ratio derived

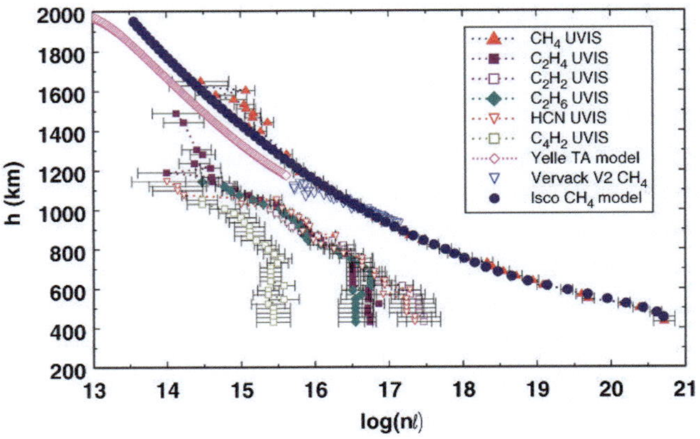

Fig. 11.4 Column densities (log base-10 and units of cm^{-2}) of methane and several other species versus slant path height derived from solar occultation measurements made by the Cassini UVIS experiment during the Tb flyby. A model fit to the UVIS methane profile is shown and was used to find a thermospheric temperature of about 140 K increasing to a bit less than 150 K in the exosphere. Methane profiles from INMS Ta data (Waite et al. 2005) and from a pre-Cassini Voyager-based occultation (Vervack et al. 2004) are shown for comparison (From Shemansky et al., Science, 308, 978, 2005; reprinted with permission from AAAS)

near an altitude of 1,000 km from UVIS data is ≈6%, which is about twice the INMS mixing ratio. However, the INMS value is representative of the northern hemisphere, whereas the UVIS occultation on Tb took place near −36° latitude in the southern hemisphere. The thermospheric temperatures derived from the UVIS occultation data were about 140–150 K.

11.2.3 Structure of the Exosphere and Atmospheric Escape

The exosphere and its link to atmospheric escape will mainly be covered in Chapters 10 and 15, but a brief discussion of this important topic is also provided here. Titan's exobase is located near 1400–1450 km according to De La Haye et al. (2007a)'s analysis of the IMMS density profiles for Ta, Tb, and T5. De La Haye et al. found that model fits to the density profiles for altitudes above about 1500 km required temperatures higher than the thermospheric temperatures derived at lower altitudes (i.e., exospheric temperatures of 175–203 K, which are 25–50 K higher the thermospheric temperatures).

De La Haye et al. suggested that in addition to the thermospheric "thermal" neutrals, a superthermal population of neutrals was also present near the exobase (with suprathermal densities of ≈4 × 10^5 cm^{-3} for N_2 and ≈10^5 cm^{-3} for CH_4, respectively). Different energy distribution functions were tried in fitting the data. The associated escape fluxes for N_2 and CH_4 were 7.7 × 10^7 N atoms cm^{-2} s^{-1} and 2.8 × 10^7 C atoms cm^{-2} s^{-1}, respectively. In order to sustain these

Table 11.1 Neutral Composition of Titan's Thermosphere near 1000–1100 km: Mixing Ratios

Species	Lebonnais	Wilson and Atreya	INMSTa	INMSW07	INMSglob–Cui-unc	INMSglob–Cui-corr	INMSglob-Magee	UVISTb	Vuitton
CH_4	7.100%	9.400%	2.19 ± 0.0002%	1.550%	2.20 ± 0.01%	2.20 ± 0.01%	2.45 ± .366%	5.700%	
H_2	0.270%	0.220%	0.410 ± 0.005%	0.400%	0.390 ± 0.01%	0.390 ± 0.01%	0.409 ± .034%		
C_2H_2	0.039%	0.063%	0.019 ± 0.005%	0.025%	0.012 ± 0.01%[b]	0.0310% ± 0.01%[b]	0.0378–.0571%	0.340%	
C_2H_4	0.080%	0.110%	0.053 ± 0.008%	0.072%	[b]	[b]	0.007–0.065%	0.430%	
C_3H_4[a]	11 ppm	18 ppm	3.9 ± 0.22 ppm	11 ppm	6.3 ppm	140 ± 90 ppm	10.7–13.0 ppm		
HCN		900 ppm					192–561 ppm	5520 ppm	200 ppm
^{40}Ar			7.1 ± 0.1 ppm		11.0 ± 0.3 ppm	11.0 ± 0.3 ppm	15.2–16.1 ppm		
C_2H_6	37 ppm	197 ppm	121 ± 6 ppm	20 ppm	27 ± 1.9 ppm	73 ± 26 ppm	44–102 ppm	4480 ppm	
HC_3N		2 ppm[d]			<0.8 ppm	32 ± 7 ppm	1.32–2.02 ppm		40 ppm
C_6H_6		0.1 ppm[d]		3.3 ppm	0.9 ± 0.04 ppm		3.22–3.34 ppm		3 ppm
C_4H_2		3 ppm[d]		5.1 ppm	2.5 ± 0.1 ppm	6.4 ± 2.7 ppm	6.09–7.37 ppm	740 ppm	10 ppm
C_6H_2		0.03 ppm[d]							0.8 ppm
C_7H_4									0.3 ppm
C_7H_8					<0.13 ppm		0.006–0.08 ppm		0.2 ppm
C_8H_2		5E-4 ppm[d]							0.2 ppm
C_2N_2				3.3 ppm	1.5 ± 0.09 ppm	48 ± 8 ppm	1.72–2.83 ppm		
CH_3CN		0.9 ppm[d]			<1.48 ppm	31 ± 7 ppm[c]			3 ppm
C_2H_3CN		1 ppm[d]			<0.57 ppm	<18 ppm	0.24–0.61 ppm		10 ppm
C_2H_5CN							0.218–0.46 ppm		0.5 ppm
C_4H_3N									4 ppm
HC_5N									1 ppm
C_5H_5N									0.4 ppm
C_6H_3N									0.3 ppm
C_6H_7N									0.1 ppm
NH_3		0.04 ppm[d]			29.9 ± 7.8 ppm				6.7 ppm
CH_2NH									10 ppm

Lebonnois et al. (2001) model – 1180 km.
Wilson and Atreya (2004) model – 1180 km.
INMSglob-Magee – global average 1000–1100 km. min and max values (Magee et al., 2009).
INMS Ta (Waite et al., 2005) – 1176 km.
INMSW07 (Waite et al. 2007; z=960 km, 71° lat; just one example).
UVIS Tb (Shemansky et al., 2005); reported relative to CH_4; converted here with n_{tot}=5.8E9 cm^{-3}; 1006 km.
INMSglob – Cui –unc: uncorrected global average 1077 km (Cui et al., 2009a)
INMSglob–Cui-cor: corrected values 1077 km (Cui et al., 2009a).
Vuitton (Vuitton et al., 2006, 2007) ionospheric model; just deduced values – 1100km

[a] C_3H_4 (CH_3C_2H and $CH_2C_2H_2$).
[b] Combined C_2H_2 and C_2H_4.
[c] altitude of 1025 km.
[d] altitude of 1000 km.

superthermal populations, energy deposition rates near the exobase of about 100 eV cm^{-3} s^{-1} and 40 eV cm^{-3} s^{-1} were required for N_2 and CH_4, respectively.

The source of such exospheric energy inputs is not obvious. De La Haye et al. (2007b) investigated the possible contribution of photochemistry to the superthermal neutral population, and in particular, to the escape flux. For example, dissociation of N_2 and CH_4 either by absorption of solar photons or by electron impact generates superthermal N atoms (Lammer and Bauer 1993). The neutral products of the very large number of ion-neutral reactions that take place in the ionosphere are also superthermal (Cravens et al. 1997) and can also lead to either heating (if the reactions occur below the exobase) or to escape (if the reactions occur near or above the exobase and if the product energies are sufficiently high). De La Haye et al. (2007b) used INMS data and a two-stream superthermal neutral transport code to critically evaluate this photochemical source of superthermal neutrals to see if it could account for the empirically-derived superthermal neutral population near the exobase (De La Haye et al. 2007a). They found that the energy deposition rate associated with photochemical reactions is about a factor of 100 too low to account for the observed superthermal populations and generated only about 10% of the escape fluxes deduced empirically.

Other processes have also been suggested for exospheric heating and for escape, including the sputtering of neutrals associated with impacting magnetospheric ions or with locally produced pick-up ions (Lammer and Bauer 1993; Michael et al. 2005; Michael and Johnson 2005; Shematovich et al. 2003).

A related puzzle is the energy source needed to power the 10^9 cm^{-2} s^{-1} methane escape fluxes discussed earlier if low eddy diffusion coefficient values are adopted (as supported by the argon measurements) (see Cui et al. 2009a, and earlier discussion in this chapter). The required energy inputs near the exobase are high compared with magnetospheric energy sources. Strobel (2008) suggested that hydrodynamic escape can explain the high methane escape flux, and the required heat input is supplied by upward heat conduction.

11.2.4 Interpretation of the Structure of the Neutral Atmosphere, Energy Balance, and Small-Scale Structure

The thermal structure of the upper atmosphere is determined by the balance between heating, cooling, and heat transport either dynamically or via thermal conduction. Yelle (1991) described some of the key features of the thermospheric energy balance and demonstrated that the main cooling agent was collisional excitation of HCN ro-vibrational levels followed by emission of infrared radiation.

Neutral heating in the thermosphere results from deposition of energy from external sources, including absorption of solar radiation in the thermosphere and deposition of energy associated with the precipitation of energetic electrons or ions from Saturn's magnetosphere. Dynamical heating can also in principle make a contribution – both from the large-scale dynamics (e.g., adiabatic heating) and from small-scale processes such as gravity wave breaking. Not all the atmospheric energy deposition ends up as neutral heat (some energy is radiated away). The fraction of deposited energy ending up as heat is known as the heating efficiency. Calculating the heating efficiency as a function of altitude involves a detailed (and uncertain) assessment of processes by which absorbed energy ends up as heat. To consider just one example, an extreme ultraviolet solar photon absorbed by an N_2 molecule can ionize it producing an N_2^+ ion plus a photoelectron with energy equal to the photon energy minus the ionization potential of N_2. The N_2^+ ion then participates in the ion-neutral chemistry, as will be discussed later in the chapter. Most of the N_2^+ ions react with CH_4 and form CH_3^+ ions that then mostly react with CH_4 producing $C_2H_5^+$. This chain of reactions continues until an ion and a thermal electron combine in a dissociative recombination reaction, thus producing fast neutrals that contribute to the neutral heating rate. Each reaction in the chain is exothermic and the reaction products carry kinetic energy, most of which ends up as neutral heat. Reaction products are often vibrationally-excited (or even electronically excited) and this energy is either lost to the atmosphere via radiative de-excitation or contributes to heating if a quenching collision takes place. Photoelectrons typically have energies of several eV or more, and they undergo a sequence of collisions on their way to being thermalized. Collisions that end up creating excited molecules that subsequently radiate the collision energy do not contribute to heating, but other electron impact processes contribute to heating. For example, ionization leading to chemistry including electron-ion recombination reactions contributes to heating and excitations leading to metastable species that are later quenched, which also contributes to the neutral heating rate.

Heating efficiency calculations have been undertaken for several planets. The efficiencies for the thermospheres of Venus and Mars are quite low (10–20% for most altitudes) due to efficient CO_2 processes (Fox and Dalgarno 1981). The cooling rate is also high on Venus and Mars due to the CO_2. Terrestrial heating efficiencies are about 50% near the peak of the solar EUV energy deposition profile (Torr et al. 1980) and for Jupiter values of ≈50% have been calculated (Waite et al. 1983). De La Haye et al. (2008a) used Cassini data for the Ta and T5 encounters and modeled energy absorption and deposition processes, and calculated altitude-dependent efficiencies for Titan. The mean thermospheric heating efficiency was about 25% in good agreement with the value found by Fox and Yelle (1991). The peak dayside column heating rate was found to be 3×10^{10} ergs cm^{-2} s^{-1} above 990 km. De La Haye et al. (2008b) also used a one-dimensional thermal model including heat conduction (only important near 1,000 km) and HCN cooling (the main cooling process for Titan's thermosphere) and found temperatures close to 150 K, in agreement with INMS values.

Variations in temperature with latitude and/or local time can drive (via pressure gradients) thermospheric winds as shown by the global general circulation models of Müller-Wodarg et al. (2000, 2003, 2008) and Bell et al. (2009) as discussed in an earlier section. Wind speeds of about 100 m/s are typical as mentioned earlier. The dynamical processes contribute to the thermal balance via dynamical terms in the energy equation (e.g, adiabatic heating and cooling). Müller-Wodarg et al. (2000) also included a neutral energy equation with heating rates from solar radiation, although they used a heating efficiency of 50%.

The Cassini data also revealed the existence of waves in the upper atmosphere. Müller-Wodarg et al. (2006) analyzed the periodic variations seen in the INMS density measurements and showed that these are consistent with waves with vertical wavelengths of 170 to 360 km and amplitudes of 4–12% of the background values. This translates into temperature amplitudes of 5–10 K. The wave amplitudes do not grow strongly with altitude implying that they are at least partially damped. This provides a constraint on the wave frequency and Müller-Wodarg et al. (2006) determined that the wave period is of the order of hours. These derived characteristics imply that heating associated with the waves is comparable to the solar heating rate and the waves should deposite a specific momentum comparable to that in a solar-driven circulation. Though the magnitudes appear to be significant, wave heating and acceleration are difficult to specify precisely and these terms have yet to be included in global energy balance or dynamical models of the upper atmosphere.

11.2.5 Observed Variations of Minor Neutral Composition with Altitude, Latitude, Longitude, and Local Time

A large variety of neutral species have been observed in the upper atmosphere of Titan in addition to the major species N_2, CH_4, and H_2. Ground-based observations and the Voyager 1 encounter with Titan revealed an atmosphere that was rich in photochemically-produced species. Minor molecular constituents detected in the upper atmosphere by the Voyager UVS occultation experiment included C_2H_2, C_2H_4, HCN and/or HC_3N (Vervack et al. 2004). The EUV occultation experiments undertaken by the UVIS instrument on Cassini detected these species as well as C_2H_6 and C_4H_2 (Shemansky et al. 2005), and provided altitude profiles (Fig. 11.4). Many observations of hydrocarbon and nitrogen-bearing species in the stratosphere have also been made, mostly using mid-infrared spectral features. The Cassini CIRS instrument observed the following species in Titan's stratosphere: C_2H_2, C_2H_4, C_2H_6, CH_3C_2H, C_3H_8, C_4H_2, C_6H_6, HC_3N, C_2N_2, and CO_2.

In situ measurements made by the INMS in its closed source mode have provided a new source of information on minor neutral species. A measured neutral mass spectrum for the Ta encounter (altitude of about 1200 km) is shown in Fig. 11.5 (Waite et al. 2005). The detected species during this and/or other flybys include: N_2, $^{29}N_2$, CH_4, $^{13}CH_4$, ^{40}Ar, C_2H_2, C_2H_4, C_2H_6, C_3H_4, C_4H_2, C_6H_6, HC_3N, NH_3 and C_2N_2. The presence of several "minor" hydrocarbon and nitrogen-bearing species, in addition to the major species already discussed, are evident in this spectrum. Interpretation of such mass spectra are complicated by the presence of multiple mass peaks for even a single species due to the dissociative ionization of the neutrals by the filament electrons and due to the possible presence of different isotopes. For example, ambient molecular nitrogen is associated with a peak in the spectrum at mass 28, as expected, but N_2 also makes a contribution to the mass 14 peak due to dissociative ionization. This mass 14 signal competes with the signal from ambient atomic nitrogen. The mass 29 peak is due to an isotope effect ($^{15}N^{14}N$ in the atmosphere). Methane generates a signal at many mass numbers including 16 (CH_4^+), 15 (CH_3^+), 14 (CH_2^+), …. The overlapping cracking patterns must be disentangled in order for densities to be extracted. A couple of different procedures have been used (e.g., Cui et al. 2009a; MaGee et al. 2009). Another complication in the analysis of INMS closed source data is that some species can be absorbed on the antechamber walls and then be desorbed at a later time and this is not fully understood yet (c.f., Waite et al. 2004; Cui et al. 2009a; MaGee et al. 2009).

Minor neutral abundances measured by the INMS during the Ta encounter near 1200 km are listed in Table 11.1. Waite et al. (2007) reported on INMS densities for 7 passes and these are included in Table 11.1. Waite et al. (2007) made the case, using both neutral and ion measurements from INMS, that the complex ion-neutral chemistry initiated by absorption of solar photons and energetic particles in the upper atmosphere plays the key role in the formation of aromatic hydrocarbon rings such as benzene. For example, ion chemistry can produce $C_6H_7^+$ ions that in turn produce benzene via dissociative recombination. Figure 11.6 shows the benzene profile reported by Waite et al. (2007). Chemical compounds like benzene act as intermediates in the growth of heavy species and ultimately lead to the formation of the aerosols populating Titan's haze layer.

Cui et al. (2009a) and MaGee et al. (2009), with more flybys to work with, have taken a more global approach to the INMS data analysis. Cui et al. (2009a), in order to deal with the

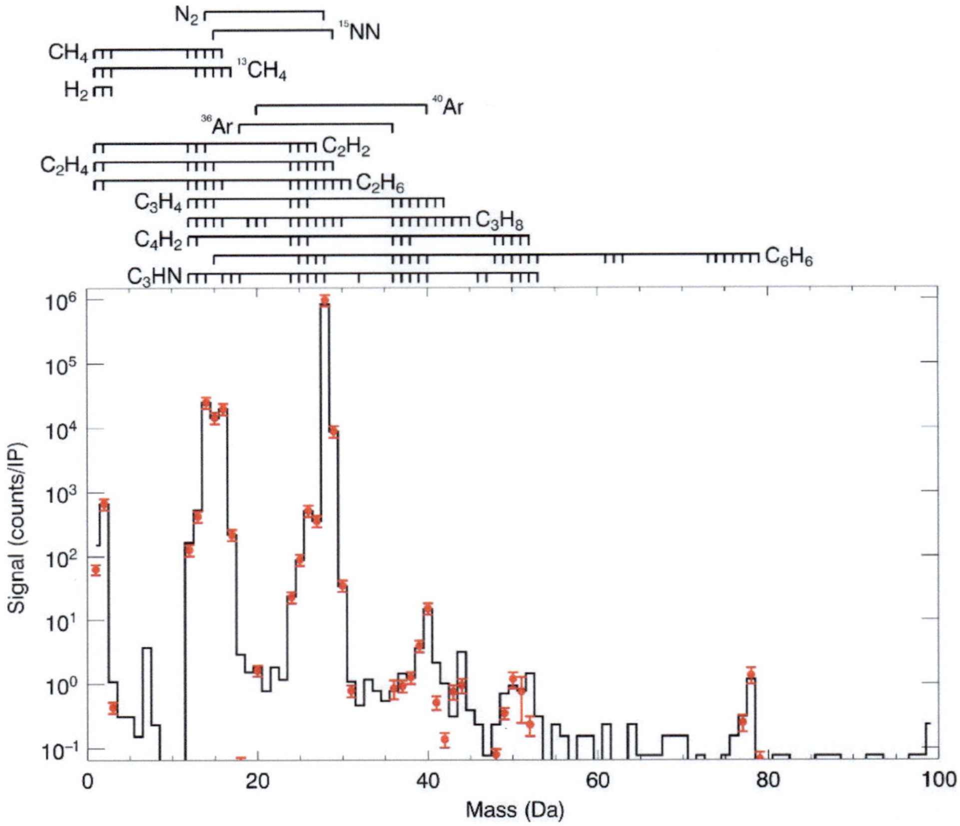

Fig. 11.5 Mass spectrum obtained at closest approach of the Ta pass by the closed source of the INMS. Instrument count rate versus mass number (charge to mass ratio in units of Daltons). Contributions at different mass peaks associated with several neutral species are indicated (From Waite et al., Science, 308, 982, 2005; reprinted with permission from AAAS)

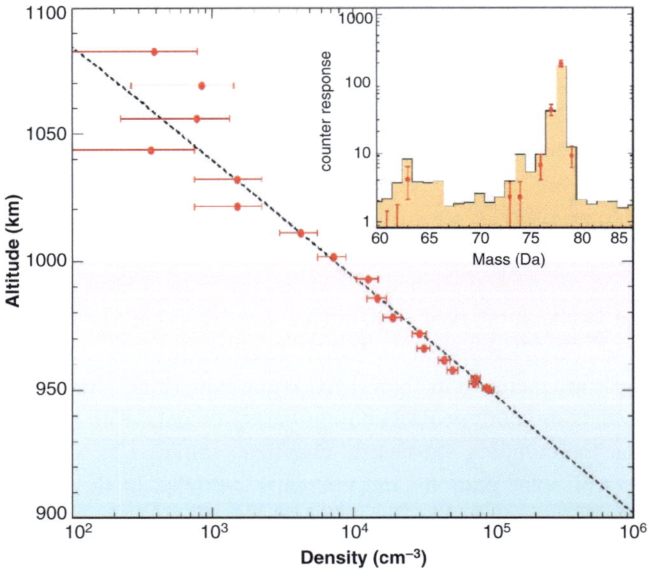

Fig. 11.6 This figure shows the altitude distribution of benzene as measured by INMS on Titan flyby T16 with statistical uncertainty as indicated by the red dots and error bars, respectively. The dashed red line indicates the molecular diffusive scale height of a molecule the mass of benzene in Titan's atmosphere. Data in the blue region were averaged to produce the mass spectrum (instrument count rate versus mass/charge in Daltons) shown in inset. Fits to the benzene spectrum using the NIST database are shown along with estimated uncertainty in red (From Waite et al., Science, 316, 870, 2005; reprinted with permission from AAAS)

cracking pattern problem, developed a fitting technique based on singular value decomposition (SVD) to determine the relative contributions of the parent species to the counting rate at each value of M/Z. The importance of correctly accounting for contributions from multiple species depends on the species under consideration. Some molecules, such as N_2, CH_4, H_2, and C_6H_6, strongly dominate in a single channel and determination of their densities is relatively straightforward. However, the mass ranges from 12–17, 24–30, 36–42, and 48–53 amu contain contributions from many species and fitting of all components simultaneously has many advantages. An example from Cui et al. of the contribution of various molecules to the signals in these mass ranges is shown in Fig. 11.7.

An additional complication discussed by Cui et al. is found in the time behavior of the signals from the minor constituents, which show a clear asymmetry between inbound and outbound densities. Moreover, the asymmetry seems to be strongly correlated with the chemical activity of the molecule (also discussed in Vuitton et al. (2008)). Inert molecules such as ^{40}Ar and C_2H_6 (see figures in the Vuitton et al. paper) display no inbound/outbound asymmetry, but for more reactive molecules the asymmetry is pronounced. Both Cui et al. (2009a) and Vuitton et al. (2008) interpreted the time delays and the correlation with chemical activity as evidence that reactive molecules are interacting with the walls of the INMS antechamber. Several types of interactions are

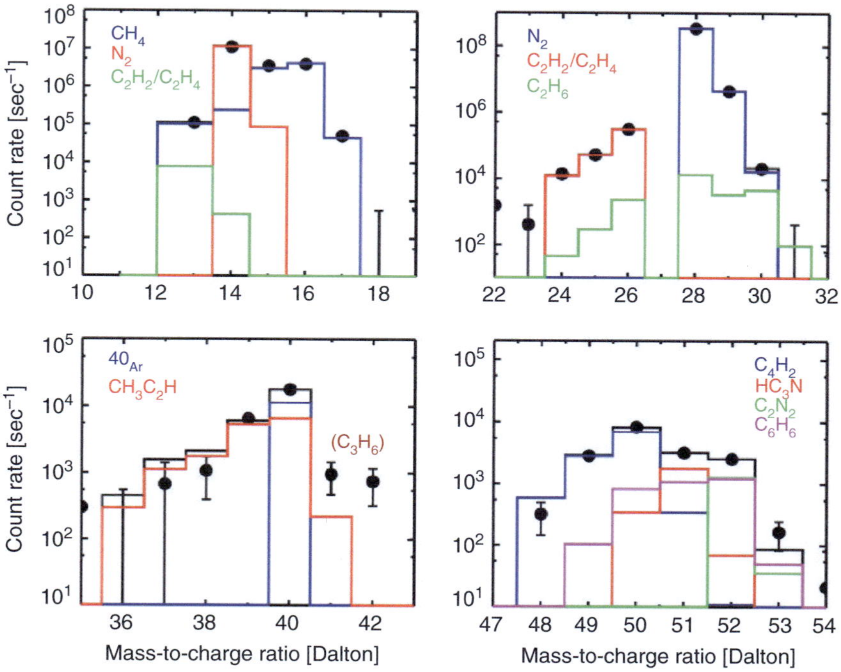

Fig. 11.7 INMS mass spectra for the T30 flyby averaged between altitudes of 960–980 km for 4 parts of the spectrum. Solid circles are measurements with 1-sigma error bars and the histograms are the contributions of each neutral species in the deconvolution procedure (from Cui et al. (2009a))

possible. Vuitton et al. (2008) suggested that chemistry on the walls of the antechamber synthesizes C_6H_6 from C_6H_5 and H and C_7H_8 from C_6H_5 and CH_3. Cui et al. (2009a) assumed that the time delay was caused by adsorption and subsequent desorption of the reactive molecules without modification by chemistry on the walls. Cui et al. developed a scheme to try to correct for these effects. Table 11.1 includes two columns for the Cui et al. neutral abundances – one column for uncorrected densities and the other that includes their corrections. Cui et al. (2009a) excluded m/z = 27 from their analysis because of concern over contamination by the very strong signal at m/z = 28.

Cui et al. also examined latitudinal variations of several neutral species measured by the INMS in the northern hemisphere (where INMS-pointing Titan passes have been concentrated up to 2008 in the Cassini mission). The ^{40}Ar mixing ratio is independent of latitude to within about ±25%. The C_2H_2, C_2H_4, C_4H_2, C_6H_6, and CH_3C_2H mixing ratios in the polar region (80° N) were observed to be about a factor of about 1.5 to 2 less than in the equatorial region (20° N. latitude). This behavior might depend on the source location for complex hydrocarbons as well as on the dynamics. The dynamical models of Müller-Wodarg et al. (2008) and Bell et al. (2009) both predicted meridional winds driven by the hotter equatorial region, and such winds should result in polar enhancements of lighter species such as methane and polar depletions (as observed) of heavier minor species such as benzene. Clearly, these observations call for sophisticated dynamical models that include sources, sinks, and dynamics for minor species.

MaGee et al. (2009) also used INMS data from a large number of flybys over 4 years to obtain abundances of minor neutral species, and these are listed in Table 11.1 as minimum and maximum values. MaGee et al. noted that the minimum mixing ratios they report (and which are reproduced in our Table 11.1) are those derived from mass spectra with high pressure background corrections, while maximum values are those without the background correction. Rather than fitting all species simultaneously, Magee et al. adopted a sequential approach in which the densities of the most abundant species are determined first, followed by subtraction of the associated signals from the spectrum. This process was continued until only 4 species remained, whose densities were determined by simultaneous fitting. Magee et al. also adopted a novel approach to the determination of the goodness of fit. Rather than the traditional chi-squared metric, Magee et al. defined a new metric in terms of the measurements and associated errors but also including information from a reference spectrum. Magee et al. did not address the time delay of the signal or correct for wall effects for the reactive minor species, though they do include a correction for wall effects in order to subtract the NH_3 contribution from m/z = 17 to estimate the $^{13}CH_4$ density. As an example of their results, they reported an average relative HCN abundance of 377 ppm at 1000 km. Two species not included in Table 11.1 but with measured values given by Magee et al. are C_3H_6 (with a mixing ratio range of 0.73–4.4 ppm) and C_3H_8 (0.35–4.9 ppm).

The Cassini UVIS experiment also provided information on several minor neutral species in the lower thermosphere as shown in Fig. 11.4 for the Tb flyby. Other UVIS occultation data is now being analyzed. Methane is depleted by a factor of about 2 for altitudes above 850 km in the sunlit atmosphere (in comparison with the lower atmosphere) and the higher order hydrocarbons show a similar depletion above 700 km. Abundances of these species at 1,000 km are also summarized in Table 11.1. The UVIS-derived information for altitudes below 1,000 km will also be discussed in Chapter 10.

Two recognizable transition regions are evident in the UVIS occultation vertical profiles (see Fig. 11.4) of higher order hydrocarbon species: (a) 600–800 km (b) ~1000 km. At the first transition (a), the vertical column density profiles show a transition from very large scale heights (of order H ~ 1000 km) to much smaller scale heights comparable to the methane scale height (H ~ 60 km). At the second transition (b), even smaller scale heights are found. The abundances of hydrocarbon species measured during the Tb occultation (acetylene, ethylene, ethane, …) all appear to be lower than the relative abundances measured by the INMS (Table 11.1). This methane difference can perhaps be explained as a latitudinal variation (higher abundances in the southern hemisphere). The postulated precursors to aerosol production, benzene and dicyanodiacetylene (c.f., Liang et al. 2007; Waite et al. 2007) were not detected in the UVIS results (Liang et al. 2007). The INMS experiment (Fig. 11.6 here) shows benzene distributed with a scale height of a diffusively separated 78 amu species (Waite et al. 2007). Projecting this distribution down to 900 km indicates that the UVIS experiment should have observed benzene in absorption.

11.2.6 Interpretation and Theoretical Considerations for the Minor Neutral Composition – Chemistry

The rich complements of ion and neutral species in Titan's upper atmosphere are closely coupled through the chemistry of the upper atmosphere. This is an extraordinary and rapidly developing subject. Because of space limitations and because investigations into this subject are still in their early stages we provide only a brief overview here, surveying some results from post-Cassini models.

The presence of methane in the atmosphere results in very complex chemical processes operating throughout the atmosphere of Titan including the thermosphere and ionosphere. Understanding these processes and how energy inputs from the Sun and magnetosphere drive chemistry leading to increasingly large and complex hydrocarbon and nitrogen-bearing compounds, and ultimately to aerosols, requires modeling. Most photochemical models of Titan have focused on altitudes below 1,000 km and have not explicitly included ion-neutral chemistry, although they have generally contained complex chemical schemes (e.g., Yung et al. 1984; Toublanc et al. 1995; Lara et al. 1996; Lebonnois et al. 2001). A number of purely ionospheric models including ion chemistry (and merely adopting neutral composition) were also constructed (and are discussed elsewhere in this chapter).

Complex hydrocarbon formation starts out in almost all the above models with the breaking of a bond in the methane molecule via photodissociation by solar radiation:

$$h\nu + CH_4 \rightarrow CH_3 + H \text{ or}$$
$$\rightarrow {}^1CH_2 + H_2 \text{ (or } H+H\text{)or} \rightarrow CH + H_2 + H \qquad (11.5)$$

(c.f., Wilson and Atreya 2004; De La Haye et al. 2008a; Lavvas et al. 2008a; and references therein). Excited methylene (${}^1CH_2$) then reacts with the methyl radical to produce ethylene:

$${}^1CH_2 + CH_3 \rightarrow C_2H_4 + H. \qquad (11.6)$$

Once ethylene is formed, it drives acetylene (C_2H_2) formation via photolysis (again with solar photons). Similarly, reactions involving CH_3 lead to ethane (C_2H_6) formation, and reaction of acetylene with ${}^1CH_2$ leads to C_3H_3 formation, which leads to the formation of other hydrocarbon species and a whole chain of subsequent reactions (c.f., Wilson and Atreya 2004).

The formation of nitrile and other nitrogen-bearing compounds (e.g., HCN, CH_3CN, ….) requires the breaking of the strong N_2 bond, which can only happen with photons with wavelengths below ≈100 nm. Atomic nitrogen can react with CH_3 radicals (from methane dissociation) to produce H_2CN, which in turn reacts with atomic hydrogen to produce HCN (Yung et al. 1984). Ionospheric chemistry is important for the formation of nitrile/nitrogen compounds, especially above the stratosphere. For example, ionization or dissociative ionization of N_2 (yielding N_2^+ and N^+ ions, respectively) results in formation of nitrogen-bearing species via reactions starting with CH_4. For example, N^+ ions react with methane to produce $HCNH^+$, the dissociative recombination of which forms HCN.

The availability of Cassini INMS ionospheric data (e.g., Cravens et al. 2006; Waite et al. 2007) is now making the quantitative assessment of these chemical pathways possible (De La Haye et al. 2008a; Vuitton et al. 2006, 2007, 2008; Krasnopolsky 2009). Neutral and ionospheric chemical pathways lead to the formation of ethylene and hydrogen cyanide as shown in a schematic (Fig. 11.8). In addition to the purely neutral pathway to C_2H_4 formation, ionization of N_2 leads to C_2H_4 production as shown in the schematic. N_2^+ is produced by solar photons or fast electrons, and in turn produces CH_3^+ and then (via reactions with CH_4) $C_2H_5^+$ ions, which in turn produce ethylene via dissociative recombination (De La Haye et al. 2008a). An example of altitude profiles of

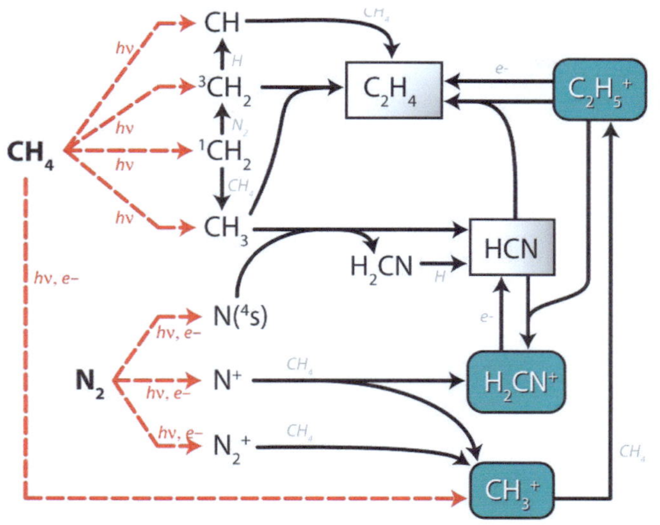

Fig. 11.8 Schematic of chemical pathways leading to the formation of ethylene (C_2H_4) and hydrogen cyanide (HCN) in Titan's atmosphere and ionosphere (from De La Haye et al. (2008a))

a minor hydrocarbon species is provided in Fig. 11.9, which includes ethylene density profiles from several models and from Ta INMS closed source neutral data (De La Haye et al. 2008a).

Lavvas et al. (2008a, b) constructed the first comprehensive photochemical models constrained by Cassini observations. The models include only neutral chemistry but incorporate a sophisticated description of haze production by photochemistry and in addition couple the chemistry and aerosol calculations to thermal structure calculations. Chemical calculations are based on an extensive reaction list (524 reactions) derived primarily from an extensive review of laboratory literature. Comparison of model predictions with Cassini measurements reveals good agreement in many cases, suggesting that chemical pathways are understood, but also some serious disagreements, implying that further work on the chemistry is needed. Model results are quite good for the main, stable hydrocarbons. The agreement between models and observations is adequate for C_2H_2, C_2H_4, C_2H_6 and C_3H_8. (There is an apparent disagreement between the models and the C_3H_8 mole fraction was based on preliminary analysis of the INMS data, but further, more sophisticated analysis of the INMS spectra by Cui et al. (2009a) determines an upper limit of 5 ppm at 1,000 km, which is perfectly consistent with the models.) The predicted C_4H_2 mole fraction, however, is only ~2 ppm, whereas the analysis of Cui et al. (2009a) finds a value of 64 ppm. The predicted abundance of C_6H_6 is far below the measured value. As discussed below, this is a reflection of the importance of ionospheric chemistry, which is not included in Lavvas et al. (2008a, b). Results for N-bearing species are also mixed. The predicted mole fraction for HCN near 1,000 km is 10^{-3}, while Magee et al. determine a value of 3×10^{-4}. CH_3CN, C_2H_5N, and C_2H_3CN are in good agreement with the values derived by Vuitton et al. (2006), but the mole fractions of HC_3N and NH_3 are under-predicted and CH_2NH over-predicted. One cannot help but suspect that the under-predicted molecules may have a significant contribution from ion chemistry. This remains as an interesting problem for future modeling investigations. Lavvas et al. (2008a, b) also examined recombination of H on aerosol surfaces as a source of H_2, utilizing the laboratory results of Sekine et al. (2008a, b). Including this process brings the calculated H_2 mole fraction into agreement with INMS measurements.

Lavvas et al. (2008a, b) also constructed a comprehensive model for the haze distribution. One unique result of this model is the prediction of significant haze production in the upper atmosphere through formation of hydrocarbon-nitrile co-polymers. The most important process is the reaction of CN with C_2H_3CN. There is a strong peak in aerosol production from this source near 850 km with significant production to above 1,000 km. It is therefore likely that this process contributes to the high altitude haze discovered by the CAPS instrument (Coates et al. 2007a). Further discussion of the high altitude haze can be found in Waite et al. (2007) and Lavvas et al. (2009).

Krasnopolsky (2009) constructed comprehensive models for the chemistry of Titan's entire atmosphere, including coupling between ion and neutral chemistry. As with Lavvas et al. (2008a, b) the main stable hydrocarbons are well produced. Agreement with many of the minor species is also quite good. The model predicted mole fractions of C_4H_2, C_6H_6, C_3H_4, and C_3H_8 that are all fairly close to measured values and the agreement between model and data for the nitriles is quite impressive. The predicted mole fractions of CH_3CN, C_2H_3CN, C_2H_5CN, C_5H_5N, CH_2NH, and HC_3N are consistent with those derived from the measured ion densities by Vuitton et al. (2007). As in Lavvas et al. (2008a, b), the model overpredicted the abundance of HCN by a factor of ~3.

De La Haye et al. (2008a) also investigated coupled ion-neutral chemistry for altitudes above 600 km. The models were also one-dimensional, but they tracked diurnal variations driven by the changing solar zenith angles and assuming solid body rotation. De La Haye et al. (2008a) predicted significant diurnal variations for C_3H_4 and HCN and this may play a role in explaining the discrepancy between models and data mentioned above. The stable hydrocarbons showed little diurnal variations, as expected, and predicted values are consistent with observations. The De La Haye et al. (2008a) models adopted fixed mole fractions at the lower boundary of 600 km which allows an unspecified flux into or out of the upper atmosphere. The ionospheric aspects of the De La Haye et al. (2008a) model are discussed in Section 11.3.6.

Hörst et al. (2008) examined the O chemistry in Titan's atmosphere and pointed out that the flux of O^+ molecules

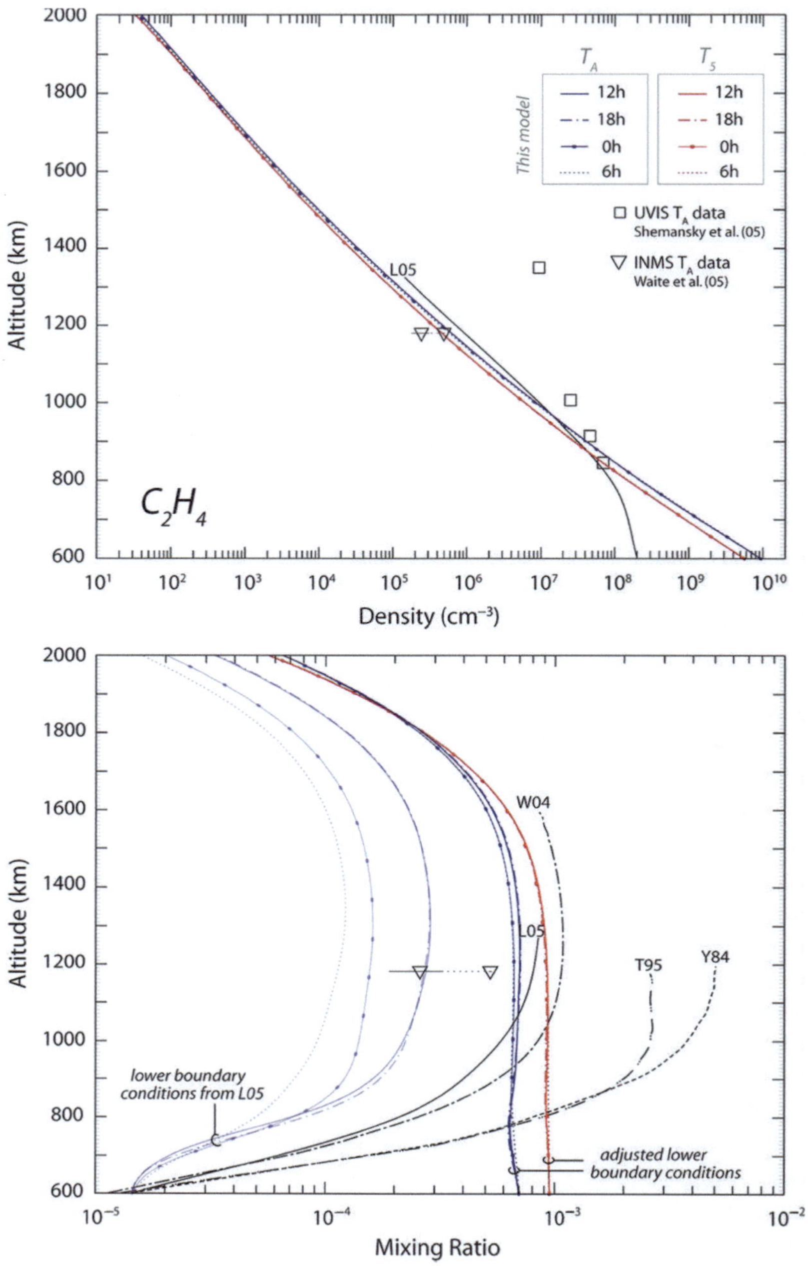

Fig. 11.9 C_2H_4 altitude profiles. Top panel: densities. Bottom panel: mixing ratios. INMS data from Ta and T5 are shown (Waite et al. 2005), and UVIS data from Ta (Shemansky et al 2005). Profiles from several models are shown. The lines labeled "this model" are from the time-dependent De La Haye et al. (2008a) model and are shown for different local times and Ta and T5 conditions. The other models are: W04 (Wilson and Atreya 2004), L05 (Lebonnais 2005), T95 (Toublanc et al. 1995), Y84 (Yung et al. 1984) (from De La Haye et al. (2008a))

from Saturn's magnetosphere measured by CAPS (Hartle et al. 2006) represented a significant source of neutral O for the atmosphere. O^+ ions precipitating into the atmosphere are neutralized near an altitude of 1000 km (also see Cravens et al. 2008a and Chapter 17). Atomic oxygen reacts with CH_3 to produce H_2CO, which is quickly photolyzed to produce CO. Previously, the only explanation for the large CO abundance was that it was a remnant of a much larger primordial CO reservoir (Wilson and Atreya 2004). Detailed models based on the O^+ hypothesis and the CAPS fluxes match the observed abundance of CO. An influx of H_2O is also required to match the observed abundances of CO_2 and H_2O. The required rates are consistent with that expected from the flux of micrometeorites into the atmosphere. CO, with a mole fraction of 50 ppm, is the fourth most abundant species in Titan's atmosphere. The connection with O^+ precipitation illustrates the very large effect that the magnetospheric interaction can have on the atmosphere as a whole.

Benzene (C_6H_6) production on Titan has been investigated by Vuitton et al. (2008) and by Waite et al. (2007). Benzene was first detected in Titan's stratosphere with mid-IR spectroscopy with a mole fraction of 6×10^{-10} (Coustenis et al. 2003). Attempts to explain the presence of C_6H_6 as a result of photochemistry focused on three-body reactions in the stratosphere (Wilson et al. 2003; Lebonnais 2005; Lavvas et al. 2009). Wilson et al. (2003) considered production of C_6H_6 by ionospheric chemistry, but found this to be a small source. Overall, three-body reactions in the stratosphere appeared to adequately explain the observed C_6H_6 abundance. The situation was overturned by the discovery of a C_6H_6 mole fraction of 1–10 ppm at 1,000 km by the INMS where the range reflects the pass-to-pass variation in derived abundance (Waite et al. 2007; Vuitton et al. 2008). The large C_6H_6 mole fraction measured by INMS indicates that the primary source of C_6H_6 is in the thermosphere rather than the stratosphere (Waite et al. 2007; Vuitton et al. 2008). At the low pressures in the thermosphere, three-body reactions are not important, and different chemical pathways for C_6H_6 production must be found. Waite et al. (2007) suggested that C_6H_6 might in fact be synthesized by ionospheric chemistry. Supporting this is the observation that $C_6H_7^+$ is present in Titan's ionosphere since recombination of this ion could produce C_6H_6. However, estimates of the strength of the recombination source by Waite et al. (2007) are about a factor of 10 smaller than required to explain the observed C_6H_6 densities, similar to the earlier estimates by Wilson et al. (2003).

Vuitton et al. (2008) conducted a more comprehensive analysis of the C_6H_6 data and coupled that with detailed models for the ion and neutral chemistry, and showed that the observed abundance can be explained by ionospheric chemistry. Photolysis of C_6H_6 is extremely rapid, resulting in production of C_6H_5 (phenyl) and H. Vuitton et al. (2008) show that the calculated C_6H_5 density is larger than the C_6H_6 density. The INMS is probably measuring a combination of C_6H_5 and C_6H_6 because heterogenous reactions on the instrument walls can convert the C_6H_5 radical into C_6H_6. C_6H_5 in the atmosphere can also react with H and reform C_6H_6, so photolysis does not necessarily represent a loss of C_6H_6, but C_6H_5 can also react with other radicals (including itself), producing more complex ring molecules. These more complex ring molecules may represent an interesting component of the Titan haze (Vuitton et al. 2008). Reaction of C_6H_5 with CH_3 produces C_7H_8 (toluene), also detected by the INMS, although it is not clear if the detected C_7H_8 is synthesized in the atmosphere or the instrument (Vuitton et al. 2008). The ionospheric model used by Vuitton et al. (2008) is discussed in more detail in Section 11.3.6.

11.3 The Structure and Composition of the Ionosphere

11.3.1 Review of Basic Processes Relevant to the Ionosphere

The structure of Titan's ionosphere can be described reasonably accurately with the fluid conservation equations: continuity equation for densities, momentum equation for bulk flow velocity (or ion fluxes), and energy equations for pressure (or temperature). These equations as applied to ionospheric plasmas are discussed in many standard textbooks (cf. Schunk and Nagy 2000; Kelley 1989; Cravens 1997). Plasmas differ from neutrals in that "free" electrons and ions respond to electric and magnetic fields via the Lorentz force. Charge particles also create electric and magnetic fields via electrical charge imbalances and electrical currents (i.e., Maxwell's equations). Ionospheric plasmas are quasi-neutral, meaning that the net charge density is extremely small so that the total positive ion density equals the electron density plus the density of negative ions for singly charged ions (the sum of n_s equals $n_e + n_-$ where n_s is the density of ion species s and n_- is the total negative ion density).

The continuity equation for each ion species includes not just the effects of transport but also sources and sinks of that species. At Titan, primary ion production is due to photoionization by solar radiation and/or due to impact ionization associated with particle precipitation. The main chemical sink of plasma as a whole is dissociative recombination. The fluid momentum equation for each species or for the plasma as a whole needs to include terms for pressure gradient forces, the Lorentz force, collisional/friction forces between ion species and neutrals, and the gravitational force. The energy equation should include terms for heat conduction, bulk heat transport, heating (e.g., due to Coulomb collisions of thermal plasma with suprathermal electrons or frictional heating) and cooling (e.g., due to electron impact excitation of vibrational states of CH_4).

Often, a single-fluid description of a plasma is useful and the momentum equation can be put into the following "magnetohydrodynamic" (MHD) form (c.f., Cravens 1997):

$$\rho\left(\frac{\partial \mathbf{u}}{\partial t}+\mathbf{u}\cdot\nabla\mathbf{u}\right)=-\nabla\left(p_e+p_i\right)+\mathbf{J}\times\mathbf{B} + \rho\mathbf{g}-\rho\nu_{in}\left(\mathbf{u}-\mathbf{u}_n\right) \quad (11.7)$$

where **u** is the plasma flow velocity, ρ is the mass density of the plasma (equal to the sum of $m_s n_s$ over all ion species s where m_s is the mass of ion species s), **B** is the magnetic field, **J** is the current density, ν_{in} is the ion-neutral momentum transfer collision frequency, and $\mathbf{u}_n$ is the neutral flow velocity. The thermal pressures for electrons and ions are $p_e = n_e k_B T_e$ and $p_i = n_i k_B T_i$, respectively, where n_i is the total ion density, k_B is Boltzmann's constant, and T_e and T_i are the electron and ion temperatures, respectively. Eq. 11.7 essentially states that a parcel of plasma accelerates (i.e. flows) in response to the net force on it. The magnetic field is usually determined in a MHD description with the magnetic induction equation (Faraday's law plus a generalized Ohm's law).

Ampere's law (minus the displacement current) can be used to rewrite the Lorentz force term solely in terms of the magnetic field:

$$\mathbf{J} \times \mathbf{B} = -\nabla \left({B^2}/{2\mu_0} \right) + \frac{1}{\mu_0} \mathbf{B} \cdot \nabla \mathbf{B} \quad (11.8)$$

The quantity $B^2/2\mu_0$ is "magnetic pressure", p_B, which can be grouped with the thermal pressure in Eq. 11.7. For planar geometries, the total pressure ($p_e + p_i + p_B$) tends to be constant. For example, at Venus, the magnetic pressure in the magnetic barrier region above the ionopause has been shown to balance the thermal pressure of the relatively cold and dense ionospheric plasma located below the ionopause (c.f., Luhmann and Cravens1991). The ion-neutral frictional force term becomes increasingly important with decreasing altitude and this severely limits plasma flow velocities in the lower ionosphere (Keller et al. 1994). At higher altitudes where collisions become less important, ionospheric flow speeds can be high and a transition to magnetospheric conditions takes place. In general, dynamical models are needed to describe the ionospheric /magnetospheric plasma above 1,400–1,500 km (e.g., Cravens et al. 1998; Kabin et al. 1999; Ma et al. 2006, 2007). At lower altitudes, dynamics can usually be neglected in the determination of ion densities and photochemical or time-dependent models suffice. The net source of each ion species is set equal to zero in photochemical models and the resulting set of coupled algebraic equations are solved for the densities.

11.3.2 Sources of Titan's Ionosphere

The ionosphere of Titan results from ionization of neutrals associated with energy deposition by solar photons or by energetic electrons and ions from Saturn's magnetosphere. Details of energy deposition processes and the magnetosphere-Titan interaction are discussed in Chapter 16 but some aspects of this topic specifically relevant to the structure and composition of the ionosphere are reviewed here. Ion production from photoionization by solar radiation can be calculated using photon fluxes in the EUV and soft x-ray portions of the solar spectrum. Tobisca et al. (2000) provided information on the solar flux for a range of solar activities. Photoabsorption and photoionization cross sections are also needed, particularly for N_2 and CH_4 (c.f., Schunk and Nagy 2000; Keller et al. 1992). The absorption of photons by the atmosphere needs to be taken into account and at Titan this is affected by the high degree of atmospheric sphericity (Müller-Wodarg et al. 2000; Cravens et al. 2004).

Ion production associated with the major neutral species can be represented by the reactions:

$$h\nu + N_2 \rightarrow N_2^+ + e \rightarrow N^+ + N + e \quad (11.9)$$

$$h\nu + CH_4 \rightarrow CH_4^+ + e \rightarrow CH_3^+ + H + e \rightarrow CH_2^+ + H_2 + e \quad (11.10)$$

Photoelectrons produced by these reactions also ionize neutrals for electron energies exceeding the relevant ionization potentials (i.e., so-called secondary ionization). These suprathermal electrons also lose (i.e., deposit) energy in ways other than ionization, such as by heating thermal electrons via Coulomb collisions (Gan et al. 1992; Galand et al. 2006) or by exciting dayglow emissions (Gan et al. 1993; Ajello et al. 2007). N_2^+ production rates (both primary and secondary) from photoionization for conditions (and neutral densities) appropriate for the Cassini T18 flyby near closest approach were shown by Robertson et al. (2009). Figure 11.10 (from Cravens et al. 2008a) shows two N_2^+ production rate profiles for dayside conditions. Secondary ionization is especially important in the bottomside ionosphere, below the peak, where the more energetic photons (including soft x-rays) are absorbed.

The most abundant ion species in Titan's ionosphere (Cravens et al. 2006) are not the "primary" ion species initially produced (e.g., N_2^+ and CH_4^+) but are species generated by ion-neutral chemistry (e.g., $C_2H_5^+$ and $HCNH^+$). Exothermic ion-neutral reactions take place rapidly and produce new ion species with lower ionization potentials. Eventually, "terminal" ion species are created and these ions dissociatively recombine with thermal electrons, regenerating the neutral atmosphere. Ionospheric models predicting the ion composition need

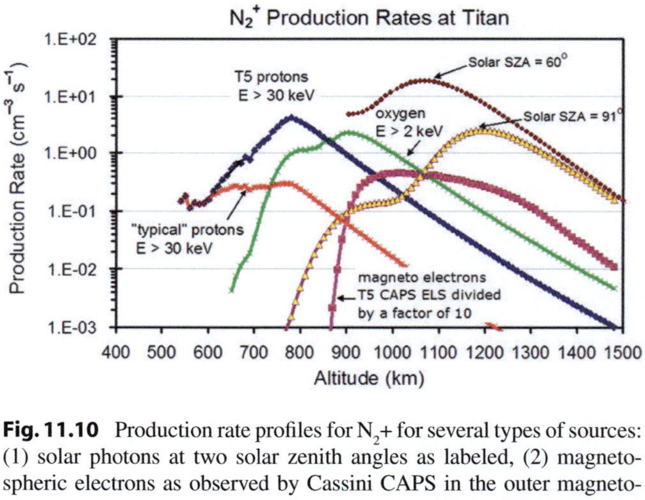

Fig. 11.10 Production rate profiles for N_2+ for several types of sources: (1) solar photons at two solar zenith angles as labeled, (2) magnetospheric electrons as observed by Cassini CAPS in the outer magnetosphere for T5, (3) magnetospheric ions adopting MIMI (Magnetospheric Imaging Instrument) observations for T5 and assuming the ions are protons (only for energies greater than 30 keV), (4) magnetospheric ions assuming the T5 MIMI fluxes are all oxygen ions, (5) magnetospheric ions assuming that MIMI measurements are protons and for typical magnetospheric conditions (from Cravens et al. (2008a); copyright 2008 American Geophysical Union–further reproduction or electronic distribution is not permitted)

to include all relevant chemical source and loss rates in the relevant continuity equations. Titan's ionospheric chemistry is discussed again in Section 11.3.6.

Ion production on the nightside obviously cannot be due to solar radiation. However, fast electrons and ions from the outer magnetosphere of Saturn near Titan can ionize the neutral gas if they can access the atmosphere. Reactions analogous to Eqs. 11.9 and 11.10 apply in this case, albeit with energetic ions and electrons rather than with photons. A more detailed discussion of electron and ion precipitation can be found in Chapter 16. The magnetic topology near Titan strongly affects the transport of electrons from Saturn's magnetosphere into Titan's atmosphere due to the small electron gyroradii (c.f., Gan et al. 1993). Lower energy protons ($E < 30$ keV roughly) are also sensitive to the magnetic topology (Ledvina et al. 2005) and, hence, to the details of the interaction of Titan with the surrounding external plasma flow (for a more detailed dicussion of energy deposition processes at Titan see Chapter 16; Hartle et al. 2006; Szego et al. 2007). On the other hand, energetic heavy ions (e.g., O^+ ions) with energies in excess of about 2 keV or protons with energies in excess of ≈30 keV have gyroradii exceeding Titan's radius and can precipitate into the atmosphere relatively unimpeded by induced magnetic fields near Titan. Ion production rates due to both solar radiation and magnetospheric particles are shown in Fig. 11.10 (Cravens et al. 2008a). It is evident that energetic ion precipitation contributes to thermal ion production mainly below ≈900 km. Both ion and electron precipitation are known to be quite variable due to the variable magnetospheric particle populations that Titan encounters (e.g., Krimigis et al 2005; Young et al. 2005; Coates et al. 2007a, 2008). Ion and electron precipitation also contribute to energy deposition in general (and not just to atmospheric ionization), including auroral emission and ionospheric heating (e.g., Michael and Johnson 2005), but this has not been systematically studied.

11.3.3 Observed Variations of the Total Ion Density, the Electron Density and Electron Temperature

Voyager 1 radio occultation observations (Bird et al. 1997) indicated the presence of an ionosphere near the terminator with a peak at ≈1180 km and a peak electron density of 2.4 × 10^3 cm^{-3}. Radio occultation measurements of the ionosphere were also made by Cassini at the S, X and Ka bands. During the Cassini prime mission four sets of occultation measurements were obtained, namely during the T12, 14, 27 and 31 flybys of Titan. Given the Sun–Earth–Saturn orbital geometry, occultations at Titan all take place within a few degrees of the dusk or dawn terminators. Figure 11.11 shows electron density profiles obtained from several occultations and also shows dawn and dusk averages of many profiles.

All the Cassini radio occultation measurements indicate that the maximum density occurs near 1,200 km. Theoretical models prior to Cassini successfully reproduce the observed peak altitude and magnitude of this peak, based solely on ionization by solar EUV radiation and the resulting secondary ionization by photoelectrons (e.g., Fox and Yelle 1997; Keller et al. 1998; Cravens et al. 2004). The comprehensive MHD model results of Ma et al. (2006, 2007) demonstrated unambiguously that chemical equilibrium conditions hold below about 1,400 km, which is important to keep in mind for the discussion that follows. The processes controlling the height and magnitude of this main peak are reasonably well understood at this time.

The shoulder, or ledge, in the electron densities located below the main peak is most likely associated with the shoulder in the solar production rate as calculated by Cravens et al. (2005) or Robertson et al. (2009) (see Fig. 11.10). Energy deposition and ionization by solar soft x-rays (and the harder EUV photons) are important for the formation of this ledge. Solar x-ray fluxes are known to be quite variable in both intensity and time (e.g., Lean et al. 2003), consistent with the observed changing densities in this 800–1,100 km chemical equilibrium region.

The Cassini prime mission provided eight occultation observations and two of these, T31 entry and exit, indicated the presence of a very significant second electron density peak in the 500 km altitude region. The fact that only two out of the eight occultations saw such a peak indicates that the ionization source must be intermittent. In a very recent paper Cravens et al. (2008a) showed that energetic ion precipitation, consistent with Cassini magnetospheric particle observations, is associated with significant electron-ion production rates, in the 500–700 km region. Here again, given that chemical equilibrium conditions prevail, production rates are directly related to electron densities. The energy and composition of the precipitating particles determines the deposition height and production efficiency, while the ion chemistry controls the effective recombination rates. Therefore, the task of calculating quantitative electron density values is an extremely difficult task at this time, given all the uncertainties in these parameters, and only order of magnitude estimates are feasible; the range of peak densities suggested by Cravens et al. (2008a) is 0.4–1.6 × 10^3 cm^{-3}. Energetic electron precipitation is also a possible source of ionization at these altitudes, although given the small gyroradii of precipitating electrons, there must be "magnetic connection" to the low altitude ionosphere from the region(s) where such electrons have been observed.

It was also suggested by Molina-Cuberos et al. (2001) that meteoric impact will cause low altitude ionization at Titan. These authors recently published the results of their model calculations, which considered metallic ion chemistry resulting from the ablation of meteoroids and the creation of

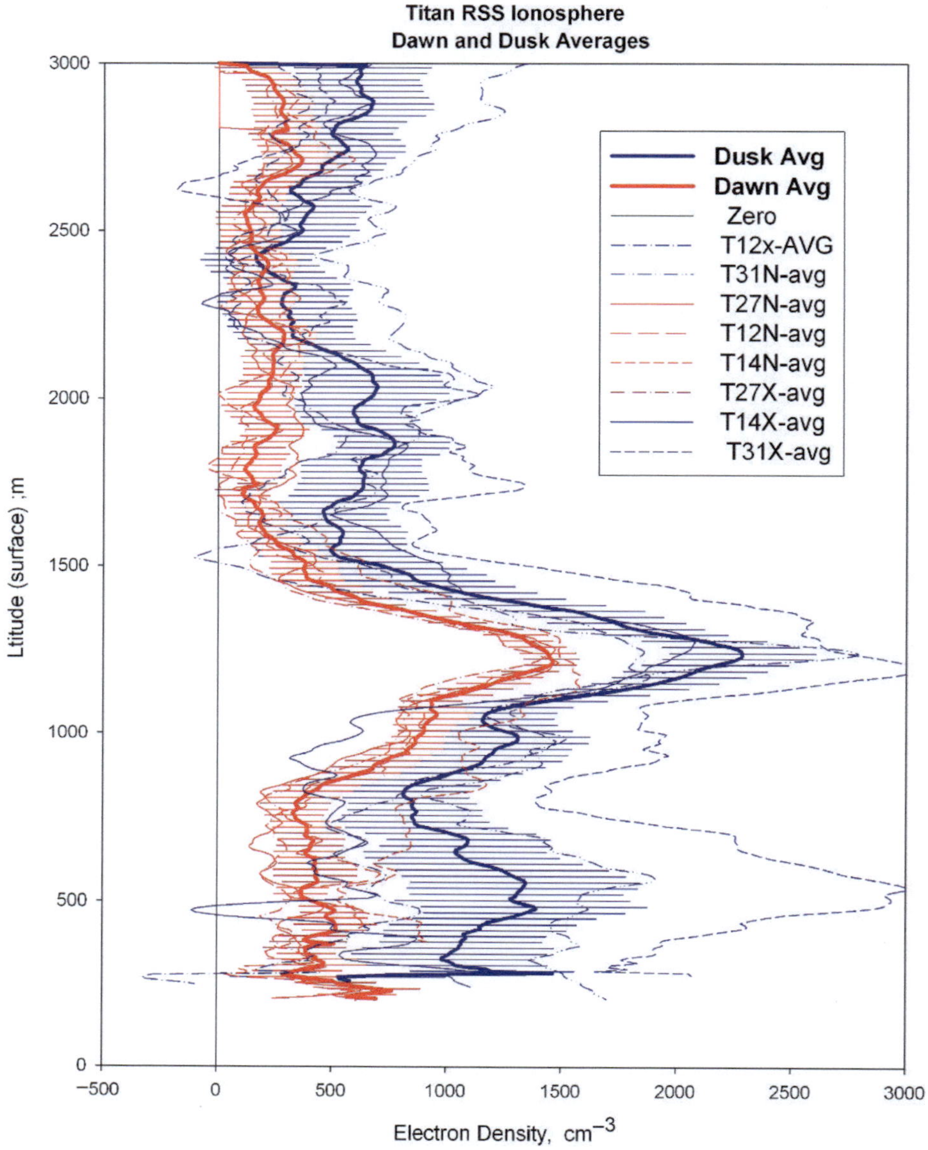

Fig. 11.11 Dawn and dusk and dawn-averaged electron density profiles, together with their cumulative averages measured by the Cassini radio occultation experiment. For T12 and T14 (southern mid-latitudes), the dusk observations are also in sunlight, and the dawn observations are in the dark (Kliore et al. 2008).

long-lived metallic ions. They predict electron densities ranging from 10^3 to 10^4 cm^{-3} in the 600–800 km altitude range. Thus this process can also fully or partially account for the electron densities observed during T31. Galactic cosmic ray ionization probably makes a contribution at the lowest altitudes (Lopez-Moreno et al. 2008).

The Cassini Ion-Neutral Spectrometer (INMS) has observed significant wave structure in the neutral atmosphere in the 1,200–1,400 km region (Waite et al. 2005; Müller-Wodarg et al. 2006). In an altitude region controlled by chemical equilibrium the ion densities follow directly the changes in the neutral densities. Thus if the wave-like features seen in the electron densities in this region above the main density peak are real and not an indication of noise in the data, it can be accounted for by the presence of the wave structure in the neutral density.

Consider the dusk to dawn drop in the average observed electron densities of Fig. 11.11. The average solar zenith angle of the dusk measurements is about 88° (see Kliore et al. 2008). Thus, these dusk densities are mainly caused by photoionization and photoelectrons, with perhaps a significantly smaller contribution due to impact ionization by magnetospheric electrons as demonstrated by model calculations (e.g., Cravens et al. 2005; Robertson et al. 2009). On the other hand, by dawn the relative importance of ionization due to solar radiation versus magnetospheric electrons must be reduced at least somewhat. The average solar zenith angle of the dawn observations is about 94°. Given the relatively long time that Titan's ionosphere is without sunlight, recombination removes a significant fraction of the ionization from the previous day if indeed the transport is from dawn to dusk.

The Langmuir Probe (LP) part of the Cassini Radio and Plasma Wave Spectrometer (RPWS) experiment also measured for the Ta encounter several plasma properties including electron density and temperature, as shown in Fig. 11.12.

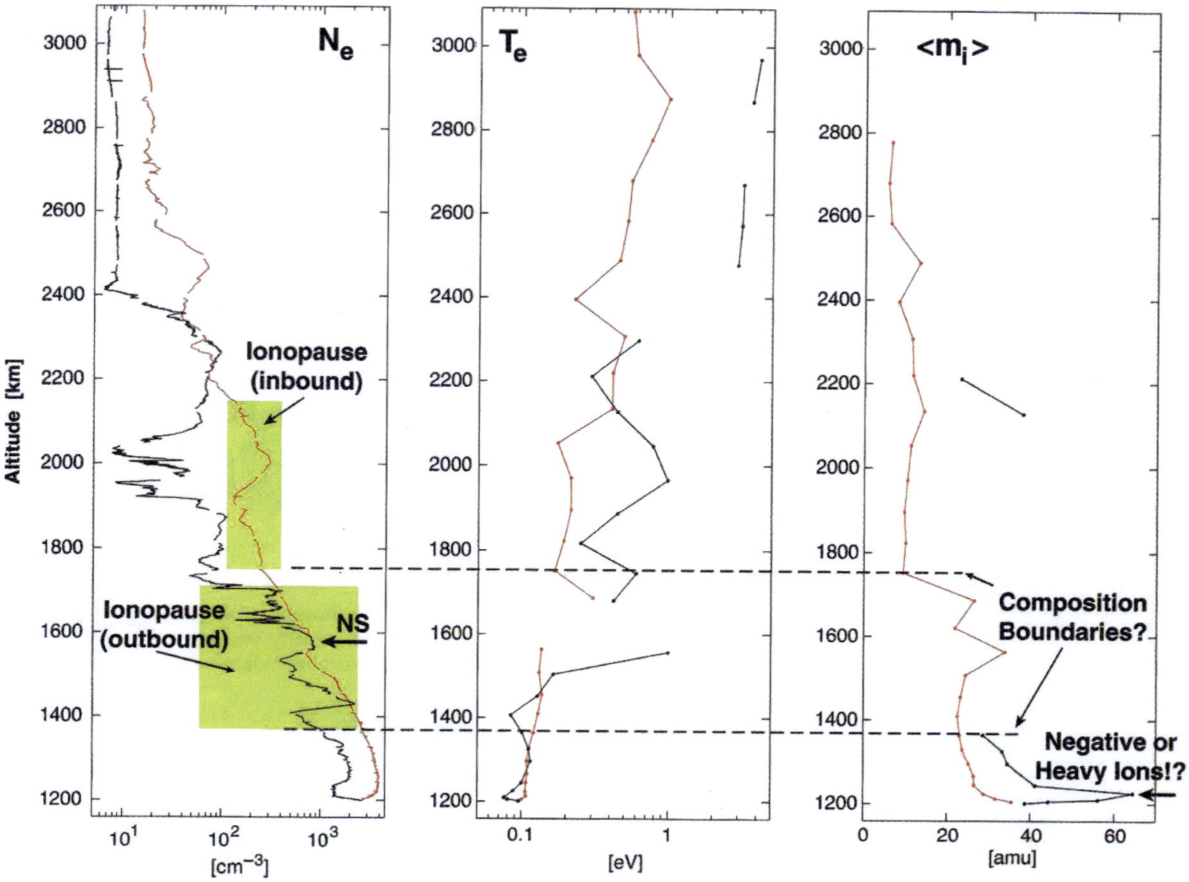

Fig. 11.12 Altitude profiles of electron density (left panel), electron temperature (middle panel), and average ion mass (right panel) for both the inbound (red lines) and outbound (black lines) Ta flyby as measured by the Cassini RPWS Langmuir probe. The "ionopause" regions for both inbound and outbound are marked (From Wahlund et al., Science, 308, 986, 2005; reprinted with permission from AAAS)

Closest approach (CA) for the Ta flyby was at a solar zenith angle of 91° (that is, at the terminator) and at an altitude of 1,175 km. Inbound was on the dayside and outbound was on the nightside, just beyond the terminator. The peak density along the spacecraft track was ≈3,000 cm^{-3} and the peak was located on the dayside. The dayside profile is rather smooth up to about 1,900 km, where Wahlund et al. (2005) noted the presence of an "ionopause" structure. This ionopause was apparently not a simple and sharp density decrease as originally defined and observed at Venus, but was defined as being where the magnetic pressure equals the thermal pressure. Wahlund et al. determined the magnetic pressure using the magnetic field strength measured by the magnetometer experiment during Ta (Backes et al. 2005). On the outbound Ta flyby, small-scale structure appears at altitudes as low as 1,400 km. Cravens et al. (2005) compared the RWPS electron density profile with the results of a photochemical model including ionization by solar radiation and ionization by precipitating magnetospheric electrons (assumed to have a 100 eV Maxwellian distribution in the magnetosphere). They concluded that solar radiation can account for most of the ionosphere on the dayside and even the ionosphere well beyond the terminator, although the magnetospheric source apparently made some contribution for Ta conditions deeper on the nightside.

The measured electron temperature in the ionosphere for altitudes between about 1,200–1,400 km for the Ta flyby was about 0.1 eV (or 1,100 K), increasing with altitude to about 0.7 eV or so (8,000 K) at 1,800 km as the ionosphere makes a transition to magnetospheric conditions. The measured ionospheric temperatures are in overall agreement with values predicted prior to Cassini (Gan et al. 1992; Roboz and Nagy 1994), as well as with a recent thermal model of the ionosphere specifically constructed for Ta conditions (Galand et al. 2006).

Ågren et al. (2009) have put together RPWS electron density and temperature data for a large number of Titan passes. Figure 11.13 shows the electron densities at the peak of the density profile versus solar zenith angle from this compilation together with a model comparison from the photochemical model of Robertson et al. (2009). The conclusion from this figure is again that solar radiation can account for the

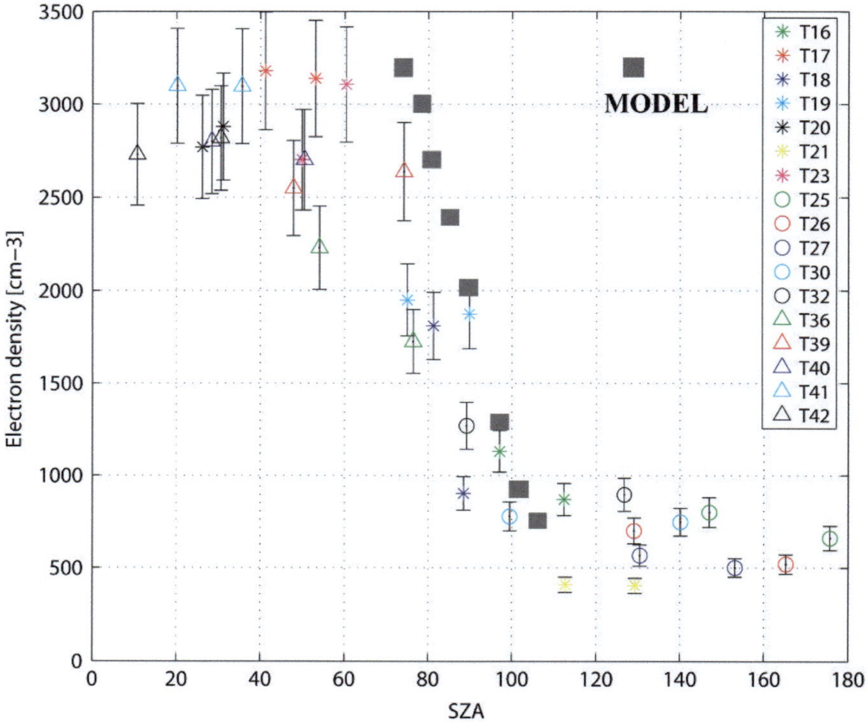

Fig. 11.13 Electron density at the peak versus solar zenith angle and altitude from the RPWS Langmuir probe for a large number of Cassini Titan flybys. Also shown are peak densities from a model of the dayside ionosphere for T18 conditions that only included ionization by solar radiation (Robertson et al. 2009) (figure adapted from Agren et al. 2009)

ionosphere for solar zenith angles out past the terminator to about 100°. The RPWS data also show an expected decrease of the peak density with solar zenith angle (Schunk and Nagy 2000). The peak nightside densities are about a factor of 3–4 less than the peak dayside densities.

The Cassini INMS in its "open source" mode (or osi mode) measured densities of ion species with mass numbers up to 99 Da. The sum of all positive ion densities at a given altitude should equal the electron density plus the sum of all negative ion densities due to quasi-neutrality. However, the INMS was configured to only measure positive ion species with masses up to 99 Da and it does not measure negative ions, which means that the total positive ion density measured by the INMS should not exactly equal the electron density measured by the RPWS Langmuir probe even if both instruments were perfectly accurate. The total ion density measured by INMS is compared in Fig. 11.14 with the RPWS electron density for the T18 pass (Robertson et al. 2009). During the T18 flyby, the spacecraft crossed the terminator from the near-nightside to the dayside. The INMS and RPWS densities agree to within ≈50% for altitudes above 1,050 km on the dayside and they also agree on the nightside for altitudes from 1050 to 1400 km. However, INMS densities are significantly lower than RPWS densities in 2 regions: (1) above 1,400 km on the inbound pass (night), and (2) below 1,050 km (i.e., times near CA). The INMS data has been corrected for spacecraft potential (values of about −1.5 V for T18), but the corrections have uncertainties and perhaps this can explain the discrepancy for this pass.

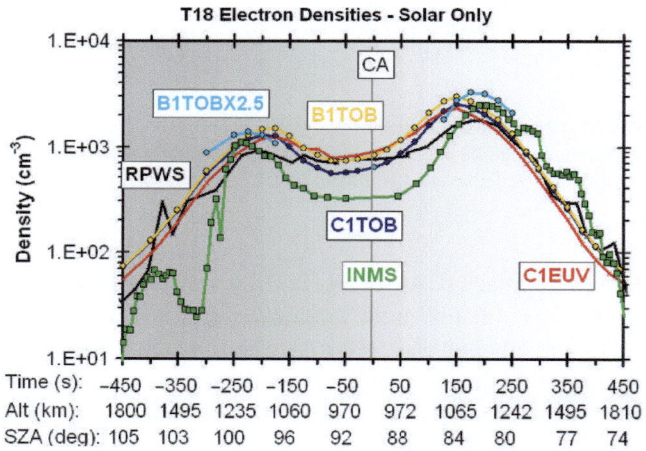

Fig. 11.14 Total ion density from the INMS, electron density from the RPWS LP, and model electron densities versus time (and solar zenith angle and altitude) for the Cassini T18 flyby. The black line is RPWS electron density data and the green line is INMS total ion density data. The other lines (*blue, red, and yellow*) are from a photochemical model in which different solar flux models and minor neutral densities are adopted (see the text) (from Robertson et al. (2009))

The INMS in its open source ion mode has a narrow field of view (≈±3°) such that for ionospheric flow speeds in excess of ≈300 m/s, depending on direction, the incident ions have their flow direction shifted out of the instrument's field of view. Dynamical models indicate that such flow speeds can occur above about 1,400–1,500 km (e.g., Cravens et al. 1998; Ledvina and Cravens 1998; Ma et al. 2004, 2006). The

RPWS Langmuir probe measurements are not subject to this limitation. The total ion densities measured by the INMS near closest approach can also be less than the RPWS electron densities for another reason; INMS does not detect ion species with m/z in excess of 99 Da and such species are apparently abundant in the chemically complex region below about 1,000 km.

An extensive comparison of CAPS, INMS, and RPWS-LP data in Titan's lower ionosphere was provided by Wahlund et al. (2009). Wahlund et al. addressed the fact that below about 1,000 km the measured electron density exceeds the total ion density from INMS, suggesting that the difference was due to ion species with mass numbers greater than 99 amu. One example of this discrepancy from Robertson et al. (2009) was shown in Fig. 11.14, but Wahlund et al. included examples from several other flybys and concluded that typically about half of all positive ions are "heavy" ion species in the lower ionosphere.

The time histories of both the INMS and RPWS density exhibit two maxima in Fig. 11.14. The figure also includes profiles from a photochemical model with 71 ion species (Robertson et al. 2009). Details of the ion composition in the model depend on the abundances of "minor" neutral species but the total ion density (i.e., electron density) is not terribly sensitive to these details because dissociative recombination rate coefficients are not too different for different ion species. The calculated densities for the two different solar flux inputs (the Solar 2000 model – Tobisca et al. 2000; and the EUVAC model – Richards and Torr 1988; Richards et al. 1994) are not too different. The EUVAC photon fluxes for wavelengths near 10 nm are somewhat greater than the Solar 2,000 fluxes and this resulted in somewhat higher densities in the bottomside ionosphere. But the densities near the peak are very close for the two solar flux cases. Apparently for T18, at least, ionization by solar radiation alone is sufficient to produce ionospheric densities as high as those observed even for solar zenith angles out past 100° for higher altitudes. Magnetospheric inputs were not needed in the model to reproduce the measured densities for T18 (Robertson et al. 2009). Figure 11.15 also illustrates this point. Model density profiles were calculated for a large set of solar zenith angles, and values along the spacecraft track (in solar zenith angle and altitude) were extracted in order to obtain time histories like those shown in Fig. 11.14. The density profiles for different spacecraft times show variations from both solar zenith angle and altitude effects. The peak densities from the model were compared with RPWS data in Fig. 11.13. Due to the "spherical" nature of Titan's atmosphere, at higher altitudes significant electron densities persist well onto the nightside. Also evident in these profiles is the ledge structure (due to absorption of higher energy EUV photons and soft x-rays) seen in the radio occultation profiles (Fig. 11.11).

It has just been shown that solar photons produce an ionosphere out past the nominal terminator, but the T5 flyby took place near SZA ≈ 130°, well beyond the reach of solar photons, even for Titan. Yet an ionosphere was observed during T5 with electron densities of $\approx 10^3$ cm^{-3}, as shown in

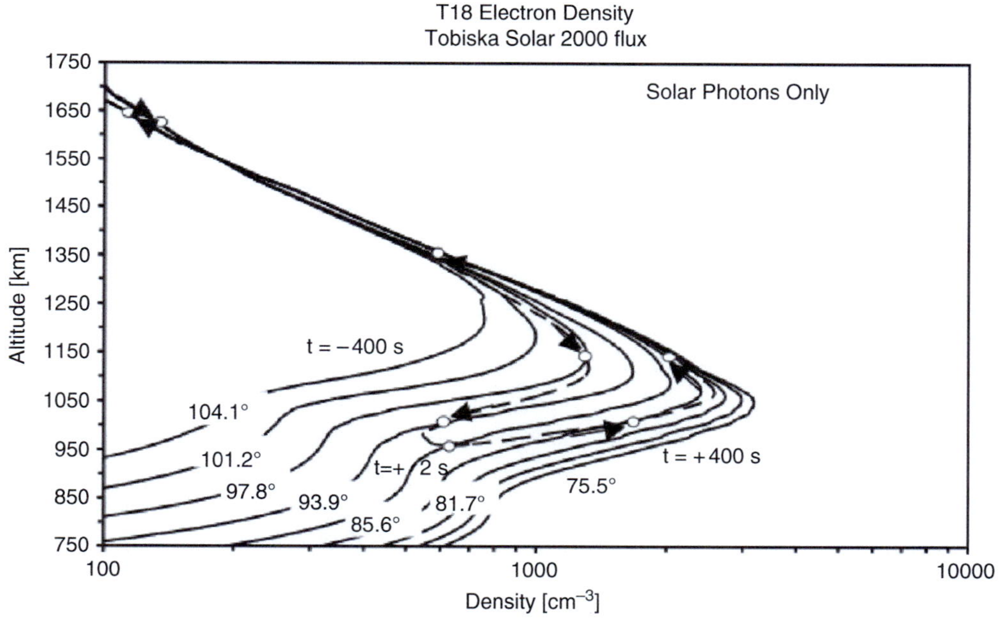

Fig. 11.15 Model electron density profiles for a large number of solar zenith angles for T18 conditions. Only solar radiation related ionization is included (no magnetospheric electrons). The spacecraft track versus altitude and solar zenith angle is shown and the densities so obtained were shown in the previous figure (for one model case) (from Robertson et al. (2009))

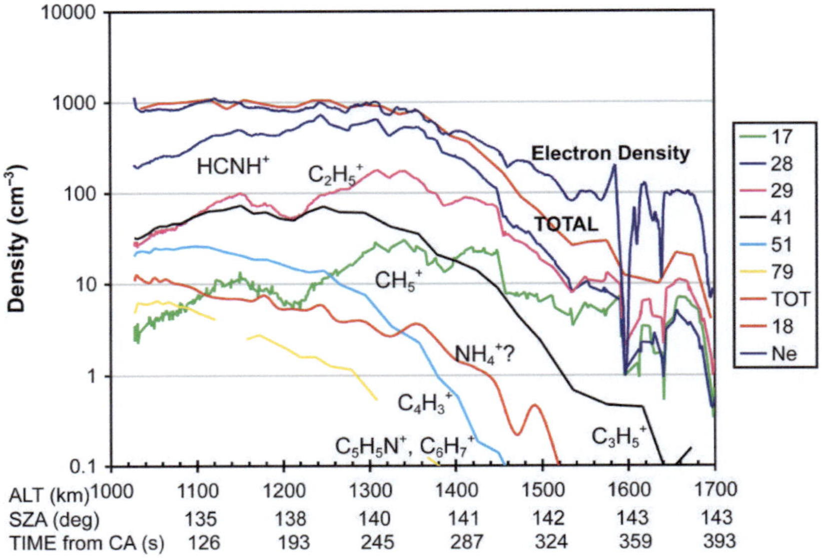

Fig. 11.16 Electron density (measured by the RPWS Langmuir probe) and total ion density (measured by the INMS) altitude profiles for the outbound leg of the T5 flyby. Density profiles for a few ion species are also shown (from Cravens et al. (2006); copyright 2006 American Geophysical Union–further reproduction or electronic distribution is not permitted)

Fig. 11.16. The total ion and electron densities measured by the INMS and RPWS experiments (Cravens et al. 2006; Ågren et al. 2009) are in excellent agreement up to an altitude of about 1,500 km, above which the the INMS densities are less than the RPWS electron densities. Perhaps the plasma flow speed is high enough (with flow direction orthogonal to the spacecraft velocity direction) such that some of the ions are not entering the narrow INMS aperture. The RPWS/LP also measured electron temperatures during T5 of about 600–1,000 K near 1,000 km as reported by Ågren et al. (2007)

Ågren et al. (2007) suggested that the T5 ionosphere was generated by electron impact ionization associated with energetic electron precipitation from Saturn's magnetosphere. They used a suprathermal electron transport code plus a relatively simple chemical model. Electron spectra in Saturn's magnetosphere were measured by the electron spectrometer (ELS) on the CAPS instrument, and for T5 the fluxes were quite high for electron energies up to several kiloelectronvolt. Agren et al. found that these fluxes were more than sufficient to produce the observed 1,000 cm^{-3} ionospheric densities but that the incident magnetospheric electron fluxes had to be reduced by a factor of about 5 to 10 from the CAPS ELS values in the magnetosphere. Cravens et al. (2008b) also modeled the T5 ionosphere using a 2-stream suprathermal electron transport code plus photochemical and time-dependent ionospheric models with 71 ion species. They found that using the CAPS magnetospheric electron fluxes resulted in a model ionospheric electron density profile in agreement with the RPWS/LP and INMS profiles only if the incident CAPS fluxes were reduced by a factor of 3 to 4. Cravens et al. demonstrated that the details of the magnetic field topology did not significantly affect the derived densities for altitudes above about 1,100 km, but did affect the profiles at lower altitudes.

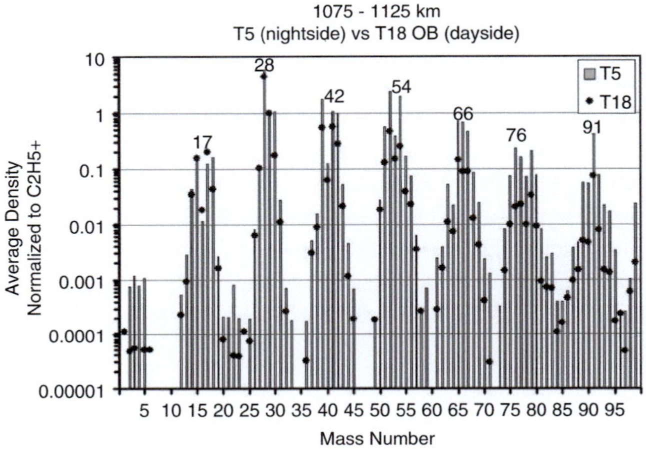

Fig. 11.17 Ion densities versus mass number for at altitude of 1100 km as measured on the dayside of Titan by the INMS on the T18 outbound (dayside) flyby and also on the T5 outbound flyby (nightside and high Northern latitude) (adapted from Robertson et al. (2009)). The densities were normalized to the m = 29 ($C_2H_5^+$) density.

11.3.4 Observed Composition of Titan's Ionosphere

It was recognized well before Cassini that the ionospheric chemistry at Titan would be complex due to the presence of hydrocarbons. The in situ ion composition measurements made by the INMS in its open source mode confirmed this expectation (Cravens et al. 2006). Figure 11.16 shows altitude/time profiles measured by INMS for several ion species, and Fig. 11.17 shows measured mass spectra at 1,100 km for both T5 (nightside and high latitude) and T18 outbound (dayside near the terminator). Fig. 11.17 shows that qualitatively the dayside and nightside ion composition are similar, but quantitative differences exist, particularly for the higher mass species that are relatively more abundant for T5 (night) than for T18 (day).

Many of the ion species observed in Titan's ionosphere were expected to be present prior to the arrival of Cassini (e.g., CH_5^+ at m = 17, $C_2H_5^+$ at m = 29, $HCNH^+$ at m = 28, C_3H_5+ at m = 42, $C_6H_7^+$ at m = 79….), but other ion species that were observed by the INMS were not expected (e.g., species at m = 30 and at m = 18, as discussed in Cravens et al. 2006; Vuitton et al. 2006, 2007). Several features are evident in Fig. 11.17. The composition is clearly organized into families separated by ≈12 mass units (i.e., the mass of carbon or nitrogen). In particular, the detailed analysis indicates these families consist mainly of species represented by $C_nH_m^+$ and $C_{n-1}H_mN^+$ where n and m are integers. It is also evident that the structure in the mass spectra appears to be continuing past m = 99 Da, the highest mass number that the INMS is capable of observing. The pre-Cassini model of Keller et al. (1998) did predict the existence of some ion species above 100 amu but not many. The measured mass spectra from T5 also show that the chemical complexity of the ionosphere increases with decreasing altitude (also expected prior to Cassini). That is, higher mass species are most abundant at the lowest altitudes sampled.

Cui et al. (2009b) analyzed INMS ion density data from many passes, organizing the data into dayside and nightside passes (Fig. 11.18). For the species shown, the greatest day-night variation is evident for CH_5^+, which is a species with a relatively short chemical lifetime (it reacts with a number of neutral species including C_2H_2, C_2H_4, H_2). On the other hand, the day–night variations for terminal (or almost terminal) ion species with long lifetimes (NH_4^+, $C_6H_7^+$, $C_2H_3CNH^+$) are rather small (or nonexistent for the last species). Cui et al. suggested that these (and similar) results imply that transport of ions from day to night by winds and/or just diurnal variations makes an important contribution to the support of the nightside ionosphere, at least for the longer-lived ion species.

The CAPS ion beam spectrometer (IBM) sensor was able to measure energy spectra of cold ionospheric plasma and these spectra were converted into low mass-resolution mass spectra by assuming that the kinetic energy of an ion is due to the spacecraft ram speed (Crary et al. 2009). Like the INMS, CAPS observed "families" of ion species, albeit at lower mass resolution. These CAPS ion spectra have been "calibrated" using simultaneously measured INMS spectra. Figure 11.19 shows one such spectrum for the altitude range 960–1,000 km. Families of ion species are present up to very high mass numbers (m > 200 amu). This topic is also discussed in Chapter 8 (Waite et al.). As discussed earlier, Wahlund et al. (2009) suggested that the presence of these heavy ion species at lower altitudes explains the fact that the measured electron density exceeds the total ion density measured by the INMS for mass numbers less than 100 amu. Wahlund et al. also discussed the conversion of heavy ion species (positively and negatively charged) into aerosols and used measured density values plus some assumptions on high-mass ion chemistry to estimate a downward mass flux of aerosol precursor species of $\approx 10^{-15}$ g cm^{-2} s^{-1} at 1,000 km.

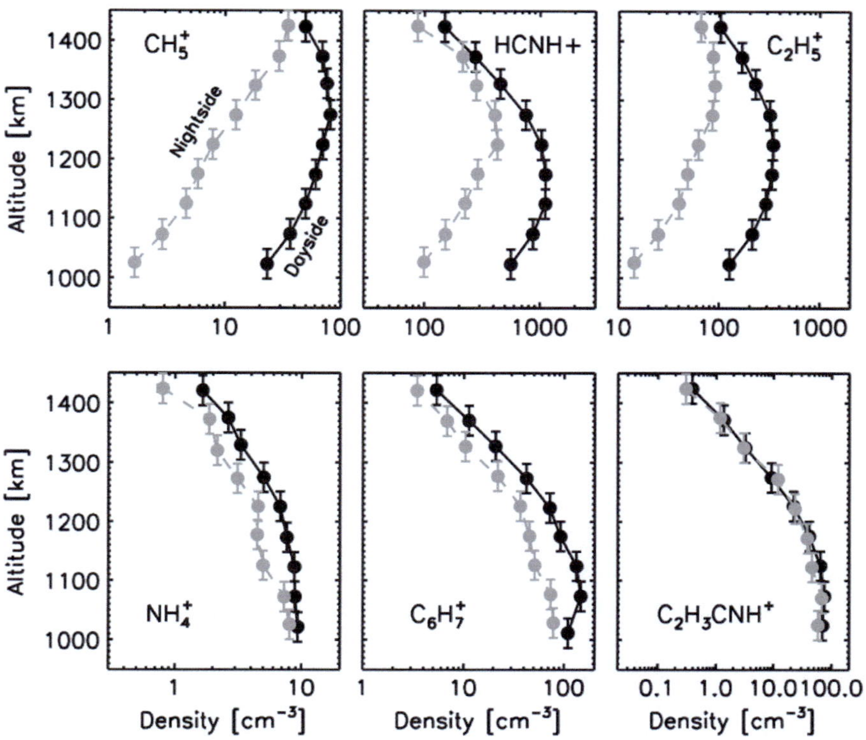

Fig. 11.18 Average ion density profiles for the dayside and the nightside as measured by the INMS for several species that differ in their chemical lifetimes (from Cui et al. (2009b))

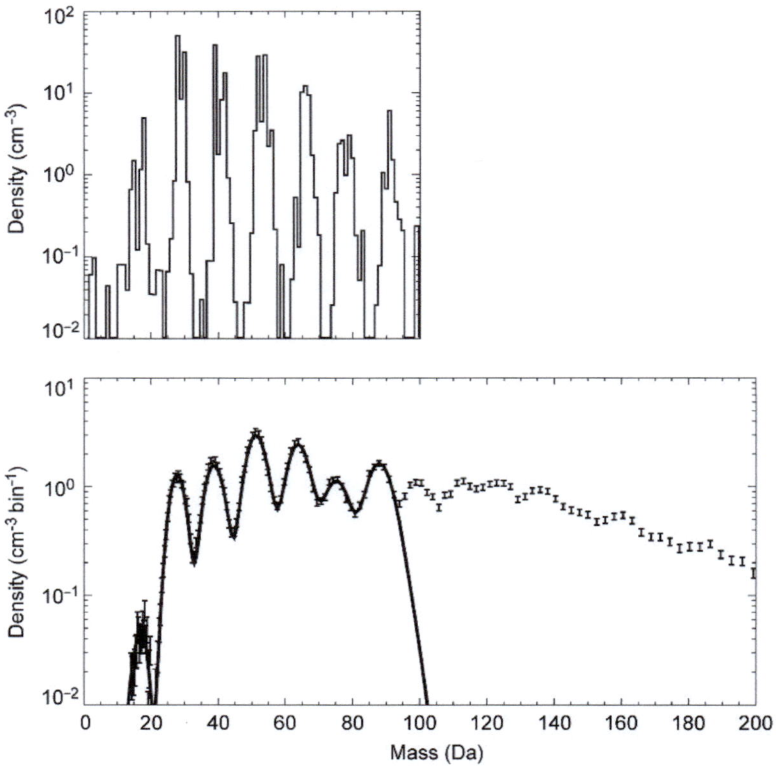

Fig. 11.19 Ion densities measured versus mass number/energy by the Cassini CAPS Ion Beam Spectrometer (IBS) at 960 km. The top panel shows the INMS mass spectrum, and the bottom panel shows CAPS IBS data for one curve and the INMS data from the top panel processed according to the CAPS IBS instrument characteristics (including lower mass resolution) (from Crary et al. (2009))

Negative as well as positive ion species were measured in Titan's ionosphere and this was a surprising discovery. The Cassini CAPS electron spectrometer (ELS) derived mass spectra from energy spectra using the ram kinetic energies of cold negative ions, albeit with low mass resolution (Coates et al. 2007b). Negative ions are known to exist in the D-region of the terrestrial ionosphere (Kelley 1989; Banks and Kockarts 1973) and are mainly associated with oxygen (O^-, O_2^- …). Titan's atmosphere is quite depleted in oxygen and so negative ions were not expected. However, some oxygen is known to be present in the stratosphere in the form of carbon monoxide (c.f., Wilson and Atreya 2004; Hörst et al. 2008). And Saturn's magnetosphere is known to have a population of oxygen ions that can precipitate into Titan's atmosphere (Horst et al. 2008; Cravens et al. 2008a) thus introducing some oxygen into the atmosphere. Species other than oxygen, and relatively abundant in Titan's atmosphere, also have rather high electron affinities (e.g., HCN) and could also be candidates for negative ions (Coates et al. 2007a).

Figure 11.20 shows a mass spectrum of negative ions as measured by the CAPS ELS at low altitudes (953 km) on the T16 flyby. Negative ion abundances are largest at lower altitudes. Another feature of the measurements is the existence of very high mass number negative ion species (up to $\approx 10^4$ amu). The total negative ion density measured during the T16 flyby near CA was about 100 cm^{-3}, which is several percent of the typical electron density at this altitude. The higher mass negative ions are probably charged aerosols (see the discussion in Waite et al. 2007). That larger ions or "grains" are negatively charged is not too surprising because grains in a thermal plasma are known to charge negatively in the absence of significant UV photon fluxes (Horanyi 1993). Negatively-charged low mass ions must result from chemistry and are likely to be species with high electron affinities. Coates et al. (2007a, 2009) suggested the following species candidates for different measured mass ranges: 10–30 amu (CN^-, NH_2^-…), 30–50 amu ($HNCN^-$, C_3H^- …), 50–80 ($C_5H_5^-$ ….).

11.3.5 Ionospheric Dynamics

To a good approximation, the combined thermal and magnetic pressure gradient forces plus gravity in an ionosphere are balanced by ion-neutral friction (c.f., Schunk and Nagy 2000), as represented in Eq. 11.7 for the single-fluid approximation. Solving Eq. 11.7 for the single-fluid plasma flow velocity yields the following simple expression:

$$\mathbf{u} - \mathbf{u}_n = \frac{1}{\rho v_{in}} \left[-\nabla \left(p_e + p_i \right) + \mathbf{J} \times \mathbf{B} + \rho \mathbf{g} \right] \qquad (11.11)$$

The ion-neutral momentum transfer collision frequency $\nu_{in} = k_{in} n_n$ is proportional to the neutral density and, hence, it increases rapidly with decreasing altitude; this strongly limits flow speeds (relative to the neutral flow speed) at lower altitudes.

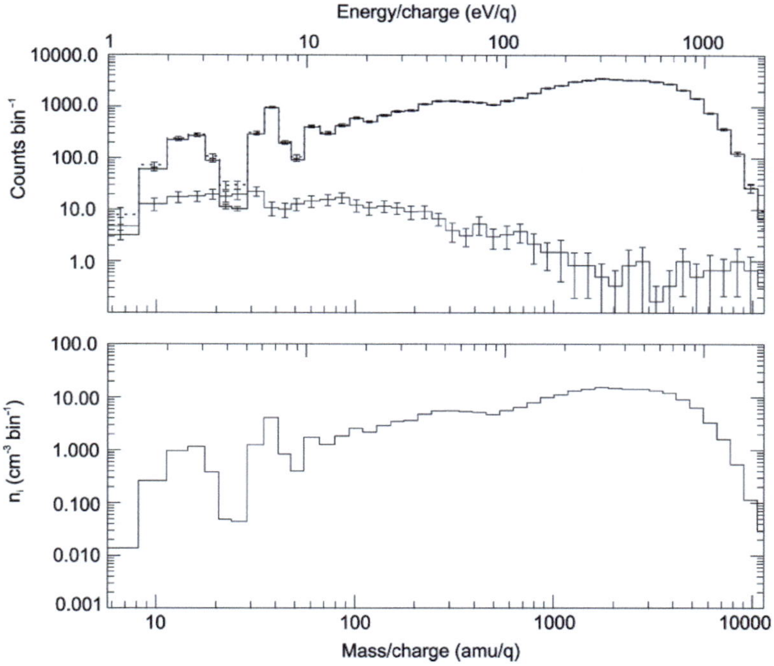

Fig. 11.20 Mass spectrum of negative ions measured by the Cassini CAPS ELS instrument on the T16 flyby at an altitude of 953 km (from Coates et al. (2007a)). Density per m/q bin is plotted in the lower panel

One-dimensional (Ip 1990; Keller et al. 1992; Keller and Cravens 1994) and two- (Cravens et al. 1998) and three-dimensional (Ledvina and Cravens 1998; Kabin et al. 1999; Ma et al. 2006, 2007; Modolo et al. 2007) numerical dynamical models of Titan's plasma environment all predict flow speeds of only a few m/s near 1,000 km, but the speeds increase rapidly with altitude. Cravens et al. (2008b) made "empirical" estimates of the flow speed (and the associated plasma transport times) versus altitude using Eq. 11.11 plus Cassini data from the T5 flyby for magnetic fields, electron densities and temperatures, and neutral densities. The collisional parameters given by Keller et al. (1992) were used. Pressure gradients were determined using the empirical pressure values at each altitude divided by characteristic lengths seen in the data. Flow speeds of u ≈5 m/s (and transport times of ≈10^4 s) were found near 1,000 km, increasing to u ≈ 1 km/s near 1,500 km (and giving a horizontal transport time of ≈500 s). Cravens et al. (2008b) compared these transport times to chemical lifetimes and found that, in general, chemistry is more important than transport in controlling ion densities for altitudes less than ≈1450 km, although different ion species had different chemical lifetimes. This is in agreement with the global MHD model calculations of Ma et al. (2006) (see Chapter 16). The plasma flows rapidly in the upper ionosphere (i.e., above 1,400 km or so) and the ionosphere begins to be "dynamically controlled" (Chapter 16). Note that for Titan (or any non-magnetic planet/body immersed in an external flowing plasma) the magnetic field plays a key role in the dynamics due to the **J** × **B** term in the momentum equation. Conversely, the distribution of the magnetic field depends on the plasma flow via the motional electric field term in the magnetic induction equation. Neutral winds can also contribute to the flow of plasma in the lower ionosphere where the ions and neutrals are collisionally coupled, and Cui et al. (2009b) suggested that this transport contributes to the maintenance of the nightside ionosphere.

11.3.6 Ionospheric Chemistry

Ionospheric chemistry at Titan has long been recognized to be complex (e.g., Atreya 1986) and many studies were made during the 10–20 years before Cassini (e.g., Keller et al. 1992; Fox and Yelle 1997; Keller et al. 1998; Wilson and Atreya 2004). A critical contribution to these chemical studies has been made by laboratory measurements of ion-neutral reaction rate coefficients (Anicich et al. 2004; Anicich 1993) and dissociative recombination reaction rate coefficients (e.g., Adams and Smith 1988; McLain et al. 2004, 2009).

Keller et al. (1998) presented a chemical flow chart illustrating key reaction paths as understood at that time. The reaction paths leading to hydrocarbon-chain building were largely accounted for in this diagram. Some nitrogen/nitrile chemistry and pathways were also included ($HCNH^+$ and $C_5H_5N^+$ were included, for example), but comparison of pre-Cassini chemical models, including the Keller et al. one, to the ionospheric ion mass spectra measured by the INMS during T5 (Cravens et al. 2006) revealed some important deficiencies in the theoretical mass spectra, particularly for the nitrile species (e.g., the models did not have sufficient density at the m = 30 mass peak). Vuitton et al. (2006, 2007) subsequently suggested a number of reactions involving neutral nitrogen-bearing species that "grab" a proton from major ion species, thus producing "new" nitrile ion species and helped to remedy the deficiencies in modeled mass spectra. Vuitton et al. used model-data comparisons to deduce the densities of minor neutral nitrile species with abundances too low to be detected by the INMS in its closed source mode (see Table 11.1). Figure 11.21 is a revised version of the Keller et al. chemical flow chart that includes the most important new pathways (but not all the chemistry) introduced by Vuitton et al. (2006, 2007).

A brief review of Titan's ionospheric chemistry is given next. The main ion species produced is N_2^+ but this species is rapidly converted first to CH_3^+ and then to $C_2H_5^+$ via the following reactions with the major species CH_4:

$$N_2^+ + CH_4 \rightarrow CH_3^+ + N_2 + H \qquad (11.12)$$

$$CH_3^+ + CH_4 \rightarrow C_2H_5^+ + H_2 \qquad (11.13)$$

In addition to reaction (Eq. 11.13), $C_2H_5^+$ is also created by reaction of CH_5^+ ions with C_2H_4, C_2H_2, or H_2. And CH_5^+ is formed by reaction of CH_4^+ (a primary species created by photoionization or electron ionization) with CH_4 or H_2. Photoionization or impact ionization of N_2 leads to N^+ production as well as to N_2^+ production. Reaction of N^+ with CH_4 produces HCN^+ and this species can then react to produce $HCNH^+$. An even more important source of $HCNH^+$ is the reaction of $C_2H_5^+$ with hydrogen cyanide (HCN). Note that an examination of the observed ionospheric mass spectrum (Fig. 11.17) indicates that high abundances are associated with mass numbers 28 ($HCNH^+$) and 29 ($C_2H_5^+$).

Higher mass hydrocarbon ion formation proceeds by a series of reactions starting with reaction of $C_2H_5^+$ with C_2H_2 and C_2H_4, and also with C_3H_4, C_4H_2, or C_2H_6. Ion species up to $C_{11}H_9^+$ are shown in Fig. 11.21, but the reaction chain should continue to even higher mass ions such as those

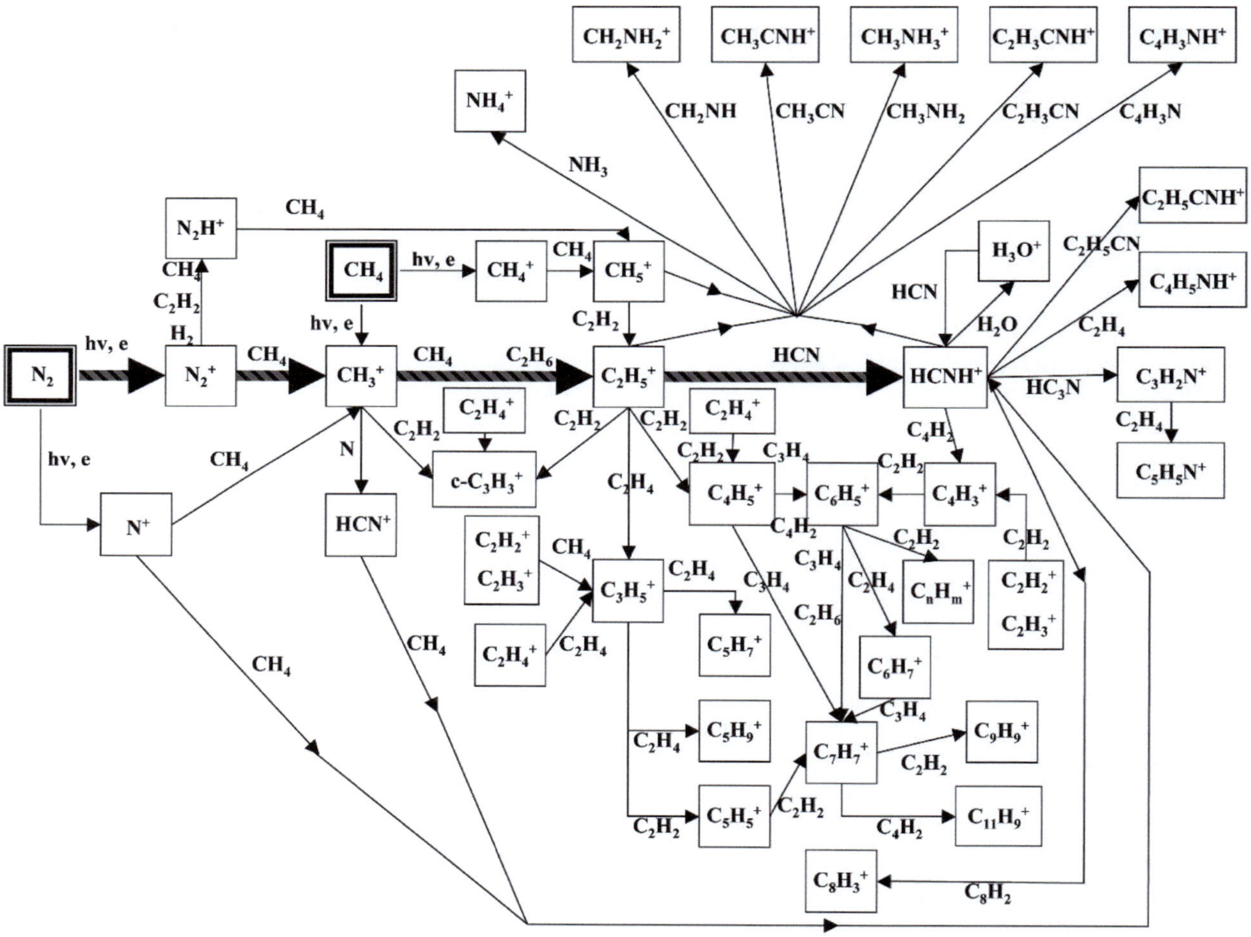

Fig. 11.21 Ionospheric chemical scheme for Titan. Adapted from Keller et al. (1998) but with the addition of some important reactions from Vuitton et al. (2006, 2007). Each of the ion species also can undergo dissociative recombination but this is mainly important further down the chemical chain for "terminal" or almost "terminal" ion species.

measured by the CAPS IBS instrument as discussed earlier. The high-mass chemistry remains quite uncertain (Carrasco et al. 2006) with reaction rates only measured for a handful of the key reactions (e.g., Ascenzi et al. 2007).

Higher mass nitrogen-bearing species are created by the reaction of major ion species such as $HCNH^+$ and $C_2H_5^+$ with neutral species such as HC_3N as first suggested by Fox and Yelle (1997) and then used by Keller et al. (1998). $C_3H_2N^+$ (m = 52) and $C_5H_5N^+$ (m = 79) are present in the ionosphere, but note that another important ion species, protonated benzene, $C_6H_7^+$, is also present at mass number 79. The Keller et al. model did not predict most of the N-bearing ion species that INMS observed on T5 (or on other passes), at least not at the relative abundances observed (Cravens et al. 2006; Vuitton et al. 2006, 2007), including species at m = 18, 30, 42, 54, 56, 66, 68 …. Vuitton et al. (2006, 2007) suggested that the missing species (mainly at even mass numbers) were N-bearing ions created mainly by proton attachment to a corresponding minor neutral species. A comparison of the Vuitton et al. (2007) photochemical model spectrum with the spectrum measured at 1,100 km by INMS during the T5 is shown in Fig. 11.22. Note that the species at m = 18 is thought to be NH_4^+ from reaction of major ion species like $HCNH^+$ and $C_2H_5^+$ with ammonia (NH_3) (also see Wilson and Atreya 2004, and Cravens et al. 2006). The m = 30 ion species was explained as being due to the reaction of species like $HCNH^+$ and $C_2H_5^+$ with CH_2NH which then produce $CH_2NH_2^+$. Similarly, m = 90 is explained by reactions with C_6H_3N to produce $C_6H_3NH^+$.

The majority of ions measured by the INMS have been identified by Vuitton et al. (2006, 2007). The identifications, shown in Fig. 11.22, are based on the chemical properties of the ions, supported by a comprehensive chemical model, incorporating the latest reaction rate data. Vuitton et al. (2007) explained that protonated closed-shell hydrocarbon ions have odd masses ($C_nH_{m+1}^+$ where n and m are integers), while ions containing a single N atom have even masses ($C_nH_mNH^+$). Thus, the surprisingly high number of ions at even masses (compared, for example, with the Keller et al. 1998, predictions) is explained as an abundance of nitrogen-bearing ions species. Moreover, Vuitton et al. (2006, 2007) demonstrated that only relatively stable species will appear with significant abundance in the spectrum and, for example, radical cations are not likely. The chemical models support

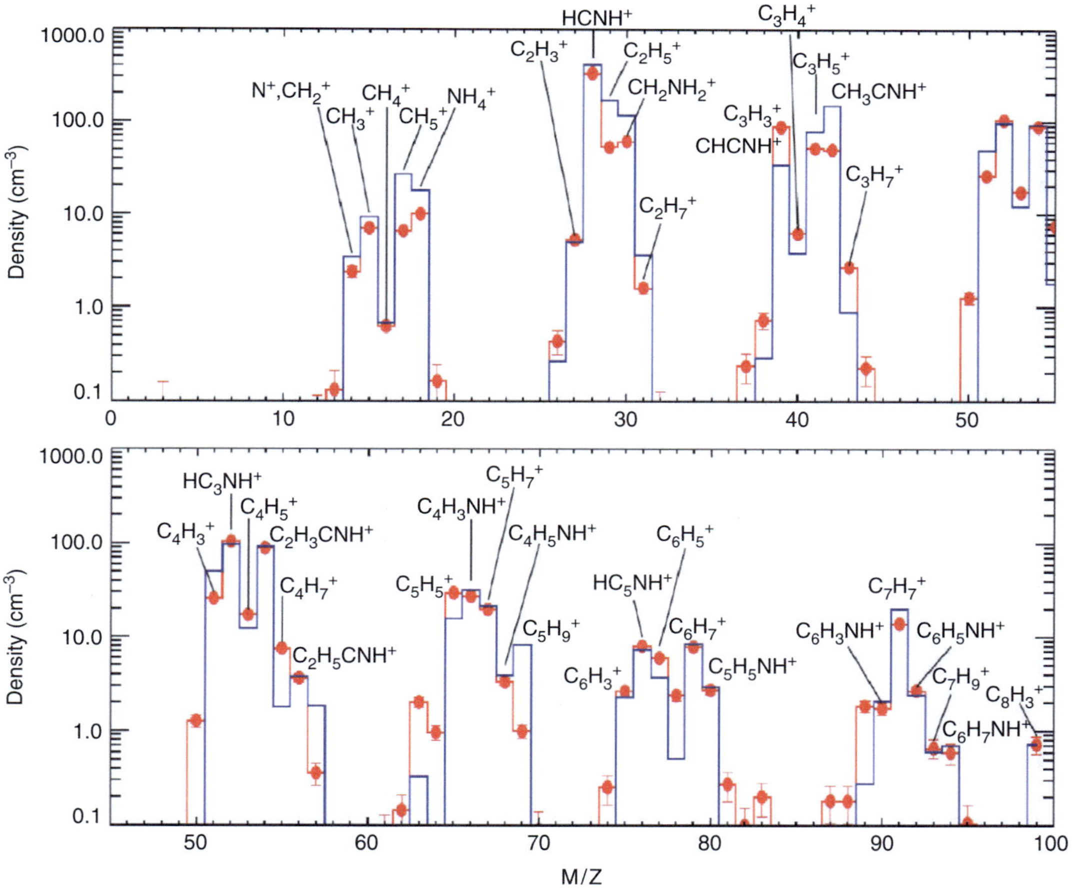

Fig. 11.22 Ion density versus mass number from a photochemical model and from the Cassini INMS for the T5 pass and 1100 km (from Vuitton et al. (2007))

these assumptions and also lead to recognition that many of the ions are simply protonated forms of the minor neutrals present in the upper atmosphere. Such species include those predicted prior to Cassini (HCN(H^+), HC_3N(H^+), C_4H_2(H^+)), …) but also new species including CH_2NH(H^+), CH_3CN(H^+), C_2H_3CN(H^+), C_2H_5CN(H^+), C_6H_2(H^+), C_8H_2(H^+), and many others. The reader is referred to Vuitton et al. (2007) for a complete list. Vuitton et al. (2006, 2007) fit the observed composition of the ionosphere primarily by adjusting the density of the minor neutral species, thereby providing a constrain on the abundance of the neutral species. Many of the neutral species are present at too low an abundance to be detected by the INMS, and in these cases analysis of the ionospheric composition provides the most sensitive means to characterize the composition of the neutral atmosphere (Vuitton et al. 2006, 2007). The mole fractions of several of the species so determined are presented in Table 11.1.

The chemical models for Titan's ionosphere rely on extensive reaction lists, based on laboratory measurements, that have been compiled over many years (Anicich et al. 2004) updated with recent measurements (Vuitton et al. 2006, 2007). Nevertheless, many uncertainties remain, especially at higher masses. Moreover, the models themselves are complicated and non-linear and the mapping from uncertainties in reaction rates and branching ratios to derived physical parameters is far from obvious. These issues have been addressed by Carrasco et al. (2007, 2008) who described a Monte Carlo method for calculating the mapping from reaction rate uncertainties to ion densites. These authors emphasize the lack of reaction rate data for heavy ion species (see Fig. 11.8 in Carrasco et al. 2008). Carrasco et al. (2008) focused on a few important uncertainties and showed that uncertainties in branching ratios alone lead to very large uncertainties in the calculated densities of many ions. This type of analysis can also identify which reaction rates have the most significant effects on the final results and therefore provide an excellent guide for laboratory investigations.

Negative ion chemistry in Titan's ionosphere has been extensively reviewed in Vuitton et al. (2009). Large particles (i.e., grains) are known to often carry an electrical charge that is proportional to its electrical potential (c.f., Horanyi 1993). The grain voltage or potential in a plasma is determined by assuming an equilibrium state and balancing all electrical current into or out of the grain (photoelectron current produced by photon absorption, electron and ion currents from the surrounding plasma, and secondary electron emission due to higher energy charged particle impacts) (see Horanyi 1993). For Titan the plasma-related currents are the most important and the potential should be given by: $V_0 \approx -2.5k_BT_e/e \approx -0.25$ V for $T_e \approx 0.1$ eV (see the discussion in Waite et al. 2007, and Coates et al. 2007a). For a spherical grain the electrical charge in terms of the potential is given by: $Q \approx 4\pi\varepsilon_0 V_0 R$, where R is the grain radius. The CAPS ELS experiment (Fig. 11.20) observed negative particles (i.e., "ions") with mass number up to $\approx 10^4$ amu, corresponding to a size of $R \approx 1.5$ nm for grain material with the density of water. The grain charge in Titan's ionosphere found this way is about $Q \approx -0.5e^-$. Large negative ions can also be produced by charge transfer reactions of large neutrals with smaller negative ions.

Now consider the origin of smaller "molecule" sized negative ions. In the terrestrial D-region oxygen with its large electron affinity attracts free electrons. In Titan's atmosphere some neutral species also have high electron affinities (e.g., HCN with an electron affinity of 1.002 eV). For example, CH_3^- can be created by electron collisions with CH_4 leading to a dissociative attachment reaction, but the threshold for this reaction is several electronvolt (Marynick and Dixon 1977) and only suprathermal electrons can accomplish this. Cravens et al. (2008a) suggested that oxygen ion precipitation into Titan's atmosphere can generate O^- ions via charge exchange collisions. Note though that most of the precipitating ions that charge exchange with neutral targets just produce fast O atoms. Once O^- ions are created in the atmosphere, further chemical reactions can transfer the electron to other higher electron affinity neutral species. The oxygen ultimately ends up as CO, as reviewed earlier in the chapter.

In the dayside terrestrial D-region ionosphere, photodetachment is also a very important loss process but not at Titan where the solar flux is 100 times less than at Earth. Ion-ion recombination is likely to be an important loss process for Titan's negative ions, and given that many positive and negative ion species have high masses in the lower ionosphere, such recombination reactions should act as an important source of large neutral species/aerosols. Cravens et al. (2008a) also suggested that dissociative attachment reactions such as suprathermal electrons plus methane giving CH_3^- might work.

Vuitton et al. (2009) presented an extensive analysis of the negative ion chemistry in Titan's upper atmosphere. These authors identify the peaks at 22 ± 4 and 44 ± 8 in the CAPS negative electron spectra as CN^- and a mix of C_3N^- and C_4H^-. This is a consequence of the high electron affinities of the cyano (CN, C_3N) or the ethynyl (C_2H, C_4H,C_6H, C_8H) groups. Vuitton et al. also suggest that the apparent peak at 82 ± 14 in the CAPS spectrum, which is of marginal statistical significance, could be C_5N^-. Vuitton et al. also construct a detailed model for the negative ion chemistry and show that dissociative electron attachment is the dominant production mechanism for negative ions through reactions such as e + HCN $\rightarrow$ CN^- + H.

This reaction is the primary source of negative ions. The H–C bond energy is a few electronvolt greater than the electron affinity of HCN (or HC_3N, C_2H_2, …) and suprathermal electrons with energies of a few electronvolt are required for the reaction. In fact, the dissociative electron detachment cross sections for reactions with HCN, HC_3N, and C_2H_2 peak at ~2.5 eV. Once negative ions are produced the charge flows to the species with the greatest electron affinity, very similar to the way the positive charge flows to the species with the greatest proton affinities (Vuitton et al. 2006, 2007). For example, C_3N^- is produced from CN^- through $CN^- + HC_3N \rightarrow C_3N^- + HCN$.

According the model calculations, this channel is slightly more important than direct dissociative attachment for the production of C_3N^-. CN^- and C_3N^- are lost primarily through reactions with radicals, principally H and CH_3. The model predicts that CN^- and C_3N^- are the dominant negative ions near 1,000 km, but heavier ions such as C_5N^-, C_4H^-, and C_6H^- become progressively more important at lower altitudes.

Many of the rates required for the chemical model are not available from laboratory measurements and were estimated by Vuitton et al. (2009). Moreover, the chemistry of heavier ions seen in the CAPS spectrum was not investigated at all because of a total lack of reaction rate data. Further progress on this topic will require specific laboratory investigations to determine the relevant rates.

11.3.7 Ionospheric Energetics

The electron (T_e) and ion (T_i) temperatures in any planetary ionosphere are expected to be equal to the neutral temperature (T_n) at low enough altitudes (i.e., thermal equilbrium) due to high electron-neutral and ion-neutral collision frequencies (Schunk and Nagy 2000). However, T_e begins to exceed both T_i and T_n at higher altitudes in most ionospheres, including Titan's (Fig. 11.12). Above about 1,000 km, the Cassini Langmuir probe measured temperatures of $T_e \approx 1{,}000$ K, greatly exceeding the observed neutral temperature of $T_n \approx 150$ K.

Prior to Cassini, Gan et al. (1992, 1993) and Roboz and Nagy (1994) constructed theoretical models of the ionospheric energetics and predicted that the electron temperatures would indeed be much higher than the neutral temperature. The electron gas is heated by Coulomb collisions with suprathermal electrons created by photoionization of neutrals by solar radiation (i.e., atmospheric photoelectrons) or by collisions with magnetospheric electrons traveling down draped magnetic fields into the ionosphere. The thermal electrons cool via Coulomb collisions with ions and also by rotationally and vibrationally exciting neutral N_2 and CH_4 (particularly methane) (Gan and Cravens 1992; Gan et al. 1992). Heat conduction is also important for the energetics and is controlled by the magnetic field topology.

Galand et al. (2006) published a post-Cassini model for the energetics with parameters tuned for the Ta flyby (and hence, the terminator region). The model included photoelectron transport and thermal electron heat conduction (Galand et al. 2006). Figure 11.23 shows calculated T_e altitude profiles for both a radial magnetic field configuration and a more realistic draped magnetic field topology. T_e is much lower for the radial magnetic field case than for the other case for two reasons: (1) heat conducts downwards more readily for the radial field case than for the mainly horizontal magnetic field case and the heat is rapidly lost by cooling in the dense lower thermosphere, and (2) the field lines in the second case are partly directed from the dayside to the nightside and thermal energy from dayside photoelectron heating is conducted towards the terminator region. The ionospheric energetics at Titan for other conditions should now be studied.

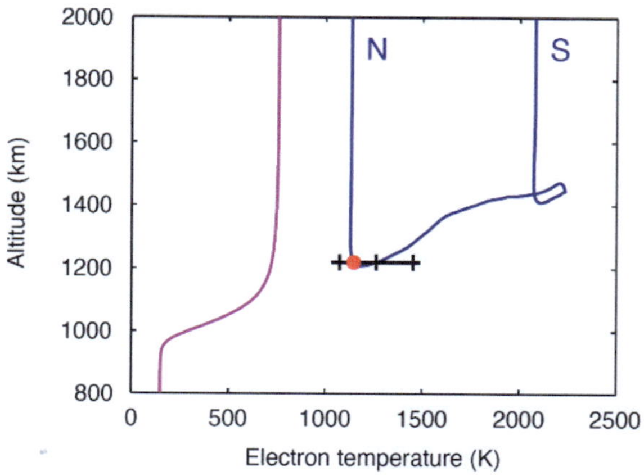

Fig. 11.23 Electron temperatures for the Ta ionosphere from a theoretical model and for two different adopted magnetic field line topologies (purple line: radial magnetic field and blue line: draped magnetic field line from a MHD model). The temperature measured at closest approach by the RPWS Langmuir probe is also shown at the data point (from Galand et al. (2006))

11.4 The Role of the Upper Atmosphere and Ionosphere for Titan Overall

11.4.1 Titan's Upper Atmosphere as an Interface Between the Magnetosphere and the Lower Atmosphere – Tranport of Energy and Momentum and Atmospheric Loss

Titan's thermosphere and ionosphere act as an important interface between the magnetosphere of Saturn (which contributes momentum and energy in the form of energetic particles and photons and magnetic field) and the lower atmosphere. The "thermal" ionosphere supplies both plasma and neutrals to the magnetosphere. The associated atmospheric loss over billions of years can affect the evolution of the whole atmosphere. The thermosphere/ionosphere region is also an important source of complex neutral and ion species which can be transported to the lower atmosphere where they contribute to haze/aerosol formation.

As alluded to earlier, an important aspect (at least for the ionosphere) of how the external plasma (Saturn's magnetospheric plasma but occasionally the solar wind plasma (Hansen. 2001)) interacts with Titan is the magnetic field induced in the ionosphere. How magnetic field lines are draped in the ionosphere affects both the dynamics of the ionosphere via the $\mathbf{J} \times \mathbf{B}$ force and the transport of suprathermal electrons and heat. As Gan et al. (1993) recognized, the supply of energetic electrons to the atmosphere, and thus available for energy deposition and ionization, depends on how long Saturnian magnetic field lines take to "move through" the atmosphere/ionosphere (c.f., Bertucci et al. 2008; Cravens et al. 2008b; Ma et al. 2006, 2007). After a couple of electron bounces in Saturn's magnetosphere, flux tubes should be depleted of electrons unless efficient replenishment processes are operating (e.g., radial diffusion). The bounce period for a few hundred electronvolts electrons is about 5–10 min.

Cassini data obtained during the T32 encounter, particularly magnetometer data, provides some insight into field line hang-up in the ionosphere (Bertucci et al. 2008). Ma et al. (2007) used a 3D MHD model to interpret Cassini data for this case. In the course of the T32 encounter, Titan moved from the magnetosphere through the magnetopause into the magnetosheath, and back. The field direction was quite different in the two regions, allowing the measured field direction to act as a "marker" of the field origin. The plasma characteristics can also help trace the history of a fluid parcel within the ionosphere (Ma et al. 2008). Apparently, in Titan's ionosphere (and in the near-wake region also), the observed fields are "fossil" fields with directions representative of an external environment from an earlier time. Bertucci et al. estimated that the field line hang-up time was at least 10 min and the Ma et al. MHD model suggests that in some places this time lag can even be hours.

The ion composition data measured by the INMS for T5 also provides some insight into magnetic field line hang-up in the ionosphere (Cravens et al. 2008b). The time/altitude profiles of ion species with short (few hundred seconds or less) chemical lifetimes (e.g., CH_4^+, CH_5^+, and to some extent $C_2H_5^+$) at an altitude of 1,200 km were observed to have small-scale structure (i.e., time scales of ≈150 s or length scales along the spacecraft track of ≈1,000 km). On the other hand, the time/altitude profiles of chemically long-lived ion species (e.g., $HCNH^+$, $CH_2NH_2^+$, NH_4^+, …) were observed to be rather smooth in their structure (see Fig. 11.16). Cravens et al. (2008b) suggested that the small-scale structure was caused by variations of the ionization rate, that in turn was caused by variations in the flux of precipitating electrons at different locations along the spacecraft track. That is, they concluded that the small-scale structures (unlike the large-scale structure) were not vertical structures associated with different energies of the precipitating electrons but were horizontal structures (or time dependent structures) associated with varying magnetospheric conditions. Agren et al's (2007) calculations of electron precipitation for T5 supports this statement because the altitude widths of the ion production rate profiles calculated for several monoenergetic beams significantly exceeded the measured extent of the short-lived ion species density structures (Fig. 11.16).

Supporting the magnetosphere-ionosphere coupling picture just described is the correlation observed between measured (by INMS) densities of short-lived ion species (e.g., CH_5^+) and low-energy suprathermal electron fluxes (measured by the CAPS ELS instrument at the same times). This is shown in Fig. 11.24. The low energy electrons observed by CAPS are secondary electrons produced by the electron impact ionization by primary electrons. The smoothness of the long-lived ion profiles for T5, unlike the short-lived ion density profiles (Cravens et al. 2006; and Fig. 11.16), implies that any temporal variations of the primary electron beam must take place over time intervals less than the lifetimes of these ion species near 1,000–1,200 km. The chemical lifetime of $HCNH^+$ near 1,100 km is ≈1000 s; hence, the field-line "hang-up" time must be less than roughly 10–20 min, at least for the T5 case. The small day-night differences in the INMS densities of high mass species shown by Cui et al. (2009b), as well as the modeling shown in that paper, suggests that long-lived species (those with smooth profiles in the T5 data) could be coming from the dayside and are not necessarily locally generated by energetic particle precipitation.

Convection electric fields induced by the interaction between the flowing magnetospheric plasma and field with

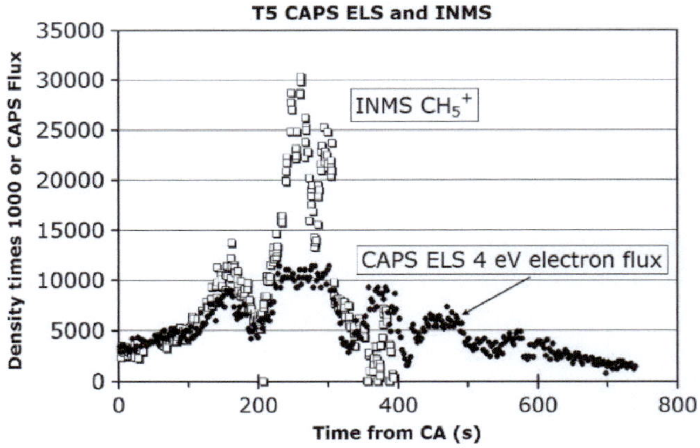

Fig. 11.24 Time history of the CH_5^+ density as measured by the INMS on T5 and the flux of 4 eV electrons measured by the CAPS ELS experiment. The values are scaled by constant factors to facilitate comparison (from Cravens et al. (2008b))

the cold plasma obstacle of Titan are mapped along magnetic field lines into the ionosphere, and cause ion plasma drifts through the neutral medium, generate electric currents, and dissipate energy via Joule heating. The flow of ionospheric currents depends strongly on the presence of a magnetic field (in Titan's case it is induced) as it creates an anisotropic ionospheric conductivity profile (Hall and Pedersen conductivities; see Schunk and Nagy 2000 and Kelley 1989). The conductivity in the terrestrial ionosphere has been extensively studied and the theory is well developed (cf. Kelley 1989). In an ionosphere electric currents perpendicular to the magnetic field arise in a region where the convective flow of ions is disturbed or partially disrupted by collisions with neutral atmospheric particles while the electrons are still drifting perpendicular to the magnetic and any applied electric field (**E** × **B** drift). This so-called dynamo region is restricted by the conditions $\Omega_e = \nu_e$ (lower boundary – about 950 km at Titan) and $\Omega_i = \nu_I$ (upper boundary – about 1,300 km at Titan). In Titan's case, the altitude region between 800 and 1,500 km is the most electrical conductive region with perpendicular to the magnetic field conductivities between 10^{-3}–10^{-1} S/m (Rosenqvist et al. 2009; see Fig. 11.25). Titan's ionosphere is therefore very conductive indeed, and would promote magnetospheric current systems to connect and in the ionosphere allow electric currents to flow perpendicular to the magnetic field. In the dynamo region further energy deposition to the thermosphere by Joule heating can take place.

The Cassini mission so far is suggesting that the thermosphere and ionosphere of Titan appear to be an important source of complex chemical species that are transported to lower regions of the atmosphere (c.f., Waite et al. 2007; Coates et al. 2008). The continuing chemistry in the lower atmosphere then results in the formation of aerosols (and tholins) as discussed elsewhere in this book (Chapters 8 and 12). The thermosphere is particularly important as a source for nitrogen-bearing species because the N_2 bond is difficult to break, but this can happen with the absorption of ionizing solar radiation and/or energetic particle precipitation.

11.4.2 Questions and Issues That Remain Concerning the Upper Atmosphere and Ionosphere of Titan – the Extended Cassini Mission

The prime Cassini–Huygens mission has taught us much over 4 years about the structure and composition of the thermosphere and ionosphere of Titan. The mission has allowed the overall vertical distribution of major neutral species and the neutral temperature in thermosphere to be determined, although a full understanding of the underlying processes remains rudimentary. Some knowledge of the horizontal distribution of neutral species (i.e., latitude, local time) has also been gained, although not to the same degree as the altitude structure. Similarly, new understanding of the sources of the ionosphere on the dayside and nightside has been achieved, and much of the chemistry for ion species with mass numbers less than 100 has been explained semiquantitatively. However, the higher mass ion chemistry is not understood (Ascenzi et al. 2007) nor is its linkage to aerosol formation understood. Negative ions with masses up to 10^4 amu have been discovered at Titan but work on the sources or sinks of such ion species is just beginning (Vuitton et al. 2009). The formation process for higher hydrocarbon species such as benzene (leading to aerosol formation) is now known to involve both neutral and ion chemistry, but again our knowledge of the

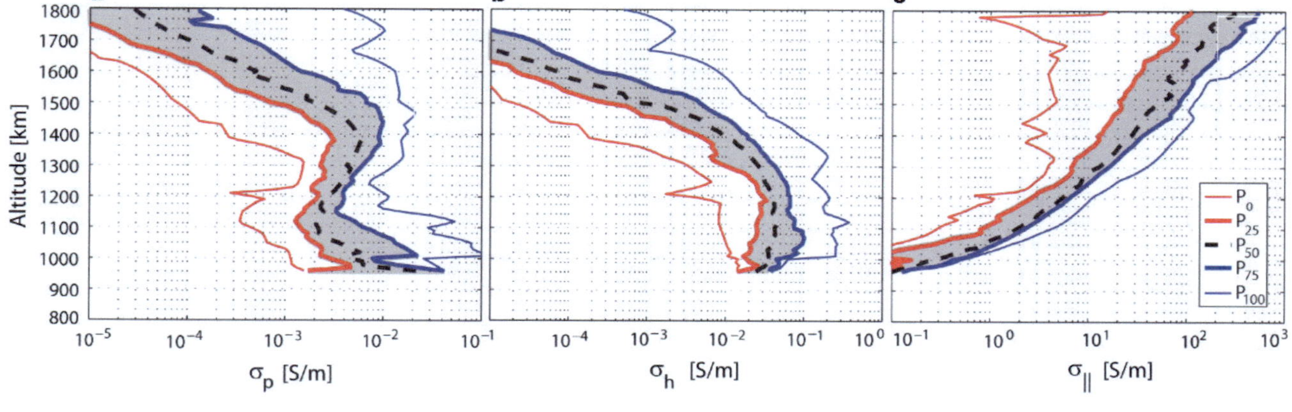

Fig. 11.25 Percentiles of the height profiles of the (**a**) Pedersen, (**b**) Hall, and (**c**) parallel conductivity observations based on 34 altitude profiles (17 flybys). The grey shaded areas show the conductivity values where 50% of all observations are. All observations are situated between the thin red and blue lines (from Rosenqvist et al. 2009))

processes involved is very sketchy. Plenty of work remains to be done during the Cassini extended mission.

A few specific outstanding issues for Titan's thermosphere and ionosphere are:

- Reconciliation of the neutral densities of species like HCN that are deduced from ion chemistry and INMS ion density measurements with densities measured by UVIS and/or predicted by theoretical models.
- Understanding the linkage between structures in measured major species and the dynamics and atmospheric escape. In particular, the trade-off in atmospheric models between eddy diffusion and methane escape is not understood.
- Ion-neutral chemistry for high mass ion species and for negative ion species is not well-understood. Effects of uncertainties in the chemical scheme and measured reaction coefficients have not been fully explored, although Carrasco et al. (2006, 2008) has started this process.
- The role of negative and positive ion chemistry for the overall Titan photochemistry and for aerosol formation is only beginning to be appreciated and more measurements and modeling work are needed to elucidate the role of the thermosphere and ionosphere.
- The role of the magnetic field topology for the transport of electrons (superthermal and thermal) and ions within the ionosphere and the linkage to the magnetosphere is beginning to be appreciated but needs more systematic study.
- A complete explanation for the high electron temperatures observed in the ionosphere is not yet available (high temperatures only for one dayside case have been explained) and no studies, data-based or modeling, have been published in the prime mission for ion temperatures.
- The lower ionospheric peak observed by the radio science experiment has not fully been linked to energetic ion precipitation, partly because the magnetospheric energetic ion composition has not been fully described and because magnetic shielding effects have not been fully explained.
- The relative roles of thermal and magnetic forcing terms and the ion-neutral collision term in the ionospheric dynamics needs further study. Explanations are needed for small-scale ionospheric structures (magnetic or plasma), including "ionopause" layers.
- Determination of bulk flow and pick-up losses of ionospheric plasma to the magnetosphere and its contribution to atmospheric escape require quantitative evaluation.

Acknowledgments The work at the University of Kansas was supported by the NASA Cassini project (grant NFP45280) via subcontract from Southwest Research Institute and by NASA Planetary Atmospheres Grant NNX07AF47G. Assistance with Figure 22 by M. Richard and P. Oswald is acknowledged, as is help with manuscript by T. Hunt-Ward.

References

Achterberg RK, Conrath BJ, Gierasch PJ, Flaser FM, Nixon CA (2008) Titan's middle-atmospheric temperatures and dynamics observed by the Cassini Composite Infrared Spectrometer, Icarus 194:263

Adams NG, Smith D (1988) Measurements of dissociative recombination coefficiencies for several polyatomic species at 300K. Chem Phys Lett 144:11

Ågren K, Wahland J-E, Modolo R, Lummerzheim D, Galand M, Müller-Wodard I, Canu P, Kurth WS, Cravens TE, Yelle RV, Waite JH Jr, Coates AJ, Lewis GR, Young DT, Bertucci C, Dougherty MK (2007) On magnetospheric electron impact ionization and dynamics in Titan's ram-side and polar ionosphere, a Cassini case study. Ann Geophys 25:2359

Ågren K, Wahlund J-E, Garnier P, Modolo R, Cui J, Galand M, Muller-Wodarg I (2009). The ionospheric structure of Titan, submitted to Planet Space Sci

Ajello JM et al (2007) Titan airglow spectra from Cassini Ultraviolet Imaging Spectrograph (UVIS): EUV analysis. Geophys Res Lett 34:L24204. doi:10.1029/2007GL031555

Ajello JM et al (2008) Titan airglow spectra from the Cassini Ultraviolet Imaging Spectrograph: FUV disk analysis. Geophys Res Lett 35:L06102. doi:10.1029/2007GL032315

Anicich VG, McEwan MJ (1997) Ion-molecule chemistry in Titan's ionosphere. Planet Space Sci 45:897

Anicich VG (1993) Evaluated bimolecular ion-molecule gas phase kinetics of positive ions for use in modeling planetary atmospheres, cometary comae, and interstellar clouds, J Chem Phys Ref Data, 22:1469

Anicich VG, Wilson P, McEwan MJ (2004) A SIFT ion-molecule study of some reactions in Titan's atmosphere. Reactions of N^+, N_2^+, and HCN^+ with CH4, C_2H_2, and C_2H_4, J Am Soc Mass Spectrum 15:1148

Ascenzi D, Cont N, Guella G, Franceschi P, Tosi P (2007) New insights into the reaction mechanisms of phenylium ions with benzene. J Phys Chem A 111:12513

Atreya SK (1986) Atmospheres and ionospheres of the outer planets and their satellites. Springer, Berlin

Backes H, Neubauer FM, Dougherty MK, Achilleos H, Andre N, Arridge CS, Bertucci C, Jones GH, Khurana KK (2005) Titan's magnetic field signature during the first Cassini encounter. Science 308:992

Banaskiewicz M, Lara LM, Rodrigo R, Lopez-Moreno JJ, Molina-Cuberos GJ (2000) A coupled model of Titan's atmosphere and ionosphere. Icarus 147

Banks PM, Kockarts G (1973) Aeronomy, Academic Press, New York

Bar-Nun A, Kleinfeld I, Kochavi E (1998) Trapping of gas-mixtures by amorphous water ice, Phys Rev B 38:7749

Bauer SJ (1987) Titan's ionosphere and atmospheric evolution. Adv. Space Res 7(5):65

Bell JM, Bougher SW, Waite JH, Ridley AJ, Bar-Nun A, Toth G, De La Haye V (2009) Simulating the global mean structure of Titan's upper atmosphere using the Titan Global Ionosphere-Thermosphere model, submitted to Planet Space Sci

Bertucci C, Achilleos N, Dougherty MK, Modolo R, Coates AJ, Szego K, Masters A, Ma Y, Neubauer FM, Garnier P, Wahlund J-E, Young DT (2008). The magnetic memory of Titan's ionized atmosphere. Science 321:1475

Bird MK et al. (1997) Detection of Titan's ionosphere from Voyager 1 radio occultation observations. Icarus 130(2):426, doi: 10.1006/icar.1997.5831

Broadfoot AL et al. (1981) Extreme ultraviolet observations from Voyager 1–encounter with Saturn. Science 212:206

Capone LA et al. (1976) The lower ionosphere of Titan. *Icarus 28*:367

Carrasco N, Dutuit O, Thissen R, Banaszkiewicz M, Pernot P (2006) Uncertainty analysis of bimolecular reactions in Titan ionosphere chemistry model. Planet Space Sci . doi:10.1016/J.pss.2006.06.004

Carrasco N, Alcaraz C, Dutuit O, Pleissis R, Thissen V, Vuitton R, Yelle V, Pernot P (2008) Sensitivity of a Titan ionospheric model to the ion-molecule reaction parameters. Planet Space Sci 56:1644, doi: 10.1016/j.pss.2008.04.007

Coates AJ, Crary FJ, Lewis GR, Young DT, Waite JH Jr (2007a) Discovery of heavy negative ions in Titan's ionosphere. Geophys Res Lett 34:L22103. doi:10.1029/2007GL030978

Coates AJ, Crary FJ, Young DT, Szego K, Arridge CS, Bebesi Z, Sittler EC Jr. (2007b) Ionospheric electrons in Titan's tail: plasma structure during the Cassini T9 encounter. Geophys Res Lett 34:L24S05, doi:10.1029

Coates AJ et al. (2009) Heavy negative ions in Titan's ionosphere: Altitude and latitude dependence. Planet Space Sci (submitted)

Coustenis A, Salama A, Schulz B, Ott S, Lellouch E, Encrenaz TH, Gautier D, Feuchtruber H (2003) Titan's atmosphere from ISO mid-infrared spectroscopy. Icarus 161:383

Crary FJ, MaGee BA, Mandt KE, Waite JH, Westlake J, Young DT (2009) Heavy ions, temperatures and winds in Titan's ionosphere: combined Cassini CAPS and INMS observations, submitted to Planet Space Sci

Cravens TE (1997) Physics of solar system plasmas. Cambridge University Press, Cambridge

Cravens TE, Keller CN, Ray B (1997) Photochemical sources of non-thermal neutrals for the exosphere of Titan. Planet Space Sci 45:889

Cravens TE, Lindgren CJ, Ledvina SA (1998) A two-dimensional multifluid MHD model of Titan's plasma environment. Planet Space Sci 46:1193

Cravens TE, Vann J, Clark J, Yu J, Keller CN, Brull C (2004) The ionosphere of Titan: An updated theoretical model. Adv Space Res (COSPAR Proceedings; special issue on Planetary atmospheres, ionospheres and plasma interactions, Kallio E, Shinagawa, H (eds)), 33:212, doi:10.1016/J.asr.2003.02.012.

Cravens TE et al. (2005) Titan's ionosphere: model comparison with Cassini Ta data. Geophys Res Lett 32:12. doi:10.1029/2005 GL023249

Cravens TE et al. (2006) The composition of Titan's ionosphere. Geophys Res Lett 33:L07105. doi:10.1029/2005GL025575

Cravens TE et al. (2008a) Energetic ion precipitation at Titan. Geophys Res Lett 35, doi: 10.1029/2007GL032451

Cravens TE, Robertson IP, Waite JH Jr, Yelle RV, Vuitton V, Coates AJ, Wahlund J-E, Agren K, Richard MS, De La Haye V, Wellbrock A, Neubauer FN (2008b) Model-data comparisons for Titan's nightside ionosphere. Icarus 199:174. doi:10.1016/j.icarus.2008.09.005

Cui J, Yelle RV, Volk K (2008) The distribution and escape of molecular hydrogen in Titan's thermosphere and exosphere. J Geophys Res 113: E10004, doi: 10.1029/2007JE003032.

Cui J et al. (2009a) Analysis of Titan's neutral upper atmosphere from Cassini Ion and Neutral Mass Spectrometer measurements. Icarus 200:581

Cui J, Galand M, Yelle RV, Vuitton V, Wahlund J-E, Lavvas P, Müller-Wodarg ICF, Cravens TE, Kasprzak WT, Waite JH Jr (2009b) Diurnal variations of Titan's ionosphere. J Geophys Res 114: A06310

De La Haye V, Waite JH Jr, Johnson RE, Yelle RV, Cravens TE, Luhmann JG, Kasprzak WT, Gell DA, Magee B, Leblanc F, Michael M, Jurac S, Robertson IP (2007a) Cassini INMS data in Titan's upper atmosphere and exosphere: Observation of a suprathermal corona. J Geophys Res 112:A07309. doi: 10.1029/2006JA012222

De La Haye V, Waite JH Jr, Cravens TE, Nagy AF, Johnson RE, Lebonnois S, Robertson IP (2007b) Titan's corona: The contribution of exothermic chemistry. Icarus 191*:236*. doi: 10.1016/ j.icarus.2007.04.031

De La Haye V, Waite JH Jr, Cravens TE, Robertson IP, Lebonnois S (2008a) Coupled ion and neutral rotating model of Titan's upper atmosphere. Icarus 197:110

De La Haye V, Waite JH Jr, Cravens TE, Bougher SW, Robertson IP, Bell JM (2008b) Heating Titan's upper atmosphere, J Geophys Res 113:A11314

Fox JL (1996a) Aeronomy. In Drake WF (ed) Atomic, molecular and optical physics handbook, AIP Press, Woodbury, New York, pp 940–968

Fox JL (1996b) Hydrocarbon ions in the ionospheres of Titan and Jupiter, in Dissociative Recombination: Theory, Experiment, and Application, edited by D. Zajfman, JBA Mitchell, and B. Rowe, World Sci, New Jersey, pp. 40–46

Fox JL, Dalgarno A (1981) Ionization, luminosity, and heating of the upper atmosphere of Venus. J Geophys Res 86:629

Fox JL, Yelle RV (1991) Solar heating efficiencies in the thermosphere of Titan, Bull Amer Astronom Soc 23:1184

Fox JL, Yelle RV (1997) Hydrocarbon ions in the ionosphere of Titan. Geophys Res Lett 24:2179

Fox JL, Yeager KE (2006) Morphology of the near-terminator Martian ionosphere: a comparison of models and data. J Geophys Res 111:A10309. doi:10.1029/2006JA011697

Galand M, Lilensten J, Toublanc D, Maurice S (1999) The ionosphere of Titan: ideal diurnal and nocturnal cases. Icarus 104:92

Galand M et al. (2006) Electron temperature of Titan's sunlit ionosphere. Geophys Res Lett 33(21):L21101

Gan L, Cravens TE (1992) Electron impact cross sections and cooling rates for methane. Planet Space Sci 40:1535

Gan L, Keller CN, Cravens TE (1992) Electrons in the ionosphere of Titan. J Geophys Res 97:12137

Gan L, Cravens TE, Keller CN (1993) A time-dependent model of suprathermal electrons at Titan. In Gombosi TI (ed) Plasma environments of non-magnetic planets, Elsevier, vol 4, p 171

Hansen KC (2001) MHD simulations of the magnetospheres of Jupiter and Saturn: Application to the Cassint mission, Ph.D thesis, University of Michigan, Ann Arbor, Mich

Hartle RE et al. (2006) Initial interpretation of Titan plasma interaction as observed by the Cassini Plasma Spectrometer: comparisons with Voyager 1. Planet Space Sci 54:1211

Hidayat J, Marten A, Bezard B, Gautier D, Owen T, Matthews HE, Paulert G (1997) Millimeter and submillimeter heterodyne observations of Titan: retrieval of the vertical profile of HCN and the $^{12}C/^{13}C$ ratio. Icarus 126:170

Hidayat J, Marten A, Bezard B, Gautier D, Owen T, Matthews HE, Paulert G (1998) Millimeter and submillimeter heterodyne observations of Titan: the vertical profile of carbon monoxide in its stratosphere. Icarus 133:109

Horanyi M (1993) Dust in planetary magnetospheres. Adv Space Res 13(10):231

Horst SM, Vuitton V, Yelle RV (2008) The origin of oxygen species in Titan's atmosphere. J Geophys Res . doi:10.1029/2008JE003135

Hunten DM, Tomasko MG, Flasar FM, Samuelson RE, Strobel DF, Stevenson DJ (1984) Titan. In Gehrels T, Matthews MS (eds) Saturn, University of Arizona Press, Tucson, pp 671–759

Ip W-H (1990) Titan's upper ionosphere. Astrophys J Part 1 362(10):354

Jacovi R, Bar-Nun R (2008) Removal of Titan's noble gases by their trapping in its haze. Icarus 196:302

Kabin K, Gombosi TI, DeZeeuw DL, Powell KG, Israelevich PL (1999) Interaction of the Saturnian magnetosphere with Titan. J Geophys Res 104:2451

Kasprzak WK et al (1996) Cassini orbiter ion and neutral mass spectrometer. Proc SPIE 2803:129

Keller CN, Cravens TE (1994) One dimensional multispecies hydrodynamic models of the wakeside ionosphere of Titan. J Geophys Res 99:6527

Keller CN, Cravens TE, Gan L (1992) A model of the ionosphere of Titan. J Geophys Res 97:12117

Keller CN, Cravens TE, Gan L (1994) One dimensional multispecies magnetohydrodynamic models of the ramside ionosphere of Titan. J Geophys Res 99:6511

Keller CN, Anicich VG, Cravens TE (1998) Model of Titan's ionosphere with detailed hydrocarbon chemistry. Planet Space Sci 46:1157

Kliore AJ et al. (2008) First results from the Cassini radio occultations of the Titan ionosphere. J Geophys Res 113, doi: 10.1029/2007JA 012965

Kelley MC (1989) The Earth's ionosphere, plasma physics, and electrodynamics. Academic Press, San Diego, CA

Krasnopolsky VA (2009) A photochemical model of Titan's atmosphere and ionosphere. Icarus 201:226

Krimigis SM et al (2005) Dynamics of Saturn's magnetosphere from MIMI during Cassini's orbital insertion. Science 307(5713):1270

Lammer H, Bauer SJ (1993) Atmospheric mass loss from Titan by sputtering. Planet Space Sci 41:657. doi:10.1016/0032-0633(93)90049-8

Lara LM, Lellouch E, Lopez-Moreno JJ, Rodrigo R (1996) Vertical distribution of Titan's atmospheric neutral constituents. J Geophys Res 101:23261

Lavvas PP, Coustenis A, Vardavas IM (2008a) Coupling photochemistry with haze formation in Titan's atmosphere: Part I: model description. Planet Space Sci 56:27

Lavvas PP, Coustenis A, Vardavas IM (2008b) Coupling photochemistry with haze formation in Titan's atmosphere, Part II: Results and validation with Cassini/Huygens data. Planet Space Sci 56:67

Lean JL et al. (2003) A new model of solar EUV irradiance variability 2. Comparisons with empirical models and observations and implications for space weather. J Geophys Res 108A(2):1059, doi:10.1029/2001JA009238

Lebonnois S, Toublanc D, Hourdin F, Ranna P (2001) Seasonal variations of Titan's atmospheric composition. Icarus 152:384

Lebonnois S, Bakes ELO, McKay CP (2003) Atomic and molecular hydrogen budget in Titan's atmosphere. Icarus 161:474. doi:10.1016/ S0019-1035(02)00039-8

Lebonnois S (2005) Benzene and aerosol production in Titan and Jupiter's atmospheres: A sensitivity study. Planet Space Sci 53:486

Ledvina SA, Cravens TE (1998) A three-dimensional MHD model of plasma flow around Titan. Planet Space Sci 46:1175

Ledvina SA, Cravens TE, Kecskemety K (2005) Ion distributions in Saturn's magnetosphere near Titan. J Geophys Res 110(A6):A06211, doi:10.1029/2004JA010771

Liang M-C, Yung YL, Shemansky DE (2007) Photolytically generated aerosols in the mesosphere and thermosphere of Titan. Astrophys J Lett 661:L199

Lilensten J, Simon C, Witasse O, Dutuit O, Thissen R, Alcarez C (2005) A fast comparison of the diurnal secondary ion production in the ionosphere of Titan. Icarus 174:285

Lindal GF, Wood GE, Sweetnam DN, Eshlemon VR, Tyler GL (1983) The atmosphere of Titan: an analysis of Voyager 1 radio occultation measurements. Icarus 53:348

Liu X, Shemansky D (2008) Analysis of Cassini UVIS observations of Titan VUV emission, presented at Titan after Cassini-Huygens Symposium, Corpus Christi, TX, 6–11 July 2008

Lopez-Moreno JJ et al (2008) Structure of Titan's low altitude ionized layer from the Relaxation Probe onboard HUYGENS. Geophys Res Lett 35:L22104. doi:10.1029/2008GL035338

Luhmann JG, Cravens TE (1991) Magnetic fields in the ionosphere of Venus. Space Sci Rev 55(201):1991

Ma Y, Nagy AF, Cravens TE, Sokolov IG, Clark J, Hansen KC (2004) 3-D global model prediction for the first close flyby of Titan by Cassini. Geophys Res Lett 31:L22803, doi:10.1029/2004GL02145

Ma Y-J, Nagy AF, Cravens TE, Sokolov IU, Hansen KC, Wahlund J-E, Crary FJ, Coates AJ, Dougherty MK (2006) Comparisons between MHD model calculations and observations of Cassini flybys of Titan. J Geophys Res 111:A05207. doi:10.1029/2005JA011481

Ma Y-J et al. (2007) 3D global multispecies Hall-MHD simulation of the Cassini T9 flyby. Geophys Res Lett 34:L24S10

MaGee B, Waite JH, Mandt KE, Bell J, Westlake J, Gell DA, De la Haye V (2009) INMS derived composition of Titan's upper atmosphere: analysis methods and model comparison, submitted to Planet Space Sci

Marynick DS, Dixon DA (1977) Electron affinity of the methyl radical: Structures of CH, PNAS, 74:410, doi: 10.1073/pnas.74.2.410

Mason M, Marrero TR (1970) The diffusion of atoms and molecules. In Bates DR, Esterman I (eds) Advances in atomic and molecular physics, Academic Press, San Diego, CA, p 155

McLain JL, Poterya V, Molek CD, Babcock LM, Adams NG (2004) Flowing afterglow studies of the temperature dependencies for dissociative recombination of O_2+, CH_5+, C_2H_5+, and C_6H_7+ with electrons. J Phys Chem A 108:6706

McLain JL et al. (2009) Flowing afterglow studies of the concentration dependencies on electron recombination of protonated cyanides (RCN)H+ and their proton-bound dimer ions $(RCN)_2$H+ where R is H, CH3, AND CH3CH2 Int J Mas 282:85

Michael M, Johnson RE (2005) Energy deposition of pickup ions and heating of Titan's atmosphere. Planet Space Sci 53:1510. doi:10.1016/j.pss. 2005.08.001

Michael M, Johnson RE, Leclanc F, Liu M, Luhmann JG, Shematovich VI (2005) Ejection of nitrogen from Titan's atmosphere by magnetospheric ions and pick-up ions. Icarus 175:263. doi:10.1016/j. icarus.2004.11.004

Modolo R, Chanteur GM, Wahlund J-E, Canu P, Kurth WS, Gurnett D, Matthews AP, Bertucci C (2007). Plasma environment in the wake of Titan from hybrid simulation: a case study. Geophys Res Lett 34:L24S07, doi: 10.1029/2

Molina-Cuberos GJ et al (2001) Ionosphere layer induced by meteoric ionization in Titan's atmosphere. Planet Space Sci 49:143

Müller-Wodarg ICF, Yelle RV, Mendillo M, Young LA, Aylward AD (2000) The thermosphere of Titan simulated by a global three-dimensional time-dependent model. J Geophys Res 105:20833

Müller-Wodarg ICF, Yelle RV, Mendillo MJ, Aylward AD (2003) On the global distribution of neutral gases in Titan's upper atmosphere and its effect on the thermal structure. J Geophys Res 108:1453. doi:10.1029/2003 JA010054

Müller-Wodarg ICF et al (2006) Waves and horizontal structure in Titan's thermosphere. J Geophys Res 111:A12315. doi:10.1029/2006 JA011961

Müller-Wodarg ICF, Yelle RV, CuiJ, Waite JH (2008) Horizontal structures and dynamics of Titan's thermosphere. J Geophys Res 113:E10005, doi:10.1029/2007JE003033

Nagy AF, Cravens TE (2002) Solar system ionospheres. In Mendillo M, Nagy AF, Waite JH (eds) Atmospheres in the solar system: comparative aeronomy, geophys. monog. 130, AGU, Washington, DC, pp 39–54

Ness NF, Acuna MH, Behannon KW (1982) The induced magnetosphere of Titan. J Geophys Res 87:1369

Neubauer FM, Gurnett DA, Scudder JD, Hartle RE (1984) Titan's magnetospheric interaction. In Gehrels T, Matthews MS (eds) Saturn, University of Arizona Press, Tucson, pp 760–787

Niemann HB et al. (2005) The abundances of constituents of Titan's atmosphere from the GCMS instrument on the Huygens probe. Nature 438:779

Richards P, Torr D (1988) Ratios of Photoelectron to EUV ionization rates for aeronomic studies. J Geophys Res 93:4060

Richards PG, Fennelly JA, Torr DG (1994) EUVAC: a solar EUV flux model for aeronomic calculations. J Geophys Res 99(A5):8981

Robertson IP, Cravens TE, Waite JH Jr, Yelle RV, Vuitton V, Coates AJ, Wahlund JE, Agren K, Magee B, Mandt K, Fattig E, Richard MS (2009) Structure of Titan's ionosphere: model comparisons with Cassini Data. Planet Space Sci (in press)

Roboz A, Nagy AF (1994) The energetics of Titan's ionosphere. J Geophys Res 99(A2):2087

Rosenqvist J-E, Whahlund K, Agren R, Modolo HJ, Opgenoorth D, Strobel I, Muller-Wodarg P, Garnier, Bertucci C (2009) Titan ionosheric conductivities, submitted to Planet Space Sci (in press)

Schunk RW, Nagy AF (2000) Ionospheres: physics, plasma physics, and chemistry. In Dressler AJ, Houghton JT, Rycroft MJ (eds) Cambridge atmospheric and space science series, Cambridge University Press, Cambridge, UK, pp 104–147

Sekine Y, Imanaka H, Matsui T, Khare BN, Bakes ELO, McKay CP, Sugita S (2008a) The role of organic haze in Titan's atmospheric chemistry. I. Laboratory investigations on heterogeneous reaction of atomic hydrogen with Titan tholin, Icarus. 194:186

Sekine Y, Imanaka H, Matsui T, Khare BN, Bakes ELO, McKay CP, Sugita S (2008b) The role of organic haze in Titan's atmospheric chemistry. II. Effect of heterogeneous reaction to the hydrogen budget and chemical composition of the atmosphere, Icarus. 194:201

Shemansky DE, Stewart AIF, West RA, Esposito LW, Hallett JT, Liu XM (2005) Science 308:978

Shematovich VI, Johnson RE, Michael M, Luhmann JG (2003) Nitrogen loss from Titan. J Geophys Res 108:5086. doi:10.1029/2003JE002096

Sittler EC, Bertucci CL, Coates AJ, Cravens TE, Dandouras I, Shemansky DE (2008) Energy deposition processes in Titan's upper atmosphere, in Titan After Cassini-Huygens, to be submitted

Smith GR, Strobel DF, Broadfoot AL, Sandel BR, Shemansky DE, Holberg JB (1982) Titan's upper atmosphere – composition and temperature from the EUV solar occultation results. J Geophys Res 87:1351

Strobel DF (1985) Photochemistry of Titan, ESA Spec Publ., ESA SP-241, p 145

Strobel DF (2008) Titan's hydrodynamically escaping atmosphere. Icarus 193:588

Szego K, Bebesi Z, Bertucci C, Coates AJ, Crary F, Erdos G, Hartle R, Sittle EC, Young DT (2007) Charged particle environment of Titan during the T9 flyby. Geophys Res Lett 34:L24503

Tobisca WKT, Woods T, Eparvier FG, Viereck R, Floyd L, Bouwer D, Rottman G, White OR (2000) The SOLAR2000 empirical solar irradiance model and forecast tool. J Atmos Sol Terr Phys 62(14):1233

Toublanc D, Parisot JP, Brillet J, Gautier D, Raulin F, McKay CP (1995) Photochemical modeling of Titan's atmosphere. Icarus 113:2. doi:10.1006/icar.1995.1002

Vervack RJ, Sandel BR, Strobel DF (2004) New perspectives on Titan's upper atmosphere from a reanalysis of the Voyager 1 UVS solar occultations. Icarus 170:91. doi:10.1016/j.icarus.2004.03.005

Vuitton V, Yelle RV, Anicich VG (2006) The nitrogen chemistry of Titan's upper atmosphere revealed. Astrophys J 647:L175

Vuitton V, Yelle RV, McEwan MJ (2007) Ion chemistry and N-containing molecules in Titan's upper atmosphere. Icarus 191:722

Vuitton V, Yelle RV, Cui J (2008) Formation and distribution of benzene on Titan. J Geophys Res 113:EDO5007, doi:10.1029/2007JE002997

Vuitton V, Lavvas P, Yelle RV, Galand M, Wellbrock A, Lewis GR, Coates AJ, Wahlund J-E (2009) Negative ion chemistry in Titan's upper atmosphere. Plant Space Sci (in press), doi: 10.1016/j.pss.2009.04.004

Wahlund J-E et al. (2005) Cassini measurements of cold plasma in the ionosphere of Titan. Science 308:986

Wahlund JE et al. (2009) On the amount of heavy ions in Titan's ionosphere, submitted to Planet Space Sci

Waite JH Jr, Cravens TE, Kozyra JU, Chen RF, Atreya SK, Nagy AF (1983) Electron precipitation and related aeronomy of the Jovian thermosphere and ionosphere. J Geophys Res 88:6143

Waite JH Jr et al. (2004) The Cassini ion and neutral mass spectrometer (INMS) investigation. Space Science Rev 114(1):113

Waite JH Jr et al. (2005) Ion Neutral Mass Spectrometer (INMS) results from the first flyby of Titan. Science 308:982

Waite JH Jr, Young DT, Cravens TE, Coates AJ, Crary FJ, Magee B, Westlake J (2007) The process of tholin formation in Titan's upper atmosphere. Science 316:870. doi:10.1126/science.1139727, 2007

Wilson EH, Atreya SK (2004) Current state of modeling the photochemistry of Titan's mutually dependent atmosphere and ionosphere. J Geophys Res 109:E06002. doi:10.1029/2003JE002181

Yelle RV (1991) Non-LTE models of Titan's upper atmosphere. Astrophys J 383:380. doi:10.1086/170796

Yelle RV, Strobel DF, Lellouch E, Gautier D (1997) Engineering models for Titant's atmosphere, ESA Spec. Publ. 1177:243–256

Yelle RG, Borggren N, de la Haye V, Kasprzak WT, Niemann HB, Müller-Wodarg I, Waite JH Jr (2006) The vertical structure of Titan's upper atmosphere from Cassin Ion Neutral Mass Spectrometer measurements. Icarus 182:567

Yelle RV, Cui J, Müller-Wodarg ICF (2008) Eddy diffusion and methane escape from Titan's atmosphere. J Geophys Res 113:E10, doi:10.1029/2007JE003031

Young DT et al. (2005) Composition and dynamics of plasma in Saturn's magnetosphere. Science 307:1262

Yung YL (1987) An update of nitrile photochemistry on Titan. Icarus 72:468

Yung YL, Allen M, Pinto JP (1984) Photochemistry of the atmosphere of Titan: comparison between model and observations. Astrophys J Suppl 55:465. doi:10.1086/190963

Chapter 12
Aerosols in Titan's Atmosphere

Martin G. Tomasko and Robert A. West

Abstract Aerosols in Titan's atmosphere play important roles in the transfer of solar and thermal radiation, in Titan's heat balance, in forcing atmospheric dynamics, and as a sink for photochemical reactions. In this chapter we briefly summarize the history of our knowledge of their distribution and optical properties before concentrating in greater detail on current knowledge of their properties and roles in the physics and chemistry in Titan's atmosphere. We discuss the size, shape, optical properties, and the vertical distribution of Titan's aerosols. We discuss variations of the optical properties of the aerosols with wavelength, variations of the distribution of the aerosols over the disk of Titan, and the seasonal and long-term variations in structure. We discuss the visible and thermal opacity of the aerosols, and their roles in Titan's heat balance. We summarize the history and current state of laboratory simulations of these particles, and present our understanding of their formation and life cycles in one and two-dimensional microphysical models. We also discuss the presence, location, and variations in condensation clouds over Titan's disk. Finally, we indicate the prospects for further progress in understanding the origin, distribution, and properties of Titan's aerosols in the future.

12.1 Introduction

The presence of aerosols in Titan's atmosphere has been known since Titan's albedo spectrum was measured, and the shape of the spectrum provided the first clues to the nature of these particles. The deep methane bands at red and infrared (IR) wavelengths require several km-amagats of methane gas for their formation (Trafton 1973). Titan's geometric albedo at blue and near ultraviolet (UV) wavelengths is lower than it would be due to molecular scattering from this abundance of clean gas. Absorbing aerosols located above most of the gaseous atmosphere are required to account for Titan's low geometric albedo at short wavelengths. The strong signatures of the methane absorption bands at red and longer wavelengths imply that the optical depths of the aerosols decrease rapidly with increasing wavelength, providing constraints on the size of the aerosol particles (Podolak and Danielson 1977).

Several significant advances in understanding the properties of Titan's aerosols occurred before the Cassini–Huygens mission. The flyby of Pioneer 11 in 1980 provided measurements of the linear polarization of blue and red sunlight scattered from Titan's disk at 90° scattering angle (Tomasko and Smith 1982). Similar measurements at other wavelengths soon followed from Voyager 2 (West et al. 1983). The degrees of polarization observed at scattering angles near 90° were very large, confirming the small size of the aerosols already deduced from their rapid decrease in opacity with increasing wavelength. However, the observations by Voyager at very high phase angles (Rages et al. 1983) showed that the aerosols produced strong forward scattering in addition to high linear polarization near 90° scattering angle. Spherical particles could not reproduce these two types of observations for the same particle size distribution, and workers concluded the aerosols in Titan's atmosphere were not spherical.

This situation led West (1991) to compute single scattering phase functions, degrees of polarization, and cross sections from large fractal aggregates of small, approximately spherical "monomers". These calculations showed that for loose aggregate structures the size of each monomer could be sufficiently small to produce the strong linear polarization while the larger size of the entire aggregate could provide the strong forward scattering. West and Smith (1991) showed that such particles had great promise for explaining these two different types of Titan observations. The size of the entire aggregate particles could still be sufficiently small to give the required rapid decrease in opacity with increasing wavelength to produce the strong methane bands at red and

M.G. Tomasko (✉)
Lunar and Planetary Laboratory, University of Arizona, Tucson, AZ, 85721, USA
e-mail: mtomasko@lpl.arizona.edu

R.A. West
Jet Propulsion Laboratory, California Institute of Technology, 4800 Oak Grove Drive, Pasadena, CA 91109, USA

R.H. Brown et al. (eds.), Titan from Cassini-Huygens,

infrared wavelengths while nearly explaining the low geometric albedo at blue and near ultraviolet wavelengths.

At wavelengths of 200 nm, the aggregates are sufficiently large compared to the wavelength that the single-scattering albedo is >0.5 for any imaginary refractive index. This is too large to reproduce the small geometric albedo of Titan in the ultraviolet. Courtin et al. (1991) discussed observations of the low ultraviolet geometric albedo of Titan, and concluded that a bimodal distribution including particles with radii not greater than 0.02 μm mixed with larger particles with radii in the range of 0.1 to 0.5 μm could fit the low UV albedo as well as the methane bands in red and near infrared wavelengths. This model did not explicitly include fractal aggregates; however, the model suggested that high altitude regions dominated by monomers alone above larger aggregate particles at lower altitudes might be capable of fitting the observed variation of geometric albedo with wavelength from the UV to the infrared. Rannou et al. (1997) used Voyager observations to constrain the altitude of production and the vertical profile of the particles near the region of formation. The Voyager images of separated layers at the limb of Titan indicated the complexity of this formation region.

The decrease of Titan's geometric albedo toward blue wavelengths is partly due to the increased opacity of the aerosols toward shorter wavelengths, but it also requires that the imaginary refractive index increase toward shorter wavelengths as well. Some estimates of the variation of refractive index as well as chemical identity were provided by early attempts to produce aerosol material in the laboratory. Khare et al. (1984) produced aerosol material which they termed "tholin" from mixtures of methane and nitrogen using an electric spark, and they published the refractive index of the solid brown or orange material produced in their reaction vessels. Many modelers used their values of imaginary index as a function of wavelength or small variations from it to produce models of Titan's geometric albedo spectrum that fit the planetary observations quite well. Chemical analyses of the composition of the tholins showed a complex mixture of a great many hydrocarbons.

Titan's aerosols play important roles in many physical and chemical processes in Titan's atmosphere. The low albedos of the aerosols at blue and shorter wavelengths absorb sunlight and heat the atmosphere at high altitudes. At longer wavelengths, the aerosols become optically thinner and permit thermal infrared radiation to escape to space. In this way they have been termed to act as an "anti-greenhouse" agent (McKay et al. 1989). Their roles in absorbing and scattering solar and thermal infrared radiation give them a strong role in the heat balance of Titan and in providing the forcing for atmospheric dynamics. In addition, their formation at high altitudes where ultraviolet sunlight penetrates implies that aerosol production provides a sink for products of methane photochemistry that contain large C to H ratios. Because of their important roles in these physical and chemical processes, gaining a greater understanding of the vertical and horizontal distribution, size, shape, and composition of Titan's aerosols were major goals of the Cassini–Huygens mission.

We present and discuss the new information about Titan's aerosols in this chapter. Sections 12.2–12.4 discuss the details of aerosol size and shape, vertical distribution, and variations of optical properties with wavelength. We present variations in properties over the disk and with time in Sections 12.5 and 12.6. We discuss the roles of the aerosols in Titan's heat balance and their microphysics in Sections 12.7 and 12.8. We outline some recent laboratory simulation studies in Section 12.9 and summarize some measurements of condensation clouds in Section 12.10. A final section discusses prospects for future progress on determining the nature of Titan's aerosols, including continued Cassini and Hubble Space Telescope observations as well as possible investigations on future missions.

12.2 Aerosol Size and Shape Estimates

Constraints on the size and shape of Titan's aggregate haze particles are available from the Descent Imager/Spectral Radiometer (DISR) observations as the Huygens probe descended through the Titan atmosphere (Tomasko et al. 2008a). The DISR instrument (Tomasko et al. 2002) measured the degree of linear polarization in two passbands centered at 938 nm ("red") and 491 nm ("blue") using crossed polarizing analyzers. For small aerosols, the position angle of linearly polarized light is known to be perpendicular to the scattering plane, so the measurements made through two orthogonal analyzers suffice to measure the total intensity and the degree of linear polarization. The linear polarization of the multiply scattered sunlight headed downward at a scattering angle of 90° was 0.65 in the red channel and 0.5 in the blue channel at the beginning of the measurements near 140 km altitude. In order to produce these high degrees of linear polarization in multiply scattered light, the degree of polarization produced in single scattering must be nearly 100%. Comparisons with model calculations constrained the radii of the monomers (which determine the minimum dimensions of the aggregate particles) to be near 0.05 μm as shown in Fig. 12.1.

The DISR instrument also measured the forward scattering nature of Titan's atmospheric aerosols using an upward-looking camera with a field of view of 20° to 70° from the zenith and covering 6° in azimuth. Some images were obtained within 10° from the azimuth of the sun. The isophote contours in these images were compared with models computed with various numbers of monomers of the size necessary to reproduce the observed linear polarization measurements. The DISR observations (see Fig. 12.2) required that the

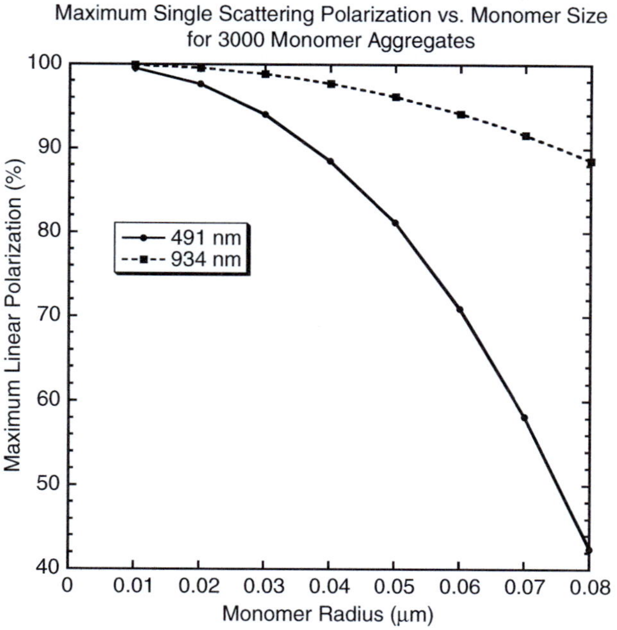

Fig. 12.1 The maximum degree of linear polarization in single scattering as a function of monomer radius at the wavelengths of the solar aureole channels of the DISR instrument is shown. In order to produce polarization of 60% in the red channel and 50% in the blue channel, Tomasko et al. (2008a) estimated that the monomer radius had to be approximately 0.05 μm or less. New models (Tomasko et al. 2009) of the DISR measurements indicate that the monomer radius is 0.04 μm.

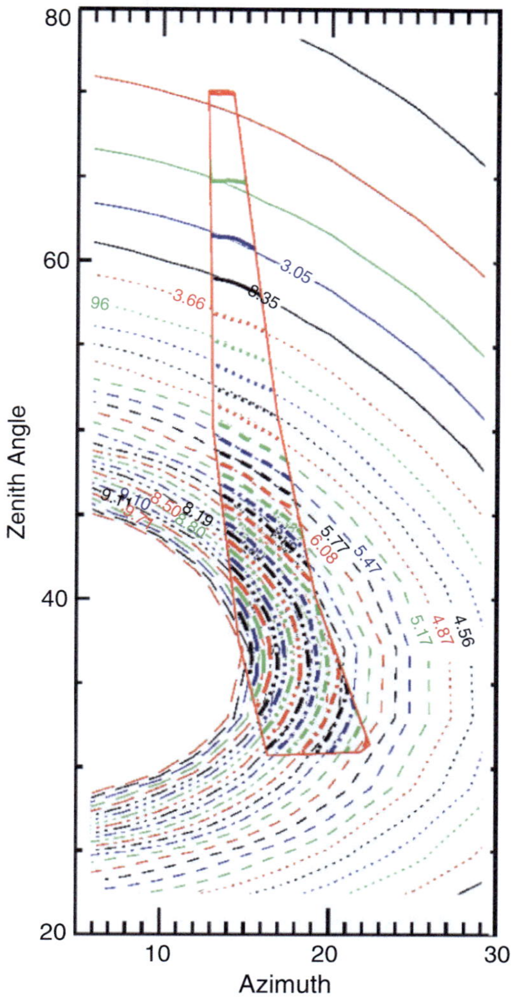

Fig. 12.2 The isophotes of intensity in the red solar aureole channel measured by the DISR instrument near the sun as measured near 60 km altitude are shown inside the red outline. The measured isophotes are compared to those of a model having 3,000 monomers of radius 0.05 μm. The model and the measurements are in good agreement.

number of monomers in an aggregate haze particle when looking upward from 60 km altitude be about 3,000 (within a factor of ~2). This seemed to be much larger than the number of monomers used in some of the earlier work with aggregate particles, but still corresponds to only about a dozen collisions between clusters of similar size found between the altitude where the aerosols are formed and the 60 km altitude where the DISR measurements were made. The radius of a spherical particle with the same volume as the 3,000 monomers in the aggregate is 0.72 μm whereas the radius of a spherical particle with the same projected area as the aggregate particle is 2.03 μm. The forward scattering nature of the haze particles is shown in Fig. 12.3 where phase functions are given for different numbers of monomers in the aggregate particles.

More complete investigations of the DISR polarization measurements suggest that the monomer size changes only very little throughout the altitude range of the DISR measurements (between 140 km and the surface). The authors assumed that the forward part of the single scattering phase functions of the aerosols remained reasonably constant in this altitude range.

It should be emphasized that computation of the single scattering properties of aggregates with such a large number of monomers is a large computational challenge. In practice, computations are made for various particles with monomer numbers up to a few hundred, and extrapolations are made to still larger particles. The approximate technique used by Tomasko et al. (2008a) used in the analysis of the DISR observations is described in an appendix to their paper.

The wavelength dependence of the cross section of the aerosols can also constrain the size of the aerosol particles. The DISR investigators fit the wavelength dependence of the aerosol extinction with power laws in three separate altitude intervals, as shown in Fig. 12.4. Notice that the slope of the power law decreases with decreasing altitude, from −2.34 above 80 km, −1.41 between 80 and 30 km altitude, to −0.97 below 30 km. Steeper extinction power laws are associated with smaller particles, e. g. slopes may reach down to −4.0 for light scattering by particles very small compared to the wavelength (Rayleigh scattering). Flatter power laws are associated with larger particles, down to zero for particles very large compared to the wavelength. Note that the extraction

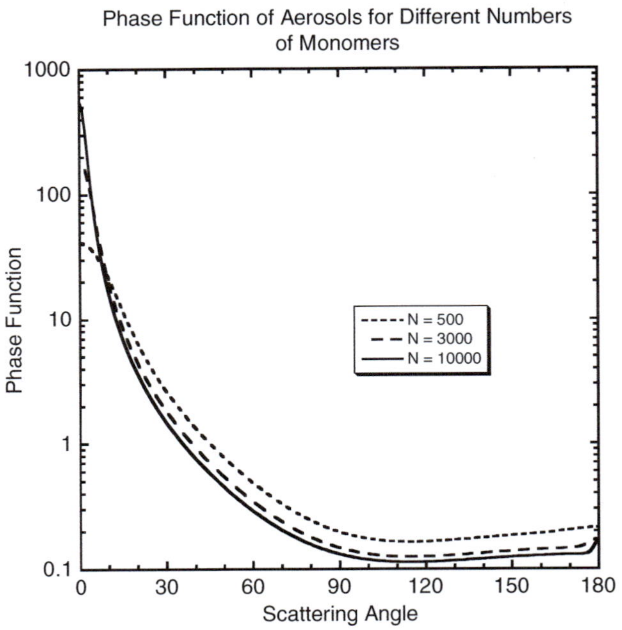

Fig. 12.3 The phase functions for aggregate particles grown by cluster-cluster collisions are shown for N different numbers of monomers each having a radius of 0.05 μm. The computations are for a wavelength of 934 nm (in the red channel of the DISR solar aureole camera).

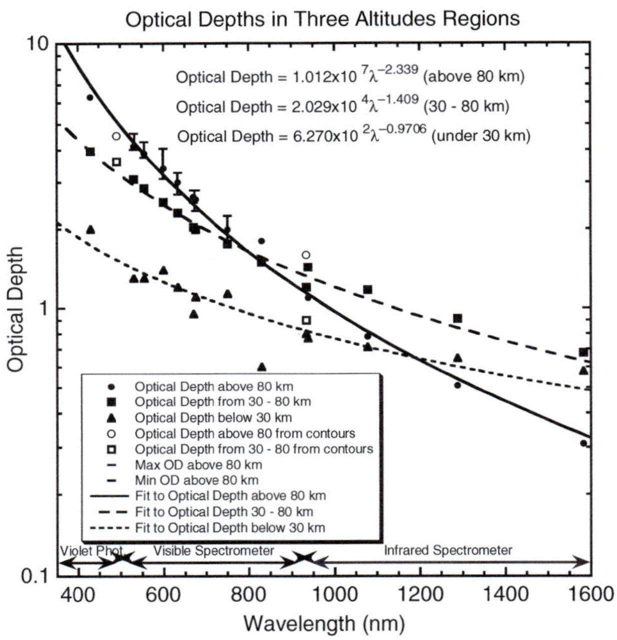

Fig. 12.4 The variation of extinction optical depth of the haze as functions of wavelength are shown for the haze particles above 80 km altitude, between 80 and 30 km altitude, and below 30 km as described by Tomasko et al. (2008a). The constraints imposed by the measurements are shown by points and are compared to power law fits shown by the lines. The slope of the variation with wavelength decreases with increasing depth into the atmosphere due to an increase in the size of the haze particles with decreasing altitude.

of particle size depends on the accuracy with which the cross section for aggregate particles can be estimated for different sized aggregates. At altitudes greater than 80 km above Titan's surface, the DISR investigators reported that the variation of cross section from blue to red matched the estimates they used for loose fractal aggregates rather well. From 80 to 30 km above Titan's surface, the observed variations in aerosol cross section between blue and red matched less well, though still to within about a factor of two. Below 30 km the variations with wavelength were rather gentile, and could be fitted with particles large compared to the wavelength.

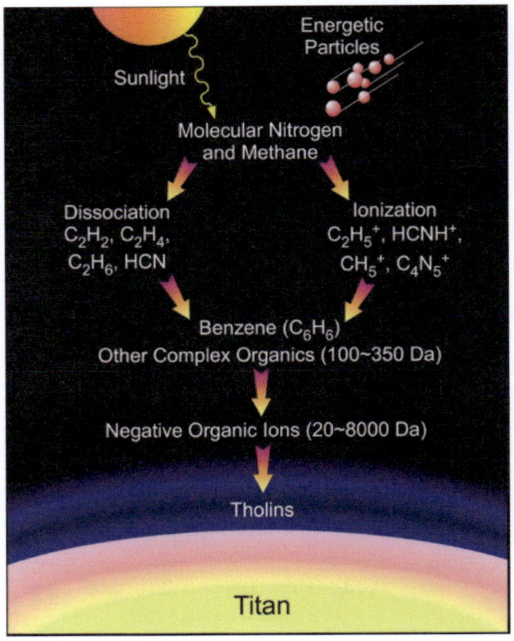

Fig. 12.5 Schematic diagram from Waite et al. (2007) showing the chemical processes leading up to the formation of tholins in Titan's upper atmosphere.

12.3 Vertical Distribution of Aerosols

The prevailing view of tholin haze formation emphasized the role of photochemical reactions in the stratosphere above about 100 km altitude (Yung et al. 1984; Wilson and Atreya 2003; Liang et al. 2007). Recent results from instruments on the Cassini orbiter call attention to ion and neutral chemistry in the high atmosphere, around 1,000 km altitude. A detailed discussion of atmospheric chemistry can be found elsewhere in this book (see chapters on stratosphere and thermosphere chemistry). Here we highlight new findings which have bearing on aerosol formation as illustrated in Fig. 12.5.

Both the Ion and Neutral Mass Spectrometer (INMS) and the Ion Beam Spectrometer (IBS) on the Cassini Plasma Spectrometer (CAPS) instrument (Young et al. 2004) detected large molecules at altitudes above 950 km in Titan's thermosphere. This section summarizes findings presented by both instruments as reported by Waite et al. 2007. Mass spectra recorded by INMS show mass peaks that extend to the mass limit of the instrument. Of particular interest is the benzene peak (C_6H_6) near 78 Da. Benzene is the smallest of the class of polycyclic aromatic hydrocarbons (PAHs) which are thought to be important for aerosol formation and are present in molecular clouds. The IBS measured positive and negative ions with charge/mass ratios up to the limit of its sensitivity near 10,000, although the number density falls rapidly above 4,000 Da.

The abundance of large negative ions was not anticipated but in retrospect seems understandable because of the electron affinities expected for PAHs and cyanoaromatics in the gas/plasma environment near the top of Titan's atmosphere. The actual mass of these clusters or proto-particles could be larger than several thousand Dalton if the charge per particle is greater than 1. Waite et al. (personal communication 2008) estimate effective radii of these large ions in the range 7.5–26 nm. These values are a factor of ten smaller than those reported by Waite et al. (2007) due to an error of a factor of 1000 in assumed mass density. These particles at 1,000 km altitude are then precursors to monomers that aggregate to form the larger haze particles observed by optical instruments at 500 km and below. However, considerable microphysical and possibly also chemical restructuring must occur over the altitude range 500–1,000 km if the monomers inferred for the optical haze are composed of aggregates of monomers having radii near 40 nm as determined from the DISR measurements (Tomasko et al. 2009).

12.3.1 Structure Above 400 km from ISS and UVIS

Ultraviolet observations of a stellar occultation provide profiles of aerosol extinction in the altitude range 300–1,000 km. The long-wave (170–190 nm) end of the Cassini Ultraviolet Imaging Spectrograph (UVIS) Far Ultraviolet (FUV) spectrograph was used to measure the extinction of light from the star λ Scorpii as it passed behind Titan's limb. An aerosol vertical profile derived from those data by Liang et al. (2007) is shown in Fig. 12.6. The profile shown in the figure was based on an assumption that the optical extinction can be calculated from Mie theory for spheres with mean radius 12.5 nm and the refractive indices for tholin given by Khare et al. 1984. Both of these assumptions are questionable, especially the assumption that the particles are spheres with radii near 12.5 nm. The important result is that an aerosol signature in the occultation data is present all the way to 1,000 km. The dip in aerosol density just below 500 km altitude is evidence for a thin haze layer in the altitude region 500–520 km.

A thin haze layer often referred to as a 'detached haze' layer was seen in Voyager images and is also seen in Cassini images from 2004 to the time of this writing (2008). Figure 12.7 provides a global view of Titan's haze and shows the change in character of the haze near 55°N latitude (from the Cassini ISS). Rages and Pollack (1983) found a prominent detached haze in Voyager images everywhere south of about 45° latitude and at altitudes 300–350 km and varying with latitude. Porco et al. (2005) report a detached haze just above 500 km altitude and almost independent of latitude below 55° latitude. These differences in altitude are too large to be attributed to measurement error. Rages and Pollack (1983) reported '... a small enhancement in the extinction at –450 km which exists at all latitudes between 75°S and 60°N'. This haze layer may have the same origin as the 510 km layer seen in Cassini images, but it is about 60 km lower. No detached haze is seen in Cassini images between 300 and 350 km and so Titan's haze must have changed in a fundamental way in the intervening years between the observations.

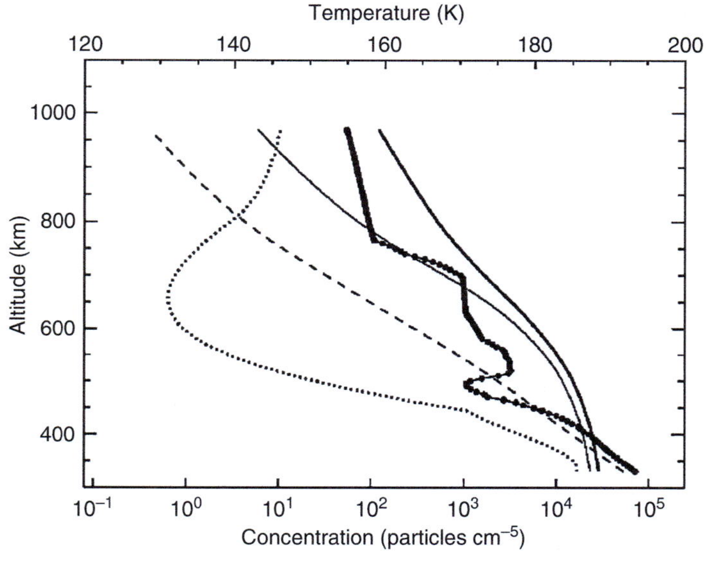

Fig. 12.6 From Liang et al. (2007) Aerosol density (filled circles) derived from the UVIS λ Sco occultation compared to the CH_4 density (dashed line) scaled by 10^{-9} (Shemansky et al. 2005a; Shemansky 2006). The increase of the mixing ratio of the UVIS aerosols through the mesosphere to at least 1,000 km implies that the production of aerosols must take place at significant rates throughout the mesosphere and thermosphere. The UVIS-derived temperature profile is shown by the dotted line. Model aerosol profiles computed by Liang et al. (2007) are shown by the thin and thick solid lines.

Fig. 12.7 Global view of Titan's stratospheric haze structure. The 'detached' haze layer near 500 Km is seen to be present at all latitudes below 55°N. Titan's spin vector (north) points to about the 11:00 o'clock position.

Fig. 12.8 Haze layers at latitudes greater than 55°N show a complex structure which changes over a three-hour period. An MPEG video showing these changes is included in the supplemental DVD.

What processes produce a stable (in both time and latitude) thin haze layer in a region of the atmosphere where the sedimentation time is short? Two hypotheses have been discussed. The layer may be a condensation region at a local temperature minimum or it may be a product of non-volatile aerosol formation originating at higher altitude. The condensate hypothesis would provide a natural explanation for the formation of a thin layer provided the location of the haze layer correspond with a local temperature minimum, and that a plausible condensate exists with the abundance and thermodynamic properties to condense at that location. A local temperature minimum is present a little below the haze location in both the in situ HASI (Huygens Atmospheric Structure Instrument) data (Fulchignoni et al. 2005) and the stellar occultation data analyzed by Sicardy et al. (2006). Liang et al. (2007) make the case for condensation of C_6N_2 throughout a broad altitude range in the mesosphere. These particles might serve as condensation nuclei for more complex non-volatile hydrocarbon and nitrile haze material. No condensate candidate has been proposed specifically for the detached layer near 510 km.

The alternate hypothesis, that microphysical processes for non-volatile aerosols originating at higher altitudes (tholins) produce the haze layer, was proposed by Lavvas et al. (2009) who point out that the haze layer coincides with a local temperature maximum just above the temperature minimum in the HASI data and in the stellar occultation (Sicardy et al. 2006) data. They show that local radiative heating by the aerosol layer could produce this feature. They propose that the particles in the detached layer have radii near 40 nm and therefore are the monomers that aggregate and sediment to lower altitude and form the larger aerosols that produce the strong forward scattering of the main haze layer. If this interpretation is correct it provides a significant new constraint on aerosol models – it identities the location where the coagulation of monomers to form aggregates occurs. Some aspects of this detail in the particle microphysics could be tested with existing observations. If the particles in the detached layer are 40-nm monomers their scattering phase functions should not be strongly forward-scattering at visible wavelengths.

Any proposal for the origin of the detached haze should also account for observations of multiple layers and time variations in the vertical structure of the haze in the region between 500 and 600 km altitude. At spatial scales better than about 6 km/pixel some Cassini Imaging Science Subsystem (ISS) images show additional layers above the main 'detached' haze layer, and some show the detached layer to consist of multiple layers. In Figs. 12.7 and 12.8 (West et al. 2008) the 'main' detached haze layer has an intensity peak near 505 km and a secondary peak near 530 km with additional fine structure up to 600 km. It is tempting to attribute multiple layers to the action of inertia-gravity waves (Porco et al. 2005) or by waves forced by tides (Walterscheid and Schubert 2006). These suggestions seem plausible but a detailed model relating wave action to intensity fluctuations remains to be constructed.

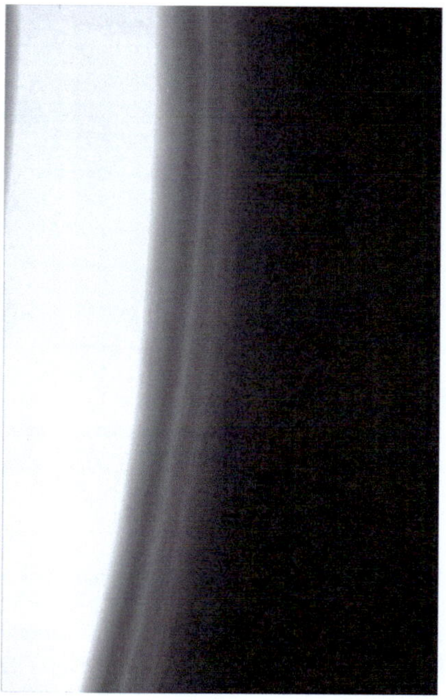

Fig. 12.9 From West et al. (2008). Cassini ISS image N1481653824 (UV3 filter). The 'detached' haze layer near 500 km altitude is composed of a main layer near 505 km altitude and additional layers at higher altitudes. This image was taken with the ISS UV3 filter. The image scale is 800 m/pixel. The phase angle is near 160°.

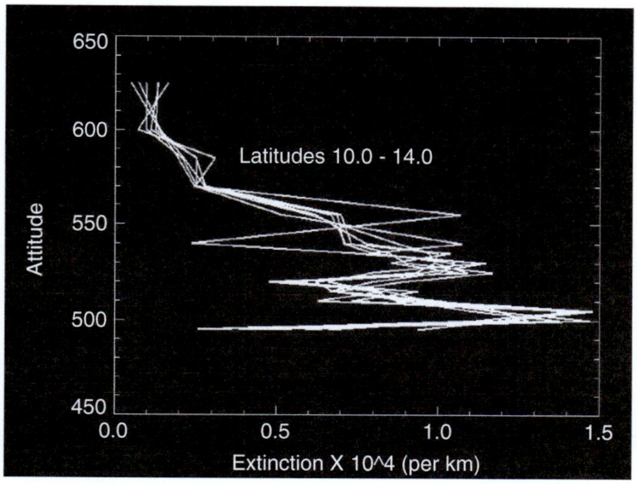

Fig. 12.10 Intensity profiles from Fig. 12.9 were used to retrieve aerosol extinction coefficient profiles for several cuts in the latitude range 10°–14°N Retrievals neglect multiple scattering (from West et al. 2008).

North of about 55°N the haze structure becomes complex and time-variable. Intensity variations occurring on time scales of tens of minutes have been observed in the northern polar vortex region (see the mpeg on the CD). One of the images from that movie sequence is shown in Fig. 12.8. High haze layers are most easily seen at high phase angles and at short wavelengths because of the strong forward scattering and larger scattering cross section at short wavelengths.

12.3.2 Direct Measurements of Titan's Aerosols by Huygens Below 150 km

At altitudes of 150 km and below, the DISR instrument aboard Huygens made direct measurements that constrain the vertical distribution of the haze aerosols. These include measurements by the upward-looking and downward-looking ultraviolet filter photometer (ULV and DLV) which covered the spectral band from 350 to 480 nm, as well as the spectral measurements between 480 and 1,600 nm looking both upward and downward from the Upward- and Downward Looking Visible Spectrometers (ULVS and DLVS) and from the Upward- and Downward Looking Infrared Spectrometers (ULIS and DLIS). All of these measurements were made at many azimuths relative to the direction to the sun. The azimuth of the probe relative to the sun was determined from onboard housekeeping data from a sun sensor as well as from the record of the modulation of the strength of the radio signal received by the Cassini orbiter from the rotating Huygens probe (Karkoschka et al. 2007). Nevertheless, the detailed tip toward the sun and in the orthogonal direction was more difficult to constrain for each observation during the entry. Some of these effects can be seen in the plots of the observations and models in Figs. 12.11 to 12.17.

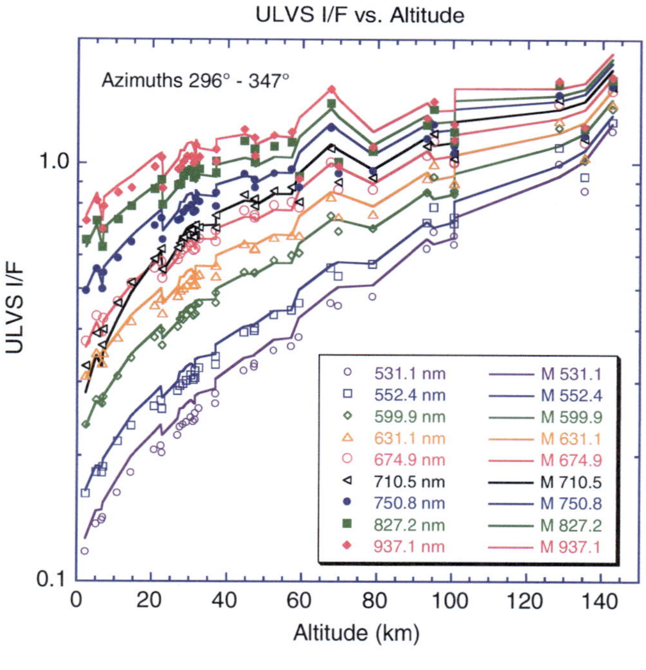

Fig. 12.11 We show the values of I/F measured by the Huygens DISR Upward-Looking Visible Spectrometer as functions of altitude for the instrument pointed at azimuths of 296° to 347° from the sun at nine continuum wavelengths. Measurements at these azimuths included the direct beam from the sun as well as the diffuse radiation from the sky. The measurements (*shown by points*) vary slightly from altitude to altitude due to variations in the exact azimuth of the instrument at different altitudes. The lines are for models computed for the instrument azimuth at each altitude. The variations of the models with altitude and with wavelength are in reasonable agreement with the observations.

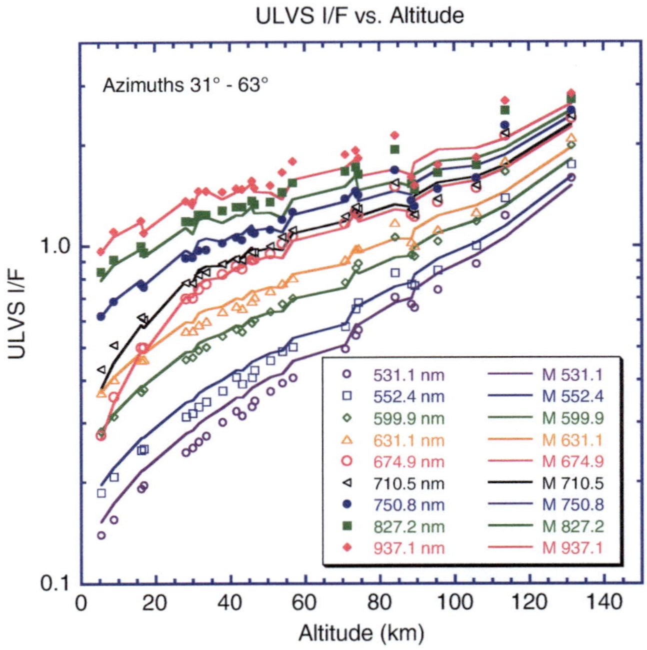

Fig. 12.12 This figure is similar to Fig. 12.11 but is for measurements made when the sun was in a different part of the field of view.

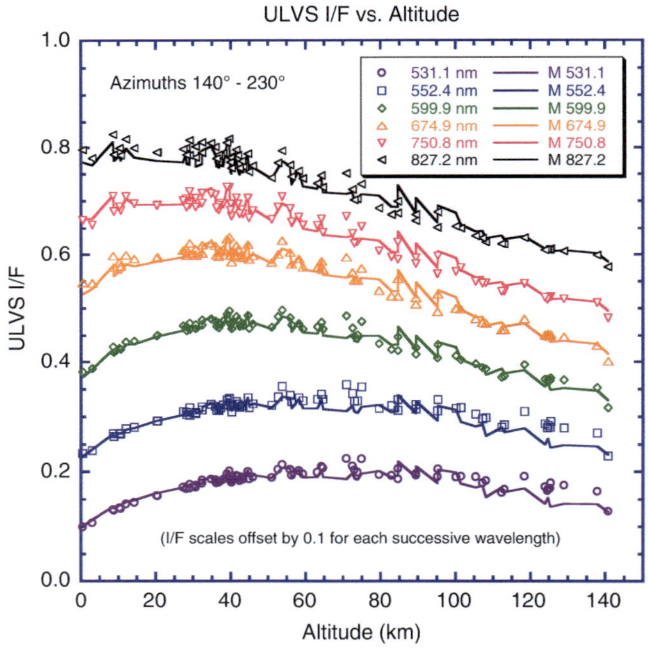

Fig. 12.13 This figure is similar to Figs. 12.11and 12.12 but is for the instrument pointed away from the sun so only diffuse light from the sky is measured. The models fit the diffuse sky brightness reasonably well also.

Despite these complications, the observations of the upward-looking instruments revealed interesting details of the vertical distribution of the haze aerosols. First, no altitude regions

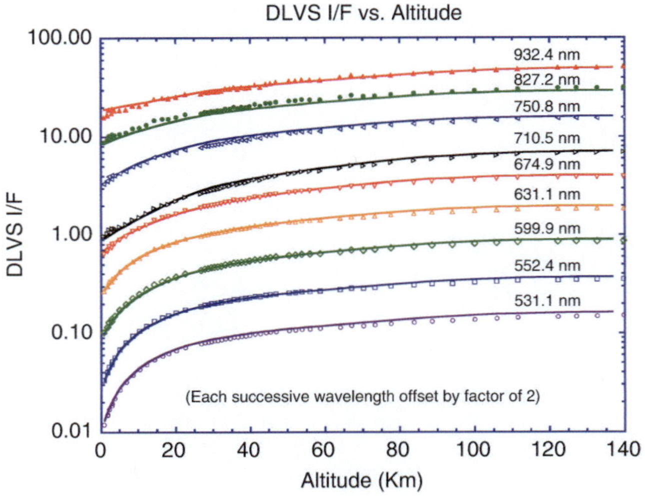

Fig. 12.14 This figure is similar to Fig. 12.13 but is for measurements made by the Huygens DISR Downward Looking Visible Spectrometer averaged over azimuth within a short time at each altitude. The models (*shown by lines*) are in good agreement with the spacecraft observations (*symbols*).

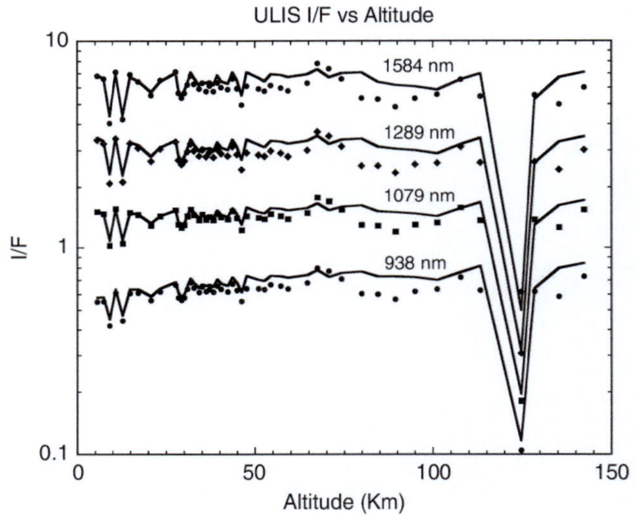

Fig. 12.15 This figure shows the measurements made by the Huygens DISR Upward-Looking Infrared Spectrometer for each 2-min cycle of the instrument during the descent. The measurements are at four continuum wavelengths as labeled. The models shown by lines are for the same set of azimuths at which the measurements were made. The low values near 120 km altitude were when the instrument slowed its rotation rate and reversed direction of rotation. The measurements here were made looking only away from the sun.

relatively clear of aerosols were seen in any altitude intervals. Second, above 80 km altitude, the vertical optical depth of the aerosols had a scale height of about 65 km. Third, between 80 and 30 km, the cumulative extinction optical depth varied linearly with altitude. Fourth, in the lowest 30 km the

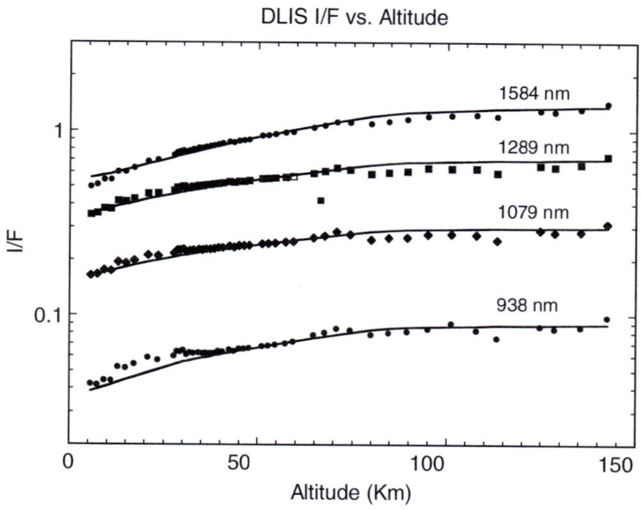

Fig. 12.16 This figure is the same as Fig. 12.15 but is for the Downward-Looking Infrared Spectrometer at the four continuum wavelengths as labeled.

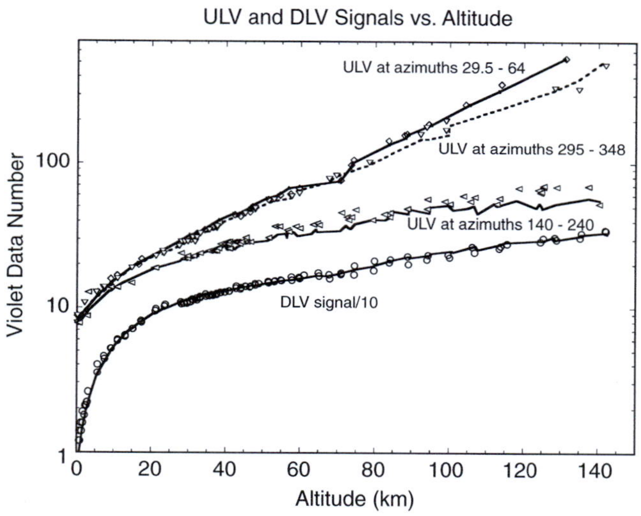

Fig. 12.17 This figure shows the data number measured by the Upward-Looking Violet (ULV) photometer at two azimuths relative to the sun (*diamonds and downward pointing triangles*) compared to models (*solid or dashed lines at the two azimuths*). The measurements of the ULV looking upward without the direct beam are shown by left-pointing triangles. The circles show the measurements of the Downward-Looking Violet (DLV) photometer averaged around all azimuths and divided by 10 to avoid overlap on the plot. The models show good agreement with the observations. The agreement with the altitude variation indicates that the vertical distribution of haze is consistent with the observations, and the agreement of the models with both the ULV and DLV measurements indicates that the variation of single scattering albedo is consistent with the measurements.

cumulative extinction optical depth again varied linearly with altitude, although at a slightly different rate than above. The variations of cumulative haze optical depth with altitude at several wavelengths are shown in Fig. 12.18.

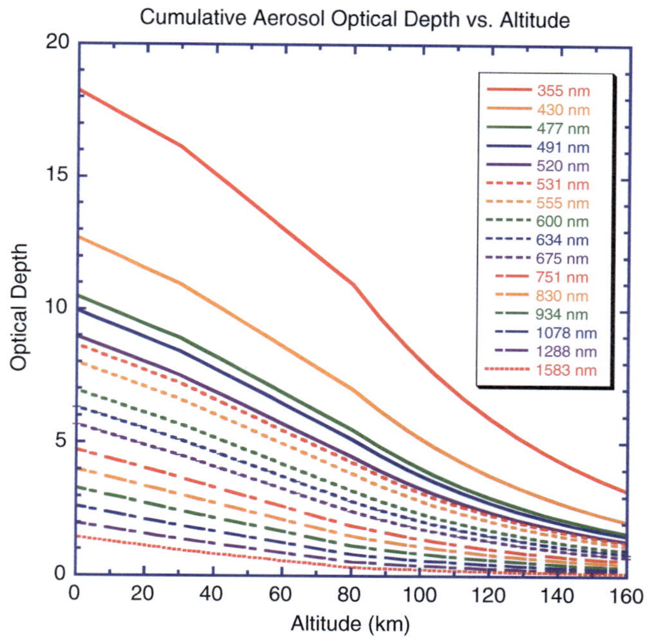

Fig. 12.18 This figure shows the variation of the extinction optical depth of the haze with altitude at several wavelengths, as labeled. No clear regions are seen in the altitude range of the DISR measurements.

12.4 Variations of the Optical Properties of Titan's Aerosols with Wavelength

12.4.1 Single Scattering Albedo

The single scattering albedo of the haze aerosols can be constrained by fitting both the upward- and downward-looking observations from the measurements during the Huygens descent. The downward minus upward looking measurements are a measure of the net flux at the altitude of the measurements. The difference in the net flux between two altitudes is due to the solar energy absorbed by the intervening layer and provides strong constraints on the single-scattering albedo. The variations in the energy absorbed with wavelength determine the variations in single scattering albedo with wavelength.

The measurements from the set of optical measurements on DISR are shown in Fig. 12.19 from Tomasko et al. 2008a. These curves result from the measurements over the complete spectral range covered by the instrument. The spectral coverage of each portion of the DISR measurements is indicated by the horizontal lines along the bottom of the figure. Note that two curves are given for the single scattering albedo as a function of wavelength. The significant difference between these curves is an indication that the single scattering albedo of the haze aerosols varies with altitude. Note that the curve for the darker aerosols refers to the higher altitudes.

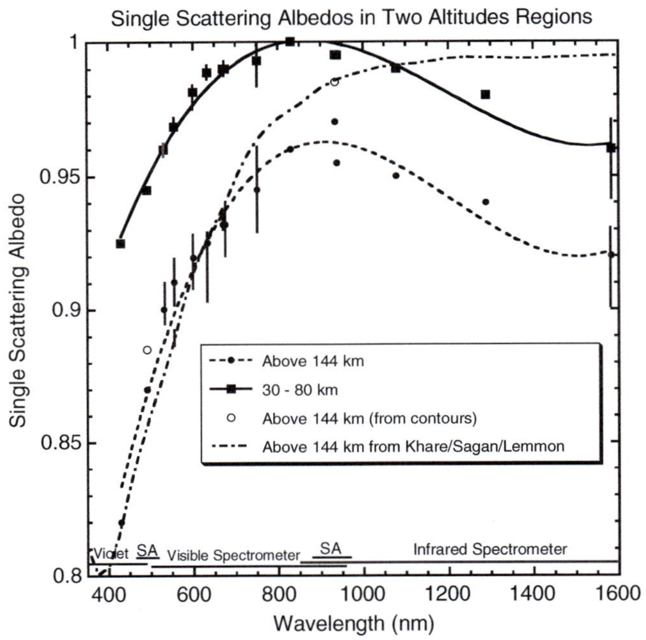

Fig. 12.19 This figure shows the variations of single scattering albedo with wavelength at altitudes above 144 km and between 30 and 80 km, as labeled from Tomasko et al. (2008a). The measurements come from different portions of the DISR data set as indicated by the horizontal lines along the bottom of the figure (Note that SA represents the two channels of the DISR solar aureole camera). The error bars approximate one sigma uncertainties in the derived values. The dotted-and-dashed line is the single scattering albedo computed for the haze aggregate particles using 1.5 times the imaginary refractive index reported by Khare et al. (1984) and shown in Fig. 12.30. The slope toward the blue is in reasonable agreement with the variation of single scattering albedo at high altitudes but does not show the decrease required for the haze aerosols longward of 900 nm.

Good fits were found for aerosols with a sharp break in the haze albedo at 80 km, as well as for a gradient in albedo with altitude above 80 km. The curve shown is for a model with fixed albedo above 144 km altitude and a linear gradient in albedo from 144 km down to 80 km. Below 80 km the data require that particles be less absorbing than those at higher altitudes.

The shapes of the curves have two interesting properties. First, the minimum absorption is near 900 nm. Toward shorter wavelengths the absorption increases rapidly. The dotted and dashed curve is for loose aggregate haze particles as determined by the solar aureole camera computed with 1.5 times the imaginary refractive index found by Khare et al. (1984). Note that the slope of the albedo to shorter wavelengths at the highest altitudes is similar. This implies that the variation of imaginary refractive index with wavelength of the haze aerosols must be similar to that measured for tholins.

Secondly, note that the albedo decreases to longer wavelengths after the peak near 900 nm. This occurs both for the aerosols above and below 80 km. This effect also occurs in the reflection spectrum of the surface measured with the DISR surface lamp (Tomasko et al. 2005). This may indicate that this feature in the surface spectrum is not due to surface hydrocarbon ices but is due to the coverage of the surface by aerosol material that has settled out of the atmosphere.

Finally, we note that for constraining the shape of the single scattering albedo curve shortward of 600 nm, the measurements used were acquired via a combination of the visible spectrometer, the blue channel of the solar aureole camera, and the ultraviolet filter photometer. Each of these measurements was a difficult challenge for the DISR instrument and resulted in significant error bars to this part of the curve. In fact, when Cassini Visual and Infrared Mapping Spectrometer (VIMS) spectra near the Huygens probe's landing site (Penteado et al., 2009) are compared with DISR haze models, some adjustment in the shape of the single scattering albedo curve is necessary to optimize the fit. This relatively small difference could be due to calibration differences in the blue portion of the VIMS and DISR instruments, but could also be due to the difficulties in retrieving this portion of the curve in the region where several of the DISR systems overlap.

12.4.2 Single Scattering Phase Functions

Uncertainties in the size of the forward peak in the single scattering phase function results in the inability to determine the true aerosol optical depth because different optical depths are retrieved for different amounts of forward scattering. The shape of the phase function requires measurements at a variety of phase angles, something that is only possible from space missions that can observe Titan over a large range of scattering angles. Observations of the shape of the backward portion of the phase function have been available from the first space missions to the Saturn system during the Pioneer 11 mission (Tomasko and Smith 1982). Pioneer observed the integrated disk over phase angles up to 90° and showed that a modest backward peak was needed for the phase functions in red and blue light. Voyager observed Titan at phase angles up to 150° and showed the need for strong forward scattering from the Titan aerosols (Rages et al. 1983).

The Huygens DISR observations that were acquired at different wavelengths and a range of phase angles during descent through Titan's atmosphere provided constraints on the phase function in several respects. First, observations within 10° of the sun constrained the forward peak at the two wavelengths of the solar aureole camera (491 and 938 nm). The visible spectrometer measured the brightness integrated over a wide field of view looking at various angles from the azimuth of the sun in both upward and downward directions. The ULVS observations near the sun and opposite to the sun constrain the side-scattering components relative to forward scattering. The downward-looking observations at a variety of azimuths relative to the sun constrain the size of the backward peak in the phase function.

The phase functions used in the DISR analysis were based on fits to phase functions computed for loose fractal aggregate particles and included the polarizing properties of the particles in addition to the intensity. Tomasko et al. (2008a) concluded that loose aggregates were necessary to permit the small size of the aggregates constituent "monomers" to produce the high degree of polarization observed. The authors concluded that the monomer size was approximately 0.05 μm in radius. A more detailed analysis of the DISR polarization analysis indicates that improved fits are actually obtained with slightly small monomers with radii of 0.04 μm (Tomasko et al. 2009).

In order to fit the strong forward scattering pattern, a large projected area of the particles is required. Assuming that the equal projected area radius of a spherical particle representing an aerosol is 2.0 μm, if each monomer has a radius of 0.05 μm, then this requires that 3,000 monomers make up an aggregate particle. A smaller radius of 0.04 μm for the monomers would imply a slightly larger number for the same projected area.

The DLVS observations as functions of azimuth angle relative to the sun show the need for modest backward peaks at wavelengths near 700 nm. At shorter wavelengths, the size of the backward peak decreases. The algorithm used to evaluate the shape of the single scattering phase function from loose fractal aggregates did not produce significant backward peaks in the wavelength range as required. The haze particles near the beginning of the probe measurements (near 150 km altitude) required the lowest single scattering albedos and the ratios of their optical depths with different wavelengths were in the best agreement with the cross section computations for our fractal algorithms. Thus, Tomasko et al. (2008a) limited the addition of the backward scattering peak to altitudes below 80 km, and used the phase functions without the backward peak from the algorithm at higher altitudes. The phase functions at lower altitudes included a smooth backward peak added to the algorithm predictions.

As the wavelength decreased from 700 nm toward blue wavelengths, the size of the backward peak in the phase functions below 80 km was decreased as required by the DLVS measurements as functions of azimuth relative to the sun. Some examples of the wavelength and altitude dependence of the phase functions required by the DISR observations is shown in Figs. 12.20 and 12.21 and can be found in the publication of the DISR analysis (Tomasko et al. 2008a).

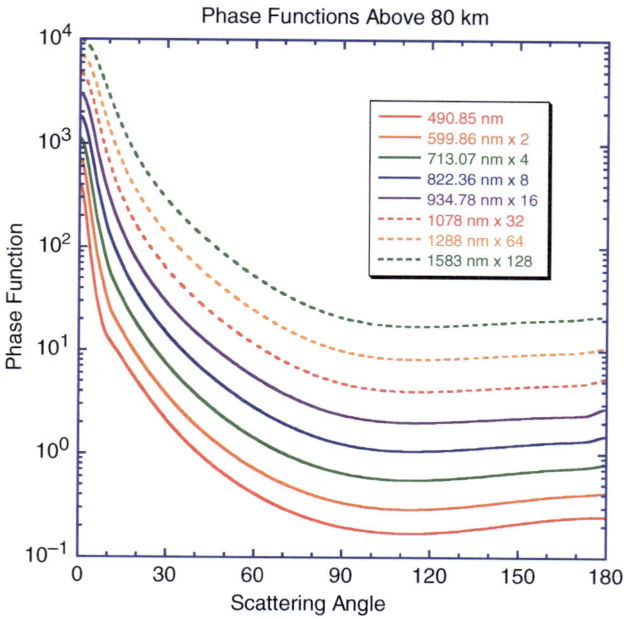

Fig. 12.20 Here we show the shapes of the single scattering phase functions for a range of wavelengths, as labeled. The curves are offset by the factors given in the legend. These phase functions are for the fractal aggregate computations used at altitudes above 80 km.

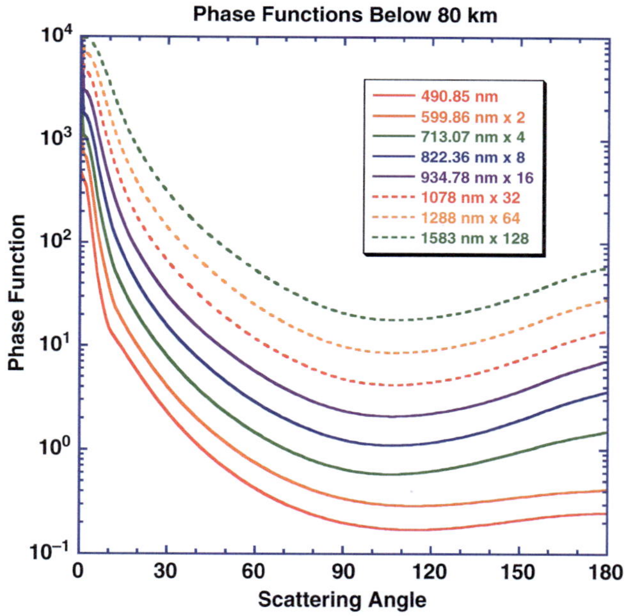

Fig. 12.21 This figure is the same as Fig. 12.20 except it is for the aerosols found at altitudes below 80 km. These curves contain enhances backscattering peaks longward of 600 nm as required by the variation in the DISR Downward-Looking Visible Spectrometer as the Huygens probe rotates.

12.4.3 Extinction Optical Depth

The variation of extinction optical depth with wavelength has been shown in Fig. 12.4 and briefly described above.

We noted that the extinction optical depth can be fitted reasonably well with power laws in wavelength, with the steepest slopes at highest altitudes and shallower slopes at lower altitudes, as would be expected for an increase in particle size with depth into the atmosphere. Above 80 km, the extinction optical depth variation with wavelength is in good agreement with the estimates for the fractal algorithm used to describe the haze aerosols, and the aerosol number density as a function of altitude can be reliably determined (see Fig. 12.22). At deeper levels in Titan's atmosphere, the agreement with the predictions for the purely fractal particles worsens, indicating a departure from the purely fractal aggregate particles thought to be present above 80 km. We know that the aerosols at these deeper levels have higher single scattering albedos. This may be caused by condensation of some hydrocarbon gasses onto the fractal particles which may fill some of the voids in the particles, and affect their total cross sections as a function of wavelength. Below 30 km, methane can condense on the particles, which might cause voids in fractal particles to collapse and the density of the aggregate to increase (i.e. to produce shallower variations of cross section with wavelength). However, this cannot happen completely, as the polarization observations in red light at deep levels requires strong linear polarization in single scattering from the haze aerosols even at these low altitudes. It may be possible that some growth in size and some collapse to denser shapes could happen below 30 km that would still permit the observation of polarization as well as optical depth with wavelength to be matched. Unfortunately, no solar aureole observations in red light close to the sun are available from DISR at altitudes below 30 km to test these ideas further.

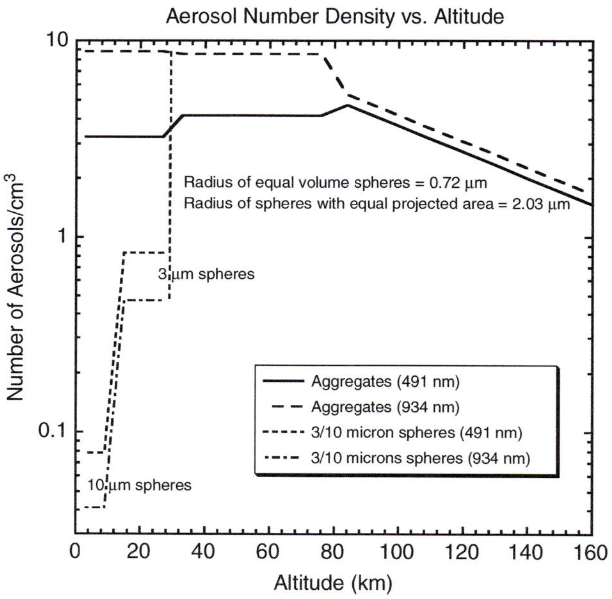

Fig. 12.22 This figure shows estimates of the aerosol number density computed from the cross sections of the fractal haze aggregates and the extinction optical depth profiles at 934 nm (*red*) and 491 nm (*blue*) wavelengths, as labeled. Above 80 km the optical depths and cross sections at the two wavelengths lead to consistent values for number density peaking near 5 particles/cc at 80 km. From 80 to 30 km the number densities derived from the red and blue wavelengths vary by a factor near two, and below 30 km the disagreement is even greater, indicating that the particles differ somewhat from the pure fractal aggregates for which the cross sections are evaluated. Perhaps this is an indication of condensation filling in some of the spaces in the particles. Below 30 km, the optical depth varies only slowly with wavelength, and the cross sections are more nearly constant in red and blue. In this case there is a relationship between the size of larger particles and the number density as indicated by the estimates for spherical particles of 3 or 10 µm radius. For larger particles the number densities in the lowest 30 km of the atmosphere could be one or even two orders of magnitude less that the value of 5/cc found near 80 km.

12.5 Variations in Haze Properties over Titan's Disk

The direct measurements by the DISR instrument on Huygens successfully determined the single scattering phase function, the single scattering albedo, and the vertical variation of cumulative optical depth as functions of wavelength between 350 and 1,600 nm at altitudes below 150 km. It is natural to ask how these properties vary over the disk of Titan, and especially with latitude.

The overall design of the Cassini–Huygens mission plan for Titan was to have the Huygens probe determine the "ground truth" at the landing site and use the remote observations of Cassini to determine the global conditions with the guidance of the Huygens observations at the landing site. Cassini VIMS spectral observations of the landing site as well as over a wide range of latitudes have been analyzed in the past few years and are under current investigation. In this regard, the Huygens observations have been helpful in measuring the absorption by methane at the long paths and low temperatures on Titan for use in the analysis of VIMS observations (Tomasko et al. 2008b). Secondly, remote observations can yield a family of solutions, in that the intensity emerging from the top of the atmosphere can be affected by the vertical variations in phase function, ground albedo, single scattering albedo, and aerosol distribution, and it is often not easy to separate these dependencies. Our goal in this section is to use as much as we can about the haze properties at the landing site and change them only as required to match the VIMS observations at other locations.

First, it should be noted that the haze model determined at the probe landing site from the Huygens measurements with no modifications predicts a spectrum seen from outside the atmosphere that is in good agreement with the VIMS observations of the landing site. This can be seen in Fig. 12.23 where the DISR haze model is used to predict what VIMS

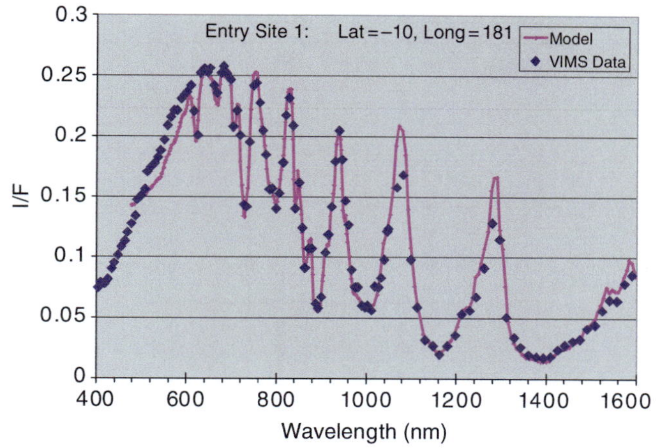

Fig. 12.23 This figure shows Cassini VIMS visible and infrared band measurements (*symbols*) near the Huygens probe landing site compared to a spectrum computed from the DISR haze model (*lines*) at the same scattering geometry. The agreement is good except for the portion shortward of 600 nm.

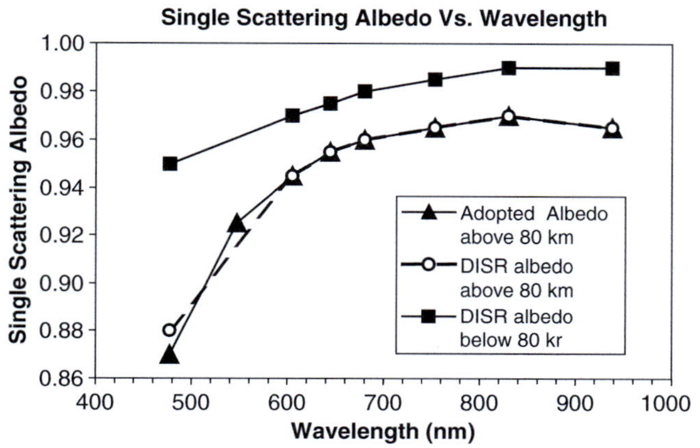

Fig. 12.24 This figure shows the variations of the single scattering albedo used in the DISR haze models above and below 80 km (*circles and squares, respectively*). Below 80 km, no adjustments to the single-scattering albedo curve are necessary to reasonably match DISR model and VIMS data. Above 80 km, a small adjustment in the slopes of the curves (*triangles and line*) significantly improves the model fit to the VIMS spectra shortward of 600 nm.

would see at the landing site. The only change to the DISR haze model needed is a small change in the shape of the haze single scattering albedo shortward of 600 nm (see Fig. 12.24) in a spectral region where DISR has the greatest difficulty measuring this quantity. The improved agreement between the haze model and the VIMS observations are shown in Fig. 12.25. We note further that adjusting only the ground albedo with no adjustment in haze properties or distribution can fit VIMS observations at other longitudes at the latitude of the probe landing.

In addition to the ground albedo, we have found only two further modifications to be necessary to fit VIMS observations over a wide range of latitudes. One is the single scattering albedo of the haze required to match the observed brightness

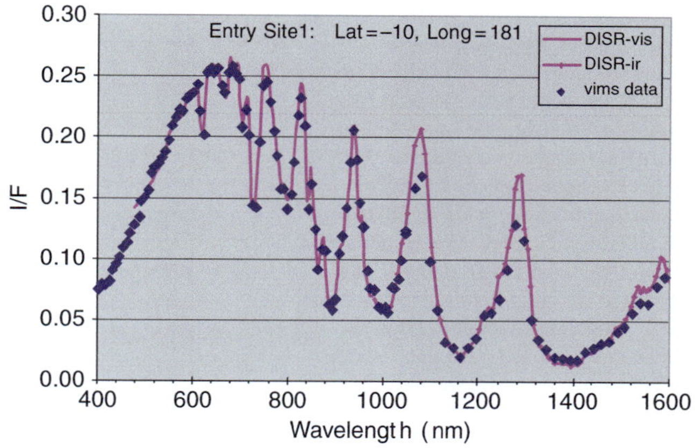

Fig. 12.25 This figure is like Fig. 12.23 except it uses the variation of single scattering albedo above 80 km as given by the triangular points and curves. The fit is quite good throughout.

near 600 nm. The haze here is sufficiently thick that the brightness observed is not much affected by the reflectivity of the ground. In the continuum between methane bands, the only way to match the brightness is to change the absorbing properties of the haze. Letting ω be the single scattering albedo of the haze in the DISR model, when the quantity (1 − ω) is multiplied by the same factor independent of wavelength the model can be adjusted to agree with the observations. Here the absorption in the haze is only adjusted at altitudes above 80 km where the DISR models show the greatest absorption. Below 80 km the absorption in the DISR model is much smaller, and adjustments here make little difference in the brightness seen outside the atmosphere.

The second adjustment is necessary to fit the deep parts of the methane absorption longward of 1,000 nm. These deep bands probe the atmosphere above 80 km. The second factor chosen multiplies the haze optical depth above 80 km. Decreasing the haze here makes the bands deeper, and increasing the factor makes the bands shallower.

It turns out that adjusting only these three parameters (the ground albedo to fit the continuum points longward of 1,000 nm, multiplying the haze absorption above 80 km by a factor, and changing the haze thickness above 80 km) is sufficient to fit VIMS spectral observations shortward of 2 µm wavelength from −60° to +50° latitude. The results are shown in Fig. 12.26 from Penteado et al. (2009). Note that the thickness of the haze shows a steep change at about −10° latitude. North of that latitude, the haze thickness is similar to that at the Huygens landing site. South of that location, the haze above 80 km is only about 60% as thick. The absorption in the haze above 80 km increases slowly with latitude by about 20% in total over the range of latitudes shown. The models require no significant changes to the haze properties or distributions below 80 km. No change in the methane mixing profile from that measured on the probe entry is required to fit the methane bands.

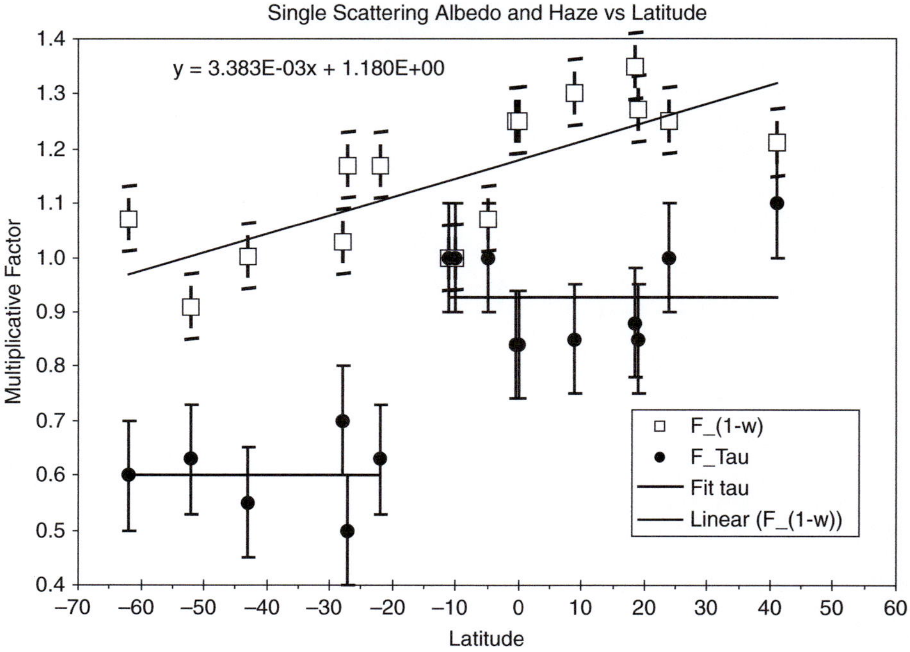

Fig. 12.26 This figure shows the factor (open squares) by which (1 – ω) (i.e., 1 minus the single scattering albedo in the DISR haze model above 80 km) must be multiplied to give good fits to VIMS spectra at latitudes varying from 60°S to 40°N. The dots show the factor by which the haze optical depth above 80 km in the DISR haze model must be multiplied to give good agreement to the VIMS spectra at the latitudes shown. North of 10°S latitude, haze parameters in the model are in good agreement with the values measured by DISR. Note that the south of 10°S latitude, the haze is about 60% as thick.

The thinner absorbing haze above 80 km combined with the smaller absorption in the haze makes the southern hemisphere appear brighter than the northern hemisphere where the absorbing haze above 80 km altitude is thicker and also more absorbing. These effects are stronger at blue wavelengths and decrease toward the red where the single scattering albedo of the haze above 80 km increases. The effects are similar to those reported in the past for the disk of Titan.

12.6 Seasonal and Long-Term Variations

Lockwood and Thompson (1979) reported on disk-integrated variations of Titan's geometric albedo that they attributed to variations in solar ultraviolet irradiance and its influence on photochemistry. An important clue to the brightness variations was provided by spatially-resolved images from the Pioneer 11 and Voyager 1 spacecraft which showed hemispheric contrast. Sromovsky et al. (1981) noted that the northern hemisphere was darker than the southern hemisphere in violet, blue and green images at the time of the Voyager observations (a little past northern spring equinox). They found four distinct latitude regions, a northern polar zone (north of latitude 55°), a northern mid-latitude zone between 10° and 55°, a southern zone between –20° and –60° and a transition region between –20° and +10°. This configuration is quite similar to what the Cassini instruments found between 2004 and 2008 (between southern summer solstice and northern spring equinox). Sromovsky et al. (1981) proposed that seasonal variations in Titan's global circulation pattern and its effect on the distribution of haze produce the hemispheric differences and dominate the variations reported by Lockwood and Thompson. The substantial phase lag of the reflectivity asymmetry relative to hemispheric asymmetry in the solar heating rate supports this idea because of the long radiative time constants of the lower atmosphere which contains most of the atmospheric mass (see Chapter 12).

Some clues to the relationship between aerosol microphysics and reflectivity asymmetry are provided by observations using the Hubble Space Telescope and the Cassini orbiter instruments. Lorenz et al. (1997) studied spatially-resolved images from the Hubble Space telescope in the period 1990–1995 (leading up to southern spring equinox). They showed that contrast in the strong methane band near 890 nm is in the opposite sense as contrast in the blue. Hemispheric asymmetry during the period 1992–1995 was the reverse of the asymmetry pattern observed during the Pioneer and Voyager epochs. Curve fitting performed by Lorenz et al. (1997) predicts that north/south hemispheric ratio observed by Cassini instruments should reach an extremum (about 0.8 at blue wavelengths) near northern spring equinox in 2009 and reverse near northern summer solstice in 2017. Lorenz et al. (1997) proposed that hemispheric contrasts are due to aerosol microphysical variations in the region above 70 km altitude and mostly below 120 km altitude driven by seasonal variations in aerosol transport by wind, in agreement with (but more specific than) the view that Sromovsky et al. (1981) put forth. Karkoschka and Lorenz (1997) derived haze aerosol radii near 0.3 μm in the northern latitudes versus 0.1 μm in the south in from 1995 HST images of Titan's shadow on Saturn. They assign these particle radii to different layers, the 'detached haze' layer at northern latitudes and the main haze layer at latitudes south of −50°. The high southern latitudes would correspond to a southern polar vortex regime.

The axis of symmetry generated by the hemispheric contrast in the haze is not aligned with Titan's spin axis. Roman et al. (2009) found a small tilt (4.1° +/− 0.3°) between the two from Cassini ISS images. The tilt is consistent with a similar offset in the axis describing symmetry of the temperature field as sensed by the Cassini CIRS data (Achterberg et al. 2008). These observations were unexpected from general circulation models of Titan and may provide an important clue to the mechanism responsible for Titan's super-rotation.

12.7 Role of Aerosols in Titan's Heat Balance

Knowledge of the haze structure and optical properties at wavelengths from the near ultraviolet to the near infrared permits computation of the net flux of sunlight as functions of altitude and solar zenith angle. These permit computation of the solar heating rate as a function of altitude averaged around circles of latitude to average over a Titan day. Further, Cassini's Composite Infrared Spectrometer (CIRS) observations at thermal infrared wavelengths can be used to obtain the thermal cooling rate as functions of altitude. These can be averaged and combined with the solar heating rates to give the net radiative heating or cooling rate as a function of altitude averaged around circles of latitude. As of the date of this book, this has only been done for the 10°S latitude of the Huygens probe entry. Figure 12.27 from Tomasko et al. (2008c) shows the results of the heating and cooling rate

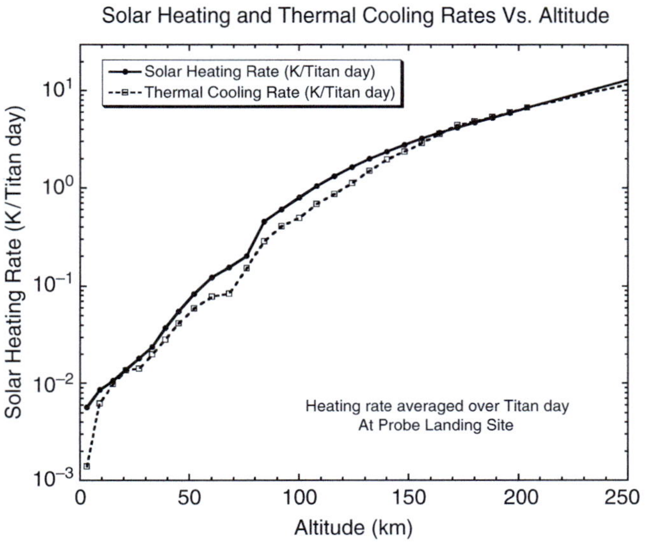

Fig. 12.27 This figure shows the solar heating rate and thermal infrared cooling rates as functions of altitude averaged around Titan in a small band at 10°S latitude near the Huygens probe landing site. The heating and cooling rates were computed using aerosol, gas composition, and temperature profiles from DISR and CIRS observations by Tomasko et al. (2008c).

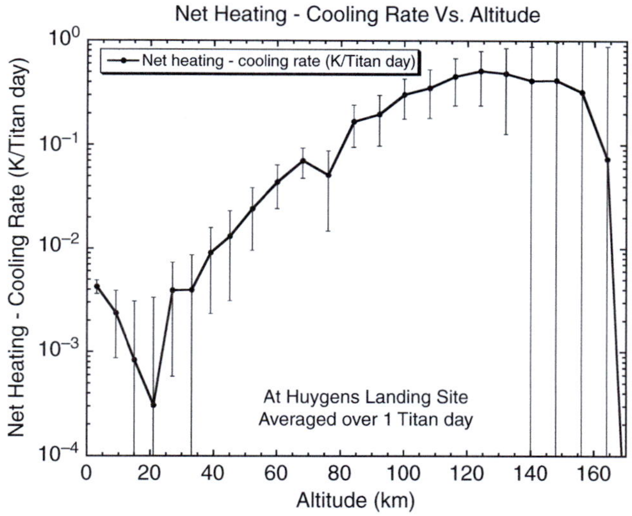

Fig. 12.28 This figure shows the net radiative heating rate (*solar heating minus infrared cooling*) as function of altitude averaged around all longitudes on Titan near 10°S latitude. Note that the net heating rate peaks near 0.5 K/Titan day near 120 km altitude and decreases toward higher and lower altitudes.

calculations for this location. Note that the solar heating rate slightly exceeds the thermal cooling rate. The excess is shown in Fig. 12.28. Because there is no evidence that the thermal structure of this region is changing, dynamical motions must redistribute this heat to other regions on Titan. These computations need to be repeated for other latitude circles to define the net radiative forcing for dynamics. We can anticipate that this will be done in the near future using the continuing observations from the Cassini orbiter.

12.8 Haze Microphysical Models

Authors attempting to determine the optical properties of Titan's haze must cope with the large number of parameters needed to describe the haze including the single-scattering phase function, albedo, cross sections, size, shape, and the variations of these properties with wavelength as well as with altitude. It is useful to apply physical constraints to decrease the degrees of freedom to a smaller number that might actually be constrained by existing observations. Among the first of these studies was that by Toon et al. (1992). These authors assumed that Titan's aerosol particles had a spherical shape, and used a one-dimensional model to combine a parameterized mass production rate, sedimentation, coagulation, and eddy diffusion to constrain aerosols properties. While limited to considering spherical particles, this study led to several interesting conclusions. The authors noted that the large variation in optical depth of the haze implied that small particles would fall out of the atmosphere slowly. They noted that eddy diffusion must increase vertical motions considerably above that produced by sedimentation to limit the vertical optical depth to reasonable values. They also noted that vertical motions of the order of 1 cm/s could cause features such as the detached haze layer and that small motions of the order of 0.05cm/s could produce significant hemispherical asymmetry. The authors pointed out that coupled two-dimensional models were likely to be required for more precise conclusions.

Shortly after this work West (1991) suggested fractal aggregates for the shape of the aerosols, and West and Smith (1991) showed that such particles could reconcile the measurements of high polarization and strong forward scattering.

In 1992, Cabane et al. (1992) published the first microphysical models that incorporated fractal aggregates. They showed that the aerosols could be divided into two regions, a region where the monomers were formed and continued to grow by the largest particles sweeping up the smaller particles followed below by a region where the monomers combined to form fractal aggregates. They found that the monomer size was relatively independent of electrical charge, mass production rate, and eddy diffusion and depended primarily on the altitude of formation since the density of gas at this altitude determined the time particles of various sizes could be suspended and grow. They also concluded that treating the monomers as spheres was a good approximation since the monomers grew by incorporating many smaller particles from the gas.

This work was followed by a paper by Cabane et al. (1993). In this paper the authors addressed the problem of how the aggregates were assembled, and how the number of monomers in the aggregates varied with altitude for different altitudes of the production region. They noted that the settling speed of the loose cluster-cluster aggregates is independent of the mass of the aggregate and is the same as the settling speed of the individual monomers. Using the suggestion of West and Smith (1991) that the aggregates each contain 8 monomers of radius 0.06 μm, the models were adjusted to give reasonable agreement with these values. They found that the production zone must be 350–400 km to give the required monomer size and an effective radius of the aggregates of 0.3 to 0.5 μm to satisfy the forward scattering constraint from the high phase angle observations of Voyager.

Assuming the charge on the particles was 30 electrons/μm, they found the monomer size to be affected little by the charge and that the monomer size depended mostly on the altitude of formation.

They noted that for production altitudes near 500 km, the number of monomers per aggregate would be larger than 500, values that seemed inconsistent with the results of West and Smith (1991). Both Toon et al. (1992) and Cabane et al. (1993) ascribed the presence of aerosols at higher altitudes (as in the detached haze) to dynamical motions.

These studies of the microphysics of fractal aggregates were followed by attempts to fit the geometric albedo of Titan with these types of particles by Rannou et al. (1995). These authors use a microphysical model for the properties and vertical distribution of the aerosols, an approximate treatment to evaluate the cross sections and asymmetry parameter of the haze particles, and a two-stream radiative transfer code to attempt to fit the geometric albedo of Titan from 0.2 to 1 μm wavelength.

In order to permit relatively accurate computations of the optical properties of the aerosols, the authors concentrated on using a production altitude of 535 km, higher than the 350 to 400 km determined from previous studies, to give monomer sizes of 0.03 μm that were sufficiently small to permit relatively good agreement with the geometric albedo down to 0.2 μm wavelength.

They varied the aerosol production rate, the electric charge of the aerosols, the imaginary refractive index, the methane abundance, and the fractal dimension of the aggregates. When the fractal computations were compared to models using spheres, they found that even compact fractal particles behaved significantly differently from spheres in that the particles reached greater optical depths at higher altitudes and had less forward scattering phase functions than the Mie particles. Both effects helped to provide the very low geometric albedo of Titan needed at short wavelengths.

They also found families of solutions where increases in the aerosol mass production rate could be compensated by changes in the methane abundance in the visible and near infrared (IR) part of the spectrum. They found that the fractal dimension of the aerosols was not well constrained by the observations, with values between 1.8 and 3 being capable of yielding fits. Nevertheless, they found that the fractal models were able to fit the low UV albedo without the ad hoc addition of very small particles used by some previous authors.

Hutzell et al. (1996) were the first to use a two-dimensional haze model together with two different estimates of the circulation on Titan to explore the effects of dynamical haze transport on the variations of Titan's reflectivity from the ultraviolet to the near infrared as functions of season and latitude. In this study, they did not have a fully coupled model in which the changing haze distributions varied the forcing for the dynamics. Nevertheless, they were able to reach several useful conclusions about the interaction of dynamics and haze and reflectivity.

These authors used two different and fixed dynamical circulations to explore the implications for the haze distribution. One dynamical circulation model was a simple Hadley cell, while the other was the result of a general circulation model in which the heating due to absorption by the haze was constant. They concluded that wind fields of the order of 1 cm/s can have significant effects on the albedo and can produce variations in the geometric albedo of Titan with time that have approximately the same magnitude as are observed by Cassini. They noted that variations due to haze production rate alone did not produce sufficiently large variations.

However, neither of the two circulation patterns they used was able to produce the observed variations of albedo with latitude that are relatively constant in each hemisphere with a relatively sharp transition at low latitude. They concluded that the Titan circulation must be broadly uniform in each hemisphere and have a sharp transition to reproduce the observations. A coupled haze-dynamical model would be needed in which the dynamical forcing changed in response to changes in the haze distribution.

Finally, the authors pointed out that the composition of the haze could also change with latitude and season, and a coupling of the chemistry that produced the haze needed also to be added to the model.

Rannou et al. (2003) used measurements of several different types including the geometric albedo spectrum from the ultraviolet to the near IR, the measurements of high linear polarization and strong forward scattering, and the profile of extinction observed at high altitudes together with a one-dimensional microphysical model for fractal aggregates to constrain the vertical profile of aerosols as well as the monomer size and number of aggregates in the haze particles. They used variations in the aerosol mass production rate, the electrical charge on the particles, the eddy diffusion profile, and the methane mixing ratio to fit the observations.

They found that increasing the eddy diffusion coefficient smoothed the aerosol profile and generally required larger aerosol mass production rates to match the methane bands. For the small values of the eddy diffusion coefficient they used, they reported an aerosol mass production rate near 0.7 $\times 10^{-14}$ g cm^{-2} s^{-1}.

The use of fractal aggregates for the shape of the aerosols easily permitted the observations of high linear polarization and strong forward scattering to be matched. They explored monomer radii between 0.04 and 0.095 μm. They excluded monomer radii greater than 0.075 mm because they did not permit fits of both the low UV albedo and the 619 nm methane band for any combination of mass production rate and eddy diffusion coefficient. It is expected that monomer radii larger than 0.075 mm would have difficulty fitting the high measured polarizations also. The authors pointed out that because the cross sections of the aggregates varied approximately with the number of monomers, the optical properties of the aggregates did not vary much with monomer number. In contrast, the optical properties of the particles depended more sensitively on monomer size.

Rannou et al. (2003) reported that one of the main difficulties was simultaneously fitting the strong methane band at 890 nm and the weaker band at 690 nm. In order to limit the depth of the 890 band to the observed value, a significant amount of haze had to be located above 80 km altitude. In the normal microphysical calculations, the optical depth per unit of altitude (the aerosol extinction) was expected to increase with depth into the atmosphere. However, once the haze optical depth was sufficiently large to fit the 890 nm band, the extinction had to decrease at low altitudes to permit photons to reach deep atmospheric levels where the 690 nm band was formed. The authors noted that several other workers reported a need for the aerosol extinction to decrease at low altitudes also.

These results must be reviewed in light of the new values of the methane absorption coefficients derived for the 890 nm methane band from the Huygens observations. Tomasko et al. (2008b) noted that the absorption coefficient for the 890 nm methane band is overestimated in the Karkoschka (1998) determination that has commonly been used in such work. With the newer lower values of the methane absorption in the 890 nm band from the Huygens observations at large paths and at low temperatures, less aerosol extinction is needed high in the atmosphere to match the observed depth of the 890 nm band than previously thought. In this case, the decrease in extinction at low altitudes to match the weak 619 nm band would be smaller than in the past.

Three relatively recent studies of microphysical processes in Titan's atmosphere have appeared since the entry of the Huygens probe into Titan's atmosphere. In one, Bar-Nun et al. (2008) compared the monomer size and aerosol density predicted from their model to that observed by the DISR instrument on Huygens. They found good agreement between the monomer size predicted by their model (0.04 μm) and the value published by Tomasko et al. (2008a) of 0.05 μm. Remarkably, recent further analysis of the DISR polarimetry in Titan's atmosphere has led to a slight revision of the monomer size to a value even closer to the value predicted by the Bar-Nun model (Tomasko et al. (2009).

A new study by Rodin et al. (2009) also investigated the microphysics of Titan's fractal aggregates and made a detailed comparison with the optical measurements made by DISR looking upward and downward in Titan's atmosphere over the visible and near IR spectral range. These authors found that a Coulomb barrier that limits the Brownian aggregation of small tholin particles possessing a single elementary charge primarily determined the monomer size of about 0.05 μm. In contrast to the earlier microphysical studies, they believed that the monomer size was controlled by the electrical charge rather than primarily by the altitude of formation.

Finally, a recent investigation by Lavvas et al. (2008) used a combination of observations by the Cassini imaging system (ISS), the UV spectrometer (UVIS), and the Huygens atmsopheric structure instrument (HASI) to determine the number density, size, and mass flux near the location of the detached haze layer at 520 km altitude. They find that aerosol radii of about 40 nm above this altitude are capable of producing the locally higher temperatures just above the location of the detached haze layer seen in the HASI temperature profile. Their aerosol profile is also consistent with the UVIS profiles. Their mass flux of 1.9 to 3.2 × 10^{-14}g $cm^{-2}s^{-1}$ is about the same as required for the main haze at lower altitudes. They suggest that the 40 nm particles are formed at altitudes up to 1,000 km by absorption of solar photons at wavelengths of <145 nm through radical and ion chemistry. They find that the lower apparent aerosol density in the detached region is due to the aggregation of the 40 nm monomers above to particles that fall faster and cause a local decrease in slant aerosol opacity at 520 km altitude. They estimate that the number of monomers in the aggregates at lower altitudes is of the order of 1,000. Their estimates of the size of the monomers and the number of monomers in the aggregates at lower altitudes are in good agreement with the measurements made by the Huygens probe (Tomasko et al. 2008a). Because the mass production above and below the detached region is the same, they believe that the real source of the haze material is due to the radical and ion chemistry at very high altitudes up to 1,000 km and not due to neutral chemistry driven by methane photolysis at lower altitudes of 300 to 500 km as been previously suggested. This work suggests that we may be beginning to understand the type of mechanisms that produce the aerosol material, the location where it is produced, and how it grows to monomers and then aggregates at lower altitudes to exhibit the sizes, shapes, number densities, and vertical distributions observed in the full suite of Cassini/Huygens observations.

12.9 Laboratory Simulations of Titan's Aerosols

Due to the difficulty in producing in the laboratory the physical and chemical conditions and the relevant time scales of the Titan atmosphere it is not obvious that laboratory studies of Titan's haze would yield meaningful results. In spite of the difficulties laboratory work has served to guide our understanding and even emboldened us to think in new ways about the optical, physical, and chemical properties of the aerosol particles.

12.9.1 *Early Tholin Production and Optical Properties*

Laboratory work on Titan haze dates back at least to 1979 (Podolak et al. 1979) when it was known that haze particles are strongly absorbing at blue wavelengths and that

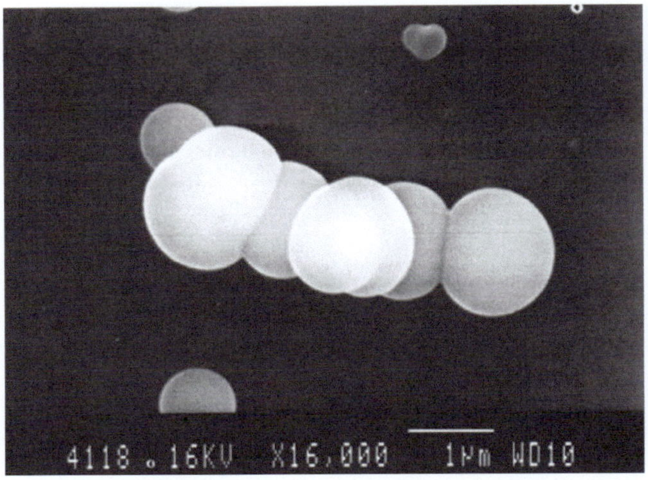

Fig. 12.29 From Bar-Nun et al. (1988). Scanning electron microscope pictures of an aggregate of polyacetylene aerosol particles, demonstrating their semiliquid nature.

photochemistry initiated by methane photolysis was likely to be the starting point for haze formation. Soon afterward analyses of Pioneer and Voyager data showed that Titan haze particles are both forward scattering and highly polarizing at middle scattering angles. Attempts to reconcile these properties with distributions of spherical particles failed, leading to the suggestion that the particles are nonspherical (West et al. 1983). Laboratory experiments by Bar-Nun et al. (1988) established that polyacetylene and polyhydrogen HCN particles form semi-liquid small spheres or deformed spheres with sticking coefficient near 1, and that these monomers naturally form aggregate particles (Fig. 12.29). However, the size of the monomers in the laboratory simulations was larger than found for Titan and the number of monomers which formed the aggregate was lower. In retrospect these differences may be due to the charge on the aerosol that might limit the growth of the monomer and was not matched in the laboratory.

Laboratory work on tholins in a variety of environments began at Cornell University in the 1970s (Sagan and Khare 1979). Khare et al. (1984) published a seminal paper on the optical constants of Titan haze analog. The values for refractive index they derived are shown in Fig. 12.30. Those values faithfully reproduce the spectral shape of the haze absorption over the visible part of the spectrum out to about 0.8 μm, but between 1 and 1.6 μm the laboratory results are not as absorbing as the DISR data indicate (Tomasko et al. 2008a).

12.9.2 *Summary of Some More Recent Laboratory Investigations*

Experimental studies of Titan haze analogs continue up to the present. Here we mention a few key results to indicate

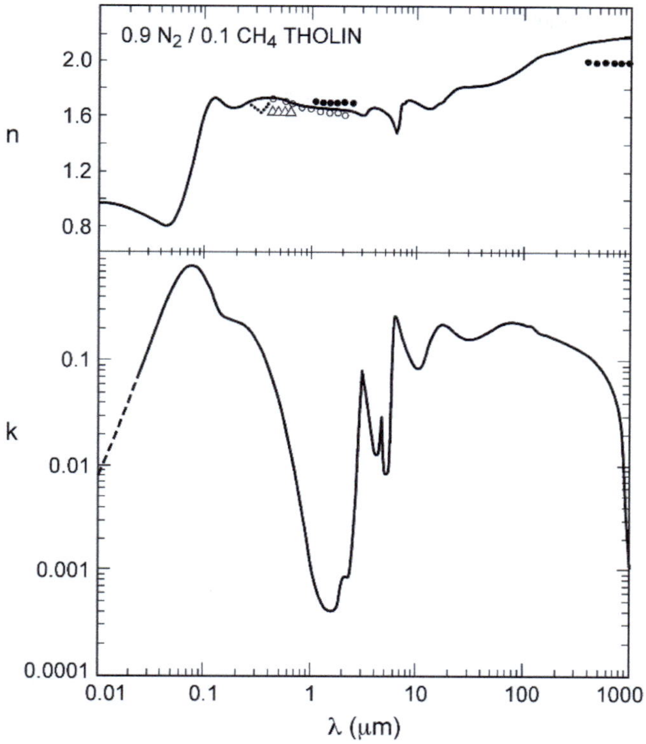

Fig. 12.30 From Khare et al. (1984). The real (n) and imaginary (k) refractive indices for laboratory tholin produced in a plasma discharge for a starting mixture of 0.9 N_2 and 0.1 CH_4.

which questions are under investigation and which groups are active in this area.

Optical constants of material derived from laboratory simulations have been reported by Ramirez et al. (2002) and by Imanaka et al. (2004). These papers explore a range of physical and chemical regimes (methane mole fraction, pressure during haze formation). Optical constants are reported in the wavelength range 200–1,000 nm. The variety of tholin materials formed under different conditions all have imaginary index k decreasing from UV to IR wavelengths but the Khare values are almost always higher than the others, sometimes by an order of magnitude. The real part of the refractive index is also highest for the Khare et al. work over most of the wavelength range.

Energy supplied by solar UV photons is likely to dominate the energetics leading to haze formation, although energetic particle bombardment from Saturn's magnetosphere can also make a contribution. The processes are more complex when one considers that photons generate photoelectrons and that ion-neutral chemistry operates on a much faster time scale than does neutral chemistry alone. Additional complexity arises from the fact that photons penetrate to different levels in the atmosphere depending on the wavelength and the dominant processes vary with altitude, with the longest wavelengths penetrating deepest and acting on previously-formed large molecules, whereas shorter wavelengths are absorbed at high altitude and operate on simple molecules as well as more complex ones. The more recent studies investigate the relative importance of the various processes. More focus is also being placed on the fate of aerosols on the surface and their possible interaction with surface water, a condition that can lead to the formation of amino acids and other organic molecules of importance in a pre-biotic or proto-biotic environment.

A substantial laboratory effort has been underway at the Laboratoire Interuniversitaire des Systèmes Atmosphériques (LISA) in Paris. Some of the work was summarized by Coll et al. (2001) who described the experiment and noted that C_4N_2 has been detected. They also reported on particle shape and size, solubility in hydrocarbons and nitriles, chemical composition, and optical behavior in the 200–900 nm wavelength range. Nguyen et al. (2008), also from LISA, showed that aerosols probably do not lead to isotopic fractionation, and they discussed hydrolysis of tholin material on the surface.

Curtis et al. (2008) investigated the role of haze as condensation nuclei for methane or ethane and found that supersaturation of ethane is more likely than that for methane. They also found results for ethane that are consistent with the observation of a polar ethane cloud (Griffith et al. 2006). Release of adsorbed methane and other volatiles from haze particles on the surface may also account for the rise in mixing ratios of those constituents observed by the Huygens Gas Chromatograph and Mass Spectrometere (GCMS) after landing (Nieman et al. 2005). Signorell and Jetzki (2007) commented on the crystalline state of methane ice particles. Adsorption aids nucleation.

Other recent results are worthy of mention. Our list is not exhaustive but it shows the range of work being done. Bernard et al. (2006) examined reflection spectra of two types of tholin classified as possible end members of relevance to surface composition. They used near-IR, visible and UV-Raman techniques. Ultraviolet Raman signatures offer the potential to identify aromatics. McGuigan et al. (2006) looked at compounds emitted by pyrolysis of tholin using a two-dimensional gas chromatograph (GC × GC) with a time-of-flight mass spectrograph. They found low-molecular-weight nitriles, alkyl substituted pyrroles, linear and branched hydrocarbons, alkyl-substituted benzenes and PAH compounds. Tran et al. (2008) found that the addition of HCN to the basic gas mixture (which includes N_2 and CH_4) does not significantly affect the optical properties of the haze. Imanaka and Smith (2007) explored the role of photochemistry stimulated by photons in the wavelength range 50–150 nm, at higher energies than used by other studies. This work is relevant to high altitudes (near 1,000 km) where Waite et al. (2007) found surprisingly complex chemistry thought to be the starting point of haze formation (see Chapter 7).

12.10 Condensation Clouds

There are two fundamentally different types of cloud systems: (1) tropospheric clouds, dominantly methane ice and probably convective, and (2) stratospheric clouds composed of trace organic ices (hydrocarbons and nitriles), probably cirrus-like and statically stable. The main source for the former appears to be the surface, and clouds are formed during atmospheric up-welling. The source for the latter is photochemistry in the mesosphere and thermosphere coupled with atmospheric down-welling due to diffusion and general circulation. Sagan and Thompson (1984) listed many photochemical products which might condense near the temperature minimum (at altitudes between 50 and 100 km). Ethane is probably a member of both of these groups (Barth and Toon 2006). We begin by discussing the first group.

12.10.1 Methane/Ethane Tropospheric Condensate Clouds

We now have observational evidence from both ground-based and spacecraft experriments for condensates near and below the temperature minimum. An east/west limb brightness asymmetry was discovered in ground-based near-infrared images and spectra (Ádámkovics et al. 2007). The asymmetry is subtle and therefore the amount of material involved is probably small. Ádámkovics et al. (207) infer that methane ice particles at an altitude of 30 km (below the temperature minimum) over the Xanadu region, are responsible for the asymmetry, and that a methane ice drizzle is present at lower altitude. The asymmetry also implies a slight cooling during the night at the altitudes where condensation occurs.

The methane 'hydrologic' cycle is a fascinating topic whose scope is largely beyond the space available here. Evidence for fluvial erosion is abundant from the widespread distribution of dendritic channels (see Chapter 4), but condensate clouds are rarely seen, especially at low latitude. The atmosphere contains a large amount of methane that may precipitate episodically and torrentially but only briefly (See Chapters 12 and 13) Another (antithetical) mode of precipitation has also been proposed: rain without clouds (Toon et al. 1988). We focus here on the observations of condensate clouds.

The first indications that condensate clouds form and dissipate in sufficient mass to be sensed with whole-disk spectrophotometry came from the ground-based observations of Griffith et al. (1998). Later, ground-based adaptive optics images were able to resolve locations of cloud fields (Brown et al. 2002; Roe et al. 2002 and subsequent papers). Observations in the years just prior to the first detailed observations by Cassini showed clouds occur predominantly at high southern latitudes (Schaller et al. 2006). Southern polar clouds appeared during the first Cassini Titan flyby in July 2004. At an image scale near 2 km/pixel the Cassini ISS images revealed south polar cloud detail and morphology similar to that for terrestrial cumulus clouds (Fig. 12.31, frames a–d). Substantial evolution of the cloud field was observed over the 4.5-h observation period. Cassini ISS observations of the cloud field in filters that sample methane bands of different strengths in the near-infrared and visible, and at nearby continuum wavelengths are consistent with methane ice clouds high in the troposphere (see Fig. 12.32). Ground-based observations are also consistent with this interpretation (Brown et al. 2002).

The latitudinal distribution of condensate clouds might provide a strong constraint on atmospheric circulation models and the methane hydrological cycle. Between 2001 and the end of 2004 almost all clouds were in the high southern latitudes. Beginning in December of 2004 ground-based

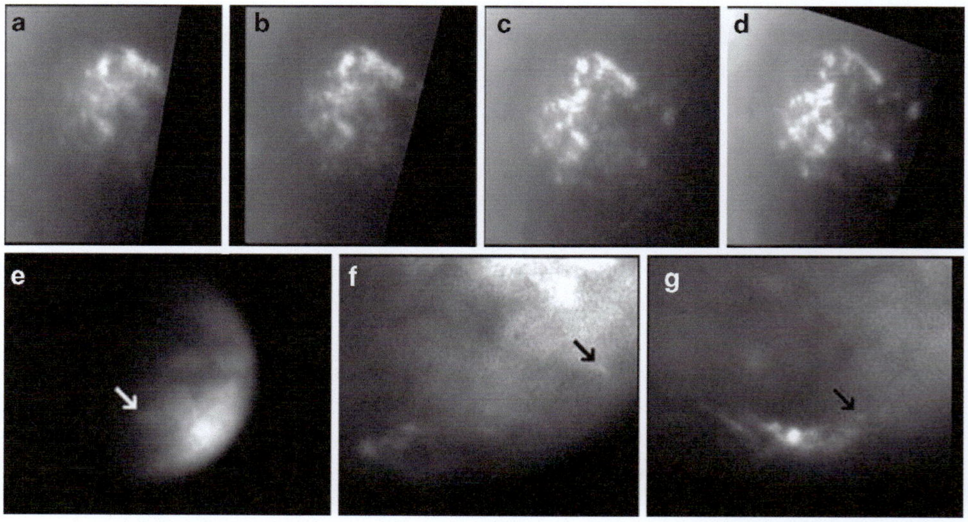

Fig. 12.31 From Porco et al. (2005). Tropospheric cloud features on Titan. Panels a–d: a sequence of four methane continuum images showing the temporal evolution over the period 05:05–09:38 of the Titan south polar cloud field on 2 July 2004. Panels e–g: three examples of discrete mid-latitude clouds (arrows) for which motions have been tracked. Coordinates are as follows. Panel e: 388 S, 818W (29 May 2004); f: 438 S, 678W (23 October 2004); g: 658 S, 1108W (25 October 2004).

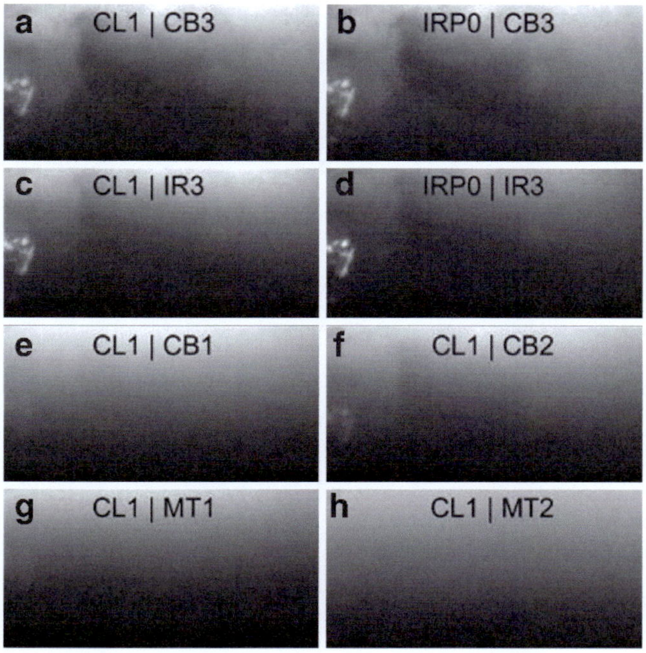

Fig. 12.32 From Porco et al. (2005) Near-simultaneous images of Titan's south polar clouds acquired on 2 July 2004 in eight filter combinations, as indicated by the legend at the top of each image. The high-contrast cloud field on the left of each image is located near the south pole. For each image, a spectral bandpass filter was combined with either a clear filter (CL1) or an infrared polarizing filter (IRP0), which helps reduce obscuration by the overlying stratospheric haze. Filters are as follows: Narrowband methane continuum images (CB3, 938 nm) without (**a**) and with (**b**) the polarizing filter. Broadband infrared (IR3, 928 nm) images, without (**c**) and with (**d**) the polarizing filter. Narrowband continuum images CB1 (635 and 603 nm; **e**) and CB2 (750 nm; **f**). (**g, h**), Methane band images in weak (MT1, 619 nm; **g**) and moderate strength (MT2, 727 nm; **h**) bands. Both clouds and apparent surface features are visible in CB3 and IR3 and to a lesser extent in CB2 and CB1, for which obscuration by stratospheric haze is stronger. In MT1, all features except the polar clouds have disappeared. Gas absorption alone is sufficient to ensure that fewer than 1% of MT1 photons reach the surface, and haze scattering reduces this further. The cloud is barely visible in MT2, which has essentially no contribution from the surface.

monitoring found only a few clouds at high southern latitude (Schaller et al. 2006; see Fig. 12.33). Clouds are also observed near latitude −40° (Turtle et al. 2009). To explain the preponderance of cloud occurrences at high southern latitude Brown et al. (2002) proposed that solar heating of the surface layers is important in initiating the chain of convective activity that results in cloud formation. Observations of surface temperature from Cassini instruments do not support this view. Brightness temperatures at south polar latitudes are cooler than elsewhere (Jennings et al. 2009). Large-scale circulation is probably the more important driver. Upwelling or a strong latitudinal temperature gradient might provide the environment conducive to convective cloud formation. Some general circulation models (GCMs) do not include the methane hydrological cycle and some do. Those that do not (Hourdin et al. 1995; Mitchell et al. 2006; Tokano et al. 1999; Richardson et al. 2007) always produce a single Hadley cell circulation, rising near the summer pole and with no causal relation to clouds at mid-latitudes. Those that do include the methane hydrological cycle (Rannou et al. 2006; Mitchell et al. 2006) have a terrestrial-like three-cell circulation. Haze aerosols could be important as cloud condensation nuclei (the contact parameter is not known) and so the transport of haze particles to locations in the troposphere could influence cloud formation. Figure 12.34 shows cloud predictions from the Rannou et al. (2006) GCM. Note the relative maxima predicted cloud cover at both poles and near 35°S latitude. The circulation model of Mitchell et al. (2006) also predicted cloud formation at the poles and at latitude −40°, provided that the methane relative humidity is near 100%. The Rannou and Mitchell models make different predictions about the future occurrence of clouds at low latitudes. This is one possible observational test, although the relevant processes are numerous and complicated, weakening the potential value of this test. The latitudinal distribution of clouds appears to be changing with time as noted by Schaller

Fig. 12.33 From Schaller et al. (2006). Titan cloud locations vs time from ground-based observations. This figure excludes all mid-latitude clouds from Roe et al. (2005a) which are all located near 40°S and are clustered in longitude near 350°W. Diamond sizes indicate the relative size of the cloud observed. Dashed line marks Titan southern summer solstice which occurred in October 2002. Dotted line marks July 2005, the time at which the south pole ceased to be the area of maximum solar insolation on Titan. Before December 2004, clouds were observed within 30° of the south pole in nearly every adaptive optics image of Titan. During the 5 months period between December 2004 and April 2005 only one image shows even a small amount of cloud activity at the pole. Images taken in the 2005–2006 Titan apparition show a resurgence of Titan's clouds at the highest latitudes yet seen.

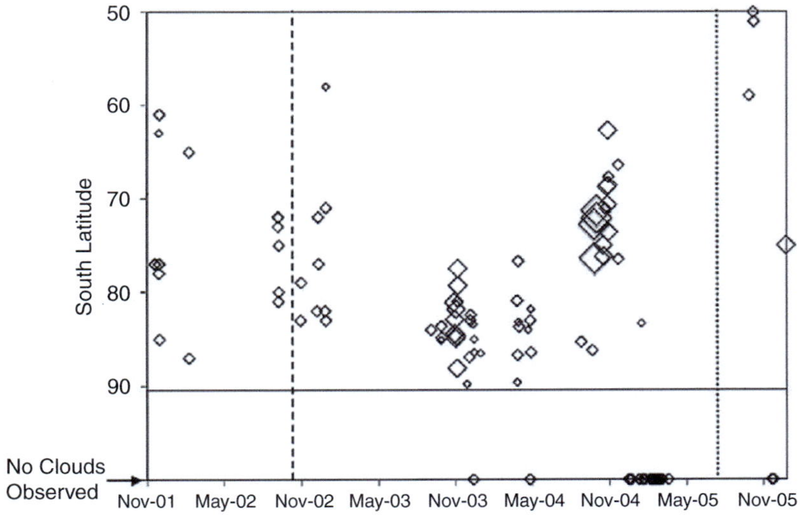

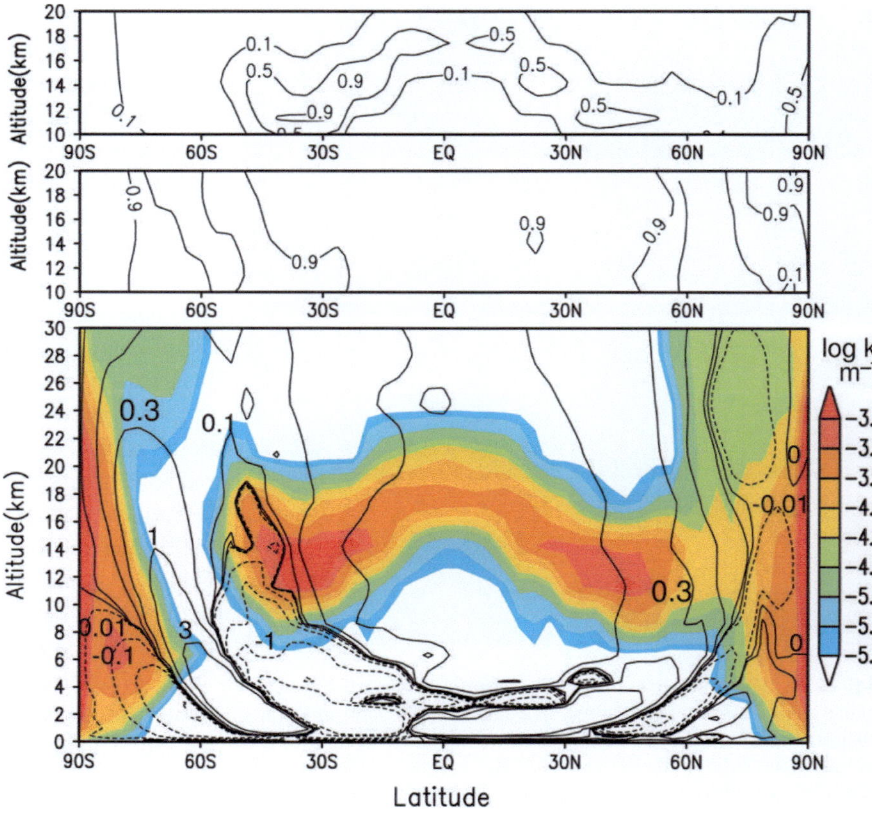

Fig. 12.34 From Rannou et al. (2006). Lower panel: modeled cloud extinction (m^{-1}) at 620 nm, averaged over one terrestrial year around the Cassini/Huygens arrival time, shown along with the stream function (10^9 kg s^{-1}) averaged over 7 years (one Titan season) before the Cassini arrival (*continuous lines denote clockwise motion*). Each result is actually the mean result of four consecutive Titan years. (*Middle panel*) The percentage of methane in drop composition (the remaining part is ethane). (*Top panel*) The fraction of time that cloud extinction exceeds a threshold of $k = 10^{-4}$ m^{-1}.

et al. (2006) from ground-based observations (Fig. 12.33) and from recent Cassini ISS observations (Turtle et al. 2009) roughly in accord with model predictions, although some details such as the abrupt cessation of south polar clouds in late 2004 are not understood.

Highly ephemeral clouds form relatively frequently near latitude −40°. These are seen in ground-based images (Roe et al. 2005a), in Cassini ISS images (Fig. 12.31 panels e–g) and Cassini VIMS images (Griffith et al 2005). The clouds form long, narrow streaks aligned in the east–west direction. This morphology would be consistent with venting from geyser or volcanic activity leading to cloud formation at an altitude where the vertical shear in zonal wind is strong, and this is the hypothesis favored by Roe et al. (2005b). Griffith et al. (2005) favor a hypothesis which incorporates the large-scale circulation as discussed above. The longitudinal distribution seen in the accumulation of Cassini ISS images from 2004 to 2008 is broader than expected if the source were a single vent, and so disfavors the volcanic origin hypothesis (Turtle et al. 2009).

12.10.2 Stratospheric Condensates

At least three additional distinct types of condensate clouds are found in the north (winter) polar vortex (north of about 55° latitude). An ethane ice cloud covering a large aerial extent has been identified in the VIMS data (Griffith et al. 2006). Griffith et al. (2006) suggest that Titan's global circulation, with downwelling in the winter pole, brings photochemically-produced ethane from the stratosphere down to a cold winter troposphere where it forms an ice cloud.

Condensate signatures have also been observed in thermal-infrared spectra from both Voyager and Cassini. Several features in the Voyager Infrared Radiometer and Interferometer Spectrometer (IRIS) spectra could not be attributed to gaseous constituents. One of them (at 478 cm^{-1}) was identified by Khanna et al. (1987) as the ν_8 band of solid C_4N_2 (dicyanoacetylene; see also Samuelson et al. 1997). It is likely that this constituent and other condensates found only in the winter polar vortex region are produced as photochemical products from lower latitudes or higher altitudes, move into the shadow region either diffusively within the vortex or are advected from outside the vortex, and are not destroyed by sunlight once in the shadow region.

The Cassini CIRS (Composite Infrared Spectrometer) instrument provided additional data which constrain compositions of candidates which produce spectral features near 221 cm^{-1} and in the range 160–190 cm^{-1}. The feature centered at 221 cm^{-1} has not been identified. The other feature might be produced by HCN ice particles with radii less than about 5 μm (Samuelson et al. 2007). Samuelson et al. (2007) extended the work of Samuelson and Mayo (1991) and Mayo and Samuelson (2005). De Kok et al. (2007) performed an analysis of the latitude and altitude distribution of the condensate spectral features. They identified four components. Two of them are widespread in latitude and are probably components of the global haze. One of these has a smaller scale height than the other. The other two components appear to be condensates based on their confined latitude and altitude distributions, reinforcing the results from Samuelson et al. (2007) and previous papers on condensates identified in thermal-infrared spectra.

Several inferences have been made regarding radii of stratospheric condensate cloud particles from far-IR spectra, using Mie theory coupled with a radiative transfer model. These results appear to be unique and suggest something about cloud growth and stability. These include particle radius determinations of: (1) 5–10 μm for C_4N_2 ice particles from the 478 cm^{-1} band (Samuelson et al. 1997), (2) 2–5 μm for a general ice cloud of unknown composition from continuum observations (Mayo and Samuelson 2005), (3) 2–5 μm for HCN ice particles from the 172 cm^{-1} band (Samuelson et al. 2007), and (4) about 2 μm for HC_3N ice particles from the 506 cm^{-1} band (Anderson et al. 2007). Anderson et al. (2009) also find evidence for ~6 μm C_4N_2 particles.

12.11 Prospects for Future Progress

To date, four types of observations have been used to study the aerosols in Titan's atmosphere. These include observations of the entire disk (one example is the geometric albedo spectrum) from the earth, observations of the spatially resolved disk from the earth using adaptive optics, observations from the Hubble Space Telescope (HST) of spectra or imaging at high spatial resolution, and observations from the Cassini/Huygens mission. Of these, only the observations from Huygens are complete and cannot be continued into the future. The observations by the Cassini orbiter by the Imaging Science System (ISS), by the Visible and Infrared Mapping Spectrometer (VIMS), and by the Composite Infrared Spectrometer (CIRS) will be continued for at least several years into the future. These observations will continue to increase our coverage over the disk of Titan. In addition, as time passes, we have the prospect of increasing the seasonal coverage of the aerosol distribution over the disk. This will materially add to our understanding of Titan's atmosphere. The spatially resolved observations from HST and from adaptive optics from the earth can also be continued to give more complete coverage of the seasonal changes on Titan.

At the current time we are just on the verge of beginning to see the results of coupled 2- and 3-dimensional models that include atmospheric circulation, aerosol production and heating, and chemistry. These theoretical studies are being increasingly constrained by the rich body of observations including seasonal coverage.

At least two other sources of information are being accumulated that are sure to aid our understanding of the roles of aerosols in Titan's atmosphere. One includes the new laboratory studies of the optical constants of "tholin" material produced in laboratory simulations. Several new studies are underway covering the material produced under wider ranges of conditions including chemistry and energetics. A second source of new information includes improved absorption coefficients for methane under conditions of long paths and low temperatures that are closer to that found on Titan. This molecule has long been used to explore the vertical structure of Titan's aerosols. With new laboratory and theoretical studies, supplemented by the direct measurements in the near infrared on the Huygens probe, these approaches will be used with greater confidence to produce results with increasing accuracy.

Finally, we also have the prospect of additional new missions to Titan. Already studies of the possible new information that could be obtained by balloons, additional probes, and spacecraft in orbit around Titan are being explored. The rich environment of Titan will without doubt continue to be the subject of new studies on many fronts for years to come.

References

Achterberg RK, Conrath BJ, Gierasch PJ, Flasar FM, Nixon CA (2008) Observation of a tilt of Titan's middle-atmospheric super-rotation. Icarus 197:549–555

Ádámkovics M, Wong MH, Laver C, de Pater I (2007) Widespread morning drizzle on Titan. Science 318:962–965

Anderson CM, Bjoraker G, Achterberg R (2007) Altitude distribution and wavenumber dependence of Titan's photochemical aerosol from Cassini/CIRS Observations. B.A.A.S. 39:530

Anderson CM, Samuelson RE, Bjoraker GL, Achterberg RK (2009) Particle sizes and abundances of HC_3N and C_4N_2 ices in Titan's stratosphere at northern high latitudes. Submitted to Icarus

Bar-Nun A, Kleinfeld I, Ganor E (1988) Shape and optical properties of aerosols formed by photolysis of acetylene, ethylene, and hydrogen cyanide. J Geophys Res 93:8383–8387

Bar-Nun A, Dimitrov V, Tomasko M (2008) Titan's aerosols: comparison between our model and DISR findings. Planet Space Sci 56:708–714

Barth EL, Toon OB (2006) Methane, ethane, and mixed clouds in Titan's atmosphere. Properties derived from microphysical modeling. Icarus 182:230–250

Bernard J-M, Quirico E, Brissaud O, Montagnac G, Reynard B, McMillan P, Coll P, Nguyen M-J, Raulin F, Schmitt B (2006) Reflectance spectra and chemical structure of Titan's tholins: application to the analysis of Cassini–Huygens observations. Icarus 185:301–307

Brown ME, Bouchez AH, Griffith CA (2002) Direct detection of tropospheric clouds near Titan's south pole. Nature 420:795–797

Cabane M, Chassefiere E, Israel G (1992) Formation and growth of photochemical aerosols in Titan's atmosphere. Icarus 96:176–189

Cabane M, Rannou P, Chassefiere E, Israel G (1993) Fractal aggregates in Titan's atmosphere. Planet Space Sci 41:257–267

Coll P, Ramirez SI, Navarro-Gonzalez R, Raulin F (2001) Chemical and optical behaviour of tholins, laboratory analogues of Titan aerosols. Adv Space Res 27:289–297

Courtin R, Wagener R, McKay CP, Caldwell J, Fricke KH, Raulin F, Bruston P (1991) UVspectroscopy of Titan's atmosphere, planetary organic chemistry and prebiological synthesis. Icarus 90:43–56

Curtis DB, Hatch CD, Hasenkopf CA, Toon OB, Tolbert MA, McKay CP, Khare BN (2008) Laboratory studies of methane and ethane adsorption and nucleation onto organic particles: application to Titan's clouds. Icarus 195:792–801

de Kok R, Irwin PGJ, Teanby NA, Nixon CA, Jennings DE, Fletcher L, Howett C, Calcutt SB, Bowles NE, Flasar FM, Taylor FW (2007) Characteristics of Titan's stratospheric aerosols and condensate clouds from Cassini CIRS far-infrared spectra. Icarus 191:223–235

Fulchignoni M et al (2005) In situ measurements of the physical characteristics of Titan's environment. Nature 438:785–791

Griffith CA, Owen T, Miller GA, Geballe T (1998) Transient clouds in Titan's lower atmosphere. Nature 395:575–578

Griffith CA, Penteado P, Baines K, Drossart P, Barnes J, Bellucci G, Bibring J, Brown R, Buratti B, Capaccioni F, Cerroni P, Clark R, Combes M, Coradini A, Cruikshank D, Formisano V, Jaumann R, Langevin Y, Matson D, McCord T, Mennella V, Nelson R, Nicholson P, Sicardy B, Sotin C, Soderblom LA, Kursinski R (2005) The evolution of Titan's mid-latitude clouds. Science 310:474–477

Griffith CA, Penteado P, Rannou P, Brown R, Boudon V, Baines KH, Clark R, Drossart P, Buratti B, Nicholson P, McKay CP, Coustenis A, Negrao A, Jaumann R (2006) Evidence for a polar ethane cloud on Titan. Science 313:1620–1622

Hourdin F, Talagrand O, Sadourny R, Courtin R, Gautier D, McKay CP (1995) Numerical simulation of the general circulation of the atmosphere of Titan. Icarus 117:358–374

Hutzell WT, McKay CP, Toon OB, Hourdin F (1996) Simulations of Titan's brightness by a two dimensional haze model. Icarus 119:112–129

Imanaka H, Smith MA (2007) Role of photoionization in the formation of complex organic molecules in Titan's upper atmosphere. Geophys Res Lett 34:L02204. doi:10.1029/2006GL028317

Imanaka H, Khare BN, Elsila JE, Bakes ELO, McKay CP, Cruikshank DP, Sugita S, Matsui T, Zare RN (2004) Laboratory experiments of Titaan tholin formed in cold plasma at various pressures: implications for nitrogen-containing polycyclis aromatic compounds in Titan haze. Icarus 168:344–366

Jennings DE, Flasar FM, Kunde VG, Samuelson RE, Pearl JC, Nixon CA, Carlson RC, Mamoutkine AA, Brasunas J, Guandique CE, Achterberg RK, Bjoraker GL, Romani PN, Segura ME, Albright SA, Elliott MH, Tingley JS, Calcutt S, Coustenis A, Courtin R (2009) Titan's surface brightness temperatures. Ap J Lett 691:L103–L105

Karkoschka E (1998) Methane, ammonia, and temperature measurements of the Jovian planets and Titan from CCD spectrophotometry. Icarus 133:134–146

Karkoschka E, Lorenz RD (1997) Latitudinal variation of aerosol sizes inferred from Titan's shadow. Icarus 125:369–379

Karkoschka E, Tomasko MG, Doose LR, See C, McFarlane EA, Schroder SE, Rizk B (2007) DISR imaging and geometry of the descent of the Huygens probe within Titan's atmosphere. Planet Space Sci 55:1896–1935

Khanna RK, Perera-Jarmer MA, Ospina MJ (1987) Vibrational infrared and Raman spectra of dicyanoacetylene. Spectrochim Acta 43A:421–425

Khare BN, Sagan C, Arakawa ET, Suits F, Calcott TA, Williams MW (1984) Optical constants of organic tholins produced in a simulated Titanian atmosphere: from X-ray to microwave frequencies. Icarus 60:127–137

Lavvas P, Yelle RV, Vuitton V (2009) The detached haze layer in Titan's atmosphere. Icarus, 201, 626–633.

Liang MC, Yung Y, Shemansky D (2007) Photolytically generated aerosols in the mesosphere and thermosphere of Titan. Astrophys J 661:L199–L202

Lockwood GW, Thompson DT (1979) A relationship between solar activity and planetary albedos. Nature 280:43–45

Lorenz RD, Smith PH, Lemmon MT, Karkoschka E, Lockwood GW, Caldwell J (1997) Titan's north-south asymmetry from HST and Voyager imaging: comparison with models and ground-based photometry. Icarus 127:173–189

Mayo LA, Samuelson RE (2005) Condensate clouds in Titan's north polar Stratosphere. Icarus 176:316–330

McGuigan M, Waite JH, Imanaka H, Sacks RD (2006) Analysis of Titan tholin pyrolysis products by comprehensive two-dimensional gas chromatography–time-of-flight mass spectrometry. J Chromatogr A 1132:280–288

McKay CP, Pollack JB, Courtin R (1989) The thermal structure of Titan's atmosphere. Icarus 80:23–53

Mitchell JL, Pierrelhumbert RT, Frierson DMW, Caballero R (2006) The dynamics behind Titan's methane clouds. Proc Nat Acad Sci USA 103:18421–18426

Nguyen M-J, Raulin F, Coll P, Derenne S, Szopa C, Cernogora G, Israël G, Bernard J-M (2008) From Titan's tholins to Titan's aerosols: Isotopic study and chemical evolution at Titan's surface. Adv Space Res 42:48–53

Niemann, HB, Atreya, SK, Bauer, SJ, Carignan, GR, Demick, JE, Frost, RL, Gautier, D, Haberman, JA, Harpold, DN, Hunten, DM, Israel, G, Lunine, JI, Kasprzak, WT, Owen, TC, Paulkovich, M, Raulin, F, Raaen, E, Way, SH (2005). The abundances of constituents of Titan's atmosphere from the GCMS instrument on the Huygens probe NATURE 438, 779–784

Penteado PF, Griffith CA, Tomasko MG, Engel S, See C, Doose L (2009) Latitudinal variations in Titan's methane and haze from Cassini VIMS observations. Icarus (in press)

Podolak M, Danielson RE (1977) Axel dust on Saturn and Titan. Icarus 30:479–492

Podolak M, Noy N, Bar-Nun A (1979) Photochemical aerosols in Titan's atmosphere. Icarus 40:193–198

Porco CC et al (2005) Imaging of Titan from the Cassini spacecraft. Nature 434:159–168

Rages K, Pollack JB (1983) Vertical distribution of scattering Hazes in Titan's upper atmosphere. Icarus 55:50–62

Rages K, Pollack JB, Smith PH (1983) Size estimates of Titan's aerosols based on Voyager high-phase-angle images. J Geophys Res 88, (NA11):8721–8728

Ramirez SI, Coll P, da Silva A, Navarro-González R, Lafait J, Raulin F (2002) Complex refractive index of Titan's aerosol analogues in the 200–900 nm domain. Icarus 256:515–529

Rannou P, Cabane M, Chassefoere E, Botet R, McKay CP, Courtin R (1995) Titan's geometric albedo: role of the fractal structure of the aerosols. Icarus 118:355–372

Rannou P, Cabane M, Botet R, Chassefiere E (1997) A new interpretation of scattered light measurements at Titan's limb. J Geophys Res 102:10977–11013

Rannou P, McKay CP, Lorenz RD (2003) A model of Titan's haze of fractal aerosols constrained by multiple observations. Planet Space Sci 51:963–976

Rannou P, Montmessin F, Hourdin F, Lebonnois F (2006) The latitudinal distribution of clouds on Titan. Science 311:201–205

Richardson MI, Toigo AD, Newman CE (2007) PlanetWRF: a general purpose, local to global numerical model for planetary atmospheric and climate dynamics. J Geophys Res 112:E09001

Rodin AV, Keller HU, Skorov YuV, Doose L, Tomasko MG (2009) Microphysical processes in Titan haze inferred from DISR/Huygens data. Icarus, (in press)

Roe HG, de Pater I, Macintosh BA, McKay CP (2002) Titan's clouds from Gemini and Keck adaptive optics imaging. Astrophys J 581:1399–1406

Roe HG, Bouchez AH, Trujillo CA, Schaller EL, Brown ME (2005a) Discovery of temperate latitude clouds on Titan. Astrophys J 618: L49–L52

Roe HG, Brown ME, Schaller EL, Bouchez AH, Trujillo CA (2005b) Geographic control of Titan's mid-latitude clouds. Science 310:477–479

Roman MT, West RA, Banfield DJ, Gierasch PJ, Achterberg RK, Nixon CA, Thomas PC (2009) Determining a tilt in Titan's north–south albedo asymmetry from Cassini images. Icarus (in press)

Sagan C, Khare BN (1979) Tholins: organic chemistry of interstellar grains and gas. Nature 277:102–108

Sagan C, Thompson WR (1984) Production and condensation of organic gases in the atmosphere of Titan. Icarus 59:133–161

Samuelson RE, Mayo LA (1991) Thermal infrared properties of Titan's stratospheric aerosol. Icarus 91:207–219

Samuelson RE, Mayo LA, Knuckles MA, Khanna RJ (1997) C_4N_2 ice in Titan's north polar stratosphere. Planet Space Sci 45:941–948

Samuelson RE, Smith MD, Achterberg RK, Pearl JC (2007) Cassini CIRS update on stratospheric ices at Titan's winter pole. Icarus 189:63–71

Schaller EL, Brown ME, Roe HG, Bouchez AH, Trujillo CA (2006) Dissipation of Titan's south polar clouds. Icarus 184:517–523

Shemansky D (2006) (needs title) in Abstracts of 36th COSPAR Scientific Assembly, 2006 July 16–23, Beijing, China (Paris: COSPAR), 2748

Shemansky DE, Stewart AIF, West RA, Esposito LW, Hallett JT, Liu XM (2005a) The Cassini UVIS stellar probe of the Titan atmosphere. Science 308:978–982

Shemansky DE, Stewart AIF, West RA, Esposito LW, Hallett JT, Liu XM (2005b) The Cassini UVIS stellar probe of the Titan atmosphere. Science 308:978–982

Sicardy B., et al. (2006) The two Titan stellar occultations of 14 November 2003. J Geophys Res 111:E11S91, doi: 10.1029/2005JE002624

Signorell R, Jetzki M (2007) Phase behavior of methane haze. Phys Rev Lett . doi:10.1103/Phys Rev Lett.98.013401

Sromovsky LA, Suomi VE, Pollack JB, Krauss RJ, Limaye SS, Owen T, Revercomb HE, Sagan C (1981) Implications of Titan's north–south brightness asymmetry. Nature 292:698–702

Tokano T, Neubauer FM, Laube, McKay CP (1999) Seasonal variation of Titan's atmospheric structure simulated by a general circulation model. Planet Space Sci 47:493–520

Tomasko MG, Smith PH (1982) Photometry and polarimetry of Titan: pioneer 11 observations and their implications for aerosol properties. Icarus 51:65–95

Tomasko MG, Buchhauser D, Bushroe M, Dafoe LE, Doose LR, Eibl A, Fellows C, McFarlane E, Prout GM, Pringle MJ, Rizk B, See C, Smith PH, Tsetsenekos K (2002) The descent imager/spectral radiometer (DISR) experiment on the Huygens entry probe of Titan. Space Sci Rev 104:469–551

Tomasko MG, Archinal B, Becker T, Bezard B, Bushroe M, Combes M, Cook D, Coustenis A, de Bergh C, Dafoe LE, Doose L, Doute S, Eibl A, Engel S, Gliem F, Grieger B, Holso K, Howington-Kraus E, Karkoschka E, Keller HU, Kirk R, Kramm R, Kuppers M, Lanagan P, Lellouch E, Lemmon M, Lunine J, McFarlane E, Moores J, Prout GM, Rizk B, Rosiek M, Rueffer P, Schroder SE, Schmitt B, See C, Smith P, Soderblom L, Thomas N, West RA (2005) Rain, winds and haze during the Huygens probe's descent to Titan's surface. Nature 438:765–778

Tomasko MG, Bezard B, Doose L, Engel S, Karkoschka E (2008b) Measurements of methane absorption by the descent imager/spectral radiometer (DISR) during its descent through Titan's atmosphere. Planet Space Sci 56:624–647

Tomasko MG, Bezard B, Doose L, Engel S, Karkoschka E, Vinatier S (2008c). Heat balance in Titan's atmosphere. Planet Space Sci 56:648–659

Tomasko MG, Doose L, Engel S, Dafoe LE, West R, Lemmon M, Karkoschka E (2008a) A model of Titan's aerosols based on measurements made inside the atmosphere. Planet Space Sci 56:669–707

Tomasko MG, Doose LR, Dafoe LE, See C (2009) Limits on the size of aerosols from measurements of linear polarization in Titan's atmosphere. Icarus (in press)

Toon OB, McKay CP, Courtin R, Ackerman TP (1988) Methane rain on Titan. Icarus 75:255–284

Toon OB, McKay CP, Griffith CA, Turco RP (1992) A physical model of Titan's aerosols. Icarus 95:24–53

Trafton LM (1973) Titan's spectrum and atmospheric composition. In Hunten DM (ed) The atmosphere of Titan. NASA SP-340, pp 17–41

Tran BN, Force M, Briggs RG, Ferris JP, Persans P, Chera JJ (2008) Titan's atmospheric chemistry: photolysis of gas mixtures containing hydrogen cyanide and carbon monoxide at 185 and 254 nm. Icarus 193:224–232

Turtle EP, Perry JE, McEwen AS, DelGenio AD, Barbara J, West RA, Fussner S, Dawson DD, Porco CC (2009) Cassini imaging of Titan's high-latitude lakes, clouds, and south-polar surface changes. Geophys Res Lett (in press)

Waite JH Jr, Young DT, Cravens TE, Coates AJ, Crary FJ, Magee B, Westlake J (2007) The process of tholin formation in Titan's upper atmosphere. Science 316:870–875

Walterscheid RL, Schubert G (2006) A vital explanation for Titan's haze layers. Icarus 183:471–478

West RA (1991) Optical properties of aggregate particles whose outer diameter is comparable to the wavelength. Appl Opt 30:5316–5324

West RA, Smith PH (1991) Evidence for aggregate particles in the atmosphere of Titan and Jupiter. Icarus 90:330–333

West RA, Lane AL, Hart H, Simmons KE, Hord CW, Coffeen DL, Esposito LW, Satao M, Pomphrey RB (1983) Voyager 2 photopolarimeter observations of Titan. J Geophys Res 88:8699–9708

West RA, Pitman KM, Martonchik JV, Dumont PJ (2008) Titan's stratospheric haze and its effect on surface contrast, Titan after Cassini–Huygens Book Symposium, Corpus Christi, TX, July 7–11

Wilson EH, Atreya SK (2003) Chemical sources of haze formation in Titan's atmosphere. Planet Space Sci 51:1017–1033

Young DT et al (2004) Cassini plasma spectrometer investigation. Space Sci Rev 114:1–112

Yung YL, Allen M, Pinto JP (1984) Photochemistry of the atmosphere of Titan: comparison between model and observations. Astrophys J Suppl Ser 55:465–506

Chapter 13
Atmospheric Dynamics and Meteorology

F. M. Flasar, K. H. Baines, M. K. Bird, T. Tokano, and R. A. West

Abstract Titan, after Venus, is the second example in the solar system of an atmosphere with a global cyclostrophic circulation, but in this case a circulation that has a strong seasonal modulation in the middle atmosphere. Direct measurement of Titan's winds, particularly observations tracking the Huygens probe at 10°S, indicate that the zonal winds are mostly in the sense of the satellite's rotation. They generally increase with altitude and become cyclostrophic near 35 km above the surface. An exception to this is a sharp minimum centered near 75 km, where the wind velocity decreases to nearly zero. Zonal winds derived from temperatures retrieved from Cassini orbiter measurements, using the thermal wind equation, indicate a strong winter circumpolar vortex, with maximum winds of 190 m s^{-1} at mid northern latitudes near 300 km. Above this level, the vortex decays. Curiously, the stratospheric zonal winds and temperatures in both hemispheres are symmetric about a pole that is offset from the surface pole by ~4°. The cause of this is not well understood, but it may reflect the response of a cyclostrophic circulation to the offset between the equator, where the distance to the rotation axis is greatest, and the seasonally varying subsolar latitude. The mean meridional circulation can be inferred from the temperature field and the meridional distribution of organic molecules and condensates and hazes. Both the warm temperatures near 400 km and the enhanced concentration of several organic molecules suggest subsidence in the north-polar region during winter and early spring. Stratospheric condensates are localized at high northern latitudes, with a sharp cut-off near 50°N. Titan's winter polar vortex appears to share many of the same characteristics of isolating high and low-latitude air masses as do the winter polar vortices on Earth that envelop the ozone holes. Global mapping of temperatures, winds, and composition in the troposphere, by contrast, is incomplete. The few suitable discrete clouds that have been found for tracking indicate smaller velocities than aloft, consistent with the *Huygens* measurements. Along the descent trajectory, the Huygens measurements indicate eastward zonal winds down to 7 km, where they shift westward, and then eastward again below 1 km down to the surface. The low-latitude dune fields seen in Cassini RADAR images have been interpreted as longitudinal dunes occurring in a mean eastward zonal wind. This is not like Earth, where the low-latitude winds are westward above the surface. Because the net zonal-mean time-averaged torque exerted by the surface on the atmosphere should vanish, there must be westward flow over part of the surface; the question is where and when. The meridional contrast in tropospheric temperatures, deduced from radio occultations at low, mid, and high latitudes, is small, ~5 K at the tropopause and ~3 K at the surface. This implies efficient heat transport, probably by axisymmetric meridional circulations. The effect of the methane "hydrological" cycle on the atmospheric circulation is not well constrained by existing measurements. Understanding the nature of the surface-atmosphere coupling will be critical to elucidating the atmospheric transports of momentum, heat, and volatiles.

13.1 Introduction

Planetary atmospheres are nonlinear dynamical systems that resist easy analysis or prediction. While theoretical studies ranging from "simple" scaling to general circulation models (GCMs) are critical tools for developing a conceptual understanding of how an atmosphere works, these must be closely tethered to observations. The range of possibilities is too rich and complex to do otherwise. The study of extraterrestrial planetary atmospheres has traditionally drawn on the much richer body of work on the terrestrial atmosphere, but Earth is only one realization. Much of the excitement in studying planetary atmospheres is to avail oneself of the large-scale natural

F.M. Flasar (✉)
NASA Goddard Space Flight Center
e-mail: f.m.flasar@nasa.gov

K.H. Baines and R.A. West
California Institute of Technology, Jet Propulsion Laboratory

M.K. Bird
Universität Bonn

T. Tokano
Universität zu Köln

R.H. Brown et al. (eds.), *Titan from Cassini-Huygens*,

laboratories that other worlds provide and examine the response of atmospheres to different sets of external factors (e.g., surface or internal rotation, solar forcing, internal heat fluxes). Trying to reconcile the observed behavior of atmospheres to different forcing factors is an important step to achieving a deeper insight into the physical processes that govern it.

In the mix of available natural planetary laboratories, Saturn's giant moon Titan offers an intriguing blend. In several aspects, it resembles Earth. Its atmosphere is primarily N_2, and its surface pressure is about 50% larger than Earth's. After N_2, the most abundant constituent is not O_2 but CH_4. Photo- and electron-impact dissociation of CH_4 and N_2 leads to the irreversible production of more complex organic molecules that either condense and precipitate or else form the photochemical smog that enshrouds Titan. There is evidence for an analog to a hydrological cycle in Titan's troposphere, involving CH_4, not H_2O. Its middle atmosphere (i.e., stratosphere and mesosphere) has strong circumpolar winds in winter, with cold polar temperatures, condensate ices, and anomalous concentrations of several gases. This is reminiscent of the ozone holes on Earth. There are differences, too. Titan is much colder than Earth, and the radiative response of its atmosphere is much longer. For this reason, thermal tides probably do not play an important role in Titan's lower and middle atmosphere, but gravitational tides may, induced by Titan's eccentric orbit about Saturn. Titan is a much slower rotator than Earth: its "day" is 15.95 terrestrial days. In this regard, it is more like Venus, another slow rotator, and it provides the second example of an atmosphere with a global cyclostrophic wind system, i.e., the atmospheric winds whip around the body in much less time that it takes the surface to rotate by 360°.

This chapter reviews the dynamic meteorology of Titan's lower and middle atmosphere, i.e., its troposphere, stratosphere, and mesosphere (Fig. 13.1), particularly drawing on the Cassini-Huygens data that have been acquired and analyzed to date. Section 13.2 briefly reviews the radiative and dynamical time-scales in Titan's atmosphere. Section 13.3 discusses Titan's temperatures and zonal winds, derived from Voyager, ground-based, and Cassini-Huygens measurements. The zonally averaged temperatures and mean zonal winds are coupled by the thermal wind equation. Meridional winds (Section 13.4) can be inferred more indirectly from the temperature field, as well as from quasi-conserved tracer gases and from the distribution of condensates; the Huygens probe also provided in situ measurements at 10°S. Section 13.5 focuses on the energy and momentum exchange between the surface and atmosphere and the structure of Titan's planetary boundary layer. Atmospheric waves, particularly gravitational tides, are the subject of Section 13.6. They are of interest because they can transport zonal momentum over large distances and, because their horizontal and vertical propagation depends on the thermal stability and

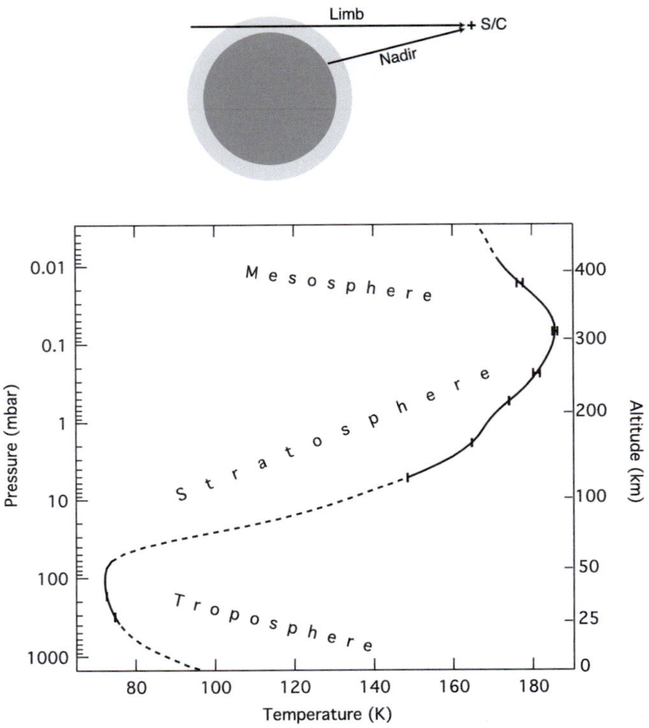

Fig. 13.1 Vertical profile of temperature at 15°S from Cassini CIRS mid- and far-infrared spectra obtained during northern winter. The *solid portions* of the profile indicate altitudes where the spectra constrain the temperature retrieval. The *upper portion* is from the spectral region of the ν_4 band of CH_4 at 7.7 μm, using both nadir- and limb-viewing geometry (illustrated at top; nadir viewing denotes the situation in which the line of sight intersects the surface); the *lower portion* is from far-infrared spectra near 100 μm, where pressure-induced N_2 absorption dominates, using only nadir-viewing geometry. The *dashed portions* of the profile are not well constrained by the spectra and are essentially initial guesses, based on temperatures retrieved from the Voyager radio occultations (after Flasar et al. 2005)

wind structure of the mean atmosphere, they are also useful probes of atmospheric structure. Section 13.7 summarizes the current understanding of Titan's general circulation, both conceptually and by the success of GCMs in simulating Titan's atmospheric behavior. Finally Section 13.8 concludes by summarizing key questions concerning Titan's meteorology and near-term prospects for addressing them. Earlier reviews of Titan's dynamic meteorology that may be of interest can be found in Hunten et al. (1984), Flasar (1998a, b), Flasar and Achterberg (2009), and Tokano (2009).

13.2 Radiative and Dynamical Time Constants

Before discussing measurements of meteorological variables on Titan and their interpretation, it is helpful to briefly discuss the notion of radiative relaxation times and dynamical turnover times, which will be used repeatedly in the ensuing discussion.

13.2.1 Radiative

Titan's atmosphere is an interesting entity: its radiative response varies by several orders of magnitude as one moves vertically through the atmosphere. The early analysis of Voyager data indicated that Titan's radiative relaxation time – the time over which its temperature relaxes to a radiative equilibrium profile from an initial disturbance, extending over an altitude that is typically on the order of a pressure scale height – was quite large in the troposphere, ~130 years (Smith et al. 1981), and decreased with altitude to a value ~1 year in the upper stratosphere near 1 mbar (Flasar et al. 1981) (see Hunten et al. 1984 and Flasar 1998b for more discussion). More recent radiative flux measurements by the Huygens Probe Descent Imager/Spectral Radiometer (DISR) have indicated a time constant on the order of 500 years in the lower troposphere (Tomasko et al. 2008; Strobel et al. 2009, Ch. 10, Fig. 10.5). Two conclusions follow. The first is that, since the radiative relaxation times in Titan's troposphere and middle atmosphere are much longer than its day (15.95 days) divided by 2π [1], one does not expect solar-driven diurnal effects, e.g., thermal tides, to be very important. The second is that radiative time scales in the upper stratosphere are much smaller than Titan's year (29.5 years) divided by 2π, so there should be a strong response to the seasonal modulation of solar heating. The measurements discussed in the following sections bear this out. Given the very long radiative time constant in the troposphere, Flasar et al. (1981) suggested that seasonal variations there would be very weak. However, this neglected any effects that seasonal variation in surface temperatures would have in coupling to the atmosphere. The heat capacity in the annual skin depth of the surface is much smaller than that corresponding to the lowest scale height of atmosphere (Tokano 2005), particularly in the absence of global oceans or widespread deep lakes. Enough solar radiation (~10%, McKay et al. 1991) makes it through the atmosphere to the surface to produce a seasonal variation in its temperature. Coupling of the surface to the atmosphere through thermally driven convection heats the lower atmosphere seasonally. GCM model simulations generally display a strong seasonal component (Section 13.7). From GCM simulations, Tokano (2005) has concluded that the thermal inertias of some plausible surface materials is low enough that a diurnal variation is possible.

[1]For a harmonic disturbance, the relevant time scale is the period divided by 2π (see, e.g., Section V of the review by Hunten et al. 1984).

13.2.2 Dynamical

For global-scale flow, the dynamical time scale is the turnover time, the horizontal scale of a circulation cell, divided by the horizontal velocity. For the axisymmetric meridional circulations thought to be important on Titan, the horizontal scale is on the order of a planetary radius (Section 13.7). The turnover times can be comparable to a season. When the dynamical time scale is longer than the radiative time scale, there can be an additional, dynamical inertia that acts to retard the atmosphere's relaxation to the radiative equilibrium state. For example, temperatures obtained by the Voyager infrared measurements indicated a north–south asymmetry at 1 mbar (Flasar and Conrath 1990). This was curious, because the Voyager season was shortly after northern equinox, and the radiative relaxation time was relatively short. Bézard et al. (1995) suggested that this might result from a hemispheric asymmetry in the opacities for solar and thermal radiation. Flasar and Conrath (1990) alternatively suggested a dynamical origin. They noted although temperatures could rapidly relax to an equilibrium configuration radiatively, they were coupled to the zonal wind fields by the thermal wind equation (Section 13.3). Hence to reach the equilibrium state angular momentum also had to be transported from the northern hemisphere to the southern. The dynamical turnover time for achieving this transport was comparable to a season on Titan, implying that the stratospheric temperatures and zonal winds would always lag the solar heating, despite the small radiative relaxation time.

13.3 Temperatures and Zonal Winds

Any study of atmospheric dynamics and meteorology is predicated on having measurements of temperatures, winds, and gaseous constituents and other tracers of motions in three dimensions. Spatially resolved observations of Titan only began with spatially resolved imaging by Pioneer 11 during its flyby in 1979 (Tomasko and Smith 1982), and in earnest with the Voyager 1 and 2 close passages in 1980 and 1981 a few months after its northern spring equinox. The Cassini orbiter and Huygens probe, observing Titan in northern winter, have provided the best resolution and global coverage to date. Images from the Hubble Space Telescope and imaging using ground-based adaptive optics have provided important information on Titan's state between the two missions, albeit at lower spatial resolution. Finally, much of what we have learned about Titan's zonal winds and their seasonal modulation before Cassini-Huygens has come from Earth-based observations, including stellar occultations, heterodyne spectroscopy, and correlation spectroscopy.

Temperature, wind, and composition fields are often cast in terms of zonal averages, i.e., averages around a latitude circle. These mean variables define the general circulation, and it is its seasonal and longer-term climatological variations that one seeks to understand. Often the coverage in longitude is not very extensive and surrogate representations must be used, e.g., retrieved variables at a specific longitude or a few longitudes. For instance, spectra from nadir-viewing observations by the Cassini Composite Infrared Spectrometer (CIRS) cover latitude and longitude extensively, and temperatures and composition retrieved from this data can be used to construct true zonal averages. On the other hand, CIRS limb-viewing observations are more sparsely distributed in longitude, and true zonal averages are not possible for these spectra. Fortunately, available evidence, including the CIRS nadir spectra, indicate that zonal variations in temperatures, derived winds, and composition on Titan are usually smaller than meridional variations (see, e.g., Teanby et al. 2008). Hence the ensuing discussion of meteorological variables in this section and the next (Section 13.4) will often be in terms of zonally averaged quantities, even though the measurements themselves have not always provided sufficient data to construct these averages.

13.3.1 Temperatures

Chapter 10 (Strobel et al. 2009) discusses the vertical structure of temperature in some detail, so comments here are brief. Figure 13.1 illustrates that the thermal structure of Titan's lower and middle atmosphere, at least at low latitudes, is remarkably reminiscent of Earth's, with a well-defined troposphere, stratosphere, and mesosphere. Titan's temperatures are much lower than Earth's, because of its greater distance form the sun, and the pressure scale height – ranging from 15 to 50 km – is much larger than on Earth (5–8 km), mainly because the gravitational acceleration on Titan–1.3 m s^{-2} at the surface – is much smaller. In fact the large scale height means that Titan's atmosphere is much more extended than most planetary atmospheres in the solar system. The 10 μbar level in the mesosphere, for example, roughly corresponds an altitude of 400 km, a sizeable fraction of Titan's 2,575-km radius. Near-infrared images of Titan from the Visual and Infrared Mapping Spectrometer (VIMS) show methane fluorescence up to ~730 km altitude (Baines et al. 2005).

The vertical profile of temperature is primarily important as a measure of the stability of an atmosphere, i.e., how stable an atmosphere is to vertical motions and whether waves can propagate and how they are refracted by the mean flow when they propagate vertically and horizontally. However, it is the lateral variations in temperature that serve as diagnostics of the departure of an atmosphere from purely radiative response and of the structure of meridional circulations (Section 13.4). Measuring horizontally resolved thermal structure has really only been achieved by the close-up reconnaissances of Voyager and Cassini-Huygens. The ingress and egress of the single Voyager 1 radio-occultation sounding were both in the equatorial region, although at nearly diametrically separated longitudes. Little variation in the troposphere and lower stratosphere was evident (Lindal et al. 1983, Lellouch et al. 1989). The Voyager thermal-infrared spectrometer (IRIS) obtained reasonable latitude coverage during the Voyager 1 flyby, but coverage in longitude was limited to two strips on the day and night sides, nearly 180° apart. Moreover, the IRIS spatial resolution was limited: spectra had to be averaged in latitude bins, typically 15° or larger, in order to obtain an adequate signal-to-noise ratio (Flasar et al. 1981). Only in the upper stratosphere~1 mbar (1 mbar = 1 hPa = 100 Pa) could temperatures be retrieved on isobars, by inverting the observed radiances in the ν_4 band of CH_4 near 1.300 cm^{-1} (7.7 μm) and assuming that the gas was uniformly distributed with latitude. Temperatures at high northern latitudes were ~12–20 K colder than at the equator; temperatures in the south were flatter between the equator and 50°, where they began to fall off toward the pole. Neither pole was well observed. Other wavelength regions near 200 cm^{-1} and 530 cm^{-1} probed the tropopause and surface, respectively. Little meridional variation was evident. Temperature variations near the tropopause were ~1 K, and the equator-to-pole contrast at the surface was ~2-3 K. The difficulty was that these differences were estimated from brightness temperatures; the unknown and heterogeneously distributed opacity sources at these wavelengths precluded direct retrieval of temperatures.

Cassini-Huygens, with a more capable array of in situ and remote-sensing instruments, has greatly extended global mapping of atmospheric temperatures at high spatial resolution. For the Cassini spacecraft, this is ensured in no small part by virtue of its being in orbit about Saturn and returning to Titan repeatedly, where its proximity allows good spatial resolution. Preliminary analysis of several radio-occultation soundings at mid and high southern latitudes and high northern latitudes, together with the descent temperature profile from the Huygens Atmospheric Structure Instrument (HASI; see Chapter 10), indicate that the meridional variation of temperatures in the troposphere is small, ~5 K at the tropopause and ~3 K near the surface. Most of the global mapping of atmospheric temperatures is from CIRS thermal-infrared spectra, but these do not permit retrieval of temperatures in the lowest 1½ scale heights (Flasar et al. 2004).

Most of the retrieved temperature maps to date have been from CIRS nadir- and limb-viewing in the mid-infrared, using the observed radiances in the ν_4 band of CH_4 as a thermometer to probe the atmosphere above the 10-mbar level. This is a classic case of going after the low-lying fruit first. The far-infrared radiances, which include pressure-induced N_2 absorption and the rotational lines of CH_4, probe lower in the atmosphere and are more affected by aerosol and condensate opacity, which require more careful analysis to retrieve temperatures. Figure 13.2 (top) depicts the meridional cross section of temperatures in Titan's middle atmosphere, retrieved from limb and nadir spectra. The vertical range over which there is information on temperature is 5 millibar –3 microbar, except in the colder regions at high northern winter latitudes, where the information content dies away at the 2 mbar level (Achterberg et al. 2008a). Given these caveats, the most dramatic aspect of the temperature field is the behavior at high northern latitudes. Below the 0.1-mbar level, the north polar region is colder than at lower latitudes. This is expected, because it is the region of polar night, and the radiative relaxation time is relatively short (Section 13.2). The temperature contrast is 20–30 K. However, higher up in the stratosphere, the high northern latitudes become the warmest part of Titan's atmosphere, with temperatures exceeding 200 K. Titan's

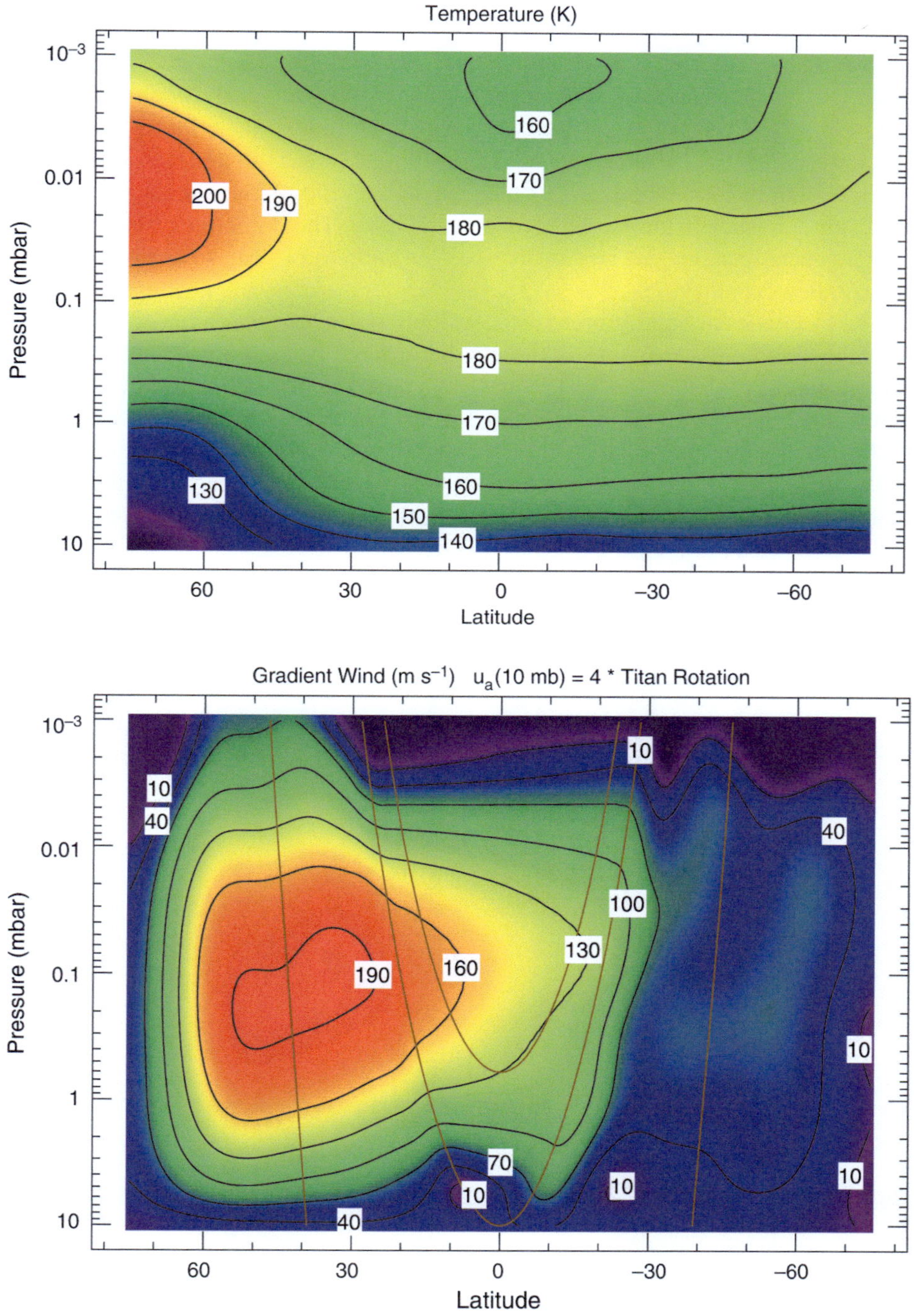

Fig. 13.2 *Top panel.* Meridional cross section of temperature (K) from CIRS limb and nadir spectra in the mid infrared. *Bottom panel.* Zonal winds (m s^{-1}) computed from the thermal wind equation (13.7) with the winds at 10 mbar set to uniform rotation at four times Titan's rotation rate. Positive winds are eastward. The *parabolic curves* correspond to surfaces parallel to Titan's rotation axis (after Achterberg et al. 2008a)

polar night does not extend to these higher altitudes. At winter solstice, its north pole is tilted by ~26.7° away from the sun, so the maximum height of shadow, in the absence of atmospheric scattering, is

$$\left(\frac{1}{\cos 26.7^{\circ}}-1\right)2575 \approx 307\ \text{km}$$

(~70 μbar). Given the thick condensate hazes that are observed at high northern latitudes during the winter and early spring, it is conceivable that radiative heating may contribute to the warm anomaly observed, but this has not been studied. It may also have a dynamical origin, and this is discussed in Section 13.4.

The figure indicates that the stratopause, the altitude of the maximum temperature marking the boundary between the stratosphere and the mesosphere, varies with latitude. It is lowest near the equator, where it is at the 70-μbar level. It rises slightly toward the south pole, but its elevation increases dramatically toward the north pole,~10 μbar. This is a far greater scale-height variation than seen on Earth (see, e.g., Andrews et al. 1987).

The longitude coverage afforded by CIRS limb soundings is limited, and nadir mapping in the middle infrared has provided most of the information on the zonal structure of temperatures, between 0.5 and 5 mbar. An unexpected result of the mapping was the discovery that the pole of symmetry of the stratospheric temperatures is tilted approximately 4° relative to the IAU definition of the polar axis, which is normal to Titan's orbit about Saturn. The spectral decomposition of the zonal temperature structure relative to the IAU pole indicated a strong wavenumber-1 component (one wavelength around the latitude circle) at most latitudes, and it was strongest at the latitudes having the strongest meridional gradients in temperature. Moreover, the phase of the wavenumber-1 component (i.e., the longitude of the warmest temperatures) remained nearly constant with latitude in each hemisphere, flipping by 180° across the equator. Since both poles are cooler than low latitudes near 1 mbar (Fig. 13.2), this strongly suggested a global tilt of the atmospheric pole of symmetry. By recalculating the zonal variances for an ensemble of new pole positions, the tilt was determined by minimizing the global variance of the temperatures from their zonal means (Achterberg et al. 2008b). Figure 13.3 displays the 1-mbar temperatures; the offset of the axis of symmetry from the IAU pole position is more evident in the northern hemisphere, where the meridional gradients are larger. In the northern hemisphere the tilt of the pole is toward the sun, but about 76°W of the solar direction. The temperatures about the tilted axis are nearly axisymmetric at most latitudes: the standard deviation of the zonal variation is not much greater than that propagating from the CIRS instrument noise. Because of the thermal wind equation, the winds tend to flow along isotherms (see Section 13.3.2.1), and thus are also centered about the offset pole. The 4° tilt is an order of magnitude greater than the recent measurement of the tilt of the rotation pole of the solid surface (Stiles et al. 2008).

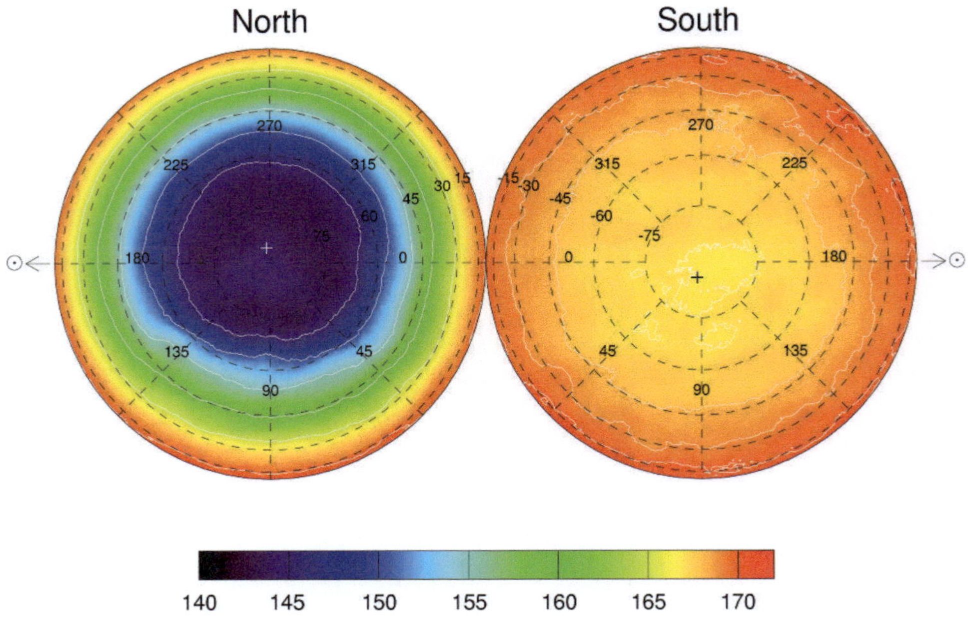

Fig. 13.3 Polar projection maps of retrieved temperatures at the 1-mbar level. The northern hemisphere is shown on the *left* and the southern hemisphere on the *right*. The color-coded temperature scale in kelvins is shown at the *bottom*. The superposed grid represents latitude and west longitude in a sun-fixed frame with the longitude of the sub-solar point at 180°W, such that the sun direction is towards the left and right edges of the figure. Temperature contours are plotted at intervals of 5 K in the northern hemisphere, and 1 K in the southern hemisphere. The fitted axis of symmetry is indicated by *white* and *black crosses* (+) in the north and south, respectively (after Achterberg et al. 2008b)

13.3.2 Zonal Winds

The likely presence of strong zonal winds on Titan had been inferred from infrared observations during the Voyager 1 flyby in November 1980. A distinct pole-to-equator latitudinal contrast in temperature was revealed, varying from $\Delta T \approx 3$ K at the surface to $\Delta T \approx 20$ K in the stratosphere (Flasar et al. 1981). The meridional gradients and the thermal wind equation implied an atmosphere that globally superrotates, analogous to that observed on Venus. To date zonal winds calculated from temperature and pressure fields have provided the most detailed picture of the global wind field. However, these suffer from ambiguities with respect to the wind direction and the need for a (often unknown) boundary condition when the zonal winds are derived from temperatures. Hence direct methods, based on observing the Doppler shifts of molecular lines, tracking discrete clouds, or tracking the descent of the Huygens probe, provide invaluable tie points for winds inferred from temperature and pressure.

13.3.2.1 Indirect Methods

The shape of the zonally averaged pressure and temperature fields is linked to the mean zonal winds. For steady, inviscid, axisymmetric flow, the balance is (see, e.g., Holton 1979; Leovy 1973):

$$\frac{-\nabla P}{\rho} - \nabla V + \left(2\Omega u + \frac{u^2}{r\cos\Lambda}\right)\mathbf{i}_\perp \approx 0 \qquad (13.1)$$

where P is pressure, ρ is atmospheric density, $\Omega = 4.56\times10^{-6}$ s^{-1} is Titan's angular rate of rotation, u is the zonal velocity, r is the radius (surface radius plus altitude), Λ is latitude, and $\mathbf{i}_\perp$ is a unit vector normal to Titan's rotation axis. V is the total potential (gravitational plus centrifugal) for a frame rotating at Titan's rotation rate; Titan is such a slow rotator that the gravitational component dominates:

$$-\nabla V \approx -g(r)\mathbf{i}_r, \qquad (13.2)$$

$\mathbf{i}_r$ is a unit vector in the radial direction from Titan's center, and g is the gravitational acceleration. The radial projection of (13.1) is the familiar hydrostatic equation:

$$\frac{1}{\rho}\left(\frac{\partial P}{\partial r}\right)_\Lambda \approx -g \qquad (13.3)$$

where the velocity terms involving u in (13.1) have been neglected in the projection, to a good approximation. The horizontal projection along lines of constant meridian yields the gradient wind relation:

$$\left(2\Omega u + \frac{u^2}{r\cos\Lambda}\right)\sin\Lambda \approx -\frac{1}{\rho r}\left(\frac{\partial P}{\partial\Lambda}\right)_V \qquad (13.4)$$

Alternatively, the gradient wind relation can be written in terms of the gravitational potential variation along isobars:

$$\left(2\Omega u + \frac{u^2}{r\cos\Lambda}\right)\sin\Lambda \approx -\frac{1}{r}\left(\frac{\partial V}{\partial\Lambda}\right)_P \qquad (13.4')$$

The balance (13.4 or 13.4') generally holds for zonally averaged variables to a good approximation when the lateral scale is large compared to a scale height. Hence, (13.1), as well as (13.3) and (13.4), can be viewed as a hydrostatic balance law in two dimensions (in height and latitude), with the centrifugal acceleration associated with the zonal winds adding vectorially to the gravitational acceleration. A possible exception is near the equator, where the left-hand side of (13.4) becomes small because of the trigonometric term. For instance, on Earth the mean lateral convergence of momentum by eddies – not included in (13.1) or (13.4) – in the intertropical convergence zone is known to be important. Usually one does not know a priori how close to the equator the gradient wind relation fails, or even if it does: (13.4) can remain valid if $(\partial P/\partial\Lambda)_V \propto \sin\Lambda$ across the equator. In practice, errors in the measured variables preclude application of (13.4) as $\Lambda\to0$, because the errors in u from the propagation of the errors on the right-hand side magnify inversely as $\sin\Lambda$ or $\sqrt{\sin\Lambda}$.

In rapidly rotating bodies, like Earth, Mars, and the outer planets, the first term on the left-hand side of (13.4, 4'), linear in u, dominates, and the equation reduces to geostrophic balance, in which the meridional gradient in pressure is balanced by the Coriolis force. For slow rotators, like Venus, the second term, quadratic in u, dominates and one has cyclostrophic balance, in which the pressure gradient is balanced by the centrifugal force associated with the zonal winds themselves. Titan is also a slow rotator, and in its middle-atmosphere the winds greatly exceed Titan's equatorial surface velocity ($\Omega a \approx 11.7$ m s^{-1}, where $a = 2{,}575$ km is Titan's radius); cyclostrophic balance is again dominant. Because the centrifugal force is quadratic in u, the direction of the zonal wind is not determined from the gradient wind relation. For the relation to hold at all, however, the pressure at constant height (or, equivalently, the potential along isobars) must decrease poleward. Hence cyclostrophic flow requires an equatorial bulge. Central flashes observed at Earth during occultations of stars by Titan have provided information on the shape of the isopycnal (i.e., constant-density) surfaces near the 0.25-mbar level (Hubbard et al. 1993; Bouchez 2003; Sicardy et al. 2006). Without too large an error, one can take these surfaces to be isobars (see Hubbard et al. 1993) and apply the gradient wind equation (13.4'). Figure 14.11 (Chapter 14; Lorenz et al. 2009) illustrates the zonal winds derived from some of these occultations.

Historically Titan's cyclostrophic winds were first inferred from temperature data using the thermal wind equation (Flasar et al. 1981; Flasar and Conrath 1990). Operating on equation (13.1) with the curl ($\nabla \times$):

$$\nabla P \times \nabla \frac{1}{\rho} + 0 + \nabla \left(2\Omega u + \frac{u^2}{r \cos \Lambda} \right) \times \mathbf{i}_\perp \approx 0 \qquad (13.5)$$

For axisymmetric flow, gradients around latitude circles vanish, and only the zonal component in (13.5) is nonzero. With the assumption of the perfect gas law and the neglect of spatial variations in the bulk composition (both valid except in the lower troposphere), this gives:

$$\frac{\partial}{\partial z_{\|}} \left(2\Omega u + \frac{u^2}{r \cos \Lambda} \right) \approx \frac{1}{P} \frac{\partial P}{\partial r} \frac{R}{r} \left(\frac{\partial T}{\partial \Lambda} \right)_P \qquad (13.6)$$

where $z_{\|}$ is the coordinate parallel to the planetary rotation axis and R is the gas constant. With (13.3), (13.6) reduces to:

$$\frac{\partial}{\partial z_{\|}} \left(2\Omega u + \frac{u^2}{r \cos \Lambda} \right) \approx -\frac{g}{T} \frac{1}{r} \left(\frac{\partial T}{\partial \Lambda} \right)_P . \qquad (13.7)$$

Unlike the gradient wind equation, the thermal wind equation (13.7) requires a boundary condition, a specification of u. Often one specifies u on a lower boundary. Zonal winds derived from (13.7) are illustrated in the bottom panel of Fig. 13.2. The salient features are the strong circumpolar wind, the polar vortex in the (northern) winter hemisphere, and the weak zonal winds in the summer hemisphere, where meridional contrasts in temperature are weaker.

For a thin atmosphere,

$$\Delta z_{\|} \approx \Delta z / \sin \Lambda_0, \qquad (13.8)$$

where Λ_0 is an average (constant) latitude. With this approximation eq. (13.7) reduces to the usual thermal wind equation:

$$\frac{\partial}{\partial r} \left[2\Omega u \sin \Lambda + \frac{u^2 \tan \Lambda}{r} \right] \approx -\frac{g}{T} \frac{1}{r} \left(\frac{\partial T}{\partial \Lambda} \right)_P , \qquad (13.9)$$

or in ln$P$ coordinates,

$$-\frac{\partial}{\partial \ln P} \left[2\Omega u \sin \Lambda + \frac{u^2 \tan \Lambda}{r} \right] \approx -\frac{R}{r} \left(\frac{\partial T}{\partial \Lambda} \right)_P . \qquad (13.9')$$

On Titan the relation (13.9, 9') fails at low latitudes, because the approximation (13.8) breaks down, and the integration of the thermal wind equation (13.7) to solve for the zonal winds must be along cylinders concentric with Titan's rotation axis (Flasar et al. 2005). For example, the 0.4-mbar isobar is approximately 130 km higher in altitude than the 10-mbar isobar. A cylindrical surface that intersects the equator at 10 mbar intersects the 0.4-mbar level at 17° latitude. These intersections (at northern and southern latitudes) lie on the parabola depicted in Fig. 13.2 tangent to the 10-mbar level. Outside this parabola, it is sufficient to apply a boundary condition at the 10-mbar level to integrate the thermal wind equation (13.7). However, cylinders intersecting the latitudes and pressure-levels within this parabola do not intersect the 10-mbar level. In this region, one must specify a boundary condition on the winds at higher altitudes, for example in the equatorial plane. However, the latter is a priori unknown. In Fig. 13.2 the winds within the 10-mbar parabola were interpolated along isobars. Although the winds at 10-mbar and lower are not well determined globally, this probably does not pose a critical problem in the winter northern hemisphere, as shown in Fig. 13.2. This is because the thermal wind equation implies that the winds increase markedly with altitude, and the zonal winds in the upper stratosphere are dominated by the thermal wind component above the 10-mbar level, i.e., the integral of the right hand side of (13.7). Since the u^2 term dominates the left-hand side of (13.7), it is not sensitive to the 10-mbar boundary condition away from the boundary. At mid and high latitudes in the southern hemisphere, the predicted thermal winds are weaker, and they are more sensitive to what is happening at the lower boundary.

Figures 13.2 and 14.11 (Chapter 14) indicate a significant seasonal variation in the stratospheric zonal winds, with the strongest velocities in the winter hemisphere. This is expected, because of the seasonal variation in temperatures (Section 13.3.1). The zonal-wind and temperature fields are coupled through the thermal wind equation, and hence changes in temperatures imply concomitant changes in the zonal winds. However, radiative processes do not transport angular momentum. It is mass motions, eddies, and waves that effect this transport and affect the nature of the seasonal variation (Sections 13.2 and 13.4).

13.3.2.2 Direct Methods

Doppler Line Shifts

One technique offering a direct determination of the wind speed and direction is to measure the differential Doppler shift of atmospheric spectral features as the field-of-view moves from east limb to west limb. Infrared heterodyne observations of Titan's ethane emission at 12 μm have been performed on three separate occasions (Kostiuk et al. 2001, 2005, 2006). Although the instrument field of view covered a significant portion of Titan's disk, the measurements consistently provided evidence for eastward winds with velocities exceeding 200 m s^{-1} at heights near 200 km (1 mbar), but with a relatively large uncertainty (Fig. 13.4). Luz et al. (2005, 2006) have performed similar observations over a wide range of wavelengths from 420 to 620 nm using a high performance spectrograph at the Very Large Telescope. Eastward wind with velocities of about 45–60 m s^{-1} were inferred at 170–200 km altitude, the primary height range of the emission features used to determine the differential Doppler shift between east and west limb. The technique has also been extended to the millimeter wavelength range, which it is sensitive to higher altitudes, depending on the specific spectral feature measured. Moreno et al. (2005) found the atmosphere to be superrotating at a speed of 160 ± 60 m s^{-1} (centered at 300 km altitude) and 60 ± 20 m s^{-1} (450 km), respectively.

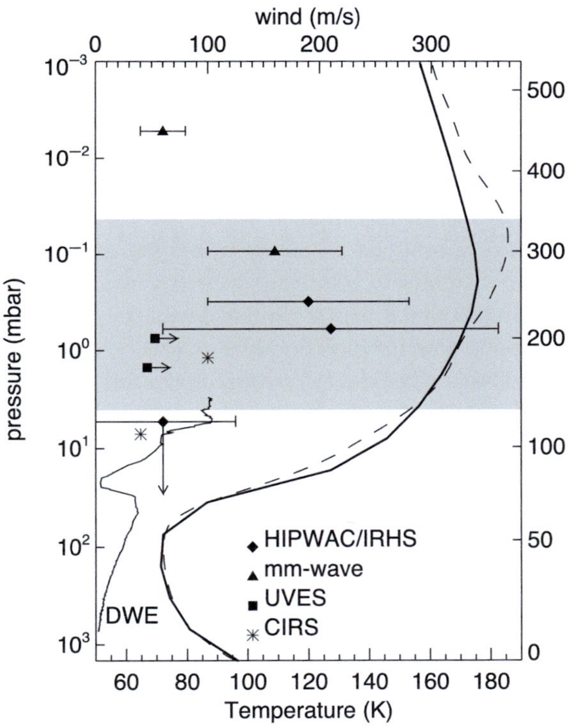

Fig. 13.4 *Upper abscissa*. Titan zonal wind velocity determined from measurements at different heights. The DWE profile from the surface to 145 km (Bird et al. 2005) is shown for comparison. In decreasing order of height the highest two data points (*triangles*) are the mm-observations from Moreno et al. (2005). The *diamonds* show results from the heterodyne IR observations of Kostiuk et al. (2001, 2005, 2006). Luz et al. (2005, 2006) reported results from broadband UVES observations (lower bounds – *squares*). Two representative points (*asterisks*) are included from the CIRS zonal wind retrieval at 15°S (Flasar et al. 2005), adjusted to be consistent with the DWE-determined wind velocity of 55 m/s at 10 mbar. *Lower abscissa*. The temperature profiles are the *solid curve* from Yelle et al. (1997) and the *dashed curve* from Flasar et al. (2005) (reproduced from Kostiuk et al. 2006)

Cloud Tracking

The measurement of winds from tracking of cloud motions has been a valuable technique on all of the giant planets and Venus. Unfortunately, Titan has not been kind to cloud trackers. Its clouds are infrequent and are typically ephemeral, lasting less than a few hours in many cases. Yet this potentially remains one of the few existing means to probe the zonal winds globally in the troposphere.

Prior to the arrival of Cassini-Huygens, disk-average ground-based observations at near infrared wavelengths in 1998 suggested the presence of reflective clouds, thought to be CH_4, covering less than 7% of Titan's disk and located at roughly 15 ± 10 km altitude (Griffith et al. 2000). Observations the following year indicated a high degree of temporal variability, with the inferred clouds dissipating in as little as 2 h. Yet they were observed over several nights. Radiative transfer modeling indicated the clouds were near 27 km altitude and covered only 0.5% of the disk, about 1% of the cover typically seen on Earth. Later observations, using adaptive optic systems that resolved Titan's disk (Brown et al. 2002; Roe et al. 2002) showed transient clouds concentrated near the south pole, which was in early summer. The authors interpreted the location and variability of the clouds as evidence of moist convection involving CH_4 condensation. Later observations in October 2004, shortly after Cassini's orbit insertion, showed an 18-fold increase in cloud brightness (Schaller et al. 2006a) near the south pole, with a factor-of-two change in brightness over 24 h. Earlier observations from 1996 to 1998, using speckle interferometry at the W. M. Keck Telescope (Gibbard et al. 2004), had also detected clouds near the south pole, supporting the notion that this was a long-term seasonal activity driven by the warm surface at the pole in late spring and early summer.

Later observations revealed clouds at temperate latitudes. With adaptive optics techniques at the Gemini North 8-m telescope, Roe et al. (2005) found tropospheric clouds clustered between 37° and 44°S on three of thirteen nights in 2003–2004. Unlike the nearly circular nature of the south polar clouds, these clouds were filamentary, extending eastward more than 1,000 km, about 30° in longitude, while only spanning a few degrees in latitude. These clouds were mostly clustered near 350°W, which suggested a geographical tie, associated perhaps with geysers or cryovolcanoes that could inject bursts of CH_4 vapor into the atmosphere. Subsequent observations by the Imaging Science Subsystem (ISS; Porco et al. 2005; Turtle et al. 2009) and the Visual and Infrared Mapping Spectrometer (VIMS; Griffith et al. 2005; Rodriguez et al. 2009) failed to corroborate the preferential location on the side of Titan facing Saturn; in fact the VIMS and ISS data indicate that the clouds in this latitude belt are nearly uniformly distributed in longitude (Rodriguez et al. 2009; Turtle et al. 2009). Other mechanisms, involving meridional transports of CH_4, have been suggested (Griffith et al. 2005).

Observations of the motions of a few well-behaved long-lived (~1 day) clouds, led to the first cloud-tracked wind velocities from ground-based observations. Tracking a temperate latitude feature on 2–3 October 2004, Roe et al. (2005) deduced a velocity of 8 m s^{-1} eastward and 3 m s^{-1} northward, which seemed consistent with the expected direction and magnitude of tidal winds (Section 13.6.1). Bouchez and Brown (2005) tracked clouds over several nights near the south pole and found no significant motion, placing 3-σ limits of 4 m s^{-1} on zonal (i.e., east–west) motion, and 2 m s^{-1} on meridional (north–south) motion.

The Cassini ISS and VIMS instruments both observed the south polar cloud during the first (distant) Titan flyby (T0) on 2 July 2004. Both sets of observations indicated sluggish eastward motions less than 4 m s^{-1} at high southern latitudes (Porco et al. 2005; Baines et al. 2005; Brown et al. 2006). VIMS observed four cloud features, the most prominent being a 700-km wide feature centered at 88°S (Baines et al. 2005). Three smaller clouds were observed within 16° of latitude of this main feature, ranging in size from 65 to 170 km in diameter. The prominent south polar cloud was observed for 13 hours, enabling a measurement of a zonal wind speed of 0.5 ± 3.3 m s^{-1} (Brown et al. 2006), slightly favoring an eastward motion. Two other cloud features, one at 74°S and the other at 78°S, showed slow zonal wind speeds of 0.9 ± 3.9 and 2.3 ± 2.0 m s^{-1}, respectively, both also consistent with eastward motions. Meridional motions were imperceptible.

The ground-based adaptive optic imagery of the south polar clouds by Schaller et al. (2006a) in October 2004, several weeks prior to the first Cassini cloud observations, also led to zonal velocity estimates. From two pairings of nights, Schaller et al. found zonal wind velocities of 2 ± 3 and 0 ± 4 m s^{-1}, consistent with the VIMS and ISS south polar cloud results. From VLT/NACO and CFHT/PUEO adaptive optics imagery acquired in 2002–2004, Hirtzig et al. (2006) measured an average south polar zonal wind speed of 3 ± 2 m s^{-1}. Subsequent observations between December 2004 and April 2005 showed the nearly complete disappearance of the south polar cloud (Schaller et al. 2006b), although VIMS observations have indicated periodic outbursts occurring almost every nine months from December 2004 through June 2007 (Rodriguez et al. 2009). These outbursts diminished in strength and persistence with time as the Titan equinox approached, perhaps consistent with the march of the summer season and the change in meridional transport by the global circulation.

ISS observations during the first Titan encounter (T0) showed the fastest cloud feature tracked so far: 34 ± 13 m s^{-1} eastward at 38°S. Eastward winds were also measured for 9 of 11 other clouds. During the third Titan flyby (Tb) in December 2004, VIMS images showed a long cloud streak extending over 25° in longitude near 41°S (Baines et al. 2005). Attempts to derive reliable zonal velocities with time lapsed observations were unsuccessful. Spectral analysis of four discrete cloud features embedded within this cloud streak by Griffith et al. (2005) showed rapid changes in cloud top altitude over 35 min, with one cloud top rising 14 km in altitude during that time, from 21 to 35 km, corresponding to updraft velocities near 10 m s^{-1}. They also concluded that high cloud centers dissipate or descend to the ambient cloud level near 10 km within an hour, consistent with the fall velocity of millimeter-sized raindrops expected for Titan (Toon et al. 1988; Lorenz 1993). Thus, temperate latitude cloud streaks can be highly dynamic and changeable, with altitudes that can vary on short timescales. Consequently, they are not typically reliable indicators of wind velocities.

Huygens Doppler Wind Experiment (DWE)

The Huygens probe has provided the most detailed vertical profile of Titan's zonal wind. After its heat shield was jettisoned and the first parachute deployed, the probe descended for nearly 150 min before impacting the surface near 10°S. During this descent phase, the lateral motion of the probe was close to that of the ambient winds. The DWE instrumentation, consisting of an atomic rubidium oscillator in the probe transmitter to assure adequate frequency stability of the radiated signal and a similar device in the orbiter receiver to maintain the high frequency stability, was implemented only on one of the radio links on the Cassini spacecraft, Channel A (2,040 MHz). Whereas the other link, Channel B (2,098 MHz), functioned flawlessly during the entire mission, the Channel A receiver was not properly configured during the probe relay sequence. All data on this link, including the probe telemetry and the planned DWE measurements, were lost. Fortunately, the primary DWE science objective, a vertical profile of winds on Titan, was largely recovered by ground-based tracking of the Channel A signal at large radio telescopes (Bird et al. 2005; Folkner et al. 2006).

During the DWE design phase it was recognized that Earth-based Doppler measurements could be combined with the orbiter Doppler measurements to reconstruct both the zonal and meridional components of the Huygens probe motion during descent (Folkner et al. 2004). Ideally, the horizontal projection of the ray paths from Huygens to Cassini and to the Earth should be perpendicular, but the 160° separation for the actual experiment geometry was still considered adequate for the calculation. A fundamental uncertainty in the Earth-based measurement was whether the received power from the Huygens carrier signal would be sufficient to support near real-time reduction of the data, or if a more extensive data processing effort, augmented with additional information from the telemetry sub-bands, would be required, as it was in the case of the ground-based detection of the Galileo Probe signal at Jupiter (Folkner et al. 1997). Despite the considerably greater distance, the probability of detecting the Huygens signal from Titan was deemed slightly more favorable than for the case of the Galileo Probe, because of the substantially higher Huygens transmitter antenna gain toward Earth. Moreover, significant residual carrier power in the Huygens signal was expected, compared to no residual carrier for the Galileo Probe. Ground-based support was solicited in a multi-facility observation proposal submitted to the National Radio Astronomy Observatory (NRAO), specifically for observation time at the Robert C. Byrd Green Bank Telescope (GBT) in West Virginia, and at eight antennas of the Very Long Baseline Array (VLBA), and to the Australia Telescope National Facility (ATNF) for observations at the Parkes Radio Telescope and several other smaller Australian antennas.

A second Earth-based experiment designed to provide ultra-precise sky positions of the Huygens probe using the Very Long Baseline Interferometry (VLBI) technique (Pogrebenko et al. 2004; Witasse et al. 2006) was conducted in parallel with the ground-based DWE observations. The results of the VLBI experiment, which enlisted a total of 17 radio telescopes in Australia, China, Japan, and the USA, are still under review.

The topocentric sky frequencies of the Huygens Channel A carrier signal recorded at GBT and Parkes are shown in Fig. 13.5. The time resolution is typically one point each 10 seconds, for which the measurement error is of the order of 1 Hz, corresponding to an uncertainty in the line-of-sight velocity of 15 cm s^{-1}. The Earth-based signal detection at GBT coincides with the initial transmission from Huygens at t_0+45^s, where t_0 is the designated start of the descent. Also indicated in Fig. 13.5 are the times of impact and loss of link to Cassini. The Huygens probe was still transmitting from Titan's surface when Titan descended below the minimum elevation limit in the terrestrial sky and the signal could no longer be tracked at the Parkes antenna.

A Doppler signature for the change from the main parachute to a smaller drogue parachute may be seen in Fig. 13.5 at t_0+15^m. At this instant, the suddenly larger descent velocity produces an abrupt decrease in the received frequency which closely follows that predicted by simulation. An additional decrease in the observed frequency comes from the rapidly decreasing zonal wind that continues through the parachute exchange event (Fig. 13.6). Similarly, a nearly discontinuous frequency increase marks the impact on the surface at 11:38:11 SCET/UTC. The magnitude of this positive jump in frequency, 26.4±0.5 Hz, reflects the abrupt change in vertical velocity from about 5 m/s to zero.

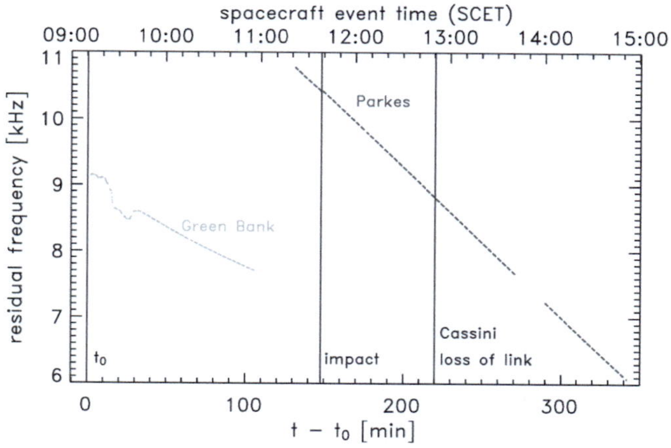

Fig. 13.5 Topocentric sky frequencies of the Huygens Channel A carrier signal recorded at the GBT and Parkes facilities. A constant value of 2,040 MHz, the nominal transmission frequency, has been subtracted from all measurements, so that the plotted data points are equivalent to the total signal Doppler shift from transmitter to receiver. The *upper time scale* is Spacecraft Event Time; the *lower scale* is minutes past the nominal start of mission at t_0=09:10:20.76 SCET/UTC. The Earth's rotation is primarily responsible for the lower recorded frequency at GBT (*near end of track*) with respect to Parkes (*near start of track*). The *small gaps* between data segments and the larger post-landing gap correspond to intervals when the radio telescopes were pointed to celestial reference sources near Titan for calibration of the simultaneously conducted VLBI experiment. A single larger gap of 26 minutes is present in the interval between the end of the observations at GBT and the start of observations at Parkes. The times of impact at $t_0+147^m\,50^s$ (11:38:11 SCET/UTC) and loss of link to Cassini at $t_0+220^m\,3^s$ (12:50:24 SCET/UTC) are indicated. The Parkes tracking pass ended at $t_0+341^m\,37^s$ (14:51:58 SCET/UTC) with the Huygens probe still transmitting from the Titan surface (after Bird et al. 2005)

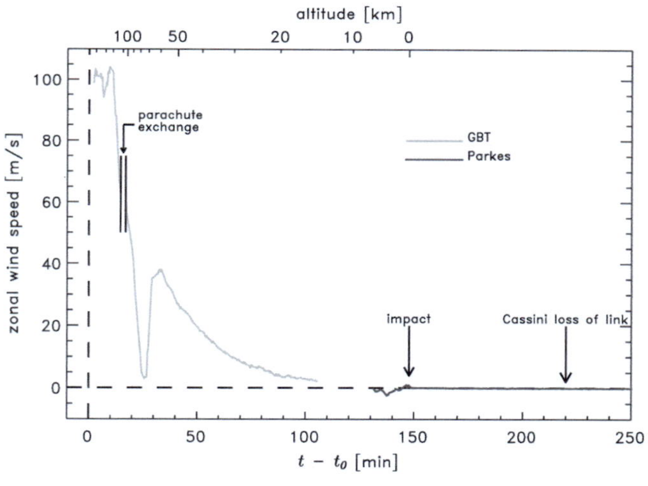

Fig. 13.6 Zonal wind velocity during the Huygens mission. The winds above 12 km are eastward (positive zonal wind), but a significant reduction in the wind speed is observed at altitudes in the interval from 60 to beyond 100 km. The interval associated with the parachute exchange, during which the probe lags the actual wind, starts at an altitude of 111 km and lasts until 104 km. A monotonic decrease in the zonal wind speed was recorded from 60 km down to the end of the GBT track at 10:56 SCET/UTC. The Parkes observations could not begin until 11:22 SCET/UTC, thereby excluding wind determinations in the height region from roughly 12 km down to 4.5 km. By this time Huygens was in a region of weak westward winds that again turned eastward about 1 km above the surface (adaptation and update of Bird et al. 2005)

With precise knowledge of the geometry, the raw measurements of sky frequency yield the line-of-sight motion of the transmitter in the Titan frame of reference. The vertical motion of the Huygens probe, which slowly decreased with decreasing altitude, was measured in situ by the HASI pressure sensors (Fulchignoni et al. 2005) and the Huygens radar altimeters. The available consolidated measurements were processed iteratively by the Huygens Descent Trajectory Working Group (DTWG) to produce a series of continually improving Huygens trajectories referenced to Titan (Atkinson et al. 2007; Kazeminejad et al. 2007). The results presented here are based on the descent velocity of the final DTWG data set #4, released in May 2007. Based on knowledge of the Huygens point of atmospheric entry following separation from Cassini, and integrating the trajectory through the entry phase with the help of accelerometer measurements (Colombatti et al. 2008), the DTWG also supplied the nominal coordinates of the Huygens probe in latitude (10.33 ± 0.17°S), longitude (196.08 ± 0.25°W), and altitude (154.8 ± 8.5 km) at the start of descent t_0.

Two key assumptions about the horizontal motion of the probe were applied to simplify the problem of determining the probe velocity during descent. The first of these is that the horizontal drift of the probe follows the horizontal wind with a negligible response time. The actual response time for the Huygens descent system is estimated to be roughly 30–40 s in the stratosphere, decreasing to 3–5 s in the lowest 10 km (Atkinson et al. 1990; Bird et al. 1997). This may be fulfilled only marginally during the early minutes of the descent and is violated briefly during the deployment of the second parachute (altitudes: 111–104 km). Barring any unanticipated parachute aerodynamics, the probe should be capable of roughly following the wind again at all altitudes below 100 km. A simple modeling analysis of the difference between the zonal wind speed and the zonal probe speed concluded that the probe was lagging the wind by up to a maximum of 7% above 40 km and that Huygens was an unreliable wind gauge only during the 2-min interval following the parachute exchange. The second assumption, that the drift in the meridional (north–south) direction is negligible, is based primarily on theoretical considerations that imply dominance of the zonal (east–west) atmospheric circulation (Flasar 1998b). The Huygens trajectory reconstructed from landmark positions determined from the DISR images below 40 km was found to have a small, but nonzero, meridional drift (Karkoschka et al. 2007). The poor projection of the north–south motion onto the line-of-sight to Earth results in a negligible effect on the observed Doppler shift and thus the zonal wind retrieval.

Small, nearly time-invariant corrections were applied for the effect of special relativity (−7.5 Hz, whereby the minus sign means a red shift), as well as the effects of general relativity associated with the Sun (18.2 Hz), Saturn (−0.7 Hz), Earth (1.4 Hz) and Titan (−0.08 Hz). A detailed discussion of these corrections may be found in Atkinson (1989). Propagation corrections to the Doppler measurements from ray refraction in the neutral and ionized intervening media (Titan, interplanetary, and Earth) have been estimated and found to be negligible. Temperature inversions, which can produce a distinct Doppler signature in the received frequency for a conventional occultation geometry with tangential ray propagation, are not effective for the case of radio propagation from Huygens to Earth because the signal propagates upward out of the atmosphere and thus parallel to any strong radial gradients. A final small correction of +10.0 Hz was applied to the absolute transmission frequency by invoking the constraint that Huygens remains stationary on Titan's surface after landing. This residual is within the error limits of the pre-launch unit-level calibration of +9.2 Hz determined for the specific DWE oscillator unit used to drive the Huygens Channel A transmitter.

The variation of the zonal wind as a function of time, derived from the data in Fig. 13.5 using an algorithm developed for the Huygens-Cassini link (Dutta-Roy and Bird 2004), but adapted for the Huygens-Earth link, is shown in Fig. 13.6. More precisely, the quantity plotted in Fig. 13.6 is the horizontal velocity of Huygens in the east direction with respect to the surface of Titan (positive value indicating the eastward direction). The time-integrated wind measurement from t_0 yields an estimate for the longitude of the Huygens landing site on Titan, 192.33 ± 0.31°W, which corresponds to an eastward drift of 3.75 ± 0.06° (165.8 ± 2.7 km) over the duration of the descent. This propagation error associated with the wind retrieval has been estimated and is found to be insignificant because the Doppler shift depends only weakly on the probe's position during the entire descent.

The variation of the zonal wind with altitude and pressure level is shown in Fig. 13.7 in comparison with the Titan engineering wind model and envelopes based on Voyager temperature data (Flasar et al. 1997). The measured profile is in rough agreement with the upper level wind speeds anticipated by the engineering model and generally eastward above 12 km altitude, within the region tracked from GBT. Assuming this local observation is representative of conditions at this latitude, the eastward wind speed profile provided the first in situ confirmation of the atmospheric superrotation anticipated from the Voyager temperature data. Moreover, the large ratio of the measured winds to Titan's equatorial rotation speed ($\Omega \cdot a \approx 11.7$ m s^{-1}, where $\Omega = 4.56 \times 10^{-6}$ s^{-1} and $a = 2{,}575$ km are Titan's rotation rate and radius, respectively) validates the condition of cyclostrophic pressure balance. Over the lower altitudes (<5 km) tracked by Parkes, Fig. 13.7 indicates that the winds had shifted westward, but again turned eastward over the lowest 1 km.

The most striking departure of the measured profile from the engineering model is the region of weak wind, sandwiched above and below by regions of strong positive and negative wind shear, where the zonal wind reaches a minimum of

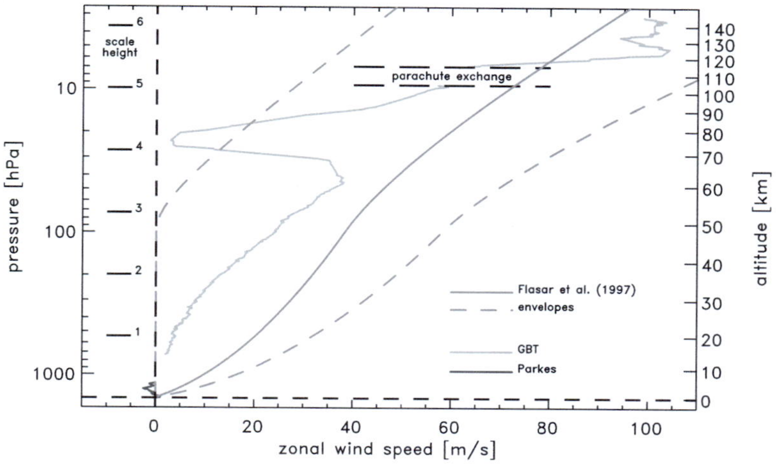

Fig. 13.7 Titan zonal wind height profile. The zonal wind derived from GBT and Parkes observations are compared with the model and envelopes proposed by Flasar et al. (1997). With the possible exception of the region above 100 km, where the wind fluctuations are greatest, the zonal flow was found to be generally weaker than those of the model. The wind shear layer in the height range between 60 and beyond 100 km was unexpected and still lacks a generally accepted explanation (adaptation and update of Bird et al. 2005)

4 m s^{-1} at about 75 km altitude. The broad minimum in zonal wind speed between 60 and 100 km, which can be seen in Fig. 13.5 as a symmetrical dip in frequency over the interval between t_0+20 m and t_0+30 m, is interpreted as a real property of Titan's atmospheric dynamics. This feature of Titan's wind profile is unlike that measured by any of the Doppler-tracked probes in the atmosphere of Venus (Counselman et al. 1980). The slow zonal wind occurs in the stratospheric region of strongest static stability, i.e., the largest vertical increase in temperature, but there is currently no generally accepted explanation of it. The thermal wind equation implies that the vertical shear in the derived zonal wind profile would be associated with strong, oppositely directed, meridional temperature gradients. Such sharp gradients might be connected with an interface region between different meridional circulation cells, but supporting observations are scarce. The slow wind layer occurs above the maximum altitude (30–40 km) for corroboration from surface feature tracking with DISR images (see below). The simultaneous VLBI tracking of Huygens had inadequate temporal resolution to verify the slow wind region (L.I. Gurvits et al., personal communication, 2009). Independent evidence in support of the strong vertical gradient in zonal wind was found in the Channel B signal amplitude measurements on the Cassini orbiter, which revealed a distinct tilt of the Huygens probe during these time intervals (Dzierma et al. 2007). The only Cassini observations that may help resolve the uncertainty in the critical height range from 60 to 100 km would be meridional temperature variations derived from either CIRS far-IR or RSS (Radio Science Subsystem) occultation data.

Descent Imager/Spectral Radiometer (DISR)

The descent trajectory of the Huygens probe has been used to derive the probe ground track and extract the implied wind speed as a function of altitude (Tomasko et al. 2005). The diverse landscape near the landing site facilitated the reconstruction of the trajectory based on a sequence of images by DISR of surface landmarks. The probe moved horizontally about 2° north of east on average from the release of the first parachute to an altitude of about 50 km (Karkoschka et al. 2007). Thereafter, the motion turned slightly to about 5° south of east between 30 and 20 km altitude (Fig. 13.8, top panel). Between 15 and 6.5 km the probe moved to the east-northeast and experienced a sharp turn towards northwest (Fig. 13.8, bottom panel). Finally its motion turned counterclockwise through almost 180° at 0.7 km altitude towards the southeast or south-southeast down to the surface. No images were available during the last 200 m of the descent, so the surface wind was constrained by the direction of the parachute, which was not visible in the surface images acquired after the landing. The wind was most likely blowing towards azimuth 160° (20° east of south) with 0.3±0.1 m s^{-1}. Thus the meridional wind was stronger than the zonal wind near the surface and the meridional wind changed direction at several altitudes. Throughout the descent the meridional wind speed varied between about ±1 m s^{-1}. The zonal winds derived from the DISR images are generally in agreement with the DWE results discussed above. In contrast to DWE, however, DISR was able to distinguish between meridional and zonal motion and could thus follow the wind direction reversals that occurred in the last minutes of the descent.

13.4 Meridional Circulations

Zonal-mean meridional and vertical motions are important for transporting angular momentum, energy, and constituents. Although eddy transports can be important, use of the transformed Eulerian-mean velocities incorporates many of these eddy fluxes into the mean circulation when motions are quasi-adiabatic and steady (Andrews et al. 1987). In Earth's middle atmosphere, the transformed Eulerian-mean circulation is equivalent to the Lagrangian circulation,

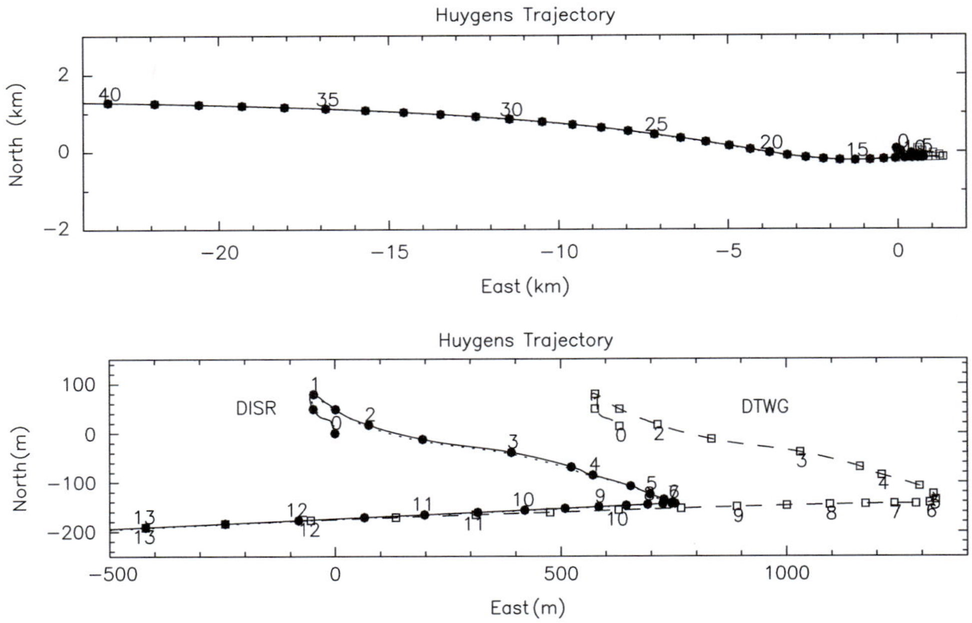

Fig. 13.8 *Top panel.* The Huygens trajectory relative to the landing site, starting at 40 km altitude. The numbers denote altitudes in km. *Solid dots* indicate the position of Huygens every full kilometer of altitude. *Bottom panel.* The same, but starting at 13 km. Trajectories derived from DISR data alone and the DTWG determination, are both shown. The *circles* and *squares* along the curves indicate altitude at 500-m intervals. The DTWG results were based on DISR data for meridional drifts but on DWE data for zonal drifts. Below 12 km, the DTWG trajectory determination is up to ~600 m further east than that from DISR. The discrepancy results from the interpolation used over a gap in the DWE data between 12 and 5 km (after Karkoschka et al. 2007)

which follows the motions of conserved quantities, to a good approximation (Dunkerton 1978). There is generally no simple diagnostic relation like that between zonally averaged temperatures or pressures and mean zonal winds. One must either obtain in-situ measurements with sufficient accuracy and coverage to determine the meridional and vertical motion fields, or else one must use variables that can serve as tracers. These include, for example, gaseous constituents with lifetimes on the order of the turnover time of the meridional circulation, temperature variations along isobars, or potential vorticity (see, e.g., Flasar 1998a for more discussion of this conserved quantity) to infer the meridional circulation.

13.4.1 Temperatures

Flasar et al. (1981) had noted that the IRIS determination of the meridional distribution of surface temperatures, which amounted to ~3 K, was not consistent with just a radiative response to solar heating. Cassini radio occultation soundings, which cover several latitudes from 74°S to 74°N, also show a contrast of only a few kelvins in the lower troposphere. The radiative solution to the annual average solar heating would imply an equator-to-pole contrast of 15 K, far larger than observed. Flasar et al. noted that a simple Hadley circulation, annually averaged, could transport sufficient heat to reduce the meridional gradient with surprisingly small meridional velocities, ~0.04 cm s^{-1}.

As noted earlier, one of the most striking features of the middle atmosphere are the elevated temperatures above the north (winter) pole at the 10-µbar level. While diabatic processes involving the thick condensate hazes over the pole may be at play, adiabatic heating associated with subsidence over the pole is likely to be substantial. This is what happens in Earth's middle atmosphere (see, e.g., Andrews et al. 1987). The winter and summer seasons are marked by a cross-equatorial circulation with ascent at high summer latitudes and subsidence at high winter latitudes, within the polar vortex. The adiabatic heating over the winter pole is critical in heating the mesosphere, and indeed the mesopause over the winter pole is markedly warmer than over the summer pole because of this. Titan GCM studies are also consistent with subsidence over the winter pole (see, e.g., Hourdin et al. 2004). The lowest-order balance is between adiabatic heating and radiative relaxation:

$$w\left(\frac{\partial T}{\partial z}+\frac{g}{C_p}\right)=-\frac{T-T_{eq}}{\tau_r}, \tag{13.10}$$

where w is the transformed Eulerian-mean vertical velocity, T is temperature, C_p is the specific heat, T_{eq} is the radiative equilibrium temperature, and τ_r is the radiative relaxation time; all quantities are zonal averages. T-T_{eq} ~ 25 K at 10 µbar seems reasonable given the contrast between the north pole and low latitudes. At a given temperature τ_r should slowly fall off with altitude, because the decrease in mass per scale height with altitude is offset to some degree by the reduced emissivity of CH_4, C_2H_2, and C_2H_6, the principal infrared gaseous coolants (Flasar et al. 1981). Using $\tau_r \sim 3\times10^7$ s, appropriate to 170 K and 1 mbar, one obtains $w \sim -1$ mm s^{-1} (Achterberg et al. 2008a). At 200 K, τ_r is a bit smaller and $|w|$ corresponding larger.

13.4.2 Gas Composition

So far, the study of the meridional distribution of gases has primarily been of the organic constituents in the stratosphere. Figure 13.9 depicts the meridional variation of several organic molecules, retrieved from Cassini and Voyager thermal-infrared spectra (Teanby et al. 2006; Coustenis et al. 2007; Coustenis and Bézard 1995). All the nitriles (HCN, HC_3N, C_2N_2) and several hydrocarbons (C_3H_4, C_4H_2, C_6H_6, C_2H_4 in the spring) exhibit enhanced concentrations at high northern latitudes. Most of the formation of these species, following the breakup of N_2 and CH_4, occurs higher up in the atmosphere. With most organics condensing in the lower stratosphere, one expects the concentration to increase with altitude, toward the source region (Chapter 10). Limb sounding at mid-infrared wavelengths with Cassini CIRS (Vinatier et al. 2007; Teanby et al. 2007) indicates that this typically is the case. Subsidence at high northern latitudes could naturally lead to an enhanced concentration of these species at the 1–10-mbar level. Polar enhancement of the more abundant organics, e.g., C_2H_6 and C_2H_2, is less, because the vertical gradients of their concentration are smaller. Teanby et al. (2009) have noted that on average the enhancements of the various constituents scale inversely with their photochemical lifetimes. In the presence of polar subsidence, this would naturally result if the rate of increase of concentration with altitude scaled inversely with the photochemical lifetime. This seems plausible, if the mean vertical profiles of individual constituents results from a competition between vertical mixing and photochemical decay: the constituents with the shortest lifetimes will have the steepest vertical gradients in their concentration. The observed enhancements of the organics and their plausible interpretation in terms of subsidence suggests an absence of strong lateral mixing of air masses within the polar vortex with those outside by eddies or waves; this isolation also occurs on Earth at the winter pole. Analysis of limb data by Teanby et al. (2008) indicates a fairly complex structure within the winter vortex (Fig. 13.10). There seem to be areas of depletion of HCN, HC_3N, and C_4H_2 within the polar vortex at 300 km altitude (~0.1 mbar). What is going on has not really been worked out, but Titan's polar vortex structure seems to be as rich in structure as Earth's ozone holes.

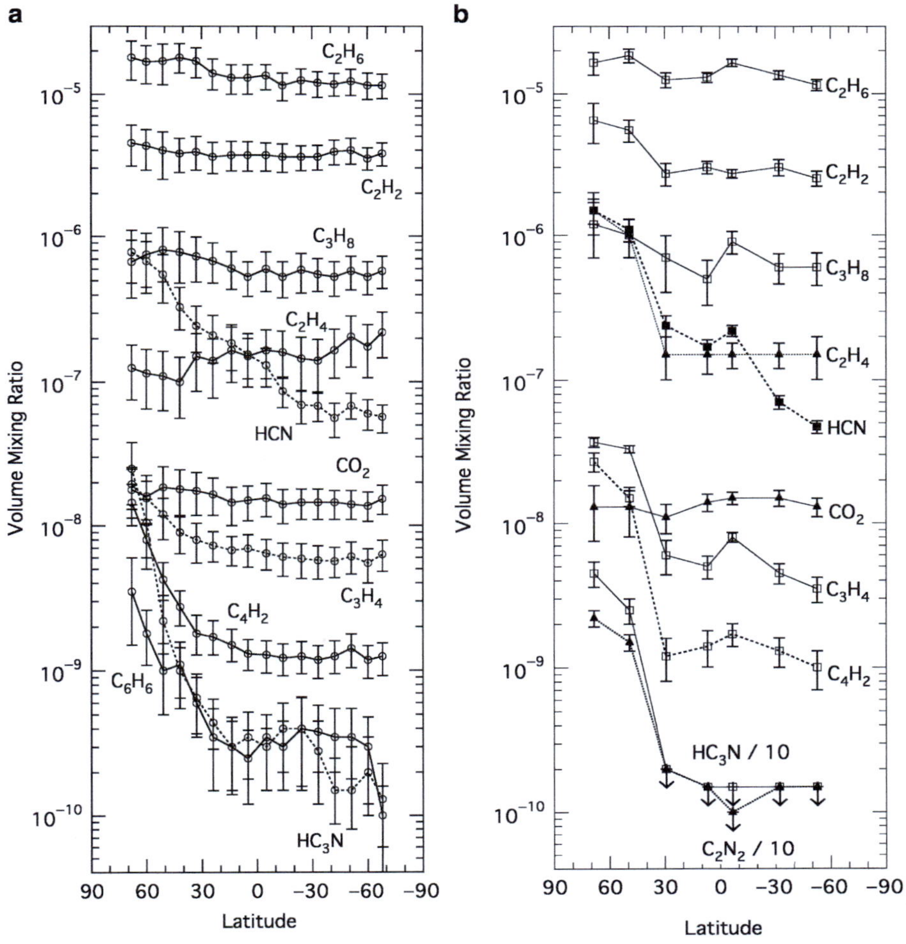

Fig. 13.9 Meridional profiles of organic gases and CO_2 from nadir-viewing observations in the mid infrared in northern spring (Coustenis and Bézard 1995) and in northern winter (Coustenis et al. 2007). The spectral emission features used to retrieve the profiles typically have maxima in their contribution functions at a level of several mbar, with a spread of several scale heights. The values of C_2H_6 shown in the figure should be multiplied by 0.7, owing to a recently improved determination of its spectroscopic constants (Vander Auwera et al. 2007). (**a**) Cassini CIRS (2004–2005): northern winter and (**b**) Voyager IRIS (1980): early northern spring. Note that theabundances of HC_3N and C_2N_2 in (**b**) have been offset by a factor of 10 for clarity (after Flasar and Achterberg 2009)

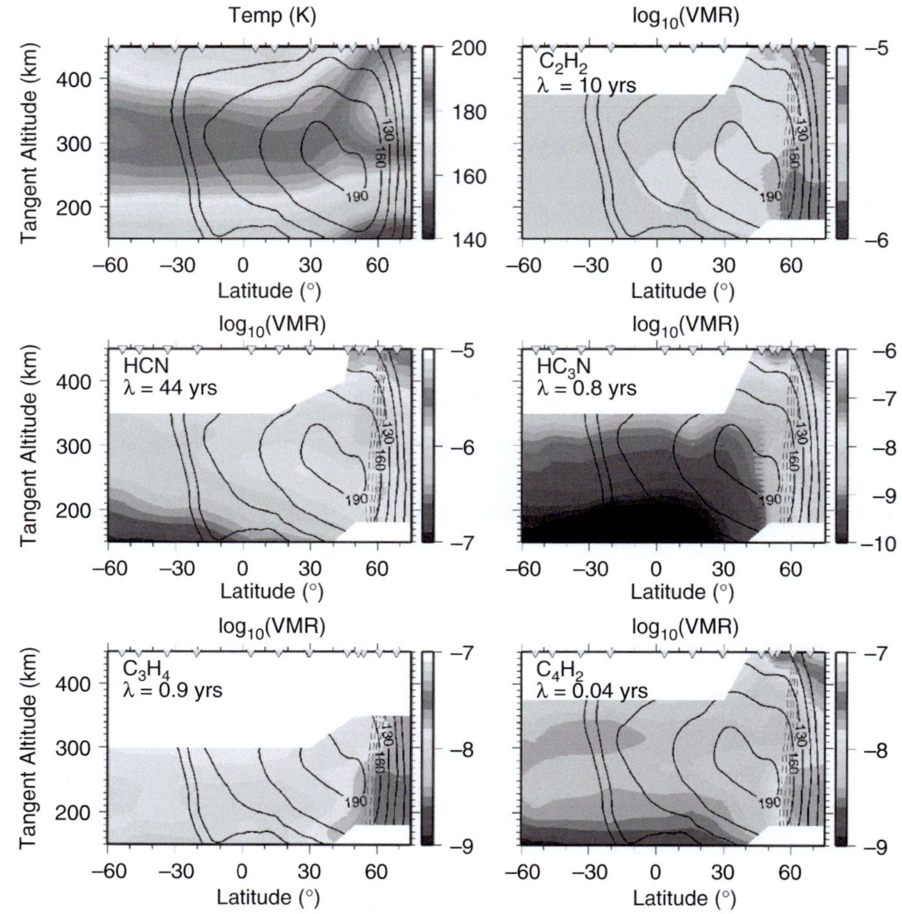

Fig. 13.10 Cross sections of temperature (*upper left panel*) and composition from CIRS mid-infrared limb sounding. Composition is given as a volume mixing ratio and the position of the observed limb profiles are denoted by the *inverted triangles* at the top of each plot. *Contours* indicate the vortex zonal wind speeds (in m s^{-1}) derived by Achterberg et al. (2008a) and *blue dashed lines* show the region with the steepest horizontal potential vorticity gradient, which indicates a dynamical mixing barrier. Altitudes with low signal-to-noise or where the atmosphere becomes opaque are not plotted. Note that southern latitudes are toward the left of each panel. VMR denotes the volume mixing ratio and λ is the photochemical lifetime at 300 km (after Teanby et al. 2008)

13.4.3 Aerosols and Condensates

That Titan has an enhanced haze layer over its winter pole, suggestive of organic ices, has been known since the Voyagers flew past in 1980–1981. The haze signature is evident in images of Titan throughout the ultraviolet, visible and near infrared, because the haze structure and sometimes also particle single scattering albedo differ significantly from those at lower latitudes. Some tentative identifications of organic ices have been made, but others remain unknown. Recent VIMS observations (Griffith et al. 2006) have detected a bright feature extending poleward of ~50° N, which is consistent with a dense C_2H_6 cloud located between 30 and 50 km altitude. Voyager IRIS and Cassini CIRS spectra both exhibit broad spectral features that are suggestive of condensates (Samuelson et al. 1997, 2007; Khanna 2005; Coustenis et al. 1999), but only at high latitudes in the hemisphere that is in winter or early spring. A particularly striking broad feature centered at 221 cm^{-1}, seen both IRIS and CIRS spectra, has yet to be identified. It may be a blend of organic ices. However, it is constricted in latitude, evident poleward of 55°N, but abruptly disappearing at 50°N and equatorward (Flasar and Achterberg 2008). This distribution does not directly provide information on meridional velocities, but it is consistent with a mixing barrier between low- and high-latitude air masses, much as occurs on Earth. It is tempting to wonder whether heterogeneous chemistry occurs on these condensates, much as it does in the terrestrial polar stratospheric clouds, but as yet there is no evidence of this.

In addition to the structure over the winter pole, there is also a more subtle hemispheric albedo asymmetry that is most noticeable in the strong methane absorption bands at 890 nm and at longer wavelengths. The boundary is displaced by about 15° from the equator. Analysis of Hubble Space Telescope (HST) images (Lorenz et al. 1997) led to the idea that hemispheric contrasts are produced by aerosol microphysical variations in the region above 70 km altitude and mostly below 120 km altitude driven by seasonal variations in aerosol transport by winds, in agreement with (but more specific than) the view that Sromovsky et al. (1981) put forth. Karkoschka and Lorenz (1997) derived haze aerosol radii near 0.3 μm in the northern latitudes versus 0.1 μm in the south from 1995 HST images of Titan's shadow on Saturn. They assign these particle radii to different layers, the 'detached haze' layer at northern latitudes and the main haze layer at latitudes south of 50ºS. Penteado et al. (2009) found that hemispheric variations in surface albedo and in haze optical depth and single scattering albedo at altitudes higher than 80 km account for hemispheric differences observed in Cassini VIMS spectra.

Roman et al. (2009) analyzed albedo patterns in 20 images obtained by the Cassini ISS instrument with the 890-nm methane filter (sensitive to haze above about 80 km altitude). These images reveal a haze hemispheric asymmetry with offset 3.8° ± 0.9° in latitude directed 79° ± 24° to the west of the sub-solar longitude. This result is very close to that derived by Achterberg et al. (2008b) from thermal data (Section 13.3.1) and indicates a zonal wind pattern that transports haze in accord with expectations from the thermal wind equation.

Hemispheric asymmetry in haze properties must be related to the meridional circulation but the details have not been worked out. The contrast boundary near the equator would suggest that two cells are operative in addition to a winter polar vortex. A fully coupled chemical, dynamical and haze microphysical model is needed to relate the haze observations to other processes. Some progress has been made in this area, using two-dimensional GCMs with parameterized eddy fluxes (see, e.g., Hourdin et al. 2004; Crespin et al. 2008).

13.4.4 In-situ Measurements

Section 13.3.2.2 has already discussed the use of DISR images to infer meridional motion during the Huygens probe descent (Fig. 13.8). Below 40 km, the meridional wind speed varied between about ±1 m s^{-1}. Near the surface the meridional wind was stronger than the zonal wind and both changed direction at several altitudes.

While the horizontal components of the atmospheric flow could be determined by the Doppler Wind Experiment and probe ground tracking, the smaller vertical component could not be separated from the trajectory by such means. However, the vertical wind along the descent trajectory of Huygens was determined from HASI accelerometer measurements combined with accurate knowledge of the atmospheric pressure and temperature profile during the Huygens descent. Mäkinen et al. (2006) solved the equation of motion of the descending probe using these data to determine the vertical wind speed. The measured vertical profile of vertical wind can be broadly subdivided into the stratosphere and troposphere (Fig. 13.11). In the entire troposphere the prevailing motion is upward, with a mean vertical speed of 5 cm s^{-1} and several wiggles are superposed. In contrast to the troposphere, the vertical flow in the stratosphere was found to be mostly downward, with maxima at 70 and 110 km altitude (up to 0.6 m s^{-1}). The vertical flow around 90 km was upward (15 cm s^{-1}). When compared with the horizontal wind profile (Bird et al. 2005; Folkner et al. 2006), the averaged zonal and vertical profiles loosely correlate with each other, the

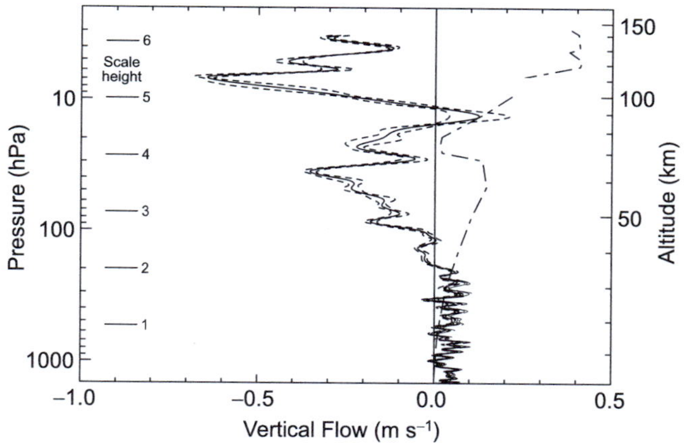

Fig. 13.11 *Solid curve*: vertical atmospheric velocity along the descent trajectory of the Huygens probe. *Dashed curves*: the error envelope. *Dot-dashed curve*: Zonal wind derived from the Huygens DWE, scaled down by a factor of 1/250 (after Mäkinen et al. 2006)

strength of the vertical wind being about 1% of that in the zonal direction.

The large vertical velocities obtained from the probe data cannot be associated with the zonal mean flow. To see this, consider the situation near 110 km (7 mbar), where Fig. 13.11 indicates a subsidence w ~ −0.5 m s^{-1}, and the implications from the heat equation balance (13.10). At this altitude $\left(\frac{\partial T}{\partial z}+\frac{g}{C_p}\right) \approx 1.6$ K km^{-1}. Flasar et al. (1981) estimated the radiative relaxation times (at 10 mbar) ~3×10^7 s–1×10^8 s, where the larger value corresponds to just gaseous cooling by C_2H_2, C_2H_6, and CH_4 and the smaller value includes cooling by aerosols in a simple model. More recent estimates of radiative cooling by the Huygens DISR experiment (see Chapter 10, Fig. 10.5) yield a larger value: $\tau_r \sim 3\times10^8$ s at 110 km. One can solve (13.10) for the departure of the atmospheric temperature from the radiative equilibrium solution, which scales linearly with τ_r. Using the smallest value, 3×10^7 s, leads to the estimate $T-T_{eq} \sim 24{,}000$ K, which is so large it can be ruled out by existing data.

Could the retrieved vertical velocities be indicative of wave motions? For the simplest case of waves propagating vertically and zonally in a mean zonal wind $\bar{u}$, the heat balance in adiabatic flow becomes:

$$w\left(\frac{\partial T}{\partial z}+\frac{g}{C_p}\right) \approx -\frac{\partial T}{\partial t}-\bar{u}\frac{\partial T}{\partial x} = i(\omega-\bar{u}k)\Delta T = \hat{\omega}\Delta T \qquad (13.11)$$

where x is the zonal (east–west) coordinate, k is the zonal wavenumber, and $T = \Delta T \exp i(kx-\omega t)$ is assumed. The quantity $\hat{\omega}$ is the (doppler-shifted) frequency of the wave relative to the background flow. For gravitational tides, which have been extensively studied (Section 13.6.1), $\hat{\omega} \sim \Omega \approx 4.6\times10^{-6}$ s^{-1}, the diurnal frequency (Tokano and Neubauer 2002;

Strobel 2006; Walterscheid and Schubert 2006), which implies a diurnal amplitude $\Delta T \sim 175$ K; this is still much too large compared to observations. Internal gravity waves may be a possibility, because they have higher frequencies. For these, $\hat{\omega}$ is bounded (see, e.g., Andrews et al. 1987) by the Brunt-Väisälä frequency,

$$\sqrt{\frac{g}{T}\left(\frac{\partial T}{\partial z}+\frac{g}{C_p}\right)} \approx 0.004\,\mathrm{s}^{-1},$$

and the Coriolis frequency, $2\Omega \sin \Lambda \sim 10^{-5}$ s^{-1}. From (13.11) $\Delta T \sim 1/\hat{\omega}$, implying smaller temperature amplitudes, so this suggestion merits further study.

13.5 Surface-Atmosphere Coupling

13.5.1 Structure of PBL

The planetary boundary layer (PBL) is the lowermost portion of the atmosphere that is affected by surface friction. The PBL lies below the "free atmosphere," discussed earlier (Section 13.3.2.1), where the gradient-wind balance typically holds for zonally averaged variables, at least away from the equator. The structure of the PBL controls the surface-atmosphere exchange of energy, momentum and matter, thus affecting meteorology and exogeneous geology. The first information on Titan's PBL came from the radio occultation experiment of Voyager 1, with a published altitude resolution of 0.5 km near the surface (Lindal et al. 1983). The retrieved temperature profiles near the equator, for an assumed N_2 atmosphere, showed a lapse rate of 1.38 K km^{-1} below about 3.5 km, close to the dry adiabatic lapse rate, and an abrupt drop to 0.9 K km^{-1} above this level. The lapse rate below 1 km in the evening profile (ingress) was slightly larger than the morning profile (egress). Hence the surface may have had a slight cooling effect on the atmosphere during the nighttime.

The first in situ investigation of the PBL structure was carried out by the Huygens probe, which landed near the equator in the morning hours (Tokano et al. 2006). Simultaneous vertical sounding of temperature and pressure with an altitude resolution of about 10 m near the surface was used to calculate the vertical profile of potential temperature, which is a measure of the static stability (the lapse rate in temperature less the dry adiabatic lapse rate) and a major classification criterion of a PBL. The potential temperature slightly decreased with altitude in the lowest 10 m, virtually stayed constant between 10 and 300 m and increased almost monotonically with altitude above 300 m. In other words, the lapse rate was superadiabatic immediately above the surface, adiabatic in the main part of the PBL and subadiabatic outside the PBL. Thus the 300-m PBL determined by Huygens was shallower than the 3.5-km thick layer indicated by the Voyager radio occultation data.

The observed vertical profile is indicative of a convective PBL in that the static stability is slightly unstable throughout the PBL. Based on the PBL depth the mean eddy diffusivity in the PBL is estimated as 7.4×10^{-3} m^2 s^{-1}, which is several orders of magnitude smaller than that in a terrestrial convective PBL. Furthermore the slope of the near-surface potential temperature gradient points to an instantaneous upward sensible (convective) heat flux of 0.02 W m^{-2} at the time and place of the probe landing. This indicates that the ground temperature was slightly higher than the air temperature immediately above the surface.

The small eddy diffusivity suggests very weak turbulence in Titan's PBL. This is also confirmed by in situ tilt measurements by the Huygens Surface Science Package (Lorenz et al. 2007), which experienced a calm descent in the lowest few kilometers. The presence of a weakly convective PBL without significant convective or stable stratification on Titan can be understood as an immediate consequence of the weak sunlight at Titan's surface (Tokano et al. 2006). The resulting small diurnal variation in the ground surface temperature can create only a minor temperature difference between the surface and atmosphere at any time, so the surface heat flux responsible for the heating or cooling of the atmosphere from below is always quite limited.

The wind speed in the PBL can be estimated as less than 0.3 m s^{-1} (Karkoschka et al. 2007) and most likely much less than this (Folkner et al. 2006; Lorenz 2006). The vertical variation of the wind direction above the surface (Section 13.3.2.2, Fig. 13.8) does not exhibit the terrestrial Ekman–spiral behavior that typifies the frictionally controlled transition between the free-atmosphere aloft in gradient-wind balance, and the surface, where the winds vanish (see, e.g., Holton 1979). GCM simulations (Tokano 2007) indicate that the sharp reversals of the wind direction observed at two altitudes between 7 km and the surface (Karkoschka et al. 2007) mark the boundary of the northward and southward branch of the Hadley circulation and the altitude of the reversal of the meridional temperature gradient (Fig 13.12). The GCM studies indicate that the force balance in Titan's PBL is one in which the instantaneous pressure gradient force is balanced by Saturn's gravitational tidal force rather than by eddy diffusion, as on Earth. Hence the forces governing the PBL on Titan appear to be substantially different from those controlling a typical terrestrial PBL.

The PBL structure at other latitudes cannot be obtained from remote sensing data with this precision, but a larger seasonal variation can be expected than at the equator.

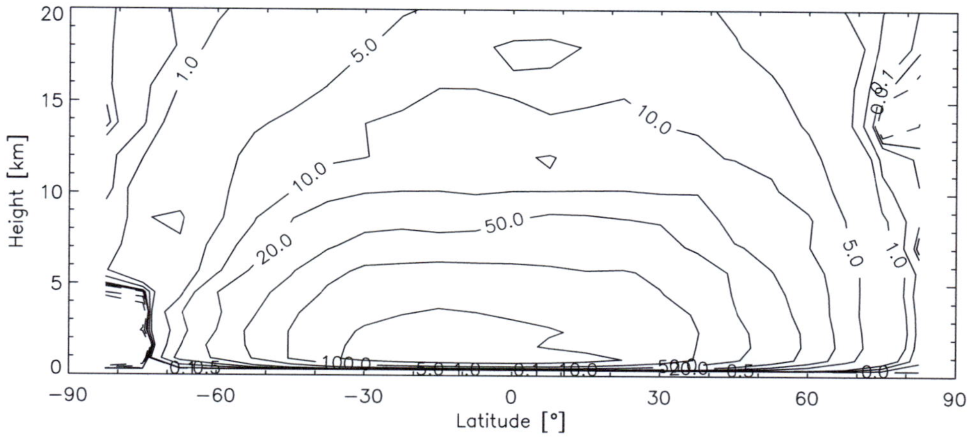

Fig. 13.12 Mass stream function (in 10^8 kg s^{-1}) of the Hadley circulation in the Huygens season from a GCM simulation (Tokano 2007). The flow is clockwise/anti-clockwise along solid/dashed streamlines. Note that southern latitudes are toward the left

13.5.2 Energy Exchange

The energy exchange at the surface importantly affects the general circulation of the atmosphere, because the diabatic heating that drives the atmospheric circulation experiences a sharp transition at the atmosphere-surface boundary. Prior to the Huygens mission the surface energy balance could only be estimated based on theoretical models. McKay et al. (1991) evaluated the various energy terms at the surface using a radiative-convective model that fits the observed vertical temperature profile. Approximately 10% of the sunlight arriving at the top of Titan's atmosphere reaches the surface, of which about a fifth is reflected back to space. To maintain thermal equilibrium at the surface, warming from absorbed solar radiation must be balanced by cooling from a net loss of thermal radiation and from convection. Most of the thermal infrared radiation emitted from the surface is radiated back to the surface by the greenhouse effect (collision-induced absorption) of the major tropospheric gases (nitrogen, methane and hydrogen), raising the surface temperature by 12 K over the effective surface temperature.

Haze and methane absorption coefficients and radiative fluxes measured during the Huygens probe descent through the atmosphere were used to calculate the net radiative fluxes (Tomasko et al. 2008). The net solar flux reaching Titan's surface nicely matches the estimate based on radiative-convective models (McKay et al. 1991), and there was a slight excess in the net heating as can be expected for low latitudes.

In addition to radiation, a small convective heat flux at the surface, corresponding to 1% of the incident solar flux at the top of the atmosphere, is required to explain the observed temperature profile (McKay et al. 1991; Lorenz and McKay 2003). This convective heat flux agrees with the inferred sensible heat flux of 0.02 W m^{-2} at the Huygens site (Tokano et al. 2006) within the estimated error. While this small heat flux usually causes merely a shallow weakly unstable PBL, it may occasionally manifest itself as convective plumes. Lorenz et al. (2005) considered the heat engine theory for convection, which assumes that steady-state convection is just strong enough that the work available from the heat engine balances the viscous dissipation opposing the convective motions. They concluded that convective events on Titan may be rare and cover only a very small area but they may be quite violent.

Given that most of Titan's surface is not covered by an ocean, particularly not at low latitudes, latent heat exchange at the surface may not be relevant for the energy balance in most regions. Numerous extended seas and lakes, however, are present near the north pole (Stofan et al. 2007) and possible seasonal evaporation of lakes (Mitri et al. 2007; Lunine and Atreya 2008) might affect the surface energy balance. This may also be the case if there is additional liquid methane in the surface material in contact with the atmosphere (Mitchell 2008).

In an early study of moist convection using a simple buoyancy parcel model with entrainment, Awal and Lunine (1994) concluded that even if moist convection contributed to the globally averaged surface energy balance the fraction of the surface area covered by updrafts would be 10^{-5}–10^{-7}, much smaller than the fraction 10^{-3} that characterizes the terrestrial intertropical convergence zone. Numerical models of moist convection (Hueso and Sánchez-Lavega 2006; Barth and Rafkin 2007) predict that convective clouds similar in shape to those observed near the south pole can develop if a warm bubble triggering moist convection is imposed near the surface and the methane humidity is higher than at the Huygens entry site. These clouds are characterized by a narrow updraft region in the core of the cloud with vertical wind speeds of typically 10–20 m s^{-1} surrounded by large downdraft regions with weaker vertical wind. The Cassini VIMS observed the evolution of clouds residing between 41°S and 61°S (Griffith et al. 2005). The sequence of cloud images indicated vigorous updrafts of up to 10 m s^{-1} in these clouds. Models also predict substantial horizontal winds converging to or diverging from the clouds. However, since there are no reliable data of methane abundance in the polar region and several other hypotheses for the formation of south polar clouds have been proposed, it is still uncertain how moist convection on Titan is triggered. Particularly, the overall small horizontal and temporal variation in the near-surface temperature might not be conducive to the development of required warm bubbles.

13.5.3 Momentum Exchange

An important side effect of atmospheric circulation is the torque exerted by the wind on the surface, by which angular momentum is exchanged between the atmosphere and surface. Since the total angular momentum in the solid body-atmosphere system is conserved except for negligible change from external torques, angular momentum exchange at the surface gives rise to a temporal change in the planetary spin rate. On Earth the alternating sequence of net acceleration and deceleration of the spin rate by varying surface winds causes an annual variation of the length-of-day (LOD) of roughly 1 ms.

Tokano and Neubauer (2005) pointed out that similar effects should be expected on Titan, because it is covered by a dense atmosphere and the tropospheric winds they predict are variable on seasonal timescales. They calculated the seasonal change in the atmospheric angular momentum using the global zonal wind field predicted by a general circulation model (GCM) as is also common practice in Earth's geodesy. The predicted seasonally exchanged angular momentum amounts to $\sim 3\times10^{25}$ kg m^2 s^{-1}, roughly comparable to that on Earth, and undergoes a semiannual cycle. This temporal change arises because at any instance the net angular momentum flux from the surface to the atmosphere in some regions and vice versa in other regions cannot be immediately balanced. The seasonal reversal of the zonal wind in the lower troposphere in response to the reversal of the meridional temperature gradient is the major driver for the angular momentum exchange.

The momentum exchange at the surface itself depends on the PBL structure. Angular momentum has to pass the planetary boundary layer, in which the vertical transport of momentum is characterized by eddy viscosity. The characteristic diffusive time scale for momentum to cross the PBL can be given as

$$t_{PBL} = \frac{H_{PBL}^{\;2}}{K_{PBL}} \qquad (13.12)$$

where H_{PBL} is the depth of the PBL and K_{PBL} is the eddy viscosity in the PBL. With H_{PBL} 300 m and $K_{PBL} = 7.4\times10^{-3}$ m^2 s^{-1} inferred at the landing site of the Huygens probe (Tokano et al. 2006), the diffusive time scale amounts to about 5×10^6 s (4 Titan days), much shorter than a Titan season. Although the eddy viscosity at other latitudes is unknown, this estimate indicates that the diffusive time scale in the PBL is short enough to allow seasonal exchange of angular momentum as predicted by Tokano and Neubauer (2005).

The major difference from Earth with respect to the seasonal angular momentum exchange is the large rotational response of Titan's solid body. The small radius and slow spin rate reduces the moment of inertia of Titan by several orders of magnitude and a seasonal LOD variation of up to 400 s is predicted, i.e., it would be 5 orders of magnitude larger than on Earth, but it also crucially depends on the yet unknown interior structure.

Repeated observations of surface landmarks by the Cassini RADAR indicate that Titan indeed rotated 0.36° per year faster than synchronous during 2004–2007 (Lorenz et al. 2008). Although the surface moment of inertia is yet unknown, the observations constrain the amount of angular momentum exchanged during this period to 2×10^{24}–7×10^{25} kg m^2 s^{-1}, thus consistent with the prediction by Tokano and Neubauer (2005). This large amount is only possible with a substantial seasonal variation in the zonal wind field, particularly of wind direction, in the lower troposphere. Conversely, the seasonal change in LOD has no influence on the atmospheric circulation.

The observations by the Cassini RADAR, however, indicate that the spin rate seems to be accelerating (Lorenz et al. 2008), while the model of Tokano and Neubauer (2005) predicted a decelerating spin rate during this period. Although an explanation for this discrepancy has not yet been given, such information may be useful in constraining the momentum exchange and may hold clues to general circulation of the troposphere that is otherwise difficult to constrain on Titan.

The global surface wind pattern relevant for the surface momentum exchange can be constrained by aeolian features, which represent a geological fingerprint of the surface winds. Dunes cover a large fraction of low latitudes of Titan and are mainly oriented in the east–west direction (Lorenz et al. 2006; Radebaugh et al. 2008). Morphologically, the dunes resemble longitudinal dunes in terrestrial deserts. Their observed properties on Titan then imply that the low-latitude surface winds are generally westerly (i.e., eastward). The Huygens DISR images and DWE tracking also indicated westerlies just above the surface (Section 13.3.2.2) at low latitudes (10°S). This is puzzling, since the equatorial surface is the least likely place for westerlies to form and not a single Titan GCM predicts persistent equatorial surface westerlies. For instance, the lower return leg of the Hadley cell on Earth tends to produce westward (i.e., subrotating) winds at low latitudes – the trade winds – because of the combination of surface friction and angular momentum conservation (Section 13.7). The mechanism for the maintenance of eastward low-latitude winds on Titan is not yet understood, but it would have a significant effect on the global angular momentum cycle as well as momentum exchange at the surface. The surface wind cannot be everywhere and always westerly since the surface would continuously decrease the atmospheric angular momentum and accelerate Titan's rotation, so the question becomes, where are the trade winds?

13.6 Waves and Their Effect on the General Circulation

Waves are of dynamical interest because of their ability to transport angular momentum and energy over large distances and because they can serve as a probe of the background flow, i.e., structure of winds and stability. Hinson and Tyler (1983) analyzed intensity scintillations in the Voyager radio occultations and in the stratosphere found them to be consistent with small-scale internal gravity waves. Analysis of waves in the Cassini radio occultations is as yet incomplete. The possibility that internal gravity waves might explain the vertical velocities derived from the Huygens descent data has been noted (Section 13.4.4). No evidence has yet been seen of thermal waves in the mapping (primarily of the middle atmosphere) by Cassini CIRS. For instance, the discovery of the pole tilt in the stratosphere at 1 mbar (Section 13.2.1) resulted from a search for zonal structure in the temperature field (Achterberg et al. 2008b). The dominant mode was the wavenumber-1 feature that corresponded to the pole tilt. When this is removed from the temperature field, the residual variance is comparable to the instrument noise. Analysis of the residuals is not complete, however, and further work is needed.

The remainder of this section discusses gravitational tides on Titan, which have been extensively studied.

13.6.1 Gravitational Tides

Atmospheric tides can be subdivided into thermal tides caused by the apparent motion of the Sun radiatively forcing the atmosphere and gravitational tides raised by an adjacent moon or planet. In the atmospheres of Earth, Mars and Venus only thermal tides are of importance. For instance, the lunar atmospheric tide on Earth causes a surface pressure variation of less than 0.1 mbar, much smaller than the pressure variation caused by weather systems (Chapman and Lindzen 1970). The gravitational tide generally becomes more relevant on outer planets because they are massive and receive little sunlight.

The gravitational tide on Titan is caused by the difference between the centrifugal force of Titan and Saturn's gravitational force, which varies spatially across the body of Titan. The tidal flow is a result of the time-dependent part of the tidal acceleration associated with Titan's orbital eccentricity (0.0292). The combination of radial and librational tidal components gives rise to a semidiurnal tide that circles Titan eastward with half of Titan's angular velocity, while the tidal potential at a given location oscillates with a period of 1 Titan day (Fig. 13.13). This tide acts equally on the entire atmosphere as well as Titan's interior. There is also a stationary and a westward propagating tidal component with wavenumber 2, but the latter one is seven times weaker than the eastward propagating or stationary component (Walterscheid and Schubert 2006). In principle, atmospheric tides can perturb Titan's external gravitational potential, both from the displacement of the atmosphere itself and also from the redistribution of mass within Titan induced by the variable atmospheric loading. In practice both effects are likely negligible (Karatekin and Van Hoolst 2006).

The influence of Saturn's gravitational tide on Titan's atmospheric dynamics was investigated by Tokano and Neubauer (2002) in the framework of a general circulation model. The most obvious effect is the surface pressure variation associated with displacements of large atmospheric masses. At a given location the surface pressure varies with a period of a Titan day by up to 1.5 mbar, thus approximately 0.1%, but there is also an eastward migration of the whole surface pressure pattern with a constant phase speed as with the tidal acceleration. This tidally caused pressure variation is superposed on the meridional surface pressure gradient caused by differential heating (Hadley circulation), which is comparable in magnitude but has almost no diurnal variation.

Another important effect is the tidal wind, which significantly distorts the general circulation. In the lower troposphere where the zonal wind is weak the tidal force is one major force that balances the pressure gradient force (Tokano 2007). Near the surface the tidal wind is characterized by rotating winds (whirls), at least at high latitudes. Near the equator the tidal wind is not very apparent since the cross-equatorial flow is relatively strong (Tokano 2008).

In most parts of the troposphere the typical speed of the tidal wind is 2 m s^{-1}, and is superposed on the mainly zonally oriented background wind. It causes a planetary wave of wavenumber 2 that travels eastward with the phase speed of the tide. Since these waves are linear and stable they do not amplify or decay.

Verification of the gravitational atmospheric tide on Titan is a difficult task and there is as yet no firm proof thereof. For instance, as noted above, there is no unambiguous

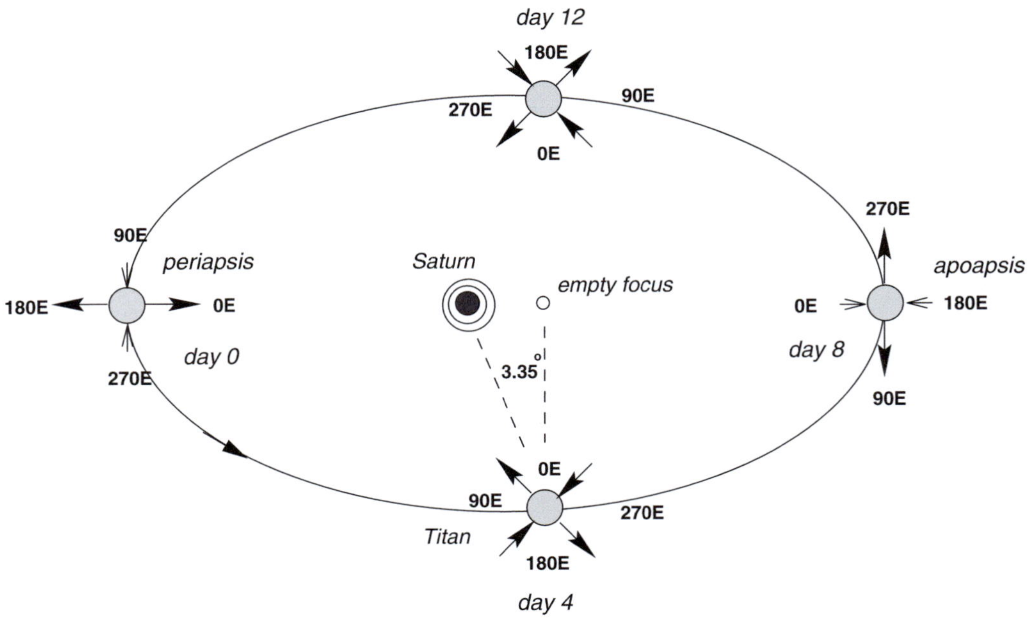

Fig. 13.13 Titan's orbital motion around Saturn and the gravitational tide. The time-dependent component of the tidal-raising force is illustrated by *arrows* for different orbital phases. The combination of radial and librational tide causes the direction of tidal forcing to move eastward on Titan. At periapsis the radial tide peaks at the sub-Saturnian and anti-Saturnian points and the librational tide disappears. The ellipticity of Titan's orbit is exaggerated in the figure. The time-independent part of the tide is not shown because it does not contribute to tidal wind (after Tokano and Neubauer 2002)

evidence in the retrieved temperatures. It is uncertain whether the meridional drift of the Huygens probe can be ascribed to tides (Tokano 2007). Point measurements of pressure and wind as performed by Huygens are generally not sufficient since spatial and temporal variations rather than particular wind speeds or directions characterize the atmospheric tides.

In addition to in situ forcing, gravitational tides also propagate vertically and affect the dynamics of the upper atmosphere (Strobel 2006). Because of the $\rho^{-1/2}$ growth in amplitude of the tidal fields with altitude, the tidal waves attain non-linear saturation amplitudes, because the local lapse rate in temperature becomes superadiabatic. The wind and temperature above 300 km undergo large oscillations with a vertical wavelength of typically 100–150 km above 500 km altitude. The amplitude of the tide is also strongly dependent on the background zonal wind speed. The oscillating vertical temperature profile in the upper atmosphere measured by Huygens (Fulchignoni et al. 2005) resembles the predicted temperature profile in the presence of tides (Strobel 2006). These tides carry significant amounts of energy and momentum vertically and are expected to provide additional heating in the range of 500–900 km by deposition of tidal energy.

Both Strobel (2006) and Walterscheid and Schubert (2006) have suggested that gravitational tides can account for the detached haze layers that have been observed above 500 km in Voyager and Cassini images (Smith et al. 1981; Porco et al. 2005), but they emphasized different mechanisms. Strobel suggested that the temperature oscillations ~ several kelvin associated with vertically propagating tides could lead to hydrocarbon condensation at discrete levels. Although there is no evidence of these amplitudes in the CIRS nadir mapping at 200 km, temperatures retrieved from stellar occultation at altitudes 300–600 km (Sicardy et al. 1999) did show fluctuations ~4–20 K in this range. Walterscheid and Schubert (2006) emphasized that vertical transport of preexisting aerosol particles by the tide could result in discrete layers. According to their calculation the more weakly forced westward propagating tide played the major role.

13.7 Titan's General Circulation

13.7.1 *Thermally Direct and Indirect Circulations*

One major component of general circulation in the lower planetary atmosphere is the thermally direct circulation (warm air rises, cold air sinks) in the meridional-vertical plane, the Hadley circulation, which results from a conversion of the available potential energy supplied by the solar heating into kinetic energy. Analytical models of the general circulation (e.g. Held and Hou 1980; Schneider 2006) show that on slowly rotating planets the Hadley circulation extends to higher latitudes than on Earth, since the smaller Coriolis parameter makes it more difficult for meridional temperature gradients to exceed the threshold for baroclinic instability. On Earth, this instability gives rise to the pattern of cyclones and anticyclones, which provide much of the eddy heat and momentum transports at mid latitudes. Although the temperature profile and atmospheric composition of Titan were unknown prior to the Voyager mission, Leovy and Pollack (1973) estimated that Titan's Hadley circulation would readily extend from the equator to the poles and the equator-to-pole temperature contrast would be tiny. This implies that a thermally indirect circulation, like the eddy-driven Ferrel circulation at mid latitudes of the Earth's troposphere, which derives its motion by converting the kinetic energy of the motion into potential energy, is absent on Titan. In fact, the role of heat transport by baroclinic eddies seems to be minimal on Titan, because the favored scale for energy conversion by the eddies, the Rossby radius of deformation, is much larger than Titan's radius (Leovy and Pollack 1973; Flasar et al. 1981; Hunten et al. 1984; Flasar 1998b).

There is little doubt that a Hadley circulation exists and transports heat meridionally in the lower atmosphere. The observed equator-to-pole thermal contrasts are much smaller than those expected if only radiative processes operated (Section 13.3.1). The need for meridional heat transport is also indicated by the imbalance seen in the radiative flux measurements by the Huygens DISR. These indicate that the solar heating rate at the landing site (10°S) is significantly larger than the thermal cooling rate in the entire lower atmosphere (Tomasko et al. 2008). Because of the expected seasonal variations in surface temperatures and the surface coupling via convection to the overlying atmosphere (Section 13.2), the Hadley cells cannot be expected to be symmetric about the equator except at limited times. When the location of maximum solar heating is off the equator the latitude of the boundary between the summer and winter Hadley cell (i.e., the intertropical convergence zone, ITCZ) has to be poleward of the latitude of maximum solar heating in order to ensure continuity of temperature and conservation of energy across this boundary (Lindzen and Hou 1988). This was indeed numerically shown by parametric GCM simulations of Williams (1988), who concluded that on Titan, which is characterized by a slow rotation and medium obliquity, the circulation should vary from the symmetric Hadley state at equinox to the solstitial-symmetric Hadley state with a single pole-to-pole cell at solstice.

Numerical modeling has provided a more complete picture of meridional circulation and its seasonal modulation, albeit with uncertainties because of the limited data available. The first Titan GCM with realistic radiative forcing and planetary parameters of Titan was developed by Hourdin et al. (1995). This 3-dimensional model is based on so-called primitive equations, which are Navier-Stokes equations of hydrodynamics in which the vertical momentum equation is replaced by the hydrostatic approximation and a few small terms in the horizontal momentum equations are neglected to ensure angular momentum conservation. The radiative forcing was calculated with the radiation model of McKay et al. (1989), which as a radiative-convective model reproduced the vertical temperature profile retrieved from the Voyager data. The predicted mean meridional circulation cell covered almost the entire globe, i.e., there is a single pole-to-pole cell with ascent near the spring and summer pole and subsidence near the autumn and winter pole. When the season approaches the equinox, a reversal of the Hadley cell takes place during which temporarily two equator-to-pole cells exist. Other GCMs that use the same radiation model (Tokano et al. 1999; Richardson et al. 2007; Friedson et al. 2008) essentially reproduce the same meridional circulation pattern and seasonal reversal. The mass stream function peaks near the surface at ~10^{10} kg s^{-1} in either model, indicating that the majority of meridional mass transport takes place in the lower troposphere. Figure 13.12 gives an illustrative example.

Given the presence of an Earth-like hydrology based on methane on Titan and the importance of the water cycle on Earth's climate, it is intuitive to expect a significant impact of methane hydrology on Titan's atmospheric circulation. However, GCMs that include methane condensation (Tokano et al. 2001; Rannou et al. 2006; Mitchell et al. 2006) have so far yielded conflicting interpretations as to the role of methane on the general circulation. Radiative effects of gaseous and condensed methane and latent heat effects have been sometimes included, but presently, models that do not include methane condensation in the troposphere seem to predict more realistic circulation patterns (Tokano 2009), indicating that either methane hydrology is not relevant for the general circulation or else it is not yet well understood.

Although eddy-driven thermally indirect circulations, like the terrestrial Ferrel cell, are unlikely on Titan, this does not mean that thermally indirect circulations do not exist, at least locally, in the middle atmosphere. The likely subsidence over the warm winter north polar stratopause, leading to an enhancement of several organic compounds at a given altitude, has already been discussed (Sections 13.4.1 and 13.4.2). Earth's warm mesopause is another example of a thermally indirect circulation, and it is likely forced by the interaction of the mean zonal flow with vertically propagating waves (see, e.g., Andrews et al. 1987). The fact that Titan's zonal wind profile (Fig. 13.2) decays with altitude above the 0.1-mbar level indicates that the flow is being damped, and waves – perhaps internal gravity waves – are a likely candidate, as they are on Earth. The zonally symmetric circulation proposed by Flasar and Conrath (1990, see Section 13.2.2) to explain the lag in cooling of the stratosphere in the summer hemisphere, as it moved into autumn, is another example of a thermally indirect flow.

13.7.2 Zonal Circulation and Superrotation

In the Earth's atmosphere the zonal wind is strongly coupled to the meridional wind via the Coriolis force and is in geostrophic wind balance. The angular momentum transport by the Hadley circulation at low latitudes and in the polar region as well as by baroclinic instability at mid latitudes basically determines the global zonal wind pattern. On the other hand, the small Coriolis parameter and the lack of baroclinic instability on Titan imply that the global zonal wind pattern is likely to be governed by another force balance and another type of angular momentum redistribution. Titan's zonal circulation is in cyclostrophic wind balance everywhere above the upper troposphere, but the winds generally decrease with decreasing altitude, and in the lower troposphere there is a transition to Earth-like geostrophic wind balance (Tokano 2007).

A fundamental question of Titan's general circulation is as to how superrotating winds can be maintained in the entire stratosphere including the equatorial region, when the maximum angular momentum/mass (m) in a purely axisymmetric circulation cannot exceed that of the equatorial surface (Hide 1969; see also Held and Hou 1980). Jets at mid or high latitudes can be generated by poleward transport of angular momentum by the mean meridional circulation, but such momentum transport would also cause easterlies at the equator as on Earth in the absence of some counter-gradient transport of momentum by eddies. Basic concepts of superrotation formation mechanism (proposed for Venus) can be broadly subdivided into those which rely on the day–night circulation, gravity waves, and mean meridional circulation with horizontal mixing (Gierasch et al. 1997).

Gierasch (1975) proposed a model to explain the maintenance of the superrotation of the Venusian atmosphere that has also been considered in the discussion of Titan's atmospheric superrotation. It involves Hadley cells with ascent at the equator, consistent with the maximum solar heating there, and subsidence at the poles. The zonal winds increase with altitude, and the specific angular momentum (m) is greatest at the equator, where the lever arm to the rotation axis is largest. (It is hard to imagine a configuration where m increases with latitude. Were it to increase with latitude along isentropes, which are nearly horizontal, inertial instability would result (see, e.g., Held and Hou 1980)). The Hadley circulation carries atmosphere with high m upward at low latitudes. The atmosphere in the upper leg of the circulation moves poleward and tends to conserve angular momentum, spinning up and forming jets at high latitudes. These jets can become barotropically unstable because of the large horizontal shear of the wind near the jet core. Barotropic eddies arising in this area efficiently transport angular momentum back towards the equator, thereby maintaining a steady balance with equatorial superrotation. There are several necessary requirements for this idealized mechanism to work: the zonal wind must be strong enough for cyclostrophic balance, vertical mixing of momentum must be slower than meridional overturning, and meridional mixing by eddies must be faster than meridional overturning.

Prior to the advent of Titan's wind data, GCMs were the only means of investigating this hypothesis for Titan's atmosphere. When a terrestrial GCM is run at Titan's slow rotation rate, all other parameters being unchanged, the Hadley circulation becomes wider in meridional extent and the zonal wind tends to superrotate, provided turbulent coupling of the atmosphere aloft with the surface is weak (Del Genio et al. 1993). This is possible when much of the solar heating occurs aloft, far removed from the surface, as it does in Venus's clouds and Titan's hazes. An analysis of the angular momentum cycle in that GCM confirmed that the Gierasch mechanism could indeed work under Titan's rotation rate. However, if Titan's Hadley circulation is not symmetric about the equator because of the seasonal migration in response to solar heating, the upwelling region of the Hadley circulation will often not be located near the equator, where the angular momentum of the atmosphere is largest. Therefore, it is insufficient to change only the rotation rate or other geophysical parameters such as planetary radius in the GCM, but importantly a realistic diabatic heating pattern as a driver of the mean meridional circulation is necessary.

The Titan GCM of Hourdin et al. (1995) predicted a slow spin-up of the stratospheric superrotation over a period of 30 Titan years (900 years), causing zonal winds of up to 120 m s^{-1} in the upper stratosphere including the equatorial region. Hourdin et al. (1995) ascribed the build-up of the superrotation to the Gierasch mechanism. In their model there was an annual mean approximate balance between the net upward transport of angular momentum by the Hadley circulation and downward diffusion of the angular momentum, although upward transport slightly prevailed during the entire simulation period. However, since the ascending branch of the Hadley circulation on Titan is located near the summer poles during most of the Titan year, angular momentum is not efficiently carried into the stratosphere except during the relatively short reversal of the Hadley circulation near the equinoxes, when the ascending branch moves across the equator.

However, the result of Hourdin et al. (1995) and the Gierasch mechanism are not universally reproduced in other Titan GCMs. Tokano et al. (1999) predicted very weak zonal winds in the entire atmosphere although the same radiation model was used and the calculated meridional circulation resembled that of Hourdin et al. (1995). The meridional heat transport by the Hadley circulation was so efficient that the meridional temperature gradient almost disappeared and was in thermal wind balance with the weak zonal wind. Similar results were also predicted by Friedson et al. (2008), using a GCM, and by Zhu and Strobel (2005), using a 2-dimensional dynamics model. Another similar GCM of Richardson et al.

(2007) predicted a similar, almost vanishing, meridional temperature gradient, but the zonal wind attained a maximum of 30 m s^{-1}, stronger than in Tokano et al. (1999) but substantially weaker than predicted by Hourdin et al. (1995) and observed by Huygens. The reason for the large discrepancies among the 3-dimensional GCMs is not understood.

Tokano and Neubauer (2002) have pointed out that the inclusion of Saturn's gravitational tide in the GCM works against the maintenance of strong superrotation, since the breaking of the tide in the stratosphere decelerates the zonal wind towards the phase speed of the tide, which is only 6 m s^{-1} at the equator. This indicates that a forcing mechanism of superrotation would have to exist that overcomes this deceleration in the upper stratosphere.

Since absorption of sunlight by the stratospheric haze is the major heat source in the stratosphere, one anticipates that seasonal transport of haze particles from hemisphere to hemisphere by the Hadley circulation would affect the heating pattern, meridional circulation and temperature. This effect was first taken into account by Rannou et al. (2004) in a 2-dimensional GCM coupled to haze microphysics and photochemistry. In these models, the effects of barotropic eddies were parameterized (Luz et al. 2003). The meridional circulation tends to accumulate the haze particles in the polar region, enhancing the thermal cooling particularly at the winter pole. This lowers the winter pole temperature, in agreement with observation, and simultaneously enhances the zonal wind in the winter hemisphere in response to a larger meridional temperature gradient. In this and subsequent model versions the predicted zonal wind and temperature in the stratosphere are generally in close agreement to the Cassini temperature and wind data (Crespin et al. 2008). However, the tropospheric wind in their GCMs is up to twice as strong as measured by Huygens. As a whole, current major Titan GCMs are faced with the peculiar situation in which one series of GCMs predicts realistic superrotation that is in balance with unrealistic, excessive tropospheric winds (Rannou et al. 2004; Crespin et al. 2008), while another series predicts realistic tropospheric winds in balance with unrealistic, negligible superrotation (Tokano et al. 1999; Tokano 2007). This may indicate the presence of additional mechanisms relevant for the superrotation that are not adequately modeled.

An adaptation of the Gierasch model has been invoked to explain the origin of the curious 4° pole tilt in the temperatures and zonal winds reported by Achterberg et al. (2008b) (Section 13.3.1). As noted above, the most favorable configuration for transporting angular moment upward by an axisymmetric circulation is with ascent at the equator. However, the thermally direct Hadley cells transport heat most efficiently with ascent below the subsolar point. Achterberg et al. speculate that the tilt represents a compromise, with the spin equator shifting toward the sun. They use a simple toy model with several frictionally coupled spherical shells of atmosphere. Imposing a vertical motion field, corresponding to a fixed meridional circulation, they follow the evolution of the spin axis of the top shell relative to the surface rotation. The model shows the spin equator of the shell tilting toward the sun, with the spin axis lagging (on seasonal time scales) the direction to the sun as observed. It predicts that the pole tilt should follow the sun, with a lag, and this should be detectable given observations over seasonal time scales. Because of the strong frictional coupling troposphere to the surface (Section 13.5), it follows that the 4° offset cannot persist at all altitudes, but must decrease through the lower atmosphere.

Another mechanism for generating equatorial superrotation involves direct forcing by waves. Forcing by thermal tides has been proposed to account for Venus's superrotation (see, e.g., Pechmann and Ingersoll 1984; Leovy 1987; Gierasch et al. 1997). Thermal tides have long been considered negligible on Titan given the relatively weak solar heating rate and the long radiative time constants compared to the diurnal cycle (Flasar 1998b; Section 13.2). More recently, Zhu (2006) performed a scale analysis of the momentum equations for Titan's atmosphere and showed that there is a balance between the zonal momentum pumping by thermal tides and frictional drag if the main solar heating layer is located near 300 km. However, no strong zonal variation in the upper stratospheric temperatures has been observed (at least about the tilted pole, Section 13.3.1) in the nadir-viewing data, which provides the most complete zonal coverage (Teanby et al. 2006; Achterberg et al. 2008b). Note that the vertical resolution of the nadir observations is comparable to a pressure scale height, H; waves with vertical wavenumbers m, such that $mH >> 1$, would be averaged out in the retrievals. Vertically propagating internal gravity waves offers another possibility. In Venus' atmosphere Hou and Farrell (1987) have proposed absorption of vertically propagating gravity waves as a mechanism for maintaining cyclostrophic winds below the clouds. Scintillations in Voyager radio-occultation profiles have led to the identification of small-scale gravity waves in Titan's stratosphere (Hinson and Tyler 1983), as noted earlier.

13.8 Key Questions and Future Prospects

Even though it is cyclostrophic, Titan's middle atmosphere exhibits a dynamical behavior that shares a lot of common ground with that of Earth, which is geostrophic. This includes:

- Strong seasonal behavior with cross-hemispheric transport
- A strong circumpolar vortex in the winter hemisphere that acts to isolate the enclosed air mass
- Anomalous concentrations of gases within the vortex
- Stratospheric condensate clouds within the vortex

Yet a lot is still elusive:

- Is heterogeneous chemistry, involving stratospheric condensates, important, as it is on Earth?
- What physical, chemical, and dynamical processes are important in the mesosphere above 400 km, a region that has had only limited sounding.
- How does the vortex break up, as it must, when the winter hemisphere moves into summer?
- What drives the global superrotation?

On Earth planetary waves disrupt the winter polar vortex as it weakens in late spring. The disruption is nonlinear, with the waves producing large distortions in the vortex itself. Planetary-scale waves have yet to be identified on Titan, so it is not obvious if its winter vortices will break up so spectacularly, or if they will simply fade away, perhaps a victim of small-scale erosion by turbulence and waves.

The meteorology of Titan's troposphere is a little more mysterious. In part, this is because it is harder to probe than the middle atmosphere via remote sensing, because of the enshrouding photochemical smog. This not only makes observations that can probe the troposphere more difficult to conceive, but can also complicate the analysis of observations that do. Terrestrial analogies are still present:

- A hydrological cycle (CH_4), although with lakes and not oceans
- Hadley circulations as a fundamental engine for redistributing heat and angular momentum
- Changes in the length of day from angular momentum transfers between the atmosphere and the surface

There are several unanswered questions in tropospheric meteorology. A few:

- Do the "lakes" of liquefied natural gas migrate seasonally?
- Does moist convection play an important role in the energy and momentum budget, as it does on Earth?
- What role does topography play in the circulation and meteorology, e.g., in forcing atmospheric waves?

There are also major differences between Titan's atmosphere and Earth's. Titan has a major gravitational tidal perturbation from Saturn's mass and proximity. Titan's circulation seems to be axisymmetric, but with significant pole tilts aloft. Is this a property of seasonally modulated cyclostrophic systems?

The highlights and questions above are not meant to be exhaustive, but they indicate two tacks that one needs to take in pursuing Titan dynamical studies. The first is that time is an important dimension to map out, in additional to spatial dimensions. Dynamical systems change, and that change is instructive, particularly if it is unanticipated. No one would have predicted, much less explained, the Antarctic ozone hole on Earth, for example, without repeated observations that extended over several years. The interaction of physics, chemistry, and dynamics turned out to be fairly complex. Titan also has an annual cycle. Planetary scientists have known of Venus's global superrotation for decades, yet there is no completely satisfactory theory of how it originated and is maintained. Titan is the second solar-system body with a superrotating atmosphere, and it brings the seasonality that Venus lacks to the analyst's table. Observing the change with seasons may provide the missing clues needed to unravel this puzzle. Titan may have longer-term cycles analogous to those on Earth, like the El Niño Southern Oscillation. Of course, this example involves ocean currents and variability in sea-surface temperatures, and oceans on Titan seem to be nonexistent, but that is not the point. What the study of Earth's atmosphere has taught everyone is that the range of physical phenomena is rich and diverse. There is no reason to think things are different on Titan. One has to show up and observe.

The second aspect is that observations different from those previously done need to be made. For a remote body, this becomes tricky: one cannot just launch a rocket from the surface or send up a tethered balloon, each laden with instruments, as has been done on Earth. But ultimately more in situ sounding will need to be done. There are limits to what remote sensing can accomplish.

Several studies of the next flagship mission to Titan, which variously include an orbiter, a balloon, and some kind of surface probe, can begin to address these issues. And the current good health of Cassini and prospects for extending its mission beyond the current Equinox Mission (ends 2010) is an important factor. Keeping in mind the length of Titan's year, it is clear that no single mission will answer all the questions that have been posed, and that a series of missions will ultimately be needed. One needs to keep a long-term view.

References

Achterberg RK, Conrath BJ, Gierasch PJ, Flasar FM, Nixon CA (2008a) Titan's middle-atmospheric temperatures and dynamics observed by the Cassini composite infrared spectrometer. Icarus 194:263–277. doi:10.1016/j.icarus.2007.09.029

Achterberg RK, Conrath BJ, Gierasch PJ, Flasar FM, Nixon CA (2008b) Observation of a tilt of titan's middle-atmospheric superrotation. Icarus 197:549–555 doi:10.1016/j.icarus.2008. 05.014

Andrews DG, Holton JR, Leovy CB (1987) Middle atmosphere dynamics. Academic, Orlando

Atkinson DH (1989) Measurement of planetary wind fields by Doppler monitoring of an atmospheric entry vehicle. Ph.D. Thesis. Washington State University, Pullman, WA

Atkinson DH, Pollack JB, Seiff A (1990) Measurement of a zonal wind profile on Titan by Doppler tracking of the Cassini entry probe. Radio Sci 25:865–882

Atkinson DH, Kazeminejad B, Lebreton J-P, Witasse O, Pérez-Ayúcar M, Matson DL (2007) The Huygens probe descent trajectory working group: organizational framework, goals, and implementation. Planet Space Sci 55:1877–1885

Awal M, Lunine JI (1994) Moist convective clouds in Titan's atmosphere. Geophys Res Lett 21:2491–2494
Baines KH, Drossart P, Momary TW, Formisano V, Griffith C, Bellucci G, Bibring J-P, Brown R.H, Buratti BJ, Capaccioni F, Cerroni P, Clark RN, Coradini A, Cruikshank DP, Jaumann R, Langevin Y, Matson DL, McCord TB, Mennella V, Nelson RM, Nicholson PD, Sicardy B, Sotin C (2005) The atmospheres of Saturn and Titan in the near-infrared: first results of Cassini/VIMS. Earth Moon Planets 96:119–147
Barth EL, Rafkin SCR (2007) TRAMS: a new dynamic cloud model for Titan's methane. Geophys Res Lett 34:L03203
Bézard B, Coustenis A, McKay CP (1995) Titan's stratospheric temperature asymmetry: a radiative origin? Icarus 113:267–276
Bird MK, Heyl M, Allison M, Asmar SW, Atkinson DH, Edenhofer P, Plettemeier D, Wohlmuth R, Iess L, Tyler GL (1997) The Huygens doppler wind experiment. In: Huygens science payload and mission, ESA SP-1177:139–162
Bird MK, Allison M, Asmar SW, Atkinson DH, Avruch IM, Dutta-Roy R, Dzierma Y, Edenhofer P, Folkner WM, Gurvits LI, Johnston DV, Plettemeier D, Pogrebenko SV, Preston RA, Tyler GL (2005) The vertical profile of winds on Titan. Nature 438:800–802
Bouchez AH (2003) Seasonal trends in Titan's atmosphere: haze, wind, and clouds. Ph. D. Thesis. California Institute of Technology. http://resolver.caltech.edu/CaltechETD:etd-10272003-092206
Bouchez AH, Brown ME (2005) Statistics of Titan's south polar tropospheric clouds. Astrophys J 618:L53–L56
Brown ME, Bouchez AH, Griffith CA (2002) Direct detection of variable tropospheric clouds near Titan's South Pole. Nature 420:795–797
Brown RH, Baines KH, Bellucci G, Buratti BJ, Capacionni F, Cerroni P, Clark RN, Coradini A, Cruikshank DP, Drossart P, Formisano V, Jaumann R, Langevin Y, Matson DL, McCord TB, Mennella V, Nelson RM, Nicholson PD, Sicardy B, Sotin C, Baugh N, Griffith C, Hansen G, Hibbitts K, Showalter MR (2006) Observations in the Saturn system during approach and orbital insertion, with Cassini's visual and infrared mapping spectrometer (VIMS). Astron Astrophys 446:707–716
Chapman S, Lindzen RS (1970) Atmospheric tides. Reidel, Dordrecht
Colombatti G, Withers P, Ferri F, Aboudan A, Ball AJ, Bettanini C, Gaborit V, Harri AM, Hathi B, Leese MR, Makinen T, Stoppato PL, Towner MC, Zarnecki JC, Angrilli F, Fulchignoni M (2008) Reconstruction of the trajectory of the Huygens probe using the Huygens atmospheric structure instrument (HASI). Planet Space Sci 56:586–600
Counselman CC III, Gourevitch SA, King RW, Loriot GB, Ginsberg ES (1980) Zonal and meridional circulation of the lower atmosphere of Venus determined by radio interferometry. J Geophys Res 85:8026–8030
Coustenis A, Bézard B (1995) Titan's atmosphere from Voyager infrared observations. IV. Latitudinal variations of temperature and composition. Icarus 115:126–140
Coustenis A, Achterberg RK, Conrath BJ, Jennings DE, Marten A, Gautier D, Nixon CA, Flasar FM, Teanby NA, Bézard B, Samuelson RE, Carlson RC, Lellouch E, Bjoraker GL, Romani PN, Taylor FW, Irwin PGJ, Fouchet T, Hubert A, Orton GS, Kunde VG, Vinatier S, Mondellini J, Abbas MM, Courtin R (2007) The composition of Titan's stratosphere from Cassini/CIRS mid-infrared spectra. Icarus 189:35–62
Coustenis A, Schmitt B, Khanna RK, Trotta F (1999) Plausible condensates in Titan's stratosphere from Voyager infrared spectra. Planet Space Sci 47:1305–1329
Crespin A, Lebonnois S, Vinatier S, Bézard B, Coustenis A, Teanby NA, Achterberg RK, Rannou P, Hourdin F (2008) Diagnostics of Titan's stratospheric dynamics using Cassini/CIRS data and the 2-dimensional IPSL circulation model. Icarus 192(2):556–571. doi:10.1016/j.icarus.2008.05.010
Del Genio AD, Zhou W, Eichler TP (1993) Equatorial superrotation in a slowly rotating GCM: implications for Titan and Venus. Icarus 101:1–17
Dunkerton T (1978) On the mean meridional mass motions of the stratosphere and mesosphere. J Atmos Sci 35:2325–2333
Dutta-Roy R, Bird MK (2004) The Huygens doppler wind experiment: a Titan zonal wind retrieval algorithm. In: Planetary probe atmospheric entry and descent trajectory analysis and science, ESA SP-544:109–116
Dzierma Y, Bird MK, Dutta-Roy R, Perez-Ayucar M, Plettemeier D, Edenhofer P (2007) Huygens probe descent dynamics inferred from channel B signal level measurements. Planet Space Sci 55:1886–1895
Flasar FM (1998a) The composition of Titan's atmosphere: a meteorological perspective. Planet Space Sci 46:1109–1124
Flasar FM (1998b) The dynamic meteorology of Titan. Planet Space Sci 46:1125–1147
Flasar FM, Achterberg RK (2009) The structure and dynamics of Titan's middle atmosphere. Phil Trans Roy Soc London 367:649–664
Flasar FM, Conrath BJ (1990) Titan's stratospheric temperature: a case for dynamical inertia? Icarus 85:346–354
Flasar FM, Samuelson RE, Conrath BJ (1981) Titan's atmosphere: temperature and dynamics. Nature 292:693–698
Flasar FM, Allison M, Lunine JI (1997) Titan zonal wind model. In: Huygens science payload and mission, ESA SP-1177:287–298
Flasar FM, Kunde VG, Abbas MM, Achterberg RK, Ade P, Barucci A, Bézard B, Bjoraker GL, Brasunas JC, Calcutt S, Carlson R, Césarsky CJ, Conrath BJ, Coradini A, Courtin R, Coustenis A, Edberg S, Edgington S, Ferrari C, Gautier D, Gierasch PJ, Grossman K, Irwin P, Jennings DE, Lellouch E, Mamoutkine AA, Marten A, Meyer JP, Nixon CA, Orton GS, Owen TC, Pearl JC, Prangé R, Raulin F, Read PL, Romani PN, Samuelson RE, Segura ME, Showalter MR, Simon-Miller AA, Smith MD, Spencer JR, Spilker LJ, Taylor FW (2004) Exploring the Saturn system in the thermal infrared: the composite infrared spectrometer. Space Sci Rev 115:169–297
Flasar FM, Achterberg RK, Conrath BJ, Gierasch PJ, Kunde VG, Nixon CA, Bjoraker GL, Jennings DE, Romani PN, Simon-Miller AA, Bézard B, Coustenis A, Irwin PGJ, Teanby NA, Brasunas JC, Pearl JC, Segura ME, Carlson RC, Mamoutkine AA, Schinder PJ, Barucci A, Courtin R, Fouchet T, Gautier D, Lellouch E, Marten A, Prangé R, Vinatier S, Strobel DF, Calcutt SB, Read PL, Taylor FW, Bowles N, Samuelson RE, Orton GS, Spilker LJ, Owen TC, Spencer JR, Showalter MR, Ferrari C, Abbas MM, Raulin F, Edgington S, Ade P, Wishnow EH (2005) Titan's atmospheric temperatures, winds, and composition. Science 308:975–978
Folkner WM, Preston RA, Border JS, Navarro J, Wilson WE, Oestreich M (1997) Earth-based radio tracking of the Galileo Probe for Jupiter wind estimation. Science 275:644–646
Folkner WM, Border JS, Lowe ST, Preston RA, Bird MK (2004) Ground-based tracking of the Huygens Probe during the Titan descent. In: Planetary probe atmospheric entry and descent trajectory analysis and science, ESA SP-544:191–196
Folkner WM, Asmar SW, Border JS, Franklin GW, Finley SG, Gorelik J, Johnston DV, Kerzhanovich VV, Lowe ST, Preston RA, Bird MK, Dutta-Roy R, Allison M, Atkinson DH, Edenhofer P. Plettemeier D, Tyler GL (2006) Winds on Titan from ground-based tracking of the Huygens probe. J Geophys Res 111:E07S02
Friedson AJ, West RA, Wilson EH, Oyafuso F, Orton GS (2009) A global climate model of Titan's atmosphere and surface. Planet Sp Sci (in press), doi: 10.1016/j.pss.2009.05.006
Fulchignoni M, Ferri F, Angrilli F, Ball AJ, Bar-Nun A, Barucci MA, Bettanini C, Bianchini G, Borucki W, Colombatti G, Coradini M, Coustenis A, Debei S, Falkner P, Fanti G, Flamini E, Gaborit V, Grard R, Hamelin M, Harri AM, Hathi B, Jernej I, Leese MR, Lehto A, Lion Stoppato PF, López-Moreno JJ, Mäkinen T, McDonnell JAM, McKay CP, Molina-Cuberos G, Neubauer FM, Pirronello V, Rodrigo R, Saggin B, Schwingenschuh K, Seiff A, Simões F, Svedhem H, Tokano T, Towner MC, Trautner R, Withers P, Zarnecki JC (2005) Titan's physical characteristics measured

by the Huygens atmospheric structure instrument (HASI). Nature 438:785–791

Gibbard SG, Macintosh B, Gavel D, Max CE, de Pater I, Roe HG, Ghez AM, Young EF, McKay CP (2004) Speckle imaging of Titan at 2 microns: surface albedo, haze optical depth, and tropospheric clouds 1996–1998. Icarus 169:429–439

Gierasch PJ (1975) Meridional circulation and the maintenance of the Venus atmospheric rotation. J Atmos Sci 32:1038–1044

Gierasch PJ, Goody RM, Young RE, Crisp D, Edwards C, Kahn R, Rider D, Del Genio A, Greeley R, Hou A, Leovy CB, McCleese D, Newman M. (1997) The general circulation of the Venus atmosphere: An assessment. In: Bougher SW, Hunten DM, Phillips RJ (eds) Venus II. University of Arizona Press, Tucson, AZ, pp 459–500

Griffith CA, Hall JL, Geballe TR (2000) Detection of daily clouds in Titan. Science 290:509–513

Griffith CA, Penteado P, Baines K, Drossart P, Barnes J, Bellucci G, Bibring J, Brown R, Buratti B, Capaccioni F, Cerroni P, Clark R, Combes M, Coradini A, Cruikshank D, Formisano V, Jaumann R, Langevin Y, Matson D, McCord T, Mennella V, Nelson R, Nicholson P, Sicardy B, Sotin C, Soderblom LA, Kursinski R (2005) The evolution of Titan's mid-latitude clouds. Science 310:474–477

Griffith CA, Penteado P, Rannou P, Brown R, Boudon V, Baines KH, Clark R, Drossart P, Buratti B, Nicholson P, McKay CP, Coustenis A, Negrao A, Jaumann R (2006) Evidence for a polar ethane cloud on Titan. Science 313:1620–1622

Held IM, Hou AY (1980) Nonlinear axially symmetric circulations in a nearly inviscid atmosphere. J Atmos Sci 37:515–533

Hide R (1969) The viscous boundary layer at the rigid bounding surface of an electrically-conducting rotating fluid in the presence of a magnetic field. J Atmos Sci 26:847–853

Hinson DP, Tyler GL (1983) Internal gravity waves in Titan's atmosphere observed by Voyager radio occultation. Icarus 54:337–352

Hirtzig M, Coustenis A, Gendron E, Drossart P, Negrão A, Combes M, Lai O, Rannou P, Lebonnois S, Luz D (2006) Monitoring atmospheric phenomena on Titan. Astron Astroph 456:761–774

Holton JR (1979) An introduction to dynamic meteorology, 2nd edn. Academic, New York

Hou AY, Farrell BF (1987) Superrotation induced by critical-level absorption of gravity waves on Venus: an assessment. J Atmos Sci 44:1049–1061

Hourdin F, Talagrand O, Sadourny R, Courtin R, Gautier D, McKay CP (1995) Numerical simulation of the general circulation of the atmosphere of Titan. Icarus 117:358–374

Hourdin F, Lebonnois S, Luz D, Rannou P (2004) Titan's stratospheric composition driven by condensation and dynamics. J Geophys Res 109:E12005. doi:10.1029/2004JE002282

Hubbard WB, Sicardy B, Miles R, Hollis AJ, Forrest RW, Nicolson IKM, Appleby G, Beisker W, Bittner C, Bode H-J, Bruns M, Denzau H, Nezel M, Riedel E, Struckmann H, Arlot JE, Roques F, Sevre F, Thuillot W, Hoffmann M, Geyer EH, Buil C, Colas F, Lecacheux J, Klots A, Thouvenot E, Vidal JL, Carreira E, Rossi F, Blanco C, Cristaldi S, Nevo Y, Reitsema HJ, Brosch N, Cernis K, Zdanavicius K, Wasserman LH, Hunter DM, Gautier D, Lellouch E, Yelle RV, Rizk B, Flasar FM, Porco CC, Toublanc D, Corugedo G (1993) The occultation of 28 Sgr by Titan. Astron Astrophys 269:541–563

Hueso R, Sánchez-Lavega A (2006) Methane storms on Saturn's moon Titan. Nature 442:428–431

Hunten DM, Tomasko MG, Flasar FM, Samuelson RE, Strobel DF, Stevenson DJ (1984) Titan. In: Gehrels T, Matthews M (eds) Saturn. University of Arizona Press, Tucson, pp 671–759

Karatekin Ö, Van Hoolst T (2006) The effect of a dense atmosphere on the tidally induced potential of Titan. Icarus 183:230–232

Karkoschka E, Lorenz RD (1997) Latitudinal variation of aerosol sizes inferred from Titan's shadow. Icarus 125:369–379

Karkoschka E, Tomasko MG, Doose LR, See C, McFarlane EA, Schröder SE, Rizk B (2007) DISR imaging and the geometry of the descent of the Huygens probe within Titan's atmosphere. Planet Space Sci 55:1896–1935

Kazeminejad B, Atkinson DH, Pérez-Ayúcar M, Lebreton J-P, Sollazzo C (2007) Huygens' entry and descent through Titan's atmosphere: methodology and results of the trajectory reconstruction. Planet Space Sci 55:1845–1876

Khanna RK (2005) Condensed species in Titans atmosphere: identification of crystalline propionitrile (C_2H_5CN, $CH_3CH_2C{\equiv}N$) based on laboratory infrared data. Icarus 177:116–121

Kostiuk T, Fast KE, Livengood TA, Hewagama T, Goldstein JJ, Espenak F, Buhl D (2001) Direct measurement of winds on Titan. Geophys Res Lett 28:2361–2364

Kostiuk T, Livengood TA, Hewagama T, Sonnabend G, Fast KE, Murakawa K, Tokunaga AT, Annen J, Buhl D, Schmülling F (2005) Titan's stratospheric zonal wind, temperature and ethane abundance a year prior to Huygens insertion. Geophys Res Lett 32:L22205

Kostiuk T, Livengood TA, Sonnabend G, Fast KE, Hewagama T, Murakawa K, Tokunaga AT, Annen J, Buhl D, Schmülling F, Luz D, Witasse O (2006) Stratospheric global winds on Titan at the time of the Huygens descent. J Geophys Res 111:E07S03

Lellouch E, Coustenis A, Gautier D, Raulin F, Dubouloz N, Frére C (1989) Titan's atmosphere and hypothesized ocean: a reanalysis of the Voyager 1 radio-occultation and IRIS 7.7 μm data. Icarus 79:328–349

Leovy CB (1973) Rotation of the upper atmosphere of Venus. J Atmos Sci 30:1218–1220

Leovy CB (1987) Zonal winds near Venus' cloud top level: an analytic model of the equatorial wind speed. J Atmos Sci 69:193–201

Leovy CB, Pollack JB (1973) A first look at atmospheric dynamics and temperature variations on Titan. Icarus 19:195–201

Lindal GF, Wood GE, Hotz HB, Sweetnam DN, Eshleman V, Tyler GL (1983) The atmosphere of Titan: an analysis of the Voyager 1 radio occultation measurements. Icarus 53:348–363

Lindzen RS, Hou AY (1988) Hadley circulations for zonally averaged heated centered off the equator. J Atmos Sci 42:2416–2427

Lorenz RD (1993) The life, death, and afterlife of a raindrop on Titan. Planet Space Sci 41:647–655

Lorenz RD (2006) Thermal interactions of the Huygens probe with the Titan environment: constraints on near-surface wind. Icarus 182:559–566

Lorenz RD, McKay CP (2003) A simple expression for vertical convective fluxes in planetary atmospheres. Icarus 165:407–413

Lorenz RD, Smith PH, Lemmon MT, Karkoschka E, Lockwood GW, Caldwell J (1997) Titan's north–south asymmetry from HST and voyager imaging: comparison with models and ground-based photometry. Icarus 127:173–189

Lorenz RD, Griffith CA, Lunine JI, McKay CP, Rennò NO (2005) Convective plumes and the scarcity of Titan's clouds. Geophys Res Lett 32:L01201

Lorenz RD, Wall S, Radebaugh J, Boubin G, Reffet E, Janssen M, Stofan E, Lopes R, Kirk R, Elachi C, Lunine J, Mitchell K, Paganelli F, Soderblom L, Wood C, Wye L, Zebker H, Anderson Y, Ostro S, Allison M, Boehmer R, Callahan P, Encrenaz P, Ori GG, Francescatti G, Gim Y, Hamilton G, Hensley G, Johnson W, Kelleher K, Muhleman D, Picardi G, Posa F, Roth L, Seu R, Shaffer S, Stiles B, Vetralla S, Flamini E, West R (2006) The sand seas of Titan: Cassini RADAR observations of longitudinal dunes. Science 312:724–727

Lorenz RD, Zarnecki JC, Towner MC, Leese MR, Ball AJ, Hathi B, Hagermann A, Ghafoor NAL (2007) Descent motions of the Huygens probe as measured by the surface science package (SSP): turbulent evidence for a cloud layer. Planet Space Sci 55: 1936–1948

Lorenz RD, Stiles BW, Kirk RL, Allison MD, Persi del Marmo P, Iess L, Lunine JI, Ostro SJ, Hensley S (2008) Titan's rotation reveals an internal ocean and changing zonal winds. Science 319:1649–1651

Lorenz RD, Brown ME, Flasar FM (2009) Seasonal change on Titan. In: Brown RH, Lebreton J-P, Waite JH (eds) Titan from Cassini-Huygens. Springer

Lunine JI, Atreya SK (2008) The methane cycle on Titan. Nature Geosci 1:159–164

Luz D, Hourdin F, Rannou P, Lebonnois S (2003) Latitudinal transport by barotropic waves in Titan's stratosphere. II. Results from a coupled dynamics-microphysics-photochemistry GCM. Icarus 166: 343–358

Luz D, Civeit T, Courtin R, Lebreton J-P, Gautier D, Rannou P, Kaufer A, Witasse O, Lara L, Ferri F (2005) Characterization of zonal winds in the stratosphere of Titan with UVES. Icarus 179:497–510

Luz D, Civeit T, Courtin R, Lebreton J-P, Gautier D, Witasse O, Kaufer A, Ferri F, Lara L, Livengood T, Kostiuk T (2006) Characterization of zonal winds in the stratosphere of Titan with UVES: 2. Observations coordinated with the Huygens probe entry. J Geophys Res 111:E08S90

Mäkinen JTT, Harri A-M, Tokano T, Savijärvi H, Siili T, Ferri F (2006) Vertical atmospheric flow on Titan as measured by the HASI instrument on board the Huygens probe. Geophys Res Lett 33:L21803

McKay CP, Pollack JB, Courtin R (1989) The thermal structure of Titan's atmosphere. Icarus 80:23–53

McKay CP, Pollack JB, Courtin R (1991) The greenhouse and antigreenhouse effects on Titan. Science 253:1118–1121

Mitchell JL (2008) The drying of Titan's dunes: Titan's methane hydrology and its impact on atmospheric circulation. J Geophys Res 113:E08105

Mitchell JL, Pierrehumbert RT, Frierson DMW, Caballero R (2006) The dynamics behind Titan's methane clouds. Proc Natl Acad Sci USA 103:18421–18426

Mitri G, Showman AP, Lunine JI, Lorenz RD (2007) Hydrocarbon lakes on Titan. Icarus 186:385–394

Moreno R, Marten A, Hidayat T (2005) Interferometric measurements of zonal winds on Titan. Astron Astrophys 437:319–328

Pechmann JB, Ingersoll AP (1984) Thermal tides in the atmosphere of Venus: comparison of model results with observations. J Atmos Sci 41:3290–3313

Penteado PF, Griffith CA, Tomasko MG, Engel S, See C, Doose L (2009) Latitudinal variations in Titan's methane and haze from Cassini VIMS observations. Icarus (in press)

Pogrebenko SV, Gurvits LI, Campbell RM, Avruch IM, Lebreton J-P, van't Klooster CGM (2004) VLBI tracking of the Huygens probe in the atmosphere of Titan. In: Planetary probe atmospheric entry and descent trajectory analysis and science, ESA SP-544:197–204

Porco CC, Baker E, Barbara J, Beurle K, Brahic A, Burns JA, Charnoz S, Cooper N, Dawson D, Del Genio AD, Denk T, Dones L, Dyudina U, Evans MW, Fussner S, Glese B, Grazier K, Helfenstein P, Ingersoll AP, Jacobson RA, Johnson TV, McEwen A, Murray CD, Neukum G, Owen WM, Perry J, Roatsch T, Spitale J, Squyres S, Thomas P, Tiscareno M, Turtle EP, Vasavada AR, Veverka J, Wagner R, West R (2005) Imaging of Titan from the Cassini spacecraft. Nature 434:159–168

Radebaugh J, Lorenz RD, Lunine JI, Wall SD, Boubin G, Reffet E, Kirk RL, Lopes RM, Stofan ER, Soderblom L, Allison M, Janssen M, Paillou P, Callahan P, Spencer C (2008) the Cassini radar team: dunes on Titan observed by Cassini radar. Icarus 194:690–703

Rannou P, Hourdin F, McKay CP, Luz D (2004) A coupled dynamics-microphysics model of Titan's atmosphere. Icarus 170:443–479

Rannou P, Montmessin F, Hourdin F, Lebonnois S (2006) The latitudinal distribution of clouds on Titan. Science 311:201–205

Richardson MI, Toigo AD, Newman CE (2007) PlanetWRF: a general purpose, local to global numerical model for planetary atmospheric and climate dynamics. J Geophys Res 112:E09001

Rodriguez S, Le Mouélic S, Rannou P, Tobie G, Baines KH, Barnes JW, Griffith CA, Hirtzig M, Pitman KM, Sotin C, Brown RH, Buratti BJ, Clark RN, Nicholson PD (2009) Global circulation as the main source of cloud activity on Titan. Nature 459: 678–682

Roe HG, de Pater I, Macintosh A, McKay CP (2002) Titan's clouds from Gemini and Keck adaptive optics imaging. Astrophys J 581: 1399–1406

Roe HG, Brown ME, Schaller EL, Bouchez AH, Trujillo CA (2005) Geographic control of Titan's mid-latitude clouds. Science 310:477–479

Roman MT, West RA, Banfield DJ, Gierasch PJ, Achterberg RK, Nixon CA, Thomas PC (2009) Determining a tilt in Titan's north–south albedo asymmetry from Cassini images. Icarus. (in press) doi: 10.1016/j.icarus.2009.04.021

Samuelson RE, Mayo LA, Knuckles MA, Khanna RK (1997) C_4N_2 ice in Titan's north polar stratosphere. Planet Space Sci 45:941–948

Samuelson RE, Smith MD, Achterberg RK, Pearl JC (2007) Cassini CIRS update on stratospheric ices at Titan's winter pole. Icarus 189:63–71

Schaller EL, Brown ME, Roe HG, Bouchez AH (2006a) A large cloud outburst at Titan's south pole. Icarus 182:224–229

Schaller EL, Brown ME, Roe HG, Bouchez AH, Trujillo CA (2006b) Dissipation of Titan's south polar clouds. Icarus 184:517–523

Schneider T (2006) The general circulation of the atmosphere. Ann Rev Earth Planet Sci 34:65–688

Sicardy B et al (1999) The structure of Titan's stratosphere from the 28 Sgr occultation. Icarus 142:357–390

Sicardy B, Colas F, Widemann T et al (2006) The two Titan stellar occultations of 14 November 2003. J Geophys Res 111:E11S91. doi: 10.1029/2005JE002624

Smith BA, Soderblom L, Beebe R, Boyce J, Briggs G, Bunker A, Collins SA, Hansen CJ, Johnson TV, Mitchell JL, Terrile RJ, Carr M, Cook II AF, Cuzzi J, Pollack JB, Danielson GE, Ingersoll A, Davies ME, Hunt GE, Masursky H, Shoemaker E, Morrison D, Owen T, Sagan C, Veverka J, Strom R, Suomi VE (1981) Encounter with Saturn: Voyager 1 imaging science results. Science 212: 163–191

Sromovsky LA, Suomi VE, Pollack JB, Krauss RJ, Limaye SS, Owen T, Revercomb HE, Sagan C (1981) Implications of Titan's north–south brightness asymmetry. Nature 292:698–702

Stiles BW, Kirk RL, Lorenz RD, Hensley S, Lee E, Ostro SJ, Allison MD, Callahan PS, Gim Y, Iess L, Persi del Marmo P, Hamilton G, Johnson WTK, West RD, the Cassini RADAR Team (2008) Determining Titan's spin state from Cassini RADAR images. Astron J 135:1669–1680

Stofan ER, Elachi C, Lunine JI, Lorenz RD, Stiles B, Mitchell KL, Ostro S, Soderblom L, Wood C, Zebker H, Wall S, Janssen M, Kirk R, Lopes R, Paganelli F, Radebaugh J, Wye L, Anderson Y, Allison M, Boehmer R, Callahan P, Encrenaz P, Flamini E, Francescetti G, Gim Y, Hamilton G, Hensley S, Johnson WTK, Kelleher K, Muhleman D, Paillou P, Picardi G, Posa F, Roth L, Seu R, Shaffer S, Vetrella S, West R (2007) The lakes of Titan. Nature 445:61–64

Strobel DF (2006) Gravitational tidal waves in Titan's upper atmosphere. Icarus 182:251–258

Strobel DF, Atreya SK, Bézard B, Ferri F, Flasar FM, Fulchignoni M, Lellouch E, Müller-Wodarg I (2009) Atmospheric composition and structure. In: Brown RH, Lebreton J-P, Waite JH (eds) Titan from Cassini-Huygens. Springer

Teanby NA, Irwin PGJ, de Kok R, Nixon CA, Coustenis A, Bézard B, Calcutt SB, Bowles NE, Flasar FM, Fletcher L, Howett C, Taylor FW (2006) Latitudinal variations of HCN, HC_3N, and C_2N_2 in Titan's stratosphere derived from Cassini CIRS data. Icarus 181:243–255

Teanby NA, Irwin PGJ, de Kok R, Nixon CA, Coustenis A, Bézard B, Calcutt SB, Bowles NE, Flasar FM, Fletcher L, Howett C, Taylor FW (2007) Latitudinal variations of HCN, HC_3N and C_2N_2 in Titan's stratosphere derived from Cassini CIRS data. Icarus 181: 364–384

Teanby NA, de Kok R, Irwin PGJ, Osprey S, Vinatier S, Gierasch PJ, Read PL, Flasar FM, Conrath BJ, Achterberg RK, Bézard B, Nixon CA, Calcutt S (2008) Titan's winter polar vortex structure revealed by chemical tracers. J Geophys Res 113:E12003

Teanby NA, Irwin PGJ, de Kok R, Nixon CA (2009) Dynamical implications of seasonal and spatial variations in Titan's stratospheric composition. Phil Trans Roy Soc London 367:697–711

Tokano T (2005) Meteorological assessment of the surface temperatures on Titan: constraints on the surface type. Icarus 173:222–242

Tokano T (2007) Near-surface winds at the Huygens site on Titan: interpretation by means of a general circulation model. Planet Space Sci 55:1990–2009

Tokano T (2008) Dune-forming winds on Titan and the influence of topography. Icarus 194:243–262

Tokano T (2009) The dynamics of Titan's troposphere. Phil Trans Roy Soc London 367:633–648

Tokano T, Neubauer FM (2002) Tidal winds on Titan caused by Saturn. Icarus 158:499–515

Tokano T, Neubauer FM (2005) Wind-induced seasonal angular momentum exchange at Titan's surface and its influence on Titan's length-of-day. Geophys Res Lett 32:L24203

Tokano T, Neubauer FM, Laube M, McKay CP (1999) Seasonal variation of Titan's atmospheric structure simulated by a general circulation model. Planet Space Sci 47:493–520

Tokano T, Neubauer FM, Laube M, McKay CP (2001) Three-dimensional modeling of the tropospheric methane cycle on Titan. Icarus 153:130–147

Tokano T, Ferri F, Colombatti G, Mäkinen T, Fulchignoni M (2006) Titan's planetary boundary layer structure at the Huygens landing site. J Geophys Res 111:E08007

Tomasko MG, Smith PH (1982) Photometry and Polarimetry of Titan: Pioneer 11 observations and their implications for aerosol properties. Icarus 51:65–95

Tomasko MG, Archinal B, Becker T, Bézard B, Bushroe M, Combes M, Cook D, Coustenis A, de Bergh C, Dafoe C, Doose LE, Douté S, Eibl A, Engel S, Gliem F, Grieger B, Holso K, Howington-Kraus E, Karkoschka E, Keller HU, Kirk R, Kramm R, Küppers M, Lanagan P, Lellouch E, Lemmon M, Lunine J, McFarlane E, Moores J, Prout GM, Rizk B, Rosiek M, Rueffer P, Schröder SE, Schmitt B, See C, Smith P, Soderblom L, Thomas N, West R (2005) Rain, winds and haze during the Huygens probe's descent to Titan's surface. Nature 438:765–778

Tomasko MG, Bézard B, Doose L, Engel S, Karkoschka E, Vinatier S (2008) Heat balance in Titan's atmosphere. Planet Space Sci 56: 648–659

Toon OB, McKay CP, Courtin R, Ackerman TP (1988) Methane rain on Titan. Icarus 75:255–284

Turtle EP, Perry JE, McEwen AS, DelGenio AD, Barbara J, West RA, Fussner S, Dawson DD, Porco CC (2009) Cassini Imaging of Titan's high-latitude lakes, clouds, and south-polar surface changes. Geophys Res Lett 36: L02204, doi: 10.1029/2008GL036186

Vander Auwera J, Moazzen-Ahmadi N, Flaud J-M (2007) Towards an accurate database for the 12 mm region of the ethane spectrum. Astrophys J 662:750–757. doi:10.1086/515567

Walterscheid RL, Schubert G (2006) A tidal explanation for the Titan haze layers. Icarus 183:471–478

Williams GP (1988) The dynamical range of global circulations–II. Climate Dynamics 3:45–84

Witasse O, Lebreton JP, Bird MK, Dutta-Roy R, Folkner, WM, Preston RA, Asmar SW, Gurvits LI, Pogrebenko SV, Avruch IM, Campbell RM, Bignall HE, Garrett MA, van Langevelde HJ, Parsley SM, Reynolds C, Szomoru A, Reynolds JE, Phillips CJ, Sault RJ, Tzioumis AK, Ghigo F, Langston G, Brisken W, Romney JD, Mujunen A, Ritakari J, Tingay SJ, Dodson RG, van't Klooster CGM, Blancquaert T, Coustenis A, Gendron E, Sicardy B, Hirtzig M, Luz D, Negrao A, Kostiuk T, Livengood TA, Hartung M, de Pater I, Adamkovics M, Lorenz RD, Roe H, Schaller E, Brown M, Bouchez AH, Trujillo CA, Buratti BJ, Caillault L, Magin T, Bourdon A, Laux C (2006) Overview of the coordinated ground-based observations of Titan during the Huygens mission. J Geophys Res 111:E07S01. doi: 10.1029/2005JE002640

Yelle RV, Strobel D, Lellouch E, Gautier D (1997) Engineering models for Titan's atmosphere. In Huygens: Science, Payload and Mission, ed. J.-P. Lebreton, Eur. Space Agency Spec. Publ. ESA SP1177, 243–256

Zhu X (2006) Maintenance of equatorial superrotation in the atmospheres of Venus and Titan. Planet Space Sci 54:761–773

Zhu X, Strobel DF (2005) On the maintenance of thermal wind balance and equatorial superrotation in Titan's stratosphere. Icarus 176:331–350

Chapter 14
Seasonal Change on Titan

Ralph D. Lorenz, Michael E. Brown, and F. Michael Flasar

Abstract Titan displays seasonal changes in the distribution of gas and hazes in its atmosphere, in the character of its methane clouds, and in its temperatures and winds. While Cassini has observed some of these changes in detail, some are observable from Earth, and the period of most rapid change may be just about to begin in the years after equinox.

14.1 Introduction

Titan is an inconstant moon, which shows changes on hourly, diurnal and seasonal timescales (e.g. see Lorenz 2008), and it was seasonal changes that were the first dynamic aspects of Titan's appearance to be noticed. Study of Titan's seasons is rewarding, but requires patience – as noted by Christiaan Huygens himself, seasons in the Saturnian system are tedious because the Saturn orbital period around the sun is some 29.5 Earth years.

In this review, we discuss observations and models of seasonal changes in Titan's haze, gas composition, temperatures and methane meteorology. For practical reasons, we discuss these aspects separately, while recognizing that all of these processes and effects are coupled. We also note that Titan's winds will change seasonally, but perforce limit our discussion on winds since there are presently few observations that constrain them.

The fundamental driver of Titan's seasons is the obliquity of Saturn (and thus the Saturn system, including Titan) of 26.7° with respect to the sun. Titan's orbit is actually inclined at 0.3° to the Saturn ring plane, and Titan's rotational pole is inclined by about 0.3° with respect to the orbit normal (Stiles et al., 2008; Lorenz et al. 2008a), but for climatological purposes these complications can be ignored. This obliquity is fractionally larger than Earth's and that of present-day Mars, and leads to very strong seasonal modulation of the illumination at high latitudes. Latitudes poleward of 70° experience a long winter night of some 12 years, while the peak daily-averaged insolation seen anywhere on Titan is at the summer pole.

Equatorial regions experience rather more constant conditions, with their strongest insolation at equinox. During the equinox season, the presence of a large primary body, Saturn, leads to the occasional condition – unique in our solar system – of an atmosphere-laden world passing into total eclipse. The eclipses themselves are rather too short (Titan's orbit has a semimajor axis of 20 saturn radii, so it is obscured for at most only about one sixtieth (1/20π) of its orbit, or about 6 h). This timescale is too short to meaningfully affect the lower atmosphere, but effects in the ionosphere may be quite profound. Eclipses occur (beginning with merely penumbral eclipses) once per Titan day over a period of about an Earth year, with the longest eclipses in the middle of the period, at equinox. The next such periods are February 2009 – January 2010, and October 2024 – November 2025. It may be noted that as well as Titan eclipses, these seasons also display Titan's shadow on Saturn.

Like the orbit of Mars, the orbit of Saturn around the sun is appreciably eccentric (e = 0.054) such that the Saturn-Sun distance varies from 9.03 AU to 10.04 AU, with a resultant change in the solar 'constant' from 16.8 Wm^{-2} to 13.6 Wm^{-2} (see Fig. 14.1). As on Earth and Mars, coincidentally, the perihelion occurs close to Southern Summer Solstice, with the result that southern summer is shorter but hotter than that in the north. This effect breaks the symmetry of the seasons and may influence long-term differences between the two hemispheres, including the distribution of volatiles on the surface. In this sense Titan may have some parallels with Mars, where the two polar caps are of rather different composition and appearance, although as with Mars it may be a challenge to disentangle the effects of asymmetric seasons from geologically-determined surface differences such as topography.

R.D. Lorenz (✉)
Applied Physics Laboratory, JHU, Laurel, MD, USA
e-mail: ralph.lorenz@jhuapl.edu

M.E. Brown
California Institute of Technology, Pasadena, CA, USA

F. Michael Flasar
Goddard Space Flight Center, Greenbelt, MD, USA

R.H. Brown et al. (eds.), Titan from Cassini-Huygens,

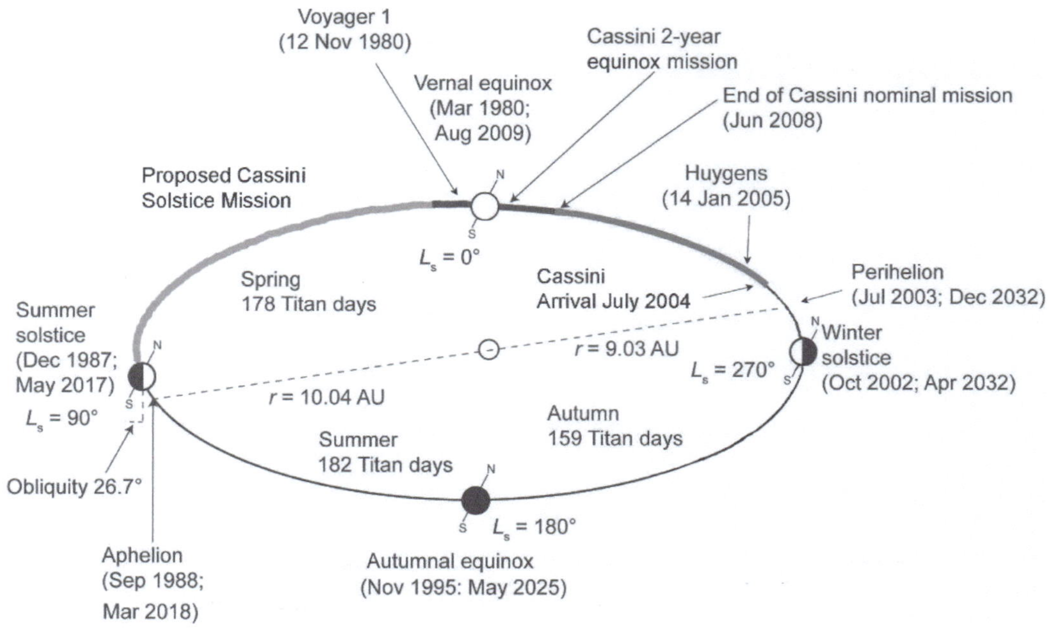

Fig. 14.1 Titan's seasons (note the unequal length) and epochs of Voyager and Cassini encounters

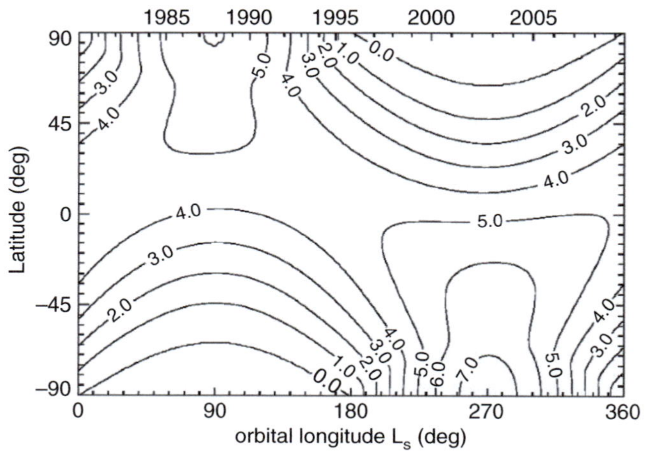

Fig. 14.2 Daily-averaged top-of-atmosphere insolation (Wm^{-2}) as a function of latitude and season. The asymmetric seasons are apparent in this plot. Most intense insolation is during southern summer ($L_s \sim 270°$, 2003)

Table 14.1 Key dates

Date	Observation/Event	Ls (deg)	Solar Lat (deg)
Sep 1979	Pioneer 11 encounter	354	−2.9
Feb 1980	Vernal equinox	360	0
Nov 1980	Voyager 1 encounter	8	+4.1
Aug 1981	Voyager 2 encounter	16	+8.0
Nov 1987	Northern summer solstice	90	+26.7
Jul 1989	28 Sgr occultation	109	+25.4
Nov 1995	Autumnal equinox	180	0
Oct 2002	Southern summer solstice	270	−26.7
Apr 2004	Cassini approach science	292	−25.0
Oct 2004	First Cassini flyby (Ta)	300	−23.5
Jan 2005	Huygens encounter	303	−22.8
Jun 2008	End nominal Cassini tour	345	−7.2
Aug 2009	Vernal equinox	360	0
Jun 2010	Cassini equinox mission ends	8	+4.1
May 2017	Northern summer solstice	90	+26.7

As with Mars climate studies, the ecliptic longitude of the Sun (Ls) is a useful climate variable for Titan, since seasons are equally-spaced along this variable – see Fig. 14.2. Table 14.1 lists the subsolar latitude and ecliptic longitudes for various pertinent dates.

Like other atmospheres, Titan's is also influenced by the solar cycle. Since much of Titan's photochemical gas and haze production is driven by ultraviolet light, the solar cycle may manifest a change in these effects with an 11-year periodicity. This effect will move in and out of phase with the seasonal forcing and may thus lead to interannual variability on Titan (i.e. some Titan years the solar maximum may coincide with e.g. southern summer solstice, while other years it will not.) Long periods of observation are required to decorrelate such effects, however, and the relationship of the Saturnian magnetosphere, which drives a not insignificant part of Titan's upper atmospheric chemistry, to the solar cycle would also need to be taken into account.

14.2 Titan's Haze

14.2.1 Observations of Titan's Main Haze

Although various observers had sketched Titan's appearance over the years, the first reproducible spatially-resolved imagery of Titan was obtained only when spacecraft reached it around 1980. However, Titan's disk-integrated brightness had been observed photometrically for some years by that time, and Titan was known to be brightening during the late 1970s, an effect initially suspected to be related to the solar cycle (Lockwood and Thompson 1979) but now known to be largely related to seasonal changes. In fact, this photometric record now spans more than a Titan year (Lockwood and Thompson 2009) and shows that the seasons do not repeat perfectly – there is interannual variability in the albedo history, with 2002 about 3% fainter at blue wavelengths than at the same season in 1973.

The Voyager encounters (Voyager 1 in November 1980, Voyager 2 in August 1981; Smith et al. 1981, 1982) showed that the optical albedo of Titan's northern hemisphere was lower than the south (Fig. 14.3). This North-South Asymmetry (NSA) was strongest at blue and green wavelengths, and declined at near-UV and red wavelengths. In blue the albedo ratio of south:north seen by Voyager 1 was 1.25: essentially the same albedo ratio (1.26) was determined in data from Pioneer 11 just over a year before.

Studies with the Hubble Space Telescope, whose resolution is approximately 0.1 arcsecond or a tenth of Titan's disk, allowed the albedo asymmetry to be monitored from the early 1990s until shortly before Cassini's arrival. The first images of Titan with HST (Caldwell et al. 1992) were at blue and green wavelengths and in the 889 nm methane band – these were acquired with the Wide-Field/Planetary Camera WFPC in 1990 while HST's primary mirror spherical aberration was unrepaired, and HST did not yet have the capability to track solar system targets, requiring that the exposures be rather short to prevent smear. Despite the resultant low signal-to-noise, and the need to deconvolve the images (at the time a somewhat novel computation procedure), it could be determined that indeed the green and blue NSA had reversed from the Voyager situation 9 years before. It was further noted that the 889 nm asymmetry was opposite in sense from that at short optical wavelengths.

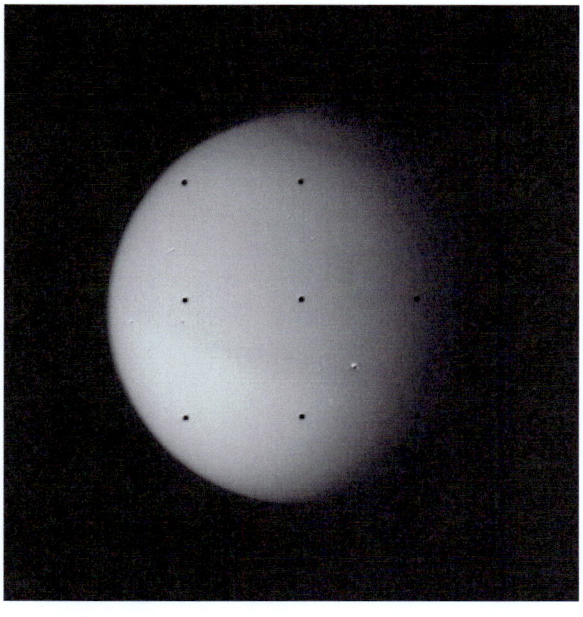

Fig. 14.3 Blue Voyager image showing the northern hemisphere of Titan in 1980 being slightly darker than the south. A dark polar collar is visible around the north pole. Black dots are reseau marks on the Voyager vidicon camera

Better WFPC images were acquired in 1992, and then after the Hubble repair WFPC-2 images were obtained in 1994, 1995, 1997, 1999, 2000 and 2002. Results from these observations (which also enabled the generation of the first near-infrared map of Titan's surface – Smith et al. 1996) were documented in a series of papers (Lorenz et al. 1997, 1999, 2001, 2004) which compare the appearance at different wavelengths over the HST period and against the appearance of Titan to Voyagers 1 and 2 and Pioneer 11. The uniform quality and filter bandpass of the WFPC-2 dataset made them particulary suitable for seasonal change monitoring.

At blue wavelengths, the deep atmosphere would be quite bright from Rayleigh scattering. However, the reddish tholin haze has a low single-scattering albedo, and large optical depth, at blue wavelengths and so it renders the atmosphere dark. Thus, for the northern hemisphere to be darker than the south means that it has a higher number density of haze particles above the bulk of the gas.

The optical depth of the haze becomes progressively lower with longer wavelength (since the haze particles have a diameter smaller than the wavelength of light). Meanwhile, the single-scattering albedo of the particles increases with wavelength, since tholin is reddish in color. Thus moving from blue towards red wavelengths the haze becomes (to draw terrestrial analogies) less like thick smoke and more like thin fog. This effect allows light to penetrate deeper into Titan's atmosphere, ultimately allowing the detection of Titan's surface, but also allowing wavelength to serve as a proxy for altitude – blue wavelengths reach an optical depth of unity high in the main haze layer (circa 80–100 km altitude), whereas red light probes down to close to the tropopause. The combination of the seasonally-changing haze distribution, and the seasonally-changing geometric aspect of Titan which exposes more of the summer hemisphere to Earth, leads to the 14.5 year cycle in disk-integrated albedo observed by Lockwood and Thompson (1979) – as shown in Fig. 14.5, the albedo (like the NSA) at yellow wavelengths varies less than that at blue.

This wavelength-as-altitude effect is more profound in the near-infrared, where absorption due to methane can vary steeply with wavelength at the edge of methane absorption bands – particularly pertinent ones of progressive depth being those at 619, 729 and 889 nm. In the near-infrared, tholin is relatively bright and the optical properties of the haze particles change only slowly with wavelength, making interpretation somewhat easier. In the methane bands, where the deep atmosphere is dark due to methane absorption, the high-altitude haze above it is bright. Thus where haze is relatively more abundant (making Titan darker in blue), the appearance in a methane band image is brighter. HST images showed (Fig. 14.4) that the NSA reversed in the 889 nm band before it did so in blue – consistent with high altitude hazes changing their distribution first.

Spatially-resolved spectra were obtained in 1997, 2000 and 2002 using the Space Telescope Imaging Spectrograph, and are presented in Lorenz et al. (2004) and Anderson et al. (2008) – see Fig 14.6. Of particular interest in these data are the rapid variation of appearance with wavelength in the near-infrared methane bands, notably between the band center at 889 nm and the window at 940 nm – at different wavelengths in this range images probe various depths in the atmosphere. A set of images from 1996 in this wavelength range was obtained with a linear ramp filter on HST and analyzed in (Young et al. 2002). Similar data (albeit of slightly lower spatial resolution) were acquired from the ground (the 2.54 m telescope on Mt. Wilson, equipped with an adaptive optics system) in 1999 with an optical-acoustical filter (Chanover et al. 2003; Anderson et al. 2004).

Groundbased adaptive optics (AO) images of Titan in the near-infrared at longer wavelengths (notably in the 2–2.3 μm window) also contain some information on the haze structure with latitude, although the emphasis in this wavelength range has been on surface mapping and tropospheric cloud detection (see Section 13.5). Gibbard et al. (1999) obtained near-infrared images of Titan with the 10 m Keck telescope in 1996 using a different technique, speckle imaging, and estimated different haze optical depths in the northern and southern limbs and at the equator at two different wavelengths – 1.6 μm (τ ~ 0.45, 0.23, 0.14 at south, equator

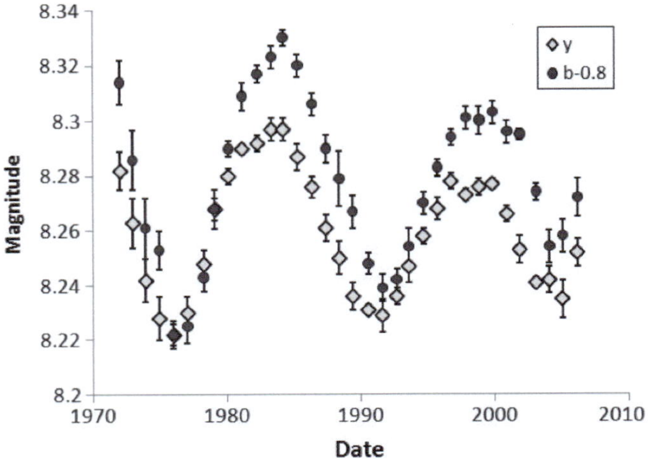

Fig. 14.5 Evolution of the disk-integrated brightness observed from Earth by Lockwood and Thompson (2009) in two visible photometer channels (*yellow and blue*). Note that this is a scale in magnitude, and thus higher on the plot means less bright, with the amplitude of the blue cycle about 10% and the yellow about 6%. Note also that the curves do not repeat, with different albedo and rate of change in 2004 and at the same Titan season in 1973

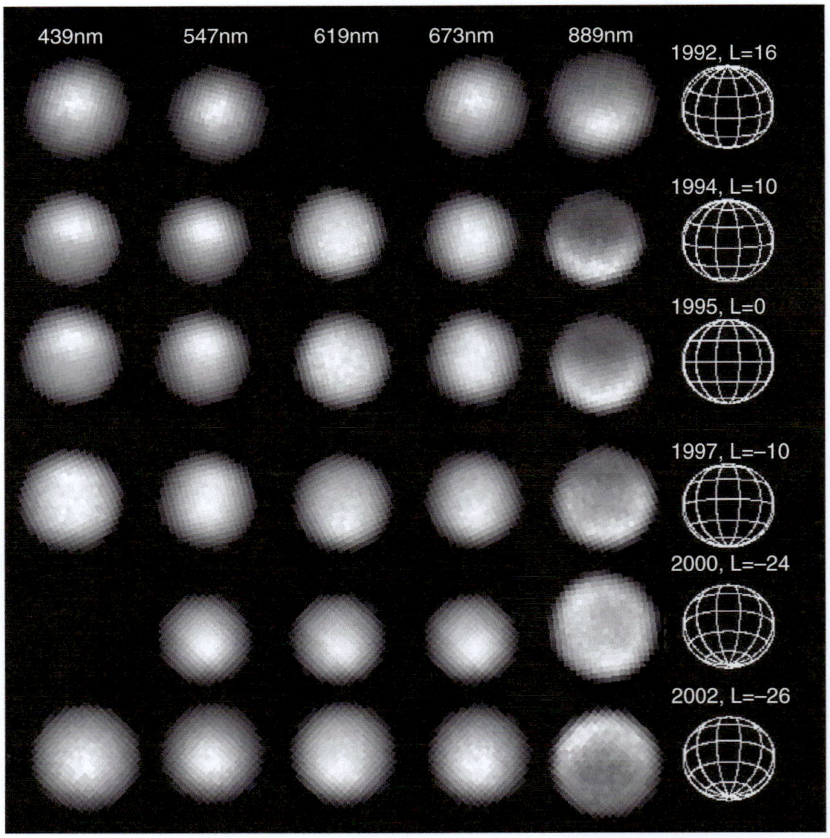

Fig. 14.4 Titan's wavelength-dependent seasonal change over a decade around Southern Spring Equinox observed by Hubble Space Telescope WF-PC and WFPC-2 images. The blue appearance (1st column) is inverted in 1994–1995 from the Voyager appearance two seasons earlier (Fig. 14.3) but the asymmetry declines and reverses by 2002. In the near-infrared methane band at 889 nm (5th column) limb-brightening is evident and the sense of the asymmetry is stronger than but reversed in sense from that at blue. The broadening and then inversion of the 'smile' is seen 1994–2002.For further discussion, see Lorenz et al. (2004)

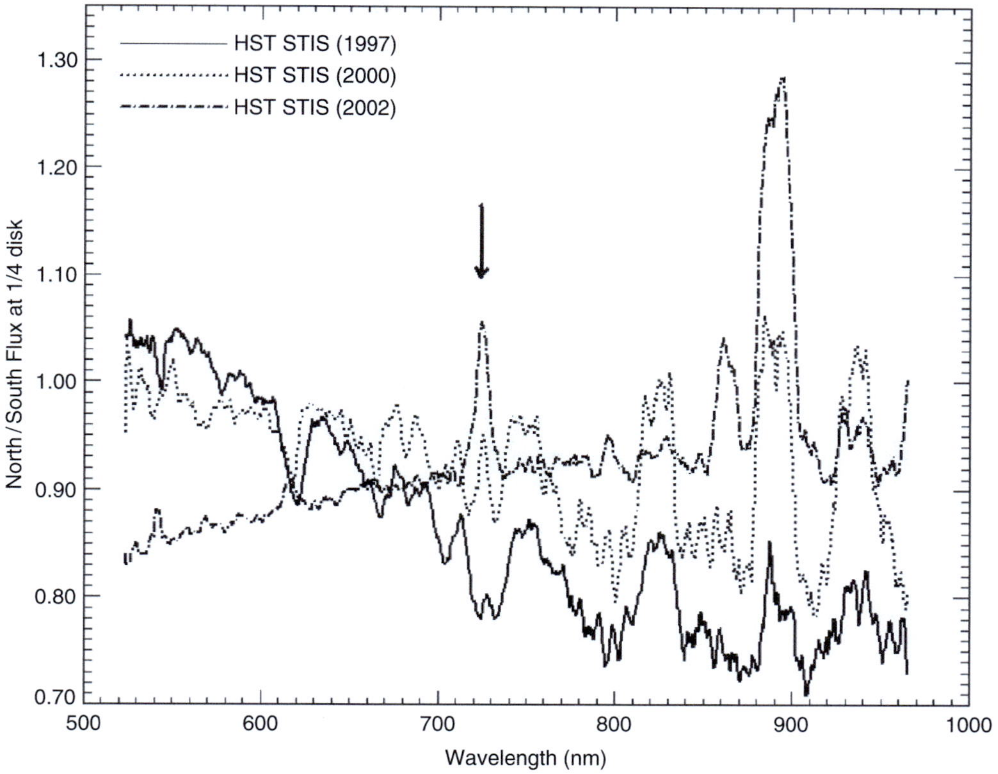

Fig. 14.6 Changing North-South Asymmetry varies strongly with near-infrared wavelength. Note in particular in the 729 nm and 889 nm methane bands. The center of the band (sensing only the highest altitudes) changes first, with the wings at intermediate altitudes and finally the continuum sensing deepest – see Lorenz et al. (2004)

and north respectively) and 2.1 μm (τ ~ 0.30, 0.15, 0.10). These optical depths, however, also include tropospheric opacity. Subsequent speckle and AO observations (Gibbard et al. 2004) show the 2-μm equatorial opacity to have remained roughly constant, while the limb values have converged over the 1996–2004 period to reach that same value (~0.18). If that trend continues, the north polar opacity may be expected to become quite large (~0.5) as the north polar regions become illuminated.

Ádámkovics et al. (2004) analyse spatially-resolved spectroscopy from Keck in 2001 in the 1.6 μm window and determine a haze opacity (from 30 km up) of 0.23 at 25°N, and 0.3 at 80°S. A challenge in assembling the pre-Cassini data into a consistent history is that different telescopes, instruments, filter bandpasses or spectral resolution have been used, and then different models have been used in the interpretation. Bouchez (2004) examined a set of spatially-resolved 2-μm spectra and found that over the 1999–2001 period the south polar tropospheric opacity had not changed, but there was a detectable increase of northern hemisphere opacity in the main haze layer, with the south remaining roughly constant (consistent with the HST imaging scenario described above).

It has been shown (see later) that advection from one hemisphere to the other by the Hadley circulation best explains the observed changes – alternative hypotheses such as particle size having been rejected by the Hubble data of the 1990s. Titan has not changed profoundly in optical appearance since Cassini's arrival in 2004 in that the Hadley circulation has likely been constant in direction. A casual inspection of Cassini images from early in the nominal mission (Fig. 14.7) shows a picture much like that seen in 2003 by HST, and indeed not grossly different from Voyager when the different illumination is taken into account. As yet no analysis of time-variation in the stratospheric haze during the Cassini epoch has been published, although it is likely that some subtle changes have occurred and can be recovered by careful analysis.

In addition to measuring densities (and thus temperatures – see next section) by the refraction of the starlight, stellar occultations can also measure haze over a small altitude range via its absorption. Occultations in 2003 indicated (Sicardy et al. 2006) that the haze distribution at 250 km altitude was close to the one predicted by Rannou et al's Global Circulation Model (GCM) studies(2004) in the southern hemisphere, but was contrary to the GCM in that a clearing of the haze north of 40°N was observed, instead of a progressive increase towards the polar hood.

If Titan's seasons are not grossly asymmetric, the 1995–2003 period following equinox suggests that the proposed

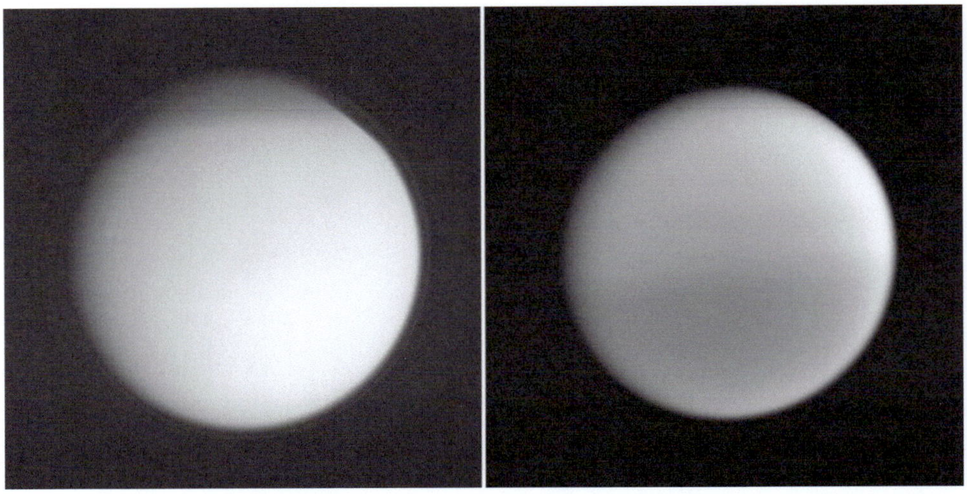

Fig. 14.7 Cassini ISS images (*North up*) acquired in September 2007 in the near-UV (*left*). The North-South asymmetry can be seen (barely – contrast in the UV is less than at blue) to be in the same sense as at Voyager (Fig. 14.3). The detached haze is visible, as is an apparent step in the height of the main haze layer. The polar hood is very apparent – and forms a complete hood rather than a ring as at Voyager. At right is an 889 nm methane band image. The contrast here is higher, and reversed compared with the UV and visible: the disk is also less limb-darkened and the detached haze is invisible

Cassini XXM (Solstice Mission) beyond about 2012 should see quite dramatic changes in Titan's optical appearance as the pole-to-pole Hadley circulation is disrupted and then reverses. The resultant haze transport will reverse the north-south asymmetry – first in the deepest methane bands at high altitudes then at progressively deeper levels in the atmosphere.

14.2.2 The Detached Haze

Voyager observations revealed the presence of a 'detached' haze layer, separated from the main haze deck by a clear gap (Fig. 14.8). This detached haze is optically thin in transmission, but can be seen near the limb, particularly at high phase angles and particularly at short (UV) wavelengths.

The detached haze is difficult to see from Earth in reflected light, since it is faint and not widely separated from the main haze. Some hint of the detached haze was noted in heavily-processed HST images by Young et al. (2004) but the altitude could not be reliably determined. An HST observation in 1995 of Titan's shadow cast onto Saturn (Karkoschka and Lorenz 1997) was interpreted to reveal the height of the detached, rather than main, haze layer.

An intriguing observation is the seasonally-varying height of the detached haze. Porco et al. (2005) determine the height to be ~500 km in late 2004 (late southern summer), whereas the Voyager determination in 1980 (northern spring equinox) by Rages and Pollack (1983) was 200–350 km. This difference has yet to be explained, and if interannual variability is not to be significant, implies a substantial change must occur between 2004 and 2009.

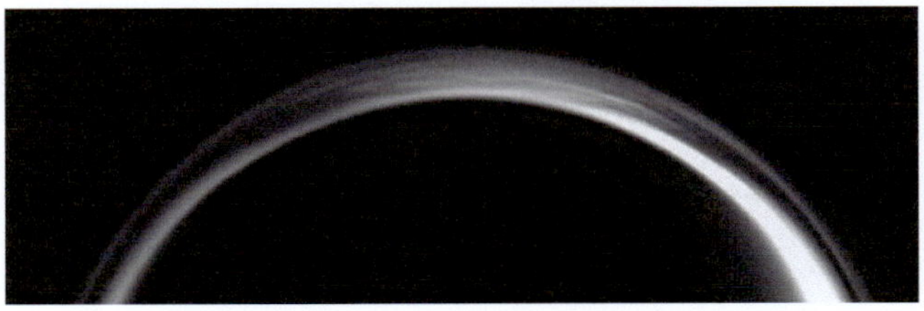

Fig. 14.8 A ten-image composite of visible-light Cassini ISS images acquired at low phase on January 16, 2006 showing details in the northern polar haze

14.2.3 Polar Hood

The dark polar hood (seen as a ring or 'polar collar' in Voyager 1 data) is a seasonal dark feature in the haze seen at about 60° latitude and above during spring and summer. It appears also to rise above the main deck and to bridge the detached haze down to the main layer.

When HST observed Titan in the early 1990s, the north polar hood was no longer present, confirming the expectation that like the north–south asymmetry, it too was a seasonal feature. It appears to be a springtime feature (it may exist during midwinter, but is then in darkness and thus largely invisible).

As Titan began to present its south pole towards Earth in the late 1990s and early 2000s, a south polar hood could be detected. HST's UV wavelength coverage was much better than Voyager's and allowed model investigation which suggested that it may be composed of rather small particles which give it a different spectral color from the main haze deck (e.g. Lorenz et al. 2006) even though the tholin

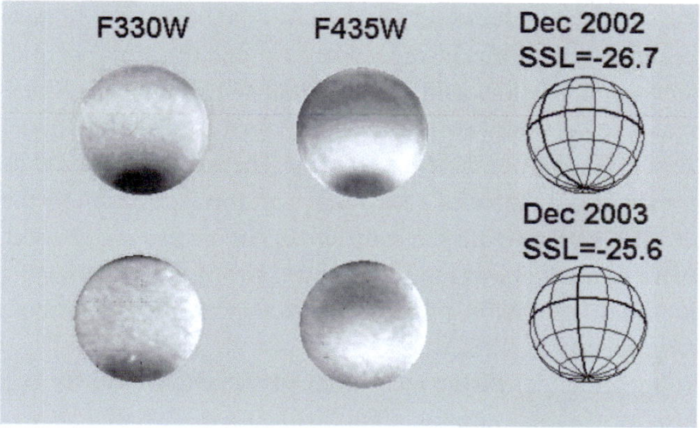

Fig. 14.9 The polar hood shown in Hubble Space Telescope ACS images (corrected for limb-darkening, and contrast-stretched) at near-UV and blue (330 and 435 nm) in 2002 and 2003, as described in Lorenz et al. (2006). The south polar hood – which had disappeared altogether by the time of Cassini's arrival – is seen to fade dramatically over only 1 year in this midsummer season, and is much more prominent in the UV than at blue

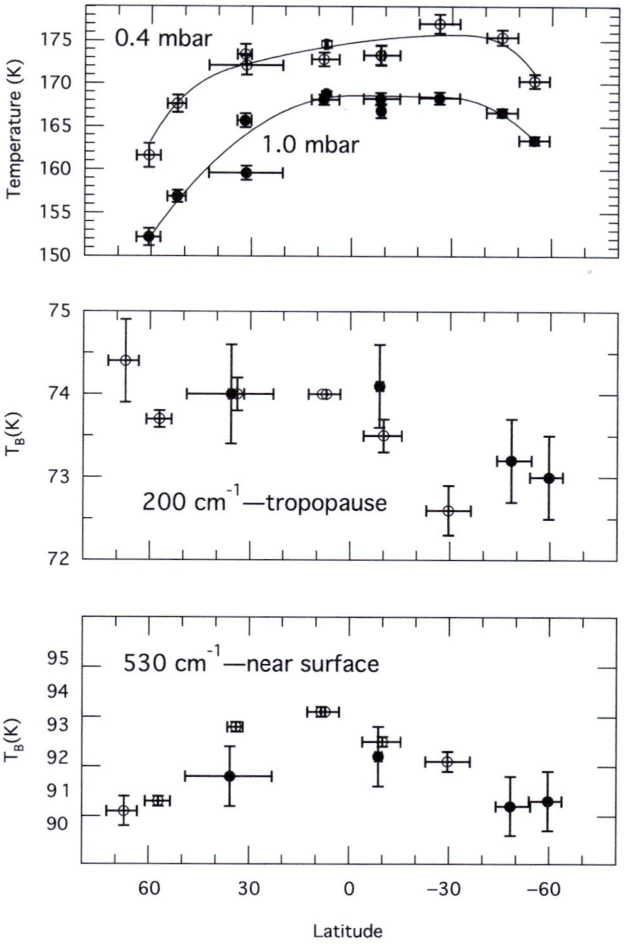

Fig. 14.10 Meridional distribution of retrieved temperatures from radiances near 1300 cm^{-1} and brightness temperatures at 530 and 200 cm^{-1}. The distribution of the latter two are normalized to an emission angle of 52.7°. Open circles are daytime data; filled circles are night-time data. The vertical bars include uncertainties from noise, calibration, and emission angle corrections. The horizontal bars denote the latitude range over which the data comprising each point extend. Adapted from Flasar et al. (1981) and Flasar and Conrath (1990)

composition may be the same. The feature is presumably associated with the descending meridional circulation over the winter pole that also leads to gas enhancements (see next section). During midsummer, the south polar hood was seen to fade (Lorenz et al. 2006 – see Fig. 14.9), and it was not detected by Cassini when it arrived in 2004 (Porco et al. 2005).

14.3 Temperatures and Zonal Winds

Titan is in an interesting dynamical regime in which the radiative response time to disturbances of order a pressure scale height is quite large in the lower troposphere (~130 terrestrial years, Smith et al. 1981), and decreases with altitude to ~1 year at 1 mbar in the upper stratosphere (Flasar et al. 1981). Hence one anticipates that the upper stratosphere (and mesosphere) will exhibit large seasonal variations in temperature, and that seasonal changes will mute as one moves downward to the troposphere. By and large this is what is observed. Figure 14.10 illustrates the temperatures derived at three altitudes from thermal-infrared spectra obtained by Voyager IRIS shortly after northern spring equinox. Temperatures near the surface and in the tropopause region show little contrast with latitude. The Voyager and Cassini radio occultations bear this out. The Voyager occultation soundings (1980) were both equatorial; those for Cassini included soundings at 74° N and S and several mid latitudes in both hemispheres but nothing near the equator. The retrieved temperature profiles exhibit a range ~4–5 K at the tropopause, and ~3 K near the surface. It is not unreasonable to expect that the meridional variations are comparable to seasonal variations and therefore can be used as an indicator of the latter. Temperatures retrieved from the thermal-infrared spectrometer on Cassini during the subsequent northern winter indicate comparable variations in latitude, but more north-south asymmetry, with the winter north pole being colder below the 0.1-mbar level (Chapter 14).

An exception to the above discussion occurs near the surface. The thermal inertia of Titan's surface is likely to be small enough to respond to the seasonal and possibly diurnal solar modulation (Tokano 2005). This raises the possibility of seasonally varying surface heating producing a planetary boundary layer coupled to the surface by thermally buoyant convection. In this situation the radiative time scales are no longer relevant, and the physics is dominated by the much shorter time scales associated with the convective eddies. Jennings et al. (2009) report surface temperatures measured in the 530 cm^{-1} window, and finds a temperature of about 94

K at low latitude, with ~91 K at the north pole and ~92 K at the south pole. They also note that dark areas (like the Belet sand sea) at low latitudes are consistently ~1 K warmer than at other equatorial longitudes.

Temperature maps have only been obtained from Voyager and Cassini observations. A more complete picture of seasonality in the stratosphere can be obtained by adding the stellar-occultation central-flash data that have been acquired in the intervening period (Hubbard et al. 1993; Bouchez 2004; Sicardy et al. 2006). The observations define the shape of the isopycnal (constant-density) surface near 0.25 mbar. Typically one assumes that the isopycnal surfaces are coincident with isobars. Since the mean zonal winds are related to the zonal mean meridional gradient in pressure by the gradient wind equation (Chapter 13), the zonal winds can be estimated. These are shown in panels b and c in Fig. 14.11 for two of the stellar occultations. The zonal winds can be similarly calculated from the Voyager (panel a) and Cassini (panel d) temperature fields, using the thermal wind equation. These require a boundary condition on the zonal wind. In the figure, zero winds were assumed at 10 mbar. The zonal winds provide a nice way to quantify seasonal variations. Although the detailed structure at each snapshot is not always understood, there is a general trend to having a strong circumpolar jet in the winter hemisphere, similar to Earth's stratosphere (Andrews et al. 1987).

In principle, direct measurements of wind motions can bear on seasonal variations, but in practice these have had limited utility so far (Chapter 13 discusses these in more detail). One method determines velocities from measuring the Doppler shifts of spectral features. This has been at visible, thermal infrared, and millimeter wavelengths using ground-based

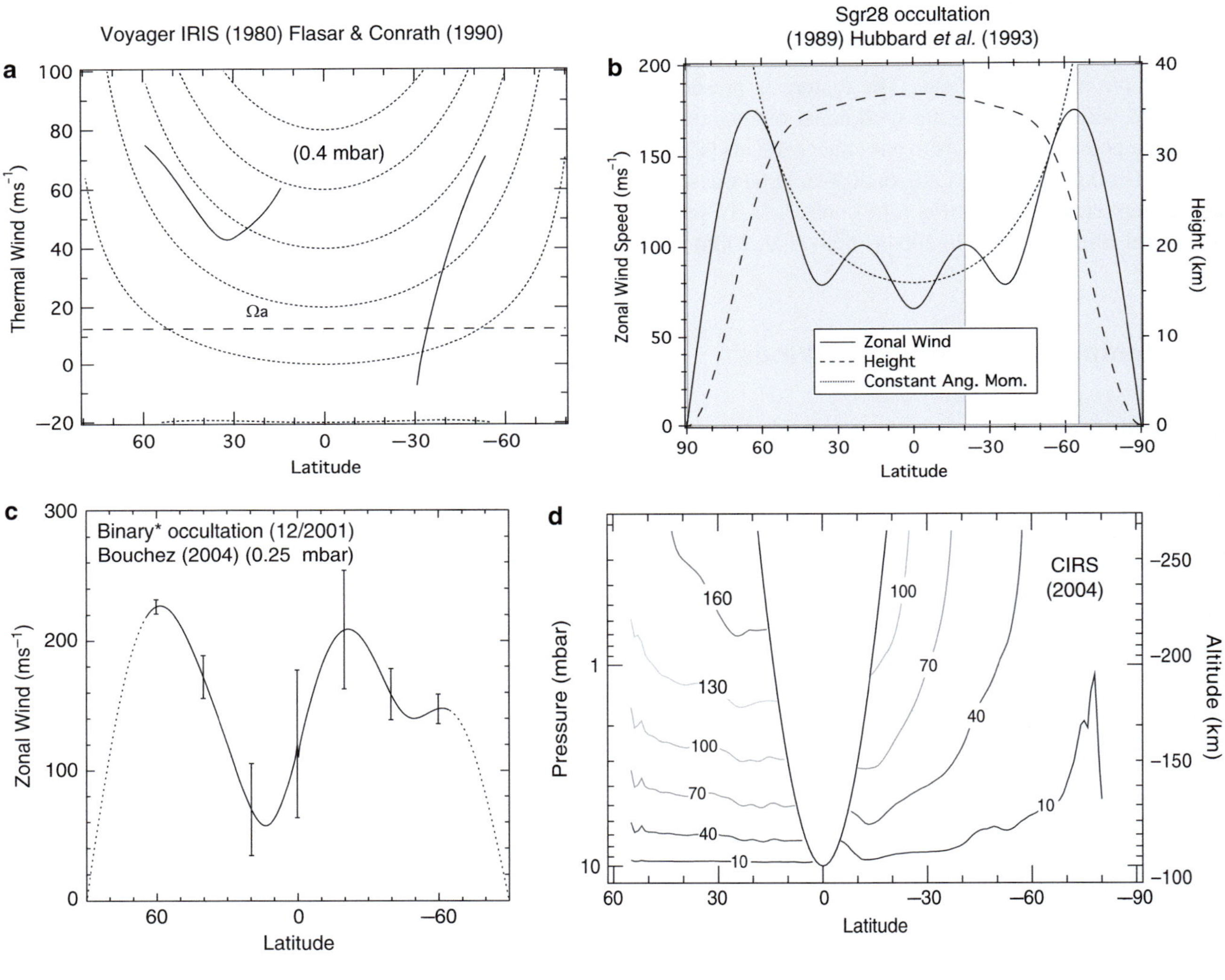

Fig. 14.11 Stratospheric zonal winds retrieved from derived temperature and pressure fields over different phases of a Titan year. (**a**) Winds at 0.4 mbar derived from temperatures retrieved from infrared spectra obtained by Voyager, assuming no motion at 10 mbar (Flasar and Conrath 1990). (**b, c**) Winds at roughly 0.25 mbar, derived from the shape of isobaric surfaces inferred from stellar occultations observed from Earth by Hubbard et al. (1993) and Bouchez (2004). (**d**) Winds derived from temperatures retrieved from Cassini CIRS spectra (Flasar et al. 2005). Except for panel d, all winds are given on a single iso-bar with latitude

telescopes (Luz et al. 2005, 2006; Kostiuk et al. 2001, 2005, 2006; Moreno et al. (2005). These observations probe the upper stratosphere and mesosphere, and they have generally indicated eastward zonal winds. Their difficulty is the limited spatial resolution: the dimension of telescopic fields of view has been on the order of 1/2 to 1 times Titan's diameter. Cloud tracking, done both from ground-based telescopes (Roe et al. 2005b; Bouchez and Brown 2005; Schaller et al. 2006a, b; Hirtzig et al. 2006) and the ISS and VIMS experiments on Cassini (Porco et al. 2005; Baines et al. 2005; Brown et al. 2006, are able to probe winds in the troposphere. However, few wind velocities have been measured, because discrete clouds are not frequently seen, and when they are, they are often ephemeral (caution must also be applied in that apparent cloud motions may not reflect the wind – clouds can evolve faster than they move, or in the case of orographic or otherwise surface-forced clouds may not appear to move at all despite the wind). Thus far winds have been measured in the south polar region (in summer), indicated weak eastward winds or no motion. A few clouds have been tracked at mid-southern latitudes (again in summer). These generally indiate eastward winds, with one determination as high as 34 ± 13 ms^{-1} (Porco et al. 2005).

A geomorphological clue about Titan's near-surface winds is the orientation pattern of the dunes which cover much of Titan's low latitudes, e.g. Lorenz and Radebaugh (2009). These indicate, contrary to model expectations, sand transport in an Eastwards (i.e. prograde) direction. If these were to represent a steady wind direction, it is difficult to understand the overall angular momentum balance (e.g. modeling by Tokano 2008) of Titan. It should be noted that the longitudinal dunes exhibit regional and local deviations from Eastwards, that such dunes form from the combination of near-orthogonal winds at different seasons, and that dunes may reflect only the strongest, rather than average, winds. Finally, it is possible that seasonal angular momentum changes in the atmosphere might cause changes in the solid body rotation of Titan's observable surface (Tokano and Neubauer 2005, which may have been observed (Stiles et al., 2008; Lorenz et al. 2008a).

14.4 Stratospheric Gases

The spatial distribution of organic gases can bear on seasonally varying meridional circulations in the middle atmosphere. The breakup of CH_4 and N_2 from photo- and electron impact dissociation tends to be high in Titan's atmosphere, and the abundances of most of the resulting organic compounds increase with altitude. Figure 14.12 illustrates the meridional variation of several organic compounds derived from observations by Voyager IRIS and Cassini CIRS. In both cases there is an enhancement of several species in the north polar region. In fact, the enhancements observed by Voyager in the northern spring are often much larger than those observed by Cassini in northern winter. Since their abundances increase with altitude (e.g. see Fig. 14.13), this is consistent with subsidence at high northern latitudes that persists from winter until early spring, probably within the polar vortex that was noted earlier. This is not unexpected. On Earth, there is a general cross equatorial circulation in the middle atmosphere, with ascent over the summer pole and descent over the winter pole within the strong circumpolar vortex (Andrews et al. 1987; Abrams et al. 1996a,b). Figure 14.13 depicts a meridional cross section of Titan's middle atmosphere, illustrating the warm temperatures and enhanced cyanoacetylene mixing ratio over the winter north pole, consistent with descent over the pole. There is also evidence in groundbased data of seasonal variation in at least one species (Roe et al. 2004).

14.5 Methane Meteorology

The dramatic images of stream networks, shorelines, and rounded pebbles on Titan taken from the Huygens probe convincingly show that Titan, like the Earth, has a surface carved by flowing fluids. Cassini imaging and radar have shown abundant lakes near the north pole, dry dunes near the equator, and a smaller number of lakes near the south pole. The central phenomenon connecting and likely explaining these disparate observations is the seasonal methane meteorological cycle on Titan – see also Chapter 7.

While for nearly two decades after Voyager it was debated whether or not Titan might support clouds or precipitation (e.g. Hunten et al. 1984 – see also Section 14.6), observations of variable cloud systems – first from ground-based observations and later from Cassini – have demonstrated that convection, condensation, and – presumably – precipitation are indeed sporadically prevalent across Titan. Understanding this methane meteorological cycle holds the key to understanding much of the diversity of features on the surface of Titan.

The first evidence for clouds on Titan came in 1995, just before Titan autumnal equinox, when Griffith et al. (1998) saw Titan brighten dramatically in surface-penetrating windows in near-infrared whole disk spectra. Radiative transfer analysis showed that this brightening was not tied to the surface but rather came from the middle of the troposphere, at the level expected for clouds caused by moist convection. The clouds that caused the brightening were estimated to have covered ~5–7% of the disk of Titan. Griffith et al. (2000) later found that a small amount of variable tropospheric cloud change was detected in 9 out of 11 nights of observations spanning from September 1993 until September

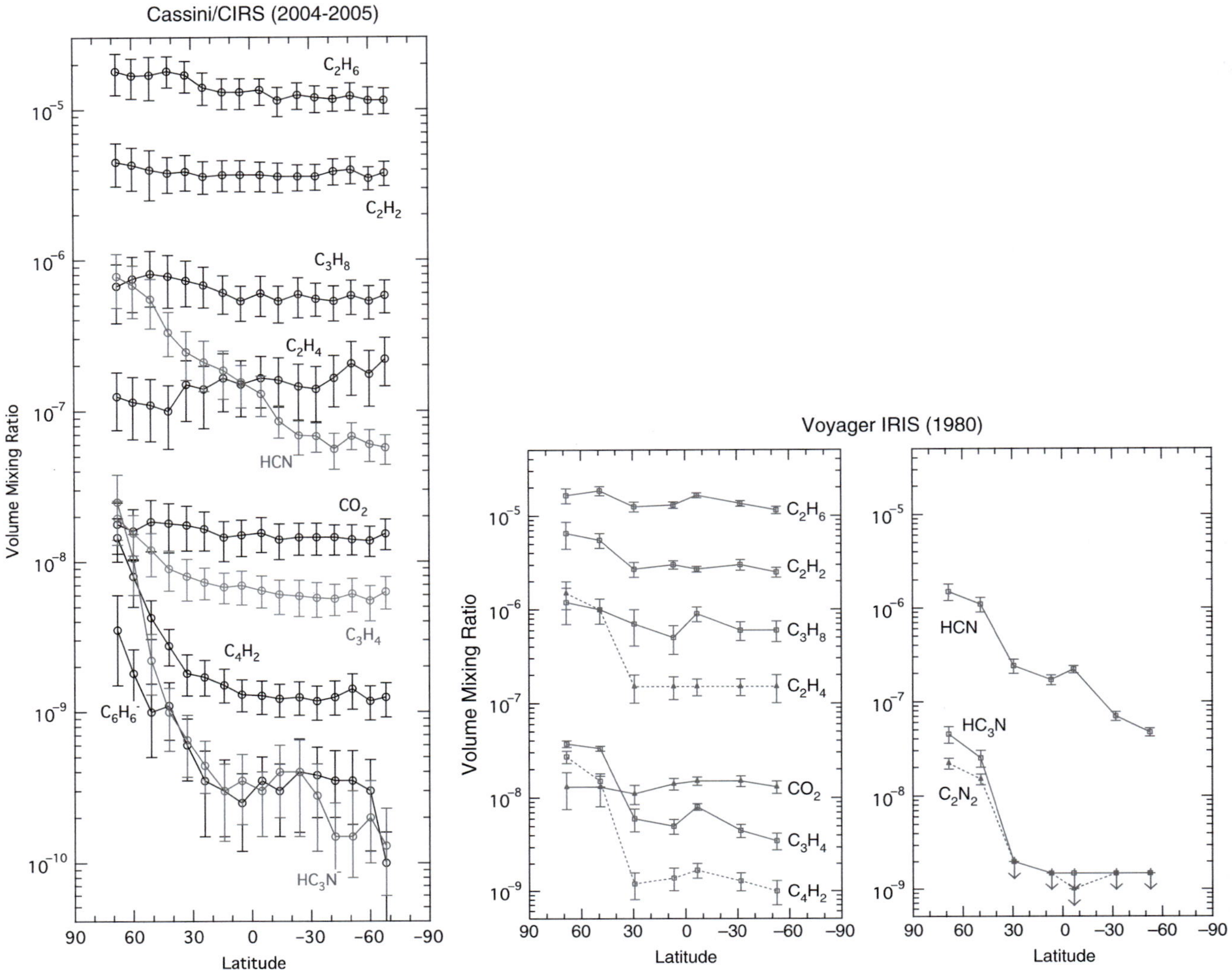

Fig. 14.12 Meridional distribution of several organic compounds from Voyager (Coustenis and Bézard 1995) and Cassini (Coustenis et al. 2007) thermal-infrared spectra.

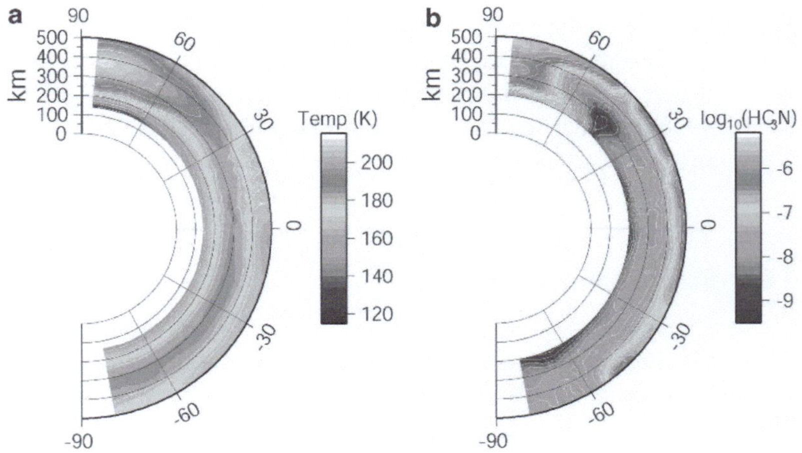

Fig. 14.13 Meridional cross section of middle-atmospheric temperatures and cyanoacetylene., retrieved from Cassini CIRS spectra obtained in early northern winter. After Teanby et al. (2007)

1999 (bracketing the 1995 equinox). These changes covered ~0.5% of the disk of Titan during this period and, for the small number of observations available, at times varied on time scales of hours or days.

The large cloud observed by Griffith et al. (1998) inspired many attempts to directly image clouds on Titan using then-new large telescope speckle interferometry and then adaptive optics systems. The first successful imaging came in December

2001, just before southern summer solistice, when Brown et al. (2002) and Roe et al. (2002) found variable clouds clustered near the south pole. This polar phenomenon had not been predicted, but Brown et al. (2002) suggested that it was possibly due to the heating of the surface during the years-long perpetual daylight of the polar summer and the subsequent initiation of convective instabilities. This hypothesis immediately drives home the realization that Titan's meteorological cycle is going to be strongly seasonally dependent, and that understanding of this cycle will require observations over a large fraction of Titan's 30 year season.

While clouds were now known to be sometimes present near the summer pole, nothing was known about their temporal and spatial distribution and variability. Bouchez and Brown (2005) first showed from 16 nights of observation in the winter of 2002 (coinciding precisely with Titan south polar solstice) that clouds were a persistent presence at the south pole of Titan, which was, at the time, the latitude of maximum daily insolation.

Photometric monitoring of Titan in the year following solstice using a 14-inch telescope and special filters designed to see photometric changes due to clouds (Fig. 14.15) found, early on, that the majority of the time the cloud cover on Titan was less than or equal to the ~1% observed by Bouchez and Brown (2005). Every several months, however, a cloud outburst covering a few percent of the disk was observed, and once in the 2 year period a cloud dwarfing all of the others was observed (Fig. 14.14). If the amount of rainfall is proportional to the cloud intensity (a reasonable but by no means certain assumption) that single large cloud accounted for the vast majority of the rainfall of the past several years (Schaller et al. 2006a). Cassini missed each of these major cloud events.

In concert with the 14-inch telescope observations, ground-based observed began obtaining snapshot observations of Titan with the adaptive optics systems at the Keck and Gemini Observatories. The long term monitoring that has been possible with the snapshot observations has led to several critical insights into the Titan cloud systems. Roe et al. (2005a) detected the first clouds seen at a location other than the south polar regions. These clouds were concentrated in a band near 40° southern latitude and first appeared about the time we had been expecting to see seasonal changes in the cloud systems. Continued long term monitoring, however, showed that these clouds were preferentially forming over a single longitude on the surface of Titan, indicative of some sort of geographic control, rather than a seasonal effect (Roe et al. 2005b). The large mid-latitude cloud outbursts seen by Roe et al. occurred between Cassini Titan encounters and so were not observed from the spacecraft (efforts to conduct distant Titan monitoring between encounters have since been introduced). Roe et al. (2005b) found that smaller mid-latitude clouds were a persistent presence apparently uncorrelated with geography. Griffith et al., (2005) analyzed high spatial resolution VIMS spectral cubes of southern mid-latitude clouds observed in Dec 2005 and found that the evolution of the cloud-top heights suggested updrafts of 2–4 m/s in larger cloud systems, and 8–10 m/s in localized spots, consistent with expected convective velocities.

The south polar cloud system was observed during Cassini's T0 encounter in early July 2004 (just after Saturn Orbit Insertion). Although a long-distance flyby (some 300,000 km) the geometry afforded good top-down views of the illuminated south polar region. A sequence of ISS images (Porco et al. 2005; Fig. 14.16) shows that the cloud system has much complex detail, and evolves during the few hours spanning the sequence. The system was also detected by VIMS (Brown et al. 2006).

In perhaps the strongest clue to date for understanding the seasonality of the hydrological cycle, Schaller et al. (2006b) found that the south polar cloud system, which was visible in

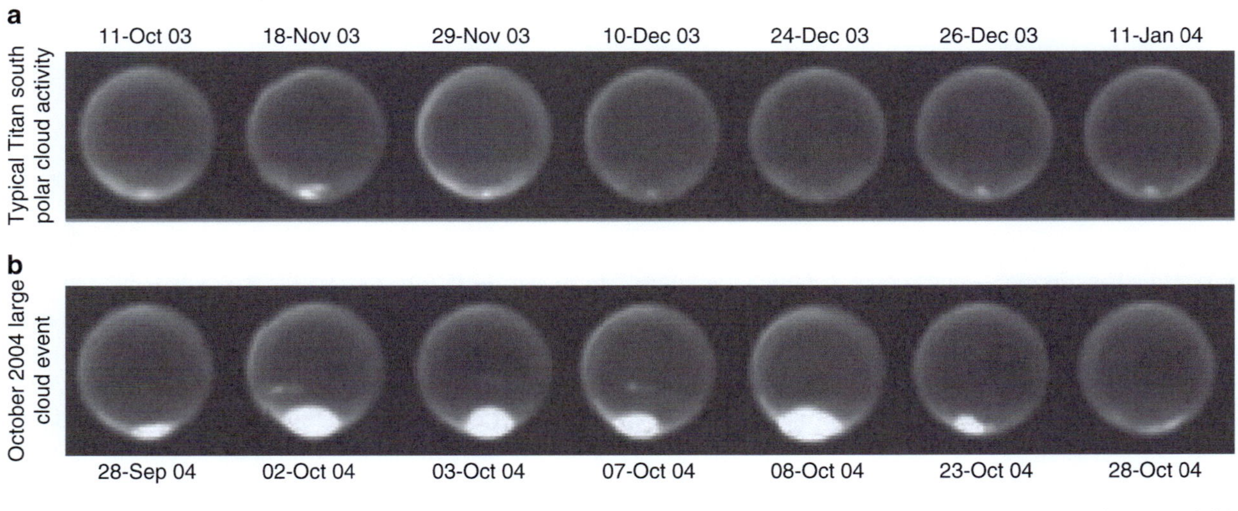

Fig. 14.14 A comparison between typical Titan cloud activity around solstice and the rare large cloud outburst of October 2004. Also visible are the 40° south clouds discovered by Roe et al. (2005a)

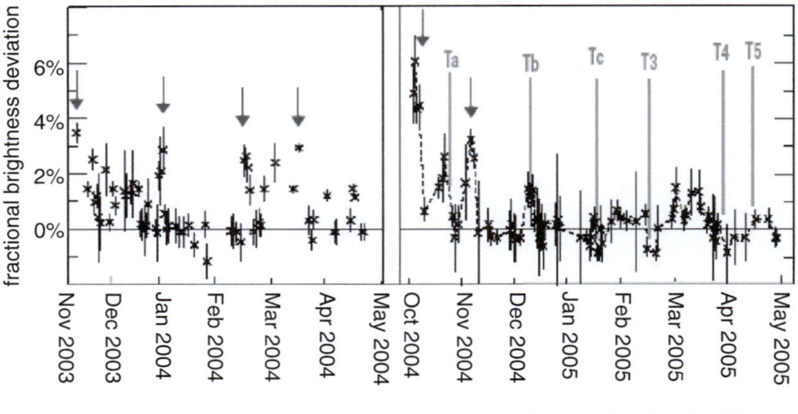

Fig. 14.15 Two year's worth of photometric monitoring of Titan. The large deviations from Titan's regular lightcurve, indicated with red arrows, correspond to large outbursts of clouds on Titan, as verified both by imaging from Keck and Gemini Observatories and from Cassini flybys. This project – now replaced with the significantly more sensitive IRTF monitoring program – first demonstrated that Titan meteorology was dynamic on timescales faster than the Cassini flybys, labeled as Ta-T5

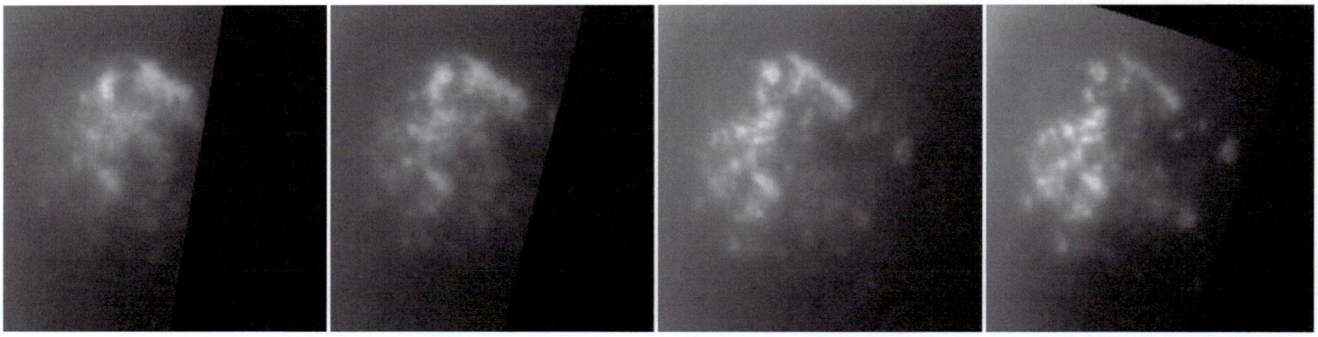

Fig. 14.16 A movie sequence of the south polar clouds. These images were taken a few hours apart, and show that the clouds around the south pole are changing on short timescales, suggesting they might be cumulus clouds with precipitation (methane rain or hail). (NASA/JPL/Space Science Institute)

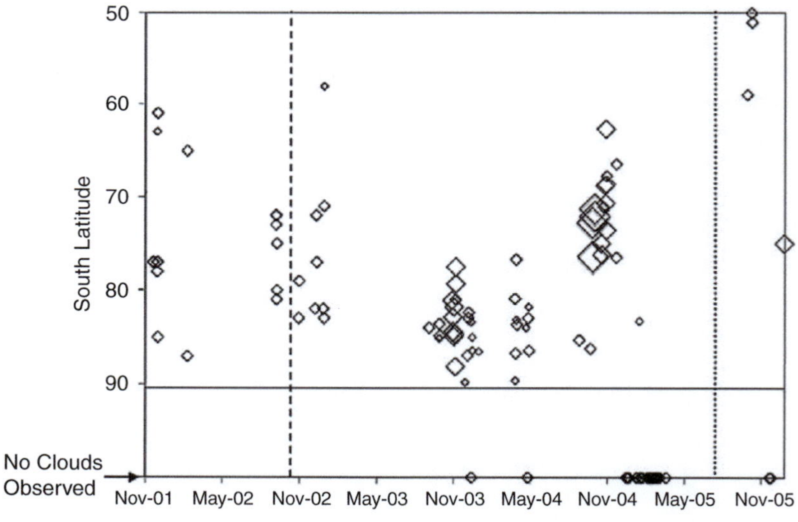

Fig. 14.17 The positions of groundbased detections of near-polar clouds since they were first observed in November 2001. The sizes of the diamonds are proportional to the size of the observed cloud. The diamonds on the lower axis show the times when no clouds were observed. Schaller et al. (2006b) found that the cloud system moved south with the solstice but then substantially dissipated soon thereafter. Continued observations showed that the major south polar system has not returned

almost every single image from 2001 until late 2004 suddenly disappeared. The disappearance coincided with the time period in which the south pole had abruptly become no longer the latitude of maximum insolation (Fig. 14.17 – see also Fig. 14.2). Cassini ISS data (Turtle et al. 2009) show a similar pattern, and indicate a few sporadic clouds at latitudes further north.

As the north pole of Titan began to come into sunlight and into view (from the Cassini perspective) starting in 2004, a large temporally and spatially unvarying scattering layer poleward of about 50° and at an altitude around 40 km began to reveal itself (see Fig 14.18). Griffith et al. (2006) used indirect evidence to suggest that this northern hood was a massive ethane cloud that had been earlier predicted by Rannou et al. (2006). The cloud is hypothesized to form when higher order hydrocarbons which were created by photolysis in the stratosphere cool and condense as they subside across the winter tropopause. The feature remains quite stable and appears as a persistent presence in the north of Titan until the present (Brown et al., submitted; Fig. 14.17). If the

Fig. 14.18 An extensive cloud system, likely of ethane ice crystals in the lower stratosphere and upper troposphere, imaged by VIMS in 2006 around Titan's North Pole. (NASA/JPL/University of Arizona)

hypothesis for the formation of this feature is correct, it is a direct tracer of circulation, showing where stratospheric air is crossing the tropopause. Following this system – and looking for its appearance in the south – as the northern winter season wanes may turn out to be an excellent diagnostic of seasonal tropospheric–stratospheric circulation.

Continued Cassini ISS and VIMS data (Turtle et al. 2009; Brown et al., submitted) show a similar pattern, are sensitive to fainter clouds than the groundbased image data and show that southern mid-latitude and polar cloud activity have continued at low-level thoughout the entire late-winter season. Continued ground-based and Cassini observations have confirmed that, after the disappearance of the major south polar clouds, the post-solstice, pre-equinox behavior of Titan has been somewhat constant. Occasional small mid-latitude clouds and occasional small south polar clouds have been observed. An even more sensitive ground-based monitoring program is now in place using same method used by Griffith et al. (1998) to detect the clouds in the first place. Schaller et al. (submitted) obtain a single full-disk spectrum of Titan on every single night that the near-infrared spectrograph is installed at the Infrared Telescope Facility (201 total nights thus far). With these spectra they can detect changes in cloud cover as small as a fraction of one percent. For most of the time period between solstice and in the years leading up to equinox there has been only modest cloud activity on Titan, although as sunlight reaches the northern polar lake region, locally-enhanced cloud activity has been detected there (Brown et al. 2009).

On the basis of the Huygens descent profile, Tokano et al. (2006) suggested drizzle might occur on Titan, and Ádámkovics et al. (2007), based on groundbased spectral imaging also suggested that drizzle might be widespread. However, Kim et al. (2008) have recently called this result into question, finding no evidence of such drizzle in similar data from the Gemini telescope. Observational evidence of drizzle is presently sparse at best.

Recently, a Cassini RADAR observation directed specifically at detecting precipitation was reported (Lorenz et al. 2008b). While optimized to see through the atmosphere, the instrument's sensitivity can be adjusted to pick up backscatter from raindrops (although in this mode the surface echo saturates the receiver, preventing the recovery of any surface information). Aerosol particles and cloud drops are much smaller than the radar wavelength, and thus – as with weather radar on earth – there is little scattering except from actual precipitation. The initial trial of this mode, yielding a negative result, took place at high northern latitude in May 2007 (T30), and placed an upper limit on the amount of drizzle that could be descending. If applied to observe methane rainstorms, of the type seen during southern midsummer at that pole, such observations would be easily able to measure the amount of rain within the cloud. It is hoped that opportunities for such observations may occur in the extended mission.

The combined ground and Cassini-based views of Titan clouds reveals that Titan is a meteorologically dynamic place with clouds changing on much faster time scales than the spacing of Cassini flybys, that major clouds frequently occur over particular spots, that major changes have occurred since the time of large cloud outbursts at southern summer solstice, and that low-level cloud activity, undetectable from the earth-based telescopes, is a nearly constant feature of Titan. that – perhaps – the several-years-long quiet period on Titan is now coming to a dramatic end.

14.6 Models of Cloud Microphysics and Dynamics

Early efforts to consider clouds on Titan were essentially as an extension of the haze, and are discussed in chapter 12 . The notion of 'rain without clouds' (Toon et al. 1988) – i.e. individual haze particles accreting a mantle of condensed ethane and then methane, became a prevalent paradigm in the absence of any data indicating discrete clouds. This was further appealing because it was believed that the haze provides surprisingly few condensation nuclei.

Consideration of clouds and rain extended beyond one dimension with the work of Samuelson et al. (1997) who considered that the polar regions might be distinct, having a larger supply of condensation nuclei and ethane. Lorenz (1993) also suggested that local saturation might occur beneath discrete clouds to form a rain shaft that would allow drops to reach the surface despite falling slowly through unsaturated air, and Lorenz et al. (2005) and Awal and Lunine (1994) have showed thermodynamic arguments to suggest that convective plumes can be somewhat energetic (with ascent rates of a few m/s) but must occupy only a small fraction of Titan's area . However, essentially one-dimensional considerations were also made since (generalizing the Huygens descent profile to much of Titan by Tokano et al. (2006) who suggested drizzle may be widespread on Titan – though Griffith et al. (2008) have noted that at least the polar and equatorial climates on Titan should be considered as quite distinct).

The first model of a storm system on Titan was that of Tokano et al. (2001a) who considered the generation of lightning on Titan, and in fact the data in that paper point to the possibility of large rain rates. Following observations of clouds on Titan, other models have been developed. Hueso and Sanchez-Lavega (2006) similarly show that terrestrial cloud dynamics models when adapted for Titan reproduce the observed scale of features, cloud top altitudes and ascent rates etc. seen on Titan. A more recent model is that by Barth and Rafkin (2007) which shows promise that the features of a variety of cloud types (stratiform as well as cumulus) may be reproduced. Barth and Rafkin (2009) have demonstrated that the cloud-top altitude is a strong function of, and thereby may serve as a proxy measurement for, methane humidity.

A complication in Titan cloud modeling is the possible role of ethane (Barth and Toon 2003) and Graves et al. (2008) have shown that evaporative cooling of drops can retard further evaporation such that individual drops may reach the ground (contrary to Lorenz 1993 who did not include this effect) even without local humidity enhancement in a rain shaft.

14.7 Models of Seasonal Change

While global circulation models have enjoyed some success at Mars, modeling Titan is fundamentally more challenging. First, the massive atmosphere has more 'memory' of its dynamical state and thus is intrinsically harder to model; indeed the model spin-up times are a computational inconvenience. Second, whereas an excellent global topography and thermal inertia dataset exists for Mars, topographic data for Titan is so far very sparse, and thermal inertia constraints even more so: Cassini will not be able to substantially address these concerns. Finally, the diverse terrestrial-like feedbacks of clouds, climate, aerosols and photochemistry make it challenging to include all the effects which influence the dynamics. A further complicating factor for Titan is the role of the gravitational tide.

Modeling efforts at Titan have generally so far been directed towards explaining various phenomena individually. Clearly a full understanding of Titan's meteorology will require the integration of a wide range of factors (in a parallel of Earth System Science).

14.7.1 Stratospheric Modeling

14.7.1.1 Haze Models

Initial attempts to explain the north-south asymmetry were largely heuristic. Smith et al. (1981) and Sromovsky et al. (1981) considered analytic descriptions of the albedo, also taking Lockwood's photometric record into account. It was noted that the asymmetry was apparently out of phase with the seasonal forcing (i.e. even though solar illumination was roughly equal in the two hemispheres at the Voyager season, one hemisphere was profoundly darker than the other).

An initial idea was that differential exposure of the two hemispheres to sunlight, and thus haze-producing ultraviolet light, might lead to different haze amounts and thus different albedos. Hutzell et al. (1993) explored this hypothesis using the McKay haze model (e.g. McKay et al. 1989; Toon et al. 1992), but found that the haze deposition timescale was too long for a seasonal effect to be important (i.e. while the seasonal forcing in the haze production would be enough in steady-state to yield the required change in haze column density, the haze column represents many Titan years-worth of photochemical production, and thus the seasonal variation is progressively diluted with depth in the atmosphere).

Hutzell et al. (1996) explored haze transport with analytically-specified meridional winds, and while quantitative agreement with observations was not obtained, the overall pattern of change was suggestive that this is a viable mechanism. There are two main effects of the Hadley circulation at a microphysical level – first there is the advection of small haze particles in the upper branch, from the summer to the winter hemisphere.

Lorenz et al. (1999) showed that a haze transport correlated with the solar latitude (i.e. incrementing the haze amount in the winter hemisphere and decrementing that in the summer by the same amount) broadly reproduced the seasonal cycle. While still analytical, this work had a more physical basis than the heuristic albedo parameterizations of the 1981 work. It further pointed out more robustly that the NSA cycle cannot completely explain the albedo changes observed from Earth – there must be some other effect due to effective area or albedo changes in both hemispheres (e.g. a change in the effective radius of Titan's haze layer – c.f. the discussion on the detached haze – or perhaps the influence of the polar hood).

Tokano et al. (1999) explored the influence of the haze cycle on Titan's circulation and temperatures, and investigated the effect of seasonal solar forcing on possible haze transport. Rannou et al. (2002) coupled haze microphysics more directly, and was able to reproduce several aspects of Titan's appearance, including the north–south asymmetry, observed limb-darkening, and perhaps most significantly the detached haze and its connection to the polar hood.

Little work has been conducted on the polar hood specifically. Yung (1987) noted the influence of the winter polar shadow on photochemistry.

14.7.1.2 Models of Seasonal Composition Changes

Attempts to model the seasonally varying distribution of organic compounds in the middle atmosphere have used two-dimensional simulations with parameterizations of eddy and wave momentum flux convergences. Until recently these simulations have had only gas distributions retrieved from Voyager IRIS spectra, which primarily provided meridional variations, and only crude information on the vertical structure. The hydrocarbon distribution was typically flat at most latitudes, except for an enhancement in the concentrations of the more reactive hydrocarbons at high winter latitudes that persisted until early spring (Fig. 14.13); the nitriles behaved similarly, except for HCN, which showed a broader range of latitudes over which its concentration increased toward the winter pole. In one of the earliest studies, Dire (2000) used a model with an array of photochemical hydrocarbon reactions coupled to a simple axisymmetric Lagrangian model with Newtonian radiative damping and Rayleigh friction. The forcing of the circulation was the seasonally varying temperature field computed from the full three-dimensional GCM of Hourdin et al (1995). Dire found that to fit the distribution of ethane, he needed a meridional circulation, but a very sluggish one. He was not able to fit the observed distribution of several other hydrocarbons, including acetylene and ethylene. Subsequent work (Lebonnois et al. 2001, 2003; Hourdin et al. 2004; Crespin et al. 2008) used a series of two-dimensional models based on the Hourdin et al. (1995) GCM, with photochemical coupling, and radiative transfer calculations that included the opacities of the organic compounds and hazes. By and large, these have had reasonable success in reproducing much of the observed spatial distributions seen by Voyager IRIS (and more recently Cassini CIRS). The models suggest that the meridional distributions were dependent on both the meridional circulations and mixing by barotropic eddies. The models predict that the circulation in the middle atmosphere is dominated by a pole-to-pole cell with ascent over the summer pole, followed by a short period after the equinox when the circulation reverses, with a brief transition period in which the ascending branch is at low latitudes and the circulation is more hemispherically symmetric. In these models the subsidence over the winter pole accounts for the enhancement in the concentrations of the organic compounds that are observed (Section 14.4).

The more complete spatial mapping provided by CIRS, both in the nadir- and limb-viewing modes, however, shows a structure that is quite complex, particularly near the north (winter) polar vortex, that the 2D models cannot adequately explain. Teanby et al (2008) have recently reported that the isoclines of several compounds, including HCN, HC_3N, C_3H_4, and C_4H_2 have an altitude-latitude structure that is tilted in the middle atmosphere, and these compounds exhibit a depletion zone near the vortex in the mesosphere. Clearly, the 2D parameterizations need to include additional transport processes to account for behavior like this. Understanding the seasonal change of the polar vortex – its buildup in the late fall and early winter, its mature phase into early spring, and its subsequent break-up – and the attendant changes in the organic concentrations will be key in elucidating these processes.

14.7.1.3 Temperature/Dynamics

Given the relatively short radiative relaxation time, the temperatures in the middle atmosphere will vary seasonally. Because the zonal winds and zonally averaged temperatures are coupled diagnostically by means of the thermal wind equation (Chapter 13), this means that the winds will also vary seasonally (Section 14.3). Angular momentum must be transported from one hemisphere to the other, in order to ensure that the strong winds are in the winter hemisphere, where it is coldest and the meridional gradient in temperature is largest. Flasar and Conrath (1990) noted that the overturning time of an axisymmetric meridional circulation is larger than the radiative relaxation time, and more comparable to a season on Titan. This "inertia" introduces a phase lag in the response of the temperature field in the middle atmosphere to the seasonal modulation of the solar heating, and, indeed, the stratospheric temperatures retrieved at northern spring equinox did exhibit some degree of asymmetry (Fig. 14.10). The simulations by Hourdin et al. (1995) indicated that eddy transports of angular momentum are also important. It is worth noting that, although these simulations did produce equatorial superrotation >100 m/s in the upper stratosphere, 3D simulations by others (Tokano et al. 1999; Richardson et al. 2007) have produced much weaker zonal winds, and the reasons for the discrepancy are not understood (Chapter 13). The angular momentum in Titan's zonal circulation likely changes seasonally (Tokano and Neubauer 2005), and the amount and timing of the changes are sensitive to model parameters such as moist convection (e.g. Mitchell 2009).

14.7.1.4 Tropospheric Modeling – Global

Since even the basics of Titan's methane meteorology were largely unknown before Cassini, there was comparatively little published work on latitudinal or seasonal variations. A notable exception is a modeling paper by Samuelson and Mayo (1997) which (while suggesting, erroneously, that moist convection could not happen on Titan) posed a balance between upward transport of methane by eddy diffusion against downward transport by raindrops. A parameter in this model is the number of condensation nuclei, which translates into the initial number of drops before they multiply by growth and breakup: the condensation nuclei were presumed to be in limited supply, formed ethane particles in turn supplied by sedimentation from above. While this assumption may not have been borne out by subsequent findings, the work is notable in that

Samuelson and Mayo suggested that the enhanced ethane flux at high latitudes resulting from the stratospheric downwelling over winter might therefore lead to a lower methane humidity near the poles, in agreement with an estimated humidity profile recovered (Samuelson et al. 1997) from Voyager 1 infrared data.

This humidity profile (varying symmetrically from a mixing ratio of about 5% at the equator to about 2% at high latitudes), together with the temperature profile recovered from the same data (93 K at the equator, dropping by 3–4 K towards both poles) constituted one of the main observables constraining the tropospheric circulation. An initial attempt to explore the effects of seasonal circulation on the methane abundance was made by Tokano et al. (2001b) who found only modest seasonal change in the methane column in a GCM with a simple condensation scheme; however, this pre-Cassini model had many unknowns, and failed even to reproduce the Samuelson data. (It should be noted that the Samuelson et al. (1997) humidity profile – for 1980 – has not been independently confirmed, and near-infrared investigations by Anderson et al. (2008) find different results, for 2000.) Tokano et al. (2001b) first proposed that large amounts of methane might be advected across the equator, leading to seasonal changes in methane amount, whereas Samuelson et al. (1997) ascribe such changes to spring polar condensation. It remains to be seen which of these (likely coupled) effects is most significant.

The humidity exerts a significant control on (but is likely only influenced slightly by) evaporation of methane from Titan's polar lakes. Stevenson and Potter (1986) suggested methane might seasonally accumulate near the poles, and that this might account for the observed symmetry in surface temperatures indicated in Voyager IRIS data. Maximum evaporation rates of several meters per year under likely Titan wind and humidity conditions have been calculated by Mitri et al. (2007), although observations have yet to show this process in action. On the other hand, lakes may be replenished with methane by rain, and Turtle et al. (2009) show evidence of dark spots near the south pole forming in 2005, in a region where cloud activity had been previously noted, suggesting perhaps that local flooding had occurred.

Models of circulation of Titan's atmosphere can generally be classified by the large-scale latitudinal structure of the circulation. The models of Rannou et al. (2006) predict a generally hemispherically symmetric circulation with northern and southern mid-latitude clouds being formed from slantwise advection of moist air from lower latitudes. In this circulation overall cloud frequency on Titan changes with time, but little latitudinal evolution appears. This model successfully reproduces the general behavior of the southern cloud systems.

Other circulation models are strongly seasonally varying and hemispherically asymmetric (Mitchell 2008; Mitchell et al. 2006; Tokano 2005). In these models the location of the mid-latitude clouds comes close to tracking, with some time lag, the seasonal latitude of maximum insolation. The location of the mid-latitude clouds marks the equivalent of the terrestrial inter-tropical convergence zone.

From the sensitivity studies performed by Tokano (2005), it appears that the essential difference between these two very different model behaviors is the presence of perennial polar haze in the Rannou et al. (2006) model. This haze prevents significant solar heating from occurring at the summer pole, causing the surface temperature maximum to be confined to within ~30° of the equator at all seasons. The muting of summer surface temperatures prevents a significant seasonal circulation asymmetry from developing.

While both classes of models do a moderately good job of reproducing the known southern cloud systems, the generally symmetric nature of the Rannou et al. (2006) model is strongly ruled out by the lack of northern mid-latitude clouds observed in Cassini VIMS data. The lack of appearance of northern mid-latitude clouds in ground-based data has long been known (Roe et al. 2005b; Schaller et al. 2006b), but, because of viewing angle and sensitivity limits, no definitive answer to the question of the existence of these clouds could be made. The VIMS data, however, with their substantially improved sensitivity and frequently excellent viewing condition conclusively show that Titan mid-latitude clouds are only present in the southern hemisphere during late northern winter (Brown et al., submitted).

The likely cause of the failure of the Rannou et al. (2006) model is the inclusion of significantly larger amounts of summer haze than actually occur on Titan. Examination of detailed haze distributions produced by the original models of Rannou et al. (2004) shows that they predict summer polar haze to by within a factor of two of the winter polar haze optical depth. While no studies have specifically focused on quantitatively examining the most recent summertime polar haze opacity, the long series of high-quality ground-based adaptive optics that were obtained starting near summer solstice (i.e. Brown et al. 2002; Roe et al. 2005a,b; Schaller et al. 2006a,b) showed significant winter haze at the north pole while showing essentially no haze over the summer southern pole. Observations at shorter wavelength could better constrain the polar haze opacities. Copious late-southern-summer observations are now available from Cassini which could quantitatively and definitively answer this question. Such an analysis should be considered a high priority.

Haze-free models like that of Mitchell et al. (2006), where the seasonal temperature maximum can travel as far as the pole itself, better reproduce the current lack of northern mid-latitude clouds. While significant winter haze is indeed known to accumulate, there is no substantial sunlight to block, so the overall effect of the circulation is small. The definitive test of these models will come, however, as Titan moves through equinox in August 2009. The cloud systems should, at this point, move through the equatorial regions (or perhaps disappear entirely) before reappearing at northern mid-latitudes and polar latitudes. While little hint of such movement can be seen in the data to March 2008 the rapid shifting should begin quite soon in these models.

One robust conclusion of the Rannou et al. simulation, however, is the existence of the winter tropopause level ethane/hydrocarbon cloud. Regardless of the details of surface thermal inertia, winter air will be subsiding, and the model appears to correctly anticipate the phenomenon. The fact that the model predicts the ethane cloud to occur northward of ~70° latitude, compared to the ~50° where it is observed is, again, likely a function of the missed details of the tropospheric circulation.

The Mitchell et al. (2006) model, in contrast, found the clouds to be strongly seasonally variable.

They found polar midsummer clouds to be common before abruptly shutting off in late summer, and they found mid-latitude clouds to be a persistent presence at a latitude related to the latitude of peak insolation. The mid-latitude clouds in their simulations are at the rising point in the Hadley cell, essentially the InterTropical Convergence Zone (ITCZ) by analogy with Earth, where the Titan tropics essentially extend from pole to pole. It should be noted, however, that the Mitchell et al. (2006) work does not accurately reproduce Huygens results.

In a recent extension of the model, Mitchell (2008) added a simple hydrological model which allowed methane to evaporate and thus disappear or to accumulate on Titan's surface. He found that, as appears to be true on Titan, the poles seasonally accumulate methane while the equator dries. The total amount of methane available appears to have a major affect on the circulation and cloud systems. It is possible that continued circulation modeling and cloud monitoring will help to understand the total reservoir available on Titan.

14.8 Conclusions

As this chapter is being written, Titan is about to enter a period of rapid seasonal change as the sun crosses the equator in 2009 into the northern hemisphere. A review of historical data and model predictions suggests the following effects may occur:

- Titan will optically darken overall over the next 5 years or so as the haze-rich northern hemisphere is presented towards the Earth.
- The meridional circulation will eventually reverse to conduct that excess haze from the north back to the south.
- The north UV polar hood will disappear by about 2013.
- This disappearance will presumably be accompanied by the breakup of the polar vortex, with concomitant changes in gas abundances and winds. A polar ethane cloud is currently present in the northern upper troposphere. Whether this will merely evaporate, or instead persist (perhaps augmented by methane condensation) remains to be seen.
- In the latter half of the decade 2010–2020, vigorous convective cloud activity (and likely precipitation) will likely occur over the northern polar regions, now known to host many hydrocarbon lakes.
- Predictions of how low- and mid-latitude tropospheric clouds will change are less certain: it will likely depend on the extent to which clouds are controlled by insolation-driven upwelling and to what extent by methane availability on the surface – the latter being rather poorly constrained. It does seem likely that northern midlatitude cloud activity will increase.

As of the time of writing, relatively little seasonal change has been documented with Cassini data alone – indeed the pace of documentation of seasonal change may have declined since the pre-Cassini era where it was one of the few topics that could be addressed with groundbased data, as described in this chapter. This slow digestion of Cassini results may be due in part to the novelty of the data, and in part due to the late summer season of Cassini's arrival, when seasonal change (apart from the abrupt termination of the south polar methane clouds) is at its slowest. In coming years, the time-base for seasonal change measurements will extend and seasonal change will become more rapid after equinox. Thus Titan's atmosphere, and efforts to document and understand it, will become more dynamic.

While observations in the Cassini Equinox and (hopefully) Solstice missions will form the centerpiece of these studies, the importance of long-term groundbased monitoring should be re-iterated, as this will be essential in bridging the gap between the end of Cassini and the start of any future Titan mission. Such groundbased monitoring includes not powerful infrared and millimeter-wave facilities for composition measurement, and adaptive optics imaging for cloud measurements, but also small-telescope monitoring of the disk-integrated albedo at a variety of wavelengths. Such measurements are within the capabilities of some amateur observers and university observatories (e.g. Lorenz et al. 2003). Small telescopes have also shown their utility in cloud monitoring (Schaller et al. 2005) and may also contribute by observations of such stellar occultation events that present themselves.

Acknowledgements RL acknowledges the support of a Cassini Data Analysis Program grant. MEB is supported by a grant from the National Science Foundation Planetary Astronomy program.

References

Abrams MC et al (1996a) ATMOS/ATLAS-3 observations of long-lived tracers and descent in the antarctic vortex in November 1994. Geophys Res Lett 23:2341–2344

Abrams MC et al. (1996b) Trace-gas transport in the arctic vortex inferred from ATMOS ATLAS-2 observations during April 1993. Geophys Res Lett 23:2345–2348

Ádámkovics M, de Pater I, Roe HG, Gibbard SG, GriffithCA (2004) Spatially-resolved spectroscopy at 1.6 microns of Titan's atmosphere and surface. Geophys Res Lett 31:L17S05

Ádámkovics M, Wong MH, Laver C, DePater I (2007) Widespread morning drizzle on Titan. Science 318:962–965

Anderson CM, Chanover NJ, McKay CP, Rannou P, Glenar DA, Hillman JJ (2004) Titan's haze structure in 1999 from spatially-resolved narrowband imaging surrounding the 0.94 Mm methane window. Geophys Res Lett 31:L17S06

Anderson CM, Young EF, Chanover NJ, McKay CP (2008) HST spectral imaging of Titan's haze and methane profile between 0.6 and 1 μm during the 2000 opposition. Icarus 194:721–745

Andrews DG, Holton JR, Leovy CB (1987) Middle atmosphere dynamics. Academic, Orlando

Awal M, Lunine JI (1994) Moist convective clouds in Titan's atmosphere. Geophys Res Lett 21:2491–2494

Baines KH, Drossart P, Momary TW, Formisano V, Griffith C, Bellucci G, Bibring J-P, Brown RH, Buratti BJ, Capaccioni F, Cerroni P, Clark RN, Coradini A, Cruikshank DP, Jaumann R, Langevin Y, Matson DL, McCord TB, Mennella V, Nelson RM, Nicholson PD, Sicardy B, Sotin C (2005) The atmospheres of Saturn and Titan in the near-infrared: first results of Cassini/VIMS. Earth Moon Planet 96:119–147

Barth EL, Rafkin SCR (2007) TRAMS: a new dynamic cloud model for Titan's methane clouds. Geophys Res Lett 34:L03203

Barth EL, Rafkin SCR (2009) Convective cloud heights as a diagnostic for methane environment on Titan. Icarus, in press

Barth EL, Toon O (2003) Microphysical modeling of ethane ice clouds in titan's atmosphere. Icarus 162:94–113

Bouchez AH (2004) Seasonal trends in Titan's atmosphere: haze, wind, and clouds. Ph. D. thesis, California Institute of Technology. URL: http://resolver.caltech.edu/CaltechETD:etd-10272003-092206

Bouchez H, Brown ME (2005) Statistics of Titan's South Polar Tropospheric Clouds. The Astrophysical Journal 618:L53–L56

Brown ME, Bouchez AH, Griffith CA (2002) Direct detection of variable tropospheric clouds near Titan's south pole. Nature 420:795–797

Brown ME, Schaller E, Roe H, Chen C, Roberts CJ, Brown RH, Baines KH, Clark RN (2009) Discovery of lake effect clouds on Titan. Geophys Res Lett 36:L01103

Brown ME, Roberts JE, and Schaller EL (2009) Clouds on Titan during the Cassini prime mission: a complete analysis of the VIMS data. Icarus, in press

Brown RH, Baines KH, Bellucci G, Buratti BJ, Capacionni F, Cerroni P, Clark RN, Coradini A, Cruikshank DP, Drossart P, Formisano V, Jaumann R, Langevin Y, Matson DL, McCord TB, Mennella V, Nelson RM, Nicholson PD, Sicardy B, Sotin C, Baugh N, Griffith C, Hansen G, Hibbitts K, Showalter M (2006) Observations in the Saturn system during approach and orbital insertion, with Cassini's Visual and Infrared Mapping Spectrometer (VIMS). Astron Astrophys 446:707–716

Caldwell JD, Smith PH, Tomasko MG, Weaver H (1992) Titan: evidence of seasonal change – a comparison of Voyager and Hubble Space Telescope images. Icarus 103:1–9

Chanover NJ, Anderson CM, McKay CP et al. (2003) Probing Titan's lower atmosphere with acousto-optic tuning. Icarus 163:150–163

Coustenis A, Bézard B (1995) Titan's atmosphere from Voyager infrared observations IV. Latitudinal variations of temperature and composition. Icarus 115:126–140

Coustenis A et al. (2007) The composition of Titan's stratosphere from Cassini/CIRS mid-infrared spectra. Icarus 189:35–62

Crespin A, Lebonnois S, Vinatier S, Bézard B, Coustenis A, Teanby NA, Achterberg RK, Rannou P, Hourdin F (2008) Diagnostics of Titan's stratospheric dynamics using Cassini/CIRS data and the 2-dimensional IPSL circulation model. Icarus 197:556–571

Dire JR (2000) Seasonal photochemical and meridional transport model for the stratosphere of Titan. Icarus 145:428–444

Flasar FM, Conrath BJ (1990) Titan's stratospheric temperature: a case for dynamical inertia? Icarus 85:346–354

Flasar FM, Samuelson RE, Conrath BJ (1981) Titan's atmosphere: temperature and dynamics. Nature 292:693–698

Flasar FM and 35 co-authors 2005. Titan's Atmospheric Temperatures, Winds, and Composition. Science 308:975–978

Gibbard SG, Macintosh B, Gavel B (1999) Titan: high resolution speckle images from the Keck Telescope. Icarus 139:189–201

Gibbard SG, de Pater I, Macintosh BA, Roe HG, Max CE, Young EF, McKay CP (2004) Titan's 2-micron surface albedo and haze optical depth in 1996–2004. Geophys Res Lett 31:L17S02

Graves SD, McKay CP, Griffith CA, Ferri F (2008) Rain and hail can reach the surface of Titan. Planet Space Sci 56(3–4):346–357

Griffith CA, Owen T, Miller GA, Geballe T (1998) Transient clouds in Titan's lower atmosphere. Nature 395:575–578

Griffith CA, Hall JL, Geballe TR (2000) Detection of daily clouds on Titan. Science 290:509–513

Griffith CA, Penteado P, Baines K, Drossart P, Barnes J, Bellucci G, Bibring J, Brown R, Buratti B, Capaccioni F, Cerroni P, Clark R, Combes M, Coradini A, Cruikshank D, Formisano V, Jaumann R, Langevin Y, Matson D, McCord T, Mennella V, Nel-son R, Nicholson P, Sicardy B, Sotin C, Soderblom LA, Kursinski R (2005) The evolution of Titan's mid-latitude clouds. Science 310: 474–477

Griffith CA, Penteado P, Rannou P, Brown R, Boudon V, Baines KH, Clark R, Drossart P, Buratti B, Nicholson P, McKay CP, Coustenis A, Negrao A, Jaumann R (2006) Evidence for a polar ethane cloud on Titan. Science 313:1620–1622

Griffith CA, McKay CP, Ferri F (2008) Titan's tropical storms in an evolving atmosphere. Astron J 687:L41–L44

Hirtzig MA, Coustenis E, Gendron P, Drossart A, Negrão M, Combes O, Lai P, Rannou S, Lebonnois, Luz D (2006) Monitoring atmospheric phenomena on Titan. Astron Astrophys 456:761–774

Hourdin F, Talagrand O, Sadourny R, Régis C, Gautier D, McKay C (1995) General circulation of the atmosphere of Titan. Icarus 117:358–374

Hourdin F, Lebonnois S, Luz D, Rannou P (2004) Titan's stratospheric composition driven by condensation and dynamics. J Geophys Res 109:E12005, doi:10.1029/2004JE002282

Hubbard WB and 45 colleagues (1993) The occultation of 28 Sgr by Titan. Astron Astrophys 269:541–563

Hueso R, Sánchez-Lavega A (2006) Methane storms on Saturn's moon Titan. Nature 442:428–431

Hunten DM, Tomasko MG, Flasar FM, Samuelson RE, Strobel DF, Stevenson DJ (1984) Titan. In Saturn, University of Arizona Press, Tucson, pp 671–759

Hutzell WT, McKay CP, Toon OB (1993) Effects of time-varying haze production on Titan's geometric albedo. Icarus 105:162–174

Hutzell WT, McKay CP, Toon OB, Hourdin F (1996) Simulations Titan's brightness by a two-dimensional haze model. Icarus 119:112–129

Jennings DE, Flasar FM, Kunde VG, Samuelson RE, Pearl JC, Nixon CA, Carlson RC, Mamoutkine AA, Brasunas JC, Guandique E, Achterberg RK, Bjoraker GL, Romani PN, Segura ME, Albright SA, Elliott MH, Tingley JS, Calcutt S, Coustenis A, Courtin R (2009) Titan's surface brightness temperatures. Astrophys J Lett 691:L103–L105

Karkoschka E, Lorenz RD (1997) Latitudinal variation of aerosol sizes inferred from Titan's shadow. Icarus 125:369–379

Kim SJ, Trafton LM, Geballe TR (2008) No evidence of morning or large-scale drizzle on Titan. Astron J 679:L53–L56

Kostiuk T, Fast KE, Livengood TA, Hewagama T, Goldstein JJ, Espenak F, Buhl D (2001) Direct measurement of winds on Titan. Geophys Res Lett 28:2361–2364

Kostiuk T, Livengood TA, Hewagama T, Sonnabend G, Fast KE, Murakawa K, Tokunaga AT, Annen J, Buhl D, Schmülling F (2005) Titan's stratospheric zonal wind, temperature and ethane abundance a year prior to Huygens insertion. Geophys Res Lett 32:L22205

Kostiuk T, Livengood TA, Sonnabend G, Fast KE, Hewagama T, Murakawa K, Tokunaga AT, Annen J, Buhl D, Schmülling F, Luz D, Witasse O (2006) Stratospheric global winds on Titan at the time of the Huygens descent. J Geophys Res 111:E07S03

Lebonnois S, Toublanc D, Hourdin F, Rannou P (2001) Seasonal variations of Titan's atmospheric composition. Icarus 152:384–406

Lebonnois S, Hourdin F, Rannou P, Luz D, Toublanc D (2003) Impact of the seasonal variations of composition on the temperature field of Titan's stratosphere. Icarus 163:164–174

Lockwood GW, Thompson DT (1979) A relationship between solar activity and planetary albedos. Nature 280:43–45

Lockwood GW, Thompson DT (2009) Seasonal photometric variability of Titan, 1972-2006, Icarus 200:616–629

Lorenz RD (1993) The life, death and afterlife of a raindrop on Titan. Planet Space Sci 41:647–655

Lorenz RD (2008) The changing face of Titan. Phys Today 61:34–39

Lorenz RD, Radebaugh J The Global Pattern of Titan's Dunes: Radar Survey from the Cassini Prime Mission (2009). Geophysical Research Letters. 36: L03202

Lorenz RD, Smith PH, Lemmon MT, Karkoschka E, Lockwood GW, Caldwell J (1997) Titan's north–south asymmetry from HST and Voyager imaging: comparisons with models and groundbased photometry. Icarus 127:173–189

Lorenz RD, Lemmon MT, Smith PH, Lockwood GW (1999) Seasonal change on Titan observed with the Hubble Space Telescope WFPC-2. Icarus 142:391–401

Lorenz RD, Young EF, Lemmon MT (2001) Titan's smile and collar: HST observations of seasonal change, 1994–2000. Geophys Res Lett 28:4453–4456

Lorenz RD, Dooley JM, West JD, Fujii M (2003) Backyard spectroscopy and photometry of Titan, Uranus and Neptune. Planet Space Sci 51:113–125

Lorenz RD, Lemmon MT, Smith PH (2004) Seasonal change in Titan's haze 1992–2002 from Hubble Space Telescope Observations. Geophys Rev Lett 31:L10702, doi:10.1029/2004GL019864

Lorenz RD, Griffith CA, Lunine JI, McKay CP, Renno NO (2005) Convective Plumes and the Scarcity of Clouds on Titan. Geophysical Review Letters 32: L01201

Lorenz RD, Lemmon MT, Smith PH (2006) Seasonal evolution of Titan's dark polar hood: midsummer disappearance observed by the Hubble Space Telescope. Monthly Notices Royal Astron Soc 369:1683–1687

Lorenz RD, Stiles B, Kirk RL, Allison M, Persi del Marmo P, Iess L, Lunine JI, Ostro SJ, Hensley S (2008a) Titan's rotation reveals an internal ocean and changing zonal winds. Science 319: 1649–1651

Lorenz RD, West RD, Johnson WTK (2008b) Cassini RADAR constraint on Titan's winter polar precipitation. Icarus 195:812–816

Luz D, Civeit T, Courtin R, Lebreton J-P, Gautier D, Rannou P, Kaufer A, Witasse O, Lara L, Ferri F (2005) Characterization of zonal winds in the stratosphere of Titan with UVES. Icarus 179: 497–510

Luz D, Civeit T, Courtin R, Lebreton JP, Gautier D, Witasse O, Kaufer A, Ferri F, Lara L, Livengood T, Kostiuk T 2006. Characterization of zonal winds in the stratosphere of Titan with UVES: 2. Observations coordinated with the Huygens Probe entry, Journal of Geophysical Research, 111, E08S90

McKay CP, Pollack JB, Courtin R (1989) The thermal structure of Titan's atmosphere. Icarus 80:23–53

Mitchell J (2008) The drying of Titan's dunes: Titan's methane hydrology and its impact on atmospheric circulation. J Geophys Res 113: E08015

Mitchell JL (2009) Coupling convectively-driven atmospheric circulation to surface rotation: evidence for active methane weather in the observed spin rate drift of Titan. Astrophys J 692:168–173

Mitchell JL, Pierrehumbert RT, Frierson DMW, Caballero R (2006) The dynamics behind Titan's methane clouds. Proc Nat Acad Sci 103:18421–18426

Mitri GL, Showman AP, Lunine JI, Lorenz RD (2007) Hydrocarbon lakes on Titan. Icarus 186:385–394

Moreno R, Marten A, Hidayat T (2005) Interferometric measurements of zonal winds on Titan. Astron Astrophys 437:319–328

Porco CC, Baker E, Barbara J, Beurle K, Brahic A, Burns JA, Charnoz S, Cooper N, Dawson DD, Del Genio AD, Denk T, Dones L, Dyudina U, Evans MW, Fussner S, Giese B, Grazier K, Helfenstein P, Ingersoll AP, Jacobson RA, Johnson TV, McEwen A, Murray CD, Neukum G, Owen WM, Perry J, Roatsch T, Spitale J, Squyres S, Thomas P, Tiscareno M, Turtle EP, Vasavada AR, Veverka J, Wagner R, West R (2005) Imaging of Titan from the Cassini spacecraft. Nature 434:159–168

Rages K, Pollack JB (1983) Vertical distribution of scattering hazes in Titan's upper atmosphere. Icarus 55:50–62

Rannou P, Hourdin F, McKay CP (2002) A wind origin for Titan's haze structure. Nature 418:853–856

Rannou P, Hourdin F, McKay CP, Luz D (2004) A coupled dynamics-microphysics model of Titan's atmosphere. Icarus 170: 443–462

Rannou P, Montmessin F, Hourdin F, Lebonnois S (2006) The latitudinal distribution of clouds on Titan. Science 311:201–205

Richardson MI, Toigo AD, Newman CE (2007) PlanetWRF: a general purpose, local to global numerical model for planetary atmospheric and climate dynamics. J Geophys Res 112:E09001

Roe HG, de Pater I, Macintosh BA, McKay CP (2002) Titan's clouds from Gemini and Keck Adaptive Optics Imaging. Astrophys J 581: 1399–1406

Roe HG, de Pater I, McKay CP (2004) Seasonal variation of Titan's stratospheric ethylene (C_2H_4) observed. Icarus 169:440–461

Roe HG, Bouchez AH, Trujillo CA, Schaller EL, Brown ME (2005a) Discovery of temperate latitude clouds on Titan. Astrophys J 618: L49–L52

Roe HG, Brown ME, Schaller E, Bouchez AH, Trujillo AC (2005b) Geographic control of Titan's mid-latitude clouds. Science 310:477–479

Samuelson RE, Mayo LA (1997) Steady-state model for methane condensation in Titan's troposphere planet. Space Sci 45:949–958

Samuelson RE, Nath NR, Borysow A (1997) Gaseous abundances and methane supersaturation in Titan's troposphere. Planet Space Sci 45:959–980

Schaller EL, Brown ME, Roe HG, Bouchez AH, Trujillo CA (2005) Cloud activity on Titan during the Cassini mission, Lunar Planet. Sci. 36 Abstract 1989

Schaller EL, Brown ME, Roe HG, Bouchez AH, Trujillo AH (2006a) A large cloud outburst at Titan's south pole. Icarus 182:224–229

Schaller EL, Brown ME, Roe HG, Bouchez AH, Trujillo CA (2006b) Dissipation of Titan's south polar clouds. Icarus 184:517–523

Sicardy B, 51 colleagues (2006) The two Titan stellar occultations of 14 November 2003. J Geophys Res 111:E11S91, doi:10.1029/2005JE002624

Smith BA, Soderblom LA, Beebe R, Boyce J, Briggs G, Bunker A, Collins SA, Hansen CJ, Johnson TV, Mitchell JL, Terrile RJ, Carr M, Cook AF II, Cuzzi J, Pollack JB, Danielson GE, Ingersoll A, Davies ME, Hunt GE, Masursky H, Shoemaker E, Morrison D, Owen T, Sagan C, Veverka J, Strom R, Suomi V (1981) Encounter with Saturn: Voyager 1 imaging results. Science 212:163–182

Smith BA, Soderblom LA, Batson R, Bridges P, Inge J, Masursky H, Shoemaker E, Beebe R, Boyce J, Briggs G, Bunker A, Collins SA, Hansen CJ, Johnson TV, Mitchell JL, Terrile RJ, Carr M, Cook AF II, Cuzzi J, Pollack JB, Danielson GE, Ingersoll A, Davies ME, Hunt GE, Morrison D, Owen T, Sagan C, Veverka J, Strom R, Suomi V (1982) A new look at the Saturn system: the Voyager 2 images. Science 215:504–537

Smith PH, Lemmon MT, Lorenz RD, Sromovsky LA, Caldwell J, Allison MD (1996) Titan's Surface, Revealed by HST Imaging. Icarus 119:336–349

Sromovsky LA, Suomi VE, Pollack JB, Kraus RJ, Limaye SS, Owen T, Revercomb HE, Sagan C (1981) Implications of Titan's north–south brightness asymmetry. Nature 292:698–702

Stevenson DJ, Potter BE (1986) Titan's latitudinal temperature distribution on seasonal cycles. Geophys Res Lett 13:93–96

Stiles B, Kirk R, Lorenz R, Hensley S, Lee E, Ostro S, Gim Y, Hamilton G, Johnson WTK, West R, Team CRADAR (2008) Estimating Titan's spin state from Cassini SAR data. Astronomical J 135: 1669–1680

Teanby NA, Irwin PGJ, de Kok R, Vinatier S, Bézard B, Nixon CA, Flasar FM, Calcutt SB, Bowles NE, Fletcher L, Howett C, Taylor FW (2007) Latitudinal variations of HCN, HC_3N and C_2N_2 in Titan's stratosphere derived from Cassini CIRS data. Icarus 181:364–384

Teanby NA, de Kok R, Irwin PGJ, Osprey S, Vinatier S, Gierasch PJ, Read PL, Flasar FM, Conrath BJ, Achterberg RK, Bézard B, Nixon CA, Calcutt SB (2008) Titan's winter polar vortex structure revealed by chemical tracers. J Geophys Res 113:E12003. doi:10.1029/2008JE003218

Tokano T (2005) Meteorological assessment of the surface temperatures on Titan: constraints on the surface type. Icarus 173:222–242

Tokano T (2008) Dune-forming winds on Titan and the influence of topography. Icarus 194:243–262

Tokano T, Neubauer FM (2005) Wind-induced seasonal angular momentum exchange at Titan's surface and its influence on Titan's length-of-day. Geophys Res Lett 32:L24203

Tokano T, Neubauer FM, Laube M, McKay CP (1999) Seasonal variation of Titan's atmospheric structure simulated by a general circulation model. Planet Space Sci 47:493–520

Tokano T, Molina-Cuberos GL, Lammer H, Stumptner W (2001a) Modelling of thunderclouds and lightning generation on Titan. Planet Space Sci 49:539–560

Tokano T, Neubauer FM, Laube M, McKay CP (2001b) Three-dimensional Modeling of the tropospheric methane cycle on Titan. Icarus 153:130–147

Tokano T, McKay CP, Neubauer FM, Atreya SK, Ferri F, Fulchignoni M, Niemann HB (2006) Methane drizzle on Titan. Nature 442:432–435

Toon OB, McKay CPR (1988) Courtin and T Ackerman, Methane Rain on Titan. Icarus 75:255–284

Toon OB, McKay CP, Griffith CA, Turco RP (1992) A physical model of Titan's aerosols. Icarus 95:24–53

Turtle EP, Perry JE, McEwen AS, DelGenio AD, Barbara J, West RA, Dawson DD, Porco CC (2009) Cassini imaging of Titans high latitude lakes, clouds, and south polar surface changes. Geophys Res Lett. 36:L02204

Young EF, Rannou P, McKay CP et al (2002) A three-dimensional map of Titan's tropospheric haze distribution based on Hubble Space Telescope imaging. Astronomical J 123:3473–3486

Young EF, Puetter R, Yahil A (2004) Direct imaging of Titan's extended Haze Layer. Geophys Res Lett 31:L17S09

Yung YL (1987) An Update of Nitrile Photochemistry on Titan. Icarus 72:468–472

Chapter 15
Mass Loss Processes in Titan's Upper Atmosphere

R.E. Johnson, O.J. Tucker, M. Michael, E.C. Sittler, H.T. Smith, D.T. Young, and J.H. Waite

Abstract Although Titan's atmospheric column density is about ten times that of the Earth's, its measured $^{15}N/^{14}N$ ratio suggests that considerable escape has occurred or that Titan's original material had a ratio closer to that of cometary materials. A number of active escape processes have been proposed: thermal escape, chemical-induced escape, slow hydrodynamic escape, pick-up ion loss, ionospheric outflow and plasma-ion-induced atmospheric sputtering. These loss processes and relevant simulations are reviewed in light of recent Cassini data.

Analysis of Cassini data collected in Titan's thermosphere and corona indicate that thermal loss of hydrogen occurs at a rate comparable to pre-Cassini estimates. This escape of hydrogen is accompanied by a significant loss of methane due to formation and precipitation of hydrocarbons (~2×10^{29} amu/s). However, there is much less agreement on the rates for the escape of methane and nitrogen. Recent estimates, ~$0.3–5 \times 10^{28}$ amu/s, are much larger than the pre-Cassini estimates and also much larger than the measured ion loss rates (~$1–5 \times 10^{26}$ amu/s). If the largest heavy molecule escape rates are assumed, a significant fraction of the present atmosphere would have been lost to space in 4 Gyr. Because understanding the nature of the active escape processes is critical, a number of data sets were used to model the methane and nitrogen escape rates. The solar heating/cooling rate and the nitrogen density profile vs altitude were used in a fluid dynamic model to extract an average net upward flux below the exobase; the altitude dependence of the diffusion of methane through nitrogen was described below the exobase allowing for upward flow and escape; the coronal structure above the exobase was simulated by plasma and photon-induced hot particle production; and measurements of the emission from excited-dissociation products were used to estimate escape. In the latter two models, hot recoils from photochemistry or plasma-ion-induced heating lead to escape, whereas in the first two models heat conduction from below was assumed to enhance Jeans escape, a process referred to as slow hydrodynamic escape. These models are compared to each other and to recent simulations of Titan's exobase region. It is found that these simulations are inconsistent with the slow hydrodynamic escape model and that the composition of the magnetospheric plasma at Titan's orbit is inconsistent with the largest carbon loss rates suggested.

R.E. Johnson and O.J. Tucker (✉)
University of Virginia, Charlottesville, VA, 22904, USA
Physics Department, NYU, NY, 10003, USA
email: rej@virginia.edu

M. Michael
Department Civil Engineering, Indian Institute of Technology, Kanpur, India

E.C. Sittler
GSFC, Greenbelt, MD, USA

H.T. Smith
JHUAPL, Laurel, MD, USA

D.T. Young and J.H. Waite
SwRI, San Antonio, TX, USA

15.1 Introduction

Titan is unique among outer solar system satellites in that it has an atmosphere with a column density about ten times that of Earth and an atmospheric mass to solid ratio comparable to that of Venus. Titan's atmosphere is not only thick, but is also extended and consists of over 95% N_2, about 2–3% CH_4 with some H_2 and other minor species. If such an atmosphere had been present on Io, Europa or Ganymede in an earlier epoch, it would have been removed by plasma trapped in the Jovian magnetosphere (Johnson 2004). Therefore, the use of Cassini spacecraft data to understand the persistence of Titan's atmosphere and to describe its present mass loss rate provides an important end point for understanding the erosion of atmospheres on other planets and satellites. It is also important for describing Titan's role in populating Saturn's magnetosphere with neutrals and plasma. The emphasis in this chapter is atmospheric escape, but mass loss to the surface will also be discussed because it is related. Loss processes are often given as a molecular flux, but to compare the relative importance of the escape processes, global average mass loss rates will also be considered.

R.H. Brown et al. (eds.), *Titan from Cassini-Huygens*,

Prior to Cassini's arrival at Saturn, modeling based on Voyager data indicated that the hydrogen escape rate was large (~1–3 × 10^{28}amu/s) (e.g., Lebonnois et al. 2003) consistent with a large loss rate for methane by precipitation as hydrocarbons, ~10–30 × 10^{28} amu/s (e.g., Wilson and Atreya 2004). However, the rates for carbon and nitrogen escape to space were estimated to be relatively small (~0.05 × 10^{28} amu/s) and dominated by atmospheric sputtering (Shematovich et al. 2003). Recent analysis of the structure of Titan's thermosphere and corona attained from Cassini data have led to substantially larger estimates of the escape rate for these heavier species (0.3–5 × 10^{28} amu/s). At the largest escape rates suggested, a mass that is about the size of the present atmosphere would have been lost to space in 4 Gyr. Although this atmospheric loss rate is still an order of magnitude smaller than the carbon precipitation rate, evaluating the proposed escape processes is critical for understanding the evolution of Titan's atmosphere.

Atoms and molecules can be lost from an atmosphere both as ions and neutrals as indicated in Fig. 15.1. A number of active escape processes have been proposed: thermal escape (Cui et al. 2008), chemically-induced escape (DeLaHaye et al. 2007b), slow hydrodynamic escape (Strobel 2008a, 2009; Yelle et al. 2008), pick-up ion loss and ionospheric outflow (Ledvina et al. 2005; Wahlund et al. 2005; Sillanpaa et al. 2006; Ma et al. 2006; Hartle et al. 2006a, b; Coates et al. 2007) and plasma-induced atmospheric sputtering (Shematovich et al. 2003; Michael et al 2005a; DeLaHaye 2007a).

The mass loss rate due to pick-up ion formation and sweeping combined with ionospheric outflow is highly variable and of the order of ~10^{25} amu/s (Wahlund et al. 2005; Hartle et al. 2006a, b). Globally averaged loss rates for neutrals have been estimated using 1D models of Cassini data. The altitude dependence of the H_2 density was used to obtain a hydrogen loss rate of ~1.6 × 10^{28} amu/s (Cui et al. 2008). Bell et al. (2009) recently obtained a similar result. Hydrogen loss implies that methane is lost by precipitation (~2 × 10^{29} amu/s using the largest rate) starting at relatively high altitudes in Titan's atmosphere (Chapter 8; Mandt et al. 2009). Three different 1D models were used to estimate a heavy molecule loss rate: ~0.3–5 × 10^{28} amu/s. The solar heating rate was used in a fluid dynamic model to extract an average net upward flux below the exobase (Strobel 2008a, 2009); the diffusion of methane through nitrogen was described below the exobase allowing for escape (Yelle et al. 2008); and the coronal structure above the exobase was simulated by presuming the observed density profiles were due to solar- and plasma-induced hot particle production (DeLaHaye et al. 2007a). In the latter, the hot recoils from photochemistry and the incident plasma produce the coronal structure. In the former models, the upward flow is assumed to be driven by heat conduction from below and referred to as slow hydrodynamic escape (Strobel 2008a). In this chapter these models are compared to each other and to recent simulations of the exobase region. The simulations and preliminary estimates of the composition of the magnetospheric plasma at Titan's orbit appear to be inconsistent with the largest methane escape rates suggested. However, because of the changing relationship between the solar and plasma illuminations (Fig. 15.1) and the plasma variability (Chapter 16), there is considerable work to be done in order to understand the extensive Cassini data set. Although the situation at Titan differs from that at Mars and Venus where the plasma and sunlight both come from, roughly, the same direction, the changing encounter geometries

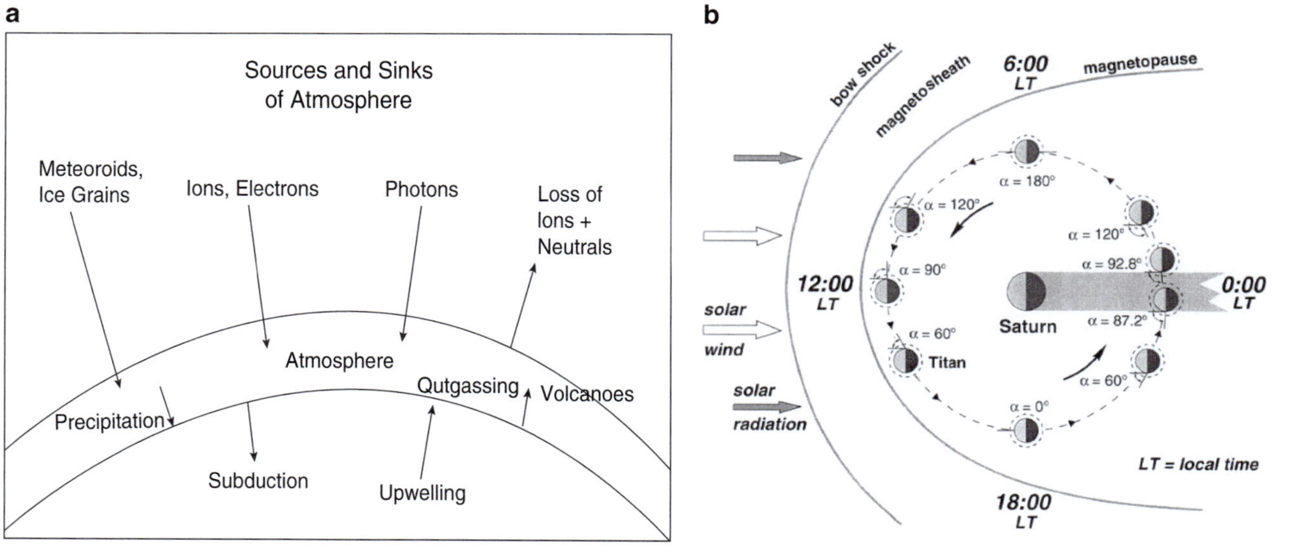

Fig. 15.1 Left panel: atmospheric sources and sinks: emphasis in this chapter is on mass loss rates due to escape and precipitation. Right panel: change in relationship between solar and plasma flux (see also Chapter 16).

at Titan can provide an opportunity to unravel the plasma- and solar-induced effects.

Below, each proposed active escape process is briefly described, followed by a short review of the pre-Cassini estimates of atmospheric mass loss, then a review of the recent mass loss calculations based on Cassini data and a summary.

15.2 Atmospheric Escape

Since the atmosphere decreases in density with increasing distance from the center of a planetary body, an altitude is eventually reached above which molecules that are moving radially outward can travel planetary scale distances with a very small probability of making a collision. At such altitudes a molecule can escape to space if its kinetic energy is greater than its gravitational binding energy and its radial velocity is outward. This region of the atmosphere is called the exosphere or the planetary corona. The lower boundary for this region, called the exobase, is defined as that altitude where the ratio of the mean free path for collisions, l_c, to the atmospheric scale height, H, is about one. In rarefied gas dynamics this ratio is the Knudsen number: $Kn = l_c/H$. It defines the transition region between where a gas that is dominated by collisions and can be modeled as a fluid, $Kn < 1$, gradually changes to a gas that should be modeled stochastically, $Kn \geq 0.1$, with the nominal exobase given by $Kn \sim 1$.

Simulations have shown that the appropriate l_c is the mean free path prior to a significant momentum transfer collision (e.g., Johnson 1994). Using realistic potentials such simulations also show that $l_c \sim 2/(n\sigma_d)$ where σ_d is momentum transfer (diffusion) cross section of the escaping particle averaged over the composition of the background gas (Johnson 1990, 1994; Johnson et al. 2000). Hot atoms and molecules escape from depths well below the nominal exobase, as seen in Fig.15.2. Therefore, the often used Chamberlain model, in which escape occurs from the exobase, is a rough approximation. Assuming hot particles move upward at random angles, the average escape depth, given as a column density is $N(r_{exo}) \approx 1.3/\sigma_d$, can be used to determine the radial position of the nominal exobase, r_{exo} (e.g., Johnson 1994; Johnson et al. 2008). This column density should be averaged over the composition, which changes with altitude, and the energy distribution of escaping molecules. It is also evident that the exobase level is different for different escaping molecules in the same atmosphere. Since the ion-neutral momentum transfer cross section, often dominated by charge exchange, is about an order of magnitude larger than the neutral–neutral cross section, effects of the plasma ions on Titan's neutral atmosphere extend well above the nominal exobase.

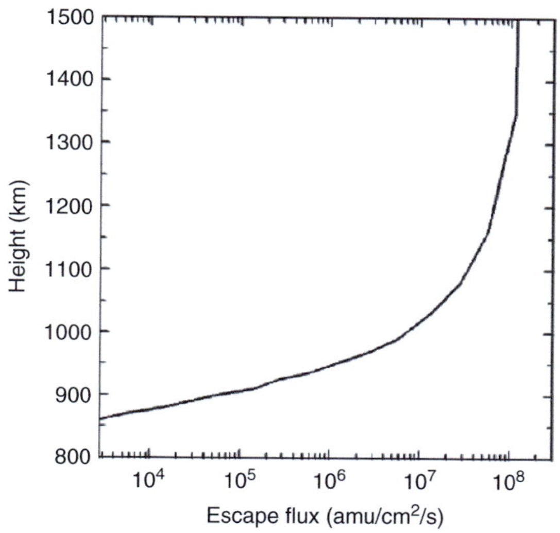

Fig. 15.2 Average upward integrated mass flux for molecular escape due to photochemistry in Titan's upper atmosphere from Shematovich et al. (2003): $\sim 1.2 \times 10^8$ amu/cm²/s at the exobase ~1500 km ($Kn \sim 1$); if the escape occurs over the hemisphere (exobase area $\sim 1.0 \times 10^{18}$ cm²) the loss rate is $\sim 1.2 \times 10^{26}$ amu/s. This figure shows the depth of origin of the escaping hot recoils, indicating the transition region, where the fluid approximation gives way to the kinetic regime, occurs well below the nominal exobase.

15.2.1 Thermal Escape

Planetary and satellite atmospheres confined by gravity are often characterized by the Jeans parameter, λ, which is the ratio of the gravitational energy to the thermal energy: $\lambda = [(GMm/r_{exo})/kT] = r_{exo}/H_{exo}$, where H_{exo} is the atmospheric scale height at the exobase and r_{exo} is the exobase radius. Since $H_{exo} = r_{exo}/\lambda$, with increasing λ the atmosphere becomes more strongly confined. In the absence of bulk flow, escape is often estimated as the fraction of upward moving atoms/molecules at the nominal exobase with velocities exceeding the escape velocity, v_{esc}. This gives the Jeans flux: $\phi = [n_{exo}\langle v\rangle/4][1 + \lambda]\exp(-\lambda)$, where n_{exo} is the density at the exobase, $n(r_{exo})$, and $\langle v\rangle$ is the mean thermal speed ($\langle v\rangle = (8kT/m\pi)^{1/2}$). In the limit that $\lambda \rightarrow 0$ in the exobase region, the atmosphere is no longer bound and blows away with a flux = $(n_{exo}\langle v\rangle/4)$ similar to the outflow from a comet. Bodies with values of $\lambda < \sim 50$ at the exobase have extended atmospheres with $r_{exo}/r_p > \sim 1.2$ (Strobel 2002) where r_p is the planetary/satellite radius. Pluto has $\lambda \sim 10$ at $r_{exo}/r_p \sim 3.5$ (Strobel 2008b), whereas Venus, Earth and Mars have $\lambda \sim (350, 130, 200)$ at $r_{exo}/r_p \sim 1.02, 1.08, 1.05$) and Titan is intermediate with $\lambda \sim 45$ at $r_{exo}/r_p \sim 1.6$.

Since the tail of the Maxwellian energy distribution is depleted by escape, the Jeans expression implies that this portion of the energy distribution is rapidly replenished. Recently, Cui et al. (2008) examined this assumption solving the collision dominated flow equation obtained in the so-called 13 moment approximation (Schunk 1975). This accounts for the perturbation in the Maxwellian energy distribution in a region in which the thermal gradients are not small. For H_2 above Titan's exobase they suggested that the upward component of the tail of the distribution is enhanced by thermal collisions, enhancing the Jeans escape rate, although that has been questioned.

15.2.2 Hydrodynamic Escape

For the lower values of λ (<~50), well above the blow-off condition, a controlled outward expansion of the atmosphere, called slow hydrodynamic escape, has been suggested to occur in planetary atmospheres. This process is assumed to be driven by solar EUV and UV heating and upward thermal conduction (McNutt 1989; Krasnopolsky 1999; Strobel 2008a; see Johnson et al. 2008 for a description). In this model of the thermosphere, the fluid equations *below* the exobase are solved under the assumptions that the escape flux, φ, is significant and hydrostatic equilibrium is a reasonable approximation (i.e., flow speed is much less than <v>). The 1D radial continuity, momentum and energy equations are scaled by the escape flux, φ, as well by λ and T. The kinetic energy of the flow is dropped from the energy and momentum equations and an optimal solution, adjusting the escape flux, is sought starting from the lower boundary up to the exobase. The scaling of the equations by φ *favors solutions with a significant escape flux*. An upward flow is extracted from the 'best' solution to the scaled equations below the exobase by matching the model to the measured density profiles. Although the flux of molecules at the exobase with speeds above the escape speed is much too small to account for the extracted escape flux, the model is continued *above* the exobase where the energy is assumed to be supplied by upward thermal conduction (Strobel 2008a, 2009). That is, thermal conduction is assumed to collisionally enhance the tail of the thermal distribution adding to the escape flux (Yelle et al. 2008; Strobel 2008a). An upper limit to the mass escape flux is obtained by equating the gravitational energy per unit time that would be carried off by the escaping particles to the heating rate $\{[GMm/r_0]\,(4\pi\varphi_{max})\} \approx \int_{ro}^{\infty} Q\, 4\pi r^2 dr$, where r_o is the lower boundary and Q is the net solar heating/ cooling rate, a quantity that varies with composition.

Due to the upward flux, adiabatic cooling occurs so that the temperature gradient is slightly negative in the thermosphere below the exobase. Above the heating maximum (or the lower boundary to the region described, whichever is highest), upward thermal conduction delivers the energy for the expansion. Since the fluid expression used for conduction begins to fail as Kn increases above about 0.1, and fails dramatically above the exobase, solutions below the exobase should be matched to a kinetic model of the exobase region and corona (e.g., Marconi et al. 1996; Tucker and Johnson 2009). Later in the chapter the results of Monte Carlo simulations of the transition region are examined. These simulations indicate that thermal conduction can not produce the proposed escape fluxes for the values of λ that are relevant to Titan's exobase region.

15.2.3 Photochemical-Induced Escape

Photochemical-induced escape occurs in response to the solar UV and EUV absorbed by molecules in the thermosphere and corona. A fraction of the energy absorbed leads to dissociations and exothermic reactions that produce atoms and molecules with energies (~eV s) that exceed the escape energy. If these events occur in the thermosphere near the exobase or in the corona above the exobase, and the products are moving radially outward, they can escape to space and contribute to atmospheric loss.

The photochemical-induced escape of N_2, CH_4, H, H_2, N, NH, HCN, CN and small hydrocarbons at Titan has been recently studied (De La Haye et al. 2007b). The escape rates were calculated for 12 species, a subset of 19 species and 47 processes examined earlier (Cravens et al. 1997). To calculate the escape fluxes, they integrated the production profiles from the exobase to 2500 km above Titan's surface. Using the 2-stream approximation for hot atoms and molecules moving through a background mixture of N_2, CH_4, and H_2, the photochemical escape rates for N and C in all forms were estimated to be relatively small, ~1.3×10^{26} amu/s and ~0.86×10^{26} amu/s respectively.

A much larger escape rate for nitrogen (~0.3×10^{28} amu/s; D. Shemansky private communication) was recently estimated based on UVIS measurements (e.g., Gustin et al. 2009). Solar occultation measurements of UV emissions during T10 and T26 produced line of sight column densities vs distance from the surface. These nitrogen emissions were interpreted as caused by photo-dissociative excitations leading to the production of hot recoils, a fraction of which have escape energies. Photodissociation of nitrogen was also the dominant heat source at ~1300 km in the model of Strobel (2008a)

15.2.4 Plasma-Induced Escape

For an unmagnetized body like Titan, the magnetospheric plasma and fields can interact with the corona and the exobase region of the thermosphere which can induce escape by a number of processes. However, the role of the plasma and fields in causing mass loss from Titan can vary considerably depending on whether Titan is in or out of Saturn's magnetospheric current sheet, where ion pressures are high ($\beta \gg 1$) and composition is dominated by heavy ions, or Titan is at higher magnetic latitudes where composition is dominated by light ions, densities and plasma pressures are low ($\beta \ll 1$) (Chapter 16) or Titan is outside the magnetosphere where solar processes dominate. Here β is the ratio of the plasma pressure to the magnetic pressure: $\beta = n_i kT_i/(B^2/2\mu_0)$ where n_i and T_i are the density and temperature of the ions, B the magnetic field and μ_0 is the permeability of space. This variation in upstream conditions can be traced to Saturn's magnetosphere being in a magnetodisk configuration at Titan's orbital position for local times away from noon where the field is more dipolar (Bertucci et al. 2008; Chapter 16). The current sheet regions are referred to as high energy input cases and the lobe like field regions are referred to as low input energy cases in Chapter 16. For the high energy case, Titan's induced magnetosphere will be more compressed and the ion gyroradii will be greater than Titan's radius, so that the energy input to Titan's ionosphere will be higher and heavy pick-up ion losses will be greater. In the low energy limit, Titan's induced magnetosphere will be less compressed and the ion gyroradii smaller since light ions dominate, so that the ion energy input to atmosphere will be lower. Magneto-hydrodynamic (MHD) and hybrid models of the interaction have been used for describing *both* the high and low energy limits. In the high energy limit, a hybrid simulation is often favored in order to account for the ion gyromotion and drifts. Since the low energy limit is more fluid like, MHD models are often favored. Another limiting case occurs when Titan is outside the magnetosphere and exposed to either the magnetosheath or the solar wind (Penz et al. 2005). In the case of the solar wind or magnetosheath, the composition is dominated by light ions but the pickup energies can be much higher due to the greater upstream speeds ~ 400 km/s (Sittler et al. 2006). Other effects also produce a highly variable ambient plasma pressure (e.g., Ma et al. 2006; Sillanpaa et al 2006).

The morphology of the induced fields and the radial scale of the exobase compared to the ion gyroradius determine the nature of magnetosphere/atmosphere interaction (e.g., Chapter 16) and the rate of atmospheric loss. Ionization in the exosphere and upper atmosphere can lead to loss by the outflow of ions from Titan's ionosphere, as is the case at Venus (e.g., Ma et al. 2008). In addition, pick-up ion formation can lead to loss if the freshly produced ions are swept down the tail of the interaction region (Chapter 16; Hartle et al. 1982; Wahlund et al. 2005; Coates et al. 2007). Such ions flowing through the corona can efficiently heat the atmosphere in the exobase region due the long-range ion–neutral interactions. Following ionization by charge exchange, energetic neutral atoms (ENAs) are formed which can escape Titan's gravity (Mitchell et al. 2005; Garnier et al. 2007, 2008; Smith et al. 2009).

A fraction of the pick-up ions and ENAs can re-impact the atmosphere with enough energy to induce heating and atmospheric sputtering (Michael et al. 2005a; Michael and Johnson 2005). This is particularly true when the pick-up ion gyroradius is of the order of the planet radius as is the case when heavy ions dominate, the high energy limit discussed above. Below we convert the loss estimates to global averages, even though there are considerable asymmetries in the plasma flux and the escape flux: e.g., pick-up ion impacts primarily occur on the Saturn side of Titan's trailing hemisphere (e.g., Hartle et al. 1982) whereas the location of the dominant photo-ionization rate changes as Titan orbits (e.g., Fig.15.1b).

The sputtering efficiency produced by the incident plasma is often given as a yield. It is the ratio of the number of escaping particles to the number of incident particles and varies inversely with the escape energy at the exobase (e.g., Johnson 1990). Sputtering of an atmosphere can occur by direct scattering of atmospheric molecules, also called knock-on, which dominates at grazing incidence. For heavy incident ions which can puncture the magnetic barrier of Titan's induced magnetosphere and penetrate the exobase, a cascade of recoils is set in motion with some having sufficient energies and the appropriate direction of motion to escape (e.g., Johnson 1994). This occurs at Titan due to the incident ambient plasma ions, such as O^+, the pickup ions, such as N_2^+ and CH_x^+, and the heavy ENAs. These cause both heating of the thermosphere near the exobase and escape as shown in simulations prior to Cassini's arrival at Titan (Shematovich et al. 2003; Michael et al. 2005a; Michael and Johnson 2005).

15.3 Simulations of the Transition Region and Escape

In the transition region the Maxwellian energy distribution can develop a non-thermal tail due the breakdown of thermal equilibrium and, more dramatically, due to the local production of hot recoils by the solar UV or the incident plasma (e.g., Fig. 15.2). In the exobase approximation described above, the fully collisional fluid approximation is used below the exobase and a ballistic, collisionless model is used above the exobase. Analytic expressions for the density vs altitude can be obtained by direct integration of the trajectories (Johnson 1990; appendix) or using the Louisville theorem (e.g., Schunk and Nagy 2000). However, the evolution of the initially energized atoms and molecules and their recoils can be described *throughout the transition region* using the Boltzmann transport equation for each molecular species or by Monte Carlo simulations.

15.3.1 Boltzmann Equation

Because solving the full set of transport equations is difficult, various approximations are used. The two stream model (Nagy et al. 1981) has been used on a number of planetary atmospheres and for analyzing the photochemistry in Titan's upper atmosphere normalized to Cassini ion-neutral mass spectrometer (INMS) data (DeLaHaye et al. 2007b). In addition, INMS H_2 data were analyzed using the collision dominated 13-moment equation, essentially the Navier–Stokes equation (Hirschfelder et al. 1964; Schunk 1975; Schunk and Nagy 2000), to describe the perturbations to the behavior of the atmosphere in a transition region (Cui et al. 2008).

A useful analytical approximation for the energy spectrum of the hot recoils has also been derived from the Boltzmann equation by ignoring spatial and temporal variations in the atmosphere (Johnson 1990, 1994). In a single component atmosphere, the number of hot recoils with energy between E and E + dE produced by a primary particle of energy E_o is given by: $\eta(E_o,E)dE \approx \beta(E_o/E^{\{2+x\}})dE$ (Johnson 1990, 1994). This applies for E » kT, where T is the atmospheric temperature, and β and x vary slowly according to the form for the interaction potential between a hot recoil and an atmospheric molecule. For a steeply varying potential, $\beta \approx 6/\pi^2$ and $x \approx 0$ has been shown to be consistent with data (Sigmund 1981; Johnson et al. 1994). In steady-state, the number of moving atoms with energy between E and E + dE produced by a source rate $\phi(t)$ of hot particles of energy E_o can be written $[\varphi(t)/\nu(E)][\delta(E_o - E) + \eta(E_o,E)]dE$. Here the delta function, $\delta(E_o - E)$, accounts for the initial recoil and $\nu(E)$ is the collision frequency between a hot molecule and a thermal molecule: $\nu(E) = [vn\sigma_d(E)]$, where v is the relative collision velocity: e.g., $v = (2E/m)^{1/2}$ with m the mass of the atmospheric recoil (Johnson 1994) and σ_d is the diffusion cross section. Monte Carlo simulations using realistic potentials have been carried out confirming this analytic model (e.g., Johnson et al. 2000; Johnson 2009). These expressions apply to the recoil production and collisional cooling of the hot atoms and molecules produced by either photo-chemistry or the incident plasma (Johnson 1994; Johnson et al. 2000). Such a model was used to fit Titan's observed coronal densities (DeLaHaye et al. 2007a).

15.3.2 Monte Carlo Simulations

Direct Simulation Monte Carlo (DSMC) calculations have also been used to describe the fate of hot recoils and escape (e.g., Michael and Johnson 2005). This is a stochastic method used to simulate a rarefied gas and is equivalent to solving the Boltzmann equation. It treats both the dynamic and stochastic nature of the gas (Bird 1994). Each species is described in terms of its phase space distribution with changes in the distributions determined by collisions between atmospheric particles (e.g., Marconi et al. 1996). Collisions with incident ions can also be included (e.g., Michael et al. 2005a; Krestyanikova and Shematovich 2006). The motions of representative particles, each assigned a weight, are calculated taking into account collisions and external forces (e.g., gravitational fields for neutrals and electro-magnetic fields for ions). DSMC is time consuming when the domain is highly collisional, but is useful for describing the transition from the collisional to collisionless regime (e.g., Marconi et al. 1996; Leblanc and Johnson 2001; Michael and Johnson 2005).

Because large numbers of particles are needed to obtain an accurate speed distribution over the transition region, approximations are often used. In exosphere approximation described above, Monte Carlo models of the phase space distributions have been used to describe the ballistic, collisionless corona. A significant improvement is referred to as the test particle method. The velocity space is divided into a hot particle population (the tail of the distribution) and a thermal background. The hot component or trace species are then tracked throughout the transition region and into the corona. These atoms and molecules move in the background atmosphere allowing collisions only with the thermal background atmosphere (e.g., Shematovich et al. 2003). This is equivalent to solving a linearized Boltzmann equation. 1D test particle simulations were compared to full DSMC simulations and the analytic model described above using a realistic interaction potential (Johnson et al. 2000). They showed that if the escape energy is well above the thermal energy, the escape rates obtained using test particle simulations compared favourably with DSMC results using realistic potentials. Both methods differed from simulations using hard sphere collisions. However, in the full DSMC model the energy spectra in the exobase region differed significantly from a Maxwellian with a power law tail, more closely resembling a kappa distribution (Johnson et al. 2000). Leblanc et al. (2002) described the effect on atmospheric sputtering when the exobase is dominated by molecules rather than by atoms.

Monte Carlo simulations have rapidly increased in complexity and are more useful than solving the Boltzmann equations when there are multiple species. The simpler test particle methods are sufficient for escape unless the response of the atmosphere is significant. In addition, test particle methods below the exobase can be coupled to DSMC models in the exobase region and above. However, in order to describe the evolution of the velocity distribution from the thermosphere across the exobase into the corona, and the proposed ability of thermal conduction to drive escape, DSMC simulations are required which span the full exobase region: Kn > ~0.1 (e.g., Marconi et al. 1996).

15.4 Estimates of Escape Flux: Pre-Cassini

Mass loss from Titan's thick and extended atmosphere induced by the ambient plasma ions, pick-up ions, ionospheric outflow and energetic re-impacting neutrals have been calculated by a number of groups. Initial estimates of the photon and electron-induced chemistry near the exobase gave very large loss rates (Strobel and Shemansky 1982), but revisions resulted in more modest rates, ~ 2×10^{26} amu/s (Shematovich et al. 2003). Initial estimates for atmospheric sputtering induced by magnetospheric ions penetrating Titan's atmosphere were also very large (Lammer and Bauer 1993). However, a direct simulation Monte Carlo (DSMC) model of the sputtering and heating using a model plasma flux consisting of magnetospheric and pick-up ions led to a few degrees increase in the exobase temperature (Michael and Johnson 2005) and a globally averaged loss rate: ~5×10^{26} amu/s (Michael et al. 2005a). Therefore, prior to the arrival of Cassini–Huygens, it was concluded that the present escape rate for carbon and nitrogen species was ~5×10^{26} amu/s and dominated by atmospheric sputtering (e.g., Shematovich et al. 2003). If this is the case, then the observed nitrogen isotope ratios must have evolved in an earlier period when the solar EUV and plasma interactions were more robust (e.g., Lammer et al. 2008; Chapter 7) or never evolved at all (Mandt et al. 2009).

Following Voyager it was also assumed that nitrogen escaping from Titan would be the dominant process for supplying Saturn's magnetosphere with heavy ions (Barbosa 1987). However, ions formed near Titan's orbit from the ejected neutrals have a high probability of being lost down Saturn's magnetotail or being lost from this region by an injection event. Therefore, the dominant source of nitrogen ions to Saturn's magnetosphere has been shown to be Enceladus (Smith et al. 2007, 2008). After reviewing the recent estimates of the escape rate, we also compare the various sources of neutral and ions for Saturn's magnetosphere.

15.5 Estimates of the Escape Flux: Cassini Data

With the many transits of Titan's exobase by Cassini, the escape processes have been re-examined. Modeling based on the Cassini data set appear to suggest larger heavy molecule loss rates than the most recent preCassini estimates. However, this modeling also indicates that the dominant heating processes for Titan's atmospheric corona and exobase region are not fully understood and can vary considerably with solar illumination angle and plasma pressure. Table 15.1 gives the principal estimates of mass loss both pre-and post Cassini. Rates are given in units of 10^{28} amu/s, which is ~0.2 of the present Titan atmospheric mass lost in 4 Gyr.

Table 15.1 Escape rates

Atmospheric component	Escape process	Loss rate
(in 10^{28} amu/s ≈ 0.2 Titan Atmospheric Masses/4 Gyr)		
Pre Cassini		
H_2[a]	Jean's escape	~1–3
CH_4[b]	Destruction/precipitation	~10–30
C, N[c]	Photochemistry and sputtering	~0.01–0.06
Based on Cassini Data		
Ions[d]	Pickup/outflow	~0.01–0.05
H_2[e]	Thermal escape	~0.8–2
CH_4[f]	Destruction/precipitation	~24
N_2/CH_4[g]	Corona fits (<~2000 km)	~0.00–0.16
N_2/CH_4[h]	Sputtering	~0.3–3.6
CH_4[i]	Hydrodynamic	~4–5
$CH_4 + H_2$[j]	Hydrodynamic	~4.5

[a] (e.g., Lebonnois et al. 2003)
[b] (Wilson and Atreya 2004)
[c] (Michael et al. 2005a; Shematovich et al. 2003; Cravens et al. 1997; Lammer and Bauer 1993; Strobel et al. 1992; Gan et al. 1992; etc.)
[d] (Hartle et al. 2008; Wei et al. 2007; Coates et al. 2007; Ma et al. 2006; Sillanpaa et al. 2006; Wahlund et al. 2005; etc.)
[e] (Cui et al. 2008 ; Garnier et al. 2008; Bell et al. 2009)
[f] (based on the H_2 data in Cui et al. 2008 and model of Wilson and Atreya 2004; Mandt et al. 2009)
[g] (DeLaHaye et al. 2007a based on kappa fits)
[h] (DeLaHaye et al. 2007a; average for 5 passes based on analytic model and coronal structure)
[i] (Yelle et al. 2008)
[j] (Strobel 2008a, 2009)

15.5.1 H_2 Escape

Data from the Cassini ion-neutral mass spectrometer (INMS) indicate a steeply varying H_2 density profile above the nominal exobase (Waite et al. 2005) as seen in Fig. 15.3a. This profile is consistent with diffusion of H_2 through the Titan atmosphere resulting in a large escape rate: a mass flux of H_2 ~1×10^{10} amu/cm²/s normalized to the surface of Titan (Cui et al. 2008; Bell et al. 2009). Assuming this rate is global (~1.6×10^{28} amu/s), then it is comparable to the pre-Cassini estimates. A large H_2 escape flux is consistent with the presence of the extended hydrogen corona imaged by the detection of energetic neutral atoms produced by charge exchange with protons (Garnier et al. 2007, 2008).

In their analysis Cui et al. 2008 extracted an exobase temperature for hydrogen that was slightly lower than that for the principal species. They interpreted the cause to be the significant escape rate. When they used their lower H_2 temperature, they estimated that the loss rate was larger than

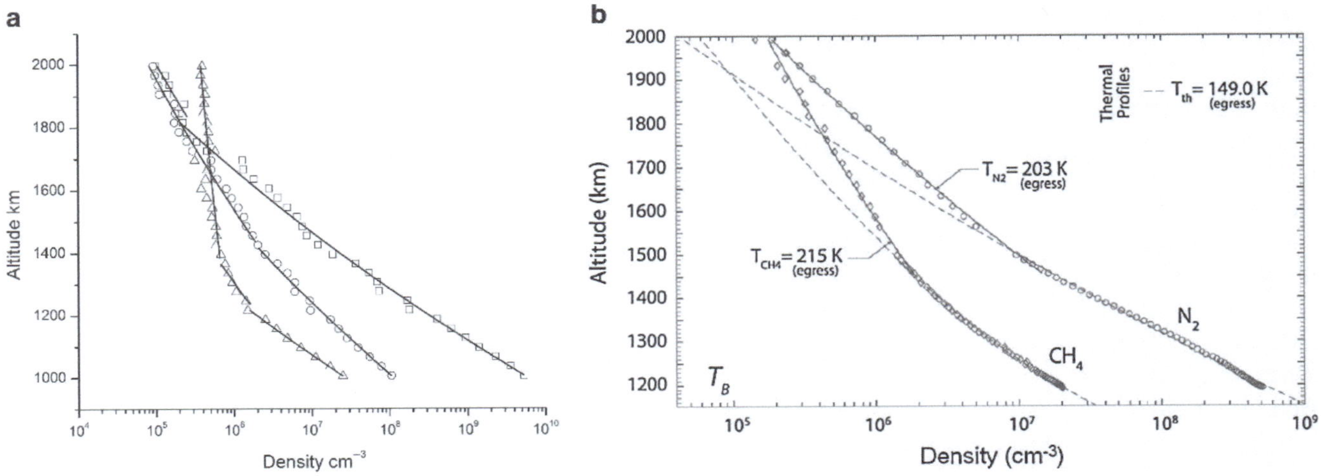

Fig. 15.3 (**a**) Density profiles for H_2 (*triangles*), CH_4 (*circles*) and N_2 (*squares*) based on an *average* of INMS ingress and egress data during 18 passes through Titan's exobase (data from Magee et al. 2009); (**b**) density vs radial distance from Titan for CH_4 and N_2 from the INMS data for Tb egress (from Fig. 6 in DeLaHaye et al. 2007a): dashed lines are the assumed isothermal fits based on data below the exobase (~1450 km); this shows a change in slope of the density profiles above the nominal exobase. The exobase temperatures *and* especially the slope of the profile above the exobase are highly variable.

that for the Jeans escape. Based on a collision-dominated flow equation, they concluded that Jeans escape was enhanced. They proposed this was due to thermal conduction in the exobase region causing a distortion in the Maxwell Boltzmann energy distribution. Bell et al (2009) find the hydrogen loss rate is consistent with Jeans escape.

15.5.2 Carbon Mass Loss by Precipitation

Since hydrogen is produced by dissociation of methane and other hydrocarbons, its loss to space implies that atmospheric methane is also lost to the surface in the form of precipitating hydrocarbon molecules and small carbonaceous grains that form in the atmosphere following dissociation of methane. Based on the H_2 escape rate and the chemistry model of Wilson and Atreya (2004), methane is lost by precipitation at a rate ~2×10^{29} amu/s (e.g., Chapter 8; Mandt et al. 2009). Therefore, even for the largest escape rates for methane and nitrogen discussed below, *atmospheric mass loss to the surface dominates mass loss by escape*.

Because the cross section for photo-absorption by methane in the UV-EUV is relatively large, the loss of methane by dissociation, reaction and precipitation affects the modeling of diffusion in Titan's upper atmosphere. That is, ignoring other effects, an upward diffusive flux is required to replace the destroyed methane: the mass flux is $\sim(m_{CH_4}/\sigma_{abs}\tau_{CH_4})[1-\exp(-\sigma_{abs}N_{CH_4})]$, where m_{CH4} is the mass of methane, τ_{CH4} is the lifetime of CH_4 exposed to the solar flux and the magnetospheric electrons above the exobase, and σ_{abs} is the average attenuation cross section of the radiation. The downward carbon flux decreases with increasing altitude (Sittler et al. 2009; Mandt et al. 2009) and the prefactor, $(m_{CH4}/\sigma_{abs}\tau_{CH4})$, is equal to the integrated methane loss rate by this process *above* the peak in the methane destruction rate. That is, assuming the destruction of methane occurs over the disc, then one fourth of the exobase area times the pre-factor must roughly equal the integrated precipitation rate, ~2×10^{29} amu/s. Methane can also be destroyed by reaction with nitrogen atoms and ions, which can dominate at high altitudes (Mandt et al. 2009). The upward diffusive flux required to replace the methane lost by this procces must be included in any model of the INMS density profiles (Chapter 8).

15.5.3 Escape of Nitrogen and Carbon: Hot Recoil Models

Based on UVIS observations (e.g., Gustin et al. 2009) the loss rate of N atoms was estimated to be ~2.0×10^{26} N/s (~0.3×10^{28} amu/s) (D. Shemansky, private communication). This is much larger than the photo-chemical loss rate of nitrogen in DeLaHaye et al. (2007b) (~1.3×10^{26} amu/s). However, it is close to the favored loss rate estimated in DeLaHaye et al. (2007a) in order to explain Titan's coronal density structure as discussed below.

INMS data for a number of early passes were examined (DeLaHaye et al. 2007a) assuming that the atmospheric structure near the exobase could be understood if the energy spectra of the molecules in the corona had a significant non-thermal, 'hot', component. That the production of hot recoils might be occurring is indicated by the UVIS results for nitrogen but was earlier suggested by the change in slope of the density vs altitude near the nominal exobase. This is seen in

Fig. 15.3b as Cassini crossed the exobase region while exiting Titan's atmosphere on orbit Tb. A change in the slope is even suggested, though more subtly, in Fig. 15.3a by the N_2 density profile vs altitude averaged over many passes through Titan's atmosphere. In addition it is clear from the five passes through the exobase examined by DeLaHaye et al (2007a), that the coronal structure varies significantly both in space and in time. This variability has been confirmed by the analysis of many additional orbits (Bell et al. 2009).

In order to simulate the densities of N_2 and CH_4 above the nominal exobase, ~1450 km, up to 2000 km above Titan's surface, expressions obtained from the Louisville theorem were used (DeLaHaye et al. 2007a). This is equivalent to a Monte Carlo ballistic trajectory model of the corona (Johnson 1990). Molecules are launched from the exobase with a chosen energy spectrum and the density vs altitude is calculated. A number of forms for the energy spectra were used and tested against Cassini data for the CH_4 and N_2 coronal structure. As shown in Fig. 15.4, results were presented for two models of the molecule energy spectrum: a thermal spectrum with a tail given by the analytic model described above and an energy spectrum described by a kappa function. They found that over the range for which they had data, the distribution of molecules in the hot corona could be better represented by a kappa distribution (Vasyliunas 1968; Jurac et al. 2002). These energy distributions are Maxwellian at the lower energies but decay as power laws at the higher energies. Such a distribution at the exobase implies that an equilibrium thermal energy distribution well below the exobase develops an enhanced tail in the exobase region. The enhanced tail was assumed to be produced by energy deposition in the transition region which they hypothesized was a result of the interaction with the local plasma. Such a smoothly varying distribution, rather than a Maxwellian with a power law tail, is also consistent with DSMC simulations of energy deposition into an atmosphere in the form of hot recoils (Johnson et al. 2000).

DeLaHaye et al. (2007a) found that for those Cassini passes through Titan's atmosphere for which a hot nitrogen corona was detected, the extracted kappa distributions exhibited tails that decayed much more steeply than the analytic recoil spectra discussed above: powers ~ 4–20. This steep decay (e.g., Fig. 15.4a) could be an artifact of the narrow range of modeled altitudes or could suggest that processes other than the production of hot recoils are responsible for the coronal structure. Although kappa fits up to only 2000 km can not give a good measure of the escape flux, they can give a measure of the amount of energy needed to form the observed hot corona. However, for four of the five exobase crossings examined, they *could not* account for the measured corona structure by assuming that the tail of the distribution was only populated by photon- and electron-induced chemical processes or by the published models of the plasma-induced heating rate. Although Yelle et al. (2008) and Strobel (2009) suggested the effect of the plasma flux is small, additional heating in the exobase region appears to be required. This must occur by energy transport from below, by horizontal transport of energy, or by larger photo- and/or plasma energy deposition rates than initially assumed.

The estimates of the required heating rate in Table 15.2 were extracted from fits to the neutral densities using both models of energy spectra at the exobase. This energy could be deposited by exothermic chemistry, by interaction with the plasma, or by vertical or horizontal heat transport. The heating rate required was found to vary considerably from a crossing of the exobase in which no hot corona was identified, hence, no hot particle production in the exobase region was required, to relatively large energy deposition rates. For the five crossings examined *average* heating rates were obtain using the kappa and the analytic model for the recoil spectrum: ~95 eV/cm³/s and ~200 eV/cm³/s respectively (Table 15.2). Similar results were found for the analysis of a much larger set of orbits using kappa functions (J.

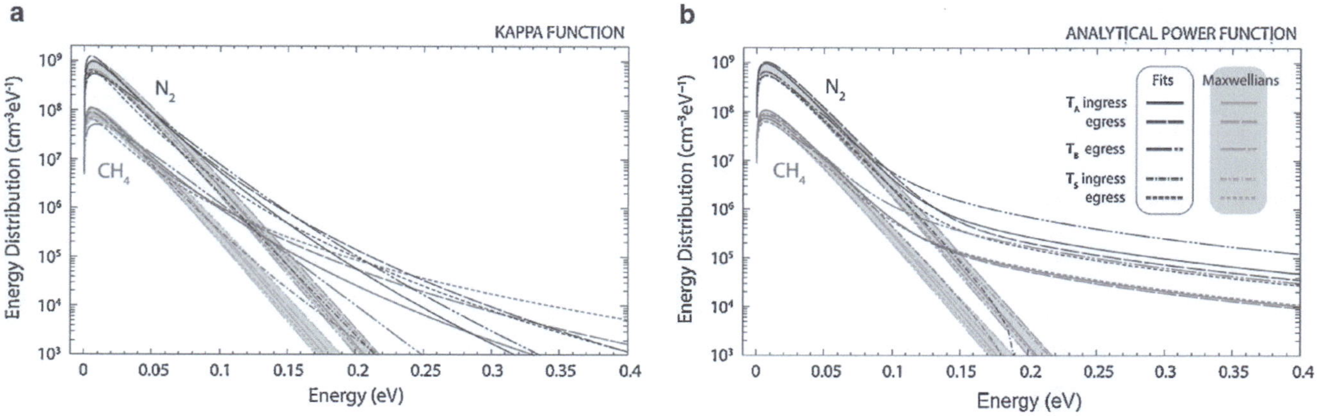

Fig. 15.4 Energy spectra at the exobase for five crossings of the exobase showing the Maxwellian (shaded profiles) and the tail of the distribution containing the hot recoils: (**a**) energy spectra are described by a kappa function and resulting densities are fit to the INMS density data in the corona; (**b**) same but with the analytic model due to hot recoils added to the Maxwellian (from DeLaHaye et al. (2007a)). Escape energies at nominal exobase (1450 km): CH_4 ~ 0.38 eV; N_2 ~ 0.67 eV. Highest altitude was 2000 km corresponding to ~12% of the escape energy.

Westlake private communication). The extracted energy deposition rates are larger than the estimated photo-induced heating plus cooling rate at that altitude (~8 eV/cm³/s; Strobel 2008a) and the pre-Cassini estimates of the globally averaged plasma-induced heating rate (~15 eV/cm³/s, Lammer et al. 1998; ~25 eV/cm³/s, Michael and Johnson 2005), but are closer to the rates obtained by assuming the incident plasma flux is dominated by low energy ions as seen in Fig. 15.5.

Table 15.2 Energy deposition and Escape rates (DeLaHaye et al. 2007a)

Titan pass	T_{exo}(K)[a]	Energy deposition (eV/cm³/s)[b]	Escape flux (10^9amu/cm²/s)[c]
TA ingress	150	82 [190]	0.02 [<18] (0.8–1.9)
TA egress	157	73 [150]	0.21 [<14] (0.7–1.5)
TB egress	149	250 [510]	0.00 [<48] (2.5–5.1)
T5 ingress	162	0 [0]	0.00 [<0.0] (0.0)
T5 egress	154	70 [130]	0.82 [<12] (0.7–1.3)
Average Global[d] (10^{28} amu/s)			0.04 [<4] (0.2–0.4)

[a]Exobase temperature extracted from INMS density vs altitude below the exobase.

[b]Energy densities extracted from best kappa fits to speed distributions describing the corona density vs altitude from 1450 to 2000 km: Sum of energy deposition in N_2 and CH_4 from Table 6; [ ] same using the analytic model for the recoil spectrum discussed earlier with x ~ 0 (Johnson 1994): data from Table 5.

[c]Flux at nominal exobase (1450 km): values from tail of kappa function fit to corona density from Table 6; [ ] analytic recoil model fit to coronal densities were described as upper limits from Table 5; () based on the energy deposition rates from the kappa and analytic model fits in column 2 scaled to recoil production and sputtering results in Michael et al (2005a): loss rate at exobase ~25. × 10^7 amu/cm²/s for energy deposition rate ~25 eV/cm³/s, or a factor of ~10^7 amu (cm/eV); for normalization of fluxes to the physical surface multiply by 2.4: i.e. exobase area 2.04 × 10^{18} cm² vs surface area 0.84 × 10^{18} cm²).

[d]Average of fluxes from column 3 assumed global.

Based on this analysis DeLaHaye et al (2007a) assumed that the plasma-induced heating associated with the magnetosphere–ionosphere interaction was more robust than pre-Cassini estimates. This might be consistent with the Cassini plasma science instrument (CAPS) observation of a low energy ion flux in the corona near the exobase (e.g., Fig. 15.6). Such ions transfer energy more efficiently to the neutrals and therefore deposit a large fraction at higher altitudes as seen in Fig. 15.5. These energies were *absent* from the early hybrid

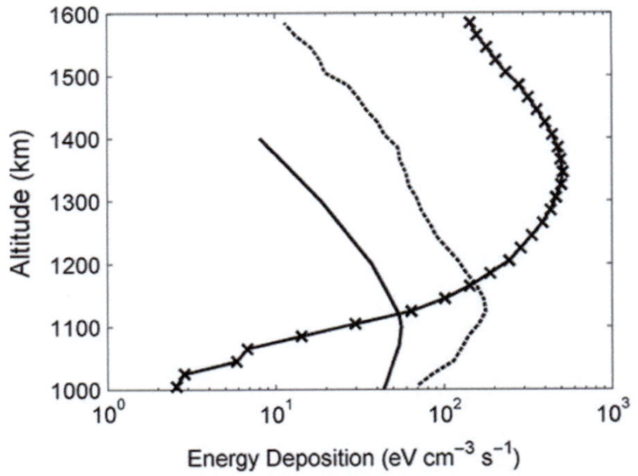

Fig. 15.5 Comparison of energy deposition rates in eV/cm³/s vs radial distance from Titan's surface: (*solid line*) the globally averaged solar EUV/UV at medium solar conditions: net heating with 25% efficiency and HCN cooling (~12% of total near exobase) (Strobel 2008a); (dashed line) the model of the incident plasma used in Michael and Johnson (2005) based on a globally averaged flux of ~5 × 10^9 eV/cm²/s at the exobase. This only includes that component of the energy going into the *production of recoils*; it also accounts for losses due to escape and to the vibrational, rotational and dissociation processes. Cooling due to emissions is not included. (*Solid line with crosses*): the same energy flux as dashed curve but the energy spectra in Fig. 15.6. This results in higher energy deposition rates near the exobase. Although the energy flux was the same the escape rate of recoils increased roughly proportional to the energy deposition rate as shown by model calculations and assumed in creating Table 15.2 (DeLaHaye et al.2007a).

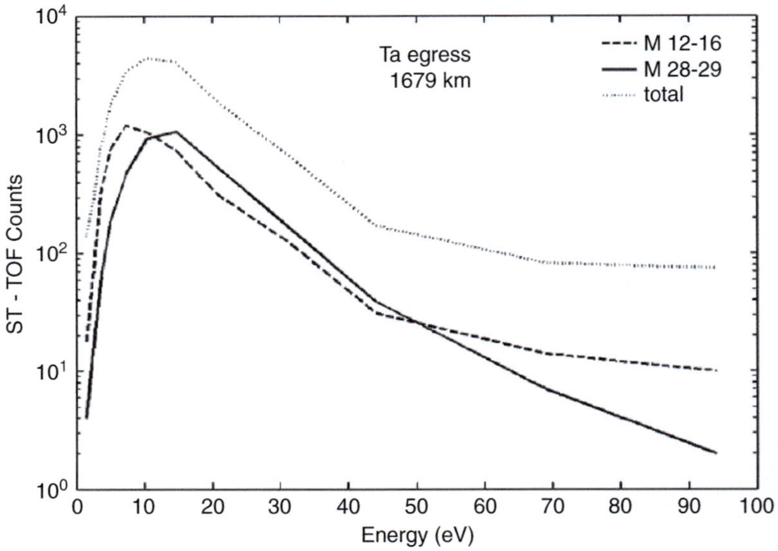

Fig. 15.6 Energy distribution for ions above Titan's exobase (~1680 km) on the ingress portion of Cassini orbit Ta obtained from CAPS ST-TOF data: results are given for the two principal heavy ions mass peaks 12–16 and 28–29 amu, as well as a total (Michael et al. 2005b). Model plasma ion energy spectra used in simulations of atmospheric sputtering (e.g., Shematovich et al. 2003) and in obtaining the dashed curve in Fig. 15.5a had a lower cut-off at ~50 eV.

simulation that was used to obtain the pre-Cassini plasma heating rate. Such energies are associated with pick-up of ions close to the exobase and/or ionospheric out flow (ions driven out of the ionosphere by the plasma pressure and scavenged by pick-up) (Wahlund et al. 2005; Ma et al. 2006; Sillanpaa et al. 2006; Hartle et al. 2006b; Chapter 16). For this reason the simulations in Michael et al. (2005a) were repeated using the same net heavy ion energy flux onto the exobase, but with an ion energy spectrum consistent with that in Fig. 15.6. Not surprisingly the energy deposition rate near the exobase increased dramatically. Although such a high rate is unlikely, the increase in the mass loss rate by sputtering is found to roughly scale with the energy deposition rate near the exobase.

The kappa energy spectra extracted from the fits to the INMS data (Fig. 15.4a) are seen to have high energy tails that decay more steeply than does the analytic model of the recoil distribution (Fig. 15.4b). Since escape rates are determined by the tail of the distribution they are very different in the two models as seen in Table 15.2. However, the tails of the distribution are not accurately tested by the atmospheric density data below 2000 km. The kappa distributions rates are heavily dominated by loss of CH_4, whereas those for the analytic model are not. It is seen that the *heating rates extracted* in the two models are much closer in size. Therefore, assuming the heating rates are due to a production of recoils, atmospheric loss rates were better estimated by scaling to the Monte Carlo model in Michael et al. (2005a). As demonstrated by the two simulations in Fig. 15.5, this scaling is valid as long as the primary recoils have energies much larger than the escape energy at the exobase (~0.38 eV for CH_4). Based on energy deposition rates extracted from the energy spectra in Fig. 15.4 (e.g., see Table 15.2) and scaling to the Monte Carlo simulations, mass loss rates obtained for the two models of the energy spectrum are within a factor of two of each other (Table 15.2): ~0.3×10^{28} amu/s and with nitrogen loss somewhat larger than that of methane. This rate is much larger than the pre-Cassini estimates. A rough *upper bound* was also obtained using the analytic model of the hot molecule spectrum. Summing the CH_4 and N_2 contributions and averaging over the five passes gives an upper bound of global mass loss rate via the heavy molecules: <~4×10^{28} amu/s. Therefore, globally averaged mass loss rates ~0.3–4×10^{28} amu/s are possible based on the analysis in DeLaHaye et al. (2007a).

15.5.4 Escape of Nitrogen and Carbon: Continuum Models

Describing Cassini data below the exobase using continuum models, Yelle et al. (2008) and Strobel (2008a, 2009) separately concluded that, in addition to H_2 loss, there is a significant escape flux of heavier molecules. These models did not include methane destruction in the thermosphere and estimated that the plasma heating rate was not important. Strobel used the slow hydrodynamic escape model described earlier to fit the HASI density and temperature profile. Three cases were examined: a lower boundary that allowed a heat flux, but no heating in the region modeled, and two different net heating/cooling rates with peaks occurring well below the exobase (e.g., Figs. 15.5 and 15.7). He concluded that the observed altitude dependence of the mass density and temperature (Fulchignoni et al. 2005) could be represented by having an upward mass flux ~5×10^{10}

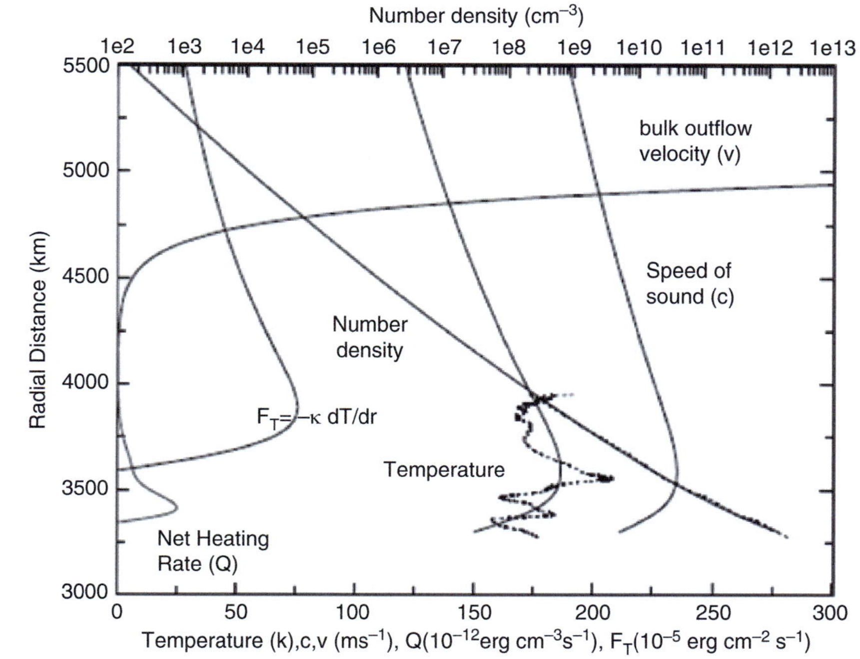

Fig. 15.7 A model for Titan's atmosphere (Strobel 2008a): n, T, c, v, and upward thermal heat conduction flux (*solid lines*) for an N_2 atmosphere with solar medium conditions: net heating due mostly to CH_4 UV heating above the lower boundary at $r_o = 3300$ km, $T_o = 158$ K, and downward thermal heat conduction flux at lower boundary = 4.0×10^{-3} erg/cm²/s. Comparison with the HASI measurements of total number density and temperature (*dashed lines*) at ~−10.3°S latitude. Assuming the model applies above the nominal exobase (here ~4300 km), the solution is valid only to ~4750 km and the mass escape rate is 4.5×10^{28} amu/s. Solar heating near the exobase contributes only ~10^{-3} of the heat conduction flux and plasma-induced heating is ignored.

amu/cm²/s with respect to Titan's surface (e.g., Fig. 15.7). Presuming this upward flux implied escape to space, the model was extended into the region above the nominal exobase with thermal conduction driving escape. Due to diffusive separation, the escape flux was estimated to be dominated by the lighter species, ~2/3 CH_4 and 1/3 H_2 by mass for a total ~5.6 × 10^{28} amu/s (Strobel 2009). However, since the expression for heat conduction was not modified above the exobase where the mean free path between collisions is large, such results are questionable.

Yelle et al. (2008) analyzed the INMS ingress data averaged over many passes for the density vs altitude of CH_4 and N_2. The changing atmospheric composition with altitude in the thermosphere (e.g., Fig. 15.3a and b) was shown to be indicative of diffusion of CH_4 through the background N_2. In order to separate eddy and molecular diffusivity, they derived an eddy diffusion coefficient from the density profile for ^{40}Ar. They then solved the coupled diffusion equations using both eddy and molecular diffusion allowing for the possibility of upward flow. Modeling the averaged profiles up to nominal exobase, they found that the CH_4 and N_2 density profiles could be described if CH_4 flows upward at a rate of ~4–5 × 10^{10} amu/cm²/s consistent with the model at mid latitudes of Muller-Wodarg et al. (2008). They also presumed this upward flux was a measure of the escape flux and pointed out that the mass flux is close to that found by Strobel (2008a). It is also close to the average upper limit for escape in DeLaHaye et al. (2007a). Recently, Bell et al. (2009) show the INMS data can be explained by a largerange of methane escape rates. Therefore, all models of the Cassini data could *in principal* be consistent (Johnson 2009).

Although the density profile below the exobase can be reasonably well described by an isothermal atmosphere, modulated by gravity waves (Waite et al. 2005; Muller-Wodarg et al. 2006; Fulchignoni et al. 2005), the continuum models yielded temperature profiles in which T decreased with increasing altitude in the exobase region consistent with upward heat flow. A 2D model of the northern hemisphere, the region primarily sampled by Cassini (Muller-Wodarg et al. 2008), also shows decreasing T profiles, primarily at latitudes below ~70°. Above that latitude the profiles are complicated resulting in significant temperature uncertainties.

15.5.5 Summary of Mass Loss: Cassini Data

Based on the UVIS measurements (D. Shemansky private communication) and the different models of the INMS and HASI data, Cassini data suggest that Titan could be experiencing an average mass loss rate to space of heavy molecules ~0.3–5 × 10^{10} amu/cm²/s. This large range of values is due to differences in the analysis of the data and is *not* related to the expected variability discussed earlier. Although all of these papers suggest the heavy molecule loss rate might be larger than the pre-Cassini estimate, there are considerable differences between the continuum and stochastic models of the data. The continuum models are reliable below the exobase. However, it is not correct to use the standard expressions for thermal conduction above the exobase in order to power escape. In the stochastic models escape is caused by the molecular physics occurring in the thermosphere and corona. Although different conceptually, the resulting estimates of the mass loss rate could be consistent. That is, if atmosphere is being removed at the top, there must be an average upward flow *below* the exobase in order to replace the material removed (Tucker and Johnson 2009). At the temperatures and densities of interest heavy molecules can be removed at the top by any number of processes: formation and precipitation of larger molecules, ionospheric outflow, pick-up ion loss, atmospheric sputtering, photo-chemistry, viscous momentum transport from H_2, transport to another region of the atmosphere, or any combination of these processes.

Strobel (2008a, 2009) and Yelle et al. (2008) suggested that photon and plasma-induced hot recoils are not required. Rather upward thermal conduction of the solar energy deposited well below the exobase would produce a non-thermal tail in energy distribution of molecules in the exobase region, and that conduction would continue to act above the exobase to accelerate a fraction of the molecules to escape energies. Although the distorted Maxwellian extracted in Fig. 15.4a is consistent with this picture, such a spectrum led to *very small* escape fluxes. Therefore, these models require a considerable enhancement in the energetic tail for the flux of heavy molecules the size of which appears to be problematic (Johnson 2009; Tucker and Johnson 2009).

The upward flow of hydrogen molecules having thermal speeds much greater than that for the heavies (~1.2 km/s at 150 K) can, in principal, transfer momentum to the heavy molecules (e.g., Chamberlain and Hunten 1987). Cui et al. (2008) modeled the viscous energy loss to the heavy background molecules. In this model viscosity and heat conduction produces a Maxwellian speed distribution with an enhanced tail. But the viscous heating rate in Fig. 15.7 of Cui et al. (2008), ~2 eV/cm³/s, falls short of that required to drive the outflow of the heavy molecules. In fact, the modeling in Cui et al. (2008) requires that H_2 *draws energy from* the background gas, enhancing the tail of the H_2 velocity distribution function and reducing the likelihood of heavy molecule escape.

15.5.6 Monte Carlo Simulations: Tests of Continuum Models

Test particle simulations (e.g., Shematovich et al. 2003) and Direct Simulation Monte Carlo (DSMC) simulations have been used to calculate the plasma-induced heating and escape at Titan (e.g., Michael et al. 2005a; Michael and Johnson 2005). These pre-Cassini results are given in Table 15.1 and Figs. 15.2 and 15.6. As mentioned earlier, the analytic model of atmospheric sputtering was tested by such simulations (Johnson 1994; Johnson et al. 2000) and was used in preliminary interpretations of Cassini INMS data in Titan's exobase region (DeLaHaye et al. 2007a).

Recently, DSMC simulations of the transition region have been carried out to test the slow hydrodynamic escape model for Titan (Tucker and Johnson 2009). Simulations of the transition region above 3900 km were carried out in 1D for both rectangular and spherical coordinates producing very similar results. The density and temperature at the lower boundary were normalized to that obtained in Strobel (2008a), Fig. 4) at 3900 km. That radial distance is well above the EUV and UV heating peaks. At an upper boundary 6200 km above the exobase, molecules were allowed to escape if they had energies greater than the escape energy, but were otherwise reflected allowing a more rapid approach to the steady state densities below 6200 km. Simulations were also carried out for an upper boundary at 8000 km showing a negligible effect on the resulting densities and temperatures. Although thermal conduction is correctly treated in the simulations, the atmosphere was found to be nearly isothermal, as seen in Fig. 15.8a, with essentially no escape. The numerical upper limit on the mass loss rate for no heating above 3900 km and a pure N_2 atmosphere is ~3 × 10^{22} amu/s as compared to the ~(4–5) × 10^{28} amu/s found in Strobel (2008a). Therefore, unlike the result in Fig. 15.7, when λ_{exo} ~ 40 at Titan's exobase *and* there is no external process causing an outflow, then no escape is expected beyond that estimated using Jean's escape.

The energy spectra of the molecules above the exobase were also examined and it was found that thermal conduction did not produce a noticeable enhanced tail (Tucker and Johnson 2009). However, when the molecules were *artificially removed* at the upper boundary of the simulations, as if by one of the loss processes discussed above (e.g., ionization and sweeping, knock-on collisions, etc.), then, not surprisingly, a decreasing temperature profile was obtained as seen in Fig. 15.8a.

Additional simulation of the exobase region and corona were carried out in which the Jeans parameter was reduced considerably in order to look for the onset of slow hydrodynamic escape. This was done by increasing the temperature at the lower boundary. Even for a Jeans parameter λ_{Jeans} ~ 12, in which the nitrogen corona becomes very extended, the actual loss rate was only about 1.5 times that of the Jeans rate calculated using the exobase temperature (Tucker and Johnson 2009).

Because CH_4 has a lower mass and is assumed to dominate the escape rate (Strobel 2008a, 2009; Yelle et al. 2008)), these simulations were then repeated for a mixed CH_4, N_2

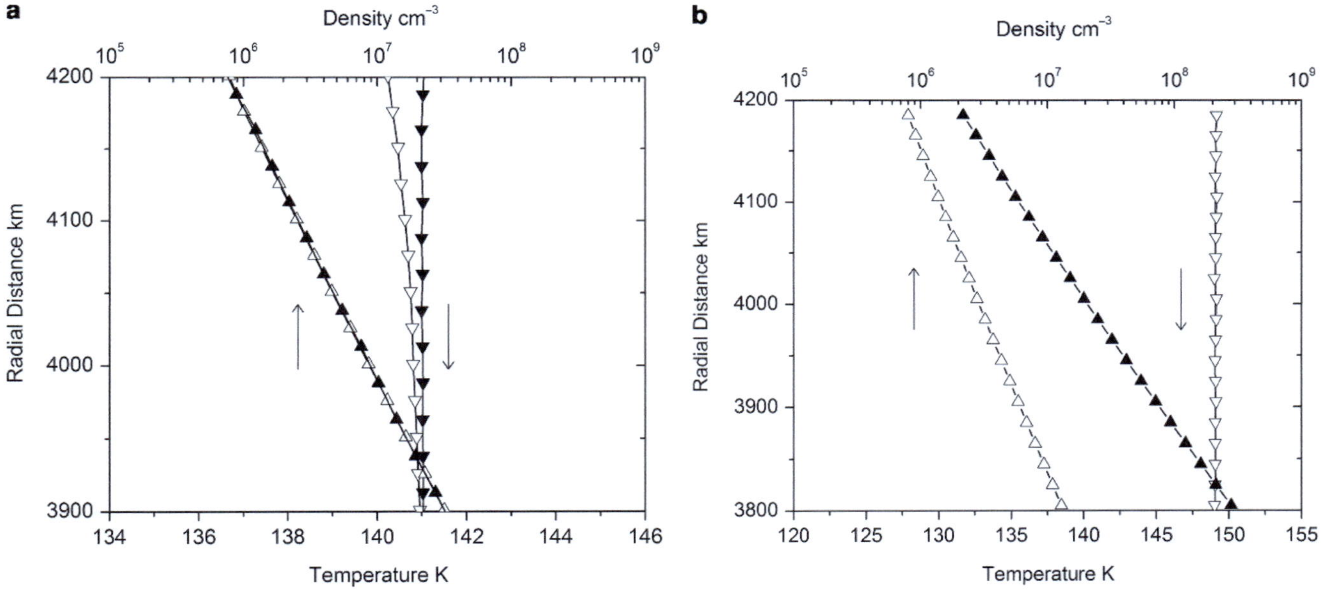

Fig. 15.8 Right panel: DSMC simulation of a nitrogen atmosphere: normalized at the lower boundary, 3800 km, to the density and temperature in Strobel (2008a) for his case of no heating above his lower boundary 3450 km. Density and temperature from two simulations are shown (Tucker and Johnson 2009); inverted triangles: no non-thermal escape process is included and T is found to be isothermal; in the second simulation molecular nitrogen is removed at the top boundary at a rate 4.5 × 10^{28} amu/s the escape rate proposed in (Strobel 2008a); the density changed little over the radial distances shown but the temperature decreases consistent with adiabatic cooling. Results are for hard sphere cross sections which gave a slightly higher thermal conductivity than that in Strobel (2008a); similar results were obtained for forward directed cross sections (Tucker and Johnson 2009). Left panel: Density and temperature for CH_4 + N_2 atmosphere obtained from a DSMC simulation normalized to the mixing ratios temperature and densities in Yelle et al. (2008). Escape is again negligible and the temperature is essentially isothermal.

atmosphere. The lower boundary at 3775 km in these simulations had a mixing ratio, temperature and density normalized to that in Yelle et al (2008). Such simulations naturally allow for diffusive separation, thermal conduction and escape. Again the DSMC simulations *did not* result in escape and the atmosphere remained isothermal as seen in Fig. 15.8b. Such results suggest that it is unlikely that thermally-driven, slow hydrodynamic escape is occurring at Titan (Tucker and Johnson 2009).

15.6 Atmospheric Loss as Plasma and Plasma Heating: Cassini

MHD and hybrid simulations (Brecht et al. 2000; Ledvina et al. 2005; Ma et al. 2006; Sillanpaa et al. 2006) have shown that ions flow through the corona and exobase region and the fields penetrate below the exobase. Due to the plasma pressure in the ionosphere, ion outflow occurs (Wahlund et al. 2005; Coates et al. 2007) similar to that observed at Io (e.g., Wilson et al. 2002) and Venus (e.g., Terada et al. 2004). Estimates of the mass loss by ionospheric outflow have been made using Cassini data (Table 15.1). For T9 Sittler et al. (2009) directly measured the ionospheric loss down Titan's induced magnetotail with values ~4 × 10^6 ions/cm^2/s for mass 17 ions and ~4 × 10^6 ions/cm^2/s for mass 29 ions with field aligned flow speeds ~10 km/s and ion temperatures T_{ion} ~ 4 eV. Such ions contribute to the low energy peak in the heavy ion energy spectra in Fig. 15.6. The mass loss rates vary by an order of magnitude ~0.1–4 × 10^{26} amu/s. This is very roughly consistent with pre-Cassini estimates of the *neutral* loss rate, but smaller than the above loss rates. These estimates of the ion loss rate are obtained more directly than the neutral loss rates.

Ionospheric outflow, pick-up ions, and the influx of the slowed and deflected ambient ions can cause heating and neutral escape via ion-neutral momentum transfer and charge exchange collisions. The escape rates in Strobel (2008a) require ~6 × 10^8 eV/cm^2/s. This is comparable to the average energy flux extracted by DeLaHaye et al. (2007a) assuming the energy deposition rates in Table 15.2 occur over a scale height. However, this energy flux is larger than the ~1 × 10^8 eV/cm^2/s carried by the ion outflow above assuming it is global. Including the enhancement in the effective area of interaction due to the gyroradii of the heavy ions, which is about a factor of two (Hartle et al. 2006a, b; Shematovich et al. 2003), the energy flux *available* in the ambient plasma and fields is much larger, ~2 × 10^{10} eV/cm^2/s (e.g., Johnson 2004). Although this indicates sufficient energy is available from the magnetosphere/ionosphere interaction, it is deposited over a broad range of altitudes. Therefore, molecular level modeling in the interaction region combined with Cassini data will be required to describe the coupling and to calculate the plasma-induced escape rate. For instance, only a fraction of this energy flux contributes to the heating of the exobase by production of recoils as seen in Fig. 15.5 (~5 × 10^9 eV/cm^2/s, Michael et al. 2005a; Michael and Johnson 2005). This flux was obtained from a hybrid model that had poor spatial resolution (~500 km) in the exobase region and produced loss rates well below the recent estimates. Therefore, the details of the plasma flow *and* the collisional coupling of the ions to the neutrals is critical. For instance, if the incident plasma has an energy spectrum such as that in Fig. 15.6, but the same energy flux used by Michael and Johnson (2005), that energy would be deposited in a much narrow region about the exobase ,as seen in Fig. 15.5, so that the required energy densities could easily be achieved.

Details of the energy deposition are given in the Chapter 16. Those authors note that the observed influx of keV oxygen ions from the magnetosphere. ~0.9 × 10^{24} O^+ ions/s impact Titan's atmosphere and corona (Hartle et al. 2006a, b), is consistent with the observation of oxygen chemistry in Titan's upper atmosphere (Hörst et al. 2008). Using a mean ion energy of ~keV, one gets an energy flux ~5.0 × 10^8 eV/cm^2/s at exobase heights. This is fortuitously close to the maximum plasma heating rate allowed by Strobel (2009) resulting in an energy deposition rate comparable to the above estimates, so the O^+ could contribute a significant fraction of the required plasma energy flux.

Sillanpaa et al. (2007) find ~10^{24} O^+/s are lost to interactions with Titan, depositing a globally averaged energy flux ~10^9 eV/cm^2/s normalized to the *exobase surface*. Estimates of the energy deposition rates by the *incoming ions* using hybrid and MHD models (Brecht et al. 2000; Ledvina et al. 2005; Sillanpaa et al. 2007) are below the rates extracted by DeLaHaye et al. (2007a). In such simulations, the ion energy deposition rate into the atmosphere is sensitive to the fields penetrating the exobase region. Therefore, modeling the ion and energetic neutral flow through the exobase region and their collisional coupling to the background neutrals is critical. For instance, ions and energetic neutrals penetrating the corona (Michael et al. 2005a), pick-up ions formed close to the exobase, and ions out-flowing from the ionosphere follow the field lines. Therefore, they not only penetrate the exobase but can cross the exobase from below. The momentum transfer from exiting ions to neutrals considerably enhances the effective sputtering efficiency and is often referred to as forward sputtering (Johnson 1990, 1994). In this way the ionospheric outflow can collisionally drag out neutrals.

Based on CAPS data (e.g., Fig. 6 and Coates et al. 2007) and data from other instruments (e.g., Wahlund et al. 2005; Sittler et al. 2008), there is a flux of low energy ions flowing through the corona that has not been incorporated into the simulations. There is also a flux of very energetic ions (»~10 keV) (Ledvina et al. 2005), but these deposit their energy at depths well below the exobase (e.g., Luna et al. 2003; Cravens et al. 2008) and do not affect escape. If atmospheric sputtering is the dominant process driving escape, which is not at all clear, it is the low energy heavy pick-up and ionospheric ions that need to be included in new simulations of Titan's exobase region. Fortunately, more detailed models of the plasma flow through the exobase region, constrained by Cassini plasma data, are now becoming available.

Based on density vs altitude profiles, such as those in Fig. 15.3b, it has been suggested that a non-thermal energizing process is occurring in the exobase region (DeLaHaye et al. 2007a). Therefore, effective scale heights for the mass density vs altitude are shown in Fig. 15.9 for a number of exobase crossings. Although there is considerable variability, possibly due to Titan's position in Saturn's plasma sheet (Chapter 16), with one exception the steepest profiles, suggestive of the highest temperatures and/or highest escape rates, are associated with the combined plasma/solar flux directions. Whereas the coronal structure can respond promptly to changes in the incident flux, time lags and horizontal transport affect the exobase structure (Muller-Wodarg et al. 2008). The full suite of the atmospheric density profiles is now being compared to the plasma flow data from the CAPS and MIMI instruments.

15.7 Titan Mass Loss: Magnetospheric Implications

Titan atmospheric mass loss is also of interest as a source of neutrals and plasma for Saturn's magnetosphere. Therefore, the composition of the ambient plasma near Titan's orbit can be used to test the atmospheric loss rate. Prior to the arrival of Cassini, hydrogen from Titan was thought to be the dominant neutral source to the magnetosphere. Although smaller mass loss rates were estimated for heavy species, Titan was still assumed to be an important, and possibly dominant, source of nitrogen for the magnetosphere. Since the present escape rates are unsettled, new models for the Titan heavy molecules tori are not yet available. Figure 15.10 shows the spatial distribution based on pre-Cassini estimates of the atmospheric sputtering rates for the nitrogen component. Based on recent estimates for the source rate, the densities shown would be far too low, but the morphology for comparable masses will be similar and the neutral densities would roughly scale by the loss rate. It was anticipated that Cassini should be able to observe pick-up ions from a neutral torus even with densities as low as those predicted in Fig. 15.10, however, such ions have not been observed.

The sources of neutrals for Saturn's magnetospheric are given in Table 15.2. Titan does not appear to be the dominant source of nitrogen (e.g., Smith et al. 2005, 2007, 2008) and is not the dominant source of mass for Saturn's magnetosphere

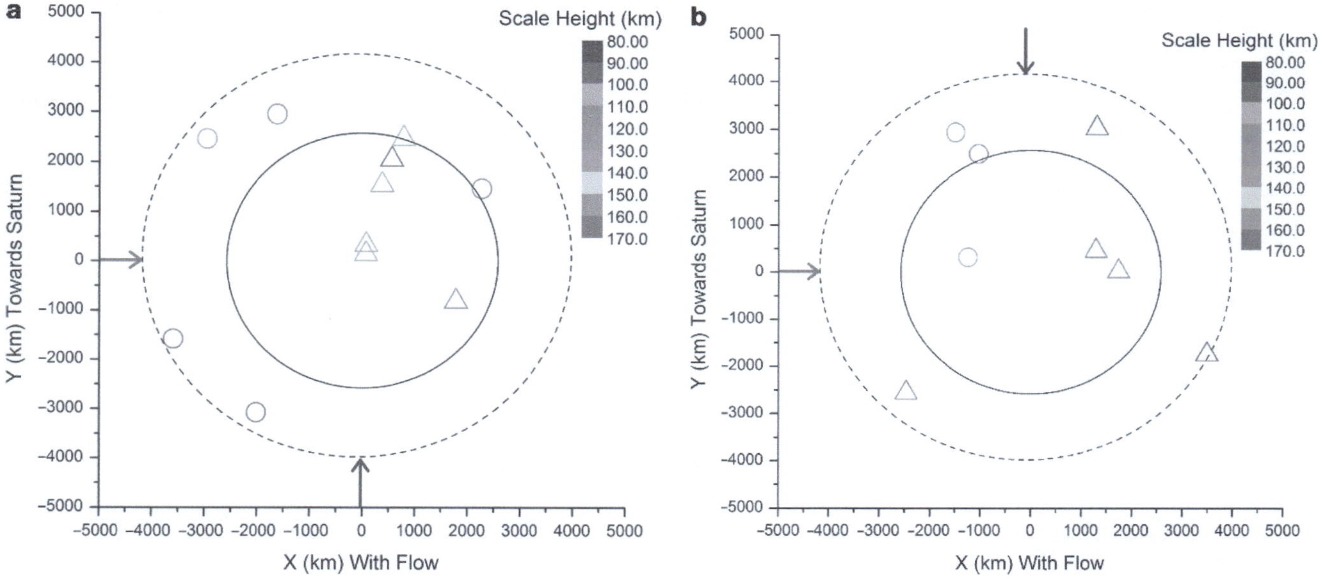

Fig. 15.9 North polar view of Titan showing the scale height in the corona vs Titan's orbital position and the position of the exobase crossing in Titan's atmosphere. Scale heights obtained from fits to the INMS data from 1500 to 2000 km (*colored triangles and circles*) represent ingress and egress points respectively: (**a**) crossings for times periods 9 am–3 pm Saturn local time (Titan on the dayside of Saturn); (**b**) from 9 pm to 3 am Saturn local time (Titan on the nightside of Saturn). Red are the steepest profiles (largest scale height) and blue are the least steep (smallest scale height). Arrows indicate directions of the external energy flux: blue, solar; red plasma: the co-rotating plasma flow which is, of course, significantly altered close to the exobase: e.g. (Sillanpaa et al. 2007) and a correlation with the actual plasma flux is needed. Solid circle indicate Titan's surface, and dashed circle indicates an exobase at 1500 km in the z plane.

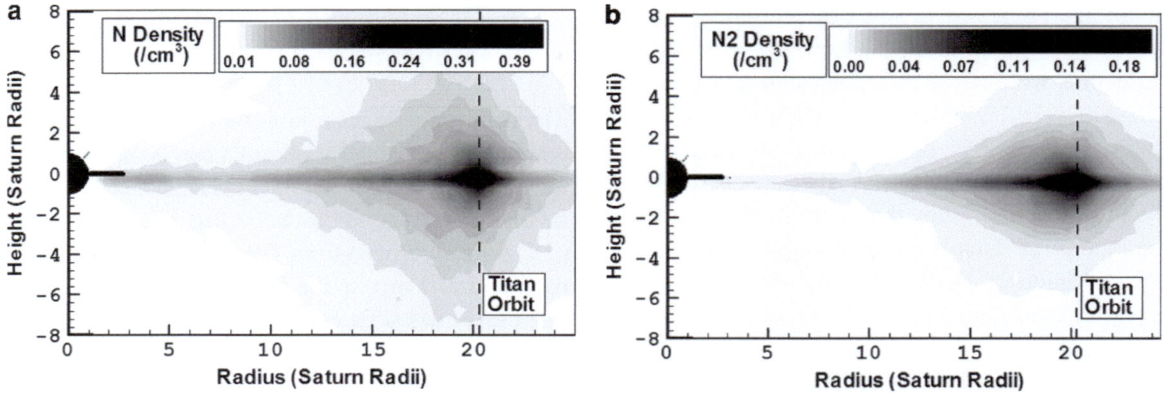

Fig. 15.10 Map of the spatial distribution of mass 14 and mass 28 (Smith et al. 2004): based on pre-Cassini estimates of the atmospheric loss rates from photochemistry (Shematovich et al. 2003) and sputtering Michael et al. 2005a): Mass 28 (~1.7×10^{26} amu/s) and Mass 14 (4.5×10^{26} s^{-1}) from a Titan nitrogen atmosphere (~ 6×10^{26} amu/s). These profiles scale roughly linearly with the source rate and, therefore can be used along with more recent estimates of the loss rates.

even for the largest proposed escape rates as seen in Table 15.2. However, if the largest methane source rate (~4–5 × 10^{28} amu/s: Yelle et al.2008; Strobel 2008a) is correct, then Titan could be the dominant source of carbon for the magnetosphere. However, Cassini measurements suggest that this is not the case.

Source	Source rate (10^{29} amu/s)	Composition[a]
Rings[b]	~6	O_2, H_2
Enceladus[c]	~1- 4	H_2O (~4% C or N species)
Titan[d]	0.03- 0.5	N_2, CH_4, H_2

[a] Molecular dissociation products would also be present for all of these species.

[b] Johnson et al. 2006b.

[c] Johnson et al. 2006a; Melin et al. 2009 ; Hansen et al. 2006 ; Waite et al. 2006.

[d] DeLaHaye et al. 2007a (N_2, CH_4); Cui et al. 2008(H_2; Strobel 2008a (CH_4,H_2); Yelle et al. 2008 (CH_4).

Since neutrals ejected from Titan continue to orbit Saturn until they are ionized, they can be detected indirectly by measuring the ambient plasma composition. Detection of Titan-generated particles is possible not only close to Titan's orbit, but over a broad region of Saturn's magnetosphere (Smith et al. 2004). Although the neutrals converted to ions close to Titan or in its wake are rapidly lost down the magnetotail or in an injection event, a background plasma has been detected in the vicinity of Titan's orbit. Outward diffusion of ions produced closer to Saturn can contribute to this plasma, but such ions are also lost prior to crossing Titan's orbit. In addition, they tend to adiabatically cool which would result in cold plasma that is not observed. Consequently, the measured phase space distributions of detected ions indicate that the ambient plasma near Titan's orbit is dominated by neutrals ionized in the outer magnetosphere (Szego et al. 2005; Sittler et al. 2006). Since any oxygen ionized in this region must be scattered to the outer magnetosphere from the Enceladus and OH tori (Johnson et al. 2006a) or from the ring atmosphere (Johnson et al. 2006b), carbon containing ions should dominate oxygen ions near Titan's orbit if the largest methane loss rates are correct.

However, the plasma upstream from Titan has been shown to contain a substantial fraction of O^+(Hartle et al. 200a, b). The ratio of oxygen to carbon containing ions for SOI was estimated to be ~1.5 (Hartle et al. 2006b). Scaling the suggested loss rates to earlier models of the neutral density from all sources (e.g., Johnson et al. 2006a, b; Smith et al. 2004, 2005, 2007), such an ion fraction is much larger than would be predicted by the highest methane source rate. Determining the present carbon loss rate from ion density measurements in the magnetosphere will require a good model of the oxygen and carbon neutral densities and ionization rates near Titan's orbit. In addition, the carbon and oxygen fractions in the ambient plasma will have to be measured for a variety of plasma conditions.

15.8 Summary

Based on Cassini data for many passes through Titan's upper atmosphere, knowledge of the thermospheric structure and its interaction of Saturn's magnetosphere are rapidly improving. Therefore, it is exciting that kinetic modeling of the present escape mechanisms, constrained by in situ data, is now possible. In this way, the differences in the loss mechanisms proposed will be resolved. However, because of the observed variability in the nature of the interaction, it is likely that analysis of the data from many passes through the thermosphere and corona will be required to unravel the atmospheric escape processes.

Although the calculated hydrogen loss rate (~0.8–2 × 10^{28} amu/s; Cui et al. 2008; Bell et al. 2009) is similar to the pre-Cassini estimate, considerable enhancements in the loss of heavy molecules have been suggested: ~0.3–4 × 10^{28} amu/s. If the largest loss rates for the heavy species persisted for ~ 4 Gyr, the net mass loss would be of the order of the present atmospheric mass. In addition, if such loss rates were primarily CH_4, then the present CH_4 inventory would be lost in <10^8 years. This rate is about an order of magnitude smaller than the principal atmospheric mass loss process, loss of methane by precipitation of hydrocarbons. However, it can have a significant affect on modeling the evolution of Titan's atmosphere (Chapters 7 and 8). For this reason accurately understanding the mechanisms driving escape (e.g., Johnson et al. 2008) is critical. The estimates of the escape rate in DeLaHaye et al (2007a) were poorly constrained since the density data covered a narrow range of altitudes in the corona. If the kappa distributions they extracted for the molecular energy spectra at the exobase are correct for describing the corona structure at all altitudes, then the heavy molecule escape rate is small, consistent with pre-Cassini estimates. Their *preferred escape rate*, obtained by scaling the extracted energy deposition rate to simulations and averaged over five passes, showed a larger heavy molecule escape rate (~0.3 × 10^{28} amu/s), but one that is much smaller than the upward flows extracted by Yelle et al. (2008), Muller-Wodarg et al. (2008) and Strobel (2008a, 2009). *However, the average upper limit* in DeLaHaye et al. (2007a), obtained using a model recoil spectrum, allowed for large escape rates. If these upper limits are correct, then a 1D analysis of the density data below the exobase would result in an upward flow such as that found in the continuum models. In that case, the three mass loss rates would be roughly consistent, although the mechanisms proposed are very different.

Although there is evidence, such as that in Fig. 15.2b, for large extensions of the corona and, possibly, concomitant large local escape rates, the lower *average* escape rates are more likely. Therefore, the continuum models likely need re-examination including both the diffusive replacement of the precipitating carbon and accurate descriptions of the transition region. In addition, good estimates of the amount and spatial distribution of the plasma-induced energy flux is required. Although, the plasma energy deposition rate in the exobase region is larger than the average EUV heating rate, an even larger plasma heating rate or more efficient coupling of the plasma to the neutrals than that in Michael et al. (2005a) would be required to obtain the largest suggested escape rates.

If the largest estimate for the upward flow of heavy molecules at the exobase is correct, but the plasma heating is inadequate to drive escape, then one of the following must be true: another non-thermal process must be active; the upward flow in the regions studied indicates transport of mass to another region of the atmosphere; the upward diffusive flux of methane is due to replacement of the methane lost by precipitation; or heat from below acts to drive escape as proposed by Yelle et al. (2008) and Strobel (2008a). However, preliminary DSMC simulations do not support the thermal conduction mechanism for the loss of carbon and nitrogen, nor does the ion composition in the ambient plasma appear to agree with the methane mass loss rates in Yelle et al (2008) and Strobel (2009). Therefore, additional molecular simulations, constrained by Cassini data, are needed and as well as a description of the morphology of the transport of material in the corona.

A detailed understanding of the present atmospheric escape rate is a precondition for understanding the origin and evolution of Titan's atmosphere and its astrobiological potential (Chapters 7–9). Therefore, it is exciting that the concepts for mass loss from Titan's atmosphere and plasma-induced heating can now be re-examined based on a wealth of Cassini data

Acknowledgements We acknowledge comments by T. Cassidy, I. Sillanpaa, J. Bell, J. Westlake, and B. Magee. Work at U. Virginia, GSFC and SwRI is supported by the Cassini Program under JPL Contract 1243218 with SwRI. Additional support at U. Virginia is provided by NASA's Cassini Data Analysis Program and NASA's Planetary Atmospheres Program.

References

Barbosa DD (1987) Titan's atomic nitrogen torus: inferred properties and consequences for the Saturnian Aurora. Icarus 72:53–61

Bell JM, Bougher SW, Waite JH Jr, Ridley AJ, Magee B, Bar-Nun A, Toth G, De La HayeV (2009) Simulating the global mean structure of Titan's upper atmosphere using the titan global ionosphere-thermosphere model. Planet Space Sci (in press)

Bertucci C; plus 11 (2008) The magnetic memory of titan's ionized atmosphere science 321:1475–1477

Bird GA (1994) DSMC procedures in a homogenous gas. In: Molecular gas dynamics and the direct simulation of gas flows. Clarendon, Oxford, UK, pp 218–256

Brecht SH, Luhmann JG, Larson DJ (2000) Simulation of the Saturnian magnetospheric interaction with Titan. J Geophys Res 105:13119–13130

Chamberlain JW, Hunten D (1987) Theory of Planetary Atmosphere. Academic, New York

Coates AJ, Crary FJ, Young DT, Szego K, Arridge CS, Bebesi Z, Sittler EC Jr, Hartle RE 4, Hill TW (2007) Geophys Res Lett 34:L24S05. doi:10.1029/2007GL030919

Cravens TE, Keller CN, Ray B (1997) Photochemical sources of non-thermal neutrals for the exosphere of Titan. Planet Space Sci 45: 889–896

Cravens TE, Robertson IP, Ledvina SA, Mitchell D, Krimigis SM, Waite JH (2008) Geophys Res Lett 35. doi: 10.1029/2007GL032451

Cui J, Yelle RV, Volk K (2008) Distribution and escape of molecular hydrogen in Titan's thermosphere and exosphere. J Geophys Res 113. doi:10.1029/2007JE003032

De La Haye V, Waite JH Jr, Cravens TE, Nagy AF, Yelle RV, Johnson RE, Lebonnois S, Robertson IP (2007b) Titan's carona: the contribution of exothermic chemistry. Icarus 191:236–250

DeLaHaye V, Waite JH Jr, Johnson RE et al. (2007a) Cassini ion and neutral mass spectrometer data in Titan's upper atmosphere and exosphere: observation of a suprathermal corona. J Geophys Res 112:A07309. doi:10.1029/2006JA0122

Fulchignoni M, Ferri F, Angrilliet F et al (2005) Titan's physical characteristics measured by the Huygens Atmospheric Structure Instrument (HASI). Nature 438:785–791

Garnier P, Dandouras I, Toublanc D, Brandt PC, Roelof EC, Mitchell DG, Krimigis SM, Krupp N, Hamilton DC, Waite H (2007) The exosphere of Titan and its interaction with the kronian magnetosphere: MIMI observations and modeling. Planetary and Space Science 55:165–173

Garnier P, Dandouras I, Toublanc D, Roelof EC, Brandt PC, Mitchell DG, Krimigis SM, Krupp N, Hamilton DC, Dutuit O, Wahlund J-E (2008)The lower exosphere of Titan: Energetic neutral atoms absorption and imaging. J Geophys Res 113, doi:10.1029/2008JA013029

Gustin J, Ajello JM, Stevens MH, Stephan AW, Stewart I, Larsen K, Esposito L, McClintock W (2009) Titan airglow spectra from Cassini UVIS: III. FUV Limb Analysis. Geophys Res Lett (in press)

Hansen CJ, Esposito L, Stewart AIF, Colwell J, Hendrix A, Pryor W, Shemansky D, West R (2006) Enceladus' water vapor plume. Science 311:1422–1425

Hartle RE, 17 authors (2006a) Preliminary interpretation of Titan plasma interaction as observed by the Cassini Plasma Spectrometer: comparisons with Voyager 1. Geophys Res Lett 33:L08201. doi:10.1029/2005GL024817

Hartle RE, 17 authors (2006b) Initial interpretation of Titan plasma interaction as observed by the Cassini plasma spectrometer: comparisons with Voyager 1. Planet Space Sci 54:1211–1224

Hartle RE, Sittler EC, Ogilvie KW, Scudder JD, Lazarus AJ, Atreya SK (1982) Titan's ion exosphere observed from Voyager 1. J Geophys Res 87:1383–1394

Hirschfelder JO, Curtiss CF, Bird RB (1964) Molecular theory of gases and liquids. Wiley, New York, 2nd corrected printing

Hörst SM, Vuitton V, Yelle RV (2008) Origin of oxygen species in Titan's atmosphere. J Geophys Res 113, doi:10.1029/2008JE003135

Johnson RE (1990) Energetic charged-particle interactions with atmospheres and surfaces, Springer, Berlin

Johnson RE (1994) Plasma-induced Sputtering of an Atmosphere. Space Sci Rev 69:215–253

Johnson RE (2004) The magnetospheric-plasma-driven evolution of satellite atmospheres. Astrophys J 609:L99–L102

Johnson RE (2009) Plasma-induced heating and escape. Proc. Royal Soc (London) 367:753–771. doi:10.1098/rsta.2008.0244

Johnson RE, Schnellenberger D, Wong MC (2000) The sputtering of an oxygen thermosphere by energetic O. J Geophys Res 105: 1659–1670

Johnson RE, Smith HT, Tucker OJ, Liu M, Tokar R (2006a) The Enceladus and OH Tori at Saturn. Astrophys J Lett 644:L137–L139

Johnson RE, Luhmann JG, Tokar RL, Bouhram M, Berthelier JJ, Sittler EC, Cooper JF, Hill TW, Smith HT, Michael M, Liu M, Crary FJ, Young DT (2006b) Production, ionization and redistribution of O_2 Saturn's ring atmosphere. Icarus 180:39

Johnson RE, Combi MR, Fox JL, Ip W-H, Leblanc F, McGrath MA, Shematovich VI, Strobel DF, Waite JH Jr (2008) Exospheres and Atmospheric Escape. Space Sci Rev 139:355–397

Jurac S, McGrath MA, Johnson RE, Richardson JD, Vasyliunas VM, Eviatar A (2002) Geophys Res Lett 29. doi:10.1029/2002GL015855

Krasnopolsky VA (1999) Hydrodynmic flow from Pluto. J Geophys Res 104:5955–5962

Krestyanikova MA, Shematovich VI (2006) Stochastic models of hot planetary and satellite coronas: a hot oxygen corona of Mars. Solar Sys Res 40:384–392

Lammer H, Bauer SJ (1993) Atmospheric mass loss from Titan by sputtering. Planet Space Sci 41:657–663

Lammer H, Stumptner W, Bauer SJ (1998) Dynamic escape of H from Titan as a consequence of sputtering induced heating. Planet Space Sci 46:1207–1213

Lammer H, Kasting JF, Chassefière E, Johnson V, Kulikov YN, Tian YN (2008) Atmospheric escape and evolution of terrestrial planets and satellites. Space Sci Rev, Space Sci Rev 139:399–436. doi 10.1007/s11214-008-9413-5

Leblanc F, Johnson RE (2001) Sputtering of the Martian atmosphere by solar wind pick-up ions. Planet Space Sci 49:645–656

Leblanc F, Johnson RE (2002) Role of molecular species in pickup Ion sputtering of the Martian atmosphere. J Geophys Res 107:5-1–5-6. doi:10.1029/2000JE001473

Lebonnois S, Bakes ELO, McKay CP (2003) Atomic and molecular hydrogen budget in Titan's atmosphere. Icarus 161:474–485

Ledvina SA, CravensTE, Kecskemety K (2005) Ion distributions in Saturn's magnetosphere near Titan. J Geophys Res 110. doi:10.1029/2004JA010771

Luna H, Michael M, Shah MB, Johnson RE, Latimer CJ, McConkey JW (2003) Dissociation of N_2 in capture and ionization collisions with fast H^+ and N^+ ions and modeling of positive ion formation in the Titan atmosphere. J Geophys Res 108. Doi:10.1029/2002JE001950

Ma Y, Nagy AF, Cravens TE, I Sokolov V, Hansen KC, Wahlund J-E, Crary FJ, Coates AJ, Dougherty MK (2006) Comparisons between MHD model calculations and observations of Cassini flybys of Titan. J Geophys Res 111. doi:10.1029/2005JA011481

Ma Y-J, Altwegg K, Breus T, Combi MR, Cravens TE, Kallio E, Ledvina SA, Luhmann JG, Miller S, Nagy AF, Ridley AJ, Strobel DF (2008) Plasma flow and related phenomena in planetary aeronomy. Space Sci Rev 139:311–353

Magee B, et Waite JH, Bell KM, Westlake JM, Gell J, De La Haye V (2009) INMS Derived Composition of Titan's Upper Atmosphere: Analysis Methods and Model Comparisons. Planet.Space Sci. (in press) doi:10.1016/j.pss.2009.06.016

Mandt KE, Waite JH Jr, Lunine JI, Magee BA (2009) Isotopic evolution of Titan's main constituents. Planet Space Sci, in press doi:10.1016/j.pss.2009.06.005

Marconi ML, Dagum L, Smyth WH (1996) Hybrid fluid mechanic/kinetic theory approach to planetary atmospheres: an example of an intermdediate mass body. Astrophys J 469:393–401

McNutt RL (1989) Models of Pluto's upper atmosphere. Geophys Res Lett 16:1225–1228

Melin H, Shemanskya DE, Liu X (2009) The distribution of hydrogen and atomic oxygen in the magnetosphere of Saturn. Icarus in press

Michael M, Johnson RE (2005) Energy deposition of pickup ions and heating of Titan's atmosphere. Planet Space Sci 53:1510–1514

Michael M, Johnson RE, Leblanc F, Liu M, Luhmann JG, Shematovich VI (2005a) Ejection of nitrogen from Titan's atmosphere by magnetospheric ions and pick-up ions. Icarus 175:263–267

Michael M, Johnson RE, Leblanc F, Smith HT, Berthelier JJ, Bouhram M, Chassefiere E, Sittler EC, Hartle RE, Coates AJ, Hill TW, Young DT (2005b) Atmospheric sputtering and heating, Spring 2005 Crete Titan/Cassini-Huygens meeting. Crete, Greece

Mitchell DG, Brandt PC, Roelof EC, Dandouras J, Krimigis SM, Mauk BH (2005) Energetic neutral atom emissions from Titan interaction with Saturn's magnetosphere. Science 308:989–992

Muller-Wodarg ICF, Yelle RV, Borggren N, Waite JH Jr (2006) Waves and horizontal structures in Titan's thermosphere. J Geophys Res 111, doi:10.1029/2006JA011961

Muller-Wodarg ICF, Yelle RV, Cui J, Waite JH Jr (2008) Horizontal structures and dynamics of Titan's thermosphere. J Geophys Res 113, CiteID E10005 10.1029/2007JE003033

Nagy AF, Cravens TE, Yee J-H, StewartAIF (1981) Hot oxygen atoms in the upper atmosphere of Venus. Geophys Res Lett 8(June):629–632

Penz T, Lammer H, Kulikov YuN, Biernat HK (2005) The influence of the solar particle and radiation environment on Titan's atmosphere evolution. Adv Sp. Res 36:241–250

Schunk RW (1975) Transport equations for aeronomy. Planet Space Sci 23:437–485

Schunk RW, Nagy AF (2000) Ionospheres: physics, plasma physics, and chemistry. In: Dressler AJ, Houghton JT, Rycroft MJ (eds) Cambridge Atmo Space Sc. Ser. Cambridge University Press, Cambridge, UK, pp 104–147

Shematovich VI, Johnson RE, Michael M, Luhmann JG (2003) Nitrogen loss from Titan. J Geophys Res 108(E8):5087. doi:10.1029/003JE002094

Sigmund P (1981) Sputtering by partclie bimbardment: theoretical concepts. In: Behrisch R (ed) Sputtering by Particle Bombardment I, Springer, Berlin, pp 9–67

Sillanpaa I, Kallio E, Janhunen P, Schmidt W, Mursula K, Vilppola J, Tanskanen P (2006) Hybrid simulation study of ion escape at Titan for different orbital positions. Adv Space Res 38:799–805

Sillanpaa I, Kallio E, Jarvinen R, Janhunen P (2007) Oxygen ions at Titan's exobase in a Voyager 1 – type interaction from a hybrid simulation J Geophys Res 112. doi:10.1029/2007JA012348, 2

Sittler EC, Thomson M, Johnson RE et al. (2007) Cassini observations of Saturn's inner plasmasphere: Saturn orbit insertion result. Planet Space Sci 54:1197–1210 (2006). Erratum: Planet Space Sci 55:2218–2220

Sittler EC et al. (2006) Energetic nitrogen ions within the inner magnetosphere of Saturn. J Geophys Res 111:A09223, doi:10.1029/2004JA010509

Sittler EC et al. (2008) Ion and neutral sources and sinks within Saturn's inner magnetosphere Cassini results: Plant Space Sci 56:3–18

Sittler EC et al. (2009) Composition of upstream flow for Titan's interaction with Saturn's magnetosphere: preliminary results. Planet Space Sci (in press)

Smith HT, Johnson RE, Shematovich VI (2004) Titan's atomic and molecular nitrogen tori 3. Geophys Res Lett 31:L16804. doi:10.1029/2004GL020580

Smith HT, Shappirio M, Sittler EC, Reisenfeld D, Johnson RE, Baragiola RA, Crary FJ, McComas DJ, Young DT (2005) Discovery of nitrogen in Saturn's inner magnetosphere. GRL 32:L14S03. doi:10.1029/2005GL022654

Smith HT, Johnson RE, Sittler EC, Shappirio M, Tucker OJ, Burger M, Crary FJ, McComas DJ, Young DT (2007) Enceladus The likely dominant nitrogen source in Saturn's magnetosphere. Icarus 188:356–366

Smith HT, Shappirio M, Johnson RE, Reisenfeld D, Sittler EC, Crary FJ, McComas DJ, Young DT (2008) Enceladus: A potential source of ammonia products and molecular nitrogen for Saturn's magnetosphere, J Geophys Res doi:10.1029/2008JA013352.

Smith HT, Mitchell DG, Johnson RE, Paranicas C (2009) Investigation of energetic proton penetration in Titan's atmosphere using the Cassini INCA instrument. Planet Space Sci doi:10.1016/j.pss.2009.03.013.

Strobel DF (2002) Aeronomic Systems on Planets, Moons, and Comets. In: Mendillo M, Nagy A, Waite H (eds) Comparative Aeronomy in the Solar System, American Geophysical Union, Geophysical Monograph Series, pp 7–22

Strobel DF (2008a) Titan's hydrodynamically escaping atmosphere. Icarus 193:588–594

Strobel DF (2008b) N_2 escape rates from Pluto's atmosphere. Icarus 193:612–619

Strobel DF (2009) Titan's hydrodynamically escaping atmosphere: escape rates and the structure of the exobase region. Icarus 202:632–641

Strobel DF, Shemansky DE (1982) EUV emissions from Titan's upper atmosphere: Voyager I encounter. J Geophys Res 87:1361–1368

Szego K et al. (2005) The global plasma environment of Titan as observed by Cassini Plasma Spectrometer during the first two close encounters with Titan. Geophys Res Lett 32:L20S05. doi:10.1029/2005GL022646

Terada N, Shinagawa H, Machida S (2004) Global hybrid model of the solar wind interaction with the Venus ionosphere: ion escape processes. Adv Space Res 33:161–166

Tucker OJ, Johnson RE (2009). Thermally driven atmospheric escape: Monte Carlo simulations for Titan's atmosphere. Planet Space Sci 11 June doi:10.1016/j.pss.2009.06.003

Vasyliunas VM (1968) A survey of low-energy electrons in the evening sector of the magnetosphere with OGO 1 and OGO 3. J Geophys Res 73:2839–2885

Wahlund J-E, 17 coauthors (2005) Cassini measurements of cold plasma in the ionosphere of Titan. Science 308:982–986

Waite JH Jr, 13 authors (2006) Cassini ion and neutral mass spectrometer: Enceladus plume composition and structure. Science 311(5766): 1419–1422

Waite JH Jr, 21 authors (2005) Ion neutral mass spectrometer results from the first flyby of Titan. Science 308:982–986

Wei HY, Russell CT, Wahlund J-E, Dougherty MK, Bertucci C, Modolo R, Ma YJ, Neubauer FM (2007) Cold ionospheric plasma in Titan's magnetotail. Geophys Res Lett 34:CiteID L24S06. doi:10.1029/2007GL030701

Wilson EH, Atreya SK (2004) Current state of modeling the photochemistry of Titan's mutually dependent atmosphere and ionosphere. J Geophys Res 109. doi:10.1029/2003JE002181

Wilson JK, Mendillo M, Baumgardner J, Schneider NM, Trauger JT, Flynn B (2002) The dual sources of Io's sodium clouds. Icarus 157:476–489

Yelle RV, Cui J, Muller-Wodarg ICF (2008) Methane escape from Titan's atmosphere. J Geophys Res 113. doi:10.1029/2007JE003031

Chapter 16
Energy Deposition Processes in Titan's Upper Atmosphere and Its Induced Magnetosphere

Edward C. Sittler, R. E. Hartle, Cesar Bertucci, Andrew Coates, Thomas Cravens, Iannis Dandouras, and Don Shemansky

Abstract Most of Titan's atmospheric organic and nitrogen chemistry, aerosol formation, and atmospheric loss are driven from external energy sources such as Solar UV, Saturn's magnetosphere, solar wind and galactic cosmic rays. The Solar UV tends to dominate the energy input at lower altitudes~1,200 km but which can extend down to ~400 km, while the plasma interaction from Saturn's magnetosphere, Saturn's magnetosheath or solar wind are more important at higher altitudes ~1,400 km, but the heavy ion plasma (O^+) ~5 keV and energetic ions (H^+) ~30 keV or higher from Saturn's magnetosphere can penetrate below 950 km. Cosmic rays with energies >1 GeV can penetrate much deeper into Titan's atmosphere with most of its energy deposited ~70 km altitude. Haze layers are observed in scattered solar photons starting at 510 km, but aerosols are broadly distributed and measured in extinction from 1,000 km downward, diffusively separated to 400 km. The induced magnetic field from Titan's interaction with the external plasma can be very complex and will tend to channel the flow of energy into Titan's upper atmosphere. Cassini observations combined with advanced hybrid simulations of the plasma interaction with Titan's upper atmosphere show significant changes in the character of the interaction with Saturn local time at Titan's orbit where the magnetosphere displays large and systematic changes with local time. The external solar wind can also drive sub-storms within the magnetosphere which can then modify the magnetospheric interaction with Titan. Another important parameter is solar zenith angle (SZA) with respect to the co-rotation direction of the magnetospheric flow which is referred to as the solar incidence-ram angle. Titan's interaction can contribute to atmospheric loss via pickup ion loss, scavenging of Titan's ionospheric plasma, loss of ionospheric plasma down its induced magnetotail via an ionospheric wind, and non-thermal loss of the atmosphere via heating and sputtering induced by the bombardment of magnetospheric keV ions and electrons. This energy input evidently drives the large positive and negative ions observed below ~1,100 km altitude with ion masses exceeding 10,000 Da. We refer to these ions as seed particles for the aerosols observed below 1,000 km altitude. These seed particles can be formed, for example, from the polymerization of acetylene (C_2H_2) and benzene (C_6H_6) molecules in Titan's upper atmosphere to form polycyclic aromatic hydrocarbons (PAH) and/or fullerenes (C_{60}). In the case of fullerenes, which are hollow spherical carbon shells, magnetospheric keV O^+ ions can become trapped inside the fullerenes and eventually find themselves inside the aerosols as free oxygen. The aerosols are then expected to fall to Titan's surface as polymerized hydrocarbons with trapped free oxygen where unknown surface chemistry can take place.

E.C. Sittler (✉) and R. Hartle
NASA/Goddard Space Flight Center, Greenbelt, MD, USA
email: Edward.C.Sittler@nasa.gov

C. Bertucci
Institute for Astronomy and Space Physics - IAFE
Ciudad Universitaria Buenos Aires, Argentina

A. Coates
Mullard Space Science Laboratory, University College London, UK

T. Cravens
University of Kansas, Kansas, USA

I. Dandouras
Centre National de la Recherche Scientifique (CNRS)
at the Centre d'Etude Spatiale des Rayonnements (CESR)
Laboratory, Toulouse, France

D. Shemansky
University of Southern California, Department of Aerospace
and Mechanical Engineering, Los Angeles, California, USA

16.1 Introduction

The deposition of energy to Titan's upper atmosphere is closely coupled to the details of Titan's interaction with Saturn's magnetosphere or magnetosheath or solar wind since the magnetic field can both insulate the atmosphere against charged particle access to it and/or can channel how this energy penetrates into its upper atmosphere. The details of the interaction will be strongly dependent upon the local configuration of the magnetosphere and as presented below can be highly variable and dependent upon Saturn Local Time (SLT). Voyager 1 was the first spacecraft to observe this interaction and therefore used to introduce discussions in this chapter.

R.H. Brown et al. (eds.), *Titan from Cassini-Huygens*,

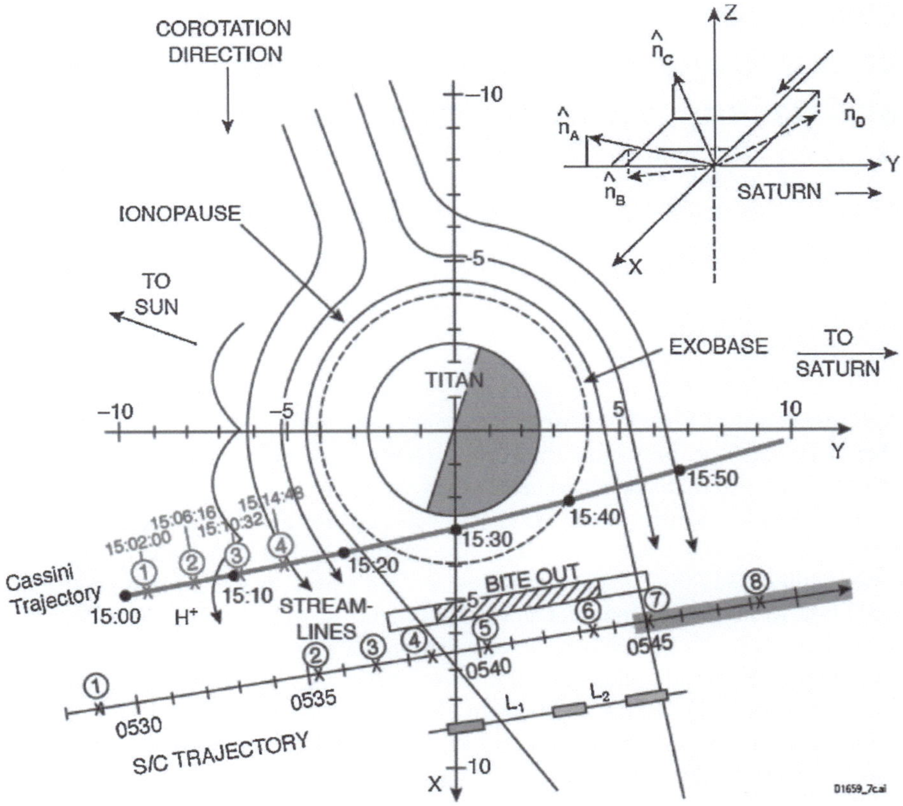

Fig. 16.1 View of Voyager 1 encounter geometry with Cassini TA encounter trajectory super-imposed. Figure is from Hartle et al. (2006b). The figure shows the coordinate system we use in this chapter and the Voyager PLS sensor alignment. The numbers along Voyager 1 trajectory (see Fig. 16.51), which is inclined ~10° relative to x–y plane, indicate location of PLS ion measurements. For TA the trajectory is inclined ~17.3° relative to x–y plane and numbers indicate position composition measurements by the IMS were made. Ion singles data was made at 4 s intervals and actuator motion was on for CAPS

The Voyager 1 (V1) encounter with Titan, on November 12, 1980, was the first close look at Titan's interaction with Saturn's magnetosphere (Bridge et al. 1981; Ness et al. 1981; Gurnett et al. 1981, 1982; Hartle et al. 1982; Neubauer et al. 1984; Sittler et al. 2005a). Magnetometer measurements revealed that Titan had no detectable intrinsic magnetic field. This meant that Saturn's rotating magnetosphere interacts directly with Titan's atmosphere/ionosphere, similar to the solar wind interaction with Venus. The Voyager 1 and Cassini TA encounter geometries are shown in Fig. 16.1 from Hartle et al. (2006b), where aspects of the plasma interaction are sketched. Voyager 1 also probed its upper ionosphere using an onboard radio science instrument (Bird et al. 1997) and probed its upper atmosphere using ultraviolet measurements of its day time airglow and solar occultation measurements during the wake pass (Broadfoot et al. 1981; Strobel and Shemansky 1982; Strobel et al. 1992). These observations clearly showed a high plasma beta $\beta >> 1$, sub-sonic and super-Alfvenic interaction with no shock observed and induced magnetosphere with draped field lines forming a magnetotail region in its wake (Ness et al. 1981; Neubauer et al. (1984). The plasma data revealed a bite-out in the $E > 500$ eV electron spectrum indicating deposition of magnetospheric electrons into Titan's upper atmosphere and presence of photoelectrons and secondaries coming into Titan's upper ionosphere along draped field lines (Bridge et al. 1981; Hartle et al. 1982). It also became clear that magnetospheric heavy ion gyro-radii >5,000 km are larger than the dimensions of Titan and that hybrid codes of the interaction would eventually have to be developed.

Development of an understanding of this interaction has extended over several decades with innovative theoretical modeling of the atmosphere and ionosphere (Yung et al. 1984; Toublanc et al. 1995; Gan et al. 1992), and the plasma interaction (Cravens et al. 1998; Brecht et al. 2000) and continued analysis of Voyager 1 data sets (Bird et al. 1997; Strobel et al. 1992; Sittler et al. 2004, 2005a). Some of this advancement was driven by preparations for the upcoming mission to Saturn and Titan by Cassini–Huygens. All the processes of the interaction, ionosphere formations, atmospheric heating and corresponding atmospheric loss can be traced to the various energy sources into Titan's upper atmosphere. In Table 16.1 the various energy sources to Titan's upper atmosphere are summarized. These estimates ignore the insulating effect of Titan's induced magnetosphere against, for example, magnetospheric electron access to Titan's ionosphere. The table does show that charged particle energy inputs can dominate all other energy input processes including solar UV. As the Cassini–Huygens mission has shown, the energy input can be traced to solar ultraviolet and soft x-rays (Cravens et al. 2005), magnetospheric plasma ion and electron precipitation (Cravens et al. 2008a,b; Hartle et al. 2006c; Coates et al. 2007a), energetic particle ion precipitation (Cravens et al. 2008a,b; Garnier et al. 2007) and precipitation from pickup ions (Michael et al. 2005; Michael and Johnson 2005). Both Sittler et al.

Table 16.1 Titan upper atmosphere energy sources

Energy source	Energy flux (erg/cm^2/s)	Global input (Watts)[d]	Comments
Plasma protons	1.6e-4	3.4e7	Magnetized
Plasma electrons	1.6e-4	3.4e7	Magnetized
Plasma heavy ions	1.5e-3	3.2e8	Unmagnetized
Energetic ions	5.0e-4 to 1.0e-2	1.05e8 to 2.0e9	$27<E_p<255$ keV[a]
energetic electrons	2.0e-4	4.0e7	$28<E_e<533$ keV[a,b]
UV airglow	1.6e-3	3.5e8	Altitude ~ 1,300 km[c]
UV ionization	1.6e-4	3.4e7	Altitude ~ 1,300 km[c]
Ohmic heating			Not yet known
GCR	1.6e-4 to 2.7e-3	3.2e7 to 5.4e8	Integrated flux
Dust	1.8e-3	1.8e8	Interplanetary dust

[a] Model by Ledvina et al. (2005) show some magnetic channeling of 50 keV protons. If heavy ions (O^+) unmagnetized. Expect energetic electrons to be more magnetized than protons.

[b] Energetic electron energy flux derived from Krupp et al. (2005).

[c] For T0 at 90° phase angle UV absorption peaked at 1,325 km, while for TB at 0° phase angle absorption peaked at 1,095 km altitude

[d] Exobase at r ~ 4,000 km and 4π area ~ 2×10^{18} cm^2.

Table 16.2 Titan Flyby geometries and magnetospheric configuration

Flyby	Magnetosphere	Mag-Ions	SRA/LT	Pass type	Energy input
V1	Dipolar	Heavy	90°/1200	Wake	High
TA, TB, T3	Dipolar	Heavy	70°/1000	Wake (close)	Medium
T5	Magnetodisk	Heavy	180°/0600	Polar and Ram	High
T9	Magnetodisk	Light	120°/0300	Wake	Low
T18	Magnetodisk	Light	110°/0200	Polar (close)	Low
T32	Magnetosheath	Light	90°/1200	Polar (close)	Medium

(2005a) for V1 and Hartle et al. (2006a,b) for TA (first close encounter with Titan by Cassini spacecraft) identified a clearing zone ~ one ion gyro-diameter above the exobase for removal of both light (~800 km) and heavy ions (~10,000 km), which resulted in much larger precipitating fluxes into Titan's atmosphere than originally foreseen. Hartle et al. (2006a,b) estimated a flux of keV O^+ ions into Titan's upper atmosphere ~ 9×10^{23} O^+/s at the exobase. This translates to an energy input ~ 3×10^{15} erg/s or energy flux ~ 1.5×10^{-3} erg/cm^2/s. Cassini INMS observations showed its neutral atmosphere composition (Waite et al. 2005) and ionosphere composition (Cravens et al. 2006) to be rich in hydrocarbons. The work by Vuitton et al. (2006, 2007) also showed the combined importance of nitrogen and hydrocarbon chemistry. Most recently, the discovery of heavy negative ions (Coates et al. 2007a) and possible insertion of magnetospheric oxygen ions to these negative ions (Sittler et al. 2009a) highlights the importance of magnetospheric input. The rich chemistry observed in Titan's upper neutral atmosphere and ionosphere underscores the importance of chemical energy and transformation of solar and magnetospheric charged particle energy input in the form of chemical potential energy and its atmospheric loss to Titan's surface (Sittler et al. 2009a).

The configuration of Saturn's magnetosphere with Saturn LT has been a big surprise and as discussed in the papers by Sittler et al. (2009b) and Bertucci et al. (2008) Titan can be exposed to upstream conditions of low energy input where ion composition is dominated by light ions and plasma pressures are low compared to magnetic field energy densities to the other extreme where heavy ions dominate and plasma pressures exceed magnetic field energy densities by orders of magnitude. Table 16.2 summarizes the geometries and Saturn magnetospheric configuration for Titan flybys selected for discussion in this chapter. The first case noted above tend to occur when Saturn's magnetosphere is in a magnetodisk configuration centered around midnight LT and Titan is below the current sheet (i.e., heavy ions are closely confined to current sheet), while the latter can also occur under similar conditions and Titan is within the current sheet region where heavy ions reside or it is near noon LT where field is dipolar and heavy ions also dominate (T25 is exception to this). In the case of low energy input Titan's induced magnetosphere will tend to be inflated and lighter pickup ions such as H^+ and H_2^+ will tend to dominate at its outer boundary, while for the high energy limit, the induced magnetosphere will be compressed and its outer boundary penetrate to ionospheric heights where heavy pickup ions such as CH_4^+ and N_2^+ will tend to dominate. For the low energy input case ion gyro-radii will be relatively low and boundaries larger in extent that multi-fluid codes could be used (although hybrid codes are still desirable but more difficult to implement even with large super computers), while for the high energy case ion gyro-radii are larger than Titan dimensions and multi-ion hybrid codes must be used.

16.1.1 Summary of Energy Input Sources to Titan's Upper Atmosphere

The various forms of energy input are shown in Fig. 16.2, which is a modified version of Fig. 7 in Waite et al. (2004). This figure shows the different layers of Titan's atmosphere from lower atmosphere to exosphere and topside ionosphere. Within the lower atmosphere is shown the presence of aerosols, while the smaller haze particles reside at higher altitudes within the thermosphere and ionosphere with lower boundary of 400 km. The exosphere's lower boundary, the exobase, is around 1,400 km altitude. To the left we show solar input to Titan's atmosphere in the form of visible light (yellow) which is strongly scattered by the haze particles and cannot penetrate down to the lower atmosphere, while blue indicates solar UV and soft x-rays, which produce Titan's ionosphere, and red indicates infrared light, which can penetrate down to Titan's surface via its narrow NIR windows. The upstream magnetospheric flow and magnetic field are also shown to the left, which is consistent with Titan being on the dusk side of Saturn. The field is shown to be vertical and undeflected on the upstream side, but as we now know the magnetospheric field can be in a magnetodisk configuration and be nearly perpendicular to the plane of the paper.

The Cassini results also identified the upstream flow to be composed of water group ions (O^+, OH^+), methane group ions (CH_n^+ with $2 \le n \le 4$), H_2^+ and H^+. When the flow and field impinges upon Titan's atmosphere-ionosphere it is mass loaded, slowed down and the field is frozen into Titan's topside ionosphere. The field then forms draped field lines often referred to having a parabolic shape and the topside ionosphere may sometimes be pulled into an extended outflow

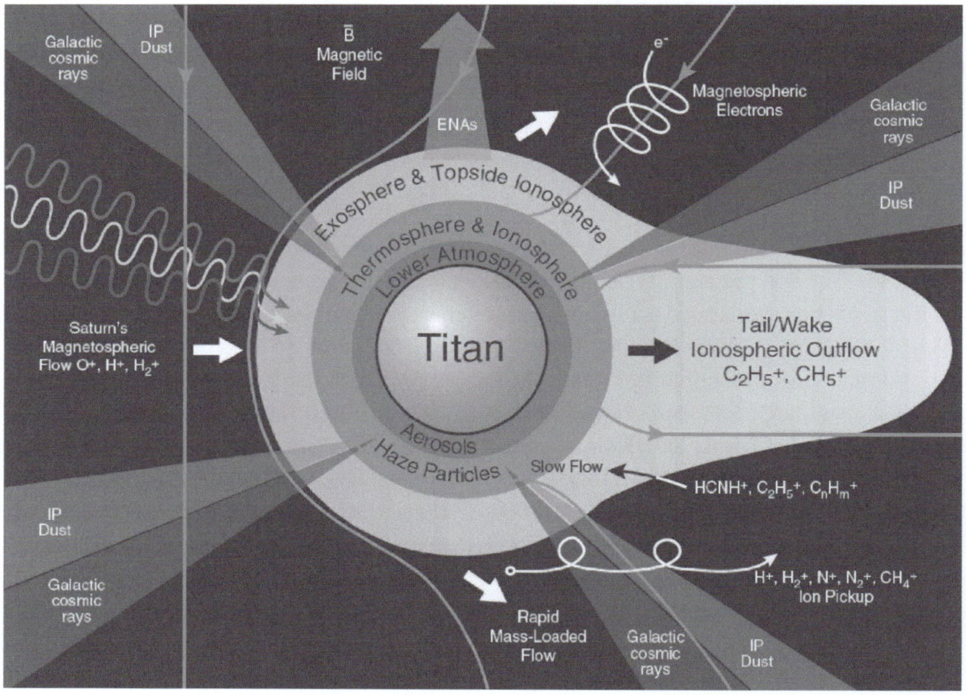

Fig. 16.2 Adapted from Waite et al. (2004). The figure shows the three main layers of Titan's atmosphere, and its illumination by solar UV (*blue rays*), solar visible light (*yellow rays*) and solar infrared light (*red rays*). The energy input from Saturn's magnetospheric interaction shows a strong interaction with Titan's atmosphere and the corresponding induced magnetosphere with draped magnetic field lines. The upstream magnetospheric flow composed of H^+, H_2^+ and O^+ ions. The energetic H^+ ions E > 50 keV and heavy keV O^+ ions can penetrate across field lines and deposit their energy directly into Titan's atmosphere. Emitted energetic neutral atoms (ENAs) from charge transfer collisions between magnetospheric energetic ions and Titan's atmosphere are shown. The magnetospheric electrons are magnetically tied to the field lines but can drift to lower altitudes via gradient and curvature drift mechanisms (Hartle et al. 1982). The magnetospheric electrons plus photoelectrons from the ionizing solar UV striking Titan's upper atmosphere produce a non-thermal electron gas within Titan's ionosphere which can then heat thermal electrons to $T_e \sim 1{,}000°K$ (Wahlund et al. 2005). The hot electrons can also produce a polarization electric field ($T_e >> T_{ION} \sim 180°K$) which will pull ions out of the ionosphere and produce an ionospheric wind as shown in the form of CH_5^+ and $C_2H_5^+$ ions. Also shown are pickup ions (PUI) from the interaction with Titan's exosphere which slows the external flow at high altitudes >1,400 km and leads to atmospheric escape of keV methane and di-nitrogen PUIs (Hartle et al. 2006a,b). The figure shows a more isotropic energy input to deeper altitudes from Galactic Cosmic Rays (GCR) with penetration below 100 km altitude (Fulchignoni et al. 2005) and Interplanetary Dust (IP) which can penetrate down to <740 km altitudes (Molina-Cuberos 2001). The IP dust flux will tend to peak on the side facing Titan's orbital motion around Saturn. In both cases ionization layers are produced deep into Titan's atmosphere where further chemistry can be driven

on the wake side forming an ionospheric wind that may be composed of the $C_2H_5^+$ and CH_5^+ ions, which dominate the higher altitudes of the ionosphere. On field lines we show magnetospheric electrons, which are highly magnetized and can only move along field lines and thus confined to higher altitudes. They can experience gradient drift around Titan (see Sittler and Hartle 1996 for Triton) and curvature drift which may allow the electrons to penetrate deeper into Titan's ionosphere (Hartle et al. 1982; Strobel et al. 1992). In contrast, the keV heavy ions in the upstream flow are not magnetized and have direct entry into Titan's lower ionosphere.

Near the bottom, pickup ions are shown, which are born at higher altitudes, where Titan's neutral exosphere extends across field lines. The figure shows pickup ions H^+, H_2^+, N^+, N_2^+ and CH_4^+, which come from the ionization of Titan's neutral H, H_2, N, N_2 and CH_4 exosphere species, respectively. The N exosphere results from keV ions bombarding Titan's upper atmosphere which then dissociate the N_2 and emit suprathermal N atoms. The N atoms can also come from UV dissociation of N_2 and bombardment by magnetospheric electrons and energetic particles. (Note, that low energy electrons produce a high neutral dissociation rate, and solar photons dissociate at and above the exobase level.) The pickup ions can mass load the flow far from Titan such that when it reaches the topside ionosphere the flow can be reduced from 120 km/s to ~5 km/s or less.

Hot plasma or energetic ions can also cross field lines but in the case of 30 keV protons Ledvina et al. (2005) found some shielding of the atmosphere by the magnetic field. In the top of the figure we show the emission of energetic neutral atoms (ENAs) which primarily result from the charge exchange of hot and energetic magnetospheric ions with Titan's atmosphere and are emitted as ENAs. Finally, we show the nearly isotropic bombardment of Titan's atmosphere by interplanetary dust (IP) particles, E ring particles and Galactic Cosmic Rays (GCR). The IP and E ring dust particles can be a source of metallic ions and water ice, respectively and can form ionization layers down to 600 km altitudes (see English et al. 1996; Lara et al. 1996; Molina-Cuberosa et al. 2001). The GCR, which have energies >1 GeV, can penetrate deep into Titan's lower atmosphere forming an ionization peak below 100 km altitude, extending this ionization all the way down to the surface for the most energetic cosmic rays (see Capone et al. 1983; Molina-Cuberos et al. 1999; Fulchignoni et al. 2005; Borucki and Whitten 2008).

In summary, this figure shows the various sources of energy to Titan's upper atmosphere and outlines the many topics to be discussed in this chapter and how the magnetospheric interaction can complicate energy input process and flow of energy to Titan's atmosphere.

16.1.2 Outline of Chapter and Relation to Other Chapters

The overall outline of the chapter is (1) The introduction. (2) Review the relevant instruments for quantifying the energy input to Titan's upper atmosphere and their observational limitations so one can best understand our later discussions and conclusions. (3) The global properties of Saturn's magnetosphere and importance of angle between solar incidence and ram direction of upstream flow or SRA. The importance of the SRA on Titan's induced magnetosphere and energy input to Titan's atmosphere is discussed. (4) In order to put the observations into proper perspective a review of the various models of Titan's interaction with Saturn's magnetosphere is given. The importance of MHD and hybrid codes are discussed and how they can complement each other. The end goal is to develop a kinetic-fluid hybrid code of the interaction which includes ionosphere physics. (5) Review radio science sounding of Titan's ionosphere and Huygens HASI results. This data gives one a global view of ionospheric structure with height and where energy is deposited into Titan's atmosphere and help focus ones discussions on energy input. (6) Reviews both Voyager 1 and Cassini observations of solar photon energy input with emphasis on FUV and EUV observations. (7) This section is the most comprehensive where Voyager 1 and Cassini observations are used to study the magnetospheric interaction and charged particle precipitation into Titan's atmosphere in the form of magnetospheric electrons, magnetospheric ions and pickup ions. This section is sub-divided into 5 sub-sections for the single Voyager 1 flyby and five of the more important Titan flybys up till now with regard to energy deposition processes and the magnetospheric interaction with Titan and is summarized in Table 16.2. (8) This section discusses the formation of ionization layers at lower altitudes ~500 km (Kliore et al. 2008) and ~70 km (Fulchignoni et al. 2005) and their possible sources by precipitating magnetospheric energetic particles, interplanetary dust at ~500 km and cosmic rays at ~70 km. (9) This section discusses the discovery of heavy negative ions within Titan's ionosphere and conversion of external energy sources into chemical energy and the possible insertion of magnetospheric oxygen into these heavy ions (Sittler et al. 2009a). These heavy ions are the likely seed population of aerosols measured at lower altitudes by UVIS (Liang et al. 2007; Shemansky et al. 2008; and Huygens Tomasko et al. 2005) and the eventual deposition of these heavy ions in the form of aerosols to Titan's surface. This oxygen will be confined to this mixture of tholins on the surface where further pre-biotic chemistry can occur from GCR and sub-surface energy sources, and (10) conclusion and outlook to the future.

16.2 Pertinent Instrument Characteristics, Observational Constraints and Observational Limitations

In order to understand Cassini observations of Titan, its interaction with Saturn's magnetosphere, solar input to Titan and the spatial and temporal properties of the energy input to Titan one needs to be aware of the relevant Cassini instrumentation, their energy or wavelength coverage and their field-of-view coverage to name a few. One must also understand their limitations. A very detailed description of all Cassini mission instruments was reported in *Space Science Reviews*, 114, 2004. For convenience, we briefly discuss the instruments here, but only those involved in the topics of this chapter. The reader can skip this section if desired and refer back to it as needed The prime instruments, listed in Table 16.3 with their characteristics, are as follows: (1) Cassini Plasma Spectrometer (CAPS) Ion Mass Spectrometer (IMS), Ion Beam Spectrometer (IBS) and Electron Spectrometer (ELS) (Young et al. 2004), (2) Cassini Ion and Neutral Mass Spectrometer (INMS) (Waite et al. 2004), (3) Cassini Magnetic Field Investigation (MAG) data (Dougherty et al. 2004), (4) Radio and Plasma Wave Science Investigation (RPWS) with Langmuir Probe (LP) (Gurnett et al. 2004), (5) the Ultraviolet Imaging Spectrograph (UVIS) instrument (Esposito et al. 2004) and (6) the Magnetosphere Imaging Instrument (MIMI) with Energetic Neutral Atom (ENA) imaging capability (INCA) (Krimigis et al. 2004). Other relevant instruments are Radio Science (Kliore et al. 2004) and Huygens HASI instrument (Fulchignoni et al. 2002, 2004).

16.2.1 CAPS IMS, IBS and ELS

Referring to Table 16.3, the CAPS IMS, IBS and ELS field-of-views (FOV) are co-aligned on a scan platform or actuator that moves in a windshield wiper motion. The spacecraft coordinates are such that $-Z_{SC}$ points along bore site of High Gain Antenna (HGA), $-Y_{SC}$ aligned along look direction of the remote sensing instruments such as UVIS and X_{SC} completes the right-hand rule. The center of the CAPS FOV and actuator scan (actuator angle $\alpha = 0°$) is along $-Y_{SC}$. Both IMS and ELS have wide but narrow fans $160° \times 8°$ so that one must scan these fans to achieve a FOV coverage~2π. The wider actuator scan is used when at higher altitudes and spatial scales of the interaction are larger, while a smaller scan is used for Titan's ionosphere where the flow is highly beamed along the ram direction of the spacecraft and one wants to increase time and thus spatial resolution of the measurements. During essentially all flybys the spacecraft ram speed is ~6 km/s, so one covers 360 km in one minute, and 21,600 km in one hour. During a wide actuator scan with actuator speed ~1 deg/s, the spacecraft moves ~1,200 km, while for the shorter actuator scan the spacecraft moves ~300 km along the spacecraft track. The geometric factor (GF) of IMS is ~100 times greater than IBS, so is best at higher altitudes where magnetospheric fluxes are relatively low ~10^5 ions/cm^2/s, while at lower altitudes within the ionosphere where the ion densities are orders of magnitude higher and the ion flux is confined to a smaller volume in velocity space (i.e., high Mach flow in spacecraft frame), one uses IBS which has a lower GF, narrower energy passband and higher angular resolution. At these lower altitudes the IMS microchannel plate (MCP) detectors will saturate due to count rates higher than they are designed to handle, one raises the minimum energy-per-charge (E/Q) from 1 V to 28 V in order to avoid the main ion mass peaks of the ionosphere. Therefore, IBS is the prime instrument for positive ions within Saturn's ionosphere. IMS can still measure high energy particles penetrating down to ionospheric heights such as observed for T5. In the case of ELS the detectors can handle count rates up to 10 MHz and in general have no restrictions in count rates since most of the ionospheric electrons are confined below 1 eV where they are measured by the LP. The ELS instrument will measure the ionospheric photoelectrons produced from the ionization of Titan's upper atmosphere by solar UV. It is important to note that CAPS IMS has the ability to separate magnetospheric O^+ ions from CH_4^+ pickup ions due to the O^- fragment produced within the IMS instrument.

16.2.2 Cassini INMS

The INMS is designed to only measure the neutral or ion density as a function of species mass. The mass coverage is from 0.5 to 8.5 and from 11.5 to 99.5 Da. For neutral measurements one normally uses the closed source which is less sensitive to spacecraft attitude. Since electron impact ionization is used to ionize the neutrals one can distinguish between species of similar mass by measuring their cracking patterns. A radio frequency quadrupole mass spectrometer is used to measure the mass spectrum. In the case of ion measurements, the open source is used which has a FOV only 6 degrees wide (full width) so that the spacecraft attitude must be oriented to accommodate this FOV requirement with spacecraft ram along $-X_{SC}$. The INMS points along the $-X_{SC}$ axis which is orthogonal to the bore sites of the various camera FOVs so they can look down for imaging of Titan if that is a requirement. Usually, when near closest approach radar is prime instrument with $-Z_{SC}$ axis pointing downward to map Titan's surface. INMS can be simultaneously accommodated with $-X_{SC}$ along spacecraft ram, and IBS-ELS can be accommodated

Table 16.3 Pertinent instrument characteristics, observational constraints and limitations

Instrument	Energy range, mass range or spectral range	FOV, actuator and pointing issues	Time resolution
CAPS-IMS	$1\ V \leq E/Q \leq 30$ kV $\Delta E/E \sim 18\%$ $1 \leq M/Q \leq 100$ amu ST $M/\Delta M \sim 8$ LEF $M/\Delta M \sim 60$	$160° \times 8°$ Fan $\Delta EL \times \Delta AZ \sim 20° \times 8°$ $-100° < \alpha < +100°$ $-Y_{SC}$ center of FOV Align along corotation direction	4 s per E sweep 6 min 24 s per AZ scan
CAPS-IBS	$1\ V \leq E/Q \leq 50$ kV $E_{RAM} \sim 0.2$ M/Q V $1 \leq M/Q < 5{,}600$ amu $\Delta E/E \sim 1.4\%$	3 Fans $1.4° \times 150°$ ΔEL X $\Delta AZ \sim 1.4° \times 1.4°$ $-25° < \alpha < +25°$ $-Y_{SC}$ center of FOV Align along spacecraft ram direction	4 s per E sweep 1 min 36 s per AZ scan
CAPS-ELS	$0.6\ eV \leq E \leq 28$ keV $\Delta E/E \sim 16.75\%$ $E_{RAM} \sim 0.2$ M/Q V $1 \leq M/Q < 5600$ amu For negative ions	$160° \times 5.24°$ Fan $\Delta EL \times \Delta AZ \sim 5.24° \times 20°$ $-100° < \alpha < +100°$ (el) $-25° < \alpha < +25°$(cations) $-Y_{SC}$ center FOVAlign along spacecraft ram direction	2 s per E sweep 6 min 24 s per AZ scan (electrons) and 1 min 36 s per AZ scan (anions)
INMS	$0.5 \leq M \leq 99.5$ Da for neutrals $\Delta M \sim 1$ and use cracking patterns and $1 \leq M \leq 100$ Da for positive ions $\Delta M \sim 1$	FOV along $-X_{SC}$ Wide FOV ±10° for closed source (neutrals) and 6° wide for open source (ions)	34 ms sample period or 200 m spatial resolution
MAG/FGM[a]	Fluxgate Vector measurement of **B** with ±40 nT range and 5 pT precision	No FOV requirements	32 vectors per second
RPWS	**E:** LFR (1–26 Hz), MFR (24 Hz to 12 kHz), HFR (3.5 kHz to 16 MHz) **B:** LFR (1–26 Hz) and MFR (24 Hz to 12 kHz)	3 E-field monopole antennas in tri-axial deployment and 10 m each Tri-axial search coil magnetic antennas and mounted on short boom	LFR: 16 s/spectrum MFR: 16 s/spectrum HFR: 0.1 s/spectrum
LP	$0.1\ 1.0 < n_e < 10^6$ el/cm^3 $100°K < T_e < 40{,}000$ K° Measures s/c potential	5-cm diameter Ti sphere 1.5 m boom Needs to view ion plasma	1 s voltage sweep time
MIMI-INCA	7 keV to 3 MeV/nucl Mass resolution: H, O	FOV $120° \times 90°$ Angular Res: $8° \times 4°$	6 s. PHA 85 s, low res 6 min, high res 23 min, full sky
MIMI-CHEMS	3–220 keV/e, $\Delta E/E \sim 3\%$ Ion species H–Fe M/Q Range 1–80 amu/e $\Delta M/M \sim 8\%$	FOV: $4° \times 159°$ 3 telescopes $53° \times 4°$ ea FOV in $(-X_{SC}, Z_{SC})$ plane	5.39 s/estep 32 E-steps/E-sweep or 2 min 52 s per E-sweep
MIMI-LEMMS	Low energy telescope: Ions:27 keV < E < 5 MeV e^-: 18 keV < E < 1 MeV High energy telescope Ions: 1.5–180 MeV/nucl e^-: 100 keV to 20 MeV	FOV:LET 15° opening HET 36°opening Actuator: 360° rotation in X_{SC}–Z_{SC} plane Angular resolution: 22.5° per angular sector and 2.81° per micro-sector	86 s per act rotation, 5.31 s per subsector and 0.66 s per micro-sector When not actuating 5.31s
UVIS	EUV: 56–118 nm FUV: 110–190 nm	EUV/FUV $64 \times (4–0.25)$ mrad	Limb scan ~8 s per scale height ~50 km
RSS	Uplink: X-band 7.2 GHz and Ka-band 34 GHz Downlink: S-band 2.3 GHz, X-band 8.4 GHz and Ka-band 32 GHz	FOV: HGA antenna pointing to Earth. Use one-way and two-way modes	USO: 2×10^{-13} for 1-s integration and 10^{-15} for 10^3 s integration
HASI	ACC: Hi-res 2–20 mg at 1–10 µg, Low-res ±20 g at ±50 mg PWA: Atm conductivity: 10^{-11}–10^{-7} /Ωm	FOV: NA	ACC: tbd PWA: 2–3 s

[a] VHM stopped working on November 21, 2005 and it is not included in this table. Its specifications can be found in Dougherty et al. (2004)

by setting actuator near +90°. This is important for ELS due to its discovery of heavy negative ions within Titan's lower ionosphere (Coates et al. 2007a).

16.2.3 Cassini Magnetometer

Cassini's dual technique magnetometer MAG consists of a vector helium (VHM) and a fluxgate (FGM) magnetometer located, respectively at the end, and halfway the spacecraft's 11 meter boom (Dougherty et al. 2004). Both VHM and FGM sensors provide high range (up to ±256 nT and ±44,000 nT, respectively) measurements of the magnetic field vectors with a maximum time resolution of 2 and 32 vectors per second respectively. Due to a malfunction, VHM stopped providing science measurements on November 18, 2005, almost one month after T8. This has required the MAG team to use more roles to correct the data for zero offsets.

Apart from the in-flight calibration rolls, the magnetometer has shown not to be sensitive to spacecraft attitude issues, and atmospheric drag effects near closest approach during Cassini's nominal mission (altitudes as low as 950 km). But also, extended mission studies indicate that the boom alignment would not be compromised at lower altitudes (at least 880km). The magnitude of the magnetic fields measured around Titan (<~ 20nT) fall into FGM's range 0 (±40 nT mode with 5 pT precision and sampling of 32 vectors per second). After the vector helium magnetometer failed the magnetic field measurements near Titan have been successfully and flawlessly carried out by FGM. With a 32 Hz sampling rate FGM has a spatial resolution of ~ 190 meters and is able to detect local proton cyclotron frequencies in fields up to ~ 1,000 nT. It cannot resolve electron cyclotron frequencies waves which have frequency ~ 140 Hz; these measurements are made by RPWS. The study of Titan's intrinsic magnetic properties and its interaction with its plasma environment is one of MAG primary science objectives. In the absence of a significant intrinsic magnetic field (Neubauer et al. 1984). Cassini MAG characterized the orientation and strength of the ambient (usually Saturn's) magnetic field which defines the orientation and structure of the moon's induced magnetosphere: is Saturn's magnetic field dipolar or does it have a magnetodisk configuration? Is it within the current sheet where heavy ions reside or at high magnetic latitudes when in magnetodisk region where fields are lobe like, plasma densities are low and composition dominated by light ions (Sittler et al. 2009b). In addition, MAG determines whether the spacecraft is magnetically connected to Titan's ionosphere for higher altitude passes (Bertucci et al. 2007). These observations will help understand the flow of charged particle energy to Titan's upper atmosphere.

16.2.4 Cassini RPWS and LP

The RPWS and LP make joint measurements of the electron component of the magnetospheric plasma and Titan's atmosphere. The plasma wave part, which can measure both electric and magnetic field components, can measure electron plasma oscillations $f_e \sim 9.1 kHz\sqrt{n_e} \sim 2{,}9$ kHz for magnetosphere and ~640 kHz for Titan's ionosphere and can measure electron cyclotron waves $f_{ge} \sim 28B(nT)\ Hz \sim 140$ Hz for magnetosphere and ~50 Hz for Titan's ionosphere. Sampling rates are ~10 Hz for plasma line measurements with spatial resolution ~600 m. The magnetic field measurements are sampled slower ~16 s per sample with spatial resolution ~ 100 km. The LP usually needs to view the ion flow in order to measure the local spacecraft potential. It is mounted on short boom ~1.5 m long. It can independently measure electron densities from ~0.1 el/cm^3 to ~10^6 el/cm^3. At the lower densities, spacecraft emitted photoelectrons can contaminate the LP measurements making it difficult to separate ambient electrons from spacecraft emitted photoelectrons. At these lower densities ELS is the best instrument for measuring electron densities. The spacecraft potential measurements will be important for measuring ionospheric winds by IBS since slight shifts in ion peak positions in E/Q spectrum will be produced by both spacecraft charging effects and atmospheric winds. Hartle et al. (2008a) estimated outward ionospheric wind speeds ~1 km/s at topside ionosphere due to polarization electric field for T_A encounter and should be detectable by IBS. This wind speed translates to an energy shift of the mass 28 peak for IBS up to ~2 eV for nominal peak energy ~5.3 eV, so sub km/s wind speeds should be detectable. This measurement has not yet been reported. The LP can also measure electron temperatures from 100°K to 40,000°K which will be important for ionospheric electron temperatures which are ~1,000°K (Wahlund et al. 2005).

16.2.5 Cassini MIMI Instrument

The Cassini MIMI instrument is composed of three main sensor heads: (1) Ion and Neutral Camera (INCA) for primarily ENA measurements, (2) Charge-Energy-Mass-Spectrometer (CHEMS) and (3) Low Energy Magnetospheric Measurements System (LEMMS). INCA can provide images of the hot magnetospheric plasma interaction with Titan and its penetration to lower altitudes, while CHEMS which uses both time-of-flight (TOF) and solid state detector (SSD) detection systems can measure the composition of the hot magnetospheric plasma for E/Q range 2.8–220 keV and LEMMS can measure the energetic particle population for both ions from 27 keV to 180 MeV and electrons from 18 keV to 20 MeV energies. LEMMS has a high energy telescope with opening angle of 30° and low energy telescope with aperture angle of 15° looking in approximately opposite directions. LEMMS has an actuator that operated during the early portion of the tour such that the telescope FOVs would scan 360° in the X_{SC}–Z_{SC} plane which is orthogonal to the $-Y_{SC}$ axis of the remote sensing instrument. But, in February 2005 the actuator presented an anomaly and it was decided to leave it fixed with the two LEMMS telescopes pointing at ~70° with respect to the spacecraft $-Z_{SC}$ axis. The CHEMS FOV is confined to three angular sectors with each ~53° wide. They are confined to the X_{SC}–Z_{SC} plane, with telescope 1 tilted toward $+Z_{SC}$, telescope 2 aligned along $-X_{SC}$ axis like INMS and telescope 3 tilted toward $-Z_{SC}$ axis. The out of plane FOV is only 4° wide. CHEMS will in general not be sensitive to ram direction ion measurements and provides best measurements when aligned along corotation direction of magnetospheric flow. It does not have the capability to separate mass 16 ions such as CH_4^+ and O^+ like CAPS. INCA, which provides images of the hot plasma interaction with Titan, has a FOV 120°×90° with center of FOV nearly co-aligned with $-Y_{SC}$ with 9.5° tilt toward $+X_{SC}$. So, it can image fast neutrals coming within a broad solid angle around the $-Y_{SC}$ axis. INCA provides nearly 4π measurements when spacecraft is rolling. Whereas the other plasma instruments are measuring local populations, INCA is performing remote-sensing measurements of the magnetosphere-exosphere interaction (except when the spacecraft is at low altitudes ~950 km, where the energetic neutral signal has been degraded by numerous collisions in the Thermosphere).

16.2.6 Cassini UVIS

The UVIS instrument is composed of a hydrogen to deuterium ratio, H/D, experiment, or Hydrogen Deuterium Absorption Cell (HDAC), a high speed photometer, HSP, and two telescope-spectrographs, covering the EUV wavelength band from 56 nm to 118 nm and one covering an FUV wavelength band from 110 nm to 190 nm. The HDAC and HSP instruments will not be addressed here. Spectrograph images from a slit are mapped onto a 2D CODACON imaging microchannel plate with one axis giving spatial information and the other spectral and spatial information. Excluding engineering pixels there are 60 spectral images obtained per exposure, covering 60 mrad×(4 pixels at 0.25 mrad/pixel in the high resolution mode of the EUV channel and 3 pixels at 0.25 mrad/pixel in the high resolution mode of the FUV channel) projected on the sky. By scanning perpendicular to slit axis, using spacecraft attitude change one can construct broader 2D spectral images. The slit look direction is along the spacecraft $-Y_{SC}$ axis like the other camera and remote sensing instruments. So, there are times when UVIS requirements are in conflict with CAPS requirements during a subset of Titan encounters. UVIS probes Titan's atmosphere from limb observations of the UV airglow and solar and stellar occultations, which give information about atmospheric composition as a function of height. Such results have been reported in the papers by Shemansky et al. (2005), Shemansky et al. (2008) and Liang et al. (2007). In Liang et al. (2007) they used stellar occultation data for wavelength band 1,850–1,900 A, where absorption by hydrocarbons and cyano species is negligible, to probe the height dependence of aerosols from the extinction of the UV in this spectral band with height.

16.2.7 Titan Observational Constraints and Limitations

The understanding of Titan's interaction with Saturn's rotating magnetosphere is very complex with non-stationary kinetic plasma processes (see Hartle et al. 2006a,b). This complexity can complicate the energy deposition into Titan's topside ionosphere and those regions not exposed to direct sunlight. Therefore, in order to make progress one must analyze the Cassini in situ and remote sensing observational data over many flybys and use innovative and advanced data analysis tools along with physically meaningful models such as the kinetic-fluid (hybrid code) model being developed by numerous groups (Modolo et al. 2007a; Ma et al. 2006; Sillanpaa et al. 2006; Lipatov et al. 2008). The encounter data and models are inter-related in a complex fashion and cannot be done quantitatively independent of each other. One requires that the models explain all relevant data sets and vice versa. In a sense they validate each other. For example, in the case of T_A flyby (Hartle et al. 2006a,b) we are able to see the pickup ions (i.e., CH_4^+ composition) within the pickup region because the CAPS field-of-view (FOV) is pointing toward Titan where the pickup ions are being born near the exobase ($\mathbf{E=-VXB}$ convection electric field points radially away from Titan), while in the case of T5 (Hartle et al. 2008b) CAPS is looking away from Titan in the pickup region and do not see the pickup ions (i.e., born far above the exobase), but rather see the ambient magnetospheric ions (i.e., O^+ composition), which precipitate into the atmosphere. So, for any particular Titan flyby we only see a piece of the ion velocity distribution function. In the case of T9 when Titan is near 0300 LT the Saturn's magnetosphere is in a magnetodisk configuration and below the current sheet. Here, the convection electric field is pointing southward and pickup ion trajectories will move below the spacecraft position which is near equatorial plane (Sittler et al. 2009c), so they may go undetected. In the case of T18 we had a north polar pass but since convective electric field was pointing downward one could not see the heavy pickup ions. The encounter geometry is usually different (i.e., upstream flow into sunlit ionosphere (TA) or upstream flow into the shadow side of ionosphere (T5)) which further complicates the data interpretation. There are times when in situ FOV requirements interfere with remote sensing or radar measurements of Titan's surface. In the case of UVIS it wants to make solar occultation or stellar occultation measurements and needs to look at Titan's limb for such observations. INCA is bore sited similar to the imaging instruments so may be in conflict with the in situ instruments. There are times when CAPS is not actuating or times when rolls occur around closest approach with good viewing going in but poor viewing going out (i.e., T18). All these complications tend to limit the quality of the data and the ability to make quantitative measurements. Data analysis techniques can be developed to mitigate against such complications. For example, in the case of CAPS data analysis for Titan flybys, one can use magnetospheric protons since their gyro-radii are relatively small which makes them act more like a fluid. This allows the CAPS ion moment programs to measure the proton flow velocity, which then allows one to measure the convective electric field $\mathbf{E}_{\mathbf{conv}} = -\mathbf{V}_{\mathbf{p}} \times \mathbf{B}$. Once one has an estimate of $\mathbf{E}_{\mathbf{conv}}$ and $\mathbf{B}$ from the magnetometer experiment, one can compute the trajectories of the heavy pickup ions such as CH_4^+. This technique has not yet been used to its full potential. As noted above the models can be used to fill in information about the interaction at a global level. INCA will give a more global picture of the interaction and the same can be said about the magnetometer which responds to the global currents of the interaction. Therefore, both INCA and MAG can allow localized in situ measurements to be connected at a global scale using the various models in use.

16.3 Global Properties of Saturn's Magnetosphere, and Importance of Solar Incidence-Ram-Angle (SRA)

With a kronocentric apoapsis which does not exceed 21 R_S, Titan is located in the confines of Saturn's magnetosphere. As a result, the nature of the plasma where the moon orbits depends on the location of the magnetopause and the properties of Saturn's outer magnetosphere. Magnetic field measurements by Voyager 1 (Ness et al. 1982) and Cassini (Backes et al. 2005; Neubauer et al. 2006) have discovered and confirmed respectively the presence of an induced magnetosphere around Titan. The orientation of an induced magnetosphere strongly depends on the directions of the external plasma flow and that of the ambient, frozen-in, magnetic field. In a simplified case where the direction of the external flow parallels the solar EUV photon flux and the magnetic field is perpendicular to them, the plane of the current sheet separating the tail lobes is perpendicular to the plane containing the field and parallels the plasma wind vectors (Israelevich et al. 1994). As a result, the induced magnetosphere rotates with the upstream field around the plasma velocity vector.

Saturn's magnetopause's stand-off distance R_{MP} is mainly controlled by the solar wind and follows a bimodal distribution with means at 21 and 27 R_S (Achilleos et al. 2008). As a result, the probability of Titan (~20 R_S) being outside Saturn's magnetopause is around 5.5% in the noon sector (SLT). This is compatible with the fact that out of 44 flybys, only one (T32) took place outside Saturn's magnetosphere (see Bertucci et al. 2008).

Figure 16.3 from Blanc et al. (2002) shows Titan's interaction as a function of local time (LT) around Saturn. The figure shows that at noon and midnight LT the flow, when in the corotation direction, is impingent upon the terminator, while at dusk LT the flow is incident upon the sunlit ionosphere of Titan, while for dawn LT the flow impinges upon the shadow side, with the sunlit ionosphere on the wake side. Here, we introduce the solar incidence-ram-angle (SRA), which is zero at dusk LT, +90° at midnight LT, 180° at dawn LT and 270° at noon LT. Encounters V1, TA, Tb and T3 have SRA ~ 270°, T5 at SRA ~ 180° and T9 and T18 at SRA ~ 120°. For T5 both sides of Titan are being illuminated by either the sun or the magnetosphere, while for dusk both sun and magnetosphere illuminate one side, with the wake receiving no input. Figure 16.4 is a modified version of the magnetospheric model by Arridge et al. (2008a) where it shows Saturn's magnetosphere in a warped magnetodisk which can appear dipolar at times near the noon meridian, while toward midnight the magnetodisk is hinged upward.

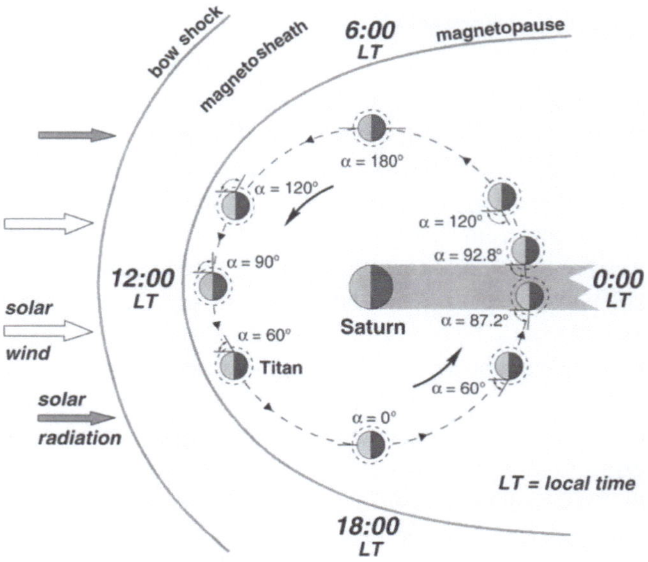

Fig. 16.3 Shows SZA variation relative to corotation direction of Saturn's magnetospheric flow as Titan moves around Saturn at different SLT positions. The figure from Blanc et al. (2002) shows nominal bow shock (BS) and magnetopause (MP) locations. The angle α is between the SZA and the corotation direction

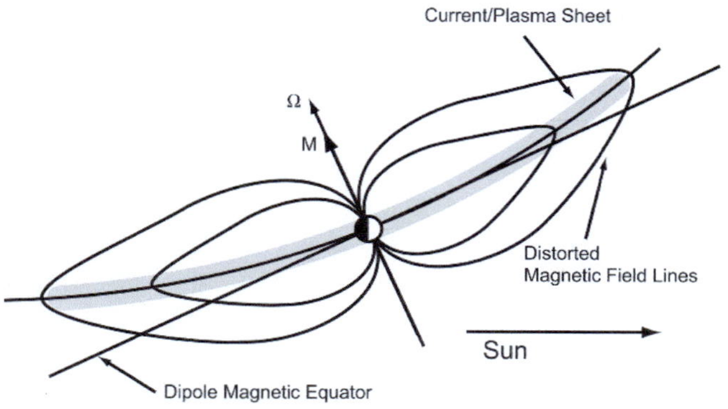

Fig. 16.4 From Arridge et al. (2009a) shows the magnetodisk geometry of Saturn's magnetosphere and Titan's position confined to Saturn's rotational equator. It shows how Saturn can be positioned below the current sheet where the heavy ions are confined

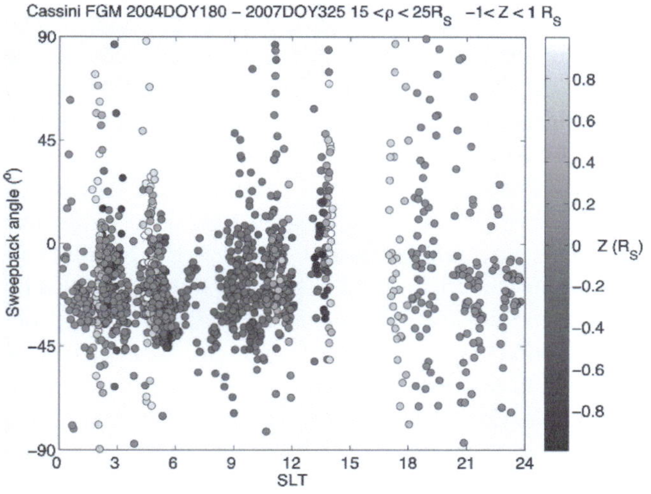

Fig. 16.5 From Bertucci et al. (2009b) shows a scatter plot of the magnetic field sweepback angle versus SLT. Near midnight LT the angle is ~−30° while near noon LT its average is ~0°. This angle is consequence of non-zero B_φ which is a result of the field being mass loaded by the rotating flow (i.e., corotation cannot be enforced by Saturn's ionosphere)

In Figs. 16.5 and 16.6 are shown the stretch angle and sweepback angle, respectively from Bertucci et al. (2009). They used 1-hour resolution magnetometer data in the KSMAG dipole-aligned cylindrical coordinate system (ρ,ϕ,Z) within Saturn's magnetosphere in order to measure the deviation of Saturn's magnetic field from a dipolar configuration. They used MAG data between day 180, 2004 (SOI) and day 325, 2007, in the range $15<\rho<25\ R_S$, and $-1<Z<0.5\ R_S$. The deviation from the dipolar field is measured by looking at the stretch and sweepback angles of the magnetic field defined as: Stretch = *atan* $(-B\rho/B_Z)$ and Sweepback = *atan* $(B\phi/B\rho)$ (Note, for Titan coordinate system used here $B_X=B_\phi$, $B_Y=-B_\rho$ and B_Z same in both frames). Large negative stretch and sweepback angles are signatures

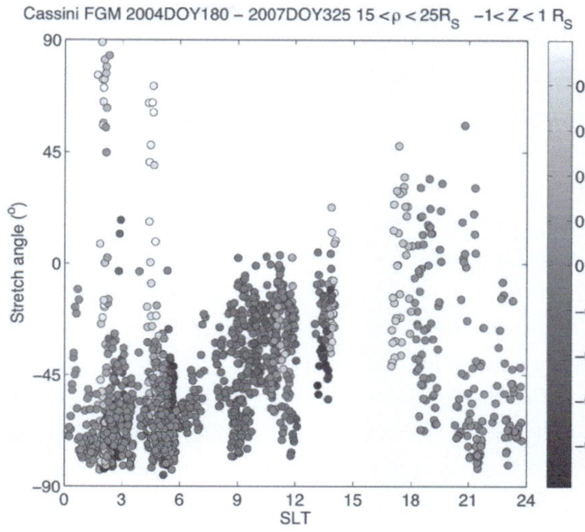

Fig. 16.6 From Bertucci et al. (2008b) shows a scatter plot of the stretch angle versus SLT. *A large negative angle* indicates magnetodisk configuration, while *small negative angles* indicate dipolar geometries. Near midnight the field is more disk like, while near noon it is more dipolar

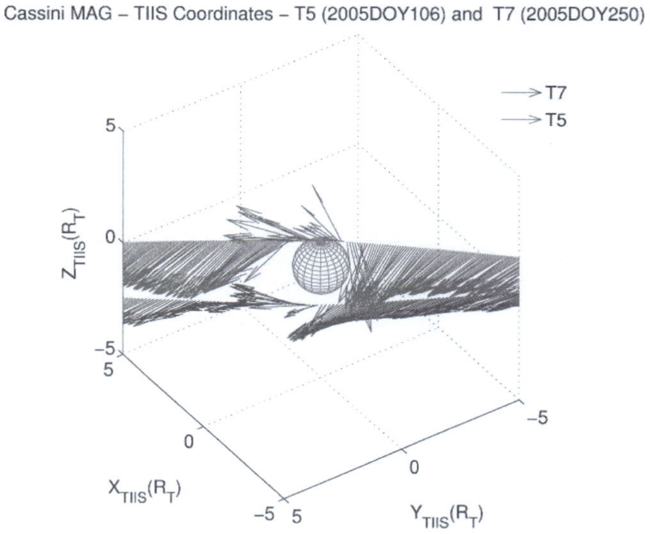

Fig. 16.7 From Bertucci et al. (2009b) shows the magnetic field vectors for T5 and T7 Cassini flybys. In both cases the magnetospheric field is nearly confined to the equatorial plane with deflections resulting from draped field lines around Titan. In a more dipolar field configuration like Voyager 1 or Cassini TA the magnetospheric field points more along – Z axis

of magnetodisk like field lines, while small values indicate dipolar like field lines. Both figures clearly show a magnetodisk like configuration near midnight local time and more dipolar near noon local time. So, with magnetodisk like field lines, the field is primarily in Saturn's equatorial plane, and when field is more dipolar, the field is perpendicular to the equatorial plane. Figure 16.7 shows magnetic field vectors for T5 and T7 from Bertucci et al. (2008), when Titan is near dawn LT and field is in equatorial plane (i.e., magnetodisk configuration).

As shown in Fig. 16.5 the stretch angle in the vicinity of Titan's orbit is usually negative, indicating that the moon, although its orbit is located on the planet's equator, is embedded in a field typical of southern magnetic latitudes. The field is more stretched and less variable in the pre-dawn and dawn magnetospheric sectors and more dipolar near noon. A certain symmetry is observed with respect to noon, although the dusk sector shows a higher variability than the pre dawn and dawn sectors. The orientation of the magnetic field is compatible with the presence of a northward warped asymmetric magnetodisk as shown in Fig. 16.55. The stretch angle provides a good estimate of the distance between Cassini and the magnetodisk's central current sheet assuming a simple Harris model (Arridge et al. 2008b). The dispersion in the stretch angle is interpreted as vertical motions of Saturn's magnetodisk (Arridge et al. 2008c). This motion seems to be partly dependent on the solar wind pressure. The more negative and less variable stretch angle values in the dawn sector (allowing comparison between flybys like T5 and T7, see Fig. 16.7) suggests that in that region Titan hardly ever interacts with Saturn's central current sheet and that the motion of the magnetodisk is not so pronounced. In the noon sector, the stronger variability suggests that the disk is more dynamic or that its thickness varies, allowing Titan to occasionally interact with the central current sheet, where the field is less stretched (see Arridge et al. 2008c). Titan being south of Saturn's magnetodisk is a seasonal effect coupled with solar wind forcing.

The sweepback angle displays a similar pattern, but less pronounced and with larger dispersion. On average, it is negative in the pre dawn, dawn, and post dusk sectors. This is also the case in the midnight sector where values are less than zero to within one standard deviation. This indicates that Titan is in the region controlled by the rotation around Saturn, rather than the tail. The near zero values in the noon sector suggest that Cassini encounters more frequently Saturn magnetodisk's central current sheet. Low variability is also observed in dusk sector. Menietti et al. (2007) found that when Titan was near midnight local time there was an enhanced occurance of SKR emissions.

Sittler et al. (2009b) looked into the composition of the magnetospheric plasma as a function of Saturn LT in the vicinity of Titan's orbit. Figure 16.8 from their paper shows fluid parameters of both protons and heavy ions and magnetic field vector data (30 min averages) for a 24 h period which includes the T9 flyby of Titan. The figure shows for the period from 0200 to 0500 hours the magnetosphere is in a magnetodisk like configuration and the spacecraft is in lobe like field lines. During this time the spacecraft is inside Titan's orbit between 15 and 17 R_S, proton densities are low and heavy ions are absent. This is also shown by the composition data displayed as insets which shows a composition dominated by protons and H_2^+ ions. For a later period, when the spacecraft is near the current sheet, the composition data shows the presence of mass 17 heavy ions. Then as the spacecraft moves to higher magnetic latitudes, where the field is more lobe like, the heavy ions start to disappear and light ions begin to dominate. This figure also shows that the ion densities are highest in the current sheets and lowest in lobe like field lines. In Fig. 16.9 we show a figure from Maurice et al. (1997) which investigated ambipolar electric fields in the magnetospheres of Jupiter and Saturn. The figure shows the latitudinal distribution of ions along magnetic field lines for Jupiter's outer magnetosphere, where it is in a magnetodisk configuration, similar to that for Saturn. The figure shows that the heavy ions are confined within a few degrees of the magnetic equator, while the protons dominate at higher magnetic latitudes. The close confinement of the heavy ions can be traced to the radial field geometry so that the centrifugal force is nearly field aligned and thus strengthens the centrifugal confinement of the plasma.

Toward latter times in Fig. 16.8, one can see the T9 encounter. Here they argue that Titan is still in near lobe like field lines but could be close to the southerly boundary of the current sheet. To some degree Titan and its extended H_2 exosphere, with the Hill sphere boundary ~ ±1 R_S from Titan, is affecting the character of the large scale magnetospheric field and plasma around Titan (Sittler et al. 2009b,c). Sittler et al. (2009c) also show observational evidence that the H^+ both upstream and downstream are in fact pickup ions and must be locally produced from Titan's H and H_2 exospheres. The composition data shown in Sittler et al. (2009c) clearly shows protons dominating the upstream flow, both inbound and outbound, and only several hours later does one see evidence of heavy ions when the spacecraft is near the current sheet. These observations show that the proton only composition for T9 was probably not a transient phenomena but a natural consequence of Saturn's magnetosphere being in a magnetodisk like configuration for T9, which was at 0300 LT and was at relatively high magnetic latitudes. Sittler et al. (2009c) also looked at T18, which was essentially at the same LT as T9 and was definitely in lobe like field lines with a composition dominated by light ions while no heavy ions were detected. As discussed in Sittler et al. (2009b,c), T9 and T18 represented a low energy limit for magnetospheric input to Titan (i.e., plasma $\beta \ll 1$). In contrast, T5 at dawn LT was in magnetodisk configuration but sufficiently close to the current sheet, where keV heavy ions had higher than average intensities and represented a high energy limit for magnetospheric ion fluxes into Titan and its induced magnetosphere (Hartle et al. 2006c).

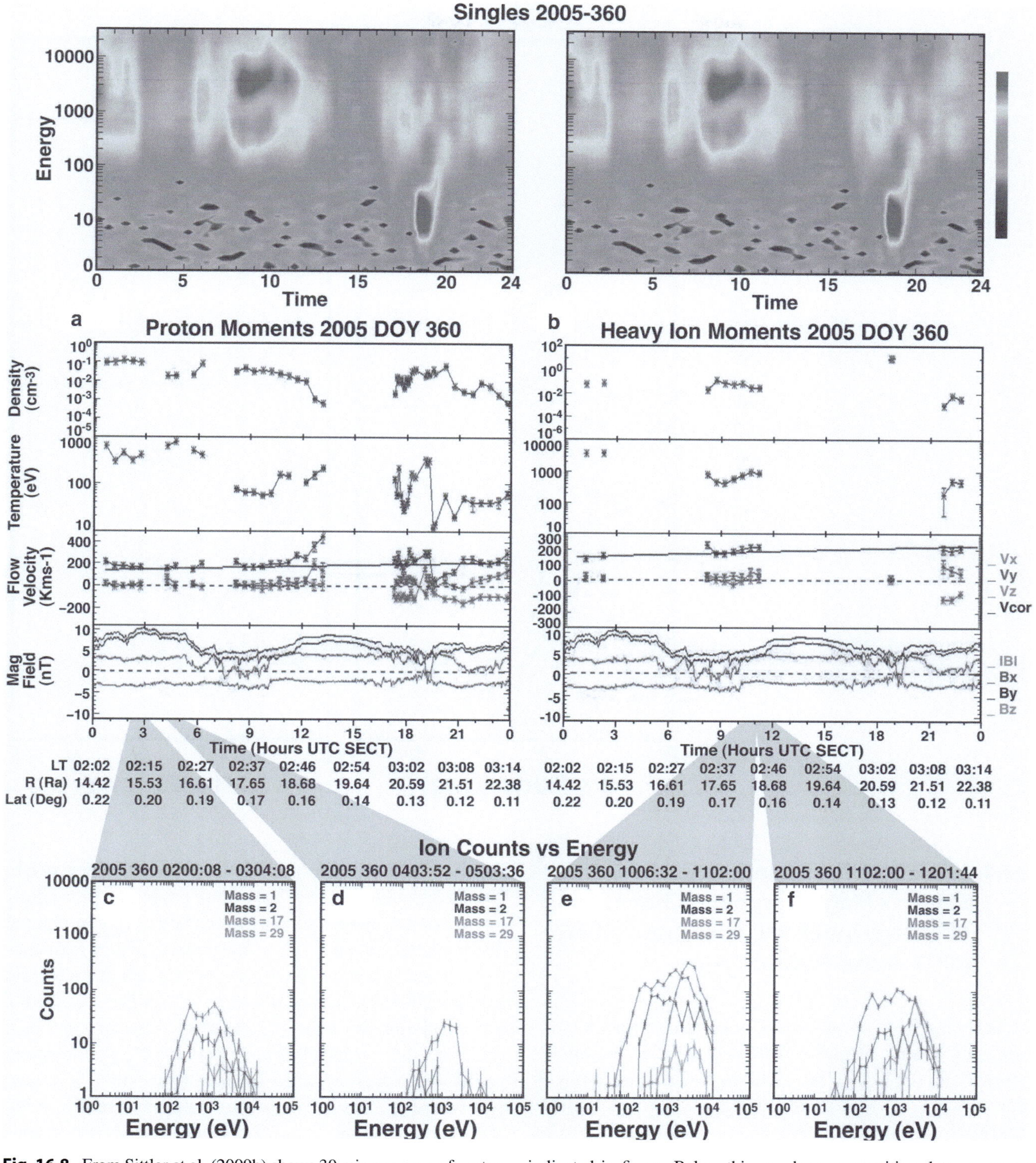

Fig. 16.8 From Sittler et al. (2009b) shows 30 min averages of proton (*left panel*) and heavy ion (*right panel*) fluid parameters DOY300 when T9 encounter occurred at a SLT ~ 0300 hours. In upper panel is energy-time spectrogram of ion counts measured by CAPS-IMS. In lower panel we show the magnetic field data in a coordinate system with x along corotation direction, y points toward Saturn (–r *direction*) and z aligned with Saturn's spin axis. Lobe like field line geometries indicated in figure. Below this we show composition data counts vs energy in log-log plot. H^+ ions in red, H_2^+ ions in blue, mass 17 ions in purple and mass 29 in green. During 0300 and 0600 hours the field is lobe-like and the composition dominated by light ions (H^+ and H_2^+), while between 1000 and 1200 hours the composition data is near edge of current sheet crossing with mass 17 ions present but as one goes to more lobe-like field lines the mass 17 peak start to disappear

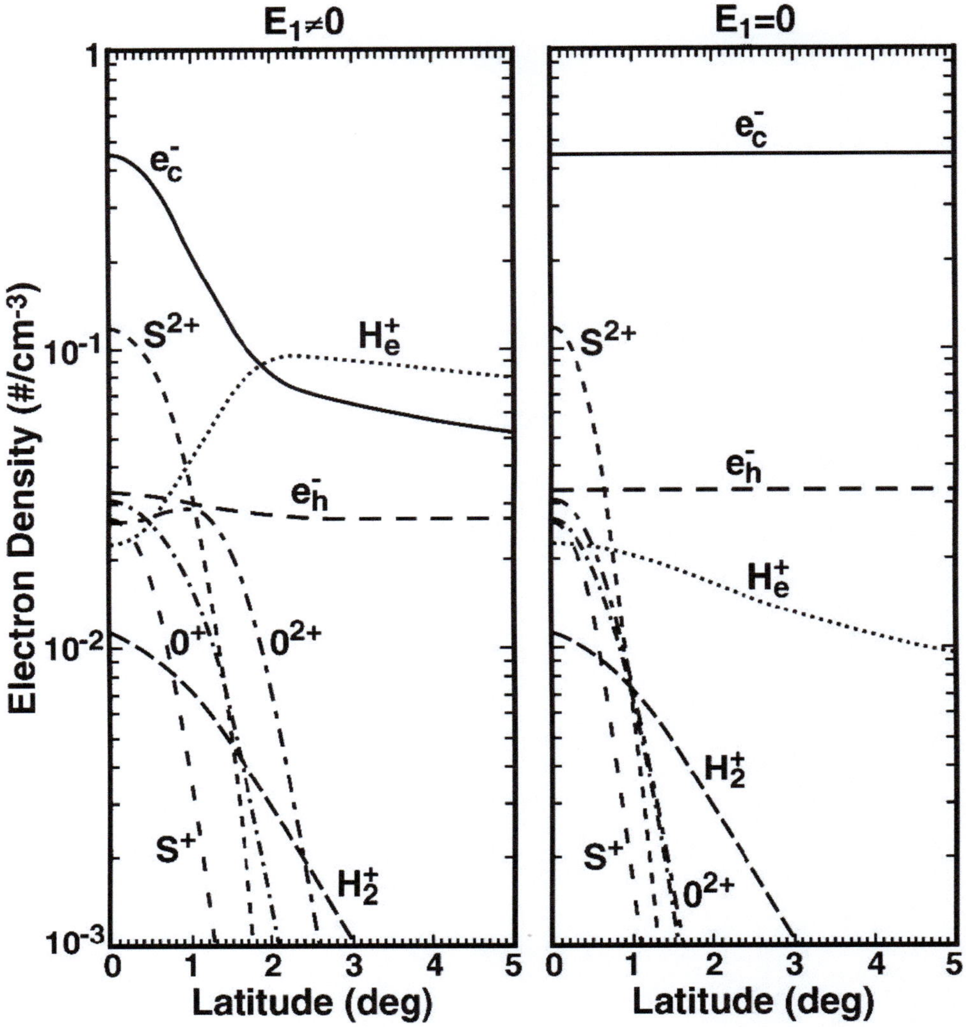

Fig. 16.9 From Maurice et al. (1997) is shown the ion density as a function of magnetic latitude for various ion species for a Jovian magnetodisk configuration. The *left figure* with non-zero ambipolar electric field is more appropriate and shows heavy ions such as S^{2+} confined within a few degrees of the magnetic equator, while protons dominate at higher latitudes

16.4 Models of Titan's Interaction with Saturn's Magnetosphere: Channeling of Energy Input to Upper Atmosphere: MHD Versus Hybrid Codes

Models have been developed to understand Titan's interaction with Saturn's magnetosphere with some earlier models being MHD (Cravens et al. 1998) to hybrid simulations by Brecht et al. (2000) and more recent 3D MHD models by Ma et al. (2006) and hybrid simulations by Modolo et al. (2007a). At higher altitudes above the exobase, where gyro-radii are larger, the hybrid simulations will work best, while at lower altitudes near and below the exobase, the flow is more mass loaded, ion gyro-radii are smaller and the interaction is more fluid like and consequently MHD codes are expected to work better (see Sittler et al. 2005a). Heavy magnetospheric ions are not magnetized and can directly deposit their energy into Titan's upper atmosphere, while light ions (H^+ and H_2^+) and electrons are more magnetized and are deflected around Titan and are not able to directly deposit their energy to Titan's upper atmosphere. This is important for atmospheric loss, since the neutral atmosphere can develop non-thermal distributions from the sputtering process of ion bombardment, which can then lead to atmospheric loss (Michael and Johnson 2005; Michael et al. 2005; De La Haye et al. 2007). Solar radiation not only ionizes Titan's atmosphere, but it also heats the thermal electron population via photoelectron production, which can then drive an ionospheric wind and corresponding loss of ionospheric plasma (Hartle et al. 2008a). We will now review past modeling and observations pertaining to this interaction, formation of induced magnetosphere and magnetic barrier around Titan, transport of energy or lack of it across this magnetic barrier and the short circuiting of the convection electric field as the plasma is mass loaded (i.e., reduce ion pickup energy) and deflection of induced magnetotail. It should also be noted that the different models discussed used different coordinate systems and we have converted their coordinate systems to be the same as ours to minimize confusion. For our discussions, X points in corotation direction, Y points toward Saturn and Z points upward and parallel to Saturn's spin axis.

16.4.1 MHD Simulations

As noted above, initial attempts were made to model Titan's interaction with Saturn's magnetosphere as observed by Voyager 1, such as the 2D quasi-multifluid MHD model by Cravens et al. (1998). Since this model is MHD they do not compute the electron properties such as its temperature and had to use the model results by Gan et al. (1992, 1993). At higher altitudes, >2,425 km, we consider their modeled electron temperatures well below Voyager 1 and Cassini observations (Hartle et al. 1982, 2006b). Figure 16.10 from Cravens et al. (1998) shows the flow inside $2R_T$ near the nose (ram direction) of the interaction slowing down considerably and a magnetic barrier forms. For this simulation the upstream flow is sub-sonic, moving at 120 km/s, super Alfvenic and plasma beta $\beta \sim 11 >> 1$. The simulation is applicable to TA as reported by Hartle et al. (2006a,b), except the plasma beta is considerably lower than for Voyager 1 (Ma et al. 2006). Figure 16.11 from Cravens et al. (1998) shows the interaction along the ram direction or nose of interaction. At around 4,500 km, the field rises over a distance of a few 100 km from 6 nT to 18 nT, while along the flanks it rose from 6 nT to 15 nT from 5,500 km to 4,800 km in radius. When this magnetic barrier forms the thermal pressure of the plasma drops with secondary peak of ionospheric pressure at around r ~ 3,750 km or altitude ~1,175 km. Both these figures show that the magnetospheric flow is diverted around Titan well above the ionosphere electron density peak and the exclusion of magnetospheric particle and field energy to lower altitudes. Overall, they found approximate pressure balance with height when plasma pressure and magnetic field pressure were added together. They had considerable magnetic field pressures below the ionospheric peak ($\beta << 1$) which must be due to their assumed magnetic diffusion coefficient and is contrary to Cassini observations where $\beta >> 1$ within the ionosphere (Wahlund et al. 2005). TA is sufficiently similar to make comparisons at ionospheric heights. In the Cravens et al. (1998) model, the electron temperature drops inside r ~ 3,800 km from $T_e \sim 1{,}000°K$ to <200°K. Focusing on the ionospheric pressure peak ~ 5×10^{-10} dynes/cm^2 in Fig. 16.18 and if one assumes $T_e \sim 1{,}000°K$ and $T_{ION} \sim 200°K$ at the peak one gets an electron density ~ 3,600 el/cm^3 at the peak and if $T_e \sim 200°K$ and $T_{ION} \sim T_e$ one gets $n_e \sim 9{,}000$ el/cm^3 at the peak. For TA at an altitude ~1,250 km Wahlund et al. (2005) observed a density max $n_e \sim 3{,}800$ el/cm^3 and $T_e \sim 1{,}000°K$, while for T_B at ~1,200 km altitude the max density $n_e \sim 3{,}200$ el/cm^3 and $110°K < T_e < 1{,}200°K$ and highly variable. So, the LP data favors the above assumption with $T_e \sim 1{,}000°K$ and $T_{ION} \sim 200°K$ at ionospheric peak. At altitudes z > 1,400 km for T_A Wahlund et al. (2005) reported $T_e \sim 10{,}000°K$ and $n_e \sim 10$ el/cm^3, while Hartle et al. (2006b) reported CAPS ELS $T_e > 10^5$ °K with $n_e \sim 1.0$ el/cm^3 at similar altitudes. At these altitudes there is no evidence of spacecraft emitted photoelectrons, so, it could be that CAPS ELS is measuring the hotter photoelectrons from solar UV and secondaries from magnetospheric electron ionization of the neutral atmosphere and the LP data from Wahlund et al. (2005) is observing the thermal population of the topside ionosphere. If so, the total electron temperature will be $T_e \sim 20{,}000$ °K with ½ the electron pressure in the hot electrons. As modeled by Gan et al. (1992) these higher thermal electron temperatures $T_{e,c} \sim 10{,}000°K$ are due to heating by ionospheric photoelectrons and magnetospheric electrons via electron–electron Coulomb collisions. From Fig. 10 in Hartle et al. (2006b) one can see a clear bite-out in the hot electrons within Titan's ionosphere down to 10 eV similar to

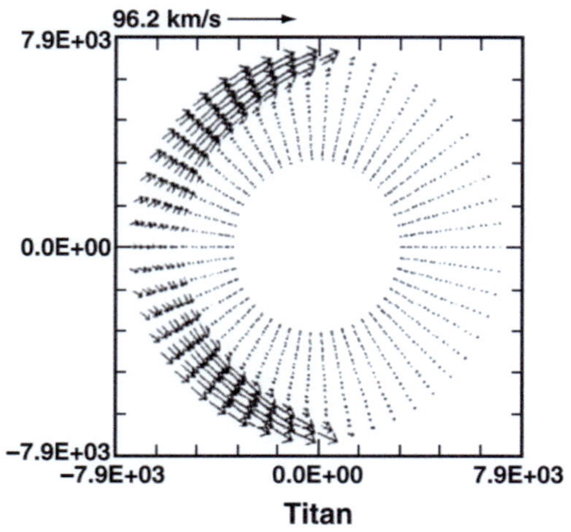

Fig. 16.10 Flow velocity vectors inside $2R_T$ of the external flow around Titan due to its interaction with Saturn's magnetosphere. Model results are from the 2D quasi-multifluid MHD model by Cravens et al. (1998)

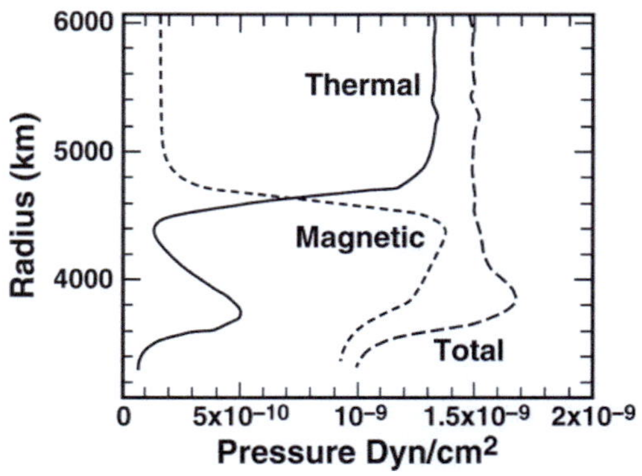

Fig. 16.11 A 1D cut of plasma and magnetic field pressure along ram direction of external flow from the Cravens et al. (1998) 2D quasi-multifluid MHD model

Voyager 1 which was at higher altitudes with biteout at higher energies E > 500 eV (Hartle et al. 1982). This loss of electrons are a combination of inelastic collisions with Titan's upper atmosphere and the gradient-curvature drift of magnetospheric electrons around Titan similar to that modeled by Sittler and Hartle (1996) for Triton (see Section 8.2 for details). Within the ionosphere the energy density of the ionosphere peaks around 400 eV/cm^3 (Wahlund et al. 2005), while in magnetosphere, for T_A and T_B, electron energy densities are ~10 eV/cm^3 (Ma et al. 2006). For Voyager 1 magnetospheric electron energy densities are ~60 eV/cm^3 and ion thermal energy densities ~600 eV/cm^3. So, there is enough energy in the upstream plasma to heat the ionosphere to its observed energy density. But, due to the magnetic barrier, it is thought that most of this energy is diverted around Titan and most of the heating and ionization come from solar UV (Cravens et al. 2005).

A more recent 3D quasi-multi-fluid MHD model (Ma et al. 2004) was presented for TA and TB by Ma et al. (2006), which showed V1 to be a high energy density encounter $\beta \sim 11$, while TA and TB were intermediate in energy density with $\beta \sim 1.2$ and 1.5, respectively. This model was able to produce a good representation of the magnetic field data and plasma parameters. Flow velocities were not available and one should consider the speeds estimated by the LP by Wahlund et al. (2005) only qualitatively accurate. Comparisons between T9 CAPS IMS observations with LP data for T9 (Modolo et al. 2007b) show this to be the case (Sittler et al. 2009c). Ion parameters were not available for the Ma et al. (2006) results, so they picked a nominal plasma temperature $T_P \sim 350$ eV using ELS electron observations as a guide with assumption $T_P = ½(T_{ION} + T_e)$ with T_{ION} T_e which is known to be violated in general. They use the chemical equilibrium model from Cravens et al. (2004) which considers SZA effects and use $n_e \sim 0.1$ el/cm^3 and $T_e \sim 100$ eV as boundary condition at end of field lines to model the effects of suprathermal electrons from the magnetosphere (two-stream model by Nagy and Banks 1970). But, locally they use the electron temperature from Gan et al. (1992). They also, self-consistently, included the photoelectrons produced from the ionization of the neutrals from incident solar UV. The solar UV and modeled electron temperature were used as the primary energy source for their ionospheric model and corresponding ion production rates with height. They also compute the conductivity of the ionosphere including electron-ion and electron-neutral collisions with the dominant neutrals N_2, CH_4 and H_2. They used ion flow speeds ~110 km/s as reported by Szego et al. (2005) and Hartle et al. (2006a,b). CAPS ion densities and temperatures have not yet been reported for these two encounters. As shown in Fig. 16.12, the model produces a nearly symmetric acceleration of the flow along the flanks, while observationally on the anti-Saturn side one sees only mass loading and pickup ions and on Saturn side one does see acceleration of the flow due to tension in the field (Hartle et al. 2006a,b). This is due to the fact that MHD cannot model finite gyro-radius effects which are known to be important for both TA and TB. The model represents the total ionosphere electron density fairly well as observed by the LP (Wahlund et al. 2005), but tends to under-estimate T_e (their model uses Gan et al. (1992) model electron temperature) when compared to the LP at altitudes below the electron density peak. In Fig. 16.13 from Ma et al. (2006) is shown a comparison between a photochemical model results with and without dynamics. The photochemical only models tend to overestimate ion densities at higher heights. The model also produces fairly high

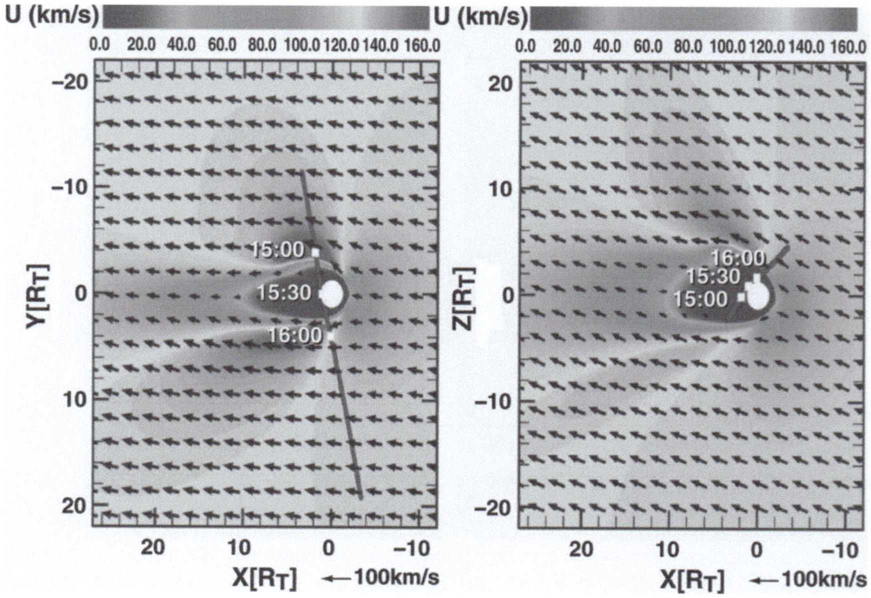

Fig. 16.12 Flow velocity vectors for TA encounter from the 3D quasi-multifluid MHD model by Ma et al. (2006). *Left panel* shows flow field in Titan orbit plane, while *right panel* shows flow field in V_X–V_Z plane showing asymmetry in north-south flow.

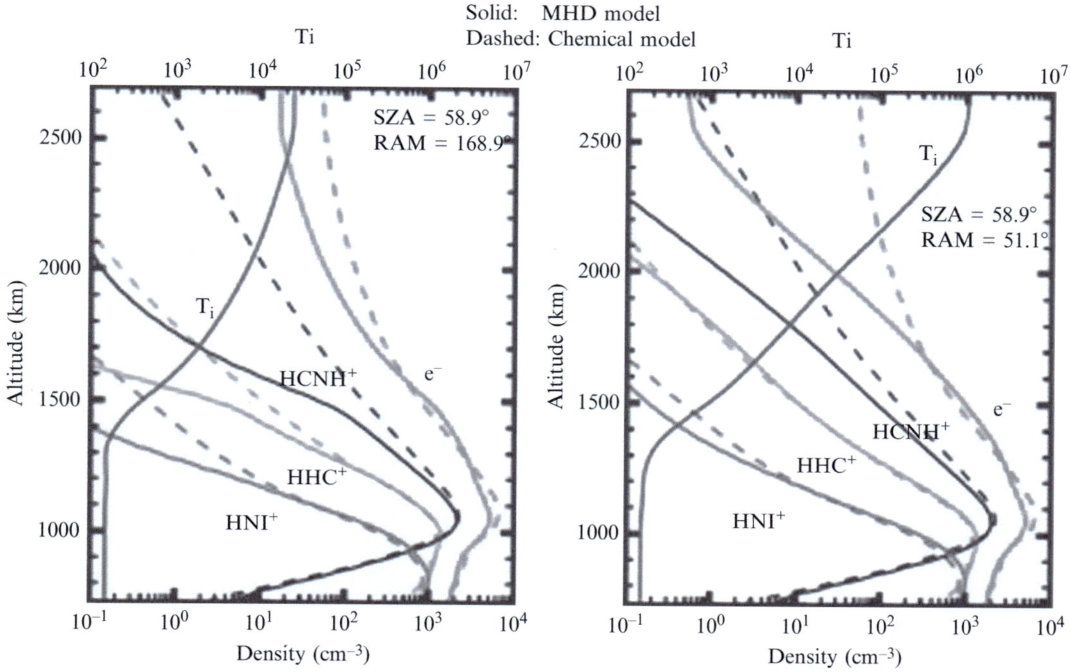

Fig. 16.13 Ion density versus height from Ma et al. (2006) model showing photochemical model results without dynamics on *left panel* and with dynamics on *right panel*

field strengths ~20 nT near the nose of the interaction and since keV magnetospheric heavy ions are unmagnetized the field may diffuse into the ionosphere more than modeled and the field pile up may be less. Up till now encounter geometries do not pass through the pile up region so field build up with height cannot be tested at this time (see Section 7.1 and discussion of ENA observations at Titan).

16.4.2 Hybrid Simulations

Brecht et al. (2000) performed the first hybrid simulation of Titan's interaction with Saturn's magnetosphere for the Voyager 1 flyby. Their simulations used 91 X 153×66 cells with minimum cell size ~500 km and spatial extent 17.5 R_T X 30 R_T X 15 R_T, 12 million particles and 20 processors using the Ames Numerical Aerodynamic Simulation Facility (NAS). With minimum cell size of 500 km they could not resolve the ionosphere, and some numerical errors can occur for protons since their gyro-radii can be smaller than cell size (see Lipatov 2002). To simplify the calculations they used the Keller et al. (1992) ionospheric model with altitude profile up to 2,500 km. They modeled the altitude profile based on CH_5^+ and $C_2H_5^+$, but in order to simplify interpretation of simulations they used a single ion model with composition of H^+, N^+ or $C_2H_5^+$, while preserving the ionosphere number density with height. $C_2H_5^+$ is used since the Keller et al. (1992) model had it as the dominant topside ionospheric ion. The ionospheric ions were used to mimic pickup ions and mass loading of the flow. They were replenished at the ion production rates given in Keller et al. (1992). The model does not include an exosphere. The equations do include the Hall term in the electron momentum equation. Electrons are treated as a fluid. The model results summarized in Fig. 16.14 shows lower magnetic fields in the pile up region with max fields between 15–20 nT and there was a definite asymmetry with mass loading on the side, where the convective electric field $\mathbf{E}_{\mathbf{conv}}$ points away from Titan (−Y direction). On this side the interaction extended further from Titan than on the Saturn side. They also found the induced magnetotail being more extended on the side where the electric field is pointing away from Titan. They concluded the observed asymmetry reported by Hartle et al. (1982) and Sittler et al. (2005a) is controlled more by photoionization of the ionosphere and not electron impact ionization. This result, like that of MHD, predicts that part of the flow energy is converted to magnetic field energy and is deflected around Titan's ionosphere so that magnetospheric electrons cannot penetrate Titan's topside ionosphere, while solar UV can. Their model results were later used by Michael et al. (2005) to estimate the energy input to the upper atmosphere by pickup N_2^+ and N^+ ions which was significant with regard to atmospheric loss.

The hybrid simulations by Sillanpaa et al. (2006) attempted to model ion escape from the ionosphere and

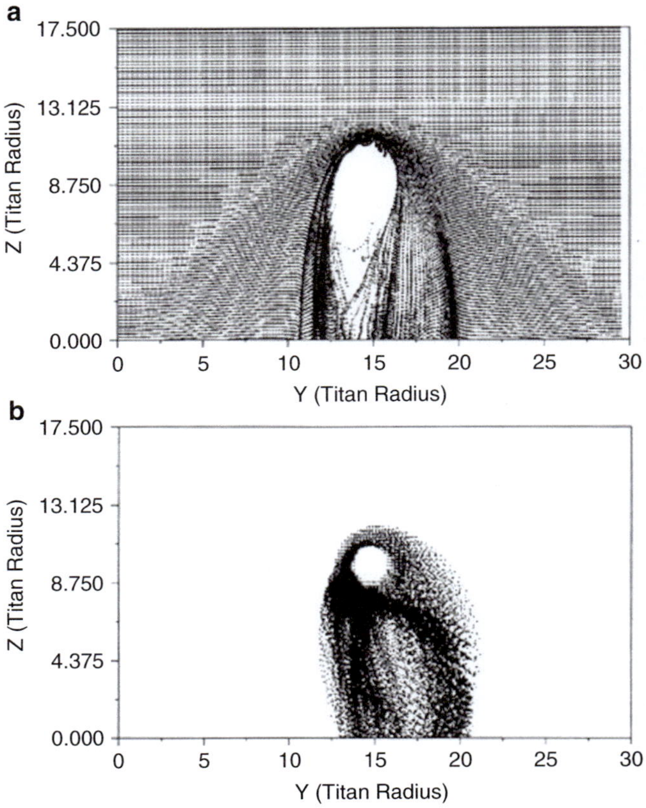

Fig. 16.14 Particle positions for plane containing convective electric field pointing away from Saturn and upstream flow velocity vector from 3D hybrid simulation by Brecht et al. (2000) for Titan's interaction with Saturn's magnetosphere for geometry similar to that for Voyager 1 flyby period. Panel A shows plasma flow ions from magnetosphere with mass 14 (assumed to be N^+) and panel B shows pickup ions modeled by ionospheric $C_2H_5^+$ ions with mass 29

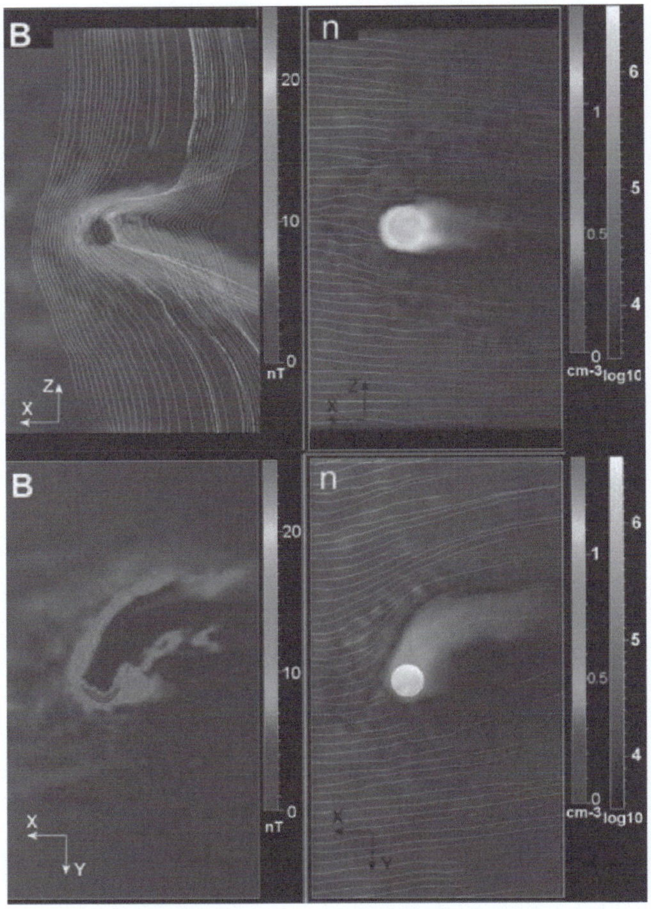

Fig. 16.15 3D hybrid model simulations by Kallio et al. (2004) which shows magnetotail deflected toward Saturn when Titan near dusk SLT. These results are similar to that modeled by Sillanpää et al. (2006)

methane exosphere as a function of Titan's LT relative to Saturn. The model approach is similar to that by Kallio et al. (2004), which was applicable to the LT = 18 hours case, and who found a significant deflection of the induced magnetotail toward Saturn (see Fig. 16.15). Sillanpaa et al. (2006)) assumed upstream parameters essentially identical to that for Voyager 1 except for the change in Saturn LT where the sunlit side varied in orientation relative to the upstream flow velocity. The upstream magnetic field was dipolar pointing along –Z direction. The escape rate of ionospheric ions was found to not be a function of LT, but the exospheric CH_4^+ ion source did change according Saturn LT although the change with LT was small. Their model did not include the importance of magnetospheric electrons and did not include the electron pressure term. In Fig. 1 of their paper this model shows a deflection of the magnetotail toward Saturn similar to that observed for Voyager 1 (Hartle et al. 1982) who assumed the upstream flow was deflected inward. For the Voyager 1 case with LT ~ 12, they see less of an asymmetry and therefore may not be in contradiction with Voyager 1 results as presented by Sittler et al. (2005a), which included a methane exosphere in their analysis. The same can be said for the more recent TA plasma results presented by Hartle et al. (2006a,b), which showed an enhancement of pickup ions on the anti-Saturn side. The Sillanpaa et al. (2006)) simulations showed a pile up of the magnetic field extending from the upstream side to the side where $\mathbf{E}_{\mathbf{conv}}$ pointed away from Titan for LT = 12. The magnetic field observations reported by Backes et al. (2005) for TA did not see a buildup of magnetic flux on this side of Titan. Voyager 1 was too far down the induced magnetotail to see this pileup of magnetic flux. It could be that most of the pickup CH_4^+ residing in the induced magnetotail of their simulations could be occurring where the flow speed is very low and pickup energies are correspondingly low. Since they only include photoionization for the exospheric methane ions and they treat their N_2^+ ionospheric ions in a similar manner, it's not surprising they behave similarly in their simulations. So, it may be what their really modeling is ionospheric ions, which one would expect to fill the induced magnetotail as observed by Sittler et al. (2009c) for T9. But, these ions have very low energies, are found to flow along **B** and gyro-radius effects are expected to be small. As argued in Sillanpaa et al. (2006)), it could be that the induced electric field resulting from the more energetic pickup ions reported in Hartle et al. (2006a,b), is sufficient to deflect the lower energy but more dense low energy CH_4^+ pickup ions toward Saturn. Their estimated ionospheric loss of N_2^+ is ~ 4.0×10^{24} ion/s (assume N_2^+ is dominant topside ionospheric ion) and exospheric ion loss of CH_4^+ is ~ 10^{25} ion/s. In Sittler et al. (2009c), using estimated tail cross-section ~4×10^{17} cm^2 (i.e., event 1 lasted 20 min and spacecraft velocity ~6 km/s), they estimated an ion loss rate ~ 3.2×10^{24} ion/s. About ½ of the event 1 ions observed by Sittler et al. (2009c) were around mass ~16–18 and the other half ~28–30. So, some of these ions could be low energy pickup ions. Combining the numbers by Sillanpaa et al. (2006)) their ion loss rate is ~ 4 times estimated by Sittler et al. (2009c).

Modolo and Chanteur (2008) performed a 3D hybrid simulation of the TA flyby. This model included an upstream composition of H^+ and O^+ and exospheric ions H_2^+, CH_4^+ and N_2^+. The upstream flow is diverted around a cavity that forms around Titan ~ exobase heights and exosphere pickup ions such as H_2^+, CH_4^+ and N_2^+ accumulates inside this cavity with deflection of plasma wake toward Saturn. A similar deflection was observed for V1.

The Modolo et al. (2007a) hybrid simulation for T9 summarized in Fig. 16.16 for CH_4^+ pickup ions, cannot be directly compared with the model calculations by Kallio et al. (2004) and Sillanpää et al. (2006), since the upstream magnetic field is in magnetodisk configuration in X–Y plane while the models by Kallio et al. (2004) and Sillanpaa et al. (2006) assumed dipolar field pointing in –Z direction. Also, for T9, the convective electric field $\mathbf{E}_{\mathbf{conv}}$ points vertically downward along –Z direction. The Modolo et al. (2007a) hybrid simulations also assumed an upstream flow more like Voyager 1 dominated by heavy ions, while Sittler et al. (2009b,c) show it to be dominated by protons. It is true that the induced magnetotail for T9 was tilted toward Saturn like that predicted by Sillanpaa et al. (2006), but the upstream magnetic field has similar alignment and probably not due to finite gyro-radius effects. Modolo et al. (2007b) are able to reproduce event 2 which may be due to that side of Titan being in shadow as they argue and less short circuiting of the upstream electric field $\mathbf{E}_{\mathbf{conv}}$. They do get an ion composition dominated by heavy ions for event 1 and light ions for event 2. Here, the larger altitude extent of the H_2 exosphere and charge exchange with upstream heavy ions explains their simulation results. If the upstream ions are H^+ the charge exchange rates are a factor of 5 smaller and electron impact ionization will dominate with H^+ the pickup ion. So, their simulated event 2 dominated by H_2^+ may not occur if they more appropriately used H^+ for the upstream ion composition. As argued in Sittler et al. (2009c), event 2 is more likely the result of draping of magnetic field lines around the induced magnetopause boundary where the upstream H^+ ions are deflected around and the scavenging of the H_2^+ exospheric ions formed inside the magnetopause where photoionization of the H_2 exosphere dominates and H_2^+ is expected to dominate over H^+ as the pickup ion. T9 is an excellent example of how the interaction details can affect the deposition of energy with height, the pickup ion composition with height and the pickup ion energy.

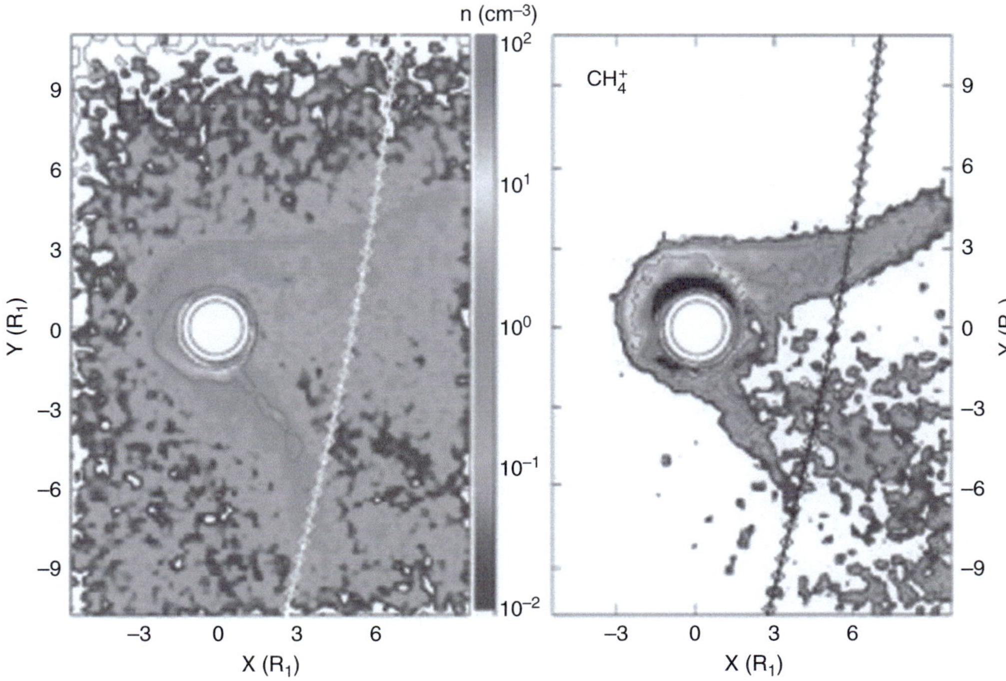

Fig. 16.16 Hybrid simulations by Modolo et al. (2007a) for CH_4^+ pickup ions showing the split tail configuration thought to apply for the T9 encounter. Their model included heavy ions for the upstream flow similar to Voyager 1 but as shown by Szego et al. (2007) and Sittler et al. (2009b,c) the composition of the upstream flow is dominated by light ions with heavy ions essentially absent

16.5 Radio Science Observations of Titan's Ionosphere and Its Height Dependence

16.5.1 Titan's Ionospheric Layer as Observed by Voyager 1

Titan's ionosphere was initially probed by Voyager 1 on Nov. 12, 1980 using the radio occultation technique (Bird et al. 1997). This initial attempt, Fig. 16.17, showed peak electron densities ~2,400 ± 1,100 el/cm^3 at an altitude z ~ 1,180 ± 150 km. Numerous models of Titan's ionosphere were developed previous to Cassini by Keller et al. (1992); Gan et al. (1992); Banaszkiewicz et al. (2000); Galand et al. (1999); Fox and Yelle (1997); Wilson and Atreya (2004) and Cravens et al. (2004). The primary ionization source they considered in all these models is solar EUV-UV and magnetospheric electron impact ionization. As shown by Voyager hot magnetospheric electrons with energies from 10 eV to 5 keV are prevalent within Saturn's outer magnetosphere where Titan is located (Sittler et al. 1983 and Maurice et al. 1996). Voyager 1 plasma observations, Fig. 16.18, reported by Bridge et al. (1981) and Hartle et al. (1982) showed a bite-out of magnetospheric electron fluxes for E > 500 eV within Titan's induced magnetotail indicating collisional loss with Titan's upper atmosphere. Since, the magnetospheric electrons are highly magnetized and the magnetotail field lines are probably draped around Titan's ionosphere, these keV electrons have access to lower altitudes via a centrifugal drift mechanism across field lines (Hartle et al. 1982). The magnetospheric electrons can then ionize Titan's topside ionosphere. Since Voyager 1 plasma instrument had lower energy cut-off of 10 eV, it could not

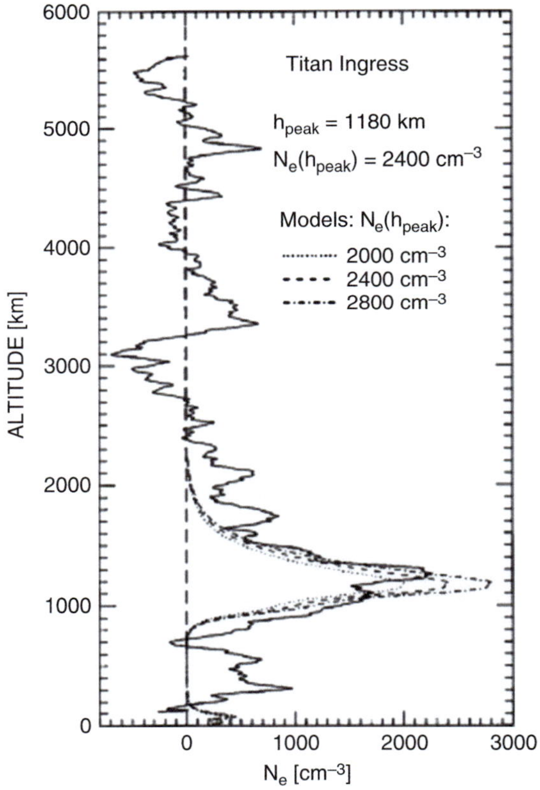

Fig. 16.17 Voyager 1 radio occultation measurement of Titan's ionospheric electron density as a function of altitude from Bird (1997). It shows peak electron density of 2,400 el/cm^3 at 1,180 km

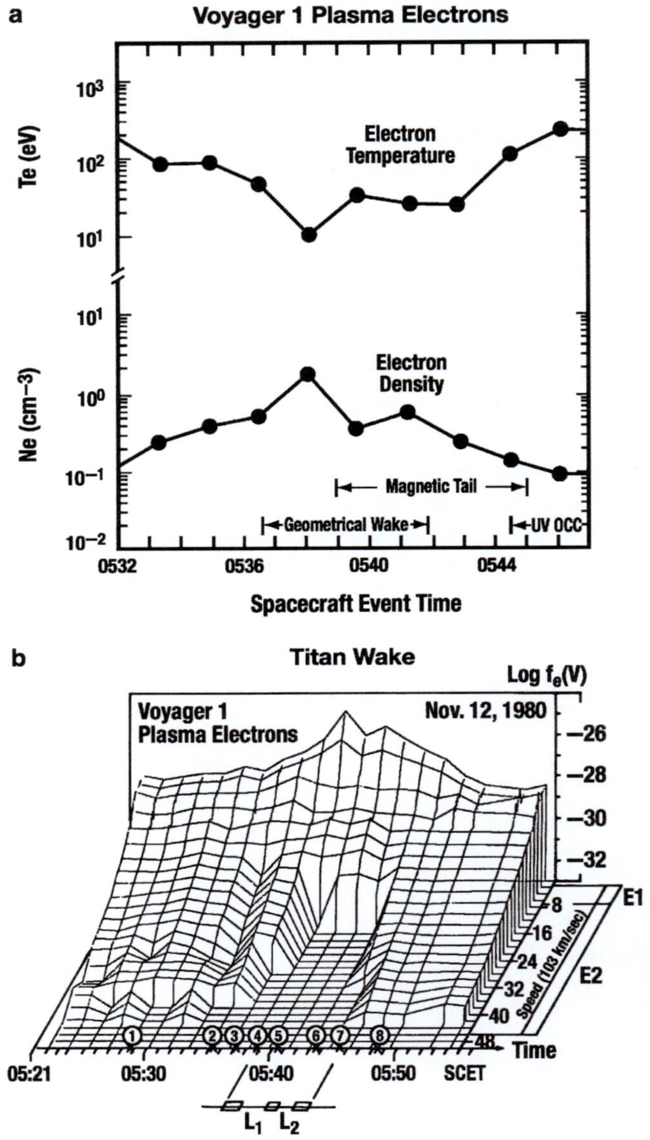

Fig. 16.18 Voyager 1 electron measurements as originally reported by Bridge et al. (1981) and Hartle et al. (1982). The *top panel* shows the total electron density and total electron temperature versus time, while the *lower panel* shows a 3D representation of the electron distribution function versus electron speed along diagonal axis and time along abscissa. The electron bite-out for E > 500 eV are clearly shown and buildup of ionospheric secondary and photoelectrons for E > 10 eV

directly measure the lower energy ionospheric electrons with $T_e < 3$ eV. Gan et al. (1992) using the two stream model by Nagy and Banks (1970) with parabolic field approximation derived an electron temperature model for the ionosphere. Outside r ~ 3,500 km (z ~ 925 km) they estimated T_e ~ 1,000°K, which then rose to T_e ~ 10,000°K or 1 eV outside r ~ 10,000 km (z ~ 7,425 km). Below r ~ 3,500 km T_e dropped to $T_e \sim T_{ION} \sim$ 180°K (i.e., collisional equilibrium with neutral gas of atmosphere). Knowledge of the thermal electron temperature $T_{e,c}$ is important for estimating recombination rates in the ionosphere (Cravens et al. 2005). It is also needed to estimate ionospheric loss via a polarization electric field ($\mathbf{E}_{\mathbf{pol}} \sim -1/n_e e \nabla_{\|} P_e$) that can drive ionospheric outflows (Hartle and Sither 2008).

As noted earlier the Gan et al. (1992) suprathermal electrons are highly magnetized, r_{ge} ~ 3–4 km, and their ionization of the neutral atmosphere with height is largely controlled by the complexities of the magnetopsheric interaction. On the ram side the magnetospheric electrons will be pushed to lower altitudes as the field lines pile up and they themselves are dragged to lower altitudes by the tension in the draped field lines. Along the flanks they will have access via centrifugal drift across horizontal field lines and once flux tubes are emptied ionization will terminate. As shown by Hartle et al. (1982) in explanation of bite-out this can happen relatively fast due to fast field aligned bounce times for keV electrons ~5 min. The plasma protons do not contribute much to atmospheric heating, but are observed to have similar temperatures as electrons. Like electrons they are magnetized with gyro-radii r_{gp} ~ 150 km, with some residual finite-gyro-radius effects at topside ionosphere height since r_{gp} ~ 2 atmospheric scale height H_{N2} ~ 70 km.

16.5.2 Cassini Radio Science Observations of Titan's Ionosphere

In Figure 11 from Cravens et al. (Chapter 11 of this book) we show a compilation of the four Cassini RSS radio occultation observations of Titan's ionosphere by Kliore et al. (2008). Radio occultation observations were made for T12 (March 20, 2006) at 14.3 S and 36.3 S latitudes, T14 (May 20, 2006) at 19.7 S and 21.3 S latitudes, T27 (March 26, 2007) at 74.6 S and 60.6 N latitudes and T31 (May 28, 2007) at 75.4 N and 74.0 S latitudes. All observations occurred near terminator with T12 and T14 observations at dawn side (atmosphere in shadow ~8 days before crossing terminator) and dusk (ionosphere on sunlit side ~8 days before crossing terminator). For T27 and T31 which are polar passes the ionosphere is on sunlit side at all times. The higher electron densities on dusk side relative to dawn side were therefore expected. The position and amplitude of the main peak is well established for all data sets and models. This includes comparison with Voyager 1 measurements by Bird et al. (1997). The shoulder at ~ 1,000 km altitude is probably due to solar soft x-rays (known to be quite variable, see Lean et al. 2003) and is similar to lower altitude shoulder in Cravens et al. (2005) solar only electron density model calculations. The reality of the other minor peaks is less certain, but as shown in Fig. 16.19 from Kliore et al. (2008) a large peak in electron density at 500 km with amplitude ~2,800 el/cm^3 is observed during egress at Titan's north pole. As shown in Cravens et al. (2008) energetic protons E>30 keV can provide peak ion production at ~750 km with residual ionization down to 500 km. The data used by Cravens et al. (2008) was a combination of Voyager plasma and T5 MIMI observations when magnetospheric energy input was higher than typical. Another possibility suggested by Molina-Cuberosa et al. (2001) and later by Kliore et al. (2008) is meteoritic impact, ablation of meteoroids and long-lived metallic ions in Titan's atmosphere with electron density peak at ~700 km with secondary shoulder at ~600 km. Peak electron densities can be as high as 10^4 el/cm^3. An earlier model by English et al. (1996) yielded similar results. Since meteor showers from comets can be time variable and is known to produce short lived Earth ionospheric layers, this could be an explanation of this sporadic lower altitude ionization layer. Galactic cosmic rays (GCR) cannot explain this lower ionization peak, since GCR are isotropic, essentially time independent and deposit their energy below 100 km (Capone et al. 1983; Molina-Cuberos et al. 1999). The observations by Kliore et al. (2008) present a challenge to future models of Titan's ionosphere.

16.6 Solar Input to Titan's Upper Atmosphere: Ionosphere Formation, Atmospheric Heating, Haze Layers and Non-Thermal Atmospheric Escape

16.6.1 Voyager 1 and Cassini UVIS Airglow Observations

In this section we discuss the various forms of solar UV and magnetospheric interactions (see item 2) with Titan's upper atmosphere that give rise to N I, II and N_2 (and recently C I) UV emission lines by collisional excitation These collision processes lead to particle, electron and photon induced fluorescence, often referred to as airglow. The fluorescence-induced excitations can be due to (1) resonance fluorescence by incident solar UV photons e.g. N I 1,200 Å, (2) direct excitation of N and dissociative excitation of N_2 by magnetospheric electrons with energies ~ 200 eV

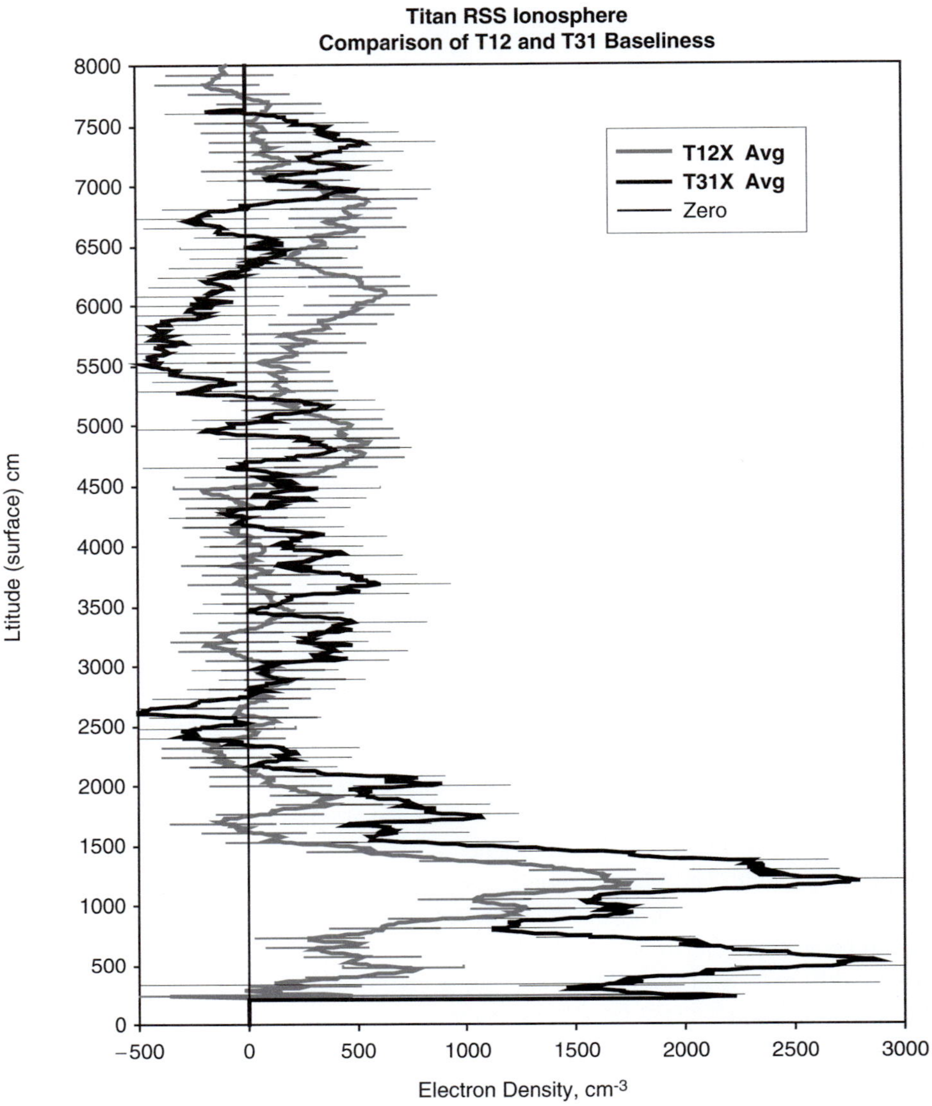

Fig. 16.19 Cassini RSS radio science occultation measurements of electron density versus height for one of the north egress occultation passes. This figure shows large ionization layer at 500 km altitude, which may be due to a meteor shower as discussed in Kliore et al. (2008)

e.g. N II 1,085 Å, (3) photoelectron production from ionization of the atmosphere by solar XUV (4–450 Å) and subsequent photoelectron (E < 20 eV) impact excitation of different neutral atmosphere constituents such as N_2 to produce the emissions from the Rydberg and Valence (RV) states of N_2 (the EUV portion of the dayglow spectrum attributed to the N_2 RV bands (b $^1\Pi_u$, b' $^1\Sigma_u^+$, c'$_4$ $^1\Sigma_u^+ \rightarrow$ X $^1\Sigma_g^+$and the FUV portion of the dayglow spectrum to the Lyman-Birge-Hopfield (LBH) band system (N_2(a $^1\Pi_g$ – X $^1\Sigma_g^+$)), (4) similar to (3) but with the predissociative ejection of fast N-atoms before RV emission, (5) photodissociative excitation (PDE) of N_2 leading to predissociation into fast N-atoms, (6) photodissociative ionization (PDI) of N_2 _>N I and N II with emission from N I, (7) Rayliegh scattering in the FUV and (8) Mie scattering off aerosols referred to as tholins in case of Titan. We first start this discussion with the Voyager 1 UVS observations back in 1980, since there has been considerable controversy with regard to their correct interpretation and only recently resolved by Cassini UVIS observations.

The Voyager 1 UVS instrument measured Titan's EUV airglow (Fig. 16.20) which was originally attributed to electron excited N_2 (Broadfoot et al. 1981; Strobel and Shemansky 1982). The N_2 Rydberg bands at 958 Å (c$'_4$ (0–0)) and 981 Å (c$'_4$ (0-1)) were the most prominent with likely excitation by magnetospheric electrons at exobase heights (Hunten et al. 1984). NI bands at 1,134 Å, 1,200 Å, 1,243 Å and 1,493 Å, and NII band at 1,243 Å were also prominent, along with N_2 RV band systems (800 Å to 1,400 Å) and LBH bands (1,260–2,300 Å) also present. At the time of Voyager it was thought all of this airglow was excited by electron impact on N_2 as modeled by Strobel and Shemansky (1982). But, then Strobel et al. (1992), after a downward revision of the Voyager UVS intensities for $\lambda < 1{,}100$ Å by Holberg et al. (1982, 1991),

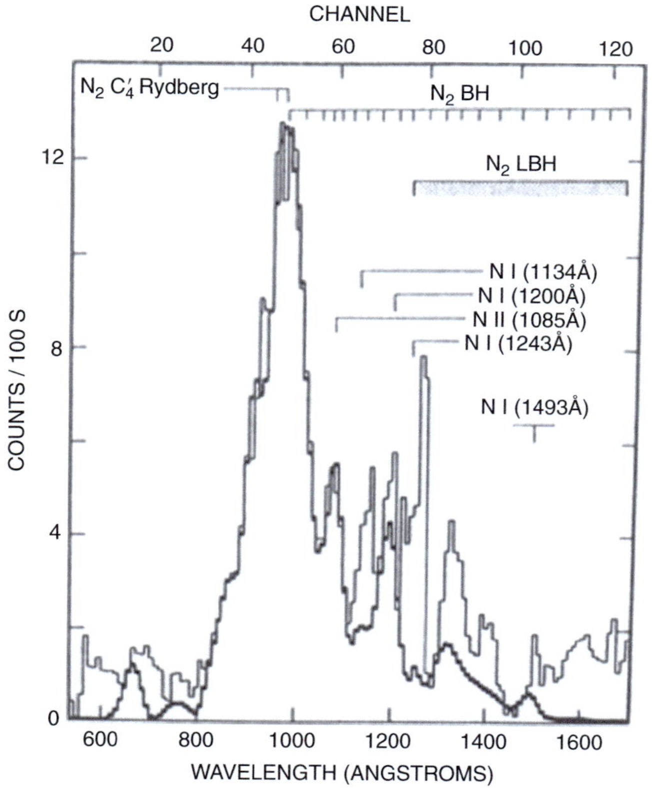

Fig. 16.20 Titan's EUV airglow observations from Voyager 1 UVS instrument as reported by Broadfoot et al. (1981), Strobel and Shemansky (1982) and Hunten et al. (1984)

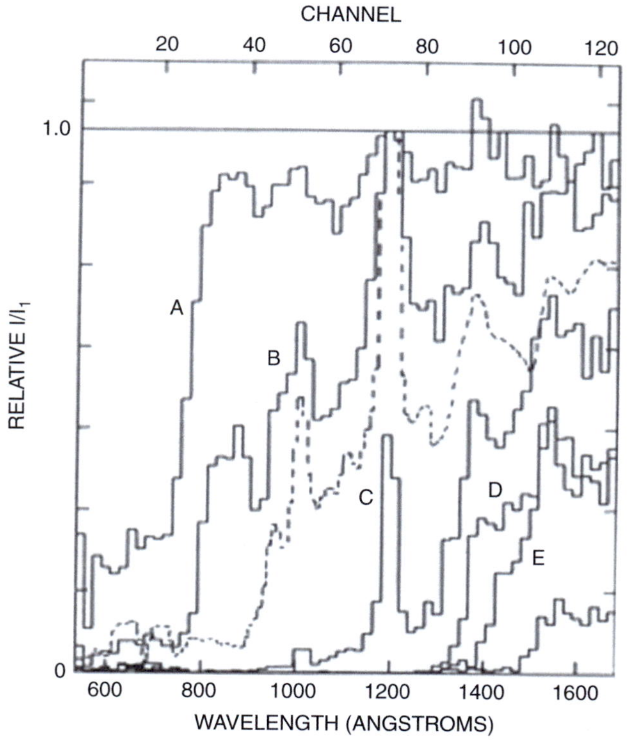

Fig. 16.21 Voyager 1 UVS occultation data showing differential absorption of sun-light as a function of altitude from Broadfoot et al. (1981)

showed that most of the N_2 LBH and N^+ (1,085 Å) was now due to solar XUV (5–500 Å) and photoelectron excitation of N_2 with only a 10% contribution by magnetospheric electrons. An exception was 40% of the N_2 c_4'-band may be due to magnetospheric electrons with remainder by photoelectrons. By downgrading the importance of the magnetospheric electrons they could provide enough of the N_2 excitation via centrifugal drift of magnetospheric electrons across magnetic field lines as originally done by Hartle et al. (1982). Some of the N_2 RV band systems (1,026–1,400 Å) could be due to H Lyman β fluorescence (1,026 Å), itself excited by electron impact on H (Hunten et al. 1984). More recently, Stevens (2001) argued that by using a stronger XUV solar irradiance and a blending of contributions from N I produced by photoelectron dissociative ionization of N_2, the c_4' (0,0) near 958 Å was misidentified and that solar excitation alone could account for the Titan EUV airglow. If true the excitation occurs at lower heights where the atmosphere is not optically thin.

As discussed in Hunten et al. (1984), the Voyager 1 UVS occultation data in Fig. 16.21 (Broadfoot et al. 1981) displayed differential absorption of sun-light as a function of altitude and wavelength in Titan's upper atmosphere for r<4,000 km (i.e., the exobase). The absorption was most prominent at shorter wavelengths ~725 Å with optical depth $\tau \sim 1$ at r ~ 3,840 km (z ~ 1,265 km) from N_2 absorption and estimated density ~2.7×10^8 N_2/cm^3 which is very close to that measured by Waite et al. (2005) and Yelle et al. (2006) using Cassini INMS data. In general, this absorption was confined between r ~ 3,800 km (z ~ 1,225 km) and r ~ 3,050 km (z ~ 475 km). Most of this absorption can be attributed to N_2, CH_4 and C_2H_2 at SZA ~ 90° (Hunten et al. 1984).

This absorption of solar XUV (note the obvious that solar UV can cross magnetic field lines and have direct access to Titan's upper atmosphere) also falls within the height range where ion production by solar UV is expected. This is shown in Fig. 16.22 from the paper by Cravens et al. (2005) with solar absorption peaking at z ~ 1,400 km for SZA 108.6° to z ~ 1,000 km for SZA 58.9°. The photoelectron production peaks at 1,200 km altitude for SZA ~ 90°, which is more like the SZA for Voyager 1 solar occultation measurements. As discussed in Section 4.0, Ma et al. (2006) 3D MHD model, which used a chemically simpler version of the photochemical model used by Cravens et al. (2004), found for z > 1,300 km lower ion densities relative to that of a photochemical only model; they attributed this difference to the importance of transport effects absent in the photochemical only models. The shoulder in the ion production below 950 km in Fig. 16.22 is attributed to solar soft x-rays which are known to be quite variable (Tobisca and Eparvier 1998; Tobisca

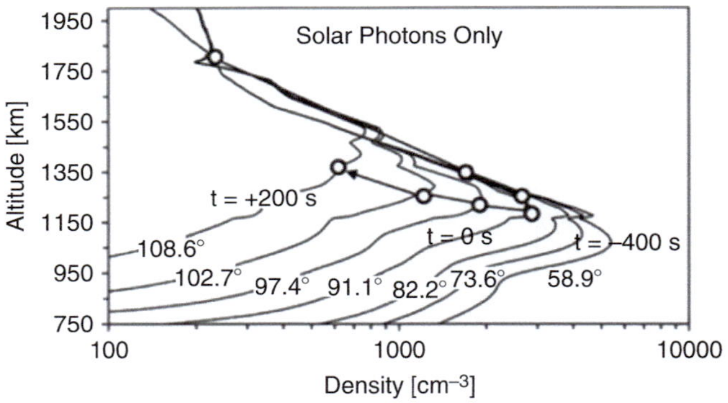

Fig. 16.22 Model predictions of Titan ionospheric electron densities as a function of altitude and solar zenith angle SZA from Cravens et al. (2005). These model results only includes the ion production from solar UV ionization and electron recombination with assumed $T_e \sim 1{,}000°K$ from T_A (Wahlund et al. 2005)

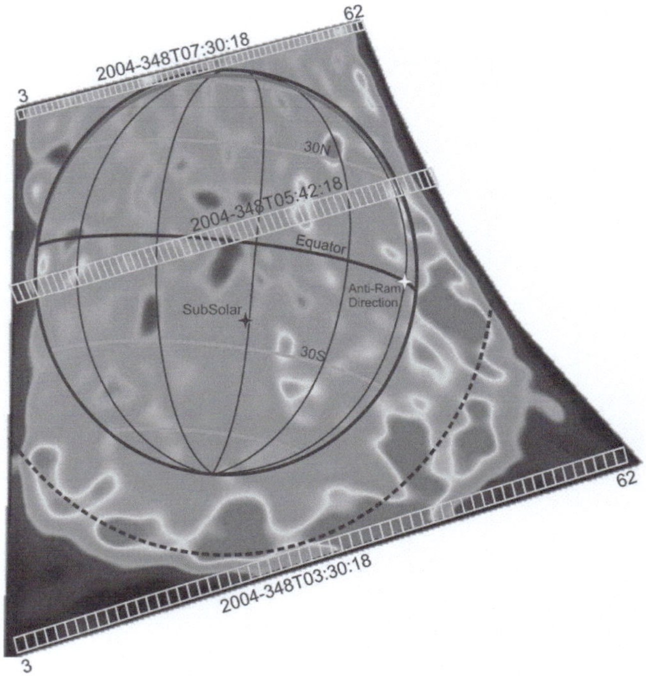

Fig. 16.23 Cassini UVIS EUV image of Titan's airglow from Ajello et al. 2007. EUV signal from 983–988 Å and 1092–1088 Å wavelength bands. Spatial pixels are simultaneously imaged through instrument entrance slit and scanned vertically upward. The accumulation period for the image was ~4 h.

et al. 2000; Lean et al. 2003). Furthermore, for altitudes less than 950 km Cravens et al. (2005) extrapolated RPWS electron temperatures $T_e \sim 1{,}000°K$ from TA (Wahlund et al. 2005), while the model by Gan et al. (1992) estimated $T_e \sim 180°K$ below 925 km. The lower T_e will increase recombination and the lower altitude shoulder may not be as pronounced as shown in Fig. 16.22.

In Fig. 16.23, from Ajello et al. (2007) is the EUV airglow (983–988 Å and 1,082–1,087 Å wavelength bands) observed by Cassini UVIS instrument before the TB flyby on 13Dec04. This image shows emission peaking at around 1,200 km altitude. In Fig. 16.24 from Gustin et al. (2009) is shown the FUV limb airglow showing solar reflectance for $\lambda > 1{,}500$ Å extending up to ~800 km in altitude. Ajello et al. (2008) attribute the solar scattering to hydrocarbon (i.e., C_2H_2, C_2H_4, and C_4H_2) absorption, dissociation, nitrile formation and subsequent solar scattering due to aerosol (tholin) scattering. The Lyα is observed at all altitudes indicating emission is also coming from a more global source of atomic H surrounding Saturn (Shemansky and Hall, 1992; Shemansky et al., 2009)

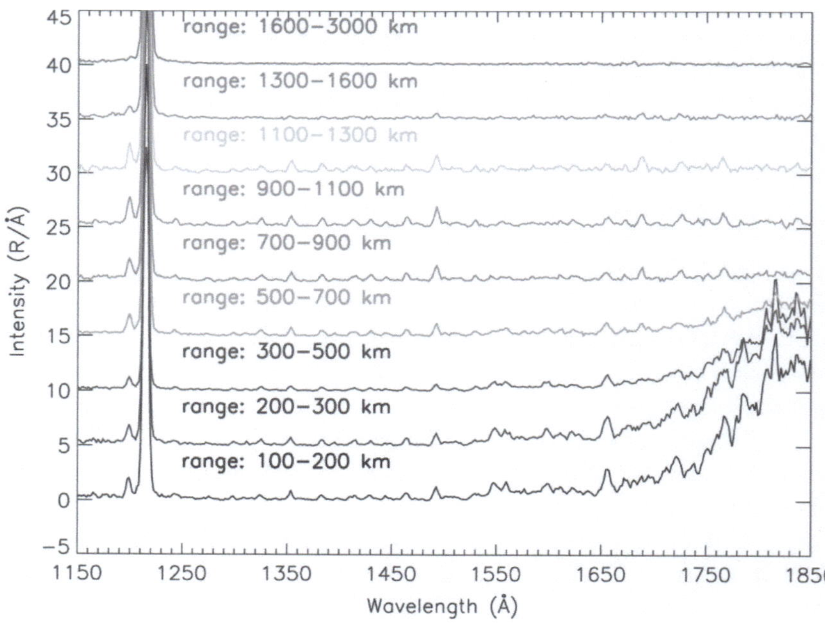

Fig. 16.24 Cassini UVIS FUV airglow limb scan from surface to exosphere from Gustin et al. 2009. Data recorded on 13Dec04 (TB flyby). Nine minimum ray height (MRH) wavelength spectra shown

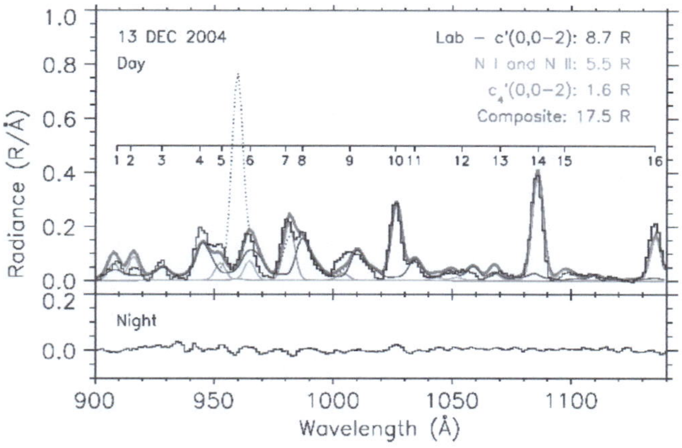

Fig. 16.25 Cassini UVIS airglow spectra from 900 to 1,200 Å from Ajello et al. (2007). Comparing this with Fig. 16.2020 it is clear that the Cassini instrument has superior spectral resolution

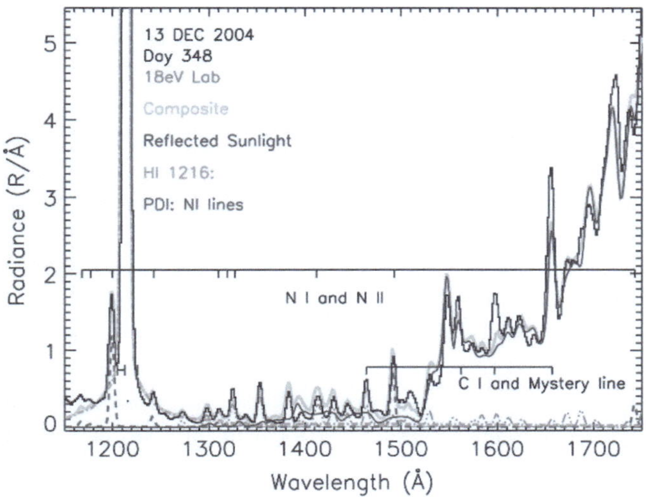

Fig. 16.26 Similar to Fig. 16.2525, but now the wavelength coverage is from 1100 to 1900 Å and comes from Ajello et al. (2008)

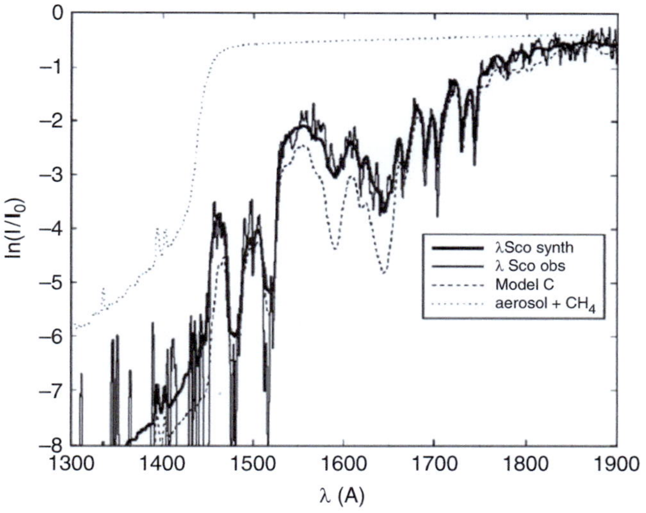

Fig. 16.27 Cassini UVIS stellar occultation observations reported by Liang et al. (2007) and showing the extinction due to aerosols

and the interstellar medium. In Fig. 16.25, from Ajello et al. (2007), is shown airglow spectrum from 900 to 1,200 Å observed by Cassini UVIS before the TB encounter where the main peak observed by Voyager 1 UVS from 950 to 1,050 Å has now been resolved (i.e., Cassini UVIS higher spectral resolution). This figure shows the absence of the c_4' (0,0) emission line at 958 Å or is very weak, but the N I multiplets at 953.2 Å and 964.5 Å are present instead, clearly resolved by the ~5.8 Å UVIS bandpass as predicted by Stevens (2001). This re-enforces the argument that this emission is primarily due to photoelectron excitation. Figure 16.26 from Ajello et al. (2008) shows the TB airglow from 1,100 to 1,900 Å with emission primarily from NI (1,200 Å), NI (1,493 Å) and N_2 LBH.(Lyman-Birge-Hopfield band system), which can be compared with the Voyager 1 UVS observation shown in Fig. 16.20. The Cassini UVIS observations show many more lines than Voyager due to its much higher spectral resolution. All of the dayglow emission observed by Cassini UVIS at high altitudes in orbit TB can be attributed to photoelectrons (Ajello et al. 2008). This interpretation is re-enforced by the absence of a nitrogen nightglow. Other orbits such as the one that occurred on May 20, 2006 show the presence of nightglow indicating the presence of soft magnetospheric electrons on other orbits. Ajello et al. (2007) estimated the UV airglow emission from N_2, N I and N II to be ~8×10^7 watts. In all cases the N_2 is dissociated into fast N atoms with PE contributing ~8×10^8 atoms/cm²/s, PDI producing ~2×10^8 atoms/cm²/s and PDE ~ 3×10^8 atoms/cm²/s. They estimated a total N_2 loss rate to be ~ 10^9 atoms/cm²/s or a global hemispherical atom production ~ 10^{27} atoms/s with an unspecified fraction of these reaching the exobase and being lost to the magnetosphere (see Ajello and Ciocca 1996).This loss of N atoms will be a source term to an atomic nitrogen torus within Saturn's outer magnetosphere and the corresponding observation of magnetospheric keV N^+ ions near Titan's orbital position (see Sittler et al. 2006a). The remainder that does not escape probably gets converted back into N_2 (Ajello et al. 2007).

In Fig. 16.27 from Liang et al. (2007) the aerosol extinction becomes evident for altitudes ~ 970 km or less and dominates all other absorbers at all wavelengths in Cassini UVIS except for CH_4 at altitudes below 400–450 km. At these lower altitudes CH_4 extinction dominates between 1,100 and 1,400 Å (Liang et al. 2007). Liang et al. (2007) have used the wavelength band 1,850–1,900 Å (SP1), which is free of hydrocarbon and cyano species, to measure aerosol scattering with height during stellar occultation by Cassini UVIS. Cassini UVIS stellar occultation observations, as shown in Fig. 16.26 have also been used to measure neutral species abundances with height from 350 to 1,000 km (Shemansky et al. 2005; Liang et al. 2007; Ajello et al. 2008). The UVIS data can be used to probe the presence of aerosols as a function of height and latitude around Titan, but have difficulty in giving the aerosol size distribution. See for example Ajello et al. 2008 who get similar results for aerosols with radius ~125 Å and 25 Å.

16.7 Magnetospheric Interaction and Charged Particle Bombardment: Electrons, Ions and Pickup Ions

In order to organize our discussions about Titan's magnetospheric interaction and resulting energy depositions we subdivide into three major sub-sections. The first Sub-section 7.1, has to do with the neutral exosphere, which extends into Saturn's magnetosphere where its constituents are ionized by the magnetospheric plasma and solar UV, producing pickup ions that mass load the plasma and slow it down. The background plasma with embedded pickup ions is deflected around Titan at high altitudes resulting in piled up magnetic field approximately centered on the ram direction. We then discuss in Sub-section 7.2 the interaction at lower altitudes where electron impact ionization of the ionosphere is potentially important, but since the electrons are magnetized they will have difficulty being transported to lower altitudes. Then, in Section 7.3 we discuss the importance of heavy magnetospheric ion precipitation and energetic charged particle precipitation into Titan's lower ionosphere. These ions are unmagnetized and can essentially move across field lines with minimal atmosphere shielding (see Ledvina et al. 2005). In doing this, we consider five Titan flybys (V1, TA, T5, T9 and T18) which have received the most attention and relate to the major topics of this section.

16.7.1 Titan's Exosphere as Source of Pickup Ions, Mass Loading of Incoming Flow, Energy Input to Upper Atmosphere and Atmospheric Loss

16.7.1.1 Voyager 1 Observations and Modeling of Titan's Exosphere

In the earlier Voyager papers by Hartle et al. (1982) and Neubauer et al. (1984) they introduced the importance of the neutral exosphere with regard to pickup ions and mass loading of the flow at heights greater than the exobase at r ~ 4,000 km. This exospheric interaction is important since it will begin the deflection of the upstream flow well above the topside ionosphere and reduce the magnetospheric input to Titan's upper atmosphere, especially with regard to magnetospheric protons and electrons which are magnetized and tied to the magnetic field lines. The exosphere model used by Hartle et al. (1982) was composed of H and N_2. During the intervening years, atmosphere models by Yung et al. (1984), Yung (1987), Toublanc et al. (1995) and Keller et al. (1998) predicted significant quantities of H_2 and CH_4 to the exosphere. In addition, models showed the presence of suprathermal nitrogen atoms N* due to electron and photon interaction with N_2 (Strobel and Shemansky 1982; Ip 1992; Strobel et al. 1992) and sputtering due to magnetospheric ion impact (Lammer and Bauer 1993; Shemantovich 1998, 1999; Shemantovich et al. 2001; Shemantovich et al. (2003); Michael et al. 2005). Cravens et al. (1997) investigated photochemical reactions that produced non-thermal source of neutrals to Titan's exosphere. Therefore, Sittler et al. (2005a) revisited the Voyager 1 flyby data using an exosphere model that now included H_2, CH_4, and N* in addition to H and N_2 (see Fig. 16.28). Yelle et al. (2006) using INMS data confirmed the presence of the N_2, CH_4 and H_2 exospheres for altitudes just above the exobase z ~ 1,400 km. The Sittler et al. (2005a) re-analysis showed that CH_4^+ pickup ions dominated mass loading the upstream flow for r < 6,000 km for Voyager 1 flyby conditions while H_2^+ displayed some slowing of the flow farther from Titan. As shown in Fig. 16.29, the Sittler et al. (2005a) model and data showed a stagnation point near ram side of interaction at r ~ 4,800 km while along the flanks a decrease in flow speed from 120 km/s to ~60 km/s at r ~ 6,000 km and to ~5 km/s at r ~ 5,580 km. From these results one can estimate an atmospheric loss rate due to pickup ions to be ~5×10^{22} ions/s. At these heights the ionosphere is not contributing to the deflection and slow down of the flow, but rather Titan's exosphere is dominating the slow

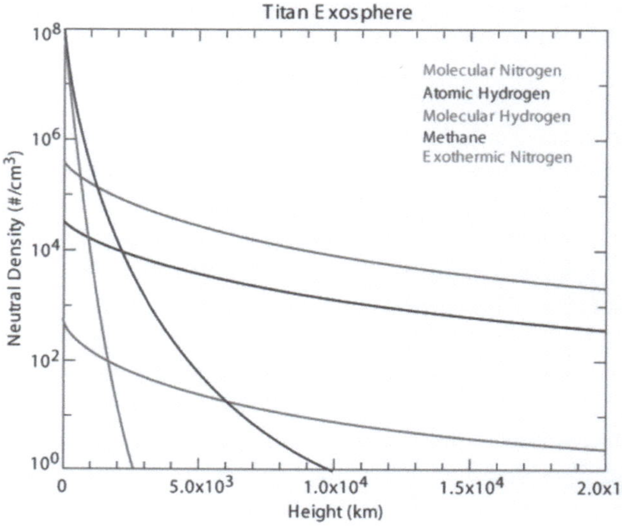

Fig. 16.28 Titan's exosphere as modeled by Sittler et al. (2005a) which includes H, H_2, CH_4, N_2 and suprathermal N from neutral sputtering due to keV magnetospheric ions bombarding Titan's upper atmosphere (Michael et al. 2005)

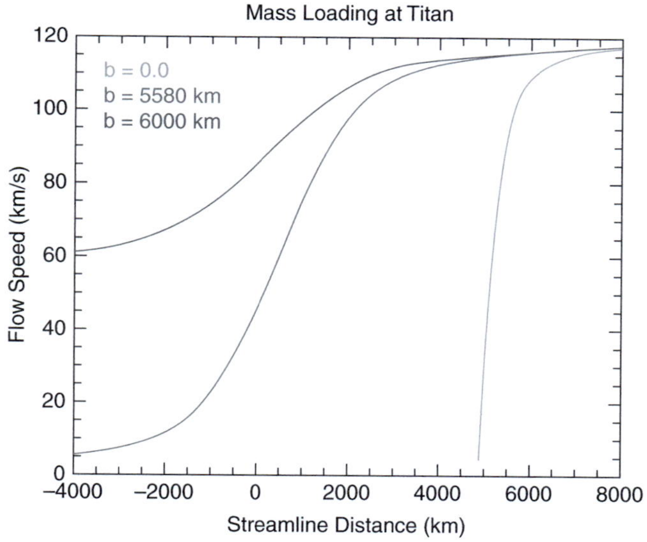

Fig. 16.29 Calculations of mass loading of the upstream flow by Titan's exosphere as modeled by Sittler et al. (2005a)

down and deflection. The primary reason for this is that the neutral exosphere can pass through the piled up magnetic field lines and extend its influence to much higher altitudes. As discussed in Sittler et al. (2009c) for T9 where light ions dominate the upstream flow, the H_2 exosphere may have a significant influence on the flow far from Titan. In addition to mass loading of the flow by the exosphere Dobe et al. (2007) presented a two-stream boundary layer model between the upstream flow and Titan's ionospheric upper boundary which further slows the magnetospheric flow which scavenges ionospheric ions away from Titan as a sink to the ionosphere.

As the hybrid simulations by Brecht et al. (2000) and modeling by Michael et al. (2005) and Michael and Johnson (2005) showed, the pickup ions on the Saturn side (Voyager 1 conditions) when $\mathbf{E}_{conv}$ is pointing into Titan, can impact Titan's upper atmosphere and via a sputtering process produce hot suprathermal N* and N_2* which can then escape Titan's atmosphere and contribute to the nitrogen torus around Saturn (Smith et al. 2004). As discussed in Sittler et al. (2006a), this nitrogen torus can contribute to the plasma nitrogen ions (Smith et al. 2005; Sittler et al. 2005b; Sittler et al. 2006a) observed by CAPS-IMS and energetic nitrogen ions observed by MIMI-CHEMS (Krimigis et al. 2005) in Saturn's magnetosphere. The magnetospheric nitrogen ions are observed to be ~5% of the heavy ion component with water group ions such as O^+ dominating the composition (Hartle et al. 2006b). Based on charge transfer collisions within the Enceladus torus (Johnson et al. 2005; Sittler et al. 2006a) have argued that these fast water group neutrals are still gravitationally bound to Saturn and then become ionized and picked up in the outer magnetosphere to form hot keV water group ions in the outer magnetosphere as observed (Hartle et al. 2006b).

16.7.1.2 CAPS Observations of Pickup Ions and Titan's Exosphere

Hartle et al. (2006a,b) showed in situ observation of exosphere pickup ions from methane and di-nitrogen exospheres. Neubauer et al. (2006) showed that the upstream magnetic field was dipolar for TA, TB and T3. As shown in Fig. 16.30 from Hartle and Sittler (2007) the heavy pickup ions will

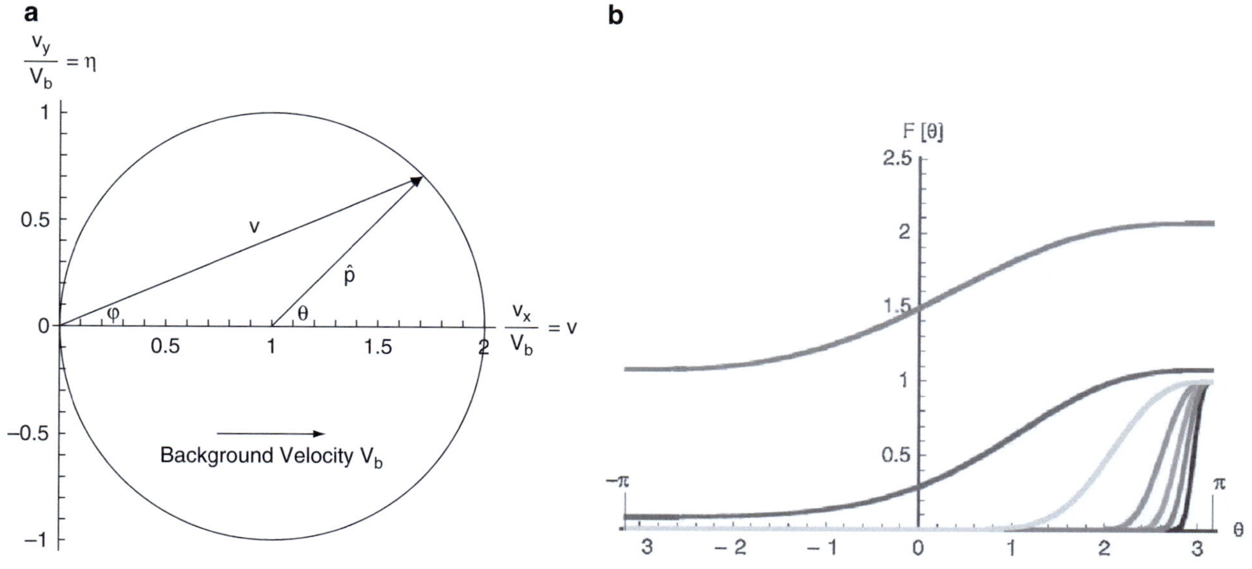

Fig. 16.30 Shows the velocity distribution function properties of pickup ions born from Titan's exosphere as modeled by Hartle and Sittler (2007). The model is 1D with $\alpha = r_g/H$ is the critical parameter. The *top panel* a shows the ring distribution and defines the angles θ and φ with upstream flow in x direction with speed V_b. Other parameters $\nu = V_x/V_b$ and $\eta = V_y/V_b$. *Right panel* b shows beam properties of pickup ions when parameter $\alpha >> 1$ (see text for details)

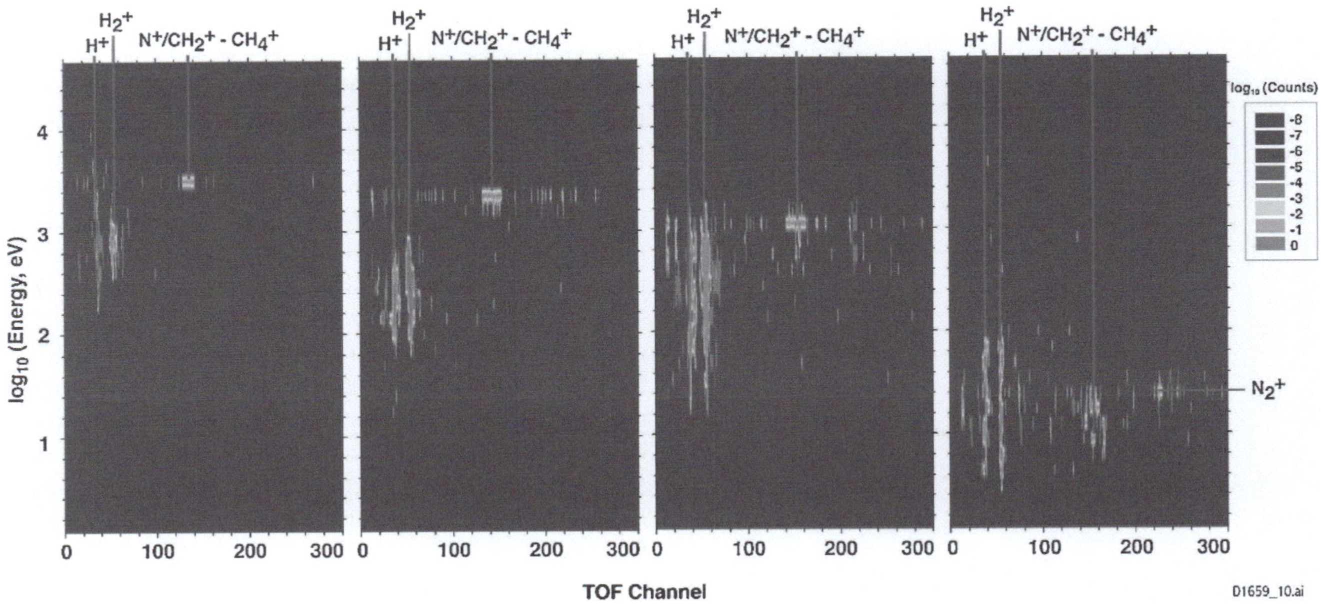

Fig. 16.31 CAPS composition data during T_A flyby as shown by Hartle et al. (2006a). The figure shows clear evidence of pickup methane ions with lack of O^- peak (i.e., O^- fragment is produced within instrument when incident ions are water group $W^+=O^+$, OH^+, H_2O^+ and H_3O^+). Theory excludes NH_2^+ ions

appear highly beamed with source points near exobase heights when the ratio of pickup ion gyro-radius and atmospheric scale $\alpha=r_g/H>>1$. This is exactly what was observed by the CAPS-IMS, shown in Fig. 16.31, during the TA flyby in the pickup ion region where $\mathbf{E}_{conv}$ points radially away from Titan. They were also able to show that the mass loading was dominated by methane ions since the O^- fragments from oxygen ions with mass = 16 were not observed (see Section 7.3 for details of O^- fragment). The ion composition data was able to rule out N^+ ions, but NH_2 was ruled out since INMS did not observe an NH_2 exosphere (Waite et al. 2005; Yelle et al. 2006; De La Haye et al. 2007); NH_2 is also not present in theoretical models of Titan's upper atmosphere (see Toublanc et al. 1995). Evidence of the N_2 exosphere was also present when the flow was highly mass loaded and ions with M/Q ~ 28 were observed. Hartle et al. (2006a,b) were not able to estimate very accurately the neutral densities of the exosphere from the TA pickup ion observations. Future modeling with spherical exosphere and more accurate models of the electric and magnetic fields from hybrid simulations may allow this to be done in the future.

16.7.1.3 Cassini ENA Imaging of Titan's Exosphere

The first MIMI-INCA energetic neutral images of Titan's interaction with Saturn's magnetosphere were first reported by Mitchell et al. (2005) for both TA and TB. They used the hydrogen energetic neutral atoms (ENA) from 20 keV to 80 keV channels. For TA they observed the classic ENA images (90° × 120° FOV), shown in Fig. 16.32 with max intensity on the downstream side relative to Titan with guiding centers coming from the right-side of Titan. This strong left/right asymmetry is a confirmation of a theoretical expectation as shown in Fig. 16.33, based on a 3D kinetic modeling of the trajectories on the ENA parent ions, which have gyroradii comparable to the Titan induced magnetotail exobase radius (Dandouras and Amsif 1999). The signal for 20 to 50 keV H ENAs peaked about 2,000 km above exobase or 3,700 km in altitude. These ions will have mean gyro-radius ~5,400 km. Garnier et al. (2007) performed a more quantitative analysis of the TA ENA data and found the ENA peak altitude of 2,000 km about a factor of 2 higher than expected; a more recent analysis by Garnier et al. (2008a) and Dandouras et al. (2008) show the peak nearer to the exobase at a lower altitude ~1,400 km. They proposed possible explanations for these observations such as field line draping and non-thermal escape of exosphere particles (De La Haye et al. 2007). Then using in situ measurements of 20–50 keV protons within the wake region, they inferred an exosphere model based on the Chamberlain formalism. Backes et al. (2005) showed significant draping of the magnetic field during the TA wake pass with significant B_X components first being negative for north lobe and positive for south lobe. Since, the gyro-radii ~5,000 km are comparable to the cross-sectional size of the induced magnetotail, the trajectories will be quite complex with guiding centers ~5,000 km to the right of the peak intensity region. The observed minimum near vertical midpoint of emission is due to reduced detection efficiency of instrument's microchannel plate (MCP).

For TB the geometry is more complex but viewing of ENAs with Titan's exosphere can be made both inbound and

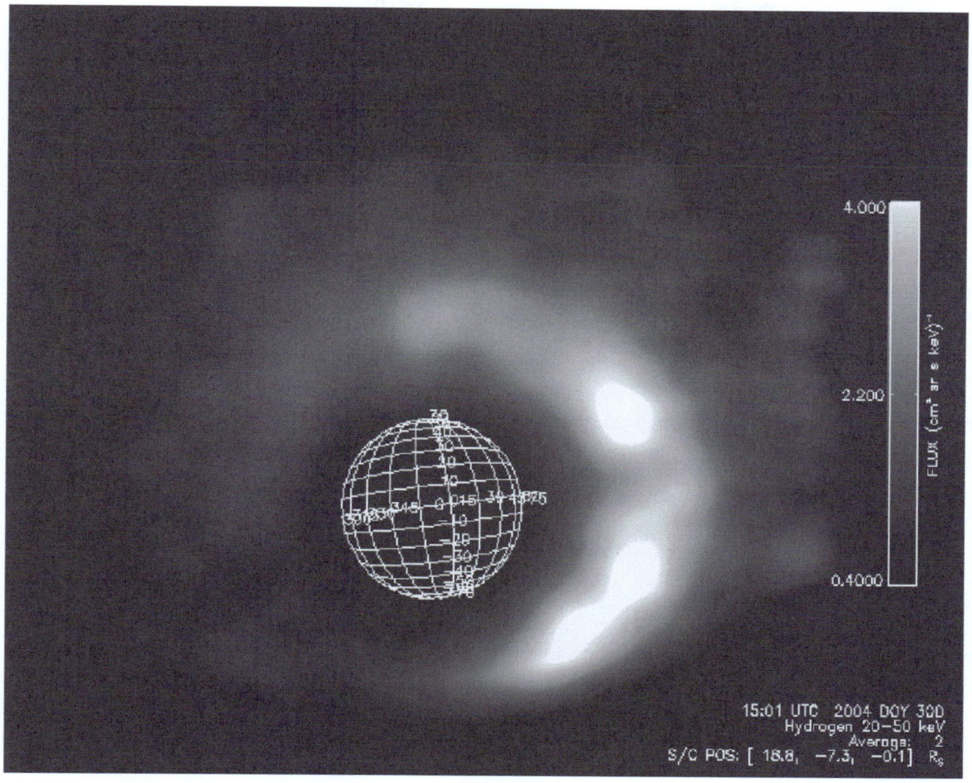

Fig. 16.32 Image of ENA 20–80 keV H atoms during the TB flyby by the MIMI-INCA instrument as reported by Mitchell et al. (2005). The lower intensity of *arc shaped feature* is due to lower detector efficiency at that location. This figure was taken during approach when spacecraft was outside Titan's orbital position and later passing through Titan's wake side ionosphere

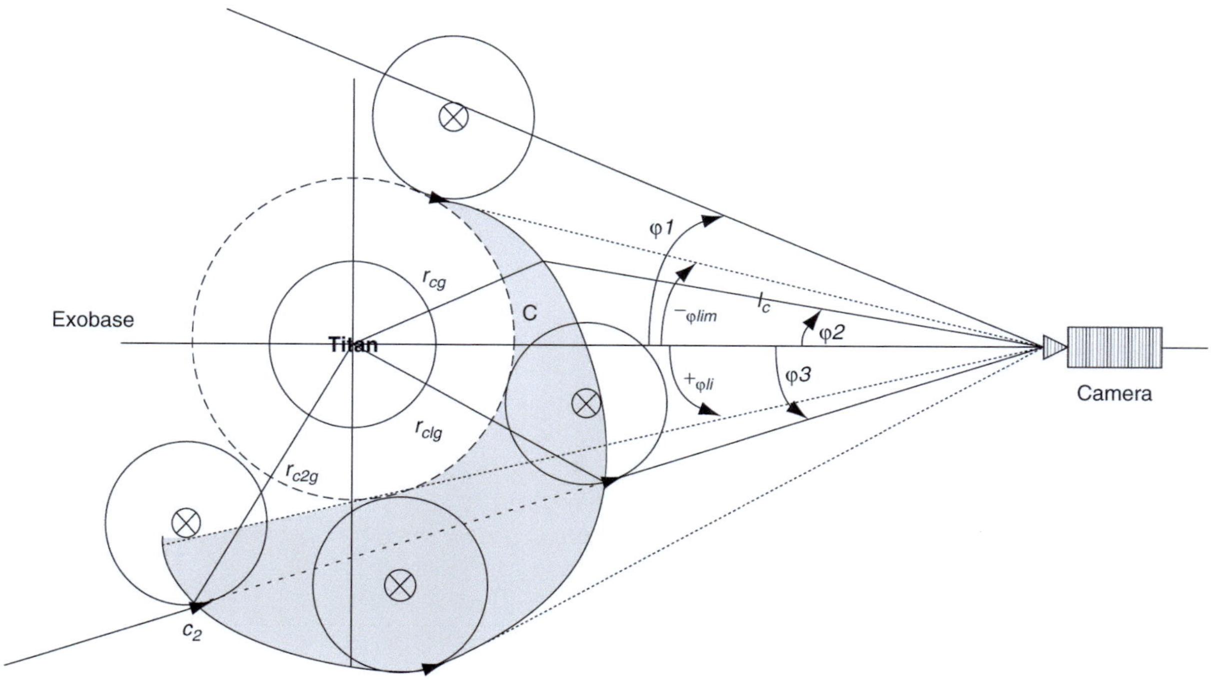

Fig. 16.33 Shows launch site of ENAs as function of ion guiding center location relative to Titan and its exosphere and why an asymmetry in ENA intensities occur as shown in Figure 16.32

outbound. As summarized in Mitchell et al. (2005) the use of unperturbed magnetic field cannot be used in the interpretation of these images, but must consider the complexities of the interaction with Titan's conductive ionosphere. Their images, shown in Fig. 16.34, shows for image 7.7.D a halo of emission around Titan at all latitudes at about exobase altitudes ~1,400 km. For image 7.7.D Cassini is outside Titan's orbit with INCA looking at Titan from the side with ram on left side and wake on right side of Titan. When Cassini is near CA, 11:35 SCET, Cassini is in Titan's wake looking at Titan in the direction of the incoming flow. This can be seen for images 7.7F to 7.7H where the emission is confined to the north since Cassini is also to the north. Here, we're probably seeing ions with their guiding centers to the Saturn side of Titan and on draped field lines. In the case of images 7.7I–7.7K the emission is more confined and to lower altitudes. Here, INCA is probably imaging the region where the magnetic field has piled up (INCA viewing of ENAs optimal for ram side of interaction) and the gyro-radii ~1,200 km are smaller, which may account for tight confinement of the observed emission with height. Reasons for the emission extending below the exobase is not as clear, but recent results on the penetration of E>30 keV protons down to 800 km altitude by Cravens et al. (2008) are consistent with this result. More recent results by Smith et al. (2009), shown in Fig. 16.35 clearly show this penetration down below 800 km. Smith et al. (2009) also performed detailed Monte Carlo calculations of the penetration into Titan's atmosphere which

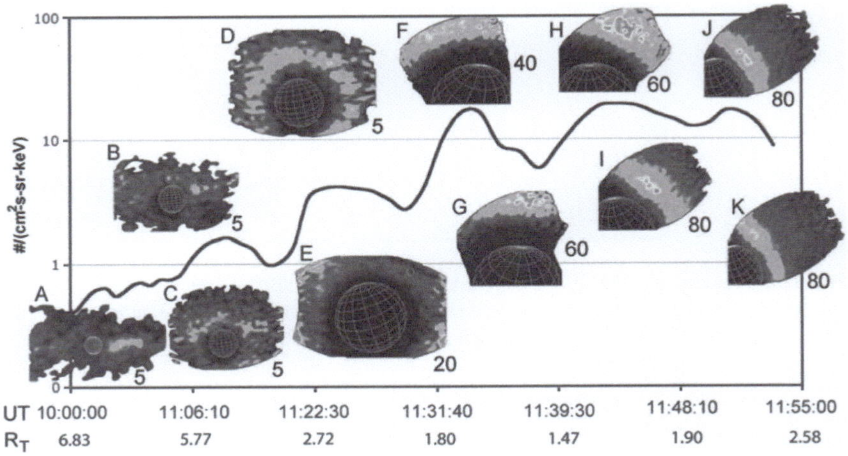

Fig. 16.34 Shows collection of ENA images for TB flyby as reported by Mitchell et al. (2005). Figures A to K increase with time and vertical scale shows gradual increase in ENA intensity as one moves from image A to image K

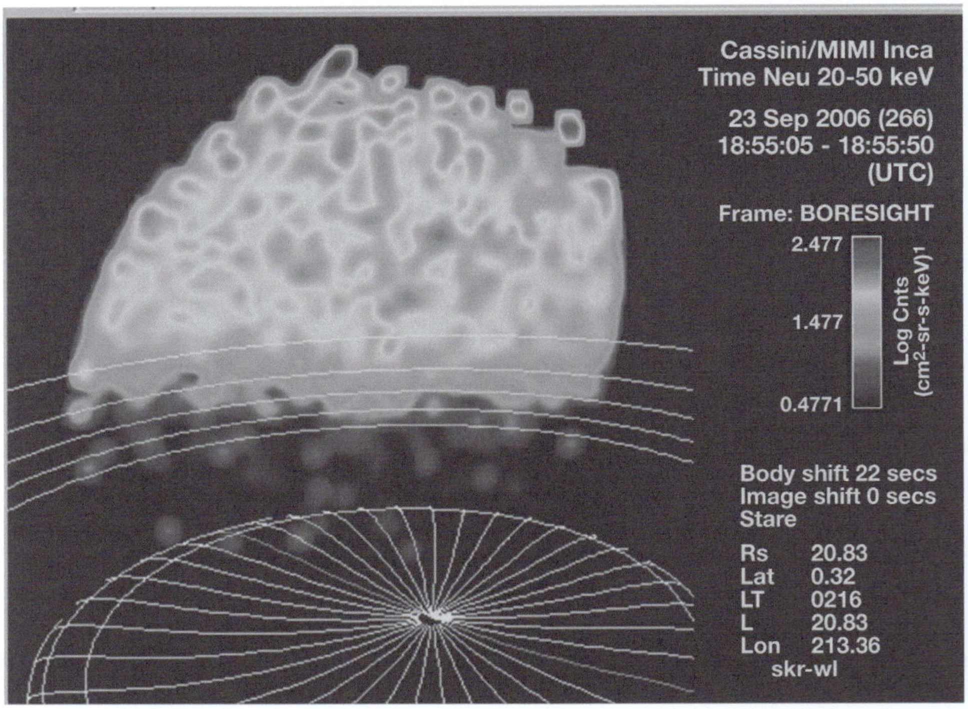

Fig. 16.35 Shows ENA image of 50 keV H atoms from Smith et al. (2009) showing the penetration of 50 keV protons down below 800 km altitude. These observations were substantiated by Monte Carlo ion neutral collisions as modeled by Smith et al. (2009)

confirmed the observations. For both TA and TB the energetic protons are directly interacting with Titan's N_2 and CH_4 exosphere with the H_2 exosphere being less important.

In summary, the ENA observations allow one to image the interaction and model the exosphere, which then allows one to estimate the energy deposition with height due to the energetic ions or hot plasma component of the magnetosphere. This can then be compared with theoretical calculations of the energy deposition with height, which can allow one to model the ionization layers observed both in situ by CAPS, INMS and RPWS-LP and remotely by RSS.

16.7.1.4 Cassini INMS Observations of Titan's Lower Exosphere

Most recently, De La Haye et al. (2007) reported evidence of a suprathermal hot component of the N_2 and CH_4 exospheres with energy inputs ~110±90 eV/cm^3/s and ~39 ±35 eV/cm^3/s, respectively for the N_2 and CH_4. From this they estimated escape fluxes of nitrogen ~7.7 ±7.1 × 10^7 N/cm^2/s and for CH_4 ~ 2.8 ±2.1 × 10^7 CH_4/cm^2/s. Yelle et al. (2006) did estimate CH_4 escape flux > 6.2 × 10^8 CH_4/cm^2/s at exobase if the eddy diffusion coefficient were <10^9 cm^2/s, but they chose not to give a final estimate in the paper. Yelle et al. (2006) did estimate an H_2 escape flux ~5.0 × 10^9 H_2/cm^2/s at the exobase. As noted above, based on Hartle et al. (2006a,b) TA estimate of O^+ into Titan's atmosphere one can estimate an energy flux into Titan's atmosphere ~5 × 10^8 eV/cm^2/s and assuming it deposits most of its energy over a scale height ~70 km, one gets energy input ~71 eV/cm^3/s to Titan's upper atmosphere. This estimate is within the error bars estimated by De La Haye et al. (2007). So, the magnetospheric keV heavy oxygen ions as reported by Hartle et al. (2006a,b) can account for the suprathermal component of N_2 and CH_4 exosphere species observed by De La Haye et al. (2007). The escape fluxes reported by De La Haye et al. (2007) will contribute ~ 1.5 ±1.4 × 10^{26} N/s and ~ 5.6 ± 4.2 × 10^{25} CH_4/s atoms and molecules to their respective tori around Saturn near Titan's orbital position. The error bars by De La Haye et al. (2007) are quite large so one can only know the N_2 and CH_4 escape fluxes within an order of magnitude.

16.7.1.5 Hydrodynamic Model and Non-Thermal Model of Titan Exospheric Escape Particles

Strobel (2008) argues that he can reproduce high atmospheric loss rates using a hydrodynamic wind model similar to that proposed by Parker (1958) for the Sun's solar wind (see chapter 15 of Titan Book, Mass Loss processes in Titan's upper atmosphere by Johnson et al. 2009 for a more detailed discussion of this topic). Strobel claims to be able to support total mass rates ~4–5 × 10^{28} amu/s. Converting the De La Haye et al. (2007) numbers into amu/s one gets loss rate ~2.1 ±2.0 × 10^{27} amu/s for N and 9.0 ±6.7 × 10^{26} amu/s for CH_4, while the H_2 escape rate from Yelle et al. (2006) is ~ 2 × 10^{28} amu/s. If one uses the estimate by Yelle et al. (2006) for methane with eddy diffusion coefficient K ~ 10^9 cm^2/s one gets total loss rate of CH_4 ~ 2 × 10^{28} amu/s. This is the high loss rate of methane that Strobel (2008) is attempting to explain. But, Strobel's model requires energy addition to the methane molecules well above the exobase where collisional transport of this energy becomes questionable. The source of the energy flux driving the hydrodynamic flow is EUV-UV absorption and NIR absorption (2.3 and 3.3 μm bands) by CH_4 at altitude ~700–900 km and EUV absorption by N_2 with peak above ~1,000 km altitude. Then heat conduction carries this heat to higher altitudes to push the H_2 and CH_4 above the gravitational well of Titan and form an escaping atmospheric wind. Strobel (2008) estimates net energy flux ~10^{-3} erg/cm^2/s to produce his hydrodynamic wind, but as estimated above the energy flux from magnetospheric O^+ ions to Titan's atmosphere at exobase heights is ~5 × 10^8 eV/cm^2/s ~ 1.4 × 10^{-3} erg/cm^2/s which is very close to the energy flux Strobel requires. Tucker and Johnson (2008) using a Monte Carlo code are unable to produce the relatively high conductive energy flux required by Strobel (2008). It's more likely the required energy comes from the precipitating magnetospheric heavy ions noted above. This is also consistent with the well known fact that bydrodynamics is not applicable in a collisionless medium. Even though the De La Haye et al. (2007) non-thermal model cannot produce the large outward fluxes predicted by Strobel (2008), there should be adequate energy input to produce the lower atmospheric loss rates reported by De La Haye et al. (2007). We emphasize that Yelle et al. (2006) and De La Haye et al. (2007) did not require such high outflows in their final analysis of the INMS data. Another consequence of Strobel's hydrodynamic wind is that a large torus of methane molecules must be present and has not yet been detected by Cassini. This will be an active area of future investigation with CAPS-IMS being the instrument most likely to detect such a methane torus in the form of pickup methane group ions.

In addition to the above analysis of atmospheric escape based primarily on the INMS data set, Garnier et al. (submitted 2008b) using INCA ENA observations and energetic particle data by LEMMS, have developed a non-thermal model of Titan's exosphere. Kappa distribution functions are used at the exobase for CH_4, N_2, and N, averaged over the TA, TB and T5 flybys. The resulting kappa parameter values, for these species, are of the order of 12 to 13. This model allows then the calculation of the escape rates from the exosphere, of the order of ~2 × 10^{22} particles per second for N_2 and CH_4. This estimate is ~1,000 times less than that estimated by De La Haye et al. (2007) and even lower than that discussed in Strobel (2008). The reason for these discrepancies is not yet known.

16.7.2 Magnetospheric Electron Energy Input Versus Solar Input to Titan's Ionosphere

16.7.2.1 Initial Cassini Observations of Titan's Ionosphere

16.7.2.1.1 Cassini TA Flyby

The first in situ measurements of Titan's ionosphere, shown in Fig. 16.36, were made by Cassini's RPWS LP during Cassini's TA and TB flybys (Wahlund et al. 2005). We concentrate on TA since TB is similar and we have CAPS published results for TA. Refer to Fig. 16.1 from Hartle et al. (2006b) for TA encounter geometry. CA occurs at 15:30:08 SCET, at altitude ~1,764 km, latitude ~41° and SZA ~ 91.1°. During ingress Cassini was in sunlight while during egress it was in shadow. These observations revealed an ionosphere during ingress at r ~ 5,000 km (15:18 SCET) with electron densities $n_{e,c}$ ~ 20 el/cm³, an ionospheric peak before closest approach (CA) having peak electron densities $n_{e,c}$ ~ 4,000 el/cm³ (15:28 SCET). An egress boundary is observed at r ~ 5,000 km (15:39:30 SCET), where one sees a sudden drop in electron density from $n_{e,c}$ ~ 100 el/cm³ to $n_{e,c}$ ~ 5 el/cm³. This outer boundary nearly coincides with what they call the mass-loading boundary (MLB), while for ingress the ionosphere resides well inside the MLB (15:14 SCET). The LP electron temperatures for TA are shown in Fig. 16.37 which show $T_{e,c}$ ~ 1,000°K at ~1,200 km altitude and $T_{e,c}$ ~ 10,000°K (T_e ~ 1 eV) at about 2,900 km altitude. These temperatures are considerably hotter than the neutral atmosphere temperatures T ~ 150°K observed by INMS for TA (Waite et al. 2005). The LP electron temperatures are close to those modeled by Gan et al. (1992) and Galand et al. (2006). The former was purely a model prediction, while the latter used TA measurements and showed the critical importance of the horizontal magnetic field in channeling electron heat transport. The heating is coming from the photoelectrons observed by ELS, which then transfer their energy to the thermal electrons via electron-electron collisions. The thermal population is cooled by electron-neutral collisions which become less important at higher altitudes above the exobase at z ~ 1,400 km. Outside the outer boundaries of the ionosphere, on the ingress and egress legs, the LP signals are dominated by spacecraft emitted photoelectrons and are not reliable; in this outer region CAPS-ELS observations shown in Fig. 16.38 as reported in Hartle et al. (2006b) are more reliable with $n_{e,h}$ ~ 0.1 el/cm³ and $T_{e,h}$ ~ 100 to 1,000 eV. In Fig. 16.36 we show the boundaries beyond which magnetospheric electrons are confined, 15:21 SCET inbound and 15:38 SCET outbound. Between 15:18 SCET and 15:21 SCET and between 15:38 SCET and 15:39:30 SCET, where the LP ionospheric and ELS magnetospheric electrons overlap, we see heating of the thermal electrons to $T_{e,c}$ ~ 10,000°K, which indicates magnetospheric electron heating of thermal ionospheric electrons. At these higher altitudes electron-neutral collisions, which can cool the thermal electrons, are less important. Within the ionosphere defined by the LP, ELS is primarily seeing photoelectrons from Titan's ionosphere with $n_{e,h}$ ~ 1.0 el/cm³ and $T_{e,h}$ ~ 20 eV near the ionospheric outer boundary at r ~ 5,000 km to max electron densities $n_{e,h}$ ~ 10 el/cm³ and $T_{e,h}$ ~ 10 eV near ionospheric density peak. Hartle et al. (2006a,b) using CAPS-IMS data observed a clearing zone of magnetospheric O^+ starting at 14:53 SCET (r ~ 12,500 km) and observed CH_4^+ pickup ions at 15:02 SCET (r ~ 11,000 km) to 15:19:04 SCET (r ~ 5,900 km) (see Figs. 16.1 and 16.31). The boundary outside of which the pickup ions are observed is shown in Fig. 16.36.

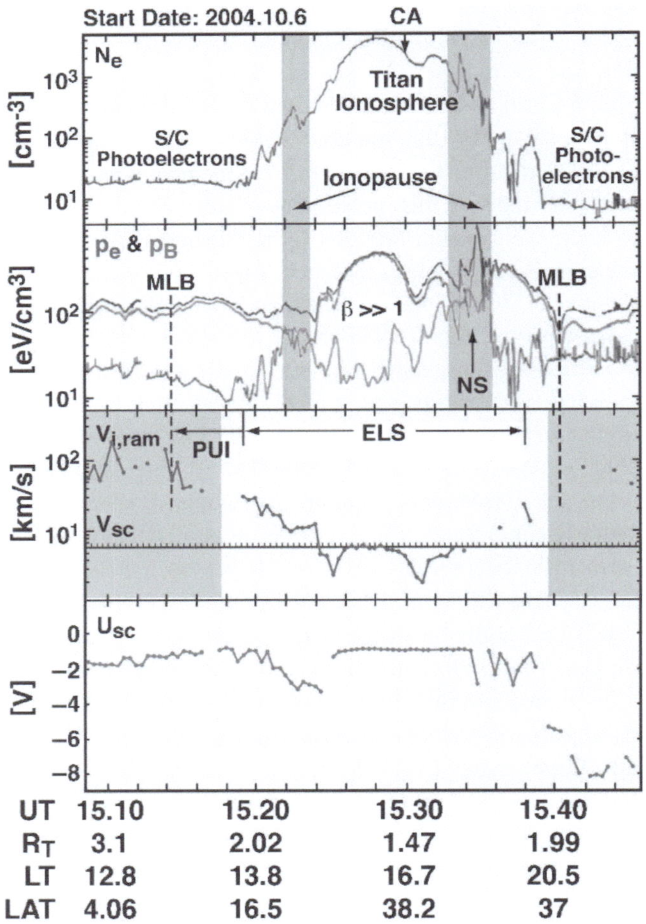

Fig. 16.36 Cassini TA flyby RPWS LP observations as reported by Wahlund et al. (2005). The top panel (**a**) shows electron density n_e, *panel* (**b**) shows electron pressure, magnetic field pressure and total pressure, panel (**c**) shows inferred ion ram speed and panel (**d**) shows spacecraft potential U_{SC}

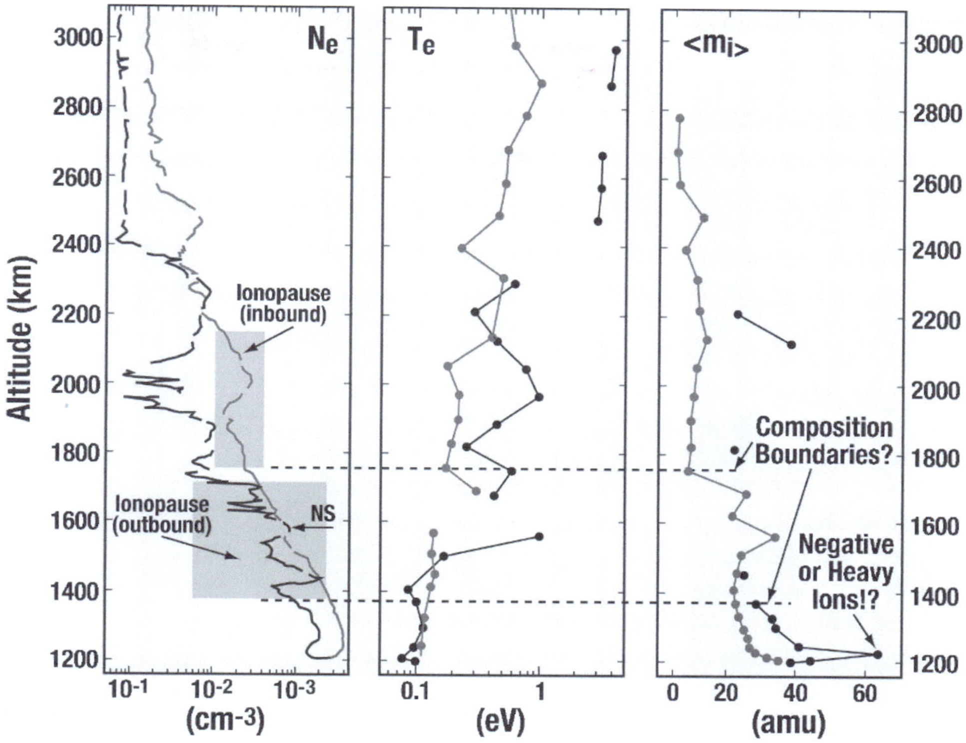

Fig. 16.37 Cassini RPWS LP results for TA flyby by Wahlund et al. (2005). *Left panel* gives thermal electron density versus altitude, *middle panel* gives thermal electron temperature versus altitude and *right panel* gives inferred ion mass

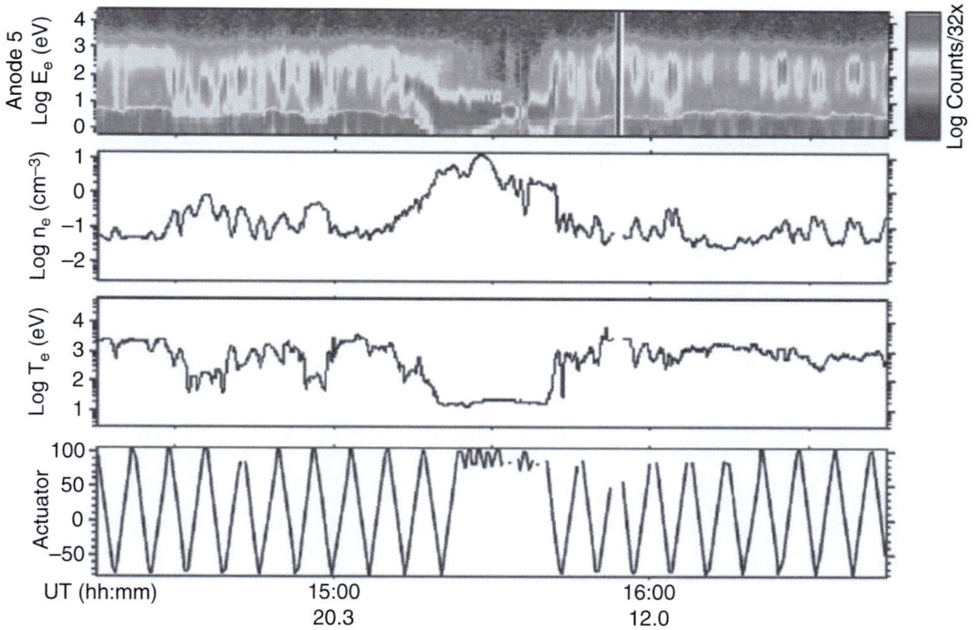

Fig. 16.38 Plots of CAPS ELS electron observations for TA flyby. *Top panel* gives colorized electron energy spectrum versus time, next *two lower panels* give electron density and electron temperature for E>1 eV and *bottom panel* gives actuator angles as function of time (i.e., CAPS works in windshield wiper mode for 2π steradian FOV coverage). These electron observations represent either magnetospheric electrons or ionospheric suprathermal electrons as photoelectrons or secondary electrons

The draping of magnetic field lines did not begin until 15:10 SCET (r ~ 8,400 km) (Backes et al. 2005). Therefore, the ingress mass loading region defined by CAPS was significantly more extended than that defined by the LP measurements. These results do show that the interaction confines the magnetospheric plasma to heights well above Titan's ionosphere and why solar UV and soft x-rays are the dominant source for ionizing the ionosphere for SZA less than 90° (see Cravens et al. 2005).

During egress CAPS observed a rapid exit from the ionosphere at 15:40 SCET (r ~ 5,600 km) after which the flow quickly recovered to speeds near corotation speeds. This is similar to that observed by Voyager 1 (Hartle et al. 1982). At the same time CAPS-ELS observed rapid heating across this boundary from $T_e \sim 10$ eV to $T_e \sim 1{,}000$ eV. Wahlund et al. (2005) noted that around 15:35 SCET, inside the boundary just noted above, a sharp rotation in the magnetic field occurred which indicated an electrical current within this neutral sheet crossing where they observed a localized enhancement in electron density $n_{e,c} \sim 850$ el/cm³, temperature $T_{e,c} \sim 10{,}000$°K and electron pressure $P_{e,c} \sim 900$ eV/cm³. One can also see a sharp increase in the magnetic pressure to about $P_B \sim 150$ eV/cm³ with the plasma beta $\beta \sim 10$ within this current layer. Magnetic field pile up is probably occurring within this region similar to the ingress crossing at 15:23 SCET. As discussed in a preceding paragraph, this is the same region where magnetospheric electrons have access to the ionospheric plasma where they can heat the thermal electrons to $T_{e,c} > 10{,}000$°K. This high plasma beta is a general feature of Titan's ionosphere where $P_e \sim 400$ eV/cm³ at density peak and the magnetic pressure is only $P_B \sim 20$ eV/cm³ with $\beta \sim 20$. These results are contrary to that modeled by Cravens et al. (1998), which can be traced to the magnetic diffusion rate which is related to the ionospheric conductivity. Evidently, the diffusion of the magnetic field into the ionosphere is very low. The energy flux of the magnetic field can be traced to the pointing flux $\mathbf{S} = 1/4\pi\mathbf{EXB} \sim (B^2/4\pi)\mathbf{V} \sim 2.4 \times 10^{-3}$ ergs/cm²/s (B ~ 5 nT and V ~ 120 km/s). Within the ionosphere egress outer boundary, where pileup is thought to occur, the magnetic field energy density $P_B \sim 150$ eV/cm³ $\sim 2.4 \times 10^{-10}$ erg/cm³; since flow speeds are <5 km/s within this region (see Hartle et al. 2006a,b), most of the magnetic field energy flux must be deflected around the topside ionosphere outer boundary. Since, the magnetospheric electrons are tied to the field, gyro-radii ~3–4 km, most of this energy does not penetrate into the ionosphere where they would otherwise contribute to the ionization.

Since INMS did not provide ion measurements for TA (Waite et al. 2005), Cravens et al. (2005) had to model Titan's ionosphere for TA using the LP electron density and temperature measurements and CAPS-ELS measurements available at the time. There are CAPS-IBS positive ion measurements of the ionosphere of lower mass resolution than that provided by INMS, but those results have not yet been published. This model included the ionizing effects of solar EUV-FUV and solar soft x-rays. As they discussed, the solar input was generally in late 2004 with F10.7 proxy indices <100. For TA F10.7 was ≈138. For their Solar 1 case they used SOLAR2000 irradiance model with F10.7≈85 (Tobisca et al. 2000). They also used the part of the solar soft x-ray spectrum described in Maurellis et al. (2000). For their Solar 2 case they enhanced the Solar 1 solar spectrum by 1.6 for F10.7≈138 (TA value). To model the magnetospheric electrons ($n_{e,h} \sim 0.1$ el/cm³ and $T_{e,h} \sim 100$ eV) they used the two-stream model by Nagy and Banks (1970) which was later applied by Gan et al. (1992) to Titan. They also investigated a series of $T_{e,h}$ from 25 eV to 200 eV. When modeling

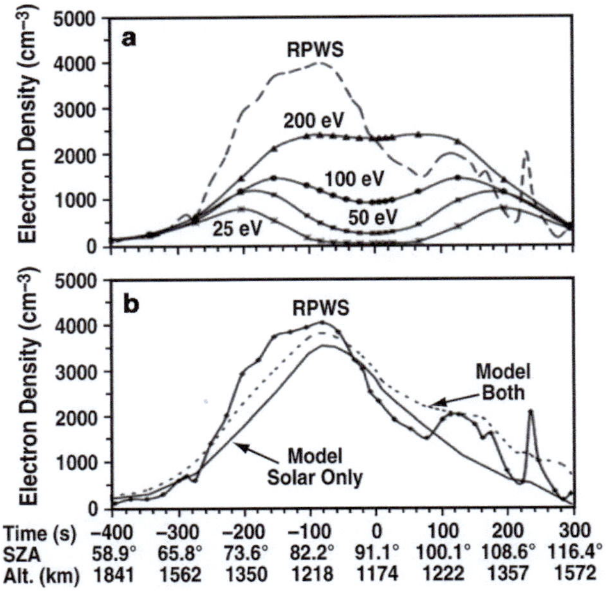

Fig. 16.39 Panel a is Cassini TA flyby observed RPWS ionospheric electron densities and model determined electron densities due to ion production from magnetospheric electrons precipitating into Titan's upper atmosphere assuming parabolic draped magnetic field geometry from Cravens et al. (2005). Panel b is the same, but here is shown RPWS observed electron density versus time and model electron densities from Cravens et al. (2005) using only ion production from solar UV or ion production from solar UV and observed magnetospheric electrons

the electron interaction they used parabolic geometry for the magnetic field, similar to that used by Gan et al. (1992) to model draped magnetic field lines. The results of Cravens et al. (2005) calculations are summarized in Fig. 16.39. The results show that solar input dominates except for periods when in shadow where magnetospheric electrons may be more important. The results shown in Fig. 16.39 are confined inside t ~ +300 s (15:35 SCET) where magnetospheric fluxes are undetectable and ionospheric photoelectrons dominate. We will return to this point below when we discuss T5 flyby results.

16.7.2.1.2 Cassini T5 Flyby

During the T5 flyby, the INMS measured ion composition in Titan's ionosphere (Cravens et al. 2006). As discussed in Sittler et al. (2009b) and Hartle et al. (2008b) T5 represents a high energy input case for the magnetospheric plasma which is dominated by 5–10 keV water group ions. Here, Saturn's magnetosphere is in its magnetodisk configuration and Titan must be close to the magnetic and centrifugal equator of the magneosphere where the heavy ion plasma dominates and hot magnetospheric plasma is also present (see Cravens et al. 2008a). In addition, as shown in Fig. 16.7, the encounter geometry was a north polar pass with wake

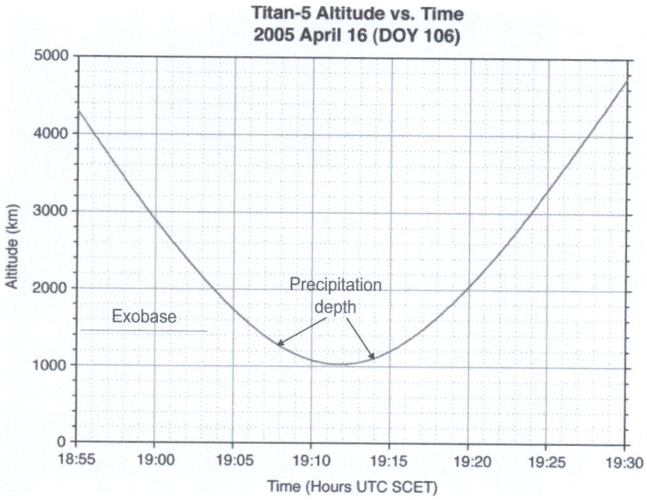

Fig. 16.40 Plot from Cravens et al. (2009) of Cassini altitude versus time for T5 flyby with exobase height indicated

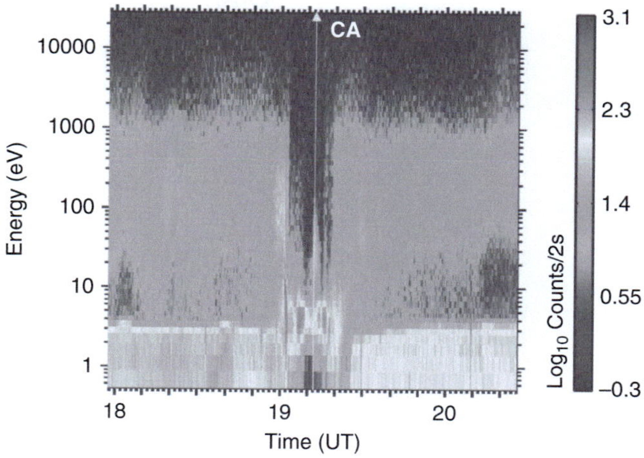

Fig. 16.41 Energy-time CAPS ELS electron energy spectrum for T5 flyby from Hartle et al. (2006c). Time of closest approach (CA) is indicated

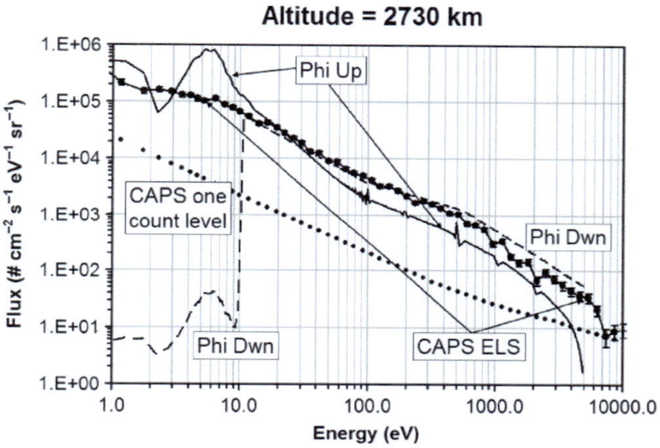

Fig. 16.42 Plot of CAPS ELS magnetospheric electron fluxes for altitude ~2,730 km from Cravens et al. (2009)

side in sunlight and ram side in shadow (i.e., SRA ~ 180°). Closest approach occurs at 19:11:48 SCETI as shown in Fig. 16.40 from Hartle et al. (2006c). Therefore, this encounter provides the unique opportunity to study the ionization of Titan's ionosphere by magnetospheric electrons with less complication from solar input. Furthermore, the shadow side is where the upstream flow impacts the upper atmosphere and magnetic field pile up occurs; however, the magnetic field is not completely shielded from the ionosphere, allowing penetration of the field and magnetospheric electrons into Titan's ionosphere. The ionospheric ion composition reported by Cravens et al. (2006) was rich in hydrocarbons $C_nH_m^+$ with (m,n) as high as (6,7) and nitrogen-bearing species such as $CH_2NH_2^+$ (Vuitton et al. 2006). Ion and electron densities were as high as ~1,000 el/cm^3, with dominant ions having mass M/Q = 17, 18, 28, 29, 41 and 51. The details of understanding the ion composition is given in Chapter 11 on ionospheric composition by Cravens et al. Here we note, that previous models were mostly based on photochemical models, while in this case magnetospheric electrons may dominate the ionization in the shadow zone.

As discussed in Cravens et al (2006), using experience with the ionospheres of Venus and Mars (Nagy and Cravens 2002), transport of ionospheric flow from dayside to night side normally occurs, but in Titan's case the magnetospheric flow during T5 is from night side to dayside, so the Venus Mars analogy may not be working. The analysis of the T5 LP data by Ågren et al. (2007) does suggest a day-night flow ~ 1-3 km/s over the poles on the night side of the interaction. The more recent paper by Cravens et al. (2009) shows the CAPS-ELS observations for T5 (Fig. 16.41). Outside Titan's ionosphere there are significant magnetospheric electron fluxes from 100 eV to 2 keV; a detailed energy spectrum from Cravens et al. (2009) is shown in Fig. 16.42. Similar to the boundary observed for protons in Hartle et al. (2006c), the magnetospheric electrons are confined outside z ~ 2,200 km inbound (19:03 SCET) and z ~ 1,750 km outbound (19:18:40 SCET). See Fig. 16.40 showing altitude versus time. Within this region, as reported in Cravens et al. (2009), there are <5 eV ionospheric photoelectrons on the sunlit side, similar to that reported by Coates et al. (2007b) for the T9 flyby, but on the nightside, secondary electrons of similar energy <5 eV are observed. The ionospheric main peak as defined by the LP measurements reported by Ågren et al. (2007) in Fig. 16.43 is confined between 19:02:30 SCET and 19:22 SCET. Inbound the magnetospheric electrons penetrate slightly across the ionospheric outer boundary, while outbound, when in shadow they do penetrate into the region defined by Ågren et al. (2007) as the exo-ionosphere. It is also in this region centered on 19:19 SCET that the magnetic field is piled up and

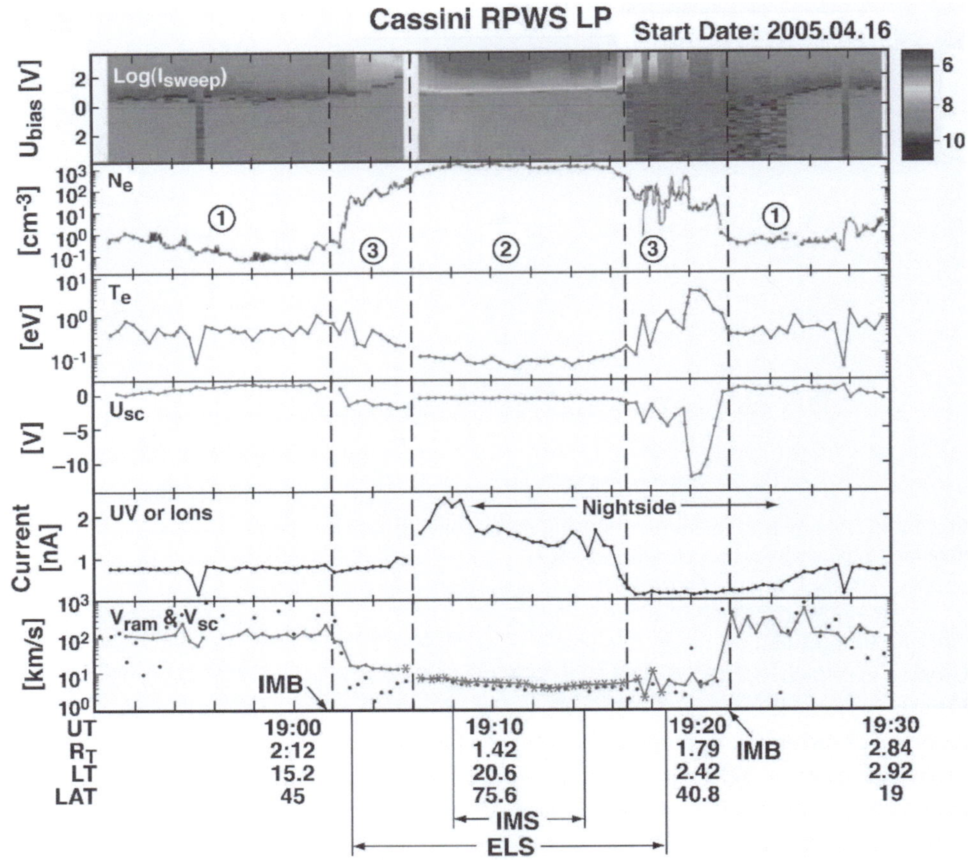

Fig. 16.43 Plot of Cassini RPWS LP thermal electron properties for T5 as reported by Ågren et al. (2007). *Top panel* shows the observed LP current versus sweep voltage on probe, *second, third* and *fourth panel* down gives thermal electron density, thermal electron temperature and spacecraft potential. *Fifth panel* down gives positive current due ions or photoemission electron current from solar UV, and *bottom panel* gives inferred ion speed

the field is able to penetrate into the topside ionosphere of Titan. As reported in Hartle et al. (2006c) the ions within this pile up region are compressed and heated. This probably explains the heating of the colder ionospheric electrons from $T_{e,c} \sim 1{,}000°K$ to $T_{e,c} \sim 10{,}000°K$ with peak values as high as $T_{e,c} \sim 50{,}000°K$ ($T_{e,c} \sim 5$ eV) and relatively high electron densities $n_{e,c} \sim 100$ el/cm^3. A similar region was observed inbound by Ågren et al. (2007), but the densities were lower $n_{e,c} \sim 50$ el/cm^3 and the mean thermal electron temperature $T_{e,c} \sim 5{,}000°K$. Here, photoelectron heating is probably more important with electron-neutral collisional cooling at exospheric heights being less important ($z > 1{,}400$ km).

Ågren et al. (2007) performed a detailed analysis of the LP data for T5 (Fig. 16.43). The CAPS-ELS electron densities in upstream magnetospheric flow is $n_{e,h} \sim 0.1$ el/cm^3. The peak ionospheric electron densities, $n_{e,c} \sim 1{,}000$ el/cm^3, measured by the LP are very close to that observed by INMS (Cravens et al. 2009). But most importantly, they developed a simplified ionospheric model with only electron impact ionization. The model of the neutral atmosphere only included the most important species N_2, CH_4, HCN, C_2H_4 and HC_3N. The N_2 and CH_4 were derived from INMS TA observations reported by Yelle et al. (2006) and C_2H_4 and HC_3N from photochemical model by Toublanc et al. (1995). They used the 3-D General Circulation Model (GCM) by Müller-Wodarg et al. (2003) to infer the HCN altitude dependence. This GCM model uses solar EUV as the energy source, but in a more suitable model precipitating electrons and ions must be used. The HCN is important since it represents an important cooling mechanism for Titan's thermosphere (Yelle 1991). For magnetospheric electrons they used CAPS-ELS fluxes similar to that shown in Fig. 16.42 (This ELS spectrum from Cravens et al. 2009, is four times less intense than reported in Ågren et al. 2007 due to a re-calibration of the ELS). To estimate the ion production they used the aurora model by Lummerzheim (1987) and Lummerzheim and Lilensten (1994). For the electron transport and energy loss they used 36 eV per ion formation from Rees (1963) and a nitrogen-methane atmosphere transport code from Galand et al. (2006). Both yield similar results shown in Figs. 7 and 9 of their paper. For the relevant recombination coefficients they used values from Keller et al. (1998) and the LP estimated electron temperatures $T_{e,c} \sim 700°K$. Their final results used the CAPS-ELS energy spectrum noted above and yielded results shown in Fig. 16.44. But when compared to observed electron densities they were an order of magnitude too high. In order to get good agreement they had to down grade the magnetospheric electron fluxes by a factor of 10. Some of this over-estimate can be explained by an older calibration of the CAPS-ELS as noted above. Therefore, the

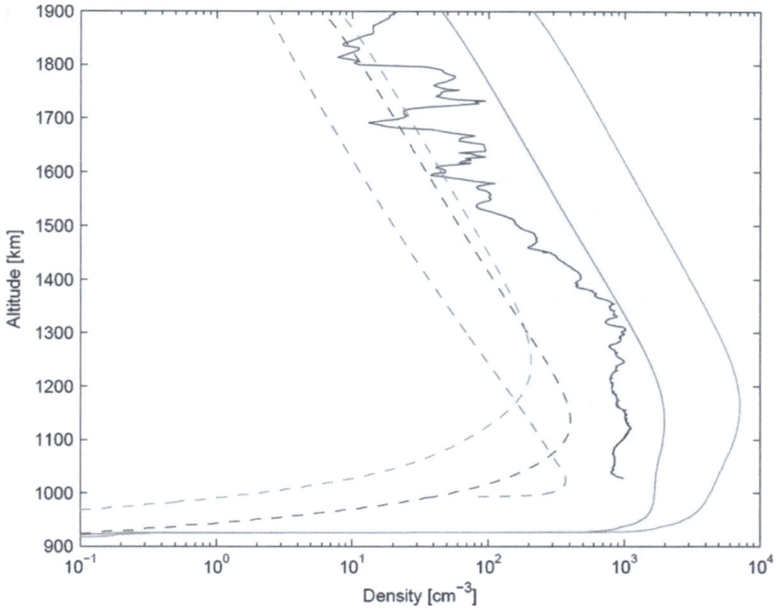

Fig. 16.44 Results of simplified model of ion production of Titan's upper atmosphere for the T5 flyby with magnetospheric electrons dominating the ionization. The model also includes atmosphereic cooling and recombination to construct model electron densities versus height. Model and observed electron densities shown

over-estimate is ~factor of 2.5. The remaining over-estimate is probably caused by not using parabolic geometry for draped magnetic field lines and by not considering the limited access of the magnetospheric electrons across draped field lines.

The CAPS-ELS observations show that the hotter 10–100 eV electrons cannot penetrate the magnetic barrier. The keV electrons can centrifugal drift across field lines as first suggested by Hartle et al. (1982) and later applied by Strobel et al. (1992). Also, as shown by Sittler and Hartle (1996) for Triton, the magnetospheric electrons will gradient B drift orthogonal to the magnetic field around the ionosphere and not penetrate inside. If draped field lines are horizontal then centrifugal drift will transport them across field lines, but will readily be emptied by inelastic collisions. So, the effective flux of electrons is limited by the rate they can centrifugal drift across the field lines (Strobel et al. 1992). To see this better we refer to Eqs. 16.1 and 16.2 for gradient and centrifugally drift from Sittler and Hartle (1996) and ½ bounce period for keV magnetospheric electrons in Eq. 16.3.

$$\vec{V}_G = \frac{E(eV)\sin^2 \alpha_{PA}}{B(nT)\vec{L}_G(km)} 10^3 \, km/s \qquad 16.1$$

$$\vec{V}_C = \frac{2E(eV)\cos^2 \alpha_{PA}}{B(nT)r(km)} \hat{y} 10^3 \, km/s \qquad 16.2$$

$$\tau_{B,e}/2 \sim 2.7 R_S L/v_e \sim 173\,s @ 1\,keV \qquad 16.3$$

In the case of gradient drift in Eq. 16.1 the scale length $\mathbf{L}_G \sim 100$ km, E(eV) ~ 1,000 eV, pitch angle $\alpha_{PA} \sim 45°$, B(nT) ~ 10 nT one gets $V_G \sim 500$ km/s. In the case of centrifugal drift the radius of curvature r ~ 4,000 km one gets $V_C \sim 25$ km/s. If one assumes keV electrons at ram side they will gradient drift ~πr(km) ~ 12,560 km to the wake side and be lost down the tail if not lost to the atmosphere. During most of the bounce period there will be very little gradient or centrifugal drift, except near Titan where the gradient and curvature scale lengths are small enough. It takes t ~πr(km)/v_e ~ 0.67 s for keV electron to pass through draped field lines and will move ~ 335 km horizontally for each ½ bounce period due to gradient drift and ~17 km vertically per ½ bounce period and will take ~12,560/670 km ~ 18 $\tau_{B,e}$ to drift around Titan and will move vertically ~18 × 34 km ~ 612 km. It takes the keV electrons ~18 × 173s × 2 ~ 6,200 s ~ 1 h 44 min to drift around Titan. Considering the total path length for the electrons they may drift vertically ~300 km or 5 scale heights before being lost. So, they may penetrate down to 1,100 km altitude, which is consistent with the CAPS-ELS data.

At this point it is useful to consider the fate of the keV electrons and their interaction with Titan's upper atmosphere by considering the two stream model developed by Gan et al. (1992) for Titan. In Fig. 16.45 we show one of their simulations of electrons on draped field lines at z ~ 1,400 km and case when at z ~ 1,100 km. In both cases they preserve electron fluxes on one end at upstream value; They then added the results for electrons moving in opposite directions. This was done to model the Voyager 1 observations reported by Hartle et al. (1982). Near Titan the keV electron fluxes are highly depleted, while photoelectrons and secondary electrons at energies <20 eV are enhanced. We then see an enhancement of the keV electrons moving away from Titan; these electrons are backscattered from Titan's N_2 and CH_4 neutral exosphere at 10% level. In reality, the electrons are

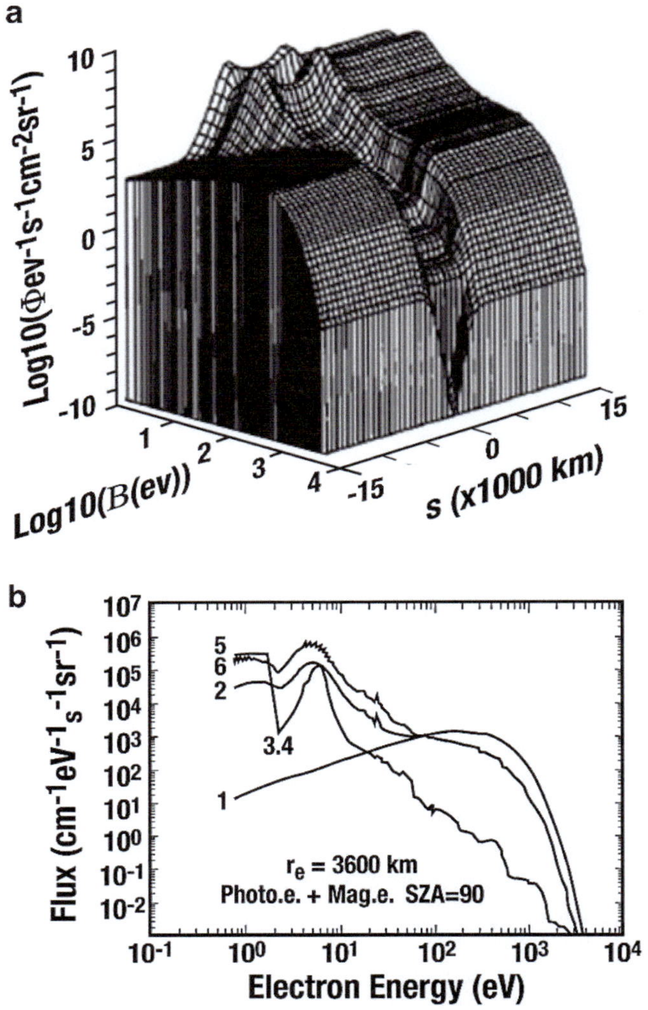

Fig. 16.45 Two-stream model by Gan et al. (1992) of magnetospheric electrons interacting with Titan's upper atmosphere. Panel (**a**) shows 3D plot of electron flux versus electron energy and field aligned distance relative to equatorial plane. Panel (**b**) shows electron flux versus energy at various positions along a magnetic field line

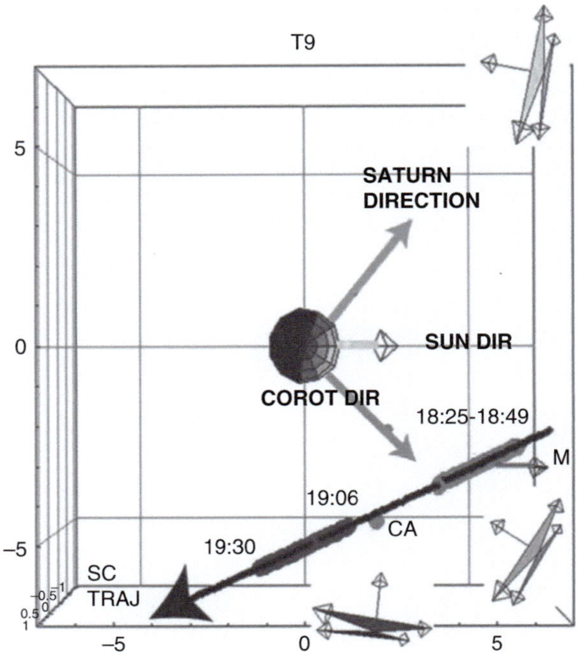

Fig. 16.46 Cassini T9 flyby encounter geometry as given by Szego et al. (2007). Corotation direction, solar direction and Saturn direction are indicated along with spacecraft track and indication of events 1 and 2. Figure also shows CAPS IMS FOV at different points along the Cassini trajectory

not replaced at the end of each field line and are eventually depleted to background levels as observed (see Cravens et al. 2008b). Finally, as discussed in the next section one must also consider heavy keV plasma ion precipitation and energetic ion precipitation to explain the lower ionosphere that can be probed in situ down to 950 km altitude and radio science down to 500 km altitude (Kliore et al. 2008).

16.7.2.1.3 Cassini T9 and T18 Flybys

As discussed above and shown in Sittler et al. (2009b,c) and Bertucci et al. (2007) the T9 flyby represents the low energy limit for magnetospheric input to Titan's upper atmosphere. T9 occurred when Titan was near 0300 LT relative to Saturn and the magnetosphere was in a magnetodisk configuration (Bertucci et al. 2007). Initial plasma ion and electron papers describing the T9 flyby was published by Szego et al. (2007) and Coates et al. (2007b). The observations showed a very complex interaction. The geometry of the flyby is shown in Fig. 16.46 from Szego et al. (2007). This was a distant wake pass ~10,000 km from Titan. In Fig. 16.47 we show the ion fluid parameters from Sittler et al. (2009c). The electron fluid parameters can be found in Fig. 4 in Sittler et al. (2009c). Figure 16.47 also shows one minute averages of the vector magnetic field. The ion flow angle relative to the local magnetic field direction is shown in Fig. 16.48. Figure 16.49 shows the composition measurements from the CAPS-IMS for T9. The data shows a proton only upstream flow near the corotation direction with speed ~200 km/s, but most importantly at right angles to the local magnetic field direction. This alignment is also maintained during the outbound leg and is preserved as both **V** and **B** vary direction. Sittler et al. (2009c) interpret as pickup H^+ and H_2^+ within Titan's extended H and H_2 neutral exosphere with Hill sphere dimensions ~±55,000 km. The proton temperature is $T_p \sim 200$ eV is ~ the local pickup energy which further supports the notion one is seeing pickup ions. The electron temperature $T_{e,h} \sim 100$ eV is close to the proton temperature and they combine to give plasma $\beta \sim 0.13 \ll 1$. Upstream of Titan one sees anti-correlation between V_X and B_X which could indicate linearly polarized Alfven waves propagating along **B** with

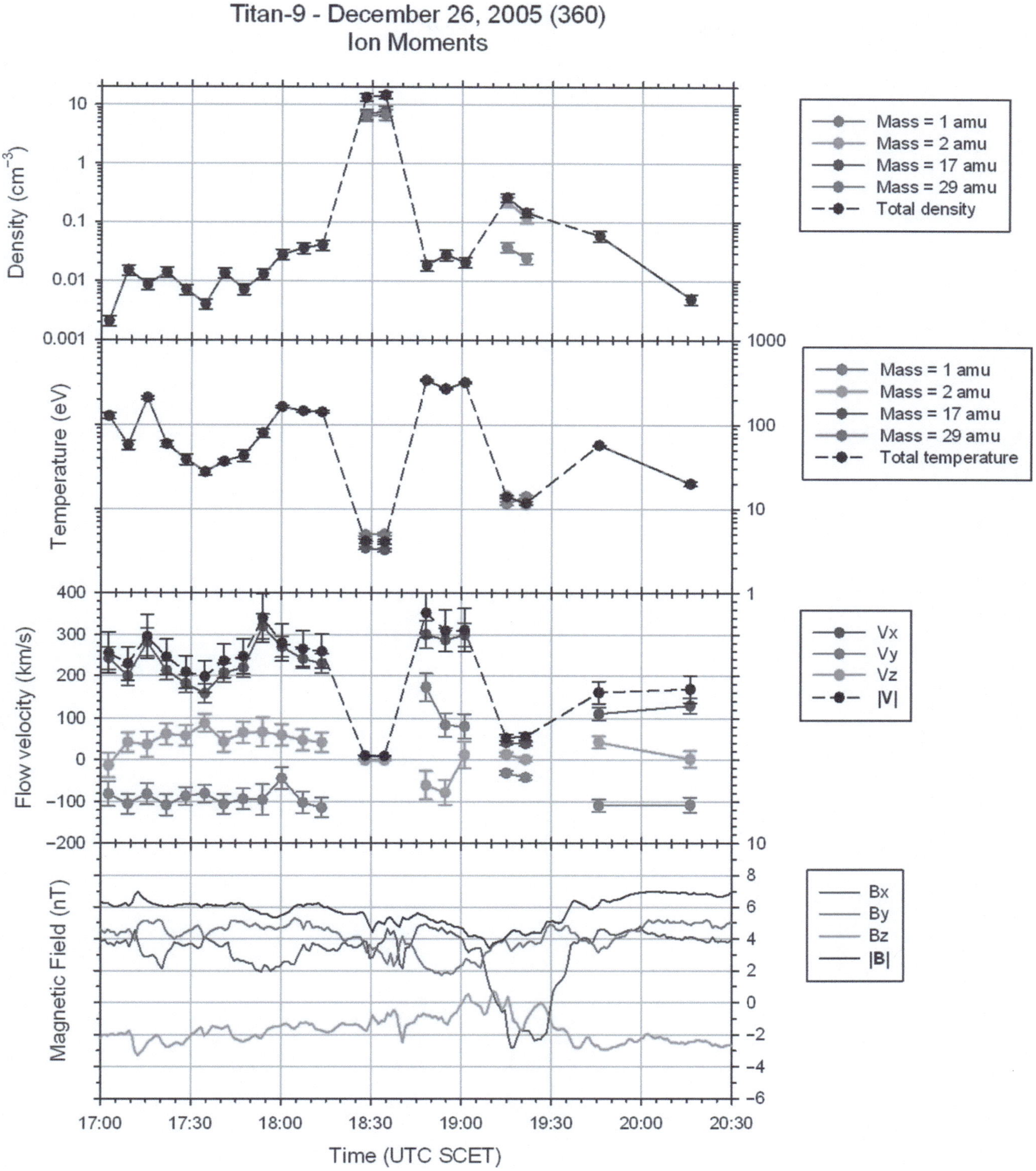

Fig. 16.47 CAPS ion fluid parameters and magnetic field vector amplitudes as a function of time from Sittler et al. (2009c)

$\delta\mathbf{V} = \pm\delta\mathbf{B}/\sqrt{4\pi\rho}$, ($\delta B_X \sim -1$ nT, $n_P \sim 0.01$ ions/cm^3 and using proton composition one get $\delta V_X \sim +200$ km/s). Here, δV_X is observed to be ~ +100 km/s which may indicate the waves are not purely Alfvenic The flow ram energy is negligible relative to the magnetic field pressure. The data also shows positive V_Z for the flow which indicates that Saturn's magnetodisk is moving upward which is consistent with Titan being below the current sheet as observed (Bertucci et al. 2007). Since the proton gyro-radii are small they may not reach the ionosphere and there are no upstream heavy ions

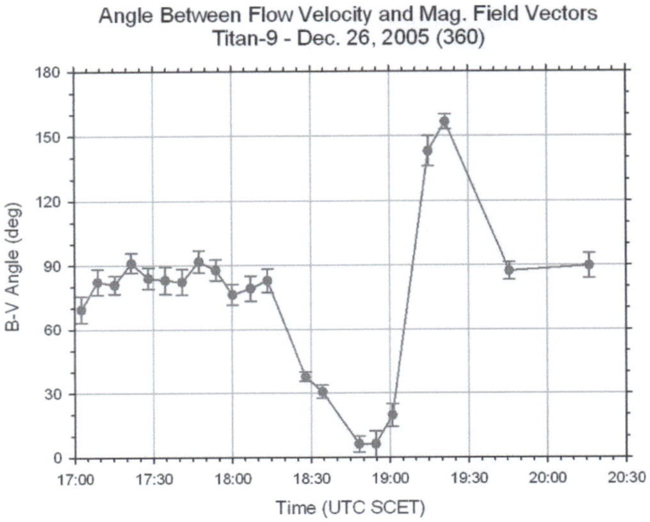

Fig. 16.48 Angle between ion flow velocity and local magnetic field vector versus time for T9 flyby from Sittler et al. (2009c)

which can puncture through this induced magnetopause boundary and reach Titan's ionosphere. Therefore, the only remaining ionization source is solar UV and soft x-rays. This is reinforced by the ionospheric photoelectron observations reported by Coates et al. (2007b) for event 1 and shown in Fig. 16.50 No evidence of secondary electrons from the precipitation of magnetospheric ion-electron populations were reported by Coates et al. (2007b).

Hartle et al. (2008a) and Sittler et al. (2009c) predicted an ionospheric wind using TA ionospheric electron measurements by Wahlund et al. (2005). The topside ionospheric electrons with $T_{e,c} > 1{,}000°K$ are significantly hotter than the ionospheric ions $T_{ion} \sim 200°K$ and are thus energetically able to produce a polarization electric field that accelerates the dominant ions from the topside ionosphere to escape speeds ~ 1 km/s with ion flux ~10^7 ions/cm^2/s. At these heights the magnetospheric and ionospheric photoelectrons can effectively heat the thermal electrons via electron-electron Coulomb collisions and at these higher heights the

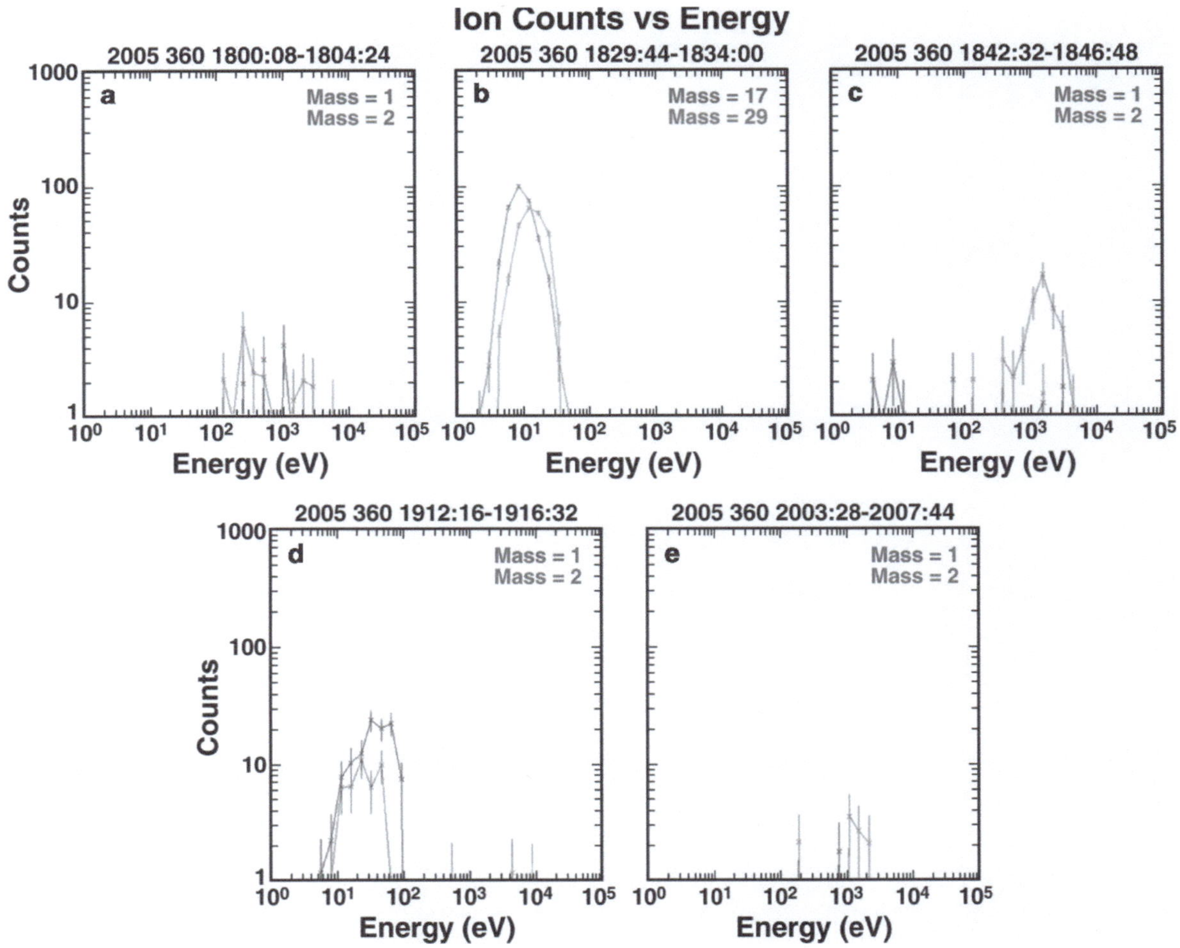

Fig. 16.49 CAPS IMS composition plots for T9 by Sittler et al. (2009c). Different colors are used for mass 1 (*red*), mass 2 (*blue*), mass 17 (*purple*) and mass 29 (*green*) ions. The plot is in format of ion counts versus ion energy in eV

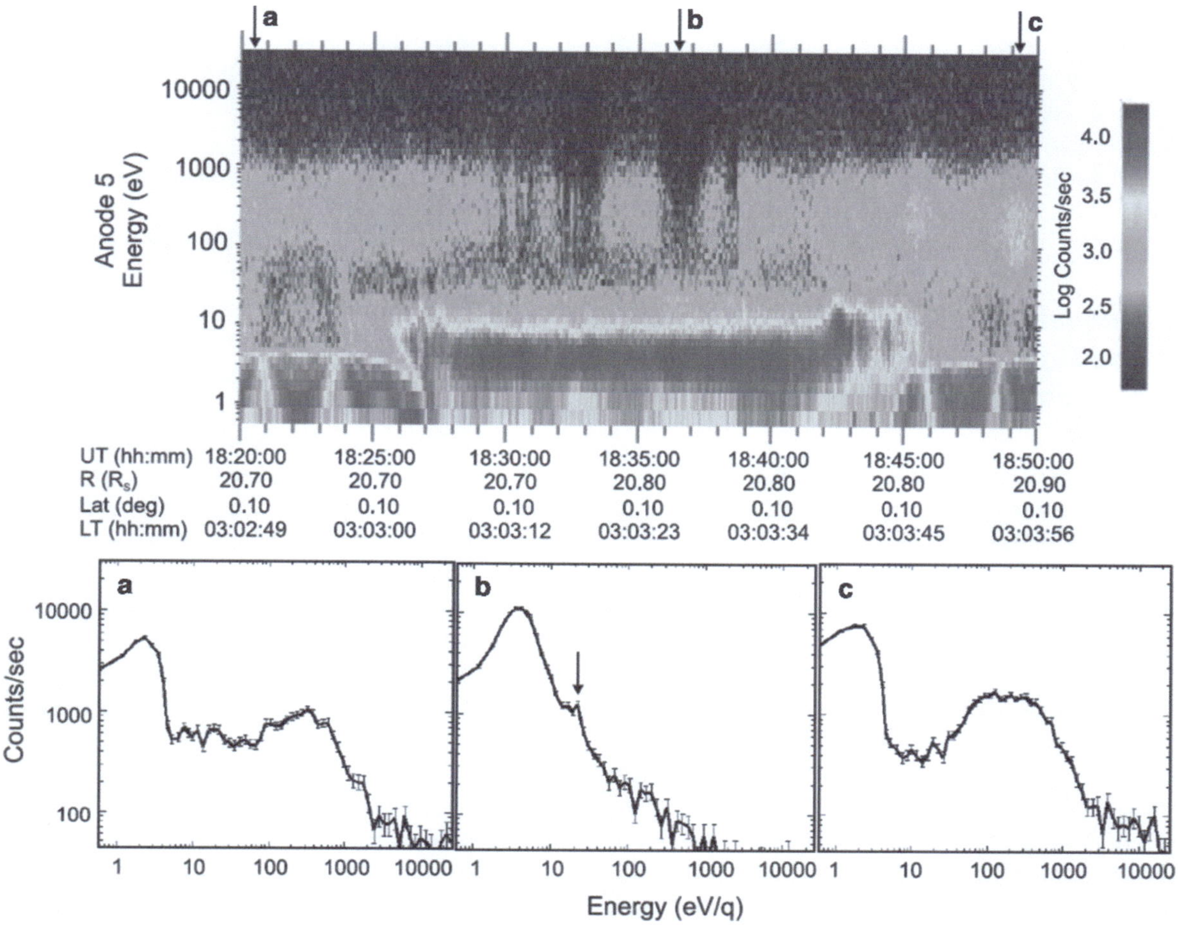

Fig. 16.50 CAPS ELS electron energy-time color spectrogram (*top panel*) and electron energy spectrum cuts at various times (*bottom panels*) for T9 from Coates et al. (2007b)

electron-neutral collisions are not sufficient to cool the electrons, so they are heated to $T_{e,c} > 1{,}000°K$ as observed (Wahlund et al. 2005). This ionospheric wind should be most important on the sunlit side and primarily move along field lines away from Titan where solar UV is the dominant energy source. This ionospheric wind will tend to move along draped field lines into Titan's induced magnetotail as observed by Voyager 1 (Hartle et al. 1982). As reported in Sittler et al. (2009c), one sees during event 1 an ionospheric wind moving within 30° magnetic field lines within Titan's induced magnetotail (see Figs. 16.47–16.49) with total ion density ~10 ions/cm^3, ion flow speed ~8 km/s; and ion temperature T_{ion} ~4 eV. The ion densities are very close to those estimated from the LP data by Modolo et al. (2007a). The composition data shows this plasma dominated by mass ~17 and mass ~29, which is consistent with CH_5^+ and $C_2H_5^+$, respectively. If the ionospheric ion temperatures are ~150°K, they must experience considerable heating before being observed by Cassini. The flow speeds are considerably higher than that estimated at the topside ionosphere by Hartle et al. (2008a) and Sittler et al. (2009c); however, the upward fluxes times the respective escape areas are expected to be similar in the topside ionosphere and the tail. The photoelectrons from the ionosphere will also tend to escape along field lines, they have comparable pressures as the thermal electrons previously discussed and therefore may provide the further acceleration of the ionospheric wind at the higher altitudes sampled during T9 event 1. Wei et al. (2007) using magnetometer and LP data also concluded that ionospheric escape must be occurring for event 1. During event 2 the ion composition is H^+ and H_2^+ which is consistent with photoionization of Titan's H_2 exosphere and lack of penetration of magnetospheric electrons. But, Sittler et al. (2009c) found for a single ion composition mass spectrum clear evidence for a slowly moving mass 16–17 ion population during event 2 (see Figure 9b of Sittler et al. 2009c). One explanation is pickup methane ions from Titan's exosphere are being observed. But, one cannot rule out observance of ionospheric ions CH_5^+, but the mass 29 ionospheric ions $C_2H_5^+$ are not observed. If ionospheric ions, then scavenging of the ionosphere is occurring and the Dobe et al. (2007) two-stream model would apply. Here, the flow is nearly aligned with the field similar to that for event 1. The light ions, H^+ and H_2^+, for event 2 are moving away from Titan with speed ~30 km/s, ion density ~0.5 ion/cm^3, T_{ion} ~30 eV and the magnetic field orientation consistent with draped field lines that are connected with Titan (Bertucci et al. 2007).

Intermediate to events 1 and 2 one sees field aligned protons moving with speeds ~300 km/s, temperature $T_P \sim 500$ eV and density $n_P \sim 0.02$ ions/cm³. The reason why the flow is field aligned is not known at this time, but could be serendipitous since main component of flow is along the corotation direction and the field is twisted primarily in the corotation or x direction.

16.7.3 Magnetospheric Heavy Ion and Suprathermal Ion Energy Input to Titan's Upper Atmosphere

16.7.3.1 Voyager and Cassini Plasma Observations of Ion Precipitation with Titan's Upper Atmosphere

Sittler et al. (2004, 2005a), in a re-analysis of Voyager 1 Titan flyby plasma data, observed finite gyro-radius effects in the high energy component originally identified by Hartle et al. (1982) as a heavy ion component with rotational flow speed ~120 km/s and temperature T ~ 2.9 keV. In the moving frame these heavy ions were identified as either N^+ or O^+ have gyro-radii ~5,600 km. Along the Voyager 1 trajectory, Fig. 16.1, are shown the position of spectra 1 to 8 in Fig. 16.51. As can be seen during the inbound pass an attenuation of this high energy component for arrival directions coming from the corotation direction occurred when spacecraft was within 10,000 km from Titan's exobase or a gyro-diameter above the exobase. Here, the ion guiding centers are on the Titan side of the spacecraft. Shortly after this the low energy component started to disappear when within a gyro-diameter for protons. During the outbound pass, the low energy component reappeared first and then the high energy component. The recovery on the outbound pass was within a gyro-radius since now the guiding centers were on the side further from Titan. So, in this sense, Titan acted as an ion mass spectrometer, clearly showing that the low energy component was protons and the high energy component was heavy ions. These observations showed for the first time the finite gyro-radius effects of Saturn's magnetosphere with Titan. These observations also showed the presence of a much larger clearing zone of magnetospheric plasma than one would estimate using a fluid approximation and effectively increases the ion energy flux to Titan's upper atmosphere. These observations also showed that the keV heavy ions are essentially unmagnetized and can easily cross field lines and deposit their energy directly to Titan's upper atmosphere. If they were essentially deflected around the obstacle one would not see the observed attenuation of the heavy ion fluxes.

Then for TA, Hartle et al. (2006a,b) observed a similar clearing zone around Titan, with the heavy ion component being swept away when within a gyro-diameter of the exobase. CAPS-IMS observations of the magnetospheric plasma ion composition before the TA flyby, shown in Fig. 16.52, indicate a heavy ion component at keV energies that is dominated by water group ions (O^+, OH^+ and H_2O^+). The signature for water group ions is the O^- fragment in the ion composition energy spectra of Fig. 16.52. Also shown is the presence of H^+ and H_2^+ ions from 100 eV to a few keV. The O^- fragment or 'finger print' occurs when the O^+ ion is accelerated inside the IMS by -14.8 kV and passes through a carbon foil where it charge exchanges to a negative oxygen ion ($O^+ - > O^-$). Then due to the electric fields within the instrument, the O^- fragment has a shorter TOF

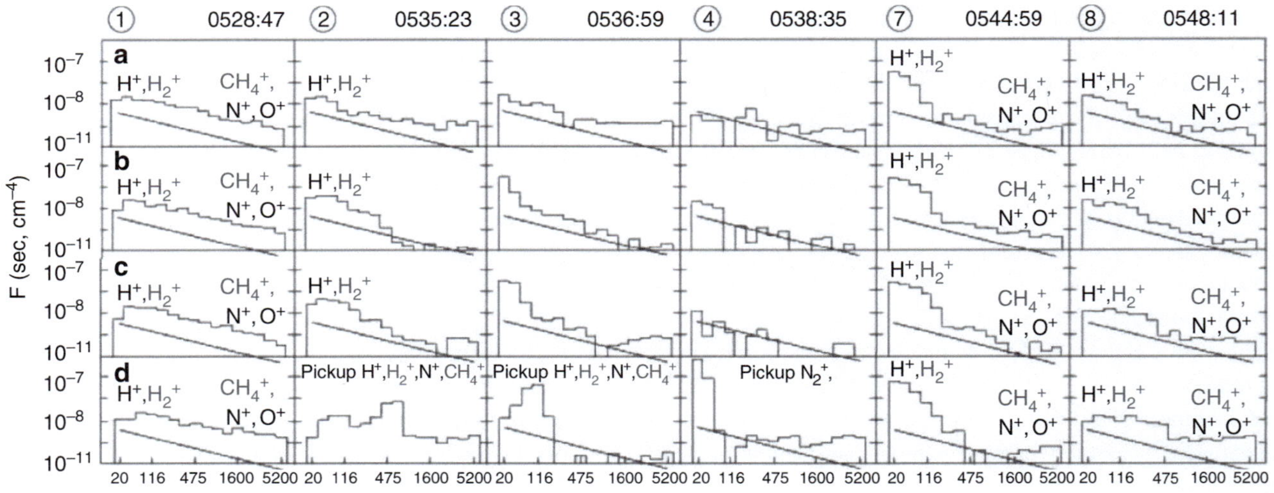

Fig. 16.51 Voyager 1 PLS ion spectra from both Sittler et al. (2005a) and Hartle et al. (2006b). Figure shows approximate position of light (H^+, H_2^+) ions and heavy (CH_4^+, N^+, O^+) ions in energy for the six ion spectra displayed

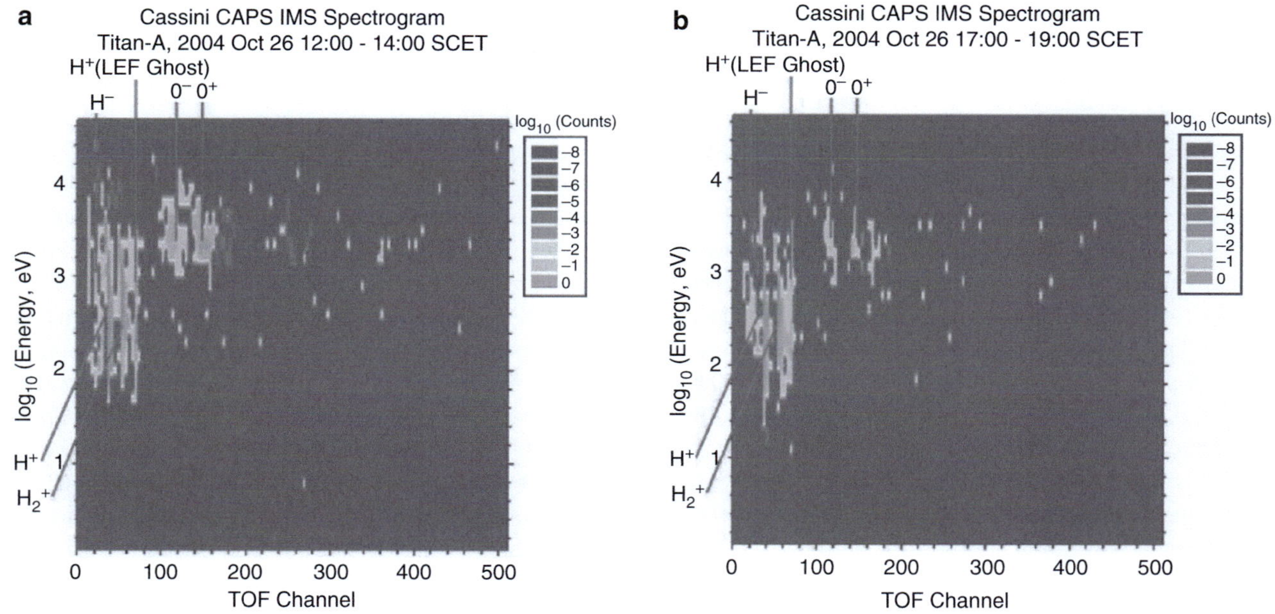

Fig. 16.52 Energy-TOF color spectrograms of the magnetospheric ions as measured by Cassini CAPS IMS both before (panel **a**) and after (panel **b**) of the TA flyby. Presence of O^- fragment peak indicates presence of water group ions with Enceladus as the likely source of the plasma

than say its neutral fragment ($O^+ - > O^0$) exiting the instrument's carbon foil. If the incident ion was N^+ or CH_X^+, one would not see this O^- fragment. This is a very powerful feature of the CAPS-IMS. Since, the water group ions have keV energies, their origin must be the Enceladus torus (see Sittler et al. 2005b, 2006b, 2008) where they charge exchange and become fast neutrals in Saturn's outer magnetosphere where they are eventually ionized and attain keV pickup energies (Johnson et al. 2005; Sittler et al. 2006a). Hartle et al. (2006b) was then able to estimate a precipitation rate of oxygen ions to Titan's upper atmosphere ~9.5×10^{23} O^+/s. Then if the mean energy of the magnetospheric oxygen ions is ~2 keV, the total energy input to Titan's upper atmosphere is ~3×10^{15} ergs/s. This translates to an energy flux ~ 1.4×10^{-3} ergs/cm²/s to Titan's atmosphere at exobase heights r ~ 4,125 km.

As shown above, Hartle et al. (2006a,b) also detected CH_4^+ pickup ions as ion beams coming from Titan during inbound pass when $\mathbf{E}_{\mathbf{conv}}$ was pointed away from Titan (see Section 16.7.1). On the Saturn side of Titan $\mathbf{E}_{\mathbf{conv}}$ points toward Titan and these pickup ions can deposit their pickup energies into Titan's upper atmosphere. Hartle et al. (2006a,b) did not estimate a quantitative energy flux from the pickup ions, which is deferred to a later study. One can estimate a pickup ion loss rate ~5×10^{22} ions/s from the Sittler et al. (2005a) model calculations. But, as modeled by Michael et al. (2005) using hybrid simulations by Brecht et al. (2000), they included the energy flux from the pickup ions as an additional energy source for sputtering loss of non-thermal N and N_2 neutrals from Titan and formation of a nitrogen torus around Saturn (Smith et al. 2004; Sittler et al. 2006a). The estimated source of nitrogen to this torus is ~4.5×10^{25} N/s (Michael et al. 2005) or about 6×10^{26} amu/s which is much less than the loss rates ~4–5×10^{28} amu/s Strobel (2008) can produce from his hydrodynamic loss mechanism. This non-thermal escape of neutrals due to ion sputtering can be in the form of power law distributions for the non-thermal recoils and dissociation of N_2 (Michael et al. 2005 and also see De La Haye et al. 2007). As noted in Michael and Johnson (2005), forward scattering along the flanks may be important, but to be sure, a 3D model of the flow is required. The total energy input estimated by Michael and Johnson (2005), including deflected magnetospheric heavy ions and molecular pickup N_2^+ ions is ~4.8×10^{-3} erg/cm²/s at 1,200 km altitude which is a few hundred km below the exobase. In an earlier paper, Lammer et al. (1998) modeled an energy flux of 1.2×10^{-3} erg/cm²/s due to sputtering of N^+ ions. The pickup N_2^+ ions dominate the energy input in the Michael and Johnson (2005) calculations. Their estimate is about a factor ~3.4 greater than that estimated from the precipitation flux of O^+ by Hartle et al. (2006b). The advantage of this energy deposition mechanism with regard to atmospheric loss is that it occurs just below the exobase, while solar EUV-UV deposits most of its energy at lower altitudes ~800–900 km, where the atmosphere is more collisional and deeper in the gravitational well of Titan. The nitrogen and oxygen ion energy fluxes discussed here are expected to change with Titan orbital positions and magnetospheric variations. The magnitude of these variations should become clear as future Cassini observations become available.

16.7.3.2 T5 Ion Energy Deposition Results

Cassini's T5 encounter provides an opportunity to quantify magnetospheric ion and electron precipitation with Titan's upper atmosphere since its upstream side is in shadow and one can more easily separate solar ionizing radiation from magnetospheric charged particle ionization effects. Also, as discussed in Hartle et al. (2006c) the flyby geometry allows one to quantify the depth of the penetration from close in when in the wake (inbound) and near the ram side of the interaction (outbound) as shown in Fig. 16.7. The figure also shows T5 to be a north polar pass. As discussed in Sittler et al. (2009b), Hartle et al. (2006c) and Cravens et al. (2008, 2009), Titan was probably near the magnetospheric current sheet near dawn LT, where heavy ions are prevalent and energy input from the magnetosphere to Titan was at its high energy limit.

For T5, we show in Fig. 16.53 from Hartle et al. (2006c) the ion singles count rates in an energy time spectrogram format where all 8 angular sectors of the CAPS-IMS have been added together. This figure only shows signal above 50 eV so the ionospheric plasma is not visible, but the region of low flux is where the spacecraft passes through Titan's ionosphere. The vertical lines define the lower boundary of penetration at 19:08 SCET (inbound) and 19:14 SCET (outbound). Referring to Fig. 16.42 this corresponds to altitudes z ~ 1,200 km inbound and z ~ 1,100 km outbound. At energies between 100 eV to 500 eV are light ions H^+ and H_2^+, while from 5 keV to 10 keV are oxygen ions (i.e., O^- peak definitely present). The higher intensity fluxes for the O^+ ions outbound could be due to the spacecraft being nearer the ram side of the interaction where the field has piled up and the ion fluxes are intensified within this region. This pile up in magnetic flux is shown in Hartle et al. (2006c) and Ågren et al. (2007). At lower energies the protons are confined to higher altitudes which could partially be due to their being more magnetized. In Fig. 16.54 from Hartle et al. (2006c) we show the decrease in heavy ion fluxes with height and its energy dependence. As expect, the energy peak moves to lower energy with decreasing altitude, reaching fluxes near the background at z ~ 1,100 km. As discussed in Cravens et al. (2008) for oxygen energies <100 keV most of the oxygen will be neutralized with O^- fraction ~5%. The CAPS-IMS can only observe positive ions, so even if these ions can penetrate below 1,100 km, CAPS-IMS cannot see them. But, if there are negative ions from charge transfer collisions CAPS-ELS should be able to see these ions at depths below that observed for the magnetospheric electrons. So, this might set an upper limit for O^- fragments exiting Titan's upper atmosphere. In order for CAPS-IMS to view these ions their trajectories must be nearly tangent with Titan's atmosphere. The equivalent vertical depth $\Delta z \sim H \log_e(1+2^{1.5}*\mathrm{sqrt}(r/H)) \sim 201$ km relative to 1,100 km or max depth ~900 km. The penetration could be deeper since we have not corrected for $O^+ -> O^0$ or O^-, but we over-estimate

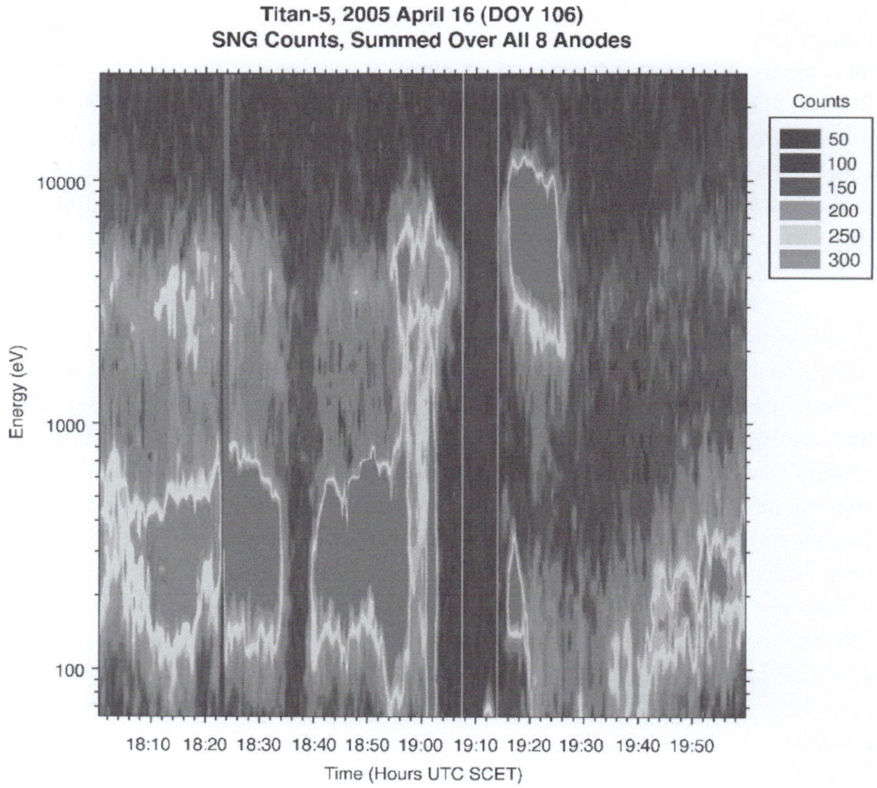

Fig. 16.53 Energy-time color spectrogram of CAPS IMS singles data for T5 flyby as reported by Hartle et al. (2006c). All eight angular sectors have been added together to enhance signal for display

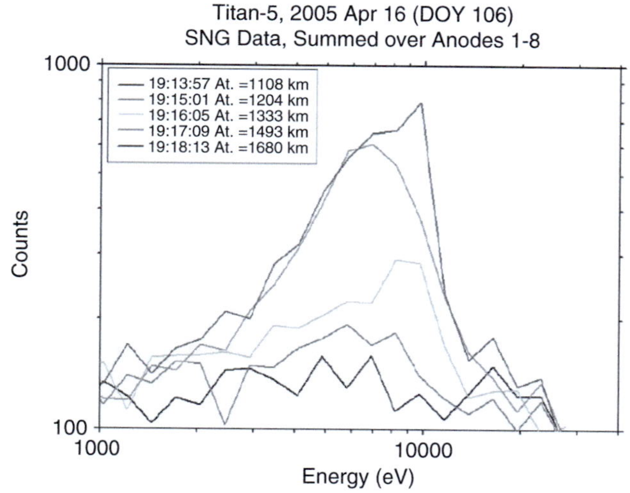

Fig. 16.54 From Hartle et al. (2006c) showing the decrease of the CAPS-IMS ion count energy spectrum with decreasing height. Note spectrum moving to lower energies with decreasing height

the penetration since not all trajectories are at normal incidence. For E>2 keV incident O^+ ions Cravens et al. (2008), Fig. 16.55, get peak ion production at ~900 km (E ~ 5 keV O^+ ions) with secondary shoulder at ~800 km (E ~ 50 keV O^+ ions) (Fig. 2 from Cravens et al. (2008) show penetration depth versus incident ion energy), and they have neglected straggling, which will cause them to over-estimate the penetration depth. Cravens et al. (2008) used the MIMI-LEMMS observations for E ~ 27 keV to 4 MeV. For energies less than 10 keV Cravens et al. used Voyager PLS observations for E<5 keV and extrapolated to MIMI energies using a power law. They then made the assumption of all protons or all oxygen ions for their ion production curves in Fig. 16.55 using stopping powers for protons and oxygen ions in a nitrogen atmosphere. Considering the uncertainties, both results by Hartle et al. (2006c) and Cravens et al. (2008) are in very good agreement.

In Fig. 16.55 from Cravens et al. (2009) the E>30 keV protons ion production peaks around 750 km altitude. But, it should be emphasized this applies to T5 when charged particle fluxes are higher than normal, while for more normal conditions very little ionization occurs below 900 km. For reference they also show the ion production from magnetospheric electrons using CAPS-ELS observations and solar UV ionization for SZA ~ 60° and 91°. The SZA ~ 91° is more applicable to atmosphere in shadow for T5. Here, the peak occurs around 1,200 km in altitude, while T5 CAPS-ELS ion production can dominate photoionization at altitudes >900 km. The electron penetration may not be as deep since they are highly magnetized and may be excluded by Titan's induced magnetosphere. The observations reported by Hartle et al. (2006c) support this conclusion. Even though significant penetration of the magnetic field in the ionosphere does occur, there is also a gradient in B to lower values inside z ~ 1,400 km (see Ågren et al. 2007, Fig. 5), which will cause the keV electrons to gradient drift around Titan and not penetrate to lower altitudes (see Sittler and Hartle 1996).

The absorption of ENAs in the lower exosphere has been studied in detail by Garnier et al. (2008a). It is shown there that, in Titan's lower exosphere/upper thermosphere, charge exchange collisions with exospheric neutrals is the dominant loss mechanisms for ENAs, and that all the other mechanisms are comparatively negligible. The calculated limit between the optically thin (statistically less than one collision) and the optically thick regime is situated, for a ~30 keV ENA, at an altitude of ~1,500 km, i.e. close to the exobase. Below this altitude an ENA can undergo several successive charge exchange collisions with exospheric neutrals, flip-flopping from neutral to ion and vice versa, and only a few eV are "lost" in each such collision. Figure 16.32, from Garnier et al. 2008a, shows the final energy of an initially 30 keV ENA originating from Titan's exosphere, when it reaches the imager which is situated well above the exobase. Between ~1,000 and ~1,500 km altitude the ENA can suffer multiple

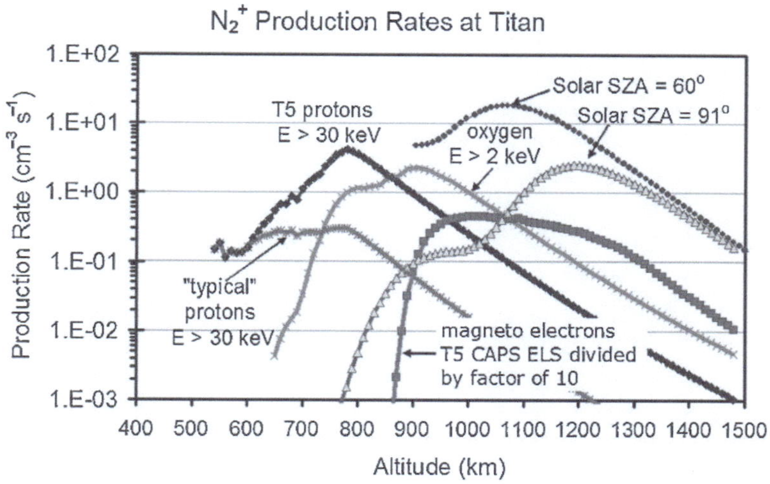

Fig. 16.55 Ion production rate for N_2^+ ions with Titan's upper atmosphere as modeled by Cravens et al. (2008) for various magnetospheric keV incident ions such as protons at E>30 keV and oxygen ions E>2 keV They also show the ion production due to solar UV and magnetospheric electrons

collisions, but it can still emerge with adequate energy (several keV) to be detected by an imager situated at higher altitudes. Below ~1,000 km, however, the collision frequency increases drastically and the accumulative loss of energy is such that the ENA can be considered as absorbed.

Smith et al. (2009) observed E > 30 keV protons penetrating down to heights ~800 km as shown in Fig. 16.35. They performed Monte Carlo calculations of the energetic protons colliding with Titan's upper atmosphere. These results showed minimal energy loss, until they reached altitudes ~800 km where their simulations showed a sudden drop in energy. This is consistent with the observations shown in Fig. 16.35 and the rule of thumb that particles lose most of their energy near the end of the track through matter. Finally, as discussed in Cravens et al. (2009), these energetic ions can drive most of the ionization observed at altitudes <800 km as observed by Kliore et al. (2008), but may not be sufficient to explain the transient ionization layer observed down to 500 km altitude.

16.8 Meteoric Ionization and Cosmic Rays: Energy Deposition at Lower Ionosphere and Thermosphere

16.8.1 Dust Particle Produced Ionization Layers

Micrometeoroid impact, atmospheric ablation and atmospheric ionization are not new when it comes to the Earth (Hughes 1978). This mechanism for atmospheric ionization of Titan's upper atmosphere was initially suggested by Ip (1990) for the case of water ice particles and stony particles, although Samuelson et al. (1983) first suggested meteoritic ablation of water ice dust particles as an important source of OH for making CO_2. The meteoroids primarily ablate at the altitude ionization occurs (Molina-Cuberosa et al. 2001). Grard (1992) made an estimate based on an Earth analogy and predicted a thin ionization layer~500 km in altitude with very high electron densities ~10^5 el/cm^3. Molina-Cuberosa et al. (2001) argue that this model is incorrect since Titan's atmosphere is more extended, three body recombination reactions are more important at these lower altitudes (i.e., 500 km) and not included in the Grard (1992) calculations and they violate the upper limits set by the Voyager radio science observations (Bird et al. 1997). English et al. (1996) estimated the energy deposition from ablation with height assuming the meteoroids are primarily composed of water ice. They showed that the water deposition rate peaking around 700 km altitude with 10% levels at 500 km and 900 km altitude. The English et al. (1996) calculations were then used by Lara et al. (1996) to estimate the production of CO and CO_2 in Titan's upper atmosphere assuming the dust particles are a source of water and OH to Titan's upper atmosphere. More recently, Hörst et al. (2008) showed that the reaction $OH + CH_3$ used by the previous photochemical models such as Lara et al. (1996) does not produce CO as previously thought, but rather H_2O (see Wong et al. 2002). They do show that direct injection of oxygen as reported by Hartle et al. (2006a,b) can produce CO directly via the reaction $O(^3P) + CH_3$ (Wong et al. 2002). Hörst et al. (2008) then show that primordial CO is not required to explain observations of CO (Lutz et al. 1983), while OH is still required to produce the observed CO_2 with $CO + OH -> CO_2 + H$ as the primary reaction (Samuelson et al. 1983).

More recently, Molina-Cuberosa et al. (2001) have developed a comprehensive model of Titan's ionospheric layers produced by micrometeoroids impacting Titan's upper atmosphere with an assumed metallic composition as observed for Earth oriented meteor showers (Hughes 1978). The meteor showers can be time dependent (Grebowsky et al. 1998), which seems to be required by the Cassini radio science observations by Kliore et al. (2008) as previously discussed. Evidence for interplanetary meteoritic dust near Saturn's orbit was observed by the Pioneer 10 and 11 dust detector as reported by Humes (1980) and more recently by the Cassini dust detector or CDA (Hillier et al. 2007a). Molina-Cuberos et al. (2001), using the Grün et al. (1985) flux model of interplanetary dust they estimated a mass flux at 10 AU ~ 6.6×10^{-18} g/cm^2/s. Ip (1984) estimated a mass flux ~ 4×10^{-18} to 2.5×10^{-17} g/cm^2/s and Cuzzi and Estrada (1998) ~ 4.5×10^{-17} g/cm^2/s. Then when they take into account gravitational focusing ~ 2.8 for Saturn plus Titan they get radial velocities (i.e., dust particle trajectories highly eccentric) away from the Sun ~ 18 km/s at Titan. Titan's orbital motion around Saturn ~5.6 km/s provides for a dust particle influx greater on the leading side ~2×10^{-17} g/cm^2/s that is an order of magnitude greater than that on the trailing side (English et al. 1996). Molina-Cuberosa et al. (2001) estimated a range of inflowing speeds ~12 to 24 km/s. Based on these model parameters with influx speed ~18 km/s and a dust particle composition of oxygen 61.7%, silicon 24.2%, magnesium 8.2% and iron 5.9% by number they estimated a neutral and ion production rate with height in Fig. 1 of their paper and shown here in Fig. 16.56. They estimate a total production rate of neutrals and ions is 1,230 cm^{-2} s^{-1} with peak ~660 km ± 50 km. The ablation process will produce ions and can be modeled using an ion production probability parameter $\beta_I \sim k_i \times (v - v_i)^2 v^{0.8}$ as used by Molina-Cuberosa et al. (2001) with k_i determined from laboratory measurements, v is inflow speed of dust particles and $v_i^* = 16.7$ (O), 9.4 (Fe), 11.1 (Mg) and 11.0 km/s (Si) from Jones (1997). This relation shows it is easier to ionize Fe and hardest to ionize O. The model by Molina-Cuberosa et al. (2001) shows

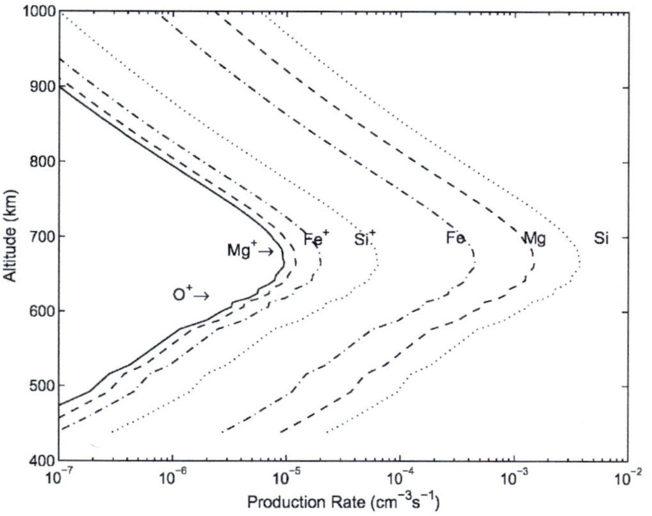

Fig. 16.56 Ion production versus altitude with Titan's atmospheree from Interplanetary dust as modeled by Molino-Cuberas et al. (2001). Here, the micrometeorites were assumed to have metal composition such as Mg, Si and Fe with some O Intermixed

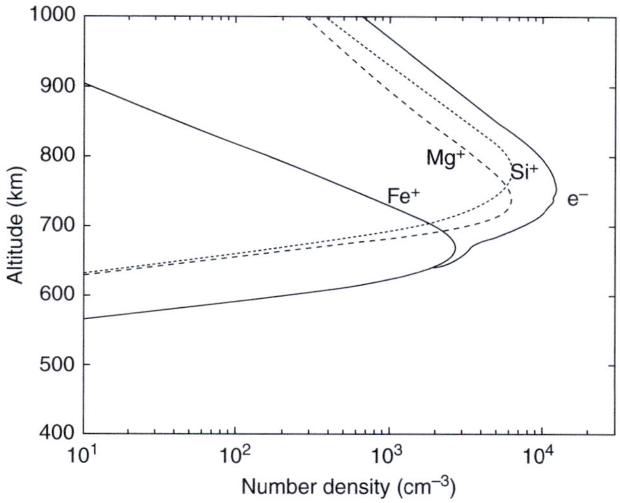

Fig. 16.57 Ion and electron densities as a function of altitude within Titan's atmosphere as modeled by Molino-Cuberas et al. (2001) using their estimated ion production rates shown in Figure 16.56

Si^+ production greater than Fe^+ production due primarily to its greater assumed abundance. The process of estimating ion and electron densities as a function of height is more difficult. They explored three model types. Model I used the most simplistic approach by only including the ion production process noted above and radiative recombination reactions with electrons which yielded at peak electron density $n_e \sim 3{,}000$ el/cm³ at 660 km altitude. When they included cluster formation due to three body reaction collisions ($Mg^+ + N_2 + N_2 -> Mg^+N_2 + N_2$) and the fact that the recombination rate of the metallic cluster is very fast ($Mg^+N_2 + e -> Mg + N_2 + h\nu$) they find a significant erosion of the predicted electron densities for heights <600 km with higher peak ~ 680 km and lower electron densities $n_e \sim 1{,}350$ el/cm³. Finally, in their model III shown in Fig. 16.57, when they include diffusive transport, charge exchange reactions between metals and non-metallic ions (ionization by ablation is small) and photoionization from solar UV (also small for heights <800 km), the ionization peak moves to ~740 km with higher peak electron density $n_e \sim 10^4$ el/cm³.

Srama et al. (2006) presented comprehensive measurements of the dust environment within Saturn's magnetosphere using the Cassini dust detector or CDA. These observations, confined inside ~15 R_S, showed an extended tenuous dust ring or E ring out to 15 R_S which can extend more than a Saturn radius off the ring plane. From their Fig. 18 we inferred a radial dependence in dust particle number density $N_d(r) \approx N_{d0}(r_0)(r_0/r)^{3.8}$ with $N_{d0}(r_0) \sim 10^{-3}$ part/m³ and $r_0 \sim 6\ R_S$. From this relation we estimate at Titan's orbit an E ring dust particle number density $N_d(20) \sim 10^{-5}$ part/m³. The dust particle size is ~1 μm in diameter with composition dominated by water ice (Hillier et al. 2007b; Postberg et al. 2008). These dust particles are believed to be of Saturn origin (i.e., Enceladus (Spahn et al. 2006a)). Spahn et al. (2006b) also presented evidence for hyper-velocity impacts on the moons of Saturn where E ring particles (ERP) impacts contribute to inner region and interplanetary dust particle (IDP) impacts tend to dominate outer E ring contributions. If these dust particle trajectories are not too eccentric one would expect impact speeds less than 6 km/s and more isotropic in Titan's frame of reference. Since we know these dust particles are primarily water ice with size ~1 μm diameter (Srama et al. 2006) one can estimate a total influx of water to Titan's atmosphere ~ 2×10^{23} H_2O/s. Considering the uncertainties this influx of water is comparable but lower than the estimate of keV O^+ ions into Titan's upper atmosphere ~9×10^{23} O^+/s as estimated by Hartle et al. (2006a,b). At these lower entry speeds ~6 km/s, ionization of the atmosphere by ablation processes is not expected to be important but can be an important source of water to Titan's atmosphere.

16.8.2 Galactic Cosmic Ray Ionization Layer

Capone et al. (1983) was the first post-Voyager encounter model of galactic cosmic ray (GCR) ionization of Titan's atmosphere. Because of their very high energies ~1 GeV or greater, they can penetrate down to depths below 70 km in altitude and produce an electron density peak at 90 km with peak densities ~2,150 el/cm³ (Molina-Cuberos et al. 1999). Capone et al. (1980), pre-Voyager found electron peak at a range of altitudes with electron densities ~1,800 el/cm³, while Borucki et al. (1987) with limit of no aerosols found electron densities ~1,600 el/cm^3 at 95 km. Due to Titan's dense atmosphere, it's the only planetary moon that can stop GCR particles in its atmosphere and develop ionization layers deep in its atmosphere. Ionizing UV and magnetospheric electrons are confined to altitudes >400 km and cannot penetrate to the great depths <100 km altitude that can be achieved by GCR. Pre-Cassini values of ion production rates with height from Molina-Cuberos et al. (1999) are shown in Fig. 16.58 and for reference their predicted ion and electron densities with height are shown in Fig. 16.59. The GCR protons and alphas produce ionization at higher altitudes >200 km as they are slowed down, while at lower altitudes where the more energetic GCR penetrate, a cascade shower is expected to develop where they decay to secondary protons, neutrons and charged and neutral pions ($\pi^{\pm}$, π^0). The charged pions with lifetimes ~3×10^{-8} s decay rapidly into charged muons ($\mu^{\pm}$) which do not interact strongly with the atmosphere and have fairly long lifetimes ~2×10^{-6} s and can thus penetrate to greater depths. The charged muons then decay to electron-positron pair, a neutrino and anti-neutrino. The electron/positron pair can then provide further ionization of the atmosphere. The neutrinos do not interact and leave the atmosphere as lost energy (see Capone et al. 1983). They usually assume the alpha particles breakup into two protons and two neutrons almost immediately once they enter the atmosphere. The lifetime of the neutral pions π^0 is only 10^{-15} sec and rapidly decay into two energetic gamma rays which decay into electron-positron pair production, subsequent bremsstrahlung and Compton scattering which give rise to high energy electrons for further ionization of the atmosphere (Capone et al. 1983; Molina-Cuberos et al. 1999). All models (Capone et al. 1983; Borucki et al. 1987; Molina-Cuberos et al. 1999) are based on the earlier work by O'Brien (1969, 1970, 1971, 1972) and O'Brien and McLaughlin (1970).

The cascade shower process as described in Capone et al. (1983) produced N, $N_2(A^3\Sigma_u^+)$, N_2^+, N^+, Ar^+, CH_4^+ and CH_3^+. Approximately 50% of N atoms are in excited state $N(^2D)$ and 50% in ground state $N(^4S)$ (Oran et al. 1975). At the time of their paper it was felt that significant mixing ratios of Ar ~ 11% could exist, but the Huygens GCMS results by Niemann et al. (2005) showed ^{36}Ar to be $(2.8\pm0.3)\times10^{-7}$ and ^{40}Ar $(4.32\ \pm0.1)\times10^{-5}$ and thus not important. The results by Capone et al. (1983) did show that Ar played a small role in the nitrogen hydrocarbon chemistry in Titan's lower atmosphere. At the time of the Capone et al. (1983) paper the importance of $N_2(A^3\Sigma_u^+)$ with hydrocarbons and N organics was unknown, but they did stress the importance of dissociation of N_2 to atomic N by GCR. The reaction of excited $N(^2D)+CH_4 - >HCN+H_2+H$ produces HCN and $N(^4S)$ atoms in ground state can react with radical CH_3 to also make HCN. But these reactions are favored at higher altitudes where dissociation of the N_2 by magnetospheric electrons can occur, while at lower altitudes where GCR are more important the N atoms will interact with H_2 to produce the amino radical NH_2. The NH_2 through 3-body interactions can make hydrazine N_2H_4 and can also interact with CH_3 to make methylamine CH_3NH_2. The CH_3 radical is formed at higher altitudes from C_2H_2 (Allen et al. 1980). The methylamine can be dissociated by $\lambda<2{,}000$ Å UV, which is expected to be absorbed at higher altitudes by aerosols, CH_4 and C_2H_2. Therefore, the methylamine may remain intact at altitudes <200 km. If so their primary loss mechanism is freezing out by the 70° cold trap and will have abundances ~50 ppbv at ~70 km altitude. One of the main uncertainties in the model by Capone et al. (1983) is the abundance of H_2.

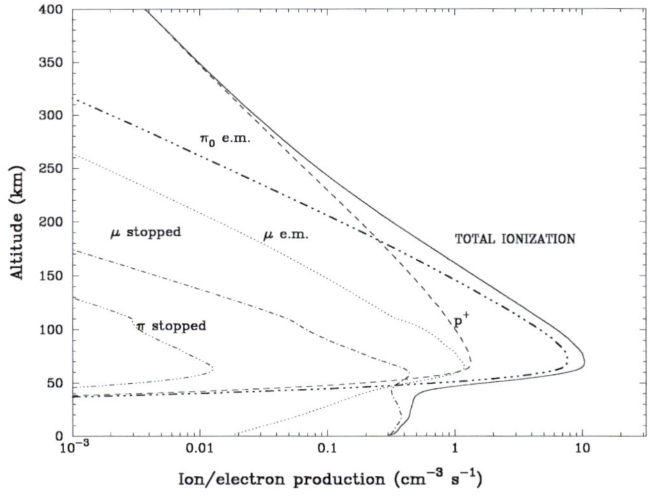

Fig. 16.58 Ion and electron production rates versus altitude for Titan's atmosphere due to Galactic Cosmic Rays as modeled by Molino-Cuberas et al. (1999)

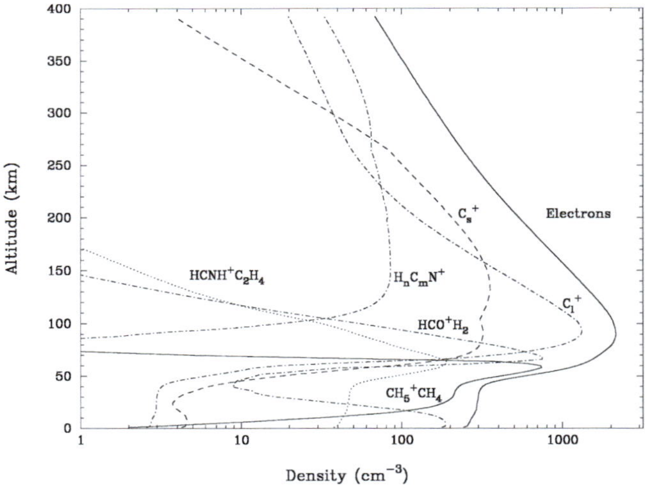

Fig. 16.59 Ion and electron densities in Titan's atmospheree as predicted by the Molino-Cuberas et al. (1999) modeled using the ion production rates due to GCRs as shown in Figure 16.58

Molina-Cuberos et al. (1999) noted that the earlier models by Capone et al. (1980, 1983) and Borucki et al. (1987) suffered from large uncertainties in Titan atmospheric models. They stressed the importance of more accurate models by Toublanc et al. (1995) and Lara et al. (1996) in estimating electron density altitude profiles and the importance of water entering Titan's atmosphere as modeled by Lara et al. (1996) and detected by ISO as reported by Coustenis et al. (1998). The presence of CO plays an important role in the model by Molina-Cuberos et al. (1999) with HCO^+H_2 being one of the dominant ions at ~70 km as shown in Fig. 16.59. Cluster cations below 70 km altitude such as $HCNH^+C_3H_4$ and $CH_5^+CH_4$ are also produced in the model by Molina-Cuberos et al. (1999). The Borucki et al. (1987) model did estimate presence of $HCNH^+$ and NH_4^+, but they had to assume presence of ammonia in the atmosphere for which Molina-Cuberos et al. (1999) said there was no observational evidence for the ammonia in Titan's atmosphere. The Borucki et al. (1987) paper also concluded that negative clusters were unlikely; if present they could absorb the electrons and thus reduce the electron densities in Titan's lower atmosphere. The recent discovery of negative heavy ions in Titan's upper atmosphere by Coates et al. (2007b), suggests that such calculations should be revisited and investigate the importance of negative ions in Titan's lower atmosphere.

Observational evidence of a lower ionization layer due to GCR was detected by Huygens HASI experiment (Fulchignoni et al. 2005) shown in Fig. 16.60 between 40 km and 140 km altitude with electrical conductivity peak at 60 km altitude. Borucki and Whitten (2008) presented electron densities, electron conductivities, and ion conductivities as a function of height for heights less than 150 km and compared them to the more recent analysis of the Huygens HASI conductivity data by Hamelin et al. (2006). Their model included the ionization by GCR and photoemission from aerosols observed by Tomasko et al. (2005). The ionization from GCR was updated from their earlier work in Borucki et al. (1987, 2006). A free parameter in their model calculations was the photoemission threshold energy for aerosols which they varied from 6 eV to 7.8 eV. Their electron conductivities were orders of magnitude greater than the Hamelin et al. (2006) results for threshold energies <7.2 eV and even at the higher threshold energies of 7.8 eV their electron conductivities were orders of magnitude greater than the Hamelin et al. (2006) results for heights >60 km. They did get fairly good agreement with the ion conductivities. They speculated that the aerosols could be coated with a material that does not photoemission at wavelengths that penetrate to lower depths, such as methane and ethane molecules which in their gas phase at higher altitudes will absorb the wavelengths that can produce the photoemission in the methane and ethane coatings. Finally, the Borucki et al. (2006) results showed that significant reductions in electron conductivities can occur if large concentrations of aerosol embryos of a few angstroms radius were present. Borucki and Whitten (2008) said that such embryos, which could not be detected by Tomasko et al. (2005), could provide the lower electron conductivities above the peak (two orders of magnitude reduction) as required by the Hamelin et al. (2006) results. In order to reduce the electron conductivity below the peak as required by the Hamelin et al. (2006) results, Borucki and Whitten (2008) speculated that a small abundance of negative molecular ions (10^{-11} mole fraction) below the peak could provide the required attenuation. In summary the Huygens HASI results have shown evidence of GCR ionization in Titan's lower atmosphere, but that earlier models cannot explain the

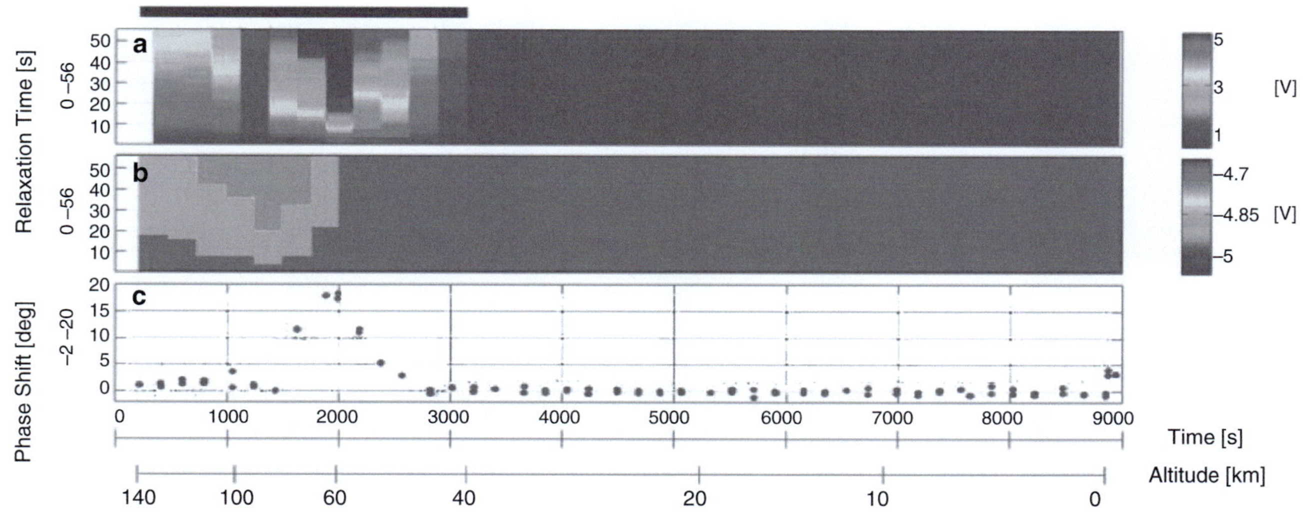

Fig. 16.60 Huygens HASI conductivity probe measurements showing ionization layer at 60 km altitude as reported by Fulchignoni et al. (2005). They interpreted layer as being due to GCRs

observations unless undetectable embryo aerosols and small amounts of negative molecules are introduced into the modeling process. The unexpected discovery of heavy negative ions in Titan's upper atmosphere by Coates et al. (2007b) underscores the possible importance of negative ions in Saturn's lower atmosphere.

16.9 Heavy Ion Formation and Aerosol Production

16.9.1 Observations of Heavy Ions by Cassini

The presence of extremely heavy and complex negative ions high in Titan's ionosphere, in addition to complex neutrals and positive ions is one of the key findings of the Cassini prime mission. Initially, high time resolution observations from CAPS-ELS in Titan's ionosphere during the TA encounter, revealed a surprising population of negative ions. These were subsequently observed on 15 other Titan encounters, and were analyzed by Coates et al. (2007b) (see also Waite et al. 2007). As shown in Fig. 16.61, these ions were observed near closest approach, and were narrowly confined to the ram direction, and contained distinct peaks. As the spacecraft flies through Titan's cold ionosphere the spacecraft velocity (~6 km/s) provides effectively a mass spectrometer for cold, ionospheric ions; assuming singly charged ions the conversion is $m_{amu} = 5.32E_{ev}$ (Coates et al. 2007b).

During TA, the observed maximum energy of the negative ions was ~60eV, corresponding to ~320 amu/q; in other encounters, notably T16 and shown in Fig. 12 from Waite et al. (Chapter 8), masses as high as 10,000 amu/q are observed. On the various encounters the ions were observed in rough mass groups at 10–30, 30–50, 50–80, 80–110, 110–200, (200–500, 500+) amu/q. Suggested mass identifications are given in Table 16.4. It was suggested that these ions may in fact be multiply charged in this region of high electron density, therefore the actual mass may be up to 5 times higher, providing a possible formation process for aerosols (Coates et al. 2007b, Waite et al. 2007).

Negative ions were not anticipated this high in Titan's atmosphere and were not included in pre-Cassini chemical schemes (e.g. Wilson and Atreya 2004) so this is an important observation requiring new chemical models (e.g. Vuitton et al. 2009, Waite et al. 2008).

The relationship between these ions and the heavy positive ion population was discussed by Waite et al. (2007, 2008). See Fig. 5 in Tomasko and West (Chapter 12), Fig. 9 in Waite et al. (Chapter 8), and Fig. 19 from Cravens et al. (Chapter 11) for summary of related observations. They suggested that nitrogen and methane in Titan's high atmosphere would be acted

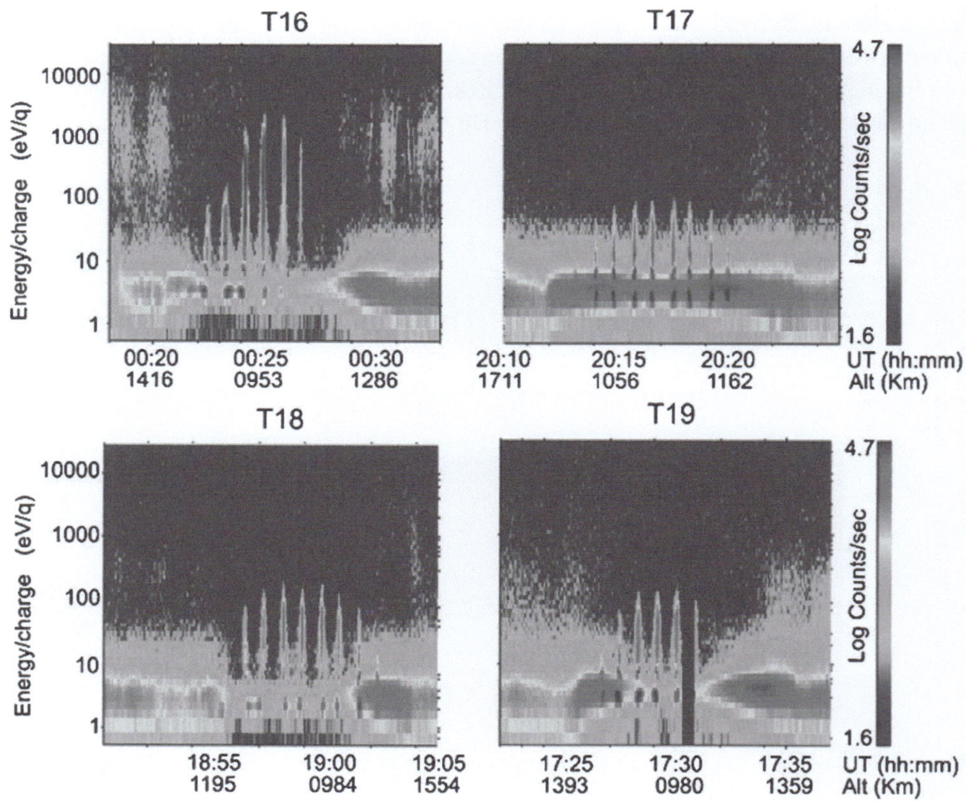

Fig. 16.61 CAPS ELS observations of heavy negative ions with Titan's ionosphere for Titan flybys TA, T16, T17, T18 and T19 as originally reported by Coates et al. (2007b). The figure shows the ions narrowly confined around the ram direction of the Cassini spacecraft indicating they are super-sonic in the spacecraft frame and thus ions and not electrons. In some cases they extend up to several keV or 10,000 amu/charge in ion M/Q

on by sunlight and magnetospheric particles forming heavier but relatively simple species by dissociation and ionization processes. Eventually this could lead to the observed benzene and other heavier ions seen by the INMS and CAPS-IBS positive ion instruments up to ~350 amu/q. The process would continue to form heavier positive and negative ions. They suggested that these may be the tholins postulated by Sagan and Khare (1979). Eventually, the large compounds would become aerosols and drift down towards Titan's surface. This may be the chain of processes by which space physics at the top of the atmosphere eventually affects the surface of Titan. This idea was supported by the more extensive observations of negative ions by Coates et al. (2007b).

Recently, further analysis of negative ion signatures has extended the number of encounters where negative ions were observed to 22 during Cassini's prime mission (Coates et al. 2009). This allows a systematic study which shows (Coates et al. 2008) that the higher mass negative ions are observed preferentially :

- At low altitudes, with the highest mass (~10,000 amu/q) at 950km, Cassini's lowest altitude so far
- At high Titan latitudes
- In the region of the terminator

On the basis of these results, Coates et al. (2008) suggested that the formation of high mass negative ions is more efficient, and/or their destruction is less efficient when sunlight is highly attenuated (Table 16.4).

16.9.2 The Formation of Fullerenes and PAHs in Titan's Upper Atmosphere

As discussed in the papers by Coates et al. (2007b, 2008), and Waite et al. (2007, 2008) the heavy positive and negative ions could be evidence for the formation of polycyclic aromatic hydrocarbons (PAH) in Titan's ionosphere with heavier ions being observed at lower altitudes with 950 km being the lowest altitude so far. These issues are further discussed in the paper by Sittler et al. (2009a) where they make reference to the review article by Bohme (1992) with regard to their presence in interstellar and circumstellar environments. As discussed in Sittler et al. (2009a), for the lower temperatures in Titan's upper atmosphere T ~ 150°K, fullerences can be produced from the polymerization of acetylene in Titan's ionosphere, formation of long carbon chains, which then form polycyclic carbon rings followed by formation of fullerences (see Hunter 1994, Thaddeus 1994 and Thaddeus 1995). At higher temperatures fullerences can also be produced by low pressure benzene-oxygen flames in the laboratory (Gerhardt et al. 1987; Richter and Howard 2007). X-ray crystallography measurements have shown fullerenes to be hollow carbon cages (see Fig. 16.62) made of 60 carbon atoms C_{60} (720 amu) which have spherical shape or 70 carbon atoms C_{70} (840 amu) with ellipsoidal shape (Krätschmer, et al. 1990). The measured diameters are ~ 7 Å with effective mass density $\rho_F \sim 1.8$ gm/cm^3 (i.e., r ~ 5.4 Å for ρ ~ 2.25 gm/cm^3 for outer shell made of carbon). This should be compared to aerosol radii $r_A \sim 12.5$ nm by Liang et al. (2007) and $r_A \sim 260$ nm by Waite et al. (2007). The higher number by Waite et al. (2007) was inadvertently over-estimated (Waite 2008, private communication). Both C_{60} and C_{70} are known to be very stable (Kroto et al. 1985; Kroto et al. 1991) and are found to be the end product of forest fires (Richter et al. 2000; Calcote and Keil 1990; Frencklach 2002). In the case of PAHs they can be formed in the laboratory using low pressure acetylene-oxygen flames (see Gerhardt et al. 1987) for which acetylene is known to be more abundant in Titan's

Table 16.4 Possible negative ion identifications (after Coates et al. 2007)

Mass group (amu/q)	Possible identification
10–30	CN^-, NH_2^-, O^-
30–50	NCN^-, $HNCN^-$, C_3H^-
50–80	$C_5H_5^-$, C_6H^-, $C_6H_5^-$
80–110	Polyyines, high order nitriles, PAHs, cyano-aromatics
110–200	
200–500	
500–10,000	

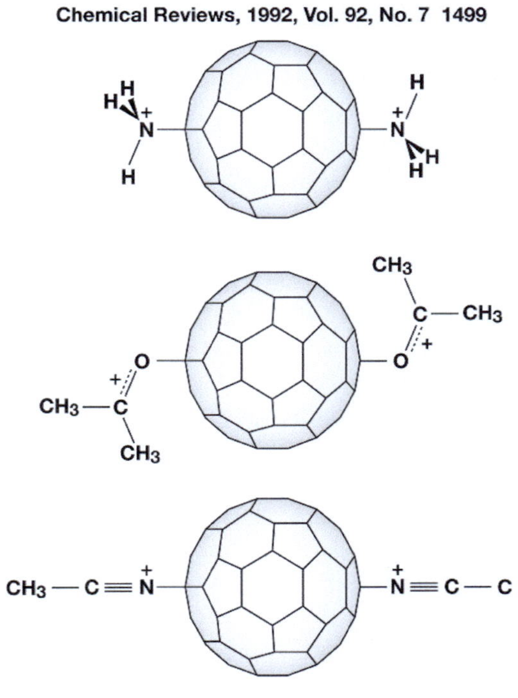

Fig. 16.62 Shows the cage structure for C_{60} fullerenes as described in the review article by Bohme (1992)

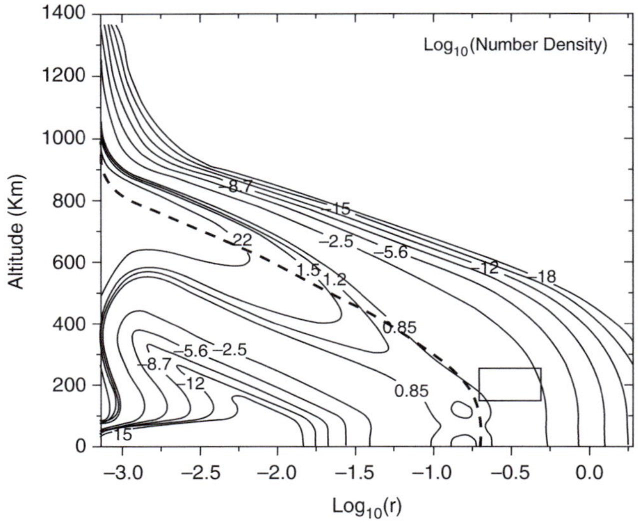

Fig. 16.63 Contour plots of log base 10 of aerosol number density versus log base 10 of aerosol radius in microns versus altitude in Titan's atmospheree by Lavvas et al. (2008a,b)

upper atmosphere ionosphere than benzene (Shemansky et al. 2005; Waite et al. 2007). The main difference between fullerenes and PAHs is that the former has no hydrogen (C_{60}), while PAHs have hydrogen ($C_{2n}H_x$). As discussed in both Waite et al. (2007) and Sittler et al. (2009a) ion-neutral collision reaction rates can be very high with cross-sections approaching geometric cross-sections of the molecules and ions. Laboratory experiments also show that oxidation can compete with PAH formation with formation of CO and CO_2. There is a "zipper" mechanism for which oxidation can keep going to make large PAHs and fullerenes. Laboratory measurements by Yang et al. (1987) and Wang et al. (1991) have shown C_{60} to have high electron affinity (2.6-2.9 eV), and since electron attachment for PAHs occur for electron affinities >1 eV (Moustefaoui et al. 1998), fullerene negative ions are allowed. But, in some cases fullerenes can be stripped of electrons to form +1 and +2 cations (Weiske et al. 1991). As noted previously, Titan's upper atmosphere and ionosphere are relatively cold with neutral and ion temperatures ~150°K (Waite et al. 2005, 2007). Electron temperatures can be quite hot with thermal electron temperatures T e,c ~ 1,000°K (Wahlund et al. 2005) and hot component photoelectron temperatures T e,h ~ 50,000°K (Coates et al. 2007a), while also being bombarded by hot keV magnetospheric heavy oxygen ions (Hartle et al. 2006a, b; Hartle et al. 2006c) and energetic protons and oxygen ions (Mitchell et al. 2005 and Cravens et al. 2008). Cravens et al. (2008) and Hartle et al. (2006c) have shown that the energetic protons and keV oxygen ions can penetrate below 950 km where the heavy positive and negative ions are observed by Cassini (Coates et al. 2007b; Waite et al. 2007). So, there is plenty of free energy to form either fullerenes and/or PAHs in Titan's upper atmosphere. Since ion-neutral and neutral-neutral (i.e., free radical C_2H) collisions dominate formation of long carbon chains from which fullerences form, low temperature reactions will dominate with the free energy sources ionizing the neutrals and making the free radicals which drive the "photochemistry". High electron temperatures will also affect charge state of the ions and recombinations rates, so indirectly will affect the reaction rates. So, the key to the formation of fullerenes and PAHs within Titan's upper atmosphere is the combined presence of methane and the high energy input to the atmosphere in the form of solar UV and soft x-rays and magnetospheric ions and electrons and at lower altitudes by GCR (Sittler et al. 2009a).

16.9.3 Role of Oxygen Input from Magnetosphere and Micro-Meteorites

As discussed in Sittler et al. (2009a) the role of oxygen can be quite important if fullerene formation occurs. For example, oxygen is an important ingredient of all amino acids attributed to life and is therefore of exobiology importance. Secondly, Titan's atmosphere is a highly reducing atmosphere with the presence of free oxygen highly unlikely (Raulin and Owen 2003). But, the presence of oxygen into Titan's upper atmosphere is not a new phenomena dating back to the discovery of CO at the 60 ppm level in Titan's atmosphere by Lutz et al. (1983). This was later followed by the discovery of CO_2 in Titan's atmosphere at the 1.5 ppb level by Samuelson et al. (1983), who showed that water entry from above may be required in order to explain the presence CO_2 which had a relatively short lifetime ~ $5x10^4$ years. In both cases the oxygen is locked up in the CO and CO_2 which are fairly stable molecules (CO – >C+O+11.17 eV; CO_2 – >CO+O+5.46 eV; CO_2 – >C+O+C+16.63 eV). Samuelson et al. (1983) suggested that oxygen entry into Titan's upper atmosphere may have been produced by the entry and ablation of micrometeorites composed of water ice. It was argued that the CO was probably primordial in origin but the recent model results by Hörst et al. (2008) showed that this was not required due to the entry of keV oxygen ions to Titan's upper atmosphere (Hartle et al. 2006a,b). Models of Titan's atmosphere after the discoveries of CO and CO_2 required an inward flux of oxygen. In the case of Yung et al. (1984) they required oxygen influx ~6.1×10^5 O atoms/cm²/s, while that by Toublanc et al. (1995) used a downward flux ~1.5×10^6 O atoms/cm²/s. Then English et al. (1996) proposed a micrometeorite ablation model for water entry with peak deposition rate at ~700 km altitude in Lara et al. (1996). The English et al. (1996) model showed peak flux ~3×10^6 mol/cm²/s. The discovery of water in Titan's atmosphere by Coustenis et al. (1998) with mole fraction ~8 ppb at 400 km altitude ended any doubts about the presence of water in Titan's upper atmosphere. The Coustenis et al. (1998) results inferred an influx rate of 0.8-2.8×10^6 mol/cm²/s. Finally, as discussed numerous times in this chapter, Hartle et al. (2006a,b) presented the first observational evidence of magnetospheric keV oxygen ions into Titan's upper atmosphere with influx rate ~1.1×10^6 O atoms/cm²/s at the exobase. Then, as discussed above, the CDA observations of E ring dust particles by Srama et al. (2006) within Saturn's outer magnetosphere allows one to estimate an influx rate of water molecules into Titan's upper atmosphere ~2.5×10^5 water mol/cm²/s at the exobase. The presence of negative oxygen ions in Titan's upper atmosphere can be inferred by the mass 16 peak in Fig. 3 from Coates et al. (2007b) or Fig. 12 from Waite et al. (Chapter 8). This might be expected due to the high electro negativity of oxygen atoms. In the case of magnetospheric oxygen ions and E ring dust particles, the source of the water and oxygen can be traced back to Enceladus (Johnson et al. 2005; Sittler et al. 2006a).

16.9.4 Trapping of Free Oxygen and Hydroxyl Ions in Seed Particles

Sittler et al. (2009a) making note of laboratory experiments where keV ions (in lab frame kinetic energy) can become trapped inside fullerenes, suggested that keV oxygen ions bombarding Titan's upper atmosphere could become trapped inside the fullerenes forming within Titan's upper atmosphere. The laboratory data is quite extensive where in most cases noble gas ions such as He, Ne, Ar, Xe and Kr were used. Weiske et al. (1991) found that $C_{60}+He->C_xHe$ $(46<x<60)$ for incident He^+ at 6 keV (30%), 8 keV (34%), 10 keV (14%) and 16 keV (<1%) with fractions produced given in parenthesis. The equivalent center of mass energies for incident He ions would be 33 eV, 44 eV, 56 eV and 89 eV, respectively. For incident O^+ ions the center of mass energy would range from 132 eV to 356 eV. (As keV O^+ ions lose energy within Titan's upper atmosphere and their energies drop below 400 eV they can become trapped with the fullerenes with higher probability. This trapping will occur deeper in the atmosphere as the O^+ incident energy increases.) In case of Argon center of mass energies used were ~450 eV, while for Kr center of mass energies were ~1,230 eV (Caldwell et al. 1991, 1992). The higher center of mass energies can cause fragmentation of the fullerene with ejection of carbon atoms. Higher center of mass collisions up to 1 keV for O^+ have yet to be done. In some cases more than one He atoms was trapped inside the fullerene. Caldwell et al. (1991) did similar experiments using also Ne and Ar and found end products $C_{60}He^+$, $C_{60}He_2^+$, $C_{70}He^+$, $C_{84}He^+$, $C_{55}Ne^+$ and $C_{55}Ar^+$. Xe and Kr were not captured due to their larger size than the space inside fullerenes. These results are consistent with x-ray crystallography measurements of fullerene dimensions and size within their hollow cage. From these laboratory results two mechanisms were invoked. Mechanism 1 was the "low pressure limit" where the ion was injected directly into the fullerene and mechanism 2 or "high pressure limit" where a collision induced fracture of the carbon cage occurred first and then He^+ capture occurred. The first mechanism probably applies for Titan where the oxygen ion or OH ion are directly injected into the fullerene. Although, most of the oxygen ions will form CO and CO_2 once thermalized, the direct entry mechanism does not have this problem and the cross-section ~the fullerene geometric cross-section $\sigma_F \sim \pi r_F^2 \sim 4\times10^{-15}$ cm² is quite large and if one uses the results by Coates et al. (2007b) for mean negative ion mass ~4,000 amu (i.e., embryo aerosol) which gives 4,000/720~6 cages, radius $r_A \sim 7$ Å and geometric cross-section $\sigma_A \sim \pi r_A^2 \sim 1.5\times10^{-14}$ cm². This oxygen or hydroxyl ions trapped inside the fullerenes could be viewed as free oxygen trapped within the aerosols which eventually settle onto the surface where unknown surface chemistry can occur. Then, with these aerosols on the surface and if a hot spot were to occur due to cryovolcanism below the surface and liquid water is present would it be energetically possible for this oxygen to allow the formation of amino acids within the surface which can be the building blocks of life? Combining this with the presence of liquid water could pre-biotic chemistry occur? This is the question raised in the Sittler et al. (2009a) paper.

16.9.5 Transport to Lower Atmosphere and Surface as Aerosols

It is important to estimate the mass flux of the heavy ions observed in Titan's ionosphere down to the surface to assess whether they can account for the aerosols observed by Tomasko et al. (2005) using the Huygens DISR imaging system and that expected from theoretical models. This will then allow one to estimate the surface accumulation and what fraction of the aerosols contain free oxygen if a significant fraction of the aerosols are composed of fullerenes in addition to PAHs. As the Coates et al. (2007b) paper showed the number of heavy negative ions increased with depth in the atmosphere with the highest masses and ion number densities at the minimum altitude probed by Cassini of 950 km. Therefore, it is not unreasonable to think that such increases will continue below 950 km. In the joint papers by Lavvas et al. (2008a,b) they developed a photochemical model of Titan's atmosphere which extended from the surface to the upper atmosphere z~1,400 km altitude. In the Lavvas et al. (2008a,b) papers they derived the size distribution of the haze particles with height using a 1D diffusive transport code with source and sink terms for the size distribution of particles. The results of their calculations are shown here in Fig. 16.63. This figure shows the size distribution at different heights and if one moves along their dashed curve of max number densities one gets ~30 part/cm^{-3} and size $r_A \sim 10^{-3}$ microns at 950 km. Then as discussed in Sittler et al. (2009a) the Lavvas et al. (2008b) results give particle mass ~5,000 amu, which is very close to that observed by Coates et al. (2007b) at 950 km. Near the surface Lavvas et al. (2008b) estimate particles sizes ~0.5 μm with number densities ~7 part/cm^3.

In the paper by Borucki and Whitten (2008) they modeled the haze particles for heights less than 150 km. Their results are shown in Fig. 16.64. They assumed three mass fluxes 4×10^{-14} kg/m^2/s for case B, 4×10^{-13} kg/m^2/s for case C and 10^{-12} kg/m^2/s for case D and assume mass flux conservation and do not use diffusion terms since they are essentially unknown. Using Stokes Equation from Roe et al. (2002), they estimated fall velocity $\sim5\times10^{-6}$ m/s for case B and $\sim4\times10^{-5}$ m/s for case D. The particle sizes were r ~ 0.45 micron for case B, 0.95 microns for case C and r ~ 1.3 micron for case D. The particle abundances were $\sim6\times10^6$ part/m^3 ~ 6 part/cm^3, which is very close to that estimated by Lavvas et al. (2008b). In Sittler et al. (2009a) they estimated a fall velocity applicable to the higher altitudes probed by Cassini which includes simple frictional drag from molecular collisions on the heavy ions. The expression is as follows:

$$V_{drift} = \frac{M_{seed}\,g(z)}{N_{N2}M_{N2}w_{N2}\sigma_{seed}} \qquad 16.4$$

with

$$\sigma_{seed,j} = \pi r_{seed,j}^2 \qquad 16.5$$

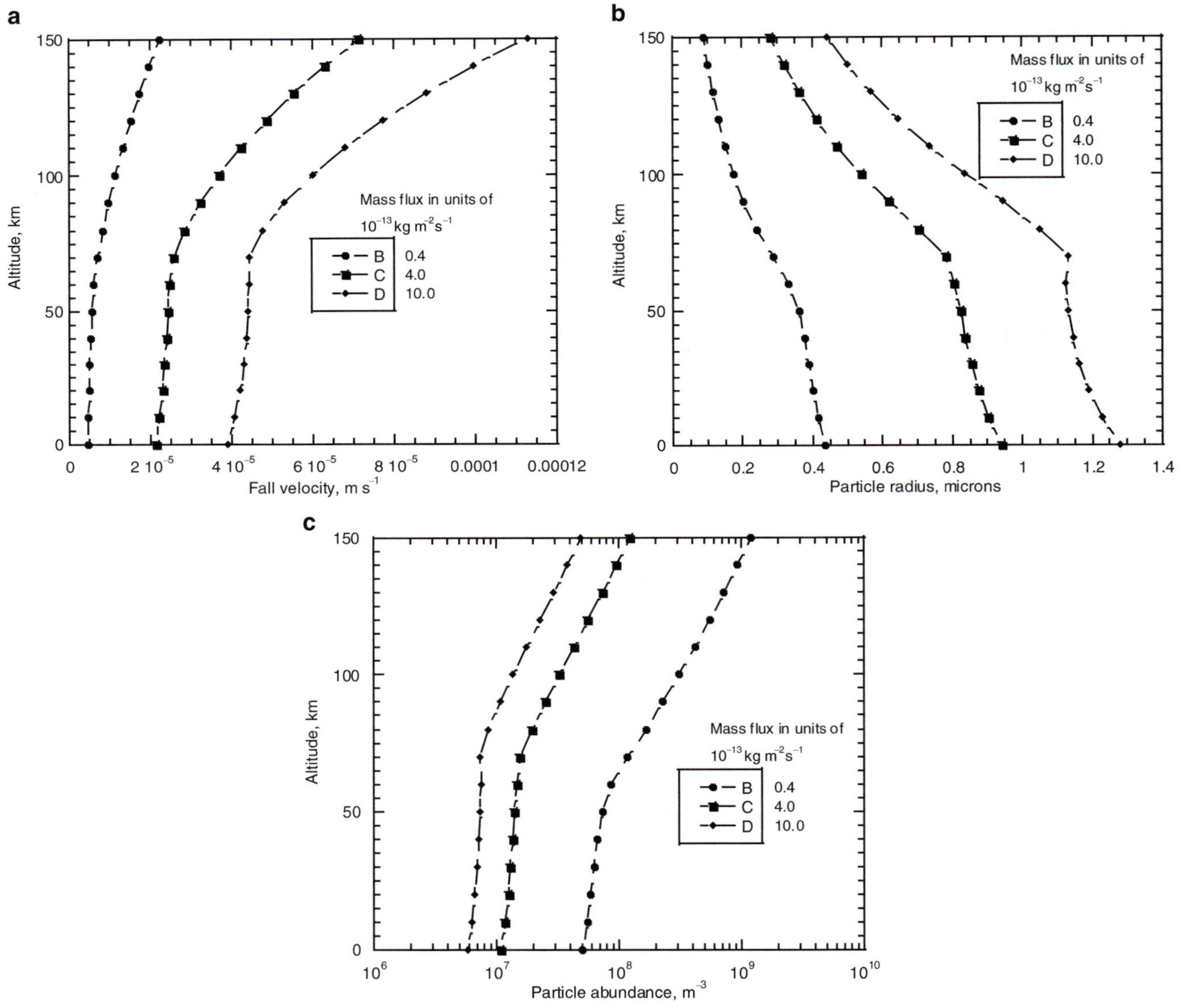

Fig. 16.64 Plot of aerosol fall velocity (m/s) versus altitude in panel (**a**), aerosol radius (microns) versus altitude in panel (**b**) and aerosol number density (#/m^3) versus altitude in panel (**c**) from Borucki and Whitten (2008a,b). The altitude range is from 0 to 150 km

Where M_{seed} is the mass of the heavy ion they refer to as seed particles for the aerosols (or embryo aerosols), g(z) is the force of gravity which is a function of height z, N_{N2} is the number density of N_2 molecules, $M_{N2}=28$ amu is the mass of N_2, w_{N2} is the thermal speed of the N_2 molecules and σ_{seed} is the geometric cross-section of the seed particles. Sittler et al. (2009a) used this expression to estimate the heavy ion fall speed at 950 km which they found to be ~10 m/s for fullerenes and ~0.4 m/s for PAHs. Assuming fullerenes and the ion number densities and mean mass of the heavy ions ~4,000 amu they estimated a downward mass flux $\Phi \sim 2.7\times10^{-14}$ kg/m²/s, which is approximately equal to the lower limit $\Phi \sim 4\times10^{-14}$ kg/m²/s used by Borucki and Whitten (2008) which was constrained by the Tomasko et al. (2005) aerosol observations. If the mass density flux increase below 950 km as expected, then the heavy ions observed above 950 km may be sufficient to account for all the aerosols observed by Tomasko et al. (2005) down near the surface. If one uses the mass flux rate ~4×10^{-14} kg/m²/s and a surface mass density ~420 kg/m³ it will take ~300 Myrs to accumulate one meter thick of aerosols. If one uses the higher mass fluxes of Borucki and Whitten (2008) the accumulation times could be as short as ~10 Myrs.

In Sittler et al. (2009a) they also considered the effect of heavy negative ion drift and formation of an ambipolar electric field but found this to be a small effect for the heavy ions which provided an effective loss rate of mass 28 ions ~5×10^6 ions/cm²/s.

Using UVIS stellar occultation data, Liang et al. (2007) observed the haze layer down to 400 km altitude and assuming a scattering radius $r_A \sim 12.5$ nm for aerosols, they estimated particle density for "tholin" particles to be ~10^4 part/cm³ at 950 km and ~10^6 part/cm³ at 400 km altitude. As discussed in Sittler et al. (2009a), such particle sizes could be in agreement with the Coates et al. (2007b) heavy negative ion results if they are flat PAHs with radii $r_{seed} \sim 3$ nm. Since the Liang et al. (2007) results are not very sensitive to particle size they could also be seeing 3 nm size particles. If Coates et al. (2007b) are really seeing fullerenes then they may be too small ($r_A \sim 7$ Å) for UVIS to see them and Liang et al. (2007) are seeing particles too big to be observed by CAPS-ELS. If a significant fraction of the haze particles observed by UVIS are charged, then we have a conflict and charge neutrality would become a serious problem when the Langmuir Probe data is also considered (Wahlund, private communication, 2009). The Liang et al. (2007) results also show that the haze particles have similar scale height as the acetylene which one might expect if the acetylene is the source of the "tholin" particles and we know that acetylene can polymerize and make fullerences, while benzene is probably needed to make PAHs.

Sittler et al. (2009a), using the influx rate of O^+ keV of 1.1×10^6 O/cm²/s from Hartle et al. (2006b), a seed cross-section $\sigma_{seed} \sim 4.5\times10^{-14}$ cm² and $N_{seed} \sim 200$ ions/cm³ an implantation rate ~10^{-5} O implanted seed/cm³/s. They then estimate the volume to make one of the aerosols observed by Tomasko et al. (2005) to be $\Delta V_{ionosphere} \sim 1.9\times10^6$ cm³ one gets ~19 O atoms/s implanted within each aerosol. If the seed particle fall rate is ~3 m/s and one uses a column density ~100 km the O^+ can penetrate they estimate ~6.3×10^5 O/aerosol. They then estimated ~2×10^9 fullerenes per aerosol, so that there are ~0.05% of the fullerenes with oxygen atoms. But, these are lower limits since the heavy negative ion densities and masses are expected to be higher below 950 km. They then showed that their estimated mass flux of free oxygen to the surface is 0.1% of that incident onto the upper atmosphere from the magnetosphere and therefore the bulk of the oxygen is going into the formation of CO and CO_2 as one might expect. If Coates et al. (2007b) are in fact seeing PAHs then they can account for all the aerosols if one confines their observations above 950 km. If the numbers increase below 950 km as expected, then one might have too many aerosols which would also violate observations. If the heavy ions are PAHs then the trapping of oxygen within them is less certain. Whatever the outcome, the possibility of free oxygen in Titan's surface raises very interesting exobiology possibilities.

16.10 Conclusion and Future Outlook

The Cassini mission has revealed new and unexpected phenomena within Saturn's magnetosphere and Titan itself that have influenced our understanding of how energy from above is deposited into Titan's atmosphere. The global Saturn magnetosphere LT configuration has had an important impact onto the nature of the interaction and whether the magnetosphere deposits this energy at high, medium and low energy limits. At noon LT the TA, TB and T3 flybys showed the magnetospheric field to be dipolar in direction with magnetospheric heavy ion energy fluxes into Titan at the medium to high energy limits with solar ram angle SRA ~ 90°. At dawn-midnight local times both T9 and T18 flybys showed a magnetodisk configuration with light ions dominating the composition, the energy input was at its low energy limit with solar-ram angle SRA ~ 120°. Here, the upstream flow does not have easy access to the sunlit atmosphere which is inclined toward the wake, while the upstream flow impinges upon the shadow side. Here ion gyro-radii are smaller and a multi-fluid interaction may be more applicable. Near dawn local time we have the T5 flyby where again the field has a magnetodisk configuration with heavy ions dominating the ion composition with energies between 5–10 keV and the energy input is at the high energy limit. The solar-ram angle SRA ~ 180° with the flow incident upon the Titan's nightside atmosphere. Here, magnetospheric electrons have direct entry to Titan's atmosphere and a transition from photoelectrons on dayside ionosphere to secondary electrons on nightside ionosphere is observed.

Yelle et al. (2008) estimated relatively high atmospheric loss rates of methane ~10^{27} mol/s. But this estimate was based on an estimate of the eddy diffusion coefficient which does tend to have high uncertainties. The hydrodynamic model by Strobel (2008) was an attempt to explain this perceived high atmospheric loss rate but has a fundamental problem of transporting energy from lower heights to above the exobase where neutral particle collisions are expected to be negligible. This model will also require a large torus of methane gas and its radicals such as CH_3 to form around Saturn. The methane molecules become ionized and seen as hot keV pickup methane ions CH_n^+ ($1 \leq n \leq 4$) by the CAPS IMS. This search is now ongoing (see Fig. 3 in Waite et al. (Chapter 8).

The energy input drives a multiple of processes such as ion production, photoelectron and secondary electron emission, upper atmosphere heating, ionosphere heating, rich neutral and ion hydrocarbon and nitrile chemistry, ionospheric loss via an ionospheric wind and scavenging, and atmospheric loss via non-thermal heating and emission of fast neutrals from sputtering processes. Various energy sources were identified such as solar UV and soft x-rays, magnetospheric ion and electron precipitation, pickup ion precipitation, interplanetary dust, E ring dust and Galactic Cosmic Rays.

It was also pointed out that the details of the magnetospheric interaction had a significant effect upon how energy was deposited into Titan's atmosphere so we reviewed both 2D and 3D MHD models which include the ionosphere physics. This gave both an historical evolution of ideas about the interaction and many of the subtle but important features of the interaction. Hybrid codes which are more accurate at higher altitudes did not arrive until later and up till now have not included the ionospheric physics very accurately. So, in this sense the MHD models are ahead of the hybrid codes. Eventually one will need to develop kinetic-fluid hybrid codes, which includes the ionosphere.

The ionosphere was probed remotely by both Voyager 1 and Cassini radio science experiments with both showing a main ionization peak at 1,200 km altitude with solar input being prime source. The Cassini radio science experiment identified a transient lower peak at 500 km either due to energetic ions or interplanetary dust from meteor showers. Huygens HASI instrument detected an ionization layer at 60 km altitude, which is probably due to GCR. But, here the observations cannot yet be explained by present theoretical models of ionization in Titan's lower atmosphere which can also have contributions from photoemission from aerosol outer surfaces.

The discovery of heavy negative ions was highlighted with their production driven by the external energy sources reviewed in this chapter. The heavy ions can also be viewed as sources of stored chemical energy. The heavier negative ions are felt to be either fullerenes (no hydrogen C_n) or PAHs (with hydrogrn C_mH_x) or both, for which fullerenes have a high electron affinity and can form negative ions as observed. Fullerences can be formed from the polymerization of acetylene molecules, while PAHs probably arise from the polymerization of benzene. Furthermore, ion neutral collision cross-sections are known to be very high and can approach geometric cross-sections of the molecules and ions. The possible implantation of magnetospheric keV O^+ into the negative ions if fullerenes were present and the estimated mass flux to surface from various publications were reviewed.

With regard to the future, the analysis of the past Titan flybys are starting to reveal new unexpected features of Titan's interaction with Saturn's magnetosphere and the extended mission will provide new opportunities to detect a wider range of magnetospheric energy input, SLT and SRA parameters. We will also observe how the various energy input processes evolve as Saturn and Titan progress from southern summer to equinox and observe the effects of Enceladus if its output of water into Saturn's magnetosphere varies with time. Chemistry changes with Titan's upper atmosphere may also evolve. The Cassini mission has provided many surprises and one looks forward to new discoveries with regard Titan's interaction with Saturn's magnetosphere.

Acknowledgements We acknowledge assistance by Joseph Ajello with regard to the section on UVIS observations at Titan. Robert Kilgore from TRAX International Corporation at Goddard Space Flight Center performed all the graphic arts suppert for the chapter's figures. Work at GSFC is supported by the Cassini Program under JPL Contract 1243218 with SwRI. Additional support at GSFC is provided by NASA's Cassini Data Analysis Program (CDAP).

References

Achilleos N, Arridge CS, Bertucci C et al (2008) Large-scale dynamics of Saturn's magnetopause: observations by Cassini. J Geophys Res Space Phys 113:A11209

Ågren K, Wahland J-E, Modolo R, Lummerzheim D, Galand M, Muller-Wodard I, Canu P, Kurth WS, Cravens TE, Yelle RV, Waite Jr JH, Coates AJ, Lewis GR, Young DT, Bertucci C, Dougherty MK (2007) On magnetospheric electron impact ionization and dynamics in Titan's ram-side and polar ionosphere, a Cassini case study. Ann Geophys 25:2359–2369

Ajello JM, Ciocca M (1996) Fast nitrogen atoms from dissociative excitation of N_2 by electron impact. J Geophys Res 101(18):953

Ajello JM, Stevens MH, Stewart I, Larsen K, Esposito L, Colwell J, McClintock W, Holsclaw G, Gustin J, Pryor W (2007) Titan airglow spectra from Cassini ultraviolet imaging spectrograph (UVIS): EUV analysis. *Geophys Res* Lett 34:L24204. doi:10.1029/2007GL031555

Ajello JM, Gustin J, Stewart I, Larsen K, Esposito L, Pryor W, McClintock W, Stevens MH, Malone CP, Dziczek D (2008) Titan airglow spectra from Cassini ultraviolet imaging spectrograph (UVIS): FUV disk analysis. Geophys Res Lett 35:L06102. doi:10.1029/2007GL032315

Allen M, Pinto JP, Yung YL (1980) Titan: aerosol photochemistry and variations related to the sunspot cycle. Astrophys J 242:L125–L128

Arridge CS, Khurana KK, Russell CT et al (2008a) Warping of Saturn's magnetospheric and magnetotail current sheets. J Geophys Res Space Phys 113:A08217

Arridge CS, Russell CT, Khurana KK et al (2008b) Saturn's magnetodisc current sheet. J Geophys Res Space Phys 113:A04214

Arridge CS, Andre N, Achilleos N et al (2008c) Thermal electron periodicities at 20R(S) in Saturn's magnetosphere. Geophys Res Lett 35:L15107

Backes H, Neubauer FM, Dougherty MK, Achilleos N, Andre N, Arridge CS, Bertucci C, Jones GH, Khurana KK, Russell CT, Wennmacher A (2005) Titan's magnetic field signature during the first Cassini encounter. Science 308:992–995. doi:10.1126/science.1109763

Banaszkiewicz M, Lara LM, Rodrigo R, et al. (2000) The upper atmosphere and ionosphere of Titan: a coupled model. In Planetary ionospheres and magnetospheres, pp 1547–1550

Bertucci C, Neubauer FM, Szego K, Wahlund J-E, Coates AJ, Dougherty MK, Young DT, Kurth WS (2007) Structure of Titan's mid-range magnetic tail: Cassini magnetometer observations during the T9 flyby. Geophys Res Lett 34:L24S02. doi:10.1029/2007GL030865

Bertucci C, Achilleos N, Dougherty MK et al (2008) The magnetic memory of Titan's ionized atmosphere. Science 321:1475–1478

Bertucci C, Sinclair B, Achilleos N, Hunt P, Dougherty MK, Arridge CS (2009) The variability of Titan's magnetic environment. Planet Space Sci (in press)

Bird MK, Dutta-Roy R, Asmar SW, Rebold TA (1997) Detection of Titan's ionosphere from Voyager 1 radio occultation observations. Icarus 130:426–436

Blanc M et al (2002) Magnetospheric and plasma science with Cassini-Huygens. Space Sci Rev 104:253–346

Bohme DK (1992) PAH and Fullerene ions and ion/molecule reactions in interstellar and circumstellar chemistry. Chem Rev 92:1487–1508

Borucki WJ, Whitten RC (2008) Influence of high abundances of aerosols on the electrical conductivity of the Titan atmosphere. Planet Space Sci 56:19–26

Borucki WJ, Levin Z, Whitten RC, Keesee RG, Capone LA, Summers AL, Toon OB, Dubach J (1987) Predictions of the electrical conductivity and charging of the aerosols in Titan's atmosphere. Icarus 72:604–622

Borucki WJ, Whitten RC, Bakes ELO, Barth E, Tripathi S (2006) Predictions of the electrical conductivities and charging of aerosols in Titan's atmosphere. Icarus 181:527–544

Brecht SH, Luhmann JG, Larson DJ (2000) Simulation of the Saturnian magnetospheric interaction with Titan. J Geophys Res 105:13, 119–13,130

Bridge HS et al (1981) Plasma observations near Saturn: initial results from Voyager 1. Science 212:217

Broadfoot AL et al (1981) Extreme ultraviolet observations from Voyager 1 – encounter with Saturn. Science 212:206

Calcote HF, Keil DG (1990) The role of ions in soot formation Pure Appl Chem 62:815

Caldwell KA, Giblin DE, Hsu CS, Cox D, Gross ML (1991) Endohedral Complexes of fullerene radical cations. J Am Chem Soc 113:8519–8521

Caldwell KA, Giblin DE, Gross ML (1992) High-energy collisions of fullerene radical cations with target gases - capture of the target gas and charge stripping of C-60(.+), C-70(.+), and C-84(.+). J Am Chem Soc 114:3743–3756

Capone LA, Dubach J, Whitten RC, Prasad SS, Santhanam K (1980) Cosmic ray synthesis of organic molecules in Titan's atmosphere. Icarus 44:72–84

Capone LA, Dubach J, Prasad SS, Whitten RC (1983) Galactic cosmic rays and N_2 dissociation on Titan. Icarus 55:73–82

Coates AJ, Crary FJ, Young DT, Szego K, Arridge CS, Bebsi Z, Sittler EC Jr, Hartle RE, Hill TW (2007a) Ionospheric electrons in Titan's tail: plasma structure during the Cassini T9 encounter. Geophys Res Lett 34:L24S05. doi:10.1029/2007GL030919

Coates AJ, Crary FJ, Lewis GR, Young DT, Waite Jr JH, Sittler EC Jr (2007b) Discovery of heavy negative ions in Titan's ionosphere. Geophys Res Lett 34:L22103. doi:10.1029/2007GL030978

Coates AJ, Lewis GR, Wellbrock A, Young DT, Crary FJ, Waite Jr JH (2009) Heavy negative ions in Titan's ionosphere: altitude and latitude dependence. Presented at Titan after Cassini-Huygens symposium. Corpus Christi, Texas, 6–11 July 2008

Coustenis A, Salama A, Lellouch E, Encrenaz T, Bjoraker GL, Sameulson RE, de Graauw T, Feuchtgruber H, Kessler MF (1998) Evidence for water vapor in Titan's atmosphere from ISO/SWS data. Astron Astrophys 336:L85–L89

Cravens TE, Keller CN, Ray B (1997) Photochemical sources of non-thermal neutrals for the exosphere of Titan. Planet Space Sci 45, No. 8:889–896

Cravens TE, Lindgren CJ, Ledvina SA (1998) A two-dimensional multifluid MHD model of Titan's plasma environment. Planet Space Sci 46, No. 9/10:1193–1205

Cravens TE, Vann J, Clark J et al (2004) The ionosphere of Titan: an updated theoretical model. 2nd World Space Congress/34th COSPAR Scientific Assembly, Oct. 10–19, 2002, Houston, TX, Planet. Atmos Ionosph Plasma Int Adv Space Res 33:212–215

Cravens TE, Robertson IP, Clark J, Wahlund J-E, Waite Jr JH, Ledvina SA, Niemann HB, Yelle RV, Kasprzak WT, Luhmann JG, McNutt RL, Ip W-H, De La Haye V, Muller-Wodarg I, Young DT, Coates AJ (2005) Titan's ionosphere: model comparison with Cassini Ta data. Geophys Res Lett 32:L12108. doi:10.1029/2005GL023249

Cravens TE, Robertson IP, Waite Jr JH, Yelle RV, Kasprzak WT, Keller CN, Ledvina SA, Niemann HB, Luhmann JG, McNutt RL, Ip W-H, De La Haye V, Mueller-Wodarg I, Wahlund J-E, Anicich VG, Vuitton V (2006) Composition of Titan's ionosphere. Geophys Res Lett 33:L07105

Cravens TE, Robertson IP, Ledvina SA, Michell D, Krimigis SM, Waite Jr JH (2008) Energetic ion precipitation at Titan. Geophys Res Lett 35:L03103

Cravens TE, Robertson IP, Waite Jr JH, Yelle RV, Vuitton V, Coates AJ, Wahlund J-E, Ågren K, Richard MS, De La Haye V, Wellbrock A, Neubauer FM (2009) Model-data comparisons for Titan's nightside ionosphere. Icarus 199:174–188

Cuzzi JN, Estrada PR (1998) Compositional evolution of Saturn's rings due to meteoroid bombardment. Icarus 132:1–32

Dandouras J, Amsif A (1999) Production and imaging of energetic neutral atoms from Titan's exosphere: a 3-D model. Planet Space Sci 47:1355–1369

Dandouras I, Garnier P, Mitchell DG, Roelof EC, Brandt PC, Krupp N, Krimigis SM (2008) Titan's exosphere and its interaction with Saturn's magnetosphere. Phil Trans R Soc A. doi:10.1098/rsta.2008.0249

De La Haye V, Waite Jr JH, Johnson RE, Yelle RV, Cravens TE, Luhmann JG, Kasprzak WT, Gell DA, Magee B, Leblanc F, Michael M, Jurac S, Robertson IP (2007) Cassini ion and neutral mass spectrometer data in Titan's upper atmosphere and exosphere: observations of a suprathermal corona. J Geophys Res 112:A07309

Dobe Z, Szego K, Quest KB, Vtali D Shapiro, Hartle RE, Sittler EC Jr (2007) Nonlinear evolution of modified two-stream instability above ionosphere of Titan: Comparison with the data of the Cassini plasma spectrometer. J Geophys Res 112:A03203. doi:10.1029/2006JA011770

Dougherty MK et al (2004) The Cassini magnetic field investigation. Space Sci Rev 114:331–383

English MA, Lara LM, Lorenz RD, Ratcli P, Rodrigo R (1996) Ablation and chemistry of meteoric materials in the atmosphere of Titan. Adv Space Res 17:157–160

Esposito LW et al (2004) The Cassini ultraviolet imaging spectrograph investigation. Space Sci Rev 115:299–361

Fox JL, Yelle RV (1997) Hydrocarbon ions in the ionosphere of Titan. Geophys Res Lett 24:2179–2182

Frencklach M (2002) Reaction mechanism of soot formation in flames. Phys Chem Chem Phys 4:2028–2037

Fulchignoni M et al (2002) The characterization of Titan's atmospheric physical properties by the Huygens atmospheric structure instrument (HASI). Space Sci Rev 104:395–431

Fulchignoni M, Aboudan A, Angrilli F et al (2004) A stratospheric balloon experiment to test the Huygens atmospheric structure instrument (HASI). Planet Space Sci 52:867–880

Fulchignoni M et al (2005) In situ measurements of the physical characteristics of Titan's environment. Nature 438:785–791

Galand M, Lilensten J, Toublanc D, Maurice S (1999) The ionosphere of Titan: Ideal diurnal and nocturnal cases. Icarus 140:92–105

Galand M et al (2006) Electron temperature of Titan's sunlit ionosphere. Geophys Res Lett 33:L21101. doi:10.1029/2006GL027488

Gan L, Keller CN, Cravens TE (1992) Electrons in the ionosphere of Titan. J Geophys Res 97:12,137–12,151

Gan L, Cravens TE, Keller CN (1993) A time-dependent model of suprathermal electrons at Titan. Plasma Environ Non-Magnet Planets 4:171. (Proceedings of COSPAR colloquium held in Ann Arbor, MI, August, 1992)

Garnier P, Dandouras I, Toublanc D, Brandt PC, Roelof EC, Mitchell DG, Krimigis SM, Krupp N, Hamilton DC, Waite H (2007) The exosphere of Titan and its interaction with the kronian magnetosphere : MIMI observations and modeling. Planet Space Sci 55:165. doi:10.1016/j.pss.2006.07.006

Garnier P, Dandouras I, Toublanc D, Roelof EC, Brandt PC, Mitchell DG, Krimigis SM, Krupp N, Hamilton DC, Dutuit O, Wahlund JE (2008a) The lower exosphere of Titan: energetic neutral atoms absorption and imaging. J Geophys Res 113:A10216. doi:10.1029/2008JA013029

Garnier P, Dandouras I, ToublancD, Roelof EC, Brandt PC, Mitchell DG, Krimigis SM, Krupp N, Hamilton DC, Waite H, Wahlund JE (2008b) A non thermal model for the extended Titan exosphere. Icarus (submitted)

Gerhardt Ph, Löffler S, Homann KH (1987) Polyhedral carbon ions in hydrocarbon flames. Chem Phys Lett 137(4):306–310

Grard R (1992) The significance of meteoric ionization for the propagation of lightning spherics in the atmosphere of Titan. ESA Sci Publ 338:125–128

Grebowsky JM, Goldberg RA, Pesnell WD (1998) Do meteor showers significantly perturb the ionosphere? J Atmos Terr Phys 60:607–615

Grün E, Zook HA, Fechtig H, Giese RH (1985) Collisional balance of the meteoritic complex. Icarus 62:244–272

Gurnett DA, Kurth WS, Scarf FL (1981) Plasma waves near Saturn: initial results from Voyager 1. Science 212:235–239

Gurnett DA, Scarf FL, Kurth WS (1982) The structure of Titan's wake from plasma wave observations. J Geophys Res 87:1395–1403

Gurnett DA et al (2004) The Cassini radio and plasma wave investigation. Space Sci Rev 114:395–463

Gustin J, Ajello J, Stevens M, Stewart I, Stephan A, Esposito L (2009) Titan airglow spectra from Cassini UVIS: III. FUV Limb Analysis. Geophys Res Lett (in review)

Hamelin M et al (2006) Titan's atmospheric electricity from measurements of the PWA-HASI instrument on board the Huygens probe. Titan's atmospheric electricity-PWA-HASI Team-EUROPLANET-Berlin

Hartle RE, Sittler EC Jr (2007) Pckup ion phase space distributions: Effects of atmospheric spatial gradients. J Geophysic Res 112:A07104. doi:10.1029/2006JA012157

Hartle RE, Sittler EC (2008) Comparisons of selected atmospheric escape mechanisms on Venus, Mars and Titan. European Planetary Science Congress 3: 2008-A-00453

Hartle RE, Sittler EC Jr, Ogilvie KW, Scudder JD, Lazarus AJ, Atreya SK (1982) Titan's ion exosphere observed from Voyager 1. J Geophys Res 87:1383

Hartle RE, Sittler EC Jr, Neubauer FM, Johnson RE, Smith HT, Crary F, McComas DJ, Young DT, Coates AJ, Simpson D, Bolton S, Reisenfeld D, Szego K, Berthelier JJ, Rymer A, Vilppola J, Steinberg JT, Andre N (2006a) Preliminary interpretation of Titan plasma interaction as observed by the Cassini plasma spectrometer: comparisons with Voyager 1. Geophys Res Lett 33:L08201. doi:10.1029/2005GL024817

Hartle RE, Sittler EC Jr, Neubauer FM, Johnson RE, Smith HT, Crary F, McComas DJ, Young DT, Coates AJ, Simpson D, Bolton S, Reisenfeld D, Szego K, Berthelier JJ, Rymer A, Vilppola J, Steinberg JT, Andre N (2006b) Initial interpretation of Titan plasma interaction as observed by the Cassini plasma spectrometer: comparisons with Voyager 1. Planet Space Sci 54:1211

Hartle RE, Sittler EC Jr, Shappirio MD, Johnson RE, Tucker OJ, Luhmann JG, Ledvina SA, Cooper JF, Coates AJ, Szego K, Burger MH, Simpson DG, Crary F, Young DT (2006c) Saturn's magnetosphere ion erosion by Titan: penetration and loss of water group ions in upper atmosphere. J Geophys Res (manuscript in preparation)

Hartle RE, Sittler EC Jr, Lipatov AS (2008a) Ion escape from the ionosphere of Titan. Geophys Res Abst Vol. 10, EGU2008-A-09860, 2008. SRef-ID: 1607-7962/gra/EGU2008-A-09860, EGU General Assembly

Hartle RE, Sittler EC Jr, Lipatov AS, Bertucci C, Coates AJ, Szego K, Shappirio MD, Simpson DG (2008b) Effects of pickup ions on Titan's interaction with Saturn's magnetosphere during the T9 Flyby, 2008 Joint Assembly, P31A-06

Hillier JK, Green SF, McBride N, Schwanethal JP, Postberg F, Srama R, Kempf S, Moragas-Klostermeyer G, McDonnell JAM, Grün E (2007b) The composition of Saturn's E ring. Mon Not R Astron Soc 377:1588–1596

Hillier JK, Green SF, McBride N, Altobelli N, Postberg F, Kempf S, Schwanethal J, Srama R, McDonnell JAM, Grün E (2007a) Interplanetary dust detected by the Cassini CDA chemical analyser. Icarus 190:643–654

Holberg JB, Forrester WT, Shemansky DE (1982) Voyager absolute far-ultraviolet spectrophotometry of hot stars. Astrophys J 257:656–671

Holberg JB, Ali B, Carone TW, Polidan RS (1991) Absolute far-ultraviolet spectrometry of hot subluminous stars from Voyager. Astrophys J 375:716–721

Hörst SM, Vuitton V, Yelle RV (2008) The origin of oxygen species in Titan's atmosphere. J Geophys Res 113:E10006

Hughes DW (1978) Meteors. In: McDonnell JAM (ed) Cosmic dust. Wiley, Chichester, pp 123–184

Humes DH (1980) Results of Pionner 10 and 11 meteoroid experiments: interplanetary and near-Saturn. J Geophys Res 85:5841–5852

Hunten DM, Tomasko MG, Flasar FM, Samuelson RE, Strobel DF, Stevenson DJ (1984) Titan. In: Saturn. Gehrels T, Matthews MS (eds). 671–759

Hunter JM, Fye JL, Roskamp EJ and Jarrold MF (1994) Isomerization of pure carbon cluster ions: From rings to fullerences, *Molecules and Grains in Space, AIP Conference Proceedings 312, Mont Sainte-Odile, France 1993, editor Iréne Nenner*, 571–588

Ip W. –H., (1984) Ring torque of Saturn from interplanetary meteoroid impact Icarus 60:547–552

Ip W-H (1990) Meteoroid ablation processes in Titan's atmosphere. Nature 345:511–512

Ip W-H (1992) The nitrogen tori of Titan and Triton. Adv Space Res 12(8):73

Israelevich PL, Neubauer FM, Ershkovich AI (1994) The induced magnetosphere of comet Halley – interplanetary magnetic-field during GIOTTO encounter. J Geophys Res Space Phys 99:6575–6583

Johnson RE, Liu M, Sittler EC Jr (2005) Plasma-induced clearing and redistribution of material embedded in planetary magnetospheres. Geophys Res Lett 32:L24201. doi:10.1029/2005GL024275

Jones W (1997) Theoretical and observational determinations of the ionization coefficient of meteors. Mon Not R Astron Soc 288:995–1003

Kallio E, Sillanpää I, Janhunen P (2004) Titan in subsonic and supersonic flow. Geophys Res Lett 31:L15703

Keller CN, Cravens TE, Gan L (1992) A model of the ionosphere of Titan. J Geophys Res 97:12117–12135

Keller CN, Anicich VG, Cravens TE (1998a) Model of Titan's ionosphere with detailed hydrocarbon chemistry. Planet Space Sci 46:1157

Keller CN, Anicich VG, Cravens TE (1998b) Model of Titan's ionosphere with detailed hydrocarbon ion chemistry. Planet Space Sci 46:1157–1174

Kliore AJ et al (2004) Cassini radio science. Space Sci Rev 115:1

Kliore AJ, Nagy AF, Marouf EA, French RG, Flasar FM, Rappaport NJ, Anabttawi A, Asmar SW, Kahann DS, Barbinis E, Goltz GL, Fleischman DU, Rochblatt DJ (2008) First results from the Cassini radio occultations of the Titan ionosphere. J Geophys Res 113:A09317

Krätschmer W, Lamb LD, Fostiropoulos K, Huffman DR (1990) Solid C_{60}: a new form of carbon. Nature 347:354–358

Krimigis SM et al (2004) Magnetospheric imaging instrument (MIMI) on the Cassini mission to Saturn/Titan. Space Sci Rev 114:233–329

Krimigis SM et al (2005) The dynamic Saturn magnetosphere: first results from Cassini/MIMI. Science 307:1270–1273

Kroto HW, Heath JR, O'Brien SC, Curl RF, Smalley RE (1985) C_{60}: Buckminsterfullerene. Nature 318:162–163

Kroto HW, Allaf AW, Balm SP (1991) C_{60}: buckminsterfullerene. Chem Rev 91:1213–1235

Krupp N, Lagg A, Woch J, Krimigis SM, Livi S, Mitchell DG, Roelof EC, Paranicas C, Mauk BH, Hamilton DC, Armstrong TP, Dougherty MK (2005) The Saturnian plasma sheet as revealed by energetic particle Measurements. Geophys Res Lett 32:L20S03. doi:10.1029/2005GL022829

Lammer H, Bauer SJ (1993) Atmospheric mass loss from Titan by sputtering. Planet Space Sci 41:657–663

Lammer H, Stumpter W, Bauer SJ (1998) Dynamic escape of H from Titan as a consequence of sputtering induced heating. Planet Space Sci 46:1207–1213

Lara LM, Lellouch E, López-Moreno JJ, Rodrigo R (1996) Vertical distribution of Titan's atmospheric neutral constituents. J Geophys Res 101:E10, 23,261–23,283

Lavvas PP, Coustenis A, Vardavas IM (2008a) Coupling photochemistry with haze formation in Titan's atmosphere: Part I: model description. Planet Space Sci 56:27–66

Lavvas PP, Coustenis A, Vardavas IM (2008b) Coupling photochemistry with haze formation in Titan's atmosphere, Part II: results and validation with Cassini/Huygens data. Planet Space Sci 56:67–99

Lean JL et al (2003) A new model of solar EUV irradiance variability 2. Comparisons with empirical models and observations and implications for space weather. J Geophys Res 108:A(2), 1059. doi:10.1029/2001JA009238

Ledvina SA, Cravens TE, Kecskemety K (2005) Ion distributions in Saturn's magnetosphere near Titan. J Geophys Res Space Phys 110:A06211

Liang M-C, Yung YL, Shemansky DE (2007) Photolytically generated aerosols in the mesosphere and thermosphere of Titan. Astrophys J Lett 661:L199–L202

Lipatov AS (2002) The Hybrid multiscale simulation technology. Springer

Lipatov A, Sittler EC Jr, Hartle RE (2008) Titan's plasma environment for T9 encounter: 3D hybrid simulation and comparison with observations, 2008 Joint Assembly, P41A-06

Lummerzheim D (1987) Electron transport and optical emissions in the aurora. Ph.D. Thesis, University of Alaska, Fairbanks

Lummerzheim D, Lilensten J (1994) Electron transport and energy degradation in the ionosphere: evaluation of the numerical solution. Comparison with laboratory experiments and auroral observations. Ann Geophys 12:1039–1051

Lutz B, de Bergh C, Owen T (1983) Titan: discovery of carbon monoxide in its atmosphere. Science 220:1374–1375

Ma Y, Nagy AF, Cravens TE, Sokolov IG, Clark J, Hansen KC (2004) 3-D global model prediction for the first close flyby of Titan by Cassini. Geophys Res Lett 31:L22803. doi:10.1029/2004GL02145

Ma Y, Nagy AF, Cravens TE, Sokolov IV, Hansen KC, Wahlund J-E, Crary FJ, Coates AJ, Dougherty MK (2006) Comparisons between MHD model calculations and observations of Cassini flybys of Titan. J Geophys Res 111:A05207. doi:10.1029/2005JA011481

Maurellis AN, Cravens TE, Gladstone GR, Waite Jr JH, Acton LW (2000) Jovian X-ray emission from solar X-ray scattering. Geophys Res Lett 27:1339

Maurice S, Sittler EC Jr, Cooper JF, Mauk BH, Blanc M, Selesnick RS (1996) Comprehensive analysis of electron observations at Saturn: Voyager 1 and 2. J Geophys Res 101:15211–15232

Maurice S, Blanc M, Prange R, Sittler EC Jr (1997) The magnetic-field-aligned polarization electric field and its effects on particle distribution in the magnetospheres of Jupiter and Saturn. Planet Space Sci 45(11):1449–1465

Menietti JD, Groene JB, Averkamp TF, Hospodarsky GB, Kurth WS, Gurnett DA, Zarka P (2007) Influence of Saturnian moons on Saturn kilometric radiation. J Geophys Res 112:A08211. doi:10.1029/2007JA012331

Michael M, Johnson RE (2005) Energy deposition of pickup ions and heating of Titan's atmosphere. Planet Space Sci 53:1510–1514

Michael M, Johnson RE, Leblanc F, Liu M, Luhmann JG, Shemantovich VI (2005) Ejection of nitrogen from Titan's atmosphere by magnetospheric ions and pickup ions. Icarus 175:263–267

Mitchell DG, Brandt PC, Roelof EC, Dandouras J, Krimigis SM, Mauk BH (2005) Energetic neutral atom emissions from Titan Interaction with Saturn's magnetosphere. Science 308:989

Modolo R, Chanteur GM (2008) A global hybrid model for Titan's interaction with the kronian plasma: application to the Cassini Ta flyby. J Geophys Res. doi:10.1029/2007JA012453

Modolo R, Wahlund J-E, Bostrom R, Canu P, Kurth , D. Gurnett, Lewis GR, Coates, Far plasma wake of Titan from the RPWS observations: a case study (2007a) Geophys Res Lett 34:L24S04. doi:10.1029/2007GL030482

Modolo R, Chanteur GM, Wahlund J-E, Canu P, Kurth WS, GurnettD, Matthews AP, Bertucci C (2007b) Plasma environment in the wake of Titan from hybrid simulation: a case study. Geophys Res Lett 34:L24S07. doi:10.1029/2007GL030489

Molina-Cuberos GJ, López-Moreno JJ, Rodrigo R, Lara LM, O'Brien K (1999) Ionization by cosmic rays of the atmosphere of Titan. Planet Space Sci 47:1347–1354

Molina-Cuberosa GJ, Lammer H, Stumptner W, Schwingenschuh K, Rucker HO, López-Moreno JJ, Rodrigo R, Tokano T (2001) Ionospheric layer induced by meteoric ionization in Titan's atmosphere. Planet Space Sci 49:143–153

Moustefaoui T, Rebrion-Rowe C, Jean-Luc Le Garrec, Rowe BR, Mitchell JBA (1998) Low temperature electron attachment to polycyclic aromatic hydrocarbons. Faraday Discuss 109:71–82

Müller-Wodarg ICF, Yelle RV, Mendillo MJ, Aylward AD (2003) On the global distribution of neutral gases in Titan's upper atmosphere and its effect on the thermal structure. J Geophys Res 108:1453. doi:10.1029/2003JA010054

Nagy AF, Banks PM (1970) Photoelectron fluxes in the ionosphere. J Geophys Res 75:6260–6270
Nagy AF, Cravens TE (2002) Solar system ionospheres. In: Mendillo M, Nagy AF, Waite JH (eds) Atmospheres in the solar system: comparative aeronomy, geophys. monograph, vol 130. AGU, Washington DC, pp 39–54
Ness NF, Acuna MH, Lepping RP, Connerney JEP, Behannon KW, Burlaga LF, Neubauer FM (1981) Magnetic field studies by Voyager 1: preliminary results at Saturn. Science 212:211
Ness NF, Acuna MH, Behannon KW et al (1982) The induced magnetosphere of Titan. J Geophys Res Space Phys 87:1369–1381
Neubauer FM, Gurnett DA, ScudderJD, Hartle RE (1984) Titan's magnetospheric interaction. In: Gehrels T, Matthews MS (eds) Saturn, pp 760–787. University of Arizona Press, Tucson
Neubauer FM et al (2006) Titan's near magnetotail from magnetic field and electron plasma observations and modeling: Cassini flybys, TA, TB and T3. J Geophys Res 111:A10220. doi:10.1029/2006JA011676
Niemann HB et al (2005) The abundance of constituents of Titan's atmosphere from the GCMS instrument on the Huygens probe. Nature. doi:10.1038/nature04122
O'Brien K (1969) Extra-nuclear hadron cascade calculation using Passow's approximation. Nucl Instrum Methods 72:93–98
O'Brien K (1970) Calculated cosmic ray ionization in the lower ionosphere. J Geophys Res 75:4357–4359
O'Brien K (1971) Cosmic-ray propagation in the atmosphere. Nuovo Cimento 3A:521–547
O'Brien K (1972) Propagation of muons underground and the primary cosmic ray spectrum below 40 TeV. Phys Rev D5:597–605
O'Brien K, McLaughlin JE (1970) Calculation of dose and dose-equivalent rates to man in the atmosphere from galactic cosmic rays. Hasl-228. Health and Safety Laboratories, U.S Atomic Energy Commission, New York
Oran ES, Julienne PS, Strobel DF (1975) The aeronomy of odd nitrogen in the thermosphere. J Geophys Res 80:3068–3076
Parker EN (1958) Dynamics of the interplanetary gas and magnetic fields. Astrophysical J 128:664–676
Postberg F, Kempf S, Hillier JK, Srama R, Green SF, McBride N, Grün E (2008) The E-ring in the vicinity of Enceladus II. Probing the moon's interior – the composition of E-ring particles. Icarus 193:438–454
Raulin F, Owen TC (2003) Organic chemistry and exobiology on Titan. Space Sci Rev 104:377–394
Rees MH (1963) Physics and chemistry of the upper atmosphere. Cambridge University Press, Cambridge
Richter, H. and J. B. Howard, Formation of polycyclic aromatic hydrocarbons and their growth to soot—a review of chemical reaction pathways, *Progress in Energy and Combustion Science*, **26**, 565–608, 2000.
Roe HG, de Pater I, MacIntosh BA, McKay CP (2002) Titan's clouds from Gemini and Keck adaptive optics imaging. Astrophys J 581:1399–1406
Sagan C, Khare BN (1979) Tholins – organic chemistry of interstellar grains and gas. Nature 277:102–107
Samuelson RE, Maguire WC, Hanel RA, Kunde VG, Jennings DE, Yung Y-L, Aikin AC (1983) CO_2 on Titan. J Geophys Res 88:8709
Shemansky DE, and Hall DT (1992) The distribution of atomic hydrogen in the magnetosphere of Saturn, *J. Geophys. Res.*, **97**, 4143–4161
Shemansky DE, Stewart AIF, West RA, Esposito LW, Hallett JT, Liu XM (2005) Science 308:978
Shemansky DE, Liang M, Yung Y (2008) Titan atmospheric structure from 2000 km to 300 km: models compared to Cassini UVIS observations. Presented at Titan after Cassini-Huygens symposium. Corpus Christi, Texas, 6–11 July 2008
Shemansky, DE, X. Liu and H. Melin (2009) The Saturn hydrogen plume, *Planet. Space Sci.*, in press
Shemantovich VI (1998) Kinetic modeling of suprathermal nitrogen atoms in the Titan's atmosphere: I. Sources. Solar System Res 32:384
Shemantovich VI (1999) Kinetic modeling of suprathermal nitrogen atoms in the Titan's atmosphere: II. Escape flux due to dissociation processes. Solar System Res 33:32
Shemantovich VI, Tully C, Johnson RE (2001) Suprathermal nitrogen atoms and molecules in Titan's corona. Adv Space Res 27:1875–1880
Shemantovich VI, Johnson RE, Michael M, Luhmann JG (2003) Nitrogen loss from Titan. J Geophys Res 108:E085087, E6, 1–11
Sillanpaa I, Kallio E, Janhunen P, Schmidt W, Mursula J Vilpolla, Tanskanen P (2006) Hybrid simulation study of ion escape at Titan for different orbital positions. Adv Space Res 38:799–805
Sittler EC Jr, Andre N, Blanc M, Burger M, Johnson RE, Coates A, Rymer A, Reisenfeld D, Thomsen MF, Persoon A, Dougherty M, Smith HT, Baragiola RA, Hartle RE, Chornay D, Shappirio MD, Simpson DG, McComas DJ, Young DT (2008) Ion and neutral sources and sinks within Saturn's inner magnetosphere: Cassini results, PSS, 56:3–18
Sittler EC Jr, Hartle RE (1996) Triton's ionospheric source: electron precipitation or photoionization. J Geophys Res 101:10,863
Sittler EC Jr, Ogilvie KW, Scudder JD (1983) Survey of low energy plasma electrons in Saturn's magnetosphere: Voyager 1 and 2. J Geophys Res 88:8847
Sittler EC Jr, Hartle RE, Vinas AF, Johnson RE, Smith HT, Mueller-Wodarg I (2004) Titan interaction with Saturn's magnetosphere: mass loading and ionopause location. In: Proceedings of the international conference TITAN from discovery to encounter. ESA Special Publication 1278, p 377, ESTEC. Noordwijk, Netherlands
Sittler EC Jr, Hartle RE, Vinas AF, Johnson RE, Smith HT, Mueller-Wodarg I (2005a) Titan interaction with Saturn's magnetosphere: Voyager 1 results revisited. J Geophys Res 110:A09302. doi:10.1029/2004JA010759
Sittler EC Jr, Thomsen M, Chornay D, Shappirio MD, Simpson D, Johnson RE, Smith HT, Coates AJ, Rymer AM, Crary F, McComas DJ, Young DT, Reisenfeld D, Dougherty M, Andre N (2005b) Preliminary results on Saturn's inner plasmasphere as observed by Cassini: comparison with Voyager. Geophys Res Lett 32:L14S07. doi:10.1029/2005GL022653
Sittler EC Jr, Johnson RE, Smith HT, Richardson JD, Jurac S, Moore M, Cooper JF, Mauk BH, Michael M, Paranicus C, Armstrong TP, Tsurutani B (2006a) Energetic nitrogen ions within the inner magnetosphere of Saturn. J Geophys Res 111:A09223. doi:10.1029/2004JA010509
Sittler EC Jr et al (2006b) Cassini observations of Saturn's inner plasmasphere: Saturn orbit insertion results. Planet Space Sci 54
Sittler EC Jr, Ali A, Cooper JF, Hartle RE, Johnson RE, Coates AJ, Young DT (2009a) Heavy ion formation in Titan's ionosphere: magnetospheric introduction of free oxygen and source of Titan's searosols? Planet Space Sci, in press
Sittler EC Jr, Hartle RE, Cooper JF, Lipatov AS, Johnson RE, Bertucci C, Coates AJ, Szego K, Shappirio M, Simpson DG, Tokar R, Young DT (2009b) Saturn's magnetosphere and properties of upstream flow at Titan: preliminary results. Planet Space Sci (submitted)
Sittler EC Jr, Hartle RE, Lipatov AS, Cooper JF, Bertucci C, Coates AJ, Szego K, Johnson RE, Shappirio M, Simpson DG, Wahlund J-E (2009c) Saturn's magnetospheric interaction with Titan as defined by Cassini encounters T9 and T18: new results. Planet Space Sci (submitted)
Smith HT, Johnson RE, Shemantovich V (2004) Titan's atomic and molecular nitrogen tori. Geophys Res Lett 31:116804. doi:10.1029/2004GL020580
Smith HT, Shappirio M, Sittler EC, Reisenfeld D, Johnson RE, Baragiola RA, Crary FJ, McComas DJ, Young DT (2005) Discovery of nitrogen in Saturn's inner magnetosphere. Geophys Res Lett 32:L14S03. doi:10.1029/2005GL022654
Smith HT, Mitchell D, Johnson RE, Paranicus C (2009) Energetic particle deposition in Titan's atmosphere, presented at Titan after

Cassini-Huygens symposium. Corpus Christi, Texas, 6–11 July 2008
Smith HT, Mitchell DG, Johnson RE, Paranicas CP (2009) Investigation of energetic proton penetration in Titan's atmosphere using the Cassini INCA instrument. Planet. Space Sci doi:10.1016/j.pss.2009.03.013
Spahn F, Albers N, Hörning M, Kempf S, Krivov AV, Makuch M, Schmidt J, Seiß M, Sremcevic M (2006a) E ring dust sources: Implications from Cassini's dust measurements. Planet Space Sci 54:1024–1032
Spahn F et al (2006b) Cassini dust measurements at Enceladus and implications for the origin of the E ring. Science 311:1416–1418
Srama R, Kempf S, Moragas-Klostermeyer G, Helfert S, Ahrens TJ, Altobelli N, Auer S, Beckmann U, Bradley JG, Burton M, Dikarev VV, Economou T, Fechtig H, Green SF, Grande M, Havnes O, Hillier JK, Horanyi M, Igenbergs E, Jessberger EK, Johnson TV, Krüger H, Matt G, McBride N, Mocker A, Lamy P, Linkert D, Linkert G, Lura F, McDonnell JAM, Möhlmann D, Morfill GE, Postberg F, Roy M, Schwehm GH, Spahn F, Svestka J, Tschernjawski V, Tuzzolino AJ, Wäsch R, Grün E (2006) In situ dust measurements in the inner Saturnian system. Planet Space Sci 54:967–987
Stevens M (2001) The EUV airglow of Titan: production and loss of NB_{2B} $c'B_{4B}(0)$-X. J Geophys Res 106:3685
Strobel DF (2008) Titan's hydrodynamically escaping atmosphere. Icarus 193:588–594
Strobel DF, Shemansky DE (1982) EUV emission from Titan's upper atmosphere: Voyager 1 encounter. J Geophys Res 87:1361–1368
Strobel DF, Summers ME, Zhu X (1992) Titan's upper atmosphere: structure and ultraviolet emissions. Icarus 100:512–526
Szego K et al (2005) The global plasma environment of Titan as observed by Cassini plasma spectrometer during the first two close encounters with Titan. Geophys Res Lett 32:L20S05. doi:10.1029/2005GL022646
Szego K, Bebesi Z, Bertucci C, Coates AJ, Crary F, Erdos G, Hartle R, Sittler EC Jr, Young DT (2007) Charged particle environment of titan during the T9 flyby. Geophys Res Lett 34:L24S03. doi:10.1029/2007GL030677
Thaddeus P (1994) On the large organic molecules in the interstellar gas, *Molecules and Grains in Space, AIP Conference Proceedings 312, Mont Sainte-Odile, France 1993, editor Irene Nenner*, 711–731
Thaddeus P (1995) Carbon chains and the diffuse interstellar bands, *A.G.G.M. Tielens and T.P. Snow (eds.), The Diffuse Interstellar Bands, Kluwer Academic Publishers, Printed in the Netherlands*, 369–378
Tobisca WK, Eparvier FG (1998) EUV97: improvements to EUV irradiance modeling in the soft X rays and FUV. Solar Phys 177:147
Tobisca WK, Woods T, Eparvier FG, Viereck R, Floyd L, Bouwer D, Rottman G, White OR (2000) The SOLAR2000 empirical solar irradiance model and forecast tool. J Atmos Sol Terr Phys 62(14):1233
Tomasko M et al (2005) Rain, winds and haze during Huygens probe's descent to Titan's surface. Nature 438(8):765–778
Toublanc D, Parisot JP, Gautier D, Raulin F, McKay CP (1995) Photochemical modeling of Titan's atmosphere. Icarus 113:2
Tucker OJ, Johnson RE (2008) A DSMC model of heating in Titan's upper atmosphere. Presented at Titan after Cassini-Huygens symposium. Corpus Christi, Texas, 6–11 July 2008
Vuitton V, Yelle RV, Anicich VG (2006) The nitrogen chemistry of Titan's upper atmosphere revealed. The Astrophys J 647:L175–L178
Vuitton V, Yelle RV, McEwan MJ (2007) Ion chemistry and N-containing molecules in Titan's upper atmosphere. Icarus 191:722–742
Vuitton V, Yelle RV, Lavvas P (2009) Composition and chemistry of Titan's thermosphere and ionosphere. Phil Trans R Soc A 367:729–741
Wahlund J-E, Bostrom R, Gustafsson G, Gurnett DA, Kurth WS, Pedersen A, Averkamp TF, Hospodarsky GB, Persoon AM, Canu P, Neubauer FM, Dougherty MK, Eriksson AI, Morooka MW, Gill R, Andre M, Eliasson L, Muller-Wodarg I (2005) Cassini measurements of cold plasma in the ionosphere of Titan. Science 308:986–989
Waite JH et al (2004) The Cassini ion and neutral mass spectrometer (INMS) Investigation. Space Sci Rev 114:113–231
Waite JH et al (2005) Ion neutral mass spectrometer results from the first flyby of Titan. Science 308:982–986
Waite JH Jr, Young DT, Cravens TE, Coates AJ, Crary FJ, Magee B, Westlake J (2007) The Process of Tholin Formation in Titan's Upper Atmosphere. Science 316:870 (11 May 2007). doi:10.1126/science.1139727
Waite JH Jr, Young DT, Coates AJ, Crary FJ, Magee BA, Mandt KE, Westlake JH (2008) The source of heavy organics and aerosols in Titan's atmosphere, submitted to organic matter in space. Proceedings IAU Symposium No. 251
Wang S, Buseck PR (1991) Packing of C60 molecules and related fullerenes in crystals – a direct view. Chem Phys Lett 182:1
Wei HY, Russell CT, Wahlund J-E, Dougherty MK, Bertucci C, Modolo R, Ma YJ, Neubauer FM (2007) Cold ionospheric plasma in Titan's magnetotail. Geophys Res Lett 34:L24S06. doi:10.1029/2007GL030701
Weiske T, Bohme DK, Schwarz H (1991) Injection of helium atoms into doubly and triply charged C_{60} cations. J Phys Chem 95:8451–8452
Wilson EH, Atreya SK (2004) Current state of modeling the photochemistry of Titan's mutually dependent atmosphere and ionosphere. J Geophys Res 109:E06002. doi:10.1029/2003JE002181
Wong A-S, Morgan CG, Yung YL, Owen T (2002) Evolution of CO on Titan. Icarus 155:382–392. doi:10.1006/icar.2001.6720
Yang SH, Pettiette CL, Conceicao J, Cheshnovsky O, Smalley RE (1987). Chem Phys Lett 139:233
Yelle RV (1991) Non-LTE models of Titan's upper atmosphere. Astrophys J 383:380–400
Yelle RG, Borggren N, de la Haye V, Kasprzak WT, Niemann HB, Muller-Wodarg I, Waite JH Jr (2006) The vertical structure of Titan's upper atmosphere from Cassin ion neutral mass spectrometer measurements. Icarus 182:567–576
Yelle RV, Cui J, Müller-Wodarg ICF (2008) Methane escape from Titan's atmosphere. J Geophys Res, 113:E10003. doi:10.1029/2007JE003031
Young DT et al (2004) Cassini plasma spectrometer investigation. Space Sci Rev 114:1–112
Young DT et al (2005) Composition and dynamics of plasma in Saturn's magnetosphere. Science 307:1262
Yung YL (1987) An update of nitrile photochemistry on Titan. Icarus 72:468
Yung YL, Allen M, Pinto JP (1984) Photochemistry of the atmosphere of Titan: comparison between model and observations. Astrophys J Suppl 55:465

Chapter 17
Titan in the Cassini–Huygens Extended Mission

C. J. Hansen, J. H. Waite, and S. J. Bolton

17.1 Titan in the Cassini–Huygens Extended Mission

At the culmination of the Primary Mission Cassini began a 2-year extended orbital phase called the Cassini Equinox Mission. The sun crosses Saturn's equatorial plane on 11 August 2009. With 2 extra years the discoveries and deficiencies in the Primary Mission can be addressed, as well as new opportunities presented by the equinox crossing. This chapter covers the state of knowledge and outstanding questions at the end of the Primary Mission, and outlines the opportunities in the Equinox Mission. Key questions are identified and the strategies by which they will be addressed are discussed. Specific geometries and scientific goals of individual Titan flybys are described. The chapter ends with a brief description of the proposed Cassini Solstice Mission, which may extend spacecraft operation to 2017.

17.1.1 Overview

The Cassini–Huygens extended mission, called the Equinox Mission (EM), began July 1, 2008. There will be 26 flybys of Titan in the 2-year extended mission, numbered T45 to T70. Tables 17.1 and 17.2 give the flyby dates and important geometric data for each of the Titan flybys in the primary and extended mission phases, respectively. The extended mission tour of the Saturnian system was designed to fill in gaps in the Primary Mission (PM), either in terms of scientific investigations or in missing geometric conditions, and to respond to discoveries made in the PM.

The overall cadence of the Cassini equinox mission is driven by a balance of different disciplinary goals. Cassini's orbit inclination at the end of the primary mission is 74.8°. Figure 17.1 shows the inclination profile punctuated by the EM Titan flybys. High inclination orbits for the first ~9 months of the EM enable viewing of stellar occultations for ring science, auroral observations, and polar passes of Enceladus. Good views of Titan's south polar region occur early while the region is still well-illuminated. Titan flybys flip from outbound to inbound node crossings between T51 and T52, changing the local solar time of the flybys from ~10 to 22 h, respectively. This sets up the orbit orientation to satisfy the Titan scientific requirements for encounters in the dusk sector of Saturn's magnetosphere. The ring science requirement to observe the equinox with the spacecraft 15° above Saturn's equatorial plane sets the slow pace from T52 to T62 of returning the spacecraft to an equatorial orbit for more icy satellite flybys and ansa-to-ansa ring occultations. T52, T53 and T54 were designed for high quality occultations of the sun and earth by Titan's atmosphere. Missing from the primary mission, thus deliberately designed into the extended mission tour, were wake passages and dusk encounters, achieved in T63 to T70. The tour is described in detail in Buffington et al. 2008.

One of the prime drivers in designing the extended mission tour was simply to have numerous Titan flybys. The decision early in the project development to eliminate the spacecraft scan platform means that multiple experiments cannot be carried out simultaneously, thus any given Titan flyby must be dedicated to just a few of Cassini's dozen instruments.

Considering the 2-year EM combined with the 4-year primary mission we also have the opportunity to observe seasonal changes. One Saturn year is 29.47 earth years; when Cassini–Huygens arrived at Saturn the season was winter in the northern hemisphere. Within the extended mission the sun crosses Saturn's equatorial plane on August 11, 2009, and the northern hemisphere of Titan will experience the onset of spring. Figure 17.2 shows the "calendar" of Titan

C.J. Hansen(✉)
Jet Propulsion Laboratory, California Institute of Technology, 4800 Oak Grove Dr. Pasadena, 91109-8099, CA, USA
e-mail: Candice.j.Hansen@jpl.nasa.gov

J.H. Waite, and S.J. Bolton
Southwest Research Institute, 6220 culebra Rd., San Antonio, TX 78238-5166, USA

R.H. Brown et al. (eds.), *Titan from Cassini-Huygens*,

Table 17.1 Titan flybys in the primary mission. Values are given for closest approach (C/A) to Titan. C/A time is formatted Day-of-year Thours:minutes:seconds. "LST" is Local Solar Time

Flyby	Orbit	Date	C/A Time	Altitude (km)	LST	Lat	Lon (West)	Phase angle
T0	0	3-Jul-04	184T09:29:00	341,500	5.76	−65.1	333.17	67.0
Ta	A	26-Oct-04	300T15:20:33	1,174	11.00	38.8	88.5	90.9
Tb	B	13-Dec-04	348T11:38:13	1,192	10.47	59.1	84.2	101.5
Tc	C	13-Jan-05	014T11:12:00	60,003	10.47	0.1	251.8	92.9
T3	3	15-Feb-05	046T06:54:21	1,579	10.33	29.9	68.9	98.5
T4	5	1-Apr-05	091T19:55:12	2,404	5.27	33.4	118.7	66.4
T5	6	16-Apr-05	106T19:11:46	1,027	5.27	74.0	272.3	127.2
T6	13	22-Aug-05	234T08:53:37	3,660	5.00	−58.5	102.7	43.1
T7	14	7-Sep-05	250T07:50:26	1,075	4.93	−67.0	308.1	84.5
T8	17	28-Oct-05	301T03:58:09	1,353	9.27	1.2	246.1	104.8
T9	19	26-Dec-05	360T18:54:15	10,411	3.00	−0.2	110.4	67.1
T10	20	15-Jan-06	015T11:41:27	2,043	8.47	0.1	250.8	120.5
T11	21	27-Feb-06	058T08:25:19	1,812	1.13	0	107.4	92.5
T12	22	19-Mar-06	078T00:05:57	1,949	6.40	0.1	250.6	148.0
T13	23	30-Apr-06	120T20:53:31	1,856	23.13	0.1	106.3	120.7
T14	24	20-May-06	140T12:18:12	1,879	4.40	0.4	249.9	163.0
T15	25	2-Jul-06	183T09:12:19	1,906	21.20	−0.4	105.5	147.9
T16	26	22-Jul-06	203T00:25:13	950	2.40	85.2	316.2	105.3
T17	28	7-Sep-06	250T20:12:04	1,000	2.27	23	56.74	44.7
T18	29	23-Sep-06	266T18:58:49	960	2.27	71.0	357.0	89.9
T19	30	9-Oct-06	282T17:23:24	980	2.20	60.7	357.5	80.9
T20	31	25-Oct-06	298T15:58:07	1,030	2.13	7.5	44.34	25.3
T21	35	12-Dec-06	346T11:41:31	1,000	2.00	43.9	265.0	123.9
T22	36	28-Dec-06	362T10:05:22	1,297	1.93	40.2	357.6	61.6
T23	37	13-Jan-07	013T08:34:00	1,000	1.93	30.5	357.9	53.1
T24	38	29-Jan-07	029T07:15:55	2,631	1.87	32.8	330.1	72.4
T25	39	22-Feb-07	053T03:10:59	1,000	13.8	30.3	16.25	161.2
T26	40	10-Mar-07	069T01:47:22	981	13.8	31.7	357.9	149.5
T27	41	26-Mar-07	085T00:21:52	1,010	13.73	41.1	358	144.0
T28	42	10-Apr-07	100T22:58:00	991	13.67	50.4	358.1	137.2
T29	43	26-Apr-07	116T21:32:52	981	13.67	59.7	358.4	129.6
T30	44	12-May-07	132T20:08:14	959	13.60	68.9	358.9	121.4
T31	45	28-May-07	148T18:51:27	2,299	13.60	76.8	360	114.3
T32	46	13-Jun-07	164T17:47:57	965	13.53	84.5	0.9	106.9
T33	47	29-Jun-07	180T16:59:46	1,932	13.53	8.1	294.9	95.6
T34	48	19-Jul-07	200T00:39:58	1,332	18.80	1.3	244.7	34.0
T35	49	31-Aug-07	243T06:34:25	3,324	11.53	63.5	110.6	87.0
T36	50	2-Oct-07	275T04:49:50	973	11.47	−59.9	109.0	67.5
T37	52	19-Nov-07	323T00:52:51	999	11.40	−22.1	117.4	51.3
T38	53	5-Dec-07	339T00:07:37	1,298	11.40	−78.9	175.1	69.7
T39	54	20-Dec-07	354T22:56:41	970	11.33	−70.1	176.6	61.3
T40	55	5-Jan-08	005T21:26:24	1,014	11.33	−12.2	130.4	37.1
T41	59	22-Feb-08	053T17:39:08	1,000	11.20	−34.7	151.7	30.1
T42	62	25-Mar-08	085T14:36:12	999	11.13	−27.1	156.4	21.3
T43	67	12-May-08	133T10:09:59	1,001	11.00	17.1	137.2	34.8
T44	69	28-May-08	149T08:33:21	1,400	10.93	12.4	150.1	23.1

flybys, compared to the equivalent earth date. Voyager flybys took place on earth-equivalent dates March 23 and April 3.

The scientific objectives for Titan in both the primary and extended mission phases have been divided informally into disciplines proceeding from Titan's interior out to its exosphere: interior structure, surface science, atmospheric investigations, and study of the interface of Titan's upper atmosphere with Saturn's magnetosphere. This chapter is divided into those four categories. For each we discuss how the new state of knowledge from data acquired in the primary mission drove

Table 17.2 Titan flybys in the Equinox Mission. Values are given for closest approach (C/A) to Titan. "LST" is Local Solar Time

Flyby	Orbit	Date	C/A Time	Altitude (km)	LST	Lat	Lon (West)	Phase angle
T45	78	31-July-08	213T02:13:11	1,614	11.0	−43.31	195.40	48.74
T46	91	3-Nov-08	308T17:35:23	1,100	10.5	−3.51	339.91	170.90
T47	93	19-Nov-08	324T15:56:28	1,023	10.5	−21.56	177.60	28.05
T48	95	5-Dec-08	340T14:25:45	960	10.5	−10.28	178.78	25.01
T49	97	21-Dec-08	356T12:59:52	970	10.4	−43.85	236.81	83.55
T50	102	7-Feb-09	038T08:50:52	960	10.2	−33.67	306.32	135.80
T51	106	27-Mar-09	086T04:43:37	960	10.0	−30.27	234.80	86.56
T52	108	4-Apr-09	094T01:47:48	4,150	22.0	−2.81	176.29	151.50
T53	109	20-Apr-09	110T00:20:46	3,600	22.0	−7.74	177.61	148.40
T54	110	5-May-09	125T22:54:16	3,244	22.0	−13.78	177.78	145.60
T55	111	21-May-09	141T21:26:42	965	22.0	−22.03	177.77	141.30
T56	112	6-June-09	157T20:00:01	965	22.0	−32.10	177.96	134.90
T57	113	22-June-09	173T18:32:36	955	22.0	−42.17	178.12	127.70
T58	114	8-July-09	189T17:04:04	965	22.0	−52.25	178.38	120.20
T59	115	24-July-09	205T15:34:04	955	22.0	−62.31	179.05	112.50
T60	116	9-Aug-09	221T14:03:54	970	21.9	−72.49	179.98	104.40
T61	117	25-Aug-09	237T12:51:39	970	21.8	−19.34	237.12	85.79
T62	119	12-Oct-09	285T08:36:25	1,300	21.7	−61.47	68.81	99.34
T63	122	12-Dec-09	346T01:03:15	4,850	17.0	33.20	114.78	124.20
T64	123	28-Dec-09	362T00:17:00	9,55	17.0	81.77	172.17	85.90
T65	124	12-Jan-10	012T23:10:37	1,073	17.0	−81.92	358.35	95.15
T66	125	28-Jan-10	028T22:28:50	7,490	17.0	−52.75	297.07	69.11
T67	129	5-Apr-10	095T15:50:39	7,462	21.0	0.0	240.35	73.00
T68	131	20-May-10	140T03:24:21	1,400	16.0	−49.09	116.61	112.40
T69	132	5-June-10	156T02:26:28	2,044	16.0	87.29	9.36	87.26
T70	133	21-June-10	172T01:27:18	880	16.0	83.52	171.67	82.43

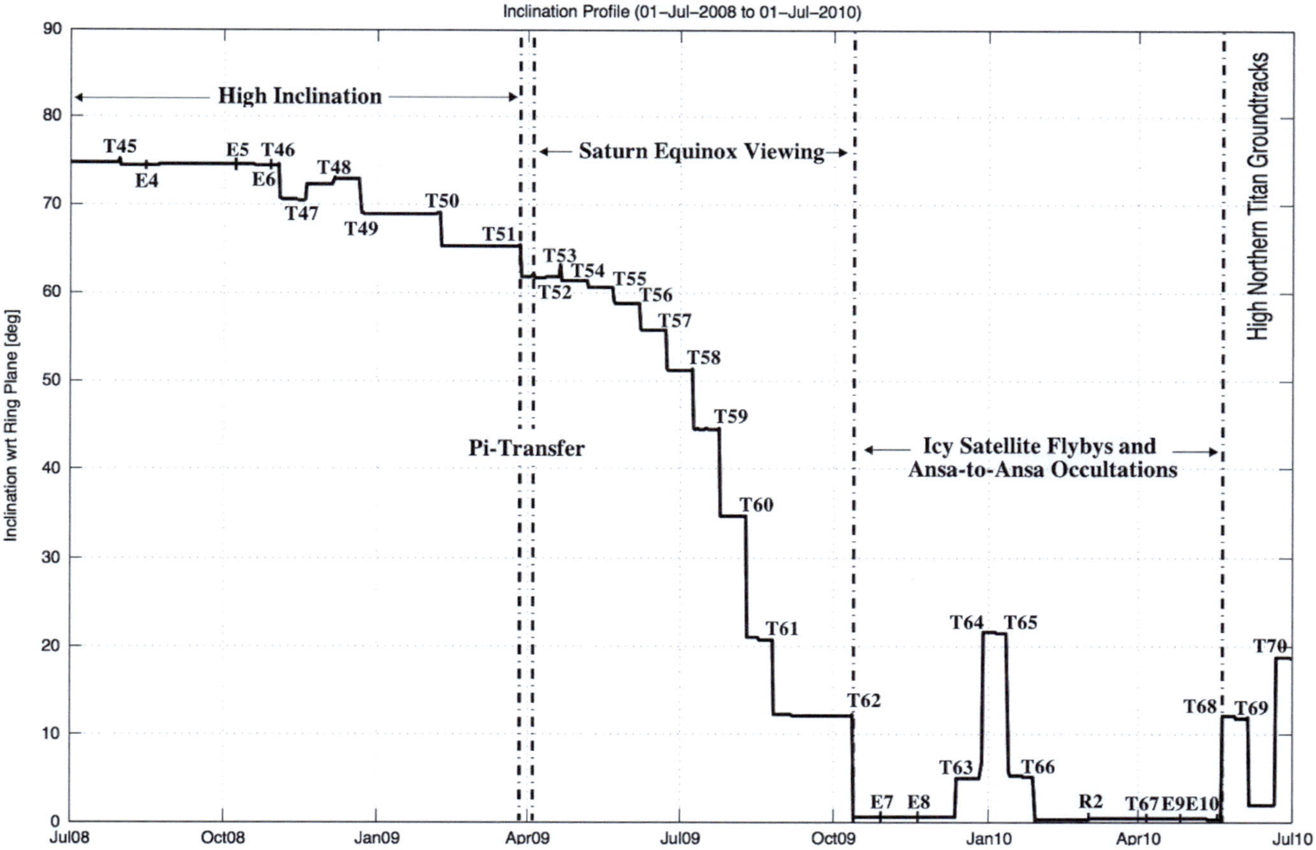

Fig. 17.1 Orbit inclination profile in the EM. Cassini's orbit inclination is 74.8° at the end of the primary mission. Titan flybys are used to shape the inclination profile and to rotate the long axis of Cassini's elliptical orbit around Saturn. This figure was originally published in Buffington et al. 2008.

Prime Mission Titan Activity	Extended Mission Titan Activity	Voyager Flyby Dates	Possible SM

January

M	T	W	T	F	S	S
				1	2	3
4	5	6	7	8	9	10
11 SOI	12	13	14	15 TA	16 TB	17 H
18 T4	19	20 T4&T5	21	22	23	24
25 T6&T7	26	27 T8	28	29 T9	30 T10	31 T11

February

M	T	W	T	F	S	S
1 T12	2 T13	3 T14	4 T15	5 T16	6	7 T17&T18
8 T18&T19	9	10 T21	11 T22&T23	12 T24	13 T25&T26	14 T27&T28
15 T29&T30	16 T31&T32	17 T33	18 T34	19 T35	20 T36	21
22 T37&T38	23 T39&T40	24	25 T41	26 T42	27	28 T43&T44

March

M	T	W	T	F	S	S
1 EOM	2 T45	3	4	5 T46	6 T47	7 T48&T49
8	9 T50	10 T51	11 T52&T53	12 T54&T55	13 T56&T57	14 T58&T59
15 T60&T61	16	17 T62	18	19 T63	20 T64&T65	21 6&Equinox
22	23 T67&V1	24	25 T68&T69	26 T70&T71	27	28
29 T72	30	31 T73				

April

M	T	W	T	F	S	S
			1 T74	2	3 T75&V2	4 T76
5 T77	6	7	8 T78	9	10	11 T79
12	13	14 T80	15 T81&T82	16	17	18 T83
19 T84	20	21 T85	22	23 T86	24 T87	25 T88
26	27	28 T89	29 T90	30		

May

M	T	W	T	F	S	S
					1 T91	2
3 T92&T93	4	5 T94	6 T95	7 T96	8 T97	9
10 T98	11 T99	12 T100	13 T101	14 T102	15 T103	16 T104
17 T105	18 T106	19	20 T107	21 T108	22 T109	23 T110
24	25 T111	26	27 T112	28	29	30 T113
31						

June

M	T	W	T	F	S	S
	1 T114	2	3 T115&T11(	4 T117	5 T118	6
7 T119	8 T120	9 T121	10 T122	11 T123	12	13 T124
14 T125	15	16	17	18 T126	19	20
21 Solstice	22	23	24	25	26	27
28	29	30				

Fig. 17.2 Titan flybys in an earth-equivalent Saturn calendar. After a 6 year mission Cassini observations of Titan cover just the northern winter/southern summer season. Titan flybys are shown on earth-equivalent dates during the primary mission (*light green*) and the EM. (*dark green*). The proposed SM flybys (*yellow*) will go to summer solstice, equivalent to just half of an earth year. Voyager flybys occurred on earth-equivalent dates of March 23 and April 3

the identification of new objectives in the extended mission, the required geometries for the extended mission observations, the flybys that will achieve the new scientific goals, and the anticipated state of knowledge at the end of the extended mission. The final section addresses tasks and geometries remaining to be achieved in the Solstice Mission, a proposed extension of Cassini operation that goes out to 2017.

17.2 Interior Structure

There are only a few ways that Cassini can probe the deep interior of Titan. These are the gravitational field and the figure of Titan, and magnetic fields of internal origin. Results from the PM are covered in Chapter 4. At the end of the primary mission two fundamental questions regarding Titan's interior remain:

- Does Titan have an intrinsic or induced magnetic field?
- Does Titan have an internal liquid layer / ocean?

17.2.1 Internal Magnetic Field

Titan could have an intrinsic magnetic field due to internal dynamo currents or an induced time-varying magnetic field due to currents driven by Saturn's magnetic field: the energy source of a dynamo is located in the interior of the body, while an induced field is generated in response to the penetration of an external time-varying field into the moon's body, provided that its interior contains conducting elements. Observational evidence shows the presence of an induced field, but whether the conduction in Titan's ionosphere has a contribution from the conducting fluid in the interior is unknown (Backes et al. 2005). The first step is to determine whether an intrinsic field exists and what its contribution is, followed by observations that allow separation of an interior induced field from the ionosphere's induced field.

In the primary mission the closest Titan flybys were at an altitude of 950 km. The presence of Titan's ionosphere and the distance and geometries of the close flybys relative to the orientation of Saturn's magnetic field masked any signature of an intrinsic Titan field. Small fields of internal origin would be easiest to detect in spatial regions near Titan that are not reached by field lines of Saturn's magnetic field because of efficient shielding by high electric conductivity in Titan's ionosphere. It was thus a high priority to go lower than 900 km, below Titan's ionosphere, on one of the flybys in the EM. The induced field is produced most effectively when Titan is on the dawn side of Saturn because the corotation plasma carrying Saturn's field impacts Titan on the night side, thus to see the intrinsic field a dusk pass is preferred. Shielded regions occur on the dayside of Titan because photoionization is by far the most important mechanism of ionization, thus this low flyby needed to be a dayside pass in the dusk quadrant of Saturn's magnetosphere where the corotation plasma comes from the same direction as the extreme ultraviolet (EUV) flux from the sun. Figure 17.3 illustrates the desired dropoff in Saturn's B field below 900 km.

Flyby T70 has the desired geometry, is at an altitude of 880 km, and will be flown in an attitude that minimizes torque on the spacecraft. One of the reasons that primary mission flybys did not go lower than 950 km was due to concerns that the pressure of the atmosphere on the spacecraft would introduce a bias in the spacecraft attitude that the spacecraft thrusters could not overcome. This would result in loss of pointing control and if it lasted long enough would put the spacecraft in safe mode. The spacecraft attitude in T70 has been selected to be the most stable available relative to the atmospheric flow. The most important magnetometer and Langmuir probe data are not very sensitive to spacecraft orientation. Unfortunately however Cassini Plasma Spectrometer (CAPS), Magnetospheric Imaging Instrument (MIMI) and Ion and Neutral Mass Spectrometer (INMS) data will not be optimal.

17.2.2 Internal Ocean

To carry out a gravity science flyby the spacecraft High Gain Antenna is pointed to earth. The telecom link is maintained throughout the flyby in order to detect subtle changes in the Doppler shift that can be attributed to Titan's gravity field.

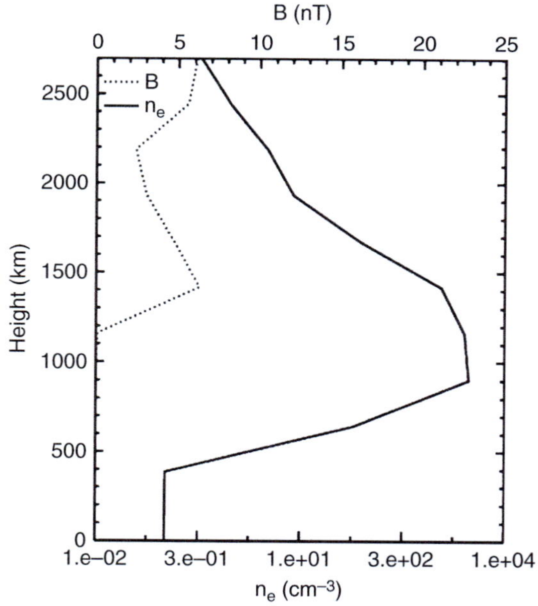

Fig. 17.3 Internal Dynamo or Induced Field. The ideal geometry for detection of an internal magnetic field requires a spacecraft dayside flyby altitude of <900 km, so that Saturn's magnetic field is blocked by Titan's ionosphere. In this plot B is the magnitude of Saturn's magnetic field, decreasing to zero as the number of electrons (n_e) in Titan's ionosphere increases. This figure was originally published by Heiko Backes (2004). Model computations assumed Saturn's plasma inflow along the positive x-axis under the solar conditions corresponding to flyby T34 of Cassini's primary mission. Incident plasma conditions were considered Voyager-like.

The Love number, k_2, is indicative of whether or not there is a layer of liquid in the interior decoupling the surface from the interior, although this result is model-dependent. The original plan in the primary mission to look for the existence of an internal liquid layer called for four flybys to measure the Love number k_2: two equatorial passes – one when Titan was at perikrone and one at apokrone, and two inclined passes – also at perikrone and apokrone. Changes in Titan's figure due to tidal and centrifugal forces would be most different and detectable between these geometries. The appropriate four primary mission flybys (T11, T22, T33 and T38) were allocated to Radio Science (RSS). Later it was found that the Deep Space Network (DSN) viewing conditions on earth for T38 were not adequate and T38 was reassigned to the Visible and Infrared Mapping Spectrometer (VIMS) for surface observations. In the extended mission it was a tour requirement to have a replacement for the T38 flyby. T45 has the required geometry, including the earth / DSN phasing needed. T68 is a backup in the event of problems at the DSN station.

A k_2 value of 0.04 is consistent with no liquid layer, while k_2 between 0.33 and 0.5, indicative of larger tidal deformation, corresponds to the presence of a liquid layer. (The exact value depends on the thickness of the ice crust over the liquid layer.) To be certain we can differentiate between the no-liquid-layer case and an internal ocean at the 2-sigma or 95%

level, the uncertainty in k_2 must be <0.1 (Rappaport et al. 2008). After three flybys in the primary mission the k_2 uncertainty of 0.2 precluded a definitive answer. One challenge to data interpretation was that the spacecraft experienced unanticipated torques due to Titan's atmosphere even at the relatively high 1400 km altitude of the T22 flyby. Another reason it has been so difficult to determine the value of k_2 may be that Titan's interior is not divided neatly into concentric shells, that the mass distribution in the interior is more complex (see Chapter 4). Simulations show that T45, the fourth flyby, will bring the uncertainty in k_2 down to 0.12, thus addressing the goal of determining the existence of a liquid layer (Rappaport et al. 2008).

In the meantime events have overtaken these EM plans. The mismatch of surface features from their predicted locations discovered in overlapping Radar swaths revealed that Titan is not rotating synchronously (Stiles et al. 2008). The discovery by the Radar team that Titan's obliquity is 0.3° and that the spin rate is slightly non-synchronous implies that an internal liquid layer is likely. Such a layer decouples the ice crust from the rest of the planet and thus makes it easier for exchange of angular momentum between the super-rotating atmosphere and the surface to change the observed rotation (Tokano and Neubauer 2005; Lorenz et al. 2008b). While the long-term average rotation is synchronous there are seasonal variations about that rate. Polar precession was also measured (Stiles et al. 2008) although libration was not included in the rotation solution. Data on surface positions after the equinox will be key in constraining libration and confirming model predictions of the coupling of the winds with the surface in response to the seasonal zonal wind changes.

17.3 Surface Science

As Cassini approached Saturn the only known surface feature on Titan was Xanadu (Smith et al. 1996). Now, at the culmination of the primary mission, we know that Titan has geology that rivals the complexity of earth. We now know that Titan has dunes, lakes of liquid ethane and methane, eroded drainage channels, craters, mountains and evidence for cryovolcanism (see Chapters 5 and 6). Maps show features, and features have names, evidence of the progress in exploration that has been made by the Cassini–Huygens mission.

17.3.1 Coverage and Resolution

One of the most important aspects of understanding a body's geologic evolution is surface coverage at an adequate resolution to see processes at work. Only the most general of planetary structures such as continents, basins, polar caps, or large craters can be studied with 10–100 km resolution. Below 10 km it becomes possible to identify regional provinces such as mountains, tectonic ridges, volcanic constructs and medium size craters. To identify geological features such as river channels, dunes, lakes, or cryovolcanic flow with certainty resolution better than 1 km is best. It is important to note that even though a surface feature may be quite large, its correct identification could require high resolution data. For example a channel might be identifiable in moderate resolution images, but to determine whether it is a tectonic rift or has been fluvially eroded requires high resolution. The peril of prematurely interpreting planetary evolution with incomplete surface coverage is well illustrated by the history of Mars exploration.

Three of Cassini's instruments are used to image the surface: Radar, the Imaging Science Subsystem (ISS) and VIMS. The lack of articulation on the Cassini spacecraft means that Radar data must be acquired separately from ISS and VIMS, which are bore sighted with respect to each other but orthogonal to the Radar. By the culmination of the primary mission Radar had achieved 22% coverage of the surface at 500 m resolution, as illustrated in Fig. 17.4a. Figure 17.4b shows the additional 8% surface coverage Radar will achieve by the end of the EM. The hazy atmosphere limits ISS resolution to ~1 km other than under ideal viewing conditions. Figure 17.5a illustrates ISS coverage and resolution at the end of the primary mission. As the sun moves toward the northern hemisphere higher northern latitudes will be imaged. Much additional coverage to be achieved in the EM is focused on filling in gaps, particularly on the trailing side (hemisphere centered on 270W longitude), and latitudes below ~40S before the equinox. It is difficult to get coverage of the leading (hemisphere centered on 90W longitude) and trailing sides of Titan on the asymptotes of ~equatorial flybys (i.e. approach and departure), because within 15° of these longitudes the spacecraft will be on a trajectory that will impact Saturn's rings. Furthermore, to get the closest approach at 90^0 or 270^0 longitude puts the spacecraft on an escape trajectory.

VIMS acquires only sparse high resolution coverage in the PM and EM, partially due to the limited number of flybys with adequate illumination at closest approach. Ideally the spacecraft should be at a phase angle <45° when it is within 50,000 km of the surface. Figure 17.6 shows range vs. phase angle for all the flybys in the EM and how few meet this criteria. It will be a Solstice Mission goal to design a tour with flybys that are at low phase angle near closest approach. Figure 17.7 illustrates the coverage and resolution VIMS will have achieved by the culmination of the PM and EM at better than 10 km resolution. VIMS has mapped about 60% of the surface at better than 20 km resolution.

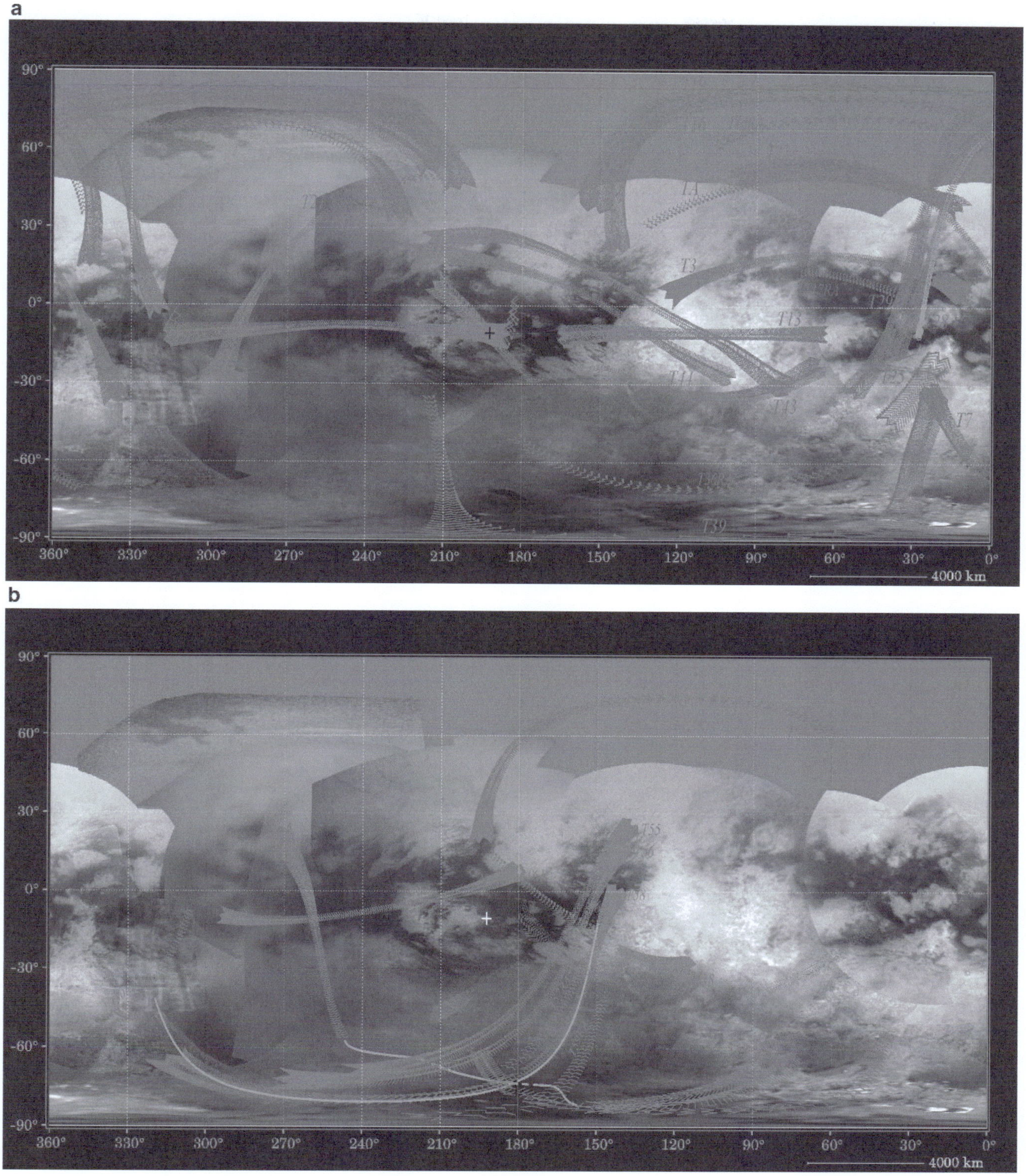

Fig. 17.4 Radar coverage in the PM and the EM. (**a**) High resolution (better than 500 m) Radar coverage in the primary mission was 22% of Titan's surface. The authors thank Yanhua Anderson and Ralph Lorenz for providing this figure. (**b**) Additional EM high resolution Radar coverage is 8% of Titan's surface.

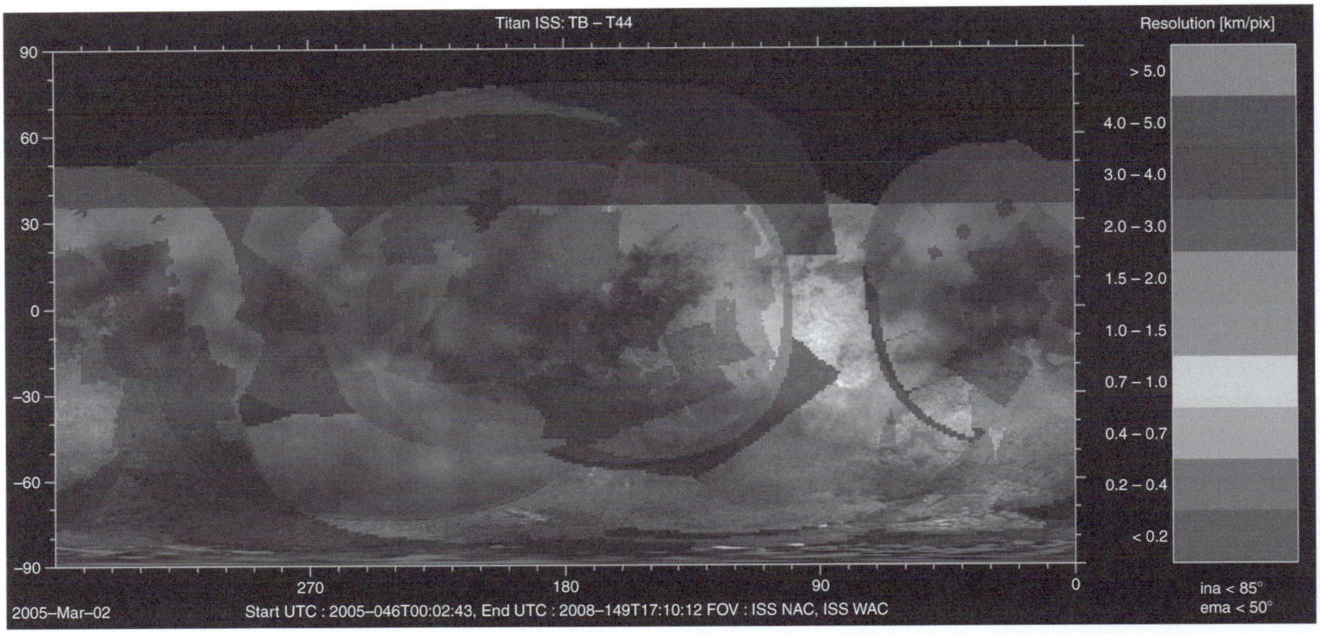

Fig. 17.5 ISS coverage and resolution in the primary mission. Acquiring coverage at leading (90W) and trailing (270W) longitudes and in the southern hemisphere are important goals for the EM, to fill in regions imaged only at low resolution in the PM. More northern hemisphere coverage will be acquired as the subsolar point moves north. Atmospheric haze limits ISS resolution to ~1 km even when the spatial scale is less than 1 km. The authors thank Elizabeth Turtle for providing this figure.

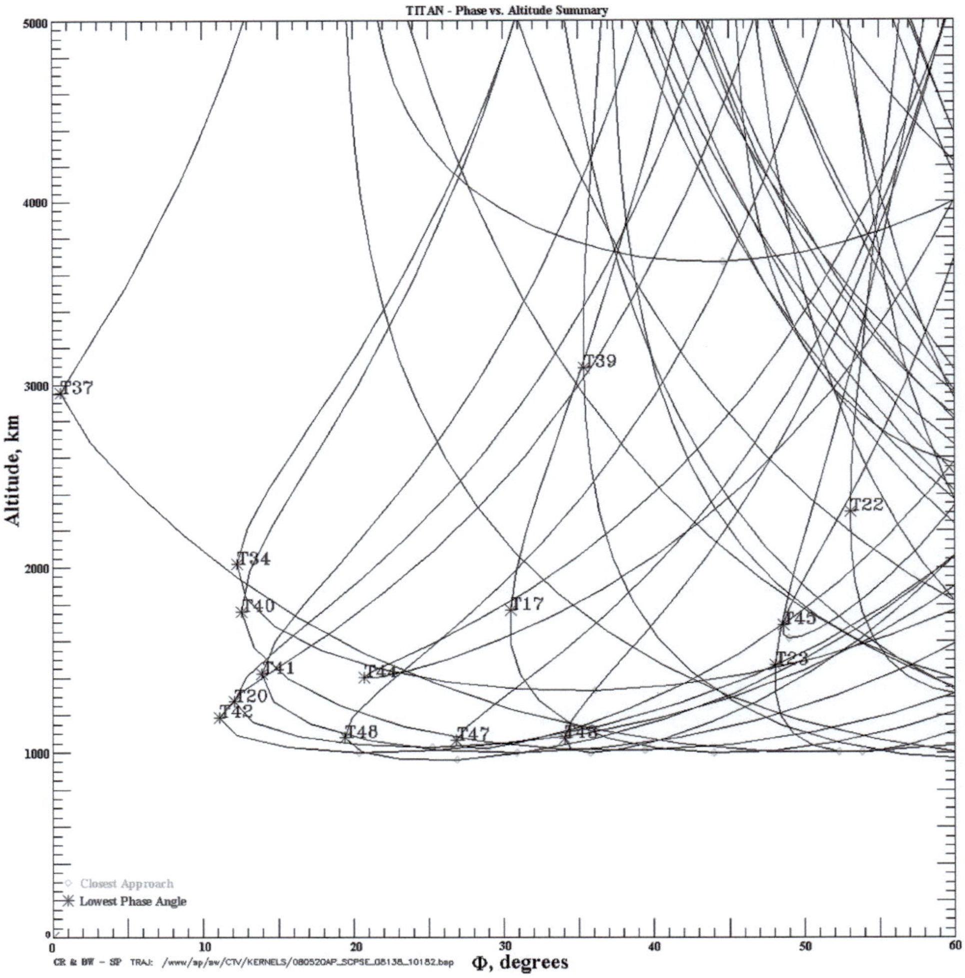

Fig. 17.6 Lighting conditions in the EM for composition measurements by VIMS. Ideal lighting conditions for VIMS are bounded by phase angles less than 45° at altitudes less than 5,000 km. The authors thank Chris Roumeliotis for providing this figure.

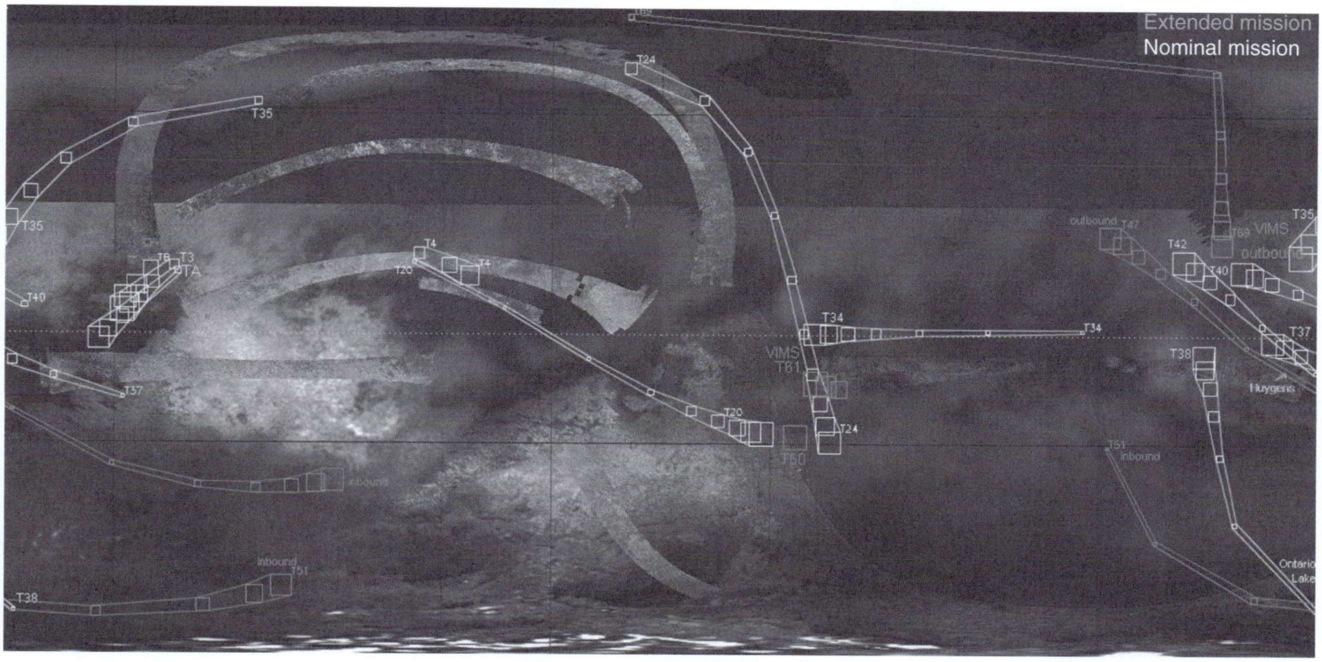

Fig. 17.7 VIMS coverage in the PM and EM at resolution better than 10 km. VIMS collects narrow swaths of data near closest approach, so it takes numerous flybys to build up high resolution coverage. This plot shows swaths collected at 5 km/pixel scale and higher, along with low resolution ISS near-global coverage and radar swaths. The authors thank Stephane Le Mouelic at the University of Nantes for providing this figure.

17.3.2 Targeted Scientific Investigations in the EM

EM scientific goals are directed at understanding processes identified in the PM, described in detail in Chapter 5. Table 17.3 lists specific targets for Titan flybys dedicated to surface scientific observations. In the EM we will continue to fill in coverage of Titan's surface at radar and optical wavelengths in order to:

- Map surface units, topography, stratigraphy – especially areas previously imaged only at low resolution; characterize latitudinal trends; Map surface composition.
- Investigate cryovolcanism; Look for changes that may be evidence of ongoing volcanic activity.
- Study eolian processes, dunes.
- Improve characterization of methane cycle timescales; Monitor seasonal change such as lakes filling and drying up.

17.3.2.1 Surface units

Major surface units identified in the primary mission by Cassini–Huygens are the ~equatorial dune fields (Lorenz et al. 2006; Radebaugh et al. 2008b), the lakes and lakebeds at high southern and northern latitudes (Stofan et al. 2007; Brown et al. 2008, Turtle et al. 2008), mountain chains (Radebaugh et al. 2008a) and relatively high terrain cut by fluvial channels draining into dark plains (Soderblom et al. 2007a; Barnes et al. 2007; Lorenz et al. 2008a). The small number of craters is consistent with a young surface modified by erosional and depositional processes (Porco et al. 2005). Tectonic activity is implied by Titan's young surface and chains of mountains, but a definitive picture of tectonic processes has not emerged from data collected in the PM. In the EM the T50 Radar swath crosses mountains identified in lower resolution VIMS data. Filling gaps in global coverage will complete the inventory of surface units and further the investigation of their latitudinal distribution, which also gives us insight into Titan's meteorology. T66 and T67 fill in major gaps in ISS and VIMS coverage on the trailing hemisphere southern and equatorial regions. Likewise, T55, T56, and T57 Radar swaths cross the southern trailing hemisphere.

17.3.2.2 Surface Composition

The powerful combination of VIMS and Radar data for interpretation of surface units has been amply illustrated (Chapter 6; Brown et al. 2006; Soderblom et al. 2007b). High resolution

Table 17.3 Targets of flybys assigned to surface science investigations

Flyby	Instrument(s)	Target coverage
T47	ISS, VIMS	Global- and regional-scale images of leading hemisphere, including Hotei Arcus, Huygens landing site
T48	Radar	Swath across Xanadu, Tui Regio
T49	Radar	Ontario Lacus (altimetry), SAR before (southern high latitudes) and after
T49	VIMS, ISS	South Xanadu, Hotei Arcus
T50	Radar	Possible mountains
T51	ISS, VIMS	High-resolution imaging of the south-polar region, including Ontario Lacus
T53	ISS, VIMS	Trailing hemisphere at high southern latitudes
T54	ISS, VIMS	High-resolution imaging of trailing hemisphere at high southern latitudes
T55	Radar	Southern hemisphere swath, cross Shangri-La dunes
T56	Radar	Southern hemisphere swath, cross Tortola Facula, then Shangri-La dunes
T57	Radar	Southern hemisphere, western edge of Xanadu
T58	Radar	Western edge of Xanadu, Ontario Lacus (SAR)
T60	Radar	Altimetry swath, Xanadu to south pole
T60	ISS, VIMS	High-resolution, low-phase-angle imaging of western Senkyo
T61	Radar	Near-equatorial swath covers Dilmun, Aderi, and Belet
T61	ISS, VIMS	Western Senkyo at low phase angles
T64	Radar	North polar swath over lakes – stereo and/or seasonal change
T65	Radar	Ontario Lacus (SAR)
T66	ISS, VIMS	Fill trailing hemisphere gap, south midlatitude
T67	ISS, VIMS	Fill trailing hemisphere gap, equatorial region
T69	ISS, VIMS	Best north polar coverage

Radar swaths identifying a specific terrain type (such as the dunes) are correlated with spectral types in the VIMS global data set, enabling global mapping of a variety of surface units (Barnes et al. 2007). The T47 ground track goes over the Huygens landing site at low phase angle. In situ data from the surface from the Huygens probe will be combined with VIMS data from above to develop an atmospheric scattering model. The model will then be applied globally to refine analysis of VIMS data. VIMS spectral data shows the presence of 9 major compositional units (Barnes et al. 2007) and with the improved atmospheric model more are certain to emerge.

Another method for acquiring surface composition data is through the Radio Science Subsystem (RSS) bistatic scattering measurements. The quasi-specular surface echo is observed by bouncing the radio signal from the spacecraft off Titan's surface to the earth. These measurements give the surface dielectric constant and with it constraints on the composition, along with surface roughness, at the location of the reflection. Results to-date are consistent in some of the locations probed with the presence of liquid hydrocarbons (Marouf et al. 2006). This observation will be carried out again on T46, T51 and T52.

17.3.2.3 Cryovolcanism

Titan's young surface age requires some mode of resurfacing to hide ancient craters. Cryovolcanism is a candidate surface modification process and its presence has been suggested by both Radar (Elachi et al. 2005) and VIMS (Sotin et al. 2005) data. The evidence is based largely on surface feature morphology and odd spectral characteristics. Lobate flows around Ganesa Macula are consistent with morphology expected from cryovolcanic activity (Lopes et al. 2007). Hotei Arcus seems to have changed its reflectivity twice over the course of the PM (Nelson et al. 2009) and Tui Regio is bright in the VIMS data in the 5 micron atmospheric spectral window (Barnes et al. 2006). Are spectral and morphological characteristics adequate to declare that cryovolcanism is in fact the responsible process? When did the cryovolcanism occur? Is it extant today? These are questions that will be addressed in the EM. One test is whether a feature appears cryovolcanic in both Radar and Optical Remote Sensing (ORS) data. For example, Tortola Facula was initially thought to be a cryovolcanic feature from VIMS data (Sotin et al. 2005), but Radar data shows nothing to support that interpretation. A Radar swath will cross Tui Regio on T48, one of the areas identified by VIMS as spectrally distinct. Definitive evidence of current cryovolcanism as compared to morphological data indicative of past volcanism could come from detection of changes on the surface. The T56 Radar swath will cross Tortola Facula again, to compare to T43 data. Hotei Arcus is targeted by VIMS and ISS on T47 and T49. Radar radiometry data has the potential to detect thermal anomalies.

17.3.2.4 Dunes

Wind is an important agent for landscape evolution on Titan, evidenced by the wide equatorial expanse of dunes

(Elachi et al. 2005; Lorenz et al. 2006; Barnes et al. 2008). Surface color from the multispectral data of VIMS allows us to extrapolate Radar dune identification to broad regions of Titan (Soderblom et al. 2007b). Questions remain: why does the orientation of the dunes defy expectations from models of low-latitude trade winds? What are the dunes composed of? Dunes are concentrated in the equatorial region – what does that tell us about the long-term meteorology of Titan and the distribution of methane humidity? More high resolution coverage will provide additional insight into dune origins from their morphologies and orientations. Radar swaths on T55 and T56 cross the Shangri-La dunes. The T61 Radar swath crosses the dunes of the Belet sand sea.

17.3.2.5 Lakes and river channels

Dark features with morphologies consistent with lakes were identified during the PM in high northern latitude Radar swaths (Stofan et al. 2007) and at high southern latitudes by ISS (McEwen et al. 2005). These features are quite likely to be lakes of liquid methane and/or ethane however other interpretations are possible. Features with similar morphology but that are Radar-bright are interpreted as dry lakebeds. A transitional class termed "granular" is also identified (Hayes et al. 2008). All three categories are identified so far only at high latitudes (above 55°). Lakes are darker and larger in size above 65N, which may be indicative of the presence of deeper liquid. Detection of seasonal change consistent with lakes evaporating as the season in the north turns to summer would allow us to definitively conclude that the lakes are indeed filled to varying levels with liquid. Surface permeability and the role of a subsurface alkanofer can also be studied by observing the long-term evolution of the lakes. Sparse coverage in the south as illumination wanes may limit our ability to document similar trends, however some evidence of changes in the south polar lakes has been detected (Turtle et al. 2009). Only one lake has been identified with certainty: Ontario Lacus. VIMS spectra of Ontario Lacus in the southern hemisphere show that it is filled with liquid ethane and methane (Brown et al. 2008). Seasonal control of the lakes is a model that has been proposed in order to explain the difference between the number of lakes in the south and north polar regions (Hayes et al. 2008).

In the EM it is urgent to get ISS and VIMS coverage of the south polar region while the surface is still illuminated, before the onset of the long winter night. The largest lake in the southern hemisphere, Ontario Lacus, will be the focus of four flybys. On T49 Radar will collect altimetry data across the lake. ISS and VIMS will image Ontario Lacus on T51. Radar SAR swaths will be collected on T58 and T65. Changes in size of the lake will refine seasonal models of precipitation and evaporation. In the northern hemisphere the Radar swath on T64 will be used to look for changes in the size of the lakes, which will bolster arguments that these are bodies of liquid. ISS will also continue to observe the northern lakes and seas, monitoring them for changes. VIMS will image the northern lakes in daylight for the first time on T69.

River channels are observed at all latitudes, not just in the polar regions (see Chapters 5 and 7). The distribution of channels and lakes suggests two different methane cycles – one in which lakes are filled by rainfall in the polar regions seasonally, and the other operative over a centuries-long timescale, capable of eroding channels but not of filling long-lasting lakes (Lunine and Atreya 2008). Observations of Titan's surface will continue to yield key insights into Titan's meteorology.

17.4 Atmospheric Investigations

Titan has a thick nitrogen atmosphere with dynamic meteorology and complex methane-based minor constituent chemistry. In the PM initial Cassini–Huygens results were focused on basic atmospheric structure and composition, then evolved to elucidation of dynamic processes as more data were acquired (see Chapters 7, 10, 12 and 13). Huygens returned in situ data including pressure, temperature, and composition during its descent to the surface (see Chapter 10) on January 14, 2005. Orbiter instruments probe Titan's atmosphere remotely from the surface up to as high as 3,500 km. Above 950 km in situ instruments on Cassini collect data (described in Section 17.5). Figure 17.8 illustrates the approximate altitudes of Titan's troposphere, stratosphere, mesosphere, thermosphere and exosphere (see Chapter 10 for a discussion of the uncertainties of these boundaries).

Atmospheric scientific objectives include study of

- Atmospheric properties from the surface to the exosphere
- Composition and chemistry; sources and sinks of minor constituents
- Titan's weather – temperatures, global winds, precipitation and the methane cycle
- Aerosols – formation/destruction/influence on atmospheric properties

All of these are a function of time and space, so it is only with data acquired in numerous locations over a long time span that we can understand the whole dynamical picture, including seasonal changes as Titan goes through equinox, especially in the polar regions.

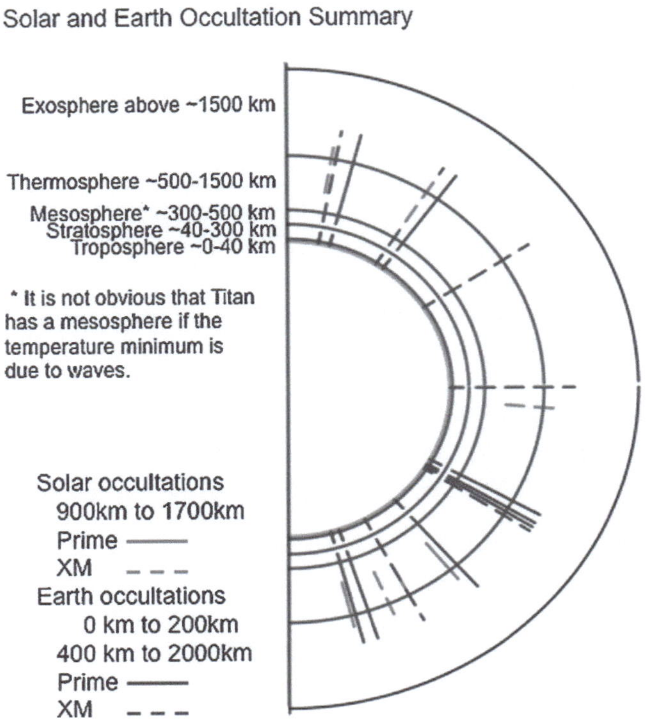

Fig. 17.8 Occultations of the sun and earth by Titan's atmosphere. Solid lines are occultations acquired in the PM, and dashed lines are occultations planned for the EM, listed in Table 17.4. Blue lines are occultations of the earth as seen from the spacecraft. The radio science signal probes from the surface to 200 km, then the ionosphere, from 400 km to 2000 km. Red lines are occultations of the sun by Titan's atmosphere as seen from the spacecraft, that probe from the surface with VIMS or 850 km altitude with UVIS, up to ~1,700 km. Altitudes of atmospheric regions (troposphere, stratosphere, etc.) vary with time and latitude and should not be used as rigorous definitions. Tracks of the sun and earth behind the atmosphere are not radial as depicted here. The authors thank Trina Ray for providing this figure.

17.4.1 Primary Mission Observations

Cassini observes Titan's atmosphere on every Titan flyby typically starting and ending at a range of ~300,000 km. On both the dayside and the nightside ORS instruments carry out scans across the limb and/or body, or point at one location to collect long integration data and monitor clouds. Within 2 h of closest approach the activities have more variety (and may be focused on other scientific discipline investigations such as surface mapping). Atmospheric investigations that may take place in these 4 h around closest approach include high resolution scans of the limb, and/or occultations of the earth, sun or stars. In situ atmospheric observations acquired at the spacecraft closest approach to Titan are discussed in Section 17.5.

The temperature and pressure from the surface up through the stratosphere is probed by the Radio Science Subsystem (RSS) using 3 wavelengths (S, X, and Ka band). The spacecraft sends radio waves to earth at these wavelengths, as the earth is occulted by the atmosphere of Titan. The signal received at the DSN is attenuated by Titan's atmosphere. The RSS measurement is sensitive to the neutral atmosphere from the surface to 200 km altitude, and to the ionosphere from 400 to 2000 km. Four occultations of the earth by Titan's atmosphere were observed during the PM in northern winter at 8 mid- and high-latitude locations: 74S, 69S, 53S, 34S, 33S, 31S, 53N, and 73N.

Cassini's Composite Infrared Spectrometer (CIRS) far infrared spectra are used to map temperatures and derive wind fields from the upper part of the troposphere through the stratosphere, to an altitude of ~400 km (Flasar et al. 2005; Achterberg et al. 2008). CIRS thermal emission spectra in the mid-infrared reveal the composition of the troposphere and stratosphere, including higher order hydrocarbons. Nitriles are enriched at latitudes above 60N, consistent with a confining circumpolar vortex and the downwelling of organic-rich air (Flasar et al. 2005). Temperature structure and composition together are used to derive atmospheric dynamics (e.g. Teanby et al. 2007).

The Ultraviolet Imaging Spectrograph (UVIS) observes ultraviolet light (1) reflected from the aerosols in Titan's atmosphere, (2) emitted from atomic and molecular species, and (3) spectrally absorbed during occultations of the sun and occultations of stars by Titan's atmosphere. Titan's reflectance spectrum is best fit with scattering from a combination of tholins and hydrocarbons (Ajello et al. 2008). At extreme ultraviolet wavelengths (EUV) UVIS detects nitrogen, CH_4 and other hydrocarbons from 850 to 3,500 km altitude with solar occultations. Observations of stellar occultations in the far ultraviolet (FUV) wavelength range determine abundance and vertical structure of aerosols, hydrocarbons, and N_2 through the range of 300 to 1,200 km (Shemansky et al. 2005).

VIMS observations of occultations of the sun by Titan's atmosphere yield the mixing ratio of CO, CH_4, CO_2, HCN and aerosols as a function of altitude. VIMS observations of thermal emission on the nightside of Titan enabled quantitative analysis of the behavior of CO in the atmosphere (Baines et al. 2006). ISS and VIMS mapped clouds in the troposphere (Griffith et al. 2006; Turtle et al. 2008). A cluster of convective clouds was imaged over the south pole in July 2004 (Porco et al. 2005). Mid-latitude clouds, particularly around 40S, have been imaged intermittently. Thin clouds are visible at 60N. VIMS observed a large cloud system in 2006 reaching from the north pole down to 62N (Barnes et al. 2006). ISS and VIMS have also imaged multiple high altitude haze layers in the stratosphere, with the most prominent at ~500 km altitude. Aerosols are re-distributed seasonally from one pole to the other, forming a polar hood. Nightside observations looking for lightning have not detected any (yet).

With these observations PM data have given us a picture of Titan's current meteorology. Most of the year solar forcing leads to a hemisphere-to-hemisphere Hadley cell, possibly with more earth-like symmetric equator-to-pole Hadley cells around equinox (see Chapter 13). Titan has a methane cycle similar to earth's hydrological cycle. The zonal wind field has been mapped and we have a global view of the distribution of minor species. The next step is to look at Titan's longer-term meteorology, seasonal changes, and implications for its climate on >1,000 year timescales.

17.4.2 Equinox Mission Scientific Investigations

In the EM the major scientific questions to be addressed are:

What are sources and sinks of higher order hydrocarbons?

How and when does the polar vortex form? Polar hood?

What is the process of aerosol formation?

What about rainfall? Will Cassini detect a methane rainstorm?

What changes in the clouds and weather take place as the season evolves?

The EM extends the temporal coverage of the primary mission to the beginning of spring. This is a time of change for the atmosphere as the dominant solar heating moves from the southern to the northern hemisphere. As in the primary mission every flyby in the EM begins and ends with remote sensing coverage. VIMS and ISS look for clouds and lightning. Near closest approach CIRS maps the thermal field and composition via limb scans. Occultations of the earth, sun and stars by Titan's atmosphere allow RSS, VIMS, and UVIS to probe the structure and vertical compositional profile of the atmosphere.

17.4.2.1 Atmospheric Structure and Distribution of Hydrocarbons

Occultations of the sun and the earth by Titan's atmosphere in the PM and EM are listed in Table 17.4. Stellar occultations in the PM and EM are listed in Table 17.5. Occultation altitudes and latitudes are illustrated in Figs. 17.8 (solar and earth) and 17.9 (stellar) for the PM and the EM. Occultations that occur at the same latitude at different times will provide data on seasonal change. CIRS high resolution limb scans will be acquired on T53, T54, T62, T66, T67, and T70.

Table 17.4 Occultations of the sun and earth by Titan's atmosphere in the PM and EM. Latitudes refer to sub-raypoint at minimum altitude. In the case of earth occultations and VIMS solar occultations this is at the surface. See Fig. 17.8

Flyby	Type	Ingress latitude	Egress latitude
Primary mission			
T12	Earth	29S	49S
T14	Earth	31S	32S
T27	Earth	72S	54N
T31	Earth	75S	75N
T10	Solar	62S	51S
T26	Solar	76S	Not observed
Equinox mission			
T46	Earth	33N	33S
T52	Earth	57N	0
T57	Earth	79N	Not observed
T53	Solar	60N	3S
T58	Solar	80N	Not observed
T62	Solar	10S	75S

Table 17.5 Occultations of stars by Titan's atmosphere in the PM and EM. The second column identifies the occultation path of the star that is illustrated in Fig. 17.9

Flyby	Occultation number in Fig. 17.9	Star	Latitude at minimum altitude (300 km)	Latitude at maximum altitude (1,200 km)
Tb	1	Alpha Virginis	51.1N	32.6N
Tb	2	Lambda Scorpii	36.0S	35.8S
T7	3	54 Alpha Pegasi	49.5N	50.9N
T13	4	Beta Orionis	17.7S	14.9S
T16	5	Alpha Virginis	71.5S	71.7S
T21	6	Alpha Eridani	35.9S	35.1S
T23	7	Eta Ursae Majoris	2.6S	8.8S
T35	8	Sigma Sagittarii	32.4S	35.0S
T40	9	Alpha Lyrae	54.3N	50.7N
T41	10	31 Eta Canis Majoris	48.2S	36.4S
T41	11	Epsilon Canis Majoris	9.0S	4.2S
T41	12	Epsilon Canis Majoris	28.7S	24.0S
T47	13	Eta Ursae Majoris	0.9N	0.7N
T47	14	Beta Canis Majoris	56.4N	53.6N
T48	15	Epsilon Canis Majoris	18.4N	20.1N
(distant)	16	Alpha Eridani	37.6N	35.0N
T53	17	Alpha Eridani	39.0N	37.8N
T56	18	Eta Ursae Majoris	45.1S	35.6S
T58	19	Eta Ursae Majoris	13.6S	6.1S
T70	20	Alpha Virginis	20.8S	31.0S

17.4.2.2 Clouds and Haze

In the EM we begin a distant Titan monitoring campaign that will continue throughout the rest of the Cassini mission. Low resolution data are sufficient to detect and locate clouds and large-scale haze structures. These observations provide isolated

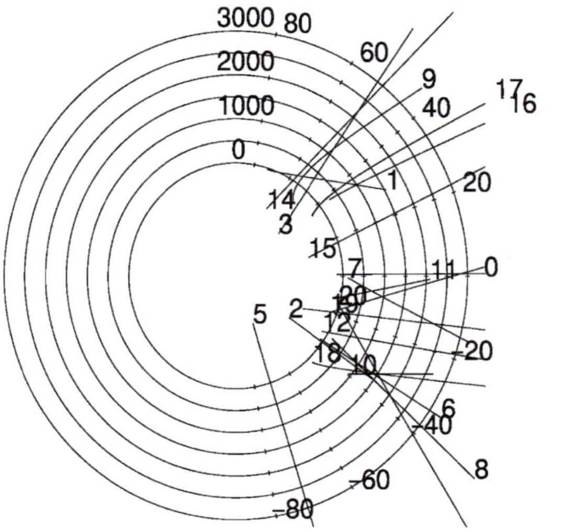

Fig. 17.9 Occultations of stars by Titan's atmosphere. Both PM and EM occultations of stars by Titan's atmosphere are shown, labeled with the number cross-referenced to Table 17.5. Altitudes are shown from 0 to 3,000 km and latitudes are labeled from +80 to −80. Occultations are rarely radial, as can be seen by the slanted track as various altitudes are crossed. The number labeling each occultation track is shown at the beginning of the track, thus differentiating between the ingress and the egress of the star, going behind or coming out from behind Titan's atmosphere, respectively. The authors thank Robert West for providing this figure.

snapshots, and are intended to improve our understanding of how often there are clouds on Titan, how quickly they appear and dissipate, where they're appearing as the seasons change, how fast and in what direction the winds blow, and also how the haze evolves with the season. These distant observations will be much more frequent than Titan flybys and will be more likely to detect sporadic cloud appearances. Ground based and Cassini observations have shown clouds changing over hours and days. Will we see a change in altitude of detached haze layer as the season changes? It was 150 km lower during Voyager observations (Rages and Pollack 1983; Porco et al. 2005). As discussed in Chapter 14, the nominal mission took place in a period of relatively slow change. The Hubble Space Telescope record of the late 1990s however shows that changes in the haze will be more rapid in the years following equinox, as the pole-to-pole Hadley circulation reverses (Lorenz et al 2004). The evolution of the UV-dark north polar hood will be observed – the Hubble record showed that the south polar hood decayed rapidly after the southern summer solstice.

17.4.2.3 Atmospheric Dynamics

Zonal winds are inferred from the temperature fields observed by CIRS (see Chapter 13). Changes in the winter polar vortex will be observed in the EM. Subsidence in the winter enhances the concentration of organics. At what point does this process begin to reverse? What is the time scale for the atmosphere to go from a single pole-to-pole Hadley cell to two hemispheric Hadley cells back to a single pole-to-pole (reversed in direction) Hadley cell? Global Circulation Models have been developed (Tokano 2008), and will be tested against Cassini data.

17.5 Titan's Thermosphere and Ionosphere, and the Interaction of Titan's Upper Atmosphere with Saturn's Magnetosphere

Titan has a weak magnetic field induced as a result of Saturn's magnetospheric plasma and energetic particles interacting directly with Titan's upper atmosphere and ionosphere. Exceptions can occur when Titan's orbit is near local solar noon where under conditions of strong solar wind dynamic pressure Titan's atmosphere can interact directly with the solar wind or Saturn's magnetosheath. Since at Titan's orbit at 20 Rs the magnetic field of Saturn is relatively weak and the gyroradii of the interaction of the magnetospheric ions can be of the order of the radius of Titan (depending on ion mass and plasma flow speed), a very asymmetric interaction is created where ions flow into Titan at Titan longitudes on the Saturn facing flank and away from Titan on the interaction flank away from Saturn. The interaction results in energy deposition in Titan's upper atmosphere from magnetospheric plasma and energetic ions that depend on Titan's longitude and Titan's position in its orbit. The orbital position is especially important in defining the energetic ion input due to the asymmetry with respect to solar local time that exists in Saturn's current sheet. For a more complete illustration of this interaction see Fig. 17.10.

Although the heating of the thermosphere is predominantly due to solar extreme ultraviolet and x-ray radiation, the magnetospheric interaction provides an appreciable, and not yet well-determined amount of energy to the upper atmosphere. Furthermore, molecular hydrogen, methane, and molecular nitrogen escape from the upper atmosphere. Jean's escape dominates the loss of molecular hydrogen from the atmosphere whereas molecular nitrogen and methane are primarily lost by the heating of the upper atmosphere from the magnetospheric interaction (via sputtering processes, and if ionized by ion pickup). See Chapters 15 and 16 for further discussion and references. The complex interaction geometries and the dimensionality of the parameter space that must be sampled require a large number of flybys under varying positions and conditions in order to fully characterize the interaction (see Figs. 17.11 and 17.12).

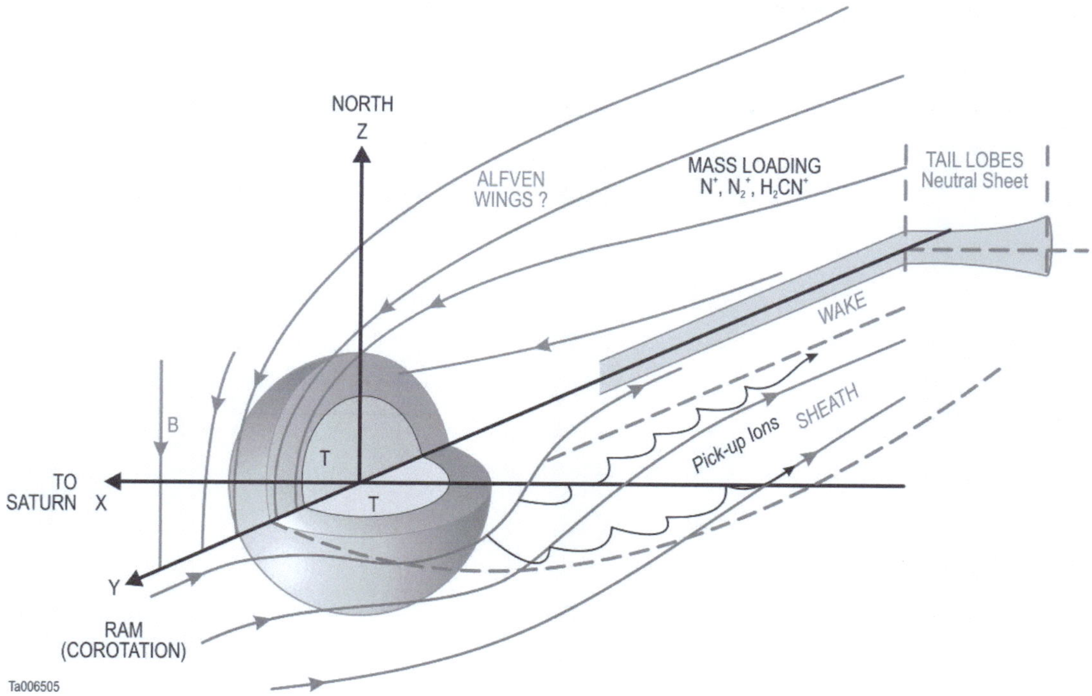

Fig. 17.10 This figure illustrates in three dimensions the interaction of Titan with Saturn's magnetosphere. The effects of the large gyroradius of the interacting ions, as compared to the size of Titan, are illustrated by the cycloidal motion of the particles. The gyroradius is schematic and can range from several Titan radii to relatively small sizes depending on ion mass and the flow speed of the underlying plasma. The rotation of Titan is in the opposite direction of ram since Saturn's spin rate is much faster than Titan's rotation.

17.5.1 Primary Mission Achievements

The basic upper atmosphere structure (see Chapter 10) and magnetospheric interaction have been characterized in the primary mission. The deposition of energy in the upper atmosphere and the resulting escape of the atmosphere (Chapter 16) have been characterized in several geometries and under a sparce set of conditions, including at least one instance of Titan being encountered by Cassini while Titan was in Saturn's magnetosheath (T32).

Four very interesting results emerge from the data available from the primary mission that require further investigation: (1) the discovery of the chemical complexity of the upper atmosphere – organic poly-aromatic compounds and associated ions having measurable densities with masses exceeding 300 Da, and most surprisingly negative ions with masses that exceed 10,000 Daltons – larger than an insulin molecule (see Chapter 8), (2) the large influx of energetic ions that deposits energy deep within Titan's atmosphere, (3) the large influx of oxygen-rich plasma from Saturn's magnetosphere, and (4) the unexpected orientation of the induced magnetic field produced by the interaction of Titan with Saturn's magnetosphere. All of these processes can produce significant effects on the chemistry and composition of the atmosphere – the significance of which is determined by the amount of energy and oxygen that is input. This input varies considerably temporally and spatially. Due to the comparable scale lengths of the ions and the radius of Titan the spatial variation is quite complicated so the EM must explore all longitudes of the interaction region at all positions of Titan within its orbit – a tall bill to fill.

17.5.2 EM Major Scientific Goals

The four important findings from the PM, described in the previous paragraph, will be the focus of EM investigations. The best spacecraft orientation for sampling the upper atmosphere has INMS pointed in the ram direction. At closest approach this can be achieved with either Radar or ORS pointed at Titan, however this compatible geometry only lasts for a few minutes. "Full" flybys leave INMS pointed in the ram direction throughout the +/–15 min around closest approach. "Half" flybys orient the INMS in the ram direction either on approach or departure. Typical altitudes at closest approach range from 950 to 1,400 km. Figure 17.13 illustrates the parameter space covered in local time, latitude, and sector of Saturn's magnetosphere on each of the INMS flybys in the PM and EM.

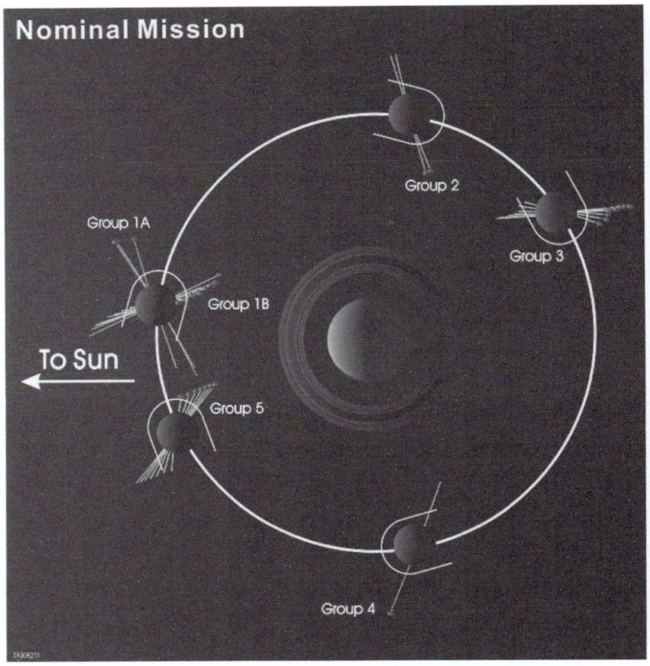

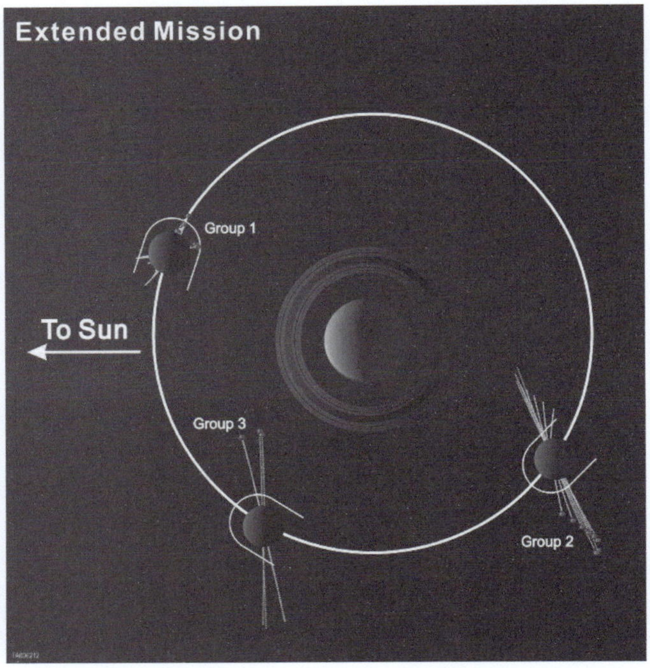

Fig. 17.11 Titan's interaction with Saturn's magnetosphere and the solar wind during Cassini's Primary Mission. This figure illustrates Cassini flybys of Titan for passes with closest approach altitudes below 1,500 km during the Primary (Nominal) Mission tour. Due to geometric constraints governed by orbital mechanics, these flybys fall into distinct groups. The average position of the groups is depicted by Titan, where the actual position of Titan during each flyby is within approximately +/− 3° in Saturn longitude (inside the radius of the Titan images depicted as the group average). Groups 1A, 1B and 5 are on Saturn's dayside, and experience the highest level of solar wind interaction. Groups 2, 3 and 4 are on the dawn, night and dusk side respectively. The yellow lines show the trajectory of Cassini for each of the flybys, with the gray dots showing the outbound leg. The white arc around Titan represents the bow shock of the interaction between Saturn's magnetosphere and Titan's atmosphere. The authors thank Greg Fletcher for providing this figure. Group 1A – Ta, Tb and T8. Group 1B – T35, T36, T37, T38, T39, T40, T41, T42, T43 and T44. Group 2 – T5 and T7. Group 3 – T16, T17, T18, T19, T20, T21, T22, T23 and T24. Group 4 – T34. Group 5 – T25, T26, T27, T28, T29, T30, T31, and T32.

Fig. 17.12 Titan's interaction with Saturn's magnetosphere and the solar wind during Cassini's Equinox Mission. This figure illustrates Cassini flybys of Titan for passes with closest approach altitudes below 1,500 km during the Equinox Mission. Due to geometric constraints governed by orbital mechanics, these flybys fall into distinct groups. The average position of the groups is depicted by Titan, where the actual position of Titan during each flyby is within approximately +/− 3° in Saturn longitude (inside the radius of the Titan images depicted as the group average). Groups 1 and 3 are on Saturn's dayside, and experience the highest level of solar wind interaction. Group 2 is on the night/dusk side. The yellow lines show the trajectory of Cassini for each of the flybys, with the gray dots showing the outbound leg. The white arc around Titan represents the bow shock of the interaction between Saturn's magnetosphere and Titan's atmosphere. The authors thank Greg Fletcher for providing this figure. Group 1 – T45, T46, T46, T47, T48, T49, T50 and T51. Group 2 – T52, T53, T54, T55, T56, T57, T58, T59, T60, T61, T62 and T67. Group 3 – T63, T64, T65, T66, T67, T68, T69, T70.

17.5.2.1 Complex Organic Formation in Titan's Upper Atmosphere

EM goals are to (a) disentangle effects of solar, magnetospheric and wave forcing on trace constituents; (b) study upper atmosphere chemistry, tholin formation; and (c) investigate complex ion composition. CAPS and INMS will have evaluated the latitude and altitude dependence of the complex ion and neutral formation process by the end of the EM. There will also be one opportunity to view seasonal variation at northern high latitudes. A first order understanding of the contribution of energetic ion precipitation to the macromolecule formation process is also anticipated.

Planned INMS and CAPS flybys

(a) Full flybys: T48, T50, T57, T59
(b) Half flybys: T51, T64, T65
(c) Coverage for several minutes at closest approach: T38, T44, T47, T49, T55, T56, T60, T61

17.5.2.2 Energetic Ion Precipitation

The key objective for the EM is to understand effects associated with the position of Titan within the current sheet and how the Saturn solar local time of the orbit affects the process. MIMI will characterize the energetic ion and electron environment of Titan to first order, and conduct correlative investigations of the energetic particle input to the upper

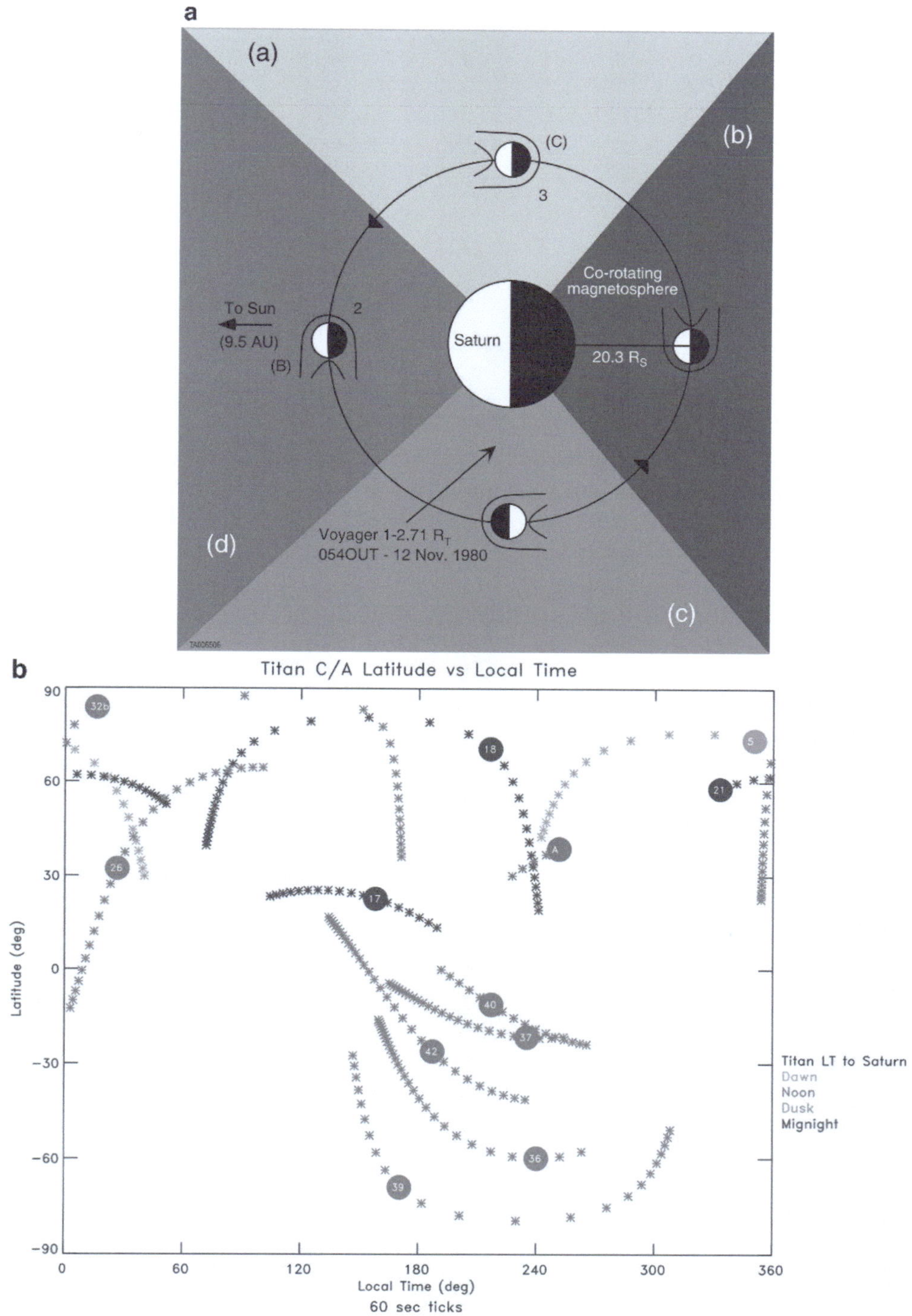

Fig. 17.13 Coverage of all sectors of the magnetosphere, latitude and local time are plotted for each INMS flyby. The authors thank Greg Fletcher for providing these figures. (**a**) Sectors of Saturn's magnetosphere referenced in 17.13b and 17.13c. (**b**) Primary mission INMS flybys' locations in Titan latitude vs. local time.

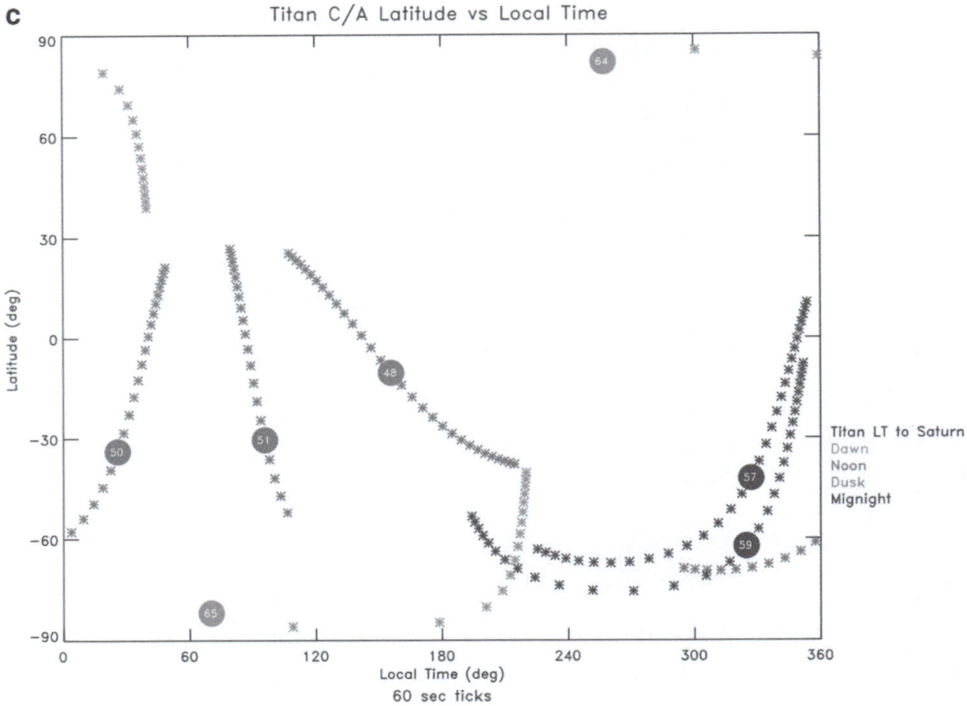

Fig. 17.13 (**c**) Extended mission INMS flybys' locations in Titan latitude vs. local time.

atmosphere and ionosphere with INMS measurements of the atmospheric chemistry and ionosphere, radio occultation data which probes the ionosphere, and CAPS measurements of the heavy negative-ion population in the upper atmosphere with a particular focus on the secondary ionospheric peak below 1,000 km altitude, and its dependence on the intensity of energetic ions and electrons. We also expect to complete our investigation of the outer reaches of Titan's exosphere, since the creation of energetic neutral atoms through charge exchange with the exospheric neutrals is a more sensitive probe of those neutrals than any other technique on the Cassini spacecraft.

Planned Magnetosphere and Plasma Science (MAPS) flybys:

T63 – dedicated to CAPS for wake passage

T65 – close downstream encounter at dusk

T68 – close downstream encounter at dusk

T70 – "Go Low" flyby for MAG to investigate induced magnetic field

17.5.2.3 Oxygen Plasma Injection into Atmosphere

In the EM we want to (a) understand effects associated with the Saturn solar local time of the orbit, and (b) sample different magnetospheric ram angles at closest approach to Titan. CAPS will assess the input of ram oxygen-rich plasmas into the nose of the interaction region and correlate this with the INMS ion and neutral composition. Planned MAPS flybys are the same as listed above.

17.5.2.4 Geometry of the Induced Magnetic Field

MAG will obtain a more accurate upper limit on Titan's weak intrinsic magnetic field as compared to the Voyager 1 magnetometer measurements (Neubauer et al. 1984) and, if there is indeed an intrinsic field, investigate its origin. With an unprecedented closest approach altitude of 880 km, we shall try to answer these questions with data from Cassini's T70 flyby. Once the intrinsic permanent field is measured, we will be able to measure any extra contribution resulting from the magnetic induction linked to the presence of conducting elements in Titan's crust and interior. Unfortunately the geometry of the Cassini encounters during the extended mission is not ideal for these studies, but hopefully flybys in the early morning sector of Saturn's magnetosphere will help answer this question.

MAG will also constrain the region downstream from Titan where its plasma escapes as a consequence of its interaction with Saturn's magnetosphere. Flybys like T9 (Barnes et al. 2007 and others within the T9 special issue of GRL, volume 34 2007) provided us with a picture of the extent of Titan's magnetotail and the way in which charged particles from Titan escape along the draped magnetic field lines, but more detailed magnetic field and plasma observations are needed beyond three Titan radii downstream from the moon. During the EM, the T63 flyby will give us the opportunity to do that. The study of Titan's flow side induced magnetosphere and the pressure balance between the magnetic barrier and the collisional ionosphere will tell us about how easily the external field from Saturn penetrates Titan's atmosphere and gives origin to an induced field with a very asymmetric geometry.

To investigate the geometry of the induced magnetic field in the EM the approach is likewise to (a) understand effects associated with the Saturn solar local time of the orbit and (b) sample different magnetospheric ram angles at closest approach to Titan. Planned MAPs flybys are the same as listed above.

17.6 The Solstice Mission

The Cassini spacecraft will have a substantial amount of fuel at the end of the EM. Planetary protection requires the disposal of the spacecraft at the end of its mission to prevent any contamination of Enceladus and/or Titan. In between the end of the EM and the end-of-mission (EOM) disposal of the spacecraft, the Project has proposed an extended scientific phase termed the "Cassini Solstice Mission". The Solstice Mission (SM) will extend from the end of the EM in July 2010 through the Saturn solstice on May 23, 2017, and then enter into an end-of-life Juno-like set of high inclination proximal orbits culminating with the spacecraft entering into Saturn's atmosphere.

The advantages of observing Titan through half of its year, illustrated in Fig. 17.14, are significant for understanding the yearly meteorological cycle, as discussed in Chapter 14. Weather and winds change seasonally in response to changes in insolation. The methane cycle can be monitored both in the atmosphere, and as evidenced by changes in liquid level in the lakes. Furthermore, doubling the length of time over which Cassini observes Titan increases the probability of identifying changes due to other geologic processes, e.g. aeolian modification, fluvial erosion, or cryovolcanism. Higher order hydrocarbon distribution in Titan's atmosphere is known to change seasonally as evidenced by overall brightness changes in the atmosphere and polar hood, and the Cassini Solstice Mission will be able to observe this process.

There are 56 Titan flybys in the 7-year SM, as shown in Fig. 17.2. There will be more opportunities to continue to fill in surface coverage at high resolution and to react to discoveries made in the PM and EM. The SM also provides the chance to achieve trajectory geometries that eluded the PM and EM, the prime example of which is low phase angle illumination at closest approach. Overall the EM tour does not provide good illumination for VIMS and ISS. Most flybys are "dark", with high phase angles at closest approach. This situation is remedied in the proposed SM.

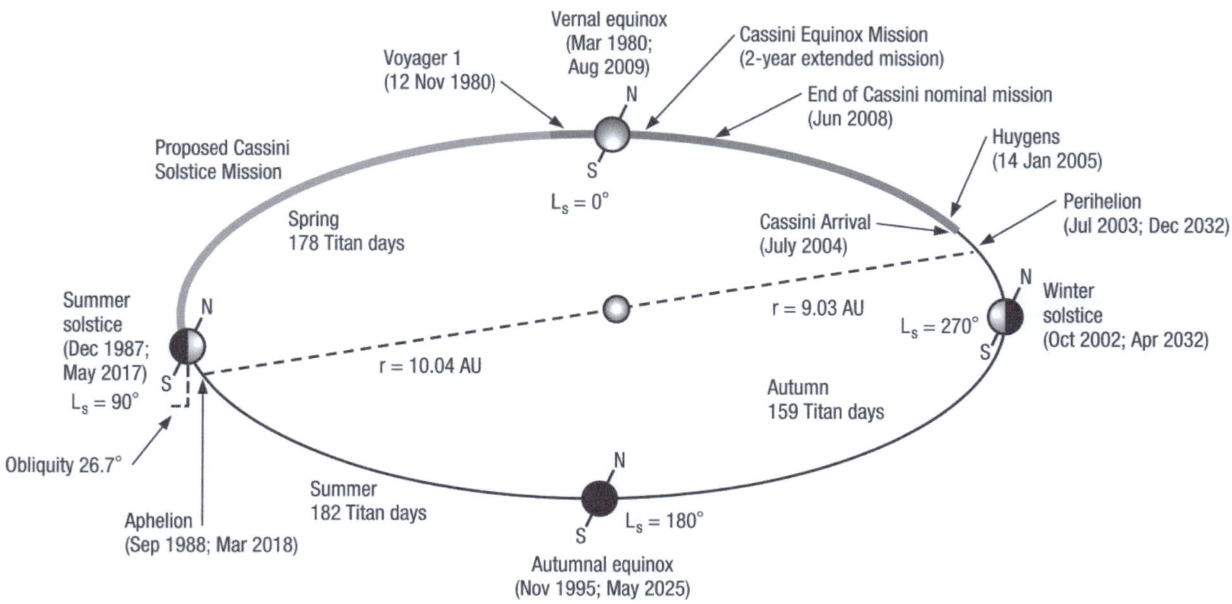

Fig. 17.14 Solstice mission. A Saturn year is 29.47 earth years. The entire primary, equinox, and solstice phases of the Cassini mission cover just half of a Saturn/Titan year. The authors thank Ralph Lorenz for providing this figure.

Specific SM Titan scientific objectives are to:

- Observe seasonal changes in the methane/hydrocarbon cycle of lakes, clouds, aerosols, and their seasonal transport.
- Observe seasonal changes in the high latitude atmosphere (specifically the temperature structure and formation and breakup of the winter polar vortex).
- Characterize changes in surface temperature.
- Observe Titan's plasma interaction as it goes from south to north of Saturn's solar-wind-warped magnetodisk from one solstice to the next.
- Determine inner and crustal structure; liquid mantle, crustal mass distribution, rotational state of the surface with time, investigate icy shell topography and viscosity.
- Further characterize the types, composition, distribution, and ages of surface units and materials, most notably lakes (i.e. filled vs. dry, depth; liquid vs. solid, composition; polar vs lake basin origin).
- Resolve current inconsistencies in atmospheric density measurements in which different techniques have yielded different results. Accurate density data are critical for any future Titan mission that uses the atmosphere for aerobraking or aerocapture.

Interior Structure. The mismatch of surface features in Radar data has been attributed to nonsynchronous rotation, arguing for the presence of an interior liquid layer (allowing slippage of the crust) and for seasonal variation in zonal winds. The nonsynchroneity during the nominal mission acted in the opposite sense to the polar precession, yielding a net drift in apparent longitude that is quite small. However if the Tokano and Neubauer (2005) zonal wind model is correct, during the EM the drift will become rather rapid as these two effects will add together as shown in Chapter 4. Such an acceleration would be convincing support of both seasonal changes in atmospheric angular momentum and of a thin ice crust decoupled from the interior. The additional data acquired during the EM may also allow higher-order changes in rotation state (such as libration) to be detected.

Surface Science. The polar regions are particularly interesting as they are the sites likely to change the most. Titan investigators have requested that the SM tour design include at least two polar excursions (preferably three) with as many Titan flybys as possible.

Atmospheric Investigations. Occultations of the earth, sun and stars by Titan's atmosphere, distributed in latitude but heavily weighted to the high latitudes, will elucidate changes in temperature, haze, and trace constituents in the atmosphere. CIRS will continue to map winds and hydrocarbon transport as the season progresses. A brand new observation will take place to resolve the ongoing discrepancy between INMS and AACS determinations of the atmospheric density at 950 km altitude, which currently differ by a factor of three. One of the SM flybys will be assigned to the navigation team and will be flown to an altitude of 950 km with Cassini's HGA pointed to earth to maintain radio contact through the time that the spacecraft is in the atmosphere. The navigation team will measure the effect on the trajectory that results from atmospheric drag on the spacecraft, enabling a third derivation of density.

Thermosphere/Magnetosphere interaction. In the SM MAPS measurements will benefit from a good mix of Titan flybys relative to Saturn's magnetic field, including several dayside (upstream) dusk flybys, and to be near dawn at solstice. Another mid-tail crossing and another low altitude flyby like T70 for MAG at dawn (vs. dusk) would be very useful to continue to refine internal and induced field measurements.

In addition to achieving basic Titan scientific goals, an over-arching consideration for all SM decisions will be to leave a legacy of data that are important for planning the next Titan mission. Table 17.6 shows the tentative plan for the

Table 17.6 Titan flybys in the Solstice Mission. Prime instruments and major science emphasis have been selected for the hours around closest approach

Flyby	Altitude (km)	Prime instruments	Science emphasis
T71	1,005	CAPS, INMS, RADAR	Mid southern latitude dawn side pass
T72	8,124	CIRS, VIMS, ISS	New surface territory, highest southern latitude
T73	7,921	CAPS, CIRS	Composition, aerosols, thermal map
T74	3,640	CAPS	Upstream pass
T75	9,996	CAPS	Wake crossing
T76	1,862	CIRS, VIMS	Belet sand sea
T77	1,383	RADAR	Xanadu
T78	5,941	CIRS, VIMS, UVIS, CAPS	Composition, aerosols, southern vortex, wake crossing
T79	3,763	CAPS	Excellent upstream pass
T80	29,331	ISS	Mid / high southern trailing hemisphere
T81	31,172	ISS	High southern leading hemisphere and Ontario Lacus on terminator
T82	3,844	CIRS	Composition, aerosols, thermal map

Table 17.6 (continued)

Flyby	Altitude (km)	Prime instruments	Science emphasis
T83	990	INMS, RADAR	Northern lakes change detection
T84	990	RADAR, INMS	Global shape
T85	990	CIRS, VIMS	Northern lakes
T86	990	CIRS, INMS	Northern lakes
T87	990	INMS, NAV	Atmospheric density
T88	1,164	CIRS, VIMS	Temperature at equator
T89	1,500	RSS, CAPS	Gravity field, flank encounter
T90	1,302	CIRS, VIMS	Tui Regio, Xanadu
T91	990	RADAR, CAPS	Global shape, altimetry, stereo of small northern lakes
T92	990	RADAR, INMS, VIMS	Stereo (with T91)
T93	990	ISS, VIMS	High northern leading latitudes, Ontario Lacus outbound
T94	990	VIMS, ISS, CIRS	High northern lakes composition
T95	990	ISS, RADAR, INMS	High northern lakes coverage
T96	990	ISS, VIMS, CIRS	High northern lakes, western Xanadu
T97	3,087	VIMS, CIRS	Dunes
T98	2,500	RADAR	Ontario Lacus change detection
T99	1,612	RSS, CAPS	Gravity field
T100	990	CIRS, VIMS, INMS	Atmospheric structure
T101	2,515	RSS	Occultation at high latitude for polar vortex, bistatic for surface properties
T102	3,288	RSS	Occultation at high latitude for polar vortex, bistatic for surface properties
T103	4,810	UVIS	Solar occultation at high latitude for polar vortex, stellar occ near equator
T104	990	INMS, RADAR, ISS	Southern edge of Kraken Mare
T105	990	VIMS, CIRS	Kraken Mare composition
T106	990	RSS	Bistatic for Kraken Mare properties
T107	990	INMS, NAV	Atmospheric density
T108	990	RADAR, INMS	Southern edge of Ligela
T109	1,200	VIMS, CIRS	Punga Mare, Sinlap
T110	2,270	VIMS, CIRS	Northern lakes composition
T111	2,722	VIMS, CIRS	Xanadu
T112	10,964	CIRS	Composition, aerosols, thermal map
T113	1,036	MAG	Intrinsic magnetic field
T114	11,907	ISS, CIRS	Hotei, Xanadu
T115	3,830	CIRS	Composition, aerosols, thermal map
T116	990	UVIS, VIMS	Solar occultation
T117	990	RSS	Occultation for seasonal variation
T118	990	UVIS, INMS	Atmospheric density
T119	990	Not assigned	
T120	990	CIRS, RADAR, INMS	Small southern lakes, global shape
T121	990	VIMS, RADAR	Tui Regio, Hotei
T122	1,679	RSS	Gravity field
T123	1,766	VIMS, CIRS	Hotei
T124	1,585	RSS	Bistatic for northern lake properties
T125	3,197	ISS, VIMS, CIRS	Hotei
T126	990	INMS, RADAR	Northern lakes change detection

Titan flybys in the SM. Developing a model of temporal changes in the structure of the atmosphere will be key to designing and flying future aerial craft such as balloons. The SM will allow us to quantify the lifetime of lakes for future missions that may send probes to land on or in the lakes.

Acknowledgements The Titan Orbiter Science Team (TOST) worked many long hours to put together a balanced scientific plan for the PM and EM. The authors acknowledge the contributions of the following members of TOST: Frank Crary (CAPS), Don Mitchell (MIMI), William Kurth and George Hospodarsky (RPWS), Hunter Waite and Greg Fletcher (INMS), Fritz Neubauer and Cesar Bertucci (MAG), Michael Flasar and Connor Nixon (CIRS), Elizabeth Turtle and Alfred McEwen (ISS), Robert West (UVIS), Bonnie Burratti and Christophe Sotin (VIMS), Ralph Lorenz (Radar), Arv Kliore (RSS). We thank Trina Ray, Doug Equils, Kim Steadman, Jo Pitesky, Dave Mohr and Chris Roumeliotis for invaluable technical support in the definition and execution of Titan activities in the EM. This effort was carried out at the Jet Propulsion Laboratory, California Institute of Technology, under a contract with the National Aeronautics and Space Administration.

References

Achterberg RK, Conrath BJ, Gierasch PJ, Flasar FM, Nixon CA (2008) Titan's middle-atmospheric temperatures and dynamics observed by the Cassini Composite Infrared Spectrometer. Icarus 194:263–277

Ajello JM, Gustin J, Stewart I, Larsen K, Esposito L, Pryor W, McClintock W, Stevens MH, Malone CP, Dziczek D (2008) Titan airglow spectra from the Cassini ultraviolet imaging spectrograph: FUV disk analysis. Geophys Res Lett 35:L06102. doi:10.1029/2007GL032315

Backes H (2004) Titan's interaction with the Saturnian magnetospheric plasma. PhD dissertation. Institut fuer Geophysik und Meteorologie, University of Cologne, Germany

Backes H, Neubauer FM, Dougherty MK, Achilleos N, Andre N, Arridge CS, Bertucci C, Jones GH, Khurana KK, Russell CT, Wennmacher A (2005) Titan's magnetic field signature during the first Cassini encounter. Science 308:992–995

Baines KH, Drossart P, Lopez-Valverde MA, Atreya SK, Sotin C, Momary TW, Brown RH, Buratti BJ, Clark RN, Nicholson PD (2006) On the discovery of CO nighttime emissions on Titan by Cassini/VIMS: derived stratospheric abundances and geological implications. Planet Space Sci 54:1552–1562

Barnes JW, Brown RH, Radebaugh J, Buratti BJ, Sotin C, Le Mouelic S, Rodriguez S, Turtle E, Perry J, Clark R, Baines KH, Nicholson PD (2006) Cassini observations of flow-like features in western Tui Regio. Titan Geophys Res Lett 33:L16204. doi:10:1029/2006GL026843

Barnes JW, Radebaugh J, Brown RH, Wall S, Soderblom L, Lunine J, Burr D, Sotin C, Le Mouelic S, Rodriguez S, Buratti BJ, Clark R, Baines KH, Jaumann R, Nicholson PD, Kirk RL, Lopes R, Lorenz RD, Mitchell K, Wood CA (2007) Near-infrared spectral mapping of Titan's mountains and channels. J Geophys Res 112:E11006. doi:10.1029/2007JE002932

Barnes JW, Brown RH, Soderblom L, Sotin C, Le Mouelic S, Rodriguez S, Jaumann R, Beyer RA, Buratti BJ, Pitman K, Baines KH, Clark R, Nicholson PD (2008) Spectroscopy, morphometry, and photoclinometry of Titan's dunefields from Cassini/VIMS. Icarus 195:400–414

Bertucci C, Neubauer FM, Szego K, Wahlund J-E, Coates AJ, Dougherty MK, Young DT, Kurth WS (2007) Structure of Titan's mid-range magnetic tail: Cassini magnetometer observations during the T9 flyby. GRL 34:doi: 10.1029/2007GL030865

Brown RH, 24 co-authors (2006) Observations in the Saturn system during approach and orbital insertion, with Cassini's Visual and Infrared Mapping Spectrometer (VIMS). Astron Astrophys 446:707–716

Brown RH, Soderblom LA, Soderblom JM, Clark RN, Jaumann R, Barnes JW, Sotin C, Buratti B, Baines KH, Nicholson PD (2008) The identification of liquid ethane in Titan's Ontario Lacus. Nature 454:607–610

Buffington B, Strange N, Smith J (2008) Overview of the Cassini extended mission trajectory. AIAA-2008-6752

Elachi C, Wall S, Allison M, Anderson Y, Boehmer R, Callahan P, Encrenaz P, Flamini E, Franceschetti G, Gim Y, Hamilton G, Hensley S, Janssen M, Johnson W, Kelleher K, Kirk R, Lopes R, Lorenz R, Lunine J, Muhleman D, Ostro S, Paganelli F, Picardi G, Posa F, Roth L, Seu R, Shaffer S, Soderblom L, Stiles B, Stofan E, Vetrella S, West R, Wood C, Wye L, Zebker H (2005) First views of the surface of Titan from the Cassini Radar. Science 308:970–974

Flasar M, 45 co-authors (2005) Titan's atmospheric temperatures, winds, and composition. Science 308:975–978

Griffith CA, Penteado P, Rannou P, Brown R, Boudon V, Baines K, Clark R, Drossart P, Buratti B, Nicholson P, Jaumann R, McKay CP, Coustenis A, Negrao A (2006) Evidence for ethane clouds on Titan from Cassini VIMS observations. Science 313:1620–1622

Hayes A, Aharonson O, Callahan P, Elachi C, Gim Y, Kirk R, Lewis K, Lopes R, Lorenz R, Lunine J, Mitchell K, Mitri G, Stofan E, Wall S (2008) Hydrocarbon lakes on Titan: distribution and interaction with a porous regolith. Geophys Res Lett 35:L09204. doi:10.1029/2008GL033409

Lopes RMC, 44 co-authors (2007) Cryovolcanic features on Titan's surface as revealed by the Cassini Titan Radar Mapper. Icarus 186:395–412

Lorenz RD, Lemmon MT, Smith PH (2004) Seasonal change in Titan's haze 1992–2002 from Hubble space telescope observations. Geophys Rev Lett 31:L10702. doi:10.1029/2004GL019864

Lorenz RD, Wall S, Radebaugh J, Boubin G, Reffet E, Janssen M, Stofan E, Lopes R, Kirk R, Elachi C, Lunine J, Mitchell K, Paganelli F, Soderblom L, Wood C, Wye L, Zebker H, Anderson Y, Ostro S, Allison M, Boehmer R, Callahan P, Encrenaz P, Ori GG, Francescetti G, Gim Y, Hamilton G, Hensley S, Johnson W, Kelleher K, Muhleman D, Picardi G, Posa F, Roth L, Seu R, Shaffer S, Stiles B, Vetrella S, Flamini E, West R (2006) The sand seas of Titan: Cassini RADAR observations of longitudinal dunes. Science 312:724–727

Lorenz RD, Lopes RM, Paganelli F, Lunine JI, Kirk RL, Mitchell KL, Soderblom LA, Stofan ER, Ori GG, Myers M, Miyamoto H, Radebaugh J, Stiles B, Wall SD, Wood CA (2008a) Fluvial channels on Titan: initial Cassini RADAR observations. Planet Space Sci 56:1132–1144

Lorenz RD, Stiles BW, Kirk RL, Allison MD, Persi del Marmo P, Iess L, Lunine JI, Ostro SJ, Hensley S (2008b) Titan's rotation reveals an internal ocean and changing zonal winds. Science 319:1649–1651

Lunine J, Atreya S (2008) The methane cycle on Titan. Nature Geosci 1:159–164

Marouf E, Flasar M, French R, Kliore A, Nagy A, Rappaport N, McGhee C, Schinder P, Simpson R, Anabtawi A, Asmar S, Barbinis E, Fleischman D, Goltz G, Kahan D, Kern A, Rochblatt D (2006) Evidence for likely liquid hydrocarbons on Titan's surface from Cassini Radio Science Bistatic scattering observations. AGUFM. P11A.07M. AGU Fall Meeting. Abstract P11A–07

McEwen A, Turtle E, Perry J, Dawson D, Fussner S, Collins G, Porco C, Johnson T, Soderblom L (2005) Mapping and monitoring the surface of Titan. Bull Am Astron Soc 37:Abstract 53.04

Nelson RM, 29 coauthors (2009) Saturn's Titan: surface change, ammonia, and implications for atmospheric and tectonic activity. Icarus 199:429–441

Neubauer F, Gurnett DA, Scudder JD, Hartle RE (1984) Titan's magnetospheric interaction. In Gehrels T, Matthews MS (eds) Saturn, Univ. of Ariz. Press, Tucson, pp 760–787

Porco CC, 36 coauthors (2005) Imaging of Titan from the Cassini spacecraft. Nature 434:159–168

Radebaugh J, Kirk RL, Lopes RM, Stofan ER, Valora P, Lunine JI, Lorenz RD, the Cassini Radar team (2008a) Mountains on Titan as evidence of global tectonism and erosion. LPSC XXXIX:2206

Radebaugh J, Lorenz RD, Lunine JI, Wall SD, Boubin G, Reffet E, Kirk RL, Lopes RM, StofanER, Soderblom L, Allison M, Janssen M, Paillou P, Callahan P, Spencer C, the Cassini Radar team (2008b) Dunes on Titan observed by Cassini Radar. Icarus 194:690–703

Rages K, Pollack JB (1983)Vertical distribution of scattering hazes in Titan's upper atmosphere. Icarus 55:50–62

Rappaport NJ, Iess L, Wahr J, Lunine JI, Armstrong JW, Asmar SW, Tortora P, Di Benedetto M, Racioppa P (2008) Can Cassini detect a subsurface ocean in Titan from gravity measurements? Icarus 194:711–720

Shemansky DE, Stewart AIF, West RA, Esposito LW, Hallett JT, Liu X (2005) The Cassini UVIS stellar probe of the Titan atmosphere. Science 308:978–982

Smith PH, Lemmon MT, Lorenz RD, Sromovsky LA, Caldwell JJ, Allison MD (1996) Titan's surface, revealed by HST imaging. Icarus 119:336–349

Soderblom LA, 27 co-authors (2007b) Correlations between Cassini VIMS spectra and RADAR SAR images: implications for Titan's surface composition and the character of the Huygens probe landing site. Planet Space Sci 55:2025–2036

Soderblom LA, Tomasko MG, Archinal BA, Becker TL, Bushroe MW, Cook DA, Doose LR, Galuszka DM, Hare TM, Howington-Kraus E, Karkoschka E, Kirk RL, Lunine JI, McFarlane EA, Redding BL, Rizk B, Rosiek MR, See C, Smith PH (2007a) Topography and geomorphology of the Huygens landing site on Titan. PSS 55:2015–2024

Sotin C, 26 co-authors (2005) Release of volatiles from a possible cryovolcano from near-infrared imaging of Titan. Nature 435:786–789

Stiles BW, Kirk RL, Lorenz RD, Hensley S, Lee E, Ostro SJ, Allison MD, Callahan PS, Gim Y, Iess L, Persi del Marmo P, Hamilton G, Johnson WTK, West RD, the Cassini Radar team (2008) Determining Titan's spin state from Cassini radar images. Astron J 135:1669–1680

Stofan ER, Elachi C, Lunine JI, Lorenz RD, Stiles B, Mitchell KL, Ostro S, Soderblom L, Wood C, Zebker H, Wall S, Janssen M, Kirk R, Lopes R, Paganelli F, Radebaugh J, Wye L, Anderson Y, Allison M, Boehmer R, Callahan P, Encrenaz P, Flamini E, Francescetti G, Gim Y, Hamilton G, Hensley S, Johnson WTK, Kelleher K, Muhleman D, Paillou P, Picardi G, Posa F, Roth L, Seu R, Shaffer S, Vetrella S, West R (2007) The lakes of Titan. Nature 445:61–64

Teanby NA, Irwin PGJ, de Kok R, Vinatier S, Bezard B, Nixon CA, Flasar FM, Calcutt SB, Bowles NE, Fletcher L, Howett C, Taylor FW (2007) Vertical profiles of HCN, HC_3N, and C_2H_2 in Titan's atmosphere derived from Cassini/CIRS data. Icarus 186: 364–384

Tokano T (2008) Dune-forming winds on Titan and the influence of topography. Icarus 194:243–262

Tokano T, Neubauer F (2005) Wind-induced seasonal angular momentum exchange at Titan's surface and its influence on Titan's length-of-day. GRL 32, doi: 10.1029/2005GL024456

Turtle EP, Perry JE, McEwen AS, DelGenio AD, Barbara J, West RA, Dawson DD, Porco CC (2009) Cassini imaging of Titan's high-latitude lakes, clouds, and south-polar surface changes. Geophys. Res. Letters Volume 36 doi: 10.1029/2008 GL 036/86

Turtle EP, Perry JE, McEwen AS, West RA, DelGenio AD, Barbara J, Dawson DD, Porco CC (2008) Cassini imaging observations of Titan's high-latitude lakes. LPSC XXXIX:Abstract #1952

Chapter 18
Titan Beyond Cassini–Huygens

Michele K. Dougherty, Athena Coustenis, and Ralph D. Lorenz

Abstract This chapter reviews the unanswered science questions which remain after the Cassini-Huygens nominal tour as well as the many new questions which has arisen following new discoveries which have been made. Further missions to the Titan system which have been studied are described, in particular that of the most recent study, the Titan Saturn System Mission.

18.1 Introduction

As preceding chapters have revealed, our understanding of the Titan environment has greatly increased thanks to the Cassini–Huygens observations from the first 4 years of orbital tour. Below we briefly summarize the knowledge we have gained, as well as detail the unresolved science questions which are as yet unanswered by Cassini–Huygens. We also describe new questions which have arisen from the discoveries that have been made. Mission studies (Lorenz 2009) carried out both in the past and more recently are also described in order to reveal the rich exploration concepts and ideas which drive our desire to resolve the many questions linked to Titan and its importance in the solar system. The chapter ends with details of the joint NASA-ESA study for a Titan Saturn System Mission whose concept we hope will follow as the next but one large outer planetary Flagship/ L-class mission.

M.K. Dougherty (✉)
Space and Atmospheric Physics, Imperial College London, SW7 2AZ, UK
email: m.dougherty@imperial.ac.uk

A. Coustenis
Laboratoire d'Etudes Spatiales et d'Instrumentation en Astrophysique, Paris-Meudon Observatory, 92195 Meudon France

R.D. Lorenz
Applied Physics Laboratory, Johns Hopkins University, Laurel, MD 20723USA

18.2 Unresolved Science Questions

18.2.1 Our Pre-Cassini–Huygens Understanding

Our pre-Cassini–Huygens knowledge of Titan and ground-based observations have been reviewed in detail in Chapter 2; here we briefly summarize the main issues. Our initial knowledge of Saturn's largest moon Titan arose from ground-based observations of its hazy atmosphere which generated a debate as to whether it consisted of a thin layer of methane or rather a dense shield of methane and nitrogen. A key characteristic of Titan is its large collection of organics, with the first discovery being that of the existence of methane and follow-on laboratory experiments suggesting that methane photolysis could result in solid organic aerosols or tholins. The Voyager 1 Titan flyby confirmed the presence of methane in a thick background atmosphere of nitrogen and in addition, many complex hydrocarbons and nitriles were observed as well as a thick organic haze. This thick haze has made resolving detailed surface or atmospheric features extremely difficult, and in fact only large surface features, such as the North–South atmospheric asymmetry and cloud systems were observed prior to Cassini–Huygens.

A single flyby of Titan by the Voyager 1 spacecraft revealed that the corotating Saturnian plasma interacted with the atmosphere of Titan rather than an internal magnetic field (see Chapter 16) which leads to the formation of an induced magnetosphere with field line draping, in a very similar way to what is observed at Venus. This also enabled an upper limit for a possible internal dynamo field to be derived. Hence, early remote sensing observations of Titan, as well as the Voyager 1 flyby data raised many more questions than were answered. These include: what is the nature of the hidden surface, what are the sources which resupply methane to the atmosphere and what is the nature of Titan's interior?

R.H. Brown et al. (eds.), *Titan from Cassini-Huygens*,

18.2.2 Post-Cassini Science Results

18.2.2.1 Cassini's Prime Mission

The combination of multiple flybys of Titan by the Cassini spacecraft, as well as the Huygens probe descent through Titan's atmosphere and its landing on Titan's surface, has resulted in a considerable increase in our understanding of the Titan system; including new information concerning its surface, atmosphere and the interaction which occurs with its surrounding plasma environment. The many discoveries which have been made are described in detail in the preceding chapters of this book and they will simply be briefly referred to here.

Titan's Interior Structure

In order to probe the internal structure of any planetary body, magnetic field and gravity observations are required in order to resolve the existence of any internal dynamo field, induced field or internal liquid layer or ocean. From Cassini flybys during the prime mission, the spacecraft flyby distances of ≥950 km resulted in any internal or intrinsic signatures being masked by ionospheric signatures from Titan. As Chapter 17 describes, a close flyby during the Cassini spacecraft's extended mission will potentially enable induced versus intrinsic signatures to be resolved. In order to diagnose the presence of a liquid layer, which decouples the interior from the surface, an understanding of Titan's gravity field is required. Upcoming Cassini gravity observations may also aid in constraining the mass distribution and possibly crustal rigidity of Titan.

The rotational dynamics of Titan also allow an understanding to be gained of its interior, and radar imagery is particularly suited to the determination of the rotation period. Recent work detailed in Chapter 4, has revealed significant changes of the rotation period of Titan, with the implication being that the crust is decoupled from the interior by a liquid water ocean. A possible detection of the Schumann resonance (Chapter 4) from the Huygens probe descent through Titan's atmosphere raises the possibility of utilizing measurements of this extremely low frequency wave and its resonance to probe the interior. There are specific Titan flybys in the extended mission that will enable resolution of some of these unresolved questions and further interior-specific flybys will be planned in the continued extension after that, the Cassini Solstice mission, allowing for additional study of these questions.

Surface Science

Our knowledge of Titan's surface has been revolutionized by the Cassini–Huygens observations. Since the orbital tour began, discoveries have been made of vast equatorial dune fields of complex organics, drainage channels likely to be carved by liquid ethane/methane, lakes, craters and mountains as well as evidence for cryovolcanism (see Chapters 5 and 6). Recently the first unequivocal detection of liquid ethane in Ontario Lacus has also been made. There remains a large fraction of the Titan surface yet to be imaged, particularly at high resolution. Furthermore, while near-infrared spectroscopy has demonstrated that Titan has a compositionally diverse surface, the confident identification of surface materials in most cases is hampered by the relatively few wavelengths covered by both Cassini's instrumentation and the transparent methane windows.

In order to understand and properly interpret the geological evolution of Titan, as complete a surface coverage as is possible at high resolution is required to resolve the various geological features. The main areas of focus center on mapping unseen regions, resolving previously imaged areas at higher spatial resolution, characterizing differences in latitude and mapping surface composition. The potential cyrovolcanic features observed to date will be investigated for any evidence of ongoing activity. The equatorial dune fields, lakes and erosion channels will be studied for insight into their formation, orientation and implications this has for meteorology. Seasonal changes will be monitored in order to better constrain timescales of the methane cycle.

Atmospheric Science

Titan's dense atmosphere has been observed by Cassini to vary seasonally in temperature, dynamical behavior and composition. There is also a winter polar structure which is comparable to the ozone hole at the Earth. During the prime mission, atmospheric structure and composition have been resolved and in situ observations from Huygens resulted in measurements of pressure, temperature and composition during its descent. Chapters 10 and 13 detail these new findings and in summary, we now have a far better understanding of the temperature, wind fields and composition of the troposphere and stratosphere; as well as of atmospheric dynamics. In addition abundance, vertical structure, isotopic and mixing ratios of aerosols and atmospheric gases from the stratosphere through to the thermosphere has been inferred at different locations over Titan's disk. Cloud structures in the troposphere have been mapped as well as higher atmospheric layers. These observations have yielded a better understanding of the current meteorology at Titan.

The main focus of the extended mission and the Cassini Solstice mission will be the dynamical processes taking place within the atmosphere, all of which evolve with time. This will be particularly important as the season changes and Titan goes through equinox. The main science questions which will be tackled include; gaining an understanding of the cloud and weather changes which may occur as the Titan season evolves, resolving the sources and sinks of the higher order hydrocarbons already detected, gaining insight into the formation of the polar vortex as well as the formation of aerosols and conducting a search for methane rainfall.

Upper Atmosphere and Its Interaction with Saturn's Magnetosphere

The type of interaction which occurs with Saturn's magnetospheric plasma is similar to that which arises at Venus. Saturn's plasma and magnetic field interact with the upper regions of Titan's ionized atmosphere. On occasion when near solar local noon, Titan can interact directly with the solar wind plasma as has occurred on a single Titan flyby thus far. Following the Titan flybys made during the prime mission the basic structure of the upper atmosphere, as well as that of the magnetospheric interaction, have been characterized, including an understanding gained of energy deposition and escape. Some of the most interesting and surprising results obtained (see Chapters 8, 15 and 16) include the discovery of massive organic compounds and ions; negative ions; energetic ion precipitation depositing energy deep within the atmosphere; a large influx of oxygen dominated plasma; and unexpected orientation of the induced magnetic tail arising from the Titan-Saturn magnetosphere interaction.

To best understand the plasma processes that result, one ideally needs to explore as much of the longitude space around Titan at as many Saturn local times as possible. The extended mission will go some way to filling the gaps which exist in this parameter space and the further extension after that, will continue to obtain local time and longitude coverage necessary to separate temporal and spatial drivers. The science focus will be driven by the surprising discoveries already made. This will enable insight to be gained into the complex formation of organics which occurs in the upper atmosphere as well as a better understanding to be achieved as to what drives the injection of oxygen-rich plasma into the atmosphere. The geometry of the induced magnetic tail at Titan will be better constrained as well as a more accurate upper limit resolved on any weak intrinsic magnetic field.

18.2.2.2 Extended Mission and Cassini Solstice Mission science

The Cassini extended mission tour has already begun and will continue until mid-2010, while plans for a further extension after that (to 2017), the Cassini Solstice mission, are well under way. The details of the type of Titan observations which will result from these extensions have been described in detail in the previous chapters. In summary, this will allow us an opportunity to observe seasonal changes at Titan, obtain more detailed and complete surface coverage, better understand temporal changes in the atmosphere, as well as observe the plasma interaction between Titan and its environment from one solstice to the next.

18.2.3 Unresolved Science Issues During and After the Cassini–Huygens Era

Cassini–Huygens discoveries have transformed our understanding of Titan, its interaction with its environment as well as its potential for having the necessary ingredients for the formation of life. The recent discoveries confirm that Titan is rich in organic compounds, possibly contains a subsurface ocean, has liquid ethane on its surface and has the energy sources required to drive chemical evolution. In order to gain an understanding of the complex interactions which arise between Titan's atmosphere, its surface, its interior and the external plasma environment, a focused and long term observation campaign over a range of temporal and spatial scales is required.

Titan is similar to the Earth on many different fronts; it has a large abundance of organic material, it has an active hydrological cycle (with methane replacing water); it has clouds and should have rain. Its landscape is remarkably like that of the Earth, with dunes, liquid carved channels, mountain ridges and polar lakes. The main atmospheric constituent is nitrogen and the atmosphere varies on a seasonal basis as well as having a polar structure similar to the ozone hole at the Earth. Confirmation from the Cassini–Huygens observations of this remarkable similarity between the Earth and Titan, as well as its astrobiological potential, are the main drivers for a return to the Titan system in the foreseeable future. In the next section, a description is given of the variety of mission studies which have been carried out on potential space missions to Titan, as well as the most recent joint study between NASA and ESA for a Titan orbiter, balloon and lander (known as the Titan Saturn System Mission, or TSSM).

Cassini–Huygens will leave many unresolved questions (TSSM Final Report 2008) including; is methane outgassing from the interior or ice crust; are the lakes fed predominantly by rain or rather by underground methane-ethane aquifiers; how often do heavy methane rains occur; did the surface previously support seas of methane compared to the smaller lakes which exist today (see Fig. 18.1). Other key questions concern the age of the surface features which have been discovered; are cryovolcanism and tectonism still active or are we simply observing remnants of an active past. Ammonia, which recent observations suggest should be present, has

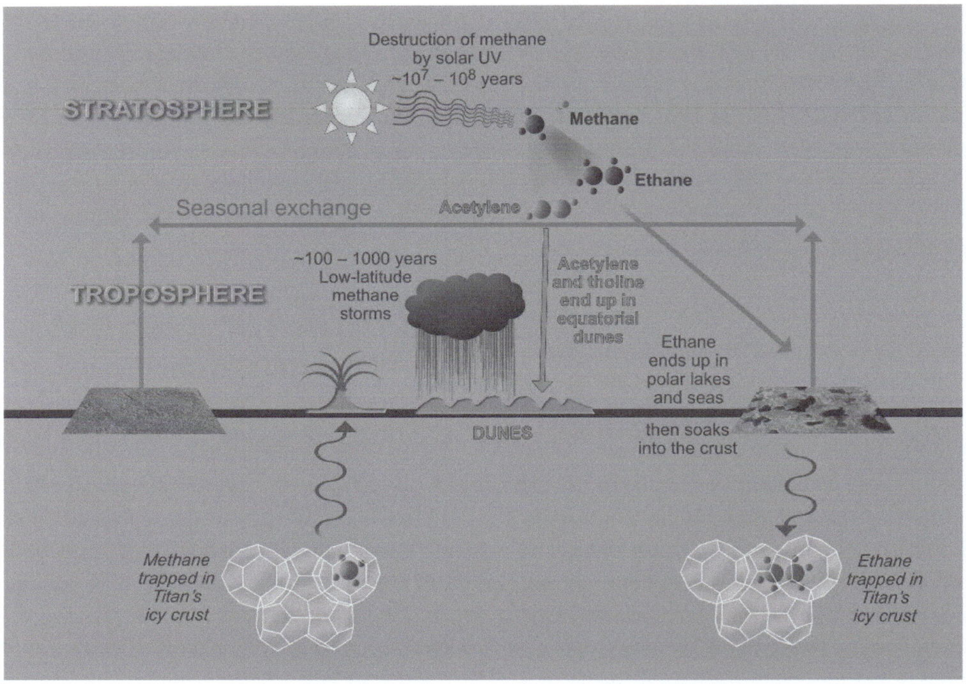

Fig. 18.1 A schematic view of the methane cycle on Titan (TSSM Final Report 2008)

still not been observed and the presence of an intrinsic magnetic field of some kind has yet to be detected. Atmospheric seasonal changes and escape of atmospheric constituents still remain topics for which a great deal remains to be revealed. Finally, the altitude range from 400–900 km will be largely unexplored following on from the Cassini–Huygens mission and much about atmospheric dynamics could be learnt from probing this region.

The main science drivers which will act as impetus for a future mission to Titan are described below and are separated into three over-riding objectives; exploring Titan as an Earth-like system; examining Titan's organic inventory and astrobiological potential; and Titan's origin and evolution.

18.2.3.1 Exploring Titan as an Earth-Like System

From the understanding we have gained from the Cassini–Huygens discoveries, it is clear that Titan presents several similarities to the Earth. It has a dense nitrogen rich atmosphere in which complex meteorological processes routinely occur, although with methane instead of water being the driving liquid. Titan's geology is also very similar to that of the Earth and even though Titan's surface is an icy one, the geological processes occurring there are very Earth-like, though with different materials. In addition there appears to be a liquid ocean, which could be mainly water (mixed with ammonia), beneath Titan's icy surface.

Being able to understand Titan as a system, and the coupling processes which occur between the diverse physical mechanisms which arise; such as geology, hydrology, meteorology and aeronomy will enable us to place Titan into context in the solar system; as well as compare it with our understanding of Venus, Mars and the Earth. Is Titan at a stage in its evolution which the Earth has already passed through or is yet to come? Why does Titan have an atmosphere but not a strong dynamo magnetic field, when Ganymede, one of Jupiter's moons which is very similar in size and density, has no atmosphere but does have an internal magnetic field?

18.2.3.2 Titan's Organic Inventory and Astrobiological Potential

Titan is extremely rich in organic molecules, both on its surface and in its atmosphere. Transient episodes of melting of the surface water ice would expose atmospheric organic molecules which had been deposited on the surface to aqueous alteration, leading potentially to chemical reactions which may replicate those which occurred on pre-biotic Earth. No other body in our solar system has this type of chemistry taking place and being able to understand the chemical cycles which generate or destroy these organics will contribute to our understanding of the origin of life within our Solar System. Some of the questions which require answers, in order to further our understanding, are what is the complexity of the organic chemistry in the atmosphere, on the surface, in the lakes and in the possible subsurface water ocean? In addition, it is important to determine how this organic inventory may differ to that which we know arises in meteorites and comets.

18.2.3.3 Titan's Origin and Evolution

The overall density of Titan requires that it contain nearly equal proportions of rock and ice. If we have knowledge of the extent of the differentiation of its interior that will help constrain temperatures in the early Saturn nebula. Cassini is expected to make gravity and magnetic field observations which will aid in obtaining a first order understanding of the internal structure of Titan. However, in order to properly constrain what is a very complex problem, many more observations, ideally from an orbiter, will be required to best resolve the internal structure, presence of an ocean and the rotational dynamics. A better understanding of the evolution of Titan's atmosphere can be derived by resolving the escape processes which occur via in-situ sampling of the plasma and energetic particle environment at Titan. This will reveal what the exchange of energy and material is between Titan and its surrounding plasma environment.

18.3 Future Missions

The exotic environment of Titan has prompted a number of imaginative ideas throughout the history of planetary exploration (Lorenz 2009). These mission concepts can be conveniently divided into several epochs separated by the encounters with Titan of missions that flew and thus prompted reassessment of the goals and opportunities at this moon. The studies from each of these epochs will be briefly described in turn, culminating in the present set of post-Cassini mission studies.

18.3.1 The Pre-Cassini Era

This era begins in the early 1970s well prior to the flyby of Titan by Voyager 1 in 1980. Titan entry probes were considered as early as 1973 paralleling the Galileo and Pioneer Venus probes which were under development. A Martin-Marietta study in 1975–1976 (Marietta 1976) was perhaps the first detailed examination of Titan mission options, and at the time it was known that Titan had an atmosphere and was interesting from an exobiological point of view. An important feature of this study is that any mission had to be tolerant of wide uncertainties in surface composition and atmospheric pressure. One concept considered was the "penetrobe", a combination of probe and penetrator with the separation of the penetrator timed to depend on the atmospheric density profile encountered. In addition, although the concept of aerocapture was not fully developed at this stage, the low entry speed and heat loads of arrival from Titan orbit were noted. The study included the idea of a dedicated remote-sensing orbiter and a Viking-like parachute-borne lander. In this sense, the study rather anticipates the mission concepts still explored some 30 years later. At around this time balloon exploration of Titan was also advocated, with a solar montgolfiere suggested (Blamont 1978).

In the years around the Voyager encounter, the paradigm for future Titan exploration included Saturn system studies (in an analog for the Galileo Jupiter mission) with a leading concept being a Saturn orbiter with two probes, one for Saturn and one for Titan (Roberts and Wright 1979). This study was notable in that it essentially defined what was to become the Huygens probe. A subsequent study (Friedlander 1984) guided by the same science objectives that shaped Cassini, considered a wide range of mission possibilities and devoted particular attention to balloons and entry probes. It was recognized around this time that the Titan atmosphere might be the key to efficient exploration via aerocapture or gravity assist (Menees 1983).

A mission named Cassini was proposed to ESA in response to a call for ideas in 1982, which was more fully explored by a joint European/US working group. With the onset of the Cassini–Huygens mission as a joint venture between NASA, ESA and ASI in 1990, Titan mission studies fell into abeyance for several years with only a handful of ad-hoc studies which examined submarines, depth probes, balloon concepts and an aerial robot.

Once Cassini–Huygens was en-route to the Saturn system, scientific priorities focused on the need to understand surface chemistry for which the Huygens probe was not well equipped. A NASA appointed working group suggested the following prioritized order (Chyba et al. 1999) for post Cassini–Huygens exploration of Titan: Surface, sub-surface and atmosphere; with the concepts emphasizing in-situ exploration vehicles rather than orbiters. Numerous other smaller studies focused on future Titan exploration in this time period including a more detailed examination of aerocapture (Lockwood, 2003), and with the Cassini findings revealing the complexity of the Titan environment and a broad set of scientific drivers arising across a range of disciplines, the argument for a combination of orbiter and in-situ elements became stronger. The final precursor to the "modern era" of mission studies was a short study commissioned by NASA to answer the question of whether any worthwhile Titan or Enceladus mission could be performed in the $650M to $1B framework. This study (Reh et al. 2007) determined that no worthwhile missions were achievable for this budget target, although single in-situ platforms at Titan were close to coming in budget but without an orbiter in support, would be limited in their ability to return science data to Earth.

18.3.2 The Present Epoch

In Section 18.2.3 we have reviewed some of the major questions which remain outstanding even after the Cassini–Huygens exploration of Titan and the Saturnian system. These questions were the basis for the science objectives which motivated the present studies for future missions to Titan. The science goals are to provide information on such aspects as; the composition of the surface and the geographic distribution of the various organic constituents; on the methane cycle and the methane reservoirs; on the ages of the surface features, and in particular on whether cryovolcanism and tectonism are actively ongoing or are relics of a more active past; on the presence or absence of ammonia, of a magnetic field and of a sub-surface ocean; on the chemistry which drives complex ion formation in the upper atmosphere; and on the *agnostosphere* (a large altitude range in the atmosphere, from 400–900 km in altitude, which will remain poorly explored after Cassini). In addition, much remains to be understood about seasonal changes of the atmosphere at all levels, and the long-term escape of constituents to space.

With these questions in mind, and in view of the apparent absence of small viable missions to Titan, the stage was set for a NASA Titan Flagship study in 2007, whose task was to develop the scientific rationale for a mission in the 2015–2020 timeframe with a budget target of $3B and descope options to $2B. At the same time a European consortium, with US collaborators, responded to ESA's 2007 Cosmic Vision 2015–2025 Call with a Titan Enceladus Mission proposal (TandEM) which was then subsequently selected for further study by ESA. These two studies will be described below, followed by the results of their merging to form the TSSM for which an international science and technical team was formed with the goal of a focused cost-effective joint NASA-ESA mission to the Titan system.

18.3.2.1 NASA Studies

The NASA Flagship study (Leary et al. 2007; Lorenz et al. 2008; Lockwood et al. 2008) considered a variety of mission architectures but quickly converged to a triple-platform mission (of orbiter, lander and balloon) package on a single Atlas 551 launch vehicle. An important constraint on the mission was that radioisotope power sources determined to be available were 100W class units (MMRTG's and ASRG's) with the cold, thick Titan atmosphere making long-lived instrument packages impossible without radioisotope power. The mission featured a long-lived lander and a montgolfiere balloon with mass allocations of some 900 and 600 kg respectively, which would be released for direct entry about 1 month prior to Titan arrival. The orbiter would use aerocapture to brake into orbit around Titan, allowing the 1,800 kg orbiter (including ~170 kg of instrument and consumables) 4 years of Titan operations. Initially an elliptical orbit would allow in-situ measurements of the ionosphere with the periapsis over the south pole and after several weeks the orbit would be circularized to 1,700 km for mapping. Towards the end of the mission, the periapsis would be lowered, this time over the north pole to perform aerosampling.

In order to bridge the global and local scales sampled in turn by an orbiter and a lander, a montgolfière balloon was included to perform a regional survey. A 1-year lifetime was assumed, with a float altitude of about 10 km, at which zonal winds should permit the balloon to circumnavigate Titan one or two times. With early Cassini data revealing the diversity of Titan's surface, a preliminary landing site for the lander was chosen to be a large equatorial region covered in organic rich sand dunes, known as Belet. Lander operations were costed for a 1-year mission although its power source of 1 ASRG's would allow it to operate near-indefinitely. The possibility of augmenting lander science with a small lander-launched airplane was also advocated.

18.3.2.2 ESA Studies

The ESA TandEM proposal (Coustenis et al. 2008) originally submitted to ESA's Cosmic Vision call in June 2007 was one of the missions chosen for further study. The original proposal was for an L-class mission (budget within €650 million) to explore in-situ both Titan and Enceladus in collaboration with NASA and the Canadian Space Agency. The proposal directly addresses several of the key scientific questions highlighted in the ESA Cosmic Vision 2015–2025 call. The mission concept was built around an orbiter, a Titan aerial platform, Titan mini-probes and Enceladus penetrators/landers carrying a large variety of scientific instruments. The baseline mission was for two, moderately-sized spacecraft to be launched around 2020; these spacecraft would use chemical propulsion and radioisotope power, while orbit insertion options such as aerocapture and aerobraking would be part of the assessment phase. The spacecraft were envisaged to consist of a Titan-Enceladus Orbiter (which would carry the Enceladus in-situ package, which consisted mainly of penetrators) and a carrier for the Titan in-situ elements (a montgolfière and up to three mini-probes).

The orbiter would first go into orbit around Saturn and then, using cycler orbits, make multiple flybys of Enceladus as well as deliver two penetrators to prime target sites on its surface. The orbiter would then go into orbit around Titan and once the initial orbiter phase was completed, the orbit would be circularized to prepare its support of the Titan probes which would arrive shortly afterwards. These probes would separate from their carrier and enter Titan's atmosphere via aerocapture. The montgolfière would use RTGs for both its heat and electric source of energy and could stay aloft for months to years. It could also deliver surface probes and at end of mission would land at a prime target location.

18.3.2.3 Joint Studies – The Titan Saturn System Mission

The joint studies recently completed by a combined US and European science and technical definition team describe a collaborative effort between NASA and ESA (TSSM Final Report 2008) to explore the Titan system. The goals of this mission build on the scientific momentum gained from the Cassini–Huygens era and will extend our understanding of the seasonal variations by arriving earlier in the annual cycle, see Fig. 18.2. The orbiter, lake lander and montgolfière which are described below, will observe Titan (and to a lesser extent Enceladus and Titan's magnetosphere) at global, regional and local scales with instruments able to achieve fundamentally new science.

The baseline architecture will integrate the orbiter and in-situ elements into a single flight system to launch on an Atlas V 551 vehicle followed by thrusting with a highly efficient SEP stage enabling a shorter trip time to the Saturn system (9 years approximately) as well as an additional 300 kg of mass margin. An orbiter, lake lander and montgolfière constitute the baseline mission. As described earlier, a better understanding of the nature of the Titan environment requires direct sampling of the surface as well as portability on the surface. In addition it is crucial to be able to globally map Titan from orbit.

The orbiter, initially placed into orbit around Saturn, will make 7 flybys of Enceladus and at least 16 of Titan. During this phase, there will be multiple opportunities to observe Saturn, several of the icy moons and the complex interaction between Titan and Saturn's magnetosphere. Titan Orbit Insertion will be carried out at the end of the Saturn tour phase and capture into an elliptical orbit will be followed by a 2-month aerobraking and aerosampling phase in which direct analysis will be made of Titan's atmosphere to altitudes well below that which Cassini has been

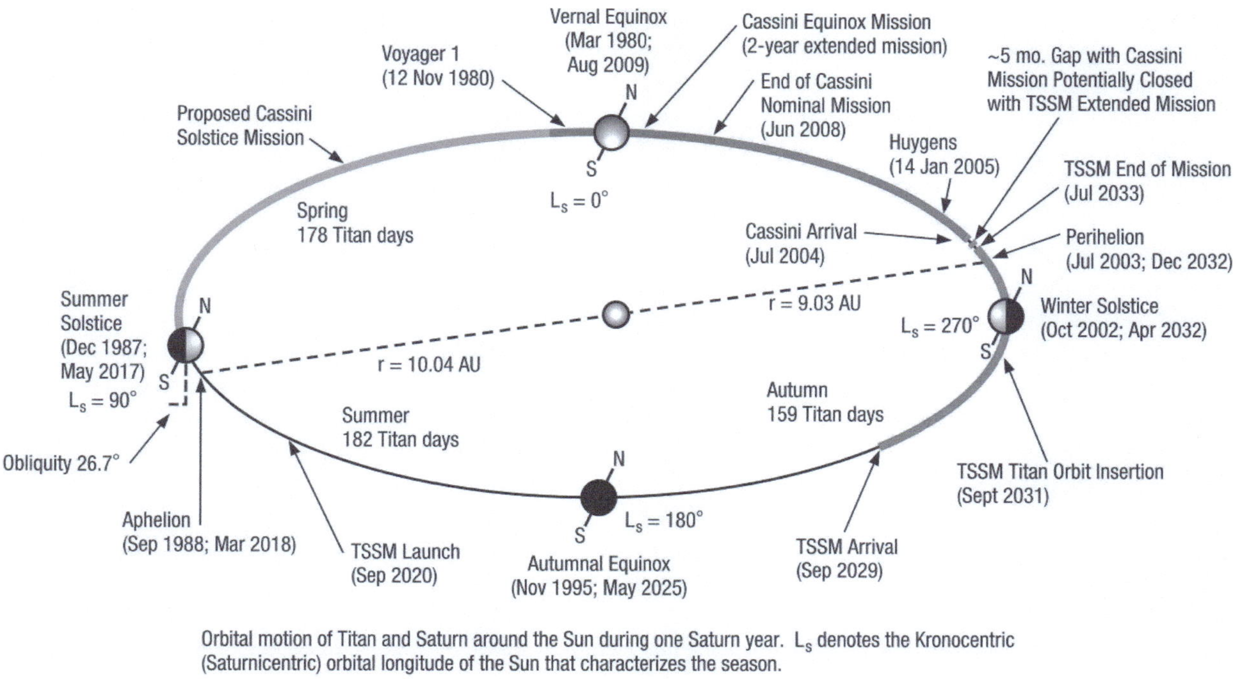

Fig. 18.2 The seasonal cycle on Titan (TSSM Final Report 2008)

capable of (as low as 600 km from the surface). The orbiter would eventually settle into a circular 1,500 km orbit around Titan, at 85° inclination polar orbit for the 24 month orbital science phase. At this altitude the orbiter will circle Titan approximately 5 times each Earth day thereby enabling a global view of the surface to be obtained. During the Saturn tour phase the orbiter will deliver the in-situ elements, the montgolfiere and lander, (TSSM in-situ elements, 2008) and then collect and relay their data back to Earth. The TSSM notional orbiter payload, as determined during the 2008 study, is composed of 7 instruments: a high-resolution imager and spectrometer (operating in the near-infrared), a penetrating radar and altimeter, a polymer mass spectrometer, thermal-infrared and sub-mm spectrometers, a magnetometer and plasma package and a radio science experiment including an accelerometer. These instruments are designed to address the science goals described earlier in this section and in Section 18.2.3.

The montgolfière will be delivered to ~20°N where the winds are the strongest (on the order of 1–2 m/s) and the baseline design has it floating around mid-latitudes at about 10 km altitude, carried by winds and heated by MMRTGs. At least one circumnavigation around Titan is expected during its prime mission of 6 months (although a 12 month mission will be planned for). In the current configuration of the in-situ TSSM mission, the balloon will carry 7 instruments (including atmospheric structure and meteorology instruments, cameras, an imaging spectrometer, an aerosol analyzer, a radar sounder, and a magnetometer). The atmospheric environment on Titan makes the possibility of operating a balloon extremely robust. Several of the newly discovered Cassini–Huygens geomorphological features (such as dunes, fluvial systems, dry lakes, mountains and potential cryovolcanic sites) are found in the region that will be investigated by the floating balloon, which also bridges the gap between the small-scale resolution observations from the lander and those (rather coarse-resolution) from the orbiter.

The short-lived (9-h nominal lifetime), battery-operated lander is designed for a wet landing in Kraken Mare at ~72°N, however just as was the case with Huygens this lander will also be capable of landing on a solid surface should the mission develop towards an equatorial dry landing. During the lander descent under parachute and below 130 km in altitude (expected to be of about 6 h in duration), the payload will be operational as well as on the surface (for a period of 3–4 h). The atmosphere above Titan's north polar regions are expected to be enhanced in gases and aerosols at the time of the TSSM probe descent, so that two completely different atmospheric environments will be studied with the two in-situ elements. The lake area was determined to be the ideal landing site because sample injection requires only a tube to equalize pressure and a membrane to control flow. The lake site is also more interesting because the solubility of organics and noble gases in liquid ethane and methane as measured in the laboratory is high enough that the lakes are a natural collection system for the global organic inventory, and for noble gases which inform about origins. Another advantage of being near the poles, is that a magnetometer system on the lake lander will be able to better complement the measurements by the montgolfière in the equatorial region. The main instrumentation foreseen on the lander (besides cameras, an atmospheric structure and electrical properties instrument) is consequently a chemical analyzer which will make direct sampling of the atmosphere and the surface and will yield the chemical composition and isotopic measurements of the air and liquid it will be in contact with. In addition, a surface properties package will perform measurements of the surface.

The combination of orbiting and in-situ elements on Titan is a powerful opportunity that is critical for synergistic investigations – synthesis of data from these carefully selected instrumentation suites is the path to understanding this profoundly complex body, as well as its environment, and Enceladus in particular (TSSM Joint Summary Report 2008).

18.4 Summary

What emerges from this review of Titan exploration concepts is the sheer diversity of the options available – from balloons to airplanes to landers, rovers and penetrators, with various flavors of orbiter; there is essentially no spacecraft concept that has not been advocated for Titan. The first major mission study was undertaken around the time of the Viking missions to Mars, and the present studies take place against the backdrop of Mars on which two rovers, a polar lander, and no less than three orbiters are active. It is therefore fully appropriate to consider Titan exploration as best accomplished by a combination of in-situ and orbital vehicles.

The rich organic chemistry at Titan was known to be of interest in the first studies, and appropriately continues to be an important focus. While organic material occurs in many solar system settings, it is most dominant on Titan, which is an environment where it is comparatively easy to deliver large amounts of in-situ instrumentation. The findings of Cassini–Huygens have only reinforced the interest in organics and exchange processes and other findings from Cassini–Huygens have broadened the range of scientific interests at Titan. Titan System Science planned with TSSM, will be the study of the interacting planetary interior, surface, atmosphere and space environment illustrated in Fig. 18.3. It will be a rich interdisciplinary endeavor of the future that parallels the study of our own planet.

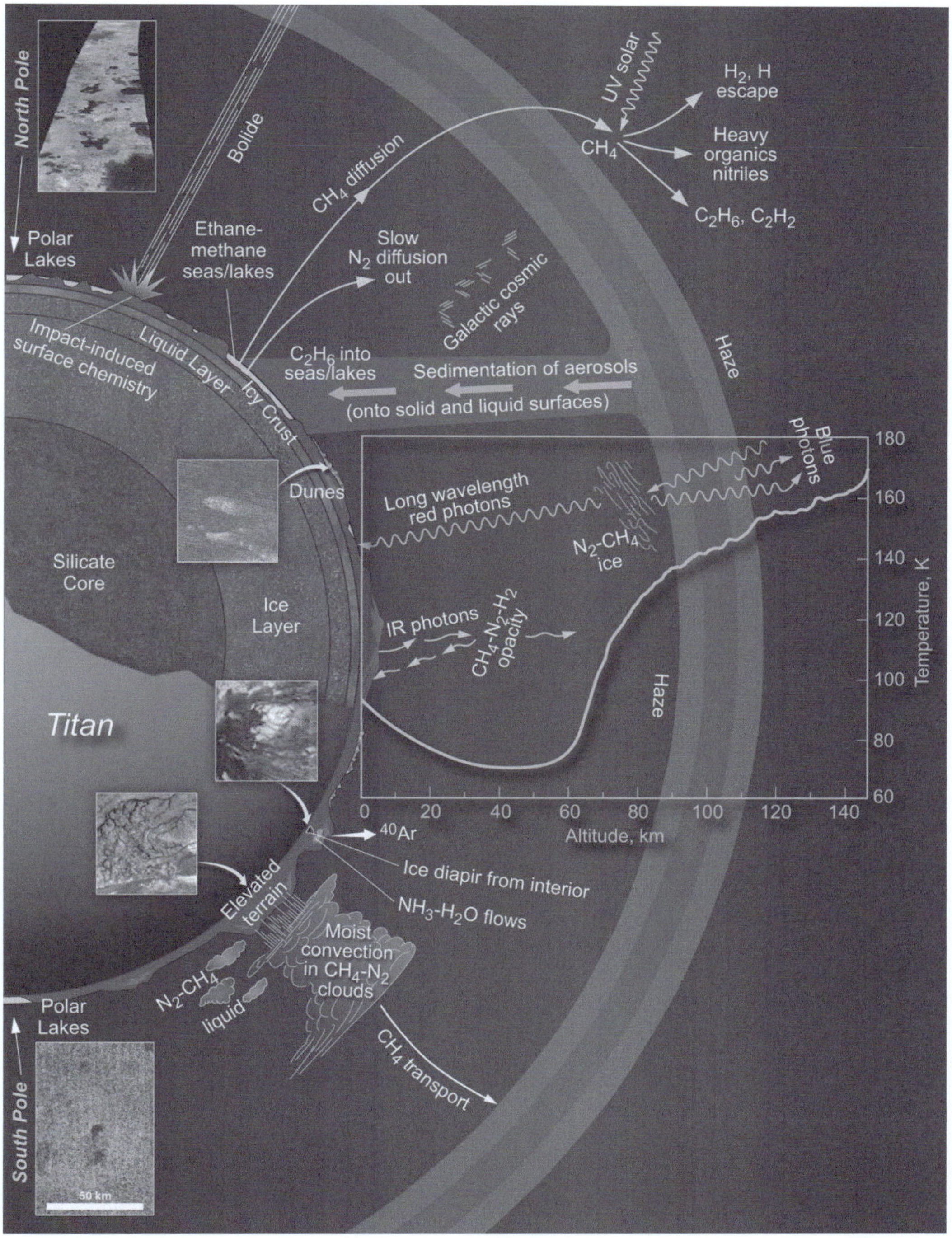

Fig. 18.3 A schematic illustration of the various inter-connections between the interior, surface, atmosphere and external environment of Titan (TSSM Final Report 2008)

References

Blamont JA (1978) Method of exploration of the atmosphere of Titan. In: Hunten DM, Morrison D (eds) The Saturn system. NASA Conference Publication 2068, pp 385–395

Chyba CF et al. (1999) Europa and Titan: preliminary recommendations of the Campaign Science Working Group on Prebiotic Chemistry in the Outer Solar System, Lunar and Planetary Science Conference, Houston

Coustenis A, Atreya S, Balint T, Brown RH, Dougherty M, Ferri F, Fulchignoni M, Gautier D, Gowen R, Griffith C, Gurvits L, Jaumann R, Langevin Y, Leese M, Lunine J, McKay CP, Moussas X, Müller-Wodarg I, Neubauer F, Owen T, Raulin F, Sittler E, Sohl F, Sotin C, Tobie G, Tokano T, Turtle E, Wahlund J-E, Waite H, Baines K, Blamont J, Dandouras I, Krimigis T, Lellouch E, Lorenz R, Morse A, Porco C, Hirtizig M, Saur J, Coates A, Spilker T, Zarnecki J, 113 co-authors (2008) TandEM: Titan and Enceladus mission. Experimental Astronomy 23:893–946

Friedlander AL (1984) Titan buoyant station. J Br Interplanet Soc 37:381–387
Leary J, Strain RD, Jones C, Lorenz R, Waite H (2007) Titan explorer NASA Flagship Mission Study. JHU Appl Phys Lab, Laurel MD
Lockwood M (2003) Titan aerocapture systems analysis. AIAA-2003-4799, AIAA/ASME/SAE/ASEE39th Joint Propulsion Conference, Huntsville, Alabama
Lockwood MK, Leary JC, Lorenz R, Waite H, Reh K, Prince J, Powell R (2008) Titan Explorer, AIAA-2008-7071, AIAA/AAS Astrodynamics Specialist Conference, Honolulu, Hawaii
Lorenz RD (2009) Titan mission studies – a historical review. JBIS:62 (in press)
Lorenz RD, Leary J, Lockwood MK, Waite JH (2008) A Titan flagship mission, STAIF Albuquerque
Martin Marietta Corporation (1976) A Titan exploration study : science, technology and mission planning options, NASA CR-137846
Menees GP (1983) Trajectory analysis of radiative heating for planetary missions with aerobraking of spacecraft, AIAA 83-0407, 21st Aerospace Sciences Meeting, Reno, NV
Reh K et al (2007) Titan Prebiotic Explorer (TiPEx) mission study final report. Jet Propulsion Laboratory, Pasadena, CA
Roberts PH, Jr., Wright JL (1979) The Saturn Orbiter Dual Probe mission concept, AAS PAPER 79-143, AAS/AIAA Astrodynamics Specialist Conference, Provincetown, MA
TSSM Final Report on the NASA Contribution to a Joint Mission with ESA, 3 November (2008) JPL D-48148, NASA Task Order NMO710851
TSSM in situ elements, ESA assessment study report, ESA-SRE (2008) 4
TSSM NASA/ESA Joint Summary Report, 15 November 2008, ESA-SRE (2008) 3, JPL D-48442, NASA Task Order NMO710851

Chapter 19
Mapping Products of Titan's Surface

Katrin Stephan, Ralf Jaumann, Erich Karkoschka, Randolph L. Kirk, Jason W. Barnes, Martin G. Tomasko, Elizabeth P. Turtle, Lucille Le Corre, Mirjam Langhans, Stéphane Le Mouélic, Ralph D. Lorenz, and Jason Perry

Abstract Remote sensing instruments aboard the Cassini spacecraft have been observed the surface of Titan globally in the infrared and radar wavelength ranges as well as locally by the Huygens instruments revealing a wealth of new morphological features indicating a geologically active surface. We present a summary of mapping products of Titan's surface derived from data of the remote sensing instruments onboard the Cassini spacecraft (ISS, VIMS, RADAR) as well as the Huygens probe (DISR) that were achieved during the nominal Cassini mission including an overview of Titan's recent nomenclature.

19.1 Introduction

The surface of Titan has long been studied with various remote sensing instruments, including those on the Hubble Space Telescope (HST) and ground-based adaptive optics systems (Coustenis et al. 2005). The first near infrared global maps of Titan were derived from images of the Hubble Space Telescope's planetary camera through the methane windows at 0.94 and 1.08 μm (Smith et al. 1996) (Fig. 19.1) giving the first indication of the diversity of Titan's surface (see Jaumann et al. 2009).

Since July 2004, the Cassini spacecraft completed 45 close flybys at Titan over its 4-year nominal mission. Onboard the Cassini spacecraft, several remote sensing instruments, that were designed to detect planetary surfaces at wavelengths from the near up to the far infrared, proved to be very successful in detecting the surface of Titan despite its dense atmosphere. Cassini investigations resulted in a wealth of mapping products displaying a great variety of Titan's surface properties. Images achieved by the Imaging Science Subsystem (ISS) (Porco et al. 2005), the Visual and Infrared Mapping Spectrometer (VIMS) (Brown et al. 2004; Sotin et al. 2005) as well as the Radio Detection and Ranging (RADAR) instrument aboard the spacecraft (Elachi et al. 2005), which complement each other in terms of wavelength, resolution and surface coverage, have provided detailed views of Titan's surface revealing an intriguing surface that appears at the same time familiar and alien.

Additionally, at 11:30 UTC on 14 January 2005, the Huygens probe successfully touched down on Titan's surface. The Huygens instruments observed the surface during the latter stages of and after its descent taking a series of the first and highest-resolution images that revealed an extraordinary world, resembling Earth in many aspects (Tomasko et al. 2002; Tomasko et al. 2005).

In this chapter we present a summary of mapping products of Titan's surface that were derived from DISR-, ISS-, VIMS- and RADAR-investigations and were achieved during the nominal Cassini mission that ended in July 2008. The coordinate system chosen for the ISS-, VIMS-, and RADAR maps presented here is the IAU "planetographic" system consisting of planetographic latitude and positive west longitude. The global maps are shown in a simple cylindrical projection centered at 0°N and 180°W, a format that matches the majority of previously published maps and in which the location of the Huygens Landing site appears in the center of the maps.

K. Stephan(✉), R. Jaumann, and M. Langhans
DLR, Institute of Planetary Research, Rutherfordstrasse 2, 12489, Berlin, Germany
email: Katrin.Stephan@dlr.de

E. Karkoschka, M.G. Tomasko, and J. Perry
Lunar and Planetary Laboratory, University of Arizona, 1629 E. University Blvd., Tucson, AZ 85721-0092, USA

R.L. Kirk
U.S. Geological Survey, Flagstaff, AZ 86011USA

J.W. Barnes
Department of Physics, University of Idaho, Moscow, ID 83844-0903, USA

E.P. Turtle and R.D. Lorenz
Johns Hopkins University, Applied Physics Lab., Laurel, MD 20723, USA

L. Le Corre and S. Le Mouélic
Laboratoire de Planétologie et Géodynamique, CNRS UMR6112, Université de Nantes, 2 rue de la Houssinière, 44072 Nantes, Cedex 3, France

R.H. Brown et al. (eds.), *Titan from Cassini-Huygens*,

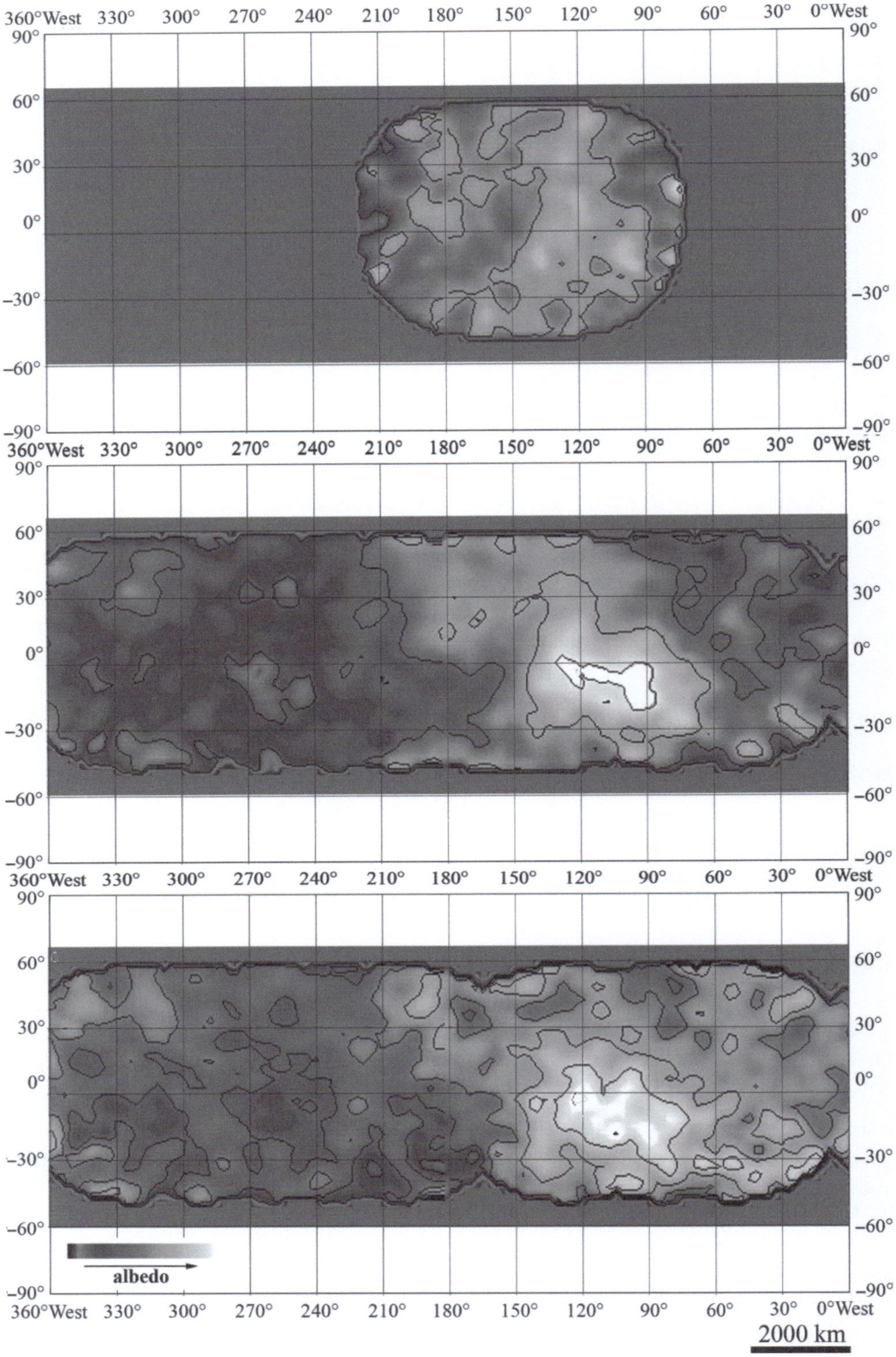

Fig. 19.1 Maps of Titan's surface showing albedo heterogeneities derived from HST observations using the filters (*top*) F673N, (*center*) F850LP, and (*bottom*) F1042M which are sensitive in the red light, as well as in the methane windows at 0.94 and 1.04 μm (adapted from Smith et al. 1996)

As a result of the numerous Cassini observations, a wealth of morphological features of the former virtually unknown surface was detected. Therefore, we also present an overview of Titan's recent nomenclature.

19.2 Huygens Image Data

Optical observations of Titan's surface exhibiting a preeminent spatial resolution so far are the images obtained by the Descent Imager/Spectral Radiometer (DISR) instrument during and after the descent of the Huygens probe (Tomasko et al. 2005), that provide invaluable ground truth to constrain observations from orbit.

The Imaging Science Subsystem (ISS) camera is potentially capable of the greatest spatial resolution in comparison to the remaining Cassini imagers, but Titan's obscuring and scattering haze limits its actual resolution on the surface to about 1 km, a value roughly similar to that available from VIMS and the RADAR imaging system. At this resolution, the bright and dark regions observed on the surface of Titan have proved difficult to interpret. In contrast, the Descent Imager/Spectral Radiometer (DISR) camera on the Huygens probe captured images of spatial resolutions between 100 m (50 m/pixel sampling) from an altitude of 50 km and 45 cm from an altitude of 220 m. In addition, the lower the probe descended, the less haze lay between the camera and the ground (Tomasko et al. 2002; Tomasko et al. 2005), providing clearer images.

The three-camera combination (High Resolution Imager HRI, Medium Resolution Imager MRI, and Side Looking Imager SLI) of the DISR instrument were designed to provide an overlapping coverage for a full 360-degree view around Huygens stretching from nadir angles between 6° and 96° (Tomasko et al. 2002). The DISR images were broadband, sensitive primarily in the near-infrared spectral range. Some 20 sets of Huygens images were planned during the descent. Because of the opacity of the haze, surface features could only be discerned in the images only below about 50 km, limiting the number of independent panoramic mosaics showing Titan's surface to 14. The loss of half of the images due to a technical problem meant that Titan's surface was not covered by systematic overlapping triplets, as expected (Tomasko et al. 2005). In the end the DISR images cover an area of about 2,500 km^2.

Figures 19.2–19.5 show different Huygens mosaics of Titan's surface at an effective wavelength of 0.77 μm combining DISR images exhibiting the highest spatial resolution. The spatial resolutions within these images (1,250 × 1,560 pixel) varies by large factors even on small scales, since some areas were imaged from low altitudes while other nearby areas were only imaged from 10 or 50 times higher altitudes. Where only low resolution was available, the

Fig. 19.2 Huygens mosaic showing a 80 × 100 km area centered on 10 km west of the Huygens landing site (HLS) constructed from images of all three DISR imagers. The spatial resolution varies across the image since some areas were imaged from low altitudes while others nearby areas were only imaged from 10 or 50 times higher altitudes. The image scale is 64 m/pixel. The dashed black line in the center of the image marks the boundary of the Huygens mosaic in Fig. 19.3

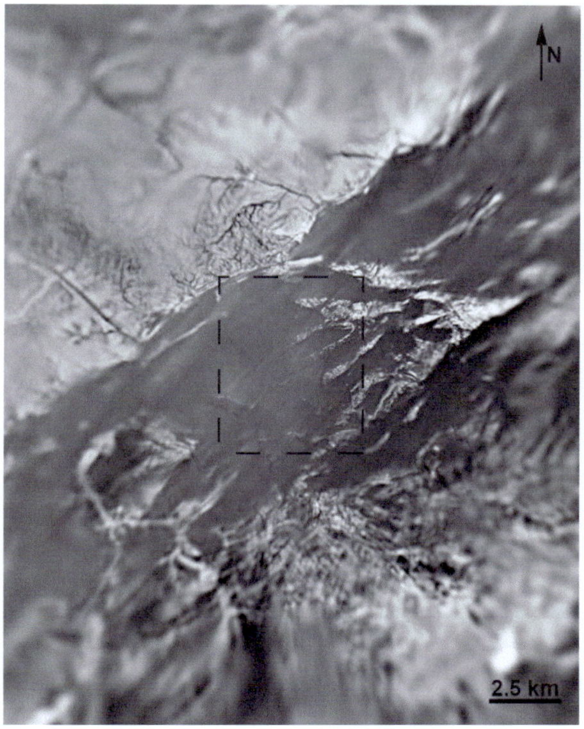

Fig. 19.3 Huygens mosaic showing a 20 × 25 km area centered on the Huygens landing site (HLS) constructed from images of all three DISR imagers. The image scale is 16 m/pixel. The dashed black line in the center of the image marks the boundary of the Huygens mosaic in Fig. 19.4

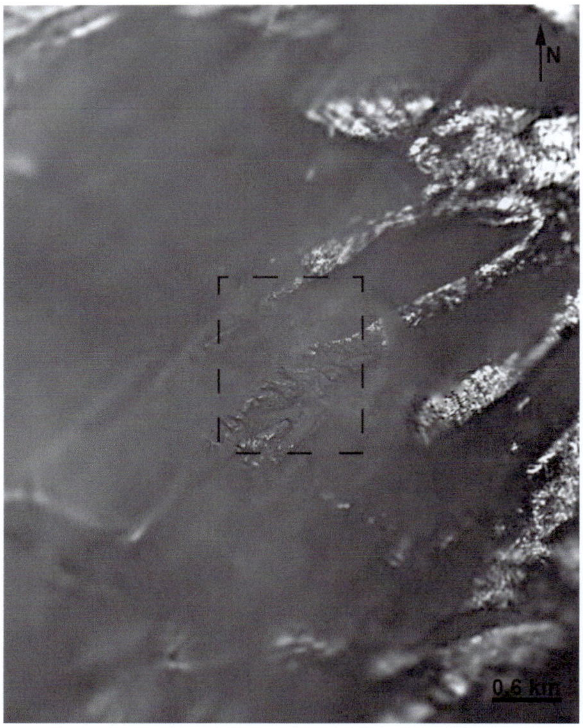

Fig. 19.4 Huygens mosaic showing a 5 × 6.25 km area centered on the Huygens landing site (HLS) constructed from images of all three DISR imagers. The spatial resolution varies across the image. The image scale is 4 m/pixel. The dashed black line in the center of the image marks the boundary of the Huygens mosaic in Fig. 19.5

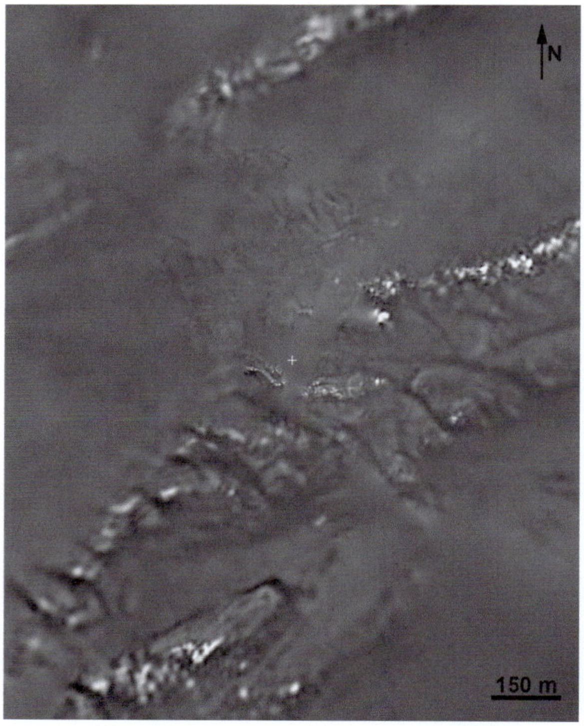

Fig. 19.5 Huygens mosaic showing a 1.25 × 1.56 km area centered on the Huygens landing site (HLS) constructed from images of all three DISR imagers. The spatial resolution varies across the image. The location of the HLS is marked by a cross of 20 m size. The image scale is 1 m/pixel

mosaic was left smooth. Thus, regions appearing to be smooth are typically regions imaged with poor resolution. The scale changes from one image to the next by a factor of four and the geometry in these images is most reliable between 500 m and 5 km and unreliable more than 10 km away from the Huygens landing site.

In the upper part of Fig. 19.2 are two parallel, dark lanes aligned east-west, about 30 and 35 km north of the landing site. Their shape matches those of two streaks seen by the Cassini RADAR, interpreted as two dunes of a large dune field extending further north. They provide the best correspondence between features seen by DISR and RADAR.

In the area directly below the Huygens probe as seen in Fig. 19.2, the boundary between the bright highlands to the north and west and dark lowland to the south is already seen. Albedo variations resemble the ones observed in highest-resolution images provided by the ISS or VIMS instruments aboard the Cassini spacecraft. Brighter regions are separated by lanes or lineaments of darker material. No obvious impact features are detectable. In the eastern part of the mosaic the images become sharper as the scale decreases and the contrast increases due to lower altitudes of the DISR cameras. More than a dozen brighter areas in that region seem to be elongated along a direction parallel with the main bright/dark boundary of that region (Tomasko et al. 2005). The area framed by the black dashed line marks the boundary of the image in Fig. 19.3 showing Titan's surface at closer range. To the left (northwest) a "coast line" is seen. Drainage channels in this bright terrain appear to flow into the ocean-like dark lowlands (Tomasko et al. 2005).

Again, the black dashed line in the center of the image marks the boundary of the next mosaic (Fig. 19.4), which shows a very dark appearing landscape. The available light is quite dim. Still, Huygens made out a surface feature that looks like a ridge dissected by a dozen darker lanes or channels. During the final part of the descent, Huygens stopped capturing images, and never got an image of the point at which it touched down the surface. This point is marked with the white cross at the center of the image in Fig. 19.5.

Additionally, the side-view imager obtained a horizontal view of the horizon. Figure 19.6 shows full 360-degree panorama of the landscape around Huygens viewed from two different altitudes i.e. 10 km (top) and 1 km (bottom) above the landing site showing a geologically very active surface. Images were colored according to spectra from the DISR down-looking visible spectrometer. The orange color of the sky and the scene on Titan is caused by the strong attenuation of blue light by Titan's haze relative to red light. A long 'coastline' separates the dark, dry 'lakebed' in the foreground from bright highland terrain that is incised by a dendritic system of channels (Tomasko et al. 2005). As the probe descended, it drifted over a plateau seen in the center of the image and was heading towards its landing site in the dark area at the bottom (Tomasko et al. 2005).

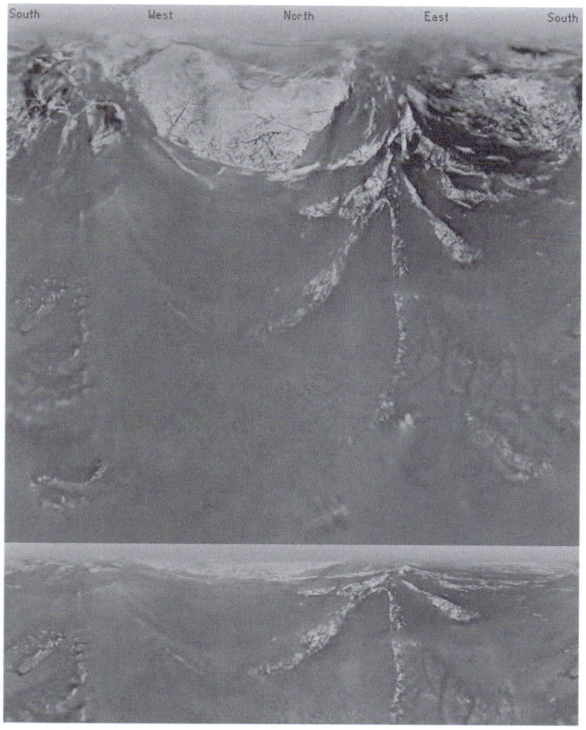

Fig. 19.6 Mercator projection views from 10 km (*top*) and 1 km (*bottom*) above the landing site. Images were colored according to spectra from the DISR down-looking visible spectrometer

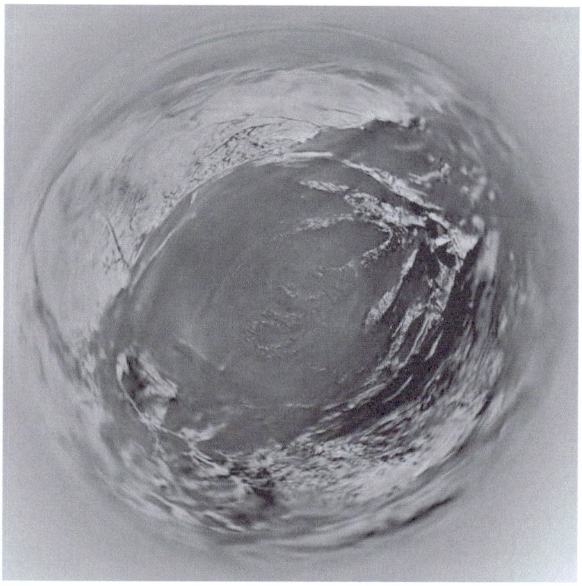

Fig. 19.7 Fish-eye view of the area imaged by Huygens, centered on the landing site

Figure 19.7 shows a fish-eye view of the area imaged by Huygens. Landforms appear distorted in this image. Figure 19.8 presents the view from Huygens after its success-

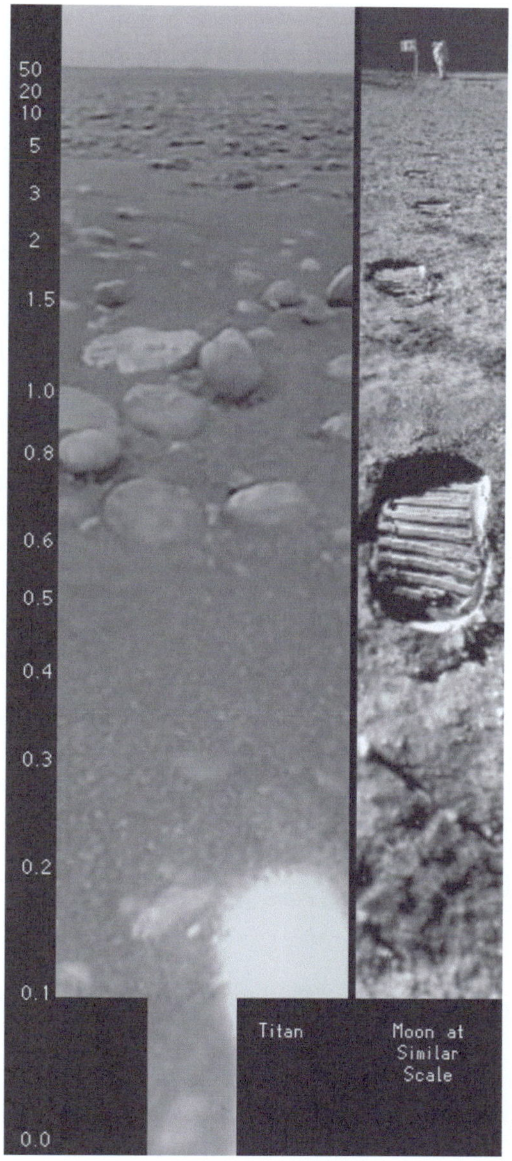

Fig. 19.8 View from Huygens after landing (*left*) and an image from the moon at similar scale for comparison (*right*). An approximate distance scale in meters is included along the left edge. The DISR image is saturated at lower right due to central illumination by the lamp

ful landing on Titan's surface in a colored view giving a better impression of the actual color of the surface (left) and an image from the moon at similar scale for comparison (right). The view is 25° wide, from south (left) to south-south-west (right). A hill in the distance appears 1° high. At the estimated distance of 60 m, it is about 1 m high. It may be part of the bright ridge seen below the cross in Fig. 19.5. Rounded stones that vary between 3 mm, the resolution limit of the imager, and 20 cm in diameter are probably composed of water ice, lie on top of a darker, finer-grained surface probably consisting of water and hydrocarbon ice as though located in the bed of a stream within the large dark lakebed

(Tomasko et al. 2005). The sun was high in the south-east, brightening the upper left side of several rocks. While only about one percent of sunlight makes it to the surface, most of the illumination comes from an area about 20 degrees around the sun due to the forward scattering nature of Titan's haze, which causes soft shadows below the largest rocks. The bright spot in the lower right corner is the illumination of the DISR surface science lamp. Black lines along the top of the two larger pebbles to the right of the "0.6" label are shadows from the lamp. The bright object next to the "0.1" label is an unfocused methane dew drop falling from the baffle of the SLI imager. It was imaged about 10 cm below the lens and about 35 cm above Titan's surface (Karkoschka and Tomasko 2009).

The method used for construction of panoramic mosaics requires the knowledge of the specific geographical location (longitude, latitude and altitude) as well as the attitude (roll, pitch and yaw) of the Huygens probe at each image. With the exception of altitude, that was provided by the Huygens Atmospheric Structure Instrument (Fulchignoni et al. 2005), none of these variables was directly measured. They were derived through an iterative image processing, providing an improved ground-track and azimuth model, which results in an upgraded trajectory, which can improve the panorama, and so on. An overview of the full data reduction process is provided by Tomasko et al. (2005) and Karkoschka et al. (2007).

19.3 Global Maps Derived from ISS Images

Imaging Titan's surface is a challenge because of almost complete obscuration at visible wavelengths by the atmosphere (Richardson et al. 2004; Turtle et al. 2008). Therefore, a narrow band pass filter at 938 nm centered in the best methane window available to ISS and infrared polarizer filters were incorporated into Cassini's Imaging Science Subsystem (ISS) (Porco et al. 2004) to take advantage of: (1) a window in methane's absorption spectrum in the near-infrared where the opacity of the atmospheric haze is lower, and (2) the high polarization of the haze at phase angles near 90° (West and Smith 1991). Despite the complications, to date the Cassini-ISS instrument has imaged almost all (~90%) of Titan's surface at resolutions better than ~10 km and a substantial fraction (~40%) of the surface at significantly better resolution, resolutions ranging from a few km down to the limit imposed by atmospheric scattering of about 1 km (Porco et al. 2004, 2005).

Most of the photons detected by ISS have been scattered by the atmospheric haze before ever reaching the surface, and at least 70–90% of the light that is reflected from the surface is widely scattered by the atmospheric haze. The ~10–30% of the light that is reflected from the surface without scattering by Titan's atmospheric haze significantly reduces the contrast of the ISS images. Therefore, more extensive data processing is needed to obtain the best detail from ISS images (McEwen et al. 2005; Perry et al. 2007).

ISS images have been processed through the following steps: (1) radiometric calibration; (2) removing of noise; (3) summation of two or three images to increase the signal-to-noise ratio (SNR); (4) division by an additional image at 619 nm acquired at the same time that shows only atmospheric haze in order to normalize brightness variations depending on illumination conditions; (5) improvement of camera-pointing information and re-projection of the images into a common map projection; (6) subtraction of 85% of a low-pass filtered image that approximates atmospheric scattering of photons reflected from the surface; (7) removal of images achieved at very high emission and incidence angles from the dataset; (8) adjustment of brightness as a function of phase angle and use of polarization filter; (9) assembly the images into mosaics and application of a seam-removal technique that preserves fine detail. For any additional and more detailed information of the ISS image processing the reader is referred to Porco et al. (2005), McEwen et al. (2007), Perry et al. (2005, 2007).

Figure 19.9 shows the resulting 938-nm albedo map of Titan's surface. The brightness variations revealed by ISS images are believed to be due to the presence of surface materials characterized by different albedos rather than topographic shading (Turtle et al. 2008). Due to the scattering of light by Titan's dense atmosphere, no topographic shading is visible in these images. The map has a scale of 4 km per pixel. Actual resolution varies greatly across the map, with the best coverage (close to the map scale) near the center and edges of the map and the worst coverage on the trailing hemisphere (centered around 270°W). Imaging coverage in the northern polar region is only just beginning to improve, and will continue to do so over the next couple of years, as Titan approaches vernal equinox in August 2009 and the north pole comes out of shadow. Large, dark and presumably liquid-hydrocarbon-filled seas are becoming visible at high latitudes.

19.4 Global Maps Derived from VIMS Observations

The Visual Infrared Mapping Spectrometer (VIMS) onboard the CASSINI spacecraft obtained new spectral data of the icy satellites of Saturn after its arrival at Saturn in June 2004 (Brown et al. 2004). VIMS operates in a spectral range from 0.35 to 5.2 µm, generating image cubes in which each pixel represents a spectrum consisting of 352 contiguous wavebands.

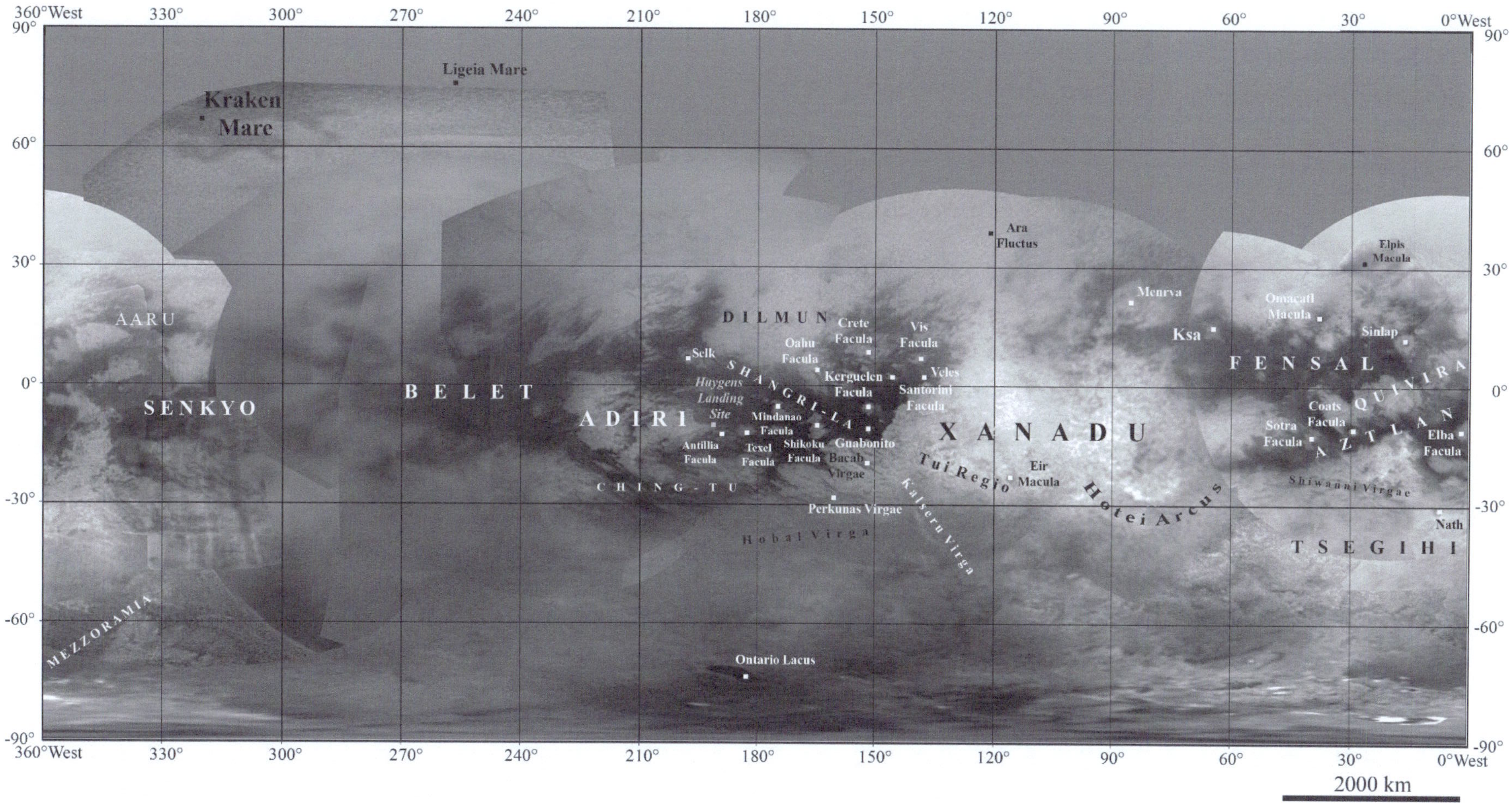

Fig. 19.9 Global map of Titan's surface produced from Cassini ISS images centered at 0.94μm including Titan's recent nomenclature. The map is displayed in a simple cylindrical projection centered at 0°N and 180°W

As an imaging spectrometer, VIMS combines the characteristics of both a spectrometer and an imaging instrument. This makes it possible to analyze the spectrum of each pixel separately and to map the spectral characteristics spatially, which is important to study the relationships between spectral information and geological and geomorphologic surface features (Jaumann et al. 2006).

Global maps presented in this chapter combine all available VIMS observations acquired during the Titan flybys TA, T4, T8, T9, T11, T13, T28, T32, T33, and T34 (from October 2004 up to July 2007) (Barnes et al. 2009).

Each VIMS observation was converted into a map projected cube according to Jaumann et al. (2006). A nearest-neighbor algorithm is used to resample the original data during the map projection process. This interpolation method does not modify the original spectral information but changes the position of the original pixel in each spectral channel due to the transformation into the new map projection (Jaumann et al. 2006).

To guarantee VIMS mosaics of high spectral and spatial quality, individual VIMS maps were selected according to the following criteria: (1) pixel ground resolution, (2) signal-to noise ratio, and (3) illumination conditions. Only VIMS cubes with an original pixel ground resolution of at least 100 km per pixel were included in the mosaics. At higher latitudes (>50°) VIMS observations usually exhibit relatively low spatial resolutions. On the contrary, equatorial and mid-latitudes regions have been observed with spatial resolutions that reach down to 1–2 km (Sotin et al. 2005) and occasional down to 500 m per pixel in nominal operation mode (Brown et al. 2004; Jaumann et al. 2008; Barnes et al. 2008). The individual maps are sorted by pixel ground resolution, and the image cube with the highest resolution is located on top of the mosaic. A map scale of 10 km per pixel was chosen in order to combine the global and local projected cubes into a mosaic. No limb observations are used in the mosaicking process. In order to guarantee the accuracy of the VIMS maps and mosaics their quality have been checked by comparison with maps of Voyager and Cassini ISS imaging data (Roatsch et al. 2006). Usually, the accuracy of maps based on lower spatial resolution VIMS data is within the limit of one VIMS pixel. If the inaccuracy exceeds one pixel, an additional registration of the VIMS data to ISS base maps has been applied (Jaumann et al. 2006).

Based on these mosaics, maps of the spectral properties for Titan's surface can be derived and attributed to geographic positions as well as to geological and geomorphologic surface features. These map-projected mosaics are the basis for all further investigations. The spectral properties of Titan's atmosphere enable the Cassini VIMS spectrometer (Brown et al. 2004) to detect the surface of Titan at the 0.94 μm wavelength corresponding to the filter used by ISS and also in seven other atmospheric "windows" between 1.08 and 5.0 μm (Barnes et al. 2007). Especially at 2 μm, the atmosphere is transparent enough and the signal-to-noise ratio (SNR) is sufficient (Jaumann et al. 2006) to map Titan's albedo features in detail (Fig. 19.10). Albedo features visible in the VIMS map in general resemble the albedo variations in ISS images (Fig. 19.9). However, the great advantage of the VIMS instrument is the ability to detect the spectral inhomogeneities of Titan's surface in more than one atmospheric window simultaneously. Thus, VIMS can be used as a multi-spectral camera providing information about the surface composition as well as geology and morphology (Jaumann et al. 2006).

The global map in Fig. 19.11 represents a mosaic of false-color composite images using VIMS channels with central wavelengths from 4.8 to 5.2 μm as red, at 2 μm as green, and 1.28 μm as blue (Barnes et al. 2007). The map illustrates the global distribution of at least two spectrally distinct types of dark material and two types of brighter material are present. The dark materials are commonly referred to as "blue" and "brown" based on their appearance in false-color spectral composites (Barnes et al. 2007). The bright spectral unit is known to be topographically high standing terrain inferring from Huygens and RADAR images. The blue spectral unit is supposed to contain water ice or to have increased water ice content while brown spectral units correlate to vast equatorial dune fields on Titan (Barnes et al. 2007; Soderblom et al. 2007).

Spectral differences on Titan's surface become obvious when using ratio combinations of VIMS channels at 1.08, 1.27, 1.59 and 2.03μm (Le Mouélic et al. 2008a). The global map in Fig. 19.12 represents a color composite image using ratios of the VIMS channels at 1.59 and 1.27 μm as red, at 2.03 and 1.27 μm as green, as well as 1.27 and 1.08 μm as blue. The map reveals a spectrally heterogeneous surface and illustrates the global distribution of at least one bright and two spectrally distinct types of dark material. In this color coding, the dark materials still appear in "blue" and "brown", but with a sharper contrast than in color composites of single bands (Fig. 19.11) (Barnes et al. 2007). The relationship between the defined spectral units and morphological features and their interpretation in terms of surface composition are described in the chapters of Jaumann et al. and Soderblom et al. (2009).

Global VIMS maps presented here are not fully corrected from photometric and atmospheric effects. It is quite hard to obtain a homogeneous VIMS map especially when numerous global VIMS observations acquired under most unequal observing conditions should be combined into mosaics as presented in Figs. 19.10–19.12. So far, only first order attempts to correct the scattering effect of aerosols in the VIMS channels were tested and proven to be useful to improve the VIMS maps (Rodriguez et al. 2006; Le Mouélic et al. 2008b).

Fig. 19.10 Global map of Titan's surface at a wavelength of 2μm produced from Cassini VIMS observations including Titan's recent nomenclature. The map is displayed in a simple cylindrical projection centered at 0°N and 180°W

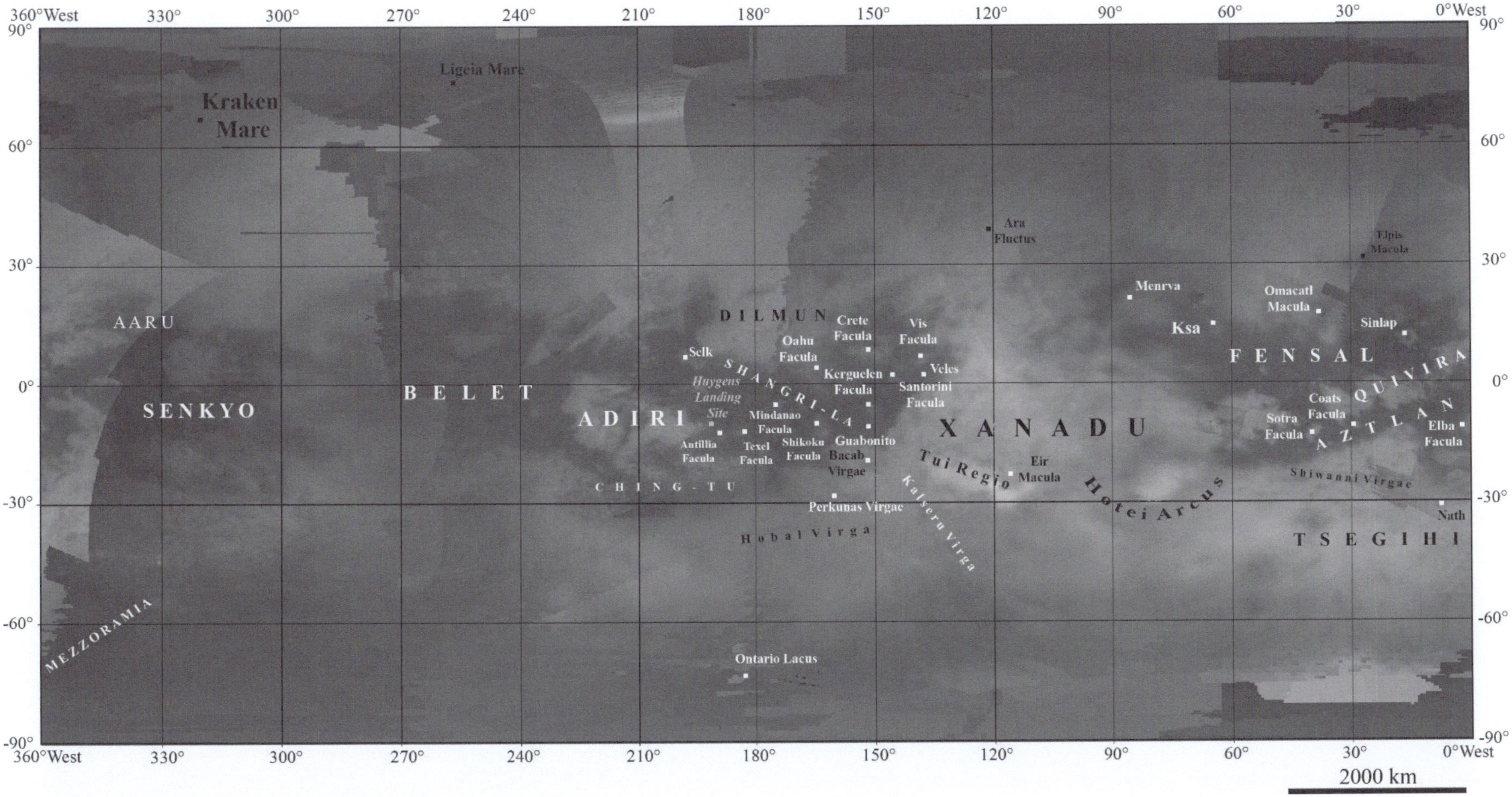

Fig. 19.11 Global VIMS map of Titan's surface displayed here as a false-color composite using the VIMS channels at 4.8–5.2 µm as red, at 2 µm as green, and 1.27 µm as blue including Titan's recent nomenclature. The map is displayed in a simple cylindrical projection centered at 0°N and 180°W

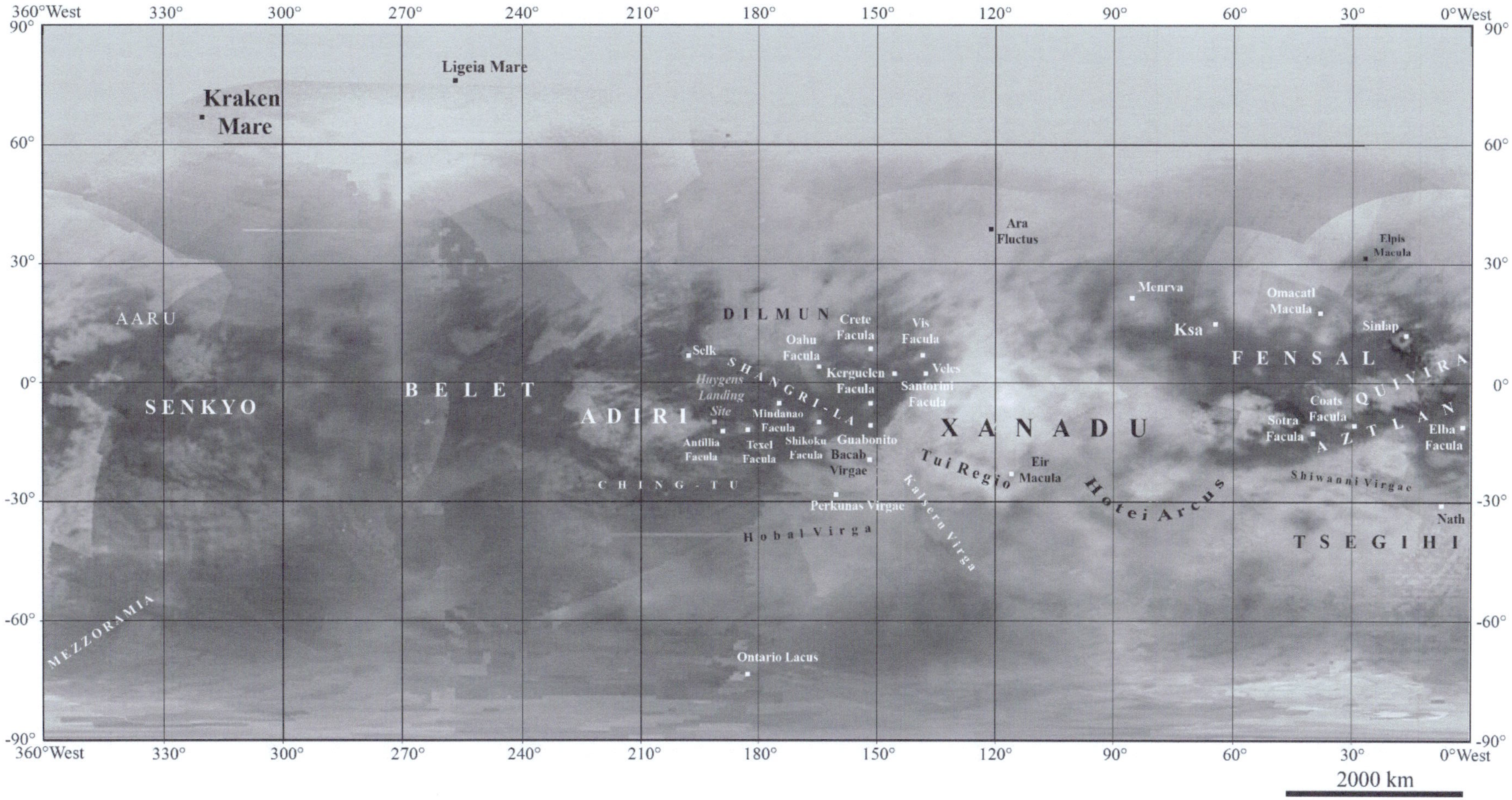

Fig. 19.12 Global VIMS map of Titan's surface displayed here as a false-color composite using ratios of the VIMS channels at 1.59 and 1.27 μm as red, at 2.03 and 1.27 μm as green, as well as 1.27 and 1.08 μm as blue. The map is displayed in a simple cylindrical projection centered at 0°N and 180°W

19.5 Global Maps Derived from RADAR Measurements

The Cassini Titan Radio Detection and Ranging (RADAR) Mapper plays a significant role in investigating the surface of Titan. This multimode radar instrument operates in the 13.8 GHz Ku-band (or 2.2 cm wavelength) and is designed to probe the optically inaccessible surface of Titan. The four different modes of the instrument (Synthetic Aperture Radar (SAR) imaging, Altimetry, Scatterometry and Radiometry) allow surface imaging as well as topographic mapping at spacecraft altitudes between 100,000 km and about 1,000 km, with resolutions for the modes ranging from hundreds of km to a few hundred meters (Elachi et al. 1991, 2004). The observations, particularly when performed in the active modes of SAR, altimetry and scatterometry, are largely unaffected by atmospheric contributions.

SAR images are obtained in the period around closest approach, and are highly elongated, with widths ranging from ~200 to ~500 km and lengths as great as 5,000 km (130° of arc around Titan). The image resolution ranges from ~300 m near closest approach to ~1.5 km at the ends of the strip, but the images are processed with a uniform sample spacing of 175 m (1/256° on Titan) in order to ensure oversampling of the data (Elachi et al. 2004). Additional images with resolutions >2 km can be obtained from higher altitude and are sometimes referred to HiSAR. Brightness in the SAR and HiSAR images is influenced by surface properties such as roughness, dielectric constant, prevalence of volume scattering, and the presence of macroscopic slopes facing toward or away from the instrument (Elachi et al. 2006). The images also contain parallax that encodes information about surface elevations, allowing stereo topographic mapping to be performed where images overlap (Kirk et al. 2009). In addition, the RADAR instrument uses multiple microwave beams to build up each strip of SAR imagery, and comparison of the signals from adjacent, overlapping beams can be used to estimate topographic relief in a single image by a technique known as "SARTopo" (Stiles et al. 2008), yielding long topographic profiles collocated with most of the images. SAR image coverage reached 27.5% (34.4% including HiSAR) at the end of the prime mission in mid-2008, and 29.5% (39%) at the time of this writing in early 2009. Figure 19.13 shows the SAR and HiSAR observations overlaid onto an ISS mosaic base.

An additional feature of this data set that is important for cartography is that the SAR image formation process is insensitive to errors in spacecraft pointing; the uncertainty in derived map coordinates comes only from the spacecraft trajectory (position and velocity) relative to Titan. Positional errors in an uncontrolled SAR mosaic will therefore be smaller than errors in an uncontrolled optical/infrared image mosaic, which are influenced by both trajectory and pointing. The errors in the reconstructed trajectories of the Cassini spacecraft relative to Titan are believed to be less than 1 km, but initial RADAR SAR mosaics showed mismatches between overlapping images that were as great as 30 km and ~10 km in a root mean square (RMS) sense (Stiles et al. 2008). These errors were found to be largely systematic in nature and were traced to inaccuracies in the pre-Cassini model of Titan's rotational state that was used to calculate ground coordinates in the latitude-longitude system of the rotating satellite. The SAR and HiSAR observations for flybys through T30 were reprocessed based on an improved model of Titan's rotation that was fit to the overlap data; images from subsequent flybys are being processed with a slightly simplified rotation model that uses the newly determined pole direction but not the time-varying rotation rate that was valid up to T30. Because the new rotation models were obtained by least-squares estimation based on the measured image features, the current mosaics (e.g., Fig. 19.13) may be considered controlled mosaics. RMS residual errors in the rotation modeling process are <1 km, as are typical mismatches in the map products. Thus, these are the most accurate maps of Titan and will be the base to which other observations will eventually be controlled.

The altimetry and scatterometry modes use the center beam of the instrument in a real aperture mode, i.e., their resolution is dictated by the angular size of the beam and is substantially poorer than that of the SAR images: from ~5 km when altimetry observations are taken near closest approach to ~150 km for the most distant scatterometry observations. The compensating advantages are the elevation information obtained when the instrument is nadir-pointed for altimetry (Zebker et al. 2009) and the much larger areal coverage and range of incidence angles for which backscatter maps are obtained in scatterometry mode (Wye et al. 2007). Figure 19.14 shows the combined SARTopo and altimetry data sets plotted on a SAR image mosaic base, whereas Fig. 19.15 presents the measurements derived from SARTopo and altimetry separately.

Figure 19.16 (top) shows the global scatterometry map. In this map, backscatter has been normalized to a uniform incidence angle by applying a model scattering function that was fitted to the full set of data (Wye et al. 2007). The data in this map, from flybys Ta through T39, cover 62.9% of Titan.

Radiometric observations are made by passively recording the microwave thermal emission of Titan during all of the other operational modes as well as during observations with altitudes up to 100,000 km. Resolution, based on the diameter of the beam footprint, thus varies widely from ~5 to ~500 km. The processing of this data set is complex and yields multiple maps that portray different aspects of Titan's surface properties (Janssen et al. 2009). Comparison of

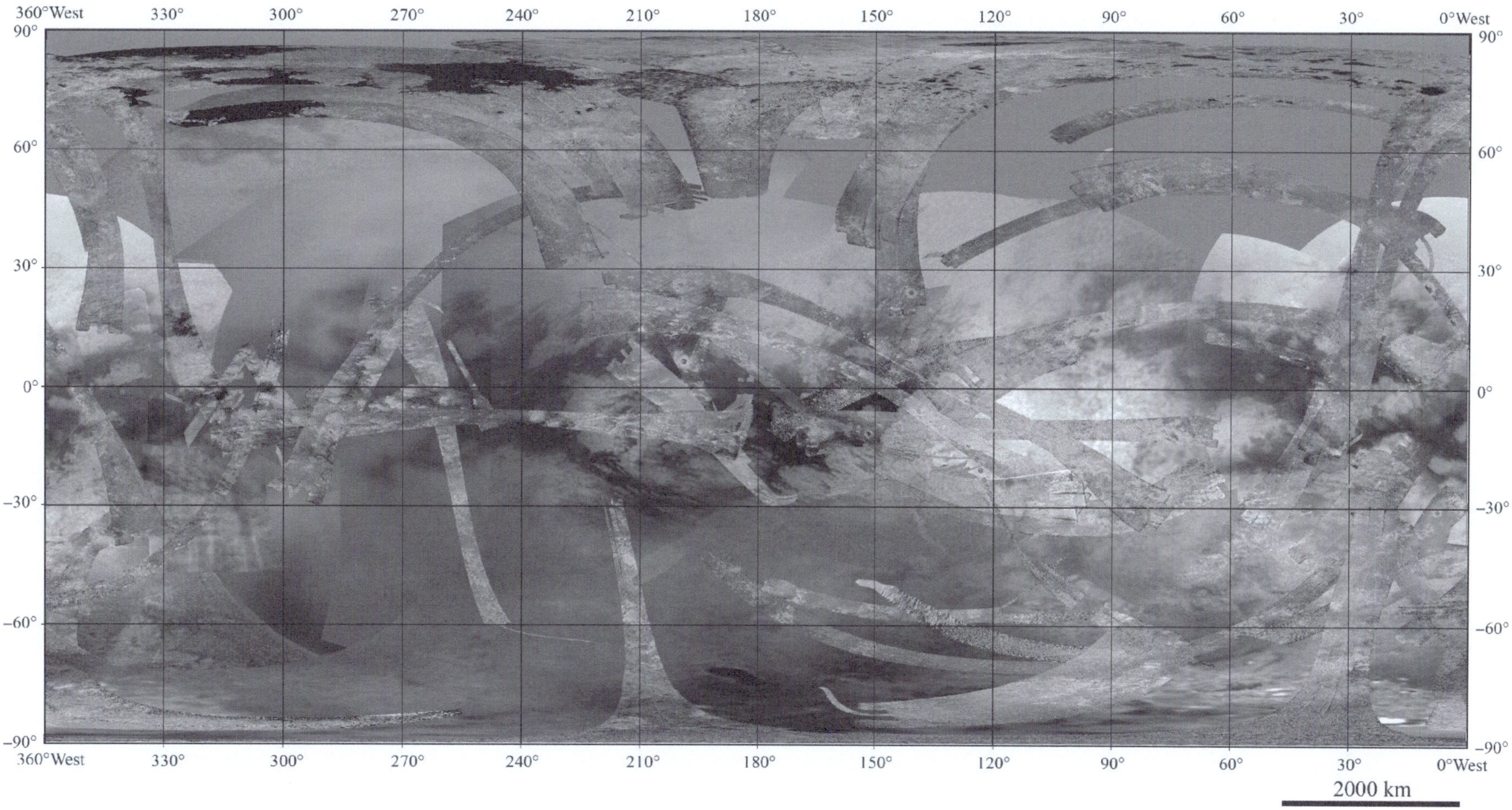

Fig. 19.13 Global map of Titan's surface showing the coverage derived from RADAR synthetic aperture (SAR) images overlaid onto the ISS map of Fig. 19.9. The map is displayed in a simple cylindrical projection centered at 0°N and 180°W. Full resolution SAR images obtained through flyby T50 (February 2009) are shown in gray, lower resolution high altitude SAR (HiSAR) observations for the same period in blue, and the base map in brown. Images have been normalized to a constant incidence angle of 35° based on the scattering model fitted by Wye et al. (2007) and scaled logarithmically. Black and dark blue correspond to -20 db, white to 5 db. Full resolution SAR coverage is 29.5% of Titan; coverage including HiSAR is 39% of Titan

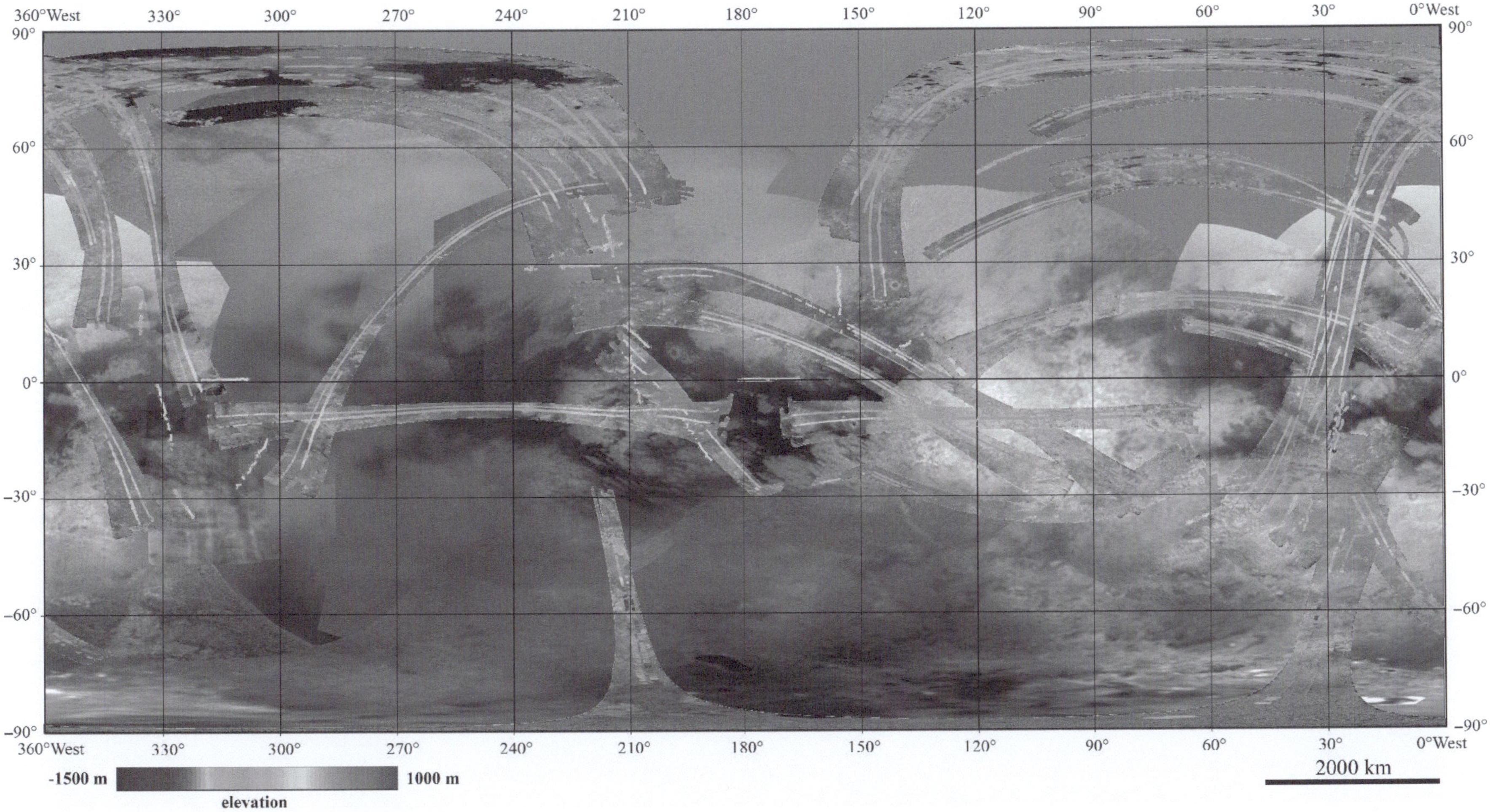

Fig. 19.14 Global map of Titan's topography as measured by the RADAR instrument. Color-coded elevations derived from altimetry and "SARTopo" (Stiles et al. 2008) are overlaid on a mosaic of RADAR SAR images (Fig. 19.13) and the ISS base map of Fig. 19.9. The map is displayed in a simple cylindrical projection centered at 0°N and 180°W

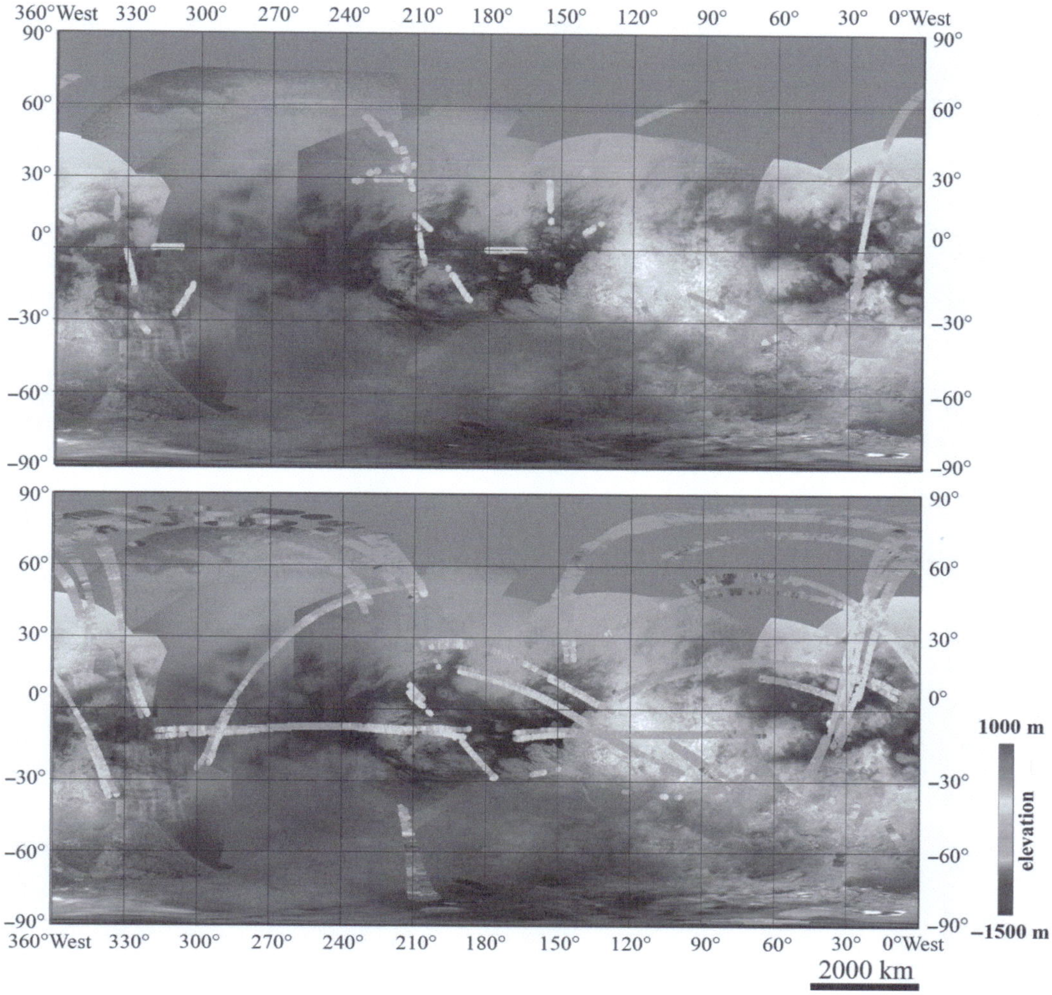

Fig. 19.15 Global map of Titan's topography as measured by the RADAR instrument similar to Fig. 19.14 but shown separately for altimetry (*top*) and "SARTopo" data (*bottom*) (Stiles et al. 2008) overlaid onto the global ISS map of Fig. 19.9. The map is displayed in a simple cylindrical projection centered at 0°N and 180°W

observations of the same ground point that were obtained with the spacecraft rotated to measure the polarization in different planes yields a map of dielectric constant (Fig. 19.17, top). This dielectric constant map is then used, along with the known incidence angle for each observation, to correct the observed brightness temperatures to normal incidence. As part of this process, the "leakage" of thermal radiation from Titan into the beam sidelobes must be estimated and removed from the observations. Because the physical temperature of Titan varies by only a few degrees, the brightness temperature map in Fig. 19.16 (bottom), which covers 98% of Titans surface (and 33% at high resolution), is a proxy for the surface thermal emissivity. Finally, the substantial departures of the emissivity from what is expected for a smooth surface of the given dielectric constant are attributed to variations in the strength of volume scattering (Fig. 19.17, bottom). Both, the measurements of the dielectric constant and of the volume scattering cover 84% of Titan's surface (A. Le Gall, pers. comm. 2009).

19.6 Thematic Maps Derived from Cassini Data

Global geological maps of Titan usually represent the result of complementary analyses of ISS and VIMS-IR images located in Titan's atmospheric windows, VIMS false-color composites (Figs. 19.11 and 19.12) and SAR swaths of the RADAR instrument (Fig. 19.13).

Figure 19.18 shows a global map illustrating the distribution of Titan's dune fields (Le Corre et al. 2008). The unit in white color on the map corresponds to the bright terrain seen in VIMS false-color composites. At high latitudes (higher than 50°), VIMS images are too low in spatial resolution to identify surface features due to the high phase angle and the presence of clouds and haze. On the contrary, brown units covering 18% of the whole Titan's surface are found mainly in equatorial regions. Dark blue units cover roughly 2% of Titan's surface. They are systematically associated with

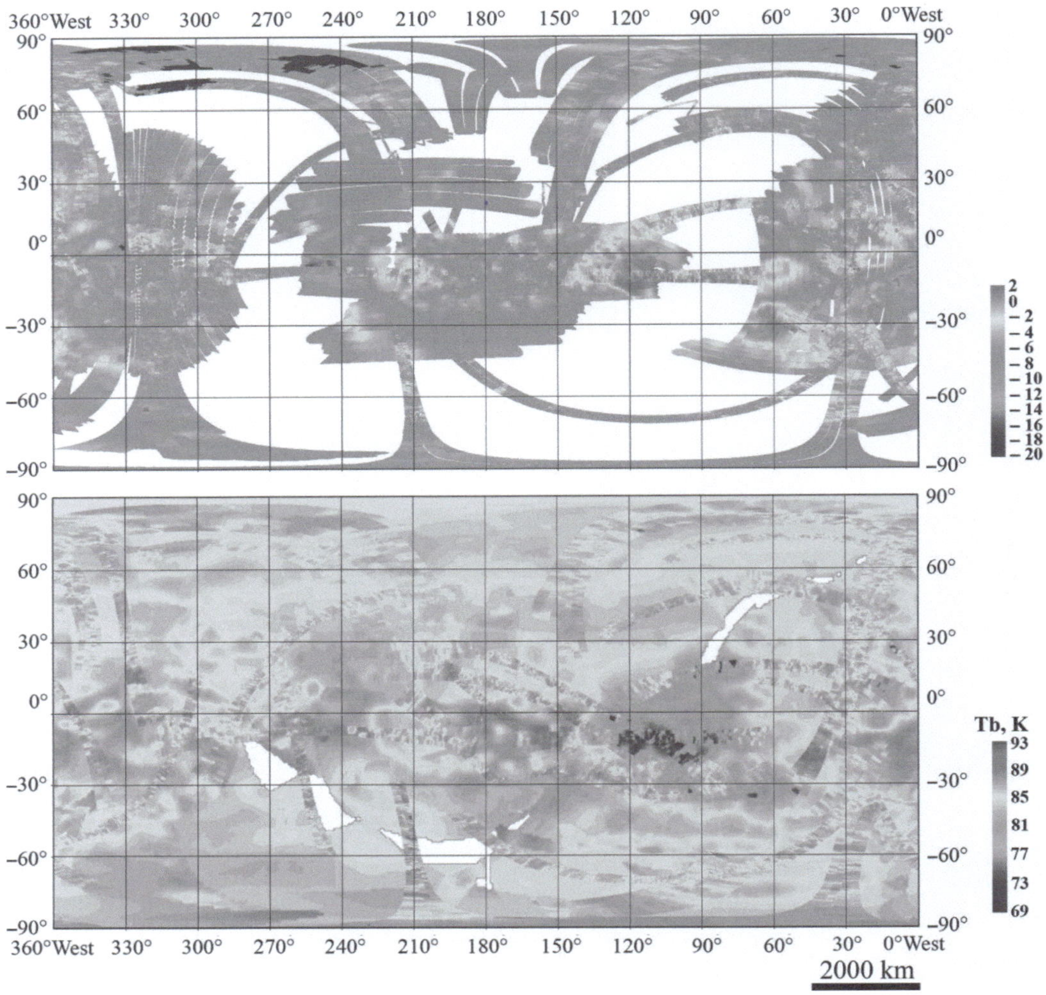

Fig. 19.16 Global maps of Titan's surface properties measured by the RADAR instrument: (1) Radar backscatter cross-section from scatterometry measurements, normalized to a constant incidence angle of 35° based on the scattering model fitted by Wye et al. (2007) and scaled logarithmically from -20 db to 2 db (*top*). Greater backscatter values correspond to rougher or more reflective surfaces or areas of enhanced subsurface volume scattering. (2) Surface brightness temperature as measured by passive radiometry, scaled to normal incidence (*bottom*) (Janssen et al. 2009). Because physical temperature is nearly constant, this is a proxy for microwave thermal emissivity. The maps are displayed in a simple cylindrical projection centered at 0°N and 180°W

bright terrains and are never found isolated within brown units. The dune fields retrieved from SAR images are highly correlated with brown infrared terrains and generally end at the boundary between infrared brown and bright units. 82% of the SAR dunes are located in brown units. Only 4.5% overlap dark blue areas. The remnant dunes, corresponding to "cat scratches" or not well defined dune fields, appear in infrared bright units as isolated patches. These dunes may form with a lower sand supply. It could account for some of the 13.5% RADAR dunes found on bright areas.

Figure 19.19 represents the resulting map of another investigation of the distribution of fluvial features on Titan derived from RADAR images that were acquired between October 2004 and April 2007 (orbits T00A–T028) and were mapped with respect to the spectral units of its surface material as seen in the false-color composites of VIMS observations (Langhans et al. 2008). Fluvial valleys could be found at all latitudes on Titan. Although, 67% of the channels could not be attributed to a specific substrate caused by missing VIMS observations with sufficient spatial resolution, channels seem to be primarily located in Titan's bright regions (27%). Numerous examples of fluvial valleys indicate that the bright spectral unit is topographically high-standing terrain compared to the dark (blue and brown) surface units. Only 3.5% and 2% of the channels are exactly located within the blue and brown unit, respectively. However, the majority (~60%) are located near, i.e., less than 100 km away from the blue spectral unit pointing to a deposition of fluvial eroded material (Jaumann et al. 2008).

Based on the complete RADAR imaging data set of the nominal Cassini mission Lorenz and Radebaugh (2009) derived a map of the orientation and extent of Titan's sand dunes (Fig. 19.20)

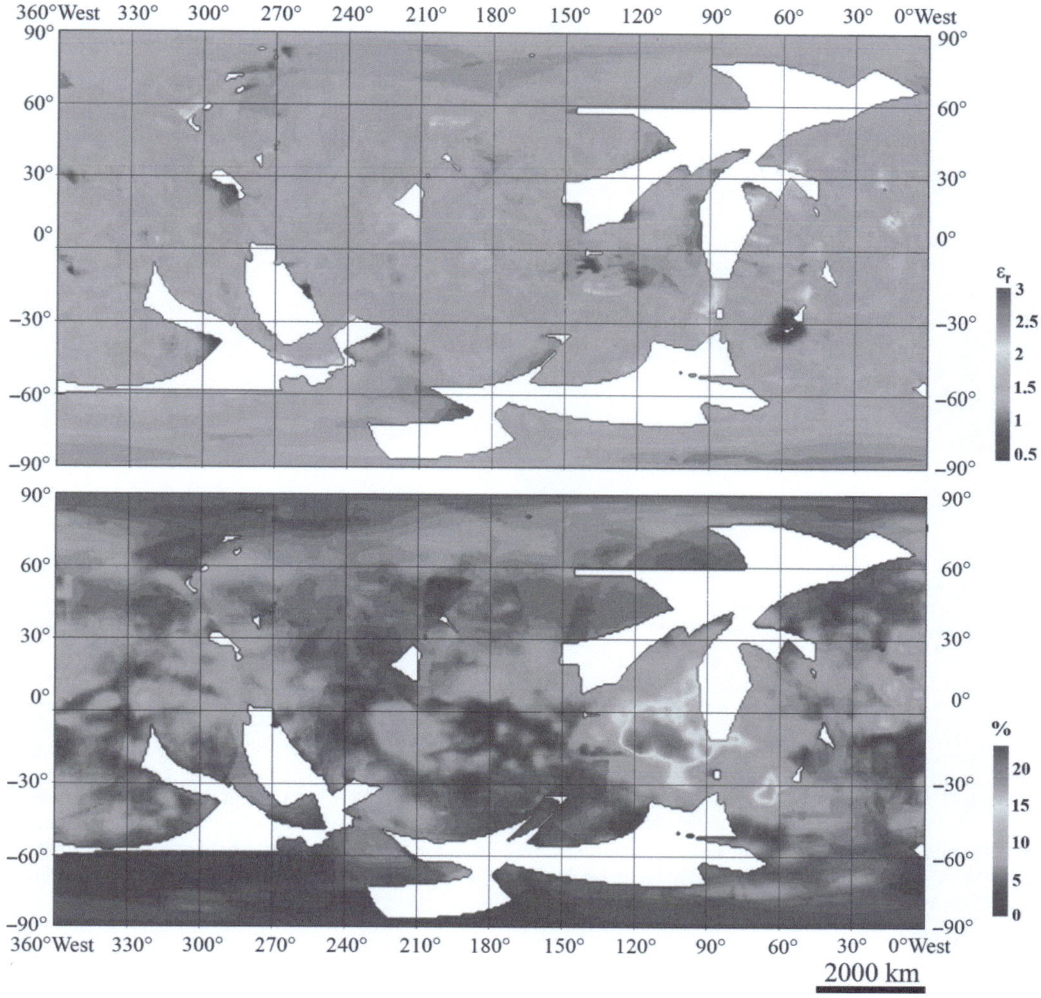

Fig. 19.17 Global maps of Titan's surface properties measured by the RADAR instrument: (1) Surface dielectric constant, inferred from the polarization of thermal emission (*top*). (2) Fraction of subsurface volume scattering (*bottom*)

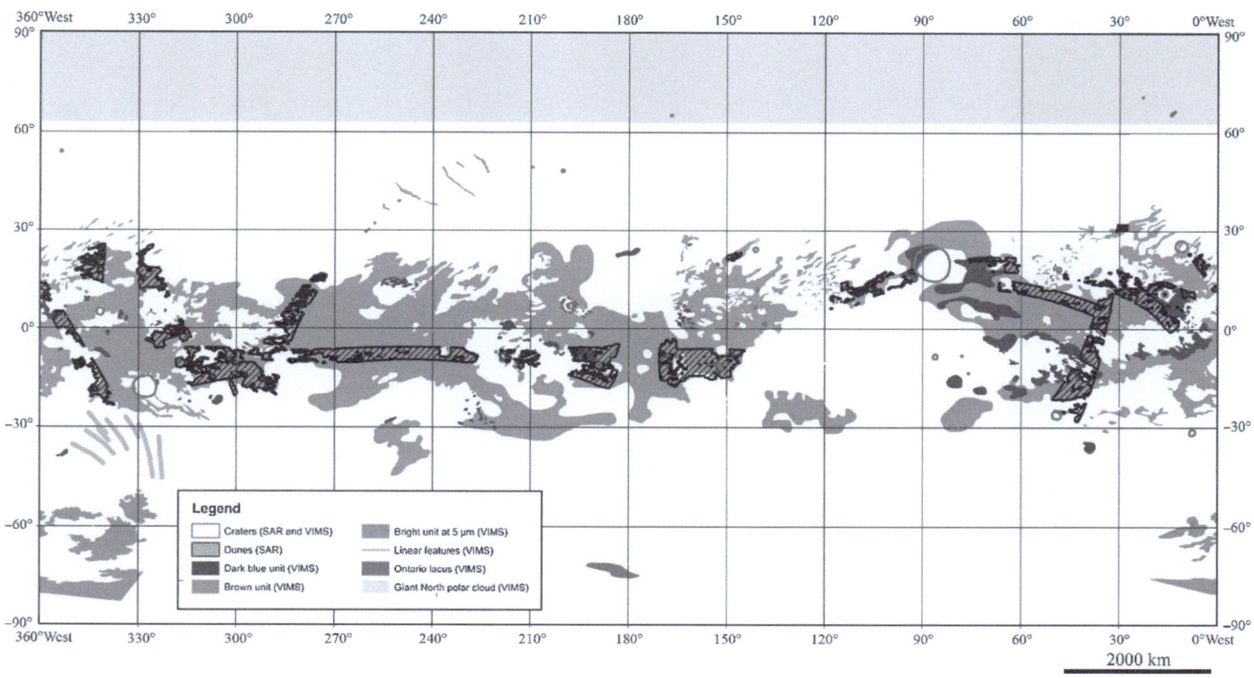

Fig. 19.18 Global map of dunes on Titan's surface derived from false-color composites of Cassini VIMS (Figs. 19.11 and 19.12) and RADAR observations (Fig. 19.13). The map is displayed in a simple cylindrical projection centered at 0°N and 180°W

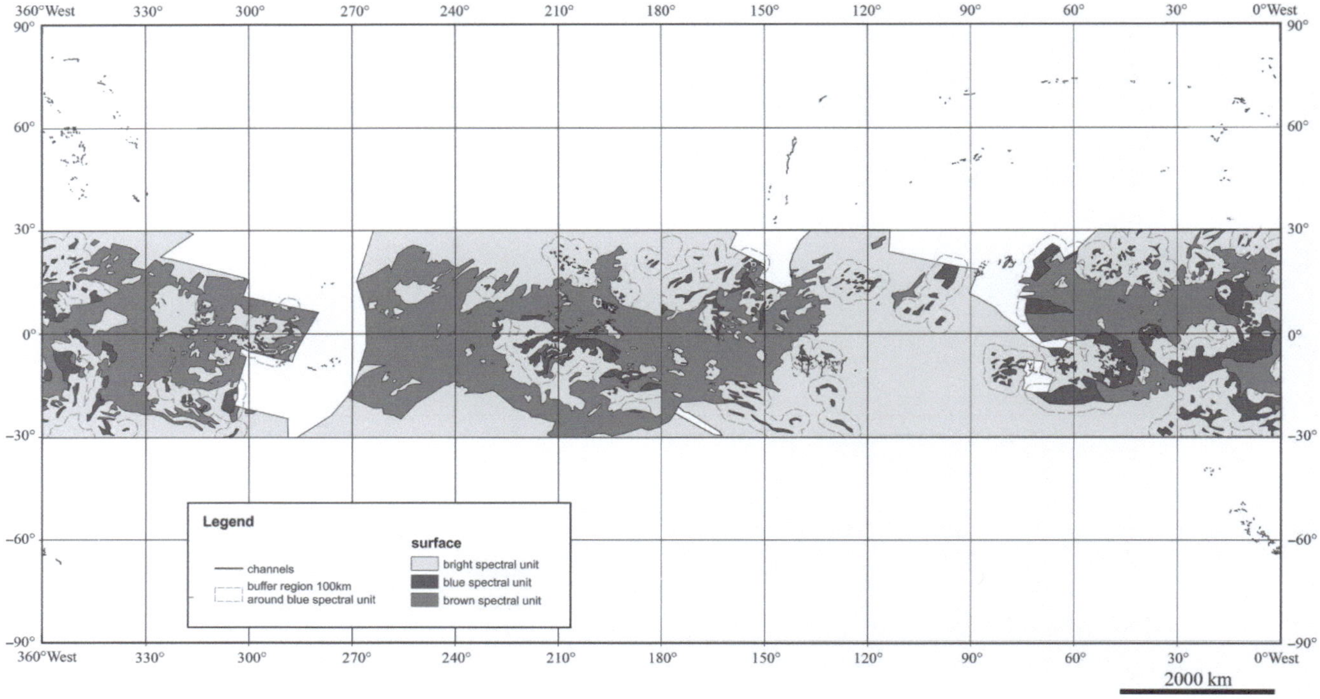

Fig. 19.19 Global map of Titan's surface displaying the distribution of channels mapped based on Cassini RADAR observations acquired during the flybys T0A up to T28 (Fig. 19.13) with respect to the spectral units derived from Cassini VIMS observations (Figs. 19.11 and 19.12). The map is displayed in a simple cylindrical projection centered at 0°N and 180°W. 27% of the mapped channels are located in the bright, 3.5% in the blue, and only 2.0% in the brown spectral unit. 67% of the channels are located in areas not mapped by VIMS with sufficient spatial resolution so far and therefore could not be associated with any of the spectral units

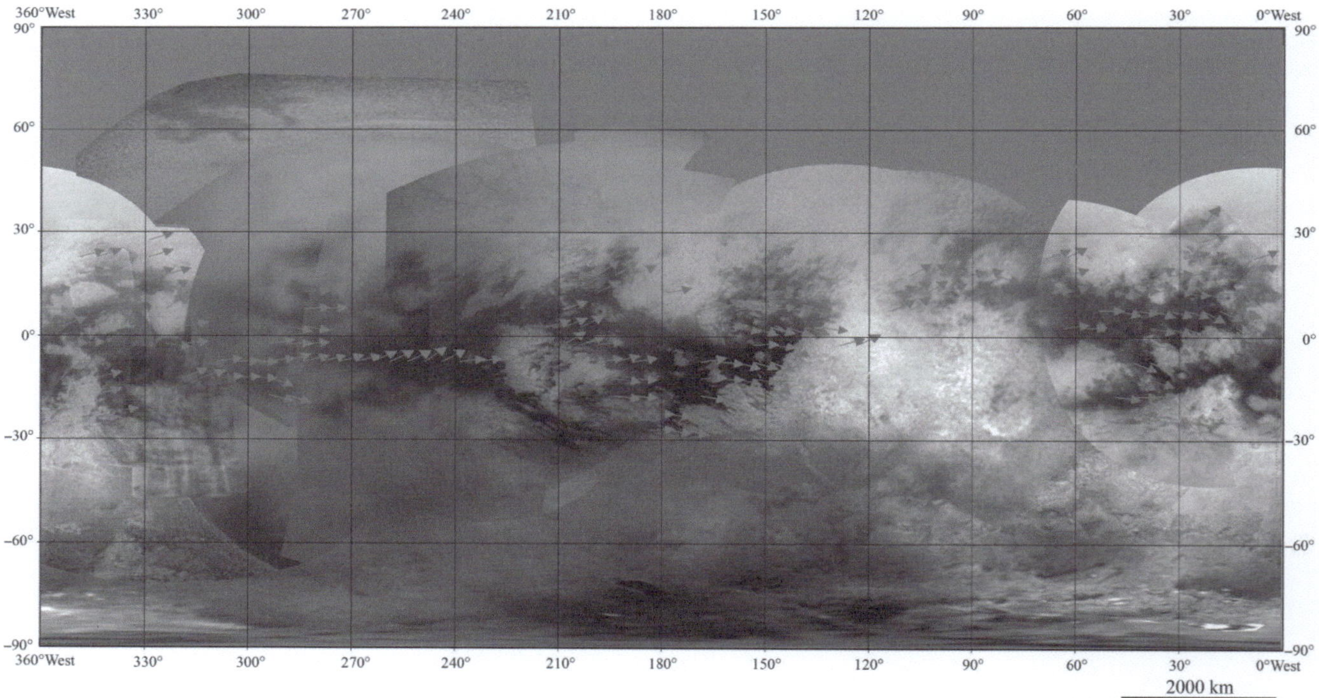

Fig. 19.20 Global dune orientation map derived from Radar data acquired during the flybys T3, T8, T13, T16, T17, T19, T21, T23, T25, T28, T29, T41, T43, and T44 overlaid. Other SAR observations (TA, T7, T18, T30, T36 and T39, generally at high latitudes) showed no observable dunes (adapted from Lorenz and Radebaugh 2009) onto the global ISS map of Fig. 19.9

A total of 16, 000 dunes have been mapped that cover ~8% of Titan's surface and are restricted to its equatorial region (±30°). The orientations of the dunes show local and regional deviations of up to about 40° from due Eastern direction. No obvious global longitudinal pattern could be observed, although some divergence with latitude is apparent. The most striking pattern is one of collimation by and divergence around bright and/or high terrain. Obstacles 100–300 m high obstruct dunes when the local slope is 1/50 or steeper, while slopes 1/200 or shallower cause dunes to thin out as they climb, or to deviate around the obstacles.

19.7 Titan's Nomenclature

As a result of the numerous observations of the remote sensing instruments aboard the Cassini spacecraft as well as the DISR camera a wealth of morphological features was detected, as Titan's surface was virtually unknown before. Consequently names of Titan's surface features have been announced and approved by the International Astronomical Union (IAU) (http://planetarynames.wr.usgs.gov/) very recently. Table 19.1 summarizes all recently named geological and/or morphological surface features. Some features were known by informal nicknames beforehand; these names are noted where appropriate.

Geological and/or morphological features on Titan include:

1. *Bright* and *dark albedo features*, that means geographic areas that can be distinguished by their specific amount of reflected light (Owen et al. 2006) that are named after sacred or enchanted places in world mythologies and literature.
2. Arc shaped features called *Arcus* (*pl.:* arcūs*)* are named after deities of happiness.
3. *Craters* on Titan are named after deities of wisdom.
4. *A Facula (pl. faculae)* represents a bright spot are named after islands on Earth that are politically independent. Groups of faculae are named after archipelagos on Earth.
5. Surface features on Titan termed *Fluctūs (pl. fluctūs)* refer to flow terrain and are named after mythological figures associated with beauty.
6. Surface features that look like a channel carved by liquid are termed *Flumen (pl.:* flumina*)*. There is one named group of flumina on Titan.
7. *Insulae* are islands within Titan's 'seas'. They are named after legendary islands.

Table 19.1 List of names of Titan's surface features approved by the International Astronomical Union (IAU) (http://planetarynames.wr.usgs.gov/)

Surface feature	Name	Named after	Central Lat.	Central Lon.	Informal nick name
	Adiri	Adiri, Melanesian paradise	–10.0	210.0	
	Dilmun	Dilmun, Sumerian heaven	15.0	170.0	
	Quivira	Quivira, legendary city in southwestern America	0.0	15.0	
	Tsegihi	Tsegihi, Navajo sacred place	–40.0	10.0	
Bright albedo features	Xanadu	Xanada, imaginary country in Coleridge's *Kubla Khan.*	–15.0	100.0	
	Aaru	Aaru, Egyptian paradise	10.0	340.0	
	Aztlan	Atzlan, mythical Aztec homeland	–10.0	20.0	Southern part of 'Lying H'
	Belet	Belet, Malay paradise	–5.0	255.0	
	Ching-tu	Ching-tu, Chinese Buddhist paradise	–30.0	205.0	
	Fensal	Fensal, Norse heavenly mansion	5.0	30.0	Northern part of 'Lying H'
	Mezzoramia	Mezzoramia, Oasis of happiness in the African dessert in an Italian legend	–70.0	0.0	
	Senkyo	Senkyo, Japanese paradise.	–5.0	320.0	
Dark albedo features	Shangri-la	Shangri-la, Tibetan paradise.	–10.0	165.0	
Arcus (*pl.* Arc s)	Hotei Arcus	Hotei, God of happiness in Japanese Buddhism	–28.0	79.0	
	Afekan	Afekan, New Guinean goddess of creation and knowledge	25.8	200.3	
	Ksa	Ksa, Lakota and Oglala god of wisdom	14.0	65.4	
	Menrva	Menvra, Etruscan goddess of wisdom	20.1	87.2	'Circus Maximus'
	Selk	Selk, Egyptian goddess of knowledge, writing, education, and reptiles	7.0	87.2	
Craters	Sinlap	Sinlap, Kachin (North Burma) wise spirit	11.3	16.0	

(continued)

Table 19.1 (continued)

Surface feature	Name	Named after	Central Lat.	Central Lon.	Informal nick name
	Antilia Faculae	Antillia, mythical Atlantic archipelago	–11.0	187.0	
	Bazaruto Facula	Bazaruto, Mozambique island	11.6	16.1	
	Coats Facula	Coats Island, Canada	–11.1	29.2	
	Crete Facula	Crete, Greek island	9.4	150.1	
	Elba Facula	Elba, Italian island	–10.8	1.2	
	Kerguelen Facula	Kerguelen Islands, French subantarctic island	–5.4	151.0	
	Mindanao Facula	Mindanao, Philippine island	–6.6	174.2	'Ireland'
	Nicobar Faculae	Nicobar Islands, Indian archipelago	2.0	159.0	
	Oahu Facula	Oahu, Hawaiian island	5.0	166.7	
	Santorini Facula	Santorini, Greek island	2.4	145.6	
	Shikoku Facula	Shikoku, Japanese island	–10.4	164.1	'Great Britain'
	Sotra Facula	Sotra, Norwegian island	–12.5	39.8	
	Texel Facula	Texel, Dutch island	–11.5	182.6	'Manhattan'
	Tortola Facula	Tortola, British Virgin Islands	8.8	143.1	'The Snail'
Facula (*pl.*: Faculae)	Vis Facula	Vis, Croatian island	7.0	138.4	
	Ara Fluctus	Ara, Armenian legendary figure	39.8	118.4	Ara Fluctus
	Leilah Fluctus	Layla, Persian goddess of beauty and chastity	50.5	77.8	Leilah Fluctus
	Rohe Fluctus	Rohe, beautiful Maori goddess, wife of Maui	47.3	37.75	Rohe Fluctus
Fluctus (*pl.* Fluct s)	Winia Fluctus	Winia, Indonesian first woman	49.0	46.0	Winia Fluctus
Flumen (*pl.* Flumina)	Elivagar Flumina	The Élivágar, a group of poisonous ice rivers in Norse mythology	19.3	78.5	
Insula (*pl.* insulae)	Mayda Insula	Mayda, legendary island in the northeast Atlantic Ocean	79.1	312.2	
Lacus (*pl.* Laci)	Abaya Lacus	Lake Abaya, Ethiopia	73.17	45.55	
	Bolsena Lacus	Lake Bolsena, Italy	75.75	10.28	
	Feia Lacus	Lake Feia, Brazil	73.7	64.41	
	Koitere Lacus	Koitere, Finland	79.4	36.14	
	Mackay Lacus	Lake Mackay, Australia	78.32	97.53	
	Mývatn Lacus	Mývatn, Iceland	78.19	135.28	
	Neagh Lacus	Lough Neagh, Northern Ireland	81.11	32.16	
	Oneida Lacus	Oneida Lake, USA	76.14	131.83	
	Ontario Lacus	Lake Ontario, USA/Canada	-72.0	183.0	
	Sotonera Lacus	Lake Sotonera, Spain	76.75	17.49	
	Sparrow Lacus	Sparrow Lake, Canada	84.3	64.7	
	Waikare Lacus	Lake Waikare, New Zealand	81.6	126.0	
	Guabonito	Guabonito, Taíno Indian sea goddess	–10.9	150.8	Guabonito
	Nath	Irish goddess of wisdom	–30.5	7.7	Nath
Large ringed features	Veles	Veles, Slavic god of wisdom	2.0	137.3	Veles
	Eir Macula	Eir, Norse goddess of healing and peace	–24.0	114.7	
	Elpis Macula	Elpis, Greek god of happiness and hope	31.2	27.0	
	Ganesa Macula	Ganesa, Hindu god of fortune and wisdom	50.0	87.3	
	Omacatl Macula	Omacatl, Aztec god of cheer and lord of banquets	17.6	37.2	
Macula (*pl.* Maculae	Polelya Macula	Polelya, Slavic god of matrimonial happiness	50.0	56.0	
Mare (*pl.* Maria)	Kraken Mare	The Kraken, Norse sea monster	68.0	310.0	
	Ligeia Mare	Ligeia, one of the Sirens, Greek monsters	79.0	248.0	
	Punga Mare	Punga, Maori supernatural being	85.1	339.7	
Regio (*pl.* Regiones)	Tui Regio	Tui, Chinese goddess of happiness, joy, and water	–20.0	130.0	
Virga (*pl.* Virgae)	Bacab Virgae	Bacab, Mayan rain god	–19.0	151.0	
	Hobal Virga	Hobal, Arabian rain god	–35.0	166.0	
	Kalseru Virga	Kalseru, Australian Aborigine rain god	–36.0	137.0	
	Perkunas Vrigae	Lithuanian supreme god, ruler of rain, thunder, and lightning	–27.0	162.0	
	Shiwanni Virgae	Shiwanni, Zuni rain god	–25.0	32.0	

8. *Laci are* believed to be filled by hydrocarbon. They are named after lakes on Earth.
9. *Large ringed features* are named after deities of wisdom in world mythology.
10. *Macula*e on Titan mark dark spots that are named after deities of happiness, peace, and harmony in world mythology.
11. Titanian *maria* are large hydrocarbon seas that are named after sea monsters in world mythology.
12. *Regiones* are distinctly different from their surroundings. They are named after deities of peace and happiness.
13. Streaks of color called *Virgae* are named after rain gods in world mythologies.

Because the exact nature of many surface features remains mysterious, a number of features have not yet received formal names and are known by nicknames. In most cases, indications of brightness and darkness refer to the infrared and RADAR images.

- '*The Sickle*': a large, dark, sickle-shaped region identified by the Hubble Space Telescope.
- '*Dog and Ball*', '*Dragon's Head*': large, dark, roughly equatorial regions identified by the European Southern Observatory's Very Large Telescope, named for their distinctive shapes.
- '*Si-Si the Cat*': a region that appears dark in radar images, named after a researcher's daughter who said it looked like a cat.

References

Barnes J, Brown RH, Soderblom L, Buratti BJ, Sotin C, Rodriguez S, Le Mouèlic S, Baines KH, Clark R, Nicholson P (2007) Global-scale surface spectral variations on Titan seen from Cassini/VIMS. Icarus 186:242–258. doi:10.1016/j.icarus.2006.08.021

Barnes J, Brown RH, Soderblom L, Sotin C, Le Mouèlic S, Rodriguez S, Jaumann R, Beyer RA, Buratti BJ, Pitman K, Baines KH, Clark R, Nicholson P (2008) Spectroscopy, morphometry, and photoclinometry of Titan's dunefields from Cassini/VIMS. Icarus 195:400–414. doi:10.1016/j.icarus.2007.12.006

Barnes JW, Soderblom J, Brown RH, Buratti BJ, Sotin C, Baines KH, Clark RN, Jaumann R, McCord TB, Nelson R, Le Mouélic S, Rodriguez S, Griffth C, Tosi F, Pitman KM, Soderblom L, Hayne P, Vixie G, Bibring J-P, Bellucci G, Capaccioni F, Cerroni P, Coradini A, Cruikshank DP, Drossart P, Formisano V, Langevin Y, Matson DL, Nicholson PD, Sicardy B (2009). VIMS Spectral Mapping Observations of Titan during the Cassini Prime Mission, PSS (submitted)

Brown RH, Baines KH, Bellucci G, Bibring J-P, Buratti BJ, Bussoletti E, Capaccioni F, Cerroni P, Clark RN, Coradini A, Cruikshank DP, Drossart P, Formisano V, Jaumann R, Langevin Y, Matson DL, McCord TB, Miller E, Nelson RM, Nicholson PD, Sicardy B, Sotin C (2004) The Cassini Visual and Infrared Mapping Spectrometer Investigation. Space Science Reviews 115:111–168

Coustenis A, Hirtzig M, Gendron E, Drossart P, Lai O, Combes M. Negrão A (2005) Maps of Titan's surface from 1 to 2.5 mm. Icarus 177, 89–105

Elachi C, Im E, Roth LE, Werner CL (1991) Cassini Titan radar mapper. Proc IEEE 79:867–880

Elachi C, Allison M, Borgarelli L, Encrenaz P, Im E, Janssen MA, Johnson WTK, Kirk RL, Lorenz RD, Lunine JI, Muhleman DO, Ostro SJ, Picardi G, Posa F, Roth LE, Seu R, Soderblom LA, Vetrella S, Wall SD, Wood CA, Zebker HA (2004) RADAR: The Cassini Titan radar mapper. Space Sci Rev 115:71–110

Elachi C, Wall S, Allison M, Anderson Y, Boehmer R, Callahan P, Encrenez P, Flamini E, Franceschetti G, Gim Y, Hamilton G, Hensley S, Janssen M, Johnson W, Kelleher K, Kirk R, Lopes R, Lorenz R, Lunine J, Muhleman D, Ostro S, Paganelli F, Picardi G, Posa F, Roth L, Seu R, Shaffer S, Soderblom L, Stiles B, Stofan E, Vetrella S, West R, Wood C, Wye L, Zebker H (2005) Cassini radar views the surface of Titan. Science 308:970–974

Elachi C, Wall S, Janssen M, Stofan E, Lopes R, Kirk R, Lorenz R, Lunine J, Paganelli F, Soderblom L, Wood C, Wye L, Zebker H, Anderson Y, Ostro S, Allison M, Boehmer R, Callahan P, Encrenez P, Flameni E, Francescetti G, Gim Y, Hamilton G, Hensley S, Johnson W, Kelleher K, Muhleman D, Picardi G, Posa F, Roth L, Seu R, Shaffer S, Stiles B, Vetrella S, West R (2006) Titan radar mapper observations from Cassini's T_A and T3 Flybys. Nature 441:709–713. doi:10.1038/nature04786

Fulchignoni M, Ferri F, Angrilli F, Ball AJ, Bar-Nun A, Barucci MA, Bettanini C, Bianchini G, Borucki W, Colombatti G, Coradini M, Coustenis A, Debei S, Falkner P, Fanti G, Flamini E, Gaborit V, Grard R, Hamelin M, Harri AM, Hathi B, Jernej I, Leese M, Lehto A, Lion Stoppado PF, Lopez-Moreno JJ, Makinen T, McDonnell JAM, McKay CP, Molina-Cuberos G, Neubauer FM, Pirronello V, Rodrido R, Saggin B, Schwingenschuh K, Seiff A, Svedhem H, Tokano T, Towner MC, Trautner R, Withers P, Zarnecki J (2005) In situ measurements of the physical characteristics of Titan's environment. Nature 438:785–791

Janssen MA, Lorenz RD, West R, Paganelli F, Stiles B, Wall SD, Callahan P, Kirk RL, Roth L, Anderson Y, and the Cassini RADAR Team (2009) Titan's surface at 2.2-cm wavelength imaged by the Cassini RADAR radiometer: Calibration and first results, Icarus 200, 222–239

Jaumann R, Stephan K, Brown RH, Buratti BJ, Clark RN, McCord TB, Coradini A, Capaccioni P, Filacchione G, Cerroni P, Baines KH, Bellucci G, Bibring JP, Combes M, Cruikshank DP, Drossart P, Formisano V, Langevin Y, Matson DL, Nelson RM, Nicholson PD, Sicardy B, Sotin C, Soderblom LA, Griffith C, Matz KD, Roatsch T, Scholten F, Porco CC (2006) High-resolution Cassini-VIMS mosaics of Titan and the icy Saturnian satellites. Planet Space Sci 54:1146–1155

Jaumann R, Brown RH, Stephan K, Barnes JW, Soderblom LA, Sotin C, Le Mouélic S, Clark RN, Soderblom J, Buratti BJ, Wagner R, McCord TB, Rodriguez S, Baines KH, Cruikshank DP, Nicholson PD, Griffith CA, Langhans M, Lorenz RD (2008) Fluvial erosion and post-erosional processes on Titan. Icarus 197:526–538

Jaumann R, Kirk R, Lorenz R, Lopes R, Stofan E, Turtle EP, Keller HU, Wood CA, Sotin C, Soderblom LA, Tomasko M (2009) Geology and surface processes on Titan, this volume

Karkoschka E, Tomasko MG (2009) Rain and dewdrops on Titan based on in situ imaging. Icarus 199:442–448 (http://dx.doi.org/10.1016/j.icarus.2008.09.020)

Karkoschka E, Tomasko MG, Doose LR, See C, McFarlane EA, Schröder SE, Rizk B (2007) DISR imaging and the geometry of the descent of the Huygens probe within Titan's atmosphere. Planet Space Sci 55(13):1896–1935

Kirk RL, Howington-Kraus E, Redding BL, Becker TL, Lee EM, Stiles BW, Hensley S, Hayes AG, Lopes RMC, Lorenz RD, Mitchell KL, Radebaugh J, Paganelli F, Soderblom LA, Stofan ER, Wall SD, Wood CA, Elachi C, and the Cassini RADAR Team (2009) Three-dimensional views of Titan's diverse surface features from Cassini RADAR stereogrammetry, Icarus, submitted

Le Corre L, Le Mouélic S, Sotin C, Barnes JW, Brown RH, Baines K, Buratti B, Clark R, Nicholson P (2008) Global map of Titan's dune fields. EPSC, Münster, Germany

Le Mouélic S, Paillou P, Janssen MA, Barnes JW, Rodriguez S, Sotin C, Brown RH, Baines K, Buratti BJ, Clark RN, Crapeau M, Encernaz PJ, Jaumann R, Geudtner D, Paganelli F, Soderblom L, Tobie G, Wall S (2008a) Mapping and interpretation of Sinlap crater on Titan using Cassini VIMS and RADAR data. J Geophys Res 113:E04003

Le Mouélic S, Barnes JW, Sotin C, Le Corre L, Brown RH, Baines K, Buratti B, Clark R, Nicholson PD (2008b) Global mapping of Titan in the infrared using an empirical decorelation between atmosphere and surface, 39th Lunar and Planetary Science Conference, Houston, 10–14 Mars, LPI Contribution No. 1391, p. 1730

Lorenz RD, Radebaugh J (2009) Global pattern of Titan's dunes: Radar survey from the Cassini prime mission. Geophys Res Lett 36:L03202. doi:10.1029/2008GL036850

Langhans M, Jaumann R, Stephan K, Brown RH, Buratti BJ, Clark R, Baine KH, Nicholson PD (2008) Analysis of fluvial features at Xanadu-Regio, Titan, observed by Cassini Radar and Cassini VIMS, EGU, Vienna, Austria, abstract

McEwen A, Turtle E, Perry J, Fussner S, Porco C, West R, Collins G, and the Cassini ISS Team (2005) Mapping global albedo patterns on Titan. Workshop on Titan after the Huygens and first Cassini encounters, Heraklion, Crete, Greece, 30 May – 3 June 2005

Owen TC, Aksnes K, Beebe R, Blue J, Brahic A, Burba GA, Smith BA, Teifel VG (2006) Nomenclature system and the very first names for one more world. LPSC XXXVIII, 1082

Perry JE, McEwen AS, Fussner S, Turtle EP, West RA, Porco CC, Knowles B, Dawson D.D and The Cassini ISS Team (2005). Processing ISS Images of Titan's Surface. LPSC XXXVI, 2312

Perry JE, Turtle EP, McEwen AS, Dawson DD, Porco CC (2007) Cassini ISS observations of Titan's Trailing Hemisphere. Ices, Oceans, and Fire: Satellites of the Outer Solar System Workshop, 6064

Porco CC, West RA, Squyres S, McEwen A, Thomas P, Murray CD, Delgenio A, Ingersoll AP, Johnson TV, Neukum G, Veverka J, Dones L, Brahic A, Burns JA, Haemmerle V, Knowles B, Dawson D, Roatsch T, Beurle K, Owen W (2004) Cassini imaging science: Instrument, characteristics and anticipated scientific investigations at Saturn. Space Sci Rev 115:363–497

Porco CC, Baker E, Barbara J, Beurle K, Brahic A, Burns JA, Charnoz S, Cooper N, Dawson DD, Del Genio AD, Denk T, Dones L, Dyudina U, Evans MW, Fussner S, Giese B, Grazier K, Helfenstein P, Ingersol AP, Jacobson RA, Johnson TV, McEwen A, Murray CD, Neukum G, Owen WM, Perry J, Roatsch T, Spitale J, Squyres S, Thomas P, Tiscareno L, Turtle EP, Vasavada AR, Veverka J, Wagner R, West R (2005) Imaging of Titan from the Cassini spacecraft. Nature 434:159–168

Richardson J, McEwen A, Lorenz RD (2004) Titan's surface and rotation – New insights from voyager images. Icarus 170:113–124

Roatsch Th, Wählisch M, Scholten F, Hoffmeister A, Matz K-D, Denk T, Neukum G, Thomas P, Helfenstein P, Porco C (2006) Mapping of the icy Saturnian satellites – first results from Cassini-ISS, PSS, 54

Rodriguez S, Le Mouélic S, Sotin C, Clénet H, Clark RN, Buratti B, Brown RH, McCord TB, Nicholson D, Baines KH, the VIMS Science Team (2006) Cassini/VIMS hyperspectral observations of the HUYGENS landing site on Titan, Planet Space Sci 54, 1510–1523

Soderblom L, Anderson J, Baines K, Barnes J, Barrett J, Brown R, Buratti B, Clark R, Cruikshank D, Elachi C, Janssen M, Jaumann R, Kirk R, Karkoschka E, Lemouelic S, Lopes R, Lorenz R, Lunine J, McCord T, Nicholson P, Radebaugh J, Rizk B, Sotin C, Stofan E, Sucharski T, Tomasko M, Wall S (2007) Correlations between Cassini VIMS spectra and RADAR SAR images: Implications for Titan's surface composition and the character of the Huygens probe landing site. Planet Space Sci 55:2025–2036. doi:10.1016/j.pss.2007.04.014

Smith PH, Lemmon MT, Lorenz RD, Sromovsky LA, Caldwell JJ, Allison MD (1996) Titan's Surface, Revealed by HST Imaging, Icarus 119:336–349

Soderblom LA, Barnes JW, Brown RH, Clark RN, Janssen MA, Marouf EA, McCord TB, Niemann HB, Tomasko MG (2009) Composition of the surface, this volume

Sotin C, Jaumann R, Buratti BJ, Brown RB, Clark RN, Soderblom LA, Baines KH, Bellucci G, Bibring J-P, Capaccioni F, Cerroni P, Coradini A, Cruikshank DP, Drossart P, Formisano V, Langevin Y, Matson DL, McCord TB, Nelson RM, Nicholson PD, Sicardy B, LeMouelic L, Rodriguez S, Stephan K, Scholz Ch (2005) Release of volatiles from a possible cryovolcano from near-infrared imaging of Titan. Nature 435:786–789

Stiles B, Kirk RL, Lorenz R, Hensley S, Lee E, Ostro S, Gim Y, Hamilton G, Johnson WTK, West R, and the Cassini RADAR Team (2008) Estimating Titan's spin state from Cassini SAR data, Astronom J 135, 1669–1680, doi:10.1088/0004-6256/135/5/1669

Tomasko MG, Buchhauser D, Bushroe M, Dafoe LE, Doose LR, Eibl A, Fellows C, McFarlane E, Prout GM, Pringle MJ, Rizk B, See C, Smith PH, Tsetsenekos K (2002) The descent imager/spectral radiometer (DISR) experiment on the Huygens entry probe of Titan. Space Sci Rev 104:469–551

Tomasko MG, Archinal B, Becker T, Bézard B, Bushroe M, Combes M, Cook D, Coustenis A, de Bergh C, Dafoe LE, Doose L, Douté S, Eibl A, Engel S, Gliem F, Grieger B, Holso K, Howington-Kraus A, Karkoschka E, Keller HU, Kirk R, Kramm R, Küppers M, Lellouch E, Lemmon M, Lunine J, McFarlane E, Moores J, Prout M, Rizk B, Rosiek M, Rüffer P, Schröder SE, Schmitt B, Smith P, Soderblom L, Thomas N, West R (2005) Rain, winds and haze during the Huygens probe's descent to Titan's surface. Nature 438(8):765–778. doi:10.1038/nature04126

Turtle EP, Perry JE, McEwen AS, West RA, DelGenio AD, Barbara J, Dawson DD, Porco CC (2008) Cassini imaging observations of Titan's high-latitude lakes. LPSC XXXIX, abstract No. 1952

West RA, Smith PH (1991) Evidence for aggregate particles in the atmospheres of Titan and Jupiter. Icarus 90:330–333

Wye LC, Zebker HA, Ostro SJ, West RD, Gim Y, Lorenz RD, and the Cassini RADAR Team (2007) Electrical properties of Titan's surface from Cassini RADAR scatterometer measurements. Icarus 188, 367–385, doi:10.1016/j.icarus.2006.12.008

Zebker HA, Gim Y, Callahan P, Hensley S, Lorenz R (2009) Analysis and interpretation of Cassini Titan radar altimeter echoes. Icarus 200(1):240–255

Appendix
The Huygens Teams, Behind the Scenes

In the pages that follow are pictures of the Huygens Operation Teams and of memorable moments during the Titan landing on 14–15 January 2005. We have also included other iconic pictures taken at key events during the development and operations of the Cassini Huygens mission over more than 25 years.

The people depicted herein, together with those presented in the appendix of the sister book, represent a subset of all of the people who have been part of the Cassini-Huygens Mission, from developing it as a candidate planetary-exploration mission in the early 1980s, to eventually being responsible for its design, construction, launch, flight and Huygens landing. It is hoped that the knowledge provided within this book will represent a small token of appreciation for the prodigious efforts of all of the people that have been involved with the Cassini-Huygens Mission over the roughly 27 years since it was proposed for a joint ESA/NASA study in 1982. It is they who have made Cassini-Huygens a shining success, and their tireless efforts will continue to bear important scientific and cultural fruits far into the future.

1. Group photo of the attendees of the Cassini-Huygens Project Science Group meeting at JPL on October 15, 2007

2. Group photo of attendees of the Huygens Critical Design Review (CDR) held at Aerospatiale, Cannes, 25–29 September 1995

The review participants included members of the following teams: ESA Huygens project team, Huygens industrial teams, JPL Cassini Project team, Huygens PI teams

3. Group Photo at the European Space Agency European Spacecraft Operations Centre (ESA/ESOC) during a Huygens Flight Operations Review in 1996

Participants included members of the Huygens Flight Operations team, Huygens Project team, Huygens Project Scientist team, Huygens Industrial team, Huygens Principal Investigator teams, JPL Flight Operations team, ESOC Public Relations team

4. Review of the Huygens mission preparation activities at the Jet Propulsion Laboratory, Pasadena, in 2002

Earl Maize (JPL), Claudio Sollazzo (JPL/HPOC), John Aiello (JPL), Shaun Standley (JPL/HPOC), Bobby Kazeminejad (ESTEC/PS team), David Allestad (JPL), Nathan Strange (JPL), Gerard Huttin (Alcatel)

5. The Huygens Flight Operations Team at ESOC

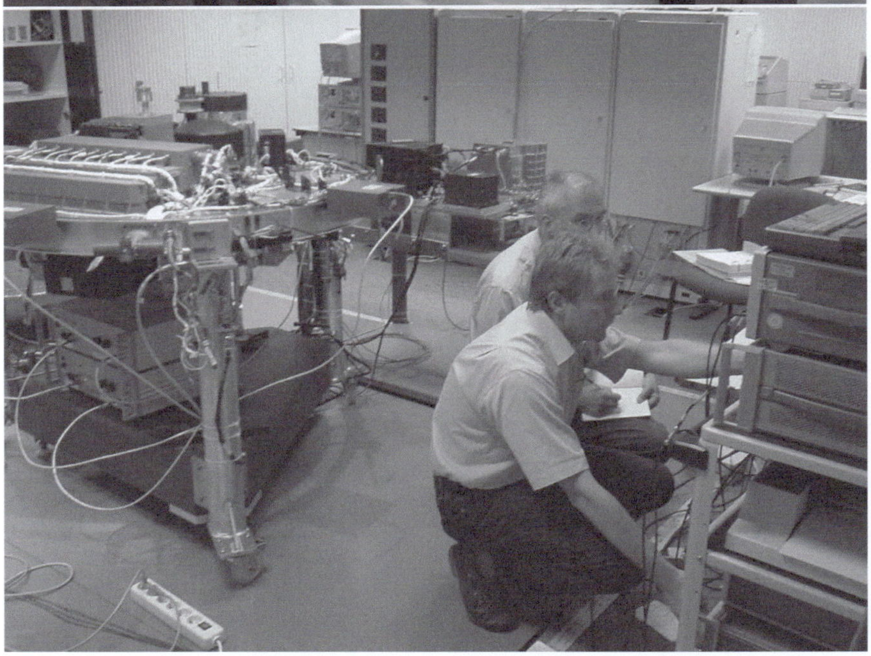

Top picture: Joe Wheadon, Martin Johnsson, Alan Smith, Martin Hermes, Dave Salt, Claudio Sollazzo, around the Huygens balloon drop test model
Bottom picture: Boris Smeds and Kevin Kewin operating the fully functional Huygens electrical model in ESOC's Huygens Probe Operations Centre (HPOC)

6. Members of the Aerosol Collector and Pyrolyser (ACP) Operations Team

Left picture: Manfred Steller, Guy Israel, Siegfried Bauer
Right picture: Hasso Niemann, Jean-Francis Brun, Guy Israel

7. Members of the Doppler Wind Experiment (DWE) Team

David Atkinson, Michael Bird, Robin Dutta-Roy, Dirk Plettemeier, Yvonne Dzierma, Michael Allison, Len Tyler

8. Members of the Descent Imager and Spectral Radiometer (DISR) Operations and Data Processing Teams

Left picture: Athena Coustenis, Laura Ellen Dafoe, Bjoern Grieger, Bernard Schmitt, Mike Bushroe, Peter Rueffer, name missing, Sylvain Doute
Right picture: Bjoern Grieger, Larry Soderblom, Bashar Rizk, Stefan Schröder, Michael Küppers

9. Members of the Gas Chromatograph Mass Spectrometer (GCMS) Operations Team

Left Picture: Eric Raaen Stan Way, Larry Frost, Hasso Niemann, Siegfried Bauer
Right Picture: George Carignan, Hasso Niemann, Jaime Demick-Montelaara, Eric Raaen In the middle back Huygens Team members Miguel Pérez Ayúcar and Thierry Blancquaert (both seen from behind) are talking to Huygens Operation Manager Claudio Sollazzo and Huygens Project Scientist Jean-Pierre Lebreton; on the far right, HASI team members Francesca Ferri and Giacomo Colombatti

10. Members of the Huygens Atmospheric Structure Instrument (HASI) Operations Team and of the Surface Science Package (SSP) Operations Team

Top Picture: John Rainnie, Ben Clark, Marcello Fulchignoni, Mark Leese, John Zarnecki, Andrew Ball, Ralph Lorenz, John Delderfield, Phil Rosenberg, Manish Patel, Martin Towner, Brijen Hathi, Francesca Ferri, Louise Patel
Bottom Picture: Francesca Ferri, Marcello Fulchignoni, Alexio Aboudan, Giacomo Colombatti, Andrew Ball, Piero Lion Stoppato, Mark Leese

11. Members of the Surface Science Package (SSP) Operations Team

Martin Towner, Phil Rosenberg, Marek Banaszkiewicz, Andrew Ball, Axel Hagermann, Manish Patel, Dave Parker, Brijen Hathi, John Rainnie

12. Chairs of the Descent Trajectory Working Group (DTWG)

Bobby Kazeminejad, David Atkinson

13. Members of the Huygens Very Long Baseline Interferometry (VLBI) Team, all affiliated (at that time) with JIVE

Cormac Reynolds, Leonid Gurvits, Sergei Pogrebenko, Ian Avruch, Arpad Szomoru, Bob Campbell, Steve Parsley

14. Members of the Jet Propulsion Laboratory Real Time Huygens Radio Tracking Team

Sami Asmar, Robert Preston, Sue Finley, Bill Folkner, Jim Border, Steve Lowe, Aseel Anabtawi. Team members not available for the picture: Garth Franklin, Jacob Gorelik, Doug Johnston and Viktor Kerzhanovich

15. Selection of the first DISR images to be shown to the world in the Huygens Science Operations Room at ESA/ESOC, Darmstadt, Germany on 14 January 2005

Top picture: Jonathan Lunine, Larry Soderblom, Laura Ellen Dafoe, Marty Tomasko, Sylvain Doute
Bottom picture: Marty Tomasko, Jocelyne Constantin-Landeau, Inigo Mascaraque

16. Presentation of the first Huygens images to the world by Marty Tomasko on 14 January 2005

17. Scenes at ESOC the day after the Huygens Landing: a gathering of Cassini-Huygens scientists, project and industry engineers, operations and media personnel, 15 January 2005

18. Members of the Huygens Data Archiving Team

19. Artistic rendering of the Huygens probe on Titan surface; the terrain surrounding of the probe has been inspired by the picture sent by DISR after landing (ESA Illustration)

Index

I

MIX
Papier aus verantwortungsvollen Quellen
Paper from responsible sources
FSC® C105338

If you have any concerns about our products,
you can contact us on
ProductSafety@springernature.com

In case Publisher is established outside the EU,
the EU authorized representative is:
Springer Nature Customer Service Center GmbH
Europaplatz 3, 69115 Heidelberg, Germany

Printed by Libri Plureos GmbH
in Hamburg, Germany